Fourth Edition

The 8088 and 8086 Microprocessors

Programming, Interfacing, Software, Hardware, and Applications

Including the 80286, 80386, 80486, and Pentium Processor Families

Walter A. Triebel
Fairleigh Dickinson University

Avtar Singh
San Jose State University

Upper Saddle River, New Jersey
Columbus, Ohio

Library of Congress Cataloging in Publication Data
Triebel, Walter A.
The 8088 and 8086 microprocessors: programming, interfacing, software, hardware, and applications/Water A. Triebel, Avtar Singh.—4th ed.
p. cm.
"Including the 80286, 80386, 80486, and Pentium processor families."
ISBN 0-13-041774-2
1. Intel 8088 (Microprocessor) 2. Intel 8086 (Microprocessor) I. Avtar Singh, 1947 June 2- II. Title.

QA76.8.I29293 T74 2003
004.165—dc21

2002029295

Editor in Chief: Stephen Helba
Assistant Vice President and Publisher: Charles E. Stewart, Jr.
Production Editor: Alexandrina Benedicto Wolf
Production Coordination: Clarinda Publication Services
Design Coordinator: Diane Ernsberger
Cover Designer: Jeff Vanik
Cover Image: Corbis Stockmarket
Production Manager: Matthew Ottenweller
Marketing Manager:

This book was set in Times Roman by The Clarinda Company. It was printed and bound by R. R. Donnelley & Sons Company. The cover was printed by Phoenix Color Corp.

Pearson Education Ltd.
Pearson Education Australia Pty. Limited
Pearson Education Singapore Pte. Ltd.
Pearson Education North Asia Ltd.
Pearson Education Canada, Ltd.
Pearson Educación de Mexico, S.A. de C.V.
Pearson Education—Japan
Pearson Education Malaysia Pte. Ltd.
Pearson Education, *Upper Saddle River, New Jersey*

10 9 8 7 6 5 4 3 2

ISBN 0-13-093081-4

To my daughter, Lindsey Triebel

—Walter A. Triebel

To my uncle Daya Singh whose timely support is the basis of my career

—Avtar Singh

Preface

Intel's 80X86 family of microprocessors is the most widely used architecture in modern microcomputer systems. The family includes both 16-bit microprocessors, such as the 8088, 8086, 80C186, 80C188, and 80286 processors, and 32-bit microprocessors, such as those of the 80386, 80486, and Pentium processor families. The 8088, which is the 8-bit bus version of the 8086, was the microprocessor used in the original IBM personal computer (PC). Many other manufacturers used the 8088 and 8086 microprocessors to make personal computers compatible with IBM's original PC. IBM's original personal computer advanced technology (PC/AT) was designed with the 80286 microprocessor. Like the PC, many other manufacturers made PC/AT-compatible personal computers, and today they are built with Pentium processor family microprocessors. Intel's 80X86 family of microprocessors is also used in a wide variety of other electronic equipment.

The 8088 and 8086 Microprocessors: Programming, Interfacing, Software, Hardware, and Applications, Fourth Edition, is a thorough study of the 8088 and 8086 microprocessors, their microcomputer system architectures, and the circuitry used in the design of the microcomputer of the original IBM PC. Written as a textbook for microprocessor courses at community colleges, four-year colleges, and universities, this book may be used in a one- or two-semester course that emphasizes both assembly language software and microcomputer circuit design.

Individuals involved in the design of microprocessor-based electronic equipment need a systems-level understanding of the 80X86 microcomputer—that is, a thorough understanding of both their software and their hardware. The first part of this book explores the software architecture of the 8088 and 8086 microprocessors and teaches the reader how to write, execute, and debug assembly language programs.

In this new edition, our coverage of software architecture and assembly language programming has been further reorganized to make the chapters shorter. Also, new material has been added on number system conversions, binary arithmetic, and combinational logic operations such as AND, OR, NOT, exclusive-OR, half- and full-adders, and half- and full-subtractors.

To successfully write assembly language programs for the 8088/8086 microprocessors, one must learn the following:

1. *Software architecture:* The internal registers, flags, memory organization and stack, and their uses from a software point of view.
2. *Software development tools:* Using the commands of the program debugger (such as DEBUG) to assemble, execute, and debug instructions and programs.
3. *Instruction set:* The function of each of the instructions in the instruction set, the permissible operand variations, and writing statements using the instructions.
4. *Programming techniques:* Basic techniques of programming, such as flowcharting, jumps, loops, strings, subroutines, and parameter passing.
5. *Applications:* The step-by-step process of writing programs for several practical applications, such as a block move routine.

All of this material is developed in detail in Chapters 2 through 7.

The software section includes many practical concepts and practical software applications. Examples are used to demonstrate practical applications such as 32-bit addition and subtraction, masking of bits, and the use of branch and loop operations to implement IF-THEN-ELSE, REPEAT-UNTIL, and WHILE-DO program structures. In addition, the various steps of the assembly language program development cycle are explored.

The study of software architecture, instruction set, and assembly language programming is closely coupled with use of the DEBUG program on the PC. That is, the line-by-line assembler in DEBUG is used to assemble instructions and programs into the memory of the PC, while other DEBUG commands are used to execute and debug the programs. The use of a practical 80X86 assembler program, the Microsoft MASM Assembler, is also covered. Using MASM and other PC-based software development tools, the student learns to create a source program; assemble the program; form a run module; and load, run, and debug a program.

The second part of the book examines the hardware architecture of microcomputers built with the 8088 and 8086 microprocessors. To understand the hardware design of an 8088- or 8086-based microcomputer system, the reader must begin by first understanding the function and operation of each of the microprocessor's hardware interfaces: memory, input/output, and interrupt. Next, the role of each of these subsystems is explored relative to overall microcomputer system operation. This material is presented in Chapters 8 through 13.

Chapter 8 examines the architecture of the 8088 and 8086 microprocessor from a hardware point of view. Included is information on pin layout, minimum and maximum mode signal interfaces, signal functions, and clock requirements. The latter part of the chapter covers the memory and input/output interfaces of the 8088/8086. This material includes extensive coverage of memory and input/output bus cycles, address maps, memory and input/output interface circuits (address latches and buffers, data bus transceivers, and address decoders), the use of programmable logic devices in implementing bus-control logic, types of input/output, and input/output instructions and programs.

This hardware introduction is followed by separate studies of the architecture, operation, devices, and typical circuit designs for the memory (Chapter 9), input/output (Chapter 10), and interrupts interfaces of the 8088/8086-based microcomputer (Chapter 11). Chapter 9 covers devices and circuits for the program storage memory (ROM, PROM, EPROM, and FLASH), data storage memory (SRAM and DRAM), and cache memory subsystems. Practical bus interface circuit and memory subsystem design techniques are also examined, including parity-checker/generator circuitry and wait-state generator circuitry.

Chapter 10 covers input/output interface circuits and LSI peripheral devices. The material on core I/O interfaces includes detailed studies of discrete parallel input/output circuits, 82C55A, 82C54, and 82C37A peripheral ICs. The chapter also explores a number of special-purpose peripheral IC devices and interfaces. For instance, serial communication and the 8250/16450 UART controllers are examined and keyboard scanning and display driving are demonstrated with the 8279 keyboard/display controller.

Chapter 11 introduces the interrupt context switching mechanism and related topics such as priority, interrupt vectors, the interrupt vector table, interrupt acknowledge bus cycle, and interrupt service routine. External hardware interrupt interface circuits are demonstrated using both discrete circuitry and the 82C59 programmable interrupt controller peripheral IC. The chapter also covers special interrupt functions such as software interrupts, the nonmaskable interrupt, reset operation, and internal interrupt processing.

The hardware design section continues in Chapter 12 with a study of the 8088-based microcomputer design used in the IBM PC. We present the circuitry used in the design of the memory subsystem, input/output interfaces, and interrupt interface on the system processor board of the PC. This chapter demonstrates a practical implementation of the material presented in the prior chapters on microcomputer interfacing techniques.

The material on hardware includes interface circuit operation, design, and troubleshooting. For example, the chapter on input/output devices explains circuits and programs for polling switches, lighting LEDs, scanning displays and keyboards, and printing characters at a parallel printer port. Moreover, Chapter 13 explores PC bus interfacing and techniques for circuit construction, testing, and troubleshooting.

The third part of the textbook provides detailed coverage of the other microprocessors of the 80X86 family: the 80286, 80386, 80486, and Pentium processors. Throughout these chapters, the focus is on how the processors' software and hardware architectures differ from those of the earlier family members. Advanced topics introduced include RISC, CRISP, and superscaler processor architectures, real-mode and protected-mode operation, burst, pipelined, and cached bus cycles, virtual memory, instruction set extensions, system control instructions, descriptors, paging, protection, multitasking, virtual 8086 mode, big and little endian data organization, clock scaling, dynamic bus sizing, address and data parity, and code and data cache memory.

Coverage of the 80486 and Pentium processor families has been further expanded in this edition. For example, new sections are included in Chapter 15 on floating-point architecture and multimedia architecture. Floating-point numbers, floating-point registers, and the floating-point instruction set are introduced relative to the 80486DX microprocessor. Material on the MMX technology, SIMD data, MMX registers, and the operation of MMX instructions is introduced relative to the Pentium Processor with MMX technology. Finally, Chapter 16 examines the newest Pentium family processors—the Pentium III processor and Pentium IV processor.

▲ SUPPLEMENTS

An extensive package of supplementary materials is available to complement the 80X86 microprocessor program offered by this textbook. It includes materials for the student and instructor for easy implementation of a practical PC-hosted laboratory program. These materials are:

1. *Instructor's Solution Manual to accompany The 8088 and 8086 Microprocessors: Programming, Interfacing, Software, Hardware, and Applications,* 4th Ed. ISBN 0-13-093082-2, Prentice Hall, Upper Saddle River, NJ 07458

   Provides the answers to all of the student exercises in the textbook as well as transparency masters for over 300 of the illustrations in the textbook. A CD-ROM is included, which contains all of the programs and executable files that are created by the student in the process of performing the 25 exercises in the laboratory manual. Based on the method identified in the exercise, the programs have been created with either the assembler in DEBUG or the Microsoft Macro Assembler.

2. *The 8088 and 8086 Microprocessors Laboratory Manual,* 4th Ed. ISBN 0-13-045231-9, Prentice Hall, Upper Saddle River, NJ 07458

   Contains 25 skill-building laboratory exercises that explore the software architecture of the 80X86 microcomputer in the PC, assembly language program development, the internal hardware of the PC, and interface circuit operation, design, testing, and troubleshooting. Also included is a CD containing all of the programs needed by the student to perform the exercises in the laboratory manual. Included are files that contain the source program, source listing, object code, and run module. These files have been produced by assembling the source program with the Microsoft Macro Assembler.

3. *PCμLAB,* Microcomputer Directions, Inc. P.O. Box 15127, Fremont, CA 94539 www.mcdlab.com (Not available through Prentice Hall)

   An easy-to-use and versatile, external hardware expansion environment for any personal computer for experimenting with microcomputer interface circuits. It extends the ISA bus external to the PC, thereby forming a bench-top laboratory text unit for building, testing, and troubleshooting interface circuits. It includes a large solderless breadboard area for working with student-constructed circuitry; a single PC/AT-compatible ISA bus slot for installation of commercially available or custom-build add-on cards; and built-in I/O devices, LEDs, switches, and a speaker. The PCμLAB also has a continuity tester and logic probe for testing circuit operation.

Walter A.Triebel
Avtar Singh

Contents

Introduction to Microprocessors and Microcomputers

▲ INTRODUCTION

In the past two decades, most of the important advances in computer system technology have been closely related to the development of high-performance 16-bit, 32-bit, and 64-bit microprocessor architectures and the microcomputer systems built with them. During this period, there has been a major change in the direction of businesses from using larger, expensive minicomputers to smaller, lower-cost microcomputers. The *IBM personal computer* (the PC, as it has become known), introduced in mid-1981, was one of the earliest microcomputers that used a 16-bit microprocessor, the 8088, as its processing unit. A few years later it was followed by another IBM personal computer, the *PC/AT* (personal computer advanced technology). This system was implemented using the more powerful 80286 microprocessor.

The PC and PC/AT quickly became cornerstones of the evolutionary process from minicomputer to microcomputer. In 1985 an even more powerful microprocessor, the 80386DX, was introduced. The 80386DX was Intel Corporation's first 32-bit member of the 8086 family of microprocessors. Availability of the 80386DX quickly lead to a new generation of high-performance PC/ATs. In the years that followed, Intel expanded its 32-bit architecture offering with the 80486 and Pentium processor families. These processors brought new levels of performance and capabilities to the personal computer marketplace. Today, Pentium IV processor-based PC/AT microcomputers represent the industry standard computer platform for the personal computer industry.

Since the introduction of the original IBM PC, the microprocessor market has matured significantly. Today, several complete families of 16- and 32-bit microprocessors are available. They all include support products such as *very large-scale integrated* (VLSI) peripheral devices, emulators, and high-level software languages. Over the same period of time, these higher-performance microprocessors have become more widely used in the design of new electronic equipment and computers. This book presents a detailed study of the software and hardware architectures of Intel Corporation's 8088 and 8086 microprocessors. An introduction to the 80286, 80386, 80486, and Pentium processors is also included.

In this chapter we begin our study with an introduction to microprocessors and microcomputers. The following topics are discussed:

1.1 The IBM and IBM-Compatible Personal Computers: Reprogrammable Microcomputers
1.2 General Architecture of a Microcomputer System
1.3 Evolution of the Intel Microprocessor Architecture
1.4 Number Systems

▲ 1.1 THE IBM AND IBM-COMPATIBLE PERSONAL COMPUTERS: REPROGRAMMABLE MICROCOMPUTERS

The IBM personal computer (the PC), shown in Fig. 1–1, was IBM's first entry into the microcomputer market. After its introduction in mid-1981, market acceptance of the PC grew by leaps and bounds, and it soon became the leading personal computer architecture. One of the important keys to its success is that an enormous amount of application software became available for the machine. Today, more than 50,000 off-the-shelf software packages are available for use with the PC, including business applications, software languages, educational programs, games, and even alternate operating systems.

Figure 1–1 Original IBM personal computer. (Courtesy of International Business Machines Corporation)

Another reason for the IBM PC's success is the fact that it offers an open system architecture. By *open system,* we mean that the functionality of the PC expands by simply adding boards into the system. Some examples of add-in hardware features are additional memory, a modem, serial communication interfaces, and a local area network interface. This system expansion is enabled by the PC's expansion bus—five card slots in the original PC's chassis. IBM defined an 8-bit expansion bus standard known as the *I/O Channel* and provided its specification to other manufacturers so that they could build different types of add-in products for the PC. Just as for software, a wide variety of add-in boards quickly became available. The result was a very flexible system that could be easily adapted to a wide variety of applications. For instance, the PC can be enhanced with add-in hardware to permit its use as a graphics terminal, to synthesize music, and even to control industrial equipment.

The success of the PC caused IBM to introduce additional family members. IBM's *PCXT* is shown in Fig. 1–2 and an 80286-based *PC/AT* is displayed in Fig. 1–3. The PCXT employed the same system architecture as that of the original PC. It was also designed with the 8088 microprocessor, but one of the floppy disk drives was replaced with a 10M-byte (10,000,000-byte) hard disk drive. The original PC/AT was designed with a 6-MHz 80286 microprocessor and defined a new open-system bus architecture called the *industry standard architecture* (ISA), which provides a 16-bit, higher-performance I/O expansion bus.

Today, Pentium IV processor-based PCs are the mainstay of the personal computer marketplace. Systems that are implemented with the Pentium IV processor no longer contain an ISA bus. They employ a new high-speed bus architecture known as the *peripheral component interface* (PCI) *bus.* PCI permits connection of high-performance I/O interfaces, such as graphics, video, and high-speed *local area network* (LAN). The PCI bus supports 32-bit and 64-bit data transfers, and its data-transfer rate is more than 10 times

Figure 1–2 PCXT personal computer. (Courtesy of International Business Machines Corporation)

Figure 1–3 PC/AT personal computer. (Courtesy of International Business Machines Corporation)

that of the ISA bus. These modern machines offer a wide variety of computing capabilities, range of performance, and software base for use in business and at home.

The IBM PC or a PC compatible is an example of a reprogrammable *microcomputer*—that is, one that is intended to run programs for a variety of applications. For example, one could use the PC with a standard application package for accounting or inventory control. In this type of application, the primary task of the microcomputer is to analyze and process a large amount of data, known as the *database.* Another user could run a word-processing software package—for example, a data input/output-intensive task—where the user enters text information that is reorganized by the microcomputer and then output to a diskette or printer. As a third example, a programmer uses a language, such as C, to write programs for a scientific application. Here the primary function of the microcomputer is to solve complex mathematical problems. The personal computer used for each of these applications is the same; the difference is in the software application that the microcomputer is running—that is, the microcomputer is simply reprogrammed to run the new application.

Let us now look at what a microcomputer is and how it differs from the other classes of computers. Evolution of the computer marketplace over the past 25 years has taken us from very large *mainframe computers* to smaller *minicomputers* and now to even smaller *microcomputers.* These three classes of computers did not originally replace each other; they all coexisted in the marketplace. Computer users had the opportunity to select the computer that best met their needs. The mainframe computer was used in an environment where it serviced a large number of users. For instance, a large university or institution would select a mainframe computer for its data-processing center where it would service hundreds of users. Mainframes are still used today to satisfy very large computer requirements.

The minicomputer had been the primary computer solution for the small, multiuser business environment. In this environment, several users connect to the system with terminals and all share the same computer system, with many of them actively working on the computer at the same time. An important characteristic of this computer system configuration is that all computational power resides at the minicomputer. The user terminals are what are known as *dumb terminals*—that is, they are not self-sufficient computers. If the minicomputer is not working, all users are down and cannot do any work at their terminals. An example of a user community that traditionally uses a minicomputer is a department at a university or a business that has a multiuser-dedicated need, such as application software development.

Managers in a department may select a microcomputer, such as the PC/AT, for their personal needs, such as word processing and database management. The original IBM PC was called a personal computer because it was initially intended to be a single-user system—that is, the user's personal computer. Several people could use the same computer, but only one at a time. Today, the microcomputer has taken over most of the traditional minicomputer user base. High-feature, high-performance microcomputer-based systems have replaced the minicomputer as a *file server.* Many users have their personal computers attached to the file server through a *local area network.* However, in this more modern computer system architecture, all users also have local computational power in their own PCs. The file server extends the computational power and system resources, such as memory available to the user, so if the file server is not operating, users can still do work with their individual personal computers.

Along the evolutionary path from mainframes to microcomputers, the basic concepts of computer architecture have not changed. Just like the mainframe and minicomputer, the microcomputer is a general-purpose electronic data-processing system intended for use in a variety of applications. The key difference is that microcomputers, such as the PC/AT, employ the newest *very large-scale integration* (VLSI) circuit technology *microprocessing unit* (MPU) to implement the system. Microcomputers, such as a Pentium III Xeon® processor-based file server, which are designed for the high-performance end of the microcomputer market, are physically smaller computer systems, outperform comparable minicomputer systems, and are available at a much lower cost.

▲ 1.2 GENERAL ARCHITECTURE OF A MICROCOMPUTER SYSTEM

The hardware of a microcomputer system can be divided into four functional sections: the *input unit, microprocessing unit, memory unit,* and *output unit.* The block diagram in Fig. 1–4 shows this general microcomputer architecture. Each of these units has a special

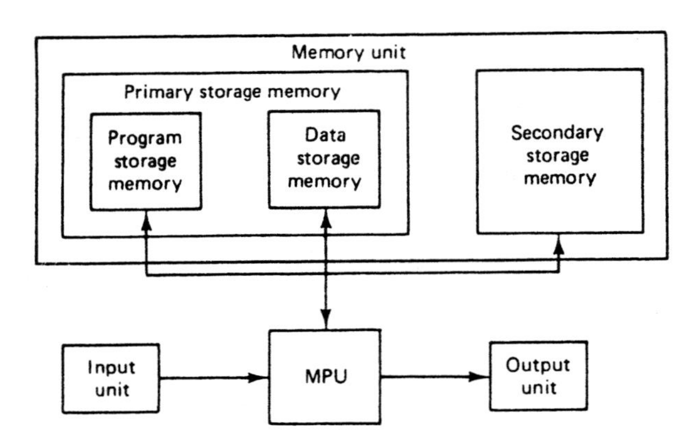

Figure 1–4 General architecture of a microcomputer system.

function in terms of overall system operation. Next we will look at each of these sections in more detail.

The heart of a microcomputer is its microprocessing unit (MPU). The MPU of a microcomputer is implemented with a VLSI device known as a *microprocessor,* or just *processor* for short. A microprocessor is a general-purpose processing unit built into a single *integrated circuit* (IC). The microprocessor used in the original IBM PC is Intel Corporation's 8088, shown in Fig. 1–5.

Earlier we indicated that the 8088 is a 16-bit microprocessor. To be more accurate, it is the 8-bit external bus version in Intel's 8086 family of 16-bit microprocessors. Even though the 8088 has an 8-bit external bus, its internal architecture is 16 bits in width and it can directly process 16-bit-wide data. For this reason, the 8088 is considered a 16-bit microprocessor.

The 8088 MPU is the part of the microcomputer that executes instructions of the program and processes data. It is responsible for performing all arithmetic operations and making the logical decisions initiated by the computer's program. In addition to arithmetic and logic functions, the MPU controls overall system operation.

The input and output units are the means by which the MPU communicates with the outside world. Input units, such as the *keyboard* on the IBM PC, allow the user to input information or commands to the MPU; for instance, a programmer could key in the lines of a BASIC program from the keyboard. Many other input devices are available for the PC; two examples include a *mouse,* for implementing a user-friendlier input interface, and a *scanner,* for reading in documents.

The most widely used output devices of a PC are the *monitor* and the *printer.* The output unit in a microcomputer is used to give the user feedback and to produce documented results. For example, key entries from the keyboard are echoed back to the monitor—that is, by looking at the information displayed on the monitor, the user can confirm that the correct entry was made. Moreover, the results produced by the MPU's processing can be either displayed or printed. For our earlier example of a BASIC program, once it is entered and corrected, a listing of the statements could be printed. Alternate output devices are also available for the microcomputer; for instance, many modern PCs are equipped with an advanced audio processing and speaker system.

Figure 1–5 Intel Corporation's 8088 microprocessor. (Courtesy of Intel Corp.)

The memory unit in a microcomputer is used to *store* information, such as number or character data. By "store" we mean that memory has the ability to hold this information for processing or for outputting at a later time. Programs that define how the computer is to operate and process data also reside in memory.

In the microcomputer system, memory can be divided into two different types: *primary storage memory* and *secondary storage memory.* Secondary storage memory is used for long-term storage of information that is not currently being used. For example, it can hold programs, files of data, and files of information. In the original IBM PC, the *floppy disk drives* represented the secondary storage memory subsystems. It had two 5¼-inch drives that used double-sided, double-density *floppy-diskette* storage media that could each store up to 360Kbytes (360,000 bytes) of data. This floppy diskette is an example of a removable media—that is, to use the diskette it is inserted into the drive and locked in place. If either the diskette is full or one with a different file or program is needed, the diskette is simply unlocked, removed, and another diskette installed.

The IBM PCXT also employed a second type of secondary storage device called a *hard disk drive.* Modern hard disk drive sizes are 5Gbytes (5000 million bytes), 10Gbytes, 20Gbytes, 40Gbytes, 60Gbytes, and 80Gbytes. Earlier we pointed out that the original IBM PCXT was equipped with a hard disk drive that could hold just 10Mbytes (10 million bytes); the hard disk drive of the original IBM PC/AT held only 20Mbytes. The hard disk drive differs from the floppy disk drive in that the media is fixed, which means that the media cannot be removed. However, being fixed is not a problem because the storage capacity of the media is so much larger. Today, desktop PCs are equipped with a hard disk drive in the 20Gbyte to 80Gbyte range.

Both the floppy diskette and hard disk are examples of read/write media—that is, a file of data can be read in from or written out to the storage media in the drive. Another secondary storage device that is very popular in personal computers today is a CD drive. Here a removable *compact disk* (CD) is used as the storage media. This media has very large storage capacity, more than 600Kbytes, but is read-only. This means you cannot write information onto a CD for storage. For this reason, it is normally used to store large programs or files of data that are not to be changed. Recently, a recordable CD (CD-R) has become available, and CD-R drives allow the PC to both read from and write to this media.

Primary storage memory is normally smaller in size and is used for the temporary storage of active information, such as the operating system of the microcomputer, the program that is currently being run, and the data that it is processing. In Fig. 1–4 we see that primary storage memory is further subdivided into *program-storage memory* and *data-storage memory.* The program section of memory is used to store instructions of the operating system and application programs. The data section normally contains data that are to be processed by the programs as they are executed (e.g., text files for a word-processor program or a database for a database-management program). However, programs can also be loaded into data memory for execution.

Typically, primary storage memory is implemented with both *read-only memory* (ROM) and *random-access read/write memory* (RAM) integrated circuits. The original IBM PC had 48Kbytes of ROM and could be configured with 256Kbytes of RAM without adding a memory-expansion board. Modern PC/ATs made with the Pentium IV processors are typically equipped with 128Mbytes of RAM.

Data, whether they represent numbers, characters, or instructions of a program, can be stored in either ROM or RAM. In the original IBM PC, a small part of the operating system and BASIC language are resident to the computer because they are supplied in ROM. By using ROM, this information is made *nonvolatile*—that is, the information is not lost if power is turned off. This type of memory can only be read from; it cannot be written into. On the other hand, data that are to be processed and information that frequently changes must be stored in a type of primary storage memory from which they can be read by the microprocessor, modified through processing, and written back for storage. This requires a type of memory that can be both read from and written into. For this reason, such data are stored in RAM instead of ROM.

Earlier we pointed out that the instructions of a program could also be stored in RAM. In fact, to run the PC *operating system* such as Windows 98®, it must be loaded into the RAM of the microcomputer. Normally the operating system, supplied on CDs, is first read from the CDs and written onto the hard disk. This is called *copying* of the operating system onto the hard disk. Once it is copied, the CD version of Windows 98 may not be used again. The PC is set up so that when it is turned on, Windows 98 is automatically read from the hard disk, written into the RAM, and then run.

RAM is an example of a volatile memory—that is, when power is turned off, the data that it holds are lost. This is why Windows 98 must be reloaded from the hard disk each time the PC is turned on.

▲ 1.3 EVOLUTION OF THE INTEL MICROPROCESSOR ARCHITECTURE

Microprocessors and microcomputers generally are categorized in terms of the maximum number of binary bits in the data they process—that is, their word length. Over time, five standard data widths have evolved for microprocessors and microcomputers: *4-bit, 8-bit, 16-bit, 32-bit,* and *64-bit.*

Figure 1–6 illustrates the evolution of Intel's microprocessors since their introduction in 1972. The first microprocessor, the 4004, was designed to process data arranged as 4-bit words. This organization is also referred to as a *nibble* of data.

The 4004 implemented a very low performance microcomputer by today's standards. This low performance and limited system capability restricted its use to simpler, special-purpose applications. A common use was in electronic calculators.

Beginning in 1974, a second generation of microprocessors was introduced. These devices, the 8008, 8080, and 8085, were 8-bit microprocessors and were designed to process 8-bit (1-byte-wide) data instead of 4-bit data. The 8080, identified in Fig. 1–6, was introduced in 1975.

These newer 8-bit microprocessors were characterized by higher-performance operation, larger system capabilities, and greater ease of programming. They were able to provide the system requirements for many applications that could not be satisfied with the earlier 4-bit microprocessors. These extended capabilities led to widespread acceptance of multichip 8-bit microcomputers for special-purpose system designs. Examples of these dedicated applications are electronic instruments, cash registers, and printers.

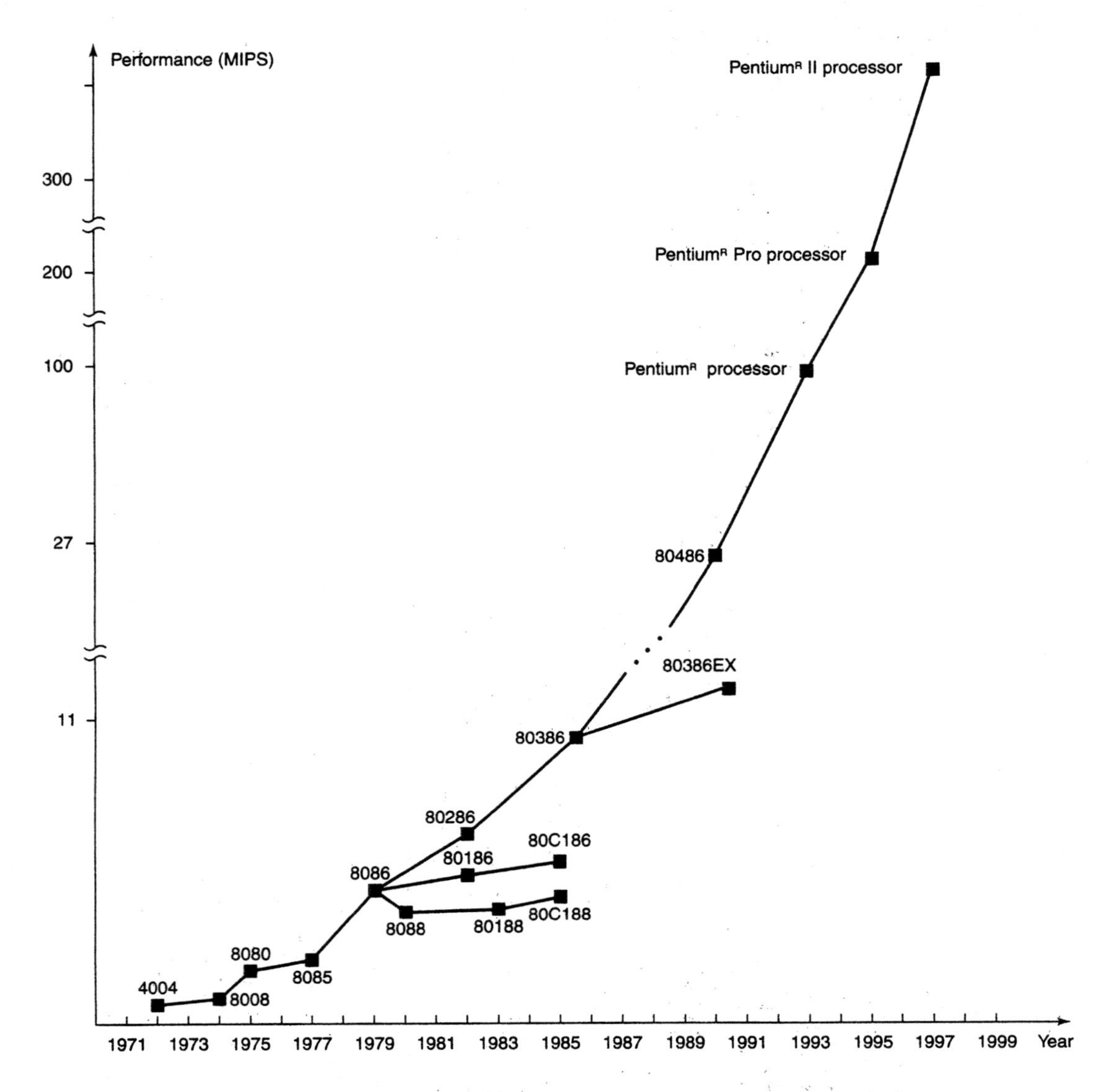

Figure 1–6 Evolution of the Intel microprocessor architecture.

In the mid-1970s, many of the leading semiconductor manufacturers announced plans for development of third-generation 16-bit microprocessors. Looking at Fig. 1–6, we see that Intel's first 16-bit microprocessor, the 8086, became available in 1979 and was followed the next year by its 8-bit bus version, the 8088. This was the birth of Intel's 8086 family architecture. Other family members, such as the 80286, 80186, and 80188, were introduced in the years that followed.

These 16-bit microprocessors provided higher performance and had the ability to satisfy a broad scope of special-purpose and general-purpose microcomputer applications.

They all have the ability to handle 8-bit, 16-bit, and special-purpose data types, and their powerful instruction sets are more in line with those provided by a minicomputer.

In 1985, Intel Corporation introduced its first 32-bit microprocessor, the 80386DX, which brought true minicomputer-level performance to the microcomputer system. This device was followed by a 16-bit external bus version, the 80386SX, in 1988. Intel's second generation of 32-bit microprocessors, called the 80486DX and 80486SX, became available in 1989 and 1990, respectively. They were followed by a yet higher-performance family, the Pentium processors, in 1993. Today, its fourth generation member—the Pentium® IV processor, represents this family.

Microprocessor Performance: MIPS and iCOMP

Figure 1–6 illustrates the 8086 microprocessor families relative to their performance. Here performance is measured in what are called *MIPS*—that is, how many million instructions they can execute per second. Today, the number of MIPS provided by a microprocessor is the standard most frequently used to compare performance. Notice that performance has vastly increased with each new generation of microprocessor. For instance, the performance identified for the 80386 corresponds to an 80386DX device operating at 33 MHz and equals approximately 11 MIPS. With the introduction of the 80486, the level of performance capability of the architecture was raised to approximately 27 MIPS. This shows that performance of the 8086 architecture was more than doubled with the introduction of the 33-MHz 80486DX microprocessor.

The MIPS used in this chart are known as Drystone V1.1 MIPS—that is, they are measured by running a test program called the *Drystone program,* and the resulting performance measurements are normalized to those of a VAX 1.1 computer (VAX 1.1 was a minicomputer manufactured by Digital Equipment Corporation). Therefore, we say that the 80486DX is capable of delivering up to 27 VAX MIPS of performance.

Intel Corporation provides another method, the iCOMP index, for comparison of the performance of its 32-bit microprocessors in a personal computer environment. In the iCOMP index chart shown in Fig. 1–7, a bar is used to represent a measure of the performance for each of Intel's MPUs. Instead of being related to the performance of a test program, such as the Dystone program, the iCOMP rating of an MPU is based on a variety of 16-bit and 32-bit MPU performance components important to the personal computer—that is, the iCOMP rating encompasses performance components that represent integer mathematics, floating-point mathematics, graphics, and video. The contribution by each of these categories is also weighted based on an estimate of their normal occurrence in widely used software applications. For this reason, iCOMP is a more broad-based rating of MPU performance for the personal computer applications.

The higher the iCOMP rating, the higher the performance offered by the MPU. Notice that the members of the 80386 family offer low performance when compared to the newer 80486 and Pentium processor families. In fact, the slowest 80386SX MPU shown in Fig. 1–7, the -20, has a performance rating of 32, whereas the fastest 80386DX, the -33, is rated at 68. Therefore, a wide range of system-performance levels can be achieved by selecting among the various members of the 80386, 80486, and Pentium processor families.

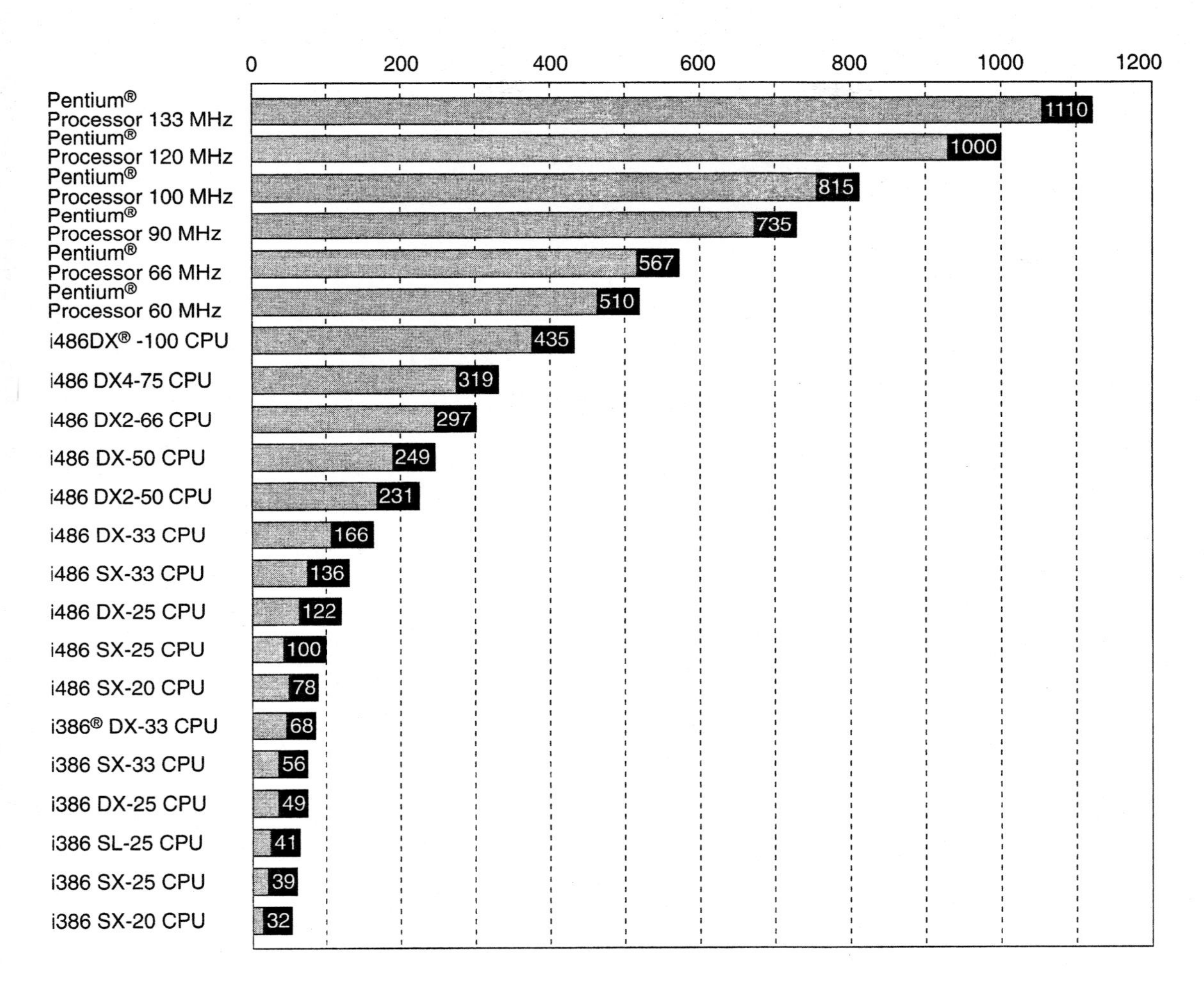

Figure 1–7 iCOMP index rating chart. (Reprinted by permission of Intel Corp. Copyright/Intel Corp. 1993.)

Transistor Density

The evolution of microprocessors is made possible by advances in semiconductor process technology. Semiconductor-device geometry decreased from about 5 microns in the early 1970s to submicron today. Smaller-device geometry permits integration of several orders of magnitude more transistors into the same-size chip and at the same time has led to higher operating speeds. Figure 1–8 shows that the 4004 contained about 10,000 transistors. Transistor density increased to about 30,000 with the development of the 8086 in 1978. With the introduction of the 80286, the transistor count further increased to approximately 140,000, and to 275,000 transistors in the 80386DX, almost doubling the transistor density. The 80486DX is the first family member with a density above the 1 million transistor level (1,200,000 transistors), and with the Pentium processor's complexity, density has risen to more than 3 million transistors.

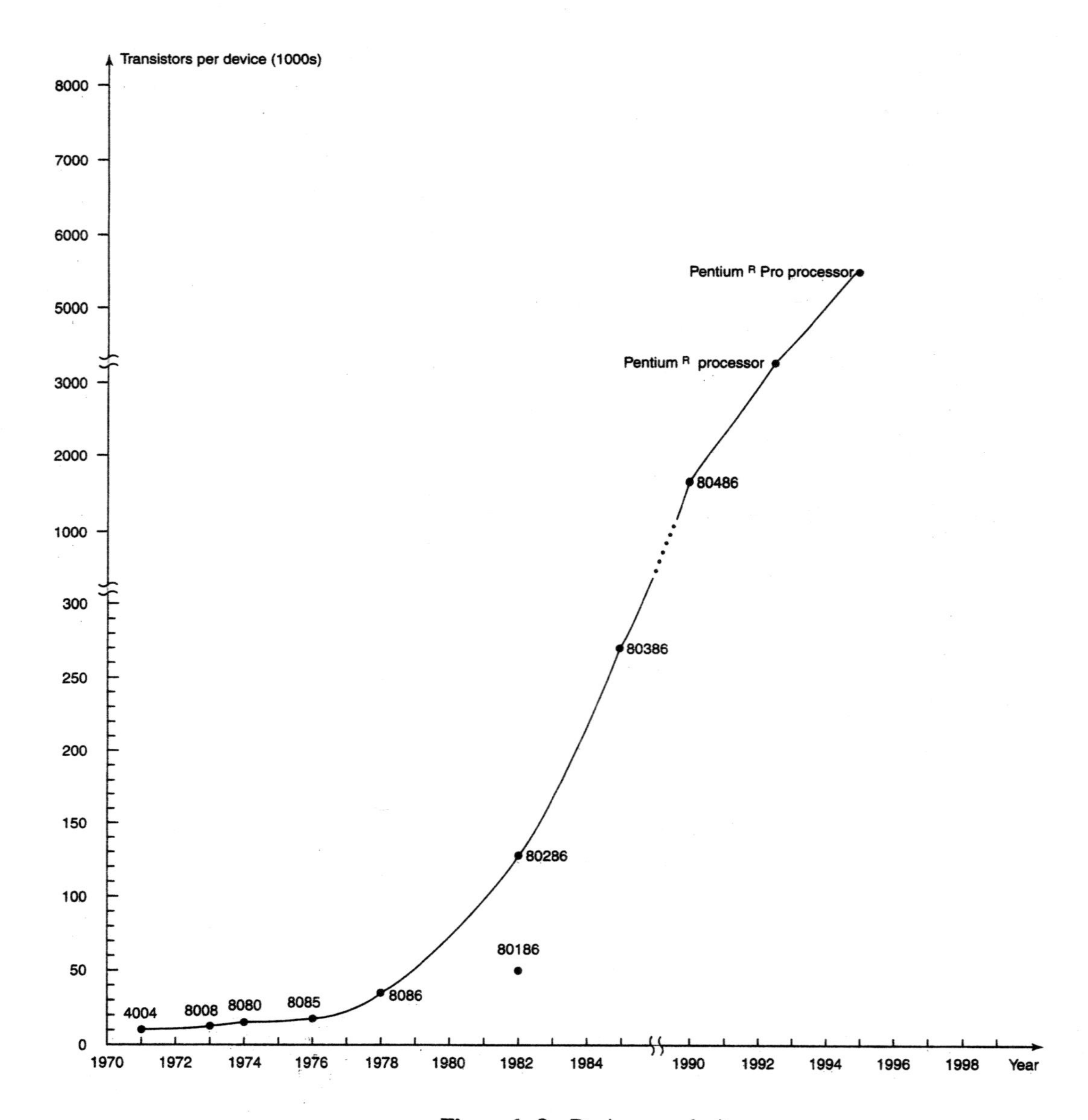

Figure 1–8 Device complexity.

Reprogrammable and Embedded Microprocessors

Microprocessors can be classified according to the type of application for which they have been designed. Figure 1–9 organizes Intel microprocessors into two application-oriented categories: *reprogrammable microprocessors* and *embedded microprocessors and microcontrollers.* Initially devices such as the 8080 were most widely used as *special-purpose microcomputers*—a system that has been tailored to meet the needs of a

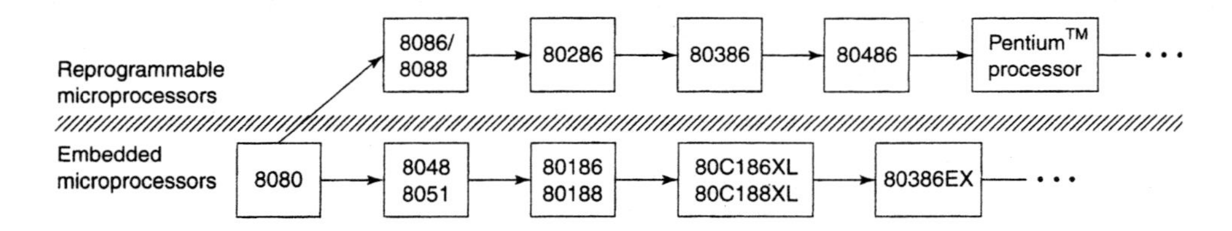

Figure 1–9 Processors for embedded control and reprogrammable applications.

specific application. These special-purpose microcomputers were used in *embedded control applications*—applications in which the microcomputer performs a dedicated control function.

Embedded control applications are further divided into those that involve primarily *event control* and those that require *data control.* An example of an embedded control application that is primarily event control is a microcomputer used for industrial process control. Here the program of the microprocessor is used to initiate a timed sequence of events. On the other hand, an application that focuses more on data control than event control is a hard disk controller interface. In this case, a block of data that is to be processed—for example, a file of data—must be quickly transferred from secondary storage memory to primary storage memory.

The spectrum of embedded control applications requires a variety of system features and performance levels. Devices developed specifically for the needs of this marketplace have stressed low cost and high integration. In Fig. 1–9, we see that highly integrated 8-bit, single-chip microcomputer devices such as the 8048 and 8051 initially replaced the earlier multichip 8080 solutions. These devices were tailored to work best as event controllers. For instance, the 8051 offers one-order-of-magnitude-higher performance than the 8080, a more powerful instruction set, and special on-chip functions such as ROM, RAM, an interval/event timer, a universal asynchronous receiver/transmitter (UART), and programmable parallel input/output ports. Today, this type of embedded control device is called a *microcontroller.*

Later, devices such as the 80C186XL, 80C188XL, and 80386EX were designed to better meet the needs of data-control applications. They are also highly integrated, but they have additional features, such as string instructions and direct-memory access channels, which better handle the movement of data. They are known as *embedded microprocessors.*

The category of reprogrammable microprocessors represents the class of applications in which a microprocessor is used to implement a *general-purpose microcomputer.* Unlike a special-purpose microcomputer, a general-purpose microcomputer is intended to run a variety of software applications—that is, while it is in use it can be easily reprogrammed to run a different application. Two examples of reprogrammable microcomputers are the personal computer and file server. Figure 1–9 shows that the 8086, 8088, 80286, 80386, 80486, and Pentium processor are the Intel microprocessors intended for use in this type of application.

Architectural compatibility is a critical need of microprocessors developed for use in reprogrammable applications. As shown in Fig. 1–10, each new member of the

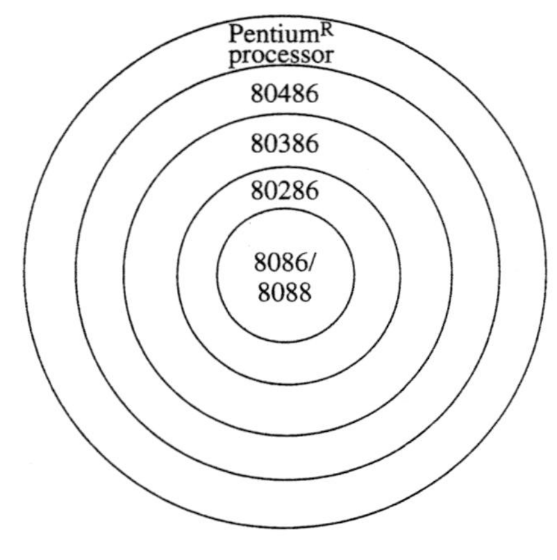

Figure 1–10 Code and system-level compatibility.

8086/8088 family provides a superset of the earlier device's architecture—that is, the features offered by the 80386 microprocessor are a superset of the 80286 architecture, and those of the 80286 are a superset of the original 8086/8088 architecture.

Actually, the 80286, 80386, 80486, and Pentium processors can operate in either of two modes: the *real-address mode* or *protected-address mode.* When in real mode, they operate like a high-performance 8086/8088. They can execute what is called the *base instruction set,* which is object code compatible with the 8086/8088. For this reason, operating systems and application programs written for the 8086 and 8088 run on the 80286, 80386, 80486, and Pentium processor architectures without modification. Further, a number of new instructions have been added in the instruction sets of the 80286, 80386, 80486, and Pentium processors to enhance their performance and functionality. We say that object code is *upward compatible* within the 8086 architecture. This means that 8086/8088 code will run on the 80286, 80386, 80486, and Pentium processors, but the reverse is not true if any of the new instructions are in use.

Microprocessors designed for implementing general-purpose microcomputers must offer more advanced system features than those of a microcontroller; for example, they need to support and manage a large memory subsystem. The 80286 is capable of managing a 1Gbyte (gigabyte) address space, and the 80386 supports 64Tbytes (64 terabytes) of memory. Moreover, a reprogrammable microcomputer, such as a personal computer, normally runs an operating system. The architectures of the 80286, 80386, 80486, and Pentium processors have been enhanced with on-chip support for operating system functions such as *memory management, protection,* and *multitasking.* These new features become active only when the MPU is operated in the protected mode. The 80386, 80486, and Pentium processors also have a special mode of operation known as *virtual 8086 mode* that permits 8086/8088 code to be run in the protected mode.

Reprogrammable microcomputers, such as those based on the 8086 family, require a variety of input/output resources. Figure 1–11 shows the kinds of interfaces that are typically implemented in a personal computer or a microcomputer system. A large family of VLSI peripheral ICs (examples are floppy disk controllers, hard disk controllers, local area network controllers, and communication controllers) is needed to support a reprogrammable microprocessor such as the 8086, 80286, 80386, 80486, and Pentium processor.

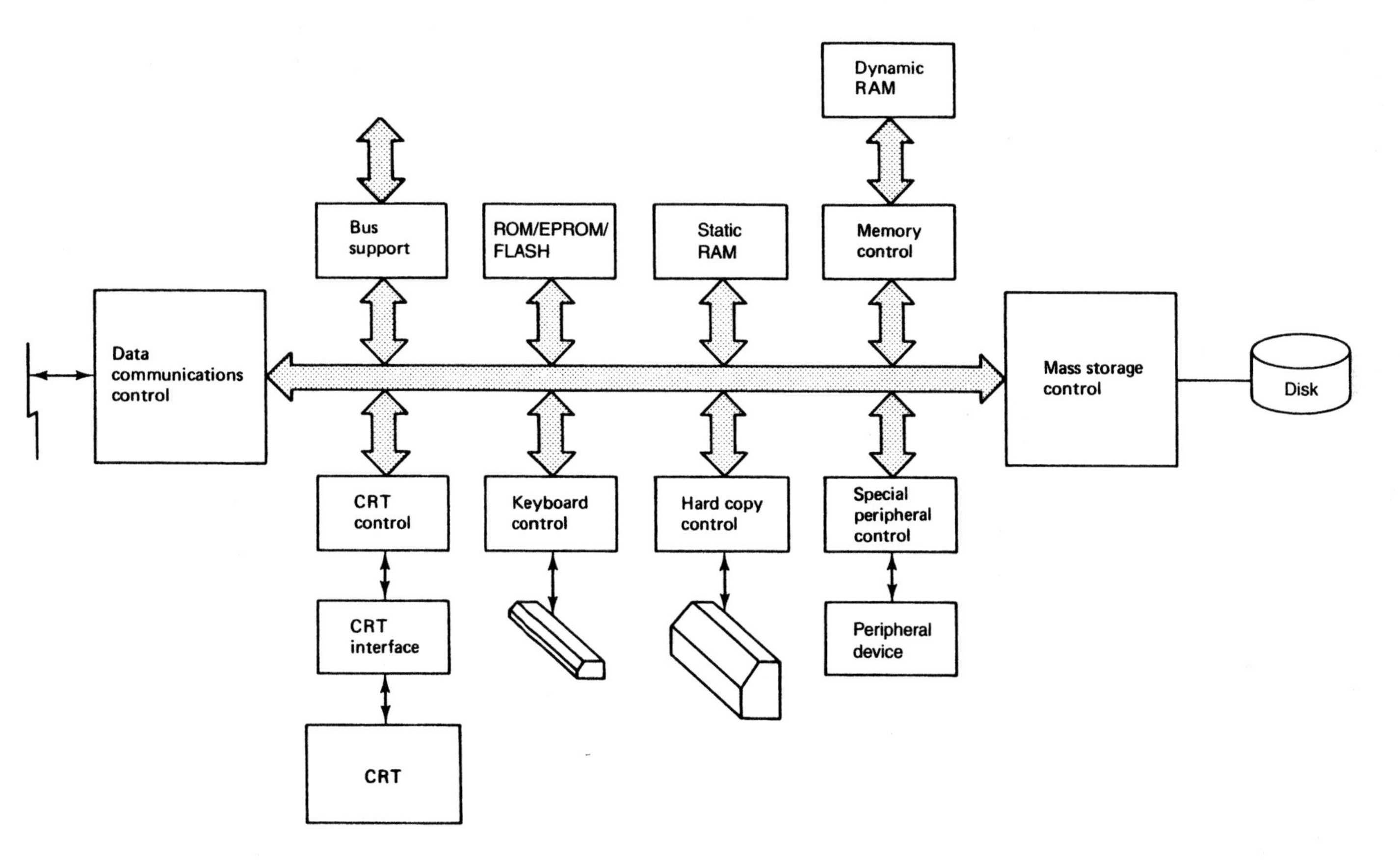

Figure 1–11 Peripheral support for the MPU.

For this reason, these processors are designed to implement a multichip microcomputer system, which can easily be configured with the appropriate set of input/output interfaces.

▲ 1.4 NUMBER SYSTEMS

For microprocessors and microcomputers, information such as instructions, data, and addresses are described with numbers. The types of numbers used are not normally the decimal numbers we are familiar with; instead, binary and hexadecimal numbers are used. We must understand how numbers are expressed in these number systems and how to convert values between them. For this reason, we shall review some basic material on number systems.

Decimal Number System

First we shall use decimal numbers to develop the general characteristics of a *number system.* Selecting a set of symbols to represent numerical values forms a number system. When doing this, we can select any group of symbols. The number of symbols used is called the *base* or *radix* of the number system.

For example, let us look at the decimal number system. Symbols 0 through 9 make up the decimal number system. These symbols are shown in Fig. 1–12(a). Here we find that 10 different symbols are used, so the base of the decimal number system is 10. Each of these symbols indicates a different numerical quantity—0 representing the smallest quantity and 9 the largest quantity.

With just the 10 basic symbols of the decimal number system, we cannot form every quantity needed in mathematics and science. For this reason, *digit notation* is used. An example of a decimal number written in digit notation is

$$735.23$$

Here the same basic symbols are used to form a larger multidigit number. This number has symbols entered into five different digit locations.

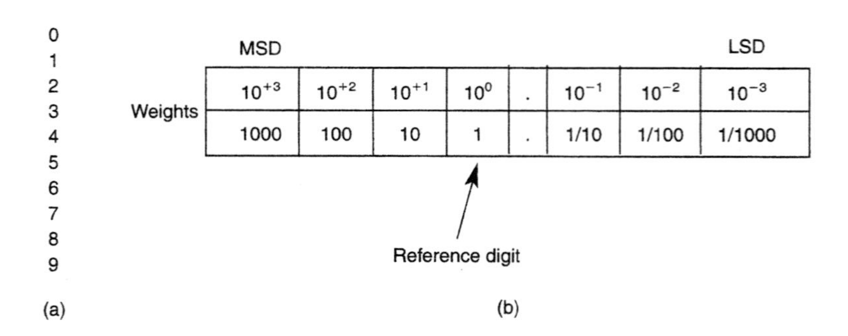

Figure 1–12 (a) Decimal number system symbols. (b) Digit notation and weights.

When digit notation is in use, the value of a symbol depends on its location in the number. The positional value of a digit is known as its *weight.* For instance, in the number 735.23 the symbol 3 occurs in two locations. Because the weights of the digits in which 3 lies are different, each takes on a different positional value.

Figure 1–12(b) shows some digit locations of the decimal number system and the corresponding weights. Note that the digit just to the left of the decimal point, the *units* digit, is used as the reference digit and its weight is 10^0 or 1.

Looking at the decimal weights, we find raising the base of the number system to a power forms them. The power to which the base 10 is raised is the + or − exponent. The value of this exponent is found by counting the number of digits to the units location. All digits to the left of the units digit are considered to have a weight with a positive exponent of the power of 10. As an example, let us look at the second digit to the left of the units digit in Fig. 1–12(b). This location has a weight of 10^{+2}, or multiplying out, we get 100. For this reason, it is called the *hundreds* digit.

Digits to the right of the reference digit have a weight with a negative exponent. For example, let us take the second digit to the right of the units location in Fig. 1–12(b). This digit has a weight of 10^{-2} or 1/100. This location is also known as the *one-hundredths* digit.

Having introduced the weight of a digit, let us now look at how it affects the value of a symbol in that location. The value of a symbol in a digit other than the units digit is found by multiplying the symbol by the weight of the location. In our earlier example, 735.23, the symbol 7 is in the hundreds digit. Therefore, it represents the quantity $7 \times 10^{+2}$ or 7(100) equals 700 instead of just 7. Likewise, the 3 in the one-hundredths digit stands for 3×10^{-2} and has a value of 3/100.

Two other terms that relate to numbers and number systems are *most significant digit* and *least significant digit.* The leftmost symbol in a number is located in the most significant digit position. This location is indicated with the abbreviation MSD. On the other hand, the symbol in the rightmost digit position is said to be in the least significant digit location or LSD.

In the number we have been using as an example, the symbol in the MSD location is 7 and its weight is 10^{+2}. Moreover, the LSD has a weight of 10^{-2} and the symbol in this location is 3.

Binary Number System

The digital electronic devices and circuits in a microcomputer system operate only in one of two states, on or off. For this reason, binary numbers instead of decimal numbers are used to describe their operation. The base of the binary number system is 2, and Fig. 1–13(a) shows that just two symbols, 0 and 1, are used to form all numbers. From an electronic circuit point of view, binary 1 normally represent a circuit input or output that is turned on. On the other hand, a 0 represents the same input or output when it is turned off.

To make a large binary number, many 0s and 1s are grouped together. The location of a symbol in a binary number is called a *bit*—a contraction for "binary digit." As an example, let us take the binary number

$$1101.001_2$$

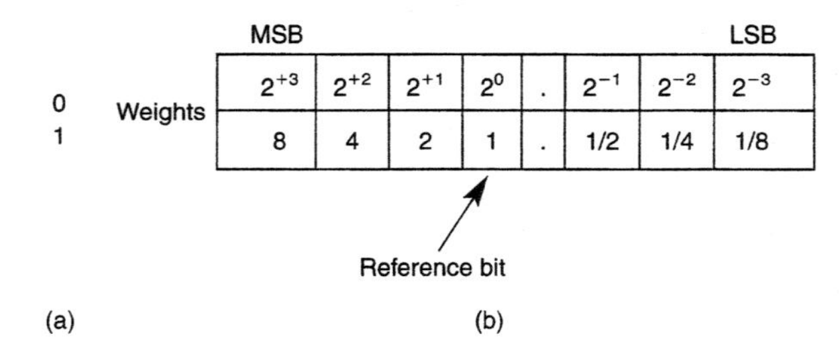

Figure 1–13 (a) Binary number system symbols. (b) Bit notation and weights.

Looking at this number, we find it has 7 bits. The number 2 written to the right and slightly below tells it is a base 2 or binary number. As with decimal numbers, each bit location has a weight, but binary weights are expressed as base 2 raised to an exponent. In Fig. 1–13(b), the weights corresponding to some binary bits are shown. Note that the weight of the reference bit is 2^0 equals 1.

Bits to the left of the 2^0 bit have weights with positive exponents, and those to the right negative exponents. For example, the weight of the third digit left of the point is 2^{+2} equals 4. If a 1 occurs in this bit location, like in the number 1101.001, it stands for a decimal value of 4.

In a binary number, the leftmost bit is referred to as the *most significant bit* (MSB) and the rightmost bit the *least significant bit* (LSB). In the number 1101.001, the MSB has a value of 1 and weight 2^{+3} equals 8. On the other hand, the least significant bit is 1 with a weight of 2^{-3} or ⅛.

Conversion between Decimal and Binary Numbers

All numbers can be expressed in both the decimal and binary number systems. Figure 1–14 lists the decimal numbers 0 through 15 along with their equivalent binary numbers. Just as one can count consecutive numbers, binary numbers can be counted as 0, 1, 10, 11, and so on. From this list, we find that the binary equivalent of decimal num-

Decimal number	Binary number
0	0
1	1
2	10
3	11
4	100
5	101
6	110
7	111
8	1000
9	1001
10	1010
11	1011
12	1100
13	1101
14	1110
15	1111

Figure 1–14 Equivalent decimal and binary numbers.

ber 0 is just binary 0. On the other hand, decimal 15 is written in binary form as 1111_2. As shown, binary numbers are often written without showing the base.

For the analysis of circuit operation, information organization, and programming digital equipment such as a microcomputer, it is important to be able to quickly convert between decimal and binary number forms. In microcomputer systems, binary numbers are normally treated in fixed lengths. For example, the 8088 has the ability to process either 8-bit-wide or 16-bit-wide data. These widths are known as a byte of data and a word of data, respectively. For instance, binary zero expressed as a byte is

$$\text{Byte-wide binary } 0 = 00000000_2$$

Address information in the 8088-based system is 20 bits long.

To find the decimal equivalent of a binary number, multiply the value in each bit of the number by the weight of the corresponding bit. After this, add the products to get the decimal number. As an example, let us find the decimal number for binary 1100_2. Multiplying bit values and weights gives

$$\begin{aligned} 1100_2 &= 1(2^{+3}) + 1(2^{+2}) + 0(2^{+1}) + 0(2^0) \\ &= 1(8) + 1(4) + 0(2) + 0(1) \\ &= 12_{10} \end{aligned}$$

This shows that 1100_2 is the binary equivalent of decimal number 12. Looking in the table of Fig. 1–14, we see that our result is correct. Expressing 12_{10} as a word-wide binary number gives

$$12_{10} = 0000000000001100_2$$

EXAMPLE 1.1

Evaluate the decimal equivalent of binary number 101.01_2.

Solution

$$\begin{aligned} 101.01_2 &= 1(2^{+2}) + 0(2^{+1}) + 1(2^{+0}) + 0(2^{-1}) + 1(2^{-2}) \\ &= 1(4) + 0(2) + 1(1) + 0(1/2) + 1(1/4) \\ &= 4_{10} + 1_{10} + 0.25_{10} \\ 101.01_2 &= 5.25_{10} \end{aligned}$$

The other conversion we must be able to perform is to express a decimal number in binary form. The binary equivalent of the integer part of a decimal number is formed by the repeated division method. Using this method, the integer decimal number is divided by 2, the quotient brought down, and the remainder written to the right. This procedure is repeated until the quotient is zero. Now, we use the remainders to form the binary number. The least significant bit of the binary number is the remainder that results from the first division or original number. Each of the remainders that follow gives the bits up to the last remainder, which gives the most significant bit. To demonstrate this method, let us convert the decimal number 12_{10} back into binary form:

2	12			
2	6	→	0	LSB
2	3	→	1	
2	1			
	0	→	1	MSB

$$12_{10} = 1100_2$$

Here we see that dividing the original number, 12, by 2 gives a quotient of 6 with 0 remainder. The quotient is brought down and the remainder written on the right. This is the LSB of the binary number for 12. Now 6 is divided once again by 2 to give a quotient of 3 and a remainder of 0. Dividing twice more, we end up with a quotient of 0 and two more remainders that are both 1. The last remainder is the MSB of the binary answer.

At this point all remainders are known; next we must form a binary number for 12_{10}. To do this, the remainders are used in the reverse order. Starting with the MSB remainder and working back toward the LSB gives 1100_2.

EXAMPLE 1.2

Convert decimal number 31_{10} to binary form. Also, express the answer as a byte-wide binary number.

Solution

2	31	→	1	LSB
2	15	→	1	
2	7	→	1	
2	3	→	1	
2	1	→	1	MSB
	0			

$$31_{10} = 11111_2$$

Expressing as a byte-wide value by filling unused more significant bits with 0s gives

$$31_{10} = 00011111_2$$

The binary equivalent of the decimal fraction part of a decimal number is found by a repeated multiplication method. Applying this method, the decimal fraction is multiplied by 2, and the integer part of that product is brought out as the binary bit. This first multiplication gives the most significant bit of the equivalent binary number. This process is repeated on the decimal fraction part of the new product until the fractional result becomes zero. Successive multiplications produce the less significant bits of the binary number. As an example, let us convert decimal fraction .8125 to its equivalent binary number:

$$2 \times .8125 \rightarrow 1 \quad \text{MSB}$$
$$2 \times .625$$
$$2 \times .25 \quad \rightarrow 1$$
$$2 \times .5 \quad \rightarrow 0$$
$$2 \times 0$$
$$.8125 = .1101_2$$

Hexadecimal Number System

The *hexadecimal number system* is important for describing microcomputer operation and programming. In fact, machine language programs, addresses, and data, which are actually binary information, are normally expressed as hexadecimal numbers. Hexadecimal numbers offer a more compact notation for representing this type of information.

The base of the hexadecimal number system is 16, and it uses numerical symbols 0 through 9 followed by letters A through F to form numbers. Letters A through F stand for numerical values equivalent to decimal numbers 10 through 15, respectively. These symbols are listed in Fig. 1–15(a).

To make a useful number, the basic hexadecimal symbols must be written in digit notation. Here the weights of the separate digits are the base 16 raised to a power. Figure 1–15(b) gives typical hexadecimal weights. Note that the weight of the reference digit is 16^0 equals 1. On the other hand, the most significant digit and least significant digit locations shown have weights of 16^{+3} and 16^{-2}, respectively. Rewriting these weights in decimal form, we get 4096 for the MSD and 1/256 for the LSD. The method used to convert directly between decimal and hexadecimal number forms is similar to that for converting binary numbers. Though it is not that commonly done in the study and analysis of microcomputer systems, this process is demonstrated here through examples.

0
1
2
3
4
5
6
7
8
9
A
B
C
D
E
F

(a)

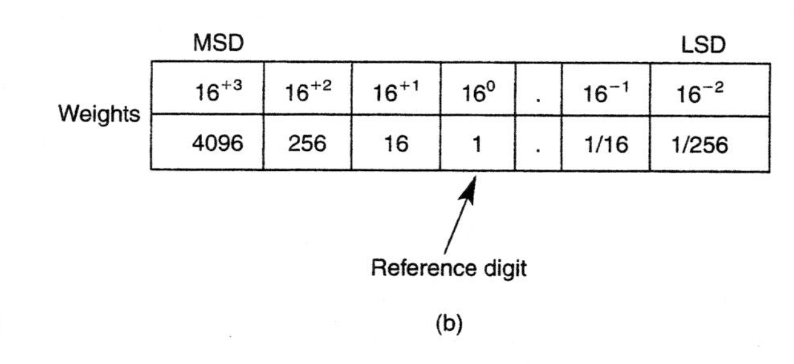

Figure 1–15 (a) Hexadecimal number system symbols. (b) Digit notation and weights.

EXAMPLE 1.3

What decimal number does $102A_{16}$ represent?

Solution

$$
\begin{aligned}
102A_{16} &= 1(16^{+3}) + 0(16^{+2}) + 2(16^{+1}) + A(16^{0}) \\
&= 1(4096) + 0(256) + 2(16) + A(1) \\
&= 4096 + 32 + 10 \\
&= 4138_{10}
\end{aligned}
$$

EXAMPLE 1.4

Convert decimal number 4138_{10} to hexadecimal form.

Solution

16	4138		
16	258 →	A	LSB
16	16 →	2	
16	1 →	0	
	0 →	1	MSB

$$4138_{10} = 102A_{16}$$

Conversion between Hexadecimal and Binary Numbers

Earlier we indicated that the importance of using hexadecimal numbers in the study of microcomputer-based systems is that they can be used to rewrite information, such as data or instructions, in a very compact way. For instance, a multibit binary number can be expressed with just a few hexadecimal digits. For this reason, it is most important to learn how to directly convert between the binary and hexadecimal number forms.

Figure 1–16(a) lists all 4-bit binary numbers and their equivalent hexadecimal number. Here we see that a 4-bit binary zero is the same as a one-digit hexadecimal zero. Moreover, each binary number that follows up through 1111_2 is the same as one of the hexadecimal numbers from 1_{16} through F_{16}. In this way, we find that 4 binary bits give a single hexadecimal digit. This fact is the basis for converting between binary and hexadecimal numbers.

The diagram in Figure 1–16(b) shows how bits of a binary number are grouped to make digits of a hexadecimal number. From this illustration, we see that the four least significant bits $2^3 2^2 2^1 2^0$ of the word-wide binary number give the least significant hexadecimal digit, 16^0. This is followed by three more groups of 4 bits for the 16^1, 16^2, and 16^3 digits. The MSD 16^3 of the hexadecimal number is formed from the four MSBs, 2^{15} through 2^{12}, of the binary number. In this way, a 16-bit binary number is rewritten with just four hexadecimal digits.

Binary number	Hexadecimal number
0000	0
0001	1
0010	2
0011	3
0100	4
0101	5
0110	6
0111	7
1000	8
1001	9
1010	A
1011	B
1100	C
1101	D
1110	E
1111	F

(a)

MSB			LSB	
$2^{15}2^{14}2^{13}2^{12}$	$2^{11}2^{10}2^{9}2^{8}$	$2^{7}2^{6}2^{5}2^{4}$	$2^{3}2^{2}2^{1}2^{0}$	Bits
16^3	16^2	16^1	16^0	Digits
MSD			LSD	

(b)

Figure 1–16 (a) Equivalent binary and hexadecimal numbers. (b) Binary bits and hexadecimal digits.

The first conversion we shall perform is to rewrite a binary number in hexadecimal notation. To do this, we start at the rightmost bit of the binary number and separate into groups each with 4 bits. After this, we replace each group of bits with its equivalent hexadecimal number.

EXAMPLE 1.5

Express the binary number 1111100100001010_2 as a hexadecimal number.

Solution

$$1111100100001010_2 = 1111\ 1001\ 0000\ 1010$$
$$= \text{F} \quad 9 \quad 0 \quad \text{A}$$
$$1111100100001010_2 = \text{F90A}_{16}$$

An H is frequently used instead of a subscript 16 to denote that a value is a hexadecimal number. For instance, using this notation the solution to Example 1.5 could also be written as F90AH. The byte-wide equivalent binary and hexadecimal numbers for decimal numbers 0 through 15 are listed in Fig. 1–17.

To write a hexadecimal number in binary form, the method just used must be reversed. That is, the value in each hexadecimal digit is simply replaced by its equivalent 4-bit number. For example, the two-digit hexadecimal number A5H is converted to an equivalent byte-wide binary number as follows:

$$\text{A5H} = 1010\ 0101 = 10100101_2$$

Decimal number	Binary number	Hexadecimal number
0	00000000	00
1	00000001	01
2	00000010	02
3	00000011	03
4	00000100	04
5	00000101	05
6	00000110	06
7	00000111	07
8	00001000	08
9	00001001	09
10	00001010	0A
11	00001011	0B
12	00001100	0C
13	00001101	0D
14	00001110	0E
15	00001111	0F

Figure 1–17 Equivalent decimal, byte-wide binary, and hexadecimal numbers.

EXAMPLE 1.6

What is the binary equivalent of the number $C315_{16}$?

Solution

$$C315_{16} = 1100\ 0011\ 0001\ 0101 = 1100001100010101_2$$

REVIEW PROBLEMS

Section 1.1

1. Which IBM personal computer employs the 8088 microprocessor?
2. What is meant by the term *open system?*
3. What is the expansion bus of the original IBM PC called?
4. What does PC/AT stand for?
5. What does ISA stand for?
6. Name the bus that replaced the ISA bus in modern PCs?
7. What is a reprogrammable microcomputer?
8. Name the three classes of computers.
9. What are the main similarities and differences between the minicomputer and the microcomputer?
10. What does VLSI stand for?

Section 1.2

11. What are the four building blocks of a microcomputer system?
12. What is the heart of the microcomputer system called?
13. Is the 8088 an 8-bit or a 16-bit microprocessor?

14. What is the primary input unit of the PC? Give two other examples of input units available for the PC.
15. What are the primary output devices of the PC?
16. Into what two sections is the memory of a PC partitioned?
17. What is the storage capacity of the standard 5¼-inch floppy diskette of the original PC? What is the storage capacity of the standard hard disk drive of the original PCXT?
18. What do ROM and RAM stand for?
19. How much ROM was provided in the original PC's processor board? What was the maximum amount of RAM that could be implemented on this processor board?
20. Why must Windows 98 be reloaded from the hard disk each time power is turned on?

Section 1.3

21. What are the standard data word lengths for which microprocessors have been developed?
22. What was the first 4-bit microprocessor introduced by Intel Corporation? Eight-bit microprocessor? Sixteen-bit microprocessor? Thirty-two-bit microprocessor?
23. Name five 16-bit members of the 8086 family architecture.
24. What does MIPS stand for?
25. Approximately how many MIPS are delivered by the 33 MHz 80486DX?
26. What is the name of the program that is used to run the MIPS measurement test for the data in Fig. 1–6?
27. What is the iCOMP rating of an 80386SX-25 MPU? An 80386DX-25 MPU?
28. Approximately how many transistors are used to implement the 8088 microprocessor? The 80286 microprocessor? The 80386DX microprocessor? The 80486DX microprocessor? The Pentium processor?
29. What is an embedded microcontroller?
30. Name the two groups into which embedded processors are categorized based on applications.
31. What is the difference between a multichip microcomputer and a single-chip microcomputer?
32. Name six 8086 family microprocessors intended for use in reprogrammable microcomputer applications.
33. Give the names for the 80386's two modes of operation.
34. What is meant by upward software compatibility relative to 8086 architecture microprocessors?
35. List three advanced architectural features provided by the 80386DX microprocessor.
36. Give three types of VLSI peripheral support devices needed in a reprogrammable microcomputer system.

Section 1.4

37. How are most significant bit and least significant bit abbreviated?
38. What is the weight of the second bit to the right of the binary point?
39. Find the symbols and weights of the most significant bit and least significant bit in the number 100111.0101_2.

40. Evaluate the decimal equivalent for each of the following integer binary numbers.
   (a) 00000110_2
   (b) 00010101_2
   (c) 01111111_2

41. What is the decimal equivalent of the minimum and maximum byte-wide integer binary numbers?

42. Convert the integer decimal numbers that follow to their equivalent byte-wide binary form.
   (a) 9
   (b) 42
   (c) 100

43. Find the word-wide binary equivalent of the decimal number 500.

44. Convert the fractional numbers that follow to their equivalent byte-wide binary form.
   (a) .5
   (b) ¼
   (c) .34375

45. Find the symbol and weight of the MSD in the hexadecimal number C8BH.

46. What is the weight of the fourth hexadecimal digit to the left of the point?

47. Convert the binary numbers that follow to hexadecimal form.
   (a) 00111001_2
   (b) 11100010_2
   (c) 0000001110100000_2

48. Evaluate the binary equivalent of each of the hexadecimal numbers that follow.
   (a) $6B_{16}$
   (b) $F3_{16}$
   (c) 02B0H

49. A byte of data read from memory in an 8088-based microcomputer is observed with an instrument to be 11000110_2. Express the data as a hexadecimal number. What is the decimal value of the data?

50. A memory display command is used to display the value of data held at a memory address in a microcomputer system. If the value is A050H, what is the value of the most significant bit? What is the value of the LSB?

51. The address output on the address bus of an 8088 microprocessor is 10000000000001011010_2. Express the address in hexadecimal form. What is its equivalent decimal value?

Software Architecture of the 8088 and 8086 Microprocessors

▲ INTRODUCTION

This chapter begins our study of the 8088 and 8086 microprocessors and their assembly language programming. To program either the 8088 or 8086 using assembly language, we must understand how the microprocessor and its memory and input/output subsystems operate from a software point of view. For this reason, in this chapter, we will examine the *software architecture* of the 8088 and 8086 microprocessors. The material that follows frequently refers only to the 8088 microprocessor, but everything that is described for the 8088 also applies to the 8086. This is because the software architecture of the 8086 is identical to that of the 8088. The following topics are covered here:

2.1 Microarchitecture of the 8088/8086 Microprocessor
2.2 Software Model of the 8088/8086 Microprocessor
2.3 Memory Address Space and Data Organization
2.4 Data Types
2.5 Segment Registers and Memory Segmentation
2.6 Dedicated, Reserved, and General-Use Memory
2.7 Instruction Pointer
2.8 Data Registers
2.9 Pointer and Index Registers
2.10 Status Register

▲ 2.1 MICROARCHITECTURE OF THE 8088/8086 MICROPROCESSOR

The microarchitecture of a processor is its internal architecture—that is, the circuit building blocks that implement the software and hardware architectures of the 8088/8086 microprocessors. Due to the need for additional features and higher performance, the microarchitecture of a microprocessor family evolves over time. In fact, a new microarchitecture is introduced for Intel's 8086 family every few years. Each new generation of processors (the 8088/8086, 80286, 80386, 80846, and Pentium processors) represents significant changes in the microarchitecture of the 8086.

The microarchitectures of the 8088 and 8086 microprocessors are similar. They both employ *parallel processing*—that is, they are implemented with several simultaneously operating processing units. Figure 2–1(a) illustrates the internal architecture of the 8088 and 8086 microprocessors. They contain two processing units: the *bus interface unit* (BIU) and the *execution unit* (EU). Each unit has dedicated functions and both operate at the same time. In essence, this parallel processing effectively makes the fetch and execution of instructions independent operations. This results in efficient use of the system bus and higher performance for 8088/8086 microcomputer systems.

The bus interface unit is the 8088/8086's connection to the outside world. By interface, we mean the path by which it connects to external devices. The BIU is responsible for performing all external bus operations, such as instruction fetching, reading and writing of data operands for memory, and inputting or outputting data for input/output peripherals. These information transfers take place over the system bus. This bus includes an 8-bit bidirectional data bus for the 8088 (16 bits for the 8086), a 20-bit address bus, and the signals needed to control transfers over the bus. The BIU is not only responsible for performing bus operations, it also performs other functions related to instruction and data acquisition. For instance, it is responsible for instruction queuing and address generation.

To implement these functions, the BIU contains the segment registers, the instruction pointer, the address generation adder, bus control logic, and an instruction queue. Figure 2–1(b) shows the bus interface unit of the 8088/8086 in more detail. The BIU uses a mechanism known as an *instruction queue* to implement a pipelined architecture. This queue permits the 8088 to prefetch up to 4 bytes (6 bytes for the 8086) of instruction code. Whenever the queue is not full—that is, it has room for at least 2 more bytes, and, at the same time, the execution unit is not asking it to read or write data from memory—the BIU is free to look ahead in the program by prefetching the next sequential instructions. Prefetched instructions are held in the first-in first-out (FIFO) queue. Whenever a byte is loaded at the input end of the queue, it is automatically shifted up through the FIFO to the empty location nearest the output. Here the code is held until the execution unit is ready to accept it. Since instructions are normally waiting in the queue, the time needed to fetch many instructions of the microcomputer's program is eliminated. If the queue is full and the EU is not requesting access to data in memory, the BIU does not

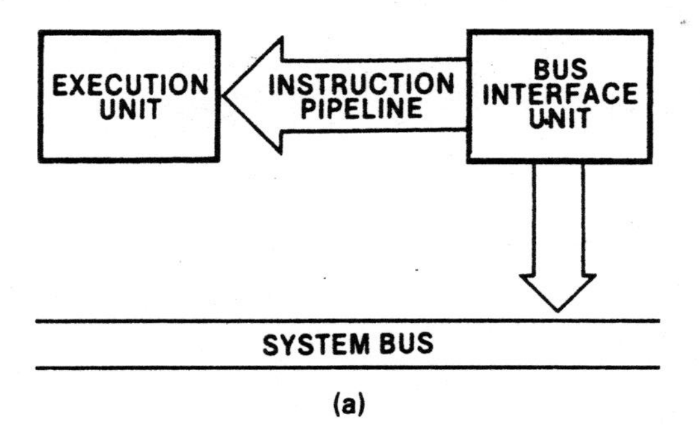

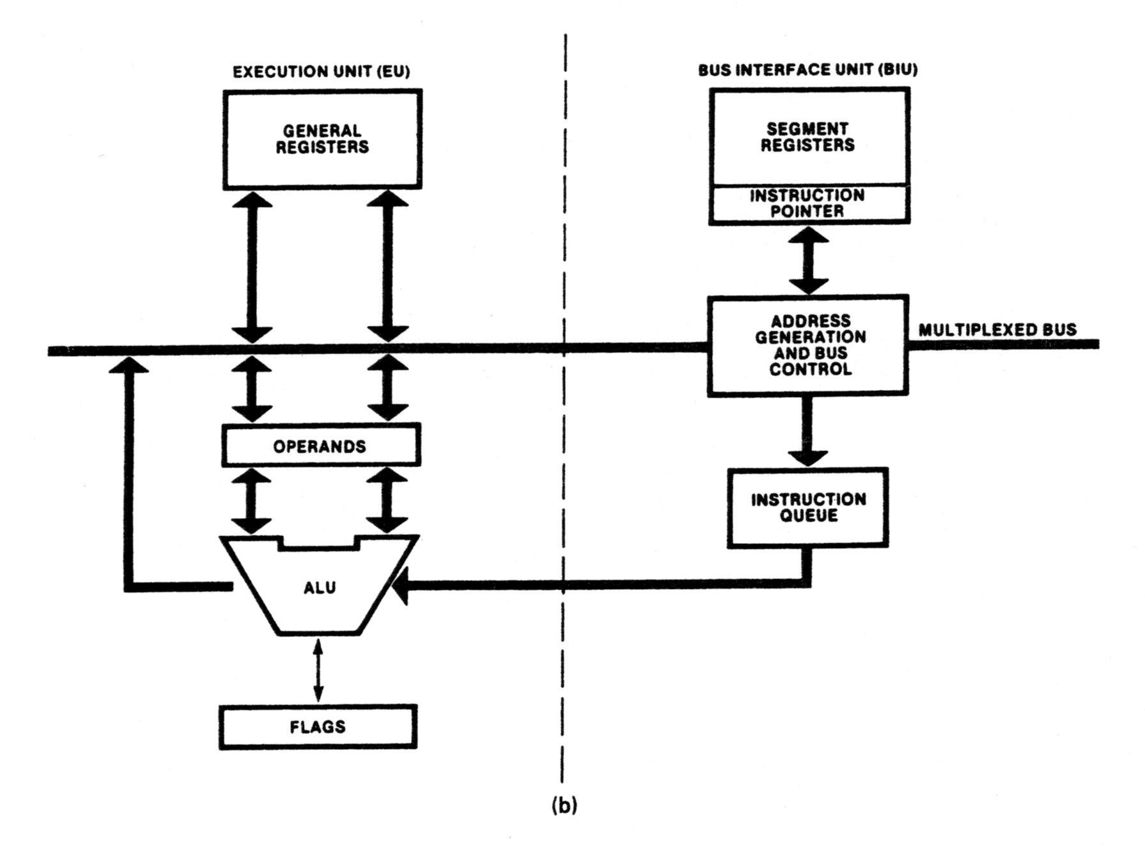

Figure 2–1 (a) Pipelined architecture of the 8088/8086 microprocessors. (Reprinted with permission of Intel Corporation, Copyright/Intel Corp. 1981) (b) Execution and bus interface units. (Reprinted with permission of Intel Corp., Copyright/Intel Corp. 1981)

need to perform any bus operations. These intervals of no bus activity, which occur between bus operations, are known as *idle states.*

The execution unit is responsible for decoding and executing instructions. Notice in Fig. 2–1(b) that it consists of the *arithmetic logic unit* (ALU), status and control flags, general-purpose registers, and temporary-operand registers. The EU accesses instructions from the output end of the instruction queue and data from the general-purpose registers or memory. It reads one instruction byte after the other from the output of the queue, decodes them, generates data addresses if necessary, passes them to the BIU and requests it to perform the read or write operations to memory or I/O, and performs the operation specified by the instruction. The ALU performs the arithmetic, logic, and shift operations required by an instruction. During execution of the instruction, the EU may test the status and control flags, and updates these flags based on the results of executing the instruction. If the queue is empty, the EU waits for the next instruction byte to be fetched and shifted to the top of the queue.

▲ 2.2 SOFTWARE MODEL OF THE 8088/8086 MICROPROCESSOR

The purpose of developing a *software model* is to aid the programmer in understanding the operation of the microcomputer system from a software point of view. To be able to program a microprocessor, one does not need to know all of its hardware architectural features. For instance, we do not necessarily need to know the function of the signals at its various pins, their electrical connections, or their electrical switching characteristics. The function, interconnection, and operation of the internal circuits of the microprocessor also may not need to be considered. What is important to the programmer is to know the various registers within the device and to understand their purpose, functions, operating capabilities, and limitations. Furthermore, it is essential that the programmer knows how external memory and input/output peripherals are organized, how information is arranged in registers, memory, and input/output, and how memory and I/O are addressed to obtain instructions and data. This information represents the software architecture of the processor. Unlike the microarchitecture, the software architecture changes only slightly from generation to generation of processor.

The software model in Fig. 2–2 illustrates the software architecture of the 8088 microprocessor. Looking at this diagram, we see that it includes 13 16-bit internal registers: the *instruction pointer* (IP), four *data registers* (AX, BX, CX, and DX), two *pointer registers* (BP and SP), two *index registers* (SI and DI), and four *segment registers* (CS, DS, SS, and ES). In addition, there is another register called the *status register* (SR), with nine of its bits implemented as status and control flags.

Figure 2–2 shows that the 8088 architecture implements independent memory and input/output address spaces. Notice that the memory address space is 1,048,576 bytes (1Mbyte) in length and the I/O address space is 65,536 bytes (64Kbytes) in length. Our concern here is what can be done with this software architecture and how to do it through software. For this purpose, we will now begin a detailed study of the elements of the model and their relationship to software.

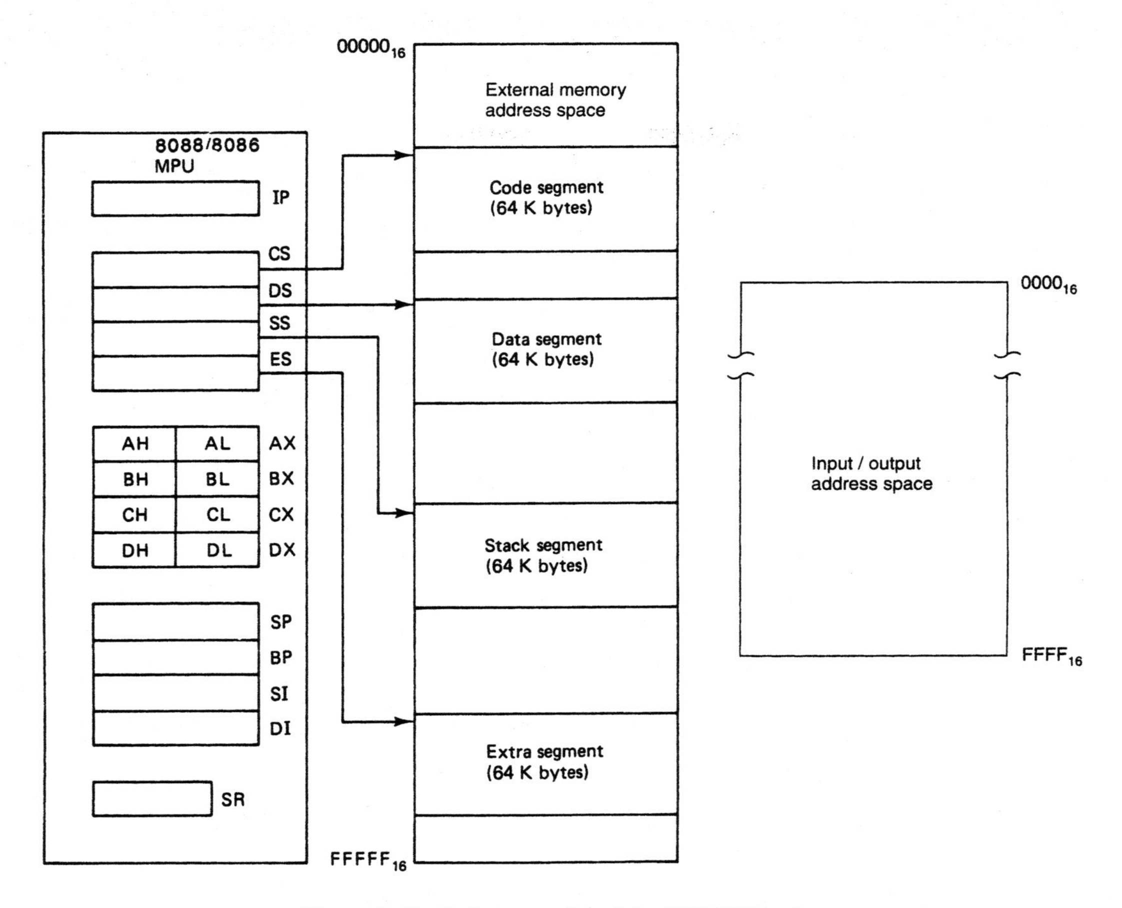

Figure 2–2 Software model of the 8088/8086 microprocessor.

▲ 2.3 MEMORY ADDRESS SPACE AND DATA ORGANIZATION

Now that we have introduced the idea of a software model, let us look at how information such as numbers, characters, and instructions is stored in memory. As shown in Fig. 2–3, the 8088 microcomputer supports 1Mbyte of external memory. This memory space is organized from a software point of view as individual bytes of data stored at consecutive addresses over the address range 00000_{16} to $FFFFF_{16}$. Therefore, memory in an 8088-based microcomputer is actually organized as 8-bit bytes, not as 16-bit words.

The 8088 can access any two consecutive bytes as a *word* of data. In this case, the lower-addressed byte is the least significant byte of the word, and the higher-addressed byte is its most significant byte. Figure 2–4(a) shows how a word of data is stored in memory. Notice that the storage location at the lower address, 00724_{16}, contains the value $00000010_2 = 02_{16}$. The contents of the next-higher-addressed storage location, 00725_{16}, are $01010101_2 = 55_{16}$. These two bytes represent the word $0101010100000010_2 = 5502_{16}$.

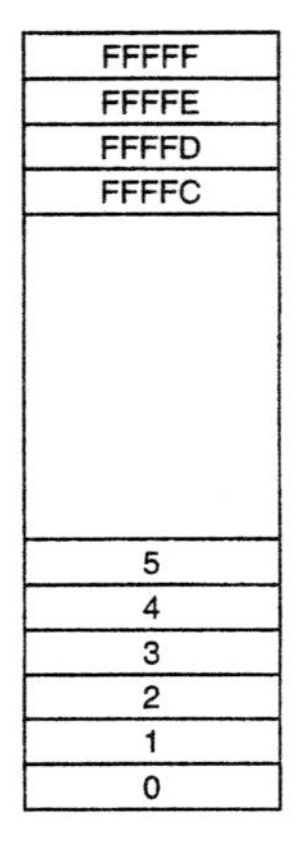

Figure 2–3 Memory address space of the 8088/8086 microprocessor.

To permit efficient use of memory, words of data can be stored at what are called even- or odd-addressed word boundaries. The least significant bit of the address determines the type of word boundary. If this bit is 0, the word is at an *even-address boundary*—that is, a word at an even-address boundary corresponds to two consecutive bytes, with the least significant byte located at an even address. For example, the word in Fig. 2–4(a) has its least significant byte at address 00724_{16}. Therefore, it is stored at an even-address boundary.

A word of data stored at an even-address boundary, such as 00000_{16}, 00002_{16}, 00004_{16}, and so on, is said to be an *aligned word*—that is, all aligned words are located at an address that is a multiple of 2. On the other hand, a word of data stored at an odd-address boundary, such as 00001_{16}, 00003_{16}, or 00005_{16} and so on, is called a *misaligned word.* Figure 2–5 shows some aligned and misaligned words of data. Here words 0, 2, 4, and 6 are examples of aligned-data words, while words 1 and 5 are misaligned words. Notice that misaligned word 1 consists of byte 1 from aligned word 0 and byte 2 from aligned word 2.

When expressing addresses and data in hexadecimal form, it is common to use the letter H to specify the base. For instance, the number $00AB_{16}$ can also be written as 00ABH.

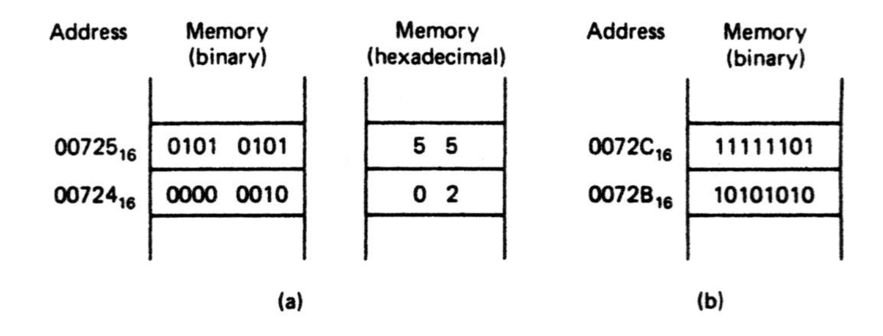

Figure 2–4 (a) Storing a word of data in memory. (b) An example.

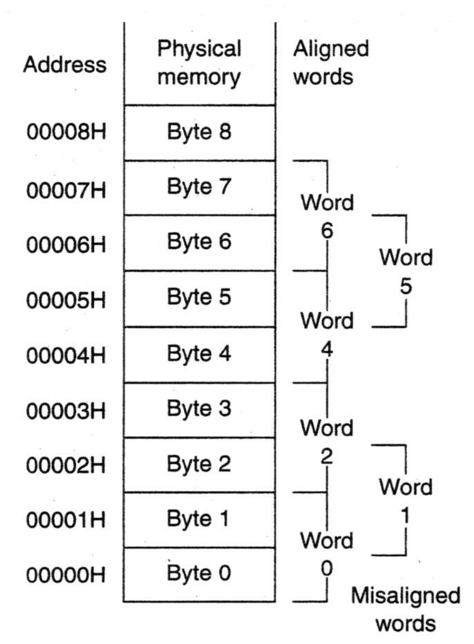

Figure 2–5 Examples of aligned and misaligned data words.

EXAMPLE 2.1

What is the data word shown in Fig. 2–4(b)? Express the result in hexadecimal form. Is it stored at an even- or odd-addressed word boundary? Is it an aligned or misaligned word of data?

Solution

The most significant byte of the word is stored at address $0072C_{16}$ and equals

$$11111101_2 = FD_{16} = FDH$$

Its least significant byte is stored at address $0072B_{16}$ and is

$$10101010_2 = AA_{16} = AAH$$

Together the two bytes give the word

$$1111110110101010_2 = FDAA_{16} = FDAAH$$

Expressing the address of the least significant byte in binary form gives

$$0072BH = 0072B_{16} = 00000000011100101011_2$$

Because the rightmost bit (LSB) is logic 1, the word is stored at an odd-address boundary in memory; therefore, it is a misaligned word of data.

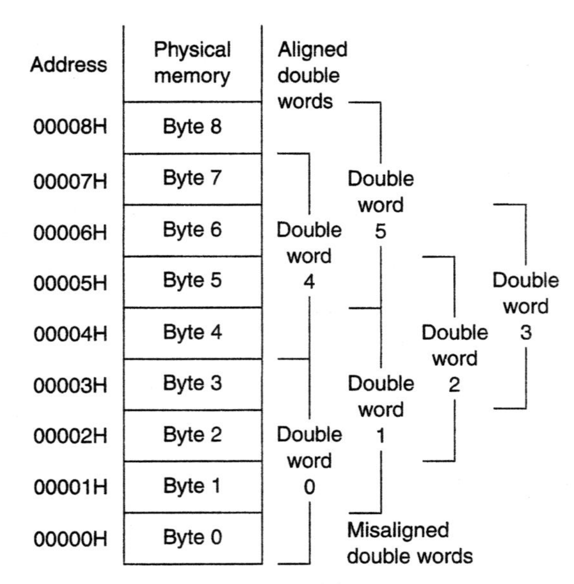

Figure 2–6 Examples of aligned and misaligned double words of data.

The *double word* is another data form that can be processed by the 8088 microcomputer. A double word corresponds to four consecutive bytes of data stored in memory; an example of double-word data is a *pointer.* A pointer is a two-word address element that is used to access data or code in memory. The word of this pointer that is stored at the higher address is called the *segment base address,* and the word at the lower address is called the *offset.*

Just like for words, a double word of data can be aligned or misaligned. An aligned double word is located at an address that is a multiple of 4 (e.g., 00000_{16}, 00004_{16}, and 00008_{16}). A number of aligned and misaligned double words of data are shown in Fig. 2–6. Of these six examples, only double words 0 and 4 are aligned double words.

An example showing the storage of a pointer in memory is given in Fig. 2–7(a). Here the higher-addressed word, which represents the segment base address, is stored

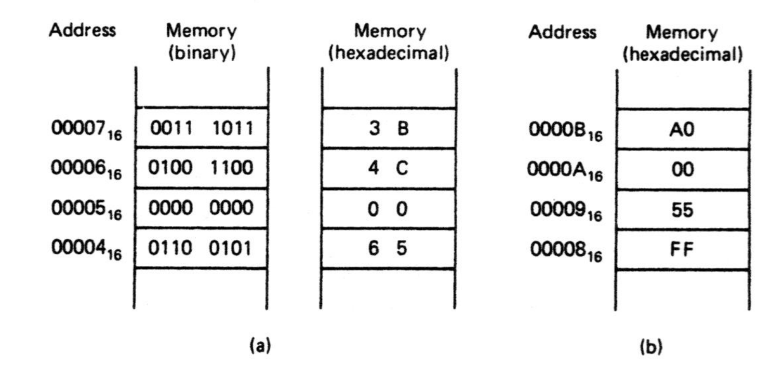

Figure 2–7 (a) Storing a 32-bit pointer in memory. (b) An example.

starting at even-address boundary 00006_{16}. The most significant byte of this word is at address 00007_{16} and equals $00111011_2 = 3B_{16}$. Its least significant byte is at address 00006_{16} and equals $01001100_2 = 4C_{16}$. Combining these two values, we get the segment base address, which equals $0011101101001100_2 = 3B4C_{16}$.

The offset part of the pointer is the lower-addressed word. Its least significant byte is stored at address 00004_{16}; this location contains $01100101_2 = 65_{16}$. The most significant byte is at address 00005_{16}, which contains $00000000_2 = 00_{16}$. The resulting offset is $0000000001100101_2 = 0065_{16}$. The complete double word is $3B4C0065_{16}$. Since this double word starts at address 00004_{16}, it is an example of an aligned double word of data.

EXAMPLE 2.2

How should the pointer with segment base address equal to $A000_{16}$ and offset address $55FF_{16}$ be stored at an even-address boundary starting at 00008_{16}? Is the double word aligned or misaligned?

Solution

Storage of the two-word pointer requires four consecutive byte locations in memory, starting at address 00008_{16}. The least-significant byte of the offset is stored at address 00008_{16} and is shown as FF_{16} in Fig. 2–7(b). The most significant byte of the offset, 55_{16}, is stored at address 00009_{16}. These two bytes are followed by the least significant byte of the segment base address, 00_{16}, at address $0000A_{16}$, and its most significant byte, $A0_{16}$, at address $0000B_{16}$. Since the double word is stored in memory starting at address 00008_{16}, it is aligned.

▲ 2.4 DATA TYPES

The preceding section identified the fundamental data formats of the 8088 as the byte (8 bits), word (16 bits), and double word (32 bits). It also showed how each of these elements is stored in memory. The next step is to examine the types of data that can be coded into these formats for processing.

The 8088 microprocessor directly processes data expressed in a number of different data types. Let us begin with the *integer data type.* The 8088 can process data as either *unsigned* or *signed integer* numbers; each type of integer can be either byte-wide or word-wide. Figure 2–8(a) represents an unsigned byte integer; this data type can be used to represent decimal numbers in the range 0 through 255. The unsigned word integer is shown in Fig. 2–8(b); it can be used to represent decimal numbers in the range 0 through 65,535.

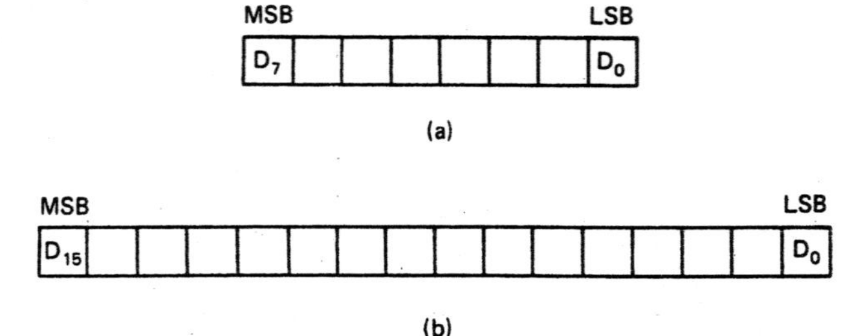

Figure 2–8 (a) Unsigned byte integer. (b) Unsigned word integer.

EXAMPLE 2.3

What value does the unsigned word integer 1000_{16} represent?

Solution

First, the hexadecimal integer is converted to binary form:

$$1000_{16} = 0001000000000000_2$$

Next, we find the value for the binary number:

$$0001000000000000_2 = 2^{12} = 4096$$

The signed byte integer and signed word integer in Figs. 2–9(a) and (b) are similar to the unsigned integer data types just introduced; however, here the most significant bit is a sign bit. A zero in this bit position identifies a positive number. For this reason, the signed integer byte can represent decimal numbers in the range +127 to −128, and the signed integer word permits numbers in the range +32,767 to −32,768, respectively. For example, the number +3 expressed as a signed integer byte is 00000011_2 (03_{16}). On the other hand, the 8088 always expresses negative numbers in 2's-complement notation. Therefore, −3 is coded as 11111101_2 (FD_{16}).

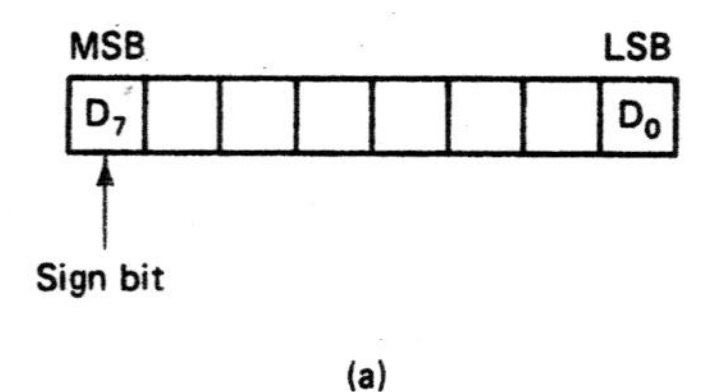

(a)

MSB
LSB
D15
D0
Sign bit

(b)

Figure 2–9 (a) Signed byte integer. (b) Signed word integer.

EXAMPLE 2.4

A signed word integer equals $FEFF_{16}$. What decimal number does it represent?

Solution

Expressing the hexadecimal number in binary form gives

$$FEFF_{16} = 1111111011111111_2$$

Since the most significant bit is 1, the number is negative and is in 2′s complement form. Converting to its binary equivalent by subtracting 1 from the least significant bit and then complementing all bits gives

$$FEFF_{16} = -0000000100000001_2$$
$$= -257$$

The 8088 can also process data that is coded as *binary-coded decimal* (BCD) *numbers.* Figure 2–10(a) lists the BCD values for decimal numbers 0 through 9. BCD data

Decimal	BCD
0	0000
1	0001
2	0010
3	0011
4	0100
5	0101
6	0110
7	0111
8	1000
9	1001

(a)

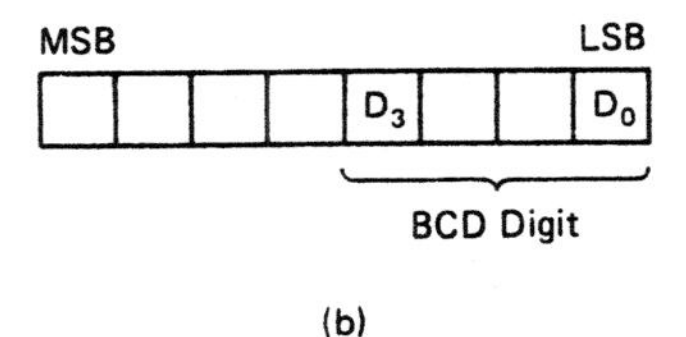

(b)

MSB LSB

D7 D4 D3 D0

BCD Digit 1 BCD Digit 0

(c)

Figure 2–10 (a) BCD numbers. (b) An Unpacked BCD digit. (c) Packed BCD digits.

can be stored in either unpacked or packed form. For instance, the unpacked BCD byte in Fig. 2–10(b) shows that a single BCD digit is stored in the four least significant bits, and the upper four bits are set to 0. Figure 2–10(c) shows a byte with packed BCD digits. Here two BCD numbers are stored in a byte. The upper four bits represent the most significant digit of a two-digit BCD number.

EXAMPLE 2.5

The packed BCD data stored at byte address 01000_{16} equal 10010001_2. What is the two-digit decimal number?

Solution

Writing the value 10010001_2 as separate BCD digits gives

$$10010001_2 = 1001_{BCD}0001_{BCD} = 91_{10}$$

Information expressed in *ASCII (American Standard Code for Information Interchange)* can also be directly processed by the 8088 microprocessor. The chart in Fig. 2–11(a) shows how numbers, letters, and control characters are coded in ASCII. For instance, the number 5 is coded as

$$H_1H_0 = 0110101_2 = 35H$$

where H denotes that the ASCII-coded number is in hexadecimal form. As shown in Fig. 2–11(b), ASCII data are stored as one character per byte.

EXAMPLE 2.6

Byte addresses 01100_{16} through 01104_{16} contain the ASCII data 01000001, 01010011, 01000011, 01001001, and 01001001, respectively. What do the data stand for?

Solution

Using the chart in Fig. 2–11(a), the data are converted to ASCII as follows:

$$(01100H) = 01000001_{ASCII} = A$$

$$(01101H) = 01010011_{ASCII} = S$$

$$(01102H) = 01000011_{ASCII} = C$$

$$(01103H) = 01001001_{ASCII} = I$$

$$(01104H) = 01001001_{ASCII} = I$$

---

	b_7	0	0	0	0	1	1	1	1
	b_6	0	0	1	1	0	0	1	1
	b_5	0	1	0	1	0	1	0	1
b_4 b_3 b_2 b_1	H_0 \ H_1	0	1	2	3	4	5	6	7
0 0 0 0	0	NUL	DLE	SP	0	@	P	'	p
0 0 0 1	1	SOH	DC1	!	1	A	Q	a	q
0 0 1 0	2	STX	DC2	"	2	B	R	b	r
0 0 1 1	3	ETX	DC3	#	3	C	S	c	s
0 1 0 0	4	EOT	DC4	$	4	D	T	d	t
0 1 0 1	5	ENQ	NAK	%	5	E	U	e	u
0 1 1 0	6	ACK	SYN	&	6	F	V	f	v
0 1 1 1	7	BEL	ETB	'	7	G	W	g	w
1 0 0 0	8	BS	CAN	(	8	H	X	h	x
1 0 0 1	9	HT	EM	)	9	I	Y	i	y
1 0 1 0	A	LF	SUB	*	:	J	Z	j	z
1 0 1 1	B	V	ESC	+	;	K	[	k	}
1 1 0 0	C	FF	FS	,	<	L	\	l	|
1 1 0 1	D	CR	GS	−	=	M	]	m	{
1 1 1 0	E	SO	RS	.	>	N	∧	n	~
1 1 1 1	F	SI	US	/	?	O	-	o	DEL

(a)

MSB LSB

D_7							D_0

ASCII Digit

(b)

Figure 2–11 (a) ASCII table. (b) ASCII digit.

▲ 2.5 SEGMENT REGISTERS AND MEMORY SEGMENTATION

Even though the 8088 has a 1Mbyte address space, not all this memory is active at one time. Actually, the 1Mbytes of memory are partitioned into 64Kbyte (65,536) *segments*. A segment represents an independently addressable unit of memory consisting of 64K consecutive byte-wide storage locations. Each segment is assigned a *base address* that identifies its starting point—that is, its lowest address byte-storage location.

Only four of these 64Kbyte segments are active at a time: the *code segment, stack segment, data segment,* and *extra segment.* The segments of memory that are active, as shown in Fig. 2–12, are identified by the values of addresses held in the 8088's four internal segment registers: *CS* (code segment), *SS* (stack segment), *DS* (data segment), and *ES* (extra segment). Each of these registers contains a 16-bit base address that points to the lowest addressed byte of the segment in memory. Four segments give a maximum of 256Kbytes of active memory. Of this, 64Kbytes are for *program storage* (code), 64Kbytes are for a *stack*, and 128Kbytes are for *data storage.*

The values held in these registers are referred to as the *current-segment register values;* for example, the value in CS points to the first word-wide storage location in the current code segment. Code is always fetched from memory as words, not as bytes.

Figure 2–13 illustrates the *segmentation of memory.* In this diagram, the 64Kbyte segments are identified with letters such as A, B, and C. The data segment (DS) register contains the value B. Therefore, the second 64Kbyte segment of memory from the top, labeled B, acts as the current data-storage segment. This is one of the segments in which data that are to be processed by the microcomputer are stored. For this reason, this part

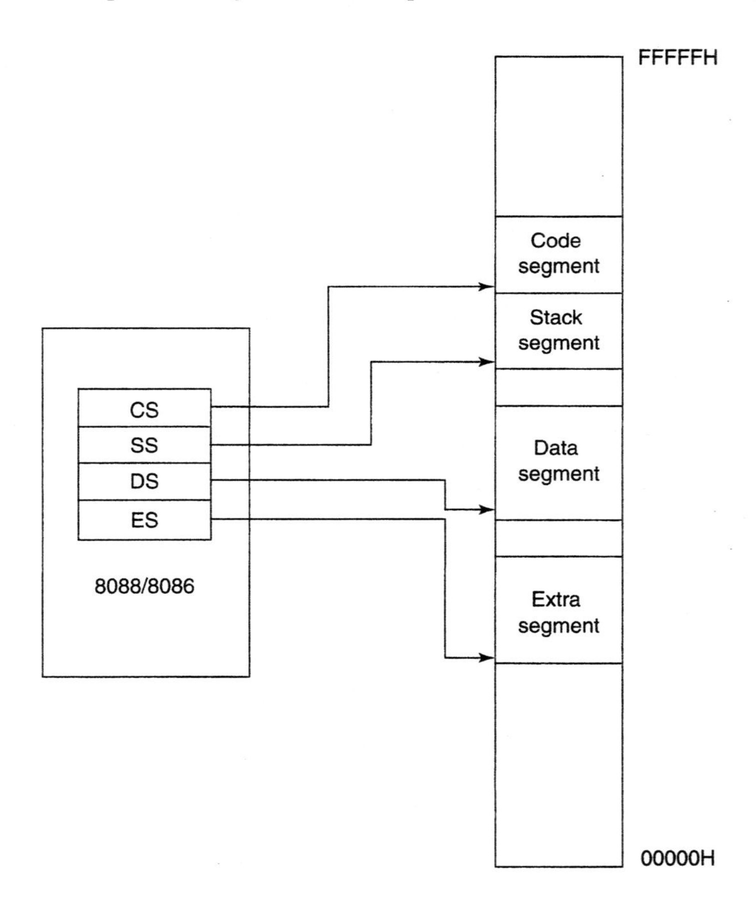

Figure 2–12 Active segments of memory.

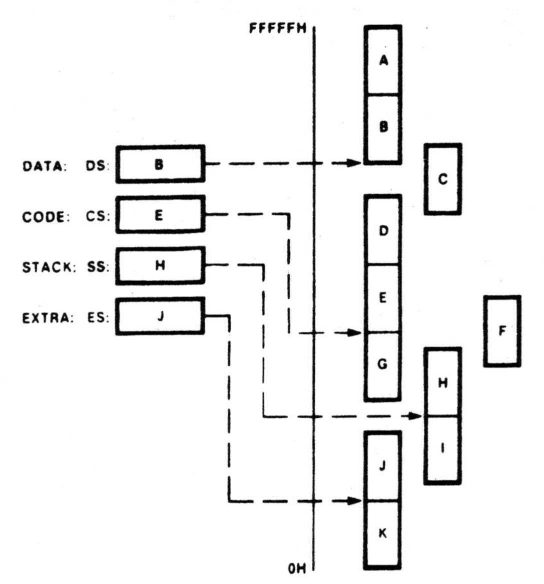

Figure 2–13 Contiguous, adjacent, disjointed, and overlapping segments. (Reprinted by permission of Intel Corp., Copyright/Intel Corp. 1979)

of the microcomputer's memory address space must contain read/write storage locations that can be accessed by instructions as storage locations for source and destination operands. CS selects segment E as the code segment. It is this segment of memory from which instructions of the program are currently being fetched for execution. The stack segment (SS) register contains H, thereby selecting the 64Kbyte segment labeled as H for use as a stack. Finally, the extra segment (ES) register is loaded with value J such that segment J of memory functions as a second 64Kbyte data storage segment.

The segment registers are said to be *user accessible.* This means that the programmer can change their contents through software. Therefore, for a program to gain access to another part of memory, one simply has to change the value of the appropriate register or registers. For instance, a new data space, with up to 128Kbytes, is brought in simply by changing the values in DS and ES.

There is one restriction on the value assigned to a segment as a base address: it must reside on a 16-byte address boundary. This is because increasing the 16-bit value in a segment register by 1 actually increases the corresponding memory address by 16; examples of valid base addresses are 00000_{16}, 00010_{16}, and 00020_{16}. Other than this restriction, segments can be set up to be contiguous, adjacent, disjointed, or even overlapping; for example, in Fig. 2–13, segments A and B are contiguous, whereas segments B and C are overlapping.

▲ 2.6 DEDICATED, RESERVED, AND GENERAL-USE MEMORY

Any part of the 8088 microcomputer's 1Mbyte address space can be implemented for the user's access; however, some address locations have *dedicated functions* and should not be used as general memory for storage of data or instructions of a program. Let us now look at these reserved, dedicated use, and general-use parts of memory.

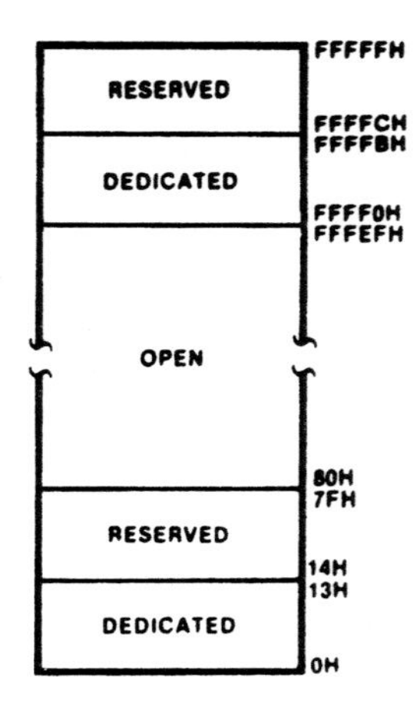

Figure 2–14 Dedicated-use, reserved, and general-use memory. (Reprinted by permission of Intel Corp., Copyright/Intel Corp. 1979)

Figure 2–14 shows the *reserved, dedicated-use,* and *general-use* parts of the 8088/8086's *address space.* Notice that storage locations from address 00000_{16} to 00013_{16} are dedicated, and those from address 00014_{16} to $0007F_{16}$ are reserved. These 128 bytes of memory are used for storage of pointers to interrupt service routines. The dedicated part is used to store the pointers for the 8088's internal interrupts and exceptions. On the other hand, the reserved locations are saved to store pointers that are used by the user-defined interrupts. As indicated earlier, a pointer is a two-word address element and requires 4 bytes of memory. The word of this pointer at the higher address is called the segment base address and the word at the lower address is the offset. Therefore, this section of memory contains up to 32 pointers.

The part of the address space labeled *open* in Fig. 2–14 is *general-use memory* and is where data or instructions of the program are stored. Notice that the general-use area of memory is the range from addresses 80_{16} through $FFFEF_{16}$.

At the high end of the memory address space is another reserved pointer area, located from address $FFFFC_{16}$ through $FFFFF_{16}$. These four memory locations are reserved for use with future products and should not be used. Intel Corporation, the original manufacturer of the 8088, has identified the 12 storage locations from address $FFFF0_{16}$ through $FFFFB_{16}$ as dedicated for functions such as storage of the hardware reset jump instruction. For instance, address $FFFF0_{16}$ is where the 8088/8086 begins execution after receiving a reset.

▲ 2.7 INSTRUCTION POINTER

The register that we will consider next in the 8088's software model shown in Fig. 2–2 is the *instruction pointer* (IP). IP is also 16 bits in length and identifies the location of the next word of instruction code to be fetched from the current code segment of memory. The IP is similar to a program counter; however, it contains the offset of the next word of instruction code instead of its actual address. This is because IP and CS are both 16 bits in length, but a 20-bit address is needed to access memory. Internal to the 8088, the offset in IP is combined with the current value in CS to generate the address of the instruction code. Therefore, the value of the address for the next code access is often denoted as CS:IP.

During normal operation, the 8088 fetches instructions from the code segment of memory, stores them in its instruction queue, and executes them one after the other. Every time a word of code is fetched from memory, the 8088 updates the value in IP such that it points to the first byte of the next sequential word of code—that is, IP is incremented by 2. Actually, the 8088 prefetches up to four bytes of instruction code into its internal code queue and holds them there waiting for execution.

After an instruction is read from the output of the instruction queue, it is decoded; if necessary, operands are read from either the data segment of memory or internal registers. Next, the operation specified in the instruction is performed on the operands and the result is written back to either an internal register or a storage location in memory. The 8088 is now ready to execute the next instruction in the code queue.

Executing an instruction that loads a new value into the CS register changes the active code segment; thus, any 64Kbyte segment of memory can be used to store the instruction code.

▲ 2.8 DATA REGISTERS

As Fig. 2–2 shows, the 8088 has four *general-purpose data registers.* During program execution, they hold temporary values of frequently used intermediate results. Software can read, load, or modify their contents. Any of the general-purpose data registers can be used as the source or destination of an operand during an arithmetic operation such as ADD or a logic operation such as AND. For instance, the values of two pieces of data, A and B, could be moved from memory into separate data registers and operations such as addition, subtraction, and multiplication performed on them. The advantage of storing these data in internal registers instead of memory during processing is that they can be accessed much faster.

The four registers, known as the *data registers,* are shown in more detail in Fig. 2–15(a). Notice that they are referred to as the *accumulator register* (A), the *base register* (B), the *count register* (C), and the *data register* (D). These names imply special functions

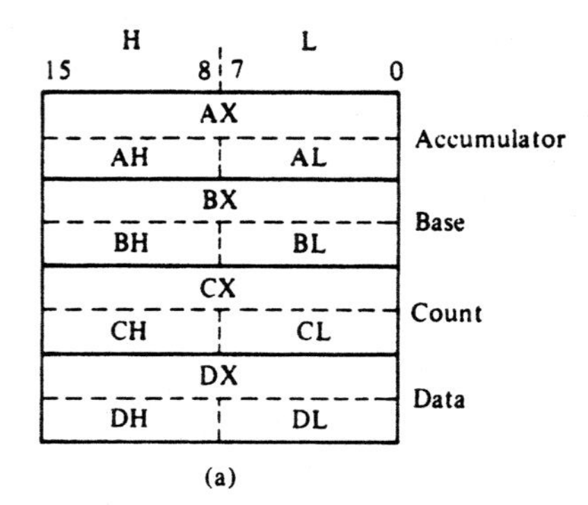

Register	Operations
AX	Word multiply, word divide, word I/O
AL	Byte multiply, byte divide, byte I/O, translate, decimal arithmetic
AH	Byte multiply, byte divide
BX	Translate
CX	String operations, loops
CL	Variable shift and rotate
DX	Word multiply, word divide, indirect I/O

(b)

Figure 2–15 (a) General-purpose data registers. (Reprinted by permission of Intel Corp., Copyright/Intel Corp. 1979) (b) Dedicated register functions. (Reprinted by permission of Intel Corp., Copyright/Intel Corp. 1979)

they are meant to perform for the 8088 microprocessor. Figure 2–15(b) summarizes these operations. Notice that string and loop operations use the C register. For example, the value in the C register is the number of bytes to be processed in a string operation. This is the reason it is given the name *count register.* Another example of the dedicated use of data registers is that all input/output operations must use accumulator register AL or AX for data.

Each of these registers can be accessed either as a whole (16 bits) for word data operations or as two 8-bit registers for byte-wide data operations. An X after the register letter identifies the reference of a register as a word; for instance, the 16-bit accumulator is referenced as AX. Similarly, the other three word registers are referred to as BX, CX, and DX.

On the other hand, when referencing one of these registers on a byte-wide basis, following the register name with the letter H or L, respectively, identifies the high byte and low byte. For the A register, the most significant byte is referred to as AH and the least significant byte as AL; the other byte-wide register pairs are BH and BL, CH and CL, and DH and DL. When software places a new value in one byte of a register, for instance AL, the value in the other byte (AH) does not change. This ability to process information in either byte location permits more efficient use of the limited register resources of the 8088 microprocessor.

Actually, some of the data registers may also store address information such as a base address or an input/output address; for example, BX could hold a 16-bit base address.

▲ 2.9 POINTER AND INDEX REGISTERS

The software model in Fig. 2–2 has four other general-purpose registers: two *pointer registers* and two *index registers.* They store what are called *offset addresses.* An offset address represents the displacement of a storage location in memory from the segment base address in a segment register—that is, it is used as a pointer or index to select a specific storage location within a 64Kbyte segment of memory. Software uses the value held in an index register to reference data in memory relative to the data segment or extra segment register, and a pointer register to access memory locations relative to the stack segment register. Just as for the data registers, the values held in these registers can be read, loaded, or modified through software. This is done prior to executing the instruction that references the register for address offset. Unlike the general-purpose data registers, the pointer and index registers are only accessed as words. To use the offset address in a register, the instruction simply specifies the register that contains the value.

Figure 2–16 shows that the two pointer registers are the *stack pointer* (SP) and *base pointer* (BP). The values in SP and BP are used as offsets from the current value of SS during the execution of instructions that involve the stack segment of memory and permit easy access to storage locations in the stack part of memory. The value in SP always represents the offset of the next stack location that is to be accessed. That is, combining SP with the value in SS (SS:SP) results in an address that points to the *top of the stack* (TOS).

BP also represents an offset relative to the SS; however, it is used to access data within the stack segment of memory. To do this, it is employed as the offset in an addressing mode called the *based addressing mode.* One common use of BP is to reference parameters that are passed to a subroutine by way of the stack. In this case, instructions are included in the subroutine that use based addressing to access the values of parameters from the stack.

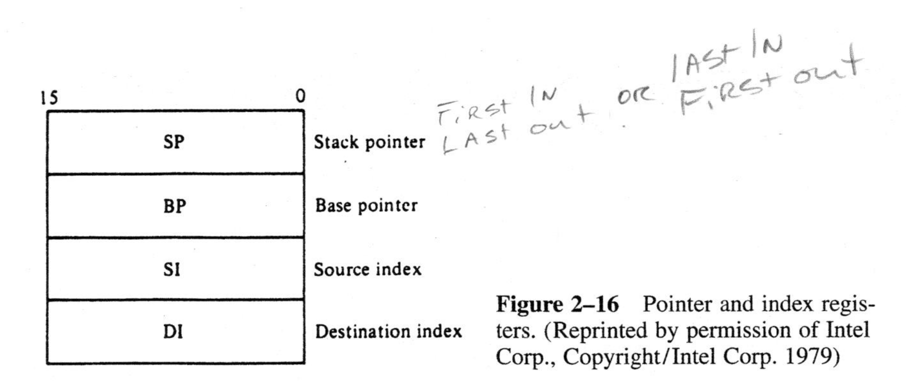

Figure 2–16 Pointer and index registers. (Reprinted by permission of Intel Corp., Copyright/Intel Corp. 1979)

The index registers are used to hold offset addresses for instructions that access data stored in the data segment of memory and are automatically combined with the value in the DS or ES register during address calculation. In instructions that involve the *indexed addressing,* the *source index* (SI) register holds an offset address that identifies the location of a source operand, and the *destination index* (DI) register holds an offset for a destination operand.

Earlier we pointed out that any of the data registers can be used as the source or destination of an operand during an arithmetic operation such as ADD, or a logic operation such as AND. However, for some operations, an operand that is to be processed may be located in memory instead of the internal register. In this case, an index address is used to identify the location of the operand in memory; for example, string instructions use the index registers to access operands in memory. SI and DI, respectively, are the pointers to the source and destination locations in memory.

The index registers can also be source or destination registers in arithmetic and logical operations. For example, an instruction may add 2 to the offset value in SI to increment its value to point to the next word-wide storage location in memory.

▲ 2.10 STATUS REGISTER

The *status register,* also called the *flags register,* is another 16-bit register within the 8088. Figure 2–17 shows the organization of this register in more detail. Notice that just nine of its bits are implemented. Six of these bits represent *status flags:* the *carry flag* (CF), *parity flag* (PF), *auxiliary carry flag* (AF), *zero flag* (ZF), *sign flag* (SF), and *overflow flag* (OF). The logic state of these status flags indicate conditions that are produced as the

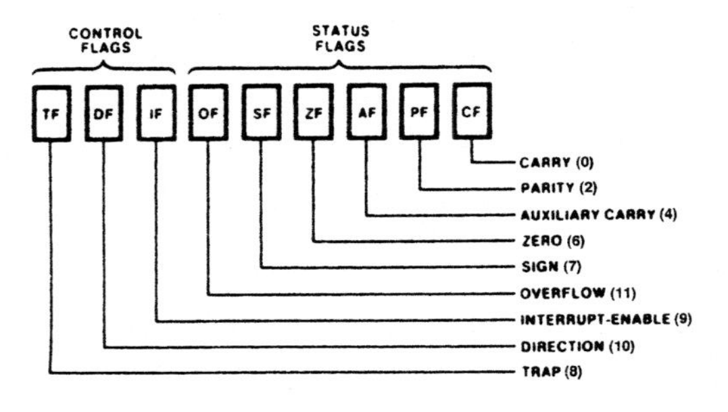

Figure 2–17 Status and control flags. (Reprinted by permission of Intel Corp., Copyright/Intel Corp. 1979)

result of executing an instruction—that is, after executing an instruction, such as ADD, specific flag bits are reset (logic 0) or set (logic 1) based on the result that is produced.

Let us first summarize the operation of these flags:

1. *The carry flag (CF):* CF is set if there is a carry-out or a borrow-in for the most significant bit of the result during the execution of an instruction. Otherwise, CF is reset.
2. *The parity flag (PF):* PF is set if the result produced by the instruction has even parity—that is, if it contains an even number of bits at the 1 logic level. If parity is odd, PF is reset.
3. *The auxiliary carry flag (AF):* AF is set if there is a carry-out from the low nibble into the high nibble or a borrow-in from the high nibble into the low nibble of the lower byte in a 16-bit word. Otherwise, AF is reset.
4. *The zero flag (ZF):* ZF is set if the result produced by an instruction is zero. Otherwise, ZF is reset.
5. *The sign flag (SF):* The MSB of the result is copied into SF. Thus, SF is set if the result is a negative number or reset if it is positive.
6. *The overflow flag (OF):* When OF is set, it indicates that the signed result is out of range. If the result is not out of range, OF remains reset.

For example, at the completion of execution of a byte-addition instruction, the carry flag (CF) could be set to indicate that the sum of the operands caused a carry-out condition. The auxiliary carry flag (AF) could also set due to the execution of the instruction. This depends on whether or not a carry-out occurred from the least significant nibble to the most significant nibble when the byte operands are added. The sign flag (SF) is also affected, and it reflects the logic level of the MSB of the result. The overflow flag (OF) is set if there is a carry-out of the sign bit, but no carry into the sign bit (an indication of overflow).

The 8088 provides instructions within its instruction set that are able to use these flags to alter the sequence in which the program is executed; for instance, a jump to another part of the program could be conditionally initiated by testing for ZF equal to logic. This operation is called *jump on zero.*

The other three implemented flag bits—the *direction flag* (DF), the *interrupt enable flag* (IF), and the *trap flag* (TF)—are *control flags.* These three flags provide control functions of the 8088 as follows:

1. *The trap flag (TF):* If TF is set, the 8088 goes into the *single-step mode* of operation. When in the single-step mode, it executes an instruction and then jumps to a special service routine that may deter·ıine the effect of executing the instruction. This type of operation is very useful ıur debugging programs.
2. *The interrupt flag (IF):* For the 8088 to recognize *maskable interrupt requests* at its interrupt (INT) input, the IF flag must be set. When IF is reset, requests at INT are ignored and the maskable interrupt interface is disabled.
3. *The direction flag (DF):* The logic level of DF determines the direction in which string operations will occur. When set, the string instruction automatically decrements the address; therefore, the string data transfers proceed from high address to low address. On the other hand, resetting DF causes the string address to be incremented—that is, data transfers proceed from low address to high address.

The instruction set of the 8088 includes instructions for saving, loading, or manipulating the flags; for instance, special instructions are provided to permit user software to set or reset CF, DF, and IF at any point in the program (e.g., just prior to the beginning of a string operation, DF is reset so that the string address automatically increments).

▲ 2.11 GENERATING A MEMORY ADDRESS

A segment base and an offset describe a *logical address* in the 8088 microcomputer system. As Fig. 2–18 shows, both the segment base and offset are 16-bit quantities, since all registers and memory locations used in address calculations are 16 bits long. However, the *physical address* that is used to access memory is 20 bits in length. The generation of the physical address involves combining a 16-bit offset value that is located in the instruction pointer, a base register, an index register, or a pointer register and a 16-bit segment base value that is located in one of the segment registers.

The source of the offset value depends on which type of memory reference is taking place. It can be the base pointer (BP) register, stack pointer (SP) register, base (BX) register, source index (SI) register, destination index (DI) register, or instruction pointer (IP). An offset can even be formed from the contents of several of these registers. On the other hand, the segment base value always resides in one of the segment registers: CS, DS, SS, or ES.

For instance, when an instruction acquisition takes place, the source of the segment base value is always the code segment (CS) register, and the source of the offset value is

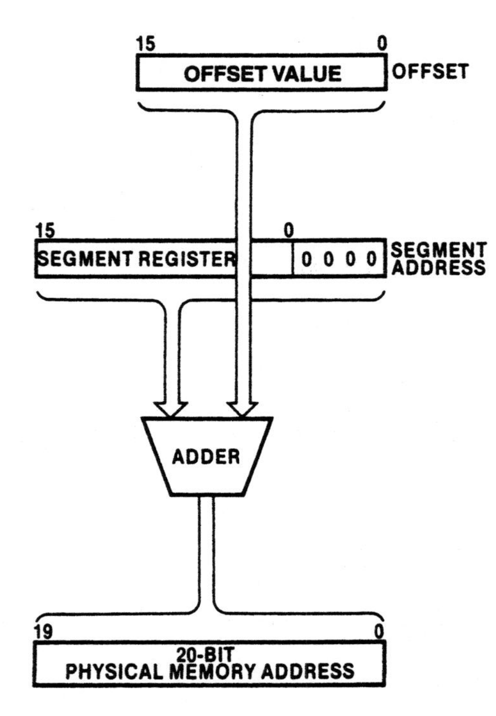

Figure 2–18 Generating a physical address. (Reprinted by permission of Intel Corp., Copyright/Intel Corp. 1981)

always the instruction pointer (IP). This physical address can be denoted as CS:IP. On the other hand, if the value of a variable is written to memory during execution of an instruction, typically the segment base value is specified by the data segment (DS) register and the offset value by the destination index (DI) register—that is, the physical address is given as DS:DI. A provision called the *segment-override prefix* is used to change the segment from which the variable is accessed; for example, a prefix could be used to make a data access occur in which the segment base is in the ES register.

Another example is the stack address that is needed when pushing parameters onto the stack. This physical address is formed from the values of the segment base in the stack segment (SS) register and offset in the stack pointer (SP) register and is described as SS:SP.

Remember that the segment base address represents the starting location of the 64Kbyte segment in memory—that is, the lowest address byte in the segment. Figure 2–19 shows that the offset identifies the distance in bytes that the storage location of interest resides from this starting address. Therefore, the lowest address byte in a segment has an offset of 0000_{16}, and the highest address byte has an offset of $FFFF_{16}$.

Figure 2–20 shows how a segment base value in a segment register and an offset value are combined to form a physical address. The value in the segment register is shifted left by four bit positions, with its LSBs filled with zeros. This gives a *segment address,* the location where the segment starts. The offset value is then added to the 16 LSBs of the shifted segment value. The result of this addition is the 20-bit physical address.

The example in Fig. 2–20 represents a segment base value of 1234_{16} and an offset value of 0022_{16}. First, let us express the segment base value in binary form. This gives

$$1234_{16} = 0001001000110100_2$$

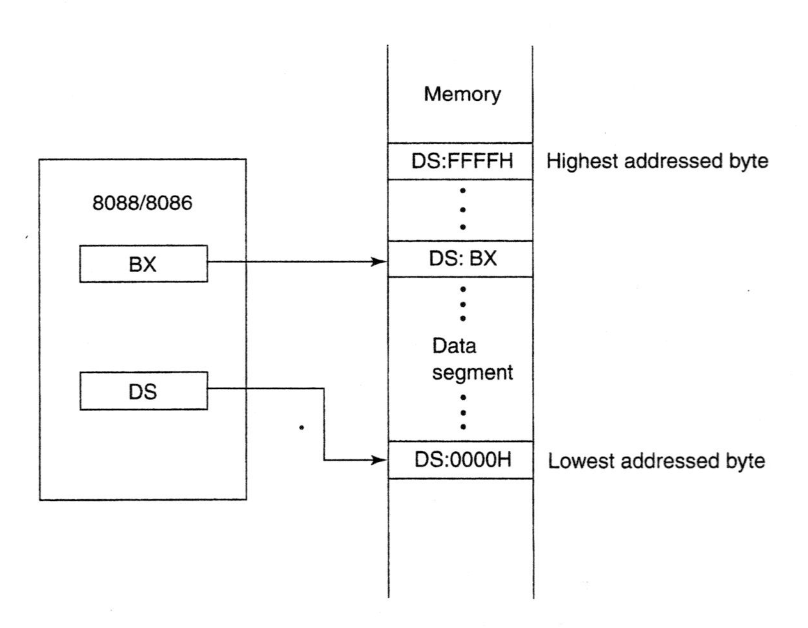

Figure 2–19 Boundaries of a segment.

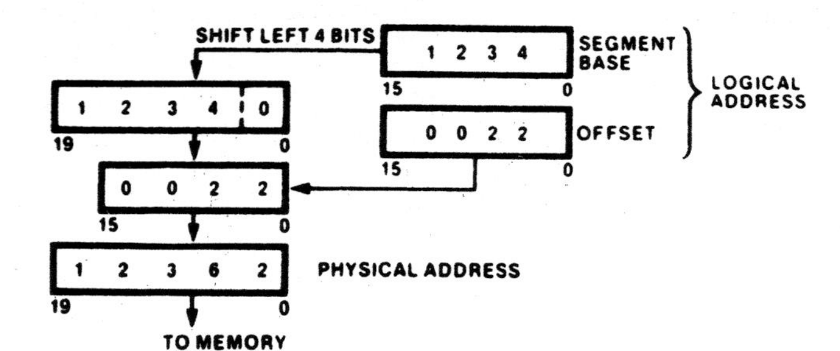

Figure 2–20 Physical address calculation example. (Reprinted by permission of Intel Corp., Copyright/Intel Corp. 1979)

Shifting left four positions and filling with zeros results in the segment address

$$00010010001101000000_2 = 12340_{16}$$

The offset in binary form is

$$0022_{16} = 0000000000100010_2$$

Adding the segment address and the offset gives

$$\begin{aligned} 00010010001101000000_2 + 0000000000100010_2 &= 00010010001101100010_2 \\ &= 12362_{16} \\ &= 12362\text{H} \end{aligned}$$

This address calculation is done automatically within the 8088 microprocessor each time a memory access is initiated.

EXAMPLE 2.7

What would be the offset required to map to physical address location $002C3_{16}$ if the contents of the corresponding segment register are $002A_{16}$?

Solution

The offset value can be obtained by shifting the contents of the segment register left by four bit positions and then subtracting from the physical address. Shifting left gives

$$002A0_{16}$$

Now subtracting, we get the value of the offset:

$$002C3_{16} - 02A0_{16} = 0023_{16}$$

Actually, many different logical addresses map to the same physical address location in memory. Simply changing the segment base value in the segment register and its corresponding offset does this. The diagram in Fig. 2–21 demonstrates this idea. Notice that segment base $002B_{16}$ with offset 0013_{16} maps to physical address $002C3_{16}$ in memory. However, if the segment base is changed to $002C_{16}$ with a new offset of 0003_{16}, the physical address is still $002C3_{16}$. We see that the physical address 002BH:0013H is equal to the physical address 002CH:0003H.

▲ 2.12 THE STACK

As indicated earlier, the *stack* is implemented in the memory of the 8088 microprocessor, and it is used for temporary storage of information such as data or addresses. For instance, when a *call instruction* is executed, the 8088 automatically pushes the current values in CS and IP onto the stack. As part of the subroutine, the contents of other registers may also be saved on the stack by executing *push instructions* (e.g., when the instruction PUSH SI is executed, it causes the contents of SI to be pushed onto the stack). Near the end of the subroutine, *pop instructions* are included to pop values from the stack back into their corresponding internal registers (e.g., POP SI causes the value at the top of the stack to be popped back into SI). At the end of the subroutine, a *return instruction* causes the values of CS and IP to be popped off the stack and put back into the internal register where they originally resided.

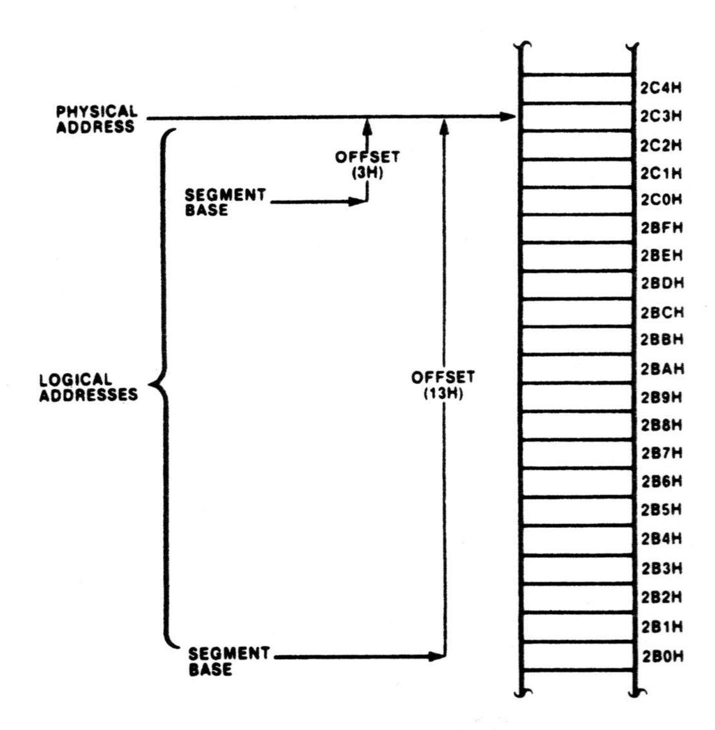

Figure 2–21 Relationship between logical and physical addresses. (Reprinted by permission of Intel Corp., Copyright/Intel Corp. 1979)

The stack is 64Kbytes long and is organized from a software point of view as 32K words. Figure 2–22 shows that the segment base value in the SS register points to the lowest address word in the current stack. The contents of the SP and BP register offset into the stack segment of memory.

Looking at Fig. 2–22, we see that SP contains an offset value that points to a storage location in the current stack segment. The address obtained from the contents of SS and SP (SS:SP) is the physical address of the last storage location in the stack to which data were pushed. This memory address is known as the *top of the stack.* At the microcomputer's startup, the value in SP is initialized to $FFFE_{16}$. Combining this value with the current value in SS gives the highest-addressed word location in the stack (SS:FFFEH)—that is, the *bottom of the stack.*

The 8088 can push data and address information onto the stack from its internal registers or a storage location in memory. Data transferred to and from the stack are word-wide, not byte-wide. Each time a word is to be pushed onto the top of the stack, the value in SP is first automatically decremented by two, and then the contents of the register are written into the stack part of memory. Therefore, the stack grows down in memory from the bottom of the stack, which corresponds to the physical address SS:FFFEH, toward the *end of the stack,* which corresponds to the physical address obtained from SS and offset 0000_{16} (SS:0000H).

When a value is popped from the top of the stack, the reverse of this sequence occurs. The physical address defined by SS and SP points to the location of the last value pushed onto the stack. Its contents are first popped off the stack and put into the specific register within the 8088; then SP is automatically incremented by two. The top of the stack then corresponds to the address of the previous value pushed onto the stack.

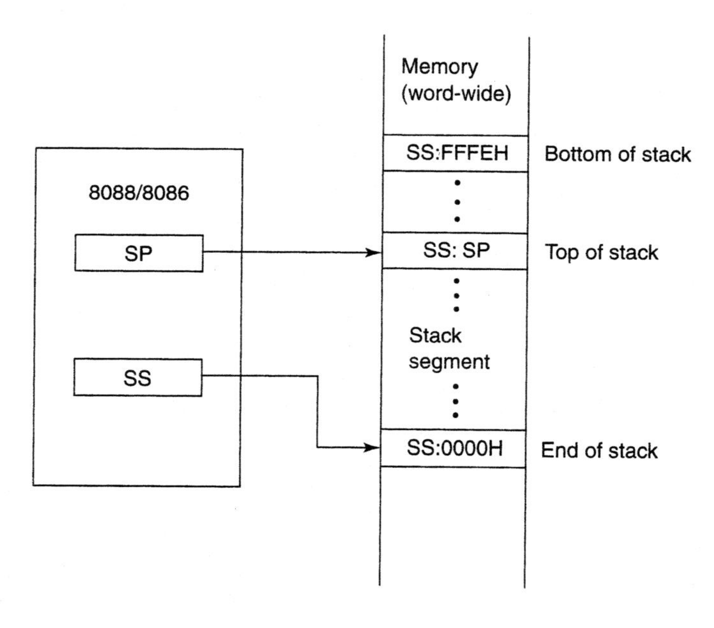

Figure 2–22 Stack segment of memory.

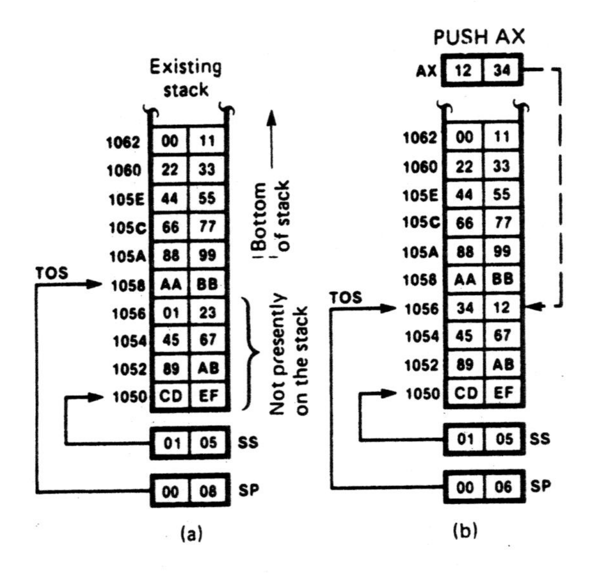

Figure 2–23 (a) Stack just prior to push operation. (Reprinted by permission of Intel Corp., Copyright/Intel Corp. 1979) (b) Stack after execution of the PUSH AX instruction. (Reprinted by permission of Intel Corp., Copyright/Intel Corp. 1979)

The example in Fig. 2–23(a) shows how the contents of a register are pushed onto the stack. Here we find the state of the stack prior to execution of the PUSH AX instruction. Notice that the stack segment register contains 105_{16}. As indicated, the bottom of the stack resides at the physical address derived from SS and offset $FFFE_{16}$. This gives the bottom-of-stack address, A_{BOS}, as

$$
\begin{aligned}
A_{BOS} &= 1050_{16} + FFFE_{16} \\
&= 1104E_{16}
\end{aligned}
$$

Furthermore, the stack pointer, which represents the offset from the beginning of the stack specified by the contents of SS to the top of the stack, equals 0008_{16}. Therefore, the current top of the stack is at physical address A_{TOS}, which equals

$$
\begin{aligned}
A_{TOS} &= 1050_{16} + 0008_{16} \\
&= 1058_{16}
\end{aligned}
$$

Addresses with higher values than that of the top of the stack, 1058_{16}, contain valid stack data. Those with lower addresses do not yet contain valid stack data. Notice that the last value pushed to the stack in Fig. 2–23(a) is $BBAA_{16}$.

Figure 2–23(b) demonstrates what happens when the PUSH AX instruction is executed. Here we see that AX initially contains the number 1234_{16}. Notice that execution of the push instruction causes the stack pointer to be decremented by two but does not affect the contents of the stack segment register. Therefore, the next stack access is to the location corresponding to address 1056_{16}. This location is where the value in AX is pushed. Notice that the most significant byte of AX, which equals 12_{16}, now resides in memory address 1057_{16}, and the least significant byte of AX, which is 34_{16}, is held in memory address 1056_{16}.

Let us next look at an example in which stack data are popped from the stack back into the registers from which they were pushed. Figure 2–24 illustrates this operation. In Fig. 2–24(a), the stack is shown to be in the state that resulted due to our prior PUSH AX example. That is, SP equals 0006_{16}, SS equals 105_{16}, the address at the top of the stack equals 1056_{16}, and the word at the top of the stack equals 1234_{16}.

Figure 2–24(b) shows what happens when the instructions POP AX and POP BX are executed in that order. Execution of the first instruction causes the 8088 to read the value from the top of the stack and put it into the AX register as 1234_{16}. Next, SP is incremented to give 0008_{16} and another read operation is initiated from the stack. This second read corresponds to the POP BX instruction, and it causes the value $BBAA_{16}$ to be loaded into the BX register. SP is incremented once more and now equals $000A_{16}$. Therefore, the new top of stack is at address $105A_{16}$.

In Fig. 2–24(b) we see that the values read out of addresses 1056_{16} and 1058_{16} remain at these locations, but now reside at locations that are above the top of the stack; therefore, they no longer represent valid stack data. If new information is pushed to the stack, these values are written over.

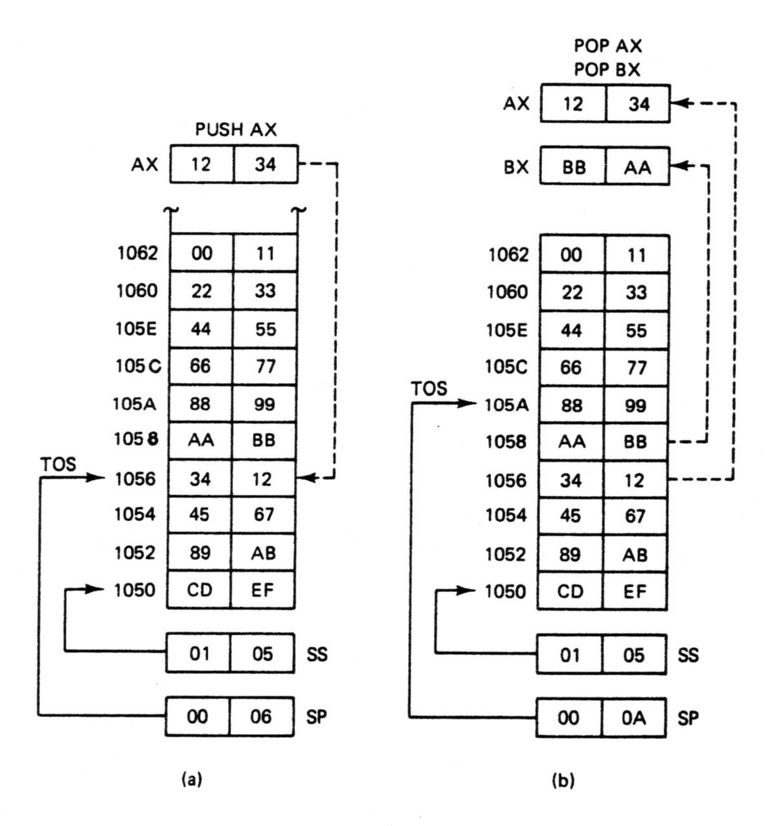

Figure 2–24 (a) Stack just prior to pop operation. (Reprinted by permission of Intel Corp., Copyright/Intel Corp. 1979) (b) Stack after the execution of the POP AX and POP BX instructions. (Reprinted by permission of Intel Corp., Copyright/Intel Corp. 1979)

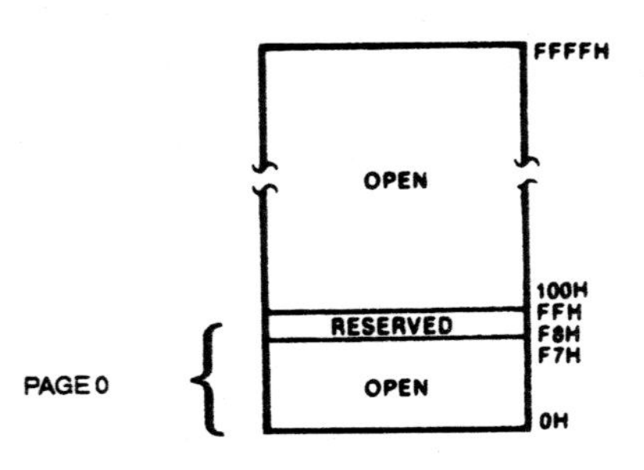

Figure 2–25 I/O address space. (Reprinted by permission of Intel Corp., Copyright/Intel Corp. 1979)

Any number of stacks may exist in an 8088 microcomputer. Simply changing the value in the SS register brings in a new stack. For instance, executing the instruction MOV SS, DX loads a new value from DX into SS. Although many stacks can exist, only one can be active at a time.

▲ 2.13 INPUT/OUTPUT ADDRESS SPACE

The 8088 has separate memory and input/output (I/O) address spaces. The *I/O address space* is the place where I/O interfaces, such as printer and monitor ports, are implemented. Figure 2–25 shows a map of the 8088's I/O address space. Notice that this address range is from 0000_{16} to $FFFF_{16}$. This represents just 64Kbyte addresses; therefore, unlike memory, I/O addresses are only 16 bits long. Each of these addresses corresponds to one byte-wide I/O port.

The part of the map from address 0000_{16} through $00FF_{16}$ is referred to as *page 0*. Certain of the 8088's I/O instructions can perform only input or output data-transfer operations to I/O devices located in this part of the I/O address space. Other I/O instructions can input or output data for devices located anywhere in the I/O address space. I/O data transfers can be byte-wide or word-wide. Notice that the eight locations from address $00F8_{16}$ through $00FF_{16}$ are specified as reserved by Intel Corporation and should not be used.

REVIEW PROBLEMS

Section 2.1

1. Name the two internal processing units of the 8088.
2. Which processing unit of the 8088 is the interface to the outside world?
3. What are the length of the 8086's address bus and data bus?
4. How large is the instruction queue of the 8088? The 8086?
5. List the elements of the execution unit.

Section 2.2

6. What is the purpose of a software model for a microprocessor?
7. What must an assembly-language programmer know about the registers within the 8088 microprocessor?

8. How many registers are located within the 8088?
9. How large is the 8088's memory address space?
10. How large is the 8086's I/O address space?

Section 2.3

11. What is the highest address in the 8088's memory address space? The lowest address?
12. Is memory in the 8088 microprocessor organized as bytes, words, or double words?
13. The contents of memory location $B0000_{16}$ are FF_{16}, and those at $B0001_{16}$ are 00_{16}. What is the data word stored at address $B0000_{16}$? Is the word aligned or misaligned?
14. What is the value of the double word stored in memory starting at address $B0003_{16}$ if the contents of memory locations $B0003_{16}$, $B0004_{16}$, $B0005_{16}$, and $B0006_{16}$ are 11_{16}, 22_{16}, 33_{16}, and 44_{16}, respectively? Is this an example of an aligned double word or a misaligned double word?
15. Show how the word $ABCD_{16}$ is stored in memory starting at address $0A002_{16}$. Is the word aligned or misaligned?
16. Show how the double word 12345678_{16} is stored in memory starting at address $A001_{16}$. Is the double word aligned or misaligned?

Section 2.4

17. List five data types processed directly by the 8088.
18. Express each of the signed decimal integers that follow as either a byte- or word-hexadecimal number (use 2's-complement notation for negative numbers).
    (a) +127
    (b) −10
    (c) −128
    (d) +500
19. How would the integer in problem 18(d) be stored in memory starting at address $0A000_{16}$?
20. How would the decimal number −1000 be expressed for processing by the 8088?
21. Express the decimal numbers that follow as unpacked and packed BCD bytes.
    (a) 29
    (b) 88
22. How would the BCD number in problem 21(a) be stored in memory starting at address $0B000_{16}$? (Assume that the least significant digit is stored at the lower address.)
23. What statement is coded in ASCII by the following binary strings?

1001110
1000101
1011000
1010100
0100000
1001001

24. How would the decimal number 1234 be coded in ASCII and stored in memory starting at address $0C000_{16}$? (Assume that the least significant digit is stored at the lower addressed memory location.)

Section 2.5

25. How large is a memory segment in the 8088 microprocessor?
26. Which of the 8088's internal registers are used for memory segmentation?
27. What register defines the beginning of the current code segment in memory?
28. What is the maximum amount of memory that can be active at a given time in the 8088 microprocessor?
29. How much of the 8088's active memory is available as general-purpose data storage memory?

Section 2.6

30. What is the dedicated use of the part of the 8088's address space from 00000_{16} through $0007F_{16}$?
31. What is the address range of the general-use part of the memory address space?
32. Which part of the 8088's memory address space can be used to store the instructions of a program?
33. What is stored at address $FFFF0_{16}$?

Section 2.7

34. What is the function of the instruction pointer register?
35. Provide an overview of the fetch and the execution of an instruction by the 8088.
36. What happens to the value in IP each time the 8088 completes an instruction fetch?

Section 2.8

37. Make a list of the general-purpose data registers of the 8088.
38. How is the word value of a data register labeled?
39. How are the upper and lower bytes of a data register denoted?
40. Name two dedicated operations assigned to the CX register.

Section 2.9

41. What kind of information is stored in the pointer and index registers?
42. Name the two pointer registers.
43. For which segment register are the contents of the pointer registers used as an offset?
44. For which segment register are the contents of the index registers used as an offset?
45. What do SI and DI stand for?
46. What is the difference between SI and DI?

Section 2.10

47. Categorize each flag bit of the 8088 as either a control flag or a flag that monitors the status due to execution of an instruction.
48. Describe the function of each status flag.

49. How does software use a status flag?

50. What does TF stand for?

51. Which flag determines whether the address for a string operation is incremented or decremented?

52. Can the state of the flags be modified through software?

Section 2.11

53. What is the word length of the 8088's physical address?

54. What two address elements are combined to form a physical address?

55. Calculate the value of each of the physical addresses that follows. Assume all numbers are hexadecimal numbers.
(a) 1000:1234
(b) 0100:ABCD
(c) A200:12CF
(d) B2C0:FA12

56. Find the unknown value for each of the following physical addresses. Assume all numbers are hexadecimal numbers.
(a) A000:? = A0123
(b) ?:14DA = 235DA
(c) D765:? = DABC0
(d) ?:CD21 = 32D21

57. If the current values in the code segment register and the instruction pointer are 0200_{16} and $01AC_{16}$, respectively, what physical address is used in the next instruction fetch?

58. A data segment is to be located from address $A0000_{16}$ to $AFFFF_{16}$. What value must be loaded into DS?

59. If the data segment register contains the value found in problem 58, what value must be loaded into DI if it is to point to a destination operand stored in memory at address $A1234_{16}$?

Section 2.12

60. What is the function of the stack?

61. If the current values in the stack segment register and stack pointer are $C000_{16}$ and $FF00_{16}$, respectively, what is the address of the current top of the stack?

62. For the base and offset addresses in problem 61, how many words of data are currently held in the stack?

63. Show how the value $EE11_{16}$ from register AX would be pushed onto the top of the stack as it exists in problem 61.

Section 2.13

64. For the 8088 microprocessor, are the input/output and memory address spaces common or separate?

65. How large is the 8088's I/O address space?

66. What name is given to the part of the I/O address space from 0000_{16} through $00FF_{16}$?

3

Assembly Language Programming

▲ INTRODUCTION

Up to this point we have studied the software architecture of the 8088/8086 microprocessor. Here we begin a detailed study of assembly language programming for the 8088/8086-based microcomputer system. This chapter introduces software and the microcomputer program, the process used to develop an assembly language program, the instruction set of the 8088/8086 microprocessor, and its addressing modes. Chapters 5 and 6 will examine the operation of the individual instructions of the instruction set. The topics covered in this chapter are as follows:

3.1 Software: The Microcomputer Program
3.2 Assembly Language Programming Development on the PC
3.3 The Instruction Set
3.4 The MOV Instruction
3.5 Addressing Modes

▲ 3.1 SOFTWARE: THE MICROCOMPUTER PROGRAM

In this section, we begin our study of 8088/8086 assembly language programming by examining software and the microcomputer program. A microcomputer does not know how to process data. It must be told exactly what to do, where to get data, what to do with the data, and where to put the results when it is done. These are the jobs of the *software* in a microcomputer system.

The sequence of commands used to tell a microcomputer what to do is called a *program.* Each command in a program is an *instruction.* A program may be simple and include just a few instructions, or it may be very complex and contain more than 100,000 instructions. When the microcomputer is operating, it fetches and executes one instruction of the program after the other. In this way, the instructions of the program guide it step by step through the task to be performed.

Software is a general name used to refer to a wide variety of programs that can be run by a microcomputer. Examples are *languages, operating systems, application programs,* and *diagnostics.* All computer systems have two types of software: *system software* and *application software.* System software represents a group of programs that enable the microcomputer to operate and is known as the operating system (OS), such as the Windows 98 operating system. The collection of programs installed on the microcomputer for the user is the application software. Examples of frequently used PC-based applications are Word, Excel, and PowerPoint

The native language of the original IBM PC is the *machine language* of the 8088 microprocessor. Programs must always be coded in this machine language before they can be executed by the microprocessor. A program written in machine language is often referred to as *machine code.* When expressed in machine code, an instruction is encoded using 0s and 1s. A single machine language instruction can take up one or more bytes of code. Even though the 8088 understands only machine code, it is almost impossible to write programs directly in machine language. For this reason, programs are normally written in other languages, such as 8088 *assembly language* or a high-level language such as *C.*

In assembly language, each of the operations that can be performed by the 8088 microprocessor is described with alphanumeric symbols instead of with 0s and 1s. A single assembly language statement represents each instruction in a program. This statement must specify which operation is to be performed and what data are to be processed. For this reason, an instruction can be divided into two parts: its *operation code (opcode)* and its *operands.* The opcode is the part of the instruction that identifies the operation that is to be performed. For example, typical operations are add, subtract, and move. Each opcode is assigned a unique letter combination called a *mnemonic.* The mnemonics for the earlier mentioned operations are ADD, SUB, and MOV. Operands describe the data that are to be processed as the microprocessor carries out the operation specified by the opcode. They identify whether the source and destination of the data are registers within the MPU or storage locations in data memory.

An example of an instruction written in 8088 assembly language is

```
ADD  AX, BX
```

This instruction says, "Add the contents of registers BX and AX together and put the sum in register AX." AX is called the *destination operand,* because it is the place where the result ends up, and BX is called the *source operand.*

An example of a complete assembly language statement is

```
START:  MOV  AX, BX ;Copy  BX into AX
```

This statement begins with the word START:. START is an address identifier for the instruction MOV AX, BX. This type of identifier is known as a *label.* The instruction is followed by ;Copy BX into AX. This part of the statement is called a *comment.* Thus a general format for an assembly language statement is

```
LABEL:  INSTRUCTION  ;Comment
```

Programs written in assembly language are referred to as *source code.* An example of a short 8088 assembly language program is shown in Fig. 3–1(a). The assembly language instructions are located toward the left. Notice that the program includes instruc-

```
TITLE BLOCK-MOVE PROGRAM

      PAGE        ,132

COMMENT *This program moves a block of specified number of bytes
         from one place to another place*

;Define constants used in this program

      N=                 16           ;Bytes to be moved
      BLK1ADDR=          100H         ;Source block offset address
      BLK2ADDR=          120H         ;Destination block offset addr
      DATASEGADDR=       2000H        ;Data segment start address

STACK_SEG           SEGMENT           STACK 'STACK'
                    DB                64 DUP (?)
STACK_SEG           ENDS
CODE_SEG            SEGMENT           'CODE'
BLOCK               PROC        FAR
      ASSUME        CS:CODE_SEG,SS:STACK_SEG

;To return to DEBUG program put return address on the stack

            PUSH  DS
            MOV   AX, 0
            PUSH  AX

;Setup the data segment address

            MOV   AX, DATASEGADDR
            MOV   DS, AX

;Setup the source and destination offset addresses

            MOV   SI, BLK1ADDR
            MOV   DI, BLK2ADDR

;Setup the count of bytes to be moved

            MOV   CX, N
;Copy source block to destination block

NXTPT:      MOV   AH, [SI]            ;Move a byte
            MOV   [DI], AH
            INC   SI                  ;Update pointers
            INC   DI
            DEC   CX                  ;Update byte counter
            JNZ   NXTPT               ;Repeat for next byte
            RET                       ;Return to DEBUG program
BLOCK             ENDP
CODE_SEG          ENDS
      END         BLOCK               ;End of program
```

(a)

Figure 3–1 (a) Example of an 8088 assembly language program. (b) Assembled version of the program.

```
Microsoft (R) Macro Assembler Version 5.10                        5/17/92 18:10:04
BLOCK-MOVE PROGRAM                                                Page     1-1

      1
      2
      3                           TITLE BLOCK-MOVE PROGRAM
      4
      5                                 PAGE         ,132
      7                           COMMENT *This program moves a block of specified number of bytes
      8                                    from one place to another place*
      9
     10
     11                           ;Define constants used in this program
     12
     13 = 0010                         N=          16           ;Bytes to be moved
     14 = 0100                         BLK1ADDR=   100H         ;Source block offset address
     15 = 0120                         BLK2ADDR=   120H         ;Destination block offset addr
     16 = 1020                         DATASEGADDR=1020H        ;Data segment start address
     17
     18
     19 0000                           STACK_SEG   SEGMENT      STACK 'STACK'
     20 0000   0040[                               DB           64 DUP(?)
     21          ??
     22                    ]
     23
     24 0040                           STACK_SEG   ENDS
     25
     26
     27 0000                           CODE_SEG    SEGMENT      'CODE'
     28 0000                           BLOCK         PROC       FAR
     29                                ASSUME      CS:CODE_SEG,SS:STACK_SEG
     30
     31                           ;To return to DEBUG program put return address on the stack
     32
     33 0000   1E                      PUSH  DS
     34 0001   B8 0000                 MOV   AX, 0
     35 0004   50                      PUSH  AX
     36
     37                           ;Setup the data segment address
     38
     39 0005   B8 1020                 MOV   AX, DATASEGADDR
     40 0008   8E D8                   MOV   DS, AX
     41
     42                           ;Setup the source and destination offset adresses
     43
     44 000A   BE 0100                 MOV   SI, BLK1ADDR
     45 000D   BF 0120                 MOV   DI, BLK2ADDR
     46
     47                           ;Setup the count of bytes to be moved
     48
     49 0010   B9 0010                 MOV   CX, N
     50
     51                           ;Copy source block to destination block
     52
     53 0013   8A 24             NXTPT:MOV   AH, [SI]                  ;Move a byte
     54 0015   88 25                   MOV   [DI], AH
     55 0017   46                      INC   SI                        ;Update pointers
     56 0018   47                      INC   DI
     57 0019   49                      DEC   CX                        ;Update byte counter
     58 001A   75 F7                   JNZ   NXTPT                     ;Repeat for next byte
     59 001C   CB                      RET                             ;Return to DEBUG program
     60 001D                           BLOCK         ENDP
     61 001D                           CODE_SEG      ENDS
     62                                END           BLOCK             ;End of program
```

(b)

Figure 3–1 (continued)

```
Segments and Groups:

                N a m e            Length      Align           Combine Class

CODE_SEG  . . . . . . . . . . . .  001D  PARA  NONE  'CODE'
STACK_SEG  . . . . . . . . . . .   0040  PARA  STACK 'STACK'

Symbols:

                N a m e            Type   Value      Attr

BLK1ADDR . . . . . . . . . . . .   NUMBER      0100
BLK2ADDR . . . . . . . . . . . .   NUMBER      0120
BLOCK . . . . . . . . . . . . . .  F PROC      0000  CODE_SEG             Length = 001D

DATASEGADDR . . . . . . . . . . .  NUMBER      1020

N . . . . . . . . . . . . . . . .  NUMBER      0010
NXTPT . . . . . . . . . . . . . .  L NEAR      0013  CODE_SEG

@CPU . . . . . . . . . . . . . .  TEXT  0101h
@FILENAME  . . . . . . . . . . .  TEXT  block
@VERSION . . . . . . . . . . . .  TEXT  510


     59 Source  Lines
     59 Total   Lines
     15 Symbols

47222 + 347542 Bytes symbol space free

     0 Warning Errors
     0 Severe Errors
```

(b)

Figure 3–1 (b) (continued)

tion statements with both a label and comment, instructions with a comment but no label, instructions without either a label or comment, and even statements that are just a comment. In fact, most statements do not have a label. An example of a statement without a label or comments is

```
MOV  DS, AX
```

On the other hand, most statements have a comment. For instance, the statement

```
INC  SI  ;Update pointers
```

has a comment, but no label. This type of documentation makes it easier for a program to be read and debugged. The comment part of the statement does not generate any machine code.

Assembly language programs cannot be directly run on the 8088. They must still be translated to an equivalent machine language program for execution by the 8088. This conversion is done automatically by running the source program through a program known as an *assembler.* The machine language output produced by the assembler is called *object code.*

Not all of the statements in the assembly language program in Fig. 3–1(a) are instruction statements. There are also statements used to control the translation process of the assembler. An example is the statement

```
DB  64 DUP(?)
```

This type of statement is known as a *directive*—that is, it supplies directions to the assembler program.

Figure 3–1(b) is the *listing* produced by assembling the assembly language source code in Fig. 3–1(a) with Microsoft's MASM macroassembler. Reading from left to right, this listing contains line numbers, addresses of memory locations, the machine language instructions, the original assembly language statements, and comments. For example, line 53, which is

```
0013 8A 24    NXTPT: MOV  AH, [SI] ;Move a byte
```

shows that the assembly language instruction MOV AH, [SI] is encoded as 8A24 in machine language and that this 2-byte instruction is loaded into memory starting at address 0013_{16} and ending at address 0014_{16}. Note that for simplicity the machine language instructions are expressed in hexadecimal notation, not in binary form. Use of assembly language makes it much easier to write a program. But notice that there is still a one-to-one relationship between assembly and machine language instructions.

EXAMPLE 3.1

What instruction is at line 58 of the program in Fig. 3–1(b)? How is this instruction expressed in machine code?

Solution

Looking at the listing in Fig. 3–1(b), we find that the instruction is

```
JNZ  NXTPT
```

and the machine code is

```
75 F7
```

High-level languages make writing programs even easier. The instructions of a high-level language are English-like statements. Source programs written in this type of language are easier to write, read, and understand. In a language such as C, high-level commands, such as FOR, IF, and WHILE, are provided. These commands no longer correspond to a single machine language statement. Instead, they implement operations that may require many assembly language statements. Again, the statements must be converted to machine code before they can be run on the 8088. The program that converts high-level-language statements to machine code instructions is called a *compiler.*

You may be asking yourself, if it is so much easier to write programs with a high-level language, why is it important to know how to program the 8088 in its assembly language? We just pointed out that if a program is written in a high-level language, such as C, it must be compiled into machine code before it can be run on the 8088. The general nature with which compilers must be designed usually results in less efficient machine code. That is, the quality of the machine code produced for the program depends on the quality of the compiler program in use. A compiled machine code implementation of a program that was written in a high-level language results in many more machine language instructions than a hand-written assembly language version of the program. This leads us to the two key benefits derived from writing programs in assembly language: first, the machine code program that results will take up less memory space than the compiled version of the program; second, it will execute faster.

Now we know the benefits of writing programs in assembly language, but we still do not know when these benefits are important. To be important, they must outweigh the additional effort necessary to write the program in assembly language instead of a high-level language. One of the major uses of assembly language programming is in *real-time applications.* By real time, we mean that the task required by the application must be completed before any other input to the program can occur that will alter its operation.

The *device service routine* that controls the operation of the PC's hard disk drive is a good example of the kind of program that might be written in assembly language. This is because it is a segment of program that must closely control the microcomputer hardware in real time. In this case, a program that is written in a high-level language probably could not respond quickly enough to control the hardware, and even if it could, operations performed with the disk subsystem would be slower. Other examples of hardware-related operations typically performed by routines written in assembly language are communication routines such as those that drive the display and printer in a personal computer and the input/output routine that scans the keyboard.

Assembly language is important not only for controlling the microcomputer system's hardware devices but also for performing pure software operations. For instance, applications frequently require the microcomputer to search through a large table of data in memory looking for a special string of characters, such as a person's name. Writing a program in a high-level language can allow the application to perform this type of operation easily; however, for large tables of data the search will take very long. Implementing the search routine through assembly language greatly improves the performance of the search operation. Other examples of software operations that may require implementation with high-performance routines derived with assembly language are *code translations,* such as from ASCII to EBCDIC, *table sort routines,* such as a bubble sort, and *mathematical routines,* such as those for floating-point arithmetic.

Not all parts of an application require real-time performance. For this reason, it is a common practice to mix, in the same program, routines developed through a high-level language and routines developed with assembly language. That is, assembly language is used to write those parts of the application that must perform real-time operations, and high-level language is used to write those parts that are not time critical. The machine code obtained by assembling or compiling the two types of program segments is linked together to form the final application program.

The compiler program can also be instructed to produce a listing that shows the equivalent assembly language statements for its machine code output. This type of output is important for understanding how the compiler implements the application program. In fact, programmers frequently use this type of output to tune the operation of an application for better performance.

▲ 3.2 ASSEMBLY LANGUAGE PROGRAM DEVELOPMENT ON THE PC

In this section, we will look at the process by which problems are solved using software. An assembly language program is written to solve a specific problem. This problem is known as the *application.* To develop a program that implements an application, the programmer goes through a multistep process. The chart in Fig. 3–2 outlines the steps in the *program-development cycle.* Let us now examine each step of the development cycle.

Describing the Problem

Figure 3–2 shows that the development cycle sequence begins by making a clear description of the problem to be solved and ends with a program that when run performs a correct solution. First the programmer must understand and describe the problem that is to be solved. A clear, concise, and accurate description of the problem is an essential part of the process of obtaining a correct and efficient software solution. This description may be provided in an informal way, such as a verbal description, or in a more formal way with a written document.

The program we used here is an example of a simple software application. Its function is to move a fixed-length block of data, called the *source block,* from one location in memory to another location in memory called the *destination block.* For the block-move program, a verbal or a written list of events may be used to describe this problem to the programmer.

On the other hand, in most practical applications, the problem to be solved is quite complex. The programmer must know what the input data are, what operations must be performed on this information, whether or not these operations need to be performed in a special sequence, whether or not there are time constraints on performing some of the operations, if error conditions can occur during the process, and what results need to be output. For this reason, most applications are described with a written document called an *application specification.* The programmers study this specification before they begin to define a software solution for the problem.

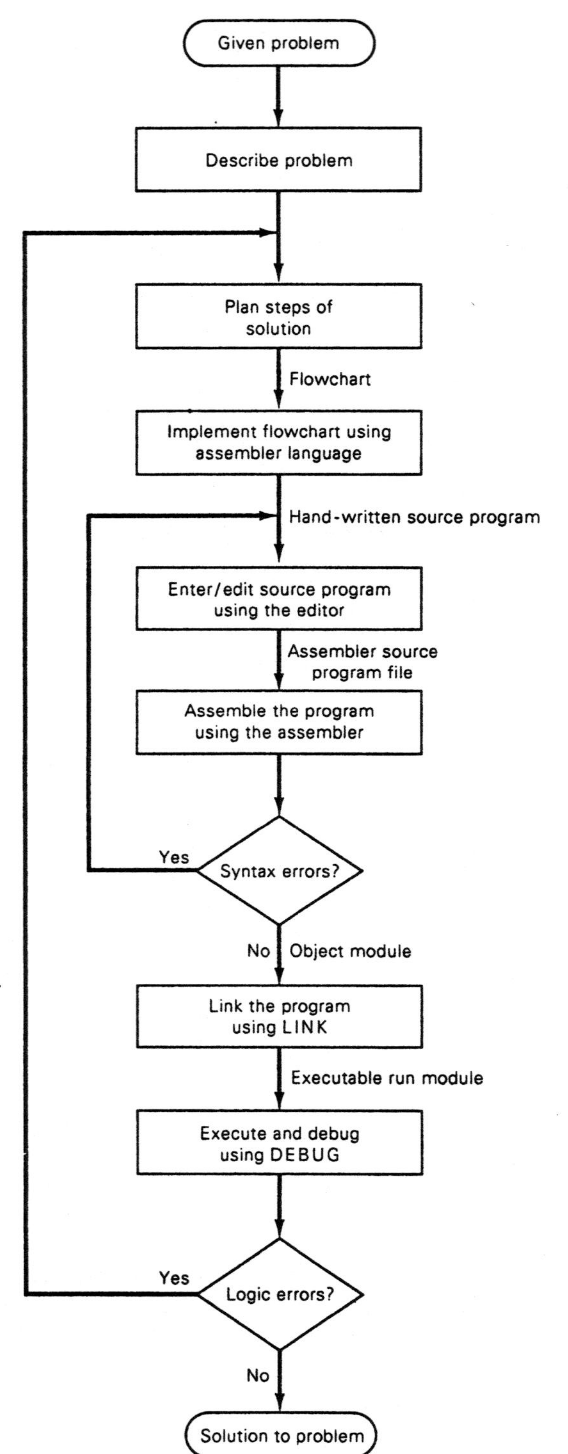

Figure 3–2 A general program development cycle.

Planning the Solution

Before writing an application program, a plan must be developed to solve the problem. Figure 3–2 shows that this is the second step in the program-development process. The decision to move to this step assumes that a complete and clear description of the problem to be solved has been provided.

The programmer carefully analyzes the application specification. Typically, the problem is broken down into a series of basic operations, which when performed in a certain sequence produce a solution to the problem. This plan defines the method by which software solves the problem. The software plan is known as the *algorithm.* Many different algorithms may be defined to solve a specific problem. However, it is important to formulate the best algorithm so that the software efficiently performs the application.

Usually, the algorithm is described with another document called the *software specification.* Also, the proposed solution may be presented in a pictorial form known as a *flowchart* in the specification. A flowchart is an outline that both documents the operations that the software must perform to implement the planned solution and shows the sequence in which they are performed. Figure 3–3(a) is the flowchart for a program that performs a block-move operation.

The flowchart identified operations that can be implemented with assembly language instructions. For example, the first block calls for setting up a data segment, initializing the pointers for the starting addresses of the source and destination blocks, and specifying the count of the number of pieces of data that are to be moved. These types of operations can be achieved by moving either immediate data, or data from a known memory location, into appropriate registers within the MPU.

A flowchart uses a set of symbols to identify both the operations required in the solution and the sequence in which they are performed. Figure 3–4 lists the most commonly used flowcharting symbols. Note that symbols are listed for identifying the beginning or end of the flowchart, input or output of data, processing functions, making a decision operation, connecting blocks within the flowchart, and connections to other flowcharts. The operation to be performed is written inside the symbol. The flowchart in Fig. 3–3(a) illustrates the use of some of these symbols. Note that a begin/end symbol, which contains the comment *Enter block move,* is used to mark the beginning of the program and another, which reads *Return to DEBUG,* marks the end of the sequence. Process function boxes are used to identify each of the tasks (initialize registers, copy source element to destination, and increment source and destination address pointers and decrement data element count) that are performed as part of the block-move routine. Arrows are used to describe the sequence (flow) of these operations as the block-move operation is performed.

The solution should be hand-tested to verify that it correctly solves the stated problem. Specifying test cases with known inputs and outputs can do this. Then, tracing through the operation sequence defined in the flowchart for these input conditions, the outputs are found and compared to the known test results. If the results are not the same, the cause of the error must be found, the algorithm is modified, and the tests rerun. When the results match, the algorithm is assumed to be correct, and the programmer is ready to move on to the next step in the development cycle. The process is called desk checking.

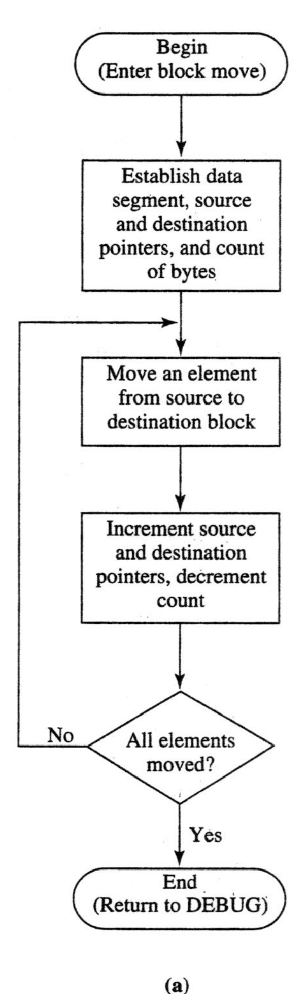

Figure 3–3 (a) Flowchart of a block-move program. (b) Block-move source program.

The flowchart representation of the planned solution is a valuable aid to the programmer when coding the solution with assembly language instructions. When a problem is a simple one, the flowcharting step may be bypassed. A list of the tasks and the sequence in which they must be performed may be enough to describe the solution to the problem. However, for complex applications, a flowchart is an important program-development tool for obtaining an accurate and timely solution.

Coding the Solution with Assembly Language

The application program is the step-by-step sequence of computer operations that must be performed to convert the input data to the required output results—that is, it is the software implementation of the algorithm. The third step of the program development

```
TITLE BLOCK-MOVE PROGRAM

      PAGE        ,132

COMMENT *This program moves a block of specified number of bytes
         from one place to another place*

;Define constants used in this program

      N=                  16          ;Bytes to be moved
      BLK1ADDR=           100H        ;Source block offset address
      BLK2ADDR=           120H        ;Destination block offset addr
      DATASEGADDR=        2000H       ;Data segment start address

STACK_SEG           SEGMENT           STACK 'STACK'
                    DB                64 DUP(?)
STACK_SEG           ENDS
CODE_SEG            SEGMENT           'CODE'
BLOCK               PROC        FAR
      ASSUME        CS:CODE_SEG,SS:STACK_SEG

;To return to DEBUG program put return address on the stack

            PUSH  DS
            MOV   AX, 0
            PUSH  AX

;Setup the data segment address

            MOV   AX, DATASEGADDR
            MOV   DS, AX

;Setup the source and destination offset addresses

            MOV   SI, BLK1ADDR
            MOV   DI, BLK2ADDR

;Setup the count of bytes to be moved

            MOV   CX, N

;Copy source block to destination block

NXTPT:      MOV   AH, [SI]            ;Move a byte
            MOV   [DI], AH
            INC   SI                  ;Update pointers
            INC   DI
            DEC   CX                  ;Update byte counter
            JNZ   NXTPT               ;Repeat for next byte
            RET                       ;Return to DEBUG program
BLOCK             ENDP
CODE_SEG          ENDS
      END         BLOCK               ;End of program
```

(b)

Figure 3–3 (continued)

cycle, as shown in Fig. 3–2, is the translation of the flowchart solution into its equivalent assembly language program. This requires the programmer to implement the operations described in each symbol of the flowchart with a sequence of assembly language instructions. These instruction sequences are then combined to form a handwritten assembly language program called the *source program.*

Two types of statements are used in the source program. First, there are the assembly language instructions. They are used to tell the microprocessor what operations are to be performed to implement the application.

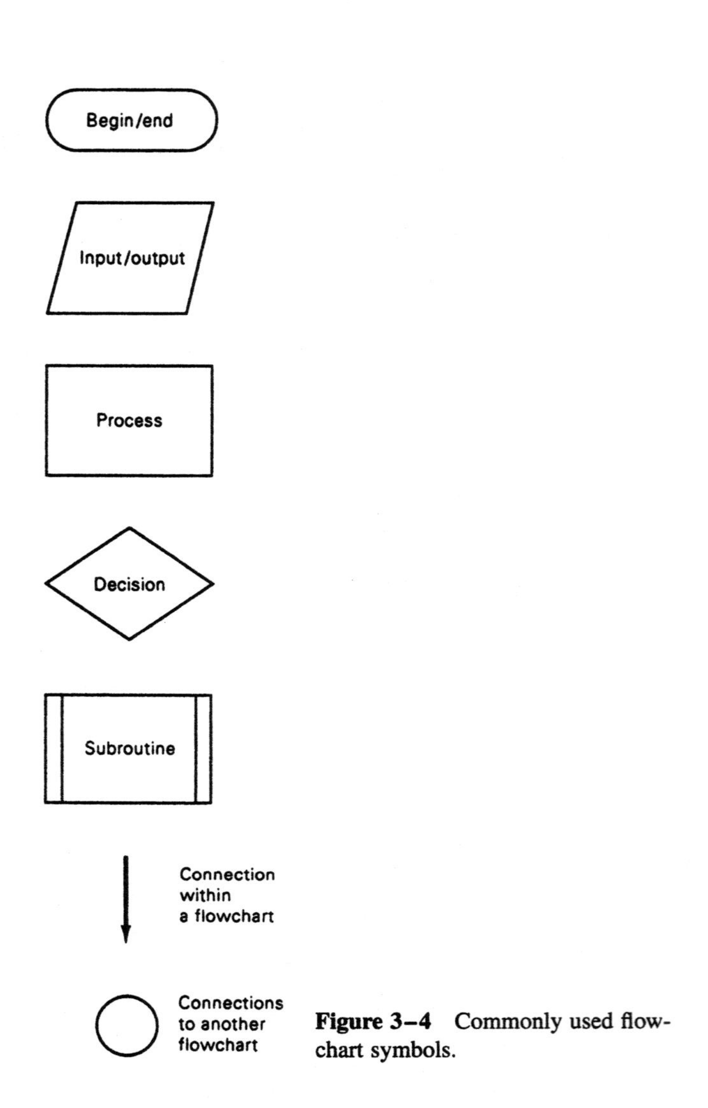

Figure 3–4 Commonly used flowchart symbols.

The assembly language program in Fig. 3–3(b) implements the block-move operation flowchart in Fig. 3–3(a). Comparing the flowchart to the program, it is easy to see that the initialization block is implemented with the assembly language statements

```
MOV  AX, DATASEGADDR
MOV  DS, AX
MOV  SI, BLK1ADDR
MOV  DI, BLK2ADDR
MOV  CX, N
```

The first two move instructions load a segment base address called DATASEGADDR into the data segment register. This defines the data segment in memory where the two blocks of data reside. Next, two more instructions are used to load SI and DI with the start offset address of the source block (BLK1ADDR) and destination block (BLK2ADDR), respectively. Finally, the count N of the number of bytes of data to be copied to the destination block is loaded into count register CX.

A source program can also contain another type of statement called a *directive,* which are instructions to the assembler program that is used to convert the assembly language program into machine code. We will discuss these statements in more detail in a later chapter. In Fig. 3–3(b), the statements

```
BLOCK  PROC  FAR
```

and

```
BLOCK  ENDP
```

are examples of modular programming directives. They mark the beginning and end, respectively, of the software procedure called BLOCK.

To do this step of the development cycle, the programmer must know the instruction set of the microprocessor, basic assembly language programming techniques, the assembler's instruction statement syntax, and the assembler's directives.

Creating the Source Program

After having handwritten the assembly language program, we are ready to enter it into the computer. This step is identified as the enter/edit source program block in the program-development cycle diagram in Fig. 3–2 and is done with a program called an *editor.* We will use the EDIT editor, which is available as part of the DOS operating system. Using an editor, each of the statements of the program is typed into the computer. If errors are made as the statements are keyed in, the corrections can either be made at the time of entry or edited at a later time. The source program is saved in a file.

Assembling the Source Program into an Object Module

The fifth step of the flowchart in Fig. 3–2 is the point at which the assembly language source program is converted to its corresponding machine language program. To do this, we use a program called an *assembler.* A program originally available from Microsoft Corporation called *MASM* is an example of an 8088/8086 assembler that runs in DOS on a PC. The assembler program reads as its input the contents of the *assembler source file;* it converts this program statement by statement to machine code and produces a machine-code program as its output. This machine-code output is stored in a file called the *object module.*

If during the conversion operation syntax errors are found—that is, violations in the rules of writing the assembly language statements for the assembler—the assembler automatically flags them. As shown in the flowchart in Fig. 3–2, before going on, the cause of each error in the source program must be identified and corrected. The corrections are made using the editor program. After the corrections are made, the source program must be reassembled. This edit-assemble sequence must be repeated until the program assembles with no error.

Producing a Run Module

The object module produced by the assembler cannot be run directly on the microcomputer. As shown in Fig. 3–2, a *LINK program* must process the module to produce an executable object module, which is known as a *run module.* The linker program converts the object module to a run module by making it address compatible with the microcomputer on which it is to be run. For instance, if our computer is implemented with memory at addresses $0A000_{16}$ through $0FFFF_{16}$, the executable machine-code output by the linker will also have addresses in this range.

There is another purpose for the use of a linker: it links different object modules to generate a single executable object module. This allows program development to be done in modules, which are later combined to form the application program.

Verifying the Solution

Now the executable object module is ready to be run on the microcomputer. Once again, the PC's DOS operating system provides us with a program, which is called DEBUG, to perform this function. DEBUG provides an environment in which we can run the program instruction by instruction or run a group of instructions at a time, look at intermediate results, display the contents of the registers within the microprocessor, and so on.

For instance, we could verify the operation of our earlier block-move program by running it for the data in the cases defined to test the algorithm. DEBUG is used to load the run module for block-move into the PC's memory. After loading is completed and verified, other DEBUG commands are employed to run the program for the data in the test case. The DEBUG program permits us to trace the operation as instructions are executed and observe each element of data as it is copied from the source to the destination block. These results are recorded and compared to those provided with the test case. If the program is found to perform the block-move operation correctly, the program-development process is complete.

On the other hand, Fig. 3–2 shows that if errors are discovered in the logic of the solution, the cause must be determined, corrections must be made to the algorithm, and then the assembly language source program must be corrected using the editor. The edited source file must be reassembled, relinked, and retested by running it with DEBUG. This loop must be repeated until it is verified that the program correctly performs the operation for which it was written.

Programs and Files Involved in the Program Development Cycle

The edit, assemble, link, and debug parts of the general program-development cycle in Fig. 3–2 are performed directly on the PC. Figure 3–5 shows the names of the pro-

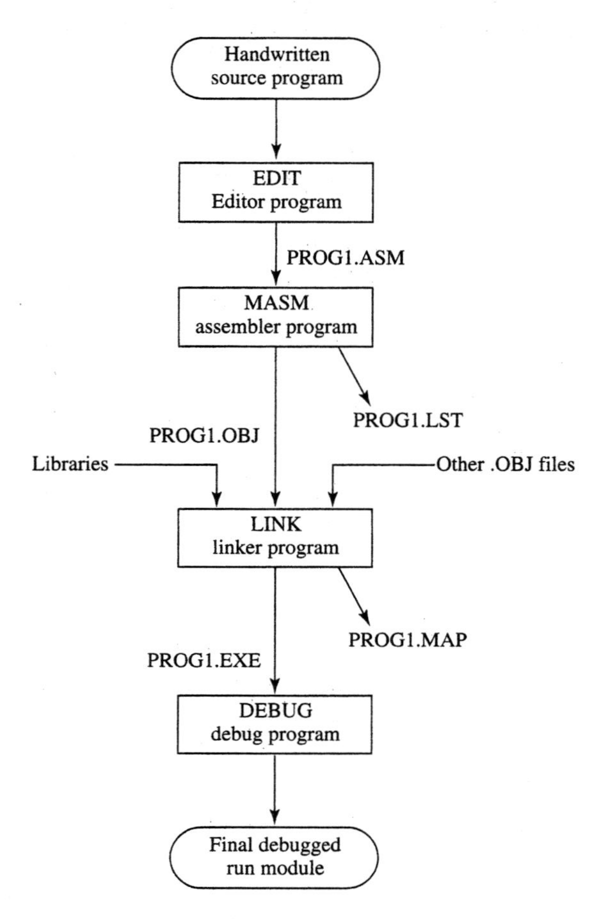

Figure 3–5 The development programs and user files.

grams and typical filenames with extensions used as inputs and outputs during this process. For example, the *EDIT program* is an editor used to create and correct assembly language source files. The program that results is shown to have the name PROG1.ASM. This stands for program 1 assembly source code.

MASM, which stands for *macroassembler,* is a program that can be used to assemble source files into object modules. The assembler converts the contents of the source input file PROG1.ASM into two output files called PROG1.OBJ and PROG1.LST. The file PROG1.OBJ contains the object code module. The PROG1.LST file provides additional information useful for debugging the application program.

Object module PROG1.OBJ can be linked to other object modules with the LINK program. For instance, programs that are available as object modules in a math library could be linked with another program to implement math operations. A library is a collection of prewritten, assembled, and tested programs. Notice that this program produces a run module in file PROG1.EXE and a map file called PROG1.MAP as outputs. The executable object module, PROG1.EXE, is run with the debugger program, called DEBUG.

Map file PROG1.MAP is supplied as support for the debugging operation by providing additional information such as where the program will be located when loaded into the microcomputer's memory.

▲ 3.3 THE INSTRUCTION SET

The microprocessor's instruction set defines the basic operations that a programmer can specify to the device to perform. The 8088 and 8086 microprocessors have the same instruction set; Fig. 3–6 contains a list of 117 basic instructions for the 8088/8086. For the purpose of discussion, these instructions are organized into groups of functionally related instructions. In Fig. 3–6, we see that these groups consist of the data transfer instructions, arithmetic instructions, logic instructions, string manipulation instructions, control transfer instructions, and processor control instructions.

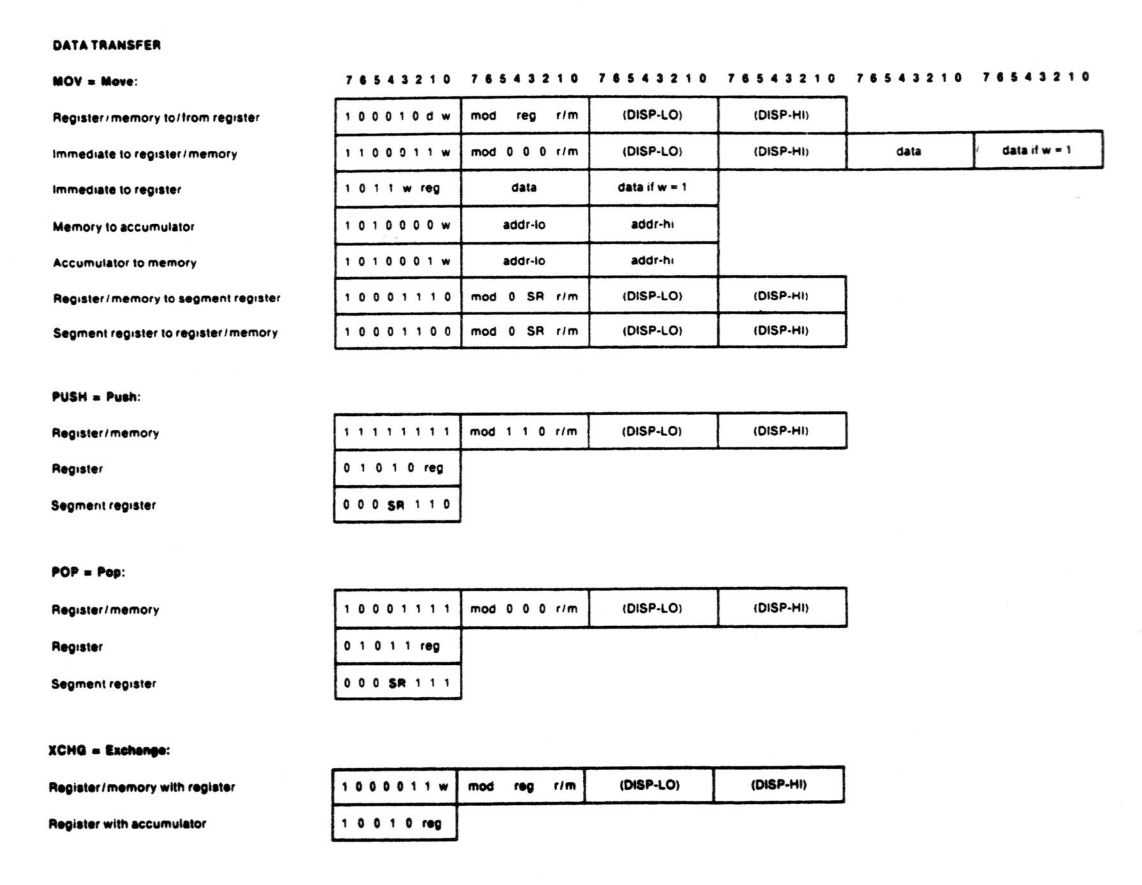

DATA TRANSFER

MOV = Move:	7 6 5 4 3 2 1 0	7 6 5 4 3 2 1 0	7 6 5 4 3 2 1 0	7 6 5 4 3 2 1 0	7 6 5 4 3 2 1 0	7 6 5 4 3 2 1 0
Register/memory to/from register	1 0 0 0 1 0 d w	mod reg r/m	(DISP-LO)	(DISP-HI)		
Immediate to register/memory	1 1 0 0 0 1 1 w	mod 0 0 0 r/m	(DISP-LO)	(DISP-HI)	data	data if w = 1
Immediate to register	1 0 1 1 w reg	data	data if w = 1			
Memory to accumulator	1 0 1 0 0 0 0 w	addr-lo	addr-hi			
Accumulator to memory	1 0 1 0 0 0 1 w	addr-lo	addr-hi			
Register/memory to segment register	1 0 0 0 1 1 1 0	mod 0 SR r/m	(DISP-LO)	(DISP-HI)		
Segment register to register/memory	1 0 0 0 1 1 0 0	mod 0 SR r/m	(DISP-LO)	(DISP-HI)		
PUSH = Push:						
Register/memory	1 1 1 1 1 1 1 1	mod 1 1 0 r/m	(DISP-LO)	(DISP-HI)		
Register	0 1 0 1 0 reg					
Segment register	0 0 0 SR 1 1 0					
POP = Pop:						
Register/memory	1 0 0 0 1 1 1 1	mod 0 0 0 r/m	(DISP-LO)	(DISP-HI)		
Register	0 1 0 1 1 reg					
Segment register	0 0 0 SR 1 1 1					
XCHG = Exchange:						
Register/memory with register	1 0 0 0 0 1 1 w	mod reg r/m	(DISP-LO)	(DISP-HI)		
Register with accumulator	1 0 0 1 0 reg					

Figure 3–6 Instruction set of the 8088/8086. (Reprinted by permission of Intel Corp. Copyright/Intel Corp. 1979)

IN = Input from:

Instruction	Byte 1	Byte 2
Fixed port	1 1 1 0 0 1 0 w	DATA-8
Variable port	1 1 1 0 1 1 0 w	

OUT = Output to:

Instruction	Byte 1	Byte 2	Byte 3	Byte 4
Fixed port	1 1 1 0 0 1 1 w	DATA-8		
Variable port	1 1 1 0 1 1 1 w			
XLAT = Translate byte to AL	1 1 0 1 0 1 1 1			
LEA = Load EA to register	1 0 0 0 1 1 0 1	mod reg r/m	(DISP-LO)	(DISP-HI)
LDS = Load pointer to DS	1 1 0 0 0 1 0 1	mod reg r/m	(DISP-LO)	(DISP-HI)
LES = Load pointer to ES	1 1 0 0 0 1 0 0	mod reg r/m	(DISP-LO)	(DISP-HI)
LAHF = Load AH with flags	1 0 0 1 1 1 1 1			
SAHF = Store AH into flags	1 0 0 1 1 1 1 0			
PUSHF = Push flags	1 0 0 1 1 1 0 0			
POPF = Pop flags	1 0 0 1 1 1 0 1			

ARITHMETIC

ADD = Add:

Instruction	7 6 5 4 3 2 1 0	7 6 5 4 3 2 1 0	7 6 5 4 3 2 1 0	7 6 5 4 3 2 1 0	7 6 5 4 3 2 1 0	7 6 5 4 3 2 1 0
Reg/memory with register to either	0 0 0 0 0 0 d w	mod reg r/m	(DISP-LO)	(DISP-HI)		
Immediate to register/memory	1 0 0 0 0 0 s w	mod 0 0 0 r/m	(DISP-LO)	(DISP-HI)	data	data if s: w=01
Immediate to accumulator	0 0 0 0 0 1 0 w	data	data if w=1			

ADC = Add with carry:

Instruction	Byte 1	Byte 2	Byte 3	Byte 4	Byte 5	Byte 6
Reg/memory with register to either	0 0 0 1 0 0 d w	mod reg r/m	(DISP-LO)	(DISP-HI)		
Immediate to register/memory	1 0 0 0 0 0 s w	mod 0 1 0 r/m	(DISP-LO)	(DISP-HI)	data	data if s: w=01
Immediate to accumulator	0 0 0 1 0 1 0 w	data	data if w=1			

INC = Increment:

Instruction	Byte 1	Byte 2	Byte 3	Byte 4
Register/memory	1 1 1 1 1 1 1 w	mod 0 0 0 r/m	(DISP-LO)	(DISP-HI)
Register	0 1 0 0 0 reg			
AAA = ASCII adjust for add	0 0 1 1 0 1 1 1			
DAA = Decimal adjust for add	0 0 1 0 0 1 1 1			

SUB = Subtract:

Instruction	Byte 1	Byte 2	Byte 3	Byte 4	Byte 5	Byte 6
Reg/memory and register to either	0 0 1 0 1 0 d w	mod reg r/m	(DISP-LO)	(DISP-HI)		
Immediate from register/memory	1 0 0 0 0 0 s w	mod 1 0 1 r/m	(DISP-LO)	(DISP-HI)	data	data if s: w=01
Immediate from accumulator	0 0 1 0 1 1 0 w	data	data if w=1			

Figure 3–6 (continued)

SBB = Subtract with borrow:

Instruction	Byte 1	Byte 2	Byte 3	Byte 4	Byte 5	Byte 6
Reg/memory and register to either	0 0 0 1 1 0 d w	mod reg r/m	(DISP-LO)	(DISP-HI)		
Immediate from register/memory	1 0 0 0 0 0 s w	mod 0 1 1 r/m	(DISP-LO)	(DISP-HI)	data	data if s: w=01
Immediate from accumulator	0 0 0 1 1 1 0 w	data	data if w=1			

DEC Decrement:

Instruction	Byte 1	Byte 2	Byte 3	Byte 4
Register/memory	1 1 1 1 1 1 1 w	mod 0 0 1 r/m	(DISP-LO)	(DISP-HI)
Register	0 1 0 0 1 reg			
NEG Change sign	1 1 1 1 0 1 1 w	mod 0 1 1 r/m	(DISP-LO)	(DISP-HI)

CMP = Compare:

Instruction	Byte 1	Byte 2	Byte 3	Byte 4	Byte 5	Byte 6
Register/memory and register	0 0 1 1 1 0 d w	mod reg r/m	(DISP-LO)	(DISP-HI)		
Immediate with register/memory	1 0 0 0 0 0 s w	mod 1 1 1 r/m	(DISP-LO)	(DISP-HI)	data	data if s: w=1
Immediate with accumulator	0 0 1 1 1 1 0 w	data				
AAS ASCII adjust for subtract	0 0 1 1 1 1 1 1					
DAS Decimal adjust for subtract	0 0 1 0 1 1 1 1					
MUL Multiply (unsigned)	1 1 1 1 0 1 1 w	mod 1 0 0 r/m	(DISP-LO)	(DISP-HI)		

ARITHMETIC	7 6 5 4 3 2 1 0	7 6 5 4 3 2 1 0	7 6 5 4 3 2 1 0	7 6 5 4 3 2 1 0	7 6 5 4 3 2 1 0	7 6 5 4 3 2 1 0
IMUL Integer multiply (signed)	1 1 1 1 0 1 1 w	mod 1 0 1 r/m	(DISP-LO)	(DISP-HI)		
AAM ASCII adjust for multiply	1 1 0 1 0 1 0 0	0 0 0 0 1 0 1 0	(DISP-LO)	(DISP-HI)		
DIV Divide (unsigned)	1 1 1 1 0 1 1 w	mod 1 1 0 r/m	(DISP-LO)	(DISP-HI)		
IDIV Integer divide (signed)	1 1 1 1 0 1 1 w	mod 1 1 1 r/m	(DISP-LO)	(DISP-HI)		
AAD ASCII adjust for divide	1 1 0 1 0 1 0 1	0 0 0 0 1 0 1 0	(DISP-LO)	(DISP-HI)		
CBW Convert byte to word	1 0 0 1 1 0 0 0					
CWD Convert word to double word	1 0 0 1 1 0 0 1					

LOGIC

Instruction	Byte 1	Byte 2	Byte 3	Byte 4
NOT Invert	1 1 1 1 0 1 1 w	mod 0 1 0 r/m	(DISP-LO)	(DISP-HI)
SHL/SAL Shift logical/arithmetic left	1 1 0 1 0 0 v w	mod 1 0 0 r/m	(DISP-LO)	(DISP-HI)
SHR Shift logical right	1 1 0 1 0 0 v w	mod 1 0 1 r/m	(DISP-LO)	(DISP-HI)
SAR Shift arithmetic right	1 1 0 1 0 0 v w	mod 1 1 1 r/m	(DISP-LO)	(DISP-HI)
ROL Rotate left	1 1 0 1 0 0 v w	mod 0 0 0 r/m	(DISP-LO)	(DISP-HI)
ROR Rotate right	1 1 0 1 0 0 v w	mod 0 0 1 r/m	(DISP-LO)	(DISP-HI)
RCL Rotate through carry flag left	1 1 0 1 0 0 v w	mod 0 1 0 r/m	(DISP-LO)	(DISP-HI)
RCR Rotate through carry right	1 1 0 1 0 0 v w	mod 0 1 1 r/m	(DISP-LO)	(DISP-HI)

Figure 3–6 (continued)

AND = And:

Reg/memory with register to either	0 0 1 0 0 0 d w	mod reg r/m	(DISP-LO)	(DISP-HI)		
Immediate to register/memory	1 0 0 0 0 0 0 w	mod 1 0 0 r/m	(DISP-LO)	(DISP-HI)	data	data if w=1
Immediate to accumulator	0 0 1 0 0 1 0 w	data	data if w=1			

TEST = And function to flags no result:

Register/memory and register	0 0 0 1 0 0 d w	mod reg r/m	(DISP-LO)	(DISP-HI)		
Immediate data and register/memory	1 1 1 1 0 1 1 w	mod 0 0 0 r/m	(DISP-LO)	(DISP-HI)	data	data if w=1
Immediate data and accumulator	1 0 1 0 1 0 0 w	data				

OR = Or:

Reg/memory and register to either	0 0 0 0 1 0 d w	mod reg r/m	(DISP-LO)	(DISP-HI)		
Immediate to register/memory	1 0 0 0 0 0 0 w	mod 0 0 1 r/m	(DISP-LO)	(DISP-HI)	data	data if w=1
Immediate to accumulator	0 0 0 0 1 1 0 w	data	data if w=1			

XOR = Exclusive or:

Reg/memory and register to either	0 0 1 1 0 0 d w	mod reg r/m	(DISP-LO)	(DISP-HI)		
Immediate to register/memory	0 0 1 1 0 1 0 w	data	(DISP-LO)	(DISP-HI)	data	data if w=1
Immediate to accumulator	0 0 1 1 0 1 0 w	data	data if w=1			

STRING MANIPULATION	7 6 5 4 3 2 1 0	7 6 5 4 3 2 1 0	7 6 5 4 3 2 1 0	7 6 5 4 3 2 1 0	7 6 5 4 3 2 1 0	7 6 5 4 3 2 1 0
REP = Repeat	1 1 1 1 0 0 1 z					
MOVS = Move byte/word	1 0 1 0 0 1 0 w					
CMPS = Compare byte/word	1 0 1 0 0 1 1 w					
SCAS = Scan byte/word	1 0 1 0 1 1 1 w					
LODS = Load byte/wd to AL/AX	1 0 1 0 1 1 0 w					
STDS = Stor byte/wd from AL/A	1 0 1 0 1 0 1 w					

CONTROL TRANS...

CALL = Call:

Direct within segment	1 1 1 0 1 0 0 0	IP-INC-LO	IP-INC-HI	
Indirect within segment	1 1 1 1 1 1 1 1	mod 0 1 0 r/m	(DISP-LO)	(DISP-HI)
Direct intersegment	1 0 0 1 1 0 1 0	IP-lo	IP-hi	
		CS-lo	CS-hi	
Indirect intersegment	1 1 1 1 1 1 1 1	mod 0 1 1 r/m	(DISP-LO)	(DISP-HI)

Figure 3–6 (continued)

JMP = Unconditional Jump:

Direct within segment	1 1 1 0 1 0 0 1	IP-INC-LO	IP-INC-HI	
Direct within segment-short	1 1 1 0 1 0 1 1	IP-INC8		
Indirect within segment	1 1 1 1 1 1 1 1	mod 1 0 0 r/m	(DISP-LO)	(DISP-HI)
Direct intersegment	1 1 1 0 1 0 1 0	IP-lo	IP-hi	
		CS-lo	CS-hi	
Indirect intersegment	1 1 1 1 1 1 1 1	mod 1 0 1 r/m	(DISP-LO)	(DISP-HI)

RET = Return from CALL:

Within segment	1 1 0 0 0 0 1 1		
Within seg adding immed to SP	1 1 0 0 0 0 1 0	data-lo	data-hi
Intersegment	1 1 0 0 1 0 1 1		
Intersegment adding immediate to SP	1 1 0 0 1 0 1 0	data-lo	data-hi
JE/JZ = Jump on equal/zero	0 1 1 1 0 1 0 0	IP-INC8	
JL/JNGE = Jump on less/not greater or equal	0 1 1 1 1 1 0 0	IP-INC8	
JLE/JNG = Jump on less or equal/not greater	0 1 1 1 1 1 1 0	IP-INC8	
JB/JNAE = Jump on below/not above or equal	0 1 1 1 0 0 1 0	IP-INC8	
JBE/JNA = Jump on below or equal/not above	0 1 1 1 0 1 1 0	IP-INC8	
JP/JPE = Jump on parity/parity even	0 1 1 1 1 0 1 0	IP-INC8	
JO = Jump on overflow	0 1 1 1 0 0 0 0	IP-INC8	
JS = Jump on sign	0 1 1 1 1 0 0 0	IP-INC8	
JNE/JNZ = Jump on not equal/not zer0	0 1 1 1 0 1 0 1	IP-INC8	

CONTROL TRANSFER (Cont'd.)	7 6 5 4 3 2 1 0	7 6 5 4 3 2 1 0	7 6 5 4 3 2 1 0	7 6 5 4 3 2 1 0	7 6 5 4 3 2 1 0	7 6 5 4 3 2 1 0
JNL/JGE = Jump on not less/greater or equal	0 1 1 1 1 1 0 1	IP-INC8				
JNLE/JG = Jump on not less or equal/greater	0 1 1 1 1 1 1 1	IP-INC8				
JNB/JAE = Jump on not below/above or equal	0 1 1 1 0 0 1 1	IP-INC8				
JNBE/JA = Jump on not below or equal/above	0 1 1 1 0 1 1 1	IP-INC8				
JNP/JPO = Jump on not par/par odd	0 1 1 1 1 0 1 1	IP-INC8				
JNO = Jump on not overflow	0 1 1 1 0 0 0 1	IP-INC8				
JNS = Jump on not sign	0 1 1 1 1 0 0 1	IP-INC8				
LOOP = Loop CX times	1 1 1 0 0 0 1 0	IP-INC8				
LOOPZ/LOOPE = Loop while zero/equal	1 1 1 0 0 0 0 1	IP-INC8				
LOOPNZ/LOOPNE = Loop while not zero/equal	1 1 1 0 0 0 0 0	IP-INC8				
JCXZ = Jump on CX zero	1 1 1 0 0 0 1 1	IP-INC8				

Figure 3–6 (continued)

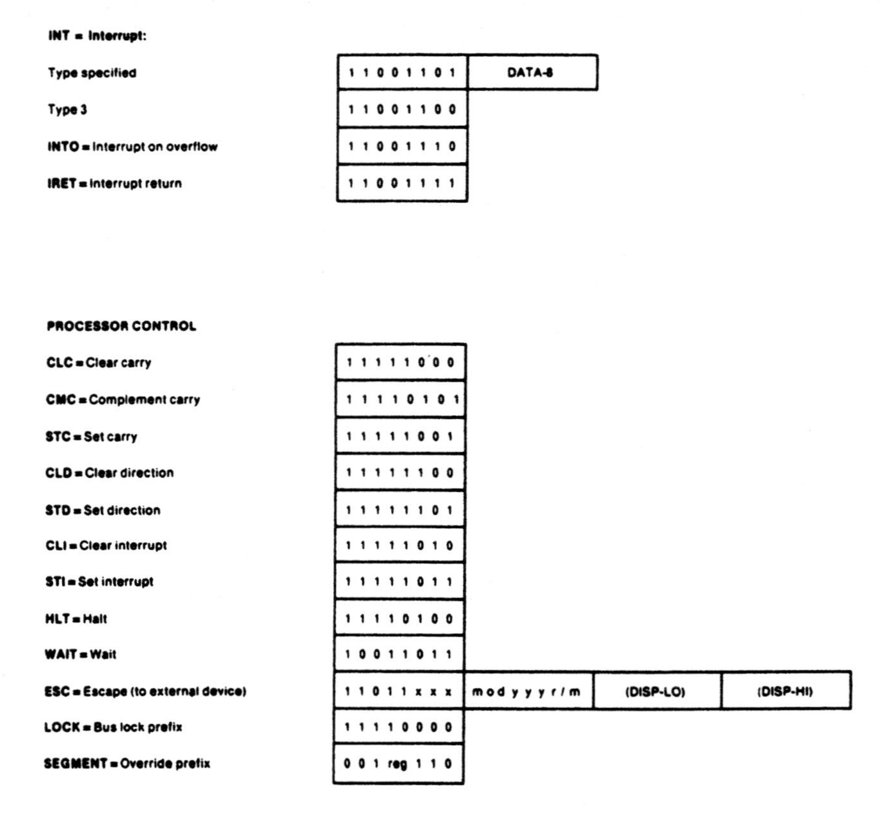

Figure 3–6 (continued)

Note that the first instruction in the data transfer group is identified as MOV (move). The wide range of operands and addressing modes permitted for use with these instructions further expands the instruction set into many more executable instructions at the machine code level. For instance, the basic MOV instruction expands into 28 different machine-level instructions.

In Chapter 5 we consider the data transfer instructions, arithmetic instructions, logic instructions, shift instructions, and rotate instructions. Advanced instructions, such as those for string manipulation and processor control, are covered in Chapter 6.

▲ 3.4 THE MOV INSTRUCTION

The move instruction is one of the instructions in the data transfer group of the 8088/8086 instruction set. The format of this instruction, as shown in Fig. 3–7(a), is written in general as

```
MOV  D, S
```

Mnemonic	Meaning	Format	Operation	Flags affected
MOV	Move	MOV D,S	(S) → (D)	None

(a)

Destination	Source
Memory	Accumulator
Accumulator	Memory
Register	Register
Register	Memory
Memory	Register
Register	Immediate
Memory	Immediate
Seg-reg	Reg16
Seg-reg	Mem16
Reg16	Seg-reg
Memory	Seg-reg

(b)

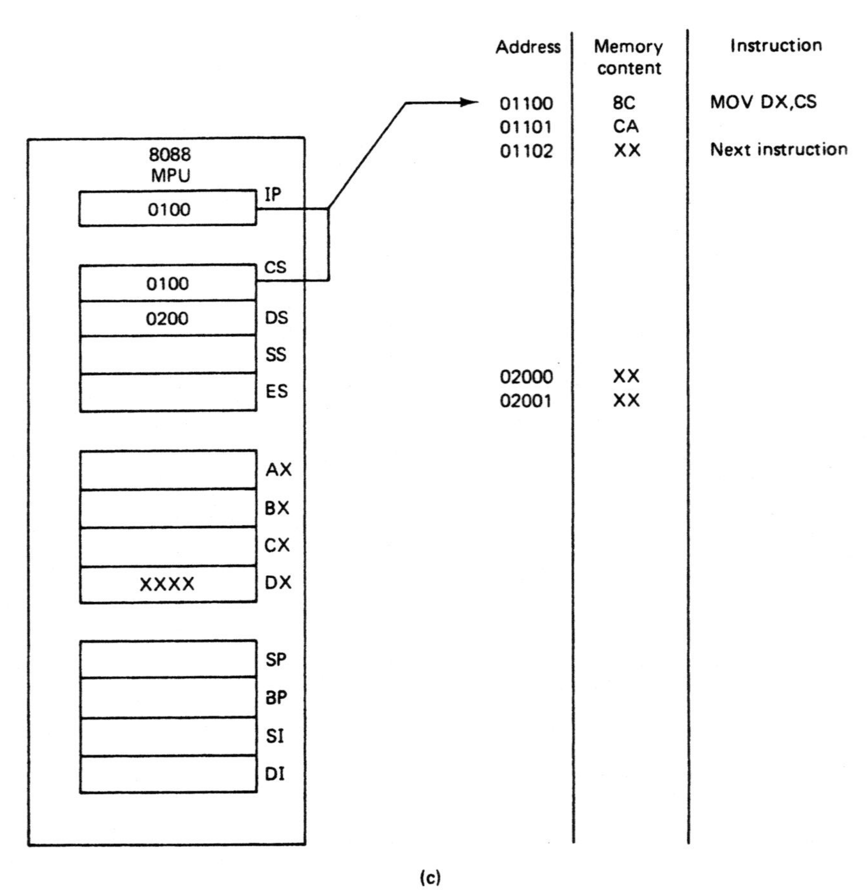

(c)

Figure 3–7

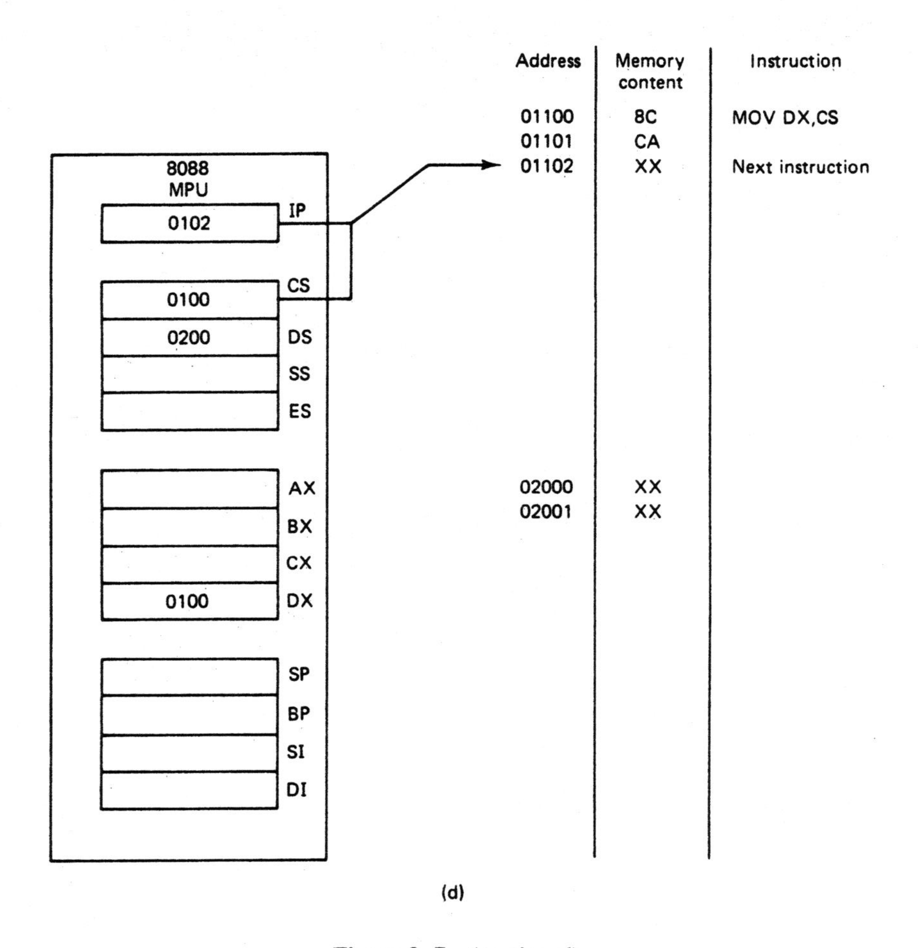

Figure 3–7 (continued)

Its operation is described in general as

$$(S) \rightarrow (D)$$

That is, execution of the instruction transfers a byte or a word of data from a source location to a destination location. These data locations can be internal registers of the 8088 and storage locations in memory. Figure 3–7(b) shows the valid source and destination variations. This large choice of source and data locations results in many different move instructions. Looking at this list, we see that data can be moved between general-purpose registers, between a general-purpose register and a segment register, between a general-purpose register or segment register and memory, or between a memory location and the accumulator.

Figure 3–7(c) shows how the instruction MOV DX,DS exists in the memory for the assumed address 01100H. To access the instruction at address 01100H, CS and IP can both be 100H. If DS contains 200H, execution of this instruction as shown in Figure 3–7(d) will place 200H in DX, and IP will increment to 102H.

▲ 3.5 ADDRESSING MODES

When the 8088 executes an instruction, it performs the specified function on data. These data, called operands, may be part of the instruction, may reside in one of the internal registers of the microprocessor, may be stored at an address in memory, or may be held at an I/O port. To access these different types of operands, the 8088 is provided with various *addressing modes.* An addressing mode is a method of specifying an operand. The addressing modes are categorized into three types: *register operand addressing, immediate operand addressing,* and *memory operand addressing.* Let us now consider in detail the addressing modes in each of these categories.

Register Operand Addressing Mode

With the *register addressing mode,* the operand to be accessed is specified as residing in an internal register of the 8088. Figure 3–8 lists the internal registers that can be used as a source or destination operand. Note that only the data registers can be accessed as either a byte or word.

An example of an instruction that uses this addressing mode is

```
MOV  AX, BX
```

Register	Operand sizes	
	Byte (Reg 8)	Word (Reg 16)
Accumulator	AL, AH	AX
Base	BL, BH	BX
Count	CL, CH	CX
Data	DL, DH	DX
Stack pointer	—	SP
Base pointer	—	BP
Source index	—	SI
Destination index	—	DI
Code segment	—	CS
Data segment	—	DS
Stack segment	—	SS
Extra segment	—	ES

Figure 3–8 Register addressing registers and operand sizes.

This stands for "move the contents of BX, which is the *source operand,* to AX, which is the *destination operand.*" Both the source and destination operands have been specified as the word contents of internal registers of the 8088.

Let us now look at the effect of executing the register addressing mode move instruction. In Fig. 3–9(a), we see the state of the 8088 just prior to fetching the instruction. Note that the logical address formed from CS and IP (CS:IP) points to the MOV AX,BX instruction at physical address 01000_{16}. This instruction is fetched into the 8088's instruction queue, where it is held waiting to be executed.

Prior to execution of this instruction, the contents of BX are $ABCD_{16}$, and the contents of AX represent a don't-care state. The instruction is read from the output side of the queue, decoded, and executed. As Fig. 3–9(b) shows, the result produced by executing this instruction is that the value $ABCD_{16}$ is copied into AX.

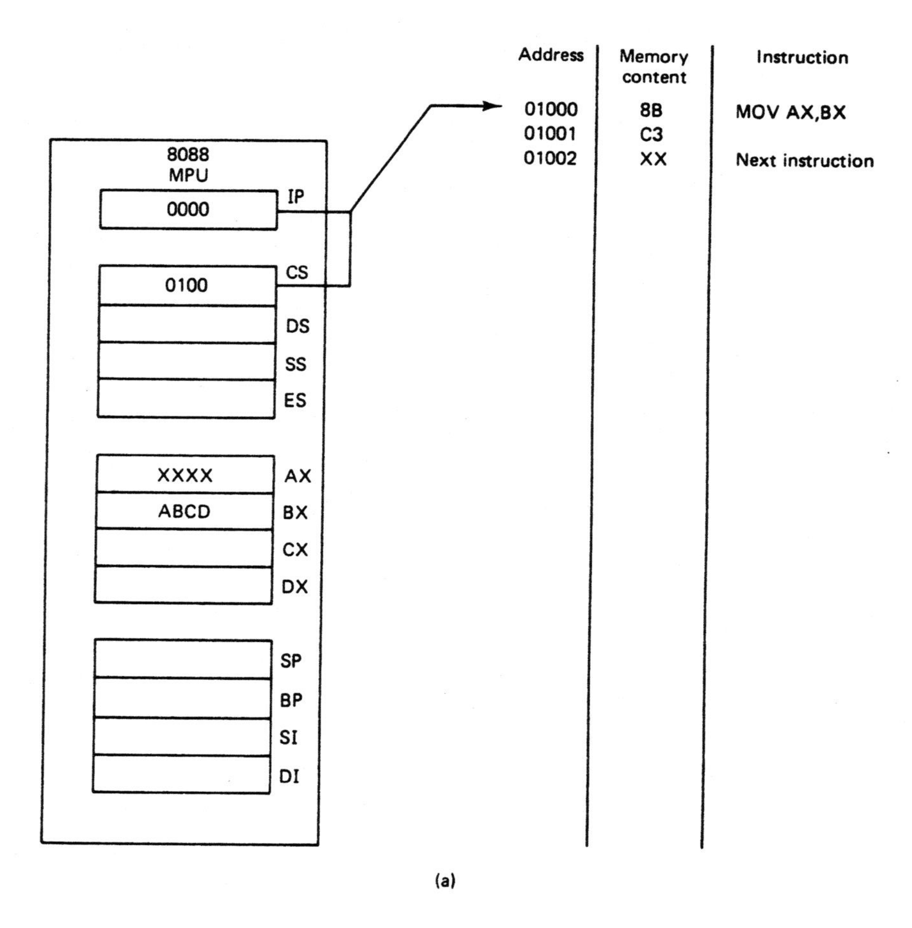

Figure 3–9 (a) Register addressing mode instruction before fetch and execution. (b) After execution.

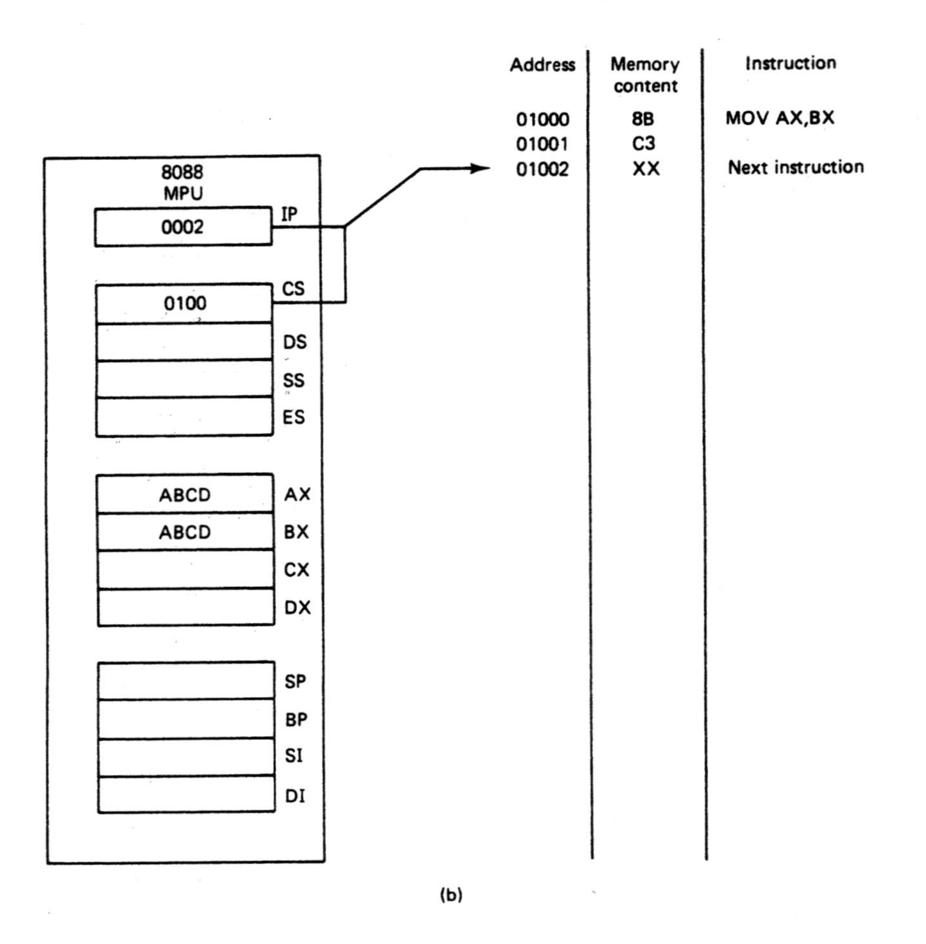

Figure 3–9 (continued)

EXAMPLE 3.2

What is the destination operand in the instruction? How large is this operand?

```
MOV  CH, AH
```

Solution

The destination operand is register CH and it specifies an 8-bit operand.

Immediate Operand Addressing Mode

If an operand is part of the instruction instead of the contents of a register or memory location, it represents what is called an *immediate operand* and is accessed using the

Opcode	Immediate operand

Figure 3–10 Instruction encoded with an immediate operand.

immediate addressing mode. Figure 3–10 shows that the operand, which can be 8 bits (Imm8) or 16 bits (Imm16) in length, is encoded as part of the instruction. Since the data are encoded directly into the instruction, immediate operands normally represent constant data. This addressing mode can only be used to specify a source operand.

In the instruction

```
MOV  AL, 15H
```

the source operand 15H (15_{16}) is an example of a byte-wide immediate source operand. The destination operand, which is the contents of AL, uses register addressing. Thus, this instruction employs both the immediate and register addressing modes.

Figures 3–11(a) and (b) illustrate the fetch and execution of this instruction. Here we find that the immediate operand 15_{16} is stored in the code segment of memory in the byte location immediately following the opcode of the instruction. This value is fetched, along with the opcode for MOV, into the instruction queue within the 8088. When it performs the move operation, the source operand is fetched from the queue, not from the memory, and no external memory operations are performed. Note that the result produced by executing this instruction is that the immediate operand, which equals 15_{16}, is loaded into the lower-byte part of the accumulator (AL).

EXAMPLE 3.3

Write an instruction that will move the immediate value 1234H into the CX register.

Solution

The instruction must use immediate operand addressing for the source operand and register operand addressing for the destination operation. This gives

```
MOV CX, 1234H
```

Memory Operand Addressing Modes

To reference an operand in memory, the 8088 must calculate the physical address (PA) of the operand and then initiate a read or write operation of this storage location. The 8088 MPU is provided with a group of addressing modes known as the *memory operand addressing modes* for this purpose. Looking at Fig. 3–12, we see that the physical address is computed from a *segment base address* (SBA) and an *effective address* (EA). SBA identifies the starting location of the segment in memory, and EA represents the offset of the operand from the beginning of this segment of memory. Earlier we showed how SBA

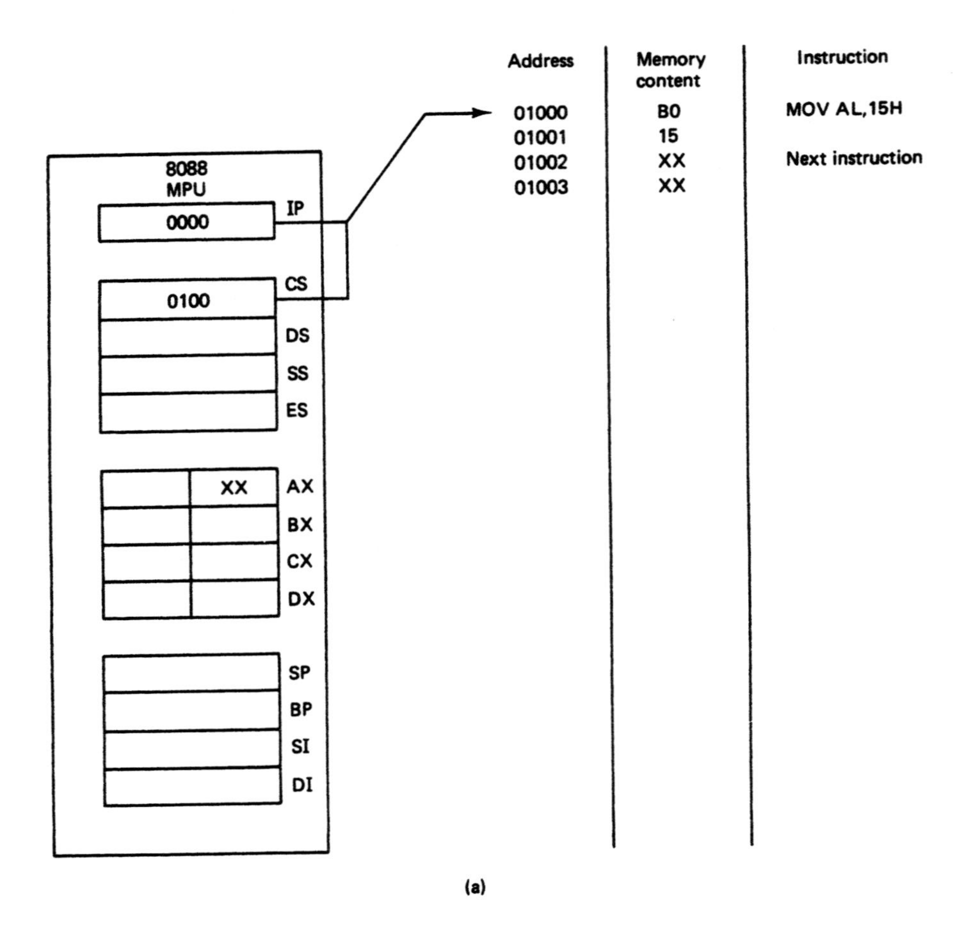

Figure 3–11 (a) Immediate addressing mode instruction before fetch and execution. (b) After execution.

and EA are combined within the 8088 to form the logical address SBA:EA and how to compute the physical address from these two values.

The value of the EA can be specified in a variety of ways. One way is to encode the effective address of the operand directly in the instruction. This represents the simplest type of memory addressing, known as the *direct addressing mode.* Figure 3–12 shows that an effective address can be made up from as many as three elements: the *base, index,* and *displacement.* Using these elements, the effective address calculation is made by the general formula

$$EA = \text{Base} + \text{Index} + \text{Displacement}$$

Figure 3–12 also identifies the registers that can be used to hold the values of the segment base, base, and index. For example, it tells us that any of the four segment registers can be the source of the segment base for the physical address calculation and that

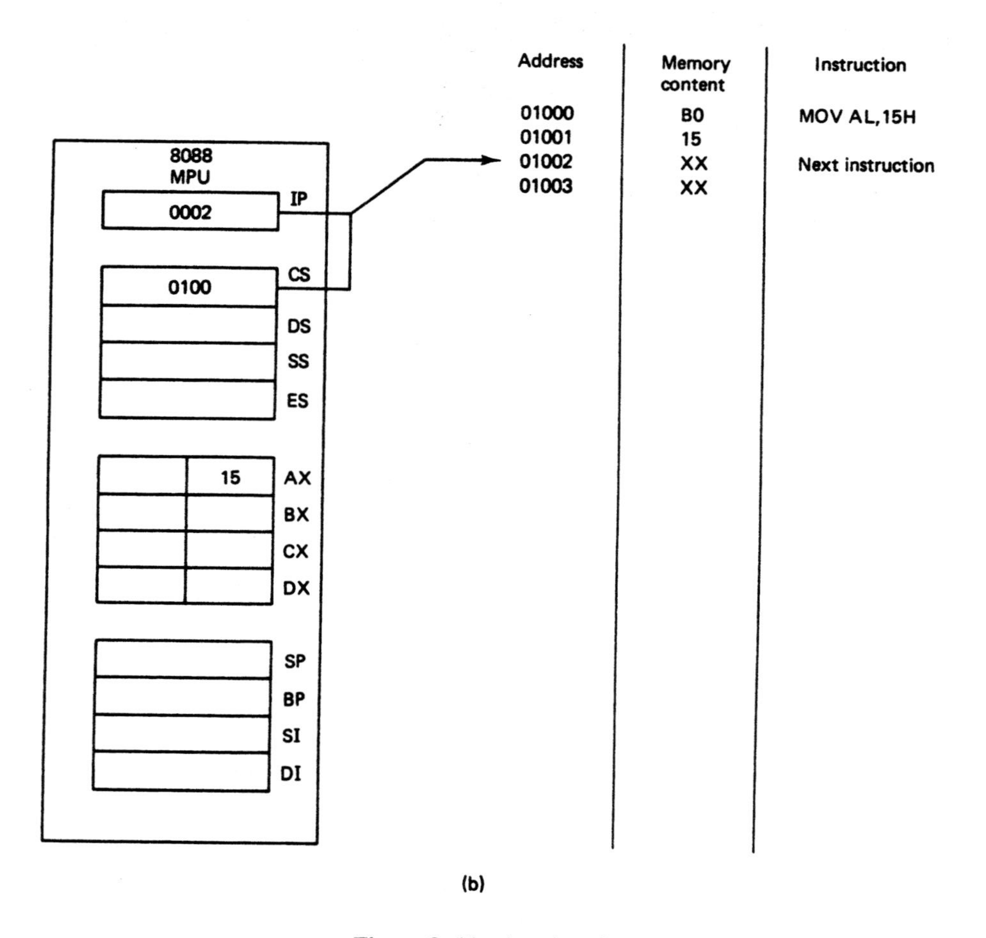

Figure 3–11 (continued)

the value of base for the effective address can be in either the base register (BX) or base pointer register (BP). Figure 3–12 also identifies the sizes permitted for the displacement.

Not all these elements are always used in the effective address calculation. In fact, a number of memory addressing modes are defined by using various combinations of these elements. They are called *register indirect addressing, based addressing, indexed addressing,* and *based-indexed addressing.* For instance, using based addressing mode, the effective address calculation includes just a base. These addressing modes provide the

PA = SBA : EA

PA = Segment base : Base + Index + Displacement

$$PA = \begin{Bmatrix} CS \\ SS \\ DS \\ ES \end{Bmatrix} : \begin{Bmatrix} BX \\ BP \end{Bmatrix} + \begin{Bmatrix} SI \\ DI \end{Bmatrix} + \begin{Bmatrix} \text{8-bit displacement} \\ \text{16-bit displacement} \end{Bmatrix}$$

Figure 3–12 Physical and effective address computation for memory operands.

PA = Segment base: Direct address

$$PA = \begin{Bmatrix} CS \\ DS \\ SS \\ ES \end{Bmatrix} : \begin{Bmatrix} \text{Direct address} \end{Bmatrix}$$

Figure 3–13 Specification of a direct memory address.

programmer with different ways of computing the effective address of an operand in memory. Next, we will examine each of the memory operand addressing modes in detail.

Direct Addressing Mode. *Direct addressing mode* is similar to immediate addressing in that information is encoded directly into the instruction. However, in this case, the instruction opcode is followed by an effective address, instead of the data. As Fig. 3–13 shows, this effective address is used directly as the 16-bit offset of the storage location of the operand from the location specified by the current value in the selected segment register.

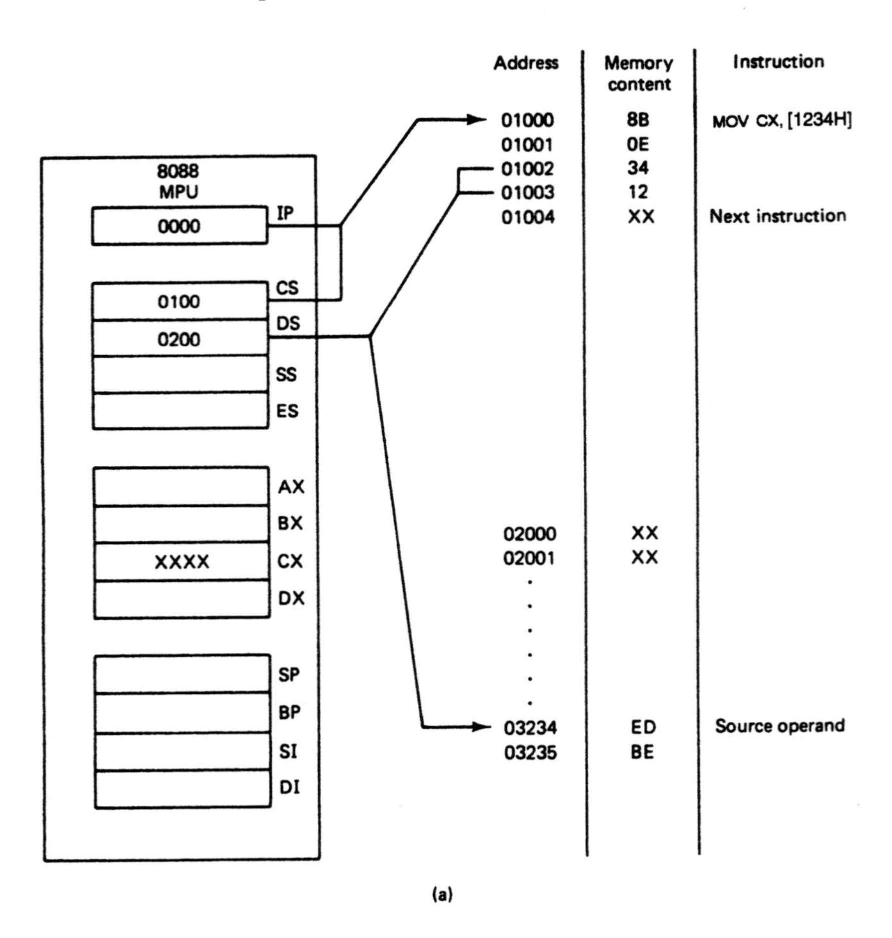

Figure 3–14 (a) Direct addressing mode instruction before fetch and execution. (b) After execution.

The default segment register is DS. Therefore, the 20-bit physical address of the operand in memory is normally obtained from logical address DS:EA. But, by using a *segment override prefix* (SEG) in the instruction, any of the four segment registers can be referenced.

An example of an instruction that uses direct addressing mode for its source operand is

```
MOV  CX, [1234H]
```

This stands for "move the contents of the memory location with offset 1234_{16} in the current data segment into internal register CX." The offset is encoded as part of the instruction's machine code.

In Fig. 3–14(a), we find that the offset is stored in the two byte locations that follow the instruction's opcode. As the instruction is executed, the 8088 combines 1234_{16} with 0200_{16} to get the physical address of the source operand as follows:

$$PA = 02000_{16} + 1234_{16}$$
$$= 03234_{16}$$

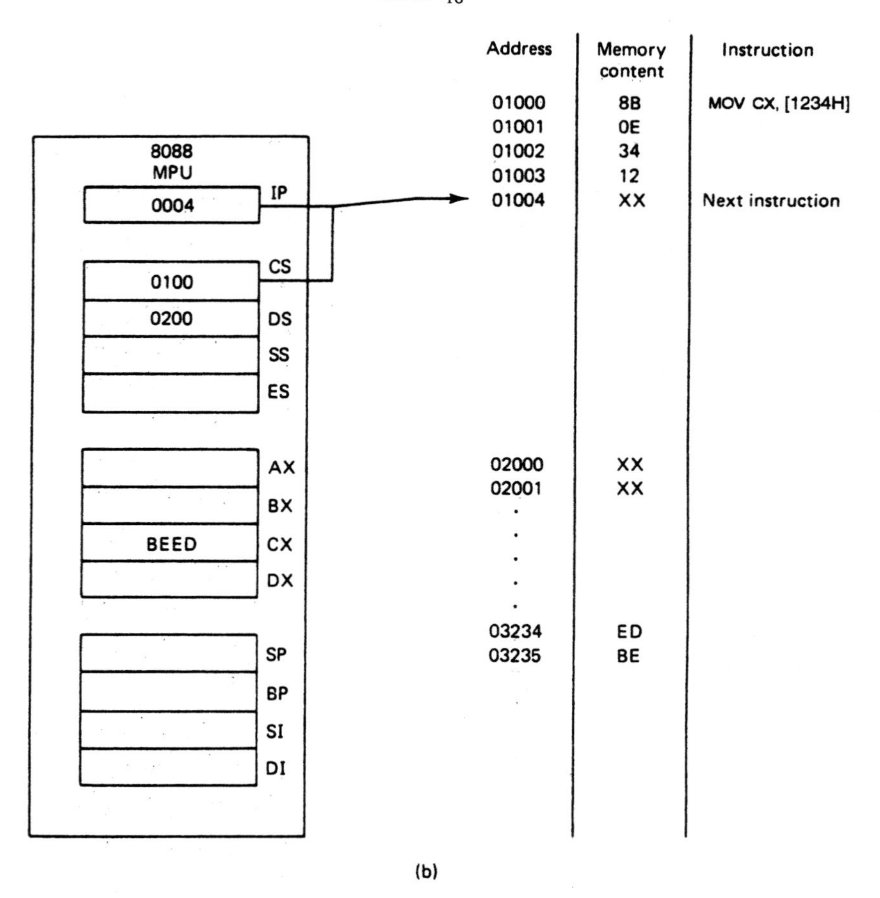

Figure 3–14 (continued)

PA = Segment base: Indirect address

$$PA = \begin{Bmatrix} CS \\ DS \\ SS \\ ES \end{Bmatrix} : \begin{Bmatrix} BX \\ BP \\ SI \\ DI \end{Bmatrix}$$

Figure 3–15 Specification of an indirect memory address.

Then it reads the word of data starting at this address, which is $BEED_{16}$, and loads it into the CX register. This result is illustrated in Fig. 3–14(b).

Register Indirect Addressing Mode. *Register indirect addressing mode* is similar to the direct addressing we just described in that an effective address is combined with the contents of DS to obtain a physical address. However, it differs in the way the offset is specified. Figure 3–15 shows that this time EA resides in either a base register or an index register within the 8088. The base register can be either base register BX or base pointer

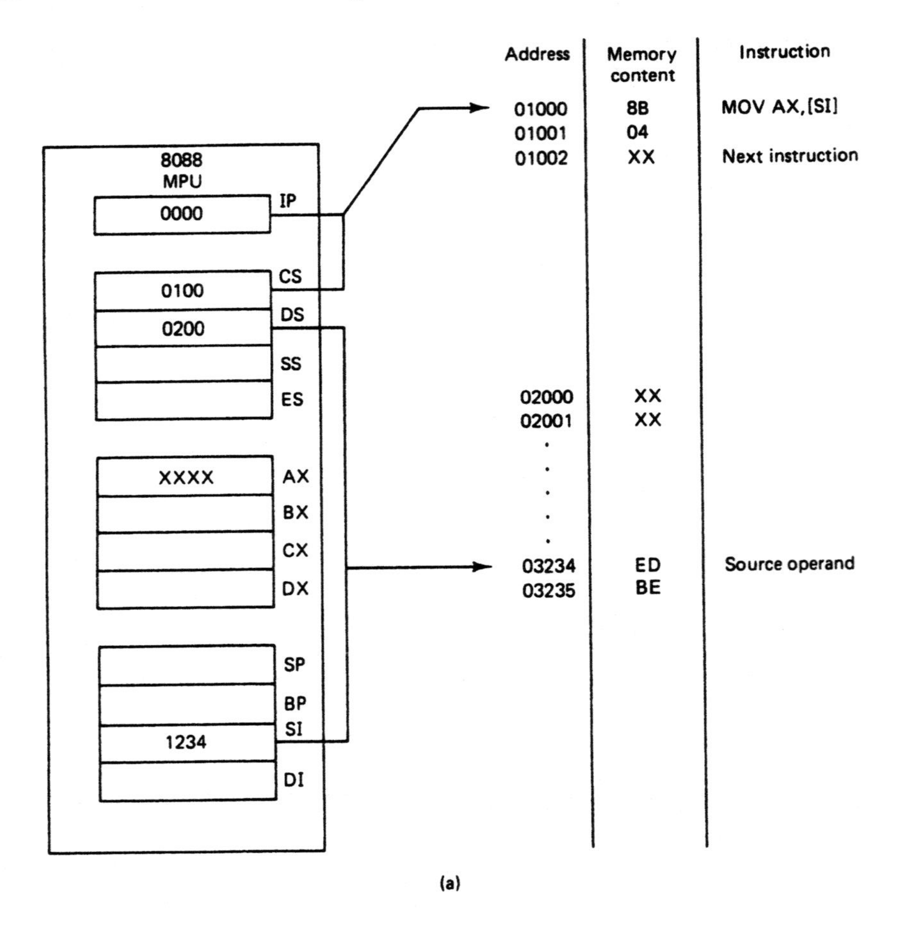

Figure 3–16 (a) Instruction using register indirect addressing mode before fetch and execution. (b) After execution.

register BP, and the index register can be source index register SI or destination index register DI. Use of a segment-override prefix permits reference of another segment register.

An example of an instruction that uses register indirect addressing for its source operand is

```
MOV  AX,[SI]
```

Execution of this instruction moves the contents of the memory location that is offset from the beginning of the current data segment by the value of EA in register SI into the AX register.

For instance, Figs. 3–16(a) on the previous page and (b) show that if SI contains 1234_{16} and DS contains 0200_{16}, the result produced by executing the instruction is that the contents of the memory location at address

$$PA = 02000_{16} + 1234_{16}$$
$$= 03234_{16}$$

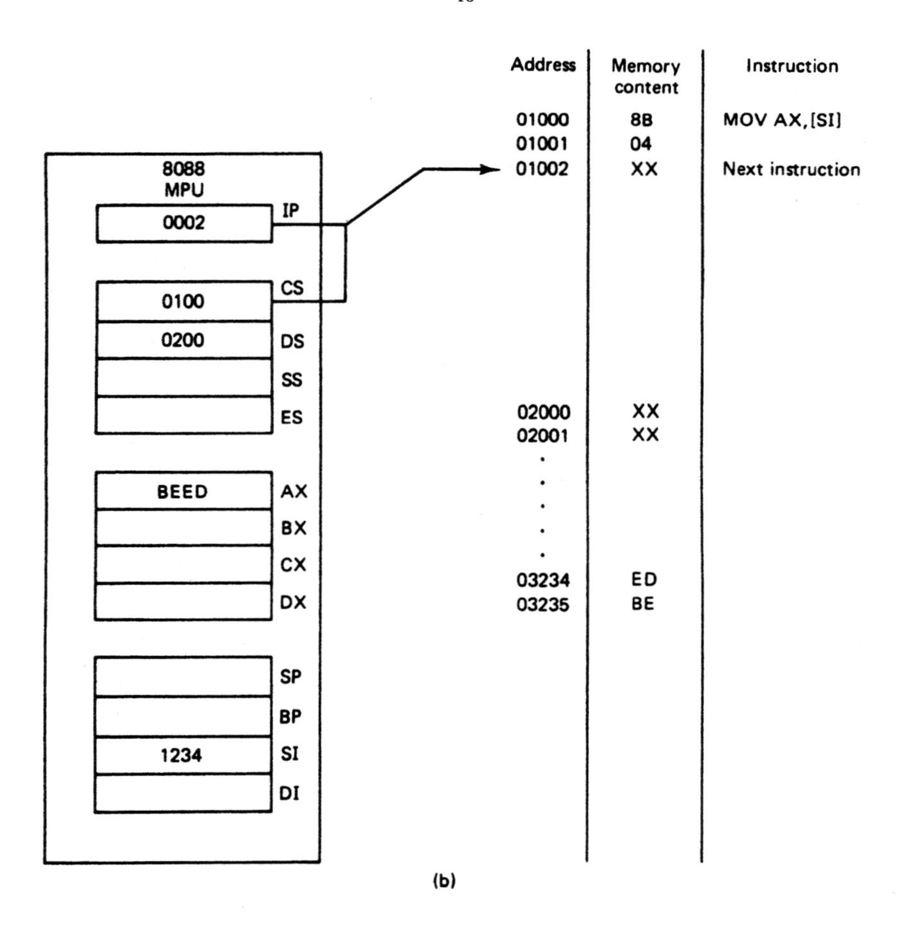

Figure 3–16 (continued)

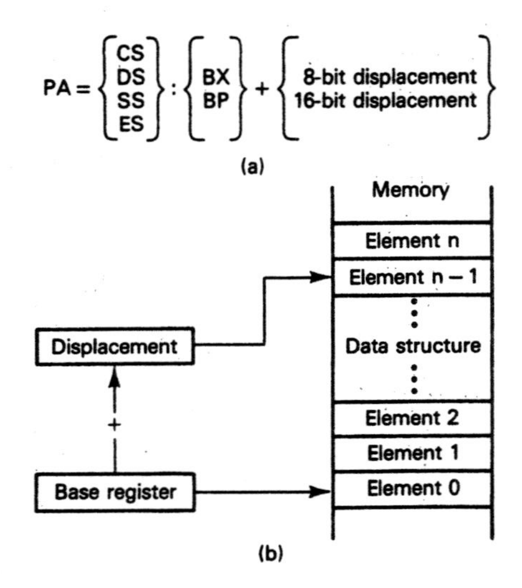

Figure 3–17 (a) Specification of a based address. (b) Based addressing of a structure of data.

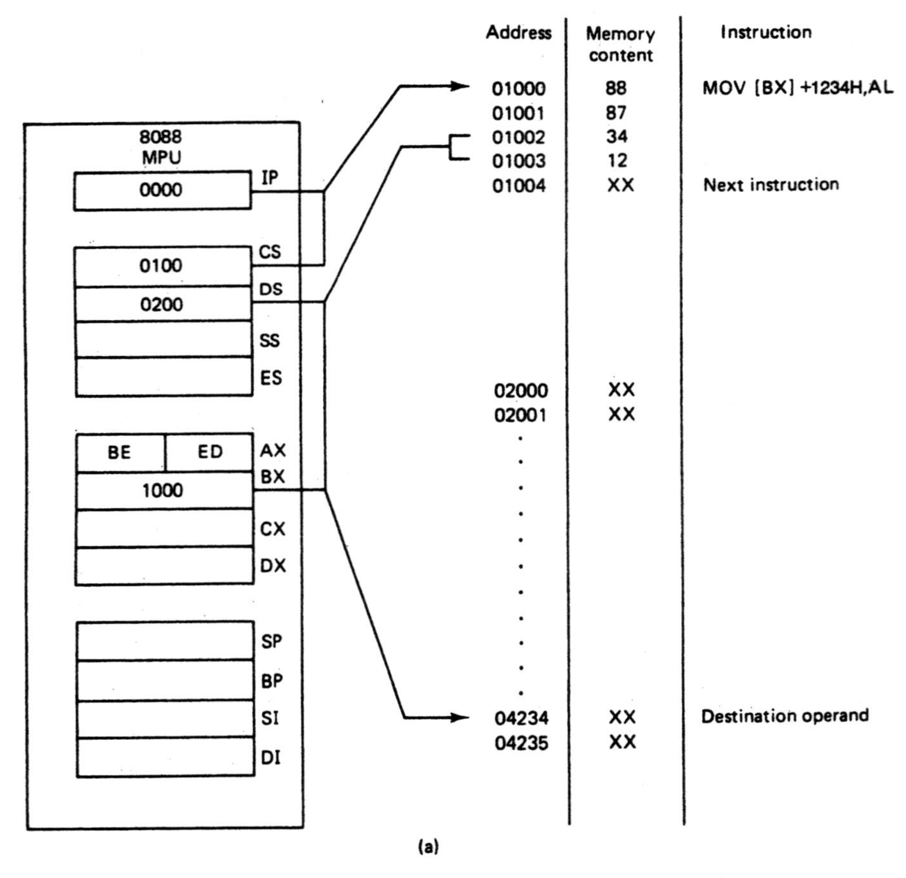

Figure 3–18 (a) Instruction using based pointer addressing mode before fetch and execution. (b) After execution.

are moved to the AX register. Notice in Fig. 3–16(b) that this value is $BEED_{16}$. In this example, the value 1234_{16} that was found in the SI register must have been loaded with another instruction prior to executing the move instruction.

The result produced by executing this instruction and that for the example for the direct addressing mode are the same. However, they differ in the way the physical address was generated. The direct addressing method lends itself to applications where the value of EA is a constant. On the other hand, register indirect addressing can be used when the value of EA is calculated and stored, for example, in SI by a previous instruction—that is, EA is a variable. For instance, the instructions executed just before our example instruction could have incremented the value in SI by two.

Based Addressing Mode. In the *based addressing mode,* the effective address of the operand is obtained by adding a direct or indirect displacement to the contents of either base register BX or base pointer register BP. The physical address calculation is shown in Fig. 3–17(a) (p. 92). Looking at Fig. 3–17(b) (p. 92), we see that the value in the base

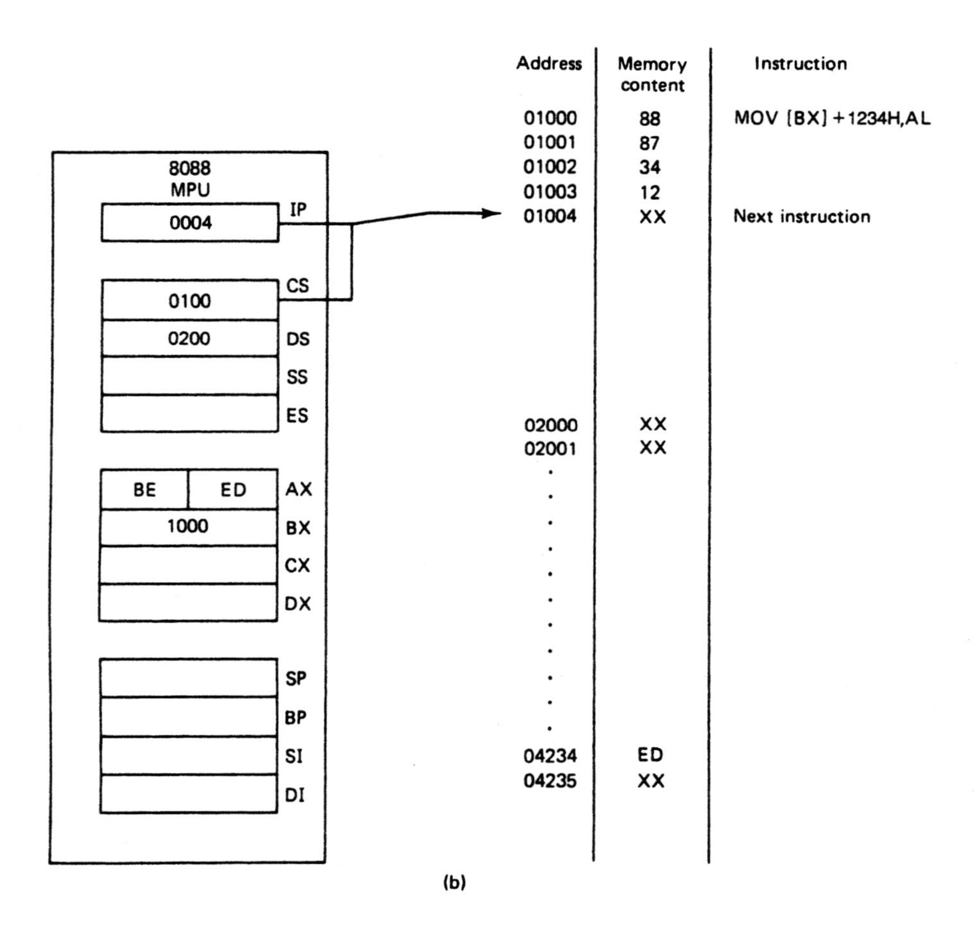

(b)

Figure 3–18 (continued)

register defines the beginning of a data structure, such as an array, in memory, and the displacement selects an element of data within this structure. To access a different element in the array, the programmer simply changes the value of the displacement. To access the same element in another similar array, the programmer can change the value in the base register so that it points to the beginning of the new array.

A move instruction that uses based addressing to specify the location of its destination operand is as follows:

```
MOV  [BX] + 1234H, AL
```

This instruction uses base register BX and direct displacement 1234_{16} to derive the EA of the destination operand. The based addressing mode is implemented by specifying the base register in brackets followed by a + sign and the direct displacement. The source operand in this example is located in byte accumulator AL.

As Figs. 3–18(a) and (b) on pgs. 92–93 show, the fetch and execution of this instruction cause the 8088 to calculate the physical address of the destination operand from the contents of DS, BX, and the direct displacement. The result is

$$PA = 02000_{16} + 1000_{16} + 1234_{16}$$
$$= 04234_{16}$$

Then it writes the contents of source operand AL into the storage location at 04234_{16}. The result is that ED_{16} is written into the destination memory location. Again, the default seg-

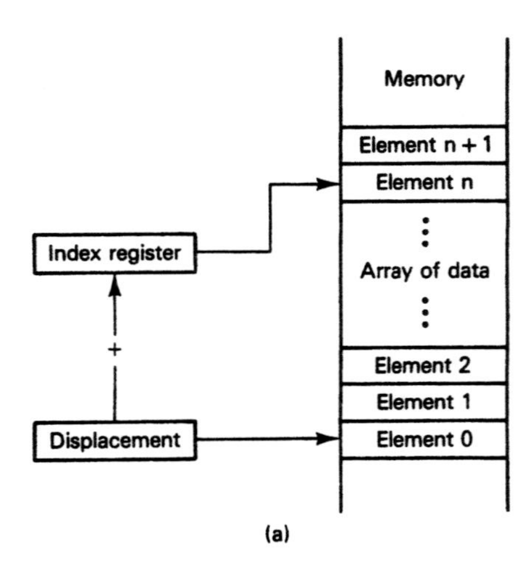

Figure 3–19 (a) Indexed addressing of an array of data elements. (b) Specification of an indexed address.

ment register for this physical address calculation is DS, but it can be changed to another segment register with the segment-override prefix.

If BP is used instead of BX, the calculation of the physical address is performed using the contents of the stack segment (SS) register instead of DS. This permits access to data in the stack segment of memory.

Indexed Addressing Mode. *Indexed addressing mode* works in a manner similar to that of the based addressing mode just described. However, as Fig. 3–19(a) shows, indexed addressing mode uses the value of the displacement as a pointer to the starting point of an array of data in memory and the contents of the specified register as an index that selects the specific element in the array that is to be accessed. For instance, for the byte-size element array in Fig. 3–19(a) on p. 94, the index register holds the value *n*. In this way, it selects data element *n* in the array. Figure 3–19(b) shows how the physical address is obtained from the value in a segment register, an index in the SI or DI register, and a displacement.

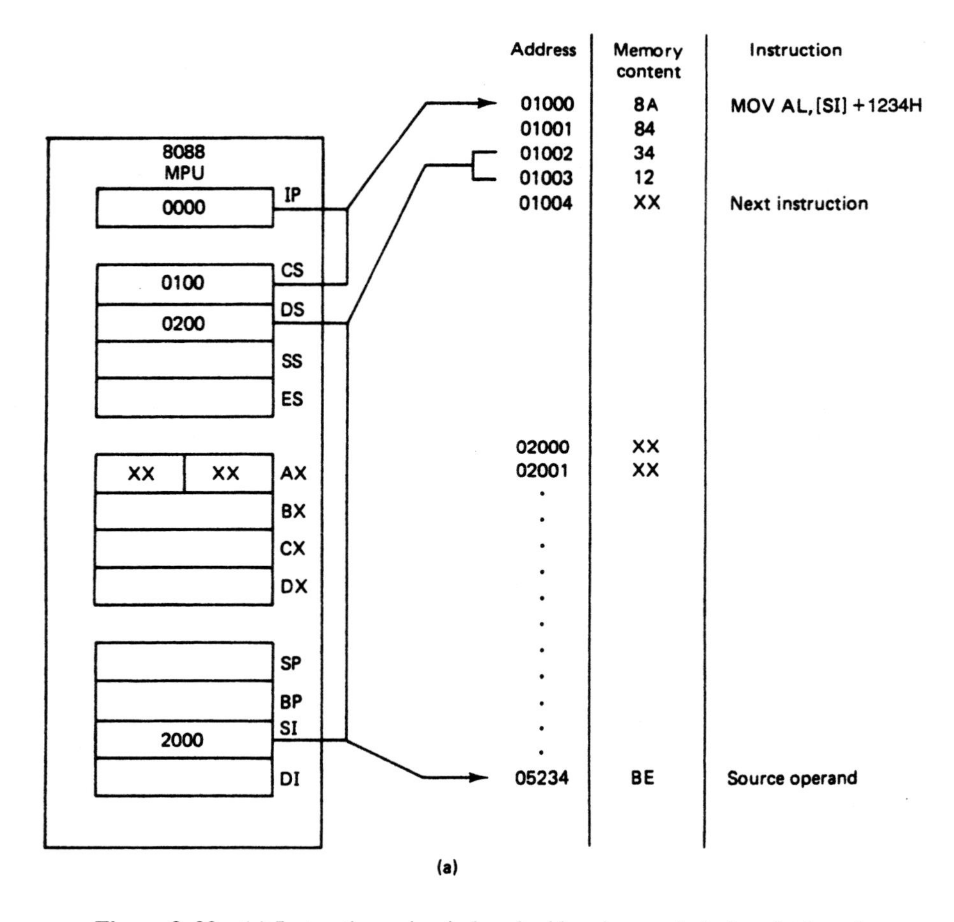

Figure 3–20 (a) Instruction using indexed addressing mode before fetch and execution. (b) After execution.

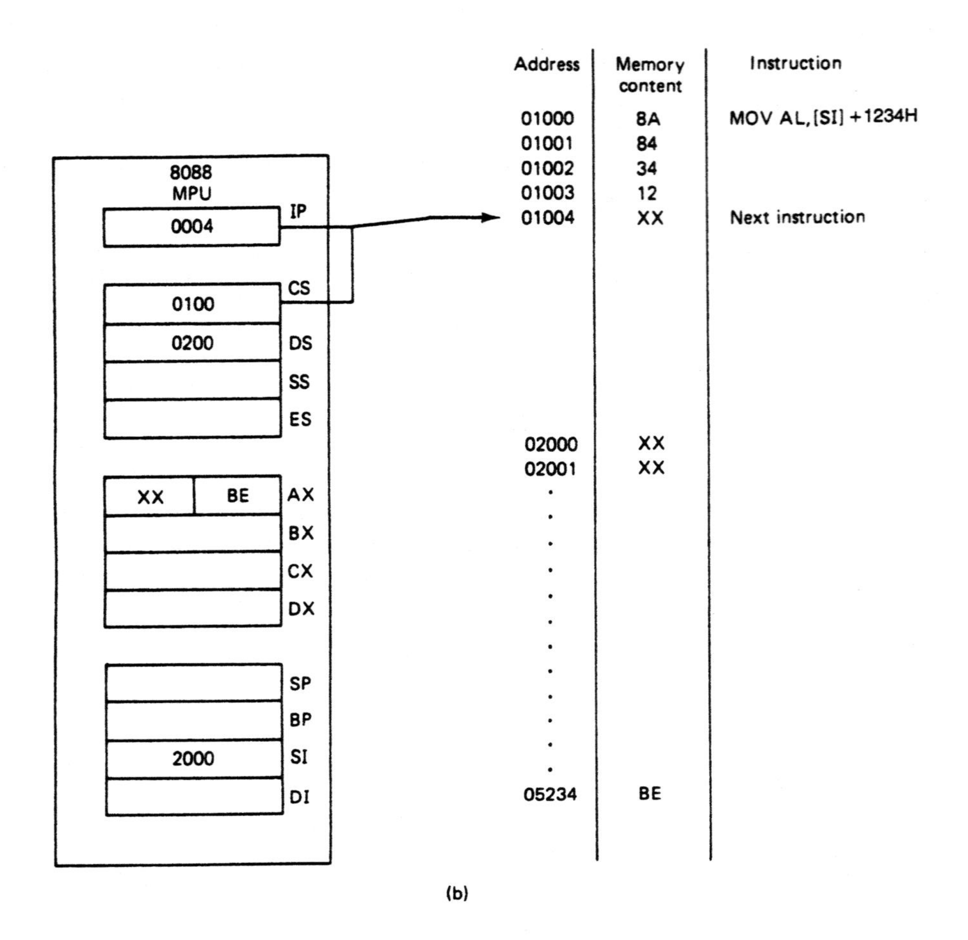

Figure 3–20 (continued)

Here is an example:

```
MOV  AL,[SI] + 2000H
```

The source operand has been specified using direct indexed addressing mode. Note that the *direct displacement* is 2000H. As with the base register in based addressing, the index register, which is SI, is enclosed in brackets. The effective address is calculated as

$$EA = (SI) + 2000H$$

and the physical address is computed by combining the contents of DS with EA.

$$PA = DS:(SI) + 2000H$$

Figures 3–20(a) and (b) (pp. 95–96) show the result of executing the move instruction. First the physical address of the source operand is calculated from the contents of DS, SI, and the direct displacement.

$$
\begin{aligned}
\text{PA} &= 02000_{16} + 2000_{16} + 1234_{16} \\
&= 05234_{16}
\end{aligned}
$$

Then the byte of data stored at this location, BE_{16}, is read into the lower byte (AL) of the accumulator register.

Based-Indexed Addressing Mode. Combining the based addressing mode and the indexed addressing mode results in a new, more powerful mode known as *based-indexed addressing mode.* This addressing mode can be used to access complex data structures such as two-dimensional arrays. Figure 3–21(a) shows how it can be used to access elements in an $m \times n$ array of data. Notice that the displacement, which is a fixed value, locates the array in memory. The base register specifies the m coordinate of the array, and the index register identifies the n coordinate. Simply changing the values in the base and

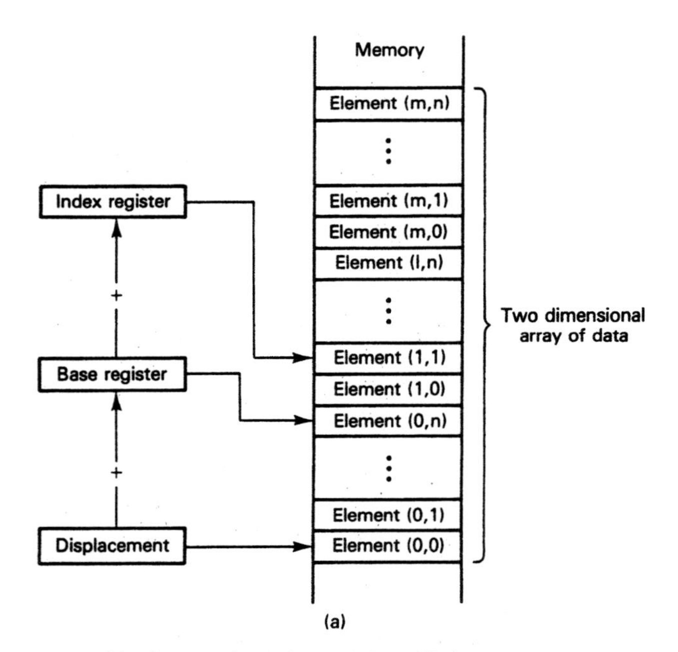

$$
\text{PA} = \begin{Bmatrix} \text{CS} \\ \text{DS} \\ \text{SS} \\ \text{ES} \end{Bmatrix} : \begin{Bmatrix} \text{BX} \\ \text{BP} \end{Bmatrix} + \begin{Bmatrix} \text{SI} \\ \text{DI} \end{Bmatrix} + \begin{Bmatrix} \text{8-bit displacement} \\ \text{16-bit displacement} \end{Bmatrix}
$$

(b)

Figure 3–21 (a) Based-indexed addressing of a two-dimensional array of data. (b) Specification of a based-indexed address.

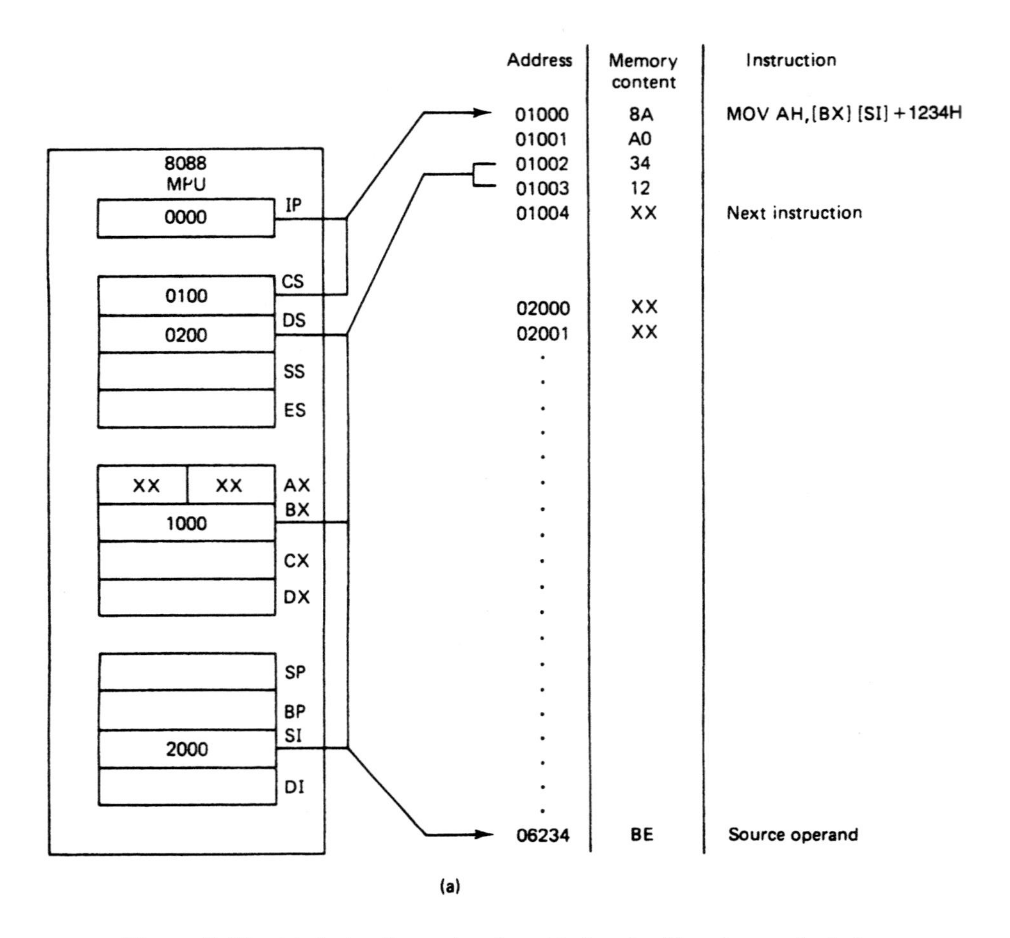

Figure 3–22 (a) Instruction using based-indexed addressing mode before fetch and execution. (b) After execution.

index registers permits access to any element in the array. Figure 3–21(b) shows the registers permitted in the based-indexed physical address computation.

Let us consider an example of a move instruction using this type of addressing.

```
MOV  AH, [BX][SI] + 1234H
```

Note that the source operand is accessed using based-indexed addressing mode. Therefore, the effective address of the source operand is obtained as

$$EA = (BX) + (SI) + 1234H$$

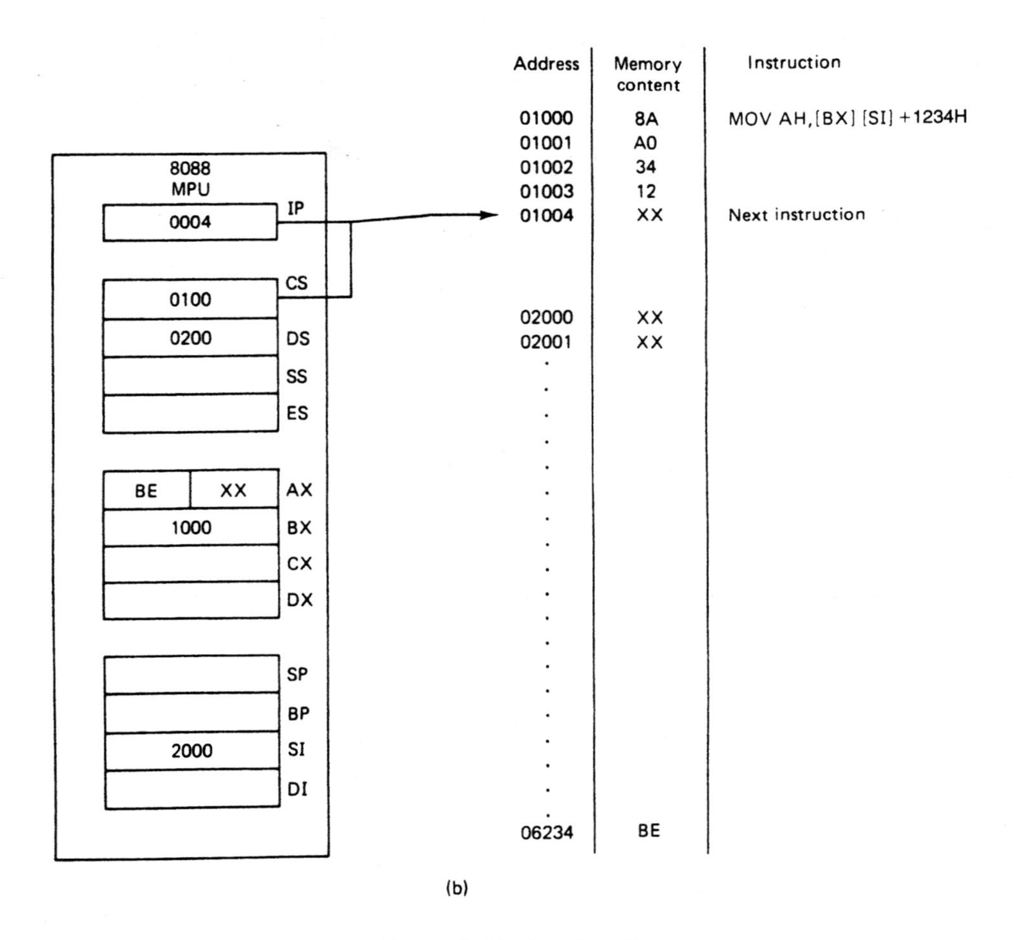

Figure 3–22 (continued)

and the physical address of the operand is computed from the current contents of DS and the calculated EA.

$$PA = DS:(BX) + (SI) + 1234H$$

Figures 3–22(a) and (b) present an example of executing this instruction. Using the contents of the various registers in the example, the address of the source operand is calculated as

$$PA = 02000_{16} + 1000_{16} + 2000_{16} + 1234_{16}$$
$$= 06234_{16}$$

Execution of the instruction causes the value stored at this location in memory to be read into AH.

REVIEW PROBLEMS

Section 3.1

1. What tells a microcomputer what to do, where to get data, how to process the data, and where to put the results when done?
2. What is the name given to a sequence of instructions used to guide a computer through a task?
3. What does OS stand for?
4. What is the native language of the 8088?
5. How does machine language differ from assembly language?
6. What does opcode stand for? Give two examples.
7. What is an operand? Give two types.
8. In the assembly language statement

```
START: ADD AX, BX ;Add BX to AX
```

   what is the label? What is the comment?
9. What is the function of an assembler? A compiler?
10. What is source code? What is object code?
11. Give two benefits derived from writing programs in assembly language instead of a high-level language.
12. What is meant by the phrase *real-time application*?
13. List two hardware-related applications that require the use of assembly language programming. Name two software-related applications.

Section 3.2

14. What document is produced as a result of the problem description step of the development cycle?
15. Give a name that is used to refer to the software solution planned for a problem. What is the name of the document used to describe this solution plan?
16. What is a flowchart?
17. Draw the flowchart symbol used to identify a subroutine.
18. What type of program is EDIT?
19. What type of program is used to produce an object module?
20. What does MASM stand for?
21. What type of program is used to produce a run module?
22. In which part of the development cycle is the EDIT program used? The MASM program? The LINK program? The DEBUG program?
23. Assuming that the filename is PROG_A, what are typical names for the files that result from the use of the EDIT program? The MASM program? The LINK program?

Section 3.3

24. How many basic instructions are in the 8088/8086 instruction set?

25. List six groups of instructions in the 8088/8086 instruction set.

Section 3.4

26. Describe the operation performed by executing the move instruction.

Section 3.5

27. What is meant by an addressing mode?

28. Make a list of the addressing modes available on the 8088.

29. What three elements can be used to form the effective address of an operand in memory?

30. Name the five memory operand addressing modes.

31. Identify the addressing modes used for the source and the destination operands in the instructions that follow.

```
(a) MOV AL, BL
(b) MOV AX, 0FFH
(c) MOV [DI], AX
(d) MOV DI,[SI]
(e) MOV [BX] + 0400H,CX
(f) MOV [DI] + 0400H,AH
(g) MOV [BX][DI] + 0400H, AL
```

32. Compute the physical address for the specified operand in each of the following instructions from problem 31. The register contents and variables are as follows: (CS) = $0A00_{16}$, (DS) = $0B00_{16}$, (SI) = 0100_{16}, (DI) = 0200_{16}, and (BX) = 0300_{16}.

(a) Destination operand of the instruction in (c)
(b) Source operand of the instruction in (d)
(c) Destination operand of the instruction in (e)
(d) Destination operand of the instruction in (f)
(e) Destination operand of the instruction in (g)

Machine Language Coding and the DEBUG Software Development Program of the PC

▲ INTRODUCTION

In this chapter, we begin by exploring how the instructions of the 8088/8086 instruction set are encoded in machine code. This is followed by a study of the software development environment provided for these microprocessors with the PC microcomputer. Here we examine the DEBUG program, which is a program-execution/debug tool that operates in the PC's *disk operating system* (DOS) environment. First we examine DEBUG's command set. Then we use these commands to load, assemble, execute, and debug programs. In Chapters 5 and 6, we learn the instructions of the 8088/8086 instruction set. The operation of many of the instructions is demonstrated by executing them with the DEBUG program. Chapter 7 completes our coverage of the 8088/8086 software development environment by exploring the use of the MASM macroassembler. The topics discussed in this chapter are as follows:

4.1 Converting Assembly Language Instructions to Machine Code
4.2 Encoding a Complete Program in Machine Code
4.3 The PC and Its DEBUG Program
4.4 Examining and Modifying the Contents of Memory
4.5 Input and Output of Data
4.6 Hexadecimal Addition and Subtraction
4.7 Loading, Verifying, and Saving Machine Language Programs

▲ 4.1 CONVERTING ASSEMBLY LANGUAGE INSTRUCTIONS TO MACHINE CODE

To convert an assembly language program to machine code, we must convert each assembly language instruction to its equivalent machine code instruction. In general, the machine code for an instruction specifies things like what operation is to be performed, what operand or operands are to be used, whether the operation is performed on byte or word data, whether the operation involves operands that are located in registers or a register and a storage location in memory, and if one of the operands is in memory, how its address is to be generated. All of this information is encoded into the bits of the machine code for the instruction.

The machine code instructions of the 8088 vary in the number of bytes used to encode them. Some instructions can be encoded with just 1 byte, others can be done in 2 bytes, and many require even more. Earlier we indicated that the maximum number of bytes an instruction might take is 6. Single-byte instructions generally specify a simpler operation with a register or a flag bit. *Complement carry* (CMC) is an example of a single-byte instruction, specified by the machine code byte 11110101_2, which equals $F5_{16}$. That is,

$$\text{CMC} = 11110101_2 = \text{F5}_{16}$$

The machine code for instructions can be obtained by following the formats used in encoding the instructions of the 8088 microprocessor. Most multibyte instructions use the *general instruction format* shown in Fig. 4–1. Exceptions to this format exist and are considered later. For now, let us describe the functions of the various bits and fields (groups of bits) in each byte of this format.

Looking at Fig. 4–1, we see that byte 1 contains three kinds of information: the *operation code* (opcode), the *register direction* (D) bit, and the *data size* (W) bit. Let us summarize the function of each of these pieces of information.

1. *Opcode field (6-bit):* Specifies the operation, such as add, subtract, or move, that is to be performed.
2. *Register direction bit (D bit):* Tells whether the register operand specified by REG in byte 2 is the source or destination operand. A logic 1 in this bit position indicates that the register operand is a destination operand, and logic 0 indicates that it is a source operand.
3. *Data size bit (W bit):* Specifies whether the operation will be performed on 8-bit or 16-bit data. Logic 0 selects 8 bits and 1 selects 16 bits as the data size.

For instance, if a 16-bit value is to be added to register AX, the six most significant bits specify the add register operation. This opcode is 000000. The next bit, D, is logic 1

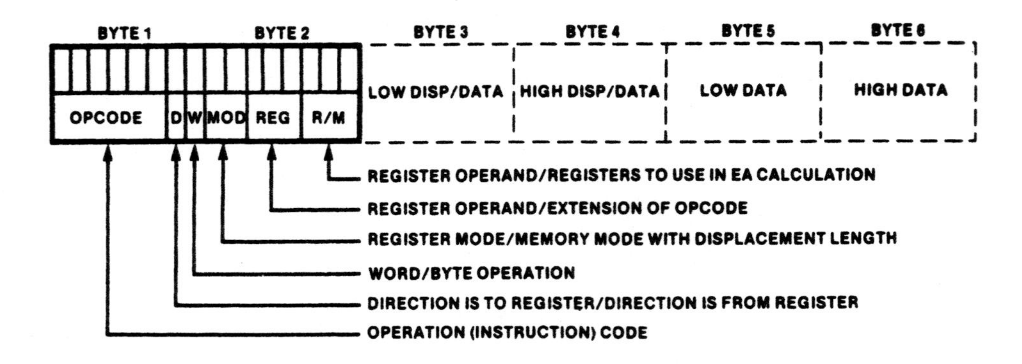

Figure 4–1 General instruction format. (Reprinted by permission of Intel Corp. Copyright/Intel Corp. 1979)

to specify that a register, AX in this case, holds the destination operand. Finally, the least significant bit, W, is logic 1 to specify a 16-bit data operation.

The second byte in Fig. 4–1 has three fields: the *mode* (MOD) *field,* the *register* (REG) *field,* and the *register/memory* (R/M) *field.* These fields are used to specify which register is used for the first operand and where the second operand is stored. The second operand can be in either a register or a memory location.

The 3-bit REG field is used to identify the register for the first operand, which is the one that was defined as the source or destination by the D bit in byte 1. Figure 4–2 shows the encoding for each of the 8088's registers. Here we find that the 16-bit register AX and the 8-bit register AL are specified by the same binary code. Note that the decision whether to use AX or AL is made based on the setting of the operation size (W) bit in byte 1.

In our earlier example, we said that the first operand, the destination operand, is register AX. For this case, the REG field is set to 000.

The 2-bit MOD field and 3-bit R/M field together specify the second operand. Encoding for these two fields is shown in Figs. 4–3(a) and (b), respectively. MOD indicates whether the operand is in a register or memory. Note that in the case of a second operand in a register, the MOD field is always 11. The R/M field, along with the W bit from byte 1, selects the register.

REG	W=0	W=1
000	AL	AX
001	CL	CX
010	DL	DX
011	BL	BX
100	AH	SP
101	CH	BP
110	DH	SI
111	BH	DI

Figure 4–2 Register (REG) field encoding. (Reprinted by permission of Intel Corp. Copyright/Intel Corp. 1979)

CODE	EXPLANATION
00	Memory Mode, no displacement follows*
01	Memory Mode, 8-bit displacement follows
10	Memory Mode, 16-bit displacement follows
11	Register Mode (no displacement)

*Except when R/M = 110, then 16-bit displacement follows

(a)

MOD = 11			EFFECTIVE ADDRESS CALCULATION			
R/M	W = 0	W = 1	R/M	MOD = 00	MOD = 01	MOD = 10
000	AL	AX	000	(BX) + (SI)	(BX) + (SI) + D8	(BX) + (SI) + D16
001	CL	CX	001	(BX) + (DI)	(BX) + (DI) + D8	(BX) + (DI) + D16
010	DL	DX	010	(BP) + (SI)	(BP) + (SI) + D8	(BP) + (SI) + D16
011	BL	BX	011	(BP) + (DI)	(BP) + (DI) + D8	(BP) + (DI) + D16
100	AH	SP	100	(SI)	(SI) + D8	(SI) + D16
101	CH	BP	101	(DI)	(DI) + D8	(DI) + D16
110	DH	SI	110	DIRECT ADDRESS	(BP) + D8	(BP) + D16
111	BH	DI	111	(BX)	(BX) + D8	(BX) + D16

(b)

Figure 4–3 (a) Mode (MOD) field encoding. (Reprinted by permission of Intel Corp. Copyright/Intel Corp. 1979) (b) Register/memory (R/M) field encoding. (Reprinted by permission of Intel Corp. Copyright/Intel Corp. 1979)

For example, if the second operand, the source operand in our earlier addition example, is to be in BX, the MOD and R/M fields are made 11 and 011, respectively.

EXAMPLE 4.1

The instruction

```
MOV  BL, AL
```

stands for "move the byte contents from source register AL to destination register BL." Using the general format in Fig. 4–1, show how to encode the instruction in machine code. Assume that the 6-bit opcode for the move operation is 100010.

Solution

In byte 1, the first six bits specify the move operation and thus must be 100010.

$$OPCODE = 100010$$

The next bit, D, indicates whether the register specified by the REG part of byte 2 is a source or destination operand. Let us say that we will encode AL in the REG field of byte 2; therefore, D is set equal to 0 for source operand.

$$D = 0$$

The last bit (W) in byte 1 must specify a byte operation. For this reason, it is also set to 0.

$$W = 0$$

This leads to

$$BYTE\ 1 = 10001000_2 = 88_{16}$$

In byte 2, the source operand, specified by the REG field, is AL. The corresponding code from Fig. 4–2 is

$$REG = 000$$

Since the second operand is also a register, the MOD field is made 11. The R/M field specifies that the destination register is BL, for which the code (from Fig. 4–3(b)) is 011. This gives

$$MOD = 11$$

$$R/M = 011$$

Therefore, byte 2 is

$$BYTE\ 2 = 11000011_2 = C3_{16}$$

Thus, the hexadecimal machine code for the instruction is given by

```
MOV  BL, AL = 88C3H
```

There are a number of ways to specify the location for the second operand located in memory. That is, any of the addressing modes supported by the 8088 microprocessor can be used to generate its address. The addressing mode is selected with the MOD and R/M fields.

Note in Fig. 4–3(b) that the addressing mode for an operand in memory is indicated by one of the three values (00, 01, or 10) in the MOD field and an appropriate R/M code. The different ways in which the operand's address can be generated are shown in the effective address calculation part of the table in Fig. 4–3(b). (These different address-calculation expressions correspond to the addressing modes introduced in Chapter 3.) For instance, if the base register (BX) contains the memory address, this fact is encoded into the instruction by making MOD = 00 and R/M = 111.

EXAMPLE 4.2

The instruction

```
ADD AX,[SI]
```

stands for "add the 16-bit contents of the memory location indirectly specified by SI to the contents of AX." Encode the instruction in machine code. The opcode for add is 000000.

Solution

To specify a 16-bit add operation with a register as the destination, the first byte of machine code will be

$$\text{BYTE 1} = 00000011_2 = 03_{16}$$

The REG field bits in byte 2 are 000 to select AX as the destination register. The other operand is in memory, and its address is specified by the contents of SI with no displacement. In Figs. 4–3(a) and (b), we find that for indirect addressing using SI with no displacement, MOD equals 00 and R/M equals 100. That is,

$$\text{MOD} = 00$$
$$\text{R/M} = 100$$

This gives

$$\text{BYTE 2} = 00000100_2 = 04_{16}$$

Thus, the machine code for the instruction is

```
ADD AX,[SI] = 0304H
```

Some of the addressing modes of the 8088 need either data or an address displacement to be coded into the instruction. These types of information are encoded using additional bytes. For instance, looking at Fig. 4–1, we see that byte 3 is needed in the encoding of an instruction if it uses a byte-size address displacement, and both byte 3 and byte 4 are needed if the instruction uses a word-size displacement.

The size of the displacement is encoded into the MOD field. For example, if the effective address is to be generated by the expression

$$(BX) + D8$$

where D8 stands for 8-bit displacement, MOD is set to 01 to specify memory mode with an 8-bit *displacement* and R/M is set to 111 to select BX.

Bytes 3 and 4 are also used to encode byte-wide immediate operands, word-wide immediate operands, and direct addresses. For example, in an instruction where direct addressing is used to identify the location of an operand in memory, the MOD field must be 00 and the R/M field, 110. The actual value of the operand's address is coded into the bytes that follow.

If both a 16-bit displacement and a 16-bit immediate operand are used in the same instruction, the displacement is encoded into bytes 3 and 4, and the immediate operand is encoded into bytes 5 and 6.

EXAMPLE 4.3

What is the machine code for the instruction

```
XOR CL, [1234H]
```

This instruction says "exclusive-OR the byte of data at memory address 1234_{16} with the byte contents of CL." The opcode for exclusive-OR is 001100.

Solution

Using 001100 as the opcode bits, 1 to denote the register as the destination operand, and 0 to denote byte data, we get

$$\text{BYTE 1} = 00110010_2 = 32_{16}$$

The REG field has to specify CL, which makes it equal to 001. In this case, a direct address has been specified for operand 2. This requires MOD = 00 and R/M = 110. Thus,

$$\text{BYTE 2} = 00001110_2 = 0E_{16}$$

To specify the address 1234_{16}, we must use byte 3 and byte 4. The least significant byte of the address is encoded first, followed by the most significant byte. This gives

$$\text{BYTE 3} = 34_{16}$$

and

$$\text{BYTE 4} = 12_{16}$$

Thus, the machine code form of the instruction is given by

```
XOR CL, 1234H = 320E3412H
```

EXAMPLE 4.4

The instruction

```
ADD [BX][DI]+1234H, AX
```

means "add the word contents of AX to the contents of the memory location specified by based-indexed addressing mode." The opcode for the add operation is 000000.

Solution

The add opcode 000000, a 0 for source operand, and a 1 for word data gives

$$\text{BYTE 1} = 00000001_2 = 01_{16}$$

The REG field in byte 2 is 000 to specify AX as the source register. Since there is a displacement and it needs 16 bits for encoding, the MOD field obtained from Fig. 4–3(a) is 10. The R/M field, also obtained from Fig. 4–3(b), is set to 001 for an effective address generated from DI and BX. This gives the second byte as

$$\text{BYTE 2} = 10000001_2 = 81_{16}$$

The displacement 1234_{16} is encoded in the next two bytes, with the least significant byte first. Therefore, the machine code that results is

```
ADD [BX][DI]+1234H, AX = 01813412H
```

As we indicated earlier, the general format in Fig. 4–1 cannot be used to encode all the instructions that can be executed by the 8088. Minor modifications must be made to this general format to encode a few instructions. In some instructions, one or more additional single-bit fields need to be added. Figure 4–4 shows these 1-bit fields and their functions. For instance, the general format of the *repeat* (REP) instruction is

```
REP = 1111001Z
```

Here bit Z is 1 or 0 depending on whether the repeat operation is to be done when the zero flag is set or when it is reset. Similarly the other two bits, S and V, in Fig. 4–4 are used to encode sign extension for arithmetic instructions and to specify the source of the count for shift or rotate instructions, respectively.

Field	Value	Function
S	0 1	No sign extension Sign extend 8-bit immediate data to 16 bits if W=1
V	0 1	Shift/rotate count is one Shift/rotate count is specified in CL register
Z	0 1	Repeat/loop while zero flag is clear Repeat/loop while zero flag is set

Figure 4–4 Additional 1-bit fields and their functions. (Reprinted by permission of Intel Corp. Copyright/Intel Corp. 1979)

The instruction set summary in Fig. 3–6 in Ch. 3 shows the formats for all of the instructions in the 8088's instruction set. This information can be used to encode any 8088 instruction.

Instructions that involve a segment register need a 2-bit field to encode which register is to be affected. This field is called the *SR field.* As shown in Fig. 3–6, PUSH and POP are examples of instructions that have an SR field. The four segment registers ES, CS, SS, and DS are encoded according to the table in Fig. 4–5.

EXAMPLE 4.5

The instruction

```
MOV  WORD PTR [BP][DI]+1234H, 0ABCDH
```

stands for "move the immediate data word $ABCD_{16}$ into the memory location specified by based-indexed addressing mode." Express the instruction in machine code.

Solution

Since this instruction does not involve one of the registers as an operand, it does not follow the general format we have been using. Figure 3–6 shows that byte 1 in an immediate data to memory move is

```
1100011W
```

Register	SR
ES	00
CS	01
SS	10
DS	11

Figure 4–5 Segment register codes.

In our case, we are moving word-size data; therefore, W equals 1. This gives

$$\text{BYTE 1} = 11000111_2 = \text{C7}_{16}$$

Again from Fig. 3–6, we find that byte 2 has the form

$$\text{BYTE 2} = \text{(MOD)000(R/M)}$$

For a memory operand using a 16-bit displacement, Fig. 4–3(a) shows that MOD equals 10, and for based-indexed addressing using BP and DI with a 16-bit displacement, Fig. 4–3(b) shows that R/M equals 011. This gives

$$\text{BYTE 2} = 10000011_2 = 83_{16}$$

Byte 3 and 4 encode the displacement with its low byte first. Thus, for a displacement of 1234_{16} we get

$$\text{BYTE 3} = 34_{16}$$

and

$$\text{BYTE 4} = 12_{16}$$

Lastly, bytes 5 and 6 encode the immediate data also with the least significant byte first. For data word ABCD_{16}, we get

$$\text{BYTE 5} = \text{CD}_{16}$$

and

$$\text{BYTE 6} = \text{AB}_{16}$$

Thus, the entire instruction in machine code is given by

```
MOV  WORD PTR  [BP][DI]+1234H, 0ABCDH = C7833412CDABH
```

EXAMPLE 4.6

The instruction

```
MOV  [BP][DI]+1234H, DS
```

says "move the contents of the data segment register to the memory location specified by based-indexed addressing mode." Express the instruction in machine code.

Solution

Figure 3–6 shows that this instruction is encoded as

10001100(MOD)0(SR)(R/M)(DISP)

The MOD and R/M fields are the same as in Example 3–6. That is,

$$MOD = 10$$

and

$$R/M = 011$$

Moreover, the value of DISP is given as 1234_{16}. Finally, Fig. 4–5 shows that to specify DS, the SR field is

$$SR = 11$$

Therefore, the instruction is coded as

$$10001100100110110011010000010010_2 = 8C9B3412_{16}$$

▲ 4.2 ENCODING A COMPLETE PROGRAM IN MACHINE CODE

To encode a complete assembly language program in machine code, we must individually encode each of its instructions. We do this by using the instruction formats in Fig. 3–6 and the information in the tables of Figs. 4–2, 4–3, 4–4, and 4–5. We first identify the general machine code format for the instruction in Fig. 3–6. After determining the format, the bit fields are evaluated using the tables of Figs. 4–2, 4–3, 4–4, and 4–5. Finally, the binary-coded instruction is expressed in hexadecimal form.

To execute a program on the PC, the machine code of the program is first stored in the code segment of memory. The bytes of machine code are stored in sequentially addressed locations in memory. The first byte of the program is stored at the lowest address, and it is followed by the other bytes in the order in which they are encoded. That is, the address is incremented by 1 after storing each byte of machine code in memory.

EXAMPLE 4.7

Encode the "block-move" program in Fig. 4–6(a) and show how it would be stored in memory starting at address 200_{16}.

```
        MOV AX,2000H   ;LOAD AX REGISTER
        MOV DS,AX      ;LOAD DATA SEGMENT ADDRESS
        MOV SI,100H    ;LOAD SOURCE BLOCK POINTER
        MOV DI,120H    ;LOAD DESTINATION BLOCK POINTER
        MOV CX,10H     ;LOAD REPEAT COUNTER
NXTPT:  MOV AH,[SI]    ;MOVE SOURCE BLOCK ELEMENT TO AH
        MOV [DI],AH    ;MOVE ELEMENT FROM AH TO DESTINATION BLOCK
        INC SI         ;INCREMENT SOURCE BLOCK POINTER
        INC DI         ;INCREMENT DESTINATION BLOCK POINTER
        DEC CX         ;DECREMENT REPEAT COUNTER
        JNZ NXTPT      ;JUMP TO NXTPT IF CX NOT EQUAL TO ZERO
        NOP            ;NO OPERATION
```

(a)

Instruction	Type of instruction	Machine code
MOV AX,2000H	Move immediate data to register	$101110000000000000100000_2 = B80020_{16}$
MOV DS,AX	Move register to segment register	$1000111011011000_2 = 8ED8_{16}$
MOV SI,100H	Move immediate data to register	$101111100000000000000001_2 = BE0001_{16}$
MOV DI,120H	Move immediate data to register	$101111110010000000000001_2 = BF2001_{16}$
MOV CX,10H	Move immediate data to register	$101110010001000000000000_2 = B91000_{16}$
MOV AH,[SI]	Move memory data to register	$1000101000100100_2 = 8A24_{16}$
MOV [DI],AH	Move register data to memory	$1000100000100101_2 = 8825_{16}$
INC SI	Increment register	$01000110_2 = 46_{16}$
INC DI	Increment register	$01000111_2 = 47_{16}$
DEC CX	Decrement register	$01001001_2 = 49_{16}$
JNZ NXTPT	Jump on not equal to zero	$0111010111110111_2 = 75F7_{16}$
NOP	No operation	$1001000_2 = 90_{16}$

(b)

Figure 4–6 (a) Block move program. (b) Machine coding of the block move program. (c) Storing the machine code in memory.

Solution

To encode this program into its equivalent machine code, we use the instruction set table in Fig. 3–6. The first instruction

```
MOV  AX, 2000H
```

Memory address	Contents	Instruction
200H	B8H	MOV AX,2000H
201H	00H	
202H	20H	
203H	8EH	MOV DS,AX
204H	D8H	
205H	BEH	MOV SI,100H
206H	00H	
207H	01H	
208H	BFH	MOV DI,120H
209H	20H	
20AH	01H	
20BH	B9H	MOV CX,10H
20CH	10H	
20DH	00H	
20EH	8AH	MOV AH,[SI]
20FH	24H	
210H	88H	MOV [DI],AH
211H	25H	
212H	46H	INC SI
213H	47H	INC DI
214H	49H	DEC CX
215H	75H	JNZ $-9
216H	F7H	
217H	90H	NOP

(c)

Figure 4–6 (continued)

is a "move immediate data to register" instruction. In Fig. 3–6, we find it has the form

1011(W)(REG)(DATA DATA IF W = 1)

Since the move is to register AX, Fig. 4–2 shows that the W bit is 1 and REG is 000. The immediate data 2000_{16} follows this byte with the least significant byte coded first. This gives the machine code for the instruction as

$$101110000000000000100000_2 = B80020_{16}$$

The second instruction

```
MOV  DS, AX
```

represents a "move register to segment register" operation. This instruction has the general format

10001110(MOD)0(SR)(R/M)

From Figs. 4–3(a) and (b), we find that for this instruction MOD = 11 and R/M is 000 for AX. Furthermore, Fig. 4–5 shows that SR = 11 for data segment. This results in the code

$$1000111011011000_2 = 8ED8_{16}$$

for the second instruction.

The next three instructions have the same format as the first instruction. In the third instruction, REG is 110 for SI and the data is 0100_{16}. This gives the instruction code as

$$101111100000000000000001_2 = BE0001_{16}$$

The fourth instruction has REG coded as 111 (DI) and the data as 0120_{16}. This results in the code

$$101111110010000000000001_2 = BF2001_{16}$$

In the fifth instruction, REG is 001 for CX, with 0010_{16} as the data. This gives its code as

$$101110010001000000000000_2 = B91000_{16}$$

The sixth instruction is a move of byte data from memory to a register. From Fig. 3–6, we find that its general format is

100010(D)(W)(MOD)(REG)(R/M)

Since AH is the destination and the instruction operates on bytes of data, the D and W bits are 1 and 0, respectively, and the REG field is 100. The contents of SI are used as a pointer to the source operand; therefore, MOD is 00 and R/M is 100. This gives the instruction code as

$$1000101000100100_2 = 8A24_{16}$$

The last MOV instruction has the same form as the previous one. However, in this case, AH is the destination and DI is the address pointer. This makes D equal to 0 and R/M equal to 101. Therefore, we get

$$1000100000100101_2 = 8825_{16}$$

The next two instructions increment registers and have the general form

01000(REG)

For the first one, register SI is incremented. Therefore, REG equals 110. This results in the instruction code as

$$01000110_2 = 46_{16}$$

In the second, REG equals 111 to encode DI. This gives its code as

$$01000111_2 = 47_{16}$$

The two INC instructions are followed by a DEC instruction. Its general form is

01001(REG)

To encode CX, REG equals 001, which results in the instruction code

$$01001001_2 = 49_{16}$$

The next instruction is a jump to the location NXTPT. Its form is

01110101(IP-INC8)

We will not yet complete this instruction because it will be easier to determine the number of bytes to be jumped after the data have been coded for storage in memory. The final instruction is NOP, and it is coded as

$$10010000_2 = 90_{16}$$

Figure 4–6(b) shows the entire machine code program.

As shown in Fig. 4–6(c), our encoded program will be stored in memory starting from memory address 200H. The choice of program-beginning address establishes the address for the NXTPT label. Note that the MOV AH, [SI] instruction, which has this label, starts at address $20E_{16}$. This location is 9 bytes back from the value in IP after fetching the JNZ instruction. Therefore, the displacement (IP-INC8) in the JNZ instruction is -9, which is $F7_{16}$ as an 8-bit hexadecimal number. Thus, the instruction is encoded as

$$0111010111110111_2 = 75F7_{16}$$

▲ 4.3 THE PC AND ITS DEBUG PROGRAM

Now that we know how to convert an assembly language program to machine code and how this machine code is stored in memory, we are ready to enter it into the PC; execute it; examine the results that it produces; and, if necessary, debug any errors in its operation. It is the *DEBUG program,* which is part of the PC's disk operating system (DOS), that permits us to initiate these types of operations from the keyboard of the microcomputer. In this section we show how to load the DEBUG program from DOS, how to use DEBUG commands to examine or modify the contents of the MPU's internal registers, and how to return to DOS from DEBUG.

Using DEBUG, the programmer can issue commands to the microcomputer in the PC. Assuming that the DOS has already been entered and that a disk that contains the DEBUG program is in drive A, DEBUG is loaded by simply issuing the command

```
C:\DOS>a:debug (↵)
```

Actually, debug can be typed in using either uppercase or lowercase characters. However, for simplicity, we will use all uppercase characters in this book

EXAMPLE 4.8

Assuming that the DOS is already running and that the DEBUG program is in the DOS directory on drive C, initiate the DEBUG program from the PC's keyboard. What prompt for command entry does the debugger display?

Solution

From DOS, DEBUG is brought up by entering

```
C:\DOS>DEBUG (↵)
```

Drive C is accessed to load the DEBUG program; DEBUG is then executed and its prompt, (-) is displayed. DEBUG is now waiting to accept a command. Figure 4–7 shows what is displayed on the screen.

The keyboard is the input unit of the debugger and permits the user to enter commands to load data, such as the machine code of a program; examine or modify the state of the MPU's internal registers and memory; or execute a program. All we need to do is type in the command and then depress the enter (↵) key. These debug commands are the tools a programmer needs in order to enter, execute, and debug programs.

When the command entry sequence is completed, the DEBUG program decodes the entry to determine which operation is to be performed, verifies that it is a valid command, and—if it is valid—passes control to a routine that performs the operation. At the completion of the operation, results are displayed on the screen and the DEBUG prompt (=) is redisplayed. The PC remains in this state until a new entry is made from the keyboard.

Six kinds of information are typically entered as part of a command: a *command letter,* an *address,* a *register name,* a *filename,* a *drive name,* and *data.* The entire command set of DEBUG is shown in Fig. 4–8. This table gives the name for each command, its function, and its general syntax. By *syntax,* we mean the order in which key entries must be made to initiate the command.

```
C:\DOS>DEBUG
-
```

Figure 4–7 Loading the DEBUG program.

Command	Syntax	Function
Register	R [REGISTER NAME]	Examine or modify the contents of an internal register
Quit	Q	End use of the DEBUG program
Dump	D [ADDRESS]	Dump the contents of memory to the display
Enter	E ADDRESS [LIST]	Examine or modify the contents of memory
Fill	F STARTING ADDRESS ENDING ADDRESS LIST	Fill a block in memory with the data in list
Move	M STARTING ADDRESS ENDING ADDRESS DESTINATION ADDRESS	Move a block of data from a source location in memory to a destination location
Compare	C STARTING ADDRESS ENDING ADDRESS DESTINATION ADDRESS	Compare two blocks of data in memory and display the locations that contain different data
Search	S STARTING ADDRESS ENDING ADDRESS LIST	Search through a block of data in memory and display all locations that match the data in list
Input	I ADDRESS	Read the input port
Output	O ADDRESS, BYTE	Write the byte to the output port
Hex Add/Subtract	H NUM1,NUM2	Generate hexadecimal sum and difference of the two numbers
Unassemble	U [STARTING ADDRESS ENDING ADDRESS]	Unassemble the machine code into its equivalent assembler instructions
Name	N FILE NAME	Assign the filename to the data to be written to the disk
Write	W [STARTING ADDRESS [DRIVE STARTING SECTOR NUMBER OF SECTORS]]	Save the contents of memory in a file on a diskette
Load	L [STARTING ADDRESS [DRIVE STARTING SECTOR NUMBER OF SECTORS]]	Load memory with the contents of a file on a diskette
Assemble	A [STARTING ADDRESS]	Assemble the instruction into machine code and store in memory
Trace	T [=ADDRESS] [NUMBER]	Trace the execution of the specified number of instructions
Go	G [=STARTING ADDRESS [BREAKPOINT ADDRESS....]]	Execute the instructions down through the breakpoint address

Figure 4–8 DEBUG program command set.

With the loading of DEBUG, the state of the microprocessor is initialized. The *initial state* depends on the DOS version and system configuration at the time the DEBUG command is issued. An example of the initial state is illustrated with the software model in Fig. 4–9. Notice that registers AX, BX, CX, DX, BP, SI, and DI are reset to zero; IP is initialized to 0100_{16}; CS, DS, SS, and ES are all loaded with 1342_{16}; and SP is loaded with $FFEE_{16}$. Finally, all the flags except IF are reset to zero. We can use the register command to verify this initial state.

Let us now look at the syntax for the *REGISTER* (R) *command.* This is the debugger command that allows us to examine or modify the contents of internal registers of the MPU. Notice that the syntax for this command is given in Fig. 4–8 as

```
R [REGISTER NAME]
```

Here the command letter is R. It is optionally followed by a register name. Figure 4–10 shows what must be entered as the Register name for each of the 8088's registers.

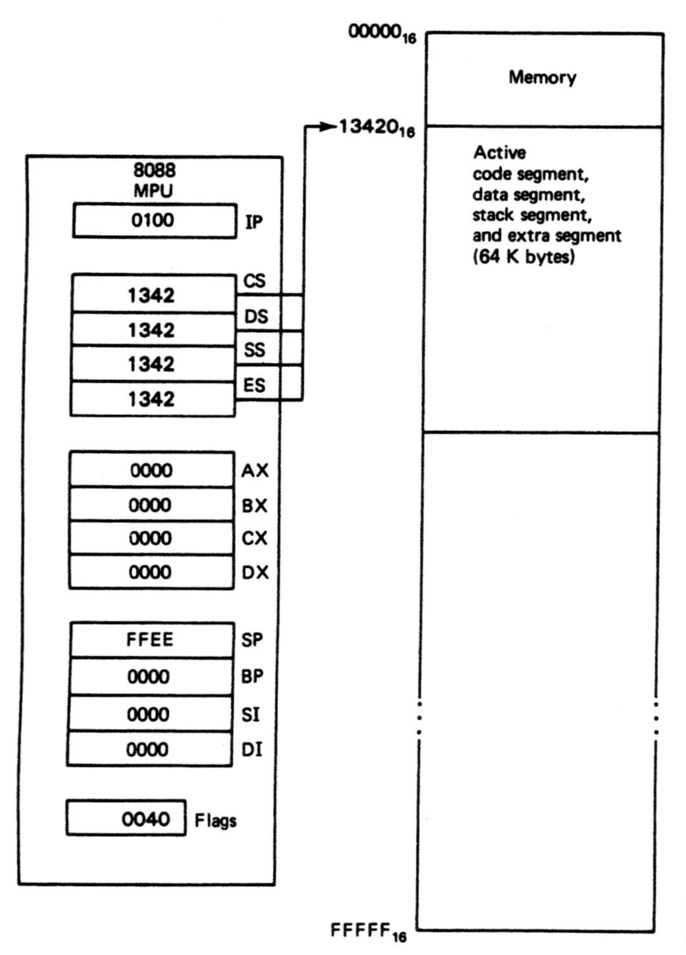

Figure 4–9 An initial state of the 8088 microprocessor.

Symbol	Register
AX	Accumulator register
BX	Base register
CX	Count register
DX	Data register
SI	Source index register
DI	Destination index register
SP	Stack pointer register
BP	Base pointer register
CS	Code segment register
DS	Data segment register
SS	Stack segment register
ES	Extra segment register
F	Flag register
IP	Instruction pointer

Figure 4–10 Register mnemonics for the R command.

An example of the command entry needed to examine or modify the value in register AX is

```
-R AX (↵)
```

Notice that brackets are not included around the register name. In Fig. 4–8 brackets are simply used to separate the various elements of the DEBUG command. They are never entered as part of the command. Execution of this Register command causes the current value in AX to be displayed as

```
AX 0000
:_
```

Here we see that AX contains 0000_{16}. The examine register command is not yet complete. Note that a colon (:) followed by the cursor is displayed. We can now either depress (↵) to complete the command, leaving the register contents unchanged, or enter a new value for AX following the colon and then depress (↵). Let us load AX with a new value of $00FF_{16}$. This is done by the entry

```
:00FF (↵)
-
```

EXAMPLE 4.9

Verify the initialized state of the 8088 by examining the contents of its registers with the Register command.

```
-R
AX=0000  BX=0000  CX=0000 DX=0000 SP=FFEE  BP=0000  SI=0000  DI=0000
DS=1342  ES=1342  SS=1342  CS=1342  IP=0100   NV UP EI PL NZ NA PO NC
1342:0100 CD21           INT     21
```

Figure 4–11 Displaying the initialized state of the MPU.

Solution

If we enter the register command without a specific register name, the debugger causes the state of all registers and flags to be displayed. That is, if we enter

```
-R (↵)
```

the information displayed is that shown in Fig. 4–11. Looking at this result, we see that all registers are initialized as expected. To verify that all flags other than IF were reset, we compare the flag settings that are listed to the right of the value for IP with the values in the table of Fig. 4–12. Note that as expected all flags but IF are in the reset state. The last line displays the machine code and assembly language statement of the instruction pointed to by the current values in CS and IP (CS:IP).

EXAMPLE 4.10

Specify PC debug commands that will cause the value in BX to be modified to $FF00_{16}$ and then verify that this new value is loaded into BX.

Solution

To modify the value in BX, all we need to do is issue the Register command with BX and then respond to the prompt :_ by entering the value $FF00_{16}$. This is done with the command sequence

```
-R BX (↵)
BX 0000
:FF00 (↵)
-
```

Flag	Meaning	Set	Reset
OF	Overflow	OV	NV
DF	Direction	DN	UP
IF	Interrupt	EI	DI
SF	Sign	NG	PL
ZF	Zero	ZR	NZ
AF	Auxiliary carry	AC	NA
PF	Parity	PE	PO
CF	Carry	CY	NC

Figure 4–12 Notations used for displaying the status flags.

```
-R BX
BX 0000
:FF00
-R BX
BX FF00
:-
-
```

Figure 4–13 Displayed information for Example 4.10.

We can verify that $FF00_{16}$ has been loaded into BX by issuing another Register command as follows:

```
-R BX (↵)
BX FF00
:_     (↵)
-
```

The displayed information for this command sequence is shown in Fig. 4–13.

The way in which the Register command is used to modify flags is different than how it is used to modify the contents of a register. If we enter the command

```
-R F  (↵)
```

the flag settings are displayed as

```
NV UP EI PL NZ NA PO NC-
```

To modify flags, just type in their new states (using the notations shown in Fig. 4–12) and depress the return key. For instance, to set the carry and zero flags, enter

```
NV UP EI PL NZ NA PO NC -CY ZR (↵)
```

Note that the new flag states can be entered in any order.

EXAMPLE 4.11

Use the Register command to set the parity flag to even parity. Verify that the flag has been changed.

Solution

To set PF for even parity, issue the Register command for the flag register and then enter PE as the new flag data. This is done with the command sequence

```
-R F (↵)
NV UP EI PL NZ NA PO NC -PE (↵)
```

```
-R F
NV UP EI PL NZ NA PO NC   -PE
-R F
NV UP EI PL NZ NA PE NC   -
-
```

Figure 4–14 Displayed information for Example 4.11.

To verify that PF has been changed to its PE state, just enter another register command for the flag register as follows:

```
-R F  (↵)

NV UP EI PL NZ NA PE NC- (↵)
```

Notice that the state of the parity flag has changed from PO to PE. Figure 4–14 shows these commands, and the displayed flag status that results.

The Register command is very important for debugging programs. For instance, it can be used to check the contents of a register or flag prior to and again just after execution of an instruction. In this way, we can tell whether or not the instruction correctly performed the required operation.

If the command that was entered is identified as being invalid, an *error message* is displayed. Let us look at an example of an invalid command entry. To do this, we repeat our earlier example in which AX was loaded with $00FF_{16}$, but in keying in $00FF_{16}$, we enter the uppercase letter O instead of zeros. The result produced by issuing this command is shown in Fig. 4–15. Here we see that a warning "Error" is displayed, and the symbol ^ is used to mark the starting location of the error in the command. To correct this error, the command is simply reentered.

We will examine one more command before going on. We now know how to invoke the DEBUG program from the DOS prompt, but we must also be able to return to the DOS when finished using DEBUG. The debugger contains a command called *QUIT* (Q) to do this. Therefore, to return to the DOS, we simply respond to the debug prompt with

```
Q (↵)
```

▲ 4.4 EXAMINING AND MODIFYING THE CONTENTS OF MEMORY

In Section 4.3 we studied the command that permitted us to examine or modify the contents of the MPU's internal registers. Here we will continue our study of DEBUG's command set with the commands used to examine and modify the contents of memory. The

```
-R AX
AX 0000
:OOFF
 ^ Error
-
```

Figure 4–15 Invalid entry.

ability to do this is essential for debugging programs. For instance, the value at an address in memory can be examined just before and just after the execution of an instruction that affects this memory location. In this way, we can verify that the instruction performs the operation correctly. This type of command also can be used to load a program into the microcomputer's memory. The complete command set of DEBUG was shown in Fig. 4–8. Six of these commands—Dump, Enter, Fill, Move, Compare, and Search—are provided for use in examining or modifying the contents of storage locations in memory. Let us now look at the operations performed with each of these commands.

DUMP Command

The *DUMP* (D) *command* allows us to examine the contents of a memory location or a block of consecutive memory locations. From Fig. 4–8, we see that the general syntax for Dump is

```
D [ADDRESS]
```

If a segment register is not specified, the value of ADDRESS entered is automatically referenced to the current value in the data segment (DS) register.

Dump can also be issued without ADDRESS. This gives the command

```
D (↵)
```

Execution of this form of the command causes the 128 consecutive bytes starting at offset 0100_{16} from the current value in DS to be displayed. If DS is initialized with 1342_{16} when DEBUG is started, issuing this command gives the memory dump shown in Fig. 4–16.

Note that 16 bytes of data are displayed per line, and only the address of the first byte is shown at the left. From Fig. 4–16 we see that the address of the first location in the first line is denoted as 1342:0100. This corresponds to the physical address

$$13420_{16} + 0100_{16} = 13520_{16}$$

The second byte of data displayed in the first line corresponds to the memory address 1342:0101, or 13521_{16}, and the last byte on this line corresponds to the memory address

```
-D
1342:0100  75 07 05 00 00 00 B2 14-4C 00 2B 04 27 0C 06 74   u.......L.+.'..t
1342:0110  46 74 05 00 00 00 31 13-B2 14 99 00 31 13 31 13   Ft....1.....1.1.
1342:0120  8A 40 8B 1E F2 39 9A 88-97 63 17 8B 1E F2 39 BA   .@...9...c....9.
1342:0130  61 46 E8 37 0B A0 5A 46-0A C0 74 2D 8B 1E AB 42   aF.7..ZF..t-...B
1342:0140  FE C8 75 0B 89 1E 5C 46-C6 06 5A 46 02 EB 1A 3B   ..u...\F..ZF...;
1342:0150  1E 5C 46 74 14 C6 06 5A-46 00 E8 17 01 75 03 E9   .\Ft...ZF....u..
1342:0160  FC 00 3C 59 75 03 E9 27-01 E8 97 01 8A C5 74 22   ..<Yu..'......t"
1342:0170  80 3E CE 09 00 75 10 80-F9 0D 75 0B E8 55 00 74   .>...u....u..U.t
-
```

Figure 4–16 Examining the contents of 128 consecutive bytes in memory.

1342:010F, or $1352F_{16}$. Note that the values of the eighth and ninth bytes are separated by a hyphen.

For all memory dumps, an ASCII version of the memory data is also displayed. It is displayed to the right of the hexadecimal data in Fig. 4–16. All bytes that result in an unprintable ASCII character are displayed as the "." symbol.

The results shown in Fig. 4–16 could be obtained with several other forms of the Dump command. One way is to enter the current value of DS, which is 1342_{16}, and an offset of 0100_{16} in the address field. This results in the command

```
D 1342:100 (↵)
```

Another way is to enter DS instead of its value with the offset. This gives

```
D DS:100 (↵)
```

In fact, the same results can be obtained by just issuing the command

```
D 100 (↵)
```

EXAMPLE 4.12

What is the physical address range of the bytes of data in the last line of data shown in Fig. 4–16?

Solution

In Fig. 4–16, we see that the first byte is at address 1342:0170. This is the physical address

$$13420_{16} + 0170_{16} = 13590_{16}$$

The last byte is at address 1342:017F, and its physical address is

$$13420_{16} + 017F_{16} = 1359F_{16}$$

EXAMPLE 4.13

What happens if we repeat the entry D (↵) after obtaining the memory dump shown in Fig. 4–16?

Solution

The contents of the next 128 consecutive bytes of memory are dumped to the display. The displayed information is shown in Fig. 4–17.

Frequently, we do not want to examine such a large block of memory. Instead, we may want to look at just a few bytes or a specific-sized block. The Dump command can also do this. This time we enter two addresses. The first address defines the starting point

```
-D
1342:0180  0B 72 0C 3C 0A 75 08 E8-4A 00 75 03 E9 D9 00 E9   .r.<.u..J.u.....
1342:0190  BF FE E8 23 40 75 02 EB-82 80 3E CE 09 00 75 15   ...#@u....>...u.
1342:01A0  80 F9 0D 75 10 E8 2C 00-74 E2 72 18 3C 0A 75 14   ...u..,.t.r.<.u.
1342:01B0  8A C5 E8 03 40 9A 49 C5-63 17 9A B3 B8 63 17 E8   ....@.I.c....c..
1342:01C0  9F 0A EB C3 9A 49 C5 63-17 9A B3 B8 63 17 E8 90   .....I.c....c...
1342:01D0  0A E9 7D FE E8 D1 01 74-66 89 2E D7 09 8A E8 24   ..}....tf......$
1342:01E0  7F 8A CB 3C 1B 74 07 3C-1D 75 3B EB 59 90 E8 B7   ...<.t.<.u;.Y...
1342:01F0  01 74 31 52 8A F0 E8 AF-01 74 38 3C 1C 75 0E C6   .t1R.....t8<.u..
-
```

Figure 4–17 Displayed information for repeat of 128-byte memory-dump command.

of the block, and the second address identifies the end of the block. For instance, if we want to examine the two bytes of data that are at offsets equal to 200(↵) and 201_{16} in the current data segment, the command is

```
D DS:200 201  (↵)
```

The result obtained by executing this command is given in Fig. 4–18.

EXAMPLE 4.14

Specify a Dump command to display the contents of the 32 bytes of memory located at offsets 0300_{16} through $031F_{16}$ in the current data segment.

Solution

The command needed to display the contents of this part of memory is

```
D 300 31F (↵)
```

and the information displayed is shown in Fig. 4–19.

Up to now, all of the data displayed with the Dump command were contained in the data segment of memory. It is also possible to examine data that are stored in the code segment, stack segment, or extra segment. To do this, we simply use the appropriate segment register name in the command. For instance, the command needed to dump the values in the first 16 bytes of the current code segment is

```
D CS:0 F (↵)
```

```
-D DS:200 201
1342:0200  06 CE
```

Figure 4–18 Displaying two bytes of data.

```
-D 300 31F
1342:0300  59 5F C3 E8 D0 3E 74 0A-E8 BF 3E E8 5C 3E 75 F3   Y_...>t...>.\>u.
1342:0310  0C 01 C3 80 3E 25 39 00-74 4A 53 52 FF 36 AB 42   ....>%9.tJSR.6.B
-
```

Figure 4–19 Displayed information for Example 4.14.

EXAMPLE 4.15

Use the Dump command to examine the 16 bytes of memory just below the top of the stack.

Solution

The top of the stack is defined by the contents of the SS and SP registers (SS:SP). Earlier we found that SP is initialized to $FFEE_{16}$ when debug is loaded. Therefore, the 16 bytes we are interested in reside at offset $FFEE_{16}$ through $FFFD_{16}$ from the current value in SS. This part of the stack is viewed with the command

```
D SS:FFEE FFFD (↵)
```

The result displayed by executing this command is shown in Fig. 4–20.

ENTER Command

The Dump command allowed us to examine the contents of memory, but we also need to be able to modify or enter information in memory—for instance, to load a machine code program. It is for this purpose that the *ENTER* (E) *command* is provided in the DEBUG program.

Figure 4–8 shows that the syntax of the Enter command is

```
E [ADDRESS] [LIST]
```

The address part of the E command is entered the same way we just described for the Dump command. If no segment name is included with the offset, the DS register is assumed. The list that follows the address is the data that get loaded into memory.

```
-D SS:FFEE FFFD
1342:FFE0                                             00 00     ..
1342:FFF0  00 00 00 00 00 00 00 00-00 00 00 00 00 00          ..............
-
```

Figure 4–20 Displayed information for Example 4.15.

As an example, let us write a command that will load five consecutive byte-wide memory locations starting at address DS:100 with the value FF_{16}. This is done with the command

```
E  DS:100 FF FF FF FF FF (↵)
```

To verify that the new values of data have been stored in memory, let us dump the contents of these locations to the display. To do this, we issue the command

```
D  DS:100 104  (↵)
```

These commands and the displayed results are shown in Fig. 4–21. Notice that the byte storage locations from address DS:100 through DS:104 now all contain the value FF_{16}.

The Enter command can also be used in a way in which it either examines or modifies the contents of memory. If we issue the command with an address but no data, the contents of the addressed storage location are displayed. For instance, the command

```
E  DS:100   (↵)
```

causes the value at this address to be displayed as follows:

```
1342:0100  FF._
```

Notice that the value at address 1342:0100 is FF_{16}.

At this point we have several options; for one, we can depress the return key. This terminates the Enter command without changing the contents of the displayed memory location and causes the debug prompt to be displayed. For another, rather than depressing return, we can depress the spacebar. Again, the contents of the displayed memory location remain unchanged, but this time the command is not terminated. Instead, the contents of the next consecutive memory address are displayed. Let us assume that this was done. Then the display reads

```
1342:0100 FF. FF._
```

Here we see that the data stored at address 1342:0101 are also FF_{16}. A third type of entry that could be made is to enter a new value of data and then depress the spacebar or return key. For example, we could enter 11_{16} and then depress the spacebar. This gives the display

```
1342:0100 FF. FF.11 FF._
```

The value pointed to by address 1342:101 has been changed to 11_{16}, and the contents of address 1342:0102, FF_{16}, are displayed. Now, depress the return key to terminate the data entry sequence.

```
-E DS:100 FF FF FF FF FF
-D DS:100 104
1342:0100   FF FF FF FF FF
-
```

Figure 4–21 Modifying five consecutive bytes in memory and verifying the change of data.

```
-E DS:100
1342:0100  FF._
1342:00FF  FF._
-
```

Figure 4–22 Using the "–" key to examine the contents of the previous memory location.

EXAMPLE 4.16

Start a data entry sequence by examining the contents of address DS:100 and then, without entering new data, depress the "-" key. What happens?

Solution

The data-entry sequence is initiated as

```
E DS:100 (↵)
```

```
1342:0100 FF._
```

Entering "-" causes the following address and data to be displayed.

```
1342:00FF FF._
```

Notice that these are the address and contents of the storage location at the address equal to one less than DS:100—that is, the previous byte storage location. This result is shown in Fig. 4–22.

The Enter command can also be used to enter ASCII data. Do this by simply enclosing the data entered in quotation marks. An example is the command

```
E DS:200 "ASCII" (↵)
```

This command causes the ASCII data for letters A, S, C, I, and I to be stored in memory at addresses DS:200, DS:201, DS:202, DS:203, and DS:204, respectively. This character data entry can be verified with the command

```
D DS:200 204 (↵)
```

Looking at the ASCII field of the data dump shown in Fig. 4–23 we see that the correct ASCII data were stored into memory. Actually, either single- or double-quote marks can be used. Therefore, the entry could also have been made as

```
E DS:200 'ASCII' (↵)
```

```
-E DS:200 "ASCII"
-D DS:200 204
1342:0200  41 53 43 49 49                                   ASCII
-
```

Figure 4–23 Loading ASCII data into memory with the Enter command.

FILL Command

Frequently, we want to fill a block of consecutive memory locations all with the same data. For example, we may need to initialize storage locations in an area of memory with zeros. To do this by entering the data address by address with the Enter command is very time-consuming. It is for this type of operation that the *FILL* (F) *command* is provided.

Figure 4–8 shows that the general form of the Fill command is

```
F [STARTING ADDRESS] [ENDING ADDRESS] [LIST]
```

Here *starting address* and *ending address* specify the block of storage locations in memory. They are followed by a *list* of data. An example is the command

```
F 100 11F 22 (↵)
```

Execution of this command causes the 32 byte locations in the range 1342:100 through 1342:11F to be loaded with 22_{16}. The fact that this change in memory contents has happened is verified with the command

```
D 100 11F (↵)
```

Figure 4–24 shows the result of executing these two commands.

EXAMPLE 4.17

Initialize all storage locations in the block of memory from DS:120 through DS:13F with the value 33_{16} and the block of storage locations from DS:140 through DS:15F with the value 44_{16}. Verify that the contents of these ranges of memory are correctly modified.

Solution

These initialization operations can be done with the Fill commands

```
F 120 13F 33 (↵)
F 140 15F 44 (↵)
```

They are then verified with the Dump command

```
D 120 15F (↵)
```

The information displayed by the command sequence is shown in Fig. 4–25.

```
-F 100 11F 22
-D 100 11F
1342:0100  22 22 22 22 22 22 22 22-22 22 22 22 22 22 22 22   """"""""""""""""
1342:0110  22 22 22 22 22 22 22 22-22 22 22 22 22 22 22 22   """"""""""""""""
-
```

Figure 4–24 Initializing a block of memory with the Fill command.

```
-F 120 13F 33
-F 140 15F 44
-D 120 15F
1342:0120  33 33 33 33 33 33 33 33-33 33 33 33 33 33 33 33   3333333333333333
1342:0130  33 33 33 33 33 33 33 33-33 33 33 33 33 33 33 33   3333333333333333
1342:0140  44 44 44 44 44 44 44 44-44 44 44 44 44 44 44 44   DDDDDDDDDDDDDDDD
1342:0150  44 44 44 44 44 44 44 44-44 44 44 44 44 44 44 44   DDDDDDDDDDDDDDDD
-
```

Figure 4–25 Displayed information for Example 4.17.

MOVE Command

The *MOVE* (M) *command* allows us to copy a block of data from one part of memory to another part. For instance, using this command, a 32-byte block of data that resides in memory from address DS:100 through DS:11F can be copied to the address range DS:200 through DS:21F with a single operation.

The general form of the Move command is given in Fig. 4–8 as

```
M  [STARTING ADDRESS]  [ENDING ADDRESS]  [DESTINATION ADDRESS]
```

Note that it is initiated by depressing the M key. After this, we must enter three addresses. The first two addresses are the *starting address* and *ending address* of the source block of data—that is, the block of data that is to be copied. The third address is the *destination starting address*—that is, the starting address of the section of memory to which the block of data is to be copied.

The command for our example, which copies a 32-byte block of data located at address DS:100 through DS:11F to the block of memory starting at address DS:200, is

```
M  100  11F 200  (↵)
```

EXAMPLE 4.18

Fill each storage location in the block of memory from address DS:100 through DS:11F with the value 11_{16}. Then copy this block of data to a destination block starting at DS:160. Verify that the block move is correctly done.

Solution

First, we fill the source block with 11_{16}. using the command

```
F  100  11F  11 (↵)
```

Next, it is copied to the destination with the command

```
M  100  11F  160  (↵)
```

Finally, we dump the complete range from DS:100 to DS:17F by issuing the command

```
D 100 17F (↵)
```

The result of this memory dump is given in Fig. 4–26. It verifies that the block move was successfully performed.

COMPARE Command

Another type of memory operation we sometimes need to perform is comparing the contents of two blocks of data to determine if they are or are not the same. This operation can be done with the *COMPARE* (C) *command* of the DEBUG program. Fig. 4–8 shows that the general form of this command is

```
C [STARTING ADDRESS] [ENDING ADDRESS] [DESTINATION ADDRESS]
```

For example, to compare a block of data located from address DS:100 through DS:11F to an equal-size block of data starting at address DS:160, we issue the command

```
C 100 10F 160 (↵)
```

This command causes the contents of corresponding address locations in each block to be compared to each other—that is, the contents of address DS:100 are compared to those at address DS:160, the contents at address DS:101 are compared to those at address DS:161, and so on. Each time unequal elements are found, the address and contents of that byte in both blocks are displayed.

Let us assume that the contents of memory are as shown in Fig. 4–26 when this command is executed. Since both of these blocks contain the same information, no data are displayed. However, if this source block is next compared to the destination block starting at address DS:120 by entering the command

```
C 100 10F 120 (↵)
```

all elements in both blocks are unequal; therefore, the information shown in Fig. 4–27 is displayed.

SEARCH Command

The *Search* (S) *command* can be used to scan through a block of data in memory to determine whether or not it contains specific value. The general form of this command as given in Fig. 4–8 is

```
S [STARTING ADDRESS] [ENDING ADDRESS] [LIST]
```

```
-F 100 11F 11
-M 100 11F 160
-D 100 17F
1342:0100  11 11 11 11 11 11 11 11-11 11 11 11 11 11 11 11   ................
1342:0110  11 11 11 11 11 11 11 11-11 11 11 11 11 11 11 11   ................
1342:0120  33 33 33 33 33 33 33 33-33 33 33 33 33 33 33 33   3333333333333333
1342:0130  33 33 33 33 33 33 33 33-33 33 33 33 33 33 33 33   3333333333333333
1342:0140  44 44 44 44 44 44 44 44-44 44 44 44 44 44 44 44   DDDDDDDDDDDDDDDD
1342:0150  44 44 44 44 44 44 44 44-44 44 44 44 44 44 44 44   DDDDDDDDDDDDDDDD
1342:0160  11 11 11 11 11 11 11 11-11 11 11 11 11 11 11 11   ................
1342:0170  11 11 11 11 11 11 11 11-11 11 11 11 11 11 11 11   ................
-
```

Figure 4–26 Displayed information for Example 4.18.

```
-C 100 10F 120
1342:0100  11  33  1342:0120
1342:0101  11  33  1342:0121
1342:0102  11  33  1342:0122
1342:0103  11  33  1342:0123
1342:0104  11  33  1342:0124
1342:0105  11  33  1342:0125
1342:0106  11  33  1342:0126
1342:0106  11  33  1342:0127
1342:0108  11  33  1342:0128
1342:0109  11  33  1342:0129
1342:010A  11  33  1342:012A
1342:010B  11  33  1342:012B
1342:010C  11  33  1342:012C
1342:010D  11  33  1342:012D
1342:010E  11  33  1342:012E
1342:010F  11  33  1342:012F
-
```

Figure 4–27 Results produced when unequal data are found with a Compare command.

When the command is issued, the contents of each storage location in the block of memory between the starting address and the ending address are compared to the data in the list. The address is displayed for each memory location where a match is found.

EXAMPLE 4.19

Perform a search of the block of data from address DS:100 through DS:17F to determine which memory locations contain 33_{16}. Assume that memory is initialized as shown in Fig. 4–26.

Solution

The Search command that must be issued is

```
S  100  17F  33 (↵)
```

Figure 4–28 shows that all addresses in the range 1342:120 through 1342:13F contain this value of data.

▲ 4.5 INPUT AND OUTPUT OF DATA

The commands studied in the previous section allowed examination or modification of information in the memory of the microcomputer, but not in its input/output address space. To access data at input/output ports, we use the *INPUT* (I) and *OUTPUT* (O) *commands*. These commands can be used to input or output data for any of the 64K byte-wide ports in the 8088's I/O address space. Let us now look at how these two commands are used to read data at an input port or write data to an output port.

The general format of the input command as shown in Fig. 4–8 is

```
I  [ADDRESS]
```

Here *address* identifies the byte-wide I/O port that is to be accessed. When the command is executed, the data are read from the port and displayed. For instance, if the command

```
I  61 (↵)
```

```
-S 100 17F 33
1342:0120
1342:0121
1342:0122
1342:0123
1342:0124
1342:0125
1342:0126
1342:0127
1342:0128
1342:0129
1342:012A
1342:012B
1342:012C
1342:012D
1342:012E
1342:012F
1342:0130
1342:0131
1342:0132
1342:0133
1342:0134
1342:0135
1342:0136
1342:0137
1342:0138
1342:0139
1342:013A
1342:013B
1342:013C
1342:013D
1342:013E
1342:013F
-
```

Figure 4–28 Displayed information for Example 4.19.

is issued, and if the result displayed on the screen is

```
4D
```

the contents of the port at I/O address 0061_{16} are $4D_{16}$.

EXAMPLE 4.20

Write a command to display the byte contents of the input port at I/O address $00FE_{16}$.

Solution

To input the contents of the byte-wide port at address FE_{16}, the command is

```
I FE (↵)
```

Figure 4–8 gives the general format of the output command as

```
O [ADDRESS] [BYTE]
```

Here we see that both the address of the output port and the byte of data that is to be written to the port must be specified. An example of the command is

```
O 61 4F (↵)
```

This command causes the value $4F_{16}$ to be written into the byte-wide output port at address 0061_{16}.

▲ 4.6 HEXADECIMAL ADDITION AND SUBTRACTION

The DEBUG program also provides the ability to add and subtract hexadecimal numbers. Both operations are performed with a single command known as the *HEXADECIMAL* (H) *command.* Figure 4–8 shows that the general format of the H command is

```
H  [NUM1]  [NUM2]
```

When executed, both the sum and difference of NUM1 and NUM2 are formed. These results are displayed as follows:

NUM1 + NUM2   NUM1 − NUM2

Both numbers and the result are limited to four hexadecimal digits.

This hexadecimal arithmetic capability is useful when debugging programs. One example of a use of the H command is the calculation of the physical address of an instruction or data in memory. For instance, if the current value in the code segment register is $0ABC_{16}$ and that in the instruction pointer is $0FFF_{16}$, the physical address is found with the command

```
H  ABC0  0FFF  (↵)

BBBF 9BC1
```

Note that the sum of these two hexadecimal numbers is $BBBF_{16}$, and their difference is $9BC1_{16}$. The sum $BBBF_{16}$ is the value of the physical address CS:IP.

The subtraction operation performed with the H command is also valuable in address calculations. For instance, a frequently used software operation jumps a number of bytes of instruction code backward in the code segment of memory. In this case, the physical address of the new location can be found by subtraction. Let us start with the physical address just found, $BBBF_{16}$, and assume that we want to jump to a new location 10_{10} bytes back in memory. First, 10_{10} is expressed in hexadecimal form as A_{16}. Then the new address is calculated as

```
H  BBBF  A  (↵)

BBC9 BBB5
```

Therefore, the new physical address is $BBB5_{16}$. Because the hexadecimal numbers are limited to four digits, physical address calculations with the H command are limited to the address range 00000_{16} through $0FFFF_{16}$.

EXAMPLE 4.21

Use the H command to find the negative of the number 0009_{16}.

Solution

The negative of a hexadecimal number can be found by subtracting it from 0. Therefore, the difference produced by the command

```
H 0 9 (↵)
0009 FFF7
```

is $FFF7_{16}$, and is the negative of 9_{16} expressed in 2's complement form.

EXAMPLE 4.22

If a byte of data is located at physical address $02A34_{16}$ and the data segment register contains 0150_{16}, what value must be loaded into the source index register such that DS:SI points to the byte storage location?

Solution

The offset required in SI can be found by subtracting the data segment base address from the physical address. Using the H command, we get

```
H 2A34 1500 (↵)
3F34 1534
```

This shows that SI must be loaded with the value 1534_{16}.

▲ 4.7 LOADING, VERIFYING, AND SAVING MACHINE LANGUAGE PROGRAMS

Up to this point we have learned how to use the register, memory, and I/O commands of DEBUG to examine or modify the contents of the (1) processor's internal registers, (2) data stored in memory, or (3) information at an input or output port. Let us now look at how we can load machine code instructions and programs into the PC's memory.

In Section 4.4 we found that the Enter command can be used to load either a single or a group of memory locations with data, such as the machine code for instructions. As an example, let us load the machine code $88C3_{16}$ for the instruction MOV BL, AL. This instruction is loaded into memory starting at address CS:100 with the Enter command

```
E CS:100 88 C3 (↵)
```

We can verify that it has been loaded correctly with the Dump command

```
D CS:100 101 (↵)
```

This command displays the data

```
1342:0100 88 C3
```

Let us now introduce another command that is important for debugging programs on the PC. It is the *UNASSEMBLE* (U) *command.* By *unassemble* we mean the process of converting machine code instructions to their equivalent assembly language source statements. The U command lets us specify a range in memory, and executing the command causes the source statements for the memory data in this range to be displayed on the screen.

Figure 4–8 shows that the syntax of the Unassemble command is

```
U [STARTING ADDRESS [ENDING ADDRESS]]
```

We can use this command to verify that the machine code entered for an instruction is correct. To do this for our earlier example, we use the command

```
U CS:100 101 (↵)
```

This command causes the starting address for the instruction to be displayed, followed by both the machine code and assembly forms of the instruction. This gives

```
1342:0100 88C3 MOV BL,AL
```

The entry sequence and displayed information for loading, verification, and unassembly of the instruction are shown in Fig. 4–29.

EXAMPLE 4.23

Use a sequence of commands to load, verify loading, and unassemble the machine code instruction 0304H. Load the instruction at address CS:200.

Solution

The machine code instruction is loaded into the code segment of the microcomputer's memory with the command

```
E CS:200 03 04 (↵)
```

Next, we can verify that it was loaded correctly with the command

```
D CS:200 201 (↵)
```

and, finally, unassemble the instruction with

```
U CS:200 201 (↵)
```

The results produced by this sequence of commands are shown in Fig. 4–30. Here we see that the instruction entered is

```
ADD  AX,[SI]
```

```
-E CS:100 88 C3
-D CS:100 101
1342:0100  88 C3
-U CS:100 101
1342:0100 88C3          MOV     BL,AL
-
```

Figure 4–29 Loading, verifying, and unassembly of an instruction.

```
-E CS:200 03 04
-D CS:200 201
1342:0200  03 04
-U CS:200 201
1342:0200 0304          ADD     AX,[SI]
-
```

Figure 4–30 Displayed information for Example 4.23.

Before going further, we will cover two more commands that are useful for loading and saving programs. They are the *WRITE* (W) *command* and the *LOAD* (L) *command.* These commands give users the ability to save data stored in memory on a diskette and to reload memory from a diskette, respectively. We can load the machine code of a program into memory with the E command the first time we use it, and then save it on a diskette. In this way, the next time the program is needed it can be simply reloaded from the diskette.

Figure 4–8 shows that the general forms of the W and L commands are

```
W  [STARTING ADDRESS  [DRIVE  STARTING SECTOR  NUMBER OF SECTORS]]
L  [STARTING ADDRESS  [DRIVE  STARTING SECTOR  NUMBER OF SECTORS]]
```

For instance, to save the ADD instruction we just loaded at address CS:200 in Example 4.23, we can issue the write command

```
W CS:200 1 10 1 (↵)
```

Note that we have selected 1 (drive B) for the disk drive specification, 10 as an arbitrary starting sector on the diskette, and an arbitrary length of 1 sector. Before the command is issued, a formatted data diskette must be inserted into drive B. Then issuing the command causes one sector of data starting at address CS:200 to be read from memory and written into sector 10 on the diskette in drive B. Unlike the earlier commands we have studied, the W command automatically references the CS register instead of the DS register. For this reason, the command

```
W 200 1 10 1 (↵)
```

will perform the same operation.

Let us digress for a moment to examine the file specification of the W command in more detail. The diskettes for a PC that has double-sided, double-density drives are organized into 10,001 sectors that are assigned sector numbers over the range 0_{16} through $27F_{16}$. Each sector is capable of storing 512 bytes of data. With the file specification in a W command, we can select any one of these sectors as the starting sector. The value of the number of sectors should be specified based on the number of bytes of data that are to be saved. The specification that we made earlier for our example of a write command selected one sector (sector number 10_{16}) and for this reason could only save up to 512 bytes of data. The maximum value of sectors that can be specified with a write command is 80_{16}.

The Load command can be used to reload a file of data stored on a diskette anywhere in memory. As an example, let us load the instruction that we just saved on a diskette with a W command at a new address (CS:300). This is done with the L command

```
L 300 1 10 1 (↵)
```

The reloading of the instruction can be verified by issuing the U command

```
U CS:300 301 (↵)
```

This command causes the display

```
1342:300 301  ADD AX,[SI]
```

EXAMPLE 4.24

Show the sequence of keyboard entries needed to enter the machine code program of Fig. 4–31 into memory of the PC. The program is to be loaded into memory starting at address CS:100. Verify that the hexadecimal machine code was entered correctly, and then unassemble the machine code to assure that it represents the source program. Save the program in sector 100 of a formatted data diskette.

Solution

We will use the Enter command to load the program.

```
E CS:100 B8 00 20 8E D8 BE 0 01 BF 20 01 B9 10 0 8A 24 88 25 46
47 49 75 F7 90  (↵)
```

Machine code	Instruction
B8H	MOV AX,2000H
00H	
20H	
8EH	MOV DS,AX
D8H	
BEH	MOV SI,100H
00H	
01H	
BFH	MOV DI,120H
20H	
01H	
B9H	MOV CX,10H
10H	
00H	
8AH	MOV AH,[SI]
24H	
88H	MOV [DI],AH
25H	
46H	INC SI
47H	INC DI
49H	DEC CX
75H	JNZ $-9
F7H	
90H	NOP

Figure 4–31 Machine code and source instructions of the block move program.

```
-E CS:100 B8 00 20 8E D8 BE 0 01 BF 20 01 B9 10 0 8A 24 88 25 46 47 49 75 F7 90
-D CS:100 117
1342:0100  B8 00 20 8E D8 BE 00 01-BF 20 01 B9 10 00 8A 24   . ....... .....$
1342:0110  88 25 46 47 49 75 F7 90                           .%FGIu..
-U CS:100 117
1342:0100 B80020        MOV     AX,2000
1342:0105 BE0001        MOV     DS,AX
1342:0105 BE0001        MOV     SI,0100
1342:0108 BF2001        MOV     DI,0120
1342:010B B91000        MOV     CX,0010
1342:010E 8A24          MOV     AH,[SI]
1342:0110 8825          MOV     [DI],AH
1342:0112 46            INC     SI
1342:0113 47            INC     DI
1342:0114 49            DEC     CX
1342:0115 75F7          JNZ     010E
1342:0117 90            NOP
-W CS:100 1 100 1
```

Figure 4–32 Displayed information for Example 4.24.

First, we verify that the machine code has been loaded correctly with the command

D CS:100 117 (↲)

Comparing the displayed source data in Fig. 4–32 to the machine code in Fig. 4–31, we see that it has been loaded correctly. Now the machine code can be unassembled by the command

U CS:100 117 (↲)

Comparing the displayed source program in Fig. 4–32 to that in Fig. 4–31, it again verifies correct entry. Finally, we save the program on the data diskette with the command

W CS:100 1 100 1 (↲)

At this time, it is important to mention that using the W command to save a program can be quite risky. For instance, if by mistake, the command is written with the wrong disk specifications, some other program or data on the diskette may be written over. Moreover, the diskette should not contain files that were created in any other way. This is because the locations of these files will not be known and may accidentally be written over by the selected file specification. Overwriting a file like this will ruin its contents. Even more important, never issue the command to the hard disk (C:). This action could destroy the installation of the operating system.

Another method of saving and loading programs is available in DEBUG, and this alternative approach eliminates the overwrite problem. We now look at how a program can be saved using a filename, instead of with a file specification.

By using the *NAME* (N) *command* along with the write command, a program can be saved on the diskette under a filename. Figure 4–8 shows that the N command is specified as

N FILE NAME

where FILE NAME has the form

NAME.EXT

Here the name of the file can be up to eight characters. On the other hand, the extension (EXT) is from zero to three characters. Neither EXE nor COM is a valid extension. Some examples of valid filenames are BLOCK, TEMP.1, BLOCK1.ASM, and BLK_1.R1.

As part of the process of using the filename command, the BX and CX registers must be updated to identify the size of the program that is to be saved in the file. The size of the program in bytes is given as

$$(BX\ CX) = \text{Number of bytes}$$

Together CX and BX specify an 8-digit hexadecimal number that identifies the number of bytes in the file. Typically, the programs we work with are small. For this reason, the upper four digits are always zero—that is, the content of BX is 0000_{16}. Just using CX permits a file to be up to 64Kbytes long.

After the name command has been issued and the CX and BX registers have been initialized, the write command is used to save the program on a diskette. The write command form is

```
W  [STARTING ADDRESS]
```

To reload the program into memory, we begin by naming the file and then simply issuing a load command with the address at which it is to start. To do this, the command sequence is

```
N  FILE NAME
L  [STARTING ADDRESS]
```

As an example, let us look at how the name command is set up to save the machine program used in Example 4.24 in a file called BLK.1 on a diskette in drive A. First, the name command

```
N  A:BLK.1  (↵)
```

is entered. Looking at Fig. 4–32, we see that the program is stored in memory from address CS:100 to CS:117. This gives a size of 18_{16} bytes. Therefore, CX and BX are initialized as follows:

```
R CX   (↵)

CX  XXXX

:18    (↵)

R BX   (↵)

BX  XXXX

:0     (↵)
```

Now the program is saved on the diskette with the command

```
W  CS:100    (↵)
```

To reload the program into memory, we simply perform the command sequence

```
N  A:BLK.1  (↵)
L  CS:100   (↵)
```

In fact, the program can be loaded starting at another address by specifying that address in the Load command.

Once saved on a diskette, the file can be changed to an executable (that is, a file with the extension .EXE) by using the DOS rename (REN) command. To do this we must first return to DOS with the command

```
Q (↵)
```

and then issue the command

```
C:\DOS> REN A:BLK.1 BLK.EXE (↵)
```

Programs that are in an executable file can be directly loaded when the DEBUG program is brought up. For our example, the program command is

```
C:\DOS> DEBUG A:BLK.EXE (↵)
```

Execution of this command loads the program at address CS:100 and then displays the DEBUG prompt. After loading, other DEBUG commands can be used to run the program.

Executable files can also be run directly in the DOS environment. This is done simply by entering the filename following the DOS prompt and then depressing the return key. Therefore, making the following entry runs program BLK.EXE

```
C:\DOS>A:BLK.EXE (↵)
```

▲ 4.8 ASSEMBLING INSTRUCTIONS WITH THE ASSEMBLE COMMAND

All the instructions we have worked with up to this point have been hand-assembled into machine code. The DEBUG program has a command that lets us automatically assemble the instructions of a program, one after the other, and store them in memory. It is called the *ASSEMBLE* (A) *command.*

The general syntax of the Assemble command as given in Fig. 4–8 is

```
A  [STARTING ADDRESS]
```

Here STARTING ADDRESS is the address at which the machine code of the first instruction of the program is to be stored. For example, to assemble the instruction ADD [BX+SI+1234H],AX and store its machine code in memory starting at address CS:100, we start with the command entry

```
A CS:100 (↵)
```

The response to this command input is the display of the starting address in the form

```
1342:0100_
```

The instruction to be assembled is typed in following this address, and when we depress the (↵) key, the instruction is assembled into machine code. It is then stored in memory, and the starting address of the next instruction is displayed. As shown in Fig. 4–33, for our example, we have

```
1342:0100 ADD [BX+SI+1234],AX (↵)
1342:0104_
```

Now we either enter the next instruction or depress the (↵) key to terminate the Assemble command.

Assuming that the assemble operation just performed was terminated by entering (↵), we can view the machine code that was produced for the instruction by issuing a Dump command. Notice that the address displayed as the starting point of the next instruction is 1342:0104. Therefore, the machine code for the ADD instruction took up 4 bytes of memory, CS:100, CS:101, CS:102, and CS:103. The command needed to display this machine code is

```
D CS:100 103 (↵)
```

Figure 4–33 shows that the machine code stored for the instruction is 01803412H.

At this point, the instruction can be executed or saved on a diskette. For instance, to save the machine code on a diskette in file INST.1, we issue the commands

```
N A:INST.1 (↵)
R CX       (↵)
:4         (↵)
R BX       (↵)
:0         (↵)
W CS:100   (↵)
```

```
-A CS:100
1342:0100 ADD [BX+SI+1234],AX
1342:0104
-D CS:100 103
1342:0100  01 80 34 12                                    ..4.
-N A:INST.1
-R CX
:4
-R BX
:0
-W CS:100
-
```

Figure 4–33 Assembling the instruction ADD [BX+SI+1234H], AX.

Now that we have shown how to assemble an instruction, view its machine code, and save the machine code on a data diskette, let us look into how a complete program can be assembled with the A command. For this purpose, we will use the program shown in Fig. 4–34(a). The same program was entered as hand-assembled machine code in Example 4.24.

We begin by assuming that the program is to be stored in memory starting at address CS:200. For this reason, we invoke the *line-by-line assembler* with the command

```
A CS:200 (↵)
```

This gives the response

```
1342:0200_
```

Now we type in the instructions of the program as follows:

```
1342:0200 MOV AX,2000 (↵)
1342:0203 MOV DS,AX   (↵)
1342:0205 MOV SI,100  (↵)
    .         .     .    .
    .         .     .    .
1342:0217 NOP         (↵)
1342:0218             (↵)
```

```
MOV   AX,2000H
MOV   DS,AX
MOV   SI,0100H
MOV   DI,0120H
MOV   CX,010H
MOV   AH,[SI]
MOV   [DI],AH
INC   SI
INC   DI
DEC   CX
JNZ   20EH
NOP
```

(a)

```
-A CS:200
1342:0200 MOV AX,2000
1342:0203 MOV DS,AX
1342:0205 MOV SI,100
1342:0208 MOV DI,120
1342:020B MOV CX,10
1342:020E MOV AH,[SI]
1342:0210 MOV [DI],AH
1342:0212 INC SI
1342:0213 INC DI
1342:0214 DEC CX
1342:0215 JNZ 20E
1342:0217 NOP
1342:0218
-
```

(b)

```
-U CS:200 217
1342:0200 B80020        MOV    AX,2000
1342:0203 8ED8          MOV    DS,AX
1342:0205 BE0001        MOV    SI,0100
1342:0208 BF2001        MOV    DI,0120
1342:020B B91000        MOV    CX,0010
1342:020E 8A24          MOV    AH,[SI]
1342:0210 8825          MOV    [DI],AH
1342:0212 46            INC    SI
1342:0213 47            INC    DI
1342:0214 49            DEC    CX
1342:0215 75F7          JNZ    020E
1342:0217 90            NOP
-
```

(c)

Figure 4–34 (a) Block move program. (b) Assembling the program. (c) Verifying the assembled program with the U command.

The details of the instruction entry sequence are shown in Fig. 4–34(b).

Now that the complete program has been entered, let us verify that it has been assembled correctly. This can be done with an Unassemble command. Notice in Fig. 4–34(b) that the program resides in memory over the address range CS:200 through CS:217. To unassemble the machine code in this part of memory, we issue the command

```
U CS:200 217  (↵)
```

The results produced with this command are shown in Fig. 4–34(c). Comparing the instructions to those in Fig. 4–34(a) confirms that the program has been assembled correctly.

The Assemble command allows us to assemble instructions involving any of the various addressing modes. For instance, the instruction we used in our earlier example

```
MOV  AX, 2000H
```

employs immediate addressing mode for the source operand. Instructions, such as

```
MOV  AX, [2000H]
```

which uses direct addressing mode for the source operand, can also be assembled into memory.

The Assemble command also supports two pseudo-instructions that can be used to assemble data directly into memory: *data byte* (DB) and *data word* (DW). An example is

```
DB  1,2,3,'JASSI'
```

With this command, the byte-size representation of numbers 1, 2, and 3 and the ASCII code for letters J, A, S, S, and I are assembled into memory.

▲ 4.9 EXECUTING INSTRUCTIONS AND PROGRAMS WITH THE TRACE AND GO COMMANDS

Once the program has been entered into the PC's memory, it is ready to be executed. The DEBUG program allows us to execute the entire program with one *GO* (G) *command* or to execute the program in several segments of instructions by using *breakpoints* in the Go command. Moreover, by using the *TRACE* (T) *command,* the program can be stepped through by executing one or more instructions at a time.

Let us begin by examining the operation of the Trace command in more detail. The Trace command provides the programmer with the ability to execute one instruction at a time. This mode of operation is also known as *single-stepping* the program; it is very useful during the early stages of program debugging. This is because the contents of registers or memory can be viewed both before and after the execution of each instruction to determine whether or not the correct operation was performed.

Figure 4–8 shows the general form of the Trace command as

```
T [=STARTING ADDRESS] [NUMBER]
```

Notice that a *starting address* is specified as part of the command and is the address of the instruction at which execution is to begin. *Number* tells how many instructions are to be executed. The equal sign before the starting address is very important. If it is left out, the microcomputer usually hangs up and will have to be restarted with a power-on reset.

If an instruction count is not specified in the command, just one instruction is executed. For instance, the command

```
T =CS:100 (↵)
```

causes the instruction starting at address CS:100 to be executed. At the end of the instruction's execution, the complete state of the MPU's internal registers is automatically displayed. At this point, other DEBUG commands can be issued—for instance, to display the contents of memory—or the next instruction can be executed.

This Trace command can also be issued as

```
T (↵)
```

In this case, the instruction pointed to by the current values of CS and IP (CS:IP) is executed. This form of the Trace command is used to execute the next instruction.

If we want to step through several instructions, the Trace command must include the number of instructions to be executed. This number is included after the address. For example, to trace through three instructions, the command is issued as

```
T =CS:100 3 (↵)
```

Again, the internal state of the MPU is displayed after each instruction is executed.

EXAMPLE 4.25

Load the instruction stored at file specification 1 10 1 on a diskette at offset 100_{16} in the current code segment. Unassemble the instruction. Then initialize AX with 1111_{16}, SI with 1234_{16}, and the word contents of memory address 1234_{16} to the value 2222_{16}. Next, display the internal state of the MPU and the word contents of address 1234_{16} to verify their initialization. Finally, execute the instruction with the Trace command. What operation is performed by the instruction?

Solution

First, the instruction is loaded from the diskette to address CS:100 with the command

```
L CS:100 1 10 1   (↵)
```

Now the machine code is unassembled to verify that the instruction has loaded correctly.

```
U 100 101           (↵)
```

Looking at the displayed information in Fig. 4–35, we see that it is an ADD instruction. Next we initialize the internal registers and memory with the command sequence

```
R AX                (↵)

AX 0000

:1111               (↵)
R SI                (↵)

SI 0000

:1234               (↵)
E DS:1234 22 22     (↵)
```

Now the initialization is verified with the commands

```
R                   (↵)
D DS:1234 1235      (↵)
```

Figure 4–35 shows that AX, SI, and the word contents of address 1234_{16} are correctly initialized. Therefore, we are ready to execute the instruction. This is done with the command

```
T =CS:100 (↵)
```

```
-L CS:100 1 10 1
-U 100 101
1342:0100 0304          ADD   AX,[SI]
-R AX
AX 0000
:1111
-R SI
SI 0000
:1234
-E DS:1234 22 22
-R
AX=1111  BX=0000  CX=0000  DX=0000  SP=FFEE  BP=0000  SI=1234  DI=0000
DS=1342  ES=1342  SS=1342  CS=1342  IP=0100   NV UP EI PL NZ NA PO NC
1342:0100 0304          ADD      AX,[SI]
-D DS:1234 1235
1342:1230              22 22                                      ""
-T =CS:100

AX=3333  BX=0000  CX=0000  DX=0000  SP=FFEE  BP=0000  SI=1234  DI=0000
DS=1342  ES=1342  SS=1342  CS=1342  IP=0102   NV UP EI PL NZ NA PE NC
1342:0102 0000          ADD      [BX+SI],AL                    DS:1234=22
-
```

Figure 4–35 Displayed information for Example 4.25

From the displayed trace information in Fig. 4–36, we find that the value 2222_{16} at address 1234_{16} was added to the value 1111_{16} held in AX. Therefore, the new contents of AX are 3333_{16}.

The Go command is typically used to run programs that are already working or to execute programs in the latter stages of debugging. For example, if the beginning part of a program is already operating correctly, a Go command can be used to execute this group of instructions and then stop execution at a point in the program where additional debugging is to begin.

The table in Fig. 4–8 shows that the general form of the Go command is

```
G [=STARTING ADDRESS [BREAKPOINT ADDRESS LIST]]
```

The first address is the *starting address* of the segment of program that is to be executed—that is, the address of the instruction at which execution is to begin. The second address, the *breakpoint address,* is the address of the end of the program segment—that is, the address of the instruction at which execution is to stop. The breakpoint address that is specified must correspond to the first byte of an instruction. A list of up to 10 breakpoint addresses can be supplied with the command.

An example of the Go command is

```
G =CS:200 217 (↵)
```

This command loads the IP register with 0200_{16}, sets a breakpoint at address CS:217, and then begins program execution at address CS:200. Instruction execution proceeds until address CS:217 is accessed. When the breakpoint address is reached, program execution is terminated, the complete internal status of the MPU is displayed, and control is returned to DEBUG.

Sometimes we just want to execute a program without using a breakpoint. This can also be done with the Go command. For instance, to execute a program that starts at offset 100_{16} in the current CS, we can issue the Go command without a breakpoint address as follows:

```
G =CS:100 (↵)
```

This command will cause the program to run to completion provided there are appropriate instructions in the program to initiate a normal termination, such as those needed to return to DEBUG. In the case of a program where CS and IP are already initialized with the correct values, we can just enter

```
G (↵)
```

However, it is recommended that the Go command always include a breakpoint address. If a Go command is issued without a breakpoint address and the value of CS and IP are not already set up or the program is not correctly prepared for normal termination, the microcomputer can lock up. This is because the program execution may go beyond the end of the program into an area of memory with information that represents invalid instructions.

EXAMPLE 4.26

In Section 4.7, we saved the block move program in file BLK.EXE on a data diskette in drive A. Load this program into memory starting at address CS:200. Then initialize the microcomputer by loading the DS register with 2000_{16}; fill the block of memory from DS:100 through DS:10F with FF_{16}, and the block of memory from DS:120 through DS:12F with 00_{16}. Verify that the blocks of memory were initialized correctly. Load DS with 1342_{16}, and display the state of the MPU's registers. Display the assembly language version of the program from CS:200 through CS:217. Use a Go command to execute the program through address CS:20E. What changes have occurred in the contents of the registers? Now execute down through address CS:215. What changes are found in the blocks of data? Next execute the program down to address CS:217. What new changes are found in the blocks of data?

Solution

The commands needed to load the program are

```
N A:BLK.EXE       (↵)
L CS:200          (↵)
```

Next we initialize the DS register and memory with the commands

```
R DS              (↵)
DS 1342
:2000             (↵)
F DS:100 10F FF   (↵)
F DS:120 12F 00   (↵)
```

The blocks of data in memory are displayed using the commands

```
D DS:100 10F      (↵)
D DS:120 12F      (↵)
```

The displayed information is shown in Fig. 4–36. DS is restored with 1342_{16} using the command

```
R DS        (↵)
DS 2000
:1342       (↵)
```

and the state of the MPU's registers is displayed with the command

```
R           (↵)
```

```
-N A:BLK.EXE
-L CS:200
-R DS
DS 1342
:2000
-F DS:100 10F FF
-F DS:120 12F 00
-D DS:100 10F
2000:0100  FF FF FF FF FF FF FF FF-FF FF FF FF FF FF FF FF   ................
-D DS:120 12F
2000:0120  00 00 00 00 00 00 00 00-00 00 00 00 00 00 00 00   ................
-R DS
DS 2000
:1342
-R
AX=0000  BX=0000  CX=0020  DX=0000  SP=FFEE  BP=0000  SI=0000  DI=0000
DS=1342  ES=1342  SS=1342  CS=1342  IP=0100   NV UP EI PL NZ NA PO NC
1342:0100 0000          ADD     [BX+SI],AL                        DS:0000=CD
-U CS:200 217
1342:0200 B80020        MOV     AX,2000
1342:0203 8ED8          MOV     DS,AX
1342:0205 BE0001        MOV     SI,0100
1342:0208 BF2001        MOV     DI,0120
1342:020B B91000        MOV     CX,0010
1342:020E 8A24          MOV     [DI],AH
1342:0212 46            INC     SI
1342:0213 47            INC     DI
1342:0214 49            DEC     CX
1342:0215 75F7          JNZ     020E
1342:0217 90            NOP
-G =CS:200 20E

AX=2000  BX=0000  CX=0010  DX=0000  SP=FFEE  BP=0000  SI=0100  DI=0120
DS=2000  ES=1342  SS=1342  CS=1342  IP=020E   NV UP EI PL NZ NA PO NC
1342:020E 8A24          MOV     AH,[SI]                           DS:0100=FF
-G =CS:20E 215

AX=FF00  BX=0000  CX=000F  DX=0000  SP=FFEE  BP=0000  SI=0101  DI=0121
DS=2000  ES=1342  SS=1342  CS=1342  IP=0215   NV UP EI PL NZ AC PE NC
1342:0215 75F7          JNZ     020E
-D DS:100 10F
2000:0100  FF FF FF FF FF FF FF FF-FF FF FF FF FF FF FF FF   ................
-D DS:120 12F
2000:0120  FF 00 00 00 00 00 00 00-00 00 00 00 00 00 00 00   ................
-G =CS:215 217

AX=FF00  BX=0000  CX=0000  DX=0000  SP=FFEE  BP=0000  SI=0110  DI=0130
DS=2000  ES=1342  SS=1342  CS=1342  IP=0217   NV UP EI PL ZR NA PE NC
1342:0217 90            NOP
-D DS:100 10F
2000:0100  FF FF FF FF FF FF FF FF-FF FF FF FF FF FF FF FF   ................
-D DS:120 12F
2000:0120  FF FF FF FF FF FF FF FF-FF FF FF FF FF FF FF FF   ................
-
```

Figure 4–36 Displayed information for Example 4.26.

Before beginning to execute the program, we will display the source code with the command

```
U CS:200 217      (↵)
```

The displayed program is shown in Fig. 4–36.

Now the first segment of program is executed with the command

```
G =CS:200 20E     (↵)
```

Looking at the displayed state of the MPU in Fig. 4–36, we see that DS was loaded with 2000_{16}, AX was loaded with 2000_{16}, SI was loaded with 0100_{16}, and CX was loaded with 0010_{16}.

Next, another Go command is used to execute the program down through address CS:215.

```
G =CS:20E 215     (↵)
```

We can check the state of the blocks of memory with the commands

```
D DS:100 10F      (↵)

D DS:120 12F      (↵)
```

From the displayed information in Fig. 4–36, we see that FF_{16} was copied from the first element of the source block to the first element of the destination block.

Now we execute through CS:217 with the command

```
G =CS:215 217     (↵)
```

and examine the blocks of data with the commands

```
D DS:100 10F      (↵)

D DS:120 12F      (↵)
```

We find that the complete source block has been copied to the destination block.

▲ 4.10 DEBUGGING A PROGRAM

In Sections 4.7, 4.8, and 4.9, we learned how to use DEBUG to load a machine code program into the PC's memory, assemble a program, and execute the program. However, we did not determine if the program when run performed the operation for which it was written. It is common to have errors in a program, and even a single error can render the program useless. For instance, if the address to which a "jump" instruction passes control is wrong, the program may get hung up. Errors in a program are also referred to as *bugs;* the process of removing them is called *debugging.*

The two types of errors that can be made by a programmer are the *syntax error* and the *execution error.* A syntax error is caused by not following the rules for coding or entering an instruction. These types of errors are typically automatically identified and signaled to the programmer with an error message. For this reason, they are usually easy to find and correct. For example, if a Dump command is keyed in as

```
D DS:100120   (↵)
```

an error condition exists. This is because the space between the starting and ending address is left out. This incorrect entry is signaled by the warning "Error" in the display; the spot where the error begins, in this case, the 1 in 120, is marked with the symbol "∧" to identify the position of the error.

An execution error is an error in the logic behind the development of the program. That is, the program is correctly coded and entered, but it still does not perform the operation for which it was written. Entering the program into the microcomputer and observing its operation can identify this type of error. Even when an execution error has been identified, it is usually not easy to find the exact cause of the problem.

Our ability to debug execution errors in a program is aided by the commands of the DEBUG program. For instance, the Trace command allows us to step through the program by executing just one instruction at a time. We can use the display of the internal register state produced by the Trace command and the memory dump command to determine the state of the MPU and memory prior to execution of an instruction and again after its execution. This information tells us whether the instruction has performed the operation planned for it. If an error is found, its cause can be identified and corrected.

To demonstrate the process of debugging a program, let us once again use the program that we stored in file A:BLK.EXE. We load it into the code segment at address CS:200 with the command

```
N A:BLK.EXE   (↵)

L 200         (↵)
```

Now the program resides in memory at addresses CS:200 through CS:217. The program is displayed with the command

```
U 200 217   (↵)
```

The program that is displayed is shown in Fig. 4–37. This program implements a block data-transfer operation. The block of data to be moved starts at memory address DS:100 and is 16 bytes in length. It is to be moved to another block of storage locations starting at address DS:120. DS equals 2000_{16}; therefore, it points to a data segment starting at physical address 20000_{16}.

Before executing the program, let us issue commands to initialize the source block of memory locations from address 100_{16} through $10F_{16}$ with FF_{16} and the bytes in the destination block starting at 120_{16} with 00_{16}. To do this, we issue the command sequence

```
F 2000:100 10F FF (↵)

F 2000:120 12F 00 (↵)
```

```
C:\DOS>DEBUG
-N A:BLK.EXE
-L 200
-U 200 217
1342:0200 B82010        MOV     AX,2000
1342:0203 8ED8          MOV     DS,AX
1342:0205 BE0001        MOV     SI,0100
1342:0208 BF2001        MOV     DI,0120
1342:020B B91000        MOV     CX,0010
1342:020E 8A24          MOV     AH,[SI]
1342:0210 8825          MOV     [DI],AH
1342:0212 46            INC     SI
1342:0213 47            INC     DI
1342:0214 49            DEC     CX
1342:0215 75F7          JNZ     020E
1342:0217 90
-F 2000:100 10F FF
-F 2000:120 12F 00
-T =CS:200 5

AX=2000  BX=0000  CX=0010  DX=0000  SP=FFEE  BP=0000  SI=0100  DI=0120
DS=1020  ES=1342  SS=1342  CS=1342  IP=0203   NV UP EI PL NZ NA PO NC
1342:0203 8ED8          MOV     DS,AX

AX=2000  BX=0000  CX=0010  DX=0000  SP=FFEE  BP=0000  SI=0100  DI=0120
DS=2000  ES=1342  SS=1342  CS=1342  IP=0205   NV UP EI PL NZ NA PO NC
1342:0205 BE0001        MOV     SI,0100

AX=2000  BX=0000  CX=0010  DX=0000  SP=FFEE  BP=0000  SI=0100  DI=0120
DS=2000  ES=1342  SS=1342  CS=1342  IP=0208   NV UP EI PL NZ NA PO NC
1342:0208 BF2001        MOV     DI,0120

AX=2000  BX=0000  CX=0010  DX=0000  SP=FFEE  BP=0000  SI=0100  DI=0120
DS=2000  ES=1342  SS=1342  CS=1342  IP=020B   NV UP EI PL NZ NA PO NC
1342:020B B91000        MOV     CX,0010

AX=2000  BX=0000  CX=0010  DX=0000  SP=FFEE  BP=0000  SI=0100  DI=0120
DS=2000  ES=1342  SS=1342  CS=1342  IP=020E   NV UP EI PL NZ NA PO NC
1342:020E 8A24          MOV     AH,[SI]                          DS:0100=FF
-D DS:120 12F
2000:0120  00 00 00 00 00 00 00 00-00 00 00 00 00 00 00 00   ................
-T 2

AX=FF00  BX=0000  CX=0010  DX=0000  SP=FFEE  BP=0000  SI=0100  DI=0120
DS=2000  ES=1342  SS=1342  CS=1342  IP=0210   NV UP EI PL NZ NA PO NC
1342:0210 8825          MOV     [DI],AH                          DS:0120=00

AX=FF00  BX=0000  CX=0010  DX=0000  SP=FFEE  BP=0000  SI=0100  DI=0120
DS=2000  ES=1342  SS=1342  CS=1342  IP=0212   NV UP EI PL NZ NA PO NC
1342:0212 46            INC     SI
-D DS:120 12F
2000:0120  FF 00 00 00 00 00 00 00-00 00 00 00 00 00 00 00   ................
-T 3

AX=FF00  BX=0000  CX=0010  DX=0000  SP=FFEE  BP=0000  SI=0101  DI=0120
DS=2000  ES=1342  SS=1342  CS=1342  IP=0213   NV UP EI PL NZ NA PO NC
1342:0213 47            INC     DI

AX=FF00  BX=0000  CX=0010  DX=0000  SP=FFEE  BP=0000  SI=0101  DI=0121
DS=2000  ES=1342  SS=1342  CS=1342  IP=0214   NV UP EI PL NZ NA PE NC
```

Figure 4–37 Program debugging.

```
1342:0214 49            DEC     CX

AX=FF00  BX=0000  CX=000F  DX=0000  SP=FFEE  BP=0000  SI=0101  DI=0121
DS=2000  ES=1342  SS=1342  CS=1342  IP=0215   NV UP EI PL NZ AC PE NC
1342:0215 75F7          JNZ       020E
-T

AX=FF00  BX=0000  CX=000F  DX=0000  SP=FFEE  BP=0000  SI=0101  DI=0121
DS=2000  ES=1342  SS=1342  CS=1342  IP=020E   NV UP EI PL NZ AC PE NC
1342:020E 8A24          MOV     AH,[SI]                        DS:0101=FF
-G =CS:20E 215

AX=FF00  BX=0000  CX=000E  DX=0000  SP=FFEE  BP=0000  SI=0102  DI=0122
DS=2000  ES=1342  SS=1342  CS=1342  IP=0215   NV UP EI PL NZ NA PO NC
1342:0215 75F7          JNZ     020E
-D DS:120 12F
2000:0120  FF FF 00 00 00 00 00 00-00 00 00 00 00 00 00 00   ................
-T

AX=FF00  BX=0000  CX=000E  DX=0000  SP=FFEE  BP=0000  SI=0102  DI=0122
DS=2000  ES=1342  SS=1342  CS=1342  IP=020E   NV UP EI PL NZ NA PO NC
1342:020E 8A24          MOV     AH,[SI]                        DS:0102=FF
-G =CS:20E 217

AX=FF00  BX=0000  CX=0000  DX=0000  SP=FFEE  BP=0000  SI=0110  DI=0130
DS=2000  ES=1342  SS=1342  CS=1342  IP=0217   NV UP EI PL ZR NA PE NC
1342:0217 90            NOP
-D DS:120 12F
2000:0120  FF FF FF FF FF FF FF FF-FF FF FF FF FF FF FF FF   ................
-
```

Figure 4–37 (continued)

The first two instructions of the program in Fig. 4–37 are

```
MOV  AX, 2000H
```

and

```
MOV  DS, AX
```

These two instructions, when executed, load the data segment register with the value 2000_{16}. In this way they define a data segment starting at the physical address 20000_{16}. The next three instructions are used to load the SI, DI, and CX registers with 100_{16}, 120_{16}, and 10_{16}, respectively. Let us now show how to execute these instructions and then determine if they perform the correct function. They are executed by issuing the command

```
T =CS:200 5   (↵)
```

To determine if the instructions that were executed performed the correct operation, we need only to look at the trace display that they produce, as shown in Fig. 4–37. Here we see that the first instruction loads AX with 2000_{16} and the second moves this value into the DS register. Also notice in the last trace displayed for this command that SI contains 0100_{16}, DI contains 0120_{16}, and CX contains 0010_{16}.

The next two instructions copy the contents of memory location 100_{16} into the storage location at address 120_{16}. Let us first check the contents of the destination block with the D command

```
D DS:120 12F (↵)
```

Looking at the dump display in Fig. 4–37, we see that the original contents of these locations are 00_{16}. Now the two instructions are executed with the command

```
T 2     (↵)
```

and the contents of address DS:120 are checked once again with the command

```
D DS:120 12F  (↵)
```

The display dump in Fig. 4–37 shows that the first element of the source block was copied to the location of the first element of the destination block. Therefore, both address 100_{16} and address 120_{16} now contain the value FF_{16}.

The next three instructions are used to increment pointers SI and DI and decrement block counter CX. To execute them, we issue the command

```
T 3 (↵)
```

Referring to the trace display in Fig. 4–37 to verify their operation, we find that the new values in SI and DI are 0101_{16} and 0121_{16}, respectively, and CX is now $000F_{16}$.

The jump instruction is next, and it transfers control to the instruction eight bytes back if CX did not become zero. It is executed with the command

```
T (↵)
```

Note that the result of executing this instruction is that the value in IP is changed to $020E_{16}$. This corresponds to the location of the instruction

```
MOV  AH, [SI]
```

In this way we see that control has been returned to the part of the program that performs the data-move operation.

The move operation performed by this part of the program was already checked; however, we must still determine if it runs to completion when the count in CX decrements to zero. Therefore, we will execute another complete loop with the Go command

```
G =CS:20E 215 (↵)
```

Correct operation is verified because the trace shows that CX has been decremented by one more and equals E. The fact that the second element has been moved can be verified by dumping the destination block with the command

```
D DS:120 12F (↵)
```

Now we are again at address CS:215. To execute the jump instruction at this location, we can again use the T command

```
T       (↵)
```

This returns control to the instruction at CS:20E. The previous two commands can be repeated until the complete block is moved and CX equals 0_{16}. Or we can use the Go command to execute to the address CS:217, which is the end of the program.

```
G =CS:20E 217 (↵)
```

At completion, the overall operation of the program can be verified by examining the contents of the destination block with the command sequence

```
D DS:120 12F (↵)
```

FF_{16} should be displayed as the data held in each storage location of the destination block.

REVIEW PROBLEMS

Section 4.1

1. How many bytes are in the general machine code instruction format?
2. Encode the following instruction using the information in Figs. 4–1 through 4–4.

   ```
   ADD AX,DX
   ```

   Assume that the opcode for the Add operation is 000000.
3. Encode the following instructions using the information in Figs. 4–1 through 4–5 and Fig. 3–6.

   ```
   (a) MOV  [DI], DX
   (b) MOV  [BX][SI], BX
   (c) MOV  DL,[BX]+10H
   ```

4. Encode the instructions that follow using the information in Figs. 4–1 through 4–6 and Fig. 3–6.

   ```
   (a) PUSH DS
   (b) ROL BL, CL
   (c) ADD AX,[1234H]
   ```

Section 4.2

5. How many bytes are required to encode the instruction MOV SI, 0100H?
6. How many bytes of memory are required to store the machine code for the program in Fig. 4–6(a)?

Section 4.3

7. What purpose does the DEBUG program serve?
8. Can DEBUG be brought up by typing the command using lowercase letters?
9. If the DEBUG command R AXBX is entered to a PC, what happens?

10. Write the Register command needed to change the value in CX to 10_{16}.
11. Write the command needed to change the state of the parity flag to PE.
12. Write a command that will dump the state of the MPU's internal registers.

Section 4.4

13. Write a Dump command that will display the contents of the first 16 bytes of the current code segment.
14. Show an Enter command that can be used to examine the contents of the same 16 bytes of memory that were displayed in problem 13.
15. Show the Enter command needed to load five consecutive bytes of memory starting at address CS:100 of the current code segment with FF_{16}.
16. Show how an Enter command can be used to initialize the first 32 bytes at the top of the stack to 00_{16}.
17. Write a sequence of commands that will fill the first six storage locations starting at address CS:100 with 11_{16}, the second six with 22_{16}, the third six with 33_{16}, the fourth six with 44_{16}, and the fifth six with 55_{16}; change the contents of storage locations CS:105 and CS:113 to FF_{16}; display the first 30 bytes of memory starting at CS:100; and then use a search command on this 30-byte block of memory to find those storage locations that contain FF_{16}.

Section 4.5

18. What DEBUG commands do I and O stand for?
19. What operation is performed by the command

```
I 123 (↵)
```

20. Write an output command that will load the byte-wide output port at I/O address 0124_{16} with the value $5A_{16}$.

Section 4.6

21. What two results are produced by the hexadecimal command?
22. How large can the numbers in an H command be?
23. The difference $FA_{16} - 5A_{16}$ is to be found. Write the H command.

Section 4.7

24. Show the sequence of commands needed to load the machine code instruction 320E3412H starting at address CS:100, unassemble it to verify that the correct instruction was loaded, and save it on a data diskette at file specification 1 50 1.
25. Write commands that will reload the instruction saved on the data diskette in problem 24 into memory at offset 400_{16} in the current code segment and unassemble it to verify correct loading.

Section 4.8

26. Show how the instruction MOV [DI], DX can be assembled into memory at address CS:100.

27. Write a sequence of commands that will first assemble the instruction ROL BL, CL into memory starting at address CS:200, and then verify its entry by unassembling the instruction.

Section 4.9

28. Show a sequence of commands that will load the instruction saved on the data diskette in problem 24 at address CS:300; unassemble it to verify correct loading; initialize the contents of register CX to $000F_{16}$ and the contents of the word memory location starting at DS:1234 to $00FF_{16}$; execute the instruction with the Trace command; and verify its operation by examining the contents of CX and the word of data stored in memory starting at DS:1234.

29. Write a sequence of commands to repeat Example 4.26; however, this time execute the complete program with one Go command.

Section 4.10

30. What is the difference between a syntax error and an execution error?

31. Give another name for an error in a program.

32. What is the name given to the process of removing errors in a program?

33. Write a sequence of commands to repeat the DEBUG demonstration presented in Section 4.10, but this time use only Go commands to execute the program.

8088/8086 Programming—Integer Instructions and Computations

▲ INTRODUCTION

Up to this point we have studied the software architecture of the 8088/8086 microprocessor, its instruction set, addressing modes, and the software development tools provided by the DEBUG program of DOS. We found that the software architectures of the 8088 and 8086 microprocessors are identical and learned how to encode assembly language instructions in machine language and how to use the debugger to enter, execute, and debug programs. In this chapter, we begin a detailed study of the instruction set of the 8088 and 8086 microprocessors by examining the data transfer, arithmetic, logic, shift, and rotate instructions. We demonstrate the use of these instructions with a number of example programs. The rest of the instruction set and some more sophisticated programming concepts are covered in Chapter 6. The following topics are presented in this chapter:

5.1 Data Transfer Instructions
5.2 Arithmetic Instructions
5.3 Logic Instructions
5.4 Shift Instructions
5.5 Rotate Instructions

▲ 5.1 DATA TRANSFER INSTRUCTIONS

The 8088 microprocessor has a group of *data-transfer instructions* that are provided to move data either between its internal registers or between an internal register and a storage location in memory. This group includes the *move byte or word* (MOV) instruction, *exchange byte or word* (XCHG) instruction, *translate byte* (XLAT) instruction, *load effective address* (LEA) instruction, *load data segment* (LDS) instruction, and *load extra segment* (LES) instruction. These instructions are discussed in this section.

The MOV Instruction

The move instruction was briefly introduce in Section 3.4 and used in Section 3.5 to demonstrate addressing modes of the 8088/8086 microprocessor. Here we examine its allowed operands and operation more thoroughly, and demonstrate its execution with the DEBUG program. Figure 5–1(a) shows that the move (MOV) instruction transfers data from a source operand to a destination operand. Earlier we found that the operands can be internal registers of the 8088 and storage locations in memory.

Figure 5–1(b) shows the valid source and destination operand variations. Note that the MOV instruction cannot transfer data directly between a source and a destination, which both reside in external memory. Instead, the data must first be moved from memory into an internal register, such as to the accumulator (AX), with one move instruction, and then moved to the new location in memory with a second move instruction.

All transfers between general-purpose registers and memory can involve either a byte or word of data. The fact that the instruction corresponds to byte or word data is designated by the way in which its operands are specified. For instance, AL or AH is used to specify a byte operand, and AX, a word operand. On the other hand, data moved between one of the general-purpose registers and a segment register or between a segment register and a memory location must always be word-wide.

Figure 5–1(a) also provides additional important information. For instance, flag bits within the 8088 are not modified by execution of a MOV instruction.

An example of a segment register to general-purpose register MOV instruction shown in Fig. 5–1(c) is

```
MOV DX, CS
```

In this instruction, the code segment register is the source operand, and the data register is the destination. It stands for "move the contents of CS into DX." That is,

$$(CS) \rightarrow (DX)$$

For example, if the contents of CS are 0100_{16}, execution of the instruction MOV DX, CS as shown in Fig. 5–1(d) makes

$$(DX) = (CS) = 0100_{16}$$

Mnemonic	Meaning	Format	Operation	Flags affected
MOV	Move	MOV D,S	(S) → (D)	None

(a)

Destination	Source
Memory	Accumulator
Accumulator	Memory
Register	Register
Register	Memory
Memory	Register
Register	Immediate
Memory	Immediate
Seg-reg	Reg16
Seg-reg	Mem16
Reg16	Seg-reg
Memory	Seg-reg

(b)

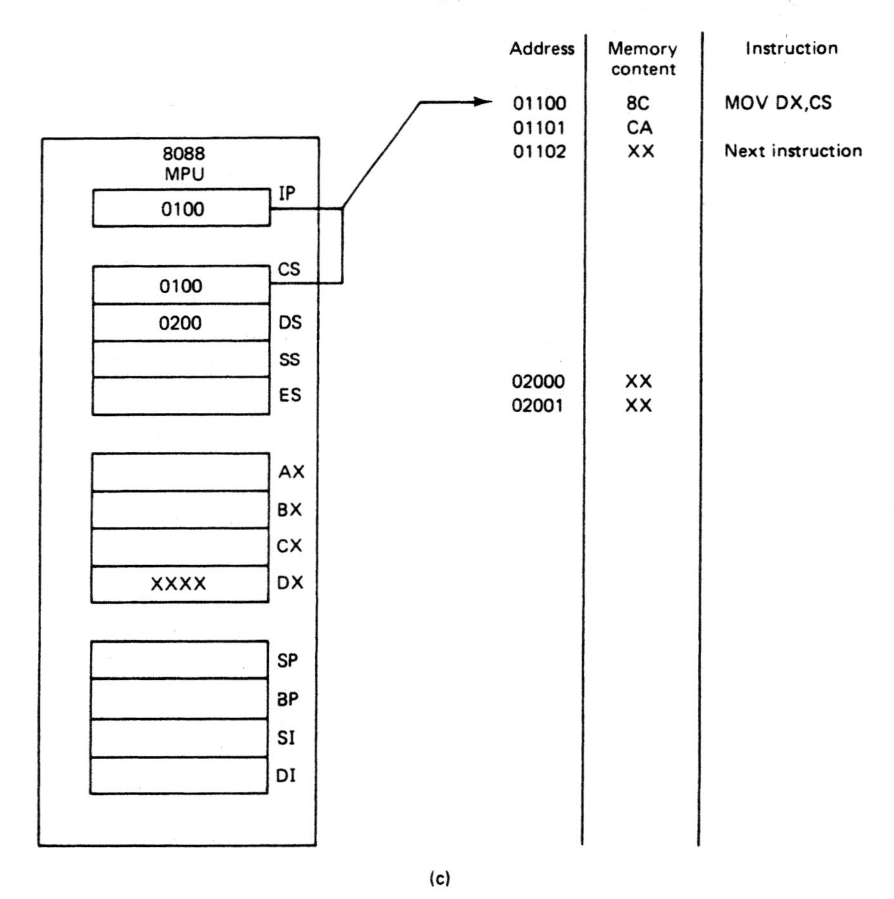

(c)

Figure 5–1 (a) MOV data transfer instruction. (b) Allowed operands. (c) MOV DX, CS before fetch and execution. (d) After execution.

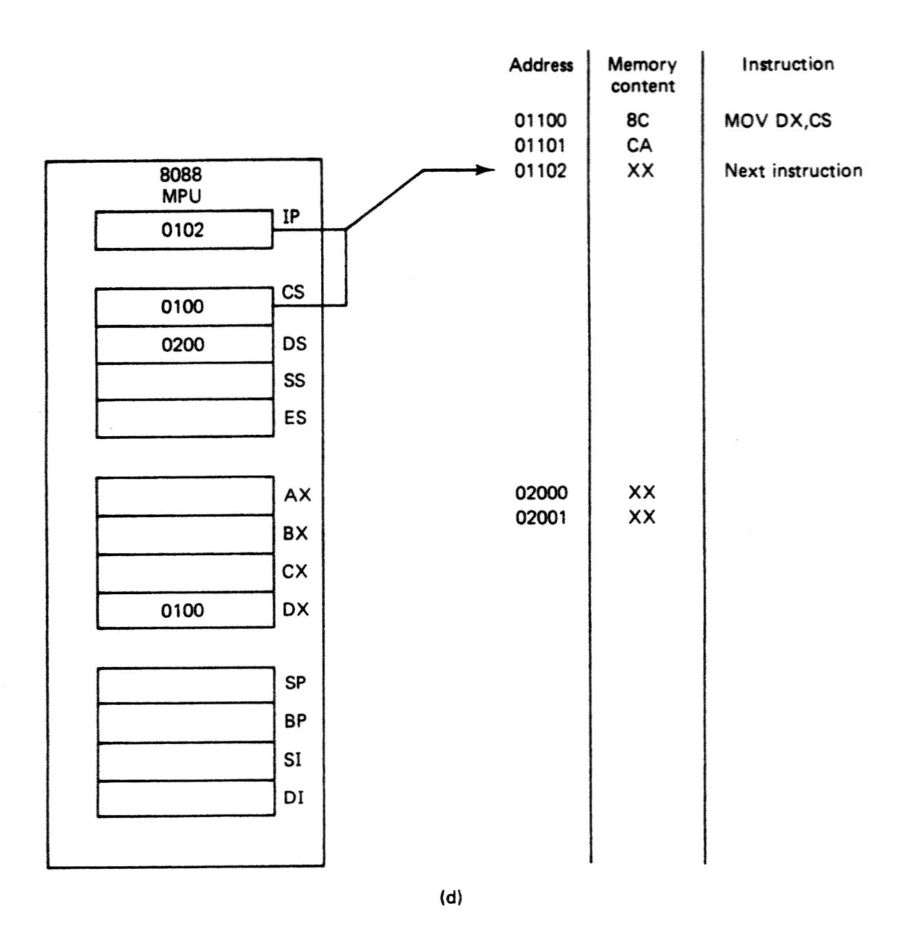

Figure 5–1 (continued)

In all memory reference MOV instructions, the machine code for the instruction includes an offset address relative to the beginning of current data segment. An example of this type of instruction is

```
MOV [SUM], AX
```

In this instruction, the memory location identified by the variable SUM is specified using direct addressing. That is, the value of the offset SUM is encoded in the two byte locations that follow its opcode.

Let us assume that the contents of DS equal 0200_{16} and that SUM equals 1212_{16}. Then this instruction means "move the contents of accumulator AX to the memory location offset by 1212_{16} from the starting location of the current data segment." The physical address of this location is obtained as

$$PA = 02000_{16} + 1212_{16} = 03212_{16}$$

Thus, the effect of the instruction is

$$(AL) \rightarrow (\text{Memory Location } 03212_{16})$$

and

$$(AH) \rightarrow (\text{Memory Location } 03213_{16})$$

EXAMPLE 5.1

What is the effect of executing the instruction

```
MOV CX, [SOURCE_MEM]
```

where SOURCE_MEM equal to 20_{16} is a memory location offset relative to the current data segment defined by data segment base address $1A00_{16}$?

Solution

Execution of this instruction results in the following:

$$((DS)0 + 20_{16}) \rightarrow (CL)$$
$$((DS)0 + 20_{16} + 1_{16}) \rightarrow (CH)$$

In other words, CL is loaded with the contents held at memory address

$$1A000_{16} + 20_{16} = 1A020_{16}$$

and CH is loaded with the contents of memory address

$$1A000_{16} + 20_{16} + 1_{16} = 1A021_{16}$$

EXAMPLE 5.2

Use the DEBUG program on the PC to verify the operation of the instruction in Example 5.1. Initialize the word storage location pointed to by SOURCE_MEM to the value $AA55_{16}$ before executing the instruction.

Solution

First, invoke the DEBUG program by entering the command

```
C:\DOS>DEBUG    (↵)
```

As Fig. 5–2 shows, this results in the display of the DEBUG prompt

-

```
C:\DOS>DEBUG
-R
AX=0000  BX=0000  CX=0000  DX=0000  SP=FFEE  BP=0000  SI=0000  DI=0000
DS=1342  ES=1342  SS=1342  CS=1342  IP=0100   NV UP EI PL NZ NA PO NC
1342:0100 OF          DB   OF
-A
1342:0100 MOV   CX,[20]
1342:0104
-R DS
DS 1342
:1A00
-E 20 55 AA
-T

AX=0000  BX=0000  CX=AA55  DX=0000  SP=FFEE  BP=0000  SI=0000  DI=0000
DS=1A00  ES=1342  SS=1342  CS=1342  IP=0104   NV UP EI PL NZ NA PO NC
1342:0104 FFF3            PUSH     BX
-Q

C:\DOS>
```

Figure 5–2 Display sequence for Example 5.2.

To determine the memory locations the debugger assigns for use in entering instructions and data, we can examine the state of the internal registers with the command

```
-R    (↵)
```

Looking at the displayed information for this command in Fig. 5–2, we find that the contents of CS and IP indicate that the starting address in the current code segment is 1342:0100 and the current data segment starts at address 1342:0000. Also note that the initial value in CX is 0000H.

To enter the instruction from Example 5.1 at location 1342:0100, we use the Assemble command

```
-A                            (↵)
1342:0100 MOV CX, [20]        (↵)
1342:0104                     (↵)
```

Note that we must enter the value of the offset address instead of symbol SOURCE_MEM and that it must be enclosed in brackets to indicate that it is a direct address.

Let us now redefine the data segment so that it starts at $1A000_{16}$. As shown in Fig. 5–2, a Register command loads the DS register with $1A00_{16}$. This entry is

```
-R DS   (↵)
DS 1342
:1A00   (↵)
```

Now we initialize the memory locations at addresses 1A00:20 and 1A00:21 to 55_{16} and AA_{16}, respectively, with the Enter command

```
-E 20 55 AA   (↵)
```

Finally, to execute the instruction, a Trace command is issued

```
-T  (↵)
```

The result of executing the instruction is shown in Fig. 5–2. Note that CX has been loaded with $AA55_{16}$.

A use of the move instruction is to load initial address and data values into the registers of the MPU. For instance, the instruction sequence in Fig. 5–3 uses immediate data to initialize the values in the segment, index, and data registers. Figure 5–1(b) shows that the immediate data cannot be directly loaded into a segment register. For this reason, the initial values of the segment base addresses for DS, ES, and SS are first loaded into AX and then copied into the appropriate segment registers. The execution of the first five instructions initializes these segment registers.

The next six instructions are used to initialize the data and index registers. First, the AX register is cleared to 0000_{16}, and then BX is also cleared by copying this value from AX to BX. Finally, CX, DX, SI, and DI are loaded with the immediate values $000A_{16}$, 0100_{16}, 0200_{16}, and 0300_{16}, respectively.

The XCHG Instruction

In our study of the move instruction, we found that it could be used to copy the contents of a register or memory location into another register or contents of a register into a storage location in memory. In all of these cases, the original contents of the source location are preserved and the original contents of the destination are destroyed. In some applications we need to exchange the contents of two registers. For instance, we might want to exchange the data in the AX and BX registers.

This could be done using multiple move instructions and storage of the data in a temporary register, such as DX. However, to perform the exchange function more efficiently, a special instruction has been provided in the instruction set of the 8088. This is the exchange (XCHG) instruction. The forms of the XCHG instruction and its allowed operands are shown in Fig. 5–4(a) and (b). Here we see that it can be used to swap data

```
MOV AX,2000H
MOV DS, AX
MOV ES, AX
MOV AX,3000H
MOV SS,AX
MOV AX,0H
MOV BX,AX
MOV CX,0AH
MOV DX,100H
MOV SI,200H
MOV DI,300H
```

Figure 5–3 Initializing the internal registers of the 8088.

Mnemonic	Meaning	Format	Operation	Flags affected
XCHG	Exchange	XCHG D,S	(D) ↔ (S)	None

(a)

Destination	Source
Accumulator	Reg16
Memory	Register
Register	Register
Register	Memory

(b)

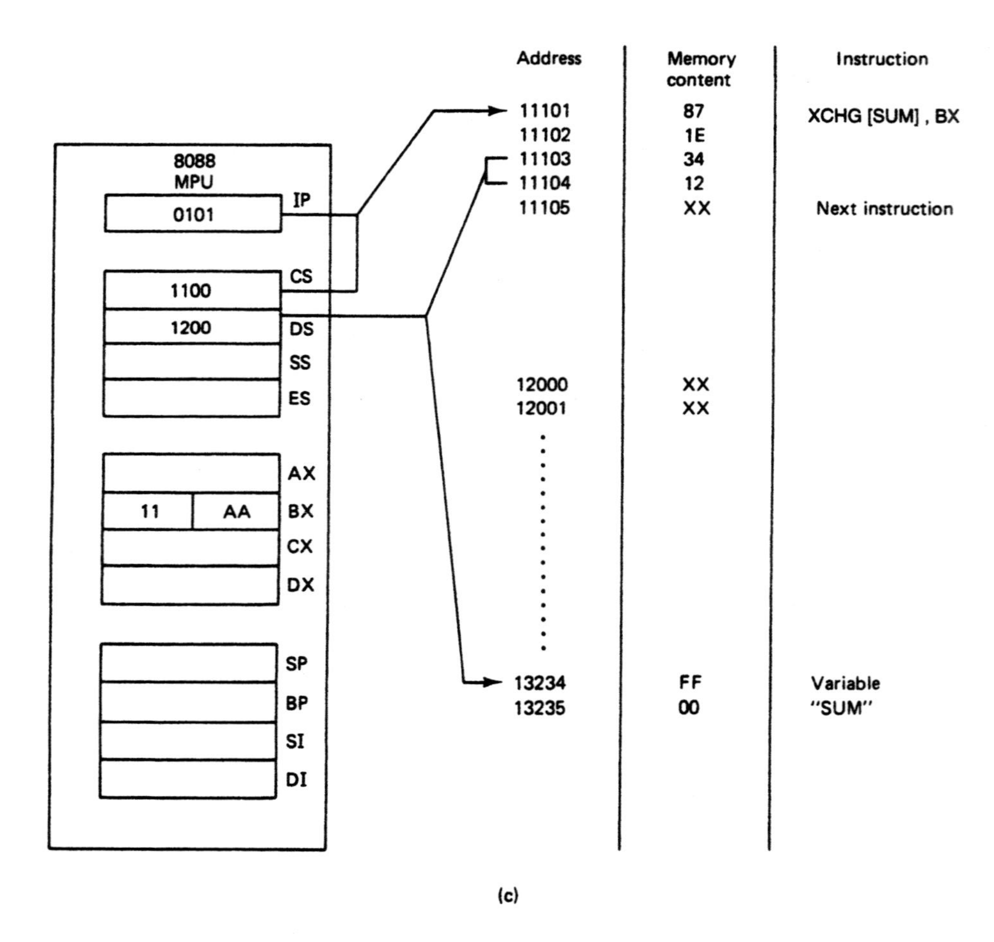

(c)

Figure 5–4 (a) XCHG data transfer instruction. (b) Allowed operands. (c) XCHG [SUM], BX before fetch and execution. (d) After execution.

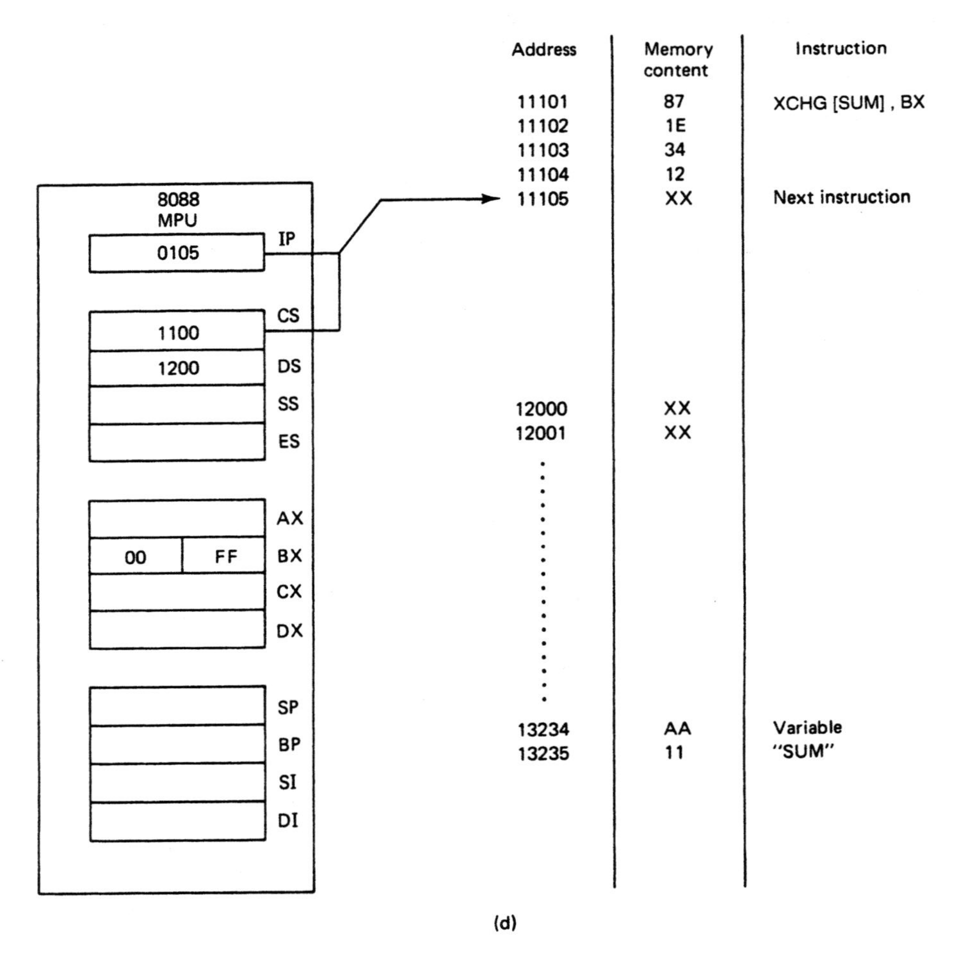

Figure 5–4 (continued)

between two general-purpose registers or between a general-purpose register and a storage location in memory. In particular, it allows for the exchange of words of data between one of the general-purpose registers, including the pointers and index registers and the accumulator (AX), exchange of a byte or word of data between one of the general-purpose registers and a storage location in memory, or between two of the general-purpose registers.

Let us consider an example of an exchange between two internal registers. Here is a typical instruction:

```
XCHG AX, DX
```

Its execution by the 8088 swaps the contents of AX with that of DX. That is,

$$(\text{AX original}) \rightarrow (\text{DX})$$
$$(\text{DX original}) \rightarrow (\text{AX})$$

or

$$(\text{AX}) \leftrightarrow (\text{DX})$$

EXAMPLE 5.3

For the data shown in Fig. 5–4(c), what is the result of executing the following instruction?

```
XCHG [SUM], BX
```

Solution

Execution of this instruction performs the operation

$$((\text{DS})0 + \text{SUM}) \leftrightarrow (\text{BX})$$

In Fig. 5–4(c), we see that (DS) = 1200_{16} and the direct address SUM = 1234_{16}. Therefore, the corresponding physical address is

$$\text{PA} = 12000_{16} + 1234_{16} = 13234_{16}$$

Note that this location contains FF_{16} and the address that follows contains 00_{16}. Moreover, BL contains AA_{16} and BH contains 11_{16}.

Execution of the instruction performs the following 16-bit swap:

$$(13234_{16}) \leftrightarrow (\text{BL})$$
$$(13234_{16}) \leftrightarrow (\text{BH})$$

As shown in Fig. 5–4(d), we get

$$(\text{BX}) = 00FF_{16}$$
$$(\text{SUM}) = 11AA_{16}$$

EXAMPLE 5.4

Use the DEBUG program to verify the operation of the instruction in Example 5.3.

Solution

The DEBUG commands needed to enter the instruction, enter the data, execute the instruction, and verify the result of its operation are shown in Fig. 5–5. Here we see that

```
C:\DOS>DEBUG
-R
AX=0000  BX=0000  CX=0000  DX=0000  SP=FFEE  BP=0000  SI=0000  DI=0000
DS=1342  ES=1342  SS=1342  CS=1342  IP=0100   NV UP EI PL NZ NA PO NC
1342:0100 OF            DB      OF
-A 1100:101
1100:0101 XCHG [1234],BX
1100:0105
-R BX
BX 0000
:11AA
-R DS
DS 1342
:1200
-R CS
CS 1342
:1100
-R IP
IP 0100
:101
-R
AX=0000  BX=11AA  CX=0000  DX=0000  SP=FFEE  BP=0000  SI=0000  DI=0000
DS=1200  ES=1342  SS=1342  CS=1100  IP=0101   NV UP EI PL NZ NA PO NC
1100:0101 871E3412      XCHG    BX,[1234]          DS:1234=0000
-E 1234 FF 00
-U 101 104
1100:0101 871E3412      XCHG    BX,[1234]
-T

AX=0000  BX=00FF  CX=0000  DX=0000  SP=FFEE  BP=0000  SI=0000  DI=0000
DS=1200  ES=1342  SS=1342  CS=1100  IP=0105   NV UP EI PL NZ NA PO NC
1100:0105 8946FE        MOV     [BP-02],AX         SS:FFFE=0000
-D 1234 1235
1200:1230              AA 11
-Q

C:\DOS>
```

Figure 5–5 Display sequence for Example 5.4.

after invoking DEBUG and displaying the initial state of the 8088's registers, the instruction is loaded into memory with the command

```
-A 1100:101                    (↵)
1100:0101 XCHG  [1234], BX     (↵)
1100:0105                      (↵)
-
```

Next, as shown in Fig. 5–5, R commands are used to initialize the contents of registers BX, DS, CS, and IP to $11AA_{16}$, 1200_{16}, 1100_{16}, and 0101_{16}, respectively, and then the updated register states are verified with another R command. Now, memory locations DS:1234H and DS:1235H are loaded with the values FF_{16} and 00_{16}, respectively, with the E command

```
-E 1234 FF 00  (↵)
-
```

Before executing the instruction, its loading is verified with an unassemble command. We see in Fig. 5–5 that it has been correctly loaded. The instruction is executed by issuing the Trace command

```
-T  (↵)
```

The displayed trace information in Fig. 5–5 shows that BX now contains $00FF_{16}$. To verify that the memory location was loaded with data from BX, we must display the data held at address DS:1234H and DS:1235H. This is done with the Dump command

```
-D 1234 1235  (↵)
1200:1230          AA 11
```

In this way, we see that the word contents of memory location DS:1234H have been exchanged with the contents of the BX register.

The XLAT Instruction

The translate (XLAT) instruction has been provided in the instruction set of the 8088 to simplify implementation of the lookup-table operation. This instruction is described in Fig. 5–6. When using XLAT, the contents of register BX represent the offset of the starting address of the lookup table from the beginning of the current data segment. Also, the contents of AL represent the offset of the element to be accessed from the beginning of the lookup table. This 8-bit element address permits a table with up to 256 elements. The values in both of these registers must be initialized prior to execution of the XLAT instruction.

Execution of XLAT replaces the contents of AL by the contents of the accessed lookup table location. The physical address of this element in the table is derived as

$$PA = (DS)0 + (BX) + (AL)$$

An example of the use of this instruction is for software code conversions. Figure 5–7 illustrates how an ASCII-to-EBCDIC conversion can be performed with the translate instruction. As shown, DS:BX defines the starting address where the EBCDIC table is stored in memory. This gives the physical address

$$PA = (DS)0 + (BX) = 03000_{16} + 0100_{16} = 03100_{16}$$

The individual EBCDIC codes are located in the table at element displacements equal to their equivalent ASCII character values. For example, the EBCDIC code $C1_{16}$, which represents letter A, is positioned at displacement 41_{16}, which equals ASCII A, from the start of the table.

Mnemonic	Meaning	Format	Operation	Flags affected
XLAT	Translate	XLAT	((AL)+(BX)+(DS)0) → (AL)	None

Figure 5–6 XLAT data transfer instruction.

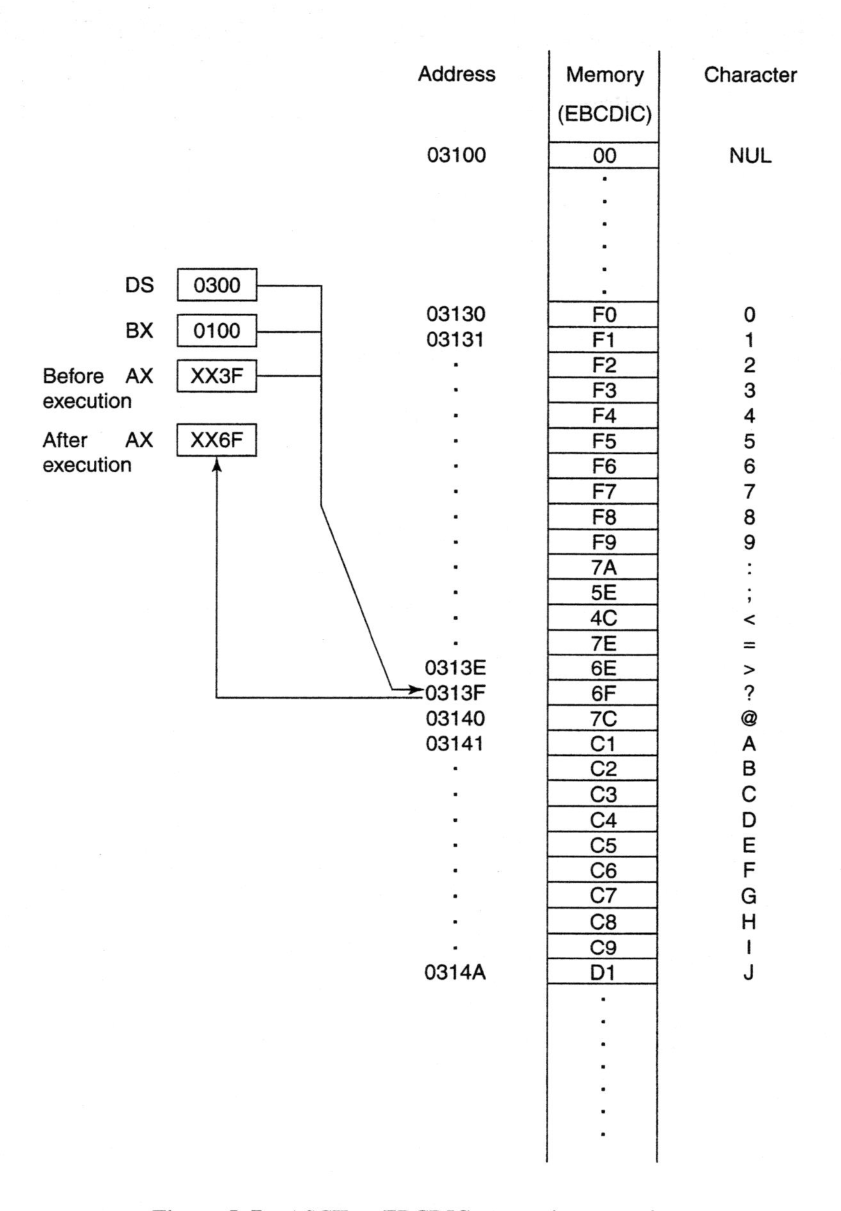

Figure 5–7 ASCII-to-EBCDIC conversion operation.

As an illustration of XLAT, let us assume (AL) = $3F_{16}$. Here $3F_{16}$ represents the ASCII symbol question mark. Execution of XLAT replaces the contents of AL by the contents of the memory location given by

$$PA = (DS)0 + (BX) + (AL)$$

$$= 03000_{16} + 0100_{16} + 3F_{16} = 0313F_{16}$$

Thus, the execution can be described by

$$(0313F_{16}) \rightarrow (AL)$$

This memory location contains $6F_{16}$ (EBCDIC code for question mark) and this value is placed in AL:

$$(AL) = 6F_{16}$$

LEA, LDS, and LES Instructions

An important type of data-transfer operation is loading a segment and a general-purpose register with an address directly from memory. Special instructions are provided in the instruction set of the 8088 to give a programmer this capability. These instructions, described in Fig. 5–8, are load register with effective address (LEA), load register and data segment register (LDS), and load register and extra segment register (LES).

Looking at Fig. 5–8(a), we see that these instructions provide programmers with the ability to manipulate memory addresses by loading either a 16-bit offset address into a general-purpose register or a 16-bit offset address into a general-purpose register together with a 16-bit segment address into either DS or ES.

The LEA instruction is used to load a specified register with a 16-bit offset address. An example of this instruction is

```
LEA SI, EA
```

When executed, it loads the SI register with an offset address value. The value of this offset is represented by the effective address EA. The value of EA can be specified by any valid addressing mode. For instance, if the value in DI equals 1000H and that in BX is 20H, then executing the instruction

```
LEA  SI, [DI + BX + 5H]
```

will load SI with the value

$$EA = 1000H + 20H + 5H = 1025H$$

That is,

$$(SI) = 1025H$$

The other two instructions, LDS and LES, are similar to LEA except that they load the specified register as well as the DS or ES segment register, respectively. That is, they are able to load a complete address pointer that is stored in memory. In this way, executing a single instruction can activate a new data segment.

Mnemonic	Meaning	Format	Operation	Flags affected
LEA	Load effective address	LEA Reg16,EA	EA → (Reg16)	None
LDS	Load register and DS	LDS Reg16,EA	EA → (Reg16) EA+2 → (DS)	None
LES	Load register and ES	LES Reg16,EA	EA → (Reg16) EA+2 → (ES)	None

(a)

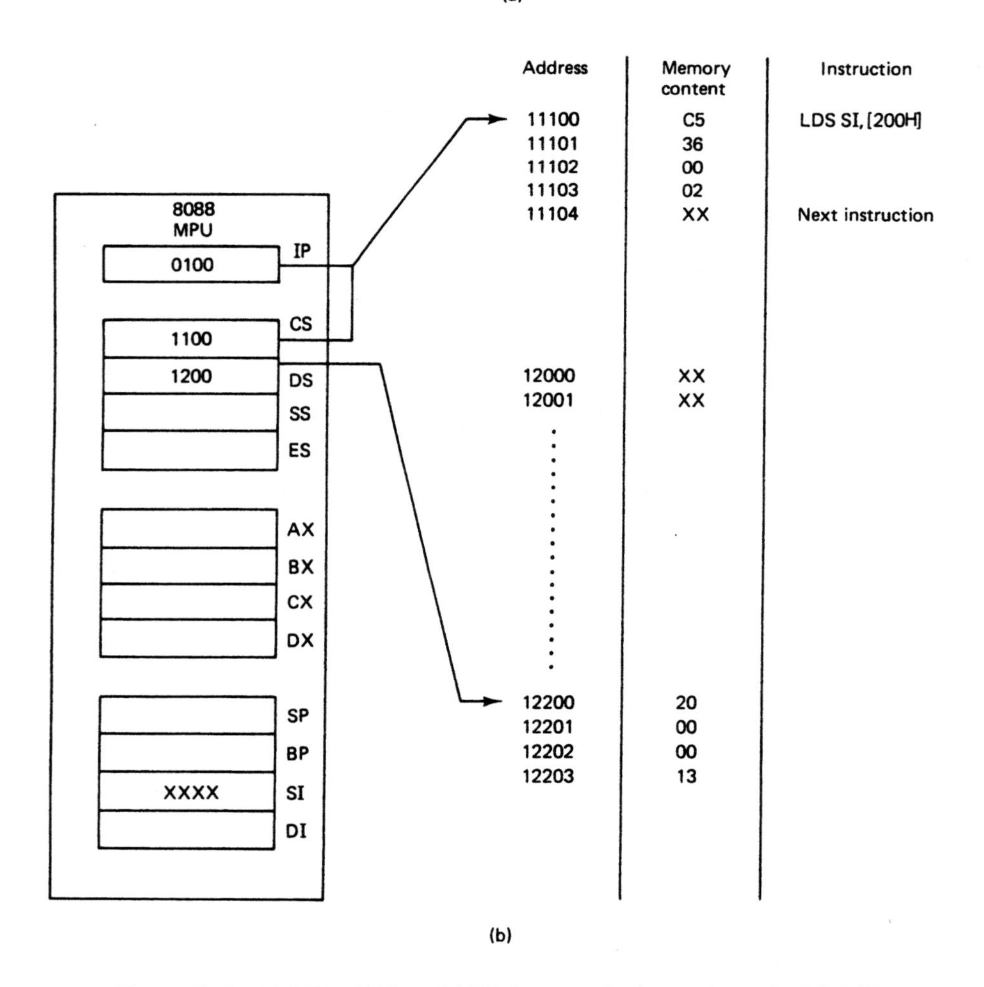

(b)

Figure 5–8 (a) LEA, LDS, and LES data transfer instructions. (b) LDS SI, [200H] before fetch and execution. (c) After execution.

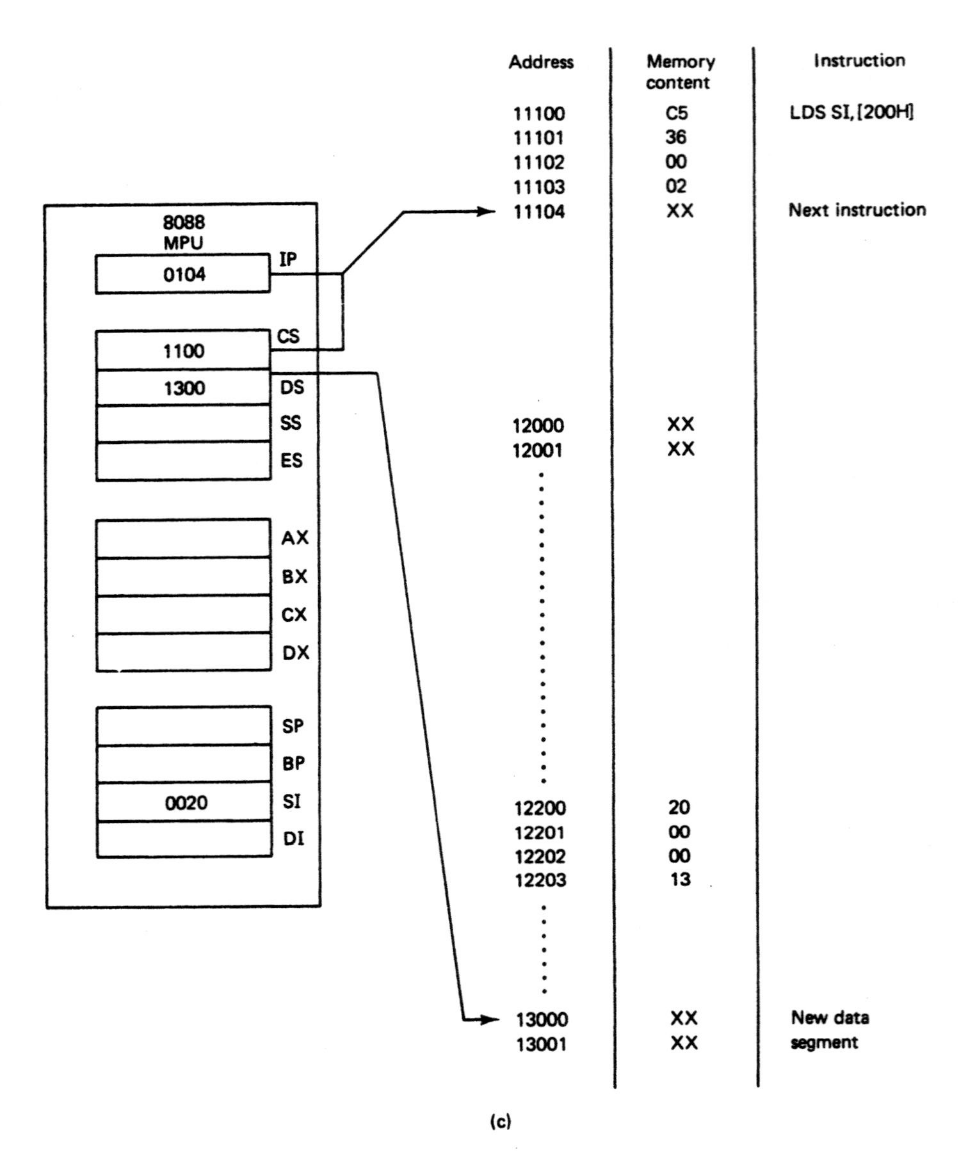

Figure 5–8 (continued)

EXAMPLE 5.5

Assuming that the 8088 is initialized as shown in Fig. 5–8(b), what is the result of executing the following instruction?

```
LDS SI, [200H]
```

Solution

Execution of the instruction loads the SI register from the word location in memory whose offset address with respect to the current data segment is 200_{16}. Figure 5–8(b) shows that the contents of DS are 1200_{16}. This gives a physical address of

$$PA = 12000_{16} + 0200_{16} = 12200_{16}$$

It is the contents of this location and the one that follows that are loaded into SI. Therefore, in Fig. 5–8(c) we find that SI contains 0020_{16}. The next two bytes—that is, the contents of addresses 12202_{16} and 12203_{16}—are loaded into the DS register. As shown, this defines a new data segment address of 13000_{16}.

EXAMPLE 5.6

Verify the execution of the instruction in Example 5.5 using the DEBUG program. The memory and register contents are to be those shown in Fig. 5–8(b).

Solution

As shown in Fig. 5–9, DEBUG is first brought up and then REGISTER commands are used to initialize registers IP, CS, DS, and SI with values 0100_{16}, 1100_{16}, 1200_{16}, and 0000_{16}, respectively. Next, the instruction is assembled at address CS:100 with the command

```
-A CS:100                       (↵)
1100:0100 LDS SI, [200]          (↵)
1100:0104                       (↵)
```

```
C:\DOS>DEBUG
-R IP
IP 0100
:
-R CS
CS 1342
:1100
-R DS
DS 1342
:1200
-R SI
SI 0000
:
-A CS:100
1100:0100 LDS   SI,[200]
1100:0104
-E 200 20 00 00 13
-T
AX=0000  BX=0000  CX=0000  DX=0000  SP=FFEE  BP=0000  SI=0020  DI=0000
DS=1300  ES=1342  SS=1342  CS=1100  IP=0104   NV UP EI PL NZ NA PO NC
1100:0104 C0             DB        C0
-Q

C:\DOS>
```

Figure 5–9 Display sequence for Example 5.6.

Before executing the instruction, we need to initialize two words of data starting at location DS:200 in memory. As shown in Fig. 5–9, this is done with an E command.

```
-E 200 20 00 00 13        (↵)
```

Then the instruction is executed with the command

```
-T                        (↵)
```

Looking at the displayed register contents in Fig. 5–9, we see that SI has been loaded with the value 0020_{16} and DS with the value 1300_{16}.

Earlier we showed how the segment registers, index registers, and data registers of the MPU can be initialized with immediate data. Another way of initializing them is from a table of data in memory. Using the LES instruction along with the MOV instruction provides an efficient method for performing register initialization from a data table. The sequence of instructions in Fig. 5–10(a) loads segment registers SS and ES, index registers SI and DI, and data registers AX, BX, CX, and DX with initial addresses and data from a table starting at offset address INIT_TABLE in the current data segment.

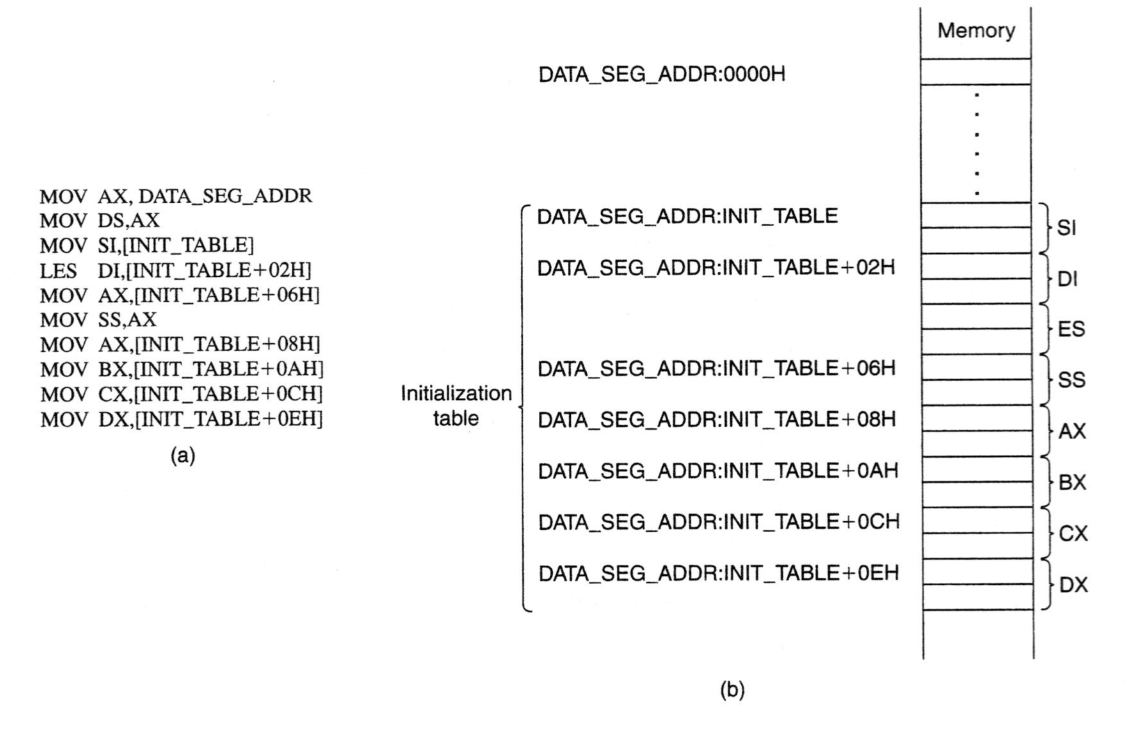

Figure 5–10 (a) Initializing the internal registers of the 8088 from a table in memory. (b) Table of initialization information.

Looking at the source addresses in the instructions, we can determine the location of each of the address or data elements in the table. For example, the 16-bit base address for register SS is held in the table at offset addresses INIT_TABLE+06H and INIT_TABLE+07H, and the word of data for DX is held at offset addresses INIT_TABLE+0EH and INIT_TABLE+0FH. Figure 5–10(b) shows that the table is 16 bytes long and spans the address range DATA_SEG_ADDR:INIT_TABLE through DATA_SEG_ADDR:INIT_TABLE+0FH. This table of information must be loaded into memory before the instruction sequence is executed.

First the program sets up a data segment starting at base address DATA_SEG_ADDR. Then, the addresses or data values are fetched from the table in memory and loaded into the appropriate registers. For instance, the third instruction reads an offset address, which is the word content of the memory location at offset address INIT_TABLE, into the SI register. Note that an LES instruction is used instead of MOV instructions to load the DI and ES registers. When this instruction is executed, the word of data for DI is loaded from table locations at offset addresses INIT_TABLE+02H and INIT_TABLE+03H, whereas the base address for DS is loaded from INIT_TABLE+04H and INIT_TABLE+05H.

▲ 5.2 ARITHMETIC INSTRUCTIONS

The instruction set of the 8088 microprocessor contains a variety of *arithmetic instructions.* They include instructions for the *addition, subtraction, multiplication,* and *division* operations. These operations can be performed on numbers expressed in a variety of numeric data formats. These formats include *unsigned* or *signed binary bytes* or *words, unpacked* or *packed decimal bytes,* or *ASCII numbers.* Remember that by *packed decimal* we mean that two BCD digits are packed into a byte-size register or a memory location. Unpacked decimal numbers are stored one BCD digit per byte. The BCD numbers are unsigned decimal numbers. ASCII numbers are expressed in ASCII code and stored one number per byte.

The status that results from the execution of an arithmetic instruction is recorded in the flags of the microprocessor. The flags that are affected by the arithmetic instructions are carry flag (CF), auxiliary flag (AF), sign flag (SF), zero flag (ZF), parity flag (PF), and overflow flag (OF). Each of these flags was discussed in Chapter 2.

For the purpose of discussion, we will divide the arithmetic instructions into the subgroups shown in Fig. 5–11.

Addition of Binary Numbers

Before examining the instructions that are provided for performing addition, let us review the topic of adding binary numbers. Adding binary numbers is governed by the following rules:

$$\begin{array}{r}0\\+\,0\\\hline 0\end{array}\qquad\begin{array}{r}0\\+\,1\\\hline 1\end{array}\qquad\begin{array}{r}1\\+\,0\\\hline 1\end{array}\qquad\begin{array}{r}1\\+\,1\\\hline 0\end{array}\ \text{\& carry}$$

Addition	
ADD	Add byte or word
ADC	Add byte or word with carry
INC	Increment byte or word by 1
AAA	ASCII adjust for addition
DAA	Decimal adjust for addition
Subtraction	
SUB	Subtract byte or word
SBB	Subtract byte or word with borrow
DEC	Decrement byte or word by 1
NEG	Negate byte or word
AAS	ASCII adjust for subtraction
DAS	Decimal adjust for subtraction
Multiplication	
MUL	Multiply byte or word unsigned
IMUL	Integer multiply byte or word
AAM	ASCII adjust for multiply
Division	
DIV	Divide byte or word unsigned
IDIV	Integer divide byte or word
AAD	ASCII adjust for division
CBW	Convert byte to word
CWD	Convert word to doubleword

Figure 5–11 Arithmetic instructions.

Looking at these additions, we see three different results. First, the addition 0 + 0 gives a sum of binary 0. On the other hand, adding 0 to a 1 results in 1. The last result is obtained by adding 1 to 1. This gives an answer of 2, but in binary 2 is written as 10. Another way of describing the answer to 1 + 1 is to say that sum is 0 and a carry of 1 to the next more significant bit. For the first three additions, there is no carry. No carry can be identified as a binary 0. Therefore, the result of 1 + 0 can be written as 01. This is read as "1 and a carry of 0" or "1 and no carry." The binary add operation is described in general as

$$A + B = S \,\&\, C_o$$

Note that the carry is identified as C_o and stands for carry-out. This type of binary addition operation is know as a half-add and is described in combinational logic by the half-adder function in Fig. 5–12(a).

In 8088/8086 assembly language, binary additions are performed with 8-bit or 16-bit binary numbers. For instance, the addition of two byte-wide numbers A and B is represented as

$$\begin{array}{r} A_7A_6A_5A_4A_3A_2A_1A_0 \\ +\ B_7\,B_6B_5B_4B_3\,B_2B_1B_0 \\ \hline S_8\,S_7\ S_6\,S_5S_4\,S_3\,S_2\,S_1S_0 \end{array}$$

Here we see that the sum of two 8-bit binary numbers can have 9 bits. The ninth bit is marked S_8 and is actually the carry out from the sum on the eighth bit.

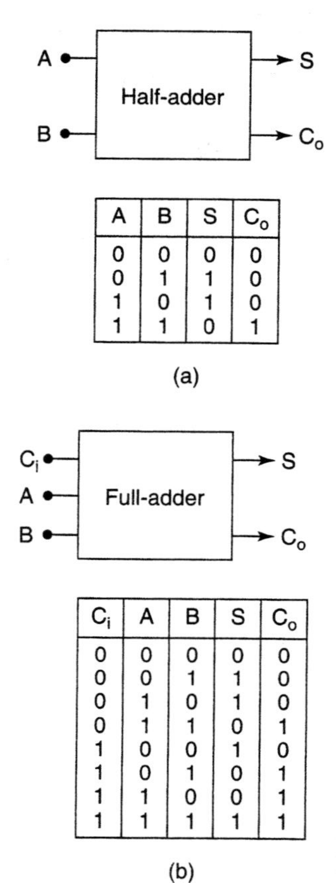

A	B	S	C_o
0	0	0	0
0	1	1	0
1	0	1	0
1	1	0	1

C_i	A	B	S	C_o
0	0	0	0	0
0	0	1	1	0
0	1	0	1	0
0	1	1	0	1
1	0	0	1	0
1	0	1	0	1
1	1	0	0	1
1	1	1	1	1

Figure 5–12 (a) Half-adder logic function. (b) Full-adder logic function.

Let us do an example to demonstrate this type of addition. Assume that

$$N1 = 170_{10} = 10101010_2$$

and

$$N2 = 238_{10} = 11101110_2$$

Then, their sum is

$$N1 + N2 = 170_{10} + 238_{10} = 408_{10}$$

Now performing the same operation by binary addition gives

```
111 111      carry
 10101010    N1
 11101110    N2
110011000    sum
```

Converting to its decimal equivalent can allow one to check this answer:

$$110011000_2 = 1(2^3) + 1(2^4) + 1(2^7) + 1(2^8) = 8 + 16 + 128 + 256 = 408_{10}$$

Looking at the addition performed for the third bit in this example, we see a more general type of binary addition. In this bit, three numbers are added. The carry-in (C_i) from the addition in the second bit is added to the sum of the A_2 and B_2. The result is the sum and a carry-out. This addition is described by the expression

$$C_i + A + B = S \ \& \ C_o$$

In combinational logic, this binary addition function is called a full-add and is represented by the full-adder logic element shown in Fig. 5–12(b).

Addition Instructions: ADD, ADC, INC, AAA, and DAA

Figure 5–13(a) shows the form of each of the instructions in the *addition subgroup;* Fig. 5–13(b) shows, the allowed operand variations for all but the INC instruction; and Fig. 5–13(c) shows the allowed operands for the INC instruction. Let us begin by looking more closely at the *add* (ADD) instruction. Note in Fig. 5–13(b) that it can be used to add an immediate operand to the contents of the accumulator, the contents of another register,

Mnemonic	Meaning	Format	Operation	Flags Affected
ADD	Addition	ADD D, S	(S) + (D) → (D) Carry → (CF)	OF, SF, ZF, AF, PF, CF
ADC	Add with carry	ADC D, S	(S) + (D) + (CF) → (D) Carry → (CF)	OF, SF, ZF, AF, PF, CF
INC	Increment by 1	INC D	(D) + 1 → (D)	OF, SF, ZF, AF, PF
AAA	ASCII adjust for addition	AAA		AF, CF OF, SF, ZF, PF undefined
DAA	Decimal adjust for addition	DAA		SF, ZF, AF, PF, CF, OF, undefined

(a)

Destination	Source
Register	Register
Register	Memory
Memory	Register
Register	Immediate
Memory	Immediate
Accumulator	Immediate

(b)

Destination
Reg16
Reg8
Memory

(c)

Figure 5–13 (a) Addition instructions. (b) Allowed operands for ADD and ADC instructions. (c) Allowed operands for INC instruction.

or the contents of a storage location in memory. It also allows us to add the contents of two registers or the contents of a register and a storage location in memory.

In general, the result of executing ADD is expressed as

$$(S) + (D) \rightarrow (D)$$

That is, the contents of the source operand are added to those of the destination operand and the sum that results is put into the location of the destination operand. The carry-out (C_o) that may occur from the addition of the most significant bit of the destination is reflected in the carry flag (CF). Therefore, this instruction performs the half-add binary arithmetic operation.

EXAMPLE 5.7

Assume that the AX and BX registers contain 1100_{16} and $0ABC_{16}$, respectively. What is the result of executing the instruction ADD AX, BX?

Solution

Executing the ADD instruction causes the contents of source operand BX to be added to the contents of destination register AX. This gives

$$(BX) + (AX) = 0ABC_{16} + 1100_{16} = 1BBC_{16}$$

This sum ends up in destination register AX. That is,

$$(AX) = 1BBC_{16}$$

A carry-out does not occur; therefore, the carry flag is reset.

$$CF = 0$$

Figures 5–14(a) and (b) illustrate the execution of this instruction.

EXAMPLE 5.8

Use the DEBUG program to verify the execution of the instruction in Example 5.7. Assume that the registers are to be initialized with the values shown in Fig. 5–14(a).

Solution

The debug sequence for this is shown in Fig. 5–15. After the debug program is brought up, the instruction is assembled into memory with the command

```
-A 1100:0100            (↵)
1100:0100 ADD AX, BX     (↵)
1100:0102               (↵)
-
```

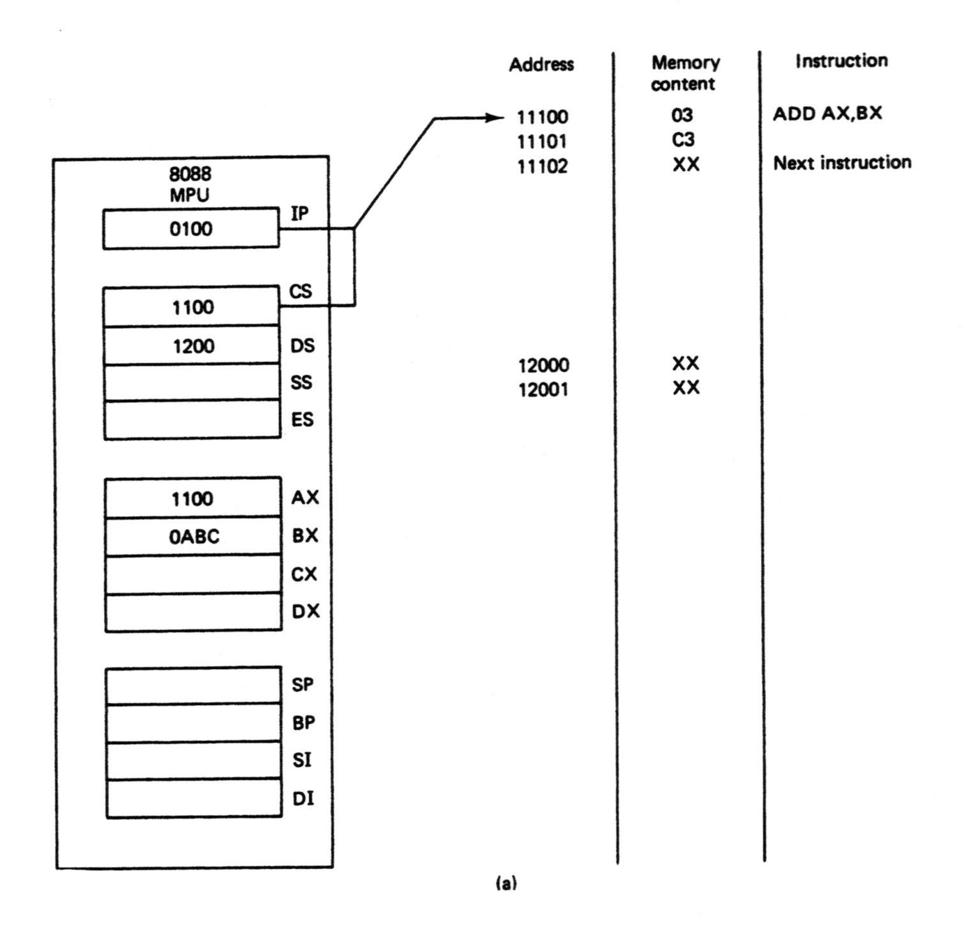

Figure 5–14 (a) ADD instruction before fetch and execution. (b) After execution.

Next, as shown in Fig. 5–15, the AX and BX registers are loaded with the values 1100_{16} and $0ABC_{16}$, respectively, using R commands.

```
-R AX        (↵)
AX 0000
:1100        (↵)
-R BX        (↵)
BX 0000
:0ABC        (↵)
```

Loading of the instruction is verified with the UNASSEMBLE command

```
-U 1100:0100 0100     (↵)
```

and is shown in Fig. 5–15 to be correct.

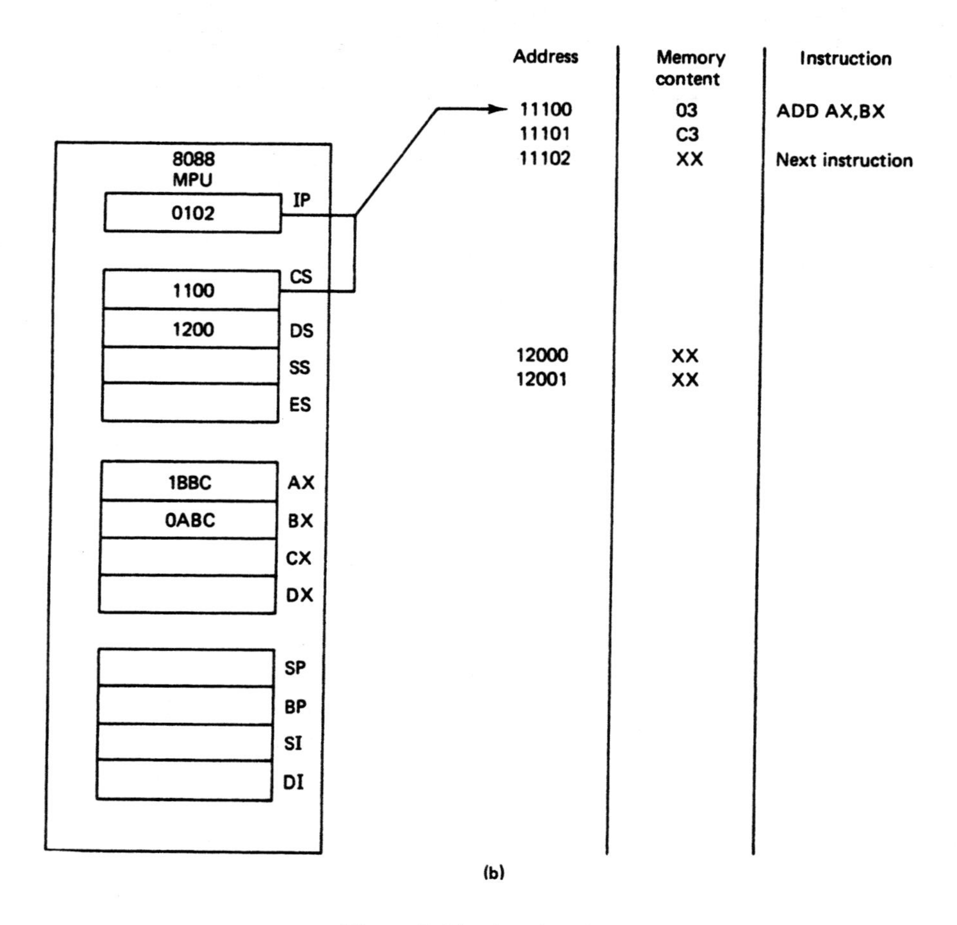

Figure 5–14 (continued)

```
C:\DOS>DEBUG
-A 1100:0100
1100:0100 ADD AX,BX
1100:0102
-R AX
AX 0000
:1100
-R BX
BX 0000
:0ABC
-U 1100:100 100
1100:0100 01D8          ADD     AX,BX
-T =1100:100

AX=1BBC  BX=0ABC  CX=0000  DX=0000  SP=FFEE  BP=0000  SI=0000  DI=0000
DS=1342  ES=1342  SS=1342  CS=1100  IP=0102   NV UP EI PL NZ NA PO NC
1100:0102 0002          ADD     [BP+SI],AL                        SS:0000=CD
-Q

C:\DOS>
```

Figure 5–15 Display sequence for Example 5.8.

We are now ready to execute the instruction with the TRACE command

```
-T =1100:0100          (↵)
```

From the trace dump in Fig. 5–15, we see that the sum of AX and BX, which equals $1BBC_{16}$, is now held in destination register AX. Also note that no carry (NC) has occurred.

The instruction *add with carry* (ADC) works similarly to ADD. But in this case, the content of the carry flag is also added— that is,

$$(S) + (D) + (CF) \rightarrow (D)$$

Here CF serves as both C_i and C_o, and the instruction performed the operation of the full-adder logic function. The valid operand combinations are the same as those for the ADD instruction. ADC is primarily used for multiword add operations.

Another instruction that can be considered part of the addition subgroup of arithmetic instructions is the *increment* (INC) instruction. As shown in Fig. 5–13(c), its operands can be the contents of a 16-bit internal register, an 8-bit internal register, or a storage location in memory. Execution of the INC instruction adds 1 to the specified operand. An example of an instruction that increments the high byte of AX is

```
INC AH
```

This instruction is typically used to increment the values of a count or address.

Figure 5–13(a) shows how the execution of these three instructions affects the earlier mentioned flags.

EXAMPLE 5.9

The original contents of AX, BL, word-size memory location SUM, and carry flag CF are 1234_{16}, AB_{16}, $00CD_{16}$, and 0_{16}, respectively. Describe the results of executing the following sequence of instructions:

```
ADD   AX, [SUM]
ADC   BL, 05H
INC   WORD PTR [SUM]
```

Solution

Executing the first instruction adds the word in the accumulator and the word in the memory location pointed to by address SUM. The result is placed in the accumulator. That is,

$$(AX) \leftarrow (AX) + (SUM) = 1234_{16} + 00CD_{16} = 1301_{16}$$

The carry flag remains reset.

Instruction	(AX)	(BL)	(SUM)	(CF)
Initial state	1234	AB	00CD	0
ADD AX, [SUM]	1301	AB	00CD	0
ADC BL,05H	1301	B0	00CD	0
INC WORD PTR [SUM]	1301	B0	00CE	0

Figure 5–16 Execution results of arithmetic instructions in Example 5.9.

The second instruction adds to the lower byte of the base register (BL) the immediate operand 5_{16} and the carry flag, which is 0_{16}. This gives

$$(BL) \leftarrow (BL) + imm8 + (CF) = AB_{16} + 5_{16} + 0_{16} = B0_{16}$$

Since no carry is generated, CF stays reset.

The last instruction increments the contents of memory location SUM by one. That is,

$$(SUM) \leftarrow (SUM) + 1_{16} = 00CD_{16} + 1_{16} = 00CE_{16}$$

These results are summarized in Fig. 5–16.

EXAMPLE 5.10

Verify the operation of the instruction sequence in Example 5.9 by executing it with the DEBUG program. A source program that includes this sequence of instructions is shown in Fig. 5–17(a), and the source listing produced when the program is assembled is shown in Fig. 5–17(b). A run module that was produced by linking this program is stored in file EX510.EXE.

Solution

The DEBUG program is brought up and at the same time the run module from file EX510.EXE is loaded with the command

```
C:\DOS>DEBUG A:EX510.EXE  (↵)
```

Next, we will verify the loading of the program by unassembling it with the command

```
-U 0 12  (↵)
```

Comparing the displayed instruction sequence in Fig. 5–17(c) to the source listing in Fig. 5–17(b), we find that the program has loaded correctly.

Notice in Fig. 5–17(c) that the instructions for which we are interested in verifying operation start at address 0D03:000A. For this reason, a Go command will be used to execute down to this point in the program. This command is

```
-G A  (↵)
```

```
TITLE   EXAMPLE 5.10
      PAGE      ,132
STACK_SEG        SEGMENT            STACK 'STACK'
                 DB                 64 DUP(?)
STACK_SEG        ENDS

DATA_SEG         SEGMENT
SUM              DW                 0CDH
DATA_SEG         ENDS

CODE_SEG         SEGMENT            'CODE'
EX510   PROC     FAR
      ASSUME  CS:CODE_SEG,   SS:STACK_SEG,   DS:DATA_SEG

;To return to DEBUG program put return address on the stack

      PUSH    DS
      MOV     AX, 0
      PUSH    AX

;Following code implements the Example 5.10

      MOV     AX, DATA_SEG        ;Establish data segment
      MOV     DS, AX

      ADD     AX, SUM
      ADC     BL, 05H
      INC     WORD PTR SUM

      RET                         ;Return to DEBUG program
EX510 ENDP

CODE_SEG      ENDS
      END     EX510
```

(a)

Figure 5–17 (a) Source program for Example 5.10. (b) Source listing produced by assembler. (c) Debug session for execution of program EX510.EXE.

Note that in the trace information displayed for this command, CS now contains $0D03_{16}$ and IP contains $000A_{16}$; therefore, the next instruction to be executed is at address 0D03:000A. This is the ADD instruction.

Now we need to initialize registers AX, BX, and the memory location pointed to by SUM (WORD PTR [0000]). We must also ensure that the CF status flag is set to NC (no carry). In Fig. 5–17(c), we find that these operations are done with the following sequence of commands:

```
-R AX           (↵)
AX 0D05
:1234           (↵)
-R BX           (↵)
BX 0000
:AB             (↵)
-R F            (↵)
NV UP EI PL NZ NA PO NC -     (↵)
-E 0 CD 00      (↵)
-D 0 1          (↵)
0D03:0000 CD 00
```

```
          TITLE   EXAMPLE 5.10

                       PAGE      ,132

0000                  STACK_SEG        SEGMENT          STACK 'STACK'
0000    40 [                           DB               64 DUP(?)
           ??
              ]

0040                  STACK_SEG        ENDS


0000                  DATA_SEG         SEGMENT
0000   00CD           SUM              DW               0CDH
0002                  DATA_SEG         ENDS

0000                  CODE_SEG         SEGMENT          'CODE'
0000                  EX510  PROC      FAR
                     ASSUME  CS:CODE_SEG, SS:STACK_SEG, DS:DATA_SEG

          ;To return to DEBUG program put return address on the stack

0000 1E                         PUSH    DS
0001 B8 0000                    MOV     AX, 0
0004 50                         PUSH    AX

          ;Following code implements the Example 5.10

0005 B8 — R                     MOV     AX, DATA_SEG     ;Establish data segment
0008 8E D8                      MOV     DS, AX

000A 03 06 0000 R               ADD     AX, SUM
000E 80 D3 05                   ADC     BL, 5H
0011 FF 06 0000 R               INC     WORD PTR SUM

0015 CB                         RET                      ;Return to DEBUG program
0016                  EX510     ENDP

0016                  CODE_SEG         ENDS

                     END      EX510

Segments and groups:

            N a m e               Size     align     combine    class

CODE_SEG . . . . . . . . . . . . . 0016     PARA      NONE       'CODE'
DATA_SEG . . . . . . . . . . . . . 0002     PARA      NONE
STACK_SEG. . . . . . . . . . . . . 0040     PARA      STACK      'STACK'

Symbols:

            N a m e               Type     Value     Attr

EX510. . . . . . . . . . . . . . . F PROC   0000      CODE_SEG          Length =0016
SUM. . . . . . . . . . . . . . . . L WORD   0000      DATA_SEG

Warning Severe
Errors  Errors
0       0
```

(b)

Figure 5–17 (continued)

Now we are ready to execute the ADD instruction. Issuing the Trace command does this.

-T (↵)

From the information displayed for this command in Fig. 5–17(c), note that the value CD_{16} has been added to the original value in AX, which was 1234_{16}, and the sum that results in AX is 1301_{16}.

```
C:DOS>DEBUG A:EX510.EXE
-U 0 12
0D03.0000 1E            PUSH    DS
0D03:0001 B80000        MOV     AX,0000
0D03:0004 50            PUSH    AX
0D03:0005 B8050D        MOV     AX,0D05
0D03:0008 8ED8          MOV     DS,AX
0D03:0C0A 03060000      ADD     AX,[0000]
0D03:000E 80D305        ADC     BL,05
0D03:0011 FF060000      INC     WORD PTR [0000]
-G A

AX=0D03  BX=0000  CX=0000  DX=0000  SP=003C  BP=0000  SI=0000  DI=0000
DS=0D05  ES-0CF3  SS=0D06  CS=0D03  IP=000A   NV UP EI PL NZ NA PO NC
0D03:000A 03060000      ADD     AX,[0000]                 DS:0000=00CD
-R AX
AX 0D05
:1234
-R BX
BX 0000
:AB
-R F
NV UP EI PL NZ NA PO NC  -
-E 0 CD 00
-D 0 1
0D05:0000  CD 00
-T

AX=1301  BX=00AB  CX=0000  DX=0000  SP=003C  BP=0000  SI=0000  DI=0000
DS=0D05  ES=0CF3  SS=0D06  CS=0D03  IP=000E   NV UP EI PL NZ AC PO NC
0D03:000E 80D305        ADC     BL,05
-T

AX=1301  BX=00B0  CX=0000  DX=0000  SP=003C  BP=0000  SI=0000  DI=0000
DS=0D05  ES=0CF3  SS=0D06  CS=0D03  IP=0011   NV UP EI NG NZ AC PO NC
0D03:0011 FF060000      INC     WORD PTR [0000]           DS:0000=00CD
-T

AX=1301  BX=00B0  CX=0000  DX=0000  SP=003C  BP=0000  SI=0000  DI=0000
DS=0D05  ES=0CF3  SS=0D06  CS=0D03  IP=0015   NV UP EI PL NZ NA PO NC
0D03:0015 CB            RETF
-D 0 1
0D05:0000  CE 00
-G

Program terminated normally
-Q

C:\DOS>
```

(c)

Figure 5–17 (continued)

Next the ADC instruction is executed with another T command.

```
-T  (↵)
```

It causes the immediate operand value 05_{16} to be added to the original contents of BL, AB_{16}, and the sum that is produced in BL is $B0_{16}$.

The last instruction is executed with one more T command, causing the SUM (WORD PTR [0000]) to be incremented by one. This can be verified by issuing the Dump command

```
-D 0 1  (↵)
```

Note that the value of SUM is identified as a WORD PTR. This assembler directive means that the memory location for SUM is to be treated as a word-wide storage location. Similarly if a byte-wide storage location, say BYTE_LOC, were to be accessed, it would be identified as BYTE PTR [BYTE_LOC].

The addition instructions we just covered can also be used to add numbers expressed in ASCII code, provided the binary result that is produced is converted back to its equivalent ASCII representation. This eliminates the need for doing a code conversion on ASCII data prior to processing them with addition operations. Whenever the 8088 does an addition on ASCII format data, an adjustment must be performed on the binary result to convert it to the equivalent decimal number. It is specifically for this purpose that the *ASCII adjust for addition* (AAA) instruction is provided in the instruction set. The AAA instruction should be executed immediately after the ADD instruction that adds ASCII data.

Assuming that AL contains the result produced by adding two ASCII coded numbers, executing the AAA instruction causes the contents of AL to be replaced by its equivalent decimal value. If the sum is greater than nine, AL contains the LSD, and AH is incremented by one. Otherwise, AL contains the sum, and AH is unchanged. Figure 5–13(a) shows that the AF and CF flags can be affected. Since AAA can adjust only data that are in AL, the destination register for ADD instructions that process ASCII numbers should be AL.

EXAMPLE 5.11

What is the result of executing the following instruction sequence?

```
ADD AL, BL
AAA
```

Assume that AL contains 32_{16} (the ASCII code for number 2), BL contains 34_{16} (the ASCII code for number 4), and AH has been cleared.

Solution

Executing the ADD instruction gives

$$(AL) \leftarrow (AL) + (BL) = 32_{16} + 34_{16} = 66_{16}$$

Next, the result is adjusted to give its equivalent decimal number. This is done by executing the AAA instruction. The equivalent of adding 2 and 4 is decimal 6 with no carry. Therefore, the result after the AAA instruction is

$$(AL) = 06_{16}$$
$$(AH) = 00_{16}$$

and both AF and CF remain cleared.

The instruction set of the 8088 includes another instruction, called *decimal adjust for addition* (DAA). This instruction is used to perform an adjust operation similar to that performed by AAA but for the addition of packed BCD numbers instead of ASCII numbers. Figure 5–13 also provides information about this instruction. Similar to AAA, DAA performs an adjustment on the value in AL. A typical instruction sequence is

```
ADD AL, BL
DAA
```

Remember that the contents of AL and BL must be packed BCD numbers—that is, two BCD digits packed into a byte. The adjusted result in AL is again a packed BCD byte.

32-Bit Binary Addition Program

As an example of the use of the addition instructions, let us perform a 32-bit binary add operation on the contents of the registers. We will implement the addition

$$(DX,CX) \leftarrow (DX,CX) + (BX,AX)$$

for the following data in the registers

$$(DX,CX) = FEDCBA98_{16}$$

$$(BX,AX) = 01234567_{16}$$

We first initialize the registers with the data using move instructions as follows:

```
MOV    DX, 0FEDCH
MOV    CX, 0BA98H
MOV    BX, 0123H
MOV    AX, 4567H
```

Next, the 16 least significant bits of the 32-bit number are added with the instruction.

```
ADD  CX, AX
```

Note that the sum from this part of the addition is in CX and the carry-out in CF.

To add the most significant 16 bits, we must account for the possibility of a carry-in from the addition of the lower 16 bits. For this reason, the ADC instruction must be used. Thus, the last instruction is

```
ADC    DX, BX
```

Execution of this instruction produces the upper 16 bits of the 32-bit sum in register DX.

Subtraction of Binary Numbers

Now that we have examined binary addition and the instructions of the 8088/8086 instruction set that perform addition operations, let us continue by reviewing how to perform binary subtraction. The basic binary subtractions are as follows:

$$\begin{array}{r} 0 \\ -0 \\ \hline 0 \end{array} \qquad \begin{array}{r} 0 \\ -1 \\ \hline 1 \end{array} \text{ \& borrow} \qquad \begin{array}{r} 1 \\ -0 \\ \hline 1 \end{array} \qquad \begin{array}{r} 1 \\ -1 \\ \hline 0 \end{array}$$

From these subtractions, we find that subtracting binary 1 from 0 requires a borrow from the more significant bit. When a 1 is borrowed, it is brought back to the less significant bit as 1 + 1, and then subtracting gives a result equal to 1. This result is expressed as 1 and a borrow of 1. The other three subtractions are performed without a borrow, or a borrow of 0. The subtraction operation is described in general as

$$A - B = D \ \& \ Br_i$$

Br_i stands for borrow-in. This form of the binary subtract operation is referred to as a half-subtract and symbolized by the combination logic function in Fig. 5–18(a).

As with addition, subtraction performed in software by the 8088/8086 microprocessor involves byte-wide or word-wide binary numbers. The subtraction of two byte-wide numbers is denoted as

$$\begin{array}{r} A_7A_6A_5A_4A_3A_2A_1A_0 \\ -\ B_7B_6B_5B_4B_3B_2B_1B_0 \\ \hline D_7D_6D_5D_4D_3D_2D_1D_0 \end{array}$$

To illustrate the subtraction process, let us do the following example:

$$N1 = 238_{10} = 11101110_2$$

and

$$N2 = 171_{10} = 10101011_2$$

Then, their difference is

$$\text{Difference} = N1 - N2 = 238_{10} - 171_{10} = 67_{10}$$

$$\begin{array}{rl} 11 & \text{Borrow} \\ 11101110 & \text{N1} \\ -10101011 & \text{N2} \\ \hline 01000011 & \text{Difference} \end{array}$$

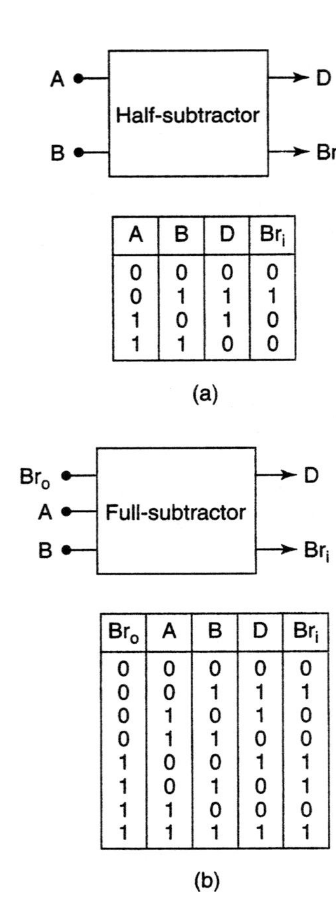

A	B	D	Br$_i$
0	0	0	0
0	1	1	1
1	0	1	0
1	1	0	0

Br$_o$	A	B	D	Br$_i$
0	0	0	0	0
0	0	1	1	1
0	1	0	1	0
0	1	1	0	0
1	0	0	1	1
1	0	1	0	1
1	1	0	0	0
1	1	1	1	1

Figure 5–18 (a) Half-subtractor logic function. (b) Full-subtractor logic function.

Checking the answer gives

$$01000011_2 = 1(2^0) + 1(2^1) + 1(2^6) = 1 + 2 + 64 = 67_{10}$$

Since A_0 is less than B_0, a borrow from A_1 must be used to do the subtraction, resulting in the difference D_0 as 1.

Moving to the A_1 bit, we find that 1 has already been borrowed to leave a 0 there. For this reason, the subtract cannot be done without borrowing. Bringing 1 back from the A_2 bit and subtracting, we obtain the difference of 1. The rest of the bits subtract without requiring any additional borrows.

Looking at the subtraction of B_1 from A_1, we notice that two borrows have been performed. First, 1 is borrowed from the A_1 bit and returned to the A_0 bit. This is called a borrow-out (Br_o) for the A_1 bit. Moreover, a 1 was borrowed from the A_2 bit and returned to the A_1 bit so that the subtraction could be performed. This type of borrow is referred to as a borrow-in (Br_i) for the A_1 bit. The subtraction performed in the A_1 bit is expressed in general as

$$A - B - Br_o = D \ \& \ Br_i$$

This binary subtract function is called a full-subtract and is represented by the full-subtractor combination logic element shown in Fig. 5–18(b).

Another way of performing binary subtraction is to use the 2's complement method. Using this method, the difference of two binary numbers is found by an addition process instead of directly through subtraction. The process requires that the value of the minuend be replaced by its 2's complement and then this value is added to the subtrahend to produce the difference. That is, the subtraction

$$(\text{Subtrahend}) - (\text{Minuend}) = \text{Difference}$$

is performed as

$$(\text{Subtrahend}) + (\text{2's complement of Minuend}) = \text{Difference}$$

Let us first review how to form the 2's complement of a binary number. To form the 2's complement of a number, first change all 1s in the number to 0s and all 0s to 1s; then 1 is added to the least significant bit. For instance, the 2's complement of the minuend in our earlier example is formed as follows:

$$N2 = 10101011_2$$

Inverting bits and adding 1 to the LSB gives

$$\text{2's complement } N2 = 01010100_2 + 1_2 = 01010101_2$$

For the earlier example, we have:

$$\begin{aligned} N1 - N2 &= 11101110_2 - 10101011_2 \\ &= 11101110_2 + \text{2's complement of } 10101011_2 \\ &= 11101110_2 + 01010101_2 \\ &= 01000011_2 \end{aligned}$$

The carry from the most significant bit is ignored. This result is identical to that obtained earlier.

Subtraction Instructions: SUB, SBB, DEC, AAS, DAS, and NEG

The instruction set of the 8088 includes an extensive set of instructions provided for implementing subtraction. As Fig. 5–19(a) shows, the subtraction subgroup is similar to the addition subgroup. It includes instructions for subtracting a source and a destination operand, decrementing an operand, and adjusting the result of subtractions of ASCII and BCD data. An additional instruction in this subgroup is negate.

The *subtract* (SUB) instruction is used to subtract the value of a source operand from a destination operand. The result of this operation in general is given as

$$(D) \leftarrow (D) - (S)$$

Mnemonic	Meaning	Format	Operation	Flags affected
SUB	Subtract	SUB D,S	(D) − (S) → (D) Borrow → (CF)	OF, SF, ZF, AF, PF, CF
SBB	Subtract with borrow	SBB D,S	(D) − (S) − (CF) → (D)	OF, SF, ZF, AF, PF, CF
DEC	Decrement by 1	DEC D	(D) − 1 → (D)	OF, SF, ZF, AF, PF
NEG	Negate	NEG D	0 − (D) → (D) 1 → (CF)	OF, SF, ZF, AF, PF, CF
DAS	Decimal adjust for subtraction	DAS		SF, ZF, AF, PF, CF OF undefined
AAS	ASCII adjust for subtraction	AAS		AF, CF OF, SF, ZF, PF undefined

(a)

Destination	Source
Register	Register
Register	Memory
Memory	Register
Accumulator	Immediate
Register	Immediate
Memory	Immediate

(b)

Destination
Reg16
Reg8
Memory

(c)

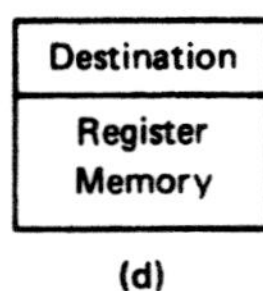

Destination
Register
Memory

(d)

Figure 5–19 (a) Subtraction instructions. (b) Allowed operands for SUB and SBB instructions. (c) Allowed operands for DEC instruction. (d) Allowed operands for NEG instruction.

The borrow-in (Br_i) that may occur from the subtraction of the most significant bit of the destination is reflected in the carry flag (CF). Therefore, this instruction performs the half-subtract binary arithmetic operation. As Fig. 5–19(b) shows, it can employ operand combinations similar to the ADD instruction.

The *subtract with borrow* (SBB) instruction is similar to SUB; however, it also subtracts the contents of the carry flag from the destination. That is,

$$(D) \leftarrow (D) - (S) - (CF)$$

CF serves as both Br_i and Br_o, and the instruction performs the operation of the full-subtractor logic function. SBB is primarily used for multiword subtract operations.

EXAMPLE 5.12

Assuming that the contents of registers BX and CX are 1234_{16} and 0123_{16}, respectively, and the carry flag is 0, what is the result of executing the following instruction?

```
SBB BX, CX
```

Solution

Since the instruction implements the operation

$$(BX) - (CX) - (CF) \rightarrow (BX)$$

we get

$$(BX) = 1234_{16} - 0123_{16} - 0_{16}$$
$$= 1111_{16}$$

Since no borrow was needed, the carry flag remains cleared.

EXAMPLE 5.13

Verify the operation of the subtract instruction in Example 5.12 by repeating the example using the DEBUG program.

Solution

As Fig. 5–20 shows, we first bring up the DEBUG program and then dump the initial state of the 8088 with a REGISTER command. Next, we load registers BX, CX, and flag CF with the values 1234_{16}, 0123_{16}, and NC, respectively. Note in Fig. 5–20 that this is done with three more R commands.

```
C:\DOS>DEBUG
-R
AX=0000  BX=0000  CX=0000  DX=0000  SP=FFEE  BP=0000  SI=0000  DI=0000
DS=1342  ES=1342  SS=1342  CS=1342  IP=0100   NV UP EI PL NZ NA PO NC
1342:0100 0F            DB       0F
-R BX
BX 0000
:1234
-R CX
CX 0000
:0123
-F F
NV UP EI PL NZ NA PO NC  -
-A
1342:0100 SBB BX,CX
1342:0102
-R
AX=0000  BX=1234  CX=0123  DX=0000  SP=FFEE  BP=0000  SI=0000  DI=0000
DS=1342  ES=1342  SS=1342  CS=1342  IP=0100   NV UP EI PL NZ NA PO NC
1342:0100 19CB          SBB      BX,CX
-U 100 101
1342:0100 19CB          SBB      BX,CX
-T

AX=0000  BX=1111  CX=0123  DX=0000  SP=FFEE  BP=0000  SI=0000  DI=0000
DS=1342  ES=1342  SS=1342  CS=1342  IP=0102   NV UP EI PL NZ NA PE NC
1342:0102 B98AFF        MOV      CX,FF8A
-Q

C:\DOS>
```

Figure 5–20 Display sequence for Example 5.13.

Now the instruction is assembled at address CS:100 with the command

```
-A                          (↵)
1342:0100 SBB BX,CX         (↵)
1342:0102                   (↵)
-
```

Before executing the instruction, we can verify the initialization of the registers and the entry of the instruction by issuing the commands

```
-R              (↵)
```

and

```
-U 100 101   (↵)
```

Looking at Fig. 5–20, we find that the registers are correctly initialized and that the instruction SBB BX,CX is correctly entered.

Finally, the instruction is executed with a TRACE command. As shown in Fig. 5–20, the result of executing the instruction is that the contents of CX are subtracted from the contents of BX. The difference, which is 1111_{16}, resides in destination register BX, and CF is still NC.

Just as the INC instruction can be used to add 1 to an operand, the *decrement* (DEC) instruction can be used to subtract 1 from its operand. The allowed operands for DEC are shown in Fig. 5–19(c).

In Fig. 5–19(d), we see that the *negate* (NEG) instruction can operate on operands in a general-purpose register or a storage location in memory. Executing this instruction causes the value of its operand to be replaced by its negative. This is actually done through subtraction—that is, the contents of the specified operand are subtracted from zero and the result is returned to the operand location. The subtraction is actually performed by the processor hardware using 2's-complement arithmetic. To obtain the correct value of the carry flag that results from a NEG operation, the carry generated by the add operation used in the 2's-complement subtraction calculation must be complemented.

EXAMPLE 5.14

Assuming that register BX contains $003A_{16}$, what is the result of executing the following instruction?

```
NEG BX
```

Solution

Executing the NEG instruction causes the 2's-complement subtraction that follows:

$$\begin{aligned}(BX) &= 0000_{16} - (BX) = 0000_{16} + \text{2's-complement of } 003A_{16}\\ &= 0000_{16} + FFC6_{16}\\ &= FFC6_{16}\end{aligned}$$

Since no carry is generated in this add operation, the carry flag is complemented to give

$$(CF) = 1$$

EXAMPLE 5.15

Verify the operation of the NEG instruction in Example 5.14 by executing it with the DEBUG program.

Solution

After starting the DEBUG program, we first initialize the contents of the BX register. This is done with the command

```
-R  BX      (↵)
BX 0000
:3A         (↵)
-
```

Next the instruction is assembled with the command

```
-A                  (↵)
1342:0100 NEG BX    (↵)
1342:0102           (↵)
```

At this point, we can verify the initialization of BX by issuing the command

```
-R BX       (↵)
BX 003A
:           (↵)
-
```

To check the assembly of the instruction, unassemble it with the command

```
-U 100 101    (↵)
1342:0100 F7DB NEG BX
-
```

Now the instruction is executed with the command

```
-T    (↵)
```

The information that is dumped by issuing this command is shown in Fig. 5–21. Here the new contents in register BX are verified as $FFC6_{16}$, which is the negative of $003A_{16}$. Also note that the carry flag is set (CY).

```
C:\DOS>DEBUG
-R BX
BX 0000
:3A
-A
1342:0100 NEG BX
1342:0102
-R BX
BX 003A
:
-U 100 101
1342:0100 F7DB      NEG   BX
-T

AX=0000  BX=FFC6  CX=0000  DX=0000  SP=FFEE  BP=0000  SI=0000  DI=0000
DS=1342  ES=1342  SS=1342  CS=1342  IP=0102   NV UP EI NG NZ AC PE CY
1342:0102 B98AFF        MOV     CX,FF8A
-Q

C:\DOS>
```

Figure 5–21 Display sequence for Example 5.15.

In our study of the addition instruction subgroup, we found that the 8088 is capable of directly adding ASCII and BCD numbers. The SUB and SBB instructions can subtract numbers represented in these formats as well. Just as for addition, the results that are obtained must be adjusted to produce the corresponding decimal numbers. In the case of ASCII subtraction, we use the *ASCII adjust for subtraction* (AAS) instruction, and for packed BCD subtraction we use the *decimal adjust for subtract* (DAS) instruction.

An example of an instruction sequence for direct ASCII subtraction is

```
SUB AL, BL
AAS
```

ASCII numbers must be loaded into AL and BL before the subtract instruction is executed. Note that the destination of the subtraction should be AL. After execution of AAS, AL contains the difference of the two numbers, and AH is unchanged if no borrow takes place or is decremented by 1 if a borrow occurs.

32-Bit Binary Subtraction Program

As an example of the use of the subtraction instructions, let us implement a 32-bit subtraction of two numbers X and Y that are stored in memory as

X = (DS:203H)(DS:202H)(DS:201H)(DS:200H)

MS byte of MS word	LS byte of MS word	MS byte of LS word	LS byte of LS word

Y = (DS:103H)(DS:102H)(DS:101H)(DS:100H)

The result of X − Y is to be saved in the place where X is stored in memory.

First, we subtract the least significant 16 bits of the 32-bit words using the instructions

```
MOV    AX, [200H]
SUB    AX, [100H]
MOV    [200H], AX
```

Next, the most significant words of X and Y are subtracted. In this part of the 32-bit subtraction, we must use the borrow that might have been generated by the subtraction of the least significant words. Therefore, SBB is used to perform the subtraction operation. The instructions to do this are

```
MOV    AX, [202H]
SBB    AX, [102H]
MOV    [202H], AX
```

These instructions used direct addressing to access the data in memory. The 32-bit subtract operation can also be done with indirect addressing with the program shown in Fig. 5–22.

Multiplication and Division Instructions: MUL, DIV, IMUL, IDIV, AAM, AAD, CBW, and CWD

The 8088 has instructions to support multiplication and division of binary and BCD numbers. Two basic types of multiplication and division instructions, for the processing of unsigned numbers and signed numbers, are available. To do these operations on unsigned numbers, the instructions are MUL and DIV. On the other hand, to multiply or divide 2's-complement signed numbers, the instructions are IMUL and IDIV.

Figure 5–23(a) describes these instructions. Note in Fig. 5.23(b) that a single byte-wide or word-wide operand is specified in a multiplication instruction. It is the source operand. As shown in Fig. 5–23(a), the other operand, which is the destination, is assumed already to be in AL for 8-bit multiplication or in AX for 16-bit multiplication.

The result of executing an MUL or IMUL instruction on byte data can be represented as

$$(AX) \leftarrow (AL) \times (8\text{-bit operand})$$

```
MOV   SI,200H      ;Initialize pointer for X
MOV   DI,100H      ;Initialize pointer for Y
MOV   AX,[SI]      ;Subtract LS words
SUB   AX,[DI]
MOV   [SI],AX      ;Save the LS word of result
MOV   AX,[SI]+2    ;Subtract MS words
SBB   AX,[DI]+2
MOV   [SI]+2,AX    ;Save the MS word of result
```

Figure 5–22 32-bit subtraction program using indirect addressing.

Mnemonic	Meaning	Format	Operation	Flags Affected
MUL	Multiply (unsigned)	MUL S	(AL) · (S8) → (AX) (AX) · (S16) → (DX),(AX)	OF, CF SF, ZF, AF, PF undefined
DIV	Division (unsigned)	DIV S	(1) Q((AX)/(S8)) → (AL) R((AX)/(S8)) → (AH) (2) Q((DX,AX)/(S16)) → (AX) R((DX,AX)/(S16)) → (DX) If Q is FF_{16} in case (1) or $FFFF_{16}$ in case (2), then type 0 interrupt occurs	OF, SF, ZF, AF, PF, CF undefined
IMUL	Integer multiply (signed)	IMUL S	(AL) · (S8) → (AX) (AX) · (S16) → (DX),(AX)	OF, CF SF, ZF, AF, PF undefined
IDIV	Integer divide (signed)	IDIV S	(1) Q((AX)/(S8)) → (AL) R((AX)/(S8)) → (AH) (2) Q((DX,AX)/(S16)) → (AX) R((DX,AX)/(S16)) → (DX) If Q is positive and exceeds $7FFF_{16}$ or if Q is negative and becomes less than 8001_{16}, then type 0 interupt occurs	OF, SF, ZF, AF, PF, CF undefined
AAM	Adjust AL for multiplication	AAM	Q((AL)/10) → (AH) R((AL)/10) → (AL)	SF, ZF, PF OF, AF,CF undefined
AAD	Adjust AX for division	AAD	(AH) · 10 + (AL) → (AL) 00 → (AH)	SF, ZF, PF OF, AF, CF undefined
CBW	Convert byte to word	CBW	(MSB of AL) → (All bits of AH)	None
CWD	Convert word to double word	CWD	(MSB of AX) → (All bits of DX)	None

(a)

Source
Reg8
Reg16
Mem8
Mem16

(b)

Figure 5–23 (a) Multiplication and division arithmetic instructions. (b) Allowed operands.

That is, the resulting 16-bit product is produced in the AX register. On the other hand, for multiplication of data words, the 32-bit result is given by

$$(DX, AX) \leftarrow (AX) \times (16\text{-bit operand})$$

where AX contains the 16 LSBs and DX the 16 MSBs.

For the division operation, again just the source operand is specified. The other operand is either the contents of AX for 16-bit dividends or the contents of both DX and AX for 32-bit dividends. The result of a DIV or IDIV instruction for an 8-bit divisor is represented by

$$(AH), (AL) \leftarrow (AX)/(8\text{-bit operand})$$

where AH contains the remainder and AL the quotient. For 16-bit division, we get

$$(DX), (AX) \leftarrow (DX, AX)/(16\text{-bit operand})$$

Here AX contains the quotient and DX contains the remainder.

EXAMPLE 5.16

The 2's-complement signed data contents of AL equal -1 and the contents of CL are -2. What result is produced in AX by executing the following instructions?

```
MUL CL
```

and

```
IMUL CL
```

Solution

As binary data, the contents of AL and CL are

$$\begin{aligned}(AL) &= -1 \text{ (as 2's complement)} = 11111111_2 = FF_{16} \\ (CL) &= -2 \text{ (as 2's complement)} = 11111110_2 = FE_{16}\end{aligned}$$

Executing the MUL instruction gives

$$\begin{aligned}(AX) &= 11111111_2 \times 11111110_2 = 1111110100000010_2 \\ &= FD02_{16}\end{aligned}$$

The second instruction multiplies the two numbers as signed numbers to generate the signed result. That is,

$$\begin{aligned}(AX) &= -1_{16} \times -2_{16} \\ &= 2_{16} = 0002H\end{aligned}$$

EXAMPLE 5.17

Verify the operation of the MUL instruction in Example 5.16 by performing the same operation with the DEBUG program.

Solution

First, the DEBUG program is loaded, and then registers AX and CX are initialized with the values FF_{16} and FE_{16}, respectively. These registers are loaded as follows:

```
-R AX        (↵)
AX 0000
:FF          (↵)
-R CX        (↵)
CX 0000
:FE          (↵)
-
```

Next, the instruction is loaded with the command

```
-A           (↵)
1342:0100 MUL CL
1342:0102  (↵)
-
```

Before executing the instruction, let us verify the loading of AX, CX, and the instruction. To do this, we use the commands as follows:

```
-R AX        (↵)
AX 00FF
:            (↵)
-R CX        (↵)
CX 00FE
:            (↵)
-U 100 101 (↵)
1342:0100 F6E1 MUL CL
-
```

To execute the instruction, we issue the T command:

```
-T           (↵)
```

The displayed result in Fig. 5–24 shows that AX now contains $FD02_{16}$, the unsigned product of FF_{16} and FE_{16}.

As Fig. 5–23(a) shows, adjust instructions for BCD multiplication and division are also provided. They are *adjust AX for multiply* (AAM) and *adjust AX for divide* (AAD).

```
C:\DOS>DEBUG
-R AX
AX 0000
:FF
-R CX
CX 0000
:FE
-A
1342:0100 MUL CL
1342:0102
-R AX
AX 00FF
:
-R CX
CX 00FE
:
-U 100 101
1342:0100 F6E1      MUL   CL
-T

AX=FD02  BX=0000  CX=00FE  DX=0000  SP=FFEE  BP=0000  SI=0000  DI=0000
DS=1342  ES=1342  SS=1342  CS=1342  IP=0102   OV UP EI NG NZ AC PE CY
1342:0102 B98AFF         MOV     CX,FF8A
-Q

C:\DOS>
```

Figure 5–24 Display sequence for Example 5.17.

The AAM instruction assumes that the instruction just before it multiplies two unpacked BCD numbers with their product produced in AL. The AAD instruction assumes that AH and AL contain unpacked BCD numbers.

The division instructions can also be used to divide an 8-bit dividend in AL by an 8-bit divisor. However, to do this, the sign of the dividend must first be extended to fill the AX register. That is, AH is filled with zeros if the number in AL is positive or with ones if it is negative. Execution of the *convert byte to word* (CBW) instruction automatically does this conversion. Note that the sign extension does not change the signed value for the data. It simply allows data to be represented using more bits.

In a similar way, the 32-bit by 16-bit division instructions can be used to divide a 16-bit dividend in AX by a 16-bit divisor. In this case, the sign bit of AX must be extended by 16 bits into the DX register. This can be done by another instruction, known as *convert word to double word* (CWD). Figure 5–23(a) shows the operations of these two sign-extension instructions.

Note that the CBW and CWD instructions are provided to handle operations where the result or intermediate results of an operation cannot be held in the correct word length for use in other arithmetic operations. Using these instructions, we can extend the value of a byte- or word-wide signed number to its equivalent signed word or double-word value.

EXAMPLE 5.18

What is the result of executing the following sequence of instructions?

```
MOV AL, 0A1H
CBW
CWD
```

Solution

The first instruction loads AL with $A1_{16}$. This gives

$$(AL) = A1_{16} = 10100001_2$$

Executing the second instruction extends the most significant bit of AL, 1, into all bits of AH. The result is

$$(AH) = 11111111_2 = FF_{16}$$

or

$$(AX) = 1111111110100001_2 = FFA1_{16}$$

This completes conversion of the byte in AL to a word in AX.

The last instruction loads each bit of DX with the most significant bit of AX. This bit is also 1. Therefore, we get

$$(DX) = 1111111111111111_2 = FFFF_{16}$$

Now the word in AX has been extended to a double word. That is,

$$(AX) = FFA1_{16}$$
$$(DX) = FFFF_{16}$$

EXAMPLE 5.19

Use an assembled version of the program in Example 5.18 to verify the results obtained when it is executed.

Solution

The source program is shown in Fig. 5–25(a). Note that this program differs from that described in Example 5.18 in that it includes the directive statements that are needed to assemble it and some additional instructions so that it can be executed using the DEBUG program.

The source file is assembled and linked to produce a run module stored in the file EX519.EXE. The source listing (EX519.LST) produced by assembling the source file, EX519.ASM, is shown in Fig. 5–25(b).

As Fig. 5–25(c) shows, the run module is loaded for execution as part of calling up the DEBUG program. This is done with the command

```
C:\DOS>DEBUG A:EX519.EXE   (↵)
```

```
TITLE   EXAMPLE  5.19

      PAGE       ,132

STACK_SEG        SEGMENT          STACK 'STACK'
                 DB               64 DUP(?)
STACK_SEG        ENDS

CODE_SEG         SEGMENT          'CODE'
EX519    PROC    FAR
      ASSUME  CS:CODE_SEG, SS:STACK_SEG

;To return to DEBUG program put return address on the stack

      PUSH    DS
      MOV     AX, 0
      PUSH    AX

;Following code implements Example 5.19

      MOV     AL, 0A1H
      CBW
      CWD

      RET                         ;Return to DEBUG program
EX519 ENDP

CODE_SEG         ENDS

      END     EX519
```

(a)

Figure 5–25 (a) Source program for Example 5.19. (b) Source listing produced by assembler. (c) Debug session for execution of program EX519.EXE.

The loading of the program can now be verified with the Unassemble command

```
-U 0 9      (↵)
```

Looking at the instructions displayed in Fig. 5–25(c), we see that the program is correct.

From the unassembled version of the program in Fig. 5–25(c), we find that the instructions in which we are interested start at address 0D03:0005. Thus, we execute the instructions prior to the MOV AL,A1 instruction by issuing the command

```
-G 5        (↵)
```

The information displayed in Fig. 5–25(c) shows that (AX) = 0000_{16} and (DX) = 0000_{16}. Moreover, (IP) = 0005_{16} and points to the first instruction in which we are interested. This instruction is executed with the command

```
-T          (↵)
```

In the trace dump information in Fig. 5–25(c), we see that AL has been loaded with $A1_{16}$ and DX contains 0000_{16}.

```
                  TITLE   EXAMPLE 5.19

                         PAGE    ,132

0000                       STACK_SEG        SEGMENT          STACK 'STACK'
0000    40 [                                DB               64 DUP(?)
          ??
             ]

0040                       STACK_SEG        ENDS


0000                       CODE_SEG         SEGMENT          'CODE'
0000                       EX519    PROC    FAR
                         ASSUME  CS:CODE_SEG, SS:STACK_SEG

                  ;To return to DEBUG program put return address on the stack

0000 1E                            PUSH     DS
0001 B8 0000                       MOV      AX, 0
0004 50                            PUSH     AX

                  ;Following code implements the Example 5.19

0005 B0 A1                         MOV      AL, 0A1H
0007 98                            CBW
0008 99                            CWD

0009 CB                            RET                       ;Return to DEBUG program
000A                       EX519   ENDP

000A                       CODE_SEG         ENDS

                         END      EX519

Segments and groups:

            N a m e                Size    align       combine class

CODE_SEG . . . . . . . . . . . . . 000A    PARA        NONE    'CODE'
STACK_SEG. . . . . . . . . . . . . 0040    PARA        STACK   'STACK'

Symbols:

            N a m e                Type    Value       Attr

EX519. . . . . . . . . . . . . . . F PROC  0000        CODE_SEG    Length =000A

Warning Severe
Errors Errors
0      0
```

(b)

Figure 5–25 (continued)

Now the second instruction is executed with the command

-T (↵)

Again looking at the trace information, we see that AX now contains the value $FFA1_{16}$ and DX still contains 0000_{16}. This shows that the byte in AL has been extended to a word in AX.

To execute the third instruction, the command is

-T (↵)

```
C:\DOS>DEBUG A:EX519.EXE
-U 0 9
0D03:0000 1E            PUSH    DS
0D03:0001 B80000        MOV     AX,0000
0D03:0004 50            PUSH    AX
0D03:0005 B0A1          MOV     AL,A1
0D03:0007 98            CBW
0D03:0008 99            CWD
0D03:0009 CB            RETF
-G 5

AX=0000  BX=0000  CX=0000  DX=0000  SP=003C  BP=0000  SI=0000  DI=0000
DS=0CF3  ES=0CF3  SS=0D04  CS=0D03  IP=0005   NV UP EI PL NZ NA PO NC
0D03:0005 B0A1          MOV     AL,A1
-T

AX=00A1  BX=0000  CX=0000  DX=0000  SP=003C  BP=0000  SI=0000  DI=0000
DS=0CF3  ES=0CF3  SS=0D04  CS=0D03  IP=0007   NV UP EI PL NZ NA PO NC
0D03:0007 98            CBW
-T

AX=FFA1  BX=0000  CX=0000  DX=0000  SP=003C  BP=0000  SI=0000  DI=0000
DS=0CF3  ES=0CF3  SS=0D04  CS=0D03  IP=0008   NV UP EI PL NZ NA PO NC
0D03:0008 99            CWD
-T

AX=FFA1  BX=0000  CX=0000  DX=FFFF  SP=003C  BP=0000  SI=0000  DI=0000
DS=0CF3  ES=0CF3  SS=0D04  CS=0D03  IP=0009   NV UP EI PL NZ NA PO NC
0D03:0009 CB            RETF
-G

Program terminated normally
-Q

C:\DOS>
```

(c)

Figure 5–25 (continued)

Then, looking at the trace information produced, we find that AX still contains $FFA1_{16}$ and the value in DX has changed to $FFFF_{16}$. This shows that the word in AX has been extended to a double word in DX and AX.

To run the program to completion, enter the command

```
-G          (↵)
```

This executes the remaining instructions, which cause the control to be returned to the DEBUG program.

▲ 5.3 LOGIC INSTRUCTIONS

The 8088 has instructions for performing the logic operations *AND, OR, exclusive-OR,* and *NOT.* Figure 5–26(a) shows the logic symbol for each of these functions and their basic operation is described in the truth table shown in Fig. 5–26(b). Note that AND function $F = A \cdot B$ is logic 1 only if both A and B are logic 1. On the other hand, the OR function $F = A + B$ is logic 1 if either A or B is logic 1.

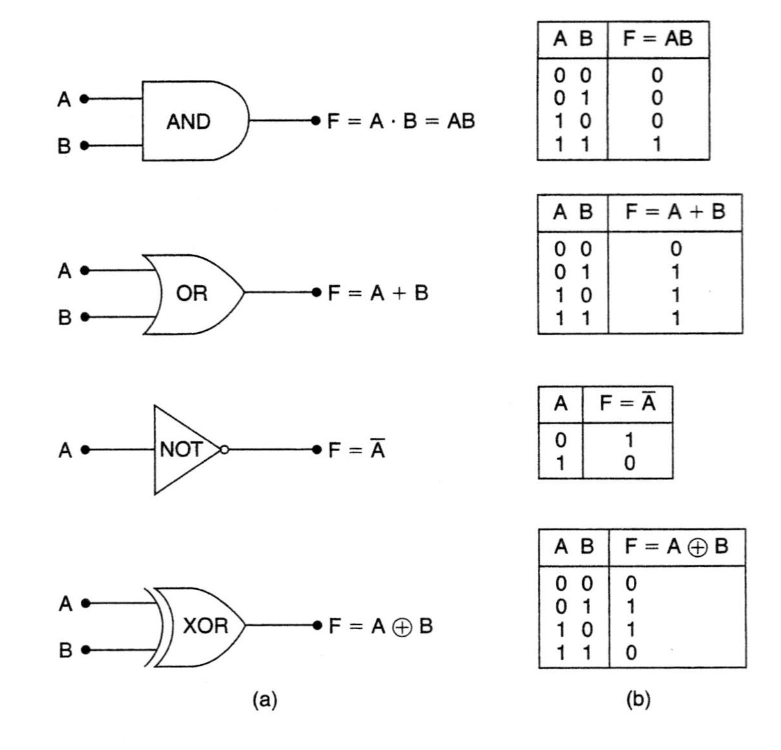

A B	F = AB
0 0	0
0 1	0
1 0	0
1 1	1

A B	F = A + B
0 0	0
0 1	1
1 0	1
1 1	1

A	F = $\overline{A}$
0	1
1	0

A B	F = A ⊕ B
0 0	0
0 1	1
1 0	1
1 1	0

Figure 5–26 (a) Logic symbols. (b) Truth tables of logic operations.

The instructions of the 8088/8086 instruction set perform these basic logic operations bit-wise on byte- and word-wide data. For example, for the byte-wide AND operation

$$A_7A_6A_5A_4A_3A_2A_1A_0 \cdot B_7B_6B_5B_4B_3B_2B_1B_0$$

The corresponding bits are ANDed to give the result.

$$A_7 \cdot B_7\ A_6 \cdot B_6\ A_5 \cdot B_5\ A_4 \cdot B_4\ A_3 \cdot B_3\ A_2 \cdot B_2\ A_1 \cdot B_1\ A_0 \cdot B_0$$

As an example, let us determine the bit-wise OR of the byte-wide values

$$A = 5A_{16} = 01011010_2$$

and

$$B = 0F_{16} = 00001111_2$$

The result is obtained as

$$A + B = 01011010_2 + 00001111_2$$

$$= 0 + 0\ 1 + 0\ 0 + 0\ 1 + 0\ 1 + 1\ 0 + 1\ 1 + 1\ 0 + 1$$

$$= 01011111_{16} = 5F_{16}$$

AND, OR, XOR, and NOT Instructions

As shown in Fig. 5–27(a), the AND, OR, and XOR instructions perform their respective logic operations bit by bit on the specified source and destination operands, the result being represented by the final contents of the destination operand. Figure 5–27(b) shows the allowed operand combinations for the AND, OR, and XOR instructions.

For example, the instruction

```
AND  AX, BX
```

causes the contents of BX to be bit-wise ANDed with the contents of AX. The result is reflected by the new contents of AX. Assuming that AX contains 1234_{16} and BX contains $000F_{16}$, the result produced by the instruction is

$$
\begin{aligned}
1234_{16} \cdot 000F_{16} &= 0001001000110100_2 \cdot 0000000000001111_2 \\
&= 0000000000000100_2 \\
&= 0004_{16}
\end{aligned}
$$

This result is stored in the destination operand and gives

$$(AX) = 0004_{16}$$

Mnemonic	Meaning	Format	Operation	Flags Affected
AND	Logical AND	AND D,S	$(S) \cdot (D) \rightarrow (D)$	OF, SF, ZF, PF, CF AF undefined
OR	Logical Inclusive-OR	OR D,S	$(S) + (D) \rightarrow (D)$	OF, SF, ZF, PF, CF AF undefined
XOR	Logical Exclusive-OR	XOR D,S	$(S) \oplus (D) \rightarrow (D)$	OF, SF, ZF, PF, CF AF undefined
NOT	Logical NOT	NOT D	$(\bar{D}) \rightarrow (D)$	None

(a)

Destination	Source
Register	Register
Register	Memory
Memory	Register
Register	Immediate
Memory	Immediate
Accumulator	Immediate

(b)

Destination
Register
Memory

(c)

Figure 5–27 (a) Logic instructions. (b) Allowed operands for the AND, OR, and XOR instructions. (c) Allowed operands for NOT instruction.

Note that the 12 most significant bits are all zeros. In this way we see how the AND instruction is used to mask the 12 most significant bits of the destination operand.

The NOT logic instruction differs from those for AND, OR, and exclusive-OR in that it operates on a single operand. Looking at Fig. 5–27(c), which shows the allowed operands for the NOT instruction, we see that this operand can be the contents of an internal register or a location in memory.

High-level languages, such as C, allow programmers to write statements that perform these bit-wise logic operations on variables. The C compiler implements the operation for these statements by applying the logic instructions of the processor's instruction set.

EXAMPLE 5.20

Describe the result of executing the following sequence of instructions:

```
MOV   AL, 01010101B
AND   AL, 00011111B
OR    AL, 11000000B
XOR   AL, 00001111B
NOT   AL
```

Here, B is used to specify a binary number.

Solution

The first instruction moves the immediate operand 01010101_2 into the AL register. This loads the data that are to be manipulated with the logic instructions. The next instruction performs a bit-by-bit AND operation of the contents of AL with immediate operand 00011111_2. This gives

$$01010101_2 \cdot 00011111_2 = 00010101_2$$

This result is placed in destination register AL:

$$(AL) = 00010101_2 = 15_{16}$$

Note that this operation has masked off the three most significant bits of AL.

The third instruction performs a bit-by-bit logical OR of the present contents of AL with immediate operand $C0_{16}$. This gives

$$00010101_2 + 11000000_2 = 11010101_2$$

$$(AL) = 11010101_2 = D5_{16}$$

This operation is equivalent to setting the two most significant bits of AL.

Instruction	(AL)
MOV AL,01010101B	01010101
AND AL,00011111B	00010101
OR AL,11000000B	11010101
XOR AL,00001111B	11011010
NOT AL	00100101

Figure 5–28 Execution results of program in Example 5.20.

The fourth instruction is an exclusive-OR operation of the contents of AL with immediate operand 00001111_2. We get

$$11010101_2 \oplus 00001111_2 = 11011010_2$$
$$(AL) = 11011010_2 = DA_{16}$$

Note that this operation complements the logic state of those bits in AL that are 1s in the immediate operand.

The last instruction, NOT AL, inverts each bit of AL. Therefore, the final contents of AL become

$$(AL) = \overline{11011010_2} = 00100101_2 = 25_{16}$$

Figure 5–28 summarizes these results.

EXAMPLE 5.21

Use the DEBUG program to verify the operation of the program in Example 5.20.

Solution

After the DEBUG program is brought up, the line-by-line assembler is used to enter the program, as shown in Fig. 5–29. The first instruction is executed by issuing the T command

```
-T    (↵)
```

The trace dump given in Fig. 5–29 shows that the value 55_{16} has been loaded into the AL register.

The second instruction is executed by issuing another T command:

```
-T    (↵)
```

Execution of this instruction causes $1F_{16}$ to be ANDed with the value 55_{16} in AL. Looking at the trace information displayed in Fig. 5–29, we see that the three most significant bits of AL have been masked off to produce the result 15_{16}.

The third instruction is executed in the same way:

```
-T    (↵)
```

```
C:\DOS>DEBUG
-A
1342:0100 MOV AL,55
1342:0102 AND AL,1F
1342:0104 OR AL,C0
1342:0106 XOR AL,0F
1342:0108 NOT AL
1342:010A
-T

AX=0055  BX=0000  CX=0000  DX=0000  SP=FFEE  BP=0000  SI=0000  DI=0000
DS=1342  ES=1342  SS=1342  CS=1342  IP=0102   NV UP EI PL NZ NA PO NC
1342:0102 241F           AND     AL,1F
-T

AX=0015  BX=0000  CX=0000  DX=0000  SP=FFEE  BP=0000  SI=0000  DI=0000
DS=1342  ES=1342  SS=1342  CS=1342  IP=0104   NV UP EI PL NZ NA PO NC
1342:0104 0CC0           OR      AL,C0
-T

AX=00D5  BX=0000  CX=0000  DX=0000  SP=FFEE  BP=0000  SI=0000  DI=0000
DS=1342  ES=1342  SS=1342  CS=1342  IP=0106   NV UP EI NG NZ NA PO NC
1342:0106 340F           XOR     AL,0F
-T

AX=00DA  BX=0000  CX=0000  DX=0000  SP=FFEE  BP=0000  SI=0000  DI=0000
DS=1342  ES=1342  SS=1342  CS=1342  IP=0108   NV UP EI NG NZ NA PO NC
1342:0108 F6D0           NOT     AL
-T

AX=0025  BX=0000  CX=0000  DX=0000  SP=FFEE  BP=0000  SI=0000  DI=0000
DS=1342  ES=1342  SS=1342  CS=1342  IP=010A   NV UP EI NG NZ NA PO NC
1342:010A 2B04           SUB     AX,[SI]                    DS:0000=20CD
-Q

C:\DOS>
```

Figure 5–29 Display sequence for Example 5.21.

It causes the value $C0_{16}$ to be ORed with the value $1F_{16}$ in AL. This gives the result $D5_{16}$ in AL.

A fourth T command is used to execute the XOR instruction,

-T (↵)

and the trace dump that results shows that the new value in AL is DA_{16}.

The last instruction is a NOT instruction and its execution with the command

-T (↵)

causes the bits of DA_{16} to be inverted. This produces 25_{16} as the final result in AL.

Clearing, Setting, and Toggling Bits of an Operand

A common use of logic instructions is to mask a group of bits of a byte or word of data. By *mask,* we mean to clear the bit or bits to zero. Remember that when a bit is ANDed with another bit that is at logic 0, the result is always 0. On the other hand, if a bit is ANDed with a bit that is at logic 1, its value remains unchanged. Thus we see that the bits that are to be masked must be set to 0 in the mask, which is the source operand, and those that are to remain unchanged are set to 1. For instance, in the instruction

```
AND  AX, 000FH
```

the mask equals $000F_{16}$; therefore, it would mask off the upper 12 bits of the word of data in destination AX. Let us assume that the original value in AX is $FFFF_{16}$. Then executing the instruction performs the operation

$$0000000000001111_2 \cdot 1111111111111111_2 = 0000000000001111_2$$

$$(AX) = 000F_{16}$$

This shows that just the lower 4 bits in AX remain intact.

The OR instruction can be used to set a bit or bits in a register or a storage location in memory to logic 1. If a bit is ORed with another bit that is 0, the value of the bit remains unchanged; however, if it is ORed with another bit that is 1, the bit becomes 1. For instance, let us assume that we want to set bit B_4 of the byte at the offset address CONTROL_FLAGS in the current data segment of memory to logic 1. This can be done with the following instruction sequence:

```
MOV  AL, [CONTROL_FLAGS]
OR   AL, 10H
MOV  [CONTROL_FLAGS], AL
```

First the value of the flags are copied into AL and the logic operation

$$(AL) = XXXXXXXX_2 + 00010000_2 = XXX1XXXX_2$$

is performed. Finally, the new byte in AL, which has bit B_4 set to 1, is written back to the memory location called CONTROL_FLAGS.

The XOR instruction can be used to reverse the logic level of a bit or bits in a register or storage location in memory. This operation is referred to as "toggling the bit."

▲ 5.4 SHIFT INSTRUCTIONS

The four shift instructions of the 8088 can perform two basic types of shift operations; the *logical shift* and the *arithmetic shift.* Moreover, each of these operations can be performed to the right or to the left. The shift instructions are *shift logical left* (SHL), *shift arithmetic left* (SAL), *shift logical right* (SHR), and *shift arithmetic right* (SAR). These instructions are used to align data, to isolate bits of a byte or word so that it can be tested, and to perform simple multiply and divide computations.

SHL, SHR, SAL, and SAR Instructions

The operation of the logical shift instructions, SHL and SHR, is described in Fig. 5–30(a). Note in Fig. 5–30(b) that the destination operand, the data whose bits are to be shifted, can be either the contents of an internal register or a storage location in memory. Moreover, the source operand can be specified in two ways. If it is assigned the value of 1, a 1-bit shift will take place. For instance, as illustrated in Fig. 5–31(a), executing

```
SHL AX, 1
```

causes the 16-bit contents of the AX register to be shifted 1 bit position to the left. Here we see that the vacated LSB location is filled with zero and the bit shifted out of the MSB is saved in CF.

On the other hand, if the source operand is specified as CL instead of 1, the count in this register represents the number of bit positions the contents of the operand are to be shifted. This permits the count to be defined under software control and allows a range of shifts from 1 to 255 bits.

An example of an instruction specified in this way is

```
SHR AX, CL
```

Mnemonic	Meaning	Format	Operation	Flags Affected
SAL/SHL	Shift arithmetic left/shift logical left	SAL/SHL D,Count	Shift the (D) left by the number of bit positions equal to Count and fill the vacated bits positions on the right with zeros	CF, PF, SF, ZF AF undefined OF undefined if count ≠ 1
SHR	Shift logical right	SHR D,Count	Shift the (D) right by the number of bit positions equal to Count and fill the vacated bit positions on the left with zeros	CF, PF, SF, ZF AF undefined OF undefined if count ≠ 1
SAR	Shift arithmetic right	SAR D,Count	Shift the (D) right by the number of bit positions equal to Count and fill the vacated bit positions on the left with the original most significant bit	SF, ZF, PF, CF AF undefined OF undefined if count ≠1

(a)

Destination	Count
Register	1
Register	CL
Memory	1
Memory	CL

(b)

Figure 5–30 (a) Shift instructions. (b) Allowed operands.

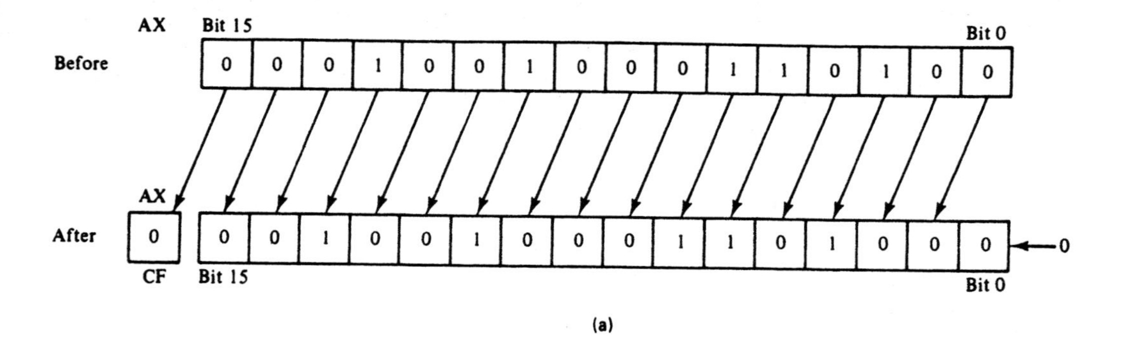

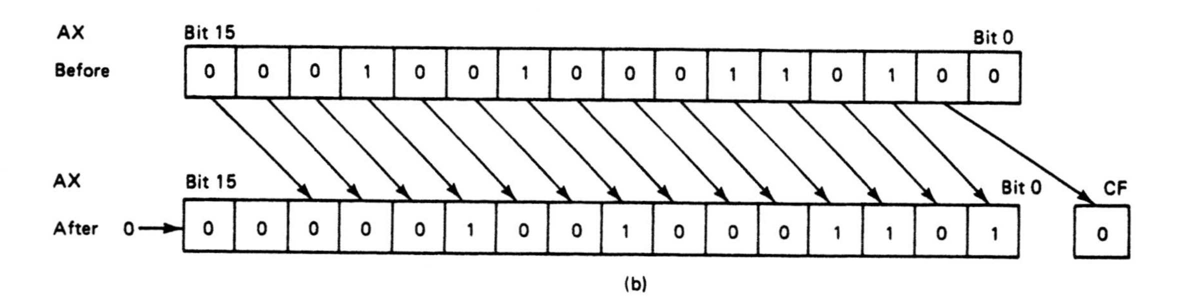

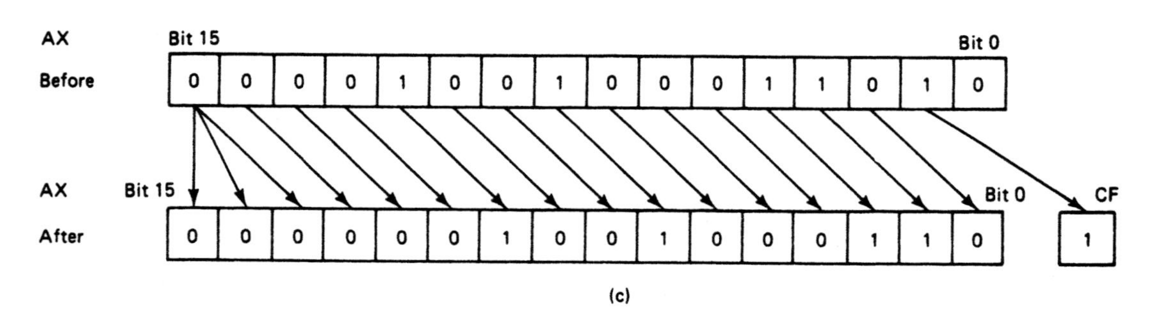

Figure 5–31 (a) Results of executing SHL AX,1. (b) Results of executing SHR AX, CL. (CL) = 02. (c) Results of executing SAR AX, CL. (CL) = 02.

Assuming that CL contains the value 02_{16}, the logical shift right that occurs is as shown in Fig. 5–31(b). Note that the two MSBs have been filled with zeros and the last bit shifted out at the LSB, which is zero, is placed in the carry flag.

In an arithmetic shift to the left, the SAL operation, the vacated bits at the right of the operand are filled with zeros, whereas in an arithmetic shift to the right, the SAR operation, the vacated bits at the left are filled with the value of the original MSB of the operand. Thus, in an arithmetic shift to the right, the original sign of the number is maintained. This operation is equivalent to division by powers of 2 as long as the bits shifted out of the LSB are zeros.

EXAMPLE 5.22

Assume that CL contains 02_{16} and AX contains $091A_{16}$. Determine the new contents of AX and the carry flag after the instruction

```
SAR AX, CL
```

is executed.

Solution

Figure 5–31(c) shows the effect of executing the instruction. Here we see that since CL contains 02_{16}, a shift right by two bit locations takes place, and the original sign bit, which is logic 0, is extended to the two vacated bit positions. Moreover, the last bit shifted out from the LSB location is placed in CF. This makes CF equal to 1. Therefore, the results produced by execution of the instruction are

$$(AX) = 0246_{16}$$

and

$$(CF) = 1_2$$

EXAMPLE 5.23

Verify the operation of the SAR instruction in Example 5.22 by executing with the DEBUG program.

Solution

After invoking the DEBUG program, we enter the instruction by assembling it with the command

```
-A                        (↵)
1342:0100 SAR AX, CL      (↵)
1342:0102                 (↵)
-
```

Next, registers AX and CL are loaded with data, and the carry flag is reset. This is done with the command sequence

```
-R AX       (↵)
AX 0000
:091A       (↵)
-R CX       (↵)
CX 0000
:2          (↵)
-R F        (↵)
NV UP EI PL NZ NA PO NC - (↵)
-
```

```
C:\DOS>DEBUG
-A
1342:0100 SAR AX,CL
1342:0102
-R AX
AX 0000
:091A
-R CX
CX 0000
:2
-R F
NV UP EI PL NZ NA PO NC  -
-T

AX=0246  BX=0000  CX=0002  DX=0000  SP=FFEE  BP=0000  SI=0000  DI=0000
DS=1342  ES=1342  SS=1342  CS=1342  IP=0102   NV UP EI PL NZ AC PO CY
1342:0102 B98AFF        MOV     CX,FF8A
-Q

C:\DOS>
```

Figure 5–32 Display sequence for Example 5.23.

Note that the carry flag was already clear, so no status entry was made.

Now the instruction is executed with the T command

```
-T    (↵)
```

Note in Fig. 5–32 that the value in AX has become 0246_{16} and a carry (CY) has occurred. These results are identical to those obtained in Example 5.22.

Isolating the Value of a Bit in an Operand

A frequent need in programming is to isolate the value of one of the bits of a word or byte of data by shifting it into the carry flag. The shift instructions may perform this operation on data either in a register or a storage location in memory. The instructions that follow perform this type of operation on a byte of data stored in memory at address CONTROL_FLAGS:

```
MOV  AL, [CONTROL_FLAGS]
MOV  CL, 04H
SHR  AL, CL
```

The first instruction reads the value of the byte of data at address CONTROL_FLAGS into AL. Next, a shift count of four is loaded into CL, and then the value in AL is shifted to the right four bit positions. Since the MSBs of AL are reloaded with zeros as part of the shift operation, the results are

$$(AL) = 0000B_7B_6B_5B_4$$

and

$$(CF) = B_3$$

In this way, we see that bit B_3 of CONTROL_FLAGS has been isolated by placing it in CF. Once this bit is in CF, it can be tested by other instructions and based on this value initiate another software operation.

▲ 5.5 ROTATE INSTRUCTIONS

Another group of instructions, the rotate instructions, are similar to the shift instructions we just introduced. This group, shown in Fig. 5–33(a), includes the *rotate left* (ROL), *rotate right* (ROR), *rotate left through carry* (RCL), and *rotate right through carry* (RCR) instructions. They perform many of the same programming functions as the shift instructions, such as alignment of data and isolation of a bit of an element of data.

ROL, ROR, RCL, and RCR Instructions

As Fig. 5–33(b) shows, the rotate instructions are similar to the shift instructions in several ways. They have the ability to rotate the contents of either an internal register or a storage location in memory. Also, the rotation that takes place can be from 1 to 255 bit positions to the left or to the right. Moreover, in the case of a multibit rotate, the number of bit positions to be rotated is specified by the value in CL. Their difference from the shift instructions lies in the fact that the bits moved out at either the MSB or LSB end are not lost; instead, they are reloaded at the other end.

Mnemonic	Meaning	Format	Operation	Flags Affected
ROL	Rotate left	ROL D,Count	Rotate the (D) left by the number of bit positions equal to Count. Each bit shifted out from the leftmost bit goes back into the rightmost bit position.	CF OF undefined if count ≠ 1
ROR	Rotate right	ROR D,Count	Rotate the (D) right by the number of bit positions equal to Count. Each bit shifted out from the rightmost bit goes into the leftmost bit position.	CF OF undefined if count ≠ 1
RCL	Rotate left through carry	RCL D,Count	Same as ROL except carry is attached to (D) for rotation.	CF OF undefined if count ≠ 1
RCR	Rotate right through carry	RCR D,Count	Same as ROR except carry is attached to (D) for rotation.	CF OF undefined if count ≠ 1

Destination	Count
Register	1
Register	CL
Memory	1
Memory	CL

(b)

Figure 5–33 (a) Rotate instructions. (b) Allowed operands.

As an example, let us look at the operation of the ROL instruction. Execution of ROL causes the contents of the selected operand to be rotated left the specified number of bit positions. Each bit shifted out at the MSB end is reloaded at the LSB end. Moreover, the content of CF reflects the state of the last bit that was shifted out. For instance, the instruction

```
ROL AX, 1
```

causes a 1-bit rotate to the left. Figure 5–34(a) shows the result produced by executing this instruction. Note that the original value of bit 15 is zero. This value has been rotated into both CF and bit 0 of AX. All other bits have been rotated one bit position to the left.

The ROR instruction operates the same way as ROL except that it causes data to be rotated to the right instead of to the left. For example, execution of

```
ROR AX, CL
```

causes the contents of AX to be rotated right by the number of bit positions specified in CL. Figure 5–34(b) illustrates the result for CL equal to four.

The other two rotate instructions, RCL and RCR, differ from ROL and ROR in that the bits are rotated through the carry flag. Figure 5–35 illustrates the rotation that takes place due to execution of the RCL instruction. Note that the value returned to bit 0 is the

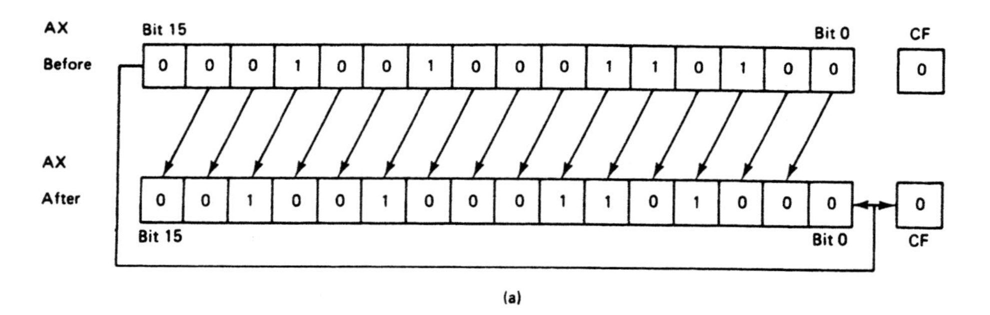

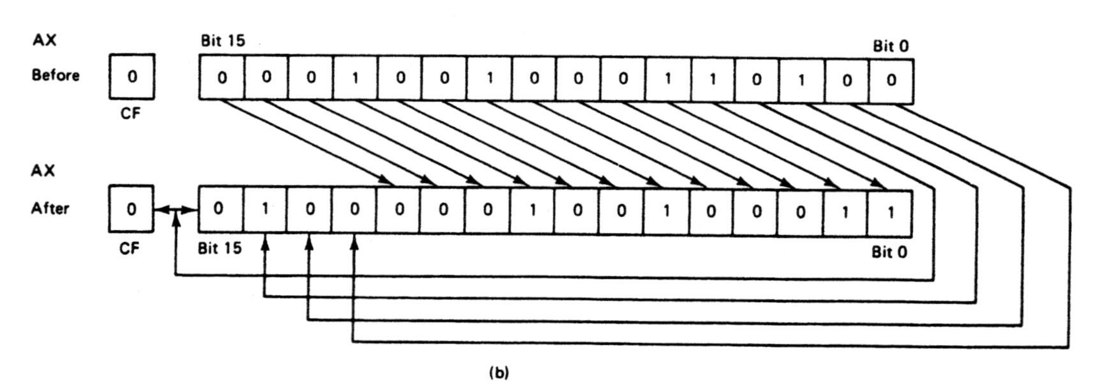

Figure 5–34 Results of executing ROL AX, 1. (b) Results of executing ROR AX, CL with (CL) = 4.

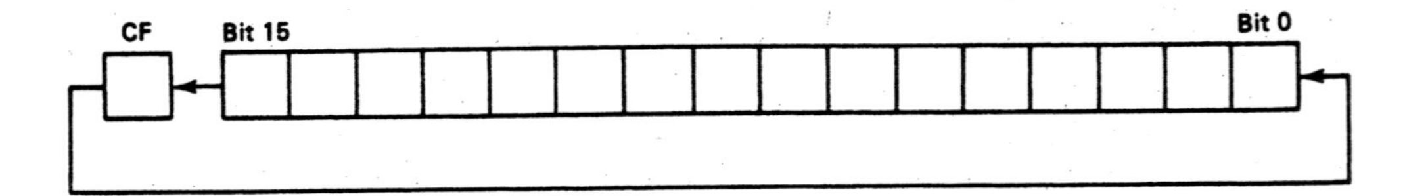

Figure 5–35 Rotation caused by execution of the RCL instruction.

prior content of CF and not bit 15. The value shifted out of bit 15 goes into the carry flag. Thus, the bits rotate through carry.

EXAMPLE 5.24

What is the result in BX and CF after execution of the following instruction?

```
RCR BX, CL
```

Assume that, prior to execution of the instruction, $(CL) = 04_{16}$, $(BX) = 1234_{16}$, and $(CF) = 0$.

Solution

The original contents of BX are

$$(BX) = 0001001000110100_2 = 1234_{16}$$

Execution of the RCR instruction causes a 4-bit rotate right through carry to take place on the data in BX. The resulting contents of BX and CF are

$$(BX) = 1000000100100011_2 = 8123_{16}$$

$$(CF) = 0_2$$

In this way, we see that the original content of bit 3, which was zero, resides in the carry flag, and 1000_2 has been reloaded from the bit-15 end of BX.

EXAMPLE 5.25

Use the DEBUG program to verify the operation of the RCR instruction in Example 5.24.

Solution

After loading DEBUG, the instruction is assembled into memory with the command

```
-A                       (↵)
1342:0100 RCR BX, CL     (↵)
1342:0102                (↵)
-
```

```
C:\DOS>DEBUG
-A
1342:0100 RCR BX,CL
1342:0102
-R BX
BX 0000
:1234
-R CX
CX 0000
:4
-R F
NV UP EI PL NZ NA PO NC -
-T

AX=0000  BX=8123  CX=0004  DX=0000  SP=FFEE  BP=0000  SI=0000  DI=0000
DS=1342  ES=1342  SS=1342  CS=1342  IP=0102   OV UP EI PL NZ NA PO NC
1342:0102 B98AFF        MOV     CX,FF8A
-Q

C:\DOS>
```

Figure 5–36 Display sequence for Example 5.25.

Next, the commands that follow load BX and CX with the initial data and clear CF:

```
-R BX          (↵)
BX 0000
:1234          (↵)
-R CX          (↵)
CX 0000
:4             (↵)
-R F           (↵)
NV UP EI PL NZ NA PO NC -     (↵)
-
```

Note that CF is already cleared (NC); therefore, no entry is made for the flag REGISTER command.

Now we can execute the instruction with the command

```
-T             (↵)
```

Looking at the trace information displayed in Fig. 5–36, we see that the new contents of BX are 8123_{16} and CF equals NC. These are the same results as obtained in Example 5.24.

Alignment of Data in Operands

An example of a software operation that can be performed with the rotate instructions is the disassembly of the two hexadecimal digits in a byte of data in memory so that they can be added. The instructions in Fig. 5–37 perform this operation. First, the byte containing the two hexadecimal digits is read into AL. Then, a copy is made in BL. Next, the four most significant bits in BL are moved to the four least significant bit locations with a rotate operation. This repositions the most significant hexadecimal digit of the

```
MOV AL,[HEX_DIGITS]
MOV BL,AL
MOV CL,04H
ROR BL,CL
AND AL,0FH
AND BL,0FH
ADD AL,BL
```

Figure 5–37 Program for disassembly and addition of two hexadecimal digits stored as a byte in memory.

original byte into the least significant digit position in BL. Now, the most significant hexadecimal digits in both AL and BL are masked off. This isolates one hexadecimal digit in the lower four bits of AL and the other in the lower four bits of BL. Finally, the two digits are added together in AL.

REVIEW PROBLEMS

Section 5.1

1. Explain what operation is performed by each of the instructions that follow.

(a) `MOV AX, 0110H`
(b) `MOV DI, AX`
(c) `MOV BL, AL`
(d) `MOV [0100H], AX`
(e) `MOV [BX+DI], AX`
(f) `MOV [DI]+4, AX`
(g) `MOV [BX][DI] +4, AX`

2. Assume that registers AX, BX, and DI are all initialized to 0000_{16} and that all the affected storage locations in data memory have been cleared. Determine the location and value of the destination operand as instructions (a) through (g) in problem 1 are executed as a sequence.

3. Write an instruction sequence that will initialize the ES register with the immediate value 1010_{16}.

4. Write an instruction that saves the contents of the ES register in memory at address DS:1000H.

5. Why does the instruction MOV CL, AX result in an error when it is assembled?

6. Describe the operation performed by each of the instructions that follow.

(a) `XCHG AX, BX`
(b) `XCHG BX, DI`
(c) `XCHG [DATA], AX`
(d) `XCHG [BX+DI], AX`

7. If register BX contains the value 0100_{16}, register DI contains 0010_{16}, and register DS contains 1075_{16}, what physical memory location is swapped with AX when the instruction in problem 6(d) is executed?

8. Assuming that $(AX) = 0010_{16}$, $(BX) = 0100_{16}$, and $(DS) = 1000_{16}$, what happens if the XLAT instruction is executed?

9. Write a single instruction that loads AX from address 0200_{16} and DS from address 0202_{16}.

Section 5.2

10. Find the binary sum in each of the following problems.
(a) 00010101 + 00011000 = ? (b) 01001111 + 11011100 = ?

11. Add the binary number 11000000 to 11011000. Convert the binary answer to hexadecimal form. Decimal form.

12. Perform the binary subtractions that follow.
(a) 00011000 − 00010101 = ? (b) 11011100 − 01001111 = ?

13. Use the 2's-complement method to subtract the binary numbers that follow.
(a) 00010000 − 00000111 = ? (b) 01111000 − 00001111 = ?

14. Using the 2's-complement method, subtract B = 00010001 from A = 00100000. Convert the answer to hexadecimal form. Decimal form.

15. What operation is performed by each of the following instructions?

(a) `ADD AX, 00FFH`
(b) `ADC SI, AX`
(c) `INC BYTE PTR [0100H]`
(d) `SUB DL, BL`
(e) `SBB DL, [0200H]`
(f) `DEC BYTE PTR [DI+BX]`
(g) `NEG BYTE PTR [DI]+0010H`
(h) `MUL DX`
(i) `IMUL BYTE PTR [BX+SI]`
(j) `DIV BYTE PTR [SI]+0030H`
(k) `IDIV BYTE PTR [BX][SI]+0030H`

16. Assume that the state of the 8088's registers and memory just prior to the execution of each instruction in problem 15 is as follows:

(AX) = 0010H
(BX) = 0020H
(CX) = 0030H
(DX) = 0040H
(SI) = 0100H
(DI) = 0200H
(CF) = 1
(DS:100H) = 10H
(DS:101H) = 00H
(DS:120H) = FFH
(DS:121H) = FFH
(DS:130H) = 08H
(DS:131H) = 00H

(DS:150H) = 02H
(DS:151H) = 00H
(DS:200H) = 30H
(DS:201H) = 00H
(DS:210H) = 40H
(DS:211H) = 00H
(DS:220H) = 30H
(DS:221H) = 00H

What result is produced in the destination operand by executing instructions (a) through (k)?

17. Write an instruction that will add the immediate value $111F_{16}$ and the carry flag to the contents of the data register DX.

18. Write an instruction that will subtract the word contents of the storage location pointed to by the base register BX and the carry flag from the accumulator.

19. Write instructions that show two different ways of incrementing the address pointer in SI by two.

20. Assuming that (AX) = 0123_{16} and (BL) = 10_{16}, what will be the new contents of AX after executing the instruction DIV BL?

21. What instruction is used to adjust the result of an addition that processed packed BCD numbers?

22. Which instruction is provided in the instruction set of the 8088 to adjust the result of a subtraction that involved ASCII-coded numbers?

23. If AL contains $A0_{16}$, what happens when the instruction CBW is executed?

24. If the value in AX is $7FFF_{16}$, what happens when the instruction CWD is executed?

25. Two byte-sized BCD integers are stored at the symbolic offset addresses NUM1 and NUM2, respectively. Write an instruction sequence to generate their difference and store it at NUM3. The difference is to be formed by subtracting the value at NUM1 from that at NUM2. Assume that all storage locations are in the current data segment.

Section 5.3

26. Perform the bit-wise logical AND operation for the binary numbers that follow:
(a) 00010101 · 00011000 = ? **(b)** 01001111 · 11011100 = ?

27. Perform the bit-wise logical OR operation for the binary numbers that follow.
(a) 00011000 + 00010101 = ? **(b)** 11011100 + 01001111 = ?

28. Perform a bit-wise logical not operation on the bits of the hexadecimal number AAAAH. Express the answer in both binary and hexadecimal notation.

29. Combine the binary numbers 11000000 and 11011000 with a bit-wise exclusive-OR operation. Convert the binary answer to hexadecimal form.

30. Describe the operation performed by each of the following instructions.

(a) `AND  BYTE PTR [0300H],  0FH`
(b) `AND  DX,  [SI]`

```
(c) OR    [BX+DI], AX
(d) OR    BYTE PTR [BX][DI]+10H, 0F0H
(e) XOR   AX, [SI+BX]
(f) NOT   BYTE PTR [0300H]
(g) NOT   WORD PTR [BX+DI]
```

31. Assume that the state of the 8088's registers and memory just prior to execution of each instruction in problem 30 is as follows:

(AX) = 5555H
(BX) = 0010H
(CX) = 0010H
(DX) = AAAAH
(SI) = 0100H
(DI) = 0200H
(DS:100H) = 0FH
(DS:101H) = F0H
(DS:110H) = 00H
(DS:111H) = FFH
(DS:200H) = 30H
(DS:201H) = 00H
(DS:210H) = AAH
(DS:211H) = AAH
(DS:220H) = 55H
(DS:221H) = 55H
(DS:300H) = AAH
(DS:301H) = 55H

What are the results produced in the destination operands after executing instructions (a) through (g)?

32. Write an instruction that when executed will mask off all but bit 7 of the contents of the data register.

33. Write an instruction that will mask off all but bit 7 of the word of data stored at address DS:0100H.

34. Specify the relation between the old and new contents of AX after executing the following sequence of instructions.

```
NOT  AX
ADD  AX, 1
```

35. Write an instruction that will toggle the logic level of the most significant bit of the value in the upper byte of the accumulator register.

36. Write an instruction sequence that will read the byte of control flags from the storage location at offset address CONTROL_FLAGS in the current data segment into register AL, mask off all but the most significant and least significant flag bits, and then save the result back in the original storage location.

37. Describe the operation that is performed by the following instruction sequence.

```
MOV  BL, [CONTROL_FLAGS]
AND  BL, 08H
XOR  BL, 08H
MOV  [CONTROL_FLAGS], BL
```

Section 5.4

38. Explain the operation performed by each of the following instructions.

(a) `SHL  DX,CL`
(b) `SHL  BYTE PTR [0400H], CL`
(c) `SHR  BYTE PTR [DI], 1`
(d) `SHR  BYTE PTR [DI+BX], CL`
(e) `SAR  WORD PTR [BX+DI], 1`
(f) `SAR  WORD PTR [BX][DI]+10H, CL`

39. Assume that the state of 8088's registers and memory just prior to execution of each instruction in problem 38 is as follows:

(AX) = 0000H	(DS:200H) = 22H
(BX) = 0010H	(DS:201H) = 44H
(CX) = 0105H	(DS:210H) = 55H
(DX) = 1111H	(DS:211H) = AAH
(SI) = 0100H	(DS:220H) = AAH
(DI) = 0200H	(DS:221H) = 55H
(CF) = 0	(DS:400H) = AAH
(DS:100H) = 0FH	(DS:401H) = 55H

What results are produced in the destination operands by executing instructions (a) through (f)?

40. Write an instruction that shifts the contents of the count register left by one bit position.

41. Write an instruction sequence that, when executed, shifts left by eight bit positions the contents of the word-wide memory location pointed to by the address in the destination index register.

42. Identify the condition under which the contents of AX would remain unchanged after execution of the instructions that follow.

```
MOV  CL, 4
SHL  AX, CL
SHR  AX, CL
```

43. If the original contents of AX, CL, and CF are 800FH, 04H, and 1, respectively, what is the content of AX and CF after executing the instruction that follows.

```
SAR  AX, CL
```

44. Describe the operation performed by the instruction sequence that follows.

```
MOV   AL, [CONTROL_FLAGS]
AND   AL, 80H
SHL   AL, 1
```

What is the result in AL after the shift is complete?

45. Write a program that will read the word of data from the offset address ASCII_DATA in the current data segment of memory. Assume that this word storage location contains two ASCII-coded characters, one character in the upper byte and the other in the lower byte. Disassemble the two bytes and save them as separate characters in the lower byte location of the word storage locations with offsets ASCII_CHAR_L and ASCII_CHAR_H in the current data segment. The upper eight bits in each of these character storage locations should be made zero. Use a SHR instruction to relocate the most significant bits.

Section 5.5

46. Describe what happens as each of the instructions that follows is executed by the 8088.

(a) `ROL  DX, CL`
(b) `RCL  BYTE PTR [0400H], CL`
(c) `ROR  BYTE PTR [DI], 1`
(d) `ROR  BYTE PTR [DI+BX], CL`
(e) `RCR  WORD PTR [BX+DI], 1`
(f) `RCR  WORD PTR [BX][DI]+10H, CL`

47. Assume that the state of the 8088's registers and memory just prior to execution of each of the instructions in problem 46 is as follows:

(AX) = 0000H
(BX) = 0010H
(CX) = 0105H
(DX) = 1111H
(SI) = 0100H
(DI) = 0200H
(CF) = 1
(DS:100H) = 0FH
(DS:200H) = 22H
(DS:201H) = 44H
(DS:210H) = 55H
(DS:211H) = AAH

$$(DS:220H) = AAH$$
$$(DS:221H) = 55H$$
$$(DS:400H) = AAH$$
$$(DS:401H) = 55H$$

What results are produced in the destination operands by executing instructions (a) through (f)?

48. Write an instruction sequence that, when executed, rotates left through carry by one bit position the contents of the word-wide memory location pointed to by the address in the base register.

49. Write a program that saves bit 5 of AL in BX as a word.

50. Repeat problem 45, but this time use a ROR instruction to perform the bit-shifting operation.

ADVANCED PROBLEMS

51. Two code-conversion tables starting with offsets TABL1 and TABL2 in the current data segment are to be accessed. Write an instruction sequence that initializes the needed registers and then replaces the contents of memory locations MEM1 and MEM2 (offsets in the current data segment) by the equivalent converted codes from the respective code-conversion tables.

52. Two word-wide unsigned integers are stored at the physical memory addresses $00A00_{16}$ and $00A02_{16}$, respectively. Write an instruction sequence that computes and stores their sum, difference, product, and quotient. Store these results at consecutive memory locations starting at physical address $00A10_{16}$ in memory. To obtain the difference, subtract the integer at $00A02_{16}$ from the integer at $00A00_{16}$. For the division, divide the integer at $00A00_{16}$ by the integer at $00A02_{16}$. Use the register indirect relative addressing mode to store the various results.

53. Write an instruction sequence that generates a byte-size integer in the memory location defined as RESULT. The value of the integer is to be calculated from the logic equation

$$(RESULT) = (AL) \cdot (NUM1) + \overline{(NUM2)} \cdot (AL) + (BL)$$

Assume that all parameters are byte-sized. NUM1, NUM2, and RESULT are the offset addresses of memory locations in the current data segment.

54. Implement the following operation using shift and arithmetic instructions.

$$7(AX) - 5(BX) - (BX)/8 \rightarrow (AX)$$

Assume that all parameters are word-sized. State any assumptions made in the calculations.

6

8088/8086 Programming—Control Flow Instructions and Program Structures

▲ INTRODUCTION

In Chapter 5 we discussed many of the instructions that can be executed by the 8088 and 8086 microprocessors. That chapter focused on instructions that performed integer computations and demonstrated their use with simple straight-line programs. In this chapter, we introduce the rest of the instruction set and at the same time cover some more complicated programming techniques. The instructions introduced here enable program control flow—that is, the ability to alter the sequence in which instructions of a program execute. The following topics are discussed in this chapter:

6.1 Flag-Control Instructions
6.2 Compare Instruction
6.3 Control Flow and the Jump Instructions
6.4 Subroutines and Subroutine-Handling Instructions
6.5 Loops and Loop-Handling Instructions
6.6 Strings and String-Handling Instructions

▲ 6.1 FLAG-CONTROL INSTRUCTIONS

The 8088 microprocessor has a set of flags that either monitors the state of executing instructions or controls options available in its operation. These flags were described in detail in Chapter 2. The instruction set includes a group of instructions that, when exe-

Mnemonic	Meaning	Operation	Flags Affected
LAHF	Load AH from flags	(AH) ← (Flags)	None
SAHF	Store AH into flags	(Flags) ← (AH)	SF, ZF, AF, PF, CF
CLC	Clear carry flag	(CF) ← 0	CF
STC	Set carry flag	(CF) ← 1	CF
CMC	Complement carry flag	(CF) ← $(\overline{CF})$	CF
CLI	Clear interrupt flag	(IF) ← 0	IF
STI	Set interrupt flag	(IF) ← 1	IF

(a)

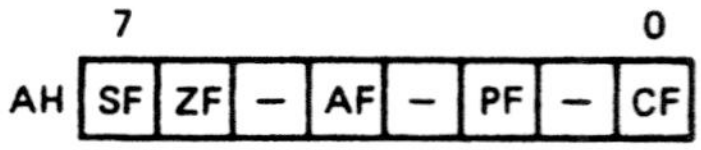

	7							0
AH	SF	ZF	–	AF	–	PF	–	CF

SF = Sign flag
ZF = Zero flag
AF = Auxiliary
PF = Parity flag
CF = Carry flag
– = Undefined (do not use)

(b)

Figure 6–1 (a) Flag-control instructions. (b) Format of the flags in AH register for the LAHF and SAHF instructions.

cuted, directly affect the state of the flags. These instructions, shown in Fig. 6–1(a), are *load AH from flags* (LAHF), *store AH into flags* (SAHF), *clear carry* (CLC), *set carry* (STC), *complement carry* (CMC), *clear interrupt* (CLI), and *set interrupt* (STI). A few more instructions exist that can directly affect the flags; however, we will not cover them until later in the chapter when we introduce the subroutine and string instructions.

Figure 6–1(a) shows that the first two instructions, LAHF and SAHF, can be used either to read the flags or to change them, respectively. Notice that the data transfer that takes place is always between the AH register and the flag register. Figure 6–1(b) shows the format of the flag information in AH. Notice that bits 1, 3, and 5 are undefined. For instance, we may want to start an operation with certain flags set or reset. Assume that we want to preset all flags to logic 1. To do this, we can first load AH with FF_{16} and then execute the SAHF instruction.

EXAMPLE 6.1

Write an instruction sequence to save the current contents of the 8088's flags in the memory location at offset MEM1 of the current data segment and then reload the flags with the contents of the storage location at offset MEM2.

Solution

To save the current flags, we must first load them into the AH register and then move them to the location MEM1. The instructions that do this are

```
LAHF
MOV  [MEM1], AH
```

```
LAHF
MOV   [MEM1],AH
MOV   AH,[MEM2]
SAHF
```

Figure 6–2 Instruction sequence for saving the contents of the flag register in a memory and loading it from another memory.

Similarly, to load the flags with the contents of MEM2, we must first copy the contents of MEM2 into AH and then store the contents of AH into the flags. The instructions for this are

```
MOV  AH, [MEM2]
SAHF
```

The entire instruction sequence is shown in Fig. 6–2.

EXAMPLE 6.2

Use the DEBUG program to enter the instruction sequence in Example 6.1 starting at memory address 00110_{16}. Assign memory addresses 00150_{16} and 00151_{16} to symbols MEM1 and MEM2, respectively. Then initialize the contents of MEMl and MEM2 to FF_{16} and 01_{16}, respectively. Verify the operation of the instructions by executing them one after the other with the TRACE command.

Solution

As Fig. 6–3 shows, the DEBUG program is called up with the DOS command

```
C:\DOS>DEBUG        (↵)
```

Now we are ready to assemble the program into memory. This is done by using the ASSEMBLE command as follows:

```
-A 0:0110                    (↵)
0000:0110 LAHF               (↵)
0000:0111 MOV [0150], AH     (↵)
0000:0115 MOV AH, [0151]     (↵)
0000:0119 SAHF               (↵)
```

Now the contents of MEM1 and MEM2 are initialized with the ENTER command

```
-E 0:0150 FF 01      (↵)
```

```
C:\DOS>DEBUG
-A 0:0110
0000:0110 LAHF
0000:0111 MOV    [0150],AH
0000:0115 MOV    AH,[0151]
0000:0119 SAHF
0000:011A
-E 0:150 FF 01
-R CS
CS 1342
:0
-R IP
IP 0100
:0110
-R DS
DS 1342
:0
-R
AX=0000  BX=0000  CX=0000  DX=0000  SP=FFEE  BP=0000  SI=0000  DI=0000
DS=0000  ES=1342  SS=1342  CS=0000  IP=0110   NV UP EI PL NZ NA PO NC
0000:0110 9F            LAHF
-T

AX=0200  BX=0000  CX=0000  DX=0000  SP=FFEE  BP=0000  SI=0000  DI=0000
DS=0000  ES=1342  SS=1342  CS=0000  IP=0111   NV UP EI PL NZ NA PO NC
0000:0111 88265001      MOV     [0150],AH                     DS:0150=FF
-T

AX=0200  BX=0000  CX=0000  DX=0000  SP=FFEE  BP=0000  SI=0000  DI=0000
DS=0000  ES=1342  SS=1342  CS=0000  IP=0115   NV UP EI PL NZ NA PO NC
0000:0115 8A265101      MOV     AH,[0151]                     DS:0151=01
-D 150 151
0000:0150  02 01
-T

AX=0100  BX=0000  CX=0000  DX=0000  SP=FFEE  BP=0000  SI=0000  DI=0000
DS=0000  ES=1342  SS=1342  CS=0000  IP=0119   NV UP EI PL NZ NA PO NC
0000:0119 9E            SAHF
-T

AX=0100  BX=0000  CX=0000  DX=0000  SP=FFEE  BP=0000  SI=0000  DI=0000
DS=0000  ES=1342  SS=1342  CS=0000  IP=011A   NV UP EI PL NZ NA PO CY
0000:011A 00F0          ADD     AL,DH
-Q

C:\DOS>
```

Figure 6–3 Display sequence for Example 6.2.

Next, the registers CS and IP must be initialized with the values 0000_{16} and 0110_{16} to provide access to the program. Also, the DS register must be initialized to permit access to the data memory locations. This is done with the commands

```
-R CS        (↵)
CS 1342
:0
-R IP        (↵)
IP 0100
:0110        (↵)
-R DS        (↵)
DS 1342
:0           (↵)
```

Before going further, let us verify the initialization of the internal registers. Displaying their state with the R command does this:

```
-R           (↵)
```

The information displayed in Fig. 6–3 shows that all three registers have been correctly initialized.

Now we are ready to step through the execution of the program. The first instruction is executed with the command

```
-T                  (↵)
```

Note from the displayed trace information in Fig. 6–3 that the contents of the status register, 02_{16}, have been copied into the AH register.

The second instruction is executed by issuing another T command

```
-T                  (↵)
```

This instruction causes the status, now in AH, to be saved in memory at address 0000:0150. The fact that this operation has occurred is verified with the D command

```
-D 150 151     (↵)
```

In Fig. 6–3, we see that the data held at address 0000:0150 is displayed by this command as 02_{16}. This verifies that status was saved at MEM1.

The third instruction is now executed with the command

```
-T                  (↵)
```

Its function is to copy the new status from MEM2 (0000:0151) into the AH register. From the data displayed in the earlier D command, we see that this value is 01_{16}. Looking at the displayed information for the third instruction, we find that 01_{16} has been copied into AH.

The last instruction is executed with another T command and, as shown by its trace information in Fig. 6–3, it has caused the carry flag to set. That is, CF is displayed with the value CY.

The next three instructions, CLC, STC, and CMC, as shown in Fig. 6–1(a), are used to manipulate the carry flag and permit CF to be cleared, set, or complemented, respectively. For example, if CF is 1 and the CMC instruction is executed, it becomes 0.

The last two instructions are used to manipulate the interrupt flag. Executing the clear interrupt (CLI) instruction sets IF to logic 0 and disables the interrupt interface. On the other hand, executing the STI instruction sets IF to 1, and the microprocessor is enabled to accept interrupts from that point on.

EXAMPLE 6.3

Of the three carry flag instructions CLC, STC, and CMC, only one is really an independent instruction—that is, the operation that it provides cannot be performed by a series of the other two instructions. Determine which one of the carry instructions is the independent instruction.

Solution

Let us begin with the CLC instruction. The clear-carry operation can be performed by an STC instruction followed by a CMC instruction. Therefore, CLC is not an independent instruction. The operation of the set-carry (STC) instruction is equivalent to the operation performed by a CLC instruction, followed by a CMC instruction. Thus, STC is also not an independent instruction. On the other hand, the operation performed by the last instruction, complement carry (CMC), cannot be expressed in terms of the CLC and STC instructions. Therefore, it is the independent instruction.

EXAMPLE 6.4

Verify the operation of the following instructions that affect the carry flag,

```
CLC
STC
CMC
```

by executing them with the DEBUG program. Start with CF set to one (CY).

Solution

After bringing up the DEBUG program, we enter the instructions with the Assemble command as

```
-A                    (↵)
1342:0100 CLC         (↵)
1342:0101 STC         (↵)
1342:0102 CMC         (↵)
1342:0103             (↵)
-
```

Next, the carry flag is initialized to CY with the R command

```
-R F    (↵)
NV UP EI PL NZ NA PO NC -CY   (↵)
```

and the updated status is displayed with another R command to verify that CF is set to the CY state. Figure 6–4 shows these commands and the results they produce.

Now the first instruction is executed with the Trace command

```
-T  (↵)
```

Looking at the displayed state information in Fig. 6–4, we see that CF has been cleared and its new state is NC.

The other two instructions are also executed with two more T commands and, as shown in Fig. 6–4, the STC instruction sets CF (CY in the state dump), and CMC inverts CF (NC in the state dump).

```
C:\DOS>DEBUG
-A
1342:0100 CLC
1342:0101 STC
1342:0102 CMC
1342:0103
-R F
NV UP EI PL NZ NA PO NC  -CY
-R F
NV UP EI PL NZ NA PO CY  -
-T

AX=0000  BX=0000  CX=0000  DX=0000  SP=FFEE  BP=0000  SI=0000  DI=0000
DS=1342  ES=1342  SS=1342  CS=1342  IP=0101   NV UP EI PL NZ NA PO NC
1342:0101 F9            STC
-T

AX=0000  BX=0000  CX=0000  DX=0000  SP=FFEE  BP=0000  SI=0000  DI=0000
DS=1342  ES=1342  SS=1342  CS=1342  IP=0102   NV UP EI PL NZ NA PO CY
1342:0102 F5            CMC
-T

AX=0000  BX=0000  CX=0000  DX=0000  SP=FFEE  BP=0000  SI=0000  DI=0000
DS=1342  ES=1342  SS=1342  CS=1342  IP=0103   NV UP EI PL NZ NA PO NC
1342:0103 8AFF          MOV     BH,BH
-Q
C:\DOS>
```

Figure 6–4 Display sequence for Example 6.4.

▲ 6.2 COMPARE INSTRUCTION

An instruction is included in the instruction set of the 8088 that can be used to compare two 8-bit or 16-bit numbers. It is the *compare* (CMP) instruction shown in Fig. 6–5(a). The compare operation enables us to determine the relationship between two numbers—that is, whether they are equal or unequal, and when they are unequal, which one is larger. Figure 6–5(b) shows that the operands for this instruction can reside in a storage location in memory, a register within the MPU, or be part of the instruction. For instance, a byte-

Mnemonic	Meaning	Format	Operation	Flags Affected
CMP	Compare	CMP D,S	(D) − (S) is used in setting or resetting the flags	CF, AF, OF, PF, SF, ZF

(a)

Destination	Source
Register	Register
Register	Memory
Memory	Register
Register	Immediate
Memory	Immediate
Accumulator	Immediate

(b)

Figure 6–5 (a) Compare instruction. (b) Allowed operands.

wide number in a register such as BL can be compared to a second byte-wide number that is supplied as immediate data.

The result of the comparison is reflected by changes in six of the status flags of the 8088. Note in Fig. 6–5(a) that it affects the overflow flag, sign flag, zero flag, auxiliary carry flag, parity flag, and carry flag. The new logic state of these flags can be used by the instructions that follow to make a decision whether or not to alter the sequence in which the program executes.

The process of comparison performed by the CMP instruction is basically a subtraction operation. The source operand is subtracted from the destination operand. However, the result of this subtraction is not saved. Instead, based on the result of the subtraction operation, the appropriate flags are set or reset. The importance of the flags lies in the fact that they lead us to an understanding of the relationship between the two numbers. For instance, if 5 is compared to 7 by subtracting 5 from 7, the ZF and CF both become logic 0. These conditions indicate that a smaller number was compared to a larger one. On the other hand, if 7 is compared to 5, we are comparing a larger number to a smaller number. This comparison results in ZF and CF equal to 0 and 1, respectively. Finally, if two equal numbers—for instance, 5 and 5—are compared, ZF is set to 1 and CF cleared to 0 to indicate the equal condition.

For example, let us assume that the destination operand equals $10011001_2 = -103_{10}$ and that the source operand equals $00011011_2 = +27_{10}$. Subtracting the source operand from the destination operand, we get

$$\begin{array}{r} 10011001_2 = -103_{10} \\ -00011011_2 = -(+27_{10}) \\ \hline 01111110_2 = +126_{10} \end{array}$$

In the process of subtraction, we get the status that follows:

1. A borrow is needed from bit 4 to bit 3; therefore, the auxiliary carry flag, AF, is set.
2. There is no borrow to bit 7. Thus, carry flag, CF, is reset.
3. Even though there is no borrow to bit 7, there is a borrow from bit 7 to bit 6. This is an indication of the overflow condition. Therefore, the OF flag is set.
4. There is an even number of 1s in the result; therefore, this sets the parity flag, PF.
5. Bit 7 of the result is zero, so the sign flag, SF, is reset.
6. The result that is produced is nonzero, which resets the zero flag, ZF.

Note that the 8-bit result of binary subtraction is not what the subtraction of the signed numbers should produce. The overflow flag having been set indicates this condition.

EXAMPLE 6.5

Describe what happens to the status flags as the sequence of instructions that follows is executed.

```
MOV  AX, 1234H
MOV  BX, 0ABCDH
CMP  AX, BX
```

Assume that flags ZF, SF, CF, AF, OF, and PF are all initially reset.

Instruction	ZF	SF	CF	AF	OF	PF
Initial state	0	0	0	0	0	0
MOV AX,1234H	0	0	0	0	0	0
MOV BX,0ABCDH	0	0	0	0	0	0
CMP AX,BX	0	0	1	1	0	0

Figure 6–6 Effect on flags of executing instructions.

Solution

The first instruction loads AX with 1234_{16}. No status flags are affected by the execution of a MOV instruction.

The second instruction puts $ABCD_{16}$ into the BX register. Again, status is not affected. Thus, after execution of these two move instructions, the contents of AX and BX are

$$(AX) = 1234_{16} = 0001001000110100_2$$

and

$$(BX) = ABCD_{16} = 1010101111001101_2$$

The third instruction is a 16-bit comparison with AX representing the destination and BX the source. Therefore, the contents of BX are subtracted from that of AX:

$$(AX) - (BX) = 0001001000110100_2 - 1010101111001101_2 = 0110011001100111_2$$

The flags are either set or reset based on the result of this subtraction. Note that the result is nonzero and positive. This makes ZF and SF equal to zero. Moreover, the overflow condition has not occurred. Therefore, OF is also at logic 0. The carry and auxiliary carry conditions occur and make CF and AF equal 1. Finally, the result has odd parity; therefore, PF is 0. Figure 6–6 summarizes these results.

EXAMPLE 6.6

Verify the execution of the instruction sequence in Example 6.5. Use DEBUG to load and run the instruction sequence provided in run module EX66.EXE.

Solution

A source program that contains the instruction sequence executed in Example 6.5 is shown in Fig. 6–7(a). This program was assembled and linked to form a run module file EX66.EXE. The source listing produced by the assembler is shown in Fig. 6–7(b).

```
TITLE    EXAMPLE 6.6

      PAGE      ,132

STACK_SEG         SEGMENT           STACK 'STACK'
                  DB                64 DUP(?)
STACK_SEG         ENDS

CODE_SEG          SEGMENT           'CODE'
EX66    PROC      FAR
      ASSUME  CS:CODE_SEG, SS:STACK_SEG

;To return to DEBUG program put return address on the stack

      PUSH    DS
      MOV     AX, 0
      PUSH    AX

;Following code implements Example 6.6

      MOV     AX, 1234H
      MOV     BX, 0ABCDH
      CMP     AX, BX

      RET                         ;Return to DEBUG program
EX66  ENDP

CODE_SEG          ENDS

      END     EX66
```

(a)

Figure 6–7 (a) Source program for Example 6.6. (b) Source listing produced by assembler. (c) Execution of the program with the DEBUG program.

To execute this program, we bring up the DEBUG program and load the file from a data diskette in drive A with the DOS command

```
C:\DOS>DEBUG A:EX66:EXE   (↵)
```

To verify its loading, the following Unassemble command can be used:

```
-U 0 D      (↵)
```

As Fig. 6–7(c) shows, the instructions of the source program are correctly displayed.

First, we execute the instructions up to the CMP instruction. This is done with the Go command

```
-G B        (↵)
```

Note in Fig. 6–7(c) that AX has been loaded with 1234_{16} and BX with the value $ABCD_{16}$.

Next, the compare instruction is executed with the command

```
-T          (↵)
```

By comparing the state information before and after execution of the CMP instruction, we find that auxiliary carry flag and carry flag are the only flags that have changed states, and they have both been set. Their new states are identified as AC and CY, respectively. These results are identical to those found in Example 6.5.

```
                         TITLE    EXAMPLE 6.6

                           PAGE      ,132

0000                          STACK_SEG        SEGMENT          STACK 'STACK'
0000      40 [                                 DB
            ??
               ]

0040                          STACK_SEG        ENDS

0000                          CODE_SEG         SEGMENT          'CODE'
0000                          EX66     PROC    FAR
                           ASSUME  CS:CODE_SEG, SS:STACK_SEG

                    ;To return to DEBUG program put return address on the stack

0000  1E                              PUSH     DS
0001  B8 0000                         MOV      AX, 0
0004  50                              PUSH     AX

                    ;Following code implements Example 6.6

0005  B8 1234                         MOV      AX, 1234H
0008  BB ABCD                         MOV      BX, 0ABCDH
000B  3B C3                           CMP      AX, BX

000D  CB                              RET                         ;Return to DEBUG program
000E                         EX66     ENDP

000E                         CODE_SEG         ENDS

                             END      EX66

Segments and groups:

             N a m e                   Size    align    combine class
CODE_SEG . . . . . . . . . . . . .     000E    PARA     NONE    'CODE'
STACK_SEG. . . . . . . . . . . . .     0040    PARA     STACK   'STACK'

Symbols:

             N a m e                   Type    Value    Attr

EX66 . . . . . . . . . . . . . . .     F PROC  0000     CODE_SEG        Length =000E

Warning Severe
Errors  Errors
0       0
```

(b)

Figure 6–7 (continued)

▲ 6.3 CONTROL FLOW AND THE JUMP INSTRUCTIONS

Earlier we pointed out that control flow relates to altering the execution path of instructions in a program. For example, a control flow decision may cause a sequence of instructions to be repeated or a group of instructions to not be executed at all. The *jump* instruction is provided in the 8088/8086 instruction set for implementing control flow operations. In the 8088 architecture, the code segment register and instruction pointer keep track of the next instruction to be fetched for execution. Thus, to initiate a change in control flow, a jump instruction must change the contents of these registers. In this way, execution continues at an address other than that of the next sequential instruction. That is, a jump occurs to another part of the program.

```
C:\DOS>DEBUG A:EX66.EXE
-U 0 D
0F50:00000 1E           PUSH   DS
0F50:0001 B80000        MOV    AX, 0000
0F50:0004 50            PUSH   AX
0F50:0005 B83412        MOV    AX,1234
0F50:0008 BBCDAB        MOV    BX,ABCD
0F50:000B 3BC3          CMP    AX,BX
0F50:000D CB            RETF
-G B

AX=1234  BX=ABCD  CX=000E  DX=0000  SP=003C  BP=0000  SI=0000  DI=0000
DS=0F40  ES=0F40  SS=0F51  CS=0F50  IP=000B   NV UP EI PL NZ NA PO NC
0F50:000B 3BC3          CMP     AX,BX
-T

AX=1234  BX=ABCD  CX=000E  DX=0000  SP=003C  BP=0000  SI=0000  DI=0000
DS=0F40  ES=0F40  SS=0F51  CS=0F50  IP=000D   NV UP EI PL NZ AC PO CY
0F50:000D CB            RETF
-G

Program terminated normally
-Q

C:\DOS>
```

(c)

Figure 6–7 (continued)

Unconditional and Conditional Jump

The 8088 microprocessor allows two different types of jump operations. They are the *unconditional jump* and the *conditional jump*. In an unconditional jump, no status requirements are imposed for the jump to occur. That is, as the instruction is executed, the jump always takes place to change the execution sequence.

The unconditional jump concept is illustrated in Fig. 6–8(a). Note that when the instruction JMP AA in part I is executed, program control is passed to a point in part III, identified by the label AA. Execution resumes with the instruction corresponding to AA. In this way, the instructions in part II of the program are bypassed—that is, they are jumped over. Some high-level languages have a GOTO statement. This is an example of a high-level language program construct that performs an unconditional jump operation.

On the other hand, for a conditional jump instruction, status conditions that exist at the time the jump instruction is executed decide whether or not the jump will occur. If the condition or conditions are met, the jump takes place; otherwise, execution continues with the next sequential instruction of the program. The conditions that can be referenced by a conditional jump instruction are status flags such as carry (CF), zero (ZF), and sign (SF) flags.

Looking at Fig. 6–8(b), we see that execution of the conditional jump instruction Jcc AA in part I causes a test to be initiated. If the conditions of the test are not met, the NO path is taken and execution continues with the next sequential instruction. This corresponds to the first instruction in part II. However, if the result of the conditional test is YES, a jump is initiated to the segment of program identified as part III, and the instructions in part II are bypassed.

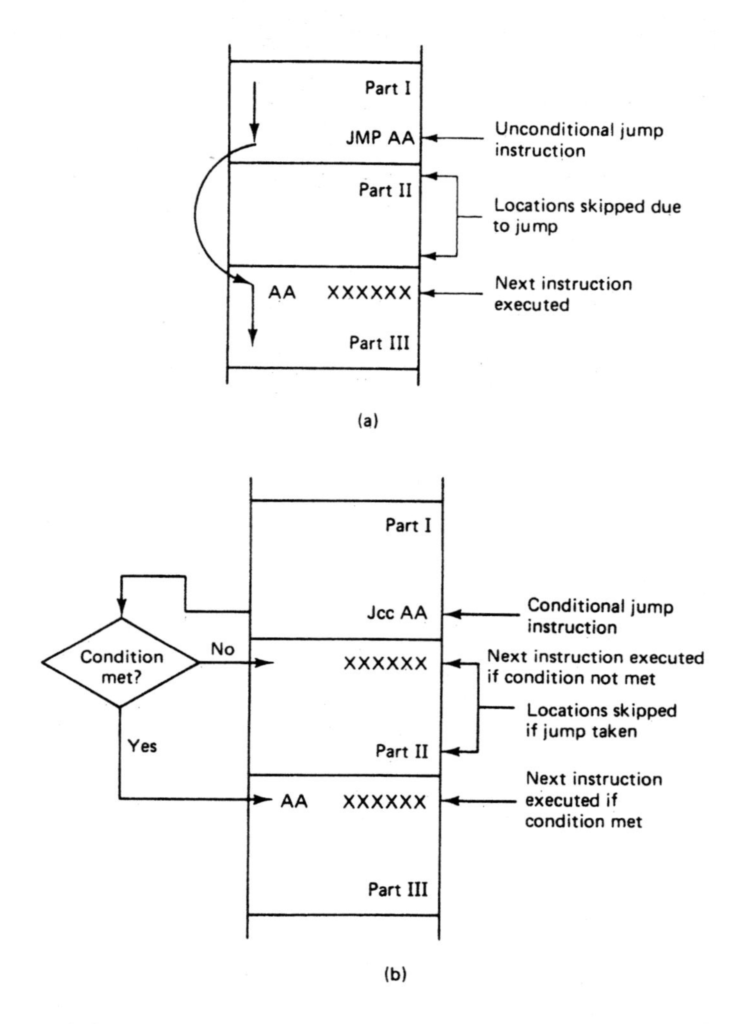

Figure 6–8 (a) Unconditional jump program sequence. (b) Conditional jump program sequence.

This type of software operation is referred to as making a *branch*—that is, a point in the program where a choice is made between two paths of execution. If the conditions specified by the jump instruction are met, program control is passed to the part of the program identified by the label. On the other hand, if they are not met, the next sequential instruction is executed.

The branch is a frequently used program structure and is sometime referred to as an *IF-THEN-ELSE* structure. By this, we mean that IF the conditions specified in the jump instruction are met, THEN control continues with the execution of the statement following the IF, ELSE program control is passed to the point identified by the label in the jump instruction.

Typically, program execution is not intended to return to the next sequential instruction after the unconditional or conditional jump instruction. Therefore, no return linkage is saved when the jump takes place.

Unconditional Jump Instruction

Figure 6–9(a) shows the unconditional jump instruction of the 8088, together with its valid operand combinations in Fig. 6–9(b). There are two basic kinds of unconditional jumps. The first, called an *intrasegment jump,* is limited to addresses within the current code segment. This type of jump is achieved by just modifying the value in IP. The second kind of jump, the *intersegment jump,* permits jumps from one code segment to another. Implementation of this type of jump requires modification of the contents of both CS and IP.

Jump instructions specified with a *Short-label, Near-label, Memptr16,* or *Regptr16 operand* represent intrasegment jumps. The Short-label and Near-label operands specify the jump relative to the address of the jump instruction itself. For example, in a Short-label jump instruction, an 8-bit number is coded as an immediate operand to specify the *signed displacement* of the next instruction to be executed from the location of the jump instruction. When the jump instruction is executed, IP is reloaded with a new value equal to the updated value in IP, which is (IP) + 2, plus the signed displacement. The new value of IP and current value in CS give the address of the next instruction to be fetched and executed. With an 8-bit displacement, the Short-label operand can only be used to initiate a jump in the range from −126 to +129 bytes from the location of the jump instruction.

On the other hand, the Near-label operand specifies a new value for IP with 16-bit immediate operand. This size of offset corresponds to the complete range of the current code segment. The value of the offset is automatically added to IP upon execution of the

Mnemonic	Meaning	Format	Operation	Affected flags
JMP	Unconditional jump	JMP Operand	Jump is initiated to the address specified by the operand	None

(a)

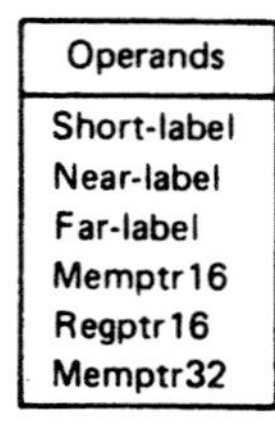

Operands
Short-label
Near-label
Far-label
Memptr16
Regptr16
Memptr32

(b)

Figure 6–9 (a) Unconditional jump instruction. (b) Allowed operands.

instruction. In this way, program control is passed to the location identified by the new IP. Consider the following example of an unconditional jump instruction:

```
JMP 1234H
```

It means jump to address 1234H. However, the value of the address encoded in the instruction is not 1234H. Instead, it is the difference between the incremented value in IP and 1234_{16}. This offset is encoded as either an 8-bit constant (Short label) or a 16-bit constant (Near label), depending on the size of the difference.

The jump-to address can also be specified indirectly by the contents of a memory location or the contents of a register, corresponding to the Memptr16 and Regptr16 operands, respectively. Just as for the Near-label operand, they both permit a jump to any address in the current code segment.

For example,

```
JMP BX
```

uses the contents of register BX for the offset in the current code segment—that is, the value in BX is copied into IP.

EXAMPLE 6.7

Verify the operation of the instruction JMP BX using the DEBUG program. Let the contents of BX be 0010_{16}.

Solution

As shown in Fig. 6–10, DEBUG is invoked, and then the Assemble command is used to load the instruction:

```
-A                        (↵)
1342:0100 JMP BX          (↵)
1342:0102                 (↵)
```

Next, BX is initialized with the command

```
-R BX                     (↵)
BX 0000
:10                       (↵)
```

Let us check the value in IP before executing the JMP instruction. This is done with another R command as

```
-R          (↵)
```

Looking at the state information displayed in Fig. 6–10, we see that IP contains 0100_{16} and BX contains 0010_{16}.

```
C:\DOS>DEBUG
-A
1342:0100 JMP BX
1342:0102
-R BX
BX 0000
:10
-R
AX=0000  BX=0010  CX=0000  DX=0000  SP=FFEE  BP=0000  SI=0000  DI=0000
DS=1342  ES=1342  SS=1342  CS=1342  IP=0100   NV UP EI PL NZ NA PO NC
1342:0100 FFE3          JMP     BX
-T

AX=0000  BX=0010  CX=0000  DX=0000  SP=FFEE  BP=0000  SI=0000  DI=0000
DS=1342  ES=1342  SS=1342  CS=1342  IP=0010   NV UP EI PL NZ NA PO NC
1342:0010 8B09          MOV     CX,[BX+DI]                        DS:0010=098B
-Q

C:\DOS>
```

Figure 6–10 Display sequence for Example 6.7.

Executing the instruction with the command

```
-T          (↵)
```

and then looking at Fig. 6–10, we see that the value in IP has become 10_{16}. Therefore, the address at which execution picks up is 1342:0010.

To specify an operand as a pointer to memory, the various addressing modes of the 8088 can be used. For instance,

```
JMP [BX]
```

uses the contents of BX as the offset address of the memory location that contains the value of IP (Memptr16 operand). This offset is loaded into IP, where it is used together with the current contents of CS to compute the "jump-to" address.

EXAMPLE 6.8

Use the DEBUG program to observe the operation of the instruction

```
JMP [BX]
```

Assume that the pointer held in BX is 1000_{16} and the value held at memory location DS: 1000 is 200_{16}. What is the address of the next instruction to be executed?

Solution

Figure 6–11 shows that first the DEBUG program is brought up and then an Assemble command is used to load the instruction. That is,

```
-A                      (↵)
1342:0100 JMP [BX]      (↵)
1342:0102               (↵)
```

```
C:\DOS>DEBUG
-A
1342:0100 JMP BX
1342:0102
-R BX
BX 0000
:1000
-E 1000 00 02
-D 1000 1001
1342:1000  00 02
-R
AX=0000  BX=1000  CX=0000  DX=0000  SP=FFEE  BP=0000  SI=0000  DI=0000
DS=1342  ES=1342  SS=1342  CS=1342  IP=0100   NV UP EI PL NZ NA PO NC
1342:0100 FF27          JMP     (BX)                          DS:1000=0200
-T

AX=0000  BX=1000  CX=0000  DX=0000  SP=FFEE  BP=0000  SI=0000  DI=0000
DS=1342  ES=1342  SS=1342  CS=1342  IP=0200   NV UP EI PL NZ NA PO NC
1342:0200 4D          DEC     BP
-Q

C:\DOS>
```

Figure 6–11 Display sequence for Example 6.8.

Next, BX is loaded with the pointer address using the R command

```
-R BX                    (↵)
BX 0000
:1000                    (↵)
```

and the memory location is initialized with the command

```
-E 1000 00 02            (↵)
```

As shown in Fig. 6–11, the loading of memory location DS:1000 and the BX register are next verified with D and R commands, respectively.

Now the instruction is executed with the command

```
-T   (↵)
```

Note from the state information displayed in Fig. 6–11 that the new value in IP is 0200_{16}. This value was loaded from memory location 1342:1000. Therefore, program execution continues with the instruction at address 1342:0200.

The intersegment unconditional jump instructions correspond to the *Far-label* and *Memptr32 operands* that are shown in Fig. 6–9(b). Far-label uses a 32-bit immediate operand to specify the jump-to address. The first 16 bits of this 32-bit pointer are loaded into IP and are an offset address relative to the contents of the code-segment register. The next 16 bits are loaded into the CS register and define the new code segment.

An indirect way to specify the offset and code-segment address for an intersegment jump is by using the Memptr32 operand. This time, four consecutive memory bytes starting at the specified address contain the offset address and the new code segment address, respectively. Just like the Memptr16 operand, the Memptr32 operand may be specified using any one of the various addressing modes of the 8088.

An example is the instruction

```
JMP DWORD PTR [DI]
```

It uses the contents of DS and DI to calculate the address of the memory location that contains the first word of the pointer that identifies the location to which the jump will take place. The two-word pointer starting at this address is read into IP and CS to pass control to the new point in the program.

Conditional Jump Instruction

The second type of jump instruction performs conditional jump operations. Figure 6–12(a) shows a general form of this instruction; Fig. 6–12(b) is a list of each of the conditional jump instructions in the 8088's instruction set. Note that each of these instructions tests for the presence or absence of certain status conditions.

For instance, the *jump on carry* (JC) instruction makes a test to determine if carry flag (CF) is set. Depending on the result of the test, the jump to the location specified by its operand either takes place or does not. If CF equals 0, the test fails and execution continues with the instruction at the address following the JC instruction. On the other hand, if CF equals 1, the test condition is satisfied and the jump is performed.

Note that for some of the instructions in Fig. 6–12(b), two different mnemonics can be used. This feature can be used to improve program readability. That is, for each occurrence of the instruction in the program, it can be identified with the mnemonic that best describes its function.

For instance, the instruction *jump on parity* (JP) or *jump on parity even* (JPE) both test parity flag (PF) for logic 1. Since PF is set to one if the result from a computation has even parity, this instruction can initiate a jump based on the occurrence of even parity. The reverse instruction JNP/JPO is also provided. It can be used to initiate a jump based on the occurrence of a result with odd instead of even parity.

In a similar manner, the instructions *jump if equal* (JE) and *jump if zero* (JZ) serve the same function. Either notation can be used in a program to determine if the result of a computation was zero.

All other conditional jump instructions work in a similar way except that they test different conditions to decide whether or not the jump is to take place. Examples of these conditions are: the contents of CX are zero, an overflow has occurred, or the result is negative.

To distinguish between comparisons of signed and unsigned numbers by jump instructions, two different names, which seem to imply the same, have been devised. They are *above* and *below*, for comparison of unsigned numbers, and *less* and *greater*, for comparison of signed numbers. For instance, the number $ABCD_{16}$ is above the number 1234_{16} if they are considered to be unsigned numbers. On the other hand, if they are treated as signed numbers, $ABCD_{16}$ is negative and 1234_{16} is positive. Therefore, $ABCD_{16}$ is less than 1234_{16}.

Mnemonic	Meaning	Format	Operation	Flags affected
Jcc	Conditional jump	Jcc Operand	If the specified condition cc is true the jump to the address specified by the operand is initiated; otherwise the next instruction is executed.	None

(a)

Mnemonic	Meaning	Condition
JA	above	CF = 0 and ZF = 0
JAE	above or equal	CF = 0
JB	below	CF = 1
JBE	below or equal	CF = 1 or ZF = 1
JC	carry	CF = 1
JCXZ	CX register is zero	(CF or ZF) = 0
JE	equal	ZF = 1
JG	greater	ZF = 0 and SF = OF
JGE	greater or equal	SF = OF
JL	less	(SF xor OF) = 1
JLE	less or equal	((SF xor OF) or ZF) = 1
JNA	not above	CF = 1 or ZF = 1
JNAE	not above nor equal	CF = 1
JNB	not below	CF = 0
JNBE	not below nor equal	CF = 0 and ZF = 0
JNC	not carry	CF = 0
JNE	not equal	ZF = 0
JNG	not greater	((SF xor OF) or ZF) = 1
JNGE	not greater nor equal	(SF xor OF) = 1
JNL	not less	SF = OF
JNLE	not less nor equal	ZF = 0 and SF = OF
JNO	not overflow	OF = 0
JNP	not parity	PF = 0
JNS	not sign	SF = 0
JNZ	not zero	ZF = 0
JO	overflow	OF = 1
JP	parity	PF = 1
JPE	parity even	PF = 1
JPO	parity odd	PF = 0
JS	sign	SF = 1
JZ	zero	ZF = 1

(b)

Figure 6–12 (a) Conditional jump instruction. (b) Types of conditional jump instructions.

Branch Program Structure—IF-THEN-ELSE

The high-level language IF-THEN-ELSE construct is expressed in general by the program structure

```
IF <condition>
THEN
     statement;
ELSE
     statement;
```

Note that if the condition tested for is satisfied, the statement associated with THEN is executed. Otherwise, the statement corresponding to ELSE is performed. Let us now look at some simple examples of how the compiler may use the conditional jump instruction to implement this software branch program structure.

One example is a branch that is made based on the flag settings that result after the contents of two registers are compared to each other. Figure 6–13 shows a program structure that implements this software operation. This program tests to confirm whether or not two values are equal.

First, the CMP instruction subtracts the value in BX from that in AX and adjusts the flags based on the result. Next, the jump on equal instruction tests the zero flag to see if it is 1. If ZF is 1, it means that the contents of AX and BX are equal and a jump is made to the ELSE path in the program identified by the label EQUAL. Otherwise, if ZF is 0, which means that the contents of AX and BX are not equal, the THEN path is taken and the instruction following the JE instruction is executed.

Similar instruction sequences can be used to initiate branch operations for other conditions. For instance, by using the instruction JA ABOVE, the branch is taken to the ELSE path if the unsigned number in AX is larger than that in BX.

Another common use of a conditional jump is to branch based on the setting of a specific bit in a register. When this is done, a logic operation is normally used to mask off the values of all of the other bits in the register. For example, we may want to mask off all bits of the value in AL other than bit 2 and then make a conditional jump if the unmasked bit is logic 1.

```
        CMP  AX, BX
        JE   EQUAL
        ---  ---          ; Next instruction if (AX) ≠ (BX)
             .
             .
EQUAL:  ---  ---          ; Next instruction if (AX) = (BX)
             .
             .
        ---  ---
```

Figure 6–13 IF-THEN branch program structure using a flag-condition test.

```
            AND   AL, 04H
            JNZ   BIT2_ONE
            ---   ---          ; Next instruction if B2 of AL = 0
                .
                .
            ---   ---
BIT2_ONE:   ---   ---          ; Next instruction if B2 of AL = 1
                .
                .
            ---   ---
```

Figure 6–14 IF-THEN branch program structure using a register-bit test.

This operation can be done with the instruction sequence in Fig. 6–14. First, the contents of AL are ANDed with 04_{16} to give

$$(AL) = XXXXXXXX_2 \cdot 00000100_2 = 00000X00_2$$

Now the content of AL is 0 if bit 2 is 0 and the resulting value in the ZF is 1. On the other hand, if bit 2 is 1, the content of AL is nonzero and ZF is 0. Remember, we want to make the jump to the ELSE path when bit 2 is 1. Therefore, the conditional test is made with a jump on not zero instruction. When ZF is 1, the THEN path is taken and the next instruction is executed, but if ZF is 0, the JNZ instruction passes control to the ELSE path and the instruction identified by the label BIT2_ONE.

Let us look at how to perform this exact same branch operation in another way. Instead of masking off all of the bits in AL, we could simply shift bit 2 into the carry flag and then make a conditional jump if CF equals 1. The program structure in Fig. 6–15 uses this method to test for logic 1 in bit 2 of AL. Notice that this implementation takes one extra instruction.

The Loop Program Structure—REPEAT-UNTIL and WHILE-DO

In many practical applications, we frequently need to repeat a part of the program many times—that is, a group of instructions may need to be executed over an over again until a condition is met. This type of control flow program structure is known as a *loop.*

```
            MOV   CL,03H
            SHR   AL, CL
            JC    BIT2_ONE
            ---   ---          ; Next instruction if B2 of AL = 0
                .
                .
            ---   ---
BIT2_ONE:   ---   ---          ; Next instruction if B2 of AL = 1
                .
                .
            ---   ---
```

Figure 6–15 IF-THEN branch program structure using an alternative register-bit test.

The high-level language *REPEAT-UNTIL* construct is an example of a typical loop operation and is expressed in general by the program structure

```
REPEAT
      Statement 1;
      Statement 2;
          .
          .
      Statement N;
UNTIL <condition>
```

Note that after statements 1 through N are performed, a conditional test is performed by the UNTIL statement. If this condition is not satisfied, control flow is returned to REPEAT and the operations performed by the statements are repeated. On the other hand, if the condition is satisfied, the loop operation is complete and control is passed to the statement following UNTIL. Since the conditional test is made at the end of the loop, it is known as a post-test. The statements in a loop that employ a post-test are always performed at least one time. A high-level language DO statement with a fixed number of iterations implements a REPEAT-UNTIL program structure. This type of loop operation may be used to repeat a computation a fixed number of times.

Figures 6–16(a) and (b) show a typical assembly language implementation of a REPEAT-UNTIL loop program structure. Here we see that the sequence of instructions from label AGAIN to the conditional jump instruction JNZ represents the loop. Note that the label for the instruction that is to be jumped to is located before the jump instruction that makes the conditional test. In this way, if the test result is true, program control returns to AGAIN and the segments of the program repeat. This continues until the condition specified by NZ is false.

Before initiating the program sequence, a parameter must be assigned to keep track of how many times the sequence of instructions has been repeated. This parameter, called the *count,* is tested each time the sequence is performed to verify whether or not it is to be repeated again. Note that register CL, which is used for the conditional test, is initialized with the value COUNT by a MOV instruction prior to entering the loop. For example, to repeat a part of a program 10 times, we begin by loading the count register CL with COUNT equal to $0A_{16}$. Then the operation of instructions 1 through *n* of the loop is performed. Next, the value in CL is decremented by 1 to indicate that the instructions in the loop are done; and after decrementing CL, ZF is tested with the JNZ instruction to see if it has reached 0. That is, the question "Should I repeat again?" is asked with software. If ZF is 0, which means that $(CL) \neq 0$, the answer is yes, repeat again, and program control is returned to the instruction labeled AGAIN and the loop instruction sequence repeats. This continues until the value in CL reaches zero to identify that the loop is done. When this happens, the answer to the conditional test is no, do not repeat again, and the jump is not taken. Instead, the instruction following JNZ AGAIN is executed.

FOR is a C language statement that can implement a REPEAT-UNTIL program structure. However, the FOR loop performs both a pretest and a post-test. That is, a test of the UNTIL condition is made prior to entry of the loop—the pretest. If the condition

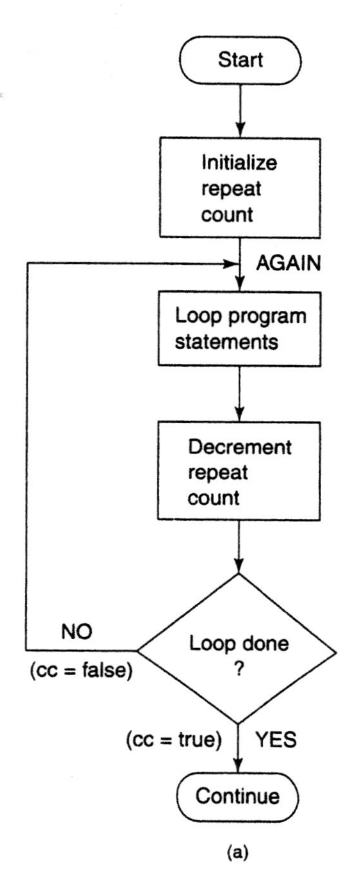

(a)

```
         MOV  CL,COUNT   ;Set loop repeat count
AGAIN:   ---  ---        ;1st instruction of loop
         ---  ---        ;2nd instruction of loop
          .    .                  .
          .    .                  .
          .    .                  .
         ---  ---        ;nth instruction of loop
         DEC  CL         ;Decrement repeat count by 1
         JNZ  AGAIN      ;Repeat from AGAIN if (CL) ≠ 00H or (ZF) = 0
         ---  ---        ;First instruction executed after the loop is
                         ;complete, (CL) = 00H, (ZF) = 1
```

(b)

Figure 6–16 (a) REPEAT-UNTIL program sequence. (b) Typical REPEAT-UNTIL instruction sequence.

is already satisfied, no iterations of the FOR loop are performed. Once the FOR loop is initiated, all remaining tests are performed at the end of the loop—the post-test. This type of loop is normally performed a fixed number of times.

Figure 6–17(a) shows another loop structure. It differs in that the conditional test used to decide whether or not the loop will repeat is made at entry of the loop instruction sequence. This type of loop operation is known as a *WHILE loop,* or sometimes as a *WHILE-DO loop.* Its program structure is expressed in general as

WHILE <expression>
Statement 1;
Statement 2;
.
.
Statement N;

Note that statements 1 through N are performed only if the conditional test performed by the WHILE statement is satisfied. For instance, it may search through a document character by character looking for the first occurrence of the ASCII code for a specific character. Since it uses a pretest, the loop may not be performed any times. However, once initiated the loop repeats until the condition is not satisfied. For this reason, the loop may repeat an unspecified number of times.

Figure 6–17(b) shows a typical WHILE-DO instruction sequence. To implement this loop, we use a conditional jump instruction to perform the pretest and unconditional jump instruction to initiate the repeat. Here the loop repeats while the conditional test finds the contents of CL not equal to zero—ZF = 0. When CL equals 0, ZF = 0, the loop is complete and control flow passes to the instruction with label NEXT.

The *no operation* (NOP) instruction is sometimes used in conjunction with loop routines. As its name implies, it performs no operation—that is, its execution does not change the contents of registers or affect the flags. However, a period of time is needed to perform the NOP function. In some practical applications—for instance, a time-delay loop—the duration it takes to execute NOP is used to extend the time-delay interval.

Applications Using the Loop and Branch Software Structures

As a practical application of the use of a conditional jump operation, let us write a program known as a *block-move program.* The purpose of this program is to move a block of *N* consecutive bytes of data starting at offset address BLK1ADDR in memory to another block of memory locations starting at offset address BLK2ADDR. We will assume that both blocks are in the same data segment, whose starting point is defined by the data segment value DATASEGADDR.

The flowchart in Fig. 6–18(a) outlines the steps followed to solve this problem. It has four distinct operations. The first operation is initialization. Initialization involves establishing the initial address of the data segment. Loading the DS register with the value DATASEGADDR does this. Furthermore, source index register SI and destination index register DI are initialized with offset addresses BLK1ADDR and BLK2ADDR, respectively. In this way, they point to the beginning of the source block and the begin-

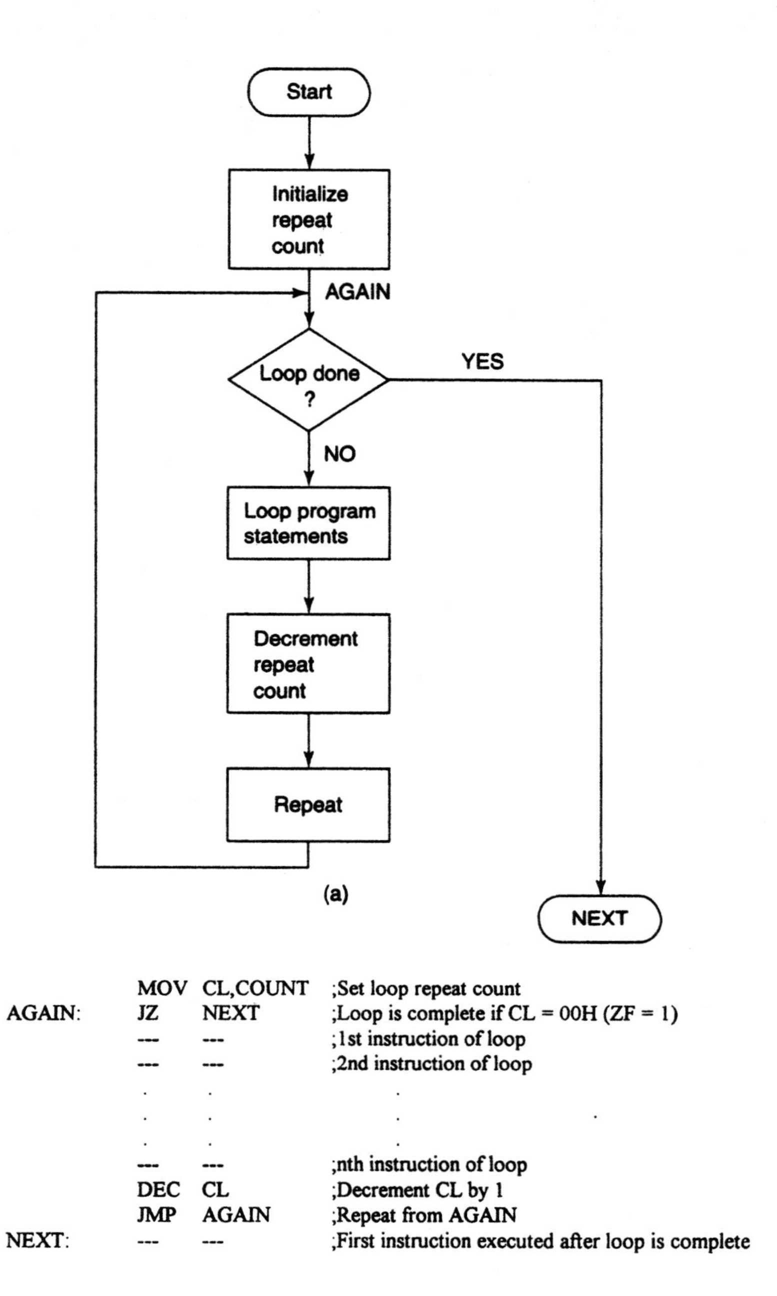

(a)

```
        MOV  CL,COUNT  ;Set loop repeat count
AGAIN:  JZ   NEXT      ;Loop is complete if CL = 00H (ZF = 1)
        ---  ---       ;1st instruction of loop
        ---  ---       ;2nd instruction of loop
        .    .          .
        .    .          .                   .
        .    .          .
        ---  ---       ;nth instruction of loop
        DEC  CL        ;Decrement CL by 1
        JMP  AGAIN     ;Repeat from AGAIN
NEXT:   ---  ---       ;First instruction executed after loop is complete
```

(b)

Figure 6–17 (a) WHILE-DO program sequence. (b) Typical WHILE-DO instruction sequence.

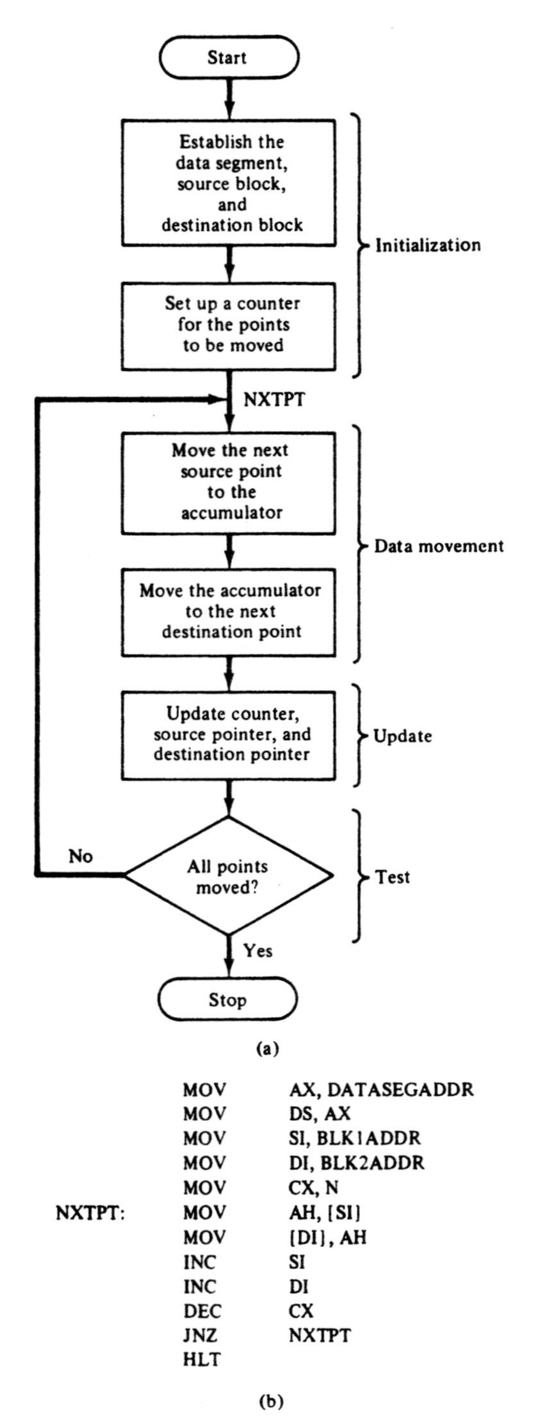

Figure 6–18 (a) Block transfer flowchart. (b) Program.

ning of the destination block, respectively. To keep track of the count of bytes transferred, register CX is initialized with *N*, the number of bytes to be moved. This leads us to the following assembly language statements:

```
MOV   AX, DATASEGADDR
MOV   DS, AX
MOV   SI, BLK1ADDR
MOV   DI, BLK2ADDR
MOV   CX, N
```

Note that DS cannot be directly loaded by immediate data with a MOV instruction. Therefore, the segment address is first loaded into AX and then moved to DS. SI, DI, and CX load directly with immediate data.

The next operation that must be performed is the actual movement of data from the source block of memory to the destination block. The offset addresses are already loaded into SI and DI; therefore, move instructions that employ indirect addressing are used to accomplish the data-transfer operation. Remember that the 8088 does not allow direct memory-to-memory moves. For this reason, AX is used as a temporary storage location for data. The source byte is moved into AX with one instruction, and then another instruction moves it from AX to the destination location. Thus, the data move is accomplished by the following instructions:

```
NXTPT:  -MOV   AH, [SI]
         MOV   [DI], AH
```

Note that for a byte move, only the higher eight bits of AX are used. Therefore, the operand is specified as AH instead of AX.

Now the pointers in SI and DI must be updated so that they are ready for the next byte-move operation. Also, the counter must be decremented so that it corresponds to the number of bytes that remain to be moved. These updates are done by the following sequence of instructions:

```
INC   SI
INC   DI
DEC   CX
```

The test operation involves determining whether or not all the data points have been moved. The contents of CX represent this condition. When its value is not 0, there still are points to be moved, whereas a value of 0 indicates that the block move is complete. This 0 condition is reflected by 1 in ZF. The instruction needed to perform this test is

```
JNZ   NXTPT
```

Here NXTPT is a label that corresponds to the first instruction in the data move operation. The last instruction in the program is a *halt* (HLT) instruction to indicate the end of the block move operation. Figure 6–18(b) shows the entire program.

EXAMPLE 6.9

The program

```
            CMP   AX, BX
            JC    DIFF2
DIFF1:      MOV   DX, AX
            SUB   DX, BX      ;(DX) = (AX) − (BX)
            JMP   DONE
DIFF2:      MOV   DX, BX
            SUB   DX, AX      ;(DX) = (BX) − (AX)
DONE:       ----
```

implements an instruction sequence that calculates the absolute difference between the contents of AX and BX and places it in DX. Use the run module produced by assembling and linking the source program in Fig. 6–19(a) to verify the operation of the program for the two cases that follow:

(a) (AX) = 6, (BX) = 2

(b) (AX) = 2, (BX) = 6

```
TITLE   EXAMPLE 6.9

        PAGE     ,132

STACK_SEG         SEGMENT            STACK 'STACK'
                  DB                 64 DUP(?)
STACK_SEG         ENDS

CODE_SEG          SEGMENT            'CODE'
EX69      PROC    FAR
        ASSUME    CS:CODE_SEG, SS:STACK_SEG

;To return to DEBUG program put return address on the stack

            PUSH    DS
            MOV     AX, 0
            PUSH    AX

;Following code implements Example 6.9

            CMP     AX, BX
            JC      DIFF2
DIFF1:      MOV     DX, AX
            SUB     DX, BX           ;(DX) = (AX) − (BX)
            JMP     DONE
DIFF2:      MOV     DX, BX
            SUB     DX, AX           ;(DX) = (BX) − (AX)
DONE:       NOP

            RET                      ;Return to DEBUG program
EX69        ENDP

CODE_SEG            ENDS

            END     EX69
```

(a)

Figure 6–19 (a) Source program for Example 6.9. (b) Source listing produced by assembler. (c) Execution of the program with DEBUG.

Solution

The source program in Fig. 6–19(a) was assembled and linked to produce run module EX69.EXE. Figure 6–19(b) shows the source listing produced as part of the assembly process.

As shown in Fig. 6–19(c), the run module is loaded as part of calling up the DEBUG program by issuing the DOS command

```
C:\DOS>DEBUG A:EX69.EXE  (↵)
```

```
                                TITLE    EXAMPLE 6.9

                                      PAGE      ,132

0000                                    STACK_SEG          SEGMENT            STACK 'STACK'
0000      40 [                                             DB                 64 DUP(?)
            ??
              ]

0040                                    STACK_SEG          ENDS

0000                                    CODE_SEG           SEGMENT            'CODE'
0000                                    EX69      PROC     FAR
                                            ASSUME  CS:CODE_SEG, SS:STACK_SEG

                                ;To return to DEBUG program put return address on the stack

0000  1E                                          PUSH     DS
0001  B8 0000                                     MOV      AX, 0
0004  50                                          PUSH     AX

                                ;Following code implements Example 6.6

0005  3B C3                                       CMP      AX, BX
0007  72 07                                       JC       DIFF2
0009  8B D0                             DIFF1:    MOV      DX, AX
000B  2B D3                                       SUB      DX, BX             ;(DX) = (AX) - (BX)
000D  EB 05 90                                    JMP      DONE
0010  8B D3                             DIFF2:    MOV      DX, BX
0012  2B D0                                       SUB      DX, AX             ;(DX) = (BX) - (AX)
0014  90                                DONE:     NOP

0015  CB                                          RET                         ;Return to DEBUG program
0016                                    EX69      ENDP

0016                                    CODE_SEG           ENDS

                                                  END      EX69

Segments and groups:

            N a m e                     Size     align    combine class

CODE_SEG . . . . . . . . . . . .        0016     PARA     NONE     'CODE'
STACK_SEG. . . . . . . . . . . .        0040     PARA     STACK    'STACK'

Symbols:

            N a m e                     Type     Value    Attr

DIFF1. . . . . . . . . . . . . .        L NEAR   0009     CODE_SEG
DIFF2. . . . . . . . . . . . . .        L NEAR   0010     CODE_SEG
DONE . . . . . . . . . . . . . .        L NEAR   0014     CODE_SEG
EX69 . . . . . . . . . . . . . .        F PROC   0000     CODE_SEG    Length =0016

Warning Severe
Errors  Errors
0       0
```

(b)

Figure 6–19 (continued)

```
C:\DOS>DEBUG A:EX69.EXE
-U 0 15
0D03:0000 1E            PUSH    DS
0D03:0001 B80000        MOV     AX,0000
0D03:0004 50            PUSH    AX
0D03:0005 3BC3          CMP     AX,BX
0D03:0007 7207          JB      0010
0D03:0009 8BD0          MOV     DX,AX
0D03:000B 2BD3          SUB     DX,BX
0D03:000D EB05          JMP     0014
0D03:000F 90            NOP
0D03:0010 8BD3          MOV     DX,BX
0D03:0012 2BD0          SUB     DX,AX
0D03:0014 90            NOP
0D03:0015 CB            RETF
-G 5

AX=0000  BX=0000  CX=0016  DX=0000  SP=003C  BP=0000  SI=0000  DI=0000
DS=0DD6  ES=0DD6  SS=0DE8  CS=0D03  IP=0005   NV UP EI PL NZ NA PO NC
0D03:0005 3BC3          CMP     AX,BX
-R AX
AX 0000
:6
-R BX
BX 0000
:2
-T

AX=0006  BX=0002  CX=0016  DX=0000  SP=003C  BP=0000  SI=0000  DI=0000
DS=0DD6  ES=0DD6  SS=0DE8  CS=0D03  IP=0007   NV UP EI PL NZ NA PO NC
0D03:0007 7207          JB      0010
-G 14

AX=0006  BX=0002  CX=0016  DX=0004  SP=003C  BP=0000  SI=0000  DI=0000
DS=0DD6  ES=0DD6  SS=0DE8  CS=0D03  IP=0014   NV UP EI PL NZ NA PO NC
0D03:0014 90            NOP
-G

Program terminated normally
-R
AX=0006  BX=0002  CX=0016  DX=0004  SP=003C  BP=0000  SI=0000  DI=0000
DS=0DD6  ES=0DD6  SS=0DE8  CS=0D03  IP=0014   NV UP EI PL NZ NA PO NC
0D03:0014 90            NOP
-R IP
IP 0014
:0
-G 5

AX=0000  BX=0002  CX=0016  DX=0004  SP=0038  BP=0000  SI=0000  DI=0000
DS=0DD6  ES=0DD6  SS=0DE8  CS=0D03  IP=0005   NV UP EI PL NZ NA PO NC
0D03:0005 3BC3          CMP     AX,BX
-R AX
AX 0000
:2
-R BX
BX 0002
:6
-T

AX=0002  BX=0006  CX=0016  DX=0004  SP=0038  BP=0000  SI=0000  DI=0000
DS=0DD6  ES=0DD6  SS=0DE8  CS=0D03  IP=0007   NV UP EI NG NZ AC PE CY
0D03:0007 7207          JB      0010
-G 14

AX=0002  BX=0006  CX=0016  DX=0004  SP=0038  BP=0000  SI=0000  DI=0000
DS=0DD6  ES=0DD6  SS=0DE8  CS=0D03  IP=0014   NV UP EI PL NZ NA PO NC
0D03:0014 90            NOP
-G

Program terminated NORMALLY
-Q

C:\DOS>
```

(c)

Figure 6–19 (continued)

Next, the loading of the program is verified with the Unassemble command

```
-U 0 15   (↵)
```

Note in Fig. 6–19(c) that the CMP instruction, which is the first instruction of the sequence that generates the absolute difference, is located at address 0D03:0005. Let us execute down to this statement with the Go command

```
-G 5   (↵)
```

Now we will load AX and BX with the case (a) data. This is done with the R commands

```
-R AX        (↵)
AX 0000
:6           (↵)
-R BX        (↵)
BX 0000
:2           (↵)
```

Next, execute the compare instruction with the command

```
-T     (↵)
```

Note in the trace information display in Fig. 6–19(c) that the carry flag is reset (NC). Therefore, no jump will take place when the JB instruction is executed.

The rest of the program can be executed by inputting the command

```
-G 14   (↵)
```

From Fig. 6–19(c), we find that DX contains 4. Executing the SUB instruction at 0D03:000B produced this result. Before executing the program for the case (b) data, the command

```
-G   (↵)
```

is issued. This command causes the program to terminate normally.

The R command shows that the value in IP must be reset to zero, and then we can execute down to the CMP instruction. This is done with the commands

```
-R IP        (↵)
IP 0014
:0           (↵)
```

and

```
-G 5         (↵)
```

Note in Fig. 6–19(c) that IP again contains 0005_{16} and points to the CMP instruction. Next, the data for case (b) are loaded with R commands. This gives

```
-R AX        (↵)
AX 0000
:2           (↵)
-R BX        (↵)
BX 0002
:6           (↵)
```

Now a T command is used to execute the CMP instruction. Notice that CY is set this time. Therefore, control is passed to the instruction at 0D03:0010.

A Go command is now used to execute down to the instruction at 0D03:0014. This command is

```
-G 14    (↵)
```

Note that DX again contains 4; however, this time it was calculated with the SUB instruction at 0D03:0012.

▲ 6.4 SUBROUTINES AND SUBROUTINE-HANDLING INSTRUCTIONS

A *subroutine* is a special segment of program that can be called for execution from any point in a program. Figure 6–20(a) illustrates the concept of a subroutine. Here we see a program structure where one part of the program is called the *main program.* In addition to this, we find a group of instructions attached to the main program, known as a subroutine. The subroutine is written to provide a function that must be performed at various points in the main program. Instead of including this piece of code in the main program each time the function is needed, it is put into the program just once as a subroutine. An assembly language subroutine is also referred to as a *procedure.*

Wherever the function must be performed, a single instruction is inserted into the main body of the program to "call" the subroutine. Remember that the logical address CS:IP identifies the next instruction to be fetched for execution. Thus, to branch to a subroutine that starts elsewhere in memory, the value in either IP or CS and IP must be modified.

After executing the subroutine, we want to return control to the instruction that immediately follows the one called the subroutine. To facilitate this return operation, return linkage is saved when the call takes place. That is, the original value of IP or IP and CS must be preserved. A return instruction is included at the end of the subroutine to initiate the *return sequence* to the main program environment. In this way, program execution resumes in the main program at the point where it left off due to the occurrence of the subroutine call.

The instructions provided to transfer control from the main program to a subroutine and return control back to the main program are called *subroutine-handling instructions.* Let us now examine the instructions provided for this purpose.

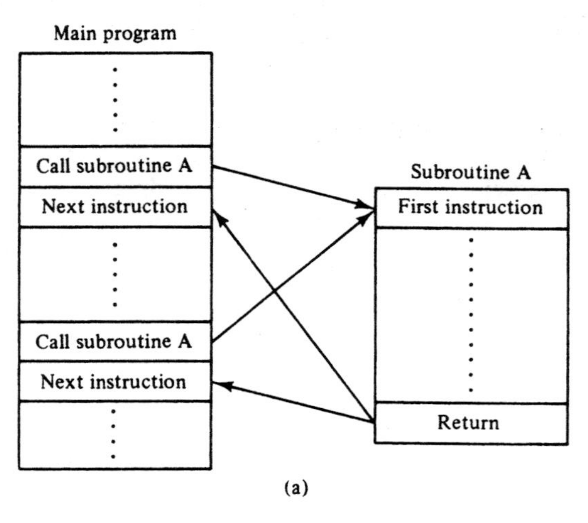

(a)

Mnemonic	Meaning	Format	Operation	Flags Affected
CALL	Subroutine call	CALL operand	Execution continues from the address of the subroutine specified by the operand. Information required to return back to the main program such as IP and CS are saved on the stack.	None

(b)

Operand
Near-proc
Far-proc
Memptr16
Regptr16
Memptr32

(c)

Figure 6–20 (a) Subroutine concept. (b) Subroutine call instruction. (c) Allowed operands.

CALL and RET Instructions

There are two basic instructions in the instruction set of the 8088 for subroutine handling: the *call* (CALL) and *return* (RET) instructions. Together they provide the mechanism for calling a subroutine into operation and returning control back to the main program at its completion. We will first discuss these two instructions and later introduce other instructions that can be used in conjunction with subroutines.

Just like the JMP instruction, CALL allows implementation of two types of operations: the *intrasegment call* and the *intersegment call*. The CALL instruction is shown in Fig. 6–20(b), and its allowed operand variations are shown in Fig. 6–20(c).

It is the operand that initiates either an intersegment or an intrasegment call. The operands Near-proc, Memptr16, and Regptr16 all specify intrasegment calls to a subroutine. In all three cases, execution of the instruction causes the contents of IP to be saved on the stack. Then the stack pointer (SP) is decremented by two. This is the push operation that was introduced in the section covering the stack in Chapter 2. The saved value of IP is the offset address of the instruction that immediately follows the CALL instruction. After saving this return address, a new 16-bit value, which is specified by the instruction's operand and corresponds to the storage location of the first instruction in the subroutine, is loaded into IP.

The types of operands represent different ways of specifying a new value of IP. Using a Near-proc operand, a subroutine located in the same code segment can be called. An example is

```
CALL  1234H
```

Here 1234H identifies the starting address of the subroutine. It is encoded as the difference between 1234H and the updated value of IP—that is, the IP for the instruction following the CALL instruction.

The Memptr16 and Regptr16 operands provide indirect subroutine addressing by specifying a memory location or an internal register, respectively, as the source of a new value for IP. The value specified is the actual offset that is loaded into IP. An example of a Regptr16 operand is

```
CALL  BX
```

When this instruction is executed, the contents of BX are loaded into IP and execution continues with the subroutine starting at the physical address derived from the current CS and the new value of IP.

By using various addressing modes of the 8088, an operand that resides in memory is used as the call to offset address. This represents a Memptr16 type of operand. For instance, the instruction

```
CALL  [BX]
```

has its subroutine offset address at the memory location whose physical address is derived from the contents of DS and BX. The value stored at this memory location is loaded into IP. Again the current contents of CS and the new value in IP point to the first instruction of the subroutine.

Note that in both intrasegment call examples the subroutine is located within the same code segment as the call instruction. The other type of CALL instruction, the intersegment call, permits the subroutine to reside in another code segment. It corresponds to the Far-proc and Memptr32 operands. These operands specify both a new offset address for IP and a new segment address for CS. In both cases, execution of the call instruction causes the contents of the CS and IP registers to be saved on the stack, and then new values are loaded into IP and CS. The saved values of CS and IP permit return to the main program from a different code segment.

Far-proc represents a 32-bit immediate operand that is stored in the four bytes that follow the opcode of the call instruction in program memory. These two words are loaded directly from code segment memory into IP and CS with execution of the CALL instruction.

On the other hand, when the operand is Memptr32, the pointer for the subroutine is stored as four consecutive bytes in data memory. The location of the first byte of the pointer can be specified indirectly by one of the 8088's memory addressing modes. An example is

```
CALL DWORD PTR [DI]
```

Here the physical address of the first byte of the 4-byte pointer in memory is derived from the contents of DS and DI.

Every subroutine must end by executing an instruction that returns control to the main program. This is the return (RET) instruction. It is described in Fig. 6–21(a) and (b). Note that its execution causes the value of IP or both the values of IP and CS that were saved on the stack to be returned back to their corresponding registers and the stack pointer to be adjusted appropriately. This is the pop operation that was discussed in Chapter 2. In general, an intrasegment return results from an intrasegment call and an intersegment return results from an intersegment call. In this way, program control is returned to the instruction that immediately follows the call instruction in program memory.

There is an additional option with the return instruction. It is that a 2-byte constant can be included with the return instruction. This constant gets added to the stack pointer after restoring the return address. The purpose of this stack pointer displacement is to provide a simple means by which the *parameters* that were saved on the stack before the call to the subroutine was initiated can be discarded. For instance, the instruction

```
RET 2
```

when executed adds 2 to SP. This discards one word parameter as part of the return sequence.

Mnemonic	Meaning	Format	Operation	Flags Affected
RET	Return	RET or RET Operand	Return to the main program by restoring IP (and CS for far-proc). If Operand is present, it is added to the contents of SP.	None

(a)

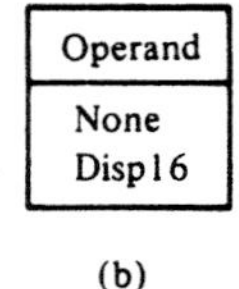

Operand
None Disp16

(b)

Figure 6–21 (a) Return instruction. (b) Allowed operands.

EXAMPLE 6.10

The source program in Fig. 6–22(a) can be used to demonstrate the use of the call and return instructions to implement a subroutine. This program was assembled and linked to produce a run module in file EX610.EXE. Its source listing is provided in Fig. 6–22(b). Trace the operation of the program by executing it with DEBUG for data (AX) = 2 and (BX) = 4.

Solution

We begin by calling up DEBUG and loading the program with the DOS command

```
C:\DOS>DEBUG A:EX610.EXE   (↵)
```

The loading of the program is now verified with the Unassemble command

```
-U 0 D   (↵)
```

Figure 6–22(c) shows that the program correctly loaded. Moreover, the CALL instruction is located at offset 0005_{16} of the current code segment. The command

```
-G 5   (↵)
```

```
TITLE    EXAMPLE 6.10

      PAGE       ,132

STACK_SEG          SEGMENT             STACK 'STACK'
                   DB                  64 DUP(?)
STACK_SEG          ENDS

CODE_SEG           SEGMENT             'CODE'
EX610    PROC      FAR
      ASSUME   CS:CODE_SEG, SS:STACK_SEG

;To return to DEBUG program put return address on the stack

      PUSH     DS
      MOV      AX, 0
      PUSH     AX

;Following code implements Example 6.10

      CALL     SUM
      RET

SUM   PROC     NEAR
      MOV      DX, AX
      ADD      DX, BX            ; (DX) = (AX) + (BX)
      RET
SUM   ENDP

EX610 ENDP
CODE_SEG       ENDS

      END      EX610
```

(a)

Figure 6–22 (a) Source program for Example 6.10. (b) Source listing produced by assembler. (c) Execution of the program with DEBUG.

```
                              TITLE   EXAMPLE 6.10

                                 PAGE    ,132

0000                               STACK_SEG      SEGMENT        STACK 'STACK'
0000      40 [                                    DB             64 DUP(?)
            ??
               ]

0040                               STACK_SEG      ENDS

0000                               CODE_SEG       SEGMENT        'CODE'
0000                               EX610    PROC  FAR
                                       ASSUME  CS:CODE_SEG, SS:STACK_SEG

                           ;To return to DEBUG program put return address on the stack

0000  1E                                   PUSH   DS
0001  B8 0000                              MOV    AX, 0
0004  50                                   PUSH   AX

                           ;Following code implements Example 6.10

0005  E8 0009 R                            CALL   SUM
0008  CB                                   RET

0009                               SUM     PROC   NEAR
0009  8B D0                                MOV    DX, AX
000B  03 D3                                ADD    DX, BX          ; (DX) = (AX) + (BX)
000D  C3                                   RET
000E  90                           SUM     ENDP

000E                               EX610   ENDP
000E                               CODE_SEG       ENDS

                                           END    EX610

Segments and groups:

            N a m e                     Size    align   combine class

CODE_SEG . . . . . . . . . . . .        000E    PARA    NONE    'CODE'
STACK_SEG. . . . . . . . . . . .        0040    PARA    STACK   'STACK'

Symbols:

            N a m e                     Type    Value   Attr

EX610. . . . . . . . . . . . . .        F PROC  0000    CODE_SEG    Length =000E
SUM  . . . . . . . . . . . . . .        N PROC  0000    CODE_SEG    Length =0005

Warning Severe
Errors  Errors
0       0
```

(b)

Figure 6–22 (continued)

executes the program down to the CALL instruction. The state information displayed in Fig. 6–22(c) shows that (CS) = $0D03_{16}$, (IP) = 0005_{16}, and (SP) = $003C_{16}$. Now let us load the AX and BX registers with R commands:

```
-R AX      (↵)
AX 0000
:2         (↵)
-R BX      (↵)
BX 0000
:4         (↵)
```

```
C:\DOS>DEBUG A:EX610.EXE
-U 0 D
0D03:0000 1E            PUSH    DS
0D03:0001 B80000        MOV     AX,0000
0D03:0004 50            PUSH    AX
0D03:0005 E80100        CALL    0009
0D03:0008 CB            RETF
0D03:0009 8BD0          MOV     DX,AX
0D03:000B 03D3          ADD     DX,BX
0D03:000D C3            RET
-G 5

AX=0000  BX=0000  CX=000E  DX=0000  SP=003C  BP=0000  SI=0000  DI=0000
DS=0F41  ES=0F41  SS=0F52  CS=0D03  IP=0005   NV UP EI PL NZ NA PO NC
0D03:0005 E80100        CALL    0009
-R AX
AX 0000
:2
-R BX
BX 0000
:4
-T

AX=0002  BX=0004  CX=000E  DX=0000  SP=003A  BP=0000  SI=0000  DI=0000
DS=0F41  ES=0F41  SS=0F52  CS=0D03  IP=0009   NV UP EI PL NZ NA PO NC
0D03:0009 8BD0          MOV     DX,AX
-D SS:3A 3B
0F52:0030                                 08 00
-T

AX=0002  BX=0004  CX=000E  DX=0002  SP=003A  BP=0000  SI=0000  DI=0000
DS=0F41  ES=0F41  SS=0F52  CS=0D03  IP=000B   NV UP EI PL NZ NA PO NC
0D03:000B 03D3          ADD     DX,BX
-T

AX=0002  BX=0004  CX=000E  DX=0006  SP=003A  BP=0000  SI=0000  DI=0000
DS=0F41  ES=0F41  SS=0F52  CS=0D03  IP=000D   NV UP EI PL NZ NA PE NC
0D03:000D C3            RET
-T

AX=0002  BX=0004  CX=000E  DX=0006  SP=003C  BP=0000  SI=0000  DI=0000
DS=0F41  ES=0F41  SS=0F52  CS=0D03  IP=0008   NV UP EI PL NZ NA PE NC
0D03:0008 CB            RETF
-G

Program terminated normally
-Q

C:\DOS>
```

(c)

Figure 6–22 (continued)

Next, the CALL instruction is executed with the T command

```
-T (↵)
```

and looking at the displayed state information in Fig. 6–22(c) shows that CS still contains $0D03_{16}$, IP has been loaded with 0009_{16}, and SP has been decremented to $003A_{16}$. This information tells us that the next instruction to be executed is the move instruction at address 0D03:0009, and a word of data has been pushed to the stack.

Before executing another instruction, let us look at what got pushed onto the stack. Issuing the memory dump command does this:

```
-D SS:3A 3B (↵)
```

Note from Fig. 6–22(c) that the value 0008_{16} has been pushed onto the stack. This is the address offset of the RETF instruction that follows the CALL instruction and is the address of the instruction to which control is to be returned at the completion of the subroutine.

Two more T commands are used to execute the move and add instructions of the subroutine. From the state information displayed in Fig. 6–22(c), we see that their execution causes the value 2_{16} in AX to be copied into DX and then the value 4_{16} in BX to be added to the value in DX. This results in the value 6_{16} in DX.

Now the RET instruction is executed by issuing another T command. Figure 6–22(c) shows that execution of this instruction causes the value 0008_{16} to be popped off the stack and put back into the IP register. Therefore, the next instruction to be executed is the one located at address 0D03:0008; this is the RETF instruction. Moreover, note that as the word is popped from the stack back into IP, the value in SP is incremented by two. After this, the program is run to completion by issuing a Go command.

PUSH and POP Instructions

Upon entering a subroutine, it is usually necessary to save the contents of certain registers or some other main program parameters. Pushing them onto the stack saves these values. Typically, these data correspond to registers and memory locations that are used by the subroutine. In this way, their original contents are kept intact in the stack segment of memory during the execution of the subroutine. Before a return to the main program takes place, the saved registers and main program parameters are restored. Popping the saved values from the stack back into their original locations does this. Thus, a typical structure of a subroutine is that shown in Fig. 6–23.

The instruction that is used to save parameters on the stack is the *push* (PUSH) instruction, and that used to retrieve them is the *pop* (POP) instruction. Note in Fig. 6–24(a) and (b) that the standard PUSH and POP instructions can be written with a general-purpose register, a segment register (excluding CS), or a storage location in memory as their operand.

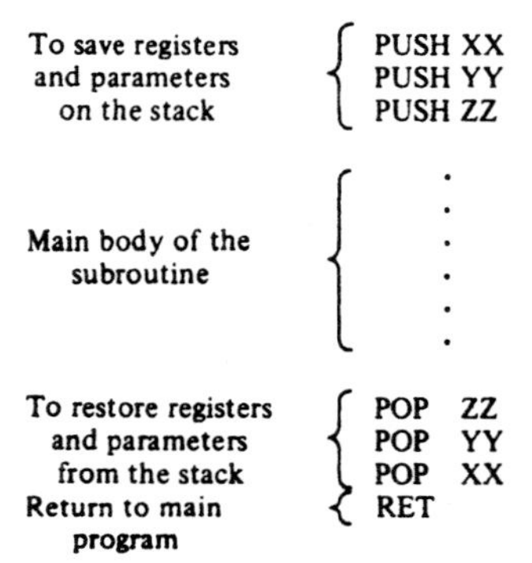

Figure 6–23 Structure of a subroutine.

Mnemonic	Meaning	Format	Operation	Flags Affected
PUSH	Push word onto stack	PUSH S	((SP)) ← (S) (SP) ← (SP)-2	None
POP	Pop word off stack	POP D	(D) ← ((SP)) (SP) ← (SP)+2	None

(a)

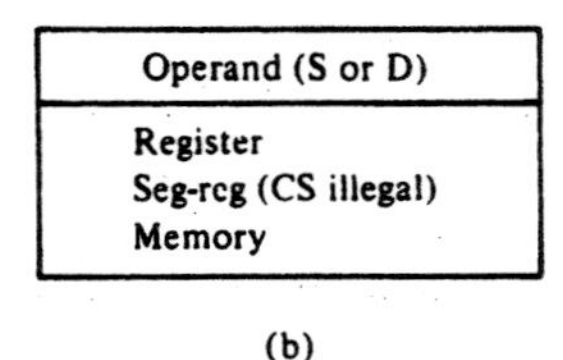

Operand (S or D)
Register Seg-reg (CS illegal) Memory

(b)

Figure 6–24 (a) PUSH and POP instructions. (b) Allowed operands.

Execution of a PUSH instruction causes the data corresponding to the operand to be pushed onto the top of the stack. For instance, if the instruction is

```
PUSH AX
```

the result is as follows:

$$((SP) - 1) \leftarrow (AH)$$
$$((SP) - 2) \leftarrow (AL)$$
$$(SP) \leftarrow (SP) - 2$$

This shows that the two bytes of AX are saved in the stack part of memory and the stack pointer is decremented by two so that it points to the new top of the stack.

On the other hand, if the instruction is

```
POP AX
```

its execution results in

$$(AL) \leftarrow ((SP))$$
$$(AH) \leftarrow ((SP) + 1)$$
$$(SP) \leftarrow (SP) + 2$$

In this manner, the saved contents of AX are restored in the register.

EXAMPLE 6.11

Write a procedure named SQUARE that squares the contents of BL and places the result in BX. Assume that this procedure is called from the main program, which is located in the same code segment.

Solution

The beginning of the procedure is defined with the directive statement

```
SQUARE PROC NEAR
```

To square the number in BL, we use the 8-bit signed multiply instruction, IMUL. This instruction requires the use of register AX for its operation. Therefore, at entry of the procedure, we must save the value currently held in AX. Pushing its contents to the stack with the instruction does this

```
PUSH AX
```

Now we load AX with the contents of BL using the instruction

```
MOV AL, BL
```

To square the contents of AL, we use the instruction

```
IMUL BL
```

which multiplies the contents of AL with the contents of BL and places the result in AX. The result is the square of the original contents of BL. To place the result in BX, use the instruction

```
MOV BX, AX
```

This completes the square operation; but before returning to the main part of the program, the original contents of AX that are saved on the stack are restored with the pop instruction

```
POP AX
```

Then a return instruction is used to pass control back to the main program:

```
RET
```

The procedure must be terminated with the end procedure directive statement that follows:

```
SQUARE ENDP
```

Figure 6–25 shows the complete instruction sequence.

```
;Subroutine:   SQUARE
;Description:  (BX) = square of (BL)

SQUARE PROC  NEAR
       PUSH AX           ;Save the register to be used
       MOV  AL,BL        ;Place the number in AL
       IMUL BL           ;Multiply with itself
       MOV  BX,AX        ;Save the result
       POP  AX           ;Restore the register used
       RET
SQUARE ENDP
```

Figure 6–25 Program for Example 6.11

EXAMPLE 6.12

Figure 6–26(a) displays a source program that can be used to demonstrate the execution of the procedure written in Example 6.11. The source listing produced when this program was assembled is given in Fig. 6–26(b), and the run module produced when it was linked is stored in file EX612.EXE. Use the DEBUG program to load this run module and verify its operation.

Solution

The DEBUG program and run module can be loaded with the command

```
C:\DOS>DEBUG A:EX612.EXE   (↵)
```

After loading is completed, the instructions of the program are unassembled with the command

```
-U 0 18   (↵)
```

The displayed information in Fig. 6–26(c) shows that the program did load correctly.

The instruction sequence in Fig. 6–26(c) shows that the part of the program whose operation is of interest starts with the CALL instruction at address 0DEC:000C. Now we execute down to this point in the program with the Go command

```
-G C   (↵)
```

Note that BL contains the number 12H, which will be squared by the subroutine.

Now, the call instruction is executed with the command

```
-T   (↵)
```

Note from the displayed information for this command in Fig. 6–26(c) that the value held in IP has been changed to 0010_{16}. Therefore, control has been passed to address 0DEC:0010, which is the first instruction of procedure SQUARE. Moreover, note that stack pointer (SP) has been decremented to the value $003A_{16}$. The new top of the stack is

```
TITLE    EXAMPLE 6.12

      PAGE      ,132

STACK_SEG          SEGMENT           STACK 'STACK'
                   DB                64 DUP(?)
STACK_SEG          ENDS

DATA_SEG           SEGMENT
TOTAL              DW                1234H
DATA_SEG           ENDS

CODE_SEG           SEGMENT           'CODE'
EX612    PROC      FAR
                     ASSUME  CS:CODE_SEG, SS:STACK_SEG, DS:DATA_SEG

;To return to DEBUG program put return address on the stack

      PUSH    DS
      MOV     AX, 0
      PUSH    AX

;Setup the data segment

      MOV     AX, DATA_SEG
      MOV     DS, AX

;Following code implements Example 6.12

      MOV     BL,12H           ;BL contents = the number to be squared
      CALL    SQUARE           ;Call the procedure to square BL contents
      RET                      ;Return to DEBUG program
EX612   ENDP

;Subroutine:      SQUARE
;Description:     (BX) = square of (BL)

SQUARE  PROC      NEAR
      PUSH    AX               ;Save the register to be used
      MOV     AL,BL            ;Place the number in AL
      IMUL    BL               ;Multiply with itself
      MOV     BX,AX            ;Save the result
      POP     AX               ;Restore the register used
      RET
SQUARE  ENDP

CODE_SEG        ENDS

      END     EX612
```

(a)

Figure 6–26 (a) Source program for Example 6.12. (b) Source listing produced by assembler. (c) Execution of the program with DEBUG.

at address 0DE7:003A. The word held at the top of the stack can be examined with the command

```
-D SS:3A 3B   (↵)
```

Note in Fig. 6–26(c) that its value is $000F_{16}$. The instruction sequence in Fig. 6–26(c) shows that this is the address of the RETF instruction. This is the instruction to which control is to be returned at the completion of the procedure.

Next, the PUSH AX instruction is executed with the command

```
-T   (↵)
```

```
                                TITLE    EXAMPLE 6.12

                                 PAGE       ,132

0000                               STACK_SEG          SEGMENT           STACK 'STACK'
0000   0040[                                          DB                64 DUP(?)
         ??
                 ]

0040                               STACK_SEG          ENDS

0000                               DATA_SEG           SEGMENT
0000   1234                        TOTAL              DW                1234H
0002                               DATA_SEG           ENDS

0000                               CODE_SEG           SEGMENT           'CODE'
0000                               EX612    PROC      FAR
                                       ASSUME  CS:CODE_SEG, SS:STACK_SEG, DS:DATA_SEG

                           ;To return to DEBUG program put return address on the stack

0000   1E                                     PUSH     DS
0001   B8 0000                                MOV      AX, 0
0004   50                                     PUSH     AX

                           ;Setup the data segment

0005   B8 —— R                                MOV      AX, DATA_SEG
0008   8E D8                                  MOV      DS, AX

                           ;Following code implements Example 6.12

000A   B3  12                                 MOV      BL,12H           ;BL contents = the number
to be squared
000C   E8 0010 R                              CALL     SQUARE           ;Call the procedure to
square BL contents
000F   CB                                     RET                       ;Return to DEBUG program
0010                               EX612      ENDP

                           ;Subroutine:       SQUARE
                           ;Description:      (BX) = square of (BL)

0010                               SQUARE     PROC     NEAR
0010   50                                     PUSH     AX               ;Save the register to be
used
0011   8A C3                                  MOV      AL,BL            ;Place the number in AL
0013   F6 EB                                  IMUL     BL               ;Multiply with itself
0015   8B D8                                  MOV      BX,AX            ;Save the result
0017   5*                                     POP      AX               ;Restore the register used
0018   C3                                     RET
0019                               SQUARE     ENDP

0019                               CODE_SEG            ENDS

                                   END        EX612
```

(b)

Figure 6–26 (continued)

and again, looking at the displayed state information, we find that SP has been decremented to the value 0038_{16}. Displaying the word at the top of the stack with the command

```
-D SS:38 39 (↵)
```

we find that it is the same as the contents of AX. This confirms that the original contents of AX are saved on the stack.

Now we execute down to the POP AX instruction with the command

```
-G 17 (↵)
```

```
Segments and Groups:

                N a m e                  Length   Align   Combine Class

CODE_SEG . . . . . . . . . . . . . .     0019     PARA    NONE    'CODE'
DATA_SEG . . . . . . . . . . . . . .     0002     PARA    NONE
STACK_SEG  . . . . . . . . . . . . .     0040     PARA    STACK   'STACK

Symbols:

                N a m e                  Type     Value    Attr

EX612  . . . . . . . . . . . . . . .     F PROC  0000     CODE_SEG        Length = 0010
SQUARE . . . . . . . . . . . . . . .     N PROC  0010     CODE_SEG        Length = 0009

TOTAL  . . . . . . . . . . . . . . .     L WORD  0000     DATA_SEG

@CPU . . . . . . . . . . . . . . . .     TEXT  0101h
@FILENAME  . . . . . . . . . . . . .     TEXT  EX612
@VERSION . . . . . . . . . . . . . .     TEXT  510

     53 Source  Lines
     53 Total   Lines
     13 Symbols

  48016 + 440523 Bytes symbol space free

      0 Warning Errors
      0 Severe  Errors
```

(b)

Figure 6–26 (continued)

Looking at the displayed state information in Fig. 6–26(c), we find that the square of the contents of BL has been formed in BX.

Next, the pop instruction is executed with the command

```
-T  (↵)
```

and the displayed information shows that the original contents of AX popped off the stack and were put back into AX. Moreover, the value in SP increments to $003A_{16}$ so that once again the return address is at the top of the stack.

Finally, the RET instruction is executed with the command

```
-T  (↵)
```

As Fig. 6–26(c) shows, this causes the value 0010_{16} to be popped from the top of the stack back into IP. Therefore, IP now equals $000F_{16}$. In this way, we see that control flow returns to the instruction at address 0DEC:000F of the main program.

At times, we also want to save the contents of the flag register, and if we save them, we will later have to restore them. These operations can be accomplished with *push flags* (PUSHF) and *pop flags* (POPF) instructions, respectively, as shown in Fig. 6–27. Note that PUSHF saves the contents of the flag register on the top of the stack. On the other hand, POPF returns the flags from the top of the stack to the flag register.

```
C:\DOS>DEBUG A:EX612.EXE
-U 0 18
0DEC:0000 1E            PUSH    DS
0DEC:0001 B80000        MOV     AX,0000
0DEC:0004 50            PUSH    AX
0DEC:0005 B8EB0D        MOV     AX,0DEB
0DEC:0008 8ED8          MOV     DS,AX
0DEC:000A B312          MOV     BL,12
0DEC:000C E80100        CALL    0010
0DEC:000F CB            RETF
0DEC:0010 50            PUSH    AX
0DEC:0011 8AC3          MOV     AL,BL
0DEC:0013 F6EB          IMUL    BL
0DEC:0015 8BD8          MOV     BX,AX
0DEC:0017 58            POP     AX
0DEC:0018 C3            RET
-G C

AX=0DEB  BX=0012  CX=0069  DX=0000  SP=003C  BP=0000  SI=0000  DI=0000
DS=0DEB  ES=0DD7  SS=0DE7  CS=0DEC  IP=000C   NV UP EI PL NZ NA PO NC
0DEC:000c E80100        CALL    0010
-T

AX=0DEB  BX=0012  CX=0069  DX=0000  SP=003A  BP=0000  SI=0000  DI=0000
DS=0DEB  ES=0DD7  SS=0DE7  CS=0DEC  IP=0010   NV UP EI PL NZ NA PO NC
0DEC:0010 50            PUSH    AX
-D SS:3A 3B
0DE7:0030                                 OF 00
-T

AX=0DEB  BX=0012  CX=0069  DX=0000  SP=0038  BP=0000  SI=0000  DI=0000
DS=0DEB  ES=0DD7  SS=0DE7  CS=0DEC  IP=0011   NV UP EI PL NZ NA PO NC
0DEC:0011 8AC3          MOV     AL,BL
-D SS:38 39
0DE7:0030                           EB 0D
-G 17

AX=0144  BX=0144  CX=0069  DX=0000  SP=0038  BP=0000  SI=0000  DI=0000
DS=0DEB  ES=0DD7  SS=0DE7  CS=0DEC  IP=0017   OV UP EI PL NZ NA PE CY
0DEC:0017 58            POP     AX
-T

AX=0DEB  BX=0144  CX=0069  DX=0000  SP=003A  BP=0000  SI=0000  DI=0000
DS=0DEB  ES=0DD7  SS=0DE7  CS=0DEC  IP=0018   OV UP EI PL NZ NA PE CY
0DEC:0018 C3            RET
-T

AX=0DEB  BX=0144  CX=0069  DX=0000  SP=003C  BP=0000  SI=0000  DI=0000
DS=0DEB  ES=0DD7  SS=0DE7  CS=0DEC  IP=000F   OV UP EI PL NZ NA PE CY
0DEC:000F CB            RETF
-G

Program terminated normally
-Q

C:\DOS>
```

(c)

Figure 6–26 (continued)

Mnemonic	Meaning	Operation	Flags Affected
PUSHF	Push flags onto stack	((SP)) ← (Flags) (SP) ← (SP)-2	None
POPF	Pop flags from stack	(Flags) ← ((SP)) (SP) ← (SP)+2	OF, DF, IF, TF, SF, ZF, AF, PF, CF

Figure 6–27 Push flags and pop flags instructions.

▲ 6.5 LOOPS AND LOOP-HANDLING INSTRUCTIONS

The 8088 microprocessor has three instructions specifically designed for implementing loop operations. These instructions can be used in place of certain conditional jump instructions and give the programmer a simpler way of writing loop sequences. The loop instructions are listed in Fig. 6–28.

The first instruction, *loop* (LOOP), works with respect to the contents of the CX register. CX must be preloaded with a count that represents the number of times the loop is to repeat. Whenever LOOP is executed, the contents of CX are first decremented by one and then checked to determine if they are equal to zero. If equal to zero, the loop is complete and the instruction following LOOP is executed; otherwise, control is returned to the instruction at the label specified in the loop instruction. In this way, we see that LOOP is a single instruction that functions the same as a decrement CX instruction followed by a JNZ instruction.

For example, the LOOP instruction sequence shown in Fig. 6–29(a) causes the part of the program from label NEXT through the instruction LOOP to repeat a number of times equal to the value stored in CX. For example, if CX contains $000A_{16}$, the sequence of instructions included in the loop executes 10 times.

Figure 6–29(b) shows a practical implementation of a loop. Here we find the block move program developed in Section 6.3 is rewritten using the LOOP instruction. Comparing this program with the one in Fig. 6–18(b), we see that the instruction LOOP NXTPT has replaced both the DEC and JNZ instructions.

Mnemonic	Meaning	Format	Operation
LOOP	Loop	LOOP Short-label	$(CX) \leftarrow (CX) - 1$ Jump is initiated to location defined by short-label if $(CX) \neq 0$; otherwise, execute next sequential instruction
LOOPE/LOOPZ	Loop while equal/ loop while zero	LOOPE/LOOPZ Short-label	$(CX) \leftarrow (CX) - 1$ Jump to location defined by short-label if $(CX) \neq 0$ and $(ZF) = 1$; otherwise, execute next sequential instruction
LOOPNE/ LOOPNZ	Loop while not equal/ loop while not zero	LOOPNE/LOOPNZ Short-label	$(CX) \leftarrow (CX) - 1$ Jump to location defined by short-label if $(CX) \neq 0$ and $(ZF) = 0$; otherwise, execute next sequential instruction

Figure 6–28 Loop instructions.

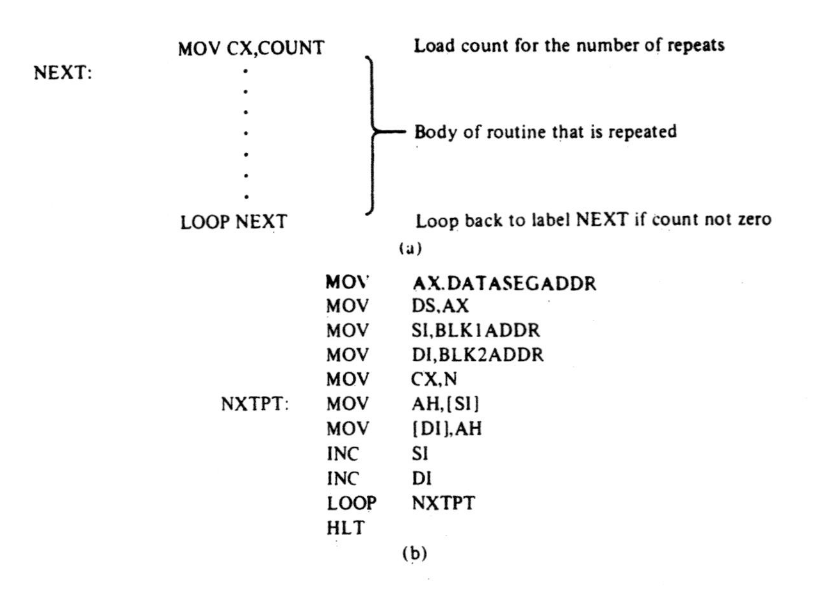

Figure 6–29 (a) Typical loop routine structure. (b) Block-move program employing the LOOP instruction.

EXAMPLE 6.13

The source program in Fig. 6–30(a) demonstrates the use of the LOOP instruction to implement a software loop operation. This program was assembled and linked to produce run module EX613.EXE. The source listing is shown in Fig. 6–30(b). Observe the operation of the loop by executing the program using DEBUG.

Solution

Looking at Fig. 6–30(c), we see that the run module is loaded with the DOS command

```
C:\DOS>DEBUG A:EX613.EXE   (↵)
```

Now the loading of the program is verified by unassembling it with the command

```
-U 0 F   (↵)
```

By comparing the instruction sequence displayed in Fig. 6–30(c) with the listing in Fig. 6.30(b), we find that the program has loaded correctly.

Also in Fig. 6–30(c), the instruction sequence of the loop whose operation is to be observed is located from address 0D03:000B through 0D03:000D. A Go command that executes the instructions down to address 0D03:000B is

```
-G B   (↵)
```

```
TITLE    EXAMPLE 6.13

         PAGE     ,132

STACK_SEG          SEGMENT             STACK 'STACK'
                   DB                  64 DUP(?)
STACK_SEG          ENDS

CODE_SEG           SEGMENT             'CODE'
EX613    PROC      FAR
         ASSUME    CS:CODE_SEG, SS:STACK_SEG

;To return to DEBUG program put return address on the stack

         PUSH      DS
         MOV       AX, 0
         PUSH      AX

;Following code implements Example 6.13

         MOV       CX, 5H
         MOV       DX, 0H
AGAIN:   NOP
         INC       DX
         LOOP      AGAIN

         RET                           ;Return to DEBUG program
EX613    ENDP
CODE_SEG           ENDS

         END       EX613
```

(a)

Figure 6–30 (a) Source program for Example 6.13. (b) Source listing produced by assembler. (c) Execution of the program with DEBUG.

Again, looking at Fig. 6–30(c), we see that these instructions initialize the count in CX to 0005_{16} and the contents of DX to 0000_{16}.

Now another Go command executes the loop down to address 0D03:000D.

```
-G D   (↵)
```

The displayed information shows that the pass count in DX has incremented by 1 to indicate that the first pass through the loop is about to be completed.

Next the LOOP instruction is executed with the Trace command

```
-T   (↵)
```

From Fig. 6–30(c), we find that the loop count CX has decremented by one, meaning that the first pass through the loop is complete, and the value in IP has been changed to $000B_{16}$. Therefore, control has returned to the NOP instruction that represents the beginning of the loop.

Now that we have observed the basic loop operation, let us execute the loop to completion with the command

```
-G F   (↵)
```

```
                                TITLE   EXAMPLE 6.13

                                     PAGE      ,132

0000                                   STACK_SEG        SEGMENT          STACK 'STACK'
0000         40 [                                       DB               64 DUP(?)
                ??
                   ]

0040                                   STACK_SEG        ENDS

0000                                   CODE_SEG         SEGMENT          'CODE'
0000                                   EX613    PROC    FAR
                                                ASSUME  CS:CODE_SEG, SS:STACK_SEG

                                ;To return to DEBUG program put return address on the stack

0000  1E                                        PUSH    DS
0001  B8 0000                                   MOV     AX, 0
0004  50                                        PUSH    AX

                                ;Following code implements Example 6.13

0005  B9 0005                                   MOV     CX, 5H
0008  BA 0000                                   MOV     DX, 0H
000B  90                               AGAIN:   NOP
000C  42                                        INC     DX
000D  E2 FC                                     LOOP    AGAIN

000F  CB                                        RET                       ;Return to DEBUG program
0010                                   EX613    ENDP
0010                                   CODE_SEG         ENDS

                                                END     EX613

Segments and groups:

                N a m e                         Size    align   combine class

CODE_SEG . . . . . . . . . . . .                0010    PARA    NONE    'CODE'
STACK_SEG. . . . . . . . . . . .                0040    PARA    STACK   'STACK'

Symbols:

                N a m e                         Type    Value   Attr

AGAIN. . . . . . . . . . . . . .                L NEAR  000B    CODE_SEG
EX613. . . . . . . . . . . . . .                F PROC  0000    CODE_SEG   Length =0010

Warning Severe
Errors  Errors
0       0
```

(b)

Figure 6–30 (continued)

The displayed information for this command in Fig. 6–30(c) shows us that at completion of the program, the loop count in CX has been decremented to 0000_{16} and the pass count in DX has been incremented to 0005_{16}.

The other two loop instructions in Fig. 6–28 operate in a similar way except that they check for two conditions. For instance, the instruction *loop while equal* (LOOPE)/*loop while zero* (LOOPZ) checks the contents of both CX and the zero flag (ZF). Each time the loop instruction is executed, CX decrements by 1 without affecting the flags, its contents are checked for 0, and the state of ZF that results from execution of the previous instruction is tested for 1. If CX is not equal to 0 and ZF equals 1, a jump is initiated to the loca-

```
C:\DOS>DEBUG A:EX613.EXE
-U 0 F
0D03:0000 1E            PUSH    DS
0D03:0001 B80000        MOV     AX,0000
0D03:0004 50            PUSH    AX
0D03:0005 B90500        MOV     CX,0005
0D03:0008 BA0000        MOV     DX,0000
0D03:000B 90            NOP
0D03:000C 42            INC     DX
0D03:000D E2FC          LOOP    000B
0D03:000F CB            RETF
-G B

AX=0000  BX=0000  CX=0005  DX=0000  SP=003C  BP=0000  SI=0000  DI=0000
DS=0DD7  ES=0DD7  SS-0DE8  CS=0D03  IP=000B   NV UP EI PL NZ NA PO NC
0D03:000B 90            NOP
-G D

AX=0000  BX=0000  CX=0005  DX=0001  SP=003C  BP=0000  SI=0000  DI=0000
DS=0DD7  ES=0DD7  SS=0DE8  CX=0D03  IP=000D   NV UP EI PL NZ NA PO NC
0D03:000D E2FC          LOOP    000B
-T

AX=0000  BX=0000  CX=0004  DX=0001  SP=003C  BP=0000  SI=0000  DI=0000
DS=0DD7  ES=0DD7  SS=0DE8  CS=0D03  IP=000B   NV UP EI PL NZ NA PO NC
0D03:000B 90            NOP
-G F

AX=0000  BX=0000  CX=0000  DX=0005  SP=003C  BP=0000  SI=0000  DI=0000
DS=0DD7  ES=0DD7  SS=0DE8  CS=0D03  IP=000F   NV UP EI PL NZ NA PE NC
0D03:000F CB
-G

Program terminated normally
-Q

C:\DOS>
```

(c)

Figure 6–30 (continued)

tion specified with the Short-label operand and the loop continues. If either CX or ZF is 0, the loop is complete, and the instruction following the loop instruction is executed.

Instruction *loop while not equal* (LOOPNE)/*loop while not zero* (LOOPNZ) works in a similar way to the LOOPE/LOOPZ instruction. The difference is that it checks ZF and CX looking for ZF equal to 0 together with CX not equal to 0. If these conditions are met, the jump back to the location specified with the Short-label operand is performed and the loop continues.

EXAMPLE 6.14

Explain what happens as the following sequence of instructions is executed.

```
        MOV    DL, 05H
        MOV    AX, 0A00H
        MOV    DS, AX
        MOV    SI, 0H
        MOV    CX, 0FH
AGAIN:  INC    SI
        CMP    [SI], DL
        LOOPNE AGAIN
```

Solution

The first five instructions are for initializing internal registers. Data register DL is loaded with 05_{16}; data segment register DS is loaded via AX with the value $0A00_{16}$; source index register SI is loaded with 0000_{16}; and count register CX is loaded with $0F_{16}$ (15_{10}). After initialization, a data segment is set up at physical address $0A000_{16}$, and SI points to the memory location at offset 0000_{16} in this data segment. DL contains the data 5_{10} and the CX register contains the loop count 15_{10}.

The part of the program that starts at the label AGAIN and ends with the LOOPNE instruction is a software loop. The first instruction in the loop increments the value in SI by 1. Therefore, the first time through the loop SI points to the memory address $0A001_{16}$. The next instruction compares the contents of this memory location with the contents of DL, 5_{10}. If the data held at $0A001_{16}$ are 5_{10}, the zero flag is set; otherwise, it is reset. The LOOPNE instruction then decrements CX (making it E_{16}) and then checks for CX = 0 or ZF = 1. If neither of these two conditions is satisfied, program control is returned to the instruction with the label AGAIN. This causes the comparison to be repeated for the examination of the contents of the next byte in memory. On the other hand, if either condition is satisfied, the loop is complete. In this way, we see that the loop is repeated until either a number 5_{10} is found or all locations in the address range $0A001_{16}$ through $0A00F_{16}$ are tested and found not to contain 5_{10}.

EXAMPLE 6.15

Figure 6–31(a) shows the source version of the program that is written in Example 6.14. This program was assembled and linked to produce run module EX615.EXE. The source listing that resulted from the assembly process is shown in Fig. 6–31(b). Verify the operation of the program by executing it with Go commands.

Solution

As Fig. 6–31(c) shows, the run module is loaded with the command

```
C:\DOS>DEBUG A:EX615.EXE   (↵)
```

The program loaded is verified with the command

```
-U 0 17   (↵)
```

Comparing the sequence of instructions displayed in Fig. 6–31(c) with those in the source listing shown in Fig. 6–31(b), we find that the program has loaded correctly.

Note that the loop that performs the memory-compare operation starts at address 0D03:0012. Let us begin by executing down to this point with the Go command

```
-G 12   (↵)
```

```
TITLE    EXAMPLE 6.15

      PAGE      ,132

STACK_SEG          SEGMENT              STACK 'STACK'
                   DB                   64 DUP(?)
STACK_SEG          ENDS

CODE_SEG           SEGMENT              'CODE'
EX615    PROC      FAR
         ASSUME    CS:CODE_SEG, SS:STACK_SEG

;To return to DEBUG program put return address on the stack

         PUSH      DS
         MOV       AX, 0
         PUSH      AX

;Following code implements Example 6.15

         MOV       DL, 5H
         MOV       AX, 0A00H
         MOV       DS, AX
         MOV       SI, OH
         MOV       CX, 0FH
AGAIN:   INC       SI
         CMP       [SI], DL
         LOOPNE    AGAIN

         RET                            ;Return to DEBUG program
EX615    ENDP
CODE_SEG           ENDS

         END       EX615
```

(a)

Figure 6–31 (a) Source program for Example 6.15 (b) Source listing produced by assembler. (c) Execution of the program with DEBUG.

The state information displayed in Fig. 6–31(c) shows that DL is loaded with 05_{16}, AX with $0A00_{16}$, DS with $0A00_{16}$, SI with 0000_{16}, and CX with $000F_{16}$.

Next, the table of data as follows is loaded with the E command

```
-E A00:0 4, 6, 3, 9, 5, 6, D, F, 9   (↵)
```

The nine values in this list are loaded into consecutive byte-wide memory locations over the range 0A00:0000 through 0A00:0008. The compare routine also checks the storage locations from 0A00:0009 through 0A00:000F. Let us dump the data held in this part of memory to verify that it has been initialized correctly. In Fig. 6–31(c), we see that this is done with the command

```
-D A00:0 F   (↵)
```

and, looking at the displayed data, we find that it has loaded correctly.

Now the loop is executed with the command

```
-G 17   (↵)
```

```
TITLE   EXAMPLE 6.15

              PAGE      ,132

0000                          STACK_SEG        SEGMENT          STACK 'STACK'
0000      40 [                                 DB               64 DUP(?)
            ??
                ]

0040                          STACK_SEG        ENDS

0000                          CODE_SEG         SEGMENT          'CODE'
0000                          EX615    PROC    FAR
                                       ASSUME  CS:CODE_SEG, SS:STACK_SEG

                  ;To return to DEBUG program put return address on the stack

0000  1E                               PUSH    DS
0001  B8 0000                          MOV     AX, 0
0004  50                               PUSH    AX

                  ;Following code implements Example 6.15

0005  B2 05                            MOV     DL, 5H
0007  B8 0A00                          MOV     AX, 0A00H
000A  8E D8                            MOV     DS, AX
000C  BE 0000                          MOV     SI, 0H
000F  B9 000F                          MOV     CX, 0FH
0012  46                      AGAIN:   INC     SI
0013  38 14                            CMP     [SI], DL
0015  E0 FB                            LOOPNE  AGAIN

0017  CB                               RET                      ;Return to DEBUG program
0018                          EX615    ENDP
0018                          CODE_SEG         ENDS

                                       END     EX615

Segments and groups:

          N a m e                      Size    align   combine class

CODE_SEG . . . . . . . . . . . . .     0018    PARA    NONE    'CODE'
STACK_SEG. . . . . . . . . . . . .     0040    PARA    STACK   'STACK'

Symbols:

          N a m e                      Type    Value   Attr

AGAIN. . . . . . . . . . . . . . .     L NEAR  0012    CODE_SEG
EX615. . . . . . . . . . . . . . .     F PROC  0000    CODE_SEG         Length =0018

Warning Severe
Errors  Errors
0       0
```

(b)

Figure 6–31 (continued)

In the display dump for this command in Fig. 6–31(c), we find that SI has incremented to the value 0004_{16}; therefore, the loop was run only four times. The fourth time through the loop SI equals four and the memory location pointed to by SI, the address 0A00:0005, contains the value 5. This value is equal to the value in DL; therefore, the instruction CMP [SI],DL results in a difference of zero, and the zero flag is set. Notice in Fig. 6–31(c) that this flag is identified as ZR in the display dump. For this reason, execution of the LOOPNZ instruction causes the loop to be terminated, and control is passed to the RETF instruction.

```
C:\DOS>DEBUG A:EX615.EXE
-U 0 17
0D03:0000 1E            PUSH    DS
0D03:0001 B80000        MOV     AX,0000
0D03:0004 50            PUSH    AX
0D03:0005 B205          MOV     DL,05
0D03:0007 B8000A        MOV     AX,0A00
0D03:000A 8ED8          MOV     DS,AX
0D03:000C BE0000        MOV     SI,0000
0D03:000F B90F00        MOV     CX,000F
0D03:0012 46            INC     SI
0D03:0013 3814          CMP     [SI],DL
0D03:0015 E0FB          LOOPNZ  0012
0D03:0017 CB            RETF
-G 12

AX=0A00  BX=0000  CX=000F  DX=0005  SP=003C  BP=0000  SI=0000  DI=0000
DS=0A00  ES=0DD7  SS=0DE9  CS=0D03  IP=0012   NV UP EI PL NZ NA PO NC
0D03:0012 46            INC     SI
-E A00:0 4,6,3,9,5,6,D,F,9
-D A00:0 F
0A00:0000  04 06 03 09 05 06 0D 0F-09 75 09 80 7C 02 54 75   .........u..|.Tu
-G 17

AX=0A00  BX=0000  CX=000B  DX=0005  SP=003C  BP=0000  SI=0004  DI=0000
DS=0A00  ES=0DD7  SS=0DE9  CS=0D03  IP=0017   NV UP EI PL ZR NA PE NC
0D03:0017 CB            RETF
-G

Program terminated normally
-Q

C:\DOS>
```

(c)

Figure 6–31 (continued)

▲ 6.6 STRINGS AND STRING-HANDLING INSTRUCTIONS

The 8088 microprocessor is equipped with special instructions to handle *string operations.* By *string* we mean a series of data words (or bytes) that reside in consecutive memory locations. The string instructions of the 8088 permit a programmer to implement operations such as to move data from one block of memory to a block elsewhere in memory. A second type of operation that is easily performed is scanning a string of data elements stored in memory to look for a specific value. Other examples are comparing the elements of two strings in order to determine whether they are the same or different, and initializing a group of consecutive memory locations. Complex operations such as these require several nonstring instructions to be implemented.

There are five basic string instructions in the instruction set of the 8088. These instructions, as listed in Fig. 6–32, are *move byte* or *word string* (MOVSB/MOVSW), *compare string* (CMPSB/CMPSW), *scan string* (SCASB/SCASW), *load string* (LODSB/LODSW), and *store string* (STOSB/STOSW). They are called the *basic string instructions* because each defines an operation for one element of a string. Thus, these operations must be repeated to handle a string of more than one element. Let us first look at the operations these instructions perform.

Mnemonic	Meaning	Format	Operation	Flags Affected
MOVS	Move string	MOVSB/MOVSW	((ES)0 + (DI)) ← ((DS)0 + (SI)) (SI) ← (SI) ± 1 or 2 (DI) ← (DI) ± 1 or 2	None
CMPS	Compare string	CMPSB/CMPSW	Set flags as per ((DS)0 + (SI)) − ((ES)0 + (DI)) (SI) ← (SI) ± 1 or 2 (DI) ← (DI) ± 1 or 2	CF, PF, AF, ZF, SF, OF
SCAS	Scan string	SCASB/SCASW	Set flags as per (AL or AX) − ((ES)0 + (DI)) (DI) ← (DI) ± 1 or 2	CF, PF, AF, ZF, SF, OF
LODS	Load string	LODSB/LODSW	(AL or AX) ← ((DS)0 + (SI)) (SI) ← (SI) ± 1 or 2	None
STOS	Store string	STOSB/STOSW	((ES)0 + (DI)) ← (AL or AX) ± 1 or 2 (DI) ← (DI) ± 1 or 2	None

Figure 6–32 Basic string instructions.

Move String—MOVSB, MOVSW

The instructions MOVSB and MOVSW perform the same basic operation. An element of the string specified by the source index (SI) register with respect to the current data segment (DS) register is moved to the location specified by the destination index (DI) register with respect to the current extra segment (ES) register. The move can be performed on a byte or a word of data. After the move is complete, the contents of both SI and DI are automatically incremented or decremented by 1 for a byte move and by 2 for a word move. The address pointers in SI and DI increment or decrement, depending on how the direction flag (DF) is set.

For example, the instruction

```
MOVSB
```

can be used to move a byte.

Figure 6–33 shows an example of a program that uses MOVSB. This program is a modified version of the block-move program shown in Fig. 6–29(b). Note that the two MOV instructions that perform the data transfer and two INC instructions that update the pointer have been replaced with one move-string byte instruction. We have also made DS equal to ES.

```
        MOV     AX,DATASEGADDR
        MOV     DS,AX
        MOV     ES,AX
        MOV     SI,BLK1ADDR
        MOV     DI,BLK2ADDR
        MOV     CX,N
        CLD
NXTPT:  MOVSB
        LOOP    NXTPT
        HLT
```

Figure 6–33 Block-move program using the move-string instruction.

Compare String and Scan String—CMPSB/CMPSW and SCASB/SCASW

The compare-strings instruction can be used to compare two elements in the same or different strings: it subtracts the destination operand from the source operand and adjusts the flags accordingly. The result of subtraction is not saved; therefore, the operation does not affect the operands in any way.

An example of a compare strings instruction for bytes of data is

```
CMPSB
```

Again, the address in SI points to the source element with respect to the current value in DS, and the destination element is specified by the contents of DI relative to the contents of ES. When executed, the operands are compared, the flags are adjusted, and both SI and DI are updated so that they point to the next elements in their respective strings.

The scan-string instruction is similar to compare strings; however, it compares the byte or word element of the destination string at the physical address derived from DI and ES to the contents of AL or AX, respectively. The flags are adjusted based on this result and DI incremented or decremented.

A program using the SCASB instruction that implements a string scan operation similar to that described in Example 6.14 is shown in Fig. 6–34. Note that we have made DS equal to ES.

Load and Store String—LODSB/LODSW and STOSB/STOSW

The last two instructions in Fig. 6–32, load string and store string, are specifically provided to move string elements between the accumulator and memory. LODSB loads a byte from a string in memory into AL. The address in SI is used relative to DS to determine the address of the memory location of the string element; SI is incremented by 1 after loading. Similarly, the instruction LODSW indicates that the word-string element at the physical address derived from DS and SI is to be loaded into AX. Then the index in SI is automatically incremented by 2.

On the other hand, STOSB stores a byte from AL into a string location in memory. This time the contents of ES and DI are used to form the address of the storage location in memory. For example, the program in Fig. 6–35 loads the block of memory locations from $0A000_{16}$ through $0A00F_{16}$ with number 5.

```
        MOV     AX,0
        MOV     DS,AX
        MOV     ES,AX
        MOV     AL,05
        MOV     DI,0A000H
        MOV     CX,0FH
        CLD
AGAIN:  SCASB
        LOOPNE  AGAIN
NEXT:
```

Figure 6–34 Block scan operation using the SCASB instruction.

```
        MOV     AX,0
        MOV     DS,AX
        MOV     ES,AX
        MOV     AL,05
        MOV     DI,0A000H
        MOV     CX,0FH
        CLD
AGAIN:  STOSB
        LOOP    AGAIN
```

Figure 6–35 Initializing a block of memory with a store string operation.

Repeat String—REP

In most applications, the basic string operations must be repeated in order to process arrays of data. Inserting a repeat prefix before the instruction that is to be repeated does this. The *repeat prefixes* of the 8088 are shown in Fig. 6–36.

The first prefix, REP, causes the basic string operation to be repeated until the contents of register CX become equal to 0. Each time the instruction is executed, it causes CX to be tested for 0. If CX is found not to be 0, it is decremented by 1 and the basic string operation is repeated. On the other hand, if it is 0, the repeat string operation is done and the next instruction in the program is executed. The repeat count must be loaded into CX prior to executing the repeat string instruction. As indicated in Fig. 6–36, the REP prefix is used with the MOVS and STOS instructions. Figure 6–37 is the memory load routine of Fig. 6–35 modified by using the REP prefix.

The prefixes REPE and REPZ stand for the same function. They are meant for use with the CMPS and SCAS instructions. With REPE/REPZ, the basic compare or scan operation repeats as long as both the count in CX is not equal to 0 and ZF is 1. The first condition, CX not equal to 0, indicates that the end of the string has not yet been reached, and the second condition, ZF = 1, indicates that the elements that were compared are equal.

The last prefix, REPNE/REPNZ, works similarly as the REPE/REPZ, except that now the operation is repeated as long as CX is not equal to 0 and ZF is 0. That is, the comparison or scanning is performed as long as the string elements are unequal and the end of the string is not yet reached.

Prefix	Used with:	Meaning
REP	MOVS STOS	Repeat while not end of string $CX \neq 0$
REPE/REPZ	CMPS SCAS	Repeat while not end of string and strings are equal $CX \neq 0$ and $ZF = 1$
REPNE/REPNZ	CMPS SCAS	Repeat while not end of string and strings are not equal $CX \neq 0$ and $ZF = 0$

Figure 6–36 Prefixes for use with the basic string operations.

```
MOV       AX,0
MOV       DS,AX
MOV       ES,AX
MOV       AL,05
MOV       DI,0A000H
MOV       CX,0FH
CLD
REPSTOSB
```

Figure 6–37 Initializing a block of memroy by repeating the STOSB instruction.

Autoindexing for String Instructions

Earlier we pointed out that during the execution of a string instruction, the address indices in SI and DI are either automatically incremented or decremented. Moreover, we indicated that the decision to increment or decrement is made based on the setting of the direction flag (DF). The 8088 provides two instructions, *clear direction flag* (CLD) and *set direction flag* (STD), to permit selection between the *autoincrement* and *autodecrement modes* of operation, as shown in Fig. 6–38. When CLD is executed, DF is set to 0. This selects autoincrement mode, and each time a string operation is performed, SI and/or DI are incremented by 1 if byte data are processed and by 2 if word data are processed.

EXAMPLE 6.16

Describe what happens as the following sequence of instructions is executed:

```
CLD
MOV  AX, DATA_SEGMENT
MOV  DS, AX
MOV  AX, EXTRA_SEGMENT
MOV  ES, AX
MOV  CX, 20H
MOV  SI, OFFSET MASTER
MOV  DI, OFFSET COPY
REPZMOVSB
```

Solution

The first instruction clears the direction flag and selects autoincrement mode of operation for string addressing. The next two instructions initialize DS with the value DATA_SEGMENT. It is followed by two instructions that load ES with the value EXTRA_SEGMENT. Then the number of repeats, 20_{16}, is loaded into CX. The next two

Mnemonic	Meaning	Format	Operation	Flags Affected
CLD	Clear DF	CLD	(DF) ← 0	DF
STD	Set DF	STD	(DF) ← 1	DF

Figure 6–38 Instructions for selecting autoincrementing and autodecrementing in string instructions.

instructions load SI and DI with offset addresses MASTER and COPY, which point to the beginning of the source and destination strings, respectively. Now we are ready to perform the string operation. Execution of REPZMOVSB moves a block of 32 consecutive bytes from the block of memory locations starting at offset MASTER in the current data segment (DS) to a block of locations starting at offset COPY in the current extra segment (ES).

EXAMPLE 6.17

The source program in Fig. 6–39(a) implements the block move operation shown in Example 6.16. This program was assembled and linked to produce a run module called EX617.EXE. The source listing that was produced during the assembly process is shown in Fig. 6–39(b). Execute the program using DEBUG and verify its operation.

```
TITLE    EXAMPLE 6.17

      PAGE      ,132

STACK_SEG        SEGMENT            STACK 'STACK'
                 DB                 64 DUP(?)
STACK_SEG        ENDS

DATA_SEG         SEGMENT            'DATA'
MASTER           DB                 32 DUP(?)
COPY             DB                 32 DUP(?)
DATA_SEG         ENDS

CODE_SEG         SEGMENT            'CODE'
EX617    PROC    FAR
         ASSUME  CS:CODE_SEG, SS:STACK_SEG, DS:DATA_SEG, ES:DATA_SEG

;To return to DEBUG program put return address on the stack

      PUSH    DS
      MOV     AX, 0
      PUSH    AX

;Following code implements Example 6.17

      MOV     AX, DATA_SEG     ;Set up data setment
      MOV     DS, AX
      MOV     ES, AX           ;Set up extra segment

      CLD
      MOV     CX, 20H
      MOV     SI, OFFSET MASTER
      MOV     DI, OFFSET COPY
REP   MOVSB

      RET                      ;Return to DEBUG program
EX617 ENDP
CODE_SEG      ENDS

      END     EX617
```

(a)

Figure 6–39 (a) Source program for Example 6.17. (b) Source listing produced by assembler. (c) Execution of the program with DEBUG.

```
                              TITLE   EXAMPLE 6.17

                                    PAGE      ,132

0000                                  STACK_SEG       SEGMENT         STACK 'STACK'
0000        40 [                                      DB              64 DUP(?)
              ??
                 ]

0040                                  STACK_SEG       ENDS

0000                                  DATA_SEG        SEGMENT         'DATA'
0000        20 [                      MASTER          DB              32 DUP(?)
              ??
                 ]

0020        20 [                      COPY            DB              32 DUP(?)
              ??
                 ]

0040                                  DATA_SEG        ENDS

0000                                  CODE_SEG        SEGMENT         'CODE'
0000                                  EX617   PROC    FAR
                                 ASSUME  CS:CODE_SEG, SS:STACK_SEG, DS:DATA_SEG, ES:DATA_SEG

                              ;To return to DEBUG program put return address on the stack

0000  1E                                      PUSH    DS
0001  B8 0000                                 MOV     AX, 0
0004  50                                      PUSH    AX

                              ;Following code implements Example 6.17

0005  B8 —— R                                 MOV     AX, DATA_SEG      ;Set up data segment
0008  8E D8                                   MOV     DS, AX
000A  8E C0                                   MOV     ES, AX            ;Set up data segment

000C  FC                                      CLD
000D  B9 0020                                 MOV     CX, 20H
0010  BE 0000                                 MOV     SI, OFFSET MASTER
0013  BF 0020 R                               MOV     DI, OFFSET COPY
0016  F3/ A4                          REP     MOVSB

0018  CB                                      RET                       ;Return to DEBUG program
0019                                  EX617   ENDP
0019                                  CODE_SEG        ENDS

                                              END     EX617

Segments and groups:

              N a m e                   Size    align   combine class

CODE_SEG . . . . . . . . . . . . .      0019    PARA    NONE    'CODE'
DATA_SEG . . . . . . . . . . . . .      0040    PARA    NONE    'DATA'
STACK_SEG. . . . . . . . . . . . .      0040    PARA    STACK   'STACK'

Symbols:

              N a m e                   Type    Value   Attr

COPY . . . . . . . . . . . . . . .      L BYTE  0020    DATA_SEG        Length =0020
EX617. . . . . . . . . . . . . . .      F PROC  0000    CODE_SEG        Length =0019
MASTER . . . . . . . . . . . . . .      L BYTE  0000    DATA_SEG        Length =0020

Warning Severe
Errors  Errors
0       0
```

(b)

Figure 6–39 (continued)

Solution

The program is loaded with the DEBUG command

```
C:\DOS>DEBUG A:EX617.EXE   (↵)
```

and verified by the Unassemble command

```
-U 0 18   (↵)
```

Comparing the displayed instruction sequence in Fig. 6–39(c) with the source listing in Fig. 6–39(b), we find that the program has loaded correctly.

First, execute down to the REPZ instruction with the Go command

```
-G 16   (↵)
```

```
C:\DOS>DEBUG A:EX617.EXE
-U 0 18
0DE7:0000 1E            PUSH    DS
0DE7:0001 B80000        MOV     AX,0000
0DE7:0004 50            PUSH    AX
0DE7:0005 B8E90D        MOV     AX,0DE9
0DE7:0008 8ED8          MOV     DS,AX
0DE7:000A 8EC0          MOV     ES,AX
0DE7:000C FC            CLD
0DE7:000D B92000        MOV     CX,0020
0DE7:0010 BE0000        MOV     SI,0000
0DE7:0013 BF2000        MOV     DI,0020
0DE7:0016 F3            REPZ
0DE7:0017 A4            MOVSB
0DE7:0018 CB            RETF
-G 16

AX=0DE9  BX=0000  CX=0020  DX=0000  SP=003C  BP=0000  SI=0000  DI=0020
DS=0DE9  ES=0DE9  SS=0DED  CS=0DE7  IP=0016   NV UP EI PL NZ NA PO NC
0DE7:0016 F3            REPZ
0DE7:0017 A4            MOVSB
-F DS:0 1F FF
-F DS:20 3F 00
-D DS:0 3F
0DE9:0000  FF FF FF FF FF FF FF FF-FF FF FF FF FF FF FF FF   ................
0DE9:0010  FF FF FF FF FF FF FF FF-FF FF FF FF FF FF FF FF   ................
0DE9:0020  00 00 00 00 00 00 00 00-00 00 00 00 00 00 00 00   ................
0DE9:0030  00 00 00 00 00 00 00 00-00 00 00 00 00 00 00 00   ................
-G 18

AX=0DE9  BX=0000  CX=0000  DX=0000  SP=003C  BP=0000  SI=0020  DI=0040
DS=0DE9  ES=0DE9  SS=0DED  CS=0DE7  IP=0018   NV UP EI PL NZ NA PO NC
0DE7:0018 CB            RETF
-D DS:0 3F
0DE9:0000  FF FF FF FF FF FF FF FF-FF FF FF FF FF FF FF FF   ................
0DE9:0010  FF FF FF FF FF FF FF FF-FF FF FF FF FF FF FF FF   ................
0DE9:0020  FF FF FF FF FF FF FF FF-FF FF FF FF FF FF FF FF   ................
0DE9:0030  FF FF FF FF FF FF FF FF-FF FF FF FF FF FF FF FF   ................
-G

Program terminated normally
-Q

C:\DOS>
```

(c)

Figure 6–39 (continued)

Looking at the displayed state information in Fig. 6–39(c), we see that CX has been loaded with 0020_{16}, SI with 0000_{16}, and DI with 0020_{16}.

Now the storage locations in the 32-byte source block that starts at address DS:0000 are initialized with the value FF_{16} using the Fill command

```
-F DS:0 1F FF   (↵)
```

and each of the 32 bytes of the destination block, which start at DS:0020, are loaded with the value 00_{16} with the Fill command

```
-F DS:20 3F 00   (↵)
```

Now a memory dump command is used to verify the initialization of memory

```
-D DS:0 3F   (↵)
```

The displayed information in Fig. 6–39(c) shows that memory has been initialized correctly.

Now execute the string move operation with the command

```
-G 18   (↵)
```

Again looking at the display in Fig. 6–39(c), we see that the repeat count in CX has been decremented to zero and that the source and destination pointers have incremented to (SI) = 0020_{16} and (DI) = 0040_{16}. Therefore, the string move instruction has executed 32 times and the source and destination addresses have been correctly incremented to complete the block transfer. The block transfer operation is verified by repeating the Dump command

```
-D DS:0 3F   (↵)
```

Note that both the source and destination blocks now contain FF_{16}.

REVIEW PROBLEMS

Section 6.1

1. Explain what happens when the instruction sequence that follows is executed.

```
LAHF
MOV   [BX+DI], AH
```

2. What operation is performed by the instruction sequence that follows?

```
MOV   AH, [BX+SI]
SAHF
```

3. What instruction should be executed to ensure that the carry flag is in the set state? The reset state?
4. Which instruction when executed disables the interrupt interface?
5. Write an instruction sequence to configure the 8088 as follows: interrupts not accepted; save the original contents of flags SF, ZF, AF, PF, and CF at the address $0A000_{16}$; and then clear CF.

Section 6.2

6. Describe the difference in operation and the effect on status flags due to the execution of the subtract words and compare words instructions.
7. Describe the operation performed by each of the instructions that follow.

(a) `CMP [0100H], AL`
(b) `CMP AX, [SI]`
(c) `CMP WORD PTR [DI], 1234H`

8. What is the state of the 8088's flags after executing the instructions in parts (a) through (c) of problem 7? Assume that the following initial state exists before executing the instructions.

(AX) = 8001H
(SI) = 0200H
(DI) = 0300H
(DS:100H) = F0H
(DS:200H) = F0H
(DS:201H) = 01H
(DS:300H) = 34H
(DS:301H) = 12H

9. What happens to the ZF and CF status flags as the following sequence of instructions is executed? Assume that they are both initially cleared.

```
MOV  BX, 1111H
MOV  AX, 0BBBBH
CMP  BX, AX
```

Section 6.3

10. What is the key difference between the unconditional jump instruction and conditional jump instruction?
11. Which registers have their contents changed during an intrasegment jump? Intersegment jump?
12. How large is a Short-label displacement? Near-label displacement? Memptr16 operand?
13. Is a Far-label used to initiate an intrasegment jump or an intersegment jump?

14. Identify the type of jump, the type of operand, and operation performed by each of the instructions that follow.

(a) `JMP   10H`
(b) `JMP   1000H`
(c) `JMP   WORD PTR [SI]`

15. If the state of the 8088 is as follows before executing each instruction in problem 14, to what address is program control passed?

$$\begin{aligned} (CS) &= 1075H \\ (IP) &= 0300H \\ (SI) &= 0100H \\ (DS{:}100H) &= 00H \\ (DS{:}101H) &= 10H \end{aligned}$$

16. Which flags are tested by the various conditional jump instructions?
17. What flag condition is tested for by the instruction JNS?
18. What flag conditions are tested for by the instruction JA?
19. Identify the type of jump, the type of operand, and operation performed by each of the instructions that follow.

(a) `JNC 10H`
(b) `JNP 1000H`
(c) `JO  DWORD PTR [BX]`

20. What value must be loaded into BX such that execution of the instruction JMP BX transfers control to the memory location offset from the beginning of the current code segment by 256_{10}?
21. The program that follows implements what is known as a *delay loop*.

```
      MOV   CX, 1000H
DLY:  DEC   CX
      NOP
      JNZ   DLY
NXT:  ---   ---
```

(a) How many times does the JNZ DLY instruction get executed?
(b) Change the program so that JNZ DLY is executed 17 times.
(c) Change the program so that JNZ DLY is executed 2^{32} times.

22. Given a number N in the range $0<N\leq5$, write a program that computes its factorial and saves the result in memory location FACT. (N! = 1*2*3*4*.....*N)
23. Write a program that compares the elements of two arrays, A(I) and B(I). Each array contains 100 16-bit signed numbers. Compare the corresponding elements of the two arrays until either two elements are found to be unequal or all elements of the arrays have been compared and found to be equal. Assume that the arrays start in the cur-

rent data segment at offset addresses $A000_{16}$ and $B000_{16}$, respectively. If the two arrays are found to be unequal, save the address of the first unequal element of A(I) in the memory location with offset address FOUND in the current data segment; otherwise, write all 0s into this location.

Section 6.4

24. What is a subroutine? What other name is used to identify a subroutine?
25. Describe the difference between a jump and call instruction.
26. Why are intersegment and intrasegment call instructions provided in the 8088/8086 instruction set?
27. What is saved on the stack when a call instruction with a Memptr16 operand is executed? A Memptr32 operand?
28. Identify the type of call, the type of operand, and operation performed by each of the instructions that follow.

(a) `CALL  1000H`
(b) `CALL  WORD PTR [100H]`
(c) `CALL  DWORD PTR [BX + SI]`

29. The state of the 8088 is as follows:

(CS) = 1075H
(IP) = 0300H
(BX) = 0100H
(SI) = 0100H
(DS:100H) = 00H
(DS:101H) = 10H
(DS:200H) = 00H
(DS:201H) = 01H
(DS:202H) = 00H
(DS:203H) = 10H

To what address is program control passed after executing each instruction in problem 28?

30. What function is performed by the RET instruction?
31. Describe the operation performed by each of the instructions that follow.

(a) `PUSH DS`
(b) `PUSH [SI]`
(c) `POP  DI`
(d) `POP  [BX + DI]`
(e) `POPF`

32. What operation is performed by the following sequence of instructions?

```
PUSH AX
PUSH BX
POP  AX
POP  BX
```

33. When must PUSHF and POPF instructions be included in a subroutine?

34. Write a segment of main program and show its subroutine structure to perform the following operations. The program is to check the three most significant bits in register DX and, depending on their setting, execute one of three subroutines: SUBA, SUBB, or SUBC. The subroutines are selected as follows:

(a) If bit 15 of DX is set, initiate SUBA.

(b) If bit 14 of DX is set and bit 15 is not set, initiate SUBB.

(c) If bit 13 of DX is set and bits 14 and 15 are not set, initiate SUBC.

If a subroutine is executed, the corresponding bit(s) of DX should be cleared and then control returned to the main program. After returning from the subroutine, the main program is repeated.

Section 6.5

35. Which flags are tested by the various conditional loop instructions?

36. What two conditions can terminate the operation performed by the instruction LOOPNE?

37. How large a jump can be employed in a loop instruction?

38. What is the maximum number of repeats that can be implemented with a loop instruction?

39. Using loop instructions, implement the program in problem 22.

40. Using loop instructions, implement the program in problem 23.

Section 6.6

41. What determines whether the SI and DI registers show an increment or a decrement during a string operation?

42. Which segment register is used to form the destination address for a string instruction?

43. Write equivalent instruction sequences using string instructions for each of the following:

(a)
```
MOV  AL, [SI]
MOV  [DI], AL
INC  SI
INC  DI
```

(b)
```
MOV  AX, [SI]
INC  SI
INC  SI
```

(c)
```
MOV  AL, [DI]
CMP  AL, [SI]
DEC  SI
DEC  DI
```

44. Use string instructions to implement the program in problem 23.

ADVANCED PROBLEMS

45. Given an array A(I) of 100 16-bit signed numbers stored in memory starting at address $A000_{16}$, write a program to generate two arrays from the given array such that one array P(J) consists of all the positive numbers and the other N(K) contains all the negative numbers. Store the array of positive numbers in memory starting at offset address $B000_{16}$ and the array of negative numbers starting at offset address $C000_{16}$.

46. Given a 16-bit binary number in DX, write a program that converts it to its equivalent BCD number in DX. If the result is bigger than 16 bits, place all 1s in DX.

47. Given an array A(I) with 100 16-bit signed integer numbers, write a program to generate a new array B(I) as follows:

$$B(I) = A(I), \text{ for } I = 1, 2, 99, \text{ and } 100,$$
$$B(I) = \text{median value of } A(I-2), A(I-1), A(I), A(I+1), \text{ and } A(I+2), \text{ for all other Is}$$

48. Write a subroutine that converts a given 16-bit BCD number to its equivalent binary number. The BCD number is in register DX. Replace it with the equivalent binary number.

49. Given an array A(I) of 100 16-bit signed integer numbers, write a program to generate a new array B(I) so that

$$B(I) = A(I), \text{ for } I = 1 \text{ and } 100$$

and

$$B(I) = 1/4\,[A(I-1) - 5A(I) + 9A(I+1)], \text{ for all other Is}$$

For the calculation of B(I), the values of A(I − 1), A(I), and A(I + 1) are to be passed to a subroutine in registers AX, BX, and CX and the subroutine returns the result B(I) in register AX.

50. Write a program to convert a table of 100 ASCII characters that are stored starting at offset address ASCII_CHAR into their equivalent table of EBCDIC characters and store them at offset address EBCDIC_CHAR. The translation is to be done using an ASCII_TO_EBCDIC conversion table starting at offset address ASCII_TO_EBCDIC. Assume that all three tables are located in different segments of memory.

7

Assembly Language Program Development with MASM

▲ INTRODUCTION

In Chapter 4, we learned how to use the DEBUG program development tool that is available in the PC's disk operating system. We found that a line-by-line assembler is included in the DEBUG program; however, this assembler is not practical to use when writing larger programs. Other assembly language-development tools, such as Microsoft's macroassembler (MASM) and the linker (LINK) programs, are available for DOS. These are the kind of software-development tools that are used to develop larger application programs. In this chapter, we will learn how to use MASM, LINK, and DEBUG to develop and debug assembly language programs for practical applications. The topics covered in the chapter are as follows:

7.1 Statement Syntax for a Source Program
7.2 Assembler Directives
7.3 Creating a Source File with an Editor
7.4 Assembling and Linking Programs
7.5 Loading and Executing a Run Module

▲ 7.1 STATEMENT SYNTAX FOR A SOURCE PROGRAM

In Chapter 3, we found that the source program is a series of assembly language and directive statements that solve a specific problem. The *assembly language statements* tell the microprocessor the operations to be performed. On the other hand, the *directive statements* instruct the assembler program on how the program is to be assembled. The structure of the assembly language statement was briefly introduced in Chapter 3. Here we continue with a detailed study of the rules that must be used when writing assembly language and directive statements for the MASM macroassembler.

Assembly Language Statement Syntax

A source program must be written using the syntax understood by the assembler program. By *syntax* we mean the rules according to which statements must be written. The general format of an assembly language statement is

```
LABEL:  OPCODE  OPERAND(S)  ;COMMENT(S)
```

Note that it contains four separate fields: the *label field, opcode field, operand field,* and *comment field.* An example is the instruction

```
START:  MOV  CX,10  ;Load a count of 10 into register CX
```

Here START is in the label field; MOV (for "move operation") is the opcode field; the operands are CX and the decimal number 10; and the comment field tells us that execution of the instruction loads a count of 10 into register CX. Note that the label field ends with a colon (:) and the comment field begins with a semicolon (;).

Not all of the fields may be present in an instruction. In fact, the only part of the format that is always required is the opcode. For instance, the instruction

```
MOV  CX, 10  ;Initialize the count in CX
```

has nothing in the label field. Other instructions may not need anything in the operand field. An example is the instruction

```
CLC  ;Clear the carry flag
```

One rule that must be followed when writing assembly language statements for MASM is that the fields must be separated by at least one blank space and, if no label is used, the opcode field must be preceded by at least one space.

Let us now look at each field of an assembly language source statement in more detail. We begin with the label field. It is used to give a *symbolic name* to an assembly language statement. When a symbolic name is given to an assembly language statement, other instructions in the program can reference the statement by simply referring to this symbol instead of the actual memory address where the instruction is stored. Consider the example,

```
        .
        .
        JMP  START
        .
        .
        .
START:  MOV CX, 10 ; Initialize the count in CX
        .
        .
```

Here execution of the jump instruction causes program execution to pass to the point in the program corresponding to label START—that is, the MOV instruction that is located further down in the program.

The label is an arbitrarily selected string of alphanumeric characters followed by a colon (:). Some examples of valid labels are START, LOOPA, SUBROUTINE_A, and COUNT_ROUTINE. As in our earlier example, the names used for labels are typically selected to help document what is happening at that point of the program.

There are some limitations on the selection of labels. For instance, the assembler recognizes only the first 31 characters of the label; moreover, the first character of the label must be a letter.

Another restriction is that *reserved symbols,* such as those used to refer to the internal registers of the processor (AH, AL, AX, etc.), cannot be used. Still another restriction is that a label cannot include embedded blanks. This is the reason that the earlier example COUNT_ROUTINE has an underscore character (_) separating the two words. Use of the underscore makes the assembler view the character string as a single label.

Each of the basic operations that can be performed by the microprocessor is identified with a three- to six-letter *mnemonic,* which is called its *operation code* (opcode). For example, the mnemonics for the add, subtract, and move operations are ADD, SUB, and MOV, respectively. It is these mnemonics that are entered into the opcode field during the writing of the assembly language statements.

The entries in the operand field tell where the data to be processed is located and how it is to be accessed. An instruction may have zero, one, or two operands. For example, the move instruction that follows has two operands:

```
MOV  AX, BX
```

Here the two operands are the accumulator (AX) register and the base (BX) register. Note that a comma separates the operands. Furthermore, assembly language instructions are written with the destination operand first. Therefore, BX is the source operand (the "move from" location) and AX is the destination operand (the "move to" location). In this way, we see that the operation performed by the instruction is to move the value held in BX into AX. The notations used to identify the processor's internal registers are shown in Fig. 7–1.

A number of addressing modes are provided for the 8088/8086 processor to help us in specifying the location of operands. Examples using each of the addressing modes are

Symbol	Register
AX	Accumulator register
AH	Accumulator register high byte
AL	Accumulator register lower byte
BX	Base register
BH	Base register high byte
BL	Base register low byte
CX	Count register
CH	Count register high byte
CL	Count register low byte
DX	Data register
DH	Data register high byte
DL	Data register low byte
SI	Source index register
DI	Destination index register
SP	Stack pointer register
BP	Base pointer register
CS	Code segment register
DS	Data segment register
SS	Stack segment register
ES	Extra segment register

Figure 7–1 Symbols used for specifying register operands.

provided in Fig. 7–2. For example, the instruction that specifies an immediate data operand simply includes the value of the piece of data or a symbol representing the value in the operand location. On the other hand, if the operand is a direct address, it is specified enclosing the value of the memory address or its symbol in brackets.

Addressing mode	Operand	Example	Segment
Register	Destination	MOV AX,15H	–
Immediate	Source	MOV AL,15H	–
Direct	Destination	MOV 15H,AX	Data
Register indirect	Source	MOV AX,[SI]	Data
		MOV AX,[BP]	Stack
		MOV AX,[DI]	Data
		MOV AX,[BX]	Data
Based	Destination	MOV [BX]+DISP,AL	Data
		MOV [BP]+DISP,AL	Stack
Indexed	Source	MOV AL,[SI]	Data
		MOV AL,[DI]	Data
Based indexed	Destination	MOV [BX] [SI]+DISP,AH	Data
		MOV [BX] [DI]+DISP,AH	Data
		MOV [BP] [SI]+DISP,AH	Stack
		MOV[BP] [DI]+DISP,AH	Stack

Figure 7–2 Examples using the addressing modes.

The comment field can be used to describe the operation performed by the instruction. It is preceded by a semicolon (;). For instance, in the instruction

```
MOV   AX, BX   ;Copy BX into AX
```

the *comment* tells us that execution of the instruction causes the value in source register BX to be copied into the destination register AX.

The assembler program ignores comments when it assembles a program. Comments are produced only in the source listing. This does not mean that comments are not important. In fact, they are very important, because they document the operation of the source program. If a program were picked up a long time after it was written—for instance, for a software update—the comments would permit the programmer to quickly understand its operation.

Directive Statement Syntax

The syntax used to write directive statements for MASM is essentially the same as that for an assembly language statement. The general format is

```
LABEL:   DIRECTIVE   OPERAND(S)   ;COMMENT(S)
```

Note that the only difference is that the instruction opcode is replaced with a *directive.* It tells the assembler which type of operation is to be performed. For example, the directive DB stands for *define byte,* and if a statement is written as

```
DB 0FFH   ;Allocate a byte location initialized to FFH
```

it causes the next byte location in memory to be loaded with the value FF_{16}. This type of command can be used to initialize memory locations with data.

Another difference between the directive statement and an assembly language statement is that directives frequently have more than two operands. For instance, the statement

```
DB   0FFH, 0FFH, 0FFH, 0FFH, 0FFH
```

causes the assembler to load the next five consecutive bytes in memory with the value FF_{16}.

Constants in a Statement

Constants in an instruction or directive, such as an immediate value of data or an address, can be expressed in any one of many data types. Some examples are the *binary, decimal, hexadecimal, octal,* and *character* data types. The first four types of data are defined by ending the number with letter B, D, H, and Q, respectively. For example, decimal number 9 is expressed in these data forms as follows:

```
1001B
   9D
   9H
  11Q
```

One exception is that decimal numbers do not have to be followed by a D. Therefore, 9D can also be written simply as 9.

Another variation is that the first digit of a hexadecimal number must always be one of the numbers in the range 0 through 9. For this reason, hexadecimal A must be written as 0AH instead of AH.

Typically, data and addresses are expressed in hexadecimal form, and the count for shift, rotate, and string instructions is more commonly expressed in decimal form.

EXAMPLE 7.1

The repeat count in CX for a string instruction is to be equal to decimal 255. Assume that the instruction that is to load the count has the form

```
MOV   CX, XX
```

where XX stands for the count, which is an immediate operand that is to be loaded into CX. Show how the instruction should be written.

Solution

Using decimal notation for XX, we write the instruction as

```
MOV   CX, 255D
```

or just

```
MOV   CX, 255
```

In hexadecimal, 255 is represented by FF. Therefore, the instruction becomes

```
MOV   CX, 0FFH
```

The numbers used as operands in assembly language and directive statements can also be *signed* (positive or negative) *numbers*. For decimal numbers, this is done by preceding them with a + or − sign. For example, an immediate count of −10 that is to be loaded into the CX register with a MOV instruction can be written as

```
MOV  CX, -10
```

However, for negative numbers expressed in binary, hexadecimal, or octal form, the 2's complement of the number must be used.

EXAMPLE 7.2

The count in a move instruction that is to load CX is to be −10. Write the instruction and express the immediate operand in binary form.

Solution

The binary form of 10_{10} is 01010_2. Forming the 2's complement by complementing each bit and adding 1, we get

$$\begin{array}{r} 10101 \\ +1 \\ \hline 10110 \end{array}$$

Therefore, the instruction is written as

```
MOV  CX, 10110B
```

Character data can also be used as an operand. For instance, a string search operation may be used to search through a block of ASCII data in memory looking for a specific ASCII character, such as the letter A. When ASCII data are used as an operand, the character or string of characters must be enclosed within double quotes. For example, if the number 1 is to be expressed as character data, instead of numeric data, it is written as "1". In a string-compare operation, the data in memory are compared to the contents of the AL register. Therefore, the character being searched for must be loaded into this register. For instance, to load the ASCII value of 1 into AL, we use the instruction

```
MOV  AL, "1"
```

A second kind of operand specifies a storage location in memory. Such operands are written using the memory-addressing modes of the microprocessor as shown in Fig. 7–2. For instance, to specify that an operand is held in a storage location that is the tenth byte from the beginning of a source block of data located in the current data segment, we can use register indirect addressing through source-index register SI. In this way, the location of the operand is specified as

10[SI] or [SI] + 10 or [SI + 10]

Before using this operand, SI must be loaded with an offset that points to the beginning of the source-data block in memory.

Certain instructions require operands that are a memory address instead of data. Two examples are the JMP instruction and the CALL instruction. Labels can be used to identify these addresses. For instance, in the instruction

```
JMP  AGAIN
```

AGAIN is a label that specifies the "jump to" address. *Attributes* may also be assigned to the label. An attribute specifies whether or not a given label is a *near, far, external,* or *internal label.*

Operand Expressions Using the Arithmetic, Relational, and Logical Operators

The operands we have used to this point have all been constants, variables, or labels. However, it is also possible to have an expression as an operand. For example, the instruction

```
MOV  AH, A+2
```

has an expression for its source operand—that is, the source operand is written as the sum of variable A and the number 2.

Figure 7–3 lists the *arithmetic, relational,* and *logical operators* that can be used to form operand expressions for use with the assembler. Expressions that are used for operands are evaluated as part of the assembly process. As the source program is assembled into an object module, the numeric values for the terms in the operand expressions are combined using the specified operators. The expression is then replaced with the resulting operand value in the final object code.

In Fig. 7–3, the operators are listed in the order of their *precedence*—that is, the order in which the assembler performs operations as it evaluates an expression. For instance, if the expression for an operand is

$$A + B * 2 / D$$

when the assembler evaluates this expression, the multiplication is performed first, the division second, and the addition third.

Using parentheses can change the order of precedence. When parentheses are in use, whatever is enclosed within them is evaluated first. For example, if we modify the example we just used as follows,

$$(A + B * 2) / D$$

the multiplication still takes place first, but now the addition and then the division follow it. Use of the set of parentheses has changed the order of precedence.

Type	Operator	Example	Function
Arithmetic	*	A * B	Multiplies A with B and makes the operand equal to the product
	/	A / B	Divides A by B and makes the operand equal to the quotient
	MOD	A MOD B	Divides A by B and assigns the remainder to the operand
	SHL	A SHL n	Shifts the value in A left by n bit positions and assigns this shifted value to the operand
	SHR	A SHR n	Shifts the value in A right by n bit positions and assigns this shifted value to the operand
	+	A + B	Adds A to B and makes the operand equal to the sum
	−	A − B	Subtracts B from A and makes the operand equal to the difference
Relational	EQ	A EQ B	Compares value of A to that of B. If A equals B, the operand is set to FFFFH and if they are not equal it is set to 0H
	NE	A NE B	Compares value of A to that of B. If A is not equal to B, the operand is set to FFFFH and if they are equal it is set to 0H
	LT	A LT B	Compares value of A to that of B. If A is less than B, the operand is set to FFFFH and if it is equal or greater than it is set to 0H
	GT	A GT B	Compares value of A to that of B. If A is greater than B, the operand is set to FFFFH and if it is equal or less than it is set to 0H
	LE	A LE B	Compares value of A to that of B. If A is less than or equal to B, the operand is set to FFFFH and if it is greater than it is set to 0H
	GE	A GE B	Compares value of A to that of B. If A is greater than or equal to B, the operand is set to FFFFH and if it is less than it is set to 0H
Logical	NOT	NOT A	Takes the logical NOT of A and makes the value that results equal to the operand
	AND	A AND B	A is ANDed with B and makes the value that results equal to the operand
	OR	A OR B	A is ORed with B and makes the value that results equal to the operand
	XOR	A XOR B	A is XORed with B and makes the value that results equal to the operand

Figure 7–3 Arithmetic, relational, and logical operators.

Figure 7–3 shows a simple expression using each of the operators and describes the function that the assembler performs for them. For example, the operand expression

A SHL n

causes the assembler to shift the value of A to the left by "n" bits.

EXAMPLE 7.3

If A = 8 and B = 5, find the value the assembler assigns to the source operand for the instruction

```
MOV   BH, (A * 4 - 2)/(B - 3)
```

Solution

The expression is evaluated as

$$\begin{aligned}(8 * 4 - 2) / (5 - 3) &= (32 - 2)/(2) \\ &= (30) / (2) \\ &= 15\end{aligned}$$

and using hexadecimal notation, we get the instruction

```
MOV   BH, 0FH
```

All the examples we have considered so far have used arithmetic operators. Let us now take an example of a relational operator. Figure 7–3 shows that there are six relational operators: *equal* (EQ), *not equal* (NE), *less than* (LT), *greater than* (GT), *less than or equal* (LE), and *greater than or equal* (GE). The example expression given for equal is

A EQ B

When the assembler evaluates this relational expression, it determines whether or not the value of A equals that of B. If they are equal, the operand is made $FFFF_{16}$; if they are unequal, the operand is made 0000_{16}.

This is true about all relational operators. If by evaluating a relational expression we find that the conditions it specifies are satisfied, the operand expression is replaced with $FFFF_{16}$; if the conditions are not met, the expression 0000_{16} replaces it.

EXAMPLE 7.4

If A = 234, B = 345, and C = 111, find the value used for the source operand in the expression

```
MOV  AX, A LE (B - C)
```

Solution

Substituting into the expression, we get

$$\begin{aligned}&234\ \text{LE}\ (345 - 111) \\ &234\ \text{LE}\ 234\end{aligned}$$

Since the relational operator is satisfied, the instruction is equivalent to

```
MOV  AX, 0FFFFH
```

The logical operators shown in Fig. 7–3 are similar to the arithmetic and relational operators; with these operators, the assembler performs the appropriate sequence of logic operations and then assigns the result to the operand.

Value-Returning and Attribute Operators

Two other types of operators are available for use with operands: the *value-returning operators* and the *attribute operators.* The operators in each group, along with an example expression and description of their function, are given in Fig. 7–4.

Type	Operator	Example	Function
Value-returning	SEG	SEG A	Assigns the contents held in the segment register corresponding to the segment in which A resides to the operand
	OFFSET	OFFSET A	Assigns the offset of the location A in its corresponding segment to the operand
	TYPE	TYPE A	Returns to the operand a number representing the type of A; 1 for a byte variable and 2 for a word variable; NEAR or FAR for the label
	SIZE	SIZE A	Returns the byte count of variable A to the operand
	LENGTH	LENGTH A	Returns the number of units (as specified by TYPE) allocated for the variable A to the operand
Attribute	PTR	NEAR PTR A	Overrides the current type of label operand A and assigns a new pointer type: BYTE, WORD, NEAR, or FAR to A
	DS:,ES:,SS:	ES:A	Overrides the normal segment for operand A and assigns a new segment to A
	SHORT	JMP SHORT A	Assigns to operand A an attribute that indicates that it is within +127 or −128 bytes of the next instruction. This lets the instruction be encoded with the minimum number of bytes
	THIS	THIS BYTE A	Assigns to operand A a distance or type attribute: BYTE, WORD, NEAR, or FAR, and the corresponding segment attribute
	HIGH	HIGH A	Returns to the operand A the high byte of the word of A
	LOW	LOW A	Returns to the operand A the low byte of the word of A

Figure 7–4 Value-returning and attribute operators.

The value-returning operators return the attribute (segment, offset, or type) value of a variable or label operand. For instance, assuming that the variable A is in a data segment, the instructions

```
MOV  AX, SEG A
MOV  SI, OFFSET A
MOV  CL, TYPE A
```

when assembled cause the 16-bit segment value for A to replace SEG A, the 16-bit offset for variable A to replace OFFSET A, and the type number of the variable A to replace TYPE A, respectively. Assuming that A is a data byte, the value 1 will be assigned to TYPE A.

The attribute operators give the programmer the ability to change the attributes of an operand or label. For example, operands that use the BX, SI, or DI registers to hold the offset to their storage locations in memory are automatically referenced with respect to the contents of the DS register. An example is the instruction

```
MOV  AX, [SI]
```

We can use the *segment override* attribute operator to select another segment register. For instance, to select the extra segment register, the instruction is written as

```
MOV  AX, ES:[SI]
```

▲ 7.2 ASSEMBLER DIRECTIVES

The primary function of an assembler program is to convert the assembly language instructions of the source program to their corresponding machine instructions. However, practical assembly language source programs do not consist of assembly language statements only; they also contain what are called *directive statements*. In this section, we will examine what a directive is, what directives are available in MASM, and how to use them in an assembly language source program.

The Directive

In Section 7.1, we introduced the syntax of a directive statement and found that it differs from the assembly language instruction statement in that it is a direction that tells the assembler how to assemble the source program instead of an instruction to be processed by the microprocessor. That is, directives are statements written in the source program but are meant only for use by the assembler program. The assembler program follows these directions during the assembling of the program, but does not produce any machine code for them.

Figure 7–5 presents a partial list of the directives provided in MASM. Notice that they are grouped into categories based on the type of operation they specify to the assembler. These categories are the *data directives, conditional directives, macro directives,* and *listing directives.* Note that each category contains a number of different directives. Here

Type	Directives		
Data	ASSUME	ENDS	NAME
	COMMENT	EQU	ORG
	DB or BYTE	=(Equal	PROC
	DD or DWORD	Sign)	PUBLIC
	DQ or GWORD	EVEN	.RADIX
	DT or TBYTE	EXTRN	RECORD
	DW or WORD	GROUP	SEGMENT
	END	INCLUDE	STRUC
	ENDP	LABEL	
Conditional	ELSE	IFDEF	IFNB
	ENDIF	IFDIF	IFNDEF
	IF	IFE	IF1
	IFB	IFIDN	IF2
Macro	ENDM	IRPC	PURGE
	EXITM	LOCAL	REPT
	IRP	MACRO	
Listing	.CREF	PAGE	TITLE
	.LALL	.SALL	.XALL
	.LFCOND	.SFCOND	.XCREF
	.LIST	SUBTTL	.XLIST
	%OUT	.TFCOND	

Figure 7–5 Assembler directives.

we will consider only the most frequently used subset of the directives in these categories. For information on those directives not covered here, the reader should consult the manual provided with the MASM macroassembler.

Data Directives

The primary purpose of the directives in the data group is to define values for constants, variables, and labels. They can also perform other functions, such as assigning a size to a variable and reserving storage locations in memory. Figure 7–6 lists the most commonly used directives for handling these types of data operations.

Directive	Meaning	Function
EQU	Equate	Assign a permanet value to a symbol
=	Equal to	Set or redefine the value of a symbol
DB or BYTE	Define byte	Define or initialize byte size variables or locations
DW or WORD	Define word	Define or initialize word size (2 byte) variables or locations
DD or DWORD	Define double word	Define or initialize double word size (4 byte) variables or locations

Figure 7–6 Commonly used data directives.

The first two data directives in Fig. 7–6 are *equate* and *equal to.* In a statement, they are written as EQU and =, respectively. Both of these directives can be used to assign a constant value to a symbol. For example, the symbol AA can be set equal to 0100_{16} with the statement

```
AA EQU 0100H
```

The value of the operand can also be assigned using the arithmetic, relational, or logic expressions discussed in Section 7.1. An example that uses an arithmetic expression to define an operand is

```
BB EQU AA+5H
```

In this statement, the symbol BB is assigned the value of symbol AA plus 5. These two operations can also be done using the = directive. This gives the statements

```
AA = 0100H
BB = AA+5H
```

Once these values are assigned to AA and BB, they can be referenced elsewhere in the program by just using symbols.

The difference between the EQU and = directives lies in the fact that the value assigned to the symbol using EQU cannot be changed, whereas when = is used to define the symbol, its value can be changed anywhere in the program.

```
AA EQU 0100H
BB EQU AA+5H
    .
    .
    .
BB EQU AA+10H   ;This is illegal
```

Here AA is set equal to 0100_{16} and BB to 0105_{16}; the value of BB cannot be changed with the third EQU statement. On the other hand, if we use = directives as follows,

```
AA = 0100H
BB = AA+5H
    .
    .
    .
BB = AA+10H  ;This is legal
```

BB is assigned the new value of AA + 10H as the third = directive is processed by the assembler.

The other three directives given in Fig. 7–6 are *define byte* (DB), *define word* (DW), and *define double word* (DD). The purpose of these directives is to define the size of variables as being byte, word, or double word in length, allocate space for them, and assign them initial values. If the initial value of a variable is not known, the DB, DW, or DD statement simply allocates a byte, word, or double word of memory to the variable name.

An example of the DB directive is

```
CC DB 7
```

Here, variable CC is defined as byte size and assigned the value 7. It is important to note that the value assigned with a DB, DW, or DD statement must not be larger than the maximum number that can be stored in the specified-size storage location. For instance, for a byte-size variable, the maximum decimal value is 255 for an unsigned number and +127 or −128 for a signed number.

Here is another example:

```
EE DB ?
```

In this case, a byte of memory is allocated to the variable EE, but no value is assigned to it. Note that the use of a ? as the operand means that an initial value is not to be assigned.

Look at another example:

```
MESSAGE DB "JASBIR"
```

Here each character in the string JASBIR is allocated a byte in memory, and these bytes are initialized with the ASCII code for the corresponding character. This is the way ASCII data are assigned to a name.

If we need to initialize a large block of storage locations with the same value, the assembler provides a way of using the byte, word, or double-word directive to repeat a value. An example is the statement

```
TABLE_A DB 10 DUP(?), 5 DUP(7)
```

This statement causes the assembler to allocate 15 bytes of memory for TABLE_A. The first ten bytes of this block of memory are left uninitialized, and the next five bytes are all initialized with the value 7. Note that use of *duplicate* (DUP) tells the assembler to duplicate the value enclosed in parentheses a number of times equal to the number that precedes DUP.

If each element of the table were to be initialized to a different value, the DB command would be written in a different way. For example, the command

```
TABLE_B  DB  0,1,2,3,4,5,6,7,8,9
```

sets up a table called TABLE_B and assigns to its ten storage locations the decimal values 0 through 9.

Segment-Control Directives

Memory of the 8088/8086-based microcomputer is partitioned into three kinds of segments: the code segment, data segments, and stack segment. The code segment is where machine-code instructions are stored, the data segments are for storage of data, and

Directive	Function
SEGMENT	Defines the beginning of a segment and specifies its kind, at what type of address boundary it is to be stored in memory, and how it is to be positioned with respect to other similar segments in memory
ENDS	Specifies the end of a segment
ASSUME	Specifies the segment address register for a given segment

Figure 7–7 Segment directives.

the stack segment is for a temporary storage location called the stack. Using the *segment-control directives* in Fig. 7–7, the statements of a source program can be partitioned and assigned to a specific memory segment. These directives can be used to specify the beginning and end of a segment in a source program and assign to them attributes such as a start address, the kind of segment, and how the segment is to be combined with other segments of the same name.

A *segment* (SEGMENT) directive identifies the beginning of a segment and its end is marked by the *end of segment* (ENDS) directive. Here is an example of a segment definition:

```
SEGA  SEGMENT  PARA PUBLIC 'CODE'
      MOV  AX, BX
        .
        .
SEGA  ENDS
```

As shown, the information between the two directive statements is the instructions of the assembly language program.

In this example, SEGA is the name given to the segment. The directive SEGMENT is followed by the operand PARA PUBLIC 'CODE.' Here PARA (*paragraph*) defines that this segment is to be aligned in memory on a 16-byte address boundary. This part of the operand is called the *align-type* attribute. The other align-type attributes that can also be used are given in Fig. 7–8 with a brief description of their function.

PUBLIC, which follows PARA in the operand of the example SEGMENT directive, defines what is called a *combine-type* attribute. It specifies that this segment is to be con-

Attribute	Function
PARA	Segment begins on a 16 byte address boundary in memory (4 LSBs of the address are equal to 0)
BYTE	Segment begins anywhere in memory
WORD	Segment begins on a word (2 byte) address boundary in memory (LSB of the address is 0)
PAGE	Segment begins on a 256 byte address boundary in memory (8 LSBs of the address are equal to 0)

Figure 7–8 Align-type attributes.

Attribute	Function
PUBLIC	Concatenates segments with the same name
COMMON	Overlaps from the beginning segments with the same name
AT [expression]	Locates the segment at the 16-bit paragraph number evaluated from the expression
STACK	The segment is part of the run-time stack segment
MEMORY	Locates the segment at an address above all other segments

Figure 7–9 Combine-type attributes.

catenated with all other segments that are assigned the name SEGA to generate one physical segment called SEGA. Other combine-type attributes are given in Fig. 7–9.

The last part of the operand in the SEGMENT statement is 'CODE,' which is an example of a *class* attribute. It specifies that the segment is a code segment. Figure 7–10 shows other segment classes.

At the end of the group of statements to be assigned to the code segment there must be an ENDS directive. Figure 7–7 shows that this statement is used to mark the end of the segment. ENDS must also be preceded by the segment name, which is SEGA in our example.

The third directive in Fig. 7–7, *assume* (ASSUME), is used to assign the segment registers that hold the base addresses to the program segments. For instance, with the statement

```
ASSUME  CS:SEGA, DS:SEGB, SS:SEGC
```

we specify that register CS holds the base address for segment SEGA, register DS holds the base address for segment SEGB, and register SS holds the base address for segment SEGC. The ASSUME directive is written at the beginning of a code segment just after the SEGMENT directive.

Figure 7–11 shows the general structure of a code segment definition using the segment control directive.

Modular Programming Directives

For the purpose of development, large programs are broken down into small segments called *modules*. Typically, each module implements a specific function and has its

Attribute	Function
CODE	Specifies the code segment
DATA	Specifies the data segment
STACK	Specifies the stack segment
EXTRA	Specifies the extra segment

Figure 7–10 Class attributes.

```
SEGA    SEGMENT    PARA PUBLIC 'CODE'
        ASSUME     CS:SEGA
        MOV        AX,BX
          .
          .
          .
SEGA    ENDS
```

Figure 7–11 Example using segment-control directives.

own code segment and data segment. However, it is common that during the execution of a module, a part of some other module may need to be called for execution or that data that resides in another module may need to be accessed for processing. To support capabilities such that a section of code in one module can be executed from another module or for data to be passed between modules, MASM provides *modular programming directives.* The most frequently used modular programming directives are listed in Fig. 7–12.

A section of a program that can be called for execution from other modules is called a *procedure.* Similar to the definition of a segment of a program, the beginning and end of a procedure must be marked with directive statements. The *procedure* (PROC) directive marks the beginning of the procedure, and the *end of procedure* (ENDP) directive marks its end.

There are two kinds of procedures: a *near procedure* and a *far procedure.* Whenever a procedure is called, the return address has to be saved on the stack. After completing the code in the called procedure, this address is used to return execution back to the point of its initiation. When a near procedure is called into operation, only the code offset address (contents of IP) is saved on the stack. Therefore, a near procedure can only be called from the same code segment. On the other hand, when a far procedure is called, both the contents of the code segment (CS) register and the code offset (IP) register are saved on the stack. For this reason, a far procedure can be called from any code segment. Depending on the kind of procedure, the *return* (RET) instruction at the end of the procedure restores either IP or both IP and CS from the stack.

If a procedure can be called from other modules, its name must be made *public* by using the PUBLIC directive. This is done before entering the procedure. Thus, the structure for defining a procedure with name SUB_NEAR will be

Directive	Function
proc-name PROC [NEAR]	Defines the beginning of a near-proc procedure
proc-name PROC FAR	Defines the beginning of a far-proc procedure
proc-name ENDP	Defines the end of a procedure
PUBLIC Symbol[.......]	The defined symbols can be referenced from other modules
EXTRN name:type[....]	The specified symbols are defined in other modules and are to be used in this module

Figure 7–12 Modular programming directives.

```
                PUBLIC  SUB_NEAR
      SUB_NEAR  PROC
                .
                .
                .
                RET
      SUB_NEAR  ENDP
```

In this example the PROC does not have either the NEAR or FAR attribute in its operand field. When no attribute is specified as an operand, NEAR is assumed as the default attribute by the assembler. In this way, we see that this procedure can be called only from other modules in the same code segment.

The structure of a far procedure that can be called from modules in other segments is

```
               PUBLIC   SUB_FAR
      SUB_FAR  PROC     FAR
               .
               .
               .
               RET
      SUB_FAR  ENDP
```

If a procedure in another module is to be called from the current module, its name must be declared external in the current procedure by using the *external reference* (EXTRN) directive. It is also important to know whether this call is to a module in the same code segment or in a different code segment. Depending on the code segments, the name of the procedure must be assigned either a NEAR or FAR attribute as part of the EXTRN statement.

The example in Fig. 7–13 illustrates the use of the EXTRN directive. Here we see that an external call is made from module 2 to the procedure SUB that resides in module 1. Therefore, the procedure SUB is defined as PUBLIC in module 1. Note that the two modules have different code segments, CSEG1 and CSEG2. Thus, the external call in module 2 is to a far procedure. Therefore, the PROC directive for SUB in module 1 and the external label definition of SUB in module 2 have the FAR attribute attached to them.

Directives for Memory Usage Control

If the machine code generated by the assembler must reside in a specific part of the memory address space, an *origin* (ORG) directive can be used to specify the starting point of that memory area. Fig. 7–14 shows that the operand in the ORG statement can be an expression. The value that results from this expression is the address at which the machine code is to begin loading. For example, the statement

```
ORG  100H
```

simply tells the assembler that the machine code for subsequent instructions is to be placed in memory starting at address 100_{16}. This directive statement is normally located at the beginning of the program.

```
         PUBLIC    SUB                         EXTRN    SUB:FAR
CSEG1    SEGMENT                     CSEG2     SEGMENT
         .                                     .
         .                                     .
SUB      PROC      FAR                         CALL     SUB
         MOV       AX,BX                       .
         .                                     .
         .                                     .
         RET                                   .
SUB      ENDP                                  .
         .                           CSEG2     ENDS
         .
         .
CSEG1    ENDS

          Module 1                              Module 2
```

Figure 7–13 An example showing the use of the EXTRN directive.

If specific memory locations must be skipped—for example, because they are in a read-only area of memory—one can use ORG directives as follows:

```
ORG   100H
ORG   $+200H
```

in which case memory locations 100_{16} to $2FF_{16}$ are skipped and the machine code of the program starts at address 300_{16}.

The End-of-Program Directive

The *end* (END) directive, also shown in Fig. 7–14, tells the assembler when to stop assembling. It must always be included at the end of the source program. Optionally, we can specify the starting point of the program with an expression in the operand field of the END statement. For instance, an END statement can be written as

```
END  PROG_BLOCK
```

where PROG_BLOCK identifies the beginning address of the program.

Directive	Function
ORG [expression]	Specifies the memory address starting from which the machine code must be placed
END [expression]	Specifies the end of the source program

Figure 7–14 ORG and END directives.

Directive for Program Listing Control

The last group of directives we will consider is called the *listing control directives.* The most widely used ones in this group are shown in Fig. 7–15. The purpose of the listing control directives is to give the programmer some options related to the way in which source program listings are produced by the assembler. For instance, we may want to set up the print output such that a certain number of lines are printed per page, or we may want to title the pages of the listing with the name of the program.

The *page* (PAGE) directive lets us set the page width and length of the source listing produced as part of the assembly process. For example, if the PAGE directive

```
PAGE  50  100
```

is encountered at the beginning of a source program, then each printed page will have 50 lines and up to 100 characters in a line. The first operand, which specifies the number of lines per page, can be any number from 10 through 255. The second operand, which specifies the maximum number of characters per line, can range from 60 to 132. The default values for these parameters are 66 lines per page and 80 characters per line. The default parameters are selected if no operand is included with the directive.

Chapter and page numbers are printed at the top of each page in a source listing. They are in the form

```
[chapter number]-[page number]
```

As the assembler produces the source listing, the page number automatically increments each time a full page of listing information is generated. On the other hand, the chapter number does not change as the listing is generated. The only way to change the chapter number is by using the directive

```
PAGE +
```

When this form of the PAGE directive is processed by the assembler, it increments the chapter count and at the same time resets the page number to 1.

The second directive in Fig. 7–15 is *title* (TITLE). When TITLE is included in a program, it causes the text in the operand field to be printed on the second line of each

Directive	Function
PAGE operand_1 operand_2	Selects the number of lines printed per page and the maximum number of characters printed per line in the listing
TITLE text	Prints "text" on the second line of each page of the listing
SUBTTL text	Prints "text" on the third line of each page of the listing

Figure 7–15 Listing control directives.

page of the source listing. Similarly, the third directive, *subtitle* (SUBTTL), prints the text included in the directive statement on the third line of each page.

Example of a Source Program Using Directives

In order to have our first experience in using directives in a source program, let us again look at the source code of the block move program. Remember that this program is written to copy a block of data from a location in memory known as the source block to another block location called the destination block. This source is repeated in Fig. 7–16. In the sections that follow, we will use this program to explore various aspects of the program

```
TITLE BLOCK-MOVE PROGRAM

      PAGE        ,132

COMMENT *This program moves a block of specified number of bytes
         from one place to another place*

;Define constants used in this program

      N=                  16          ;Bytes to be moved
      BLK1ADDR=           100H        ;Source block offset address
      BLK2ADDR=           120H        ;Destination block offset addr
      DATASEGADDR=        2000H       ;Data segment start address

STACK_SEG           SEGMENT           STACK 'STACK'
                    DB                64 DUP(?)
STACK_SEG           ENDS
CODE_SEG            SEGMENT           'CODE'
BLOCK               PROC        FAR
      ASSUME        CS:CODE_SEG,SS:STACK_SEG

;To return to DEBUG program put return address on the stack

            PUSH  DS
            MOV   AX, 0
            PUSH  AX

;Setup the data segment address

            MOV   AX, DATASEGADDR
            MOV   DS, AX

;Setup the source and destination offset addresses

            MOV   SI, BLK1ADDR
            MOV   DI, BLK2ADDR

;Setup the count of bytes to be moved

            MOV   CX, N

;Copy source block to destination block

NXTPT:      MOV   AH, [SI]            ;Move a byte
            MOV   [DI], AH
            INC   SI                  ;Update pointers
            INC   DI
            DEC   CX                  ;Update byte counter
            JNZ   NXTPT               ;Repeat for next byte
            RET                       ;Return to DEBUG program
BLOCK             ENDP
CODE_SEG          ENDS
      END         BLOCK               ;End of program
```

Figure 7–16 Example source program.

development process, such as creating a source file, assembling, linking, and debugging. For now we will just examine the directives used in the program.

The program starts with a TITLE directive statement. The text "BLOCK-MOVE PROGRAM" included in this statement will be printed on the second line of each page of the source listing. This text should be limited to 60 characters. The second and third statements in the program are also directives. The third statement is a *comment directive* and is used to place descriptive comments in the program. Note that it begins with COMMENT and is followed by the text enclosed within the delimiter asterisk (*). This comment gives a brief description of the function of the program.

There is another way of including comments in a program. This is by using a semicolon (;) followed by the text of the comment. The next line in the program is an example of this type of comment. It indicates that the next part of the program is used to define variables that are used in the program. Four "equal to" (=) directive statements follow the comment. Notice that they equate N to the value 16_{10}, BLK1ADDR to the value 100_{16}, BLK2ADDR to the value 120_{16}, and DATASEGADDR to the value 2000_{16}.

There are two segments in the program: the stack segment and the code segment. The next three directive statements define the stack segment. They are

```
STACK_SEG   SEGMENT   STACK   'STACK'
            DB     64     DUP(?)
STACK_SEG   ENDS
```

In the first statement, the stack segment is assigned the name STACK_SEG; the second statement allocates a block of 64 bytes of memory for use as stack and leaves this area uninitialized. The third statement defines the end of the stack.

The code segment is defined between the statements

```
CODE_SEG     SEGMENT     'CODE'
```

and

```
CODE_SEG     ENDS
```

Here CODE_SEG is the name we have used for the code segment. At the beginning of the code segment an ASSUME directive is used to specify the base registers for the code and stack segments. Notice that CS is the base register for the code segment, and SS is the base register for the stack segment. The instruction statements and comments that form the assembly language program follow this directive.

We also find an END directive at the end of the program. It identifies the end of the program, and BLOCK in this statement defines the starting address of the source program. Processing of this statement tells the assembler that the assembly is complete.

A PROC directive is included at the beginning of the source program, and the ENDP directive is included at the end of the program. This makes the program segment a procedure that can be used as a module in a larger program.

▲ 7.3 CREATING A SOURCE FILE WITH AN EDITOR

Now that we have introduced assembly language syntax, the directives, and an example of an assembly language program, let us continue by looking at how the source-program file is created on a PC. Source-program files are generated using a program called an *editor.* Many editors are available for this purpose. In this section we will use an editor program called EDIT, provided in the DOS of the PC. We will assume that the reader is already familiar with the DOS commands and the use of the EDIT program. For this reason, we will only briefly describe how EDIT is used to create a source-program file. For additional details, the reader should consult the appropriate DOS manual.

EDIT is a *menu-driven* text editor. That is, a simple key sequence is entered to display a *menu* of operations that can be performed. The user selects the desired operation from the menu and then depresses the return key. At this point, a *dialog box* is displayed that describes the operation to be performed. The user fills out the appropriate information by typing it in at the keyboard and then presses the return key to initiate the defined operation. These operations can also be initiated using a mouse.

Figure 7–17 shows the sequence of events that take place when a source program file is created with EDIT. As an example, we will enter instructions of the block move program in Fig. 7–16 and save it as the file BLOCK.ASM on a diskette in drive A. The sequence begins with a command to load and run the editor program. This command is

```
C:>EDIT A:BLOCK.ASM   (↵)
```

In this command, it is assumed that we are logged onto drive C and the path is set to the directory that contains the EDIT program. The EDIT program first checks to see if the file BLOCK.ASM already exists on the diskette in drive A. If it does, the file is read from the diskette, loaded into memory, and displayed on the screen. This would be the case if an existing source program were to be corrected or changed.

Looking at the flowchart, we see that the next step is to make the changes in the program. Then, the program should be carefully examined to verify that no additional errors were made during the edit process. The last step is to save the modified program in either the old file or as a file with a new name. Note that a backup file is not automatically made as part of the file-save process of EDIT.

Let us assume that the file BLOCK.ASM does not already exist and go through the sequence of events that must take place to create the source program. In this case, the same command can be used to bring up the editor. However, when it comes up, no program can be loaded, so the screen remains blank. The lines of the program are simply typed in one after the other. The TAB key is used to make the appropriate indents in the statements. For instance, to enter the instruction MOV AX, DATASEGADDR of the program in Fig. 7–16, the user depresses the TAB key once for a single indent and then types in the word "MOV." Another TAB is needed to indent again and then the operands AX, DATASEGADDR are entered. The instruction is now complete, so the user depresses the return key to position the cursor for entry of the next instruction. This sequence is repeated until the whole block-move program has been entered. Note that when entering the instruction with the label NXTPT:, the user enters the label at the left margin and then depresses TAB to indent to the position for MOV.

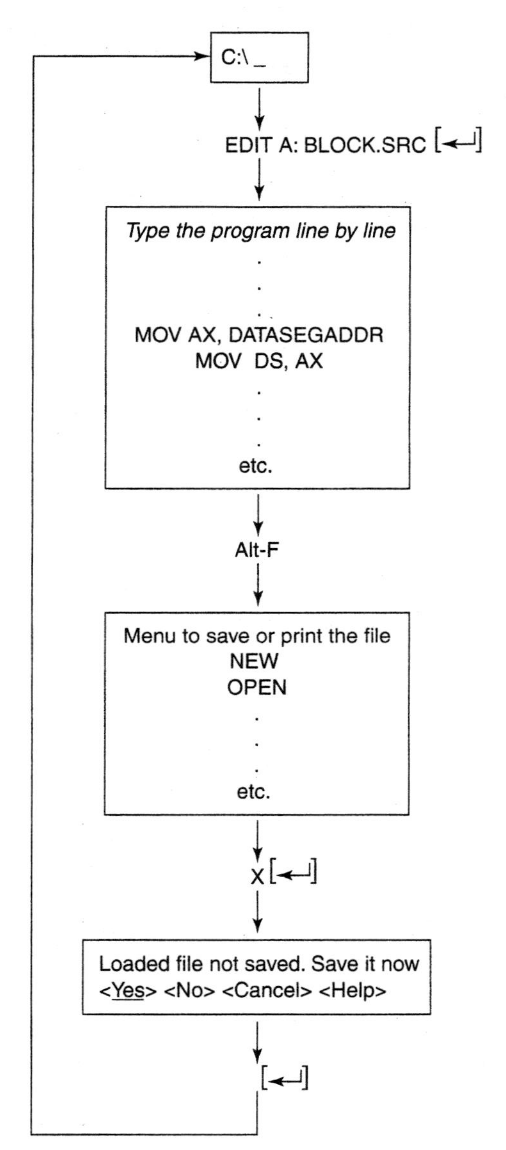

Figure 7–17 Flowchart for creating and editing of source files with EDIT.

If errors are made as the instructions of the block-move program are entered, the user can use the arrow keys to reposition the cursor to the spot that needs to be changed. Next the delete key (for characters to the right of the cursor) or backspace key (for characters to the left of the cursor) is used to remove the incorrect characters. Then, the correct character or characters are typed. This process is repeated until all corrections are made. Actually, the editing capability of EDIT is more versatile than just described. Commands are provided to move, copy, delete, find, or find and replace a character, string of characters, or block of text. For instance, we may need to change the name of an operand at all places it occurs in a large program. Instead of having to go through the program instruction by instruction to find each occurrence of the operand, this operation can be done with a single find-and-replace command.

Assuming that all corrections have been made, we are now ready to save the program. At the top of the EDIT screen is a *menu bar* with four menus: *File, Edit, Search,* and *Options.* The save operations are located under the File menu. To select this menu, hold down the ALT key and depress the F key (or click on File with the mouse). This causes a *pop-down menu* that lists the file commands to appear at the top of the screen. When the menu appears, the *New* command is highlighted. Use the arrow-down (↓) key to move the highlighted area down to the *Save As* operation. Now depress the return (↵) key to initiate the file-save operation. This displays the Save As dialog box. If a file that already existed is being edited, the file's name will automatically be filled into the dialog box. If the filename is to remain the same, simply depress the return key to save the file; if the name is to be changed, type in the new drive designator and filename information before depressing the return key. Since we have assumed that BLOCK.ASM is a new file, its name must be keyed into the spot for the filename and then the return key depressed.

At this moment, the source program has been created, verified, and saved, but we are still in the EDIT program. We are ready to exit EDIT. *EXIT* is another operation that is in the File menu. To exit EDIT, depress ALT and then F to display the File pop-down menu. Use the (↓) key to select Exit (or type X) and then depress the return key. The EDIT program terminates and the DOS prompt reappears on the screen. We have described the use of EDIT using keyboard entries to select editor operations. However, if a mouse is available, moving the mouse cursor to select the operation and then clicking on it can do these operations more conveniently.

▲ 7.4 ASSEMBLING AND LINKING PROGRAMS

Up to this point, we have studied the steps involved in writing a program, the assembly language syntax and directive statements provided in the MASM, the structure of an assembly language program, and how to create a source program using the EDIT editor. Now is the time to learn how to bring up MASM, use it to assemble a source-program file into an object-code file, and examine the other outputs produced by the assembler.

Earlier we said that an assembler is the program used to convert a file that contains an assembly language source program to its equivalent file of 8088/8086 machine code. Figure 7–18 shows that the input to the assembler program is the assembly language source program. This is the program that is to be assembled. The assembler program reads the source file and translates it into an object module file and a source listing file.

Source Listing

When the block-move source program shown in Fig. 7–16 is assembled with MASM, the source listing shown in Fig. 7–19 can be produced as an output. The leftmost column in the source listing contains the starting offset address of a machine-language instruction from the beginning of the current code segment. For instance, at an offset of 0007_{16}, we find the instruction MOV AX, DATASEGADDR. This information is followed

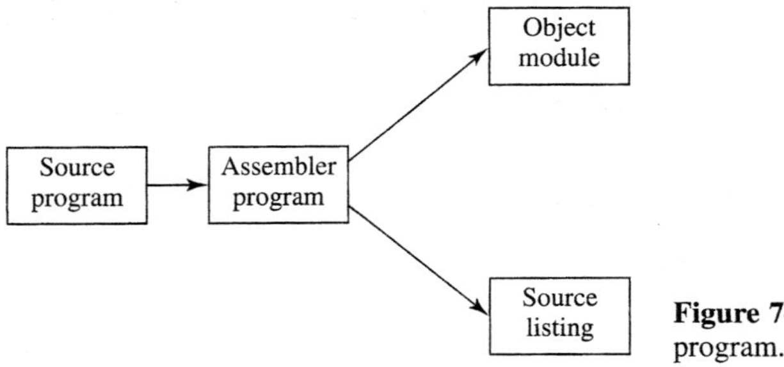

Figure 7–18 Assembling a source program.

by the bytes of machine code for the instruction, which are equal to B8 2000. The original source-code instructions and the comments are also shown in the listing.

A symbol table is also produced as part of the source listing. It is a list of all of the symbols used in the program. The symbol table for our example program is shown at the end of the source listing in Fig. 7–19. In this table, each symbol is listed along with its type, value, and attribute. The types of symbols are *label, variable, number,* and *procedure.* For example, in our example the symbol BLK1ADDR is a number and its value is 0100_{16}. For this symbol, no attribute is indicated. On the other hand, for the symbol NXTPT, which is a near-label with value 0013_{16}, the attribute is CODE_SEG.

If the assembler program identifies *syntax errors* in the source file while it is being assembled, the locations of the errors are marked in the source listing file with an *error number* and *error message.* The total number of errors is given at the end of the assembly listing. Looking at this information in the source listing for our example in Fig. 7–19, we find that no errors occurred.

Figure 7–20 shows the listing produced when four syntax errors are found during the assembly process. Notice how the errors are marked in the source listing. For instance, in the line of the source listing identified by cblock.asm(54), we find error number A2008. This error message stands for a syntax error. The error made is that the mnemonic of the instruction is spelled wrong. It should read DEC CX instead of DCR CX. The source program must first be edited to correct this error and the other three errors, and then be reassembled.

EXAMPLE 7.5

What is the meaning of the error code at line 46, identified as cblock.asm(46) in the source listing of Fig. 7–20?

Solution

Figure 7–20 shows the error number as A2006. This error means that a symbol was not defined. The undefined symbol is N.

The source listing is a valuable aid in correcting errors in the program. Since syntax errors are marked in the source listing, they can be found and corrected easily. Both the source

```
                                TITLE BLOCK-MOVE PROGRAM

                                PAGE   ,132

                          COMMENT *This program moves a block of specified number of bytes
                               from one place to another place*

                          ;Define constants used in this program

=0010                                N     =     16              ;Bytes to be moved
=0100                                BLK1ADDR=        100H       ;Source block offset address
=0120                                BLK2ADDR=        120H       ;Destination block offset addr
=2000                                DATASEGADDR=    2000H       ;Data segment start address

0000                          STACK_SEG    SEGMENT      STACK 'STACK'
0000 0040[                                 DB      64DUP(?)
    00
   ]
0040                          STACK_SEG    ENDS

0000                          CODE_SEG     SEGMENT      'CODE'
0000                          BLOCK      PROC      FAR
                                     ASSUME CS:CODE_SEG,SS:STACK_SEG

                          ;To return to DEBUG program put return address on the stack

0000 1E                              PUSH  DS
0001 B8 0000                         MOV   AX,0
0004 50                              PUSH  AX

                          ;Set up the data segment address

0007 B8 2000                         MOV   AX,DATASEGADDR
0008 8E D8                           MOV   DS,AX

                          ;Set up the source and destination offset addresses

000A BE 0100                         MOV   SI,BLK1ADDR
000D BF 0120                         MOV   DI,BLK2ADDR

                          ;Set up the count of bytes to be moved

0010 B9 0010                         MOV   CX,N

                          ;Copy source block to destination block
```

Figure 7–19 Source listing produced by assembling the block-move program.

and corresponding machine code are provided in the source listing. It also serves as a valuable tool for identifying and correcting logical errors in the writing of the program.

Object Module

The most important output produced by the assembler is the object-code file. The contents of this file are called the object module. The object module is a machine-language version of the program. Even through the object module is the machine code for the source program, it is still not an executable file that can be directly run on the microcomputer. That is, in its current state it cannot be loaded with the DEBUG program and run on the microprocessor in the PC. To convert an object module to an executable machine-code file (run module), we must process it with the linker. The LINK program performs the link operation for object code.

```
0013 8A 24                        NXTPT: MOV  AH,[SI]       ;Move a byte
0015 88 25                               MOV  [DI],AH
0017 46                                  INC  SI            ;Update pointers
0018 47                                  INC  DI
0019 49                                  DEC  CX            ;Update byte counter

001A 75 F7                               JNZ  NXTPT         ;Repeat for next byte
001C CB                                  RET                ;Return to DEBUG program
001D                              BLOCK       ENDP
001D                              CODE_SEG    ENDS
                                         END  BLOCK         ;End of program

Segments and Groups:

           N a m e                  Size     Length   Align    Combine Class

CODE_SEG . . . . . . . . . . . .             16 Bit   001D     Para    Private'CODE'
STACK_SEG. . . . . . . . . . . .             16 Bit   0040     Para    Stack  'STACK'

Procedures, parameters and locals:

           N a m e                  Type     Value    Attr

BLOCK. . . . . . . . . . . . . .    P Far    0000     CODE_SEG   Length= 001D Public
  NXTPT. . . . . . . . . . . . . .  L Near   0013     CODE_SEG

Symbols:

           N a m e                  Type     Value    Attr

BLK1ADDR . . . . . . . . . . . .             Number 0100h
BLK2ADDR . . . . . . . . . . . .             Number 0120h
DATASEGADDR. . . . . . . . . . .             Number 2000h
N. . . . . . . . . . . . . . . .    Number 0010h

      0 Warnings
      0 Errors
```

Figure 7–19 (continued)

The Link Program and Modular Programming

Let us now look into an important idea behind the use of the LINK program—that is, *modular programming*. The program we have been using as an example in this chapter is quite simple. For this reason, it can be contained easily in a single source file. However, most practical application programs are very large. For example, a source program may contain 2,000 assembly language statements. When assembled, this can result in as much as 4,000 to 6,000 bytes of machine code. For development purposes, programs of this size are frequently broken down into a number of parts called *modules* and different programmers may work on the individual modules. Each module is written independently, and then all modules are combined together to form a single executable run module as illustrated in Fig. 7–21. Note that the LINK program is the software tool used to combine the modules together. Its inputs are the object code for modules 1, 2, and 3. Execution of the linker combines these programs into a single run module.

The technique of writing larger programs as a series of modules has several benefits. First, since several programmers can work on the project simultaneously, the program is completed in a shorter period of time. Another benefit is that the smaller sizes of the

```
                              TITLE BLOCK-MOVE PROGRAM

                               PAGE   ,132

                         COMMENT *This program moves a block of specified number of bytes
                              from one place to another place*

                         ;Define constants used in this program

                                     N     =     16            ;Bytes to be moved
cblock.asm(13):error A2008: syntax error:N
=0100                                BLK1ADDR=     100H        ;Source block offset address
=0120                                BLK2ADDR=     120H        ;Destination block offset addr
=2000                                DATASEGADDR=  2000H       ;Data segment start address

0000                          STACK_SEG    SEGMENT     STACK 'STACK'
0000 0040[                                 DB       64DUP(?)
    00
   ]
0040                          STACK_SEG    ENDS

0000                          CODE_SEG     SEGMENT      'CODE'
0000                          BLOCK      PROC      FAR
                                     ASSUME CS:CODE_SEG,SS:STACK_SEG

                              ;To return to DEBUG program put return address on the stack

0000 1E                              PUSH  DS
0001 B8 0000                         MOV   AX,0
0004 50                              PUSH  AX

                              ;Set up the data segment address

0005 B8 2000                         MOV   AX,DATASEGADDR
0008 8E D8                           MOV   DS,AX

                              ;Set up the source and destination offset addresses
cblock.asm(39):error A2008:syntax error:up

000A BE 0100                         MOV   SI,BLK1ADDR
000D BF 0120                         MOV   DI,BLK2ADDR

                              ;Set up the count of bytes to be moved

                                     MOV   CX,N
cblock.asm(46):error A2006:undefined symbol:N
```

Figure 7–20 Source listing for assembly of a file with syntax errors.

modules require less time to edit and assemble. For instance, if we were not using modular programming, to make a change in just one statement, the complete program would have to be edited, reassembled, and relinked. When using modular programming, only the module containing the statement that needs to be changed can be edited and reassembled. Then the new object module is relinked with the rest of the old object modules to generate a new run module.

A third benefit derived from modular programming is that it makes it easier to reuse old software. For instance, a new software design may need some functions for which modules have already been written as part of an old design. If these modules were part of a single large program, we would need to edit them out carefully and transfer them to the new source file. However, if these segments of program exist as separate modules, we may need to do no additional editing to integrate them into the new application.

```
                               ;Copy source block to destination block

001B 8A 24                     NXTPT: MOV  AH,[SI]      ;Move a byte
001D 88 25                            MOV  [DI],AH

001F 46                               INC  SI           ;Update pointers
0020 47                               INC  DI
                                      DCR  CX           ;Update byte counter
cblock.asm(54):error A2008:syntax error:cx
0021 75 F8                            JNZ  NXTPT        ;Repeat for next byte
0023 CB                               RET               ;Return to DEBUG program
0024                           BLOCK       ENDP
0024                           CODE_SEG    ENDS
                                      END  BLOCK        ;End of program

Segments and Groups:

          N a m e                Size    Length   Align    Combine Class

CODE_SEG . . . . . . . . . . . .         16 Bit   0024     Para    Private'CODE'
STACK_SEG. . . . . . . . . . . .         16 Bit   0040     Para    Stack  'STACK'

Procedures, parameters and locals:

          N a m e                Type    Value    Attr

BLOCK. . . . . . . . . . . . . . P Far   0000     CODE_SEG   Length= 0024 Public
NXTPT. . . . . . . . . . . . . . L Near  001B     CODE_SEG

Symbols:

          N a m e                Type    Value    Attr

BLK1ADDR . . . . . . . . . . . .         Number 0100h
BLK2ADDR . . . . . . . . . . . .         Number 0120h
DATASEGADDR. . . . . . . . . . .         Number 2000h

      0 Warnings
      4 Errors
```

Figure 7–20 (continued)

Initiating the Assembly and Linking Processes

To assemble and link a program, we must first ensure that there is a path to the directories that contain the assembler and linker programs. Assuming that the source program is on a diskette, insert the diskette into drive A. The source file diskette should not be write-protected. Select drive A to be the drive to initiate the command for assembling and linking. Now the command used to initiate both assembly and linking of files with the MASM version 6.11 is given as

```
ML [options] file name [[options] file name]... [/link link-options]
```

In this command, filename(s) is (are) the file(s) to be assembled and linked. The options allow one to specify assembly variations, and the link options are for specifying link variations. There are a number of assembly options, a few of which are considered here. The option /Fl is used to generate a listing file as part of the assembly process. Similarly /Fm is used to generate a map file as part of the linking process. If no options are specified, the linker generates just the object file and the executable file.

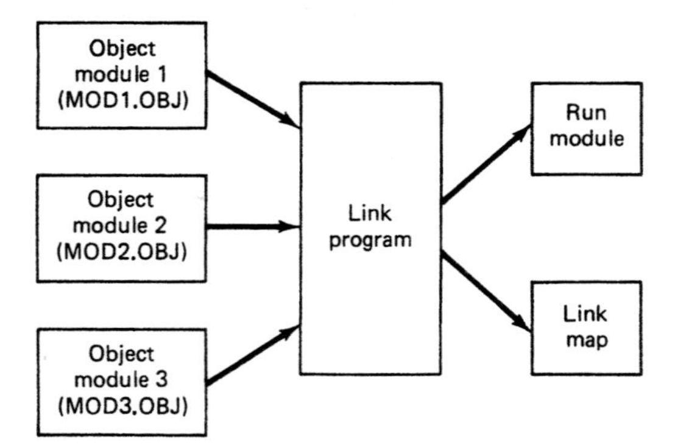

Figure 7–21 Linking object modules.

For instance, the command

```
C:\MASM611>ML  BLOCK.ASM  (↵)
```

generates BLOCK.OBJ (the object file) and BLOCK.EXE (the executable file) from the source file BLOCK.ASM. Similarly, the command

```
C:\MASM611>ML /Fl /Fm BLOCK.ASM  (↵)
```

generates files BLOCK.OBJ, BLOCK.LST, BLOCK.EXE, and BLOCK.MAP. Figure 7–22 shows the display sequence used to initiate the assembly process. Figure 7–19 shows the source listing produced by this assembly command.

There is an option, /Zm, that enables MASM5.10 compatibility. Inclusion of this option allows MASM6.11 to assemble programs that were written according to the rules of MASM version 5.10. This option is useful because it allows a programmer to use older programs with the new version of the assembler.

Figure 7–23 shows the map file generated by the above command.

```
A:\>ML/Fl/Fm BLOCK ASM
Microsoft (R) Macro Assembler Version 6.11
Copyright (C) Microsoft Corp 1981-1993. All rights reserved

Assembling:BLOCK ASM

Microsoft (R) Segmented Executable Linker Version 5.31.009 Jul 13 1992
Copyright (C) Microsoft Corp 1984-1992. All rights reserved.


Object Modules [.obj]: BLOCK.obj
Run File [BLOCK.exe]: "BLOCK.exe"
List File [nul,map]: "BLOCK.map"/m
Libraries [.lib]:
Definitions File [nul.def]:

A:\>
```

Figure 7–22 Display sequence for initiating assembly of a program.

```
Start  Stop  Length Name          Class
00000H 0003FH 00040H STACK_SEG       STACK
00040H 0005CH 0001DH CODE_SEG        CODE

Program entry point at 0004:0000
```

Figure 7–23 Link map file.

There is also an option /C that initiates the assembly process, but without the linker. This option may be used to generate object files, which can be linked separately using the link command. The command

```
C:\MASM611>LINK  (↵)
```

invokes the linker, which further asks the user to supply the names of the object modules. Figure 7–24 shows this link process for linking object modules together.

▲ 7.5 LOADING AND EXECUTING A RUN MODULE

In Chapter 4 we learned how to bring up the DEBUG program; how to use its commands; and how to load, execute, and debug the operation of a program. At that time, we loaded the machine code for the program and data with memory-modify commands. Up to this point in this chapter, we have learned how to form a source program using assembly language and assembler directive statements and how to assemble and link the program into object and run modules. Here we will load and execute the run module BLOCK.EXE that was produced in Section 7–4 for the source program BLOCK.ASM.

When the DEBUG program was loaded in Chapter 4, we did not have a run module available. For this reason, we brought up the debugger by typing in DEBUG and depressing (↵). Now that we do have a run module, we will bring up the debugger in a different way and load the run module at the same time. Issuing the command that follows does this:

```
C:\DOS>DEBUG  A:BLOCK.EXE    (↵)
```

In response to this command, both the DEBUG program and the run module BLOCK.EXE are loaded into the PC's memory. As Fig. 7–25 shows, after loading, the debug prompt "-" is displayed. Next, the register status is dumped with an R command. Note that DS is initialized with the value $11C5_{16}$.

```
A:\>LINK

Microsoft (R) Segmented Executable Linker Version 5.31.009 Jul 13 1992
Copyright (C) Microsoft Corp 1984-1992. All rights reserved.

Object Modules[.obj]: BLOCK
Run File [BLOCK.exe]:
List File [nul.map]: BLOCK
Libraries [.lib]
Definitions File [nul.def]:

A:\>
```

Figure 7–24 Display sequence to initiate linking of object files.

```
C:\DOS>DEBUG A:BLOCK.EXE
-R
AX=0000  BX=0000  CX=005D  DX=0000  SP=0040  BP=0000  SI=0000  DI=0000
DS=11C5  ES=11C5  SS=11D5  CS=11D9  IP=0000   NV UP EI PL NZ NA PO NC
11D9:0000 1E            PUSH    DS
-U CS:0 1C
11D9:0000 1E            PUSH    DS
11D9:0001 B80000        MOV     AX,0000
11D9:0004 50            PUSH    AX
11D9:0005 B80020        MOV     AX,2000
11D9:0008 8ED8          MOV     DS,AX
11D9:000A BE0001        MOV     SI,0100
11D9:000D BF2001        MOV     DI,0120
11D9:0010 B91000        MOV     CX,0010
11D9:0013 8A24          MOV     AH,[SI]
11D9:0015 8825          MOV     [DI],AH
11D9:0017 46            INC     SI
11D9:0018 47            INC     DI
11D9:0019 49            DEC     CX
11D9:001A 75F7          JNZ     0013
11D9:001C CB            RETF
-G =CS:0 13

AX=2000  BX=0000  CX=0010  DX=0000  SP=003C  BP=0000  SI=0100  DI=0120
DS=2000  ES=11C5  SS=11D5  CS=11D9  IP=0013   NV UP EI PL NZ NA PO NC
11D9:0013 8A24          MOV     AH,[SI]                          DS:0100=50
-F DS:100 10F FF
-F DS:120 12F 00
= -G =CS:13 1A

AX=FF20  BX=0000  CX=000F  DX=0000  SP=003C  BP=0000  SI=0101  DI=0121
DS=2000  ES=11C5  SS=11D5  CS=11D9  IP=001A   NV UP EI PL NZ AC PE NC
11D9:001A 75F7          JNZ     0013
-D DS:100 10F
2000:0100  FF FF FF FF FF FF FF FF-FF FF FF FF FF FF FF FF   ................
-D DS:120 12F
2000:0120  FF 00 00 00 00 00 00 00-00 00 00 00 00 00 00 00   ................
-G

Program terminated normally
-D DS:100 10F
2000:0100  FF FF FF FF FF FF FF FF-FF FF FF FF FF FF FF FF   ................
-D DS:120 12F
2000:0120  FF FF FF FF FF FF FF FF-FF FF FF FF FF FF FF FF   ................
-Q

C:\DOS>
```

Figure 7–25 Loading and executing the run module BLOCK.EXE.

Let us now verify that the program has loaded correctly by using the command

```
-U  CS:000  01C    (↵)
```

Comparing the program displayed in Fig. 7–25 to the source program in Fig. 7–16, we see that they are essentially the same. Therefore, the program has been loaded correctly.

Now we will execute the first eight instructions of the program and verify the operation they perform. To do this, we issue the command

```
-G  =CS:000  013    (↵)
```

The information displayed at the completion of this command is also shown in Fig. 7–25. Here we find that the registers have been initialized as follows: DS contains 2000_{16}, AX contains 2000_{16}, SI contains 0100_{16}, DI contains 0120_{16}, and CX contains 0010_{16}.

The Fill command is used to initialize the bytes of data in the source and destination blocks. The storage locations in the source block are loaded with FF_{16} with the command

```
-F  DS:100 10F FF     (↵)
```

and the storage locations in the destination block are loaded with 00_{16} with the command

```
-F  DS:120 12F 00     (↵)
```

Next, we execute down through the program to the JNZ instruction, address $001A_{16}$:

```
-G  =CS:013  01A     (↵)
```

To check the state of the data blocks, we use the commands

```
-D  DS:100 10F     (↵)
-D  DS:120 12F     (↵)
```

Looking at the displayed blocks of data in Fig. 7–25, we see that the source block is unchanged and that FF_{16} has been copied into the first element of the destination block.

Finally, the program is run to completion with the command

```
-G  (↵)
```

By once more looking at the two blocks of data with the commands

```
D  DS:100 10F     (↵)
D  DS:120 12F     (↵)
```

we find that the contents of the source block have been copied into the destination block.

DEBUG is used to run a program when we must either debug the operation of the program or want to understand its execution step by step. If we only want to run a program instead of observing its operation, we can use another method. The run module (.EXE file) can be executed at the DOS prompt by simply entering its name followed by (↵). This will cause the program to be loaded and then executed to completion. For instance, to execute the run module BLOCK.EXE that resides on a diskette in drive A, we enter

```
C:\DOS>A:BLOCK  (↵)
```

REVIEW PROBLEMS

Section 7.1

1. What kind of application program is MASM?
2. What are the two types of statements in a source program?
3. What is the function of an assembly language instruction?

4. What is the function of a directive?
5. What are the four elements of an assembly language statement?
6. What part of the instruction format is always required?
7. What are the two limitations on format when writing source statements for the MASM?
8. What is the function of a label?
9. What is the maximum number of characters in a label that the MASM will recognize?
10. What is the function of the opcode?
11. What is the function of operands?
12. In the instruction statement

```
SUB_A:   MOV   CL, 0FFH
```

    what is the source operand and destination operand?
13. What is the purpose of the comment field? How does an assembler process comments?
14. Give two differences between an assembly language statement and a directive statement.
15. Write the instruction MOV AX, [32728D] with the source operand expressed both in binary and hexadecimal forms.
16. Rewrite the jump instruction JMP +25D with the operand expressed both in binary and hexadecimal forms.
17. Repeat Example 7.4 with the values A = 345, B = 234, and C = 111.

Section 7.2

18. Give another name for a directive.
19. List the names of the directive categories.
20. What is the function of the data directive?
21. What happens when the statements

```
SRC_BLOCK  =  0100H
DEST_BLOCK  =  SRC_BLOCK  +  20H
```

    are processed by MASM?
22. Describe the difference between the EQU and = directive.
23. What does the statement

```
SEG_ADDR   DW   1234H
```

    do when processed by the assembler?
24. What happens when the statement

```
BLOCK_1   DB   128   DUP(?)
```

    is processed by the assembler?

25. Write a data directive statement to define INIT_COUNT as word size and assign it the value $F000_{16}$.
26. Write a data directive statement to allocate a block of 16 words in memory called SOURCE_BLOCK, but do not initialize them with data.
27. Write a directive statement that will initialize the block of memory storage locations allocated in problem 26 with the data values 0000H, 1000H, 2000H, 3000H, 4000H, 5000H, 6000H, 7000H, 8000H, 9000H, A000H, B000H, C000H, D000H, E000H, and F000H.
28. What does the statement

```
DATA_SEG   SEGMENT   BYTE MEMORY   'DATA'
```

mean?

29. Show how the segment-control directives are used to define a segment called DATA_SEG that is aligned on a word-address boundary, overlaps other segments with the same name, and is a data segment.
30. What is the name of the smaller segments in which modular programming techniques specify that programs should be developed?
31. What is a procedure?
32. Show the general structure of a far procedure called BLOCK that is to be accessible from other modules.
33. What is the function of an ORG directive?
34. Write an origin statement that causes machine code to be loaded at offset 1000H of the current code segment.
35. Write a page statement that will set up the printout for 55 lines per page and 80 characters per line and a title statement that will title pages of the source listing with "BLOCK-MOVE PROGRAM."

Section 7.3

36. What type of program is EDIT?
37. List the basic editing operations that can be performed using EDIT.
38. What are the names of the four menus in the EDIT menu bar?
39. How could a backup copy of the program in file BLOCK.ASM be created from the file menu?

Section 7.4

40. What is the input to the assembler program?
41. What are the outputs of the assembler? Give a brief description of each.
42. What is the cause of the first error statement in the source listing shown in Fig. 7–20?
43. What is the cause of the error A2008 that is located at the line identified as cblock.asm(39) in the source listing shown in Fig. 7–20?
44. Can the output of the assembler be directly executed by the 8088 microprocessor in the PC?
45. Give three benefits of modular programming.

46. What is the input to the LINK program?

47. What are the outputs of the LINK program? Give a brief description of each.

48. What name does the command entry

```
C:\MASM611>ML /Fl / BLOCK.ASM
```

assign to each of the input and output files?

49. If three object modules called MAIN.OBJ, SUB1.OBJ, and SUB2.OBJ are to be combined with LINK, write the response that must be made to the linker's prompt. Assume that all three files are on a diskette in drive A.

Section 7.5

50. Write a DOS command that will load run module LAB.EXE while bringing up the DEBUG program. Assume that the run module file is on a diskette in drive B.

Application Problems

51. Upgrade the programs written for the review problems 10, 15, 21, 27, 35, 37, 41, and 42 in Chapter 5 so that they can be assembled using the MASM and debugged using DEBUG.

52. Upgrade the programs written for the review problems 5, 22, 23, 24, 25, 26, 35, 37, 38, 43, 44, 48, and 49 in Chapter 6 so that they can be assembled using the MASM and debugged using DEBUG.

The 8088 and 8086 Microprocessors and Their Memory and Input/Output Interfaces

▲ INTRODUCTION

Up to this point in the book, we have studied the 8088 and 8086 microprocessors from a software point of view. We have covered their software architecture, instruction set, how to write, execute, and debug programs in assembly language, and found that the 8088 and 8086 were identical from the software point of view. This is not true of the hardware architectures of the 8088 and 8086 microcomputer systems. Now we begin examining the 8088 and 8086 microcomputer from the hardware point of view. In this chapter, we cover the 8088/8086's signal interfaces, memory interfaces, input/output interfaces, and bus cycles. The chapters that follow cover other hardware and interfacing aspects of these processors. This chapter includes the following topics:

8.1 8088 and 8086 Microprocessors
8.2 Minimum-Mode and Maximum-Mode Systems
8.3 Minimum-Mode Interface Signals
8.4 Maximum-Mode Interface Signals
8.5 Electrical Characteristics
8.6 System Clock
8.7 Bus Cycle and Time States
8.8 Hardware Organization of the Memory Address Space
8.9 Address Bus Status Codes

▲ 8.1 8088 AND 8086 MICROPROCESSORS

The 8086, announced in 1978, was the first 16-bit microprocessor introduced by Intel Corporation. A second member of the 8086 family, the 8088 microprocessor, followed it in 1979. The 8088 is fully software compatible with its predecessor, the 8086. The difference between these two devices is in their hardware architecture. Just like the 8086, the 8088 is internally a 16-bit MPU. However, externally the 8086 has a 16-bit data bus, and the 8088 has an 8-bit data bus. This is the key hardware difference. Both devices have the ability to address up to 1Mbyte of memory via their 20-bit address buses. Moreover, they can address up to 64K of byte-wide input/output ports.

The 8088 and 8086 are both manufactured using *high-performance metal-oxide semiconductor (HMOS) technology,* and the circuitry on their chips is equivalent to approximately 29,000 transistors. They are housed in a 40-pin dual in-line package. This package can be mounted into a socket that is soldered to the circuit board or have its leads inserted through holes in the board and soldered. The signals pinned out to each lead are shown in Figs. 8–1(a) and (b), respectively. Many of their pins have multiple functions. For example, in the pin layout diagram of the 8088, we see that address bus lines A_0 through A_7 and data bus lines D_0 through D_7 are multiplexed. For this reason, these leads are labeled AD_0 through AD_7. By *multiplexed* we mean that the same physical pin carries an address bit at one time and the data bit at another time.

EXAMPLE 8.1

At what pin location on the 8088's package is address bit A_{16} output? With what other signal is it multiplexed? What function does this pin serve on the 8086?

Solution

Looking at Fig. 8–1(a), we find that the signal A_{16} is located at pin 38 on the 8088 and that it is multiplexed with signal S_3. Figure 8–1(b) shows us that pin 38 serves the same functions on the 8086.

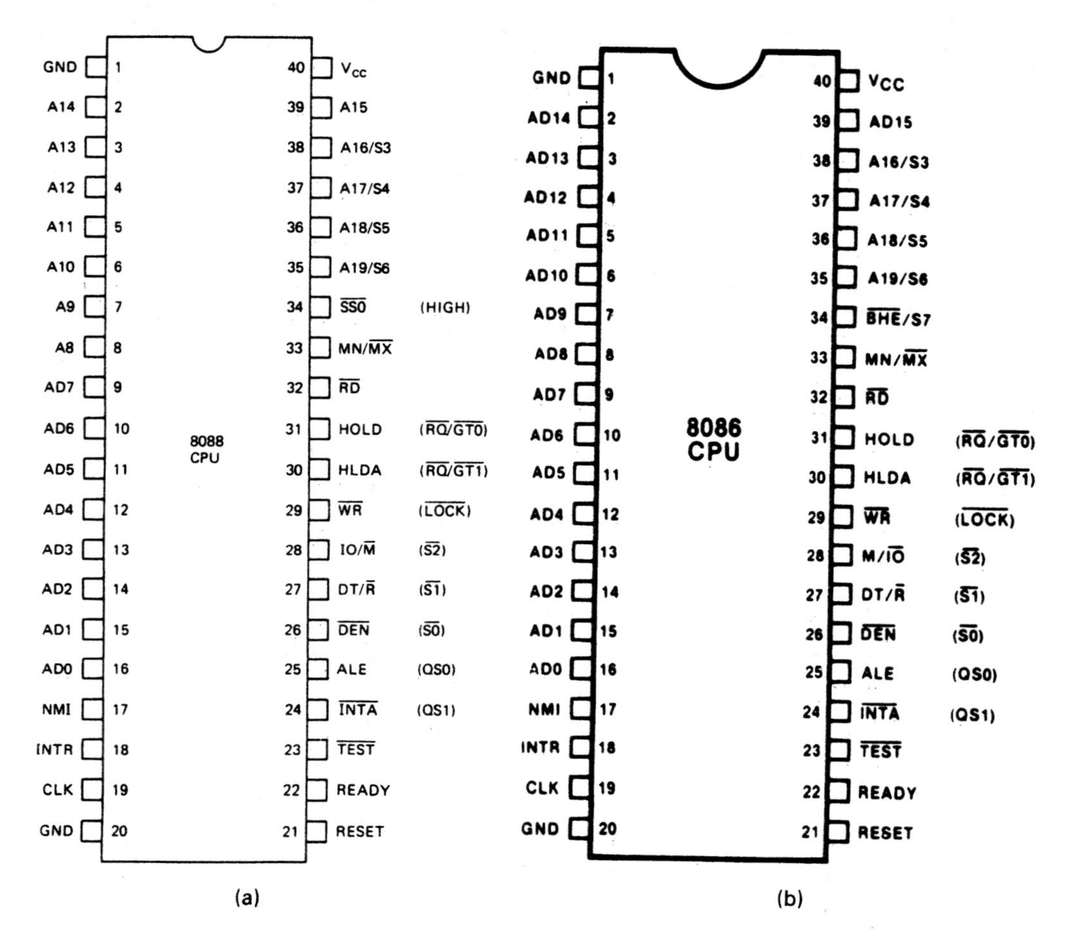

Figure 8–1 (a) Pin layout of the 8088 microprocessor. (Reprinted with permission of Intel Corporation, © 1981) (b) Pin layout of the 8086 microprocessor. (Reprinted with permission of Intel Corporation, © 1979)

▲ 8.2 MINIMUM-MODE AND MAXIMUM-MODE SYSTEMS

The 8088 and 8086 microprocessors can be configured to work in either of two modes: the *minimum mode* or the *maximum mode.* The minimum mode is selected by applying logic 1 to the MN/$\overline{MX}$ input lead. Minimum mode 8088/8086 systems are typically smaller and contain a single microprocessor. Connecting MN/$\overline{MX}$ to logic 0 selects the maximum mode of operation. This configures the 8088/8086 system for use in larger systems and with multiple processors. This mode-selection feature lets the 8088 or 8086 better meet the needs of a wide variety of system requirements.

Depending on the mode of operation selected, the assignments for a number of the pins on the microprocessor package are changed. As Fig. 8–1(a) shows, the pin functions of the 8088 specified in parentheses pertain to a maximum-mode system.

The signals of the 8088 microprocessor common to both modes of operation, those unique to minimum mode and those unique to maximum mode, are listed in Figs. 8–2(a), (b), and (c), respectively. Here we find the name, function, and type for each signal. For example, the signal $\overline{RD}$ is in the common group. It functions as a read control output and is used to signal memory or I/O devices when the 8088's system bus is set up to read in data. Moreover, note that the signals hold request (HOLD) and hold acknowledge (HLDA) are produced only in the minimum-mode system. If the 8088 is set up for maximum mode, they are replaced by the request/grant bus access control lines $\overline{RQ}/\overline{GT}_0$ and $\overline{RQ}/\overline{GT}_1$.

Common signals		
Name	Function	Type
AD7-AD0	Address/data bus	Bidirectional, 3-state
A15-A8	Address bus	Output, 3-state
A19/S6-A16/S3	Address/status	Output, 3-state
MN/$\overline{MX}$	Minimum/maximum Mode control	Input
$\overline{RD}$	Read control	Output, 3-state
$\overline{TEST}$	Wait on test control	Input
READY	Wait state control	Input
RESET	System reset	Input
NMI	Nonmaskable Interrupt request	Input
INTR	Interrupt request	Input
CLK	System clock	Input
V_{CC}	+5 V	Input
GND	Ground	

(a)

Minimum mode signals (MN/$\overline{MX}$ = V_{CC})		
Name	Function	Type
HOLD	Hold request	Input
HLDA	Hold acknowledge	Output
$\overline{WR}$	Write control	Output, 3-state
IO/$\overline{M}$	IO/memory control	Output, 3-state
DT/$\overline{R}$	Data transmit/receive	Output, 3-state
$\overline{DEN}$	Data enable	Output, 3-state
$\overline{SSO}$	Status line	Output, 3-state
ALE	Address latch enable	Output
$\overline{INTA}$	Interrupt acknowledge	Output

(b)

Maximum mode signals (MN/$\overline{MX}$ = GND)		
Name	Function	Type
$\overline{RQ}/\overline{GT1, 0}$	Request/grant bus access control	Bidirectional
$\overline{LOCK}$	Bus priority lock control	Output, 3-state
$\overline{S2}$-$\overline{S0}$	Bus cycle status	Output, 3-state
QS1, QS0	Instruction queue status	Output

(c)

Figure 8–2 (a) Signals common to both minimum and maximum modes. (b) Unique minimum-mode signals. (c) Unique maximum-mode signals.

EXAMPLE 8.2

Which pins provide different signal functions in the minimum-mode 8088 and minimum-mode 8086?

Solution

Comparing the pin layouts of the 8088 and 8086 in Fig. 8–1, we find the following:

1. Pins 2 through 8 on the 8088 are address lines A_{14} through A_8, but on the 8086 they are address/data lines AD_{14} through AD_8.
2. Pin 28 on the 8088 is the $IO/\overline{M}$ output and on the 8086 it is the $M/\overline{IO}$ output.
3. Pin 34 of the 8088 is the $\overline{SSO}$ output, and on the 8086 this pin supplies the $\overline{BHE}/S_7$ output.

▲ 8.3 MINIMUM-MODE INTERFACE SIGNALS

When minimum mode operation is selected, the 8088 or 8086 itself provides all the control signals needed to implement the memory and I/O interfaces. Figures 8–3(a) and (b) show block diagrams of a minimum-mode configuration of the 8088 and 8086, respectively. The minimum-mode signals can be divided into the following basic groups: address/data bus, status, control, interrupt, and DMA.

Address/Data Bus

Let us first look at the address/data bus. In an 8088-based microcomputer system, these lines serve two functions. As an *address bus,* they are used to carry address information to the memory and I/O ports. The address bus is 20 bits long and consists of signal lines A_0 through A_{19}. Of these, A_{19} represents the MSB and A_0 the LSB. A 20-bit address gives the 8088 a 1Mbyte memory address space. However, only address lines A_0 through A_{15} are used when accessing I/O. This gives the 8088 an independent I/O address space that is 64Kbytes in length.

The eight *data bus* lines D_0 through D_7 are actually multiplexed with address lines A_0 through A_7, respectively. For this reason, they are denoted as AD_0 through AD_7. Data line D_7 is the MSB in the byte of data and D_0 the LSB. When acting as a data bus, they carry read/write data for memory, input/output data for I/O devices, and interrupt-type codes from an interrupt controller.

Looking at Fig. 8–3(b), we see that the 8086 has 16 data bus lines instead of 8 as in the 8088. Data lines are multiplexed with address lines A_0 through A_{15} and are therefore denoted as AD_0 through AD_{15}.

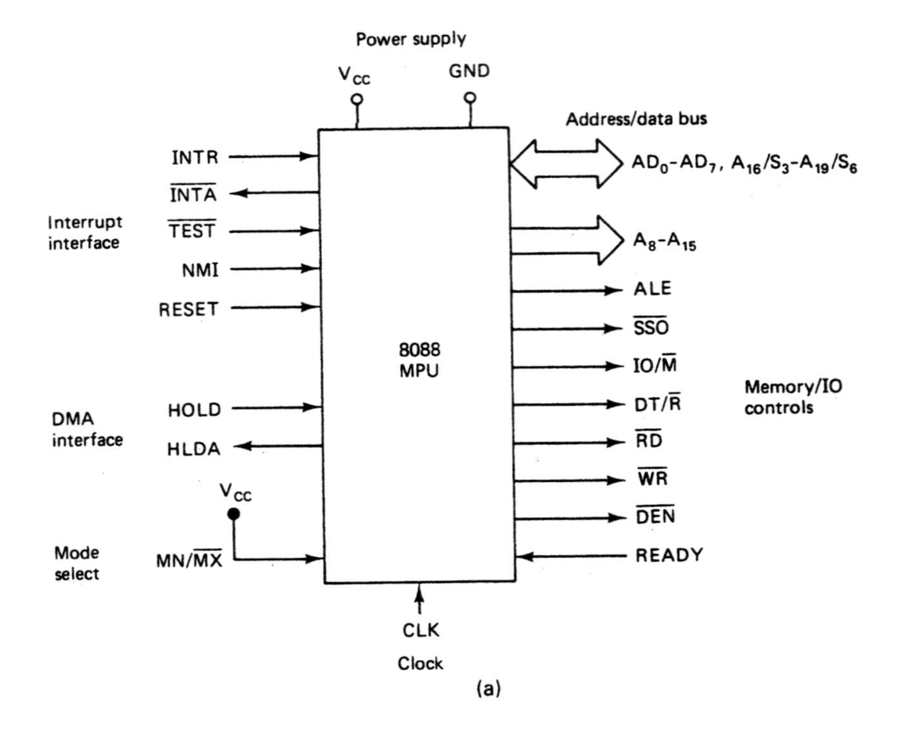

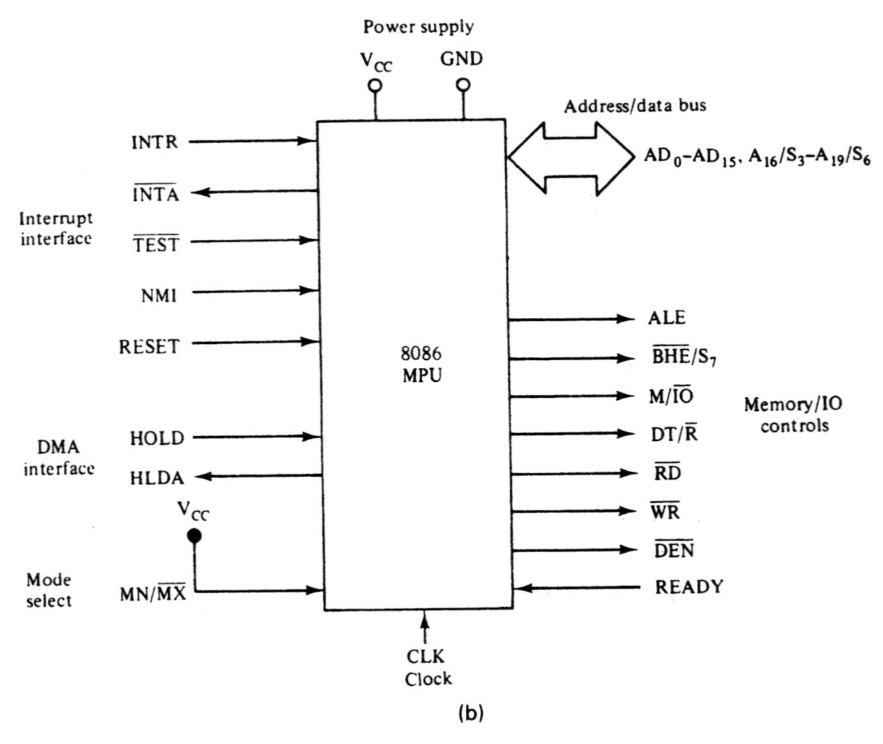

Figure 8–3 (a) Block diagram of the minimum-mode 8088 MPU. (b) Block diagram of the minimum-mode 8086 MPU.

S_4	S_3	Address Status
0	0	Alternate (relative to the ES segment)
0	1	Stack (relative to the SS segment)
1	0	Code/None (relative to the CS segment or a default of zero)
1	1	Data (relative to the DS segment)

Figure 8–4 Address bus status codes. (Reprinted with permission of Intel Corporation, © 1979)

Status Signals

The four most significant address lines, A_{19} through A_{16} of both the 8088 and 8086 are also multiplexed, but in this case with *status signals* S_6 through S_3. These status bits are output on the bus at the same time that data are transferred over the other bus lines. Bits S_4 and S_3 together form a 2-bit binary code that identifies which of the internal segment registers was used to generate the physical address that was output on the address bus during the current bus cycle. These four codes and the registers they represent are shown in Fig. 8–4. Note that the code $S_4S_3 = 00$ identifies the extra segment register as the source of the segment address.

Status line S_5 reflects the status of another internal characteristic of the MPU. It is the logic level of the internal interrupt enable flag. The status bit S_6 is always at the 0 logic level.

Control Signals

The *control signals* are provided to support the memory and I/O interfaces of the 8088 and 8086. They control functions such as when the bus carries a valid address, which direction data are transferred over the bus, when valid write data are on the bus, and when to put read data on the system bus. For example, *address latch enable* (ALE) is a pulse to logic 1 that signals external circuitry when a valid address is on the bus. This address can be latched in external circuitry on the 1-to-0 edge of the pulse at ALE.

Using the $IO/\overline{M}$ (*IO/memory*) line, $DT/\overline{R}$ (*data transmit/receive*) line, and $\overline{SSO}$ (*status output*) line, the 8088 signals which type of bus cycle is in progress and in which direction data are to be transferred over the bus. The logic level of $IO/\overline{M}$ tells external circuitry whether a memory or I/O transfer is taking place over the bus. Logic 0 at this output signals a memory operation, and logic 1 signals an I/O operation. The direction of data transfer over the bus is signaled by the logic level output at $DT/\overline{R}$. When this line is logic 1 during the data transfer part of a bus cycle, the bus is in the transmit mode. Therefore, data are either written into memory or output to an I/O device. On the other hand, logic 0 at $DT/\overline{R}$ signals that the bus is in the receive mode. This corresponds to reading data from memory or input of data from an input port.

Comparing Figs. 8–3(a) and 8–3(b), we find two differences between the minimum-mode 8088 and 8086 microprocessors. First, the 8086's memory/IO control ($M/\overline{IO}$) signal is the complement of the equivalent signal of the 8088. Second, the 8088's $\overline{SSO}$ status signal is replaced by *bank high enable* ($\overline{BHE}$) on the 8086. Logic 0 on this line is used as a memory enable signal for the most significant byte half of the data bus, D_8 through D_{15}. This line also carries status bit S_7.

The signals *read* ($\overline{RD}$) and *write* ($\overline{WR}$) indicate that a read bus cycle or a write bus cycle, respectively, is in progress. The MPU switches $\overline{WR}$ to logic 0 to signal external devices that valid write or output data are on the bus. On the other hand, $\overline{RD}$ indicates that the MPU is performing a read of data off the bus. During read operations, one other control signal, $\overline{DEN}$ (*data enable*), is also supplied. It enables external devices to supply data to the microprocessor.

One other control signal involved with the memory and I/O interface, the READY signal, can be used to insert wait states into the bus cycle so that it is extended by a number of clock periods. This signal is provided by way of an external clock generator device and can be supplied by the memory or I/O subsystem to signal the MPU when it is ready to permit the data transfer to be completed.

Interrupt Signals

The key interrupt interface signals are *interrupt request* (INTR) and *interrupt acknowledge* ($\overline{INTA}$). INTR is an input to the 8088 and 8086 that can be used by an external device to signal that it needs to be serviced. This input is sampled during the final clock period of each *instruction acquisition cycle.* Logic 1 at INTR represents an active interrupt request. When the MPU recognizes an interrupt request, it indicates this fact to external circuits with pulses to logic 0 at the $\overline{INTA}$ output.

The $\overline{TEST}$ input is also related to the external interrupt interface. For example, execution of a WAIT instruction causes the 8088 or 8086 to check the logic level at the $\overline{TEST}$ input. If logic 1 is found at this input, the MPU suspends operation and goes into what is known as the *idle state.* The MPU no longer executes instructions; instead, it repeatedly checks the logic level of the $\overline{TEST}$ input waiting for its transition back to logic 0. As $\overline{TEST}$ switches to 0, execution resumes with the next instruction in the program. This feature can be used to synchronize the operation of the MPU to an event in external hardware.

There are two more inputs in the interrupt interface: *nonmaskable interrupt* (NMI) and *reset* (RESET). On the 0-to-1 transition of NMI, control is passed to a nonmaskable interrupt service routine at completion of execution of the current instruction. NMI is the interrupt request with highest priority and cannot be masked by software. The RESET input is used to provide a hardware reset for the MPU. Switching RESET to logic 0 initializes the internal registers of the MPU and initiates a reset service routine.

DMA Interface Signals

The *direct memory access* (DMA) interface of the 8088/8086 minimum-mode microcomputer system consists of the HOLD and HLDA signals. When an external device wants to take control of the system bus, it signals this fact to the MPU by switching HOLD to the 1 logic level. For example, when the HOLD input of the 8088 becomes active, it enters the hold state at the completion of the current bus cycle. When in the hold state, signal lines AD_0 through AD_7, A_8 through A_{15}, A_{16}/S_3 through A_{19}/S_6, $\overline{SSO}$,

$IO/\overline{M}$, $DT/\overline{R}$, $\overline{RD}$, $\overline{WR}$, $\overline{DEN}$, and INTR are all put into the high-Z state. The 8088 signals external devices that it is in this state by switching its HLDA output to the 1 logic level.

▲ 8.4 MAXIMUM-MODE INTERFACE SIGNALS

When the 8088 or 8086 microprocessor is set for the maximum-mode configuration, it produces signals for implementing a *multiprocessor/coprocessor system environment.* By *multiprocessor environment* we mean that multiple microprocessors exist in the system and that each processor executes its own program. Usually in this type of system environment, some system resources are common to all processors. They are called *global resources.* There are also other resources that are assigned to specific processors. These dedicated resources are known as *local* or *private resources.*

In the maximum-mode system, facilities are provided for implementing allocation of global resources and passing bus control to other microprocessors sharing the system bus.

8288 Bus Controller: Bus Commands and Control Signals

Looking at the maximum-mode block diagram in Fig. 8–5(a), we see that the 8088 does not directly provide all the signals that are required to control the memory, I/O, and interrupt interfaces. Specifically, the $\overline{WR}$, $IO/\overline{M}$, $DT/\overline{R}$, $\overline{DEN}$, ALE, and $\overline{INTA}$ signals are no longer produced by the 8088. Instead, it outputs a status code on three signals lines, $\overline{S}_0$, $\overline{S}_1$, and $\overline{S}_2$, prior to the initiation of each bus cycle. This 3-bit *bus status code* identifies which type of bus cycle is to follow. $\overline{S}_2\overline{S}_1\overline{S}_0$ are input to the external *bus controller* device, the 8288, which decodes them to identify the type of MPU bus cycle. The block diagram and pin layout of the 8288 are shown in Figs. 8–6(a) and (b), respectively. In response, the bus controller generates the appropriately timed command and control signals.

Figure 8–7 shows the relationship between the bus status codes and the types of bus cycles. Also shown are the output signals generated to tell external circuitry which type of bus cycle is taking place. These output signals are *memory read command* ($\overline{MRDC}$), *memory write command* ($\overline{MWTC}$), *advanced memory write command* ($\overline{AMWC}$), *I/O read command* ($\overline{IORC}$), *I/O write command* ($\overline{IOWC}$), *advanced I/O write command* ($\overline{AIOWC}$), and *interrupt acknowledge* ($\overline{INTA}$).

The 8288 produces one or two of these seven command signals for each bus cycle. For instance, when the 8088 outputs the code $\overline{S}_2\overline{S}_1\overline{S}_0 = 001$, it indicates that an I/O read cycle is to be performed. In turn, the 8288 makes its $\overline{IORC}$ output switch to logic 0. On the other hand, if the code 111 is output by the 8088, it is signaling that no bus activity is to take place; the 8288 produces no command signals.

The other control outputs produced by the 8288 consist of DEN, $DT/\overline{R}$, and ALE. These three signals provide the same functions as those described for the minimum mode. Figure 8–5(b) shows that the 8288 bus controller connects to the 8086 in the same way as the 8088, and it also produces the same output signals.

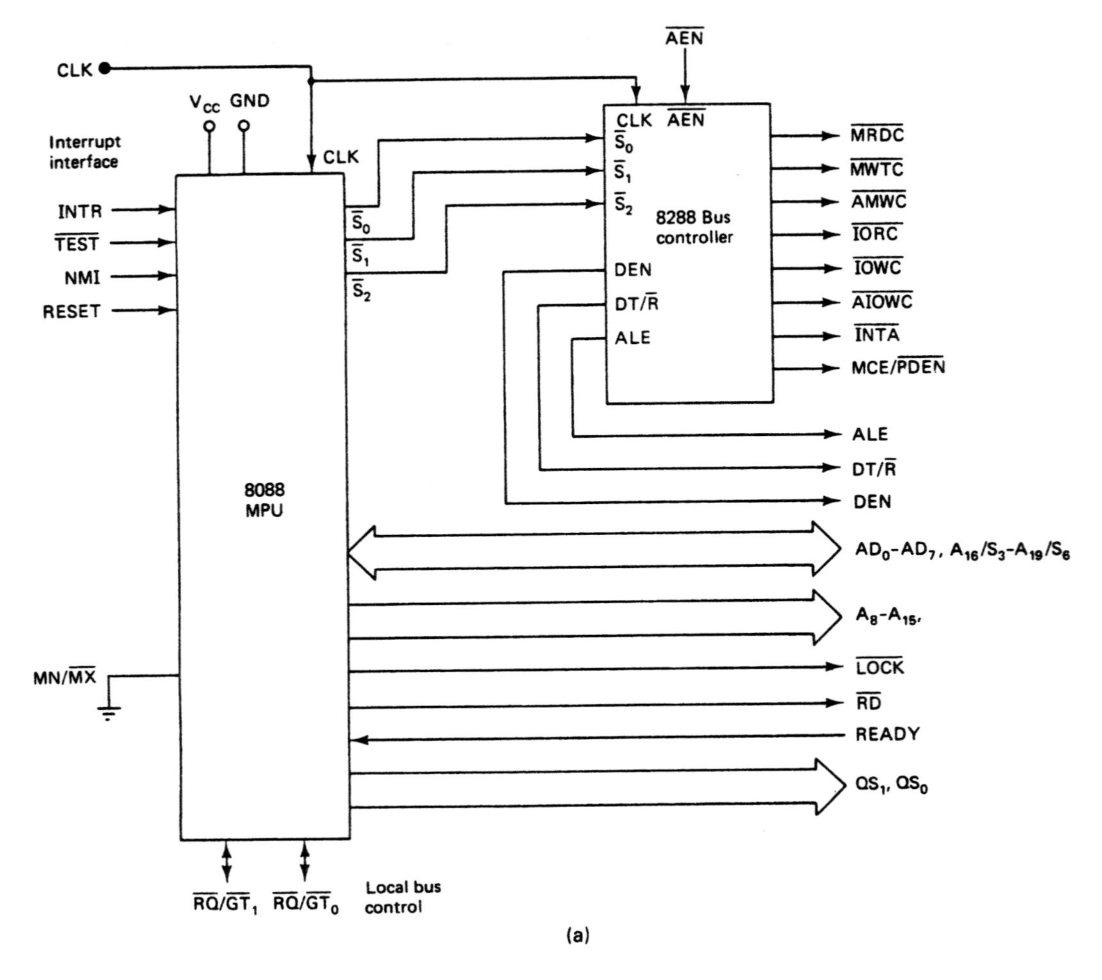

Figure 8–5 (a) 8088 maximum-mode block diagram. (b) 8086 maximum-mode block diagram.

EXAMPLE 8.3

If the bus status code $\overline{S}_2\overline{S}_1\overline{S}_0$ equals 101, what type of bus activity is taking place? Which command output is produced by the 8288?

Solution

Looking at the table in Fig. 8–7, we see that bus status code 101 identifies a read memory bus cycle and causes the $\overline{\text{MRDC}}$ output of the bus controller to be switched to logic 0.

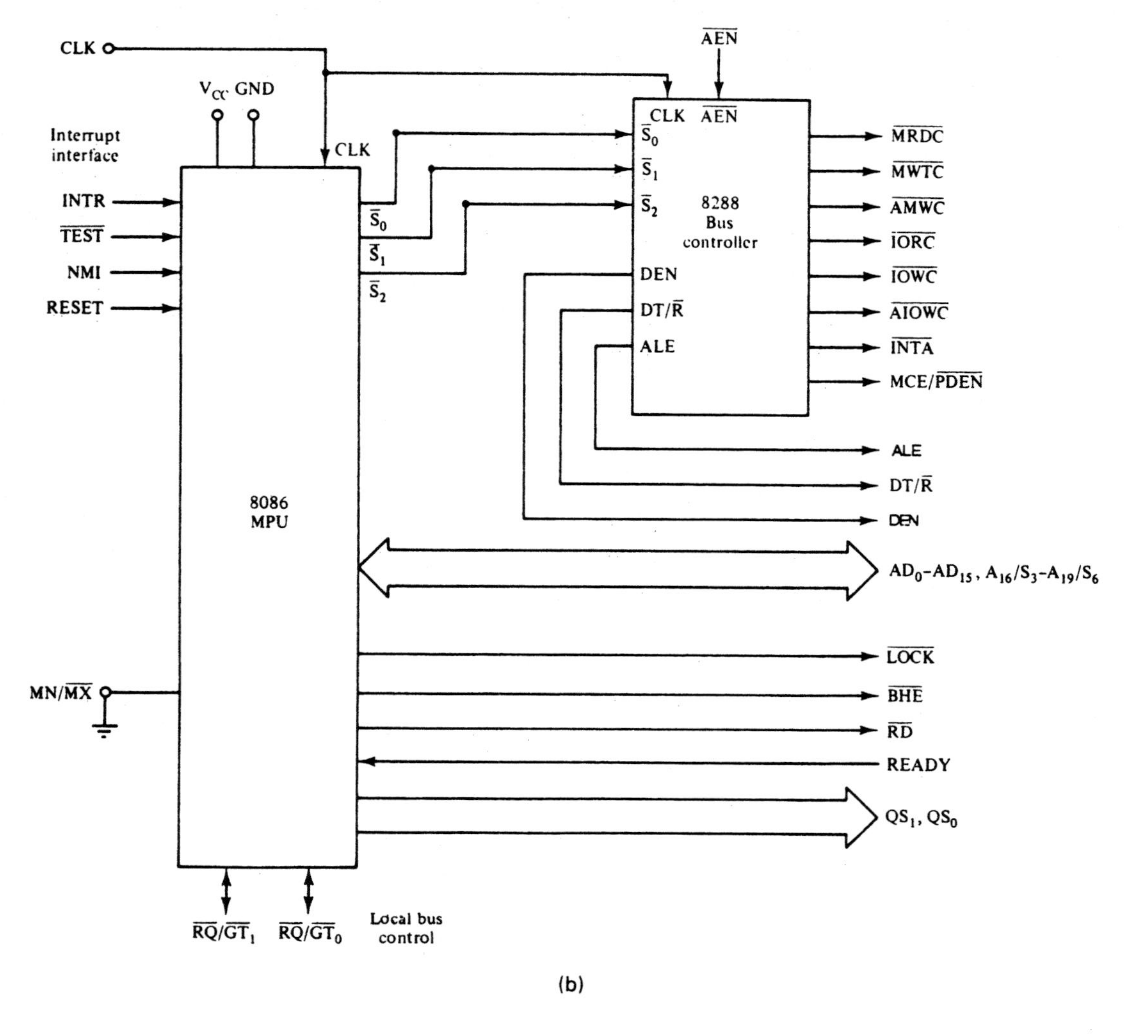

Figure 8–5 (continued)

Lock Signal

To implement a multiprocessor system, a signal called lock ($\overline{\text{LOCK}}$) is provided on the 8088 and 8086. This signal is meant to be output (logic 0) whenever the processor wants to lock out the other processors from using the bus. This would be the case when a shared resource is accessed. The $\overline{\text{LOCK}}$ signal is compatible with the *Multibus,* an industry standard for interfacing microprocessor systems in a multiprocessor environment.

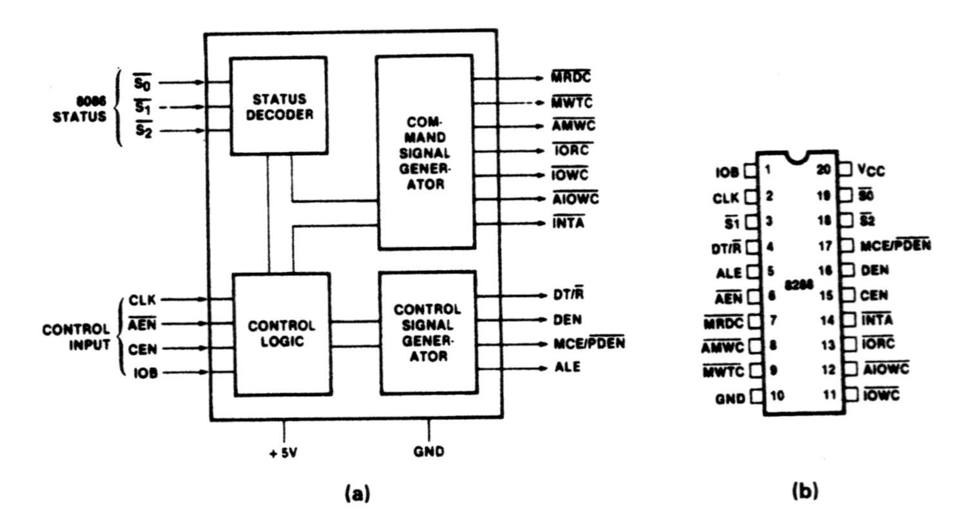

Figure 8–6 (a) Block diagram of the 8288. (Reprinted with permission of Intel Corporation, © 1979) (b) Pin layout. (Reprinted with permission of Intel Corporation, © 1979)

Queue Status Signals

Two other signals produced by the 8088 and 8086, in the maximum-mode microcomputer system, are queue status outputs QS_0 and QS_1 that form a 2-bit *queue status code,* QS_1QS_0. This code tells the external circuitry what type of information was removed from the instruction queue during the previous clock cycle. Figure 8–8 shows the four different queue status codes. Note that $QS_1QS_0 = 01$ indicates that the first byte of an instruction was taken off the queue. As shown, the fetch of the next byte of the instruction is identified by the code 11. Whenever the queue is reset due to a transfer of control, the reinitialization code 10 is output.

Status Inputs			CPU Cycle	8288 Command
$\overline{S2}$	$\overline{S1}$	$\overline{S0}$		
0	0	0	Interrupt Acknowledge	$\overline{INTA}$
0	0	1	Read I/O Port	$\overline{IORC}$
0	1	0	Write I/O Port	$\overline{IOWC}$, $\overline{AIOWC}$
0	1	1	Halt	None
1	0	0	Instruction Fetch	$\overline{MRDC}$
1	0	1	Read Memory	$\overline{MRDC}$
1	1	0	Write Memory	$\overline{MWTC}$, $\overline{AMWC}$
1	1	1	Passive	None

Figure 8–7 Bus status codes. (Reprinted with permission of Intel Corporation, © 1979)

QS1	QS0	Queue Status
0 (low)	0	No Operation. During the last clock cycle, nothing was taken from the queue.
0	1	First Byte. The byte taken from the queue was the first byte of the instruction.
1 (high)	0	Queue Empty. The queue has been reinitialized as a result of the execution of a transfer instruction.
1	1	Subsequent Byte. The byte taken from the queue was a subsequent byte of the instruction.

Figure 8–8 Queue status codes. (Reprinted with permission of Intel Corporation, © 1979)

Local Bus Control Signals

In a maximum-mode configuration, the minimum-mode HOLD and HLDA interface of the 8088/8086 is also changed. These two signals are replaced by *request/grant lines* $\overline{RQ}/\overline{GT}_0$ and $\overline{RQ}/\overline{GT}_1$. They provide a prioritized bus access mechanism for accessing the *local bus.*

▲ 8.5 ELECTRICAL CHARACTERISTICS

In the preceding sections, the pin layout and minimum- and maximum-mode interface signals of the 8088 and 8086 microprocessors were introduced. Here we will first look at the power supply ratings of these processors and then their input and output electrical characteristics.

Looking at Fig. 8–1(a), we find that power is applied between pin 40 (V_{cc}) and pins 1(GND) and 20(GND). Pins 1 and 20 should be connected together. The nominal value of V_{cc} is specified as +5 V dc with a tolerance of ±10%. This means that the 8088 or 8086 will operate correctly as long as the difference in voltage between V_{cc} and GND is greater than 4.5 V dc and less than 5.5 V dc. At room temperature (25°C), both the 8088 and 8086 draw a maximum of 340 mA from the supply.

Let us now look at the dc I/O characteristics of the microprocessor—that is, its input and output logic levels. These ratings tell the minimum and maximum voltages for the 0 and 1 logic states for which the circuit will operate correctly. Different values are specified for the inputs and outputs.

Figure 8–9 shows the I/O voltage specifications for the 8088. Notice that the minimum logic 1 (high-level) voltage at an output (V_{OH}) is 2.4 V. This voltage is specified for a test condition that identifies the amount of current being sourced by the output (I_{OH}) as −400 μA. All processors must be tested during manufacturing to ensure that under this test condition the voltages at all outputs will remain above the value of V_{OHmin}.

Symbol	Meaning	Minimum	Maximum	Test condition
V_{IL}	Input low voltage	−0.5 V	+0.8 V	
V_{IH}	Input high voltage	+2.0 V	V_{cc} + 0.5 V	
V_{OL}	Output low voltage		+0.45 V	I_{OL} = 2.0mA
V_{OH}	Output high voltage	+2.4 V		I_{OH} = −400 uA

Figure 8–9 I/O voltage levels.

Input voltage levels are specified in a similar way; except here the ratings identify the range of voltage that will be correctly identified as a logic 0 or a logic 1 at an input. For instance, voltages in the range $V_{ILmin} = -0.5$ V to $V_{ILmax} = +0.8$ V represent a valid logic 0 (lower level) at an input of the 8088.

The I/O voltage levels of the 8086 microprocessor are identical to those for the 8088 as shown in Fig. 8–9. However, there is one difference in the test conditions. For the 8086, V_{OL} is measured at 2.5 mA instead of 2.0 mA.

▲ 8.6 SYSTEM CLOCK

The time base for synchronization of the internal and external operations of the microprocessor in a microcomputer system is provided by the *clock* (CLK) input signal. At present, the 8088 is available in two different speeds. The standard part operates at 5 MHz and the 8088-2 operates at 8 MHz. On the other hand, the 8086 microprocessor is manufactured in three speeds: the 5-MHz 8086, the 8-MHz 8086-2, and the 10-MHz 8086-1. The 8284 clock generator and driver IC generates CLK. Figure 8–10 is a block diagram of this device.

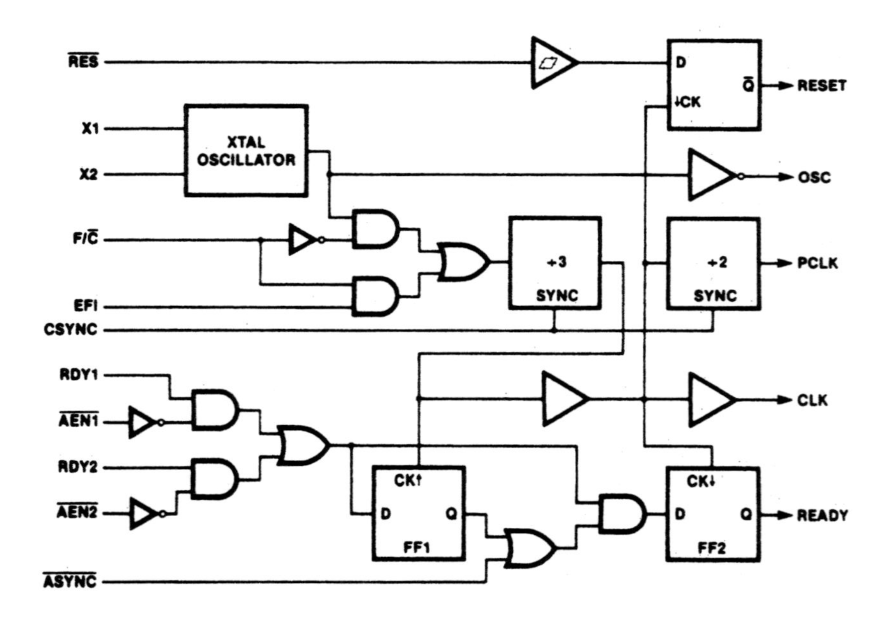

Figure 8–10 Block diagram of the 8284 clock generator. (Reprinted with permission of Intel Corporation, © 1979)

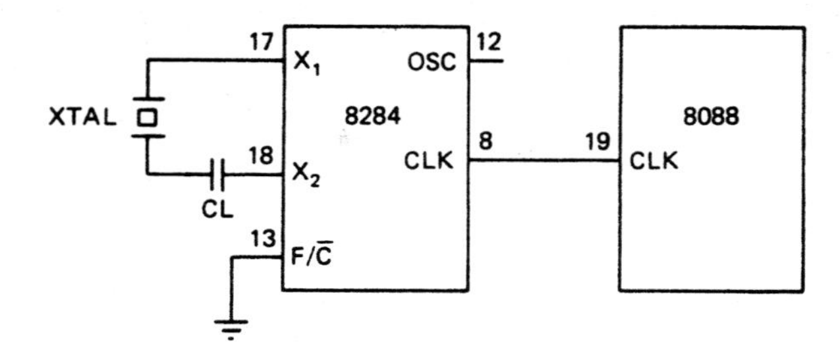

Figure 8–11 Connecting the 8284 to the 8088. (Reprinted with permission of Intel Corporation, © 1979)

The standard way in which this clock chip is used with the 8088 is to connect either a 15- or 24-MHz crystal between its X_1 and X_2 inputs. This circuit connection is shown in Fig. 8–11. Note that a series capacitor C_L is also required. Its typical value when used with the 15-MHz crystal is 12 pF. The *fundamental crystal frequency* is divided by 3 within the 8284 to give either a 5- or 8-MHz clock signal. This signal is internally buffered and output at CLK. The CLK output of the 8284 can be directly connected to the CLK input of the 8088. The 8284 connects to the 8086 in exactly the same way.

Figure 8–12 shows the waveform of CLK. Here we see that the signal is specified at metal oxide semiconductor (MOS)-compatible voltage levels and not transistor transistor logic (TTL) levels. Its minimum and maximum low logic levels are $V_{Lmin} = -0.5$ V and $V_{Lmax} = 0.6$ V, respectively. Moreover, the minimum and maximum high logic levels are $V_{Hmin} = 3.9$ V and $V_{Hmax} = V_{cc} + 1$ V, respectively. The *period* of the clock signal of a 5-MHz 8088 can range from a minimum of 200 ns to a maximum of 500 ns, and the maximum *rise* and *fall times* of its edges equal 10 ns.

Figure 8–10 shows two more clock outputs on the 8284: the *peripheral clock* (PCLK) and *oscillator clock* (OSC). These signals are provided to drive peripheral ICs. The clock signal output at PCLK is half the frequency of CLK. For instance, if an 8088 is operated at 5 MHz, PCLK is 2.5 MHz. Also, it is at TTL-compatible levels rather than MOS levels. On the other hand, the OSC output is at the crystal frequency, which is three times that of CLK. Figure 8–13 illustrates these relationships.

The 8284 can also be driven from an external clock source. The external clock signal is applied to the external frequency input (EFI). Input $F/\overline{C}$ is provided for clock source selection. When it is strapped to the 0 logic level, the crystal between X_1 and X_2 is used. On the other hand, applying logic 1 to $F/\overline{C}$ selects EFI as the source of the clock. The clock sync (CSYNC) input can be used for external synchronization in systems that employ multiple clocks.

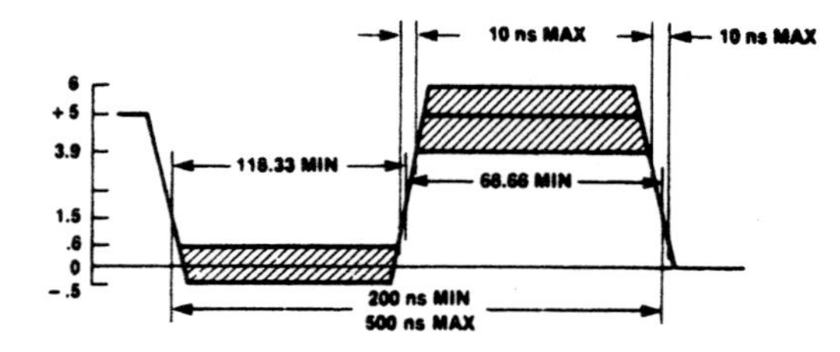

Figure 8–12 CLK voltage and timing characteristics for a 5-MHz processor. (Reprinted with permission of Intel Corporation, © 1979)

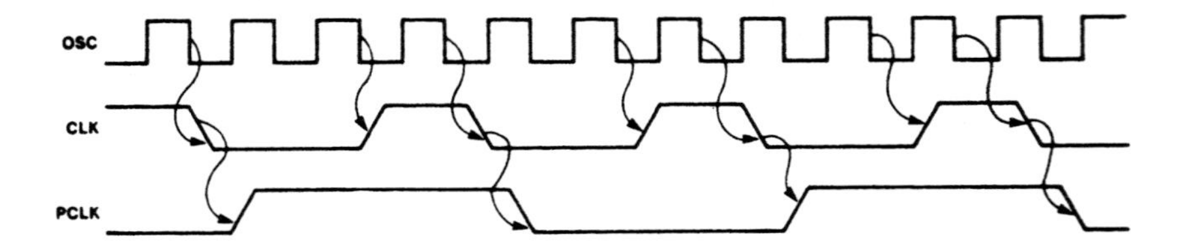

Figure 8–13 Relationship between CLK and PCLK. (Reprinted with permission of Intel Corporation, © 1979)

EXAMPLE 8.4

If the CLK input of an 8086 MPU is to be driven by a 9-MHz signal, what speed version of the 8086 must be used and what frequency crystal must be attached to the 8284?

Solution

The 8086-1 is the version of the 8086 that can be run at 9 MHz. To create the 9-MHz clock, a 27-MHz crystal must be used on the 8284.

▲ 8.7 BUS CYCLE AND TIME STATES

A *bus cycle* defines the basic operation that a microprocessor performs to communicate with external devices. Examples of bus cycles are the memory read, memory write, input/output read, and input/output write. As shown in Fig. 8–14(a), a bus cycle corresponds to a sequence of events that start with an address being output on the system bus followed by a read or write data transfer. During these operations, the MPU produces a series of control signals to control the direction and timing of the bus.

The bus cycle of the 8088 and 8086 microprocessors consists of at least four clock periods. These four time states are called T_1, T_2, T_3, and T_4. During T_1, the MPU puts an address on the bus. For a write memory cycle, data are put on the bus during state T_2 and maintained through T_3 and T_4. When a read cycle is to be performed, the bus is first put in the high-Z state during T_2 and then the data to be read must be available on the bus during T_3 and T_4. These four clock states give a *bus cycle duration* of 125 ns $\times$ 4 = 500 ns in an 8-MHz 8088 system.

If no bus cycles are required, the microprocessor performs what are known as *idle states.* During these states, no bus activity takes place. Each idle state is one clock period long, and any number of them can be inserted between bus cycles. Figure 8–14(b) shows two bus cycles separated by idle states. Idle states are performed if the instruction queue inside the microprocessor is full and it does not need to read or write operands from memory.

Wait states can also be inserted into a bus cycle. This is done in response to a request by an event in external hardware instead of an internal event such as a full queue.

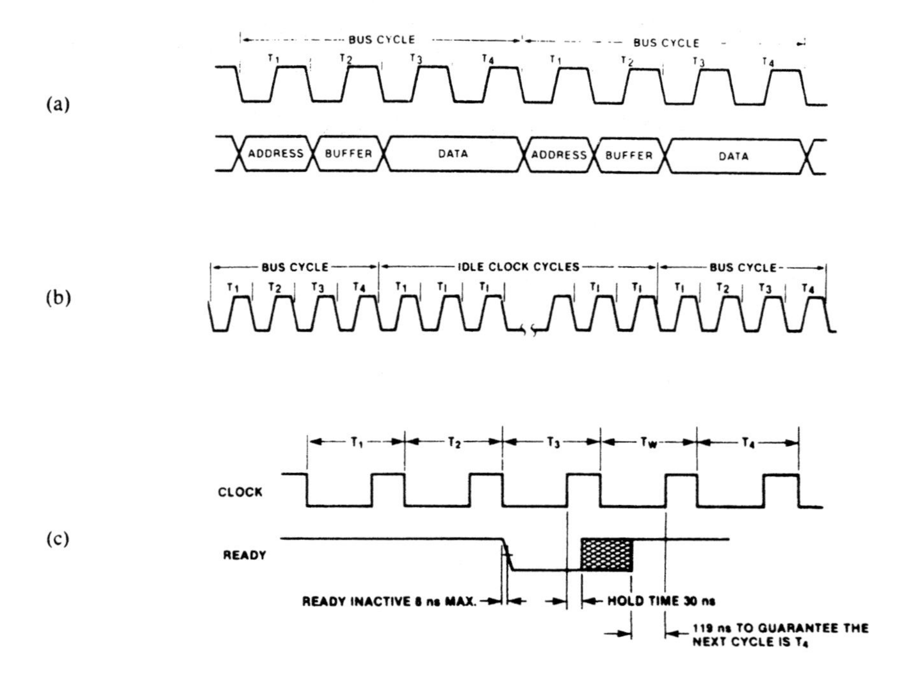

Figure 8–14 (a) Bus cycle clock periods. (Reprinted with permission of Intel Corporation, © 1979) (b) Bus cycle with idle states. (Reprinted with permission of Intel Corporation, © 1979) (c) Bus cycle with wait states. (Reprinted with permission of Intel Corporation, © 1979)

In fact, the READY input of the MPU is provided specifically for this purpose. Figure 8–14(c) shows that logic 0 at this input indicates that the current bus cycle should not be completed. As long as READY is held at the 0 level, wait states are inserted between states T_3 and T_4 of the current bus cycle, and the data that were on the bus during T_3 are maintained. The bus cycle is not completed until the external hardware returns READY back to the 1 logic level. This extends the duration of the bus cycle, thereby permitting the use of slower memory and I/O devices in the system.

EXAMPLE 8.5

What is the duration of the bus cycle in the 8088-based microcomputer if the clock is 8 MHz and two wait states are inserted?

Solution

The duration of the bus cycle in an 8-MHz system is given in general by

$$t_{cyc} = 500 \text{ ns} + N \times 125 \text{ ns}$$

In this expression N stands for the number of wait states. For a bus cycle with two wait states, we get

$$t_{cyc} = 500 \text{ ns} + 2 \times 125 \text{ ns} = 500 \text{ ns} + 250 \text{ ns}$$
$$= 750 \text{ ns}$$

▲ 8.8 HARDWARE ORGANIZATION OF THE MEMORY ADDRESS SPACE

From a hardware point of view, the memory address spaces of the 8088- and 8086-based microcomputers are organized differently. Figure 8–15(a) shows that the 8088's memory subsystem is implemented as a single 1M × 8 memory bank. Looking at the block diagram in Fig. 8–15(a), we see that these byte-wide storage locations are assigned to consecutive addresses over the range from 00000_{16} through $FFFFF_{16}$. During memory operations, a 20-bit address is applied to the memory bank over address lines A_0 through A_{19}. It is this address that selects the storage location that is to be accessed. Bytes of data are transferred between the 8088 and memory over data bus lines D_0 through D_7.

On the other hand, the 8086's 1Mbyte memory address space, as shown in Fig. 8–15(b), is implemented as two independent 512Kbyte banks: the *low (even) bank* and the *high (odd) bank*. Data bytes associated with an even address (00000_{16}, 00002_{16}, etc.) reside in the low bank, and those with odd addresses (00001_{16}, 00003_{16}, etc.) reside in the high bank.

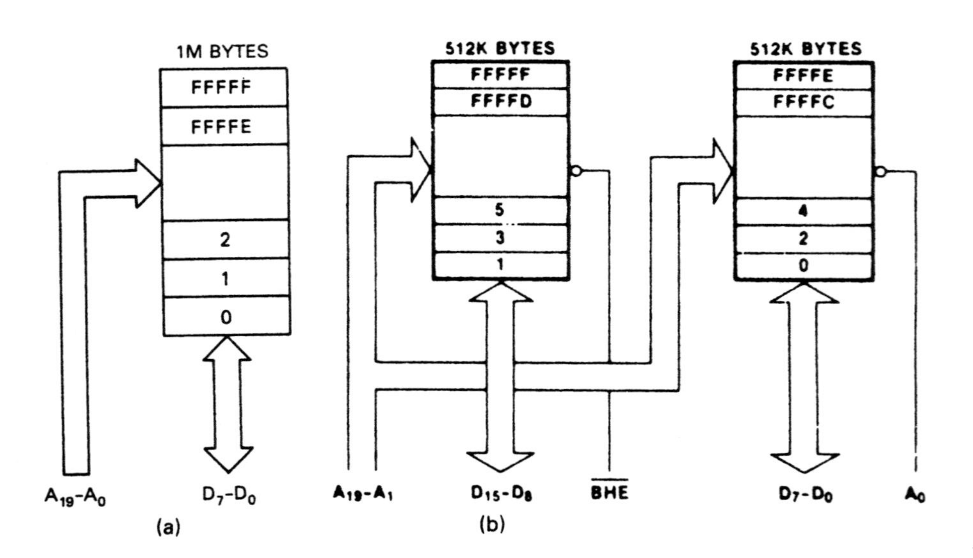

Figure 8–15 (a) 1M × 8 memory bank of the 8088. (b) High and low memory banks of the 8086. (Reprinted with permission of Intel Corporation, © 1979)

The diagram in Fig. 8–15(b) shows that for the 8086 address bits, A_1 through A_{19} select the storage location that is to be accessed. They are applied to both banks in parallel. A_0 and bank high enable ($\overline{BHE}$) are used as bank-select signals. Logic 0 at A_0 identifies an even-addressed byte of data and causes the low bank of memory to be enabled. On the other hand, $\overline{BHE}$ equal to 0 enables the high bank to access an odd-addressed byte of data. Each of the memory banks provides half of the 8086's 16-bit data bus. Notice that the lower bank transfers bytes of data over data lines D_0 through D_7, while data transfers for a high bank use D_8 through D_{15}.

We just saw that the memory subsystem of the 8088-based microcomputer system is actually organized as 8-bit bytes, not as 16-bit words. However, the contents of any two consecutive byte storage locations can be accessed as a word. The lower-addressed byte is the least significant byte of the word, and the higher-addressed byte is its most significant byte. Let us now look at how a byte and a word of data are read from memory.

Figure 8–16(a) shows how a byte-memory operation is performed to the storage location at address X. As shown in the diagram, the address is supplied to the memory

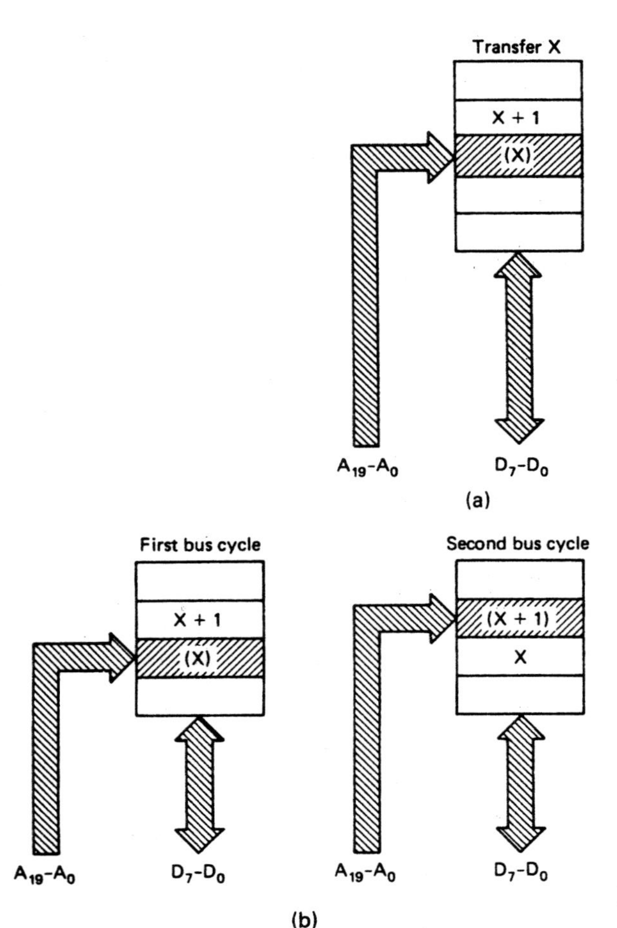

Figure 8–16 (a) Byte transfer by the 8088. (b) Word transfer by the 8088.

bank over lines A_0 through A_{19}, and the byte of data is written into or read from storage location X over lines D_0 through D_7. D_7 carries the MSB of the byte of data, and D_0 carries the LSB. This shows that a byte of data is accessed by the 8088 in one bus cycle. A memory cycle for an 8088 running at 5 MHz with no wait states takes 800 ns.

When a word of data is to be transferred between the 8088 and memory, we must perform two accesses of memory, reading or writing a byte in each access. Figure 8–16(b) illustrates how the word storage location starting at address X is accessed. Two bus cycles are required to access a word of data. During the first bus cycle, the least significant byte of the word, located at address X, is accessed. Again the address is applied to the memory bank over A_0 through A_{19}, and the byte of data is transferred to or from storage location X over D_0 through D_7.

Next, the 8088 automatically increments the address so that it now points to byte address X + 1. This address points to the next consecutive byte storage location in memory, which corresponds to the most significant byte of the word of data at X. Now a second memory bus cycle is initiated. During this second cycle, data are written into or read from the storage location at address X + 1. Since word accesses of memory take two bus cycles instead of one, it takes 1.6 ms to access a word of data when the 8088 is operating at a 5-MHz clock rate with no wait states.

The 8086 microprocessor performs byte and word data transfers differently from the 8088. Let us next examine the data transfers that can take place in an 8086-based microcomputer.

Figure 8–17(a) shows that when a byte-memory operation is performed to address X, an even-addressed storage location in the low bank is accessed. Therefore, A_0 is set to logic 0 to enable the low bank of memory and $\overline{BHE}$ to logic 1 to disable the high bank. As shown in the block diagram, data are transferred to or from the lower bank over data bus lines D_0 through D_7. Line D_7 carries the MSB of the byte, and D_0 the LSB.

On the other hand, to access a byte of data at an odd address such as X + 1 in Fig. 8–17(b), A_0 is set to logic 1 and $\overline{BHE}$ to logic 0. This enables the high bank of memory and disables the low bank. Data are transferred between the 8086 and the high bank over bus lines D_8 through D_{15}. Here D_{15} represents the MSB and D_8 the LSB.

Whenever an even-addressed word of data is accessed, both the high and low banks are accessed at the same time. Figure 8–17(c) illustrates how a word at even address X is accessed. Note that both A_0 and $\overline{BHE}$ equal 0; therefore, both banks are enabled. In this case, bytes of data are transferred from or to both banks at the same time. This 16-bit word is transferred over the complete data bus D_0 through D_{15}. The bytes of an even-addressed word are said to be aligned and can be transferred with a memory operation that takes just one bus cycle.

A word at an odd-addressed boundary is said to be unaligned. That is, the least significant byte is at the lower address location in the high memory bank. This is demonstrated in Fig. 8–17(d). Here we see that the odd byte of the word is located at address X + 1 and the even byte at address X + 2.

Two bus cycles are required to access an unaligned word. During the first bus cycle, the odd byte of the word, which is located at address X + 1 in the high bank, is accessed. This is accompanied by select signals $A_0 = 1$ and $\overline{BHE} = 0$ and a data transfer over D_8 through D_{15}. Even though the data transfer uses data lines D_8 through D_{15}, to the processor it is the low byte of the addressed data word.

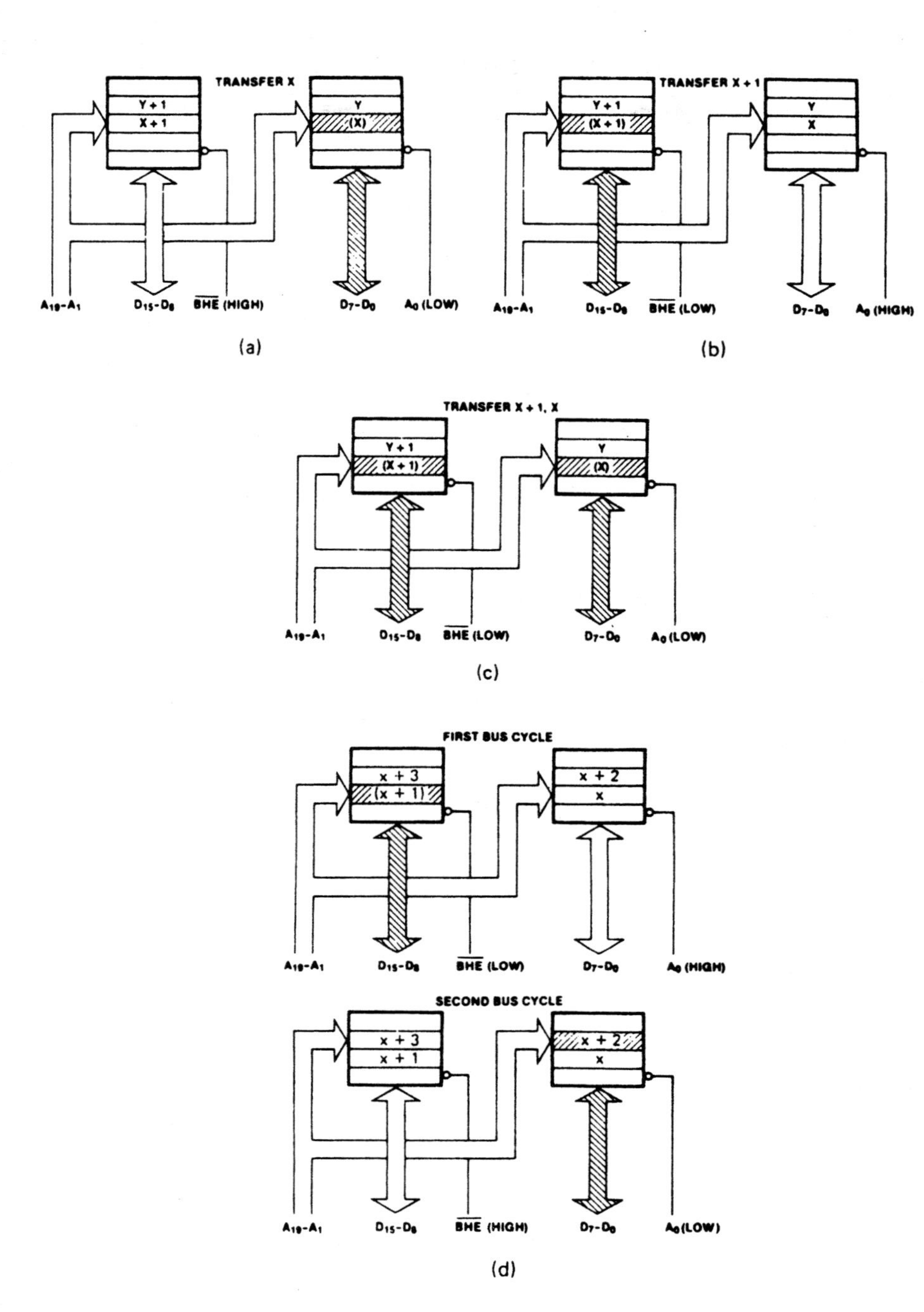

Figure 8–17 (a) Even-address byte transfer by the 8086. (Reprinted with permission of Intel Corporation, © 1979) (b) Odd-address byte transfer by the 8086. (Reprinted with permission of Intel Corporation, © 1979) (c) Even-address word transfer by the 8086. (Reprinted with permission of Intel Corporation, © 1979) (d) Odd-address word transfer by the 8086. (Reprinted with permission of Intel Corporation, © 1979)

Next, the 8086 automatically increments the address so that $A_0 = 0$. This represents the next address in memory, which is even. Then a second memory bus cycle is initiated. During this second cycle, the even byte located at X + 2 in the low bank is accessed. The data transfer takes place over bus lines D_0 through D_7. This transfer is accompanied by $A_0 = 0$ and $\overline{BHE} = 1$. To the processor, this is the high byte of the word of data.

EXAMPLE 8.6

Is the word at memory address 01231_{16} of an 8086-based microcomputer aligned or misaligned? How many bus cycles are required to read it from memory?

Solution

The first byte of the word is the second byte at the aligned-word address 01230_{16}. Therefore, the word is misaligned and requires two bus cycles to be read from memory.

▲ 8.9 ADDRESS BUS STATUS CODES

Whenever a memory bus cycle is in progress, an address bus status code S_4S_3 is output by the processor. The status code is multiplexed with address bits A_{17} and A_{16}. This two-bit code is output at the same time the data are carried over the data lines.

Bits S_4 and S_3 together form a 2-bit binary code that identifies which one of the four segment registers was used to generate the physical address that was output during the address period in the current bus cycle. The four *address bus status codes* are listed in Fig. 8–4. Here we find that code $S_4S_3 = 00$ identifies the extra segment register, 01 identifies the stack segment register, 10 identifies the code segment register, and 11 identifies the data segment register.

These status codes are output in both the minimum and the maximum modes. The codes can be examined by external circuitry. For example, they can be decoded with external circuitry to enable separate 1Mbyte address spaces for ES, SS, CS, and DS. In this way, the memory address reach of the microprocessor can be expanded to 4Mbytes.

▲ 8.10 MEMORY CONTROL SIGNALS

Earlier in the chapter we saw that similar control signals are produced in the maximum and minimum mode. Moreover, we found that in the minimum mode, the 8088 and 8086 microprocessors produce all the control signals. But in the maximum mode, the 8288 bus controller produces them. Here we will look more closely at each of these signals and their functions with respect to memory interface operation.

Minimum-Mode Memory Control Signals

In the 8088 microcomputer system shown in Fig. 8–18, which is configured for the minimum mode of operation, we find that the control signals provided to support the interface to the memory subsystem are ALE, $IO/\overline{M}$, $DT/\overline{R}$, $\overline{RD}$, $\overline{WR}$, and $\overline{DEN}$. These

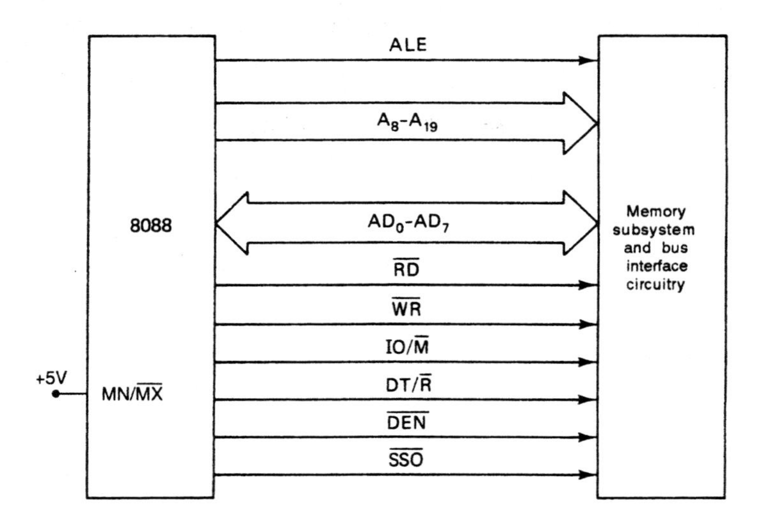

Figure 8–18 Minimum-mode 8088 memory interface.

control signals are required to tell the memory subsystem when the bus is carrying a valid address, in which direction data are to be transferred over the bus, when valid write data are on the bus, and when to put read data on the bus. For example, *address latch enable* (ALE) signals external circuitry that a valid address is on the bus. It is a pulse to the 1 logic level and is used to latch the address in external circuitry.

The *input-output/memory* (IO/$\overline{M}$) and *data transmit/receive* (DT/$\overline{R}$) lines signal external circuitry whether a memory or I/O bus cycle is in progress and whether the 8088 will transmit or receive data over the bus. During all memory bus cycles, IO/$\overline{M}$ is held at the 0 logic level. The 8088 switches DT/$\overline{R}$ to logic 1 during the data transfer part of the bus cycle, the bus is in the transmit mode, and data are written into memory. On the other hand, it sets DT/$\overline{R}$ to logic 0 to signal that the bus is in the receive mode, which corresponds to reading of memory.

The signals *read* ($\overline{RD}$) and *write* ($\overline{WR}$) identify that a read or write bus cycle, respectively, is in progress. The 8088 switches $\overline{WR}$ to logic 0 to signal memory that a write cycle is taking place over the bus. On the other hand, $\overline{RD}$ is switched to logic 0 whenever a read cycle is in progress. During all memory operations, the 8088 produces one other control signal, *data enable* ($\overline{DEN}$). Logic 0 at this output is used to enable the data bus.

Status line $\overline{SSO}$ is also part of the minimum-mode memory interface. The logic level that is output on this line during read bus cycles identifies whether a code or data access is in progress. $\overline{SSO}$ is set to logic 0 whenever instruction code is read from memory.

The control signals for the 8086's minimum-mode memory interface differ in three ways. First, the 8088's IO/$\overline{M}$ signal is replaced by the memory/input-output (M/$\overline{IO}$) signal. Whenever a memory bus cycle is in progress, the M/$\overline{IO}$ output is switched to logic 1. Second, the signal $\overline{SSO}$ is removed from the interface. Third, a new signal, *bank high enable* ($\overline{BHE}$), has been added to the interface. $\overline{BHE}$ is used as a select input for the high bank of memory in the 8086's memory subsystem. That is, logic 0 is output on this line during the address part of all the bus cycles in which data in the high-bank part of memory is to be accessed.

Maximum-Mode Memory Control Signals

When the 8088 is configured to work in the maximum mode, it does not directly provide all the control signals to support the memory interface. Instead, an external bus controller, the 8288, provides memory commands and control signals. Figure 8–19 shows an 8088 connected in this way.

Specifically, the $\overline{WR}$, IO/$\overline{M}$, DT/$\overline{R}$, $\overline{DEN}$, ALE, and $\overline{SSO}$ signal lines on the 8088 are changed. They are replaced with *multiprocessor lock* ($\overline{LOCK}$) signal, a *bus status code* ($\overline{S}_2\overline{S}_1\overline{S}_0$), and a *queue status code* (QS_1QS_0). The 8088 still does produce the signal $\overline{RD}$, which provides the same function as it did in minimum mode.

The 3-bit bus status code $\overline{S}_2\overline{S}_1\overline{S}_0$ is output prior to the initiation of each bus cycle. It identifies which type of bus cycle is to follow. This code is input to the 8288 bus controller. Here it is decoded to identify which type of bus cycle command signals must be generated.

Figure 8–20 shows the relationship between the bus status codes and the types of bus cycles produced. Also shown in this chart are the names of the corresponding command signals that are generated at the outputs of the 8288. For instance, the input code $\overline{S}_2\overline{S}_1\overline{S}_0$ equal to 100 indicates that an instruction fetch bus cycle is to take place. Since the instruction fetch is a memory read, the 8288 makes the *memory read command* ($\overline{MRDC}$) output switch to logic 0.

Another bus command provided for the memory subsystem is $\overline{S}_2\overline{S}_1\overline{S}_0$ equal to 110. This represents a memory write cycle and it causes both the *memory write command*

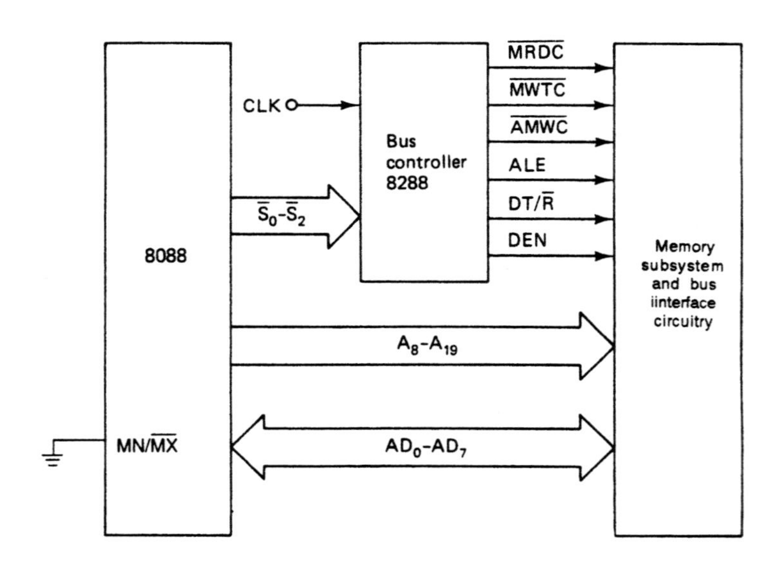

Figure 8–19 Maximum-mode 8088 memory interface.

Status Inputs			CPU Cycle	8288 Command
$\overline{S_2}$	$\overline{S_1}$	$\overline{S_0}$		
0	0	0	Interrupt acknowledge	$\overline{\text{INTA}}$
0	0	1	Read I/O port	$\overline{\text{IORC}}$
0	1	0	Write I/O port	$\overline{\text{IOWC}}$, $\overline{\text{AIOWC}}$
0	1	1	Halt	None
1	0	0	Instruction fetch	$\overline{\text{MRDC}}$
1	0	1	Read memory	$\overline{\text{MRDC}}$
1	1	0	Write memory	$\overline{\text{MWTC}}$, $\overline{\text{AMWC}}$
1	1	1	Passive	None

Figure 8–20 Memory bus cycle status codes produced in maximum mode. (Reprinted with permission of Intel Corporation, © 1979)

($\overline{\text{MWTC}}$) and *advanced memory write command* ($\overline{\text{AMWC}}$) outputs to switch to the 0 logic level.

The other control outputs produced by the 8288 are DEN, DT/$\overline{\text{R}}$, and ALE. These signals provide the same functions as those produced by the corresponding pins on the 8088 in the minimum system mode.

The two status signals, QS_0 and QS_1, form an instruction queue code. This code tells the external circuitry what type of information was removed from the queue during the previous clock cycle. Figure 8–8 shows the four different queue statuses. For instance, $QS_1QS_0 = 01$ indicates that the first byte of an instruction was taken from the queue. The next byte of the instruction that is fetched is identified by queue status code 11. Whenever the queue is reset (e.g., due to a transfer of control) the reinitialization code 10 is output. Similarly, if no queue operation occurred, status code 00 is output.

The *bus priority lock* ($\overline{\text{LOCK}}$) signal, as shown in the interface, can be used as an input to a bus arbiter. The bus arbiter is used to lock other processors off the system bus during accesses of common system resources such as *global memory* in a multiprocessor system. The READY signal is used to interface slow memory devices.

All of the memory control signals we just described for the 8088-based microcomputer system serve the same function in the maximum-mode 8086 microcomputer. However, there is one additional control signal in the 8086's memory interface, the $\overline{\text{BHE}}$. The $\overline{\text{BHE}}$ performs the same function as it did in the minimum-mode system. That is, it is used as an enable input to the high bank of memory.

▲ 8.11 READ AND WRITE BUS CYCLES

In the preceding section we introduced the status and control signals associated with the memory interface. Here we continue by studying the sequence in which they occur during the read and write bus cycles of memory.

Read Cycle

Figure 8–21 shows the memory interface signals of a minimum-mode 8088 system. Here their occurrence is illustrated relative to the four time states T_1, T_2, T_3, and T_4 of the 8088's bus cycle. Let us trace the events that occur as data or instructions are read from memory.

The *read bus cycle* begins with state T_1. During this period, the 8088 outputs the 20-bit address of the memory location to be accessed on its multiplexed address/data bus AD_0 through AD_7, A_8 through A_{15}, and multiplexed lines A_{16}/S_3 through A_{19}/S_6. Note that at the same time a pulse is also produced at ALE. The trailing edge or the high level of this pulse should be used to latch the address in external circuitry.

Also we see that at the start of T_1, signals $IO/\overline{M}$ and $DT/\overline{R}$ are set to the 0 logic level. This indicates to circuitry in the memory subsystem that a memory cycle is in progress and that the 8088 is going to receive data from the bus. Status $\overline{SSO}$ is also out-

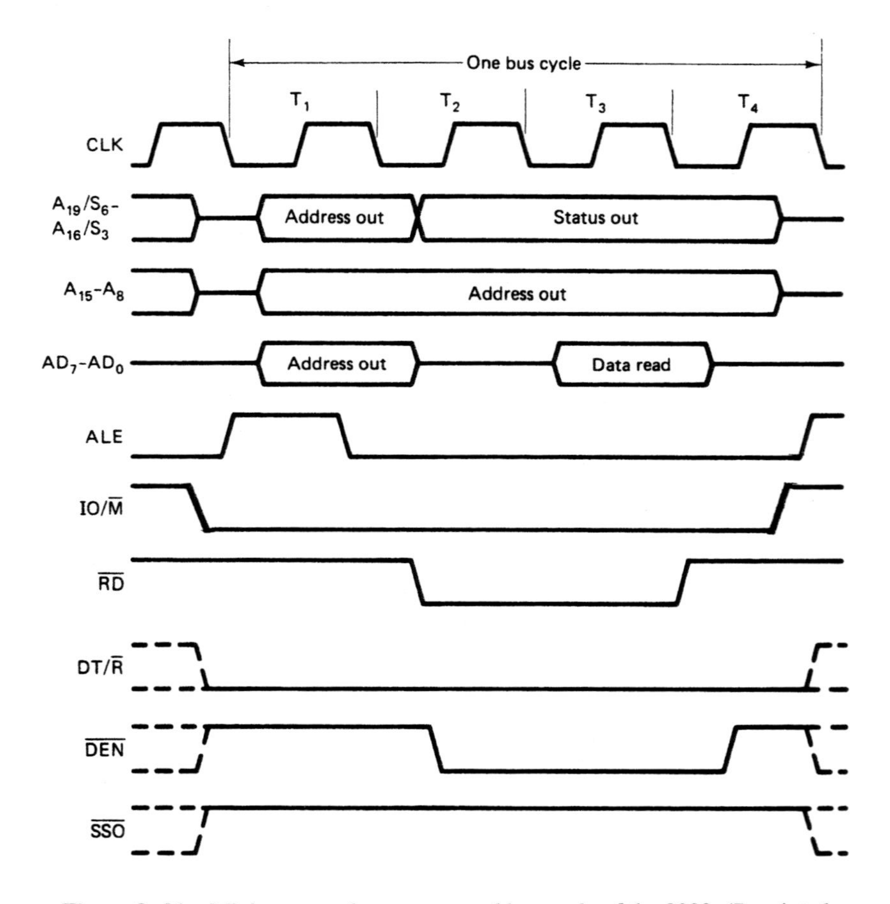

Figure 8–21 Minimum-mode memory read bus cycle of the 8088. (Reprinted with permission of Intel Corporation, © 1979)

put at this time. Note that all three of these signals are maintained at these logic levels throughout all four periods of the bus cycle.

Beginning with state T_2, status bits S_3 through S_6 are output on the upper four address bus lines A_{16} through A_{19}. Remember that bits S_3 and S_4 identify to external circuitry which segment register was used to generate the address just output. This status information is maintained through periods T_3 and T_4. The part of the address output on address bus lines A_8 through A_{15} is maintained through states T_2, T_3, and T_4. On the other hand, address/data bus lines AD_0 through AD_7 are put in the high-Z state during T_2.

Late in period T_2, $\overline{RD}$ is switched to logic 0. This indicates to the memory subsystem that a read cycle is in progress. $\overline{DEN}$ is switched to logic 0 to enable external circuitry to allow the data to move from memory onto the microprocessor's data bus.

As shown in the waveforms, input data are read by the 8088 during T_3. The memory must provide valid data during T_3 and maintain it until after the processor terminates the read operation. As Fig. 8–21 shows, it is in T_4 that the 8088 switches $\overline{RD}$ to the inactive 1 logic level to terminate the read operation. $\overline{DEN}$ returns to its inactive logic level late during T_4 to disable the external circuitry, which allows data to move from memory to the processor. The read cycle is now complete.

A timing diagram for the 8086's memory read cycle is given in Fig. 8–22(a). Comparing these waveforms to those of the 8088 in Fig. 8–21, we find just four differences; $\overline{BHE}$ is output along with the address during T_1; the data read by the 8086 during T_3 can be carried over all 16 data bus lines; $M/\overline{IO}$, which replaces $IO/\overline{M}$, is switched to logic 1

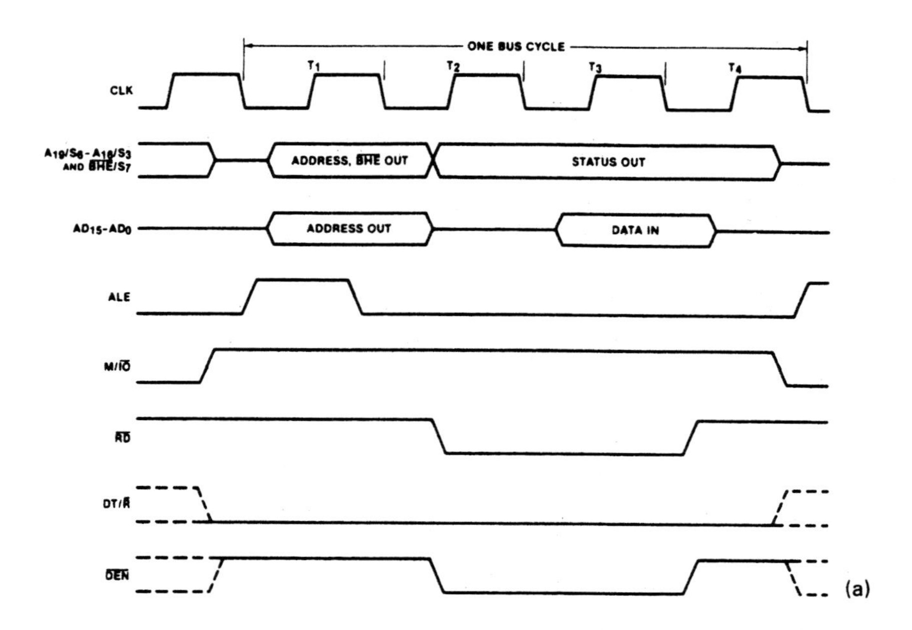

Figure 8–22 (a) Minimum-mode memory read bus cycle of the 8086. (Reprinted with permission of Intel Corporation, © 1979) (b) Maximum-mode memory read bus cycle of the 8086. (Reprinted with permission of Intel Corporation, © 1979)

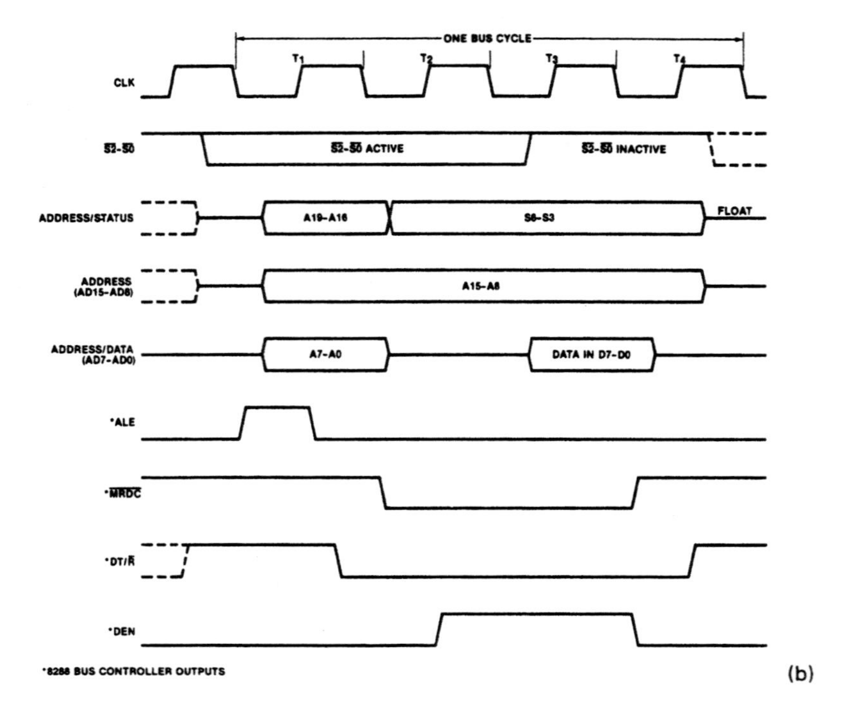

Figure 8–22 (continued)

at the beginning of T_1 and is held at this level for the duration of the bus cycle; and the $\overline{SSO}$ status signal is not produced.

Figure 8–22(b) shows a read cycle of 8-bit data in a maximum-mode 8086-based microcomputer system. These waveforms are similar to those given for the minimum-mode read cycle in Fig. 8–22(a). Comparing these two timing diagrams, we see that the address and data transfers that take place are identical. In fact, the only difference found in the maximum-mode waveforms is that a bus cycle status code, $\overline{S_2}\overline{S_1}\overline{S_0}$, is output just prior to the beginning of the bus cycle. This status information is decoded by the 8288 to produce control signals ALE, $\overline{MRDC}$, DT/$\overline{R}$, and DEN.

Write Cycle

Figure 8–23(a) illustrates the *write bus cycle* timing of the 8088 in minimum mode. It is similar to that given for a read cycle in Fig. 8–21. Looking at the write cycle waveforms, we find that during T_1 the address is output and latched with the ALE pulse. This is identical to the read cycle. Moreover, IO/$\overline{M}$ is set to logic 0 to indicate that a memory cycle is in progress and status information is output at $\overline{SSO}$. However, this time DT/$\overline{R}$ is switched to logic 1. This signals external circuits that the 8088 is going to transmit data over the bus.

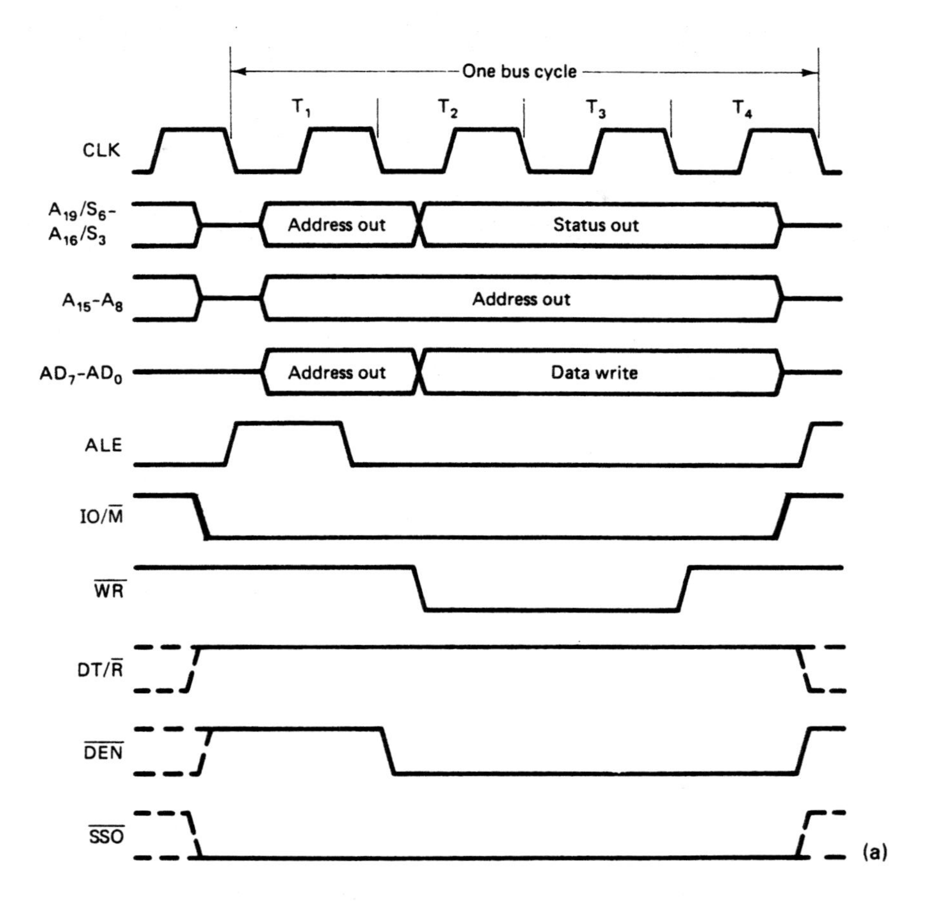

Figure 8–23 (a) Minimum-mode memory write bus cycle of the 8088. (Reprinted with permission of Intel Corporation, © 1979) (b) Maximum-mode memory write bus cycle of the 8086. (Reprinted with permission of Intel Corporation, © 1979)

As T_2 starts, the 8088 switches $\overline{WR}$ to logic 0. This tells the memory subsystem that a write operation is to follow over the bus. The 8088 puts the data on the bus late in T_2 and maintains the data valid through T_4. The writing of data into memory starts as $\overline{WR}$ becomes 0, and continues as it changes to 1 early in T_4. $\overline{DEN}$ enables the external circuitry to provide a path for data from the processor to the memory. This completes the write cycle.

Just as we described for the read bus cycle, the write cycle of the 8086 differs from that of the 8088 in four ways; again, $\overline{SSO}$ is not produced; $\overline{BHE}$ is output along with the address; data are carried over all 16 data bus lines; and finally, M/$\overline{IO}$ is the complement of the 8088's IO/$\overline{M}$ signal. The waveforms in Fig. 8–23(b) illustrate a write cycle of word data in a maximum-mode 8086 system.

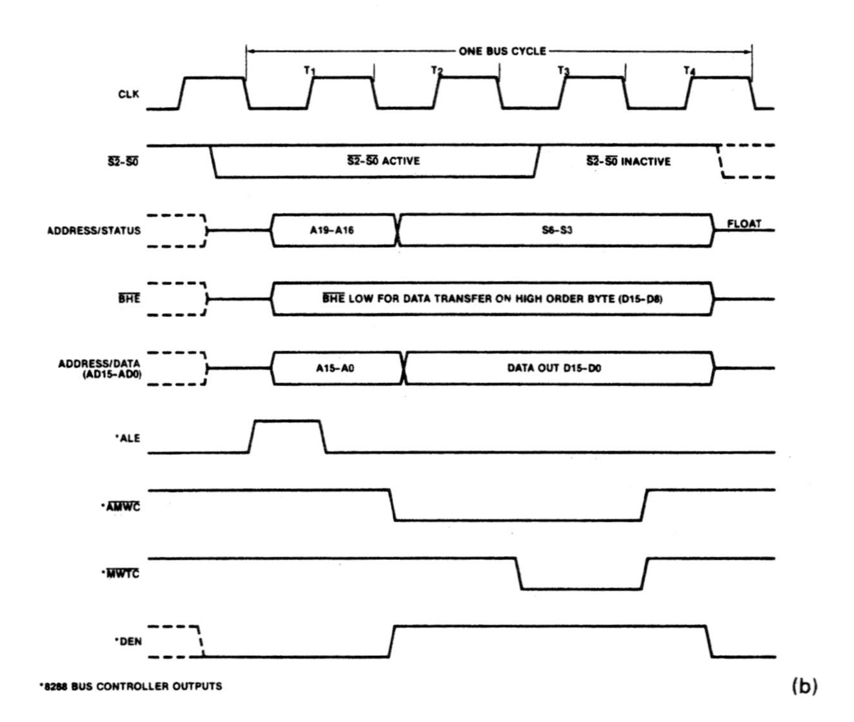

Figure 8–23 (continued)

▲ 8.12 MEMORY INTERFACE CIRCUITS

This section describes the memory interface circuits of an 8086-based microcomputer system. The 8086 system was selected instead of an 8088 microcomputer because it is more complex. Figure 8–24 shows a memory interface diagram for a maximum-mode 8086-based microcomputer system. Here we find that the interface includes the 8288 bus controller, address bus latches and an address decoder, data bus transceiver/buffers, and bank read and write control logic. The 8088 microcomputer is simpler in that the interface does not require bank write control logic because its address space is organized as a single bank.

Looking at Fig. 8–24, we see that bus status code signals $\overline{S}_2$, $\overline{S}_1$, and $\overline{S}_0$, which are outputs of the 8086, are supplied directly to the 8288 bus controller. Here they are decoded to produce the command and control signals needed to coordinate data transfers over the bus. Figure 8–20 highlights the status codes that relate to the memory interface. For example, the code $\overline{S}_2\overline{S}_1\overline{S}_0 = 101$ indicates that a data memory read bus cycle is in progress. This code makes the $\overline{MRDC}$ command output of the bus control logic switch to logic 0. Note in Fig. 8–24 that $\overline{MRDC}$ is applied to the bank read control logic.

Next let us look at how the address bus is latched, buffered, and decoded. Looking at Fig. 8–24, we see that address lines A_0 through A_{19} are latched along with control signal $\overline{BHE}$ in the address bus latch. The latched address lines A_{17L} through A_{19L} are

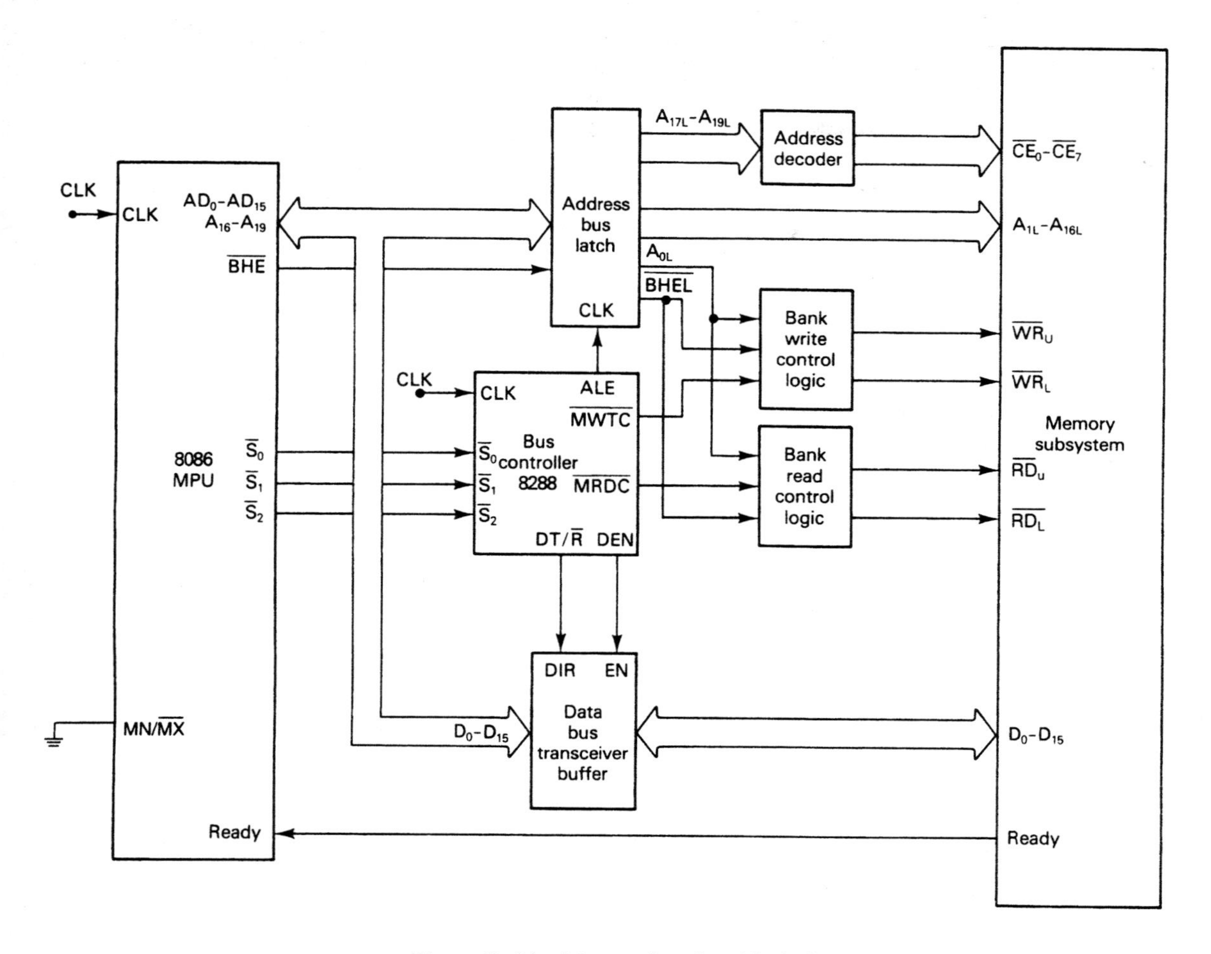

Figure 8–24 Memory interface block diagram.

decoded to produce chip enable outputs $\overline{CE}_0$ through $\overline{CE}_7$. Notice that the 8288 bus controller produces the address latch enable (ALE) control signal from $\overline{S}_2\overline{S}_1\overline{S}_0$. ALE is applied to the CLK input of the latches and strobes the bits of the address and bank high enable signal into the address bus latches. The address latch devices buffer these signals. Latched address lines A_{1L} through A_{16L} and $\overline{CE}_0$ through $\overline{CE}_7$ are applied directly to the memory subsystem.

During read bus cycles, the $\overline{MRDC}$ output of the bus control logic enables the bytes of data at the outputs of the memory subsystem onto data bus lines D_0 through D_{15}. During read operations from memory, the bank read control logic determines whether the data are read from one of the two memory banks or from both. This depends on whether a byte- or word-data transfer is taking place over the bus.

Similarly during write bus cycles, the $\overline{MWTC}$ output of the bus control logic enables bytes of data from the data bus D_0 through D_{15} to be written into the memory. The bank write control logic determines to which memory bank the data are written.

Note in Fig. 8–24 that in the bank write control logic the latched bank high enable signal $\overline{BHEL}$ and address line A_{0L} are gated with the memory write command signal $\overline{MWTC}$ to produce a separate write enable signal for each bank. These signals are denoted as $\overline{WR_U}$ and $\overline{WR_L}$. For example, if a word of data is to be written to memory over data bus lines D_0 through D_{15}, both $\overline{WR_U}$ and $\overline{WR_L}$ are switched to their active 0 logic level. Similarly the memory read control logic uses $\overline{MRDC}$, A_{0L}, and $\overline{BHEL}$ to generate $\overline{RD_U}$ and $\overline{RD_L}$ signals for bank read control.

The bus transceivers control the direction of data transfer between the MPU and memory subsystem. In Fig. 8–24, we see that the operation of the transceivers is controlled by the DT/$\overline{R}$ and DEN outputs of the bus controller. DEN is applied to the EN input of the transceivers and enables them for operation. This happens during all read and write bus cycles. DT/$\overline{R}$ selects the direction of data transfer through the devices. Note that it is supplied to the DIR input of the data bus transceivers. When a read cycle is in progress, DT/$\overline{R}$ is set to 0 and data are passed from the memory subsystem to the MPU. On the other hand, when a write cycle is taking place, DT/$\overline{R}$ is switched to logic 1 and data are carried from the MPU to the memory subsystem.

Address Bus Latches and Buffers

The 74F373 is an example of an octal latch device that can be used to implement the *address latch* section of the 8086's memory interface circuit. A block diagram of this device is shown in Fig. 8–25(a) and its internal circuitry is shown in Fig. 8–25(b). Note that it accepts eight inputs: 1D through 8D. As long as the clock (C) input is at logic 1, the outputs of the D-type flip-flops follow the logic level of the data applied to their corresponding inputs. When C is switched to logic 0, the current contents of the D-type flip-flops are latched. The latched information in the flip-flops is not output at data outputs 1Q through 8Q unless the output-control ($\overline{OC}$) input of the buffers that follow the latches is at logic 0. If $\overline{OC}$ is at logic 1, the outputs are in the high-impedance state. Figure 8–25(c) summarizes this operation.

In the 8086 microcomputer system, the 20 address lines (AD_0–AD_{15}, A_{16}–A_{19}) and the bank high enable signal $\overline{BHE}$ are normally latched in the address bus latch. The circuit configuration shown in Fig. 8–26 can be used to latch these signals. Fixing $\overline{OC}$ at the 0 logic level permanently enables latched outputs A_{0L} through A_{19L} and $\overline{BHEL}$. Moreover, the address information is latched at the outputs as the ALE signal from the bus controller returns to logic 0—that is, when the CLK input of all devices is switched to logic 0.

In general, it is important to minimize the propagation delay of the address signals as they go through the bus interface circuit. The switching property of the 74F373 latches that determine this delay for the circuit of Fig. 8–26 is called *enable-to-output propagation delay* and has a maximum value of 13 ns. By selecting fast latches—that is, latches with a shorter propagation delay time—a maximum amount of the 8086's bus cycle time is preserved for the access time of the memory devices. In this way slower, lower cost memory ICs can be used. These latches also provide buffering for the 8086's address lines. The outputs of the latch can sink a maximum of 24 mA.

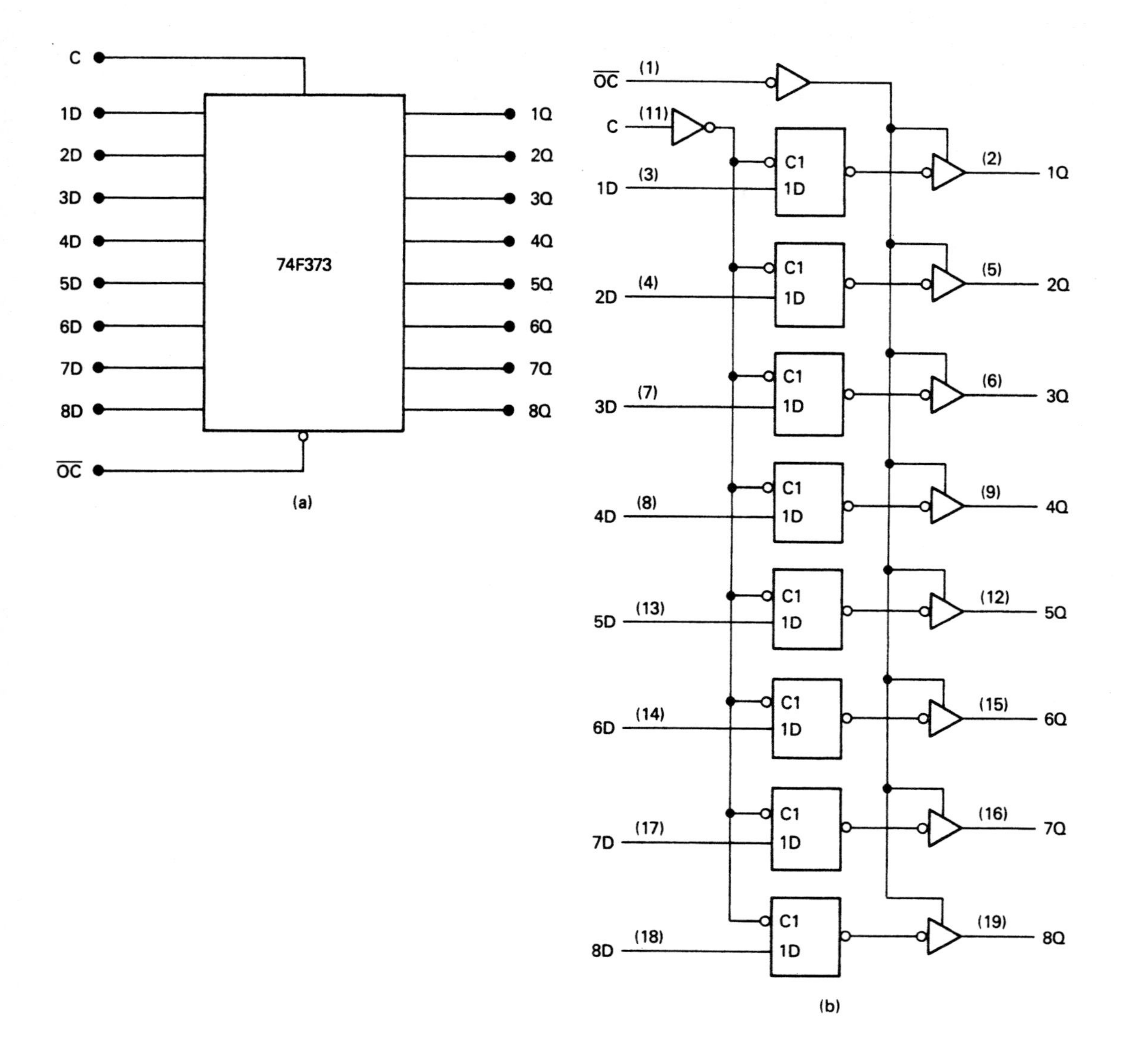

Inputs			Output
$\overline{OC}$	Enable C	D	Q
L	H	H	H
L	H	L	L
L	L	X	Q_0
H	X	X	Z

(c)

Figure 8–25 (a) Block diagram of an octal D-type latch. (b) Circuit diagram of the 74F373. (Courtesy of Texas Instruments Incorporated) (c) Operation of the 74F373. (Courtesy of Texas Instruments Incorporated).

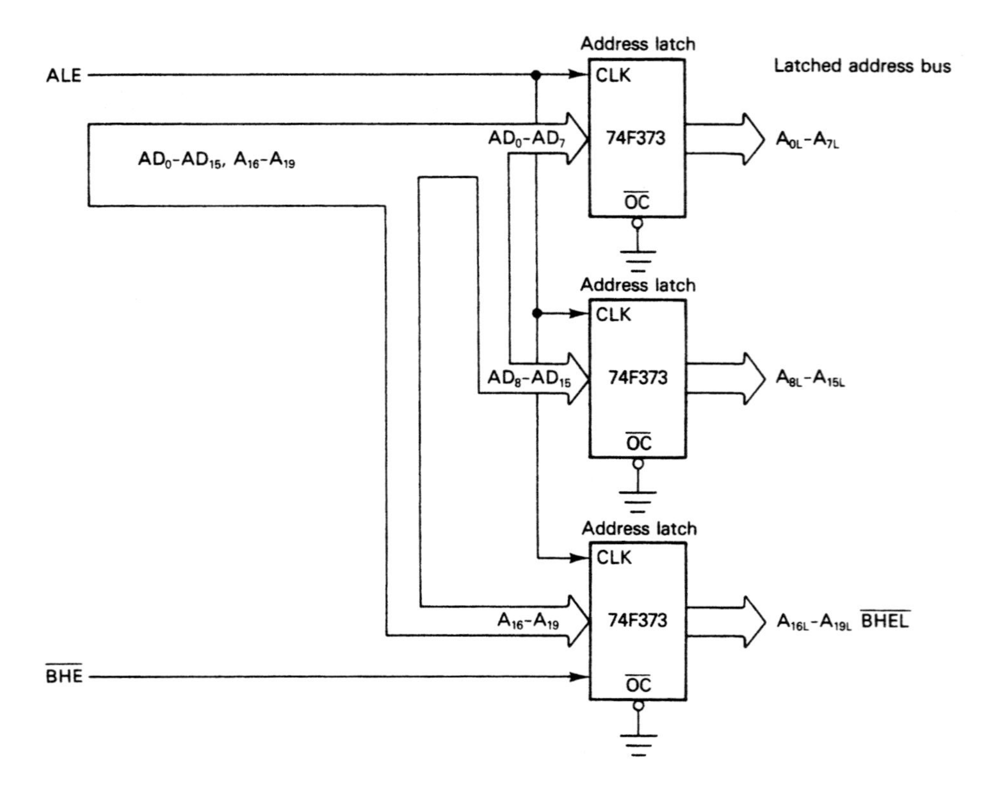

Figure 8–26 Address latch circuit.

Bank Write and Bank Read Control Logic

The memory of the 8086 microcomputer is organized in upper and lower banks. It requires separate write and read control signals for the two banks. The logic circuit in Fig. 8–27 shows how the bank write control signals, $\overline{WR_U}$ for the upper bank and $\overline{WR_L}$ for the lower bank can be generated from the bus controller signals $\overline{WRTC}$, the address bus latch signals A_{0L} and $\overline{BHEL}$. Two OR gates are used for this purpose.

Similar to the bank write control logic circuit, the bank read control logic circuit can be designed to generate $\overline{RD_U}$, the read for the upper bank of memory, and $\overline{RD_L}$, the read for the lower bank. Figure 8–28 illustrates such a circuit. Note that the circuit uses the $\overline{MRDC}$ signal from the bus controller.

Data Bus Transceivers

The *data bus transceiver* block of the bus interface circuit can be implemented with 74F245 octal bus transceiver ICs. Figure 8–29(a) shows a block diagram of this device. Note that its bidirectional input/output lines are called A_1 through A_8 and B_1 through B_8. Looking at the circuit diagram in Fig. 8–29(b), we see that the $\overline{G}$ input is used to enable

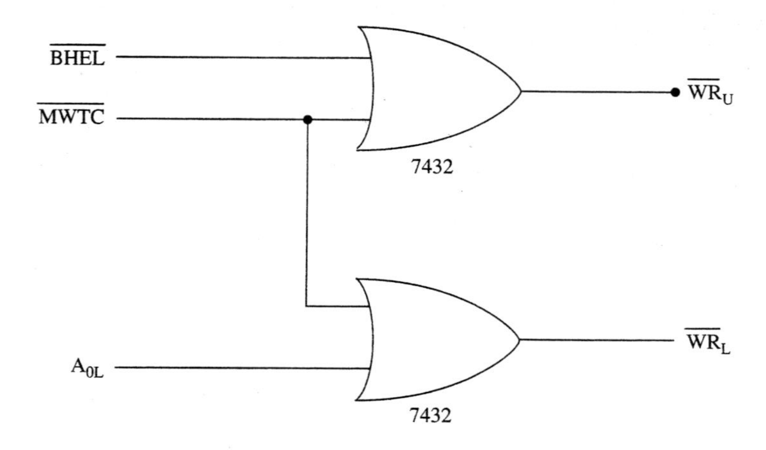

Figure 8–27 Bank write control logic.

the buffer for operation. On the other hand, the logic level at the direction (DIR) input selects the direction in which data are transferred through the device. For instance, logic 0 at this input sets the transceiver to pass data from the B lines to the A lines. Switching DIR to logic 1 reverses the direction of data transfer.

Figure 8–30 shows a circuit that implements the data bus transceiver block of the bus interface circuit using the 74F245. For the 16-bit data bus of the 8086 microcomputer, two devices are required. Here the DIR input is driven by the signal data transmit/receive ($DT/\overline{R}$), and $\overline{G}$ is supplied by data bus enable (DEN). These signals are outputs of the 8288 bus controller.

Another key function of the data bus transceiver circuit is to buffer the data bus lines. This capability is defined by how much current the devices can sink at their outputs. The I_{OL} rating of the 74F245 is 64 mA.

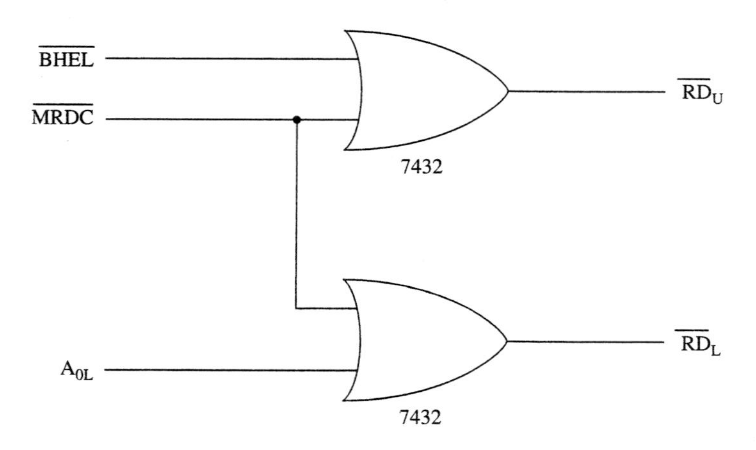

Figure 8–28 Bank read control logic.

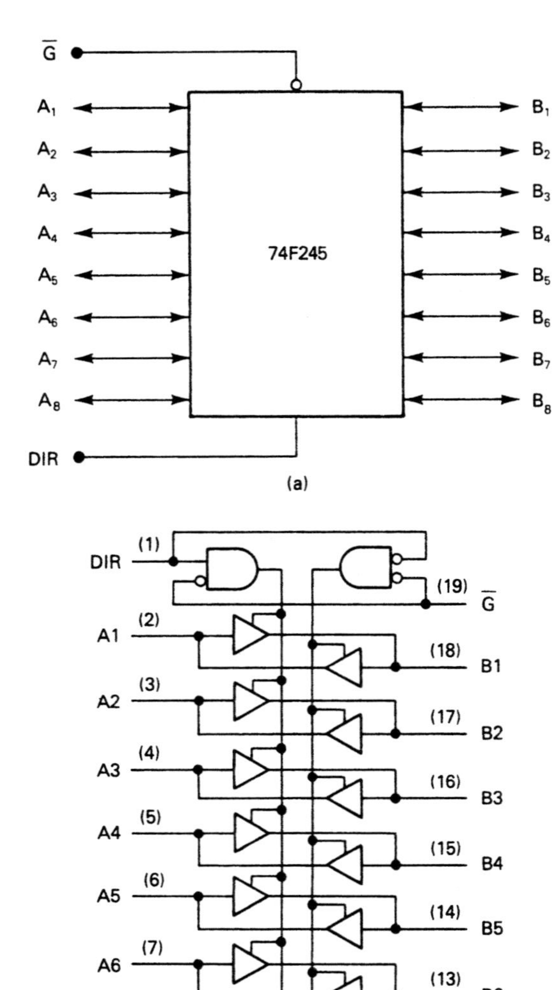

Figure 8–29 (a) Block diagram of the 74F245 octal bidirectional bus transceiver. (b) Circuit diagram of the 74F245. (Courtesy of Texas Instruments Incorporated)

Address Decoders

As shown in Fig. 8–31, the *address decoder* in the 8086 microcomputer system is located at the output side of the address latch. A typical device used to perform this decode function is the 74F139 dual 2-line to 4-line decoder. Figures 8–32(a) and (b) show a block diagram and circuit diagram for this device, respectively. When the enable ($\overline{G}$) input is at its active 0 logic level, the output corresponding to the code at the BA inputs

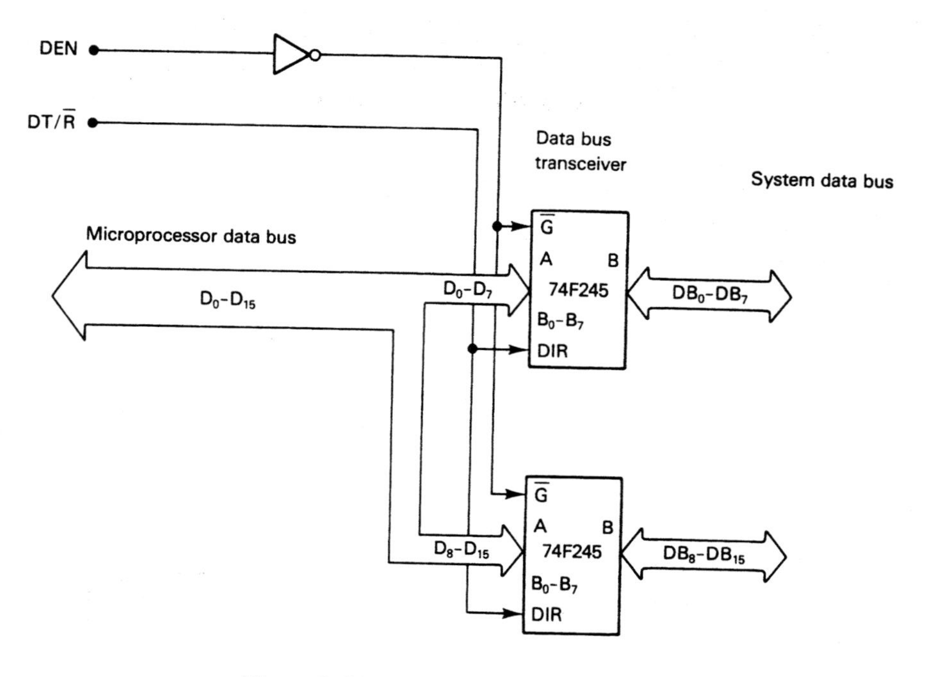

Figure 8–30 Data bus transceiver circuit.

switches to the 0 logic level. For instance, when BA = 01, output Y_1 is logic 0. The table in Fig. 8–32(c) summarizes the operation of the 74F139.

The circuit in Fig. 8–33 employs the address decoder configuration shown in Fig. 8–31. Note that address lines A_{17L} and A_{18L} are applied to the A and B inputs of the 74F139 decoder. The address line A_{19L} is used to enable one of the decoders and $\overline{A}_{19L}$, obtained using an inverter, enables the second decoder of the 74F139. Each decoder generates four chip enable (CE) outputs. Thus both decoders of the 74F139 together produce the eight outputs $\overline{CE}_0$ through $\overline{CE}_7$.

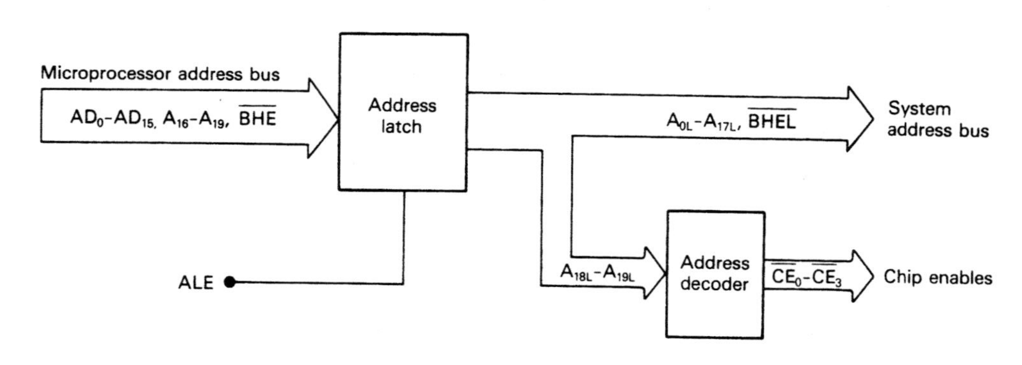

Figure 8–31 Address bus configuration with address decoding.

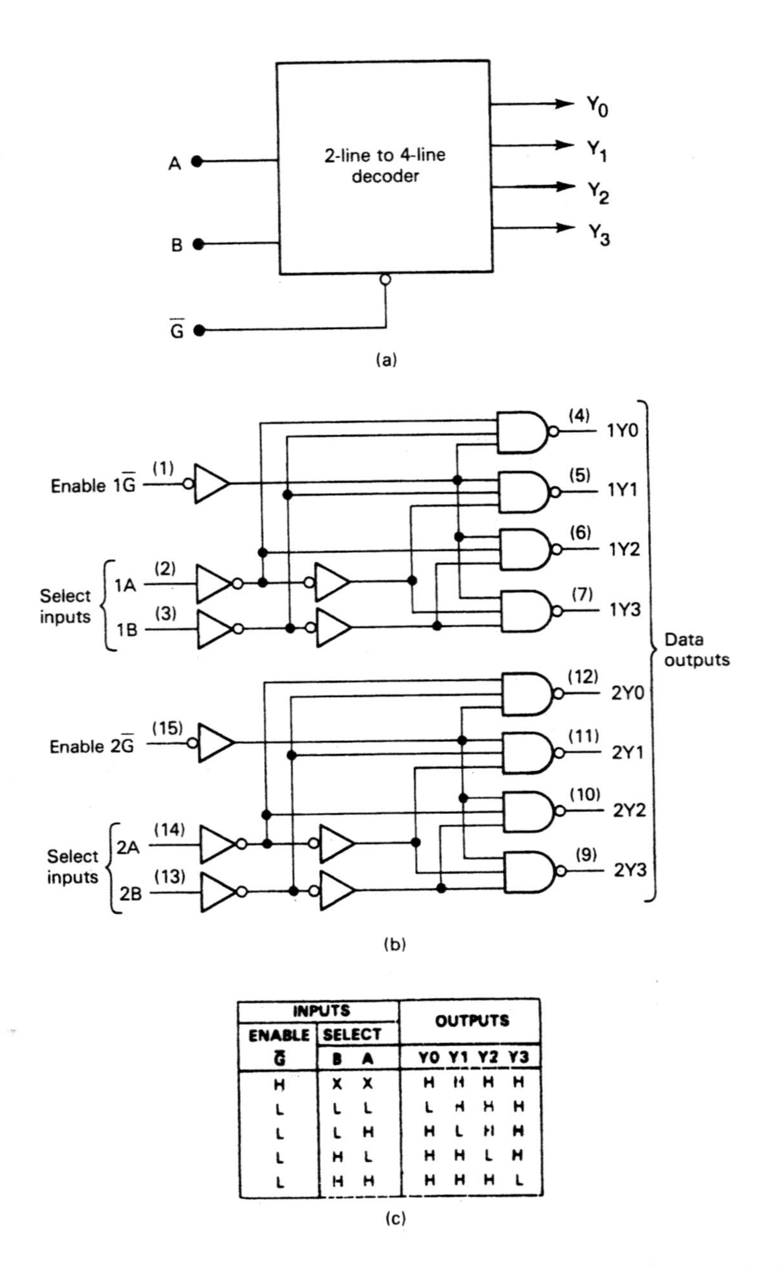

INPUTS			OUTPUTS			
ENABLE	SELECT					
$\overline{G}$	B	A	Y0	Y1	Y2	Y3
H	X	X	H	H	H	H
L	L	L	L	H	H	H
L	L	H	H	L	H	H
L	H	L	H	H	L	H
L	H	H	H	H	H	L

Figure 8–32 (a) Block diagram of the 74F139 2-line to 4-line decoder/demultiplexer. (b) Circuit diagram of the 74F139. (Courtesy of Texas Instruments Incorporated) (c) Operation of the 74F139 decoder. (Courtesy of Texas Instruments Incorporated)

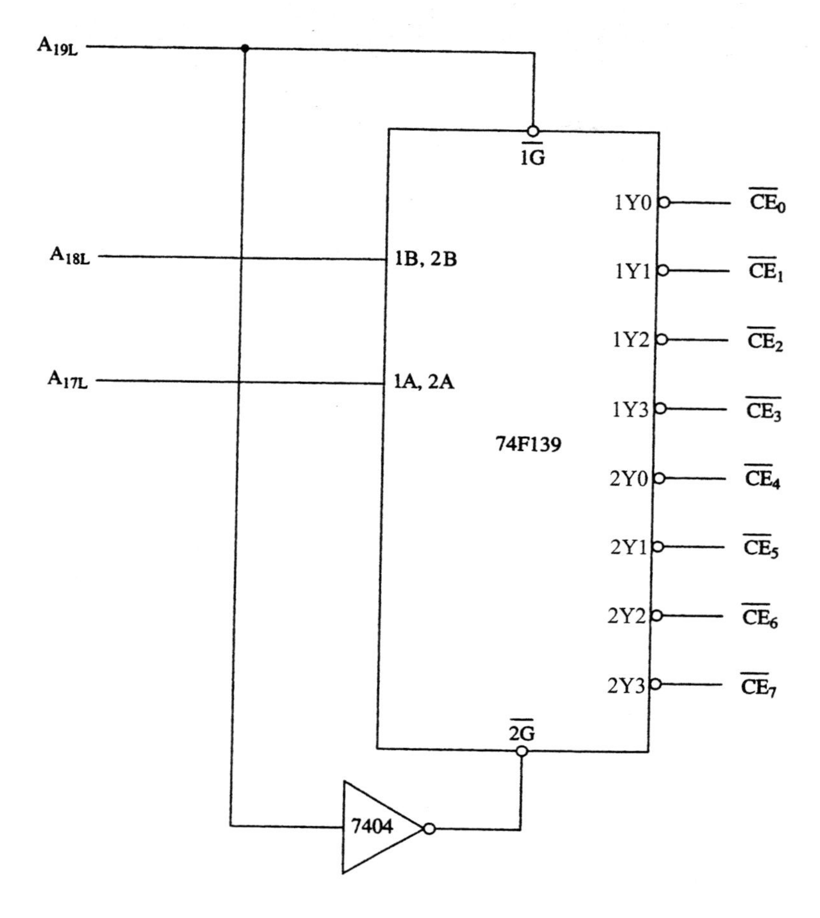

Figure 8–33 Address decoder circuit.

The block diagram of another commonly used decoder, the 74F138, is shown in Fig. 8–34(a). The 74F138 is similar to the 74F139, except that it is a single three-line to eight-line decoder. The circuit used in this device is shown in Fig. 8–34(b). Note that it can be used to produce eight $\overline{CE}$ outputs. The table in Fig. 8–34(c) describes the operation of the 74F138. Here we find that when enabled, only the output that corresponds to the code at the CBA inputs switches to the active 0 logic level.

The circuit in Fig. 8–35 uses the 74F138 to generate chip enable signals $\overline{CE_0}$ through $\overline{CE_7}$ by decoding address lines A_{17L}, A_{18L}, and A_{19L}. Connecting the enable inputs to +5V and ground permanently enables the decoder. The advantage of using the 74F138 over the 74F139 for decoding is that it does not require an extra inverter to generate eight chip enable signals.

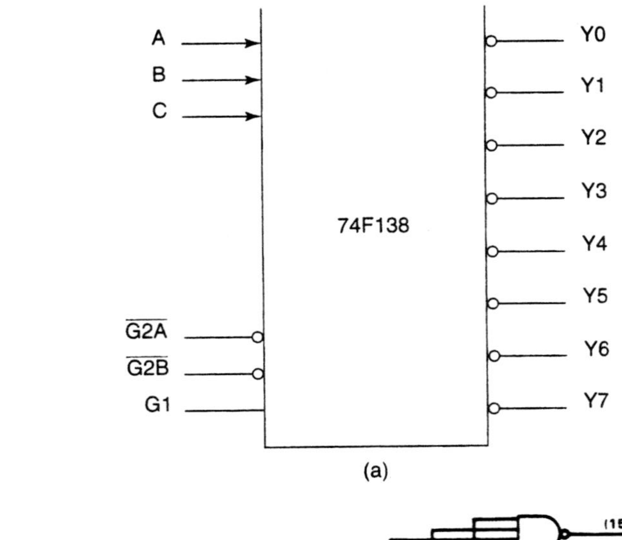

(a)

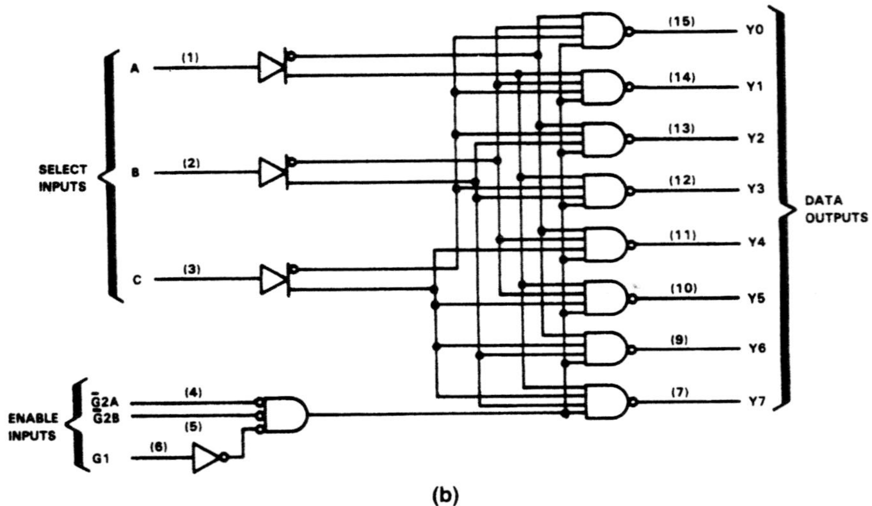

(b)

ENABLE INPUTS			SELECT INPUTS			OUTPUTS							
G1	$\overline{G2A}$	$\overline{G2B}$	C	B	A	Y0	Y1	Y2	Y3	Y4	Y5	Y6	Y7
X	H	X	X	X	X	H	H	H	H	H	H	H	H
X	X	H	X	X	X	H	H	H	H	H	H	H	H
L	X	X	X	X	X	H	H	H	H	H	H	H	H
H	L	L	L	L	L	L	H	H	H	H	H	H	H
H	L	L	L	L	H	H	L	H	H	H	H	H	H
H	L	L	L	H	L	H	H	L	H	H	H	H	H
H	L	L	L	H	H	H	H	H	L	H	H	H	H
H	L	L	H	L	L	H	H	H	H	L	H	H	H
H	L	L	H	L	H	H	H	H	H	H	L	H	H
H	L	L	H	H	L	H	H	H	H	H	H	L	H
H	L	L	H	H	H	H	H	H	H	H	H	H	L

(c)

Figure 8–34 (a) Block diagram of 74F138. (b) 74F138 circuit diagram. (Courtesy of Texas Instruments Incorporated) (c) Operation of the 74F138. (Courtesy of Texas Instruments Incorporated)

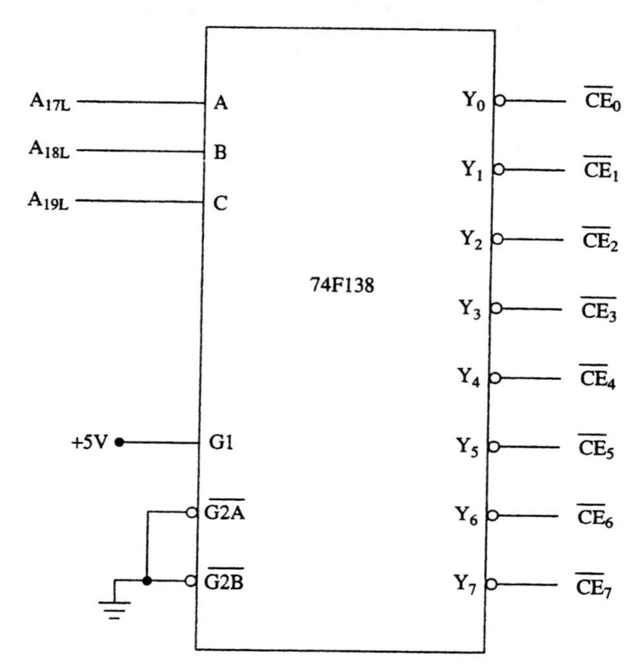

Figure 8–35 Address decoder circuit using 74F138.

▲ 8.13 PROGRAMMABLE LOGIC ARRAYS

In the last section we found that basic logic devices such as latches, transceivers, and decoders are required in the bus interface section of the 8086 microcomputer system. We showed that these functions were performed with standard logic devices such as the 74F373 octal transparent latch, 74F245 octal bus transceiver, and 74F139 two-line to four-line decoder, respectively. Today *programmable logic array* (PLA) devices are becoming very important in the design of microcomputer systems. For example, address and control signal decoding in the memory interface in Fig. 8–24 can be implemented with PLAs, instead of with separate logic ICs. Unlike the earlier mentioned devices, PLAs do not implement a specific logic function. Instead, they are general-purpose logic devices that have the ability to perform a wide variety of specialized logic functions. A PLA contains a general-purpose AND-OR-NOT array of logic gate circuits. The user has the ability to interconnect the inputs to the AND gates of this array. The definition of these inputs determines the logic function that is implemented. The process used to connect or disconnect inputs of the AND gate array is known as *programming,* which leads to the name programmable logic array.

PLAs, GALs, and EPLDs

A variety of different types of PLA devices are available. Early devices were all manufactured with the bipolar semiconductor process. These devices are referred to as *PALs* and remain in use today. Bipolar devices are programmed with an interconnect pattern by burning out fuse links within the device. In the initial state, all of these fuse links are intact. During programming, unwanted links are open-circuited by injecting a current through the fuse to burn it out. For this reason, once a device is programmed it cannot be reused. If a design modification is required in the pattern, a new device must be programmed and substituted for the original device. Since PALs are made with an older bipolar technology, they are limited to simpler functions and characterized by slower operating speeds and high power consumption.

Newer PLA devices are manufactured with the CMOS process. With this process, very complex, high-speed, low-power devices can be made. Two kinds of CMOS PLAs are in wide use today: the *GAL* and the *EPLD*. These devices differ in the type of CMOS technology used in their design. GALs are designed using *electrically erasable read-only memory* (E^2ROM) technology. The input/output operation of this device is determined by the programming of cells. These electrically programmable cells are also electrically erasable. For this reason, a GAL can be used for one application, erased, and then reprogrammed for another application. EPLDs are similar to GALs in that they can be programmed, erased, and reused; however, the erase mechanism is different. They are manufactured with *electrically programmable read only memory* (EPROM) technology. That is, they employ EPROM cells instead of E^2ROM cells. Therefore, to be erased an EPLD must be exposed to ultraviolet light. GALs and EPLDs are currently the most rapidly growing segments of the PLA marketplace.

Block Diagram of a PLA

The block diagram in Fig. 8–36 represents a typical PLA. Looking at this diagram, we see that it has 16 input leads, marked I_0 through I_{15}. There are eight output leads, labeled F_0 through F_7. This PLA is equipped with three-state outputs. For this reason, it has a chip-enable control lead. In the block diagram, this control input is marked $\overline{CE}$. The logic level of $\overline{CE}$ determines if the outputs are enabled or disabled.

When a PLA is used to implement random logic functions, the inputs represent Boolean variables, and the outputs are used to provide eight separate random logic functions. The internal AND-OR-NOT array is programmed to define a sum-of-product equation for each of these outputs in terms of the inputs and their complements. In this way,

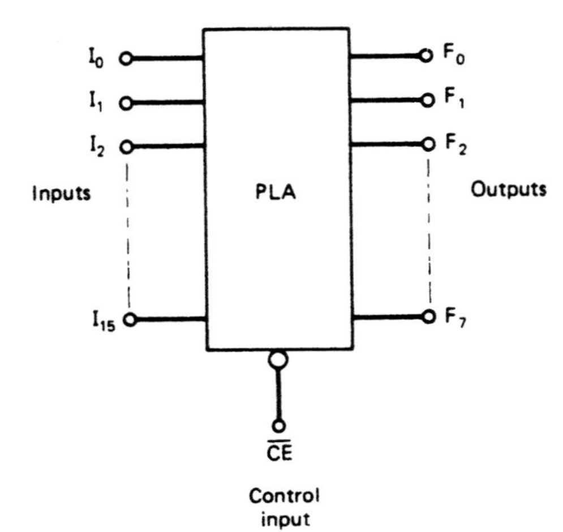

Figure 8–36 Block diagram of a PLA. (Reprinted with the permission of Walter A. Triebel)

we see that the logic levels applied at inputs I_0 through I_{15} and the programming of the AND array determine what logic levels are produced at outputs F_0 through F_7. Therefore, the capacity of a PLA is measured by three properties: the number of inputs, the number of outputs, and the number of product terms (P-terms).

Architecture of a PLA

We just pointed out that the circuitry of a PLA is a general purpose AND-OR-NOT array. Figure 8–37(a) shows this architecture. Here we see that the input buffers supply input signals A and B and their complements $\overline{A}$ and $\overline{B}$. Programmable connections in the AND array permit any combination of these inputs to be combined to form a product term. The product term outputs of the AND array are supplied to fixed inputs of the

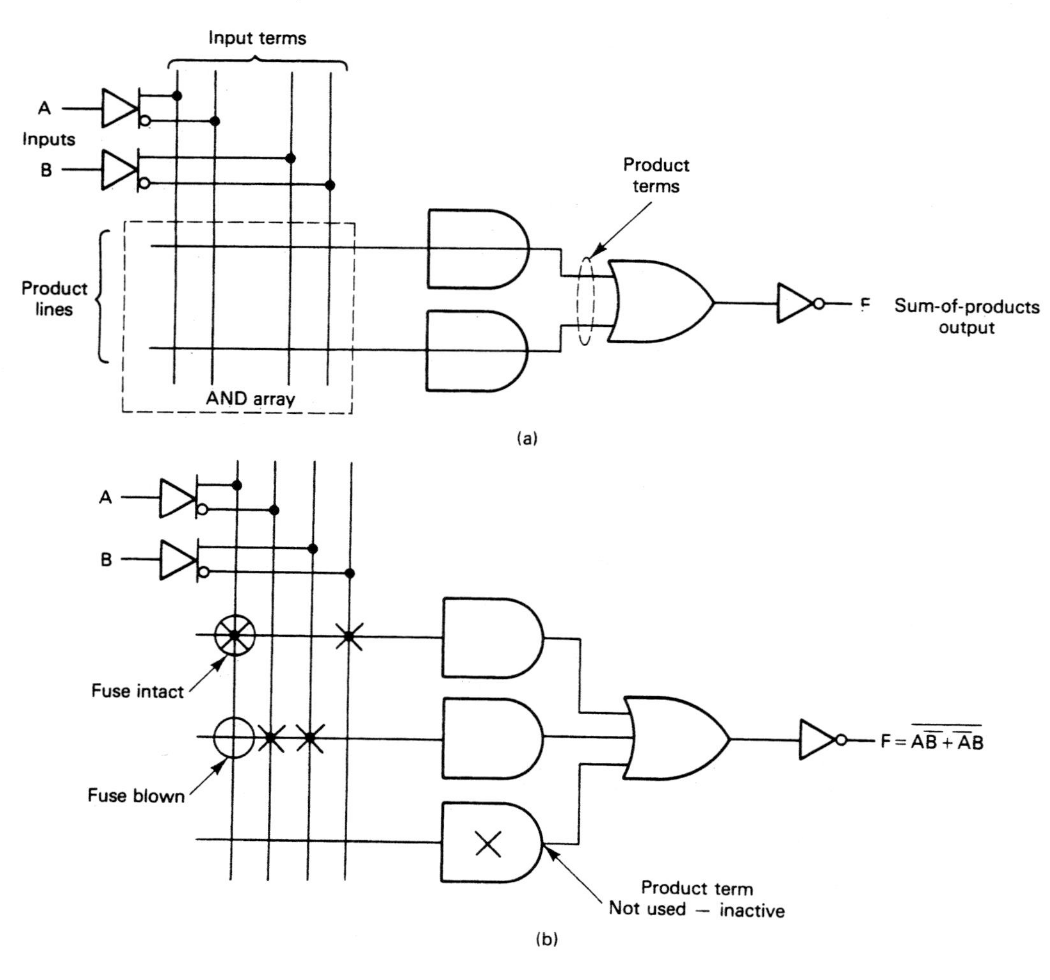

Figure 8–37 (a) Basic PLA architecture. (b) Implementing the logic function F = $\overline{(A\overline{B} + \overline{A}B)}$.

OR array. The output of the OR gate produces a sum-of-products function. Finally, the inverter complements this function.

The circuit in Fig. 8–37(b) shows how the function $F = \overline{(A\overline{B} + \overline{A}B)}$ is implemented with the AND-OR-NOT array. Notice that an X marked into the AND array means that the fuse is left intact, and no marking means that it has been blown to form an open circuit. For this reason, the upper AND gate is connected to A and $\overline{B}$ and produces the product term $A\overline{B}$. The second AND gate from the top connects to $\overline{A}$ and B to produce the product term $\overline{A}B$. The bottom AND gate is marked with an X to indicate that it is not in use. Gates like this that are not to be active should have all of their input fuse links left intact.

Figure 8–38(a) shows the circuit structure that is most widely used in PLAs. It differs from the circuit shown in Fig. 8–37(a) in two ways. First, the inverter has a programmable three-state control and can be used to isolate the logic function from the output. Second, the buffered output is fed back to form another set of inputs to the AND array. This new output configuration permits the output pin to be programmed to work as a *standard output, standard input,* or *logic-controlled input/output.* For instance, if the upper AND gate, which is the control gate for the output buffer, is set up to permanently enable the inverter, and the fuse links for its inputs that are fed back from the outputs are all blown open, the output functions as a standard output.

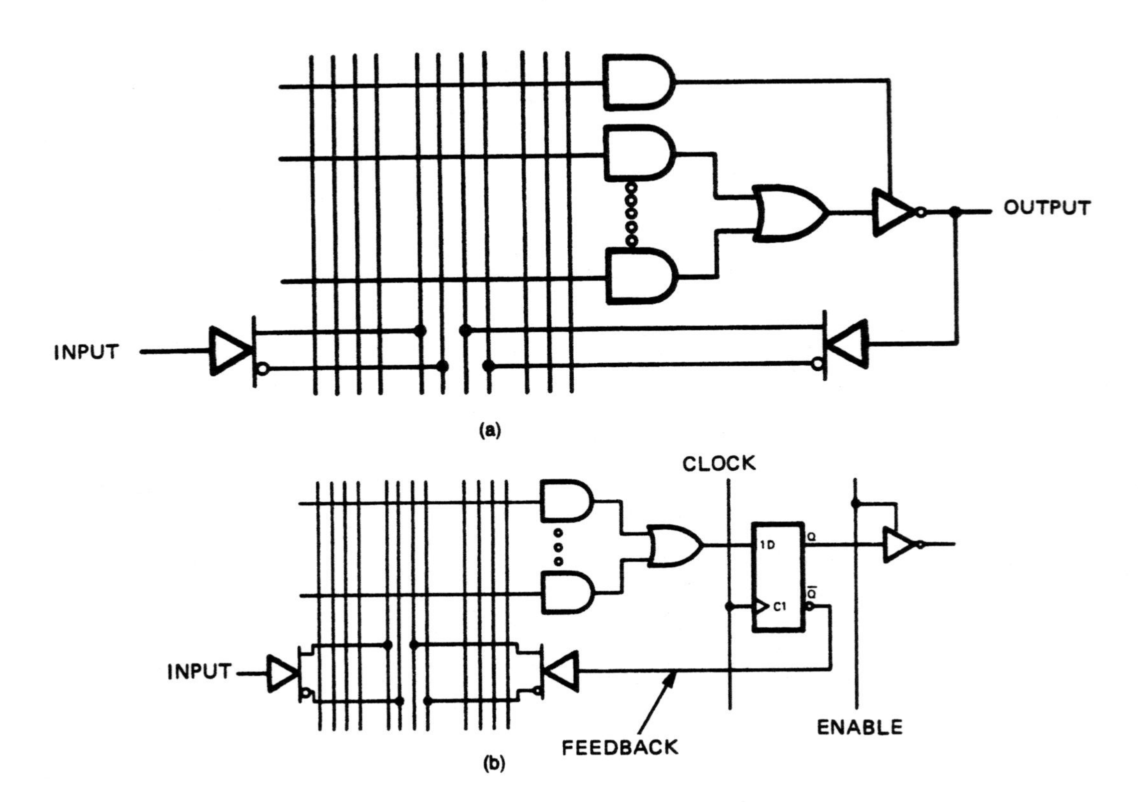

Figure 8–38 (a) Typical PLA architecture. (Courtesy of Texas Instruments Incorporated) (b) PLA with output latch. (Courtesy of Texas Instruments Incorporated)

PLAs are also available in which the outputs are latched with registers. Figure 8–38(b) shows a circuit for this type of device. Here we see that the output of the OR gate is applied to the D input of a clocked D-type flip-flop. In this way, the logic level produced by the AND-OR array is not presented at the output until a pulse is first applied at the CLOCK input. Furthermore, the feedback input is produced from the complemented output of the flip-flop, not the output of the inverter. This configuration is known as a *PLA with registered outputs* and is designed to simplify implementation of *state machine* designs.

Standard PAL™ Devices

Now that we have introduced the types of PLAs, block diagram of the PLA, and internal architecture of the PLA, let us continue by examining a few of the widely used PAL devices. A PAL, or a programmable array logic, is a PLA in which the OR array is fixed; only the AND array is programmable.

The 16L8 is a widely used PAL IC. Its internal circuitry and pin numbering are shown in Fig. 8–39(a). This device is housed in a 20-pin package, as shown in Fig. 8–39(b). Looking at this diagram, we see that it employs the PLA architecture illustrated in Fig. 8–38(a). Note that it has 10 dedicated input pins. All of these pins are labeled I. There are also two dedicated outputs, which are labeled with the letter O, and six programmable I/O lines, which are labeled I/O. Using the programmable I/O lines, the number of input lines can be expanded to as many as 16 inputs or the number of outputs can be increased to as many as eight lines.

All the 16L8's inputs are buffered and produce both the original form of the signal and its complement. The outputs of the buffer are applied to the inputs of the AND array. This array is capable of producing 64 product terms. Note that the AND gates are arranged into eight groups of eight. The outputs of seven gates in each of these groups are used as inputs to an OR gate, and the eighth output is used to produce an enable signal for the corresponding three-state output buffer. In this way, we see that the 16L8 is capable of producing up to seven product terms for each output, and the product terms can be formed using any combination of the 16 inputs.

The 16L8 is manufactured with bipolar technology. It operates from a +5V ±10% dc power supply and draws a maximum of 180 mA. Moreover, all its inputs and outputs are at TTL-compatible voltage levels. This device exhibits high-speed input-output propagation delays. In fact, the maximum I-to-O propagation delay is rated as 7 ns.

Another widely used PAL is the 20L8 device. Looking at the circuitry of this device in Fig. 8–40(a), we see that it is similar to that of the 16L8 just described. However, the 20L8 has a maximum of 20 inputs, eight outputs, and 64 P-terms. The device's 24-pin package is shown in Fig. 8–40(b).

The 16R8 is also a popular 20-pin PLA. The circuit diagram and pin layout for this device are shown in Figs. 8–41(a) and (b), respectively. From Fig. 8–41(a), we find that its eight fixed I inputs and AND-OR array are essentially the same as those of the 16L8. There is one change. The outputs of eight AND gates, instead of seven, are supplied to the inputs of each OR gate.

A number of changes have been made at the output side of the 16R8. Note that the outputs of the OR gates are first latched in D-type flip-flops with the CLK signal. They are then buffered and supplied to the eight Q outputs. Another change is that the enable

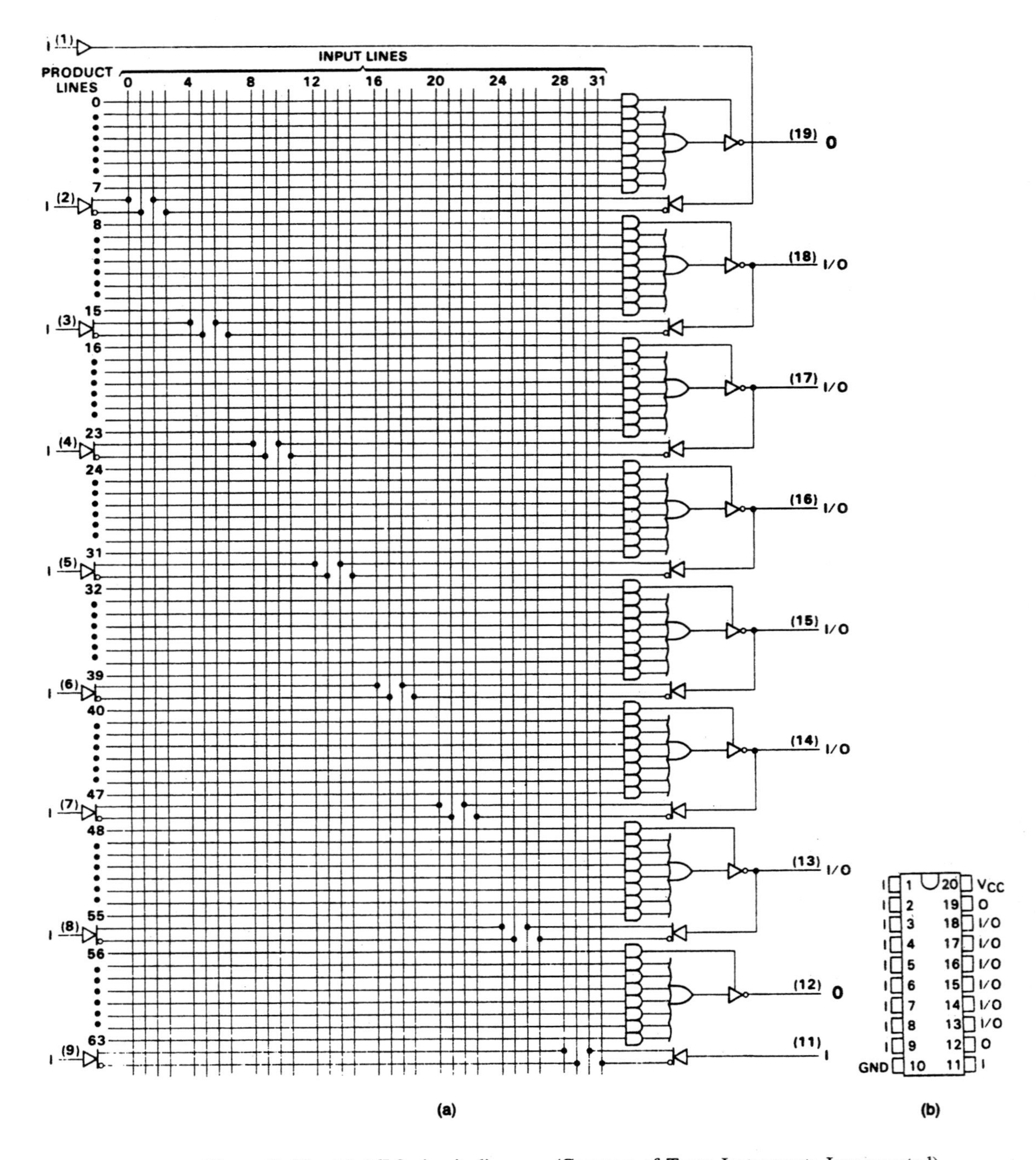

Figure 8–39 (a) 16L8 circuit diagram. (Courtesy of Texas Instruments Incorporated) (b) 16L8 pin layout. (Courtesy of Texas Instruments Incorporated)

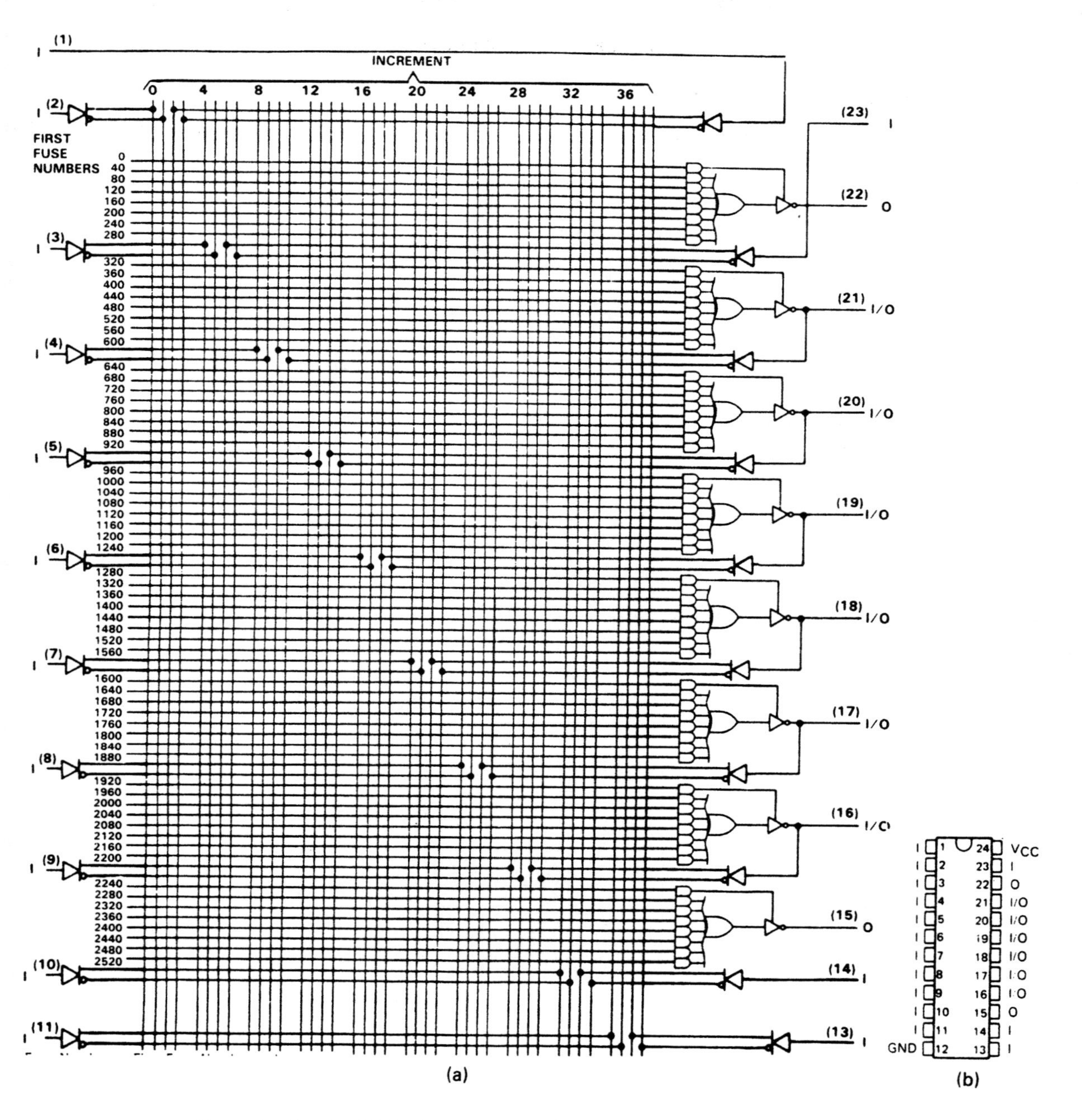

Figure 8–40 (a) 20L8 circuit diagram. (Courtesy of Texas Instruments Incorporated) (b) 20L8 pin layout. (Courtesy of Texas Instruments Incorporated)

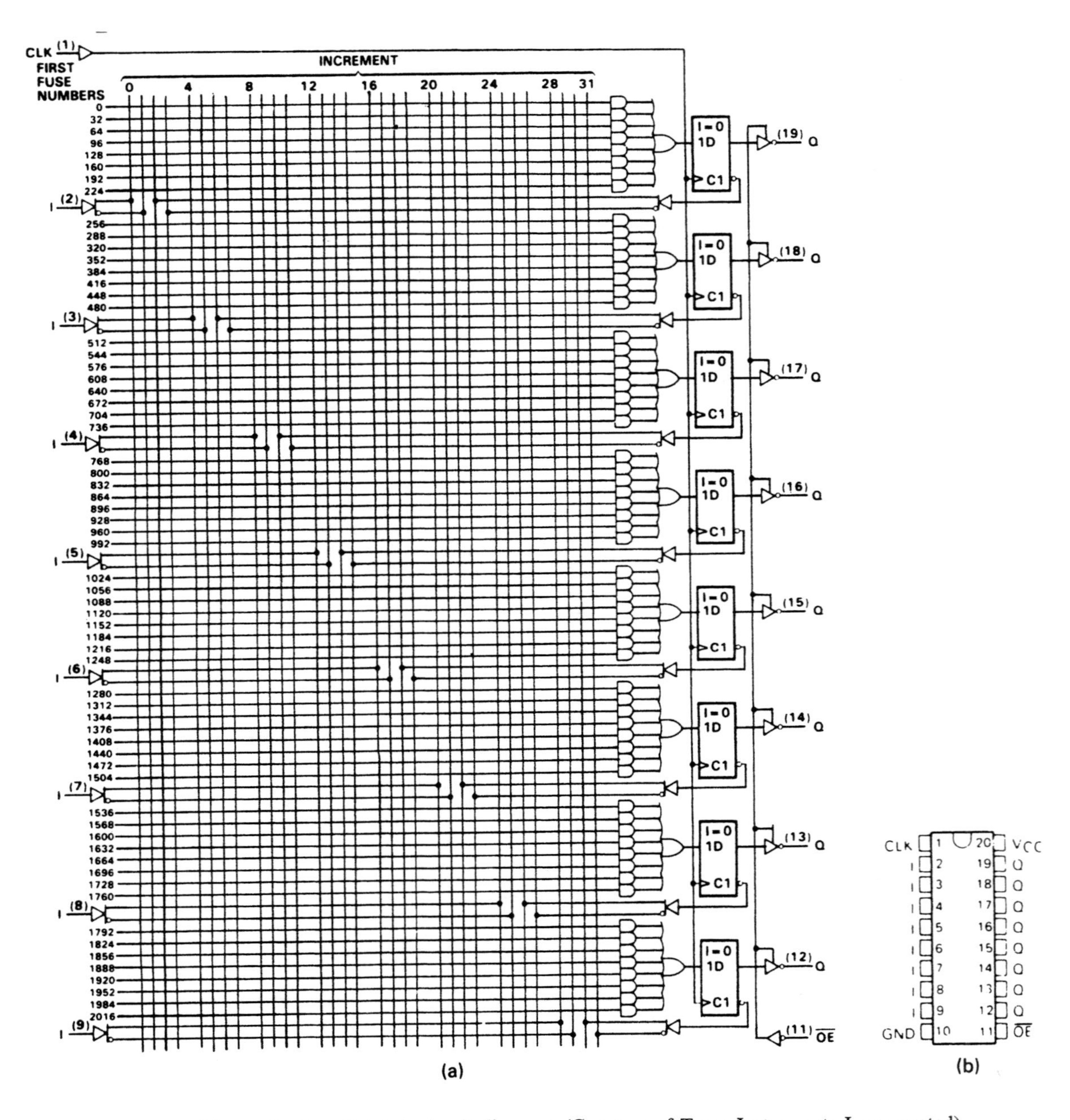

Figure 8–41 (a) 16R8 circuit diagram. (Courtesy of Texas Instruments Incorporated) (b) 16R8 pin layout. (Courtesy of Texas Instruments Incorporated)

signals for the output inverters are no longer programmable. Now the logic level of the $\overline{OE}$ control input enables all three-state outputs.

The last change is in the part of the circuit that produces the feedback inputs. In the 16R8, these eight input signals are derived from the complementary output of the corresponding latch instead of the output of the buffer. For this reason, the output leads can no longer be programmed to work as direct inputs.

The 20R8 is the register output version of the 20L8 PAL. Its circuit diagram and pin layout are given in Figs. 8–42(a) and (b), respectively.

Expanding PLA Capacity

Some applications have requirements that exceed the capacity of a single PLA IC. For instance, a 16L8 device has the ability to supply a maximum of 16 inputs, 8 outputs, and 64 product terms. Connecting several devices together can expand capacity. Let us now look at the way in which PLAs are interconnected to expand the number of inputs and outputs.

If a single PLA does not have enough outputs, two or more devices can be connected together into the configuration of Fig. 8–43(a). Here we see that the inputs I_0 through I_{15} on the two devices are individually connected in parallel. This connection does not change the number of inputs.

On the other hand, the eight outputs of the two PLAs are separately used to form the upper and lower bytes of a 16-bit output word. The bits of this word are denoted as O_0 through O_{15}. So with this connection, we have doubled the number of outputs.

When data are applied to the inputs, PLA 1 outputs the eight least significant bits of data. At the same instant PLA 2 outputs the eight most significant bits. These outputs can be used to represent individual logic functions.

Another limitation on the application of PLAs is the number of inputs. The maximum number of inputs on a single 16L8 is 16. However, additional ICs can be connected to expand the capacity of inputs. Figure 8–43(b) shows how one additional input is added. This permits a 17-bit input denoted as I_0 through I_{16}. The new bit I_{16} is supplied through inverters to the $\overline{CE}$ inputs on the two PLAs. At the output side of the PLAs, outputs O_0 through O_7 of the two devices are individually connected in parallel. To implement this connection, PLA devices with open-collector or three-state outputs must be used.

When I_{16} is logic 0, $\overline{CE}$ on PLA 1 is logic 0. This enables the device for operation, and the output functions coded for input I_0 through I_{15} are output at O_0 through O_7. At the same instant, $\overline{CE}$ on PLA 2 is logic 1 and it remains disabled. Making the logic level of I_{16} equal to 1 disables PLA 1 and enables PLA 2. Now the input at I_0 through I_{15} causes the output function defined by PLA 2 to be output at O_0 through O_7. Actually, this connection doubles the number of product terms as well as increases the number of inputs.

▲ 8.14 TYPES OF INPUT/OUTPUT

The input/output system of the microprocessor allows peripherals to provide data or receive results of processing the data. This is done using I/O ports. The 8088 and 8086 microcomputers can employ two different types of input/output (I/O): *isolated I/O* and *memory-mapped I/O*. These I/O methods differ in how I/O ports are mapped into the 8088/8086's address spaces. Some microcomputer systems employ both kinds of I/O—

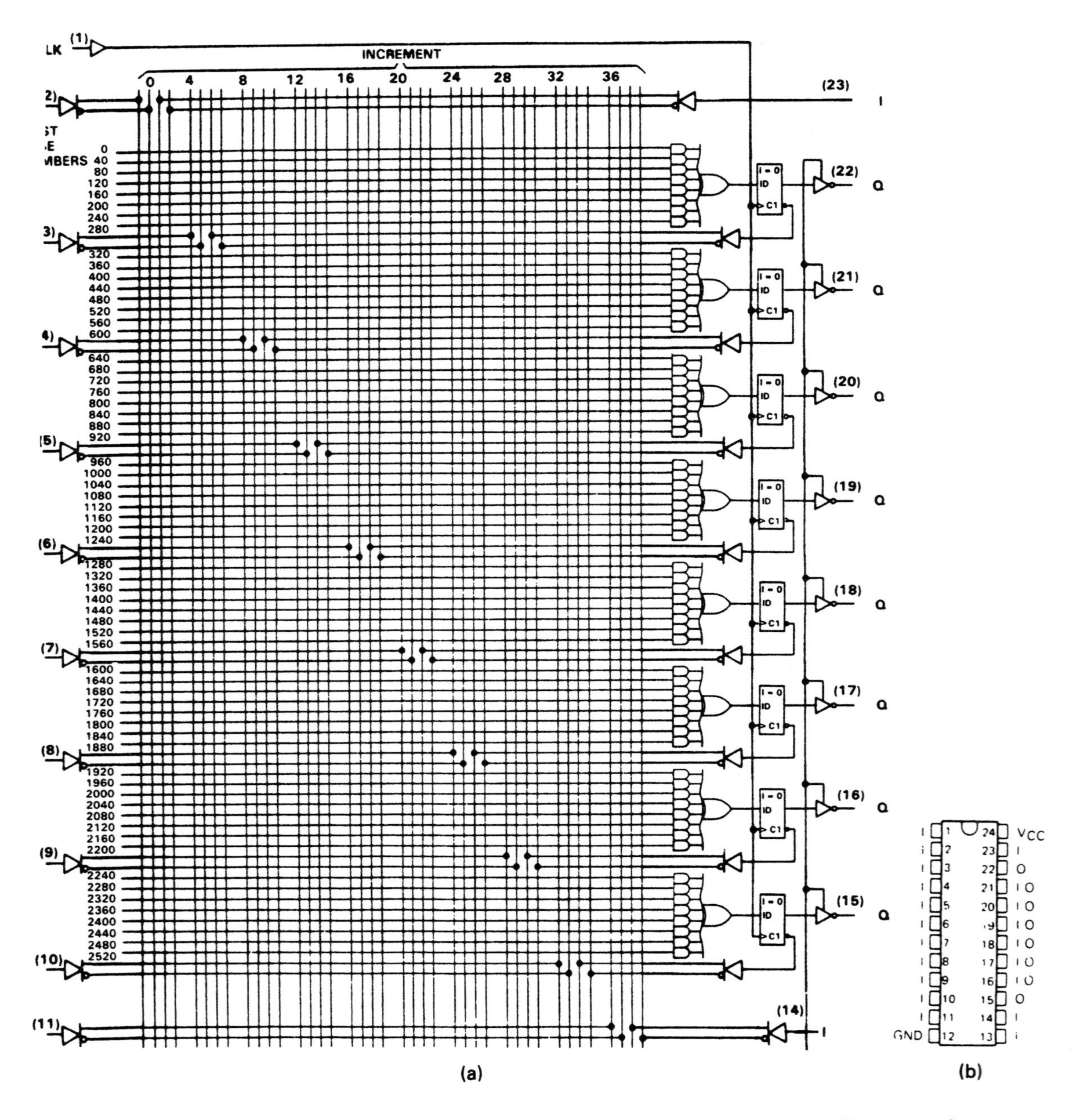

Figure 8–42 (a) 20R8 circuit diagram. (Courtesy of Texas Instruments Incorporated) (b) 20R8 pin layout. (Courtesy of Texas Instruments Incorporated)

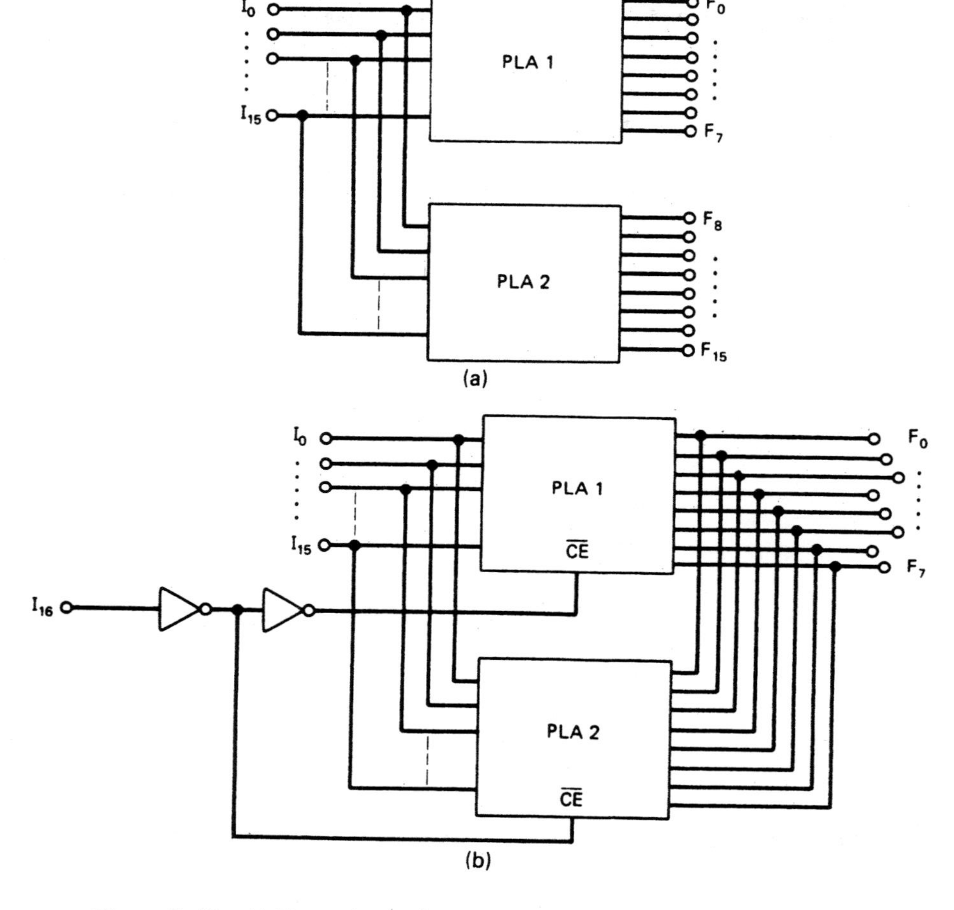

Figure 8–43 (a) Expanding output word length. (Reprinted with the permission of Walter A. Triebel) (b) Expanding input word length. (Reprinted with the permission of Walter A. Triebel)

that is, some peripheral ICs are treated as isolated I/O devices and others as memory-mapped I/O devices. Let us now look at each of these types of I/O.

Isolated Input/Output

When using isolated I/O in a microcomputer system, the I/O devices are treated separate from memory. This is achieved because the software and hardware architectures of the 8088/8086 support separate memory and I/O address spaces. Figure 8–44 illustrates these memory and I/O address spaces.

In our study of 8088/8086 software architecture in Chapter 2, we examined these address spaces from a software point of view. We found that information in memory or at

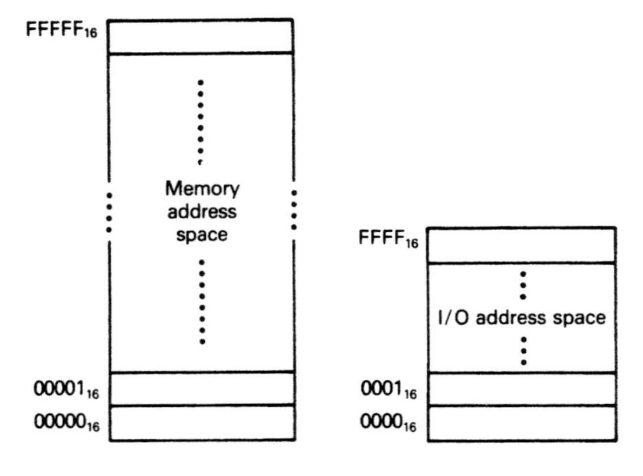

Figure 8–44 8088/8086 memory and I/O address spaces.

I/O ports is organized as bytes of data; that the memory address space contains 1M consecutive byte addresses in the range 00000_{16} through $FFFFF_{16}$; and that the I/O address space contains 64K consecutive byte addresses in the range 0000_{16} through $FFFF_{16}$.

Figure 8–45(a) shows a more detailed map of this I/O address space. Here we find that the bytes of data in two consecutive I/O addresses could be accessed as word-wide data. For instance, I/O addresses 0000_{16}, 0001_{16}, 0002_{16}, and 0003_{16} can be treated as independent byte-wide I/O ports, ports 0, 1, 2, and 3, or ports 0 and 1 may be considered together as word-wide port 0.

Note that the part of the I/O address space in Fig. 8–45(a) from address 0000_{16} through $00FF_{16}$ is referred to as *page 0.* Certain I/O instructions can only perform operations to ports in this part of the address range. Other I/O instructions can input or output data for ports anywhere in the I/O address space.

This isolated method of I/O offers some advantages. First, the complete 1Mbyte memory address space is available for use with memory. Second, special instructions have been provided in the instruction set of the 8088/8086 to perform isolated I/O input and output operations. These instructions have been tailored to maximize I/O performance. A disadvantage of this type of I/O is that all input and output data transfers must take place between the AL or AX register and the I/O port.

Memory-Mapped Input/Output

I/O devices can be placed in the memory address space of the microcomputer as well as in the independent I/O address space. In this case, the MPU looks at the I/O port as though it is a storage location in memory. For this reason, the method is known as *memory-mapped I/O.*

In a microcomputer system with memory-mapped I/O, some of the memory address space is dedicated to I/O ports. For example, in Fig. 8–45(b) the 4096 memory addresses in the range from $E0000_{16}$ through $E0FFF_{16}$ are assigned to I/O devices. Here the contents of address $E0000_{16}$ represent byte-wide I/O port 0, and the contents of addresses $E0000_{16}$ and $E0001_{16}$ correspond to word-wide port 0.

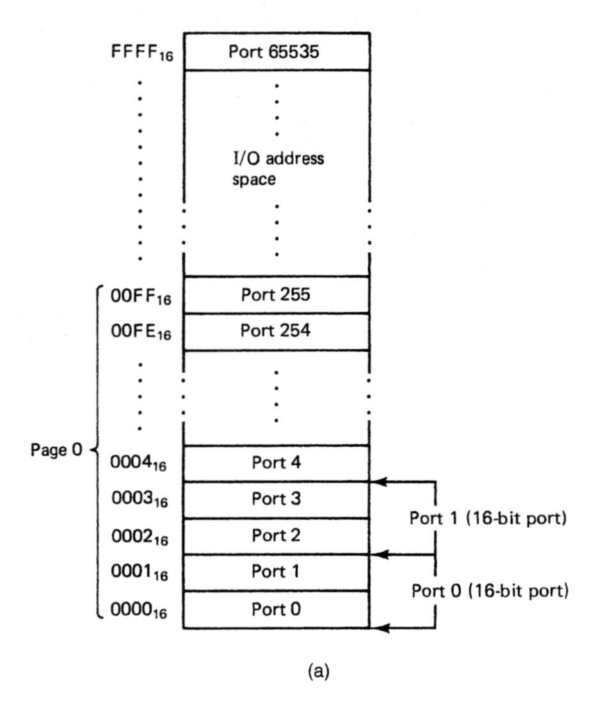

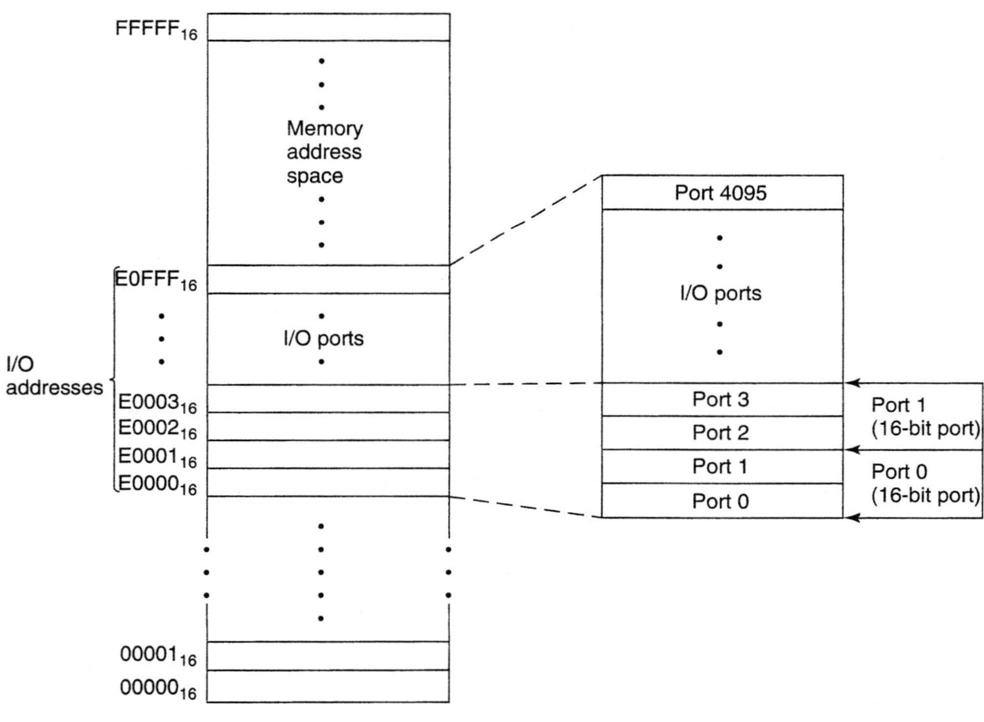

Figure 8–45 (a) Isolated I/O ports. (b) Memory-mapped I/O ports.

When I/O is configured in this way, instructions that affect data in memory are used instead of the special input/output instructions. This is an advantage in that many more instructions and addressing modes are available to perform I/O operations. For instance, the contents of a memory-mapped I/O port can be directly ANDed with a value in an internal register. In addition, I/O transfers can now take place between an I/O port and an internal register other than just AL or AX. However, this also leads to a disadvantage. That is, the memory instructions tend to execute slower than those specifically designed for isolated I/O. Therefore, a memory-mapped I/O routine may take longer to perform than an equivalent program using the input/output instructions.

Another disadvantage of using this method is that part of the memory address space is lost. For instance, in Fig. 8–45(b) addresses in the range from $E0000_{16}$ through $E0FFF_{16}$, allocated to I/O, cannot be used to implement memory.

▲ 8.15 ISOLATED INPUT/OUTPUT INTERFACE

The *isolated input/output interface* of the 8088 and 8086 microcomputers permits them to communicate with the outside world. The way in which the MPU deals with input/output circuitry is similar to the way in which it interfaces with memory circuitry. That is, input/output data transfers also take place over the multiplexed address/data bus. This parallel bus permits easy interface to LSI peripherals such as parallel I/O expanders, interval timers, and serial communication controllers. Through this I/O interface, the MPU can input or output data in bit, byte, or word (for the 8086) formats. Let us continue by looking at how an isolated I/O interface is implemented for minimum- and maximum-mode 8088 and 8086 microcomputer systems.

Minimum-Mode Interface

Let us begin by looking at the isolated I/O interface for a minimum-mode 8088 system. Figure 8–46(a) shows this minimum-mode interface. Here we find the 8088, interface circuitry, and I/O ports for devices 0 through N. I/O devices 0 through N can represent input devices such as a keyboard, output devices such as a printer, or input/output devices such as an asynchronous serial communications port. An example of a typical I/O device used in the I/O subsystem is a programmable peripheral interface (PPI) IC, such as the 82C55A. This type of device is used to implement parallel input and output ports. The circuits in the interface section must perform functions such as select the I/O port, latch output data, sample input data, synchronize data transfers, and translate between TTL voltage levels and those required to operate the I/O devices.

The data path between the 8088 and I/O interface circuits is the multiplexed address/data bus. Unlike the memory interface, this time just the 16 least significant lines of the bus, AD_0 through AD_7 and A_8 through A_{15}, are in use. This interface also involves the control signals that we discussed as part of the memory interface—that is ALE, $\overline{SSO}$, $\overline{RD}$, $\overline{WR}$, $IO/\overline{M}$, $DT/\overline{R}$, and $\overline{DEN}$.

Figure 8–46(b) shows the isolated I/O interface of a minimum-mode 8086-based microcomputer system. Looking at this diagram, we find that the interface differs from that of the 8088 microcomputer in several ways. First, the complete data bus AD_0 through

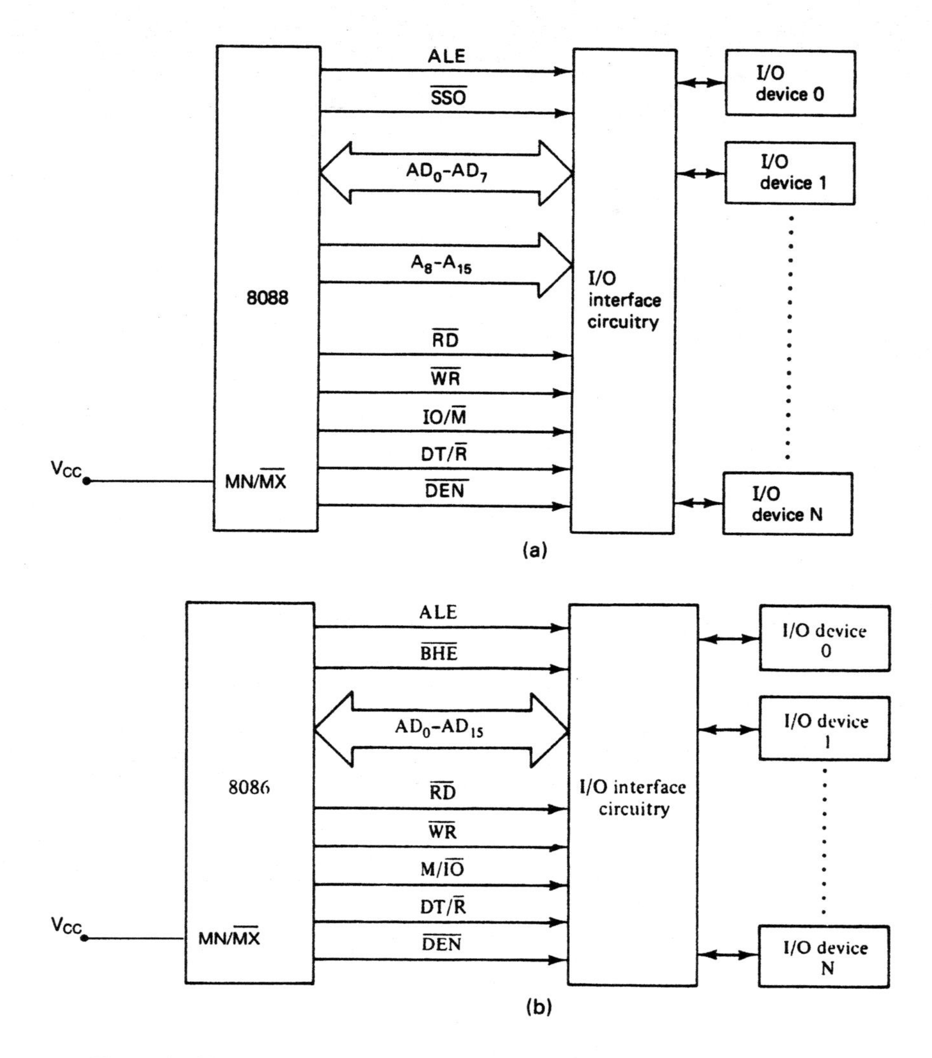

Figure 8–46 (a) Minimum-mode 8088 system I/O interface. (b) Minimum-mode 8086 system I/O interface.

AD_{15} is used for input and output data transfers; second, the M/$\overline{IO}$ control signal is the complement of the equivalent signal IO/$\overline{M}$ in the 8088's interface; and third, status signal $\overline{SSO}$ is replaced by $\overline{BHE}$.

Maximum-Mode Interface

When the 8088 is strapped to operate in the maximum mode (MN/$\overline{MX}$ connected to ground), the interface to the I/O circuitry changes. Figure 8–47(a) illustrates this configuration.

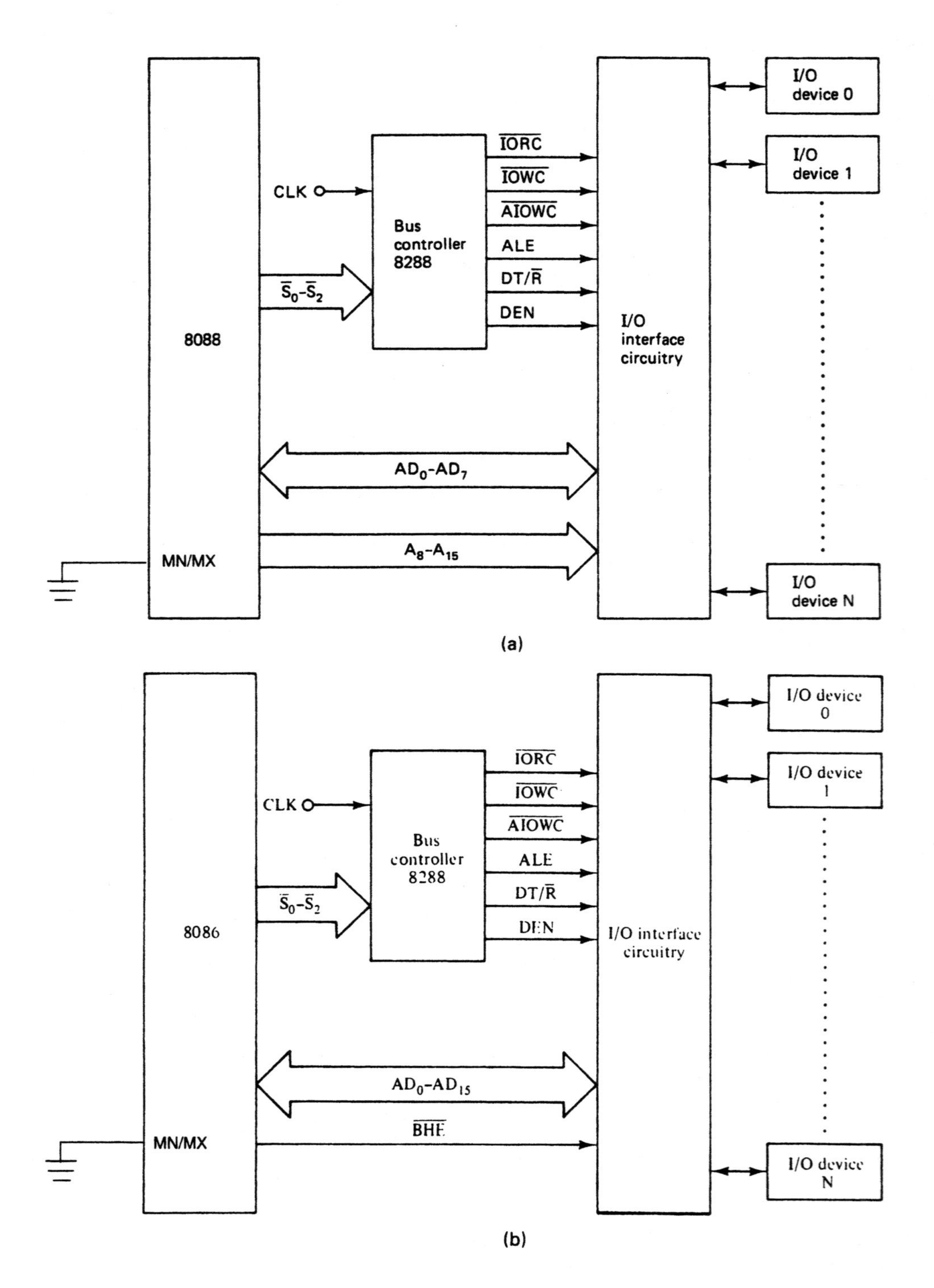

Figure 8–47 (a) Maximum-mode 8088 system I/O interface. (b) Maximum-mode 8086 system I/O interface.

Status inputs			CPU cycle	8288 command
$\overline{S}_2$	$\overline{S}_1$	$\overline{S}_0$		
0	0	0	Interrupt acknowledge	$\overline{INTA}$
0	0	1	Read I/O port	$\overline{IORC}$
0	1	0	Write I/O port	$\overline{IOWC}$, $\overline{AIOWC}$
0	1	1	Halt	None
1	0	0	Instruction fetch	$\overline{MRDC}$
1	0	1	Read memory	$\overline{MRDC}$
1	1	0	Write memory	$\overline{MWTC}$, $\overline{AMWC}$
1	1	1	Passive	None

Figure 8–48 I/O bus cycle status codes. (Reprinted with permission of Intel Corporation, Copyright/Intel Corp. 1979)

As in the maximum-mode memory interface, the 8288 bus controller produces the control signals for the I/O subsystem. The 8288 decodes bus command status codes output by the 8088 at $\overline{S}_2\overline{S}_1\overline{S}_0$. These codes tell which type of bus cycle is in progress. If the code corresponds to an I/O read bus cycle, the 8288 generates the *I/O read command output* ($\overline{IORC}$) and for an I/O write cycle it generates *I/O write command outputs* ($\overline{IOWC}$) and ($\overline{AIOWC}$). The 8288 also produces the control signals ALE, DT/$\overline{R}$, and DEN. The address and data transfer path between 8088 and maximum-mode I/O interface remains address/data bus lines AD_0 through AD_7 and A_8 through A_{15}.

Figure 8–47(b) shows the maximum-mode isolated I/O interface of an 8086 microprocessor system. There are only two differences between this interface diagram and that for the 8088 microprocessor. As in the minimum mode, the full 16-bit data bus is the path for data transfers, and the signal $\overline{BHE}$, which is not supplied by the 8088, is included in the interface.

The table in Fig. 8–48 shows the bus command status codes together with the command signals that they produce. Those for I/O bus cycles are highlighted. The MPU indicates that data are to be input (read I/O port) by code $\overline{S}_2\overline{S}_1\overline{S}_0$ = 001. This code causes the bus controller to produce control output I/O read command ($\overline{IORC}$). There is one other code that represents an output bus cycle, the write I/O port code $\overline{S}_2\overline{S}_1\overline{S}_0$ = 010. It produces two output command signals: I/O write cycle ($\overline{IOWC}$) and advanced I/O write cycle ($\overline{AIOWC}$). These command signals are used to enable data from the I/O ports onto the system bus during an input operation and from the MPU to the I/O ports during an output operation.

▲ 8.16 INPUT/OUTPUT DATA TRANSFERS

Input/output data transfers in the 8088 and 8086 microcomputers can be either byte-wide or word-wide. The port that is accessed for input or output of data is selected by an *I/O address*. This address is specified as part of the instruction that performs the I/O operation.

I/O addresses are 16 bits in length and are output by the 8088 to the I/O interface over bus lines AD_0 through AD_7 and A_8 through A_{15}. AD_0 represents the LSB and A_{15} the

MSB. The most significant address lines, A_{16} through A_{19}, are held at the 0 logic level during the address period (T_1) of all I/O bus cycles. Since 16 address lines are used to address I/O ports, the 8088's I/O address space consists of 64K byte-wide I/O ports.

The 8088 signals to external circuitry that the address on the bus is for an I/O port instead of a memory location by switching the $IO/\overline{M}$ control line to the 1 logic level. This signal is held at the 1 level during the complete input or output bus cycle. For this reason, it can be used to enable the address latch or address decoder in external I/O circuitry.

Data transfers between the 8088 and I/O devices are performed over the data bus. Data transfers to byte-wide I/O ports always require one bus cycle. Byte data transfers to a port are performed over bus lines D_0 through D_7. Word transfers also take place over the data bus, D_0 through D_7. However, this type of operation is performed as two consecutive byte-wide data transfers and takes two bus cycles.

For the 8086 microcomputer, I/O addresses are output on address/data bus lines AD_0 through AD_{15}. The logic levels of signals A_0 and $\overline{BHE}$ determine whether data are input/output for an odd-addressed byte-wide port, even-addressed byte-wide port, or a word-wide port. For example, if $A_0\overline{BHE} = 10$, an odd-addressed byte-wide I/O port is accessed. Byte data transfers to a port at an even address are performed over bus lines D_0 through D_7 and those to an odd-addressed port are performed over D_8 through D_{15}. Data transfers to byte-wide I/O ports always take place in one bus cycle.

Word data transfers between the 8086 and I/O devices are accompanied by the code $A_0\overline{BHE} = 00$ and are performed over the complete data bus, D_0 through D_{15}. A word transfer can require either one or two bus cycles. To ensure that just one bus cycle is required for the word data transfer, word-wide I/O ports should be aligned at even-address boundaries.

▲ 8.17 INPUT/OUTPUT INSTRUCTIONS

Input/output operations are performed by the 8088 and 8086 microprocessors that employ isolated I/O using special input and output instructions together with the I/O port addressing modes. These instructions, *in* (IN) and *out* (OUT), are listed in Fig. 8–49. Their mnemonics and formats are provided together with a brief description of their operations.

Note that there are two different forms of IN and OUT instructions: the *direct I/O instructions* and *variable I/O instructions.* Either of these two types of instructions can be used to transfer a byte or word of data. All data transfers take place between an I/O device and the MPU's accumulator register. For this reason, this method of performing I/O is known as *accumulator I/O.* Byte transfers involve the AL register, and word transfers the AX regis-

Mnemonic	Meaning	Format	Operation
IN	Input direct	IN Acc,Port	(Acc) ← (Port) Acc = AL or AX
	Input indirect (variable)	IN Acc,DX	(Acc) ← ((DX))
OUT	Output direct	OUT Port,Acc	(Port) ← (Acc)
	Output indirect (variable)	OUT DX,Acc	((DX)) ← (Acc)

Figure 8–49 Input/output instructions.

ter. In fact, specifying AL as the source or destination register in an I/O instruction indicates that it corresponds to a byte transfer. That is, byte-wide or word-word input/output is selected by specifying the accumulator (Acc) in the instruction as AL or AX, respectively.

In a direct I/O instruction, the address of the I/O port is specified as part of the instruction. Eight bits are provided for this direct address. For this reason, its value is limited to the address range from 0_{10} equals 00_{16} to 255_{10} equals FF_{16}. This range is referred to as page 0 in the I/O address space.

An example is the instruction

```
IN AL, 0FEH
```

As Fig. 8–49 shows, execution of this instruction causes the contents of the byte-wide I/O port at address FE_{16} of the I/O address space to be input to the AL register. This data transfer takes place in one input bus cycle.

EXAMPLE 8.7

Write a sequence of instructions that will output the data FF_{16} to a byte-wide output port at address AB_{16} of the I/O address space.

Solution

First, the AL register is loaded with FF_{16} as an immediate operand in the instruction

```
MOV  AL, 0FFH
```

Now the data in AL can be output to the byte-wide output port with the instruction

```
OUT  0ABH, AL
```

The difference between the direct and variable I/O instructions lies in the way in which the address of the I/O port is specified. We just saw that for direct I/O instructions an 8-bit address is specified as part of the instruction. On the other hand, the variable I/O instructions use a 16-bit address that resides in the DX register within the MPU. The value in DX is not an offset. It is the actual address that is to be output on AD_0 through AD_7 and A_8 through A_{15} during the I/O bus cycle. Since this address is a full 16 bits in length, variable I/O instructions can access ports located anywhere in the 64K-byte I/O address space.

When using either type of I/O instruction, the data must be loaded into or removed from the AL or AX register before another input or output operation can be performed. In the case of variable I/O instructions, the DX register must be loaded with the address. This requires execution of additional instructions. For instance, the instruction sequence

```
MOV  DX, 0A000H
IN   AL, DX
MOV  BL, AL
```

inputs the contents of the byte-wide input port at $A000_{16}$ of the I/O address space into AL and then saves it in BL.

EXAMPLE 8.8

Write a series of instructions that will output FF_{16} to an output port located at address $B000_{16}$ of the I/O address space.

Solution

The DX register must first be loaded with the address of the output port. This is done with the instruction

```
MOV  DX, 0B000H
```

Next, the data that are to be output must be loaded into AL with the instruction

```
MOV  AL, 0FFH
```

Finally, the data are output with the instruction

```
OUT  DX, AL
```

EXAMPLE 8.9

Data are to be read in from two byte-wide input ports at addresses AA_{16} and $A9_{16}$ and then output as a word to a word-wide output port at address $B000_{16}$. Write a sequence of instructions to perform this input/output operation.

Solution

We can first read in the byte from the port at address AA_{16} into AL and move it to AH. This is done with the instructions

```
IN   AL, 0AAH
MOV  AH, AL
```

Now the other byte, which is at port $A9_{16}$, can be read into AL by the instruction

```
IN  AL, 0A9H
```

The word is now held in AX. To write out the word of data, we load DX with the address $B000_{16}$ and use a variable output instruction. This leads to the following:

```
MOV  DX, 0B000H
OUT  DX, AX
```

▲ 8.18 INPUT/OUTPUT BUS CYCLES

In Section 8.15, we found that the isolated I/O interface signals for the minimum-mode 8088 and 8086 microcomputer systems are essentially the same as those involved in the memory interface. In fact, the function, logic levels, and timing of all signals other than IO/$\overline{M}$ (M/$\overline{IO}$) are identical to those already described for the memory interface in Section 8.11.

Waveforms for the 8088's *I/O input (I/O read) bus cycle* and *I/O output (I/O write) bus cycle* are shown in Figs. 8–50 and 8–51, respectively. Looking at the input and output bus cycle waveforms, we see that the timing of IO/$\overline{M}$ does not change. The 8088 switches it to logic 1 to indicate that an I/O bus cycle is in progress. It is maintained at the 1 logic level for the duration of the I/O bus cycle. As in memory cycles, the address is output together with ALE during clock period T_1. For the input bus cycle, $\overline{DEN}$ is switched to logic 0 to signal the I/O interface circuitry when to put the data onto the bus and the 8088 reads data off the bus during period T_3.

On the other hand, for the output bus cycle in Fig. 8–51, the 8088 puts write data on the bus late in T_2 and maintains it during the rest of the bus cycle. This time $\overline{WR}$ switches to logic 0 to signal the I/O system that valid data are on the bus.

The waveforms of the 8086's input and output bus cycles are shown in Figs. 8–52 and 8–53, respectively. Let us just look at the differences between the input cycle of the 8086 and that of the 8088. Comparing the waveforms in Fig. 8–52 to those in Fig. 8–50, we see that the 8086 outputs the signal $\overline{BHE}$ along with the address in T-state T_1. Remem-

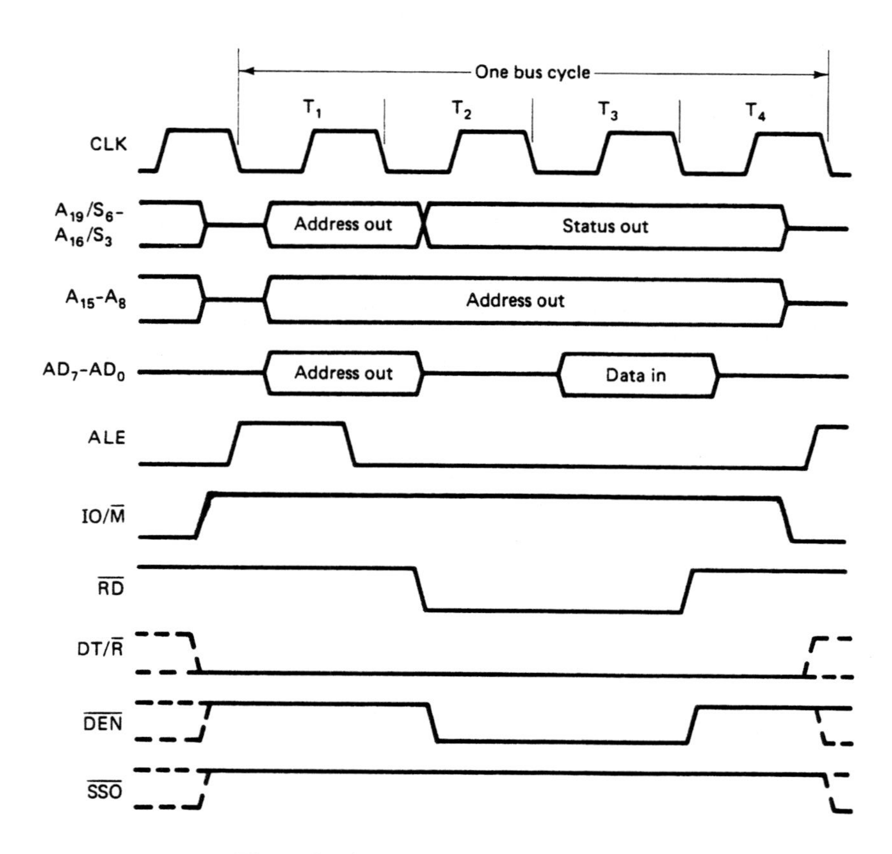

Figure 8–50 Input bus cycle of the 8088.

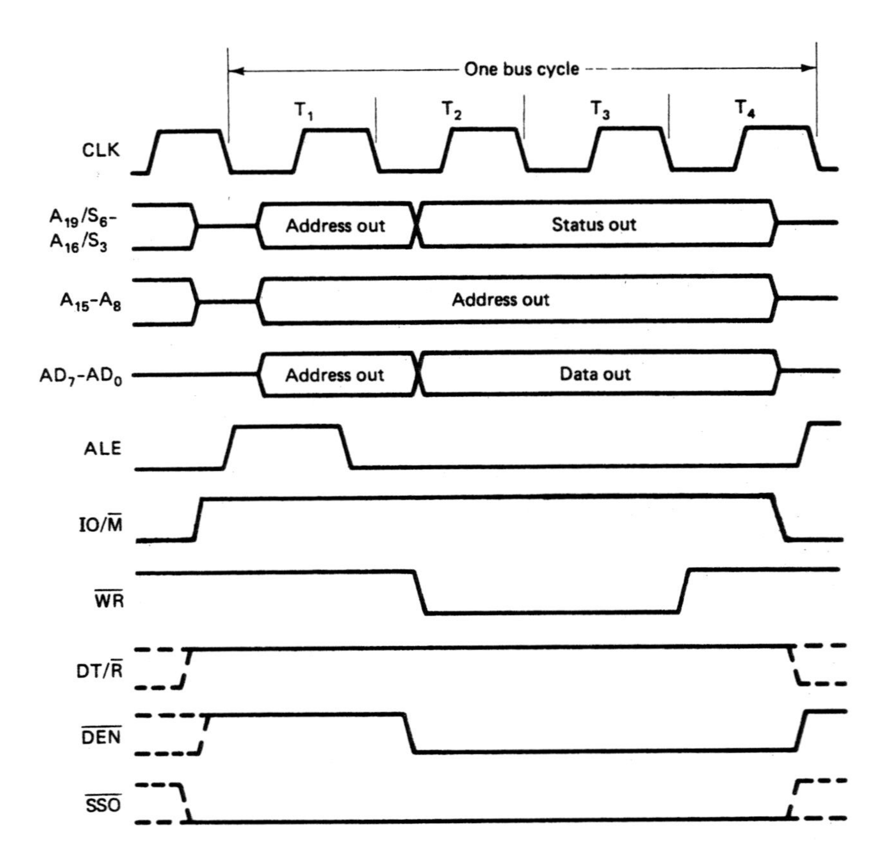

Figure 8–51 Output bus cycle of the 8088.

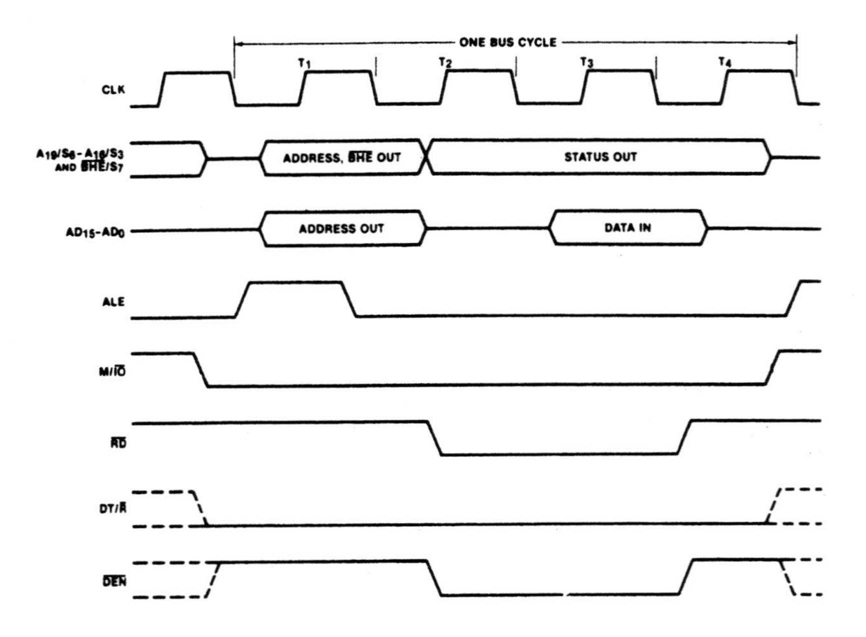

Figure 8–52 Input bus cycle of the 8086.

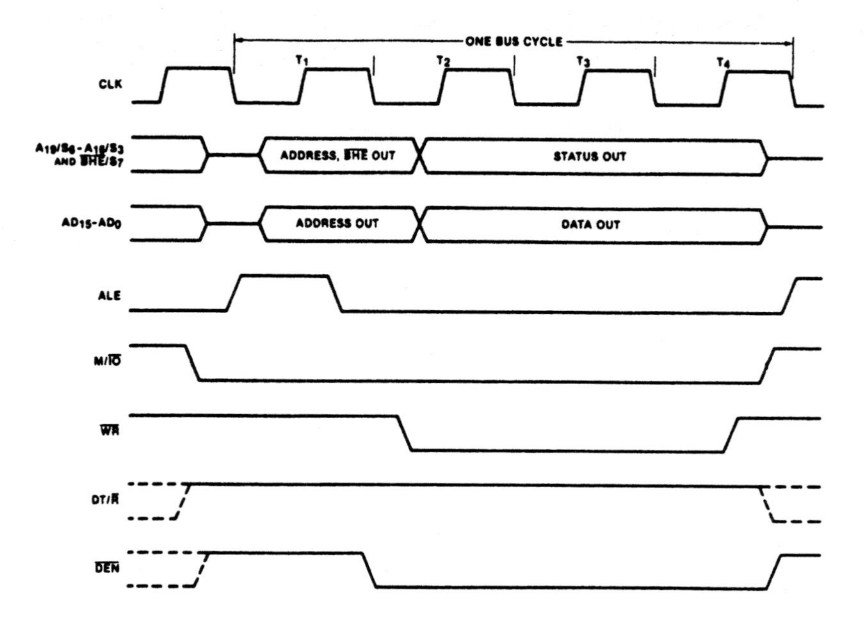

Figure 8–53 Output bus cycle of the 8086.

ber that for the 8086 microprocessor this signal is used along with A_0 to select the byte-wide or the word-wide port. Next, the 8086's data transfer path to the I/O interface is the 16-bit address/data bus, not 8 bits as in the 8088 system. Therefore, data transfers, which take place during T_3, can take place over the lower 8 data bus lines, upper 8 data bus lines, or all 16 data bus lines. Third, the 8086 outputs logic 0 on the $M/\overline{IO}$ line, while the 8088 outputs logic 1 on the $IO/\overline{M}$ line. That is, the $M/\overline{IO}$ control signal of the 8086 is the complement of that of the 8088. Finally, the 8086 does not produce an $\overline{SSO}$ output signal like the one in the 8088.

REVIEW PROBLEMS

Section 8.1

1. Name the technology used to fabricate the 8088 and 8086 microprocessors.
2. What is the transistor count of the 8088?
3. Which pin is used as the NMI input on the 8088?
4. Which pin provides the $\overline{BHE}/S_7$ output signals on the 8086?
5. How much memory can the 8088 and 8086 directly address?
6. How large is the I/O address space of the 8088 and 8086?

Section 8.2

7. How is minimum or maximum mode of operation selected?
8. Describe the difference between the minimum-mode 8088 system and maximum-mode 8088 system.

9. What output function is performed by pin 29 of the 8088 when in the minimum mode? Maximum mode?
10. Is the signal $M/\overline{IO}$ an input or output of the 8086?
11. Name one signal that is supplied by the 8088 but not by the 8086.
12. Are the signals QS_0 and QS_1 produced in the minimum mode or maximum mode?

Section 8.3

13. What are the word lengths of the 8088's address bus and data bus? The 8086's address bus and data bus?
14. Does the 8088 have a multiplexed address/data bus or independent address and data buses?
15. What mnemonic is used to identify the least significant bit of the 8088's address bus? The most significant bit of the 8088's data bus?
16. What does status code $S_4S_3 = 01$ mean in terms of the memory segment being accessed?
17. Which output is used to signal external circuitry that a byte of data is available on the upper half of the 8086's data bus?
18. What does the logic level on $M/\overline{IO}$ signal to external circuitry in an 8086 microcomputer?
19. Which output is used to signal external circuitry in an 8088-based microcomputer that valid data is on the bus during a write cycle?
20. What signal does a minimum-mode 8088 respond with when it acknowledges an active interrupt request?
21. Which signals implement the DMA interface in a minimum-mode 8088 or 8086 microcomputer system?
22. List the signals of the 8088 that are put in the high-Z state in response to a DMA request.

Section 8.4

23. Identify the signal lines of the 8088 that are different for the minimum-mode and maximum-mode interfaces.
24. What status outputs of the 8088 are inputs to the 8288?
25. What maximum-mode control signals are generated by the 8288?
26. What function does the $\overline{LOCK}$ signal serve in a maximum-mode 8088 microcomputer system?
27. What status code is output by the 8088 to the 8288 if a memory read bus cycle is taking place?
28. What command output becomes active if the status inputs of the 8288 are 100_2?
29. If the 8088 executes a jump instruction, what queue status code would be output?
30. What signals are provided for local bus control in a maximum-mode 8088 system?

Section 8.5

31. What is the range of power supply voltage over which the 8088 is guaranteed to work correctly?

32. What is the maximum value of voltage that is considered a valid logic 0 at bit D_0 of the 8088's data bus? Assume that the output is sinking 2 mA.

33. What is the minimum value of voltage that would represent a valid logic 1 at the INTR input of the 8088?

34. At what value current is V_{OLmax} measured on the 8086?

Section 8.6

35. At what speeds are 8088s generally available?

36. What frequency crystal must be connected between the X_1 and X_2 inputs of the clock generator if an 8088-2 is to run at full speed?

37. What clock outputs are produced by the 8284? What would be their frequencies if a 30-MHz crystal were used?

38. What are the logic levels of the clock waveforms applied to the 8088?

Section 8.7

39. How many clock states are in an 8088 bus cycle that has no wait states? How are these states denoted?

40. What is the duration of the bus cycle for a 5-MHz 8088 that is running at full speed and with no wait states?

41. What is an idle state?

42. What is a wait state?

43. If an 8086 running at 10 MHz performs bus cycles with two wait states, what is the duration of the bus cycle?

Section 8.8

44. How is the memory of an 8088 microcomputer organized from a hardware point of view? An 8086 microcomputer?

45. Give an overview of how a byte of data is read from memory address $B0003_{16}$ of an 8088-based microcomputer, and list the memory control signals along with their active logic levels that occur during the memory read bus cycle.

46. Give an overview of how a word of data is written to memory starting at address $A0000_{16}$ of an 8088-based microcomputer, and list the memory control signals together with their active logic levels that occur during the memory write cycle.

47. In which bank of memory in an 8086-based microcomputer are odd-addressed bytes of data stored? What bank select signal is used to enable this bank of memory?

48. Over which of the 8086's data bus lines are even-addressed bytes of data transferred and which bank select signal is active?

49. List the memory control signals together with their active logic levels that occur when a word of data is written to memory address $A0000_{16}$ in a minimum-mode 8086 microcomputer system.

50. List the memory control signals together with their active logic levels that occur when a byte of data is written to memory address $B0003_{16}$ in a minimum-mode 8086 microcomputer. Over which data lines is the byte of data transferred?

Section 8.9

51. In a maximum-mode 8088 microcomputer, what code is output on S_4S_3 when an instruction-fetch bus cycle is in progress?
52. What is the value of S_4S_3 if the operand of a pop instruction is being read from memory? Assume the microcomputer employs the 8088 in the maximum mode.

Section 8.10

53. Which of the 8088's memory control signals is the complement of the corresponding signal on the 8086?
54. What memory control output of the 8088 is not provided on the 8086? What signal replaces it on the 8086?
55. In a maximum-mode 8088-based microcomputer, what memory bus status code is output when a word of instruction code is fetched from memory? Which memory control output(s) is (are) produced by the 8288?
56. In maximum mode, what memory bus status code is output when a destination operand is written to memory? Which memory control output(s) is (are) produced by the 8288?
57. When the instruction PUSH AX is executed, what address bus status code and memory bus cycle code are output by the 8088 in a maximum-mode microcomputer system? Which command signals are output by the 8288?

Section 8.11

58. How many clock states are in a read bus cycle that has no wait states? What would be the duration of this bus cycle if the 8086 were operating at 10 MHz?
59. What happens in the T_1 part of the 8088's memory read or write bus cycle?
60. Describe the bus activity that takes place as an 8088, in minimum mode, writes a byte of data into memory address $B0010_{16}$.
61. Which two signals can be used to determine that the current bus cycle is a write cycle?
62. Which signal can be used to identify the start of a bus cycle?

Section 8.12

63. Give an overview of the function of each block in the memory interface diagram shown in Fig. 8–24.
64. When the instruction PUSH AX is executed, what bus status code is output by the 8086 in maximum mode, what are the logic levels of A_0 and $\overline{BHE}$, and what read/write control signals are produced by the bus controller?
65. What type of basic logic devices is provided by the 74F373?
66. Specify the logic levels of $\overline{BHEL}$, $\overline{MWRC}$, $\overline{MRDC}$, and A_{0L} when the 8086 in Fig. 8–24 reads a word of data from address 12340H.

67. Make a truth table, using the circuits in Figs. 8–27 and 8–28, to specify the logic levels of $\overline{RD_U}$, $\overline{RD_L}$, $\overline{WR_U}$, $\overline{WR_L}$, $\overline{BHEL}$, $\overline{MRDC}$, $\overline{MWTC}$, and A_{0L} when the processor
   (a) reads a byte from address 01234H
   (b) writes a byte to address 01235H
   (c) reads a word from address 01234H
   (d) writes a word to address 01234H

68. What logic devices are provided by the 74F245?

69. In the circuit of Fig. 8–30, what logic levels must be applied to the $\overline{DEN}$ and $DT/\overline{R}$ inputs to cause data on the system data bus to be transferred to the microprocessor data bus?

70. Make a drawing like that shown in Fig. 8–30 to illustrate the data bus transceiver circuit needed in an 8088-based microcomputer system.

71. How many address lines must be decoded to generate five chip select signals?

72. Name an IC that implements a two-line to four-line decoder logic function.

73. If the inputs to a 74F138 decoder are $G_1 = 1$, $\overline{G_{2A}} = 0$, $G_{2B} = 0$, and CBA = 101 , which output is active?

74. Make a drawing for a minimum-mode 8088-based microcomputer for which a 74F138 decoder is used to generate $\overline{MEMR}$ and $\overline{MEMW}$ from the $\overline{RD}$, $\overline{WR}$, and $IO/\overline{M}$ signals.

Section 8.13

75. What does PLA stand for?

76. List three properties that measure the capacity of a PLA.

77. What is the programming mechanism used in the PAL called?

78. What does PAL stand for? Give the key difference between a PAL and a PLA.

79. Redraw the circuit shown in Fig. 8–37(b) to illustrate how it can implement the logic function $F = \overline{(\overline{A}\,\overline{B} + AB)}$.

80. How many dedicated inputs, dedicated outputs, programmable input/outputs, and product terms are supported on the 16L8 PAL?

81. What is the maximum number of inputs on a 20L8 PAL? The maximum number of outputs?

82. How do the outputs of the 16R8 differ from those of the 16L8?

83. Use a 16L8 to decode address lines A_{17L} through A_{19L} to generate $\overline{CE_0}$ through $\overline{CE_7}$.

Section 8.14

84. Name the two types of input/output.

85. What type of I/O is in use when peripheral devices are mapped to the 8088's I/O address space?

86. Which type of I/O has the disadvantage that part of the address space must be given up to implement I/O ports?

87. Which type of I/O has the disadvantage that all I/O data transfers must take place through the AL or AX register?

Section 8.15

88. What are the functions of the 8088's address and data bus lines relative to an isolated I/O operation?

89. In a minimum-mode 8088 microcomputer, which signal indicates to external circuitry that the current bus cycle is for the I/O interface and not the memory interface?

90. List the differences between the 8088's minimum-mode I/O interface in Fig. 8–46(a) and that of the 8086 in Fig. 8–46(b).

91. What is the logic relationship between the signals $IO/\overline{M}$ and $M/\overline{IO}$?

92. In a maximum-mode system, which device produces the input (read), output (write), and bus control signals for the I/O interface?

93. Briefly describe the function of each block in the I/O interface circuit in Fig. 8–47(a).

94. In a maximum-mode 8086 microcomputer, what status code identifies an input bus cycle?

95. In the maximum-mode I/O interface shown in Fig. 8–47(a), what are the logic levels of $\overline{IORC}$, $\overline{IOWC}$, and $\overline{AIOWC}$ during an output bus cycle?

Section 8.16

96. How many bits are in the 8088's I/O address?

97. What is the range of byte addresses in the 8088's I/O address space?

98. What is the size of the 8086's I/O address space in terms of word-wide I/O ports?

99. In an 8086-based microcomputer system, what are the logic levels of A_0 and $\overline{BHE}$ when a byte of data is being written to I/O address $A000_{16}$? If a word of data is being written to address $A000_{16}$?

100. In an 8088 microcomputer system, how many bus cycles are required to output a word of data to I/O address $A000_{16}$? In an 8086 microcomputer system?

Section 8.17

101. Describe the operation performed by the instruction IN AX, 1AH.

102. Write an instruction sequence to perform the same operation as that of the instruction in problem 101, but this time use variable or indirect I/O.

103. Describe the operation performed by the instruction OUT 2AH, AL.

104. Write an instruction sequence that outputs the byte of data $0F_{16}$ to an output port at address 1000_{16}.

105. Write a sequence of instructions that inputs the byte of data from input ports at I/O addresses $A000_{16}$ and $B000_{16}$, adds these values together, and saves the sum in memory location IO_SUM.

106. Write a sequence of instructions that will input the contents of the input port at I/O address $B0_{16}$ and jump to the beginning of a service routine identified by the label ACTIVE_INPUT if the least significant bit of the data is 1.

Section 8.18

107. In the 8088's input bus cycle, during which T state do the $IO/\overline{M}$, ALE, $\overline{RD}$, and $\overline{DEN}$ control signals become active?

108. During which T state in the 8088's input bus cycle is the address output on the bus? Are data read from the bus by the MPU?

109. If an 8088 is running at 5 MHz, what is the duration of the output bus operation performed by executing the instruction OUT 0C0H, AX?

110. If an 8086 running at 10 MHz inserts two wait states into all I/O bus cycles, what is the duration of a bus cycle in which a byte of data is being output?

111. If the 8086 in problem 110 outputs a word of data to a word-wide port at I/O address $1A1_{16}$, what is the duration of the bus cycle?

9

Memory Devices, Circuits, and Subsystem Design

▲ INTRODUCTION

In the previous chapter, we began our study of the hardware architecture of the 8088- and 8086-based microcomputer systems. This included a study of the MPU's memory interface and the circuits needed to connect to a memory subsystem. We examined the memory interface signals, read and write bus cycles, hardware organization of the memory address space, and memory interface circuits. In this chapter, we continue our study of microcomputer hardware by examining the devices, circuits, and techniques used in the design of memory subsystems. For this purpose, this chapter explores the following topics:

9.1 Program and Data Storage Memory
9.2 Read-Only Memory
9.3 Random Access Read/Write Memories
9.4 Parity, the Parity Bit, and Parity-Checker/Generator Circuit
9.5 FLASH Memory
9.6 Wait-State Circuitry
9.7 8088/8086 Microcomputer System Memory Circuitry

▲ 9.1 PROGRAM AND DATA STORAGE MEMORY

Memory provides the ability to store and retrieve digital information and is one of the key elements of a microcomputer system. By digital information, we mean that instructions and data are encoded with 0s and 1s and then saved in memory. The ability to store information is made possible by the part of the microcomputer system known as the *memory unit.* In Chapter 1, we indicated that the memory unit of the microcomputer is partitioned into a *primary storage section* and *secondary storage section.* Figure 9–1 illustrates this subdivision of the memory unit.

Secondary storage memory is used for storage of data, information, and programs that are not in use. This part of the memory unit can be slow speed, but it requires very large storage capacity. For this reason, it is normally implemented with magnetic storage devices, such as the floppy disk and hard disk drive. Hard disk drives used in today's personal computers have the ability to store 10 gigabytes (Gbyte) to 80Gbytes of information.

The other part, primary storage memory, is used for working information, such as the instructions of the program currently being run and data that it is processing. This section normally requires high-speed operation but does not normally require very large storage capacity. Therefore, it is implemented with semiconductor memory devices. Most modern personal computers have 128 megabytes (128Mbytes) of primary storage memory.

Figure 9–1 shows that the primary storage memory is further partitioned into *program storage memory* and *data storage memory.* The program storage part of the memory subsystem is used to hold information such as the instructions of the program. That is, when a program is executed by the microcomputer, it is read one byte or word at a time from the program storage part of the memory subsystem. These programs can be either permanently stored in memory, which makes them always available for execution, or temporarily loaded into memory before execution. The program storage memory section does not normally contain only instructions, it can also store other fixed information such as constant data and lookup tables.

The program storage memory in a personal computer is implemented exactly this way. It has a fixed part of program memory that contains the *basic input/output system*

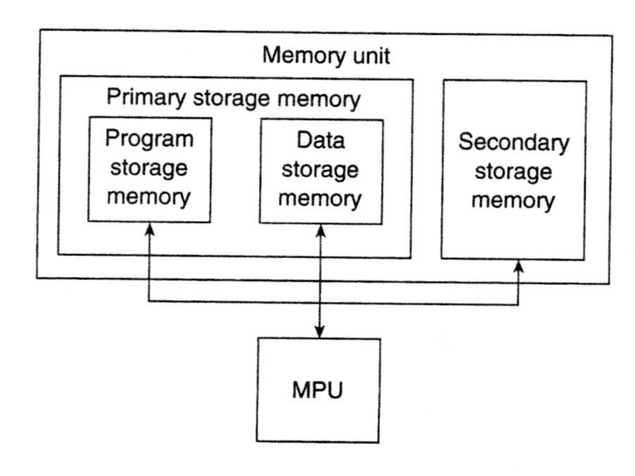

Figure 9–1 Partitioning of the microcomputer's memory unit.

(BIOS). These programs are permanently held in a read-only memory device mounted on the main processor board. Programs held this way in ROM are called *firmware* because of their permanent nature. The typically size of a BIOS ROM used in a PC today is 2 megabits (MB), which equal 256Kbytes.

The much larger part of the program storage memory in a PC is built with dynamic random access read/write memory devices (DRAMS). They may be either mounted on the main processor board or on an add-in memory module or board. Use of DRAMs allows this part of the program storage memory to be either read from or written into. Its purpose is again to store programs that are to be executed, but in this case they are loaded into memory only when needed. Programs are normally read in from the secondary storage device, stored in the program storage part of memory, and then run. When the program is terminated, the part of the program memory where it resides is given back to the operating system for reuse. Moreover, if power is turned off, the contents of the RAM-based part of the program storage memory are lost. Due to the temporary nature of these programs, they are referred to as software.

Earlier we indicated that the primary storage memory of a microcomputer is typically 128Mbytes. This number represented the total of the DRAM part of the memory subsystem and is given as the size of memory because the ROM BIOS is almost negligible when compared to the amount of DRAM. In the PC, a major part of primary storage is available for use as program storage memory.

In other microcomputer applications, such as an electronic game or telephone, the complete program storage memory is implemented with ROM devices.

Information that frequently changes is stored in the data storage part of the microcomputer's memory subsystem. For instance, the data to be processed by the microcomputer is held in the data storage part of the primary storage memory. When a program is run by the microcomputer, the values of the data can change repeatedly. For this reason, data storage memory must be implemented with RAM. In a PC, the data does not automatically reside in the data storage part of memory. Just like software, it is read into memory from a secondary storage device, such as the hard disk. Any part of the PCs DRAM can be assigned for data storage. The operating system software does this.

When a program is run, data are modified while in DRAM and writing them to the disk saves the new values. Data does not have to be numeric in form; they can also be alphanumeric characters, codes, and graphical patterns. For instance, when running a word processor application, the data are alphanumeric and graphical information.

▲ 9.2 READ-ONLY MEMORY

We begin our study of semiconductor memory devices with the *read-only memory* (ROM). ROM is one type of semiconductor memory device. It is most widely used in microcomputer systems for storage of the program that determines overall system operation. The information stored within a ROM integrated circuit is permanent—or *nonvolatile.* This means that when the power supply of the device is turned off, the stored information is not lost.

ROM, PROM, and EPROM

For some ROM devices, information (the microcomputer program) must be built in during manufacturing, and for others the data must be electrically entered. The process of entering the data into a ROM is called *programming.* As the name ROM implies, once entered into the device this information can be read only. For this reason, these devices are used primarily in applications where the stored information would not change frequently.

Three types of ROM devices are in wide use today: the *mask-programmable read-only memory* (ROM), the *one-time-programmable read-only memory* (PROM), and the *erasable programmable read-only memory* (EPROM). Let us continue by looking more closely into the first type of device, the mask-programmable read-only memory. This device has its data pattern programmed as part of the manufacturing process and is known as *mask programming.* Once the device is programmed, its contents can never be changed. Because of this fact and the cost for making the programming masks, ROMs are used mainly in high-volume applications where the data will not change.

The other two types of read-only memories, the PROM and EPROM, differ from the ROM in that the user electrically enters the bit pattern for the data. Programming is usually done with an instrument known as an *EPROM programmer.* Both the PROM and EPROM are programmed in the same way. Once a PROM is programmed, its contents cannot be changed. This is the reason they are sometimes called one-time programmable PROMs. On the other hand, exposing an EPROM to ultraviolet light erases the information it holds. That is, the programmed bit pattern is cleared out to restore the device to its unprogrammed state. In this way, the device can be used over and over again simply by erasing and reprogramming. PROMs and EPROMs are most often used during the design of a product, for early production, when the code of the microcomputer may need frequent changes, and for production in low-volume applications that do not warrant making a mask programmed device.

Figure 9–2(a) shows a typical EPROM programmer unit. Programming units like this usually have the ability to verify that an EPROM is erased, program it with new data, verify correct programming, and read the information out of a programmed EPROM. An erasing unit such as that shown in Fig. 9–2(b) can be used to erase a number of EPROM ICs at one time.

Block Diagram of a Read-Only Memory

Figure 9–3 shows a block diagram of a typical read-only memory. Here we see that the device has three sets of signal lines: the address inputs, data outputs, and control inputs. This block diagram is valid for a ROM, PROM, or EPROM. Let us now look at the function of each of these sets of signal lines.

The address bus is used to input the signals that select between the storage locations within the ROM device. In Fig. 9–3 we find that this bus consists of 11 address lines, A_0 through A_{10}. The bits in the address are arranged so that A_{10} is the MSB and A_0 is the LSB. With an 11-bit address, the memory device has $2^{11} = 2048$ unique byte-storage locations. The individual storage locations correspond to consecutive addresses over the range $00000000000_2 = 000_{16}$ through $11111111111_2 = 7FF_{16}$.

(a)

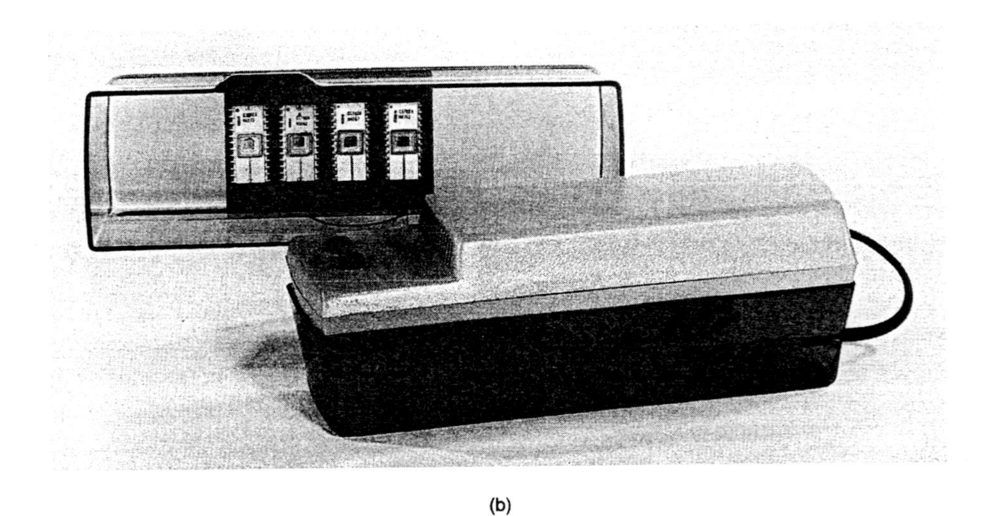

(b)

Figure 9–2 (a) EPROM programming unit. (Data I/O, Inc.) (b) EPROM erasing unit. (Ultra-Violet Products, Inc.)

Earlier we pointed out that information is stored inside a ROM, PROM, or EPROM as either a binary 0 or binary 1. Actually, 8 bits of data are stored at every address. Therefore, the organization of the ROM is described as 2048 × 8. The total storage capacity of the ROM is identified as the number of bits of information it can hold. We know 2048 bytes corresponds to 16,384 bits; therefore, the device we are describing is actually a 16K bit or 16KB ROM.

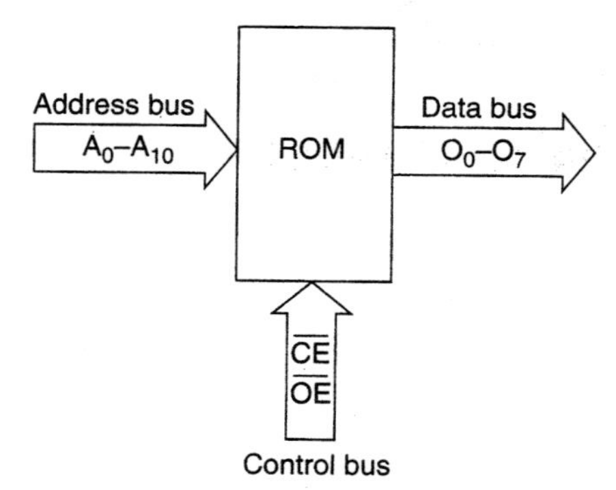

Figure 9–3 Block diagram of a ROM.

By applying the address of a storage location to the address inputs of the ROM, the byte of data held at the addressed location is read out onto the data lines. The block diagram in Fig. 9–3 shows that the data bus consists of eight lines labeled as O_0 through O_7. Here O_7 represents the MSB and O_0 the LSB. For instance, applying the address $A_{10} \ldots A_1A_0 = 10000000000_2 = 400_{16}$ will cause the byte of data held in this storage location to be output as $O_7O_6O_5O_4O_3O_2O_1O_0$.

EXAMPLE 9.1

Suppose the block diagram in Fig. 9–3 had 15 address lines and eight data lines. How many bytes of information can be stored in the ROM? What is its total storage capacity?

Solution

With 8 data lines, the number of bytes is equal to the number of locations, which is

$$2^{15} = 32{,}768 \text{ bytes}$$

This gives a total storage of

$$32{,}768 \times 8 = 262{,}144 \text{ bits}$$

The control bus represents the control signals required to enable or disable the ROM, PROM, or the EPROM device. The block diagram in Fig. 9–3 identifies two control inputs: output enable ($\overline{OE}$) and chip enable ($\overline{CE}$). For example, logic 0 at $\overline{OE}$ enables the three state outputs, O_0 through O_7, of the device. If $\overline{OE}$ is switched to the 1 logic level, these outputs are disabled (put in the high-Z state). Moreover, $\overline{CE}$ must be at logic 0 for the device to be active. Logic 1 at $\overline{CE}$ puts the device in a low-power standby mode. When in this state, the data outputs are in the high-Z state independent of the logic level of $\overline{OE}$. In this way we see that both $\overline{OE}$ and $\overline{CE}$ must be at their active 0 logic levels for the device to be ready for operation.

Read Operation

It is the role of the MPU and its memory interface circuitry to provide the address and control input signals and to read the output data at the appropriate times during the memory-read bus cycle. The block diagram in Fig. 9–4 shows a typical read-only memory interface. For a microprocessor to read a byte of data from the device, it must apply a binary address to inputs A_0 through A_{10} of the EPROM. This address gets decoded inside the device to select the storage location of the byte of data that is to be read. Remember that the microprocessor must switch $\overline{CE}$ and $\overline{OE}$ to logic 0 to enable the device and its outputs. Once done, the byte of data is made available at O_0 through O_7 and the microprocessor can read the data over its data bus.

Standard EPROM ICs

A large number of standard EPROM ICs are available today. Figure 9–5 lists the part numbers, bit densities, and byte capacities of nine popular devices. They range in size from the 16KB density (2K × 8) 2716 device, to the 4MB (512K × 8) 27C040 device. Higher-density devices, such as the 27C256 through 27C020, are now popular for system designs. In fact, many manufacturers have already discontinued some of the older devices, such as the 2716 and 2732. Let us now look at some of these EPROMs in more detail.

The 27C256 is an EPROM IC manufactured with the CMOS technology. Looking at Fig. 9–5, we find that it is a 256KB device, and its storage array is organized as 32K × 8 bits. Figure 9–6 shows the pin layout of the 27C256. Here we see that it has

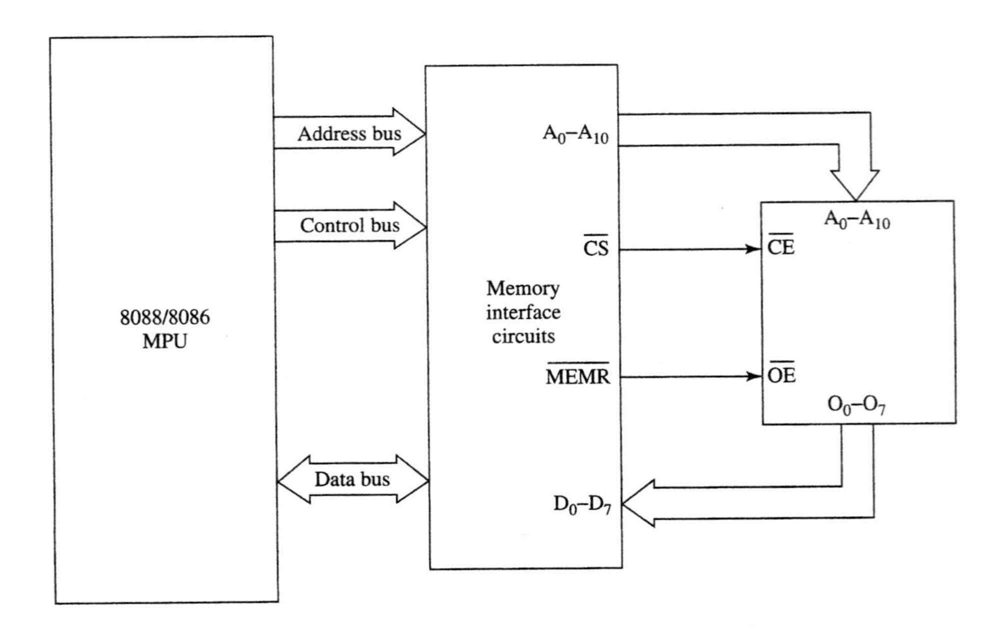

Figure 9–4 Read-only memory interface.

EPROM	Density (bits)	Capacity (bytes)
2716	16K	2K × 8
2732	32K	4K × 8
27C64	64K	8K × 8
27C128	128K	16K × 8
27C256	256K	32K × 8
27C512	512K	64K × 8
27C010	1M	128K × 8
27C020	2M	256K × 8
27C040	4M	512K × 8

Figure 9–5 Standard EPROM devices.

15 address inputs, labeled A_0 through A_{14}, eight data outputs, identified as O_0 through O_7, and two control signals $\overline{CE}$ and $\overline{OE}$.

From our earlier description of the read operation, it appears that after the inputs of the EPROM are set up, the output is available immediately; however, in practice this is not true. A short delay exists between address inputs and data outputs. This leads us to three important timing properties defined for the read cycle of an EPROM: *access time* (t_{ACC}), *chip-enable time,* (t_{CE}), and *chip-deselect time* (t_{DF}). The values of these timing properties are provided in the read-cycle switching characteristics shown in Fig. 9–7(a) and identified in the switching waveforms shown in Fig. 9–7(b).

Access time tells us how long it takes to access data stored in an EPROM. Here we assume that both $\overline{CE}$ and $\overline{OE}$ are already at their active 0 levels, and then the address is applied to the inputs of the EPROM. In this case, the delay t_{ACC} occurs before the data stored at the addressed location are stable at the outputs. The microprocessor must wait at least this long before reading the data; otherwise, invalid results may be obtained. Figure 9–7(a) shows that the standard EPROMs are available with a variety of access time ratings. The maximum values of access time are given as 170 ns, 200 ns, and 250 ns. The speed of the device is selected to match that of the MPU. If the access time of the fastest standard device is too long for the MPU, wait-state circuitry needs to be added to the interface. In this way, wait states can be inserted to slow down the memory read bus cycle.

Chip-enable time is similar to access time. In fact, for most EPROMs they are equal in value. They differ in how the device is set up initially. This time the address is applied and $\overline{OE}$ is switched to 0, then the read operation is initiated by making $\overline{CE}$ active. Therefore, t_{CE} represents the chip-enable-to-output delay instead of the address-to-output delay. Looking at Fig. 9–7(a), we see that the maximum values of t_{CE} are also 170 ns, 200 ns, and 250 ns.

Chip-deselect time is the opposite of access or chip-enable time. It represents the amount of time the device takes for the data outputs to return to the high-Z state after $\overline{OE}$ becomes inactive—that is, the recovery time of the outputs. Figure 9–7(a) shows that the maximum values for this timing property are 55 ns, 55 ns, and 60 ns.

In an erased EPROM, all storage cells hold logic 1. The device is put into the programming mode by switching on the V_{pp} power supply. Once in this mode, the address of the storage location to be programmed is applied to the address inputs, and the byte of

27C512	27C128	27C64	2732A	2716	Pin
A_{15}	V_{PP}	V_{PP}			1
A_{12}	A_{12}	A_{12}			2
A_7	A_7	A_7	A_7	A_7	3
A_6	A_6	A_6	A_6	A_6	4
A_5	A_5	A_5	A_5	A_5	5
A_4	A_4	A_4	A_4	A_4	6
A_3	A_3	A_3	A_3	A_3	7
A_2	A_2	A_2	A_2	A_2	8
A_1	A_1	A_1	A_1	A_1	9
A_0	A_0	A_0	A_0	A_0	10
O_0	O_0	O_0	O_0	O_0	11
O_1	O_1	O_1	O_1	O_1	12
O_2	O_2	O_2	O_2	O_2	13
Gnd	Gnd	Gnd	Gnd	Gnd	14

27C256

Signal	Pin	Pin	Signal
V_{PP}	1	28	V_{CC}
A_{12}	2	27	A_{14}
A_7	3	26	A_{13}
A_6	4	25	A_8
A_5	5	24	A_9
A_4	6	23	A_{11}
A_3	7	22	$\overline{OE}$
A_2	8	21	A_{10}
A_1	9	20	$\overline{CE}$
A_0	10	19	O_7
O_0	11	18	O_6
O_1	12	17	O_5
O_2	13	16	O_4
Gnd	14	15	O_3

Pin	2716	2732A	27C64	27C128	27C512
28			V_{CC}	V_{CC}	V_{CC}
27			$\overline{PGM}$	$\overline{PGM}$	A_{14}
26	V_{CC}	V_{CC}	N.C.	A_{13}	A_{13}
25	A_8	A_8	A_8	A_8	A_8
24	A_9	A_9	A_9	A_9	A_9
23	V_{PP}	A_{11}	A_{11}	A_{11}	A_{11}
22	$\overline{OE}$	$\overline{OE}/V_{PP}$	$\overline{OE}$	$\overline{OE}$	$\overline{OE}/V_{PP}$
21	A_{10}	A_{10}	A_{10}	A_{10}	A_{10}
20	$\overline{CE}$	$\overline{CE}$	$\overline{CE}$	$\overline{CE}$	$\overline{CE}$
19	O_7	O_7	O_7	O_7	O_7
18	O_6	O_6	O_6	O_6	O_6
17	O_5	O_5	O_5	O_5	O_5
16	O_4	O_4	O_4	O_4	O_4
15	O_3	O_3	O_3	O_3	O_3

Figure 9–6 Pin layouts of standard EPROMs.

Versions		$V_{CC} \pm 5\%$	27C256-120V05		27C256-135V05		27C256-150V05		27C256-1 P27C256-1 N27C256-1		27C256-2 P27C256-2 N27C256-2		27C256 P27C256 N27C256		Unit
		$V_{CC} \pm 10\%$			27C256-135V10		27C256-150V10				27C256-20 P27C256-20 N27C256-20		27C256-25 P27C256-25 N27C256-25		
Symbol	**Parameter**		**Min**	**Max**	**Min**	**Max**	**Min**	**Max**	**Min**	**Max**	**Min**	**Max**	**Min**	**Max**	
t_{ACC}	Address to output delay			120		135		150		170		200		250	ns
t_{CE}	$\overline{CE}$ to output delay			120		135		150		170		200		250	ns
t_{OE}	$\overline{OE}$ to output delay			60		65		70		75		75		100	ns
$t_{DF}^{(2)}$	$\overline{OE}$ high to output high-Z			30		35		45		55		55		60	ns
$t_{OH}^{(2)}$	Output hold from addresses, $\overline{CE}$ or $\overline{OE}$ change—whichever is first		0		0		0		0		0		0		ns

Notes:

1. A.C. characteristics tested at $V_{IH} = 2.4$ V and $V_{IL} = 0.45$ V.
   Timing measurements made at $V_{OL} = 0.8$ V and $V_{OH} = 2.0$ V.
2. Guaranteed and sampled.
3. Package Prefixes: No Prefix = CERDIP; N = PLCC; P = Plastic DIP.

(a)

Figure 9–7 (a) EPROM device timing characteristics. (Reprinted by permission of Intel Corporation; Copyright Intel Corp. 1989) (b) EPROM switching waveforms. (Reprinted by permission of Intel Corporation; Copyright Intel Corp. 1989)

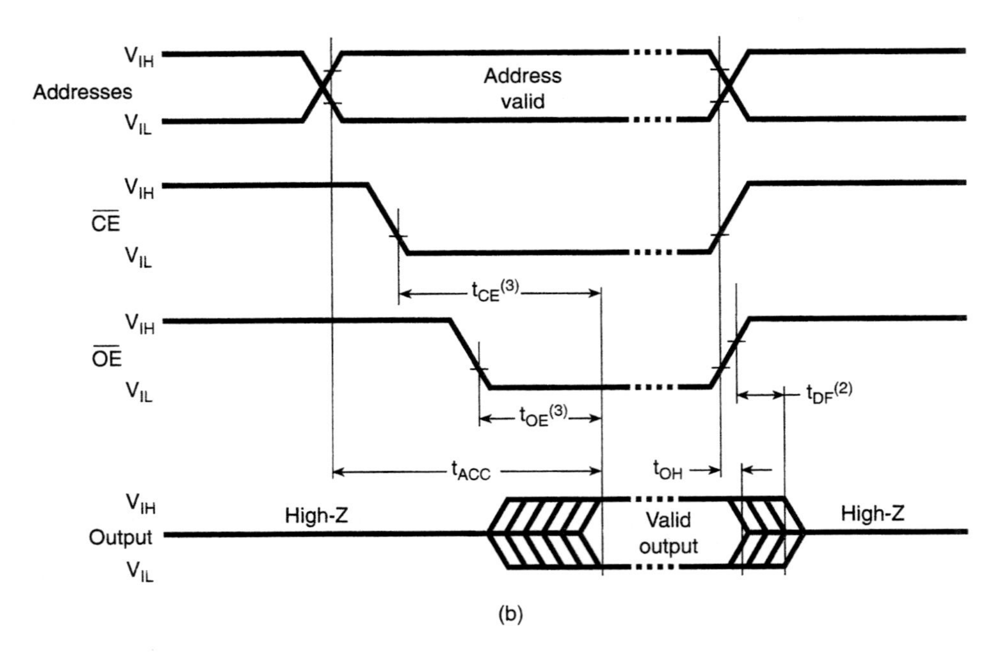

Figure 9–7 (continued)

data to be loaded into this location is supplied as inputs to the data leads. Note that the data outputs act as inputs when the EPROM is set up for programming mode of operation. Next the $\overline{CE}$ input is pulsed to load the data. Actually, a complex series of program and verify operations are performed to program each storage location in an EPROM. The two widely used programming sequences are the *Quick-Pulse Programming Algorithm* and the *Intelligent Programming Algorithm.* Flowcharts for these programming algorithms are given in Figs. 9–8(a) and (b), respectively.

Figure 9–9 presents another group of important electrical characteristics for the 27C256 EPROM. They are the device's dc electrical operating characteristics. CMOS EPROMs are designed to provide TTL-compatible input and output logic levels. Here we find the output logic level ratings are $V_{OHmin} = 3.5$ V and $V_{OLmax} = 0.45$ V. Also provided is the operating current rating of the device, identified as $I_{CC} = 30$ mA. This shows that if the device is operating at 5 V, it will consume 150 mW of power.

Figure 9–6 also shows the pin layouts for the 2716 through 27C512 EPROM devices. In this diagram, we find that both the 27C256 and 27C512 are available in a 28-pin package. A comparison of the pin configuration of the 27C512 with that of the 27C256 shows that the only differences between the two pinouts are that pin 1 on the 27C512 becomes the new address input A_{15}, and V_{pp}, which was at pin 1 on the 27C256, becomes a second function performed by pin 22 on the 27C512.

Expanding EPROM Word Length and Word Capacity

In many applications, the microcomputer system requirements for EPROM are greater than what is available in a single device. There are two basic reasons for expand-

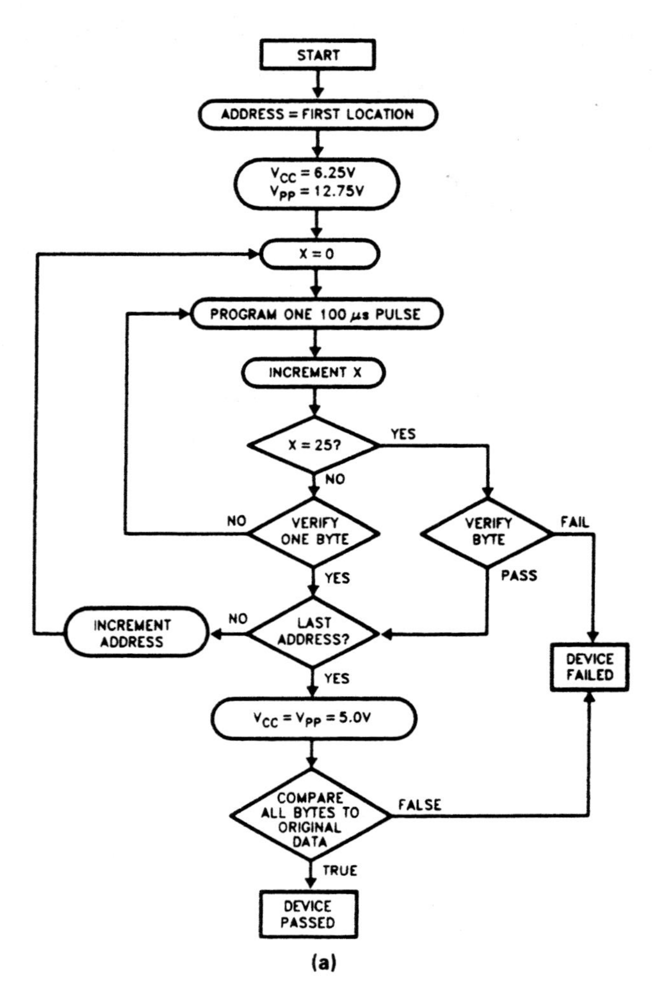

Figure 9–8 (a) Quick-Pulse Programming Algorithm flowchart. (Reprinted by permission of Intel Corporation; Copyright Intel Corp. 1989) (b) Intelligent Programming Algorithm flowchart. (Reprinted by permission of Intel Corporation; Copyright Intel Corp. 1989)

ing EPROM capacity: first, the byte-wide length is not large enough; and second, the total storage capacity is not enough bytes. Both of these expansion needs can be satisfied by interconnecting a number of ICs.

For instance, the 8086 microprocessor has a 16-bit data bus. Therefore, its program memory subsystem needs to be implemented with two 27C256 EPROMs connected, as shown in Fig. 9–10(a). Notice that the individual address inputs, chip enable lines, and output enable lines on the two devices are connected in parallel. On the other hand, the eight data outputs of each device are used to supply eight lines of the MPU's 16-bit data bus. This circuit configuration has a total storage capacity equal to 32K words or 512KB.

Figure 9–10(b) shows how two 27C256s can be interconnected to expand the number of bytes of storage. Here the individual address inputs, data outputs, and output enable lines of the two devices are connected in parallel. However, the $\overline{CE}$ inputs of the individual devices remain independent and can be supplied by different chip enable outputs,

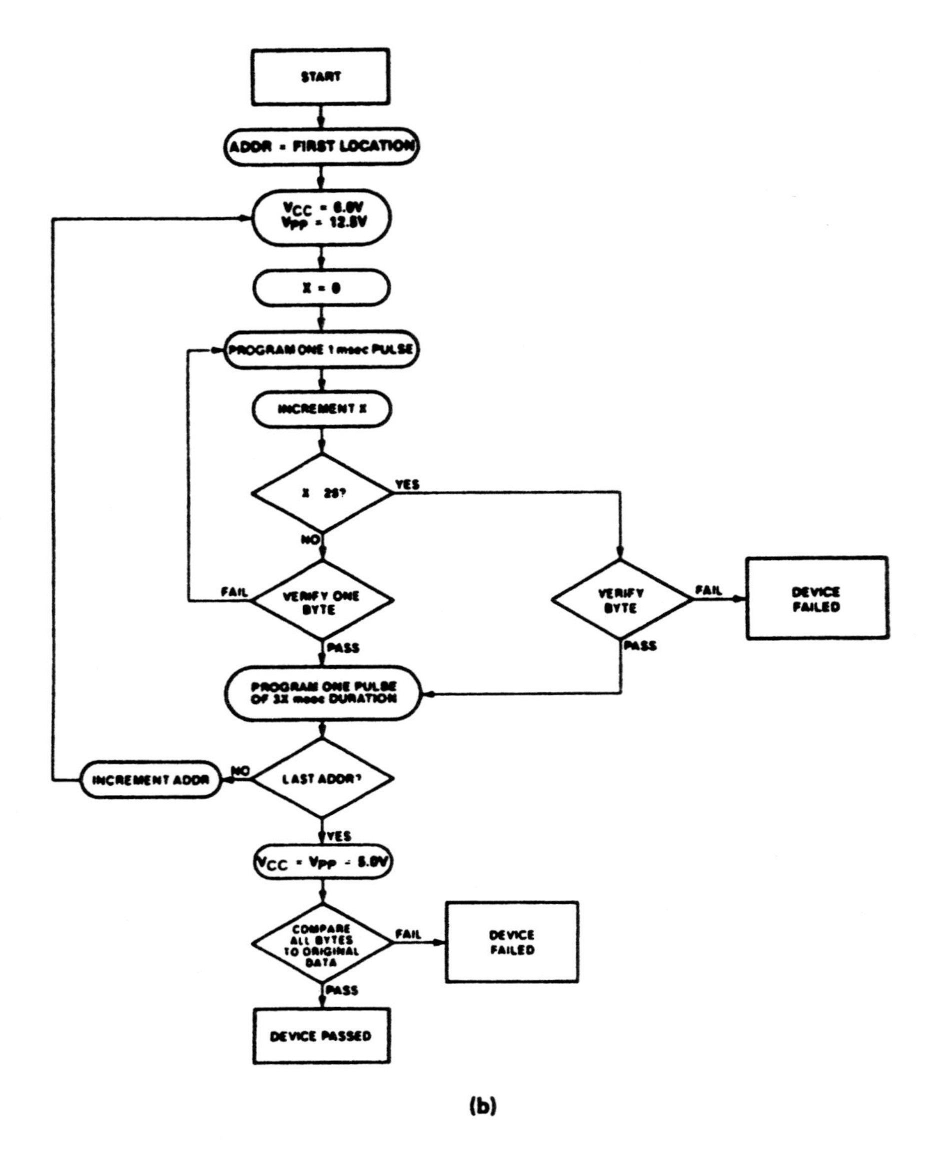

(b)

Figure 9–8 (continued)

identified as $\overline{CS}_0$ and $\overline{CS}_1$, of an address decoder circuit. In this way, only one of the two devices is enabled at one time. This configuration results in a total storage capacity of 64Kbytes or 512KB. When several EPROMs are used in an 8088-based microcomputer, they are connected in this way. To double the word capacity of the circuit in Fig. 9–10(a), this same connection must be made for each of the EPROMs.

Symbol	Parameter		Notes	Min	Typ(3)	Max	Unit	Test Conditions
I_{LI}	Input load current				0.01	1.0	μA	V_{IN} = 0V to V_{CC}
I_{LO}	Output leakage current					± 10	μA	V_{OUT} = 0V to V_{CC}
I_{PP_1}	V_{PP} read current		5			200	μA	$V_{PP} = V_{CC}$
I_{SB_1}	V_{CC} current standby	TTL	8			1.0	mA	$\overline{CE} = V_{IH}$
I_{SB_2}		CMOS	4			100	μA	$\overline{CE} = V_{CC}$
I_{CC_1}	V_{CC} current active		5, 8			30	mA	$\overline{CE} = V_{IL}$ f = 5 MHz
V_{IL}	Input low voltage (± 10% supply) (TTL)			–0.5		0.8	V	
	Input low voltage (CMOS)			–0.2		0.8		
V_{IH}	Input high voltage (± 10% supply) (TTL)			2.0		V_{CC} + 0.5	V	
	Input high voltage (CMOS)			0.7 V_{CC}		V_{CC} + 0.2		
V_{OL}	Output low voltage					0.45	V	I_{OL} = 2.1 mA
V_{OH}	Output high voltage			3.5			V	I_{OH} = –2.5 mA
I_{OS}	Output short circuit current		6			100	mA	
V_{PP}	V_{PP} read voltage		7	V_{CC} – 0.7		V_{CC}	V	

Notes:

1. Minimum D.C. input voltage is –0.5V. During transitions, the inputs may undershoot to –2.0V for periods less than 20 ns. Maximum D.C. voltage on output pins is V_{CC} + 0.5V which may overshoot to V_{CC} + 2V for periods less than 20 ns.
2. Operating temperature is for commercial product defined by this specification. Extended temperature options are available in EXPRESS and Military version.
3. Typical limits are at V_{CC} = 5V, T_A = +25°C.
4. $\overline{CE}$ is V_{CC} ± 0.2V. All other inputs can have any value within spec.
5. Maximum Active power usage is the sum $I_{PP} + I_{CC}$. The maximum current value is with outputs O_0 to O_7 unloaded.
6. Output shorted for no more than one second. No more than one output shorted at a time. I_{OS} is sampled but not 100% tested.
7. V_{PP} may be one diode voltage drop below V_{CC}. It may be connected directly to V_{CC}. Also, V_{CC} must be applied simultaneously or before V_{PP} and removed simultaneously or after V_{PP}.
8. V_{IL}, V_{IH} levels at TTL inputs.

Figure 9–9 DC electrical characteristics of the 27C256. (Reprinted by permission of Intel Corporation; Copyright Intel Corp. 1989)

▲ 9.3 RANDOM ACCESS READ/WRITE MEMORIES

The memory section of a microcomputer system is normally formed from both read-only memories and *random access read/write memories* (RAM). Earlier we pointed out that the ROM is used to store permanent information such as the microcomputer's hardware control program. RAM is similar to ROM in that its storage location can be accessed in a random order, but it is different from ROM in two important ways. First, data stored in

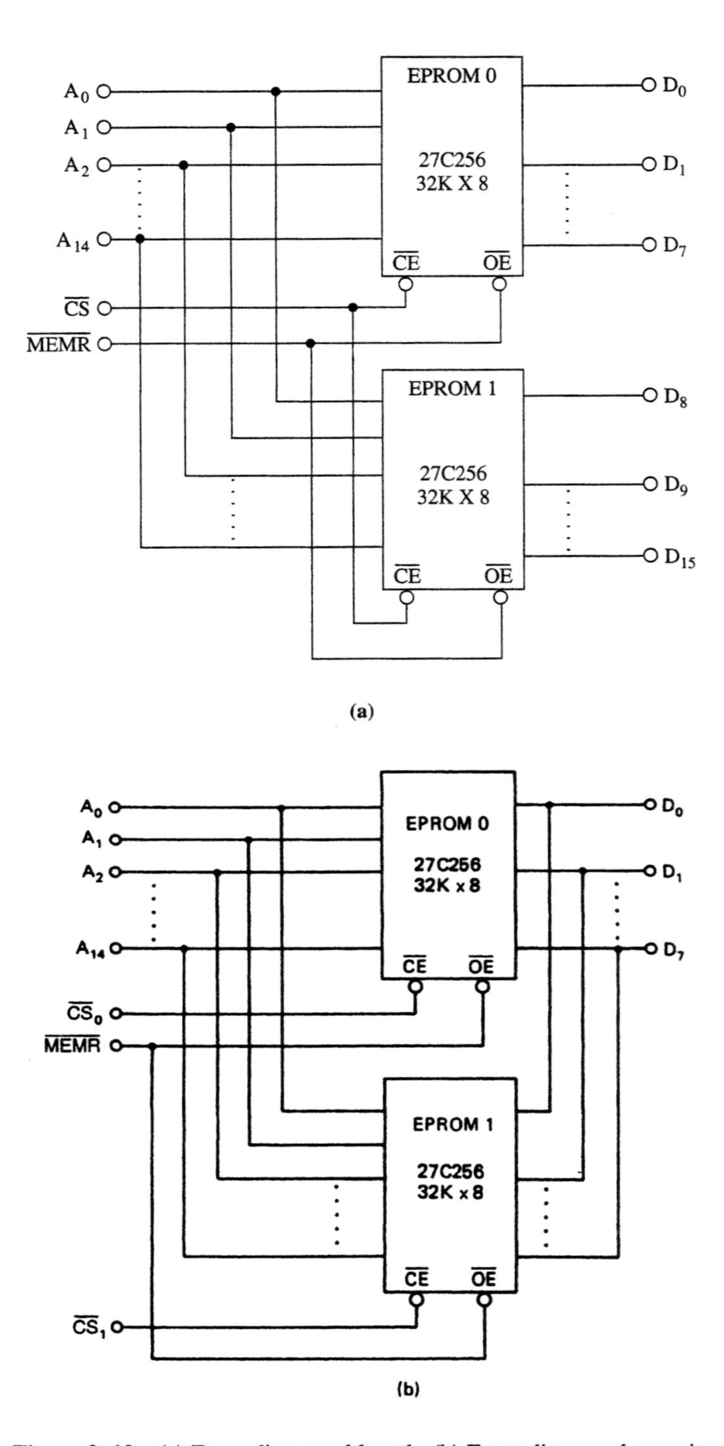

Figure 9–10 (a) Expanding word length. (b) Expanding word capacity.

RAM is not permanent in nature—that is, it can be altered. RAM can be used to save data by writing to it, and later the data can be read back for additional processing. Because of its read and write features, RAM finds wide use where data and programs need to be placed in memory only temporarily. For this reason, it is normally used to store data and application programs for execution. The second difference is that RAM is *volatile*—that is, if power is removed from RAM, the stored data are lost.

Static and Dynamic RAMs

Two types of RAMs are in wide use today: the *static RAM* (SRAM) and *dynamic RAM* (DRAM). For a static RAM, data, once entered, remain valid as long as the power supply is not turned off. On the other hand, to retain data in a DRAM, it is not sufficient just to maintain the power supply. For this type of device, we must both keep the power supply turned on and periodically restore the data in each storage location. This added requirement is necessary because the storage elements in a DRAM are capacitive nodes. If the storage nodes are not recharged within a specific interval of time, data are lost. This recharging process is known as *refreshing* the DRAM.

Block Diagram of a Static RAM

Figure 9–11 shows a block diagram of a typical static RAM IC. By comparing this diagram with the one shown for a ROM in Fig. 9–3, we see that they are similar in many ways. For example, they both have address lines, data lines, and control lines. These signal buses perform similar functions when the RAM is operated. Because of the RAMs read/write capability, data lines, however, act as both inputs and outputs. For this reason, they are identified as a bidirectional bus.

A variety of static RAM ICs are currently available. They differ both in density and organization. The most commonly used densities in system designs are the 64KB and 256KB devices. The structure of the data bus determines the organization of the SRAMs storage array. Figure 9–11 shows an 8-bit data bus. This type of organization is known as a *byte-wide* SRAM. Devices are also manufactured with by 1 and by 4 data I/O organizations. The 64KB density results in three standard device organizations: 64K × 1, 16K × 4, and 8K × 8.

The address bus on the SRAM in Fig. 9–11 consists of the lines labeled A_0 through A_{12}. This 13-bit address is what is needed to select between the 8K individual storage locations in an 8K × 8-bit SRAM IC. The 16K × 4 and 64K × 1 devices require a 14-bit and 16-bit address, respectively.

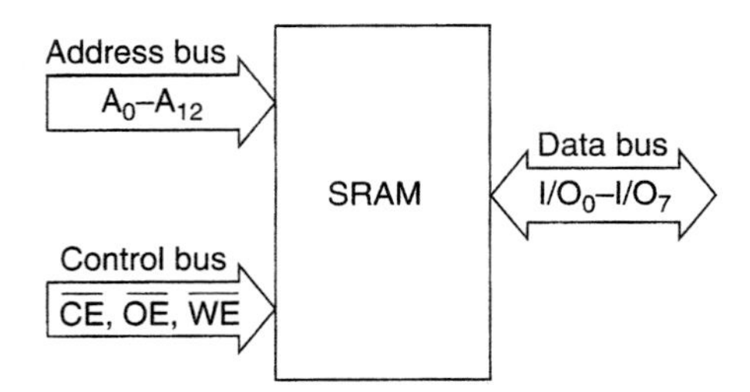

Figure 9–11 Block diagram of a static RAM.

To either read from or write to SRAM, the device must first be chip enabled. Just like for a ROM, this is done by switching the $\overline{CE}$ input of the SRAM to logic 0. Earlier we indicated that data lines I/O_0 through I/O_7 in Fig. 9–11 are bidirectional. This means that they act as inputs when writing data into the SRAM or as outputs when reading data from the SRAM. The setting of a new control signal, the *write enable* ($\overline{WE}$) input, determines how the data lines operate. During all write operations to a storage location within the SRAM, the appropriate $\overline{WE}$ inputs must be switched to the 0 logic level. This configures the data lines as inputs. On the other hand, if data are to be read from a storage location, $\overline{WE}$ is left at the 1 logic level. When reading data from the SRAM, output enable ($\overline{OE}$) must be active. Applying the active memory signal, logic 0, at this input, enables the device's three-state outputs.

A Static RAM System

Three-state data bus lines of SRAM devices allow for the parallel connection needed to expand data memory using multiple devices. For example, Fig. 9–12 shows how four 8K × 8-bit SRAMs are interconnected to form a 16K × 16-bit memory system. In this circuit, the separate $\overline{CE}$ inputs of the SRAM ICs in bank 0 are wired together and connected to a common chip-select input $\overline{CS_0}$. The same type of connection is used for the SRAMs in bank 1 using chip-select input $\overline{CS_1}$. These inputs are activated by the chip-select output of the address decoder circuit and must be logic 0 to select a bank of SRAMs for operation. The $\overline{OE}$ inputs of the individual SRAMs are connected in parallel. The combined output-enable input that results is driven by the $\overline{MEMR}$ output of the memory interface circuit and enables the outputs of all SRAMs during all memory-read bus cycles. Similarly, the write enables of all SRAMs are supplied from $\overline{MEMW}$ to write to the selected bank. Note that the memory system allows only word writes and reads.

Standard Static RAM ICs

Figure 9–13 lists a number of standard static RAM ICs. Here we find their part numbers, densities, and organizations. For example, the 4361, 4363, and 4364 are all 64KB density devices; however, they are each organized differently. The 4361 is a 64K × 1-bit device, the 4363 is a 16K × 4-bit device, and the 4364 is an 8K × 8-bit device.

The pin layouts of the 4364 and 43256A ICs are given in Figs. 9–14(a) and (b), respectively. Looking at the 4364 we see that it is almost identical to the block diagram shown in Fig. 9–11. The one difference is that it has two chip-enable lines instead of one. They are labeled $\overline{CE_1}$ and CE_2. Note that logic 0 activates one, and logic 1 activates the other. Both of these chip-enable inputs must be at their active logic levels to enable the device for operation.

EXAMPLE 9.2

How does the 43256A SRAM differ from the block diagram in Fig. 9–11?

Solution

It has two additional address inputs, A_{13} and A_{14}, and the chip-enable input is labeled $\overline{CS}$ instead of $\overline{CE}$.

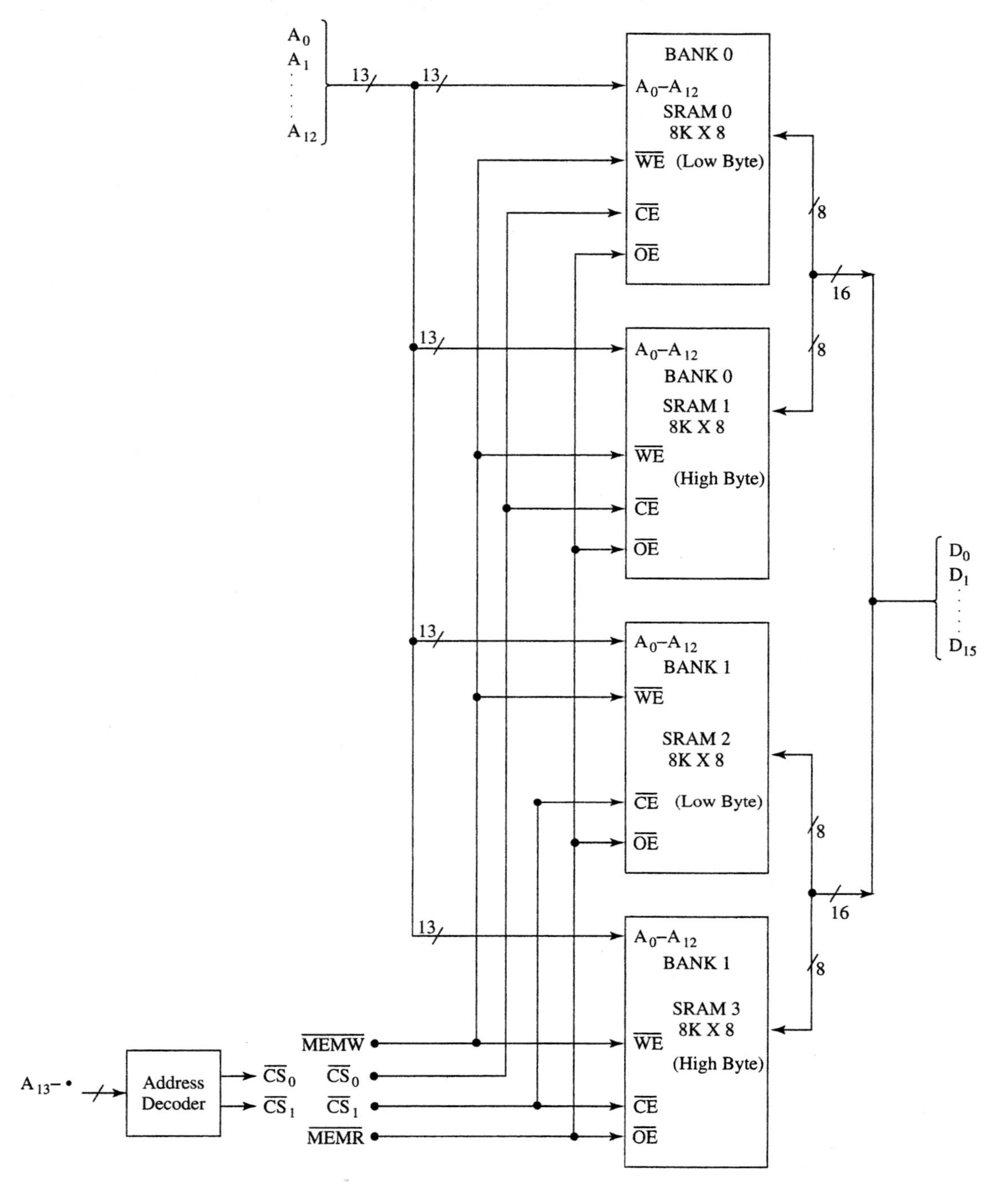

Figure 9–12 16K × 16-bit SRAM circuit.

SRAM	Density (bits)	Organization
4361	64K	64K × 1
4363	64K	16K × 4
4364	64K	8K × 8
43254	256K	64K × 4
43256A	256K	32K × 8
431000A	1M	128K × 8

Figure 9–13 Standard SRAM devices.

As Fig. 9–15 shows, the 4364 is available in four speeds. For example, the minimum read cycle and write cycle time for the 4364-10 is 100 ns. Figure 9–16 is a list of the 4364's dc electrical characteristics. Note (1) shows that the 100 ns device draws a maximum of 45 mA when operating at maximum frequency (minimum cycle time).

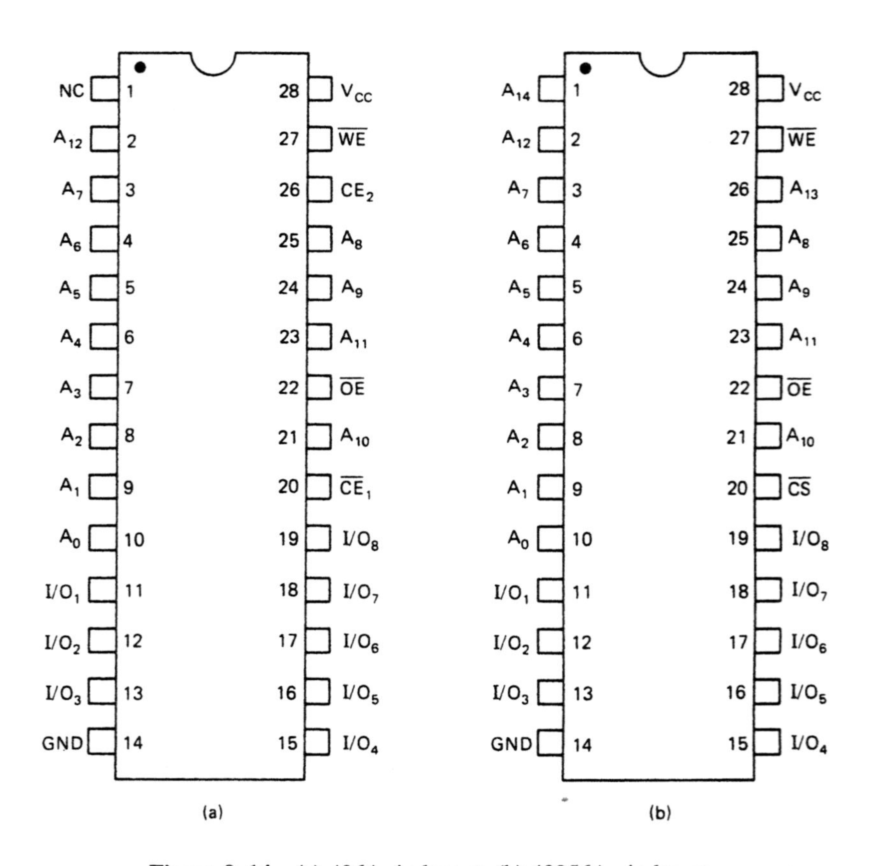

Figure 9–14 (a) 4364 pin layout. (b) 43256A pin layout.

Part number	Read/write cycle time
4364-10	100 ns
4364-12	120 ns
4364-15	150 ns
4364-20	200 ns

Figure 9–15 Speed selections for the 4364 SRAM.

SRAM Read and Write Cycle Operation

Figure 9–17 illustrates the waveforms for a typical write cycle. Let us trace the events that take place during the write cycle. Here we see that all critical timing is referenced to the point at which the address becomes valid. Note that the minimum duration of the write cycle is identified as t_{WC}. This is the 100-ns *write cycle time* of the 4364-10. The address must remain stable for this complete interval of time.

Next, $\overline{CE}_1$ and CE_2 become active and must remain active until the end of the write cycle. The duration of these pulses are identified as *$\overline{CE}_1$ to end of write time* (t_{CW1}) and *CE_2 to end of write time* (t_{CW2}). As the waveforms show, we are assuming here that they begin at any time after the occurrence of the address but before the leading edge of $\overline{WE}$. The minimum value for both of these times is 80 ns. On the other hand, $\overline{WE}$ is shown not to occur until the interval t_{AS} elapses. This is the *address-setup time* and represents the minimum amount of time the address inputs must be stable before $\overline{WE}$ can be switched to logic 0. For the 4364, however, this parameter is equal to 0 ns. The width of the write enable pulse is identified as t_{WP}, and its minimum value equals 60 ns.

Data applied to the D_{IN} data inputs are written into the device synchronous with the trailing edge of $\overline{WE}$. Note that the data must be valid for an interval equal to t_{DW} before this edge. This interval, called *data valid to end of write,* has a minimum value of 40 ns for the 4364-10. Moreover, it is shown to remain valid for an interval of time equal to t_{DH} after this edge. This *data-hold time,* however, just like address-setup time, equals 0 ns for the 4364. Finally, a short recovery period takes place after $\overline{WE}$ returns to logic 1 before the write cycle is complete. This interval is identified as t_{WR} in the waveforms, and its minimum value equals 5 ns.

The read cycle of a static RAM, such as the 4364, is similar to that of a ROM. Figure 9–18 gives waveforms of a read operation.

Standard Dynamic RAM ICs

Dynamic RAMs are available in higher densities than static RAMs. Currently, the most widely used DRAMs are the 64K-bit, 256K-bit, 1M-bit, and 4M-bit devices. Figure 9–19 lists a number of popular DRAM ICs. Here we find the 2164B, organized as 64K × 1 bit; the 21256, organized as 256K × 1 bit; the 21464, organized as 64K × 4 bits; the 421000, organized as 1M × 1 bit; and the 424256, organized as 256K × 4 bits. Pin layouts for the 2164B, 21256, and 421000 are shown in Figs. 9–20(a), (b), and (c), respectively.

Some other benefits of using DRAMs over SRAMs are that they cost less, consume less power, and their 16- and 18-pin packages take up less space. For these reasons, DRAMs are normally used in applications that require a large amount of memory. For example, most systems that support at least 1Mbyte of data memory are designed using DRAMs.

Parameter	Symbol	Limits Min	Typ	Max	Unit	Test Conditions
Input leakage current	I_{LI}			1	μA	V_{IN} = 0 V to V_{CC}
Output leakage current	I_{LO}			1	μA	$V_{I/O}$ = 0 V to V_{CC} $\overline{CE}_1 = V_{IH}$ or $CE_2 = V_{IL}$ or $\overline{OE} = V_{IH}$ or $\overline{WE} = V_{IL}$
Operating supply current	I_{CCA1}			(1)	mA	$\overline{CE}_1 = V_{IL}$, $CE_2 = V_{IH}$, $I_{I/O} = 0$, Min cycle
	I_{CCA2}		5	10	mA	$\overline{CE}_1 = V_{IL}$, $CE_2 = V_{IH}$, $I_{I/O} = 0$, DC current
	I_{CCA3}		3	5	mA	$\overline{CE}_1 \leq 0.2$ V, $CE_2 \geq V_{CC} - 0.2$ V, $V_{IL} \leq 0.2$ V, $V_{IH} \geq V_{CC} - 0.2$ V, f = 1 MHz, $I_{I/O} = 0$
Standby supply current	I_{SB}			(2)	mA	$\overline{CE}_1 \geq V_{IH}$ or $CE_2 = V_{IL}$
	I_{SB1}			(3)	mA	$\overline{CE}_1 \geq V_{CC}$–0.2 V $CE_2 \geq V_{CC}$–0.2 V
	I_{SB2}			(3)	mA	$CE_2 \leq 0.2$ V
Output voltage, low	V_{OL}			0.4	V	I_{OL} = 2.1 mA
Output voltage, high	V_{OH}	2.4			V	I_{OH} = –1.0 mA

Notes:

(1) μPD4364-10/10L: 45 mA max
μPD4364-12/12L/12LL: 40 mA max
μPD4364-15/15L/15LL: 40 mA max
μPD4364-20/20L/20LL: 35 mA max

(2) μPD4364-xx: 5 mA max
μPD4364-xxL: 3 mA max
μPD4364-xxLL: 3 mA max

(3) μPD4364-xx: 2 mA max
μPD4364-xxL: 100 μA max
μPD4364-xxLL: 50 μA max

Figure 9–16 DC electrical characteristics of the 4364.

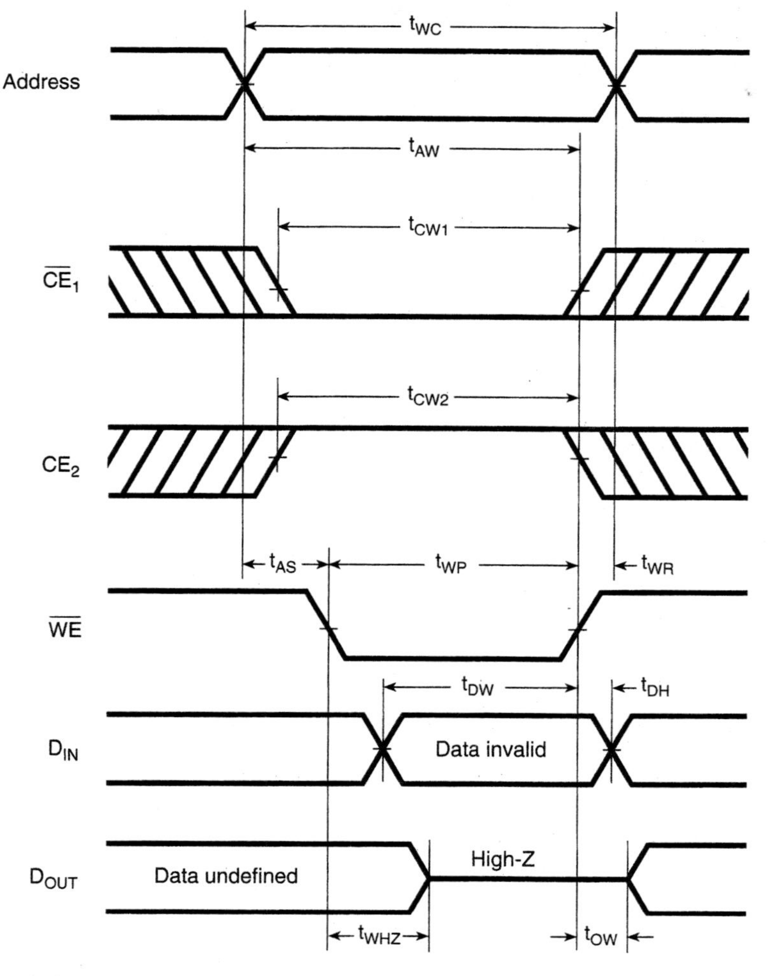

Notes:

1. A write occurs during the overlap of a low $\overline{CE}_1$ and a high CE_2 and a low $\overline{WE}$.
2. $\overline{CE}_1$ or $\overline{WE}$ [or CE_2] must be high [low] during any address transaction.
3. If $\overline{OE}$ is high the I/O pins remain in a high-impedance state.

Figure 9–17 Write-cycle timing diagram.

The 2164B is one of the older NMOS DRAM devices. Figure 9–21 presents a block diagram of the device. Looking at the block diagram, we find that it has eight address inputs, A_0 through A_7, a data input and a data output marked D and Q, respectively, and three control inputs, *row-address strobe* ($\overline{RAS}$), *column-address strobe* ($\overline{CAS}$), and *read/write* ($\overline{W}$).

The storage array within the 2164B is capable of storing 65,536 (64K) individual bits of data. To address this many storage locations, we need a 16-bit address; however,

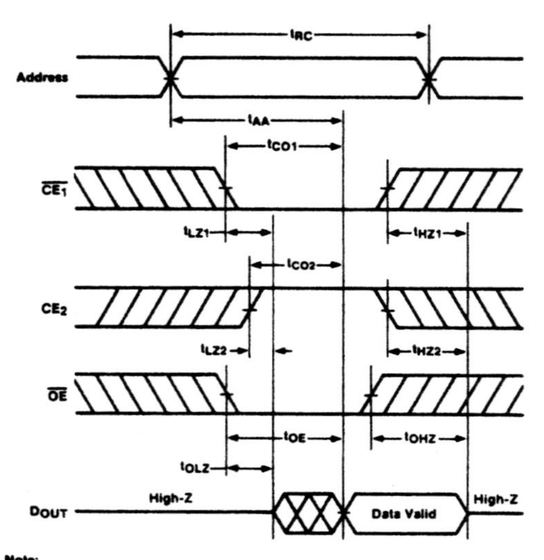

Figure 9–18 Read-cycle timing diagram.

this device's package has just 16 pins. For this reason, the 16-bit address is divided into two separate parts: an 8-bit *row address* and an 8-bit *column address*. These two parts of the address are time-multiplexed into the device over a single set of address lines, A_0 through A_7. First the row address is applied to A_0 through A_7. Then $\overline{RAS}$ is pulsed to logic 0 to latch it into the device. Next, the column address is applied and strobes $\overline{CAS}$ to logic 0. This 16-bit address selects which one of the 64K storage locations is to be accessed.

Data are either written into or read from the addressed storage location in DRAMs. Write data are applied to the D input and read data are output at Q. The logic levels of control signals $\overline{W}$, $\overline{RAS}$, and $\overline{CAS}$ tell the DRAM whether a read or write data transfer is taking place and control the three-state outputs. For example, during a write operation, the logic level at D is latched into the addressed storage location at the falling edge of either $\overline{CAS}$ or $\overline{W}$. If $\overline{W}$ is switched to logic 0 by an active $\overline{MWTC}$ signal before $\overline{CAS}$, an

DRAM	Density (bits)	Organization
2164B	64K	64K × 1
21256	256K	256K × 1
21464	256K	64K × 4
421000	1M	1M × 1
424256	1M	256K × 4
44100	4M	4M × 1
44400	4M	1M × 4
44160	4M	256K × 16
416800	16M	8M × 2
416400	16M	4M × 4
416160	16M	1M × 16

Figure 9–19 Standard DRAM devices.

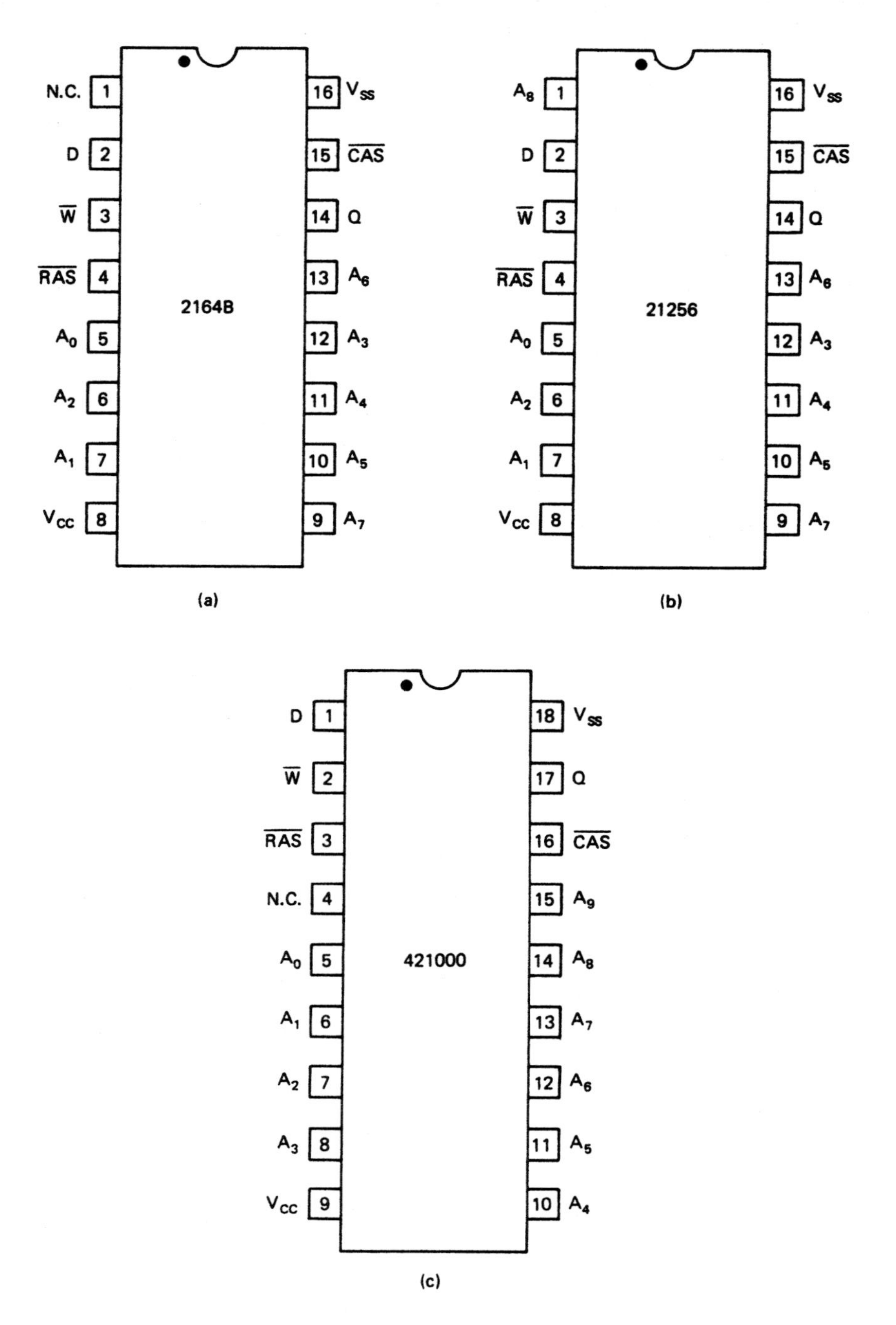

Figure 9–20 (a) 2164B pin layout. (b) 21256 pin layout. (c) 421000 pin layout.

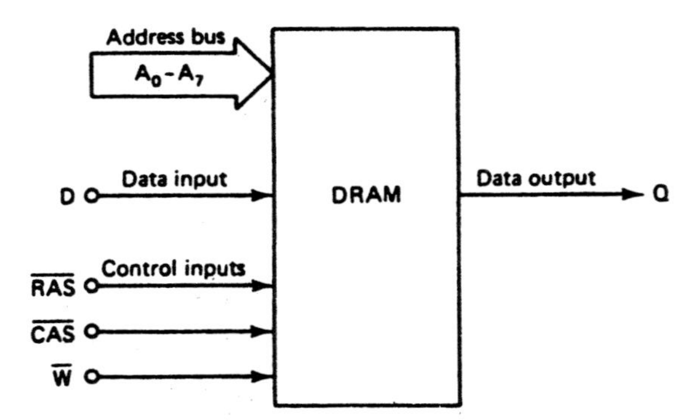

Figure 9–21 Block diagram of the 2164B DRAM.

early write cycle is performed. During this type of write cycle, the outputs are maintained in the high-Z state throughout the complete bus cycle. The fact that the output is put in the high-Z state during the write operation allows the D input and Q output of the DRAM to be tied together. The Q output is also in the high-Z state whenever $\overline{CAS}$ is logic 1. This is the connection and mode of operation normally used when attaching DRAMs to the bidirectional data bus of a microprocessor. Figure 9–22 shows how 16 2164B devices are connected to make up a 64K × 16-bit DRAM array.

The 2164B also has the ability to perform what are called *page-mode* accesses. If $\overline{RAS}$ is left at logic 0 after the row address is latched inside the device, the address is maintained within the device. Then, by simply supplying successive column addresses, data cells along the selected row are accessed. This permits faster access of memory by eliminating the time needed to set up and strobe additional row addresses.

Earlier we pointed out that the key difference between the DRAM and SRAM is that the storage cells in the DRAM need to be periodically refreshed; otherwise, they lose their data. To maintain the integrity of the data in a DRAM, each of the rows of the storage array must typically be refreshed periodically, such as every 2 ms. All the storage cells in an array are refreshed by simply cycling through the row addresses. As long as $\overline{CAS}$ is held at logic 1 during the refresh cycle, no data are output.

External circuitry is required to perform the address multiplexing, $\overline{RAS}$/$\overline{CAS}$ generation, and refresh operations for a DRAM subsystem. *DRAM-refresh controller* ICs are available to permit easy implementation of these functions.

Battery Backup for the RAM Subsystem

Even though RAM ICs are volatile, in some equipment it is necessary to make all or part of the RAM memory subsystem nonvolatile (e.g., an electronic cash register). In this application, a power failure could result in the loss of irreplaceable information about the operation of the business.

To satisfy the nonvolatile requirement, additional circuitry can be included in the RAM subsystem. These circuits must sense the occurrence of a power failure and automatically switch the memory subsystem over to a backup battery. An orderly transition must take place from system power to battery power. Therefore, when a loss of power is detected, the power-fail circuit must permit the completion of any read or write cycle that

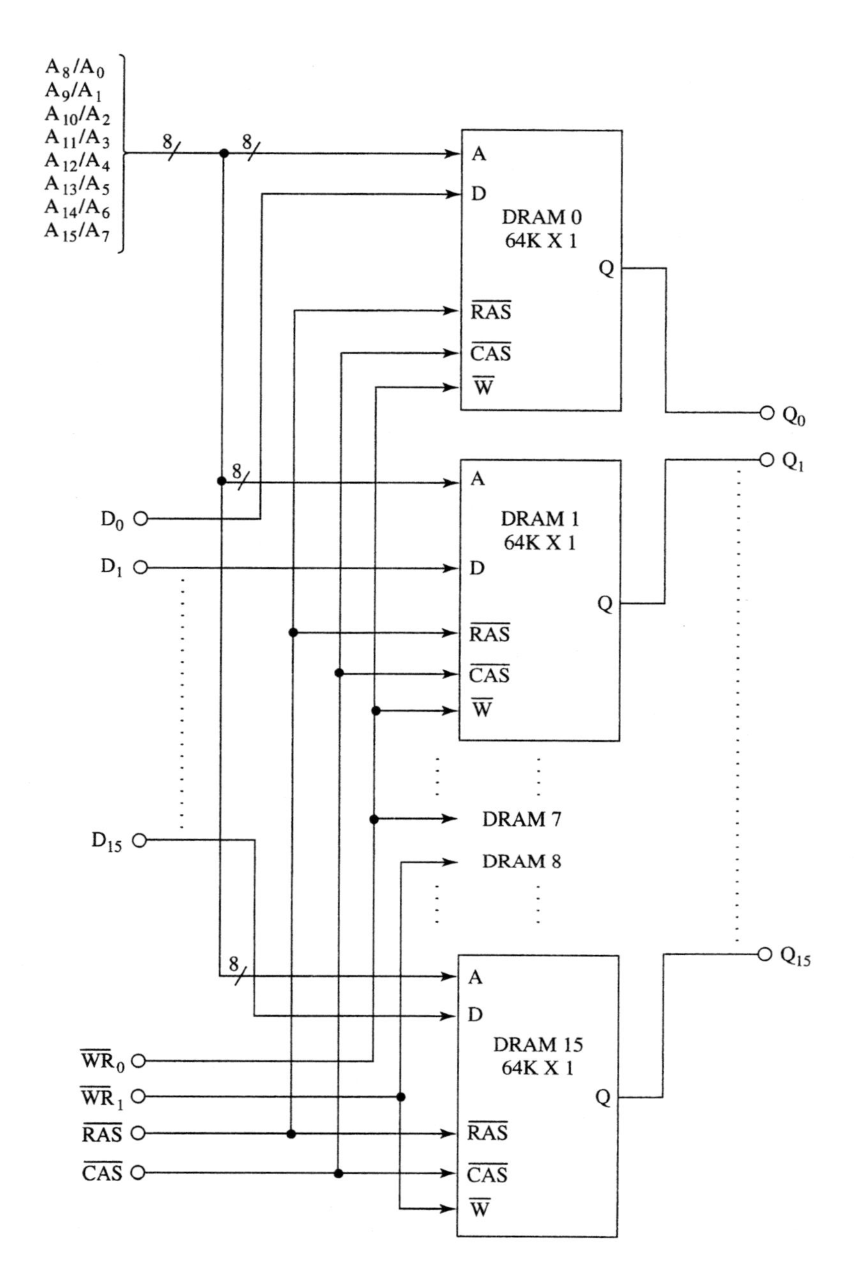

Figure 9–22 64K × 16-bit DRAM circuit.

is in progress and then lock the memory against the occurrence of additional read/write operations. The memory subsystem remains in this state until power is restored. In this way, the RAM subsystem can be made at least temporarily nonvolatile.

▲ 9.4 PARITY, THE PARITY BIT, AND PARITY-CHECKER/GENERATOR CIRCUIT

In microcomputer systems, the data exchanges that take place between the MPU and the memory must be done without error. However, problems such as noise, transient signals, or even bad memory bits can produce errors in the transfer of data and instructions. For instance, the storage location for 1 bit in a large DRAM array may be bad and stuck at the 0 logic level. This will not present a problem if the logic level of the data written to the storage location is 0, but if it is 1, the value will always be read as 0. To improve the reliability of information transfer between the MPU and memory, a *parity bit* can be added to each byte of data. To implement data transfers with parity, a *parity-checker/generator* circuit is required.

Figure 9–23 shows a parity-checker/generator circuit added to the memory interface of a microcomputer system. Note that the data passed between the MPU and memory subsystem is applied in parallel to the parity-checker/generator circuit. Assuming that the microprocessor has an 8-bit data bus, data words read from or written to memory by the

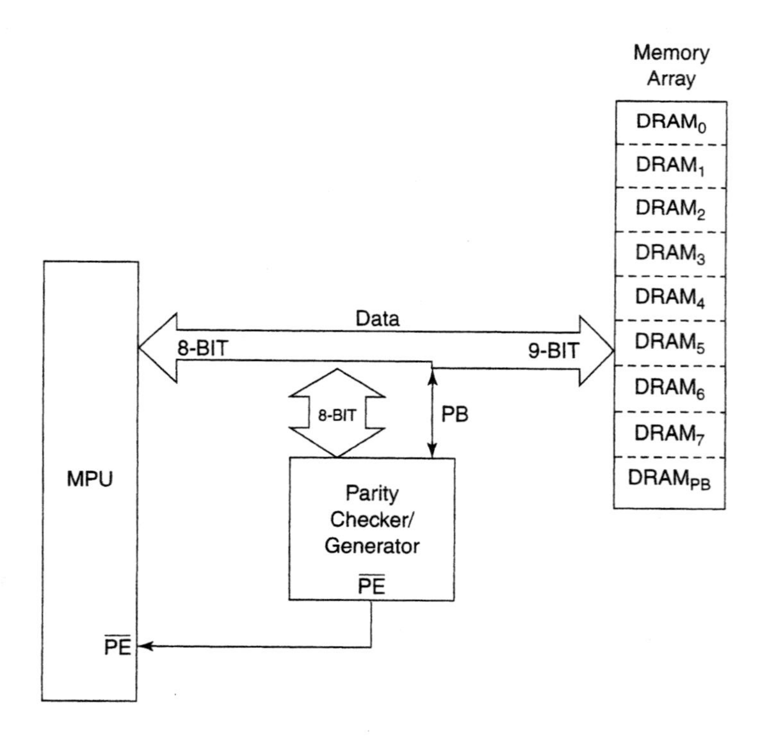

Figure 9–23 Data-storage memory interface with parity-checker generator.

MPU over the data bus are still byte-wide, but the data stored in memory is 9 bits long. The data in memory consists of 8 bits of data and 1 parity bit. Assuming that the memory array is constructed with 256K × 1-bit DRAMs, then the memory array for a memory subsystem with parity would have nine DRAM ICs instead of eight. The extra DRAM is needed for storage of the parity bit for each byte of data stored in the other eight DRAM devices.

The parity-checker/generator circuit can be set up to produce either *even parity* or *odd parity*. The 9-bit word of data stored in memory has even parity if it contains an even number of bits that are at the 1 logic level and odd parity if the number of bits at logic 1 is odd.

Let us assume that the circuit in Fig. 9–23 is used to generate and check for even parity. If the byte of data written to memory over the MPU's data bus is FFH, the binary data is 11111111_2. This byte has 8 bits at logic 1—that is, it already has even parity. Therefore, the parity-checker/generator circuit, which operates in the parity generate mode, outputs logic 0 on the parity bit line (PB), and the 9 bits of data stored in memory is 011111111_2. On the other hand, if the byte written to memory is 7FH, the binary word are 01111111_2. Since only 7 bits are at logic 1, parity is odd. In this case, the parity-checker/generator circuit makes the parity bit logic 1, and the 9 bits of data saved in memory is 101111111_2. Notice that the data held in memory has even parity. In this way, we see that during all data memory write cycles, the parity-checker/generator circuit simply checks the data that are to be stored in memory and generates a parity bit. The parity bit is attached to the original 8 bits of data to make it 9 bits. The 9 bits of data stored in memory have even parity.

The parity-checker/generator works differently when data are read from memory. Now the circuit must perform its parity-check function. Note that the 8 bits of data from the addressed storage location in memory are sent directly to the MPU. However, at the same time, this byte and the parity bit are applied to the inputs of the parity-checker/generator circuit. This circuit checks to determine whether there is an even or odd number of logic 1s in the word with parity. Again we will assume that the circuit is set up to check for even parity. If the 9 bits of data read from memory are found to have an even number of bits at the 1 logic level, parity is correct. The parity-checker/generator signals this fact to the MPU by making the parity error ($\overline{PE}$) output inactive logic 1. This signal is normally sent to the MPU to identify whether or not a memory *parity error* has occurred. If an odd number of bits are found to be logic 1, a parity error has been detected and $\overline{PE}$ is set to 0 to tell the MPU of the error condition. Once alerted to the error, the MPU can do any one of a number of things under software control to recover. For instance, it could simply repeat the memory-read cycle to see if it takes place correctly the next time.

The 74AS280 device implements a parity-checker/generator function similar to that just described. Figure 9–24(a) shows a block diagram of the device. Note that it has nine data-input lines, which are labeled A through I. In the memory interface, lines A through H are attached to data bus lines D_0 through D_7, respectively, and during a read operation the parity bit output of the memory array, D_{PB}, is applied to the I input.

The function table in Fig. 9–24(b) describes the operation of the 74AS280. It shows how the Σ_{EVEN} and Σ_{ODD} outputs respond to an even or odd number of data inputs at logic 1. Note that if there are 0, 2, 4, or 8 inputs at logic 1, the Σ_{EVEN} output switches to logic 1 and Σ_{ODD} to logic 0. This output response signals the even parity condition.

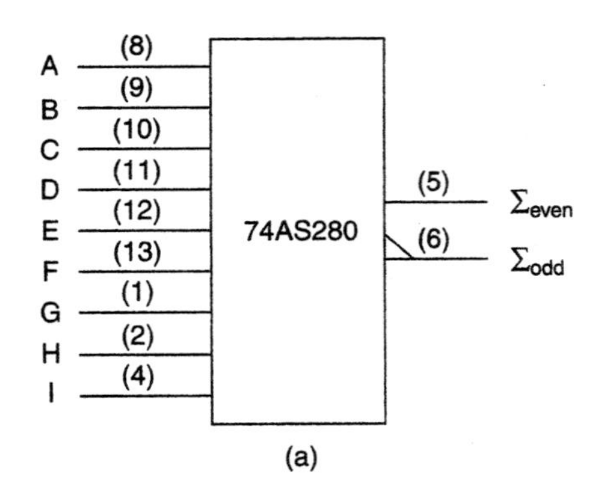

(a)

NUMBER OF INPUTS A THRU I THAT ARE HIGH	OUTPUTS	
	Σ EVEN	Σ ODD
0,2,4,6,8	H	L
1,3,5,7,9	L	H

(b)

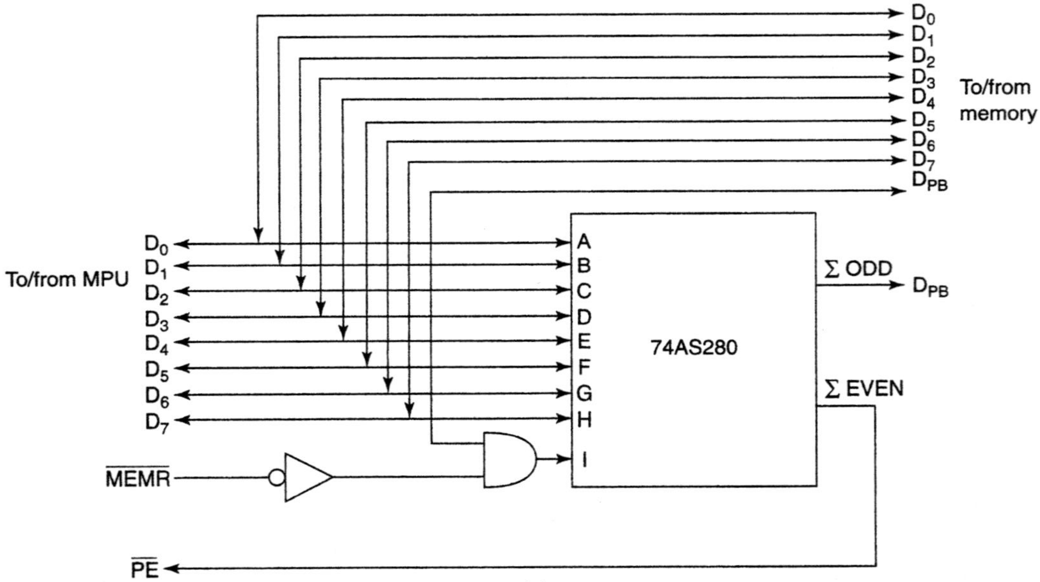

(c)

Figure 9–24 (a) Block diagram of the 74AS280. (Texas Instruments Incorporated) (b) Function table. (Texas Instruments Incorporated) (c) Even-parity checker/generator connection.

In practical applications, the Σ_{EVEN} and Σ_{ODD} outputs are used to produce the parity bit and parity error signal lines. Figure 9–24(c) is an even parity-checker/generator configuration. Note that Σ_{ODD} is used as the parity bit (D_{PB}) output that gets applied to the data input of the parity bit DRAM in the memory array. During a write operation $\overline{MEMR}$ is 0, which makes the I input 0, and therefore the parity of the byte depends only on data bits D_0 through D_7, which are applied to the A through H inputs of the 74AS280. As long as the input at A through H has an even number of bits at logic 1 during a memory write cycle, Σ_{ODD}, which is D_{PB}, is at logic 0 and the 9 bits of data written to memory retain an even number of bits that are 1, or even parity. On the other hand, if the incoming byte at A through H has an odd number of bits that are logic 1, Σ_{ODD} switches to logic 1. The logic 1 at D_{PB} along with the odd number of 1s in the original byte again give the 9 bits of data stored in memory an even parity.

Let us next look at what happens in the parity-checker/generator circuit during a memory-read cycle for the data-storage memory subsystem. When the MPU is reading a byte of data from memory, the 74AS280 performs the parity-check operation. In response to the MPU's read request, the memory array outputs 9 bits of data. They are applied to inputs A through I of the parity-checker/generator circuit. The 74AS280 checks the parity and adjusts the logic levels of Σ_{EVEN} and Σ_{ODD} to represent this parity. If parity is even as expected, Σ_{EVEN}, which represents the parity error ($\overline{PE}$) signal, is at logic 1. This tells the MPU that a valid data transfer is taking place. However, if the data at A through I has an odd number of bits at logic 1, Σ_{EVEN} switches to logic 0 and informs the MPU that a parity error has occurred.

In a 16-bit microcomputer system, such as that built with the 8086 MPU, there are normally two 8-bit banks of DRAM ICs in the data-storage memory array. In this case, a parity bit DRAM is added to each bank. Therefore, parity is implemented separately for each of the two bytes of a data word stored in memory. This is important because the 8086 can read either bytes or words of data from memory. For this reason, two parity-checker/generator circuits are also required, one for the upper eight lines of the data bus and one for the lower eight lines. Gating them together combines the parity error outputs of the two circuits and the resulting parity error signal is supplied to the MPU. In this way, the MPU is notified of a parity error if it occurs in an even-addressed byte data transfer, odd-addressed byte data transfer, or in either or both bytes of a 16-bit data transfer.

▲ 9.5 FLASH MEMORY

Another memory technology important to the study of microcomputer systems is what is known as *FLASH memory.* FLASH memory devices are similar to EPROMs in many ways, but are different in several very important ways. In fact, FLASH memories act just like EPROMs: they are nonvolatile, are read just like an EPROM, and program with an EPROM-like algorithm.

The key difference between a FLASH memory and an EPROM is that its memory cells are erased electrically, instead of by exposure to ultraviolet light. That is, the storage array of a FLASH memory can be both electrically erased and reprogrammed with new data. Unlike RAMs, they are not byte erasable and writeable. When an erase operation is performed on a FLASH memory, either the complete memory array or a large block of storage locations, not just one byte, is erased. Moreover, the erase process is complex and can take as long as several seconds. This erase operation can be followed by a write operation—a programming cycle—that loads new data into the storage location. This write operation also takes a long time when compared to the write cycle times of a RAM.

Even through FLASH memories are writeable, like EPROMs they find their widest use in microcomputer systems for storage of firmware. However, their limited erase/rewrite capability enables their use in applications where data must be rewritten, though not frequently. Some examples: implementation of a nonvolatile writeable lookup table, in-system programming for code updates, and solid state drives. An example of the use of FLASH memory as a lookup table is the storage of a directory of phone numbers in a cellular phone.

Block Diagram of a FLASH Memory

Earlier we pointed out that FLASH memories operate in a way very similar to an EPROM. Figure 9–25 shows a block diagram of a typical FLASH memory device. Let us compare this block diagram to that of the ROM in Fig. 9–3. Address lines A_0 through A_{17}, chip enable ($\overline{CE}$), and output enable ($\overline{OE}$) serve the exact same function for both devices. That is, the address picks the storage location that is to be accessed, $\overline{CE}$ enables the device for operation, and $\overline{OE}$ enables the data to the outputs during read cycles.

We also see that they differ in two ways. First, the data bus is identified as bidirectional, because the FLASH memory can be used in an application where it is written into as well as read from. Second, another control input, write enable ($\overline{WE}$), is provided. This signal must be at its active 0 logic level during all write operations to the device. In fact, this block diagram is exactly the same as that given for SRAM in Fig. 9–11.

Bulk-Erase, Boot Block, and FlashFile FLASH Memories

FLASH memory devices are available with several different memory array architectures. These architectures relate to how the device is organized for the purpose of erasing. Earlier we pointed out that when an erase operation is performed to a FLASH memory device, either all or a large block of memory storage locations are erased. The three standard FLASH memory array architectures, *bulk-erase, boot block, and FlashFile,* are shown in Fig. 9–26. In a bulk-erase device, the complete storage array is arranged as a single block. Whenever an erase operation is performed, the contents of all storage loca-

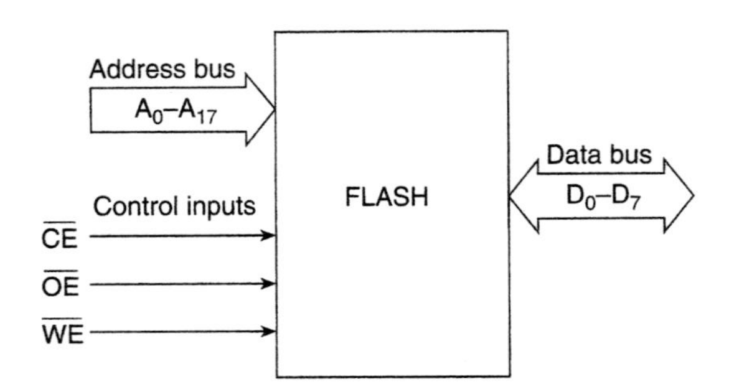

Figure 9–25 Block diagram of a FLASH memory.

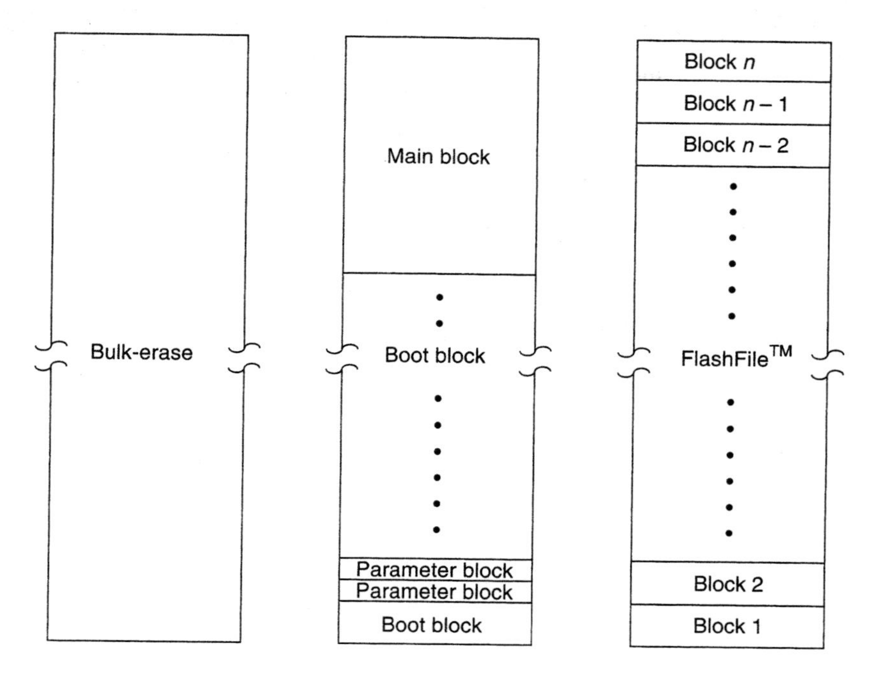

Figure 9–26 FLASH memory array architectures.

tions are cleared. This is the architecture used in the design of the earliest FLASH memory devices.

More modern FLASH memory devices employ either the boot block or FlashFile architecture for their memory array. They add granularity to the programming process. Now the complete memory array does not have to be erased. Instead, each of the independent blocks of storage locations erases separately. Note that the blocks on a boot block device are asymmetrical is size. There is one small block known as the *boot block.* This block is intended for storage of the boot code for the system. Two small blocks that are called parameter blocks follow it. Their intended use is for storage of certain system parameters, for instance, a system configuration table or lookup time. Finally, there are a number of much larger blocks of memory identified as *main blocks,* where the firmware code is stored.

Boot block devices are intended for use in a variety of applications that require smaller memory capacity and benefit from the asymmetrical blocking. One such application is known as *in-system programming.* In this type of application, the boot code used to start up the microcomputer is held in the boot block part of the FLASH memory. When the system is powered on, a memory-loading program is copied from the boot area of FLASH into RAM. Then, program execution is transferred to this program; the firmware that is to be loaded into the FLASH memory is downloaded from a communication line or external storage device such as a drive; the firmware is written into the main blocks of the FLASH memory devices; and finally the program is executed out of FLASH. In this way, we see that the FLASH memory devices are not loaded with the microcomputer's program in advance; instead, they are programmed while in the system.

FlashFile architecture FLASH memory devices differ from boot block devices in that the memory array is organized into equal-sized blocks. For this reason, it is said to be symmetrically blocked. This type of organization is primarily used in the design of high-density devices. High-density flash devices are used in applications that require a large amount of code or data to be stored (e.g., a FLASH memory drive).

Standard Bulk-Erase FLASH Memories

Bulk-erase FLASH memories are the oldest type of FLASH devices and are available in densities similar to those of EPROMs. Figure 9–27 lists the part number, bit density, and storage capacity of some of the popular devices. Note that the part numbers of FLASH devices are similar to those used for the EPROMs described earlier. The differences are that the 7 in the EPROM part number is replaced by an 8, representing FLASH, and instead of a C, which is used to identify that the circuitry of the EPROM is made with a CMOS process, an F identifies FLASH technology. Remember the 2MB EPROM was labeled 27C020; therefore, the 2MB FLASH memory is labeled as 28F020. This device is organized as 256K byte-wide storage locations.

Since FLASH memories are electrically erased, they do not need to be manufactured in a windowed package. For this reason, and the trend toward the use of surface-mount packaging, the most popular package for housing FLASH memory ICs is the plastic leaded chip carrier, or PLCC as it is commonly known. Figure 9–28(a) shows the PLCC pin layout of the 28F020. All of the devices listed in Fig. 9–27 are manufactured in this same-size package and with compatible pin layouts.

Looking at the signals identified in the pin layout, we find that the device is exactly the same as the block diagram in Fig. 9–25. To select between its 256K byte-wide storage locations, it has 18 address inputs, A_0 through A_{17}, and to support byte-wide data-read and -write transfers, it has an 8-bit data bus, DQ_0 through DQ_7. Finally, to enable the chip and its outputs and distinguish between read- and write-data transfers, it has control lines $\overline{CE}$, $\overline{OE}$, and $\overline{WE}$, respectively. As Fig. 9–28(b) shows, the device is available with read access times ranging from 70 ns for the 28F020-70 to 150 ns for the 28F020-150.

The power supply voltage and current requirements depend on whether the FLASH memory is performing a read, erase, or write operation. During read mode of operation, the 28F020 is powered by 5V $\pm10\%$ between the V_{cc} and V_{ss} pins, and it draws a maximum current of 30 mA. On the other hand, when either an erase or write cycle is taking place, 12V $\pm5\%$ must also be applied to the V_{pp} power supply input.

The 28F256, 28F512, 28F010, and 28F020 employ a bulk-erase storage array. For this reason, when an erase operation is performed to the device, all bytes in the storage

FLASH	Density (bits)	Capacity (bytes)
28F256	256K	32K × 8
28F512	512K	64K × 8
28F010	1M	128K × 8
28F020	2M	256K × 8

Figure 9–27 Standard bulk-erase FLASH memory devices.

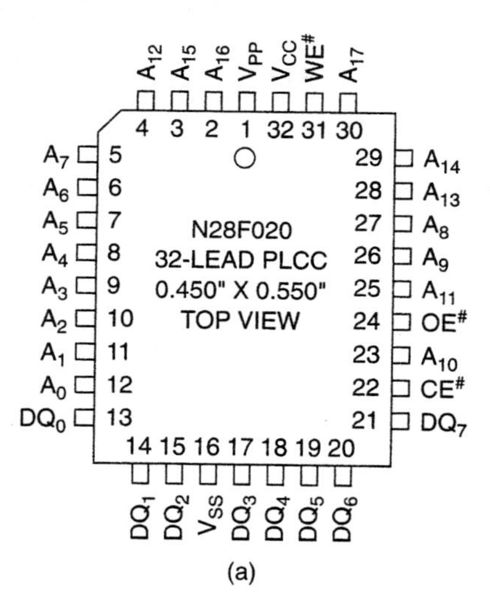

Part number	Access time
28F020-70	70 ns
28F020-90	90 ns
28F020-120	120 ns
28F020-150	150 ns

(b)

Figure 9–28 (a) Pin layout of the 28F020. (Reprinted by permission of Intel Corporation, Copyright Intel Corp. 1995) (b) Standard speed selections for the 28F020.

array are restored to FF_{16}, which represents the erased state. The method employed to erase the 28F020 FLASH memory IC is known as the *quick-erase algorithm.* A flowchart that outlines the sequence of events that must take place to erase a 28F020 is given in Fig. 9–29. This programming sequence can be performed either with a FLASH memory-programming instrument or by the software of the microprocessor to which the FLASH device is attached. Let us next look more closely at how a 28F020 is erased.

To change the contents of a memory array—that is, either erase the storage array or write bytes of data into the array—commands must be written to the FLASH memory device. Unlike an EPROM, a FLASH memory has an internal command register. Figure 9–30 lists the commands that can be issued to the 28F020. Note that they include a *read* (read memory), *set up and erase* (set up erase/erase), and *erase verification* (erase verify) commands. These three are used as part of the quick-erase algorithm process. The command register can be accessed only when +12V is applied to the V_{pp} pin of the 28F020.

Figure 9–29 also includes a table of the bus operation and command activity that takes place during an erase operation of the 28F020. From the bus operation and command columns, we see that as part of the erase process, the microprocessor (or programming instrument) must issue a set up erase/erase command, followed by an erase verify command, and then a read command to the FLASH device. It does this by executing a

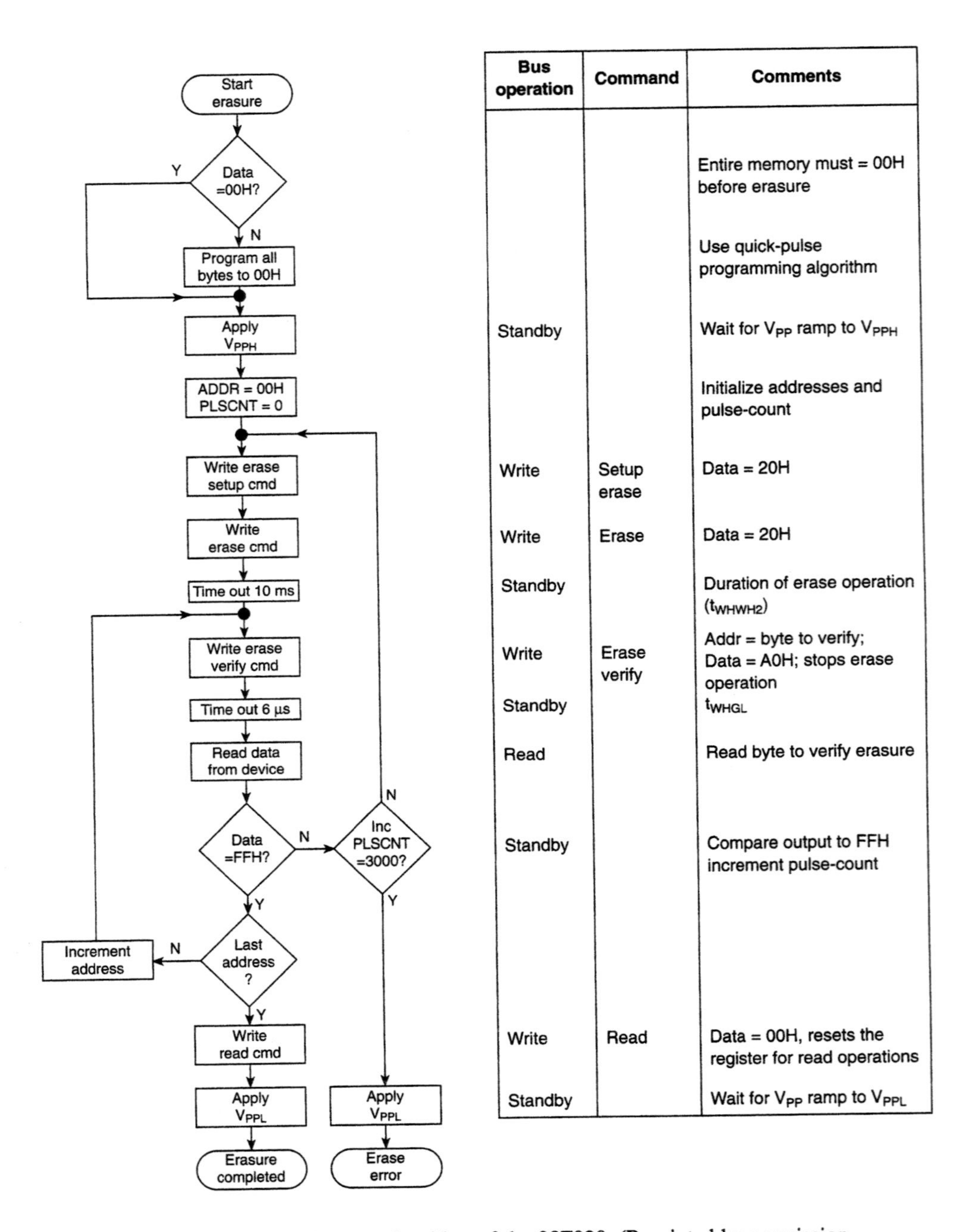

Bus operation	Command	Comments
		Entire memory must = 00H before erasure
		Use quick-pulse programming algorithm
Standby		Wait for V_{PP} ramp to V_{PPH}
		Initialize addresses and pulse-count
Write	Setup erase	Data = 20H
Write	Erase	Data = 20H
Standby		Duration of erase operation (t_{WHWH2})
Write	Erase verify	Addr = byte to verify; Data = A0H; stops erase operation
Standby		t_{WHGL}
Read		Read byte to verify erasure
Standby		Compare output to FFH increment pulse-count
Write	Read	Data = 00H, resets the register for read operations
Standby		Wait for V_{PP} ramp to V_{PPL}

Figure 9–29 Quick-erase algorithm of the 28F020. (Reprinted by permission of Intel Corporation, Copyright Intel Corp. 1995)

Command	Bus Cycles Req'd	First Bus Cycle			Second Bus Cycle		
		Operation	Address	Data	Operation	Address	Data
Read memory	1	Write	X	00H			
Read intelligent identifier codes	3	Write	X	90H	Read		
Setup erase/erase	2	Write	X	20H	Write	X	20H
Erase verify	2	Write	EA	A0H	Read	X	EVD
Setup program/program	2	Write	X	40H	Write	PA	PD
Program verify	2	Write	X	C0H	Read	X	PVD
Reset	2	Write	X	FFH	Write	X	FFH

Figure 9–30 28F020 command definitions. (Reprinted by permission of Intel Corporation, Copyright Intel Corp. 1995)

FLASH memory programming control program that causes the write of the command to the FLASH memory device at the appropriate time. Actually, all storage locations in the memory array must always be programmed with 00_{16} before initiating the erase process.

Figure 9–29 shows that two consecutive set up erase/erase commands are used in the quick-erase sequence. The first one prepares the 28F020 to be erased, and the second initiates the erase process. Figure 9–30 shows that this sequence is identified as two write cycles. During both of these bus cycles, a value of data equal to 20_{16} is written to any address in the address range of the FLASH device being erased. Once these commands have been issued, a state-machine within the device automatically initiates and directs the erase process through completion.

The next step in the quick-erase process is to determine whether or not the device has erased completely. This is done with the erase verify command. Figure 9–30 shows that this operation requires a write cycle followed by a read cycle. During the write cycle, the data bus carries the erase verify command, $A0_{16}$, and the address bus carries the address of the storage location that is to be tested, EA. The read cycle that follows is used to transfer the data from the storage location corresponding to EA to the MPU. This data is identified as EVD in Fig. 9–30. The flowchart shows that the MPU must verify that the value of data read out of FLASH is FF_{16}. This erase verify step is repeated for every storage location in the 28F020. If any storage location does not verify erasure by reading back FF_{16}, the complete erase process is immediately repeated.

After complete erasure has been verified, the software must issue a read command to the device. From Fig. 9–30, we find that it requires a single write bus cycle and is accompanied by any address that corresponds to the FLASH device being erased and a command data value of 00_{16}. Issuing this command puts the device into the read mode and readies it for read operation. Figure 9–31 outlines the *quick-pulse programming algorithm* of the 28F020. This process is similar to that just described for erasing devices; however, it uses the *set up and program* (set up program/program), *program verification* (program verify), and read (read memory) commands.

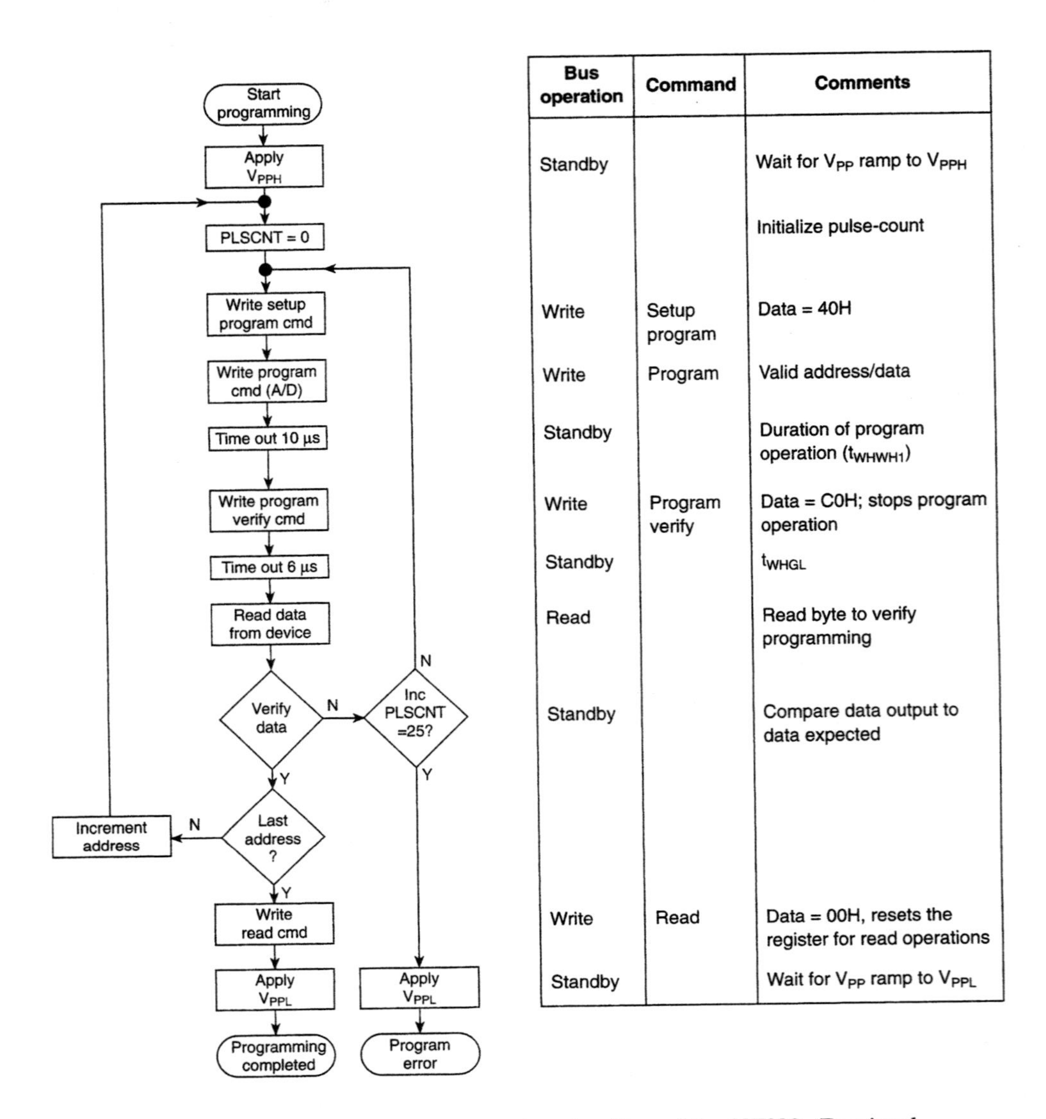

Bus operation	Command	Comments
Standby		Wait for V_{PP} ramp to V_{PPH}
		Initialize pulse-count
Write	Setup program	Data = 40H
Write	Program	Valid address/data
Standby		Duration of program operation (t_{WHWH1})
Write	Program verify	Data = C0H; stops program operation
Standby		t_{WHGL}
Read		Read byte to verify programming
Standby		Compare data output to data expected
Write	Read	Data = 00H, resets the register for read operations
Standby		Wait for V_{PP} ramp to V_{PPL}

Figure 9–31 Quick-pulse programming algorithm of the 28F020. (Reprinted by permission of Intel Corporation, Copyright Intel Corp. 1995)

Standard Boot Block FLASH Memories

Earlier we pointed out that boot block FLASH memories are designed for use in embedded microprocessor application. These newer devices are available in higher densities than bulk-erase devices. Fig. 9–32 shows the pin layouts for three compatible standard densities: the 2MB, 4MB, and 8MB devices. Notice that the corresponding devices are identified with the part numbers 28F002, 28F004, and 28F008, respectively. These devices have different densities but have a common set of operating features and capabilities. The pinout information given is for a 40-pin *thin small outline package* (TSOP).

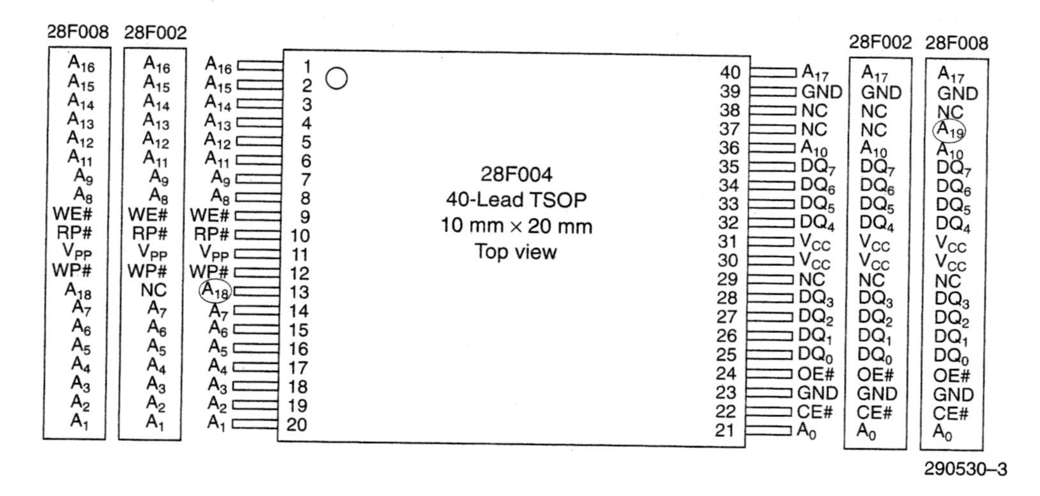

Figure 9–32 Pin-layout comparison of the TSOP 28F002, 28F004, and 28F008 ICs.

These devices offer a number of new architectural features when compared to the bulk-erase devices just described. One of the most important of these new features is what is known as *SmartVoltage.* This capability enables the device to be programmed with either a 5-V or 12-V value of V_{pp}. In fact, the device can be installed into a circuit using either value of V_{pp}. This is because the device has the ability to automatically detect and adjust its programming operation to the value of the programming supply voltage in use. These devices are available with either of two read voltage, V_{cc} supply ratings: Smart 5, which operates off a 5V V_{cc}, and Smart 3, which operates off a 3V V_{cc}.

A second important difference is that devices are available at each of these three densities that can be organized with either an 8-bit or 16-bit bus. A block diagram of such a device is shown in Fig. 9–33(a). This device is identified as a 28F004/28F400. The 28F004 device is available in the 40-pin TSOP package and only operates in the 8-bit data bus mode. The 28F400 device has 16 data lines, D_0 through D_{15}, and can be configured to operate with either an 8-bit or 16-bit data bus. This is done with the $\overline{\text{BYTE}}$ input. Logic 0 at $\overline{\text{BYTE}}$ selects byte-wide mode of operation and logic 1 chooses word-wide operation. To permit the extra data lines, the 28F400 is housed in a 56-lead TSOP.

Remember that the storage array of a boot block device is arranged as multiple asymmetrically sized, independently erasable blocks. In fact, the 28F004 has one 16Kbyte boot block, two 8Kbyte parameter blocks, three 128Kbyte main blocks, and a fourth main block that is only 96Kbytes. Two different organizations of these blocks are available, as shown in Fig. 9–33(b). The configuration on the left is known as the *top boot* (T), and that on the right is known as the *bottom boot* (B). Note that they differ in how the blocks are assigned to the address space. That is, the T version has the 16Kbyte boot block at the top of the address space (highest address), followed by the parameter blocks, and then the main blocks. The address space of the B version is a mirror image; therefore, the 16Kbyte boot block starts at the bottom of the address space (lowest address).

Another new feature introduced with the boot block architecture is that of a hardware-lockable block. In the 28F004/28F400, the 16Kbyte boot block section can be

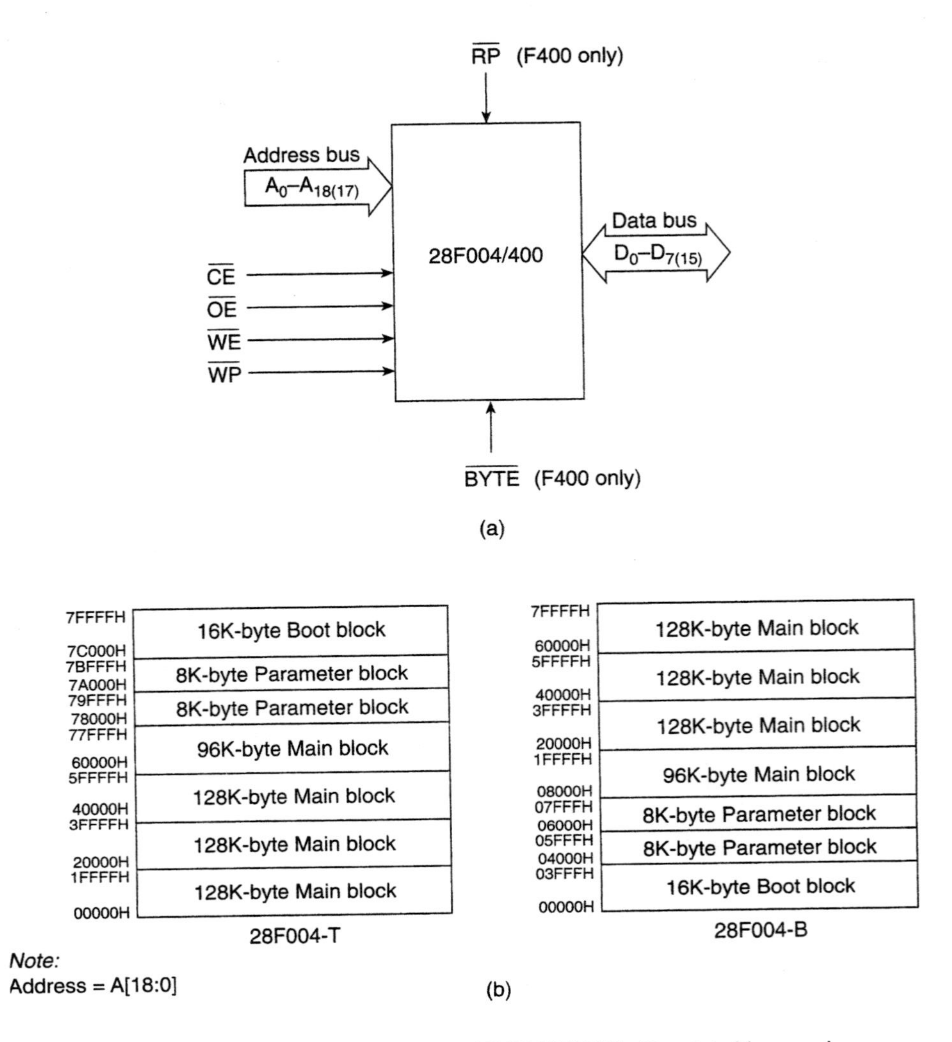

Figure 9–33 (a) Block diagram of the 28F004/28F400. (Reprinted by permission of Intel Corporation, Copyright Intel Corp. 1995) (b) Top and Bottom boot block organization of the 28F004. (Reprinted by permission of Intel Corporation, Copyright Intel Corp. 1995)

locked. If external hardware applies logic 0 to the write protect ($\overline{WP}$) input, the boot block is locked. Any attempt to erase or program this block when it is locked results in an error condition. Therefore, we say that the boot block is *write protected.* In an in-system programming application, the boot block part of the storage array typically would contain the part of the microcomputer program (boot program) that is used to load the system software into FLASH memory. For this reason, it would be locked and should remain that way.

Looking at Fig. 9–33(a), we find one more new input on the 28F004/28F400, the reset/deep power-down ($\overline{RP}$) input. This input must be at logic 1 to enable normal read, erase, and program operations. During read operations, the device can draw as much as 60 mA of current. If the device is not in use, it can be put into the deep power-down mode to conserve power. To do this, external circuitry must switch $\overline{RP}$ to logic 0. In this mode, it draws just 0.2 μA.

The last difference we will describe is that the 28F004/28F400 is equipped with what is known as *automatic erase and write.* No longer do we need to implement the complex quick-erase and quick-pulse programming algorithms in software as done for the 28F020. Instead, the 28F004/28F400 uses a *command user interface* (CUI), *status register,* and *write-state machine* to initiate an internally implemented and highly automated method of erasing and programming the blocks of the storage array.

Let us now look briefly at how an erase operation is performed. The command bus definitions of the 28F004/28F400 are shown in Fig. 9–34(a), the bit definitions of its status register are given in Fig. 9–34(b), and its erase cycle flowchart in Fig. 9–34(c). Here we see that all that needs to be done to initiate an erase operation is to write to the device a command bus definition that includes an erase setup command and an erase confirm command. These commands contain an address that identifies the block to be erased. In response to these commands, the write state machine drives a sequence that automatically programs all of the bits in this block to logic 0, verifies that they have been programmed, erases all of the bits in the block, and then verifies that each bit in the block has been erased. While it is performing this process, the write state machine status bit (WSMS) of the status register is reset to 0 to say that the device is busy. The microcomputer's software can simply poll this bit to see if it is still busy. When WSMS is read as logic 1 (ready), the erase operation is complete and all the bits in the erased block are at the 1 logic level. In this way, we see that the new programming software only has to initiate the automatic erase process and then poll the status register to determine when the erase operation is finished.

Standard FLASHFile FLASH Memories

The highest-density FLASH memories available today are those designed with the FLASHFile architecture. As pointed out earlier, they use a symmetrically sized, independently erasable organization for blocking of their storage array. Two popular devices, the 8MB 28F008S5 and the 16MB 28F016SA/SV, are intended for use in large-code storage applications and to implement solid-state mass-storage devices such as the FLASH card and FLASH drive.

A block diagram of the 28F016SA/SV FLASHFile memory device is shown in Fig. 9–35(a) and its pin layout for a *shrink small outline package* (SSOP) is given in Fig. 9–35(b). Comparing this device to the 28F004/28F400 in Fig. 9–33(a), we find many similarities. For instance, both devices have an address bus, data bus, and control signals $\overline{OE}$, $\overline{WE}$, $\overline{WP}$, $\overline{RP}$, and $\overline{BYTE}$, and they serve similar functions relative to device operation. One difference is that there are now two chip-enable inputs, labeled $\overline{CE_0}$ and $\overline{CE_1}$, instead of just one. Both of these inputs must be at logic 0 to enable the device for operation.

Another change found on the 28F016SA/SV is the addition of the ready/busy (RY/$\overline{BY}$) output. This output has been provided to further reduce the software overhead

Command	Notes	First Bus Cycle			Second Bus Cycle		
		Oper	Addr	Data	Oper	Addr	Data
Read array	8	Write	X	FFH			
Intelligent identifier	1	Write	X	90H	Read	IA	IID
Read status register	2,4	Write	X	70H	Read	X	SRD
Clear status register	3	Write	X	50H			
Word/byte write		Write	WA	40H	Write	WA	WD
Alternate word/byte write	6,7	Write	WA	10H	Write	WA	WD
Block erase/confirm	6,7	Write	BA	20H	Write	BA	D0H
Erase suspend/resume	5	Write	X	B0H	Write	X	D0H

Address
BA = Block Address
IA = Identifier Address
WA = Write Address
X = Don't Care

Data
SRD = Status Register Data
IID = Identifier Data
WD = Write Data

Notes:
1. Bus operations are defined in Tables 4 and 5.
2. IA = Identifier Address: A0 = 0 for manufacturer code, A0 = 1 for device code.
3. SRD—Data read from Status Register.
4. IID = Intelligent Identifier Data. Following the Intelligent Identifier command, two Read operations access manufacturer and device codes.
5. BA = Address within the block being erased.
6. WA = Address to be written. WD = Data to be written at location WD.
7. Either 40H or 10H commands is valid.
8. When writing commands to the device, the upper data bus [DQ_8–DQ_{15}] = X (28F400 only) which is either V_{IL} or V_{IH}, to minimize current draw.

(a)

Figure 9–34 (a) 28F004 command bus definitions. (Reprinted by permission of Intel Corporation, Copyright Intel Corp. 1995) (b) Status register bit definitions. (Reprinted by permission of Intel Corporation, Copyright Intel Corp. 1995) (c) Erase operation flowchart and bus activity. (Reprinted by permission of Intel Corporation, Copyright Intel Corp. 1995)

on the MPU during the erase and programming processes. When this output is 0, it signals that the on-chip write-state machine of the FLASH memory is busy performing an operation. Logic 1 means that it is ready to start a new operation. For the boot block devices we introduced earlier, the busy condition had to be determined through software by polling the WSMS bit of the status register. One approach for the 28F016SA/SV is that software could poll $RY/\overline{BY}$ as an input waiting for the FLASH device to be ready. On the other hand, this signal could be used as an interrupt input to the MPU. In this way, no software and MPU overhead is needed to recognize when the FLASH memory is ready to perform another operation. This is the default mode of operation and is known as *level mode.*

WSMS	ESS	ES	DWS	VPPS	R	R	R
7	6	5	4	3	2	1	0

	Notes:
SR.7 = WRITE STATE MACHINE STATUS (WSMS) 1 = Ready 0 = Busy	Write State Machine bit must first be checked to determine Byte/Word program or Block Erase completion before the Program or Erase Status bits are checked for success.
SR.6 = ERASE-SUSPEND STATUS (ESS) 1 = Erase Suspended 0 = Erase in Progress/Completed	When Erase Suspend is issued, WSM halts execution and sets both WSMS and ESS bits to "1." ESS bit remains set to "1" until an Erase Resume command is issued.
SR.5 = ERASE STATUS 1 = Error in Block Erasure 0 = Successful Block Erase	When this bit is set to "1," WSM has applied the maximum number of erase pulses to the block and is still unable to successfully verify block erasure.
SR.4 = PROGRAM STATUS 1 = Error in Byte/Word Program 0 = Successful Byte/Word Program	When this bit is set to "1," WSM has attempted but failed to program a byte or word.
SR.3 = V_{PP} STATUS 1 = V_{PP} Low Detect, Operation Abort 0 = V_{PP} OK	The V_{PP} Status bit, unlike an A/D converter, does not provide continuous indication of V_{PP} level. The WSM interrogates V_{PP} level only after the Byte Write or Erase command sequences have been entered, and informs the system if V_{PP} has not been switched on. The V_{PP} Status bit is not guaranteed to report accurate feedback bewteen V_{PPLK} and V_{PPH}.
SR.2–SR.0 = RESERVED FOR FUTURE ENHANCEMENTS	These bits are reserved for future use and should be masked out when polling the Status Register.

(b)

Figure 9–34 (continued)

The function of the RY/$\overline{\text{BY}}$ output can be configured for a number of different modes of operation under software control. Writing a device configuration code to the 28F016SA/SV does this. For instance, it can be set to produce a pulse on write or a pulse on erase or even be disabled.

The last signal line in the block diagram that is new is the 3/$\overline{5}$ input. Note that this input is implemented only on the 28F016SA IC. Logic 1 at this input selects 3.3V operation for V_{cc}, and logic 0 indicates that a 5V supply is in use. Since the 28F016SV is a SmartVoltage device, this input is not needed.

Figure 9–35(c) illustrates the blocking of the 28F016SA/SV configured for byte-wide mode of operation. Here we find that the 16MB address space is partitioned into 32 independent 64K byte blocks. Note that block 0 is in the address range from 000000_{16} through $00FFFF_{16}$ and block 31 corresponds to the range $1F0000_{16}$ through $1FFFFF_{16}$. If

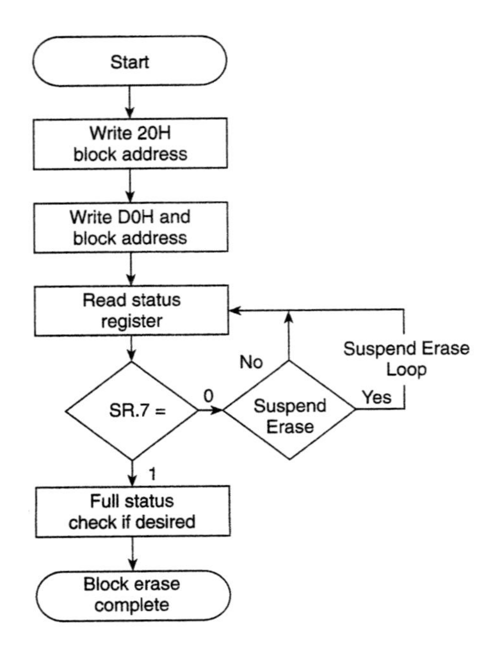

Bus Operation	Command	Comments
Write	Erase setup	Data = 20H Addr = Within block to be erased
Write	Erase confirm	Data = D0H Addr = Within block to be erased
Read		Status register data toggle CE# or OE# to update status register
Standby		Check SR.7 1 = WSM ready 0 = WSM busy
Repeat for subsequent block erasures. Full Status Check can be done after each block erase, or after a sequence of block erasures. Write FFH after the last operation to reset device to read array mode.		

(c)

Figure 9–34 (continued)

the device is strapped for word-mode operation with logic 1 at the $\overline{BYTE}$ input, there are still 32 blocks, but they are now 32K words in size.

Just as for the 28F004/28F400, the 28F016SA/SV supports block locking. However, in these devices, the 32 blocks are independently programmable as locked or unlocked. In fact, there is a separate block status register for each of the 32 blocks. This block status register contains both control and status bits related to a corresponding block. The *block-lock status* (BLS) bit, bit 6 in this register, is an example of a control bit. When it is set to logic 1 under software control, the corresponding block is configured as unlocked and write and erase operations is permitted. Changing it to logic 0 locks the block so that it cannot be written into or erased. Bit 7, *block status* (BS), is an example of a status bit that can be read by the MPU. Logic 1 in this bit means that the block is ready, and logic 0 signals that it is busy. When the write-protect ($\overline{WP}$) input is active (logic 0), write and erase operations are not permitted to those blocks marked as locked with an 0 in the BLS bit in their corresponding block status register.

Finally, the internal algorithms and hardware of the 28F016SA/SV have been expanded to improve programming performance. For instance, two 256-byte (128-word) write buffers have been added into the architecture to enable paged data writes. Moreover, the programming algorithm has been enhanced to support queuing of commands and overlapping of erase and write operations. Therefore, additional commands can be sent to a device while it is still executing a prior command. They are held in the queue until processed. The overlapping write/erase capability enables the devices to erase one block while writing data to another. All these features result in easier and faster programming for the 28F016SA/SV.

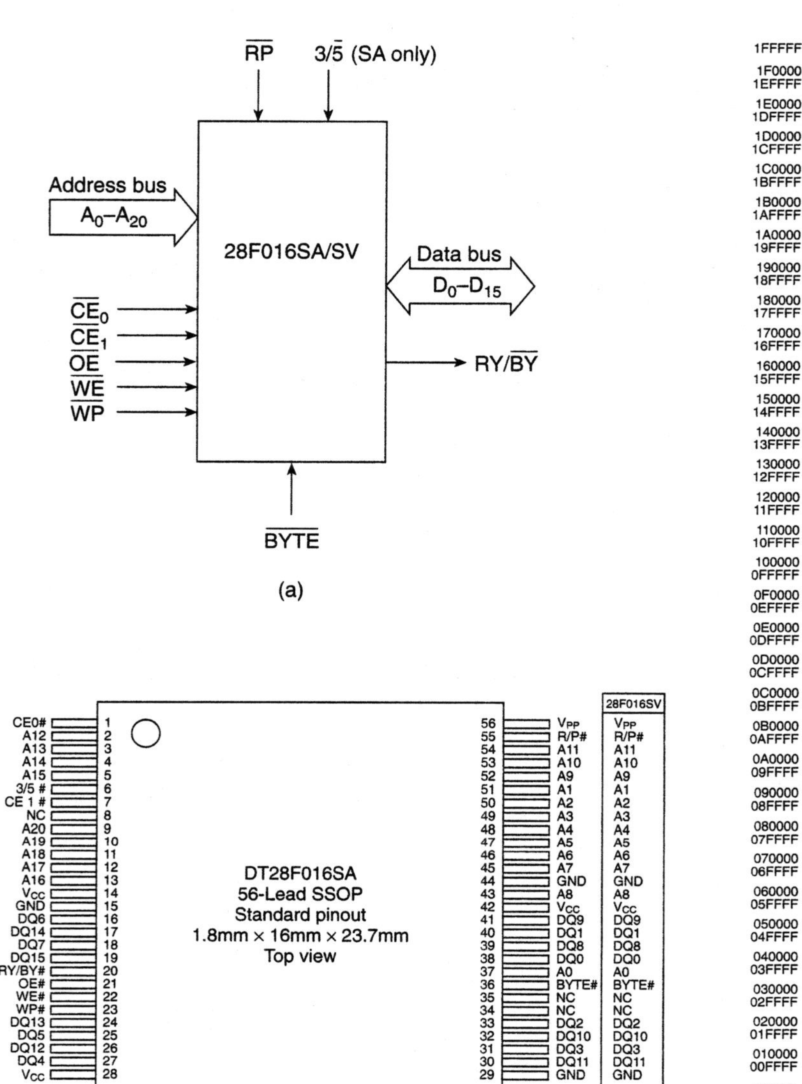

Figure 9–35 (a) Block diagram of the 28F016SA/SV FlashFile memory. (b) Pin layout. (Reprinted by permission of Intel Corporation, Copyright Intel Corp. 1995) (c) Byte-wide mode memory map. (Reprinted by permission of Intel Corporation, Copyright Intel Corp. 1995)

▲ 9.6 WAIT-STATE CIRCUITRY

Depending on the access time of the memory devices used and the clock rate of the MPU, a number of wait states may need to be inserted into external memory read and write operations. In our study of 8088/8086 bus cycles in Chapter 8, we found that the memory subsystem signals the MPU whether or not wait states are needed in a bus cycle with the logic level applied to its $\overline{READY}$ input.

The circuit that implements this function for a microcomputer system is known as a *wait-state generator.* Figure 9–36(a) shows a block diagram of this type of circuit. Note that the circuit has six inputs and just one output. The two inputs located at the top of the circuit, $\overline{CS_0}$ and $\overline{CS_1}$, are outputs of the memory chip-select logic. They could represent

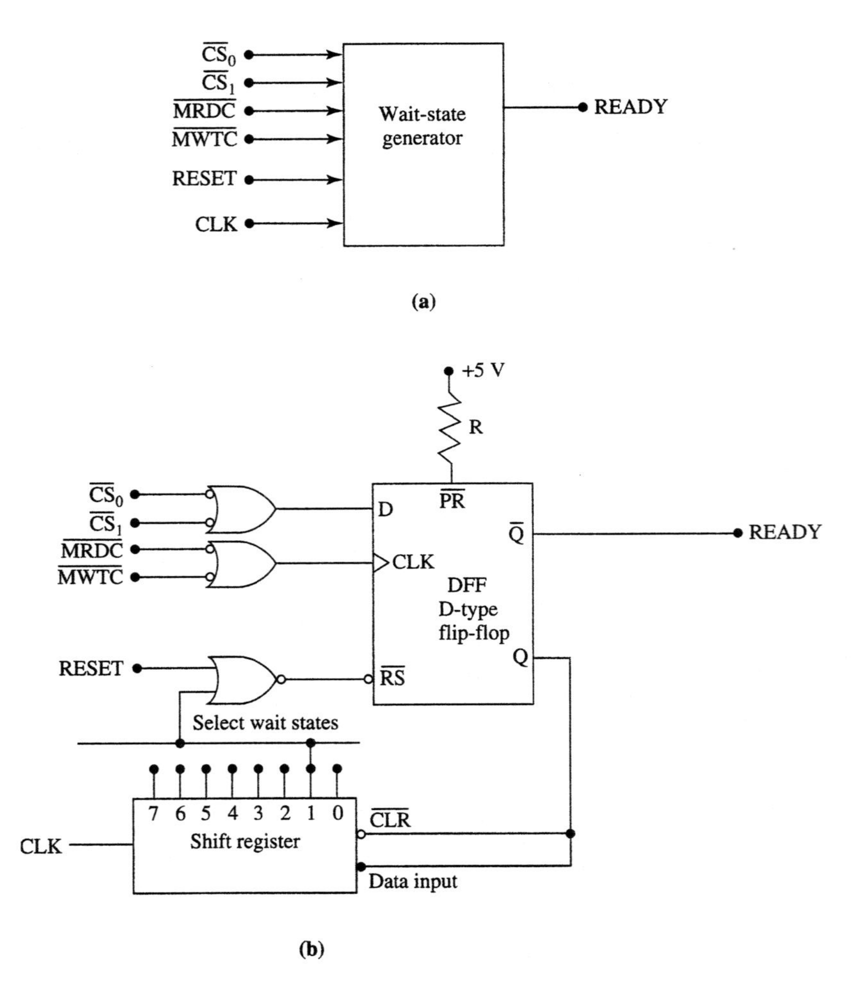

Figure 9–36 (a) Wait-state generator circuit block diagram. (b) Typical wait-state generator circuit.

the chip selects for the program storage and data-storage memory subsystems, respectively, and tell the circuit whether or not these parts of the memory subsystem are being accessed. The two middle inputs are the memory read command ($\overline{\text{MRDC}}$) and memory write command ($\overline{\text{MWTC}}$) outputs of the bus controller. They indicate that a read or write operation is taking place to the memory subsystem. The last two inputs are the system reset (RESET) signal and system clock (CLK) signals. The output is READY.

The circuit in Fig. 9–36(b) can be used to implement a wait-state generator for an 8088/8086-based microcomputer system. To design this circuit, we must select the appropriate D-type flip-flop, shift register, and gates.

Let us look briefly at how this wait-state generator circuit works. The READY output is returned directly to the READY input of the MPU. Logic 1 at this output tells the MPU that the current read/write operation is to be completed. Logic 0 means that the memory bus cycle must be extended by inserting wait states.

Whenever an external memory bus cycle is initiated, the D-type flip-flop is used to start the wait-state generation circuit. Before either $\overline{\text{CS}}_0$, $\overline{\text{CS}}_1$, or RESET becomes active, the $\overline{\text{Q}}$ output of the flip-flop is held at logic 1 and signals that wait states are not needed. The Q output, logic 0, is applied to the $\overline{\text{CLR}}$ input of the shift register. Logic 0 at $\overline{\text{CLR}}$ holds it in the reset state and holds outputs 0 through 7 all at logic 0.

Whenever a read or write operation takes place to a storage location in the external memory's address range, a logic 0 is produced at either $\overline{\text{CS}}_0$ or $\overline{\text{CS}}_1$ and a logic 0 is produced at either $\overline{\text{MRDC}}$ or $\overline{\text{MWTC}}$. The active chip-select input makes the D input of the flip-flop logic 1 and the transition to logic 0 by the read/write command signal causes the flip-flop to set. This makes the Q output switch to logic 1 and $\overline{\text{Q}}$ to logic 0. Now READY tells the MPU to start inserting wait states into the current memory bus cycle. The Q output makes both the $\overline{\text{CLR}}$ and data input of the shift register logic 1. Therefore, it is released and the logic 1 at the data input shifts up through the register synchronous with clock pulses from the MPU's system clock. When the select wait-state output becomes logic 1, it makes the $\overline{\text{RS}}$ input of the flip-flop active, thereby resetting the Q output to logic 0 and $\overline{\text{Q}}$ to logic 1. Thus, the READY output returns to logic 1 and terminates the insertion of wait states, and the MPU completes the bus cycle. The number of wait states inserted depends on how many clock periods READY remains at logic 0. Simply attaching the select wait-state line to a different output of the shift register can change this. For instance, the connection shown in Fig. 9–36(b) represents operation with two wait states.

▲ 9.7 8088/8086 MICROCOMPUTER SYSTEM MEMORY CIRCUITRY

In Chapter 8 we introduced the bus cycles, hardware organization of the memory address space, and memory interface circuits of the 8088/8086-based microcomputer system. The earlier sections of this chapter covered the types of memory devices used in the memory subsystem. Here we will show how the memory interface circuits and memory subsystem are interconnected in a simple microcomputer system, shown in Fig. 9–37(a). We will use the information we have acquired to analyze the memory circuits of this microcomputer system.

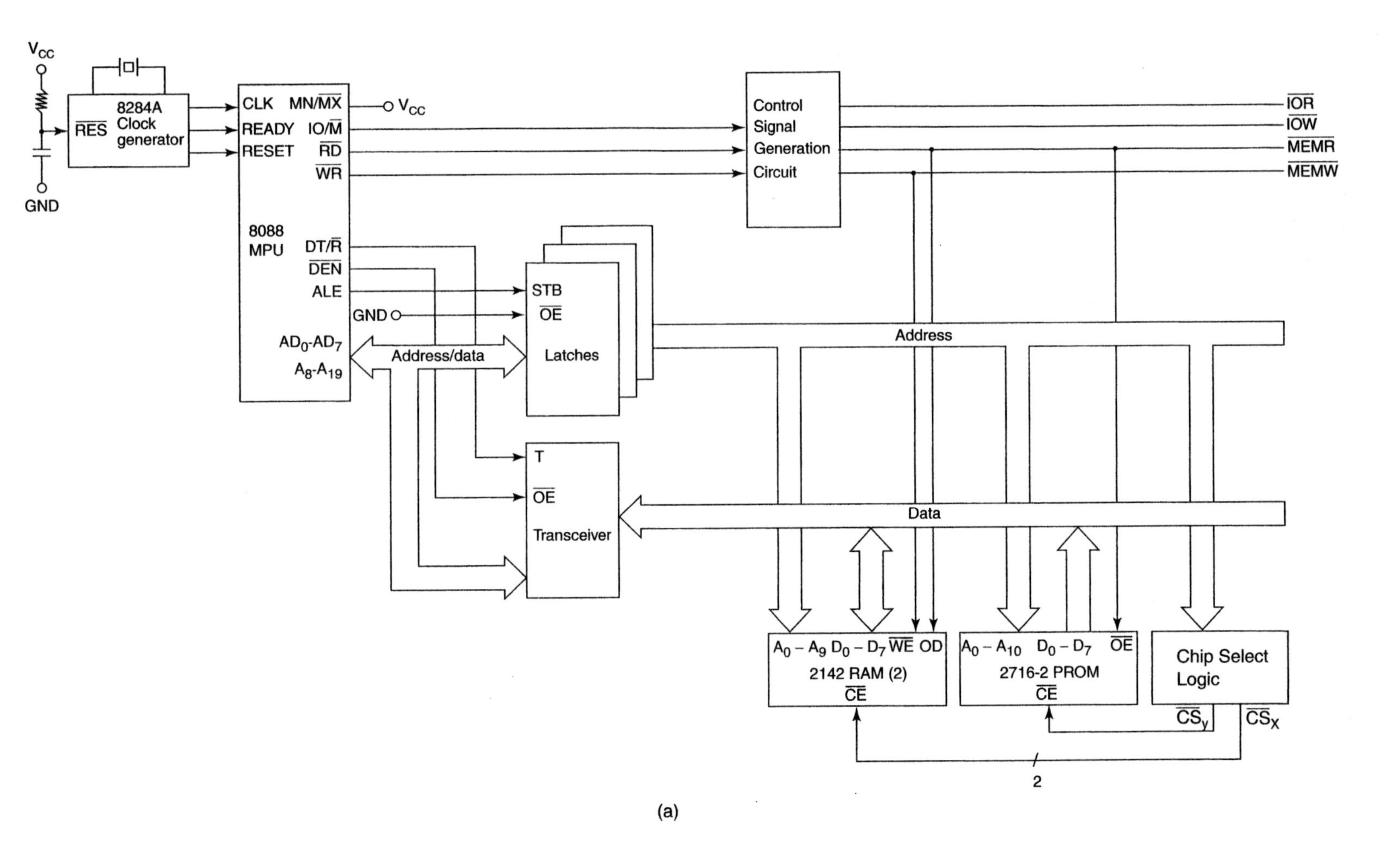

Figure 9–37 (a) Minimum-mode 8088 system memory interface. (Reprinted with permission of Intel Corporation, Copyright/Intel Corp. 1981) (b) Minimum-mode 8086 system memory interface. (Reprinted with permission of Intel Corporation, Copyright/Intel Corp. 1979) (c) Maximum-mode 8088 system memory interface. (Reprinted with permission of Intel Corporation, Copyright/Intel Corp. 1981)

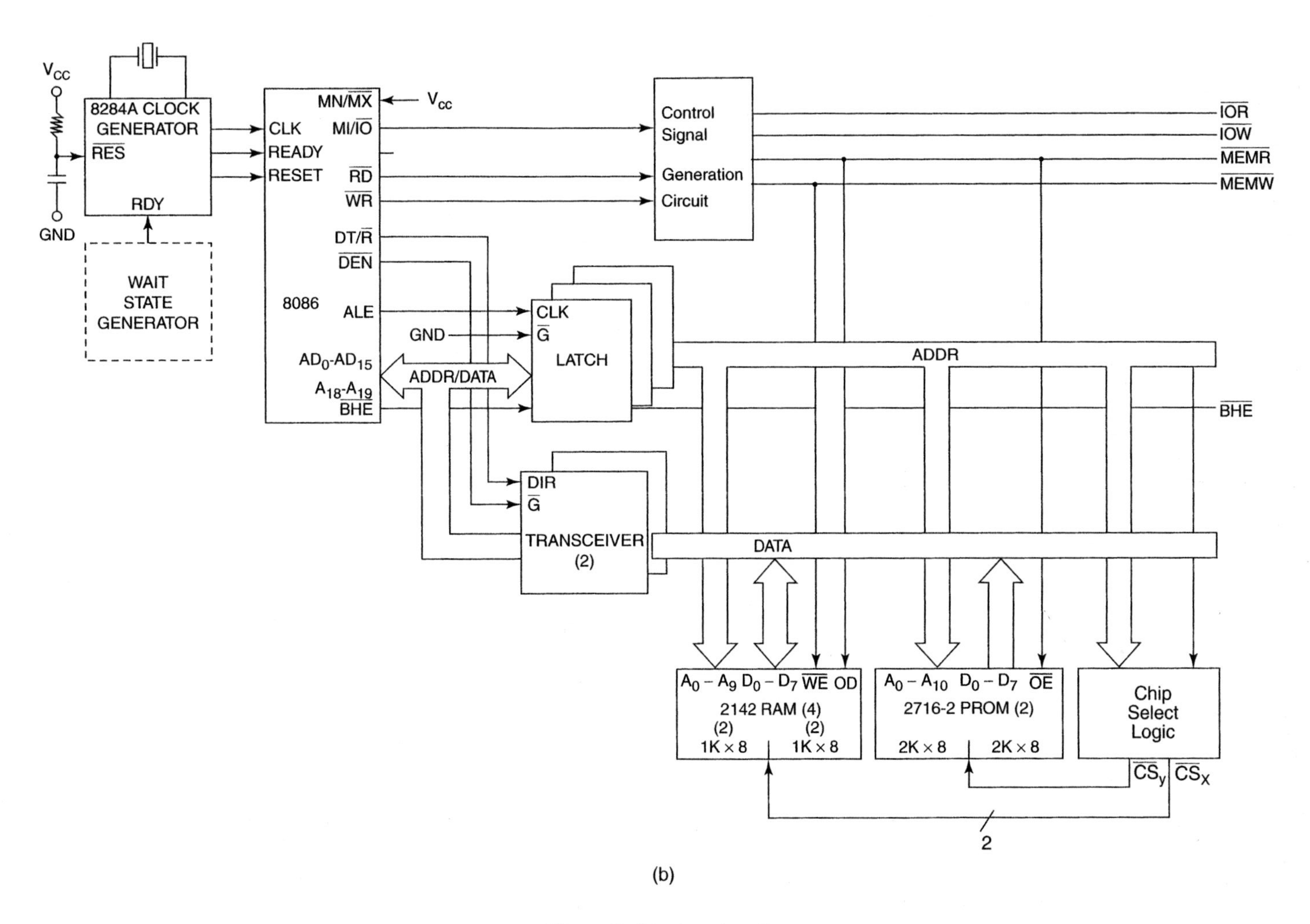

(b)

Figure 9–37 (continued)

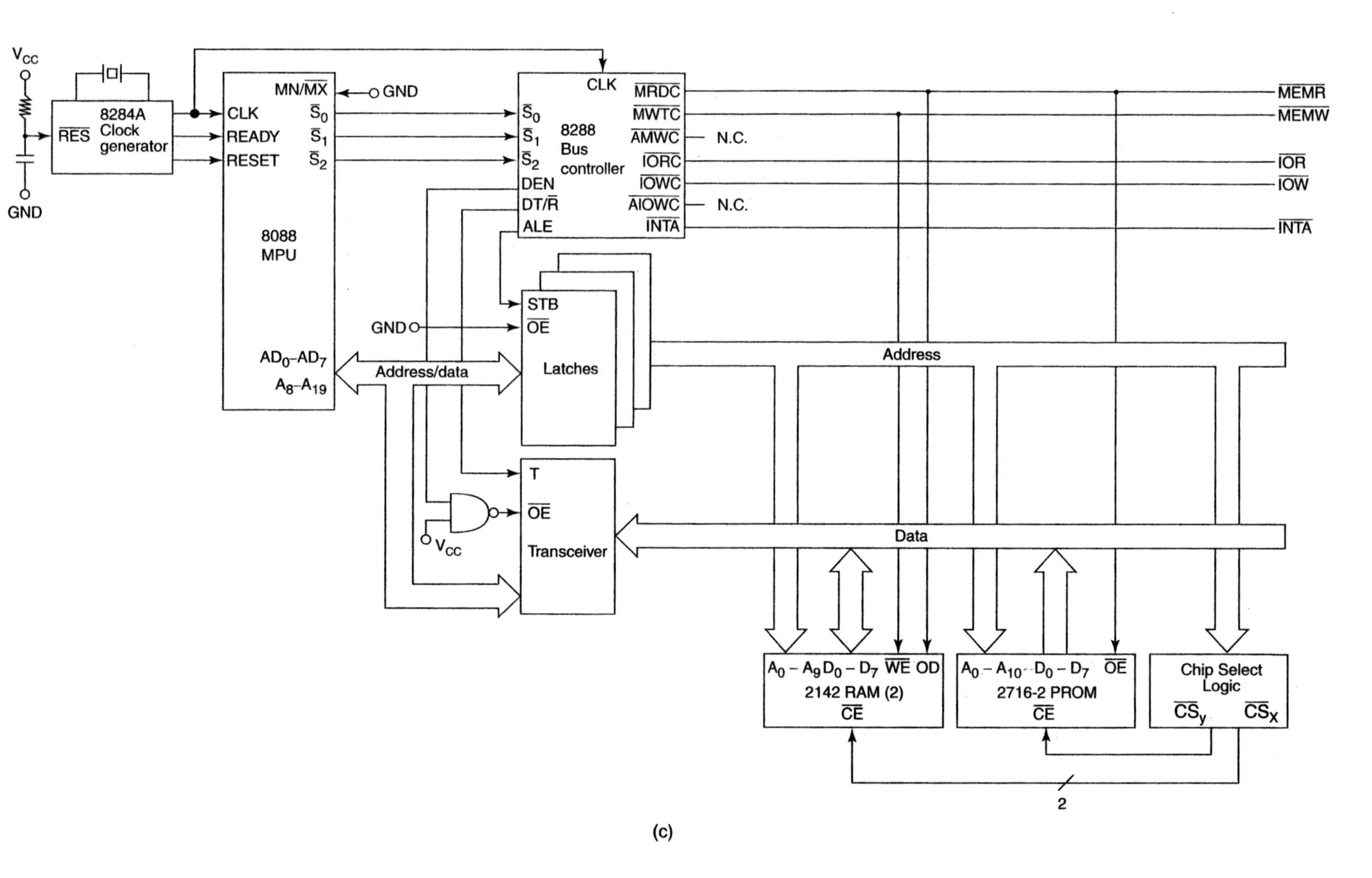

Figure 9–37 (continued)

Program Storage Memory

Earlier we found that program storage memory is used to store fixed information such as instructions of the program or tables of data. This part of the microcomputer's memory subsystem is typically implemented with nonvolatile memory devices, such as the ROM, PROM, EPROM, or FLASH memory. EPROM devices, such as the 2716, 2764, and 27C256, are organized with a byte-wide output; therefore, a single device is required to supply the 8-bit data bus of the 8088. They need to be arranged to provide a word-wide output when used in an 8086 system.

Figure 9–37(a) shows how a 2716 is connected to the demultiplexed system bus of a minimum-mode 8088-based microcomputer. This device supplies 2Kbytes of program storage memory. To select one of the 2K storage locations within the 2716, 11 bits of address are applied to address inputs A_0 through A_{10} of the EPROM. If A_0 through A_{10} of the 8088's address bus supply these inputs, the address range corresponding to program memory is from

$$A_{10}A_9 \ldots A_0 = 00000000000_2 = 00000_{16}$$

to

$$A_{10}A_9 \ldots A_0 = 11111111111_2 = 007FF_{16}$$

assuming that $A_{19} \ldots A_{11} = 0 \ldots 0$ generates the chip select signal $\overline{CS_x}$. Data outputs D_0 through D_7 of the EPROM are applied to data bus lines D_0 through D_7, respectively, of the 8088's system data bus. Data held at the addressed storage location are enabled onto the data bus by the control signal $\overline{MEMR}$ (memory read), which is applied to the $\overline{OE}$ (output enable) input of the EPROM.

In most applications, attaching several EPROM devices to the system bus expands the capacity of program storage memory. In this case, high-order bits of the 8088's address are decoded to produce chip-select signals. For instance, two address bits, A_{11} and A_{12}, can be decoded to provide four chip-select signals. Each of these chip-selects is applied to the $\overline{CE}$ (chip-enable) input of one EPROM. When an address is on the bus, just one of the outputs of the decoder becomes active and enables the corresponding EPROM for operation. By using four 2716s, the program storage memory is increased to 8Kbytes.

Now that we have explained how EPROMs are attached to the 8088's system bus, let us trace through the operation of the circuit for a bus cycle in which a byte of code is fetched from program storage memory. During an instruction acquisition bus cycle, the instruction fetch sequence of the 8088 causes the instruction to be read from memory byte by byte. The values in CS and IP are combined within the 8088 to give the address of a storage location in the address range of the program storage memory. This address is output on A_0 through A_{19} and latched into the address latches synchronously with the signal ALE. Bits A_0 through A_{10} of the system address bus are applied to the address inputs of the 2716. This part of the address selects the byte of code to be output. When the 8088 switches $\overline{RD}$ to logic 0 and $IO/\overline{M}$ to logic 0, the control signal generation circuit switches $\overline{MEMR}$ to logic 0. Logic 0 at $\overline{MEMR}$ enables the outputs of the 2716 and the byte of data at the addressed storage location is output onto system data bus lines D_0 through D_7. Early in the read bus cycle, the 8088 switches $DT/\overline{R}$ to logic 0 to signal the bus transceiver that data are to be input to the microprocessor, and later in the bus cycle $\overline{DEN}$ is switched to logic 0 to enable the transceiver for operation. Now the byte of data is passed from the system data bus onto the multiplexed address/data bus from which it is read by the MPU.

The circuit in Fig. 9–37(b) shows a similar circuit for a minimum-mode 8086 microcomputer system. Note that because of the 16-bit data bus, two octal transceivers and two EPROMs are required.

Figure 9–37(c) shows the program storage memory implementation for a maximum-mode 8088 microcomputer system. Let us look at how this circuit differs from the minimum-mode circuit of Fig. 9–37(a). The key difference in this circuit is that the 8288 bus controller is used to produce the control signals for the memory interface. Remember that in maximum mode the code output on status lines $\overline{S}_0$ through $\overline{S}_2$ identifies the type of bus cycle that is in progress. During all read operations of program memory, the 8088 outputs the instruction fetch memory bus status code, $\overline{S}_2\overline{S}_1\overline{S}_0 = 101$, to the 8288. In response to this input, the bus controller produces the memory read command ($\overline{MRDC}$) output, which is used as the $\overline{OE}$ input of the 2716 EPROM and enables it for data output.

In the maximum-mode circuit, the 8288, rather than the 8088, produces the control signals for the address latches and data bus transceiver. Notice that three address latches are again used, but this time the ALE output of the 8288 is used to strobe the memory address into these latches. ALE is applied to the STB inputs of all three latch devices in parallel. The direction of data transfer through the data bus transceiver is set by the $DT/\overline{R}$ output of the bus controller and the DEN output is used to generate the $\overline{OE}$ input of the transceiver. Since DEN, not $\overline{DEN}$, is produced by the 8288, an inverter is constructed from the NAND gate that drives $\overline{OE}$ of the transceiver.

Data Storage Memory

Information that frequently changes is stored in the data storage part of the microcomputer's memory subsystem (e.g., application programs and data). This part of the memory subsystem is normally implemented with random access read/write memory (RAM). If the amount of memory required in the microcomputer is small, for instance, less than 32Kbytes, the memory subsystem will usually be designed with static RAMs. On the other hand, systems that require a larger amount of data storage memory normally use dynamic RAMs (DRAMs), which provide larger storage capacity in the same size package. DRAMs require refresh support circuits. This additional circuitry is not warranted if storage requirements are small.

A 1Kbyte random access read/write memory is also supplied in the minimum-mode 8088-based microcomputer circuit in Fig. 9–37(a). This part of the memory subsystem is implemented with two 2142 static RAM ICs. Each 2142 contains 1K, 4-bit storage locations; therefore, they both supply storage for just 4 bits of the byte. The storage location to be accessed is selected by a 10-bit address, which is applied to both RAMs in parallel over address lines A_0 through A_9. Data are read from or written into the selected storage location over data bus lines D_0 through D_7. Of course, through software, the 8088 can read data from memory either as bytes, words, or double words. The logic level of $\overline{MEMW}$ (memory write), which is applied to the write enable ($\overline{WE}$) input of both RAMs in parallel, signals whether a read or write bus cycle is in progress. $\overline{\text{MEMR}}$ is applied to the OD (output disable) input of both RAMs in parallel. When a write cycle is in progress, $\overline{\text{RD}}$ is at logic 1, which disables the outputs of the RAMs. Now the data lines act as inputs.

Just as for program storage memory, simply attaching additional banks of static RAMs to the system bus can expand data storage memory. Once again, high-order address bits can be decoded to produce chip-select signals. Each chip-select output is applied to the chip-enable input of both RAMs in a bank and, when active, enables that bank of RAMs for operation.

Let us assume that the value of a byte-wide data operand is to be updated in memory. In this case, the 8088 must perform a write bus cycle to the address of the operand's storage location. First, the address of the operand is formed and output on the multiplexed address/data bus. When the address is stable, a pulse at ALE is used to latch it into the address latches. Bits A_0 through A_9 of the system address bus are applied to the address inputs of the 2142s. This part of the address selects the storage location into which the byte of data is to be written. Next the 8088 switches DT/$\overline{\text{R}}$ to logic 1 to signal the octal transceivers that data are to be output to memory. Later in the bus cycle, $\overline{\text{DEN}}$ is switched

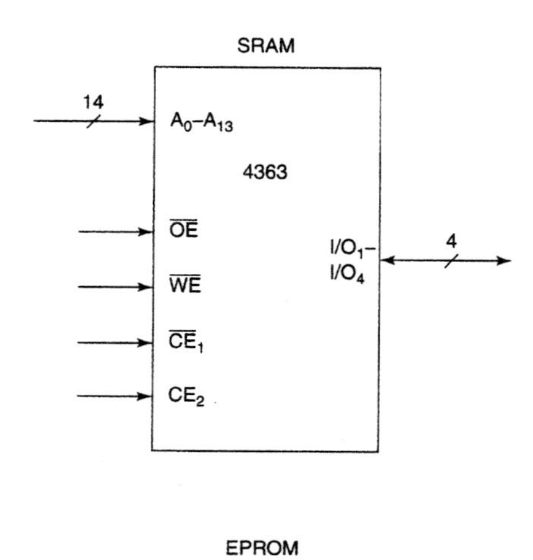

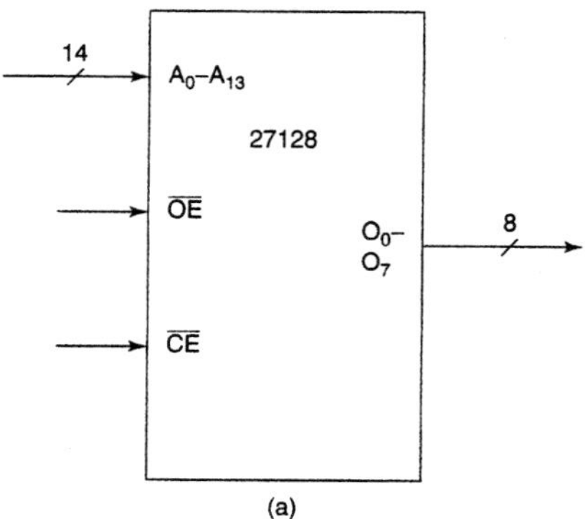

Figure 9–38 (a) Devices to be used in the system design of Example 9.3 (p. 458). (b) Memory map of the system to be designed. (c) Memory organization for the system design. (d) Address range analysis for the design of chip select signals $\overline{\text{CS}_0}$, $\overline{\text{CS}_1}$, $\overline{\text{CS}_2}$, and $\overline{\text{CS}_3}$. (e) Chip-select logic.

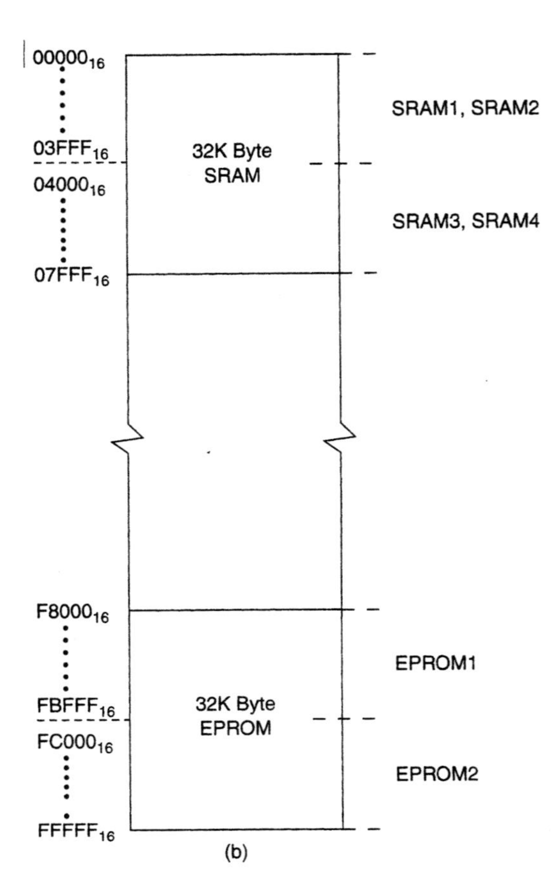

Figure 9–38 (continued)

to logic 0 to enable the data bus transceiver for operation. Now the byte of data is output on the multiplexed address/data bus and passed through the transceiver to the system data bus and data inputs of the RAMs. Finally, the byte of data is written into the addressed storage location synchronously with the occurrence of the $\overline{\text{MEMW}}$ control signal.

The data storage memory circuitry of a minimum-mode 8086 system is also shown in Fig. 9–37(b). Here we see that two banks of RAM ICs are required.

Figure 9–37(c) shows the data storage memory circuit of a maximum-mode 8088 microcomputer. Similar to our description of the program storage memory part of this circuit, the difference between the maximum-mode and minimum-mode data storage memory circuits lies in the fact that the 8288 bus controller produces the control signals for the memory and bus interface logic devices. When the 8088 is accessing data storage memory, it outputs either the read memory (101) or write memory (110) bus status code. These codes are decoded by the 8288 to produce appropriate memory control signals. For instance, the status code 110 (write memory) causes the memory write command ($\overline{\text{MWTC}}$) and advanced memory write command ($\overline{\text{AMWC}}$) outputs to become active during all write bus cycles. Figure 9–37(c) shows that $\overline{\text{MWTC}}$ or $\overline{\text{MEMW}}$ is used to drive the $\overline{\text{WE}}$ input of the 2142 SRAMs. When $\overline{\text{MWTC}}$ is at its active 0 logic level, the input

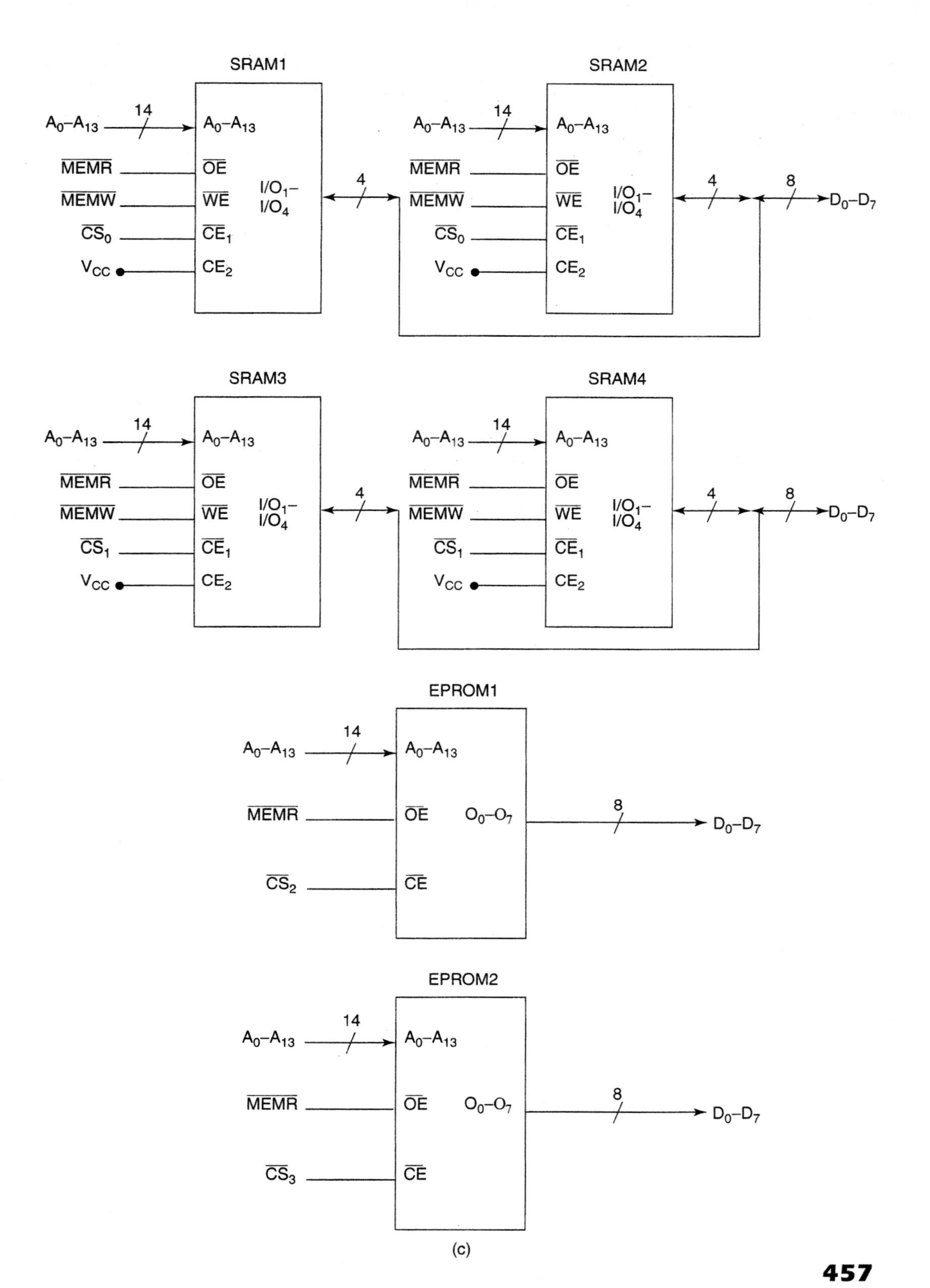

Figure 9–38 (continued)

$$
\begin{array}{l}
\qquad\qquad A_{19}\cdots\cdots\cdots\cdots A_0 \\
00000_{16} = 0000\ 0000\ 0000\ 0000\ 0000 \\
\vdots \\
03FFF_{16} = \underbrace{0000\ 0011}_{\overline{CS}_0}\ 1111\ 1111\ 1111 \\
04000_{16} = 0000\ 0100\ 0000\ 0000\ 0000 \\
\vdots \\
07FFF_{16} = \underbrace{0000\ 0111}_{\overline{CS}_1}\ 1111\ 1111\ 1111 \\
F8000_{16} = 1111\ 1000\ 0000\ 0000\ 0000 \\
\vdots \\
FBFFF_{16} = \underbrace{1111\ 1011}_{\overline{CS}_2}\ 1111\ 1111\ 1111 \\
FC000_{16} = 1111\ 1100\ 0000\ 0000\ 0000 \\
\vdots \\
FFFFF_{16} = \underbrace{1111\ 1111}_{\overline{CS}_3}\ 1111\ 1111\ 1111
\end{array}
$$

(d) **Figure 9–38** (continued)

buffers of the SRAMs are enabled for operation. On the other hand, during read bus cycles, $\overline{MRDC}$ or $\overline{MEMR}$ is used to enable the outputs of the SRAMs.

EXAMPLE 9.3

Design a memory system consisting of 32Kbytes of R/W memory and 32Kbytes of ROM memory. Use SRAM devices to implement R/W memory and EPROM devices to implement ROM memory. The memory devices to be used are shown in Fig. 9–38(a) (p. 455). R/W memory is to reside over the address range 00000_{16} through $07FFF_{16}$ and the address range of ROM memory is to be $F8000_{16}$ through $FFFFF_{16}$. Assume that the 8088 microprocessor system bus signals that follow are available for use: A_0 through A_{19}, D_0 through D_7, $\overline{MEMR}$, and $\overline{MEMW}$.

Solution

First let us determine the number of SRAM devices needed to implement the R/W memory. Since each device provides $2^{14} \times 4$ or $16K \times 4$ of storage, the number of SRAM devices needed to implement 32Kbytes of storage is

$$\text{No. of SRAM devices} = 32\text{Kbyte}/(16K \times 4) = 4$$

To provide an 8-bit data bus, two SRAMs must be connected in parallel. Two pairs connected in this way are then placed in series to implement the R/W address range, and each

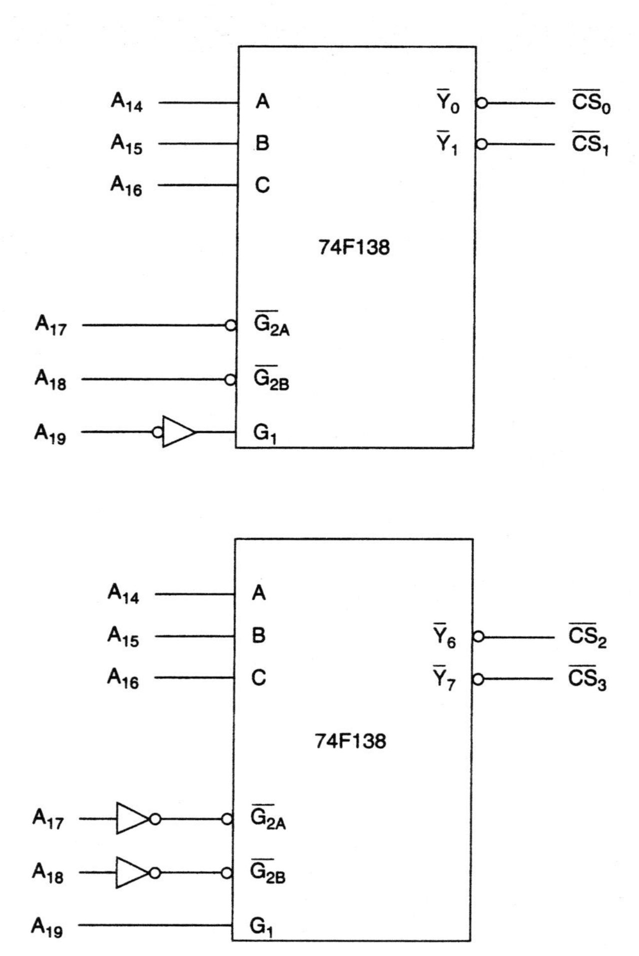

Figure 9–38 (continued)

pair implements 16Kbytes. The first pair, $SRAM_1$ and $SRAM_2$, implements the address range 00000_{16} through $03FFF_{16}$, and the second pair, $SRAM_3$ and $SRAM_4$, implements addresses 04000_{16} through $07FFF_{16}$. The memory map in Fig. 9–38(b) shows the device allocation for this implementation.

Next let us determine the number of EPROM devices that are needed to implement the ROM memory. In this case, each device provides $2^{14} \times 8$ or 16Kbytes of storage. To implement 32Kbytes of storage, the number of EPROM devices needed is

$$\text{No. of EPROM devices} = 32\text{Kbyte}/16\text{Kbyte} = 2$$

These two devices must be connected in series to implement the ROM address range and each device implements 16Kbytes of storage. As shown in the memory map in Fig. 9–38(b), the first device, $EPROM_1$, implements the address range $F8000_{16}$ through $FBFFF_{16}$. The second device, $EPROM_2$, implements the address range $FC000_{16}$ through $FFFFF_{16}$.

The memory organization based on the preceding allocation of devices is shown in Fig. 9–38(c). Notice that we have used the various 8088 system bus signals ($A_0 - A_{19}$, $D_0 - D_7$, $\overline{MEMR}$, and $\overline{MEMW}$) to draw the circuit diagram. For example, the $\overline{MEMW}$ signal is applied to the $\overline{WE}$ input of all four SRAMs in parallel, but it is not connected to the EPROMs.

The four chip select signals, $\overline{CS_0}$, $\overline{CS_1}$, $\overline{CS_2}$, and $\overline{CS_3}$, that are used in the circuit need to be produced for the appropriate address ranges. To design the circuit for generating the chip-select signals, we first analyze the address ranges as shown in Fig. 9–38(d) to determine the address bits that should be used. For instance, to generate the range represented by $SRAM_1$ and $SRAM_2$, $\overline{CS_0}$ should be active for $A_{19}A_{18}A_{17}A_{16}A_{15}A_{14} = 000000_2$. Similarly the other address ranges tell us which address bits are needed to produce the other chip-select signals. This information is used in Fig. 9–38(e) to design the chip-select logic circuit with 74F138 three-line to eight-line decoders.

REVIEW PROBLEMS

Section 9.1

1. Which part of the primary storage memory is used to store instructions of the program and fixed information such as constant data and lookup tables? Data that changes frequently?
2. What does BIOS stand for?
3. What term is used to refer to programs stored in ROM?
4. Can DRAMs be used to construct a program storage memory?

Section 9.2

5. What is meant by the term *nonvolatile memory*?
6. What does PROM stand for? EPROM?
7. What must an EPROM be exposed to in order to erase its stored data?
8. If the block diagram in Fig. 9–3 has address lines A_0 through A_{16} and data lines D_0 through D_7, what are its bit density and byte capacity?
9. Summarize the read cycle of an EPROM. Assume that both $\overline{CE}$ and $\overline{OE}$ are active before the address is applied.
10. Which standard EPROM stores 64K 8-bit words?
11. What is the difference between a 27C64A and a 27C64A-1?
12. What are the values of V_{CC} and V_{pp} for the intelligent programming algorithm?
13. What is the duration of the programming pulses used for the intelligent programming algorithm?

Section 9.3

14. What do SRAM and DRAM stand for?
15. Are RAM ICs examples of nonvolatile or volatile memory devices?

16. What must be done to maintain valid data in a DRAM device?
17. Find the total storage capacity of the circuit similar to Fig. 9–12 if the memory devices are 43256As.
18. List the minimum values of each of the write cycle parameters that follow for the 4364-10 SRAM: t_{WC}, t_{CW1}, t_{CW2}, t_{WP}, t_{DW}, and t_{WR}.
19. Give two benefits of DRAMs over SRAMs.
20. Name the two parts of a DRAM address.
21. Show how the circuit in Fig. 9–22 can be expanded to 128K × 16 bits.
22. Give a disadvantage of the use of DRAMs in an application that does not require a large amount of memory.

Section 9.4

23. What type of circuit can be added to the data storage memory interface to improve the reliability of data transfers over the data bus?
24. If in Fig. 9–23 the data read from memory is 100100100_2 , what is its parity? Repeat the same if the data is 011110000_2?
25. If the input to a 74AS280 parity-checker/generator circuit that is set up for odd parity checking and generation is IH . . . A = 111111111_2, what are its outputs?
26. What changes must be made to the circuit in Fig. 9–24(c) to convert it to an odd parity configuration?
27. Make a drawing similar to that shown in Fig 9–24(c) that can be used as the parity-checker/generator in the data storage memory subsystem of an 8086 microcomputer system. Assume that parity checking is performed independently for the upper and lower banks of the memory array and that the parity error outputs for the two banks are combined to form a single parity error signal.

Section 9.5

28. What is the key difference between a FLASH memory and an EPROM?
29. What is the key difference between the bulk-erase architecture and the boot block or FlashFile architectures?
30. What is the key difference between the boot block architecture and the FlashFile architecture?
31. What architecture is used in the 28F010 FLASH memory IC?
32. What power supply voltage must be applied to a 28F010 device when it is being erased or written into?
33. Give the names of two boot block FLASH devices.
34. What value V_{cc} power supply voltages can be applied to a Smart 5 boot block FLASH IC? What value V_{pp} power supply voltages?
35. Name the three types of blocks used in the storage array of the 28F004. How many of each is provided? What are their sizes?
36. Name two FlashFile FLASH memory devices.
37. What is the function of the RY/$\overline{BY}$ output of the 28F016SA/SV?

Section 9.6

38. What function does a wait-state generator circuit perform?

39. What output signal does the wait-state generator produce?

40. Does the circuit in Fig. 9–36(b) produce the same number of wait states for the memory subsystems corresponding to both chip selects?

41. What is the maximum number of wait states that can be produced with the circuit in Fig. 9–36(b)?

Section 9.7

42. Make a diagram showing how 2764 EPROMs can be connected to form a 16Kbyte program storage memory subsystem. Also show a 16Kword program memory subsystem.

43. If we assume that the high-order address bits in the circuits formed in problem 42 are all logic 0, what is the address range of the program memory subsystems?

44. How many 2142 static RAMs would be needed in the memory array of the circuit in Fig. 9–37(a) if the capacity of data storage memory were to be expanded to 64Kbytes?

45. How many 2716 EPROMs would be needed in the program memory array in the circuit of Fig. 9–37(a) to expand its capacity to 96K bits? If 2732s were used instead of 2716s, how many devices are needed to implement the 96K-bit program memory?

46. Repeat the design in Example 9.3 for the 8086 microprocessor system bus signals A_0 through A_{19}, D_0 through D_{15}, $\overline{MEMR}$, $\overline{MEMW}$, and $\overline{BHE}$. Use the same memory and device specifications.

10

Input/Output Interface Circuits and LSI Peripheral Devices

▲ INTRODUCTION

In Chapter 8 we introduced the input/output interface of the 8088 and 8086 microprocessors. At that time, we discussed the topics of isolated and memory-mapped I/O, minimum- and maximum-mode isolated I/O interface signals, I/O bus cycles, and I/O instructions. Here we continue our study of input/output by examining circuits and large-scale-integrated peripheral ICs that are used to implement input/output subsystems for the microcomputer systems. The following topics are covered here:

10.1 Core and Special-Purpose I/O Interfaces
10.2 Byte-Wide Output Ports Using Isolated I/O
10.3 Byte-Wide Input Ports Using Isolated I/O
10.4 Input/Output Handshaking and a Parallel Printer Interface
10.5 82C55A Programmable Peripheral Interface
10.6 82C55A Implementation of Parallel Input/Output Ports
10.7 Memory-Mapped Input/Output Ports
10.8 82C54 Programmable Interval Timer
10.9 82C37A Programmable Direct Memory Access Controller
10.10 Serial Communications Interface

▲ 10.1 CORE AND SPECIAL-PURPOSE I/O INTERFACES

In Chapter 1 we indicated that the input/output unit provides the microcomputer with the means for communicating with the outside world. For instance, the PC keyboard permits the user to input information such as programs or data for an application. The display outputs information about the program or application for the user to read. These examples represent one type of input/output function, which we will call *special-purpose I/O interfaces.* Other examples of special-purpose I/O interfaces are parallel printer interfaces, serial communication interfaces, and local area network interfaces. They are referred to as special-purpose interfaces because not all microcomputer systems employ each of these types.

The original PC is capable of supporting a variety of input/output interfaces. In fact all the interfaces just mentioned are available for the PC. Since they are special-purpose interfaces, they are all implemented in the original PC as add-on cards. That is, to support a keyboard and display interface on the PC, a special keyboard/display controller card is inserted into a slot of the PC and then the keyboard and display are attached to the card with cables connected at connectors.

In microcomputer circuit design, various other types of circuits is also classified as input/output circuitry. Parallel input/output ports, interval timers, and direct memory access control are examples of interfaces that are also considered to be part of the I/O subsystem. These I/O functions are employed by most microcomputer systems. For this reason, we will refer to them as *core input/output interfaces.*

The core I/O functions are not as visible to the user of the microcomputer; however, they are just as important to overall microcomputer function. The circuitry of the original PC contains all core microcomputer functions. For example, parallel I/O is the method used to read the settings of the DIP switches on the processor board. Also an interval time is used as part of the DRAM refresh process and to keep track of the time of day. The circuitry for these core I/O functions is built right on the PC's main processor board. They are also included as part of the MPU IC in some highly integrated processors, such as the 80C188XL and 80C186XL.

In the sections that follow, we explore the circuits and operations of both the core and special-purpose input/output functions.

▲ 10.2 BYTE-WIDE OUTPUT PORTS USING ISOLATED I/O

We start with circuits that can be used to implement parallel output ports in a microcomputer system employing isolated I/O. Figure 10–1(a) shows such a circuit for an 8088-based microcomputer, which provides eight byte-wide output ports that are implemented using 74F374 octal latches. In this circuit, the ports are labeled port 0 through port 7. These eight ports give a total of 64 parallel output lines, which are labeled O_0 through O_{63}.

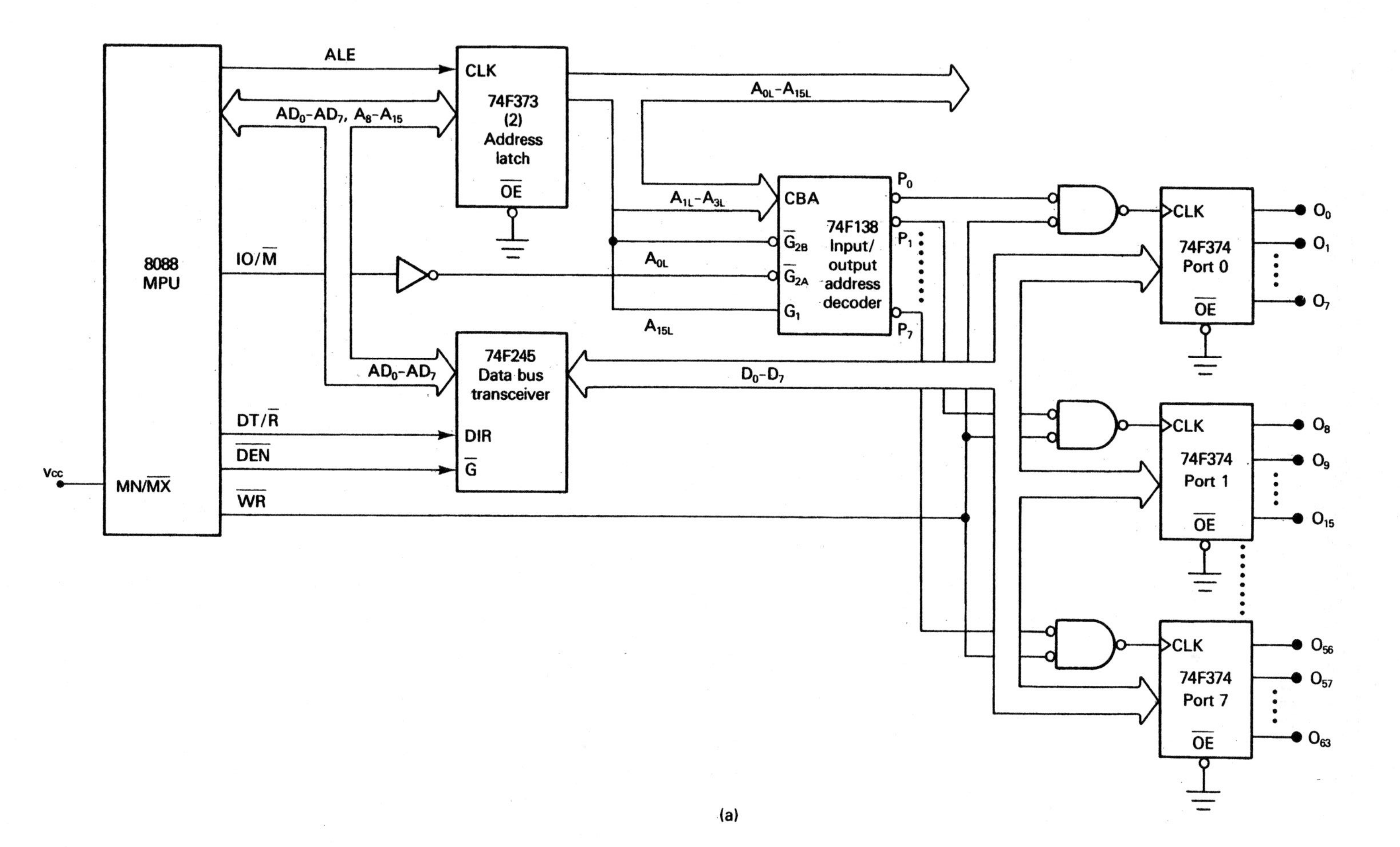

Figure 10–1 (a) Sixty-four-line parallel output circuit for an 8088-based microcomputer. (b) I/O address decoding for ports 0 through 7. (c) Sixty-four-line parallel output circuit for an 8086-based microcomputer.

I/O port	I/O address
Port 0	$1XXXXXXXXXXX0000_2$
Port 1	$1XXXXXXXXXXX0010_2$
Port 2	$1XXXXXXXXXXX0100_2$
Port 3	$1XXXXXXXXXXX0110_2$
Port 4	$1XXXXXXXXXXX1000_2$
Port 5	$1XXXXXXXXXXX1010_2$
Port 6	$1XXXXXXXXXXX1100_2$
Port 7	$1XXXXXXXXXXX1110_2$

(b)

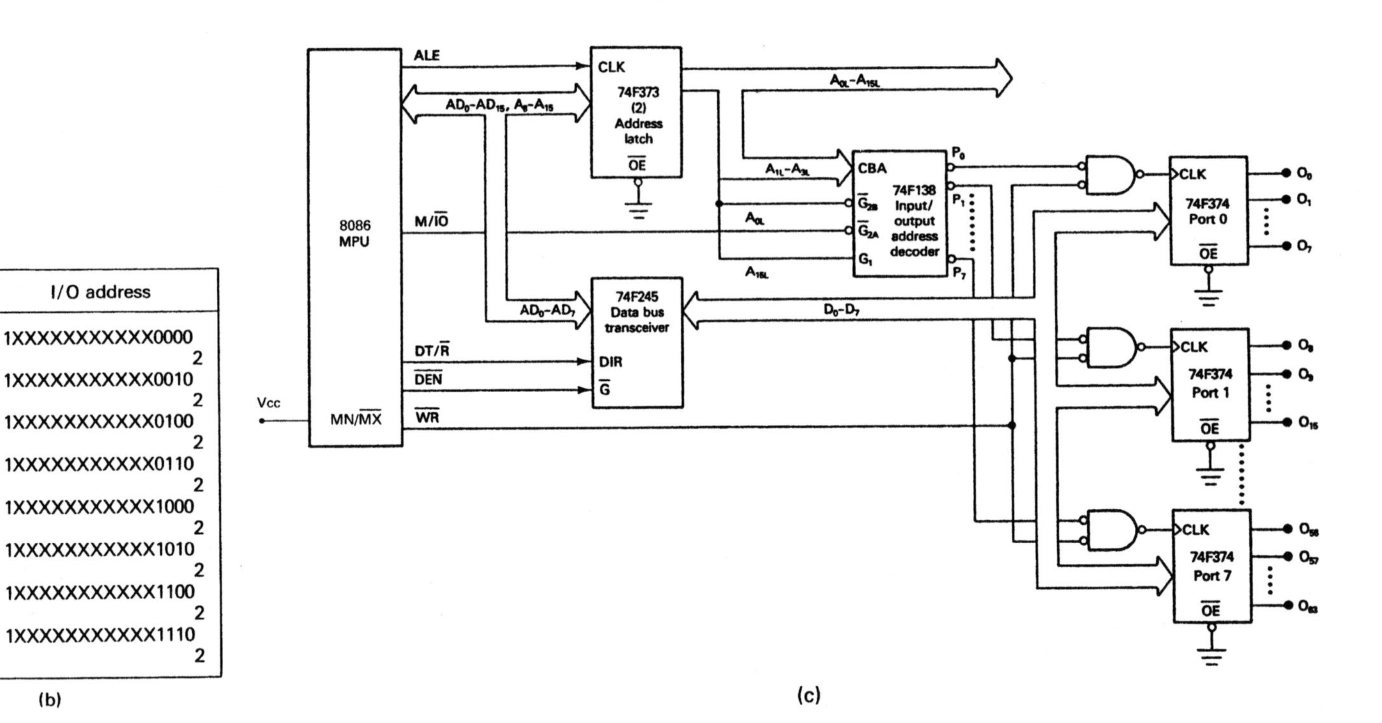

(c)

Figure 10–1 (continued)

Looking at the circuit, we see that the 8088's address/data bus is demultiplexed just as was done for the memory interface. Note that two 74F373 octal latches are used to form a 16-bit address latch. These devices latch the address A_0 through A_{15} synchronously with the ALE pulse. The latched address outputs are labeled A_{0L} through A_{15L}. Remember that address lines A_{16} through A_{19} are not involved in the I/O interface. For this reason, they are not shown in the circuit diagram.

Address/data bus lines AD_0 through AD_7 are also applied to one side of the 74F245 bus transceiver. At the other side of the transceiver, data bus lines D_0 through D_7 are shown connecting to the input side of the output latches. It is over these lines that the 8088 writes data into the output ports.

Address lines A_{0L} and A_{15L} provide two of the three enable inputs of the 74F138 input/output address decoder. These signals are applied to enable inputs $\overline{G}_{2B}$ and G_1, respectively. The decoder requires one more enable signal at its $\overline{G}_{2A}$ input, which is supplied by the complement of IO/$\overline{M}$. The enable inputs must be $\overline{G}_{2B}\overline{G}_{2A}G_1 = 001$ to enable the decoder for operation. The condition $\overline{G}_{2B} = 0$ corresponds to an even address, and $\overline{G}_{2A} = 0$ represents the fact that an I/O bus cycle is in progress. The third condition, $G_1 = 1$, is an additional requirement that A_{15L} be at logic 1 during all data transfers for this section of parallel output ports.

Note that three address lines, $A_{3L}A_{2L}A_{1L}$, are applied to select inputs CBA of the 74F138 3 line-to-8 line decoder. When the decoder is enabled, the P output corresponding to these select inputs switches to logic 0. Logic 0 at this output enables the $\overline{WR}$ signal to the clock (CLK) input of the corresponding output latch. In this way, just one of the eight ports is selected for operation.

When valid output data are on D_0 through D_7, the 8088 switches $\overline{WR}$ to logic 0. This change in logic level causes the selected 74F374 device to latch in the data from the bus. The 0 logic level at their $\overline{OE}$ inputs of the latches permanently enables the outputs. Therefore, the latched data appear at the appropriate port outputs.

The 74F245 in the circuit allows the data being output to pass from the 8088 to the output ports. Enabling the 74F245's DIR and $\overline{G}$ inputs with the DT/$\overline{R}$ and $\overline{DEN}$ signals, which are at logic 1 and 0, respectively, accomplishes this.

Note in Fig. 10–1(a) that not all address bits are used in the I/O address decoding. Here only latched address bits A_{0L}, A_{1L}, A_{2L}, A_{3L}, and A_{15L} are decoded. Figure 10–1(b) shows the addresses that select each of the I/O ports. Unused bits are shown as don't-care states. By assigning various logic combinations to the unused bits, the same port can be selected by different addresses. In this way, we see that many addresses will decode to select each of the I/O ports. For instance, if all don't-care address bits are made 0, the address of port 0 is

$$1000000000000000_2 = 8000_{16}$$

However, if these bits are all made equal to 1 instead of 0, the address is

$$1111111111110000_2 = FFF0_{16}$$

and it still decodes to enable port 0. In fact, every I/O address in the range from 8000_{16} through $FFF0_{16}$ that has its lower four bits equal to 0000_2 decodes to enable port 0. Some other examples are $8FF0_{16}$ and $F000_{16}$.

EXAMPLE 10.1

To which output port in Fig. 10–1(a) are data written when the address put on the bus during an output bus cycle is 8002_{16}?

Solution

Expressing the address in binary form, we get

$$A_{15} \ldots A_0 = A_{15L} \ldots A_{0L} = 1000000000000010_2$$

That is,

$$A_{15L} = 1$$
$$A_{0L} = 0$$

and

$$A_{3L}A_{2L}A_{IL} = 001$$

Moreover, whenever an output bus cycle is in progress, $IO/\overline{M}$ is logic 1. Therefore, the enable inputs of the 74F138 decoder are

$$\overline{G}_{2B} = A_{0L} \quad = 0$$
$$\overline{G}_{2A} = \overline{IO/\overline{M}} = 0$$
$$G_1 = A_{15L} \quad = 1$$

These inputs enable the decoder for operation. At the same time, its select inputs are supplied with the code 001. This input causes output P_1 to switch to logic 0:

$$P_1 = 0$$

The gate at the CLK input of port 1 has as its inputs P_1 and $\overline{WR}$. When valid output data are on the bus, $\overline{WR}$ switches to logic 0. Since P_1 is also 0, the CLK input of the 74F374 for port 1 switches to logic 0. At the end of the $\overline{WR}$ pulse, the clock switches from 0 to 1, a positive transition. This causes the data on D_0 through D_7 to be latched and become available at output lines O_8 through O_{15} of port 1.

EXAMPLE 10.2

Write a series of instructions that will output the byte contents of the memory address DATA to output port 0 in the circuit of Fig. 10–1(a).

Solution

To write a byte to output port 0, the address that must be output on the 8088's address bus is

$$A_{15}A_{14} \ldots A_0 = 1XXXXXXXXXXX0000_2$$

Assuming that the don't-care bits are all made logic 0, we get

$$A_{15}A_{14} \ldots A_0 = 1000000000000000_2$$
$$= 8000_{16}$$

The instruction sequence needed to output the contents of memory address DATA to port 0 is

```
MOV  DX, 8000H
MOV  AL, [DATA]
OUT  DX, AL
```

Figure 10–1(c) shows a similar output circuit for an 8086-based microcomputer system. Here again, 64 output lines are implemented as 8 byte-wide parallel ports, port 0 through port 7. Comparing this circuit to that for an 8088-based microcomputer in Fig. 10–1(a), we find just one difference, that the control signal $M/\overline{IO}$ is applied directly to the $\overline{G}_{2A}$ input of the 74F138 input/output address decoder. Since $M/\overline{IO}$ is the complement of the 8088's $IO/\overline{M}$ signal, it does not have to be inverted.

Time-Delay Loop and Blinking an LED at an Output Port

The circuit in Fig. 10–2 has an LED attached to output O_7 of parallel port 0. This circuit is identical to that shown in Fig. 10–1(a). Therefore, the port address as found in Example 10.2 is 8000H, and the LED corresponds to bit 7 of the byte of data that is written to port 0. For the LED to turn on, O_7 must be switched to logic 0, and it will remain on until this output is switched back to 1. The 74F374 is not an inverting latch; therefore, to make O_7 logic 0, we simply write 0 to that bit of the octal latch. To make the LED blink, we must write a program that first makes O_7 logic 0 to turn on the LED, delays for a short period of time, and then switches O_7 back to 1 to turn off the LED. This piece of program can run as a loop to make the LED continuously blink.

Let us begin by writing the sequence of instructions needed to initialize O_7 to logic 0. This is done as follows:

```
           MOV  DX, 8000H    ;Initialize address of port 0
           MOV  AL, 00H      ;Load data with bit 7 as logic 0
ON_OFF:    OUT  DX, AL       ;Output the data to port 0
```

After the out operation is performed, the LED will be turned on.

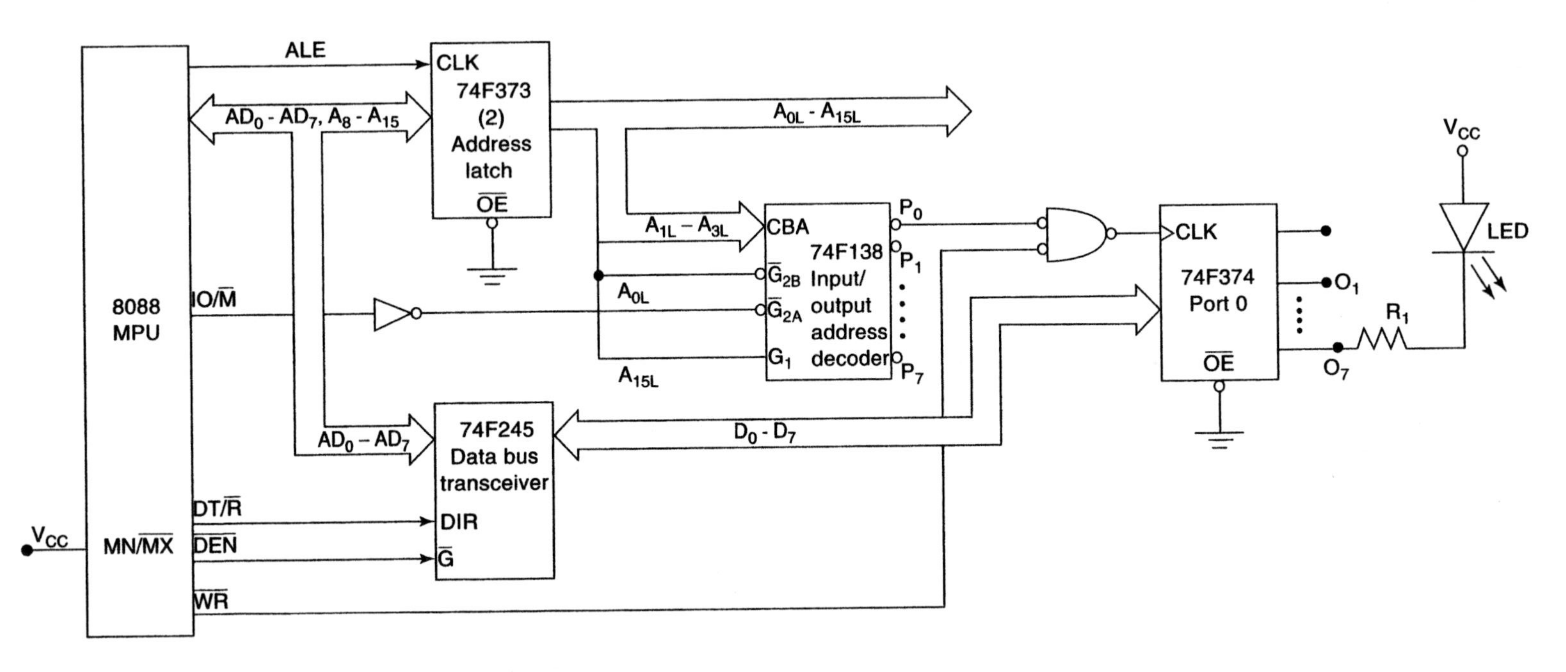

Figure 10–2 Driving an LED connected to an output port.

Next we must delay for a short period of time so as to maintain the data written to the LED. This can be done with a software loop. The following instruction sequence produces such a delay:

```
            MOV CX, 0FFFFH      ;Load delay count of FFFFH
HERE:       LOOP HERE           ;Time delay loop
```

First the count register is loaded with $FFFF_{16}$. Then the loop instruction is executed repeatedly. With each occurrence of the loop, the count in CX is decremented by 1. After 65,335 repeats of the loop, the count in CX is 0000_{16} and the loop operation is complete. These executions perform no software function for the program other than to use time, which is the duration of the time delay. By using $FFFF_{16}$ as the count, the maximum delay is obtained. Loading a smaller number in CX shortens the duration of the delay.

Next the value in bit 7 of AL is complemented to 1 and then a jump is performed to return to the output operation that writes the data to the output port:

```
XOR  AL, 80H     ;Complement bit 7 of AL
JMP  ON_OFF      ;Repeat to output the new bit 7
```

By performing an exclusive-OR operation on the value in AL with the value 80_{16}, the most significant bit is complemented to 1. The jump instruction returns control to the OUT instruction. Now the new value in AL, with MSB equal to 1, is output to port 0, and the LED turns off. After this, the time delay repeats, the value in AL is complemented back to 00_{16}, and the LED turns back on. In this way, we see that the LED blinks repeatedly with an equal period of on and off time that is set by the count in CX.

▲ 10.3 BYTE-WIDE INPUT PORTS USING ISOLATED I/O

In Section 10.2, we showed circuits that implemented eight byte-wide output ports for the 8088- and 8086-microcomputer systems. These circuits used the 74F374 octal latch to provide the output ports. Here we will examine a similar circuit that implements input ports for the microcomputer system.

The circuit in Fig. 10–3 provides eight byte-wide input ports for an 8088-based microcomputer system employing isolated I/O. Just like in the output circuit in Fig. 10–1(a), the ports are labeled port 0 through port 7; however, this time the 64 parallel port lines are inputs, I_0 through I_{63}. Note that eight 74F244 octal buffers are used to implement the ports. The outputs of the buffers are applied to the data bus for input to the MPU. These buffers have three-state outputs.

When an input bus cycle is in progress, the I/O address selects the port whose data are to be input. First A_0 through A_{15} is latched into the 74F373 address latches. This address is accompanied by logic 1 on the $IO/\overline{M}$ control line. Note that $IO/\overline{M}$ is inverted and applied to the $\overline{G}_{2A}$ input of the I/O address decoder. If during the bus cycle address bit $A_{0L} = 0$ and $A_{15L} = 1$, the address decoder is enabled for operation. Then the code $A_{3L}A_{2L}A_{1L}$ is decoded to produce an active logic level at one of the decoder's outputs. For instance, an input of $A_{3L}A_{2L}A_{1L} = 001$ switches the P_1 output to logic 0. P_1 is gated with $\overline{RD}$ to produce the $\overline{G}$ enable input for the port 1 buffer. If both $IO/\overline{M}$ and P_1 are logic 0, the $\overline{G}$ input for port 1 is switched to logic 0 and the outputs of the 74F244 are enabled.

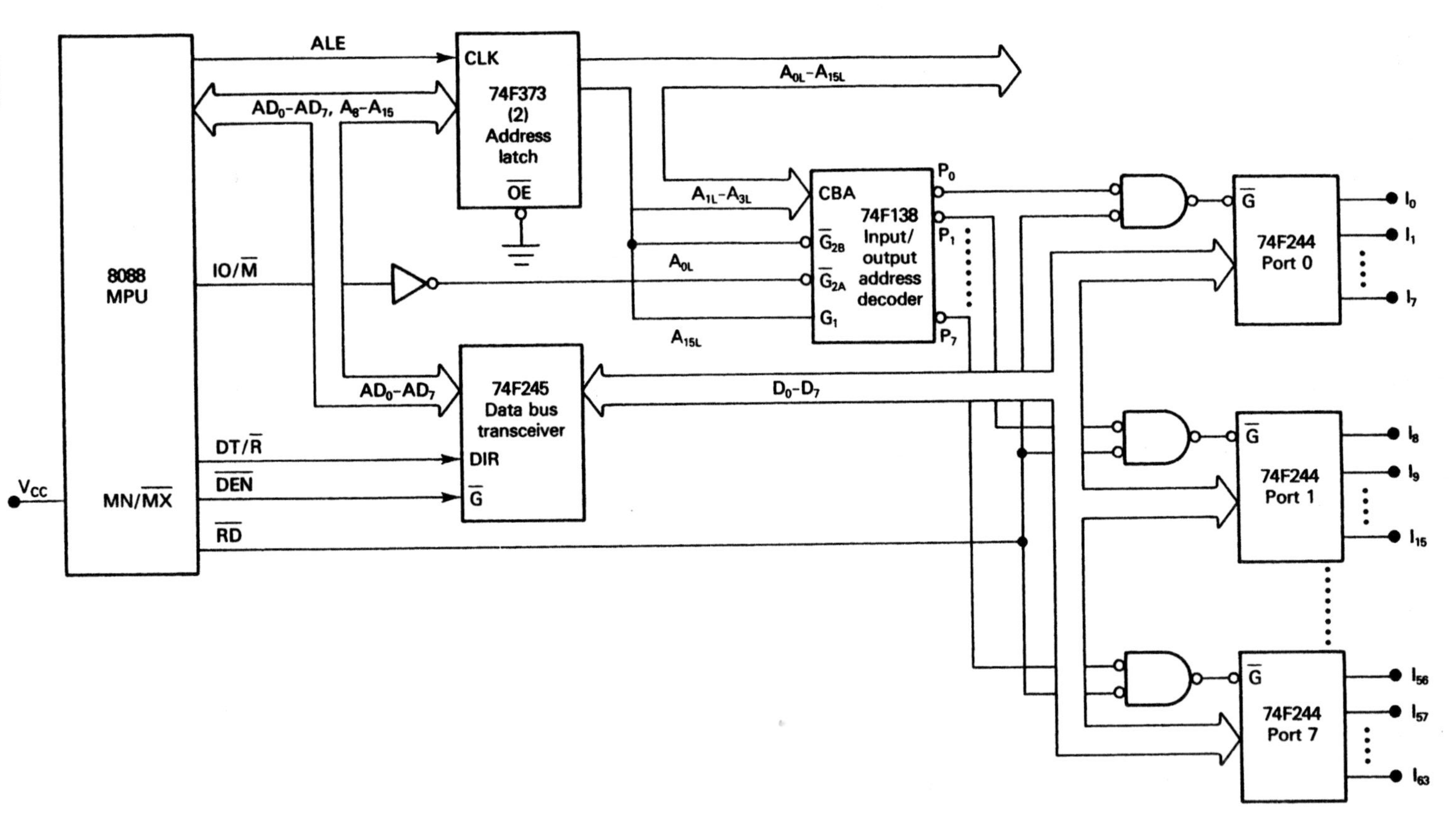

Figure 10–3 Sixty-four-line parallel input circuit for an 8088-based microcomputer.

In this case, the logic levels at inputs I_8 through I_{15} are passed onto data bus lines D_0 through D_7, respectively. This byte of data is carried through the enabled data bus transceiver to the data bus of the 8088. As part of the input operation, the 8088 reads this byte of data into the AL register.

EXAMPLE 10.3

What is the I/O address of port 7 in the circuit of Fig. 10–3? Assume all unused address bits are at logic 0.

Solution

For the I/O address decoder to be enabled, address bits A_{15} and A_0 must be

$$A_{15} = 1$$

and

$$A_0 = 0$$

To select port 7, the address applied to the CBA inputs of the decoder must be

$$A_{3L}A_{2L}A_{1L} = 111$$

Using 0s for the unused bits gives the address

$$A_{15L} \ldots A_{1L}A_{0L} = 1000000000001110_2$$
$$= 800E_{16}$$

EXAMPLE 10.4

For the circuit of Fig. 10–3, write an instruction sequence that inputs the byte contents of input port 7 to the memory location DATA_7.

Solution

In Example 10.3 we found that the address of port 7 is $800E_{16}$. This address is loaded into the DX register with the instruction

```
MOV  DX, 800EH
```

Now the contents of this port are input to the AL register by executing the instruction

```
IN  AL, DX
```

Finally, the byte of data is copied to memory location DATA_7 with the instruction

```
MOV  DATA_7, AL
```

In practical applications, it is sometimes necessary within an I/O service routine to repeatedly read the value at an input line and test this value for a specific logic level. For instance, input I_3 at port 0 in Fig. 10–3 can be checked to determine if it is at the 1 logic level. Normally, the I/O routine does not continue until the input under test switches to the appropriate logic level. This technique is known as *polling* an input. This polling technique can be used to synchronize the execution of an I/O routine to an event in external hardware.

Let us now look at how a polling software routine is written. The first step in the polling operation is to read the contents of the input port. For instance, the instructions needed to read the contents of port 0 in the circuit of Fig. 10–3 are

```
           MOV DX, 8000H
POLL_I3:   IN  AL, DX
```

A label has been added to identify the beginning of the polling routine. After executing these instructions, the byte contents of port 0 are held in the AL register. Let us assume that input I_3 at this port is the line that is being polled. Therefore, all other bits in AL are masked off with the instruction

```
AND AL, 08H
```

After this instruction is executed, the contents of AL will be either 00H or 08H. Moreover the zero flag is 1 if AL contains 00H or else it is 0. The state of the zero flag can be tested with a jump-on-zero instruction:

```
JZ POLL_I3
```

If zero flag is 1, a jump is initiated to POLL_I3, and the sequence repeats. On the other hand, if it is 0, the jump is not made; instead, the instruction following the jump instruction is executed. That is, the polling loop repeats until input I_3 is tested and found to be logic 1.

Polling the Setting of a Switch

Figure 10–4, which is similar to Fig. 10–3, shows a switch connected to input 7 of input port 0. Note that when the switch is open, input I_7 is pulled to +5 V (logic 1) through pull-up resistor R_1. When the switch is closed, I_7 is connected to ground (logic 0). It is a common practice to poll a switch like this with software waiting for it to close.

The instruction sequence that follows will poll the switch at I_7:

```
           MOV  DX, 8000H
POLL_I7:   IN   AL, DX
           SHL  AL, 1
           JC   POLL_I7
CONTINUE:
```

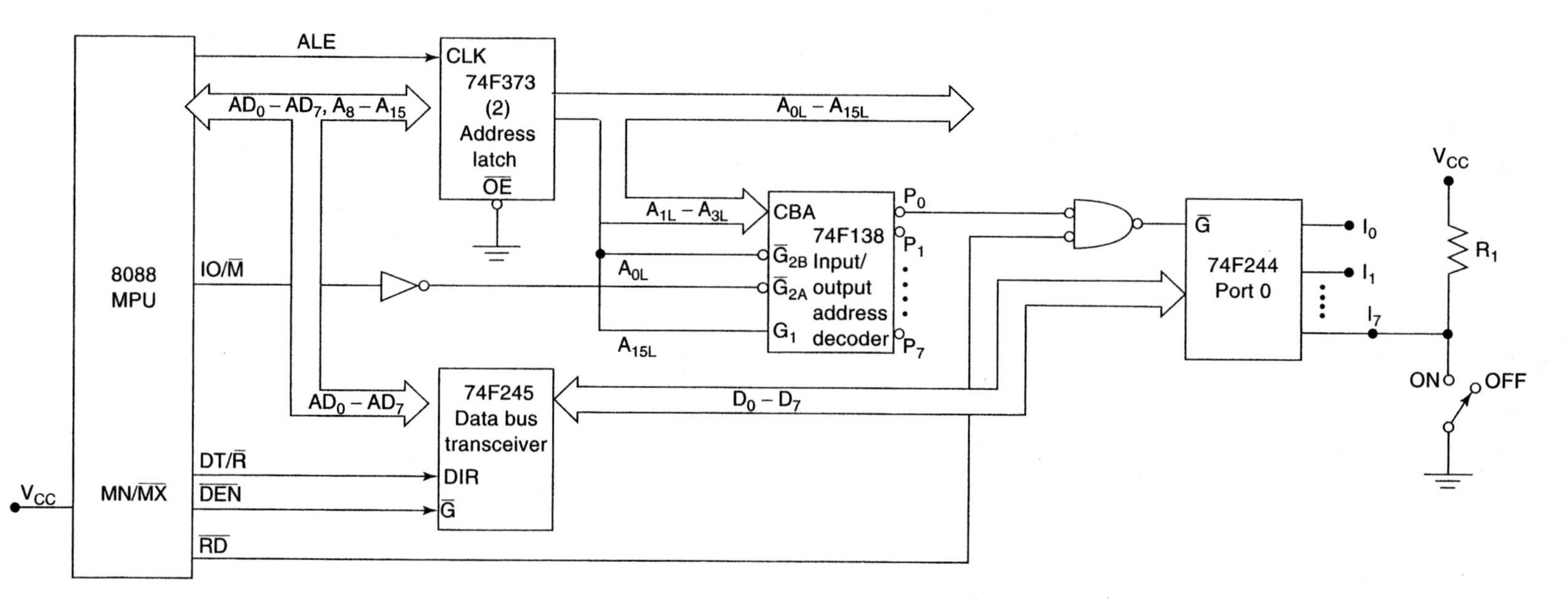

Figure 10–4 Reading the setting of a switch connected to an input port.

First, DX is loaded with the address of port 0. Then the contents of port 0 are input to the AL register. Since the logic level at I_7 is in bit 7 of the byte of data in AL, a shift left by one bit position will put this logic level into CF. Now a jump-on-carry instruction is executed to test CF. If CF is 1, the switch is not yet closed. In this case, control is returned to the IN instruction and the poll sequence repeats. On the other hand, if the switch is closed, bit 7 in AL is 0 and this value is shifted into CF. When the JC instruction detects this condition, the polling operation is complete, and the instruction following JC is executed.

▲ 10.4 INPUT/OUTPUT HANDSHAKING AND A PARALLEL PRINTER INTERFACE

In some applications, the microcomputer must synchronize the input or output of information to a peripheral device. Two examples of interfaces that may require a synchronized data transfer are a serial communications interface and a parallel printer interface. Sometimes it is necessary as part of the I/O synchronization process first to poll an input from an I/O device and, after receiving the appropriate level at the poll input, to acknowledge this fact to the device with an output. This type of synchronization is achieved by implementing what is known as *handshaking* as part of the input/output interface.

Figure 10–5(a) shows a conceptual view of the interface between the printer and a *parallel printer port.* There are three general types of signals at the *printer interface:* data, control, and status. The data lines are the parallel paths used to transfer data to the printer. Transfers of data over this bus are synchronized with an appropriate sequence of control signals. However, data transfers can only take place if the printer is ready to accept data. Printer readiness is indicated through the parallel interface by a set of signals called status lines. This interface handshake sequence is summarized by the flowchart shown in Fig. 10–5(b).

The printer is attached to the microcomputer system at a connector known as the *parallel printer port.* On a PC, a 25-pin connector is used to attach the printer. Figure 10–5(c) shows the actual signals supplied at the pins of this connector. Note that there are five status signals available at the interface, and they are called Ack, Busy, Paper Empty, Select, and Error. In a particular implementation only some of these signals may be used. For instance, to send a character to the printer, the software may test only the Busy signal. If it is inactive, it may be a sufficient indication to proceed with the transfer.

Figure 10–6(a) shows a detailed block diagram of a simple parallel printer interface. Here we find eight data-output lines, D_0 through D_7, control signal strobe ($\overline{STB}$), and status signal busy (BUSY). The MPU outputs data representing the character to be printed through the parallel printer interface. Character data are latched at the outputs of the parallel interface and are carried to the data inputs of the printer over data lines D_0 through D_7. The $\overline{STB}$ output of the parallel printer interface is used to signal the printer that new character data are available. Whenever the printer is already busy printing a character, it signals this fact to the MPU with the BUSY input of the parallel printer interface. This handshake signal sequence is illustrated in Fig. 10–6(b).

Let us now look at the sequence of events that take place at the parallel printer interface when data are output to the printer. Figure 10–6(c) is a flowchart of a subrou-

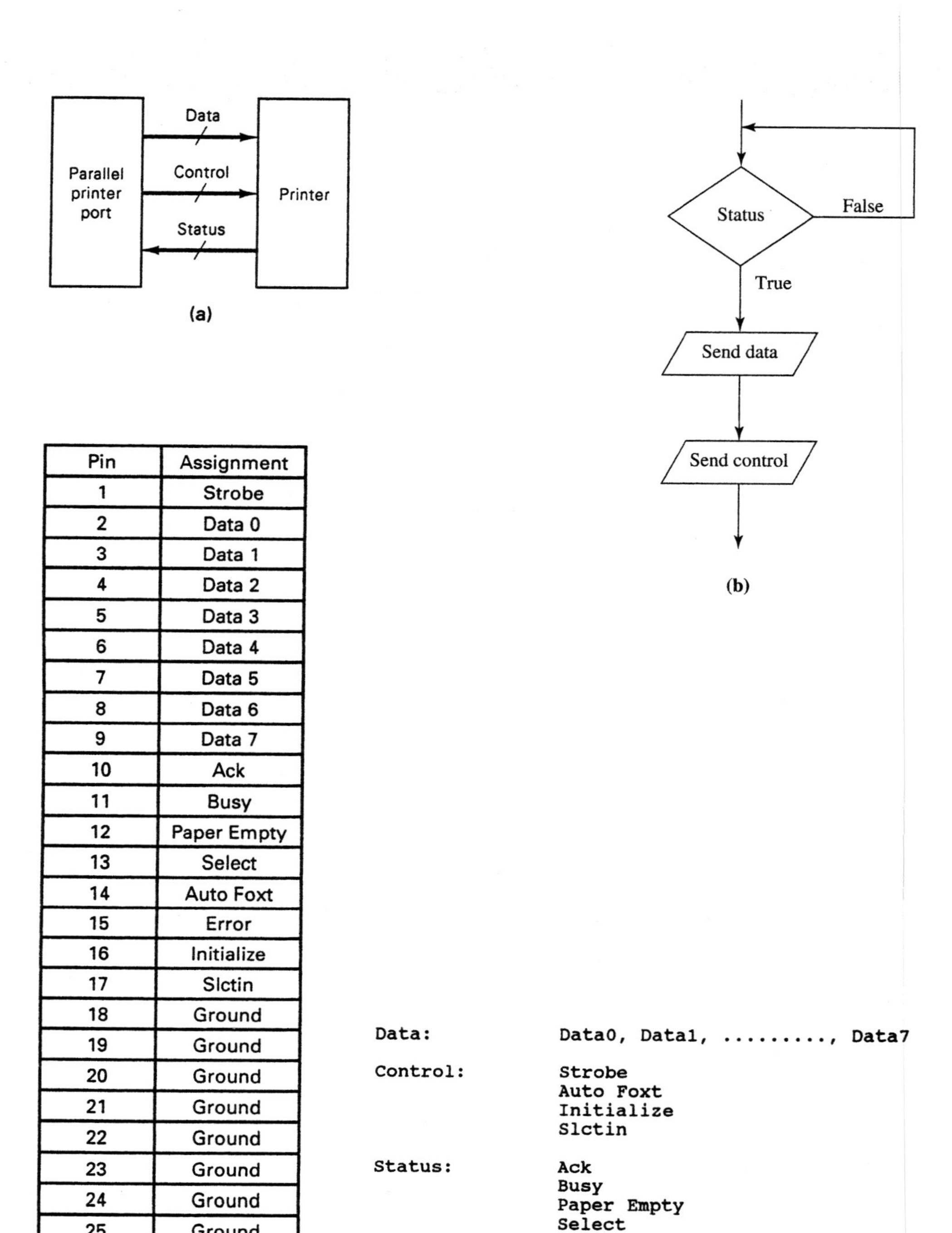

Pin	Assignment
1	Strobe
2	Data 0
3	Data 1
4	Data 2
5	Data 3
6	Data 4
7	Data 5
8	Data 6
9	Data 7
10	Ack
11	Busy
12	Paper Empty
13	Select
14	Auto Foxt
15	Error
16	Initialize
17	Slctin
18	Ground
19	Ground
20	Ground
21	Ground
22	Ground
23	Ground
24	Ground
25	Ground

Figure 10–5 (a) Parallel printer interface. (b) Flowchart showing the data transfer in a parallel printer interface. (c) Parallel printer port pin assignments and types of interface signals.

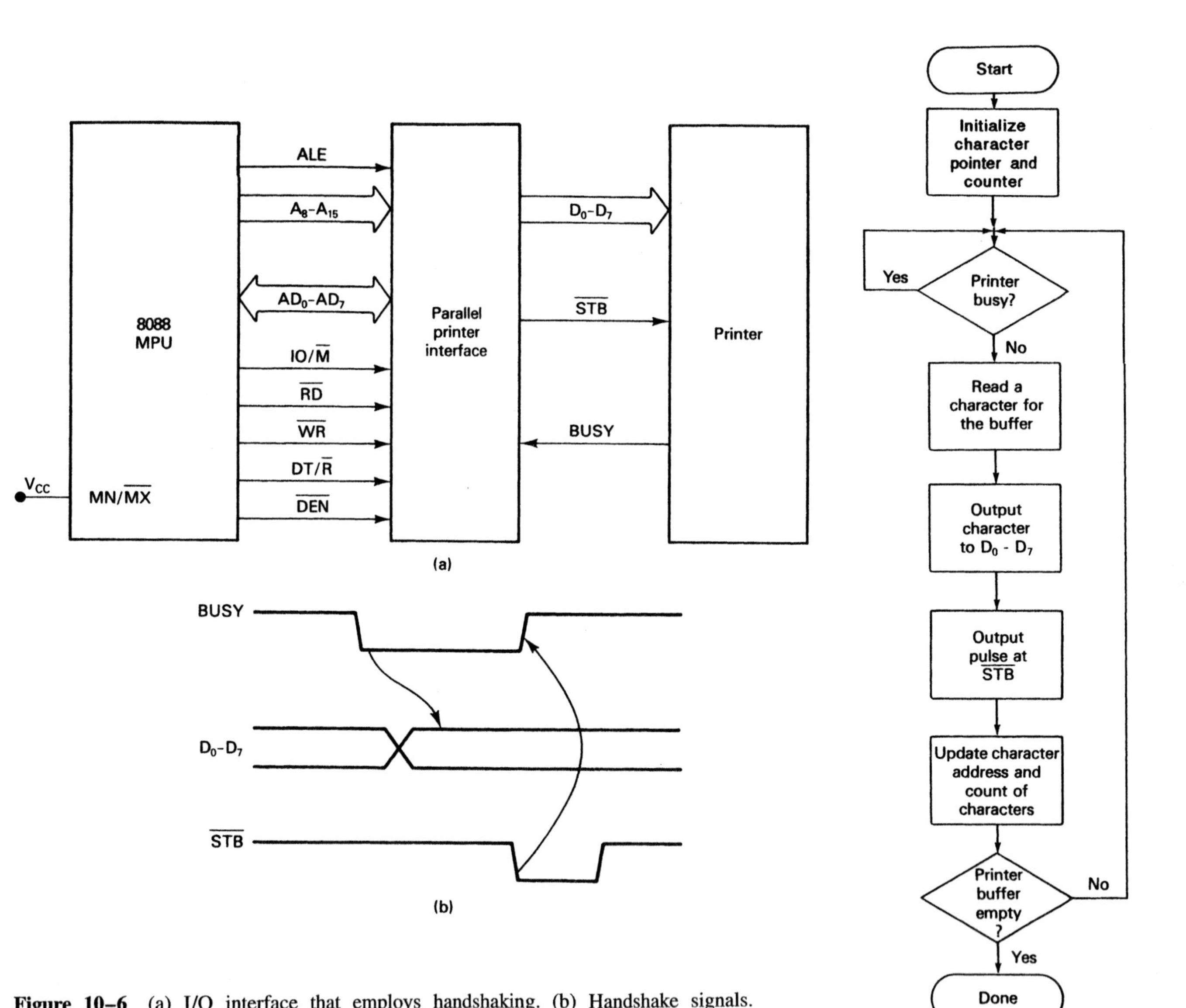

Figure 10–6 (a) I/O interface that employs handshaking. (b) Handshake signals. (c) Handshake sequence flowchart. (d) Handshaking printer interface circuit.

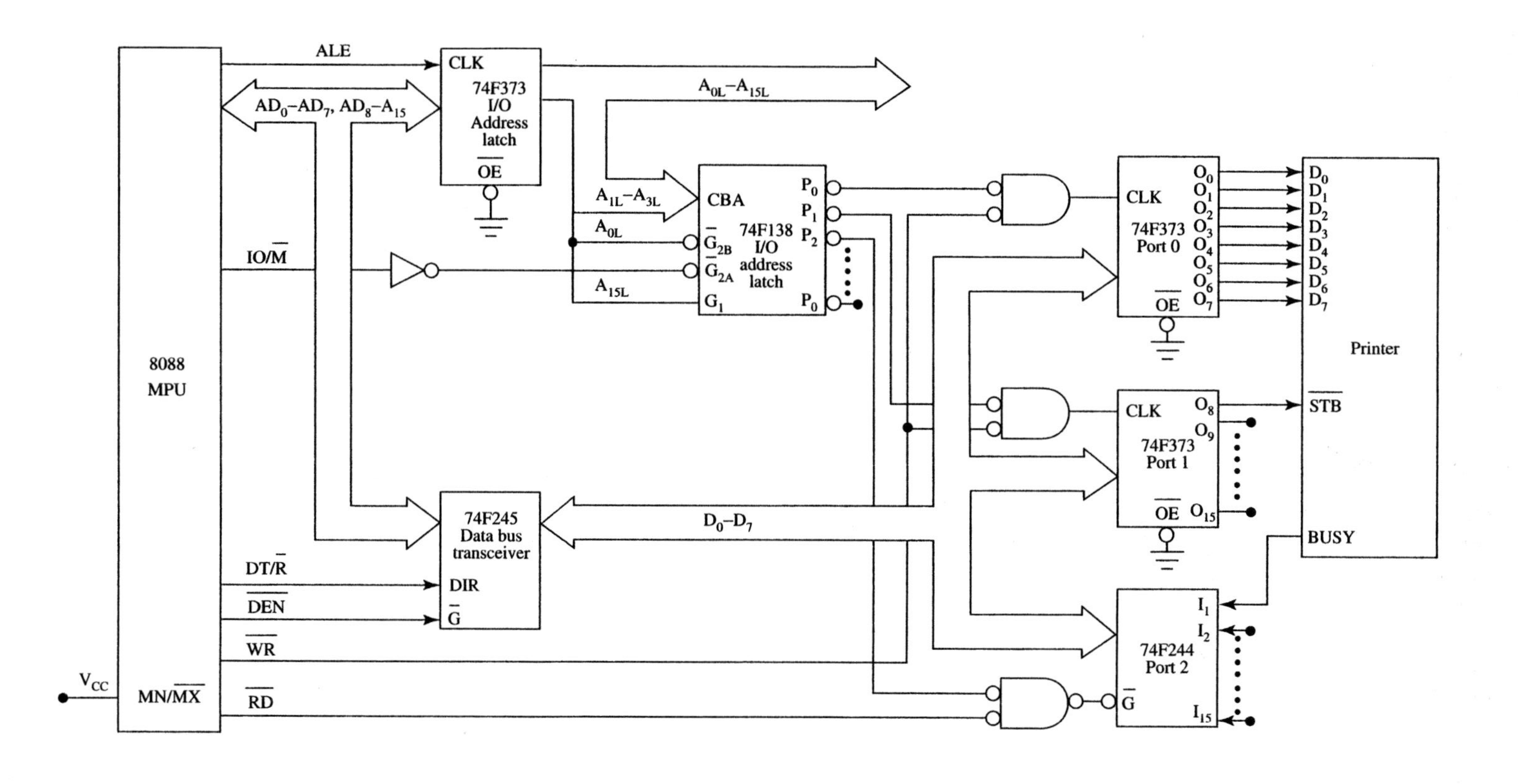

(d)

Figure 10–6 (continued)

tine that performs a parallel printer interface character-transfer operation. First the BUSY input of the parallel printer interface is tested. Note that this is done with a polling operation. That is, the MPU tests the logic level of BUSY repeatedly until it is found to be at the not-busy logic level. *Busy* means that the printer is currently printing a character. On the other hand, *not busy* signals that the printer is ready to receive another character for printing. After finding a not-busy condition, a count of the number of characters in the printer buffer (microprocessor memory) is read; a byte of character data is read from the printer buffer; the character is output to the parallel interface; and then a pulse is produced at $\overline{STB}$. This pulse tells the printer to read the character off the data bus lines. The printer is again printing a character and signals this fact at BUSY. The handshake sequence is now complete. Now the count that represents the number of characters in the buffer is decremented and checked to see if the buffer is empty. If empty, the print operation is complete. Otherwise, the character transfer sequence is repeated for the next character.

The circuit in Fig. 10–6(d) implements the parallel printer interface in Fig. 10–6(a).

EXAMPLE 10.5

What are the addresses of the ports that provide the data lines, strobe output, and busy input in the circuit shown in Fig. 10–6(d)? Assume that all unused address bits are 0s.

Solution

The I/O addresses that enable port 0 for the data lines, port 1 for the strobe output, and port 2 for the busy input are found as follows:

$$\text{Address of port 0} = 1000000000000000_2 = 8000_{16}$$

$$\text{Address of port 1} = 1000000000000010_2 = 8002_{16}$$

$$\text{Address of port 2} = 1000000000000100_2 = 8004_{16}$$

EXAMPLE 10.6

Write a program that will implement the sequence in Fig. 10–6(c) for the circuit in Fig. 10–6(d). Character data are held in memory starting at address PRNT_BUFF, and the number of characters held in the buffer is identified by the count at address CHAR_COUNT. Use the port addresses from Example 10.5.

Solution

First, the character counter and the character pointer are set up with the instructions

```
MOV  CL, CHAR_COUNT       ;(CL) = character count
MOV  SI, PRNT_BUFF        ;(SI) = character pointer
```

Next, the BUSY input is checked with the instructions

```
POLL_BUSY:    MOV  DX, 8004H      ;Keep polling till busy = 0
              IN   AL, DX
              AND  AL, 01H
              JNZ  POLL_BUSY
```

The character is copied into AL, and then it is output to port 0:

```
              MOV  AL, [SI]       ;Get the next character
              MOV  DX, 8000H
              OUT  DX, AL         ;and output it to port 0
```

Now, a strobe pulse is generated at port 1 with the instructions

```
              MOV  AL, 00H        ;STB = 0
              MOV  DX, 8002H
              OUT  DX, AL
              MOV  BX, 0FH        ;Delay for STB duration
STROBE:       DEC  BX
              JNZ  STROBE
              MOV  AL, 01H        ;STB = 1
              OUT  DX, AL
```

At this point, the value of PRNT_BUFF must be incremented, and the value of CHAR_COUNT must be decremented:

```
              INC  SI             ;Update character counter
              DEC  CL             ;and pointer
```

Finally, a check is made to see if the printer buffer is empty. If it is not empty, we need to repeat the prior instruction sequence. To do this, we execute the instruction

```
              JNZ  POLL_BUSY      ;Repeat till all characters
                                  ;have been transferred
DONE:         -
```

The program comes to the DONE label after all characters are transferred to the printer.

▲ 10.5 82C55A PROGRAMMABLE PERIPHERAL INTERFACE

The 82C55A is an LSI peripheral designed to permit easy implementation of *parallel I/O* in the 8088- and 8086-microcomputer systems. It provides a flexible parallel interface, which includes features such as single-bit, 4-bit, and byte-wide input and output ports;

level-sensitive inputs; latched outputs; strobed inputs or outputs; and strobed bidirectional input/outputs. These features are selected under software control.

A block diagram of the 82C55A is shown in Fig. 10–7(a) and its pin layout appears in Fig. 10–7(b). The left side of the block represents the *microprocessor's interface.* It includes an *8-bit bidirectional data bus* D_0 through D_7. Over these lines, commands, status information, and data are transferred between the MPU and 82C55A. These data are transferred whenever the MPU performs an input or output bus cycle to an address of a register within the device. Timing of the data transfers to the 82C55A is controlled by the *read/write control* ($\overline{RD}$ and $\overline{WR}$) signals.

The source or destination register within the 82C55A is selected by a 2-bit *register select code.* The MPU must apply this code to the *register-select inputs* A_0 and A_1 of the 82C55A. The *port A, port B,* and *port C registers* correspond to codes $A_1A_0 = 00$, $A_1A_0 = 01$, and $A_1A_0 = 10$, respectively.

Two other signals are shown on the microprocessor interface side of the block diagram. They are the *reset* (RESET) and *chip-select* ($\overline{CS}$) inputs. $\overline{CS}$ must be logic 0 during all read or write operations to the 82C55A. It enables the 82C55A's microprocessor interface circuitry for an input or output operation.

On the other hand, RESET is used to initialize the device. Switching it to logic 0 at power-up causes the internal registers of the 82C55A to be cleared. *Initialization* configures all I/O ports for input mode of operation.

The other side of the block corresponds to three *byte-wide I/O ports,* called *port A, port B,* and *port C,* and represent *I/O lines* PA_0 through PA_7, PB_0 through PB_7, and PC_0 through PC_7, respectively. These ports can be configured for input or output operation. This gives us a total of 24 I/O lines.

We already mentioned that the operating characteristics of the 82C55A could be configured under software control. It contains an 8-bit internal control register for this purpose. The *group A* and *group B control blocks* in Fig. 10–7(a) represent this register. Logic 0 or 1 can be written to the bit positions in this register to configure the individual ports for input or output operation and to enable one of its three modes of operation. The control register is write only and its contents can be modified using microprocessor instructions. A write bus cycle to the 82C55A with register-select code $A_1A_0 = 11$, and an appropriate control word is used to modify the control registers.

The circuit in Fig. 10–8 is an example of how the 82C55A can be interfaced to a microprocessor. Here we see that address lines A_0 and A_1 of the microprocessor drive the 82C55A's register-select inputs A_1 and A_0, respectively. The $\overline{CS}$ input of the 82C55A is supplied from the output of the address decoder circuit whose inputs are address lines A_2 through A_{15} and $IO/\overline{M}$. To access either a port or the control register of the 82C55A, $\overline{CS}$ must be active. Then the code A_1A_0 selects the port or control register to be accessed. The select codes are shown in Fig. 10–8.

For instance, to access port A, $A_1A_0 = 00$, $A_{15} = A_{14} = 1$, $A_{13} = A_{12} = \ldots = A_2 = 0$, which gives the port A address as

$$1100\ldots\ldots00_2 = C000_{16}$$

Similarly, it can be determined that the address of port B equals $C001_{16}$, that of port C is $C002_{16}$, and the address of the control register is $C003_{16}$.

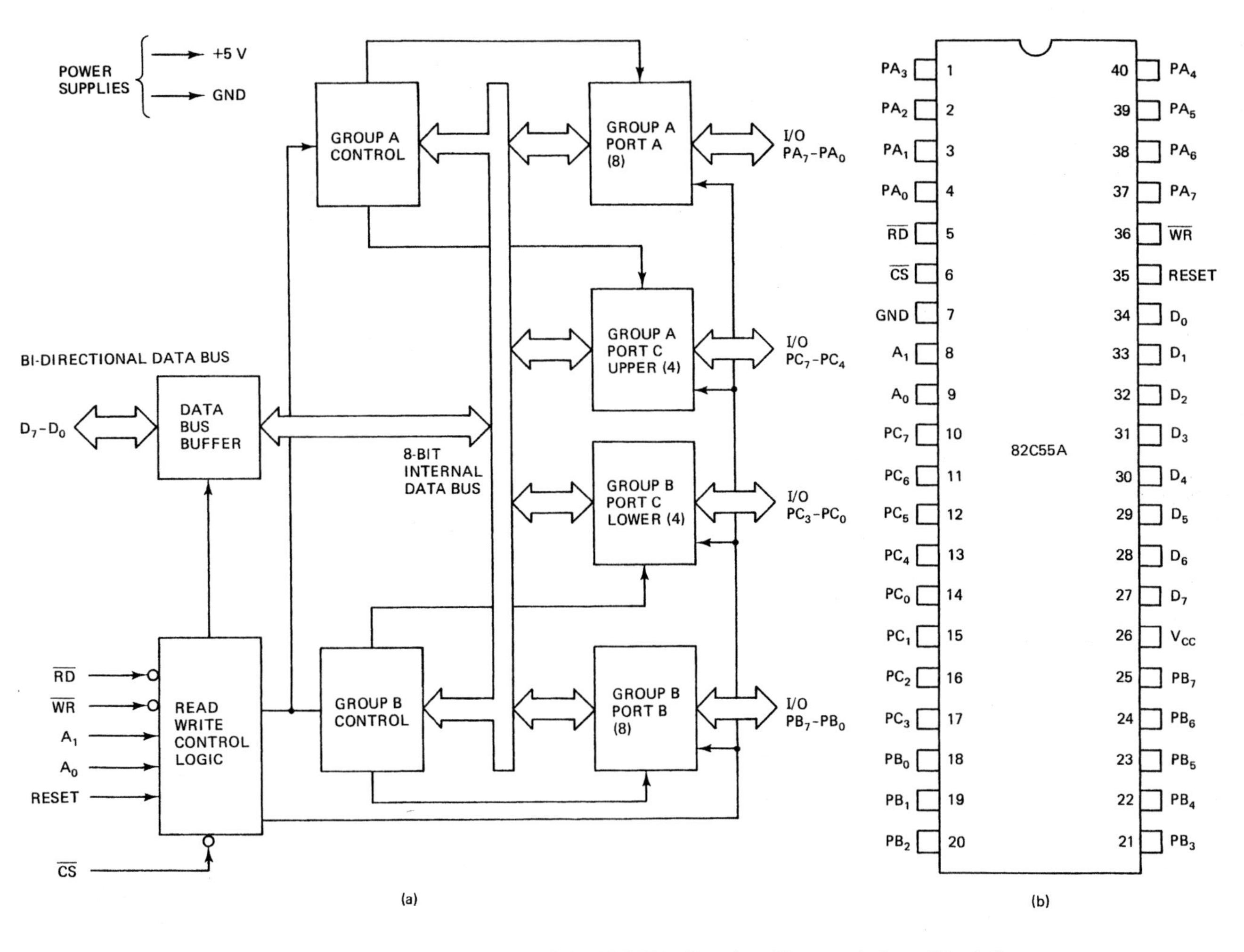

Figure 10–7 (a) Block diagram of the 82C55A. (Reprinted by permission of Intel Corporation. Copyright/Intel Corp. 1980) (b) Pin layout. (Reprinted by permission of Intel Corporation. Copyright/Intel Corp. 1980)

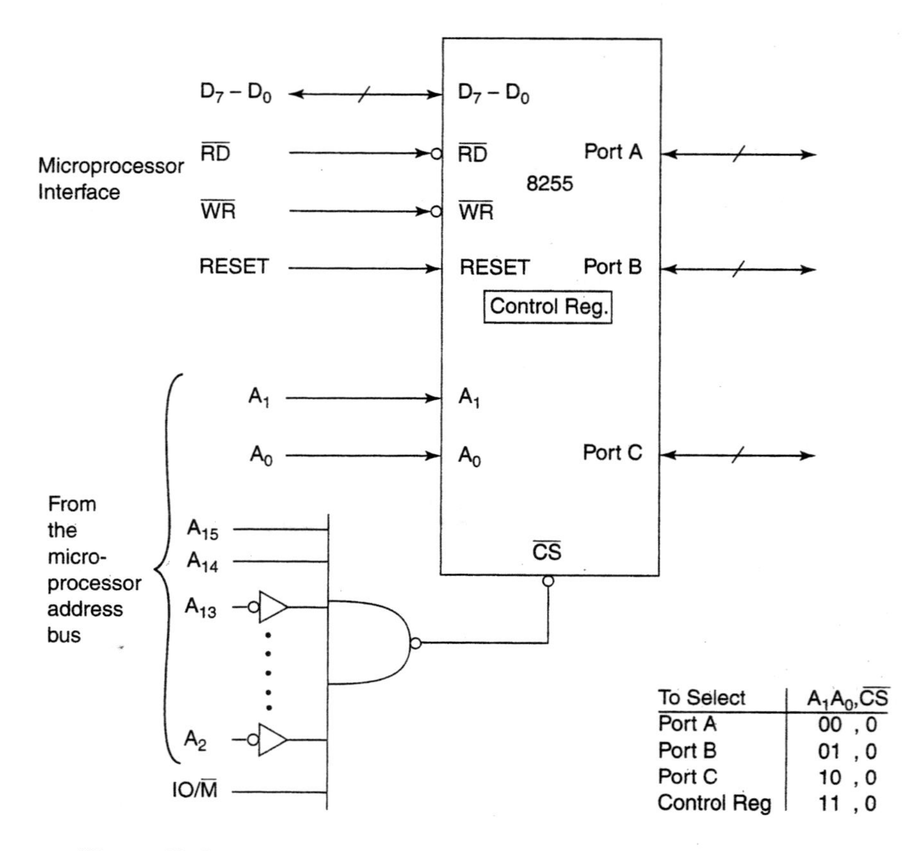

Figure 10–8 Addressing an 82C55A using the microprocessor interface signals.

The bits of the control register and their control functions are shown in Fig. 10–9. Here we see that bits D_0 through D_2 correspond to the group B control block in the diagram of Fig. 10.7(a). Bit D_0 configures the lower four lines of port C for input or output operation. Notice that logic 1 at D_0 selects input operation, and logic 0 selects output operation. The next bit, D_1, configures port B as an 8-bit-wide input or output port. Again, logic 1 selects input operation, and logic 0 selects output operation.

The D_2 bit is the mode-select bit for port B and the lower 4 bits of port C. It permits selection of one of two different modes of operation, called *mode 0* and *mode 1*. Logic 0 in bit D_2 selects mode 0, whereas logic 1 selects mode 1. These modes are discussed in detail in subsequent sections.

The next 4 bits in the control register, D_3 through D_6, correspond to the group A control block in Fig. 10–7(a). Bits D_3 and D_4 of the control register are used to configure the operation of the upper half of port C and all of port A, respectively. These bits work in the same way as D_0 and D_1 to configure the lower half of port C and port B. However, there are now two mode-select bits, D_5 and D_6, instead of just one. They are used to select between three modes of operation, *mode 0, mode 1,* and *mode 2.*

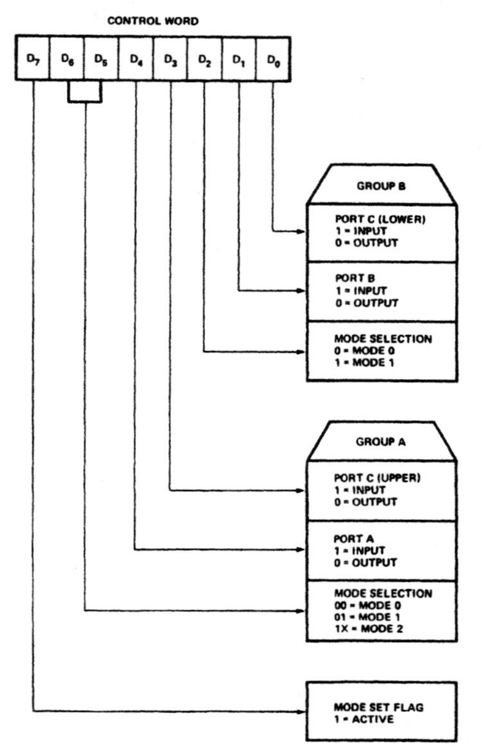

Figure 10–9 Control-word bit functions. (Reprinted by permission of Intel Corporation. Copyright/Intel Corp. 1980)

The last control register bit, D_7, is the *mode-set flag*. It must be at logic 1 (active) whenever the mode of operation is to be changed.

Mode 0 selects what is called *simple I/O operation*. By simple I/O, we mean that the lines of the port can be configured as level-sensitive inputs or latched outputs. To set all ports for this mode of operation, load bit D_7 of the control register with logic 1, bits $D_6D_5 = 00$, and $D_2 = 0$. Logic 1 at D_7 represents an active mode set flag. Now port A and port B can be configured as 8-bit input or output ports, and port C can be configured for operation as two independent 4-bit input or output ports. Setting or resetting bits D_4, D_3, D_1, and D_0 does this. Figure 10–10 summarizes the port pins and the functions they can perform in mode 0.

For example, if $80_{16} = 10000000_2$ is written to the control register, the 1 in D_7 activates the mode-set flag. Mode 0 operation is selected for all three ports because bits D_6, D_5, and D_2 are logic 0. At the same time, the zeros in D_4, D_3, D_1, and D_0 set up all port lines to work as outputs. Figure 10–11(a) illustrates this configuration.

By writing different binary combinations into bit locations D_4, D_3, D_1, and D_0, any one of 16 different mode 0 I/O configurations can be obtained. The control word and I/O setup for the rest of these combinations are shown in Fig. 10–11(b) through (p).

Pin	MODE 0	
	IN	OUT
PA_0	IN	OUT
PA_1	IN	OUT
PA_2	IN	OUT
PA_3	IN	OUT
PA_4	IN	OUT
PA_5	IN	OUT
PA_6	IN	OUT
PA_7	IN	OUT
PB_0	IN	OUT
PB_1	IN	OUT
PB_2	IN	OUT
PB_3	IN	OUT
PB_4	IN	OUT
PB_5	IN	OUT
PB_6	IN	OUT
PB_7	IN	OUT
PC_0	IN	OUT
PC_1	IN	OUT
PC_2	IN	OUT
PC_3	IN	OUT
PC_4	IN	OUT
PC_5	IN	OUT
PC_6	IN	OUT
PC_7	IN	OUT

Figure 10–10 Mode 0 port pin functions.

EXAMPLE 10.7

What is the mode and I/O configuration for ports A, B, and C of an 82C55A after its control register is loaded with 82_{16}?

Solution

Expressing the control register contents in binary form, we get

$$D_7D_6D_5D_4D_3D_2D_1D_0 = 10000010_2$$

Since D_7 is 1, the modes of operation of the ports are selected by the control word. The three least significant bits of the control word configure port B and the lower four bits of port C:

$D_0 = 0$ Lower four bits of port C are outputs.

$D_1 = 1$ Port B is an input port.

$D_2 = 0$ Mode 0 operation for both port B and the lower four bits of port C.

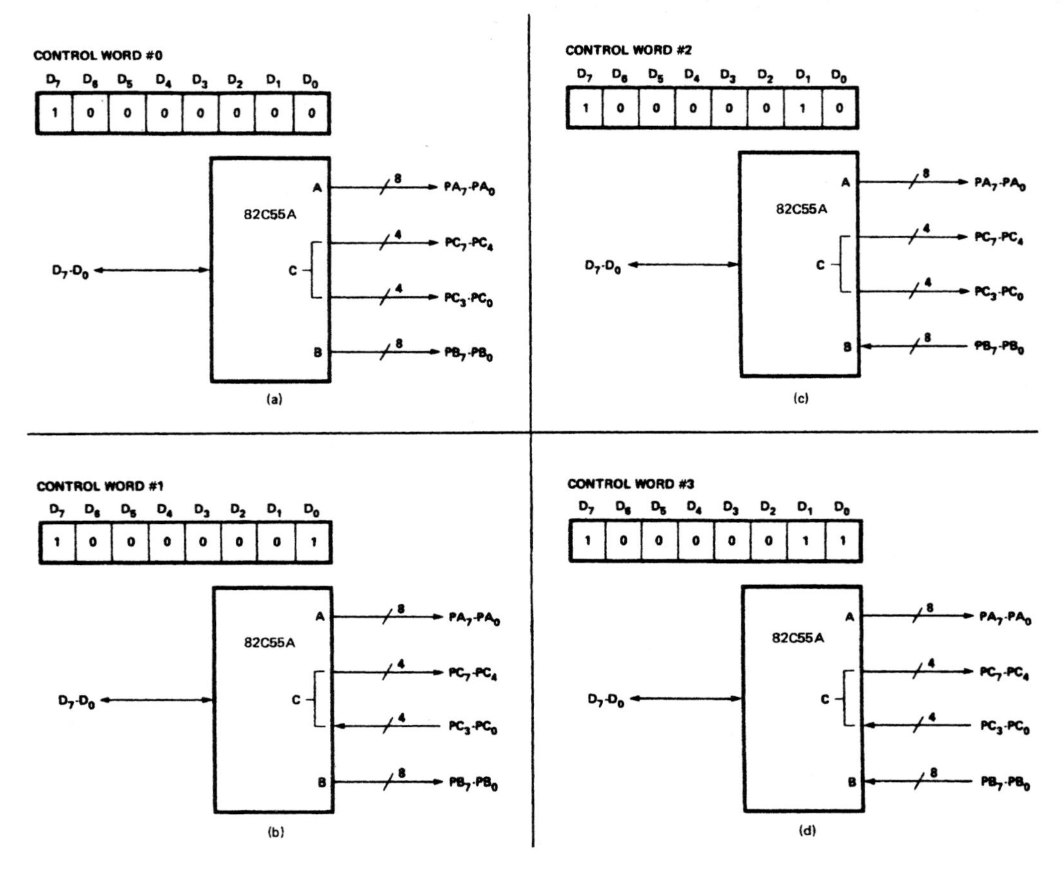

Figure 10–11 (a-p) Mode 0 control words and corresponding input/output configurations. (Reprinted by permission of Intel Corp. Copyright/Intel Corp. 1980)

The next four bits configure the upper part of port C and port A:

$D_3 = 0$	Upper four bits of port C are outputs.
$D_4 = 0$	Port A is an output port.
$D_6D_5 = 00$	Mode 0 operation for both port A and the upper part of port C.

Figure 10–11(c) shows this mode 0 I/O configuration.

Mode 1 operation represents what is known as *strobed I/O.* The ports of the 82C55A are put into this mode of operation by setting $D_7 = 1$ to activate the mode-set flag and setting $D_6D_5 = 01$ and $D_2 = 1$.

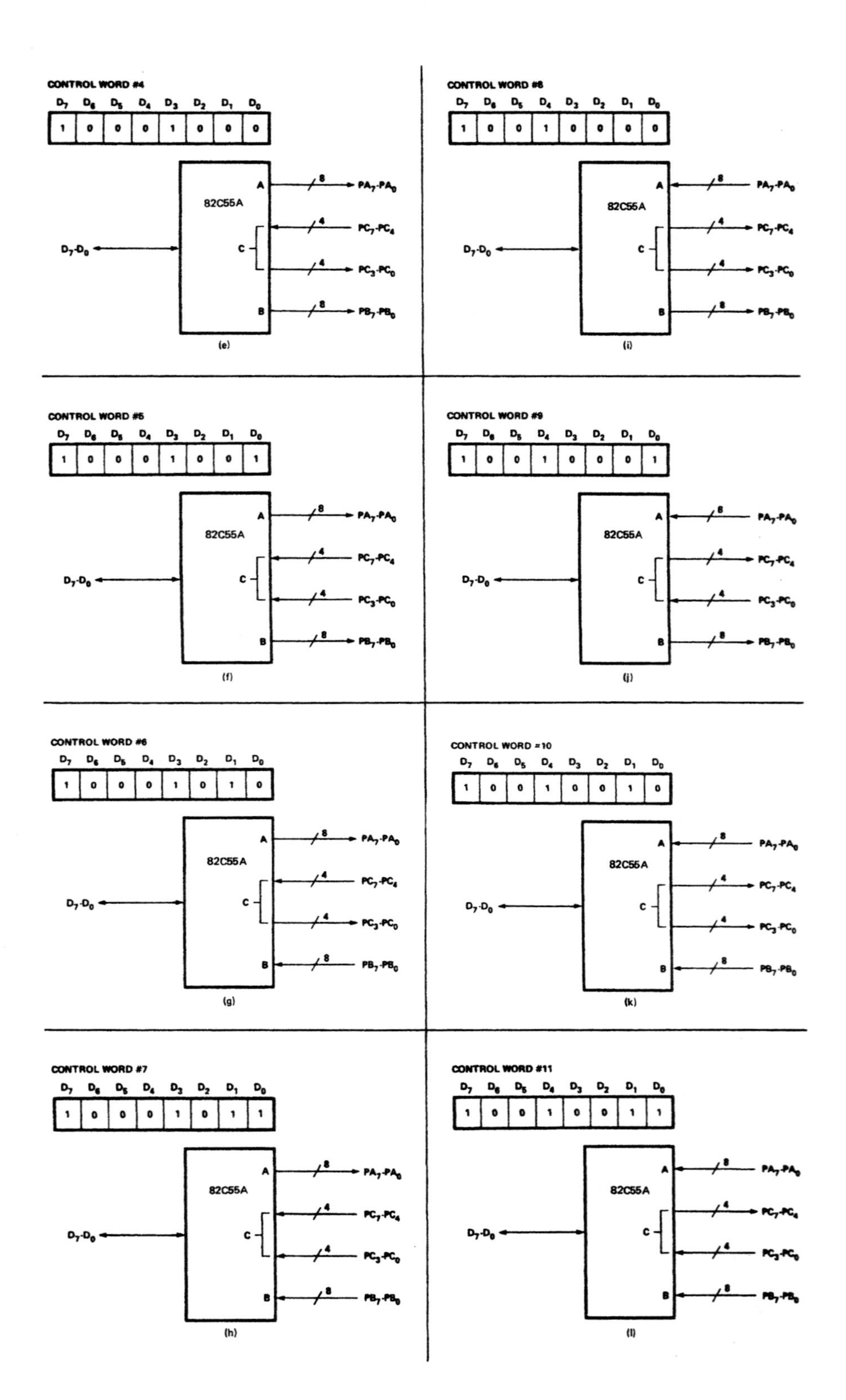

Figure 10–11 (continued)

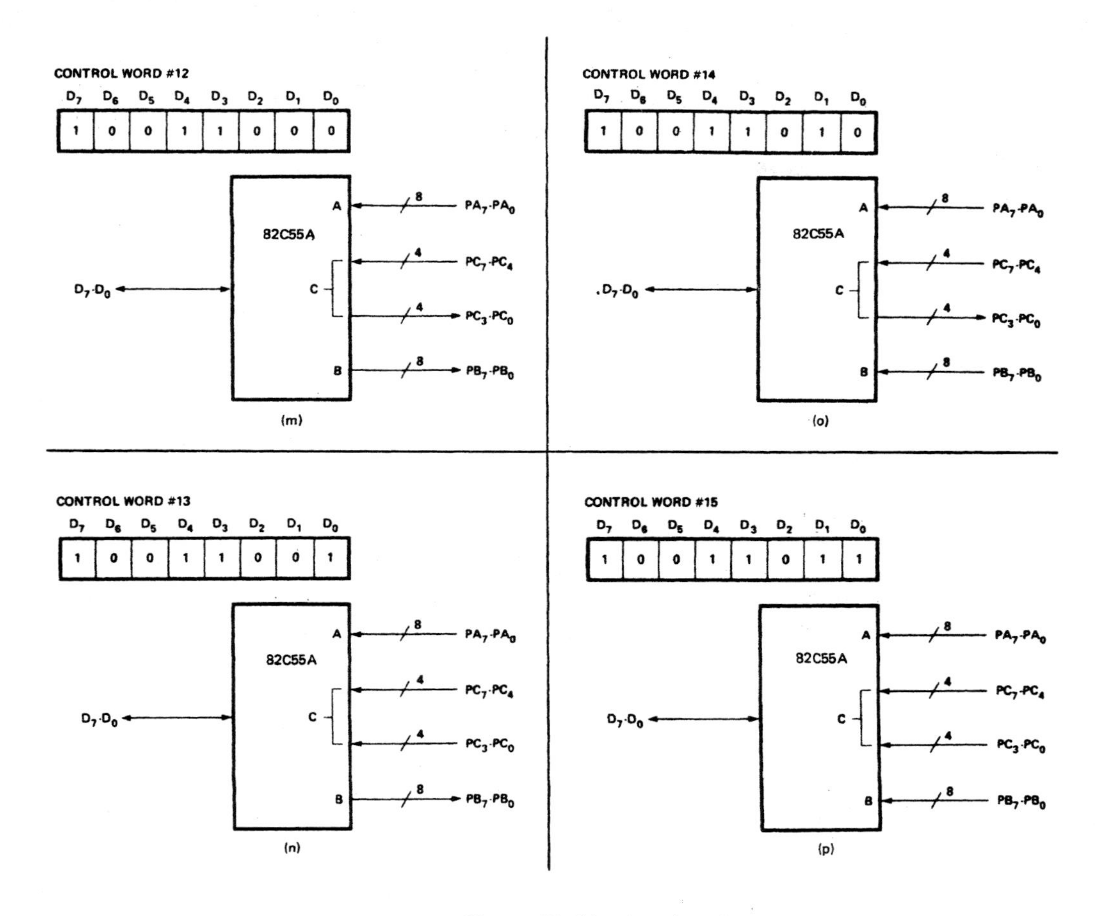

Figure 10–11 (continued)

In this way, the A and B ports are configured as two independent *byte-wide I/O ports,* each of which has a *4-bit control/data port* associated with it. The control/data ports are formed from the lower and upper nibbles of port C, respectively. Figure 10–12 lists the mode 1 functions of each pin at ports A, B, and C.

When configured in this way, data applied to an input port must be strobed in with a signal produced in external hardware. An output port in mode 1 is provided with handshake signals that indicate when new data are available at its outputs and when an external device has read these values.

As an example, let us assume for the moment that the control register of an 82C55A is loaded with $D_7D_6D_5D_4D_3D_2D_1D_0$ = 10111XXX. This configures port A as a mode 1 input port. Figure 10–13(a) shows the function of the signal lines for this example. Note that PA_7 through PA_0 form an 8-bit input port. On the other hand, the function of the upper port C lines are reconfigured to provide the port A control/data lines. The PC_4 line becomes *strobe input* ($\overline{STB}_A$), which is used to strobe data at PA_7 through PA_0 into

Pin	MODE 1	
	IN	OUT
PA_0	IN	OUT
PA_1	IN	OUT
PA_2	IN	OUT
PA_3	IN	OUT
PA_4	IN	OUT
PA_5	IN	OUT
PA_6	IN	OUT
PA_7	IN	OUT
PB_0	IN	OUT
PB_1	IN	OUT
PB_2	IN	OUT
PB_3	IN	OUT
PB_4	IN	OUT
PB_5	IN	OUT
PB_6	IN	OUT
PB_7	IN	OUT
PC_0	$INTR_B$	$INTR_B$
PC_1	IBF_B	$\overline{OBF_B}$
PC_2	$\overline{STB_B}$	$\overline{ACK_B}$
PC_3	$INTR_A$	$INTR_A$
PC_4	$\overline{STB_A}$	I/O
PC_5	IBF_A	I/O
PC_6	I/O	$\overline{ACK_A}$
PC_7	I/O	$\overline{OBF_A}$

Figure 10–12 Mode 1 port pin functions.

the input latch. Moreover, PC_5 becomes *input buffer full* (IBF_A). Logic 1 at this output indicates to external circuitry that a word has already been strobed into the latch.

The third control signal is at PC_3 and is labeled *interrupt request* ($INTR_A$). It switches to logic 1 when $\overline{STB_A} = 1$ making $IBF_A = 1$, and an internal signal *interrupt enable* ($INTE_A$) = 1. $INTE_A$ is set to logic 0 or 1 under software control by using the bit

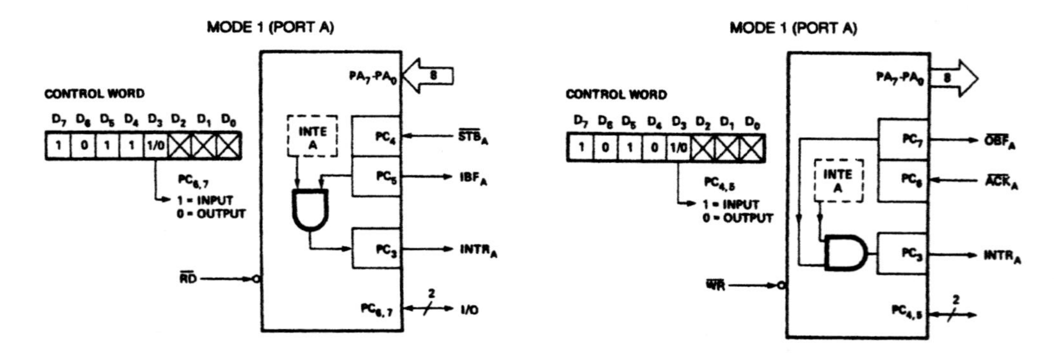

Figure 10–13 (a) Mode 1, port A input configuration. (Reprinted by permission of Intel Corporation. Copyright/Intel Corp. 1980) (b) Mode 1, port A output configuration. (Reprinted by permission of Intel Corporation. Copyright/Intel Corp. 1980)

set/reset feature of the 82C55A. This feature will be discussed later. Looking at Fig. 10–13(a), we see that logic 1 in $INTE_A$ enables the logic level of IBF_A to the $INTR_A$ output. This signal can be applied to an interrupt input of the MPU to signal it that new data are available at the input port. The corresponding interrupt-service routine reads the data, which clears $INTR_A$ and IBF_A. The timing diagram in Fig. 10–14(a) summarizes these events for an input port configured in mode 1.

As another example, let us assume that the contents of the control register are changed to $D_7D_6D_5D_4D_3D_2D_1D_0$ = 10100XXX. This I/O configuration is shown in Fig. 10–13(b). Note that port A is now configured for output operation instead of input operation. PA_7 through PA_0 make up the 8-bit output port. The control line at PC_7 is *output buffer full* ($\overline{OBF}_A$). When data have been written into the output port, $\overline{OBF}_A$ switches to the 0 logic level. In this way, it signals external circuitry that new data are available at the output lines.

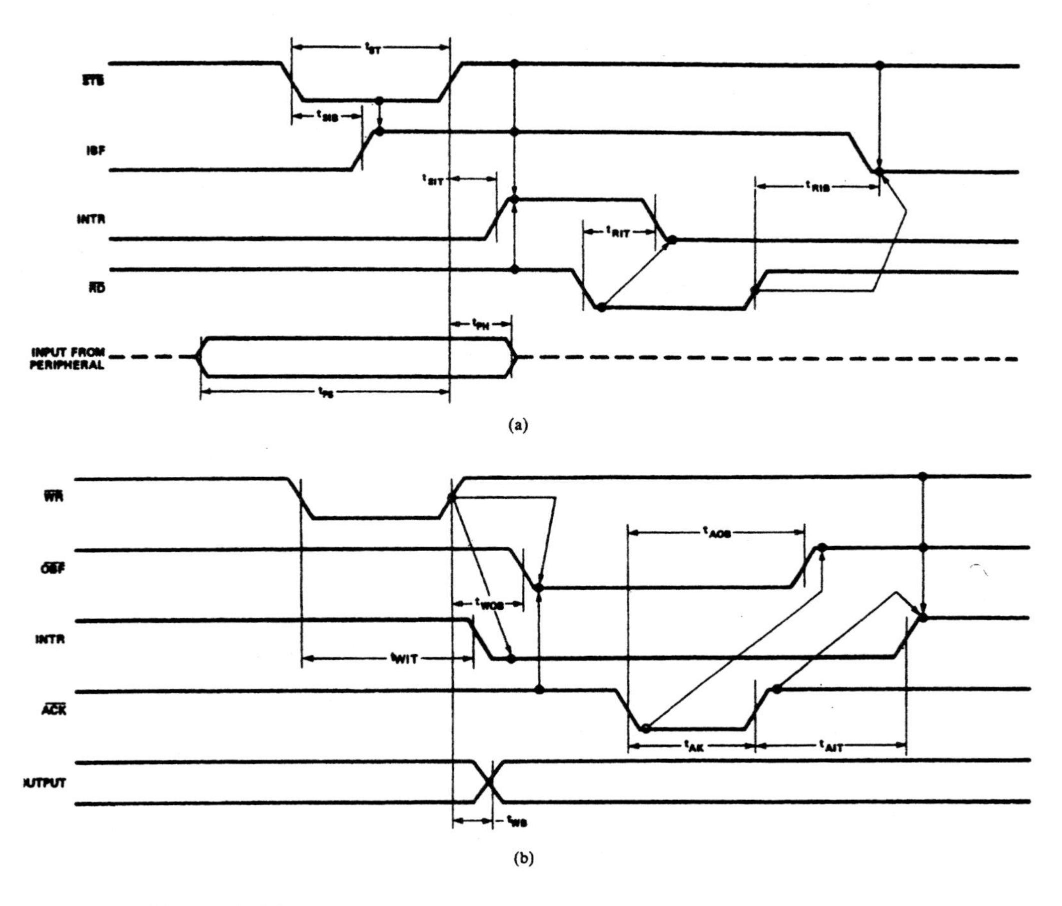

Figure 10–14 (a) Timing diagram for an input port in mode 1 configuration. (Reprinted by permission of Intel Corporation. Copyright/Intel Corp. 1980) (b) Timing diagram for an output port in mode 1 configuration. (Reprinted by permission of Intel Corporation. Copyright/Intel Corp. 1980)

Signal line PC_6 becomes *acknowledge* ($\overline{ACK_A}$), which is an input. An external device reads the data and signals the 82C55A that it has accepted the data provided at the output port by switching $\overline{ACK_A}$ to logic 0. When the $\overline{ACK_A} = 0$ is received by the 82C55A, it in turn deactivates the $\overline{OBF_A}$ output. The last signal at the control port is the *interrupt request* ($INTR_A$) output, which is produced at the PC_3 lead. This output is switched to logic 1 when the $\overline{ACK_A}$ input becomes inactive. It is used to signal the MPU with an interrupt that indicates that an external device has accepted the data from the outputs. To produce the $INTR_A$, the interrupt enable ($INTE_A$) bit must equal 1. Again $INTE_A$ must be set using the bit set/reset feature to write a 1 to PC_6. The timing diagram in Fig. 10–14(b) summarizes these events for an output port configured in mode 1.

EXAMPLE 10.8

Figures 10–15(a) and (b) show how port B can be configured for mode 1 operation. Describe what happens in Fig. 10–15(a) when the $\overline{STB_B}$ input is pulsed to logic 0. Assume that $INTE_B$ is already set to 1.

Solution

As $\overline{STB_B}$ is pulsed, the byte of data at PB_7 through PB_0 is latched into the port B register. This causes the IBF_B output to switch to 1. Since $INTE_B$ is 1, $INTR_B$ switches to logic 1.

The last mode of operation, mode 2, represents what is known as *strobed bidirectional I/O*. The key difference is that now the port works as either inputs or outputs and control signals are provided for both functions. Only port A can be configured to work in this way. The I/O port and control signal pins are shown in Fig. 10–16.

To set up this mode, the control register is set to $D_7D_6D_5D_4D_3D_2D_1D_0$ = 11XXXXXX. The I/O configuration that results is shown in Fig. 10–17. Here we find that PA_7 through PA_0 operate as an *8-bit bidirectional port* instead of a unidirectional

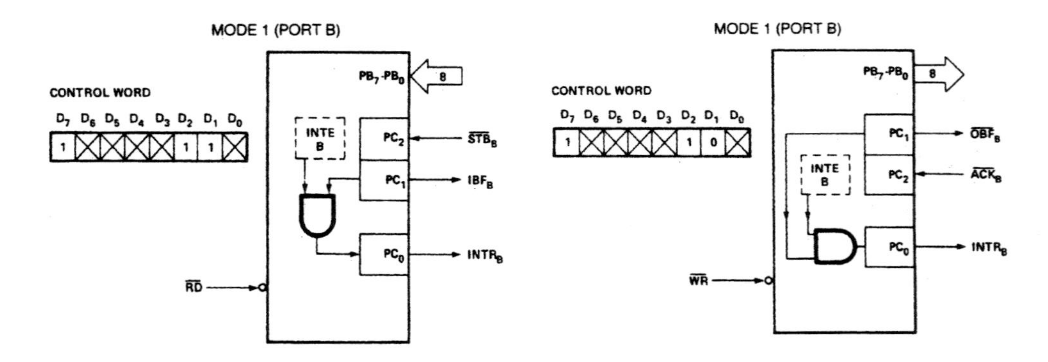

Figure 10–15 (a) Mode 1, port B input configuration. (Reprinted by permission of Intel Corporation. Copyright/Intel Corp. 1980) (b) Mode 1, port B output configuration. (Reprinted by permission of Intel Corporation. Copyright/Intel Corp. 1980)

Pin	MODE 2
	GROUP A ONLY
PA_0	↔
PA_1	↔
PA_2	↔
PA_3	↔
PA_4	↔
PA_5	↔
PA_6	↔
PA_7	↔
PB_0	—
PB_1	—
PB_2	—
PB_3	—
PB_4	—
PB_5	—
PB_6	—
PB_7	—
PC_0	I/O or $INTR_B$
PC_1	I/O or $\overline{OBF}_B$ or $\overline{IBF}_B$
PC_2	I/O or $\overline{ACK}_B$ or $\overline{STB}_B$
PC_3	$INTR_A$
PC_4	$\overline{STB}_A$
PC_5	IBF_A
PC_6	$\overline{ACK}_A$
PC_7	$\overline{OBF}_A$

MODE 0 OR MODE 1 ONLY (PB_0–PB_7)

Figure 10–16 Mode 2 port pin functions.

port. Its control signals are $\overline{OBF}_A$ at PC_7, $\overline{ACK}_A$ at PC_6, $\overline{STB}_A$ at PC_4, IBF_A at PC_5, and $INTR_A$ at PC_3. Their functions are similar to those already discussed for mode 1. One difference is that $INTR_A$ is produced by either gating $\overline{OBF}_A$ with $INTE_1$ or IBF_A with $INTE_2$.

In our discussion of mode 1, we mentioned that the *bit set/reset* feature could be used to set or reset the INTE bits. For instance, to enable $INTR_A$ for port A as output, PC_4

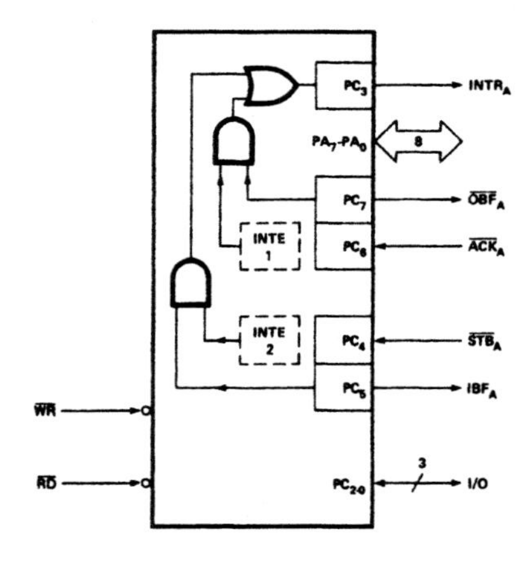

Figure 10–17 Mode 2 input/output configuration. (Reprinted by permission of Intel Corporation. Copyright/Intel Corp. 1980)

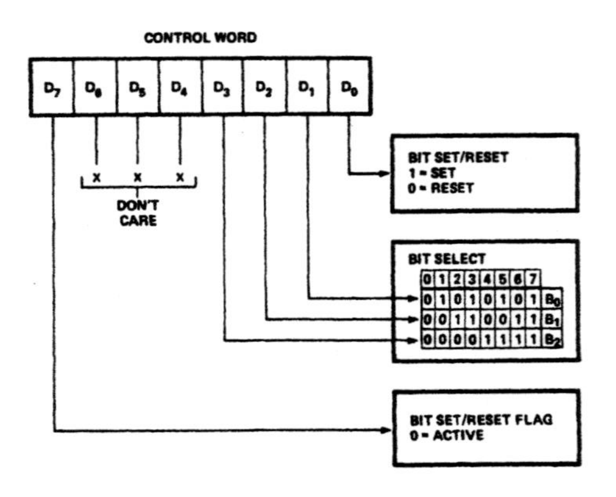

Figure 10–18 Bit set/reset format. (Reprinted by permission of Intel Corporation. Copyright/Intel Corp. 1980)

should be set to make $INTE_2$ (dotted box next to PC_4 in Fig. 10–17) logic 1. The bit set/reset feature also allows the individual bits of port C to be set or reset. To do this, we write logic 0 to bit D_7 of the control register. This resets the bit set/reset flag. The logic level that is to be latched at a port C line is included as bit D_0 of the control word. This value is latched at the I/O line of port C, which corresponds to the three-bit code at $D_3D_2D_1$.

The relationship between the set/reset control word and input/output lines is illustrated in Fig. 10–18. For instance, writing $D_7D_6D_5D_4D_3D_2D_1D_0 = 00001111_2$ into the control register of the 82C55A selects bit 7 and sets it to 1. Therefore, output PC_7 at port C is switched to the 1 logic level.

EXAMPLE 10.9

The interrupt-control flag $INTE_A$ for output port A in mode 1 is controlled by PC_6. Using the set/reset feature of the 82C55A, what command code must be written to the control register of the 82C55A to set it to enable the control flag?

Solution

To use the set/reset feature, D_7 must be logic 0. Moreover, $INTE_A$ is to be set; therefore, D_0 must be logic 1. Finally, to select PC_6, the code at bits $D_3D_2D_1$ must be 110. The rest of the bits are don't-care states. This gives us the control word

$$D_7D_6D_5D_4D_3D_2D_1D_0 = 0XXX1101_2$$

Replacing the don't-care states with the 0 logic level, we get

$$D_7D_6D_5D_4D_3D_2D_1D_0 = 00001101_2 = 0D_{16}$$

We have just described and given examples of each of the modes of operation that can be assigned to the ports of the 82C55A. It is also possible to configure the A and B ports with different modes. For example, Fig. 10–19(a) shows the control word and port

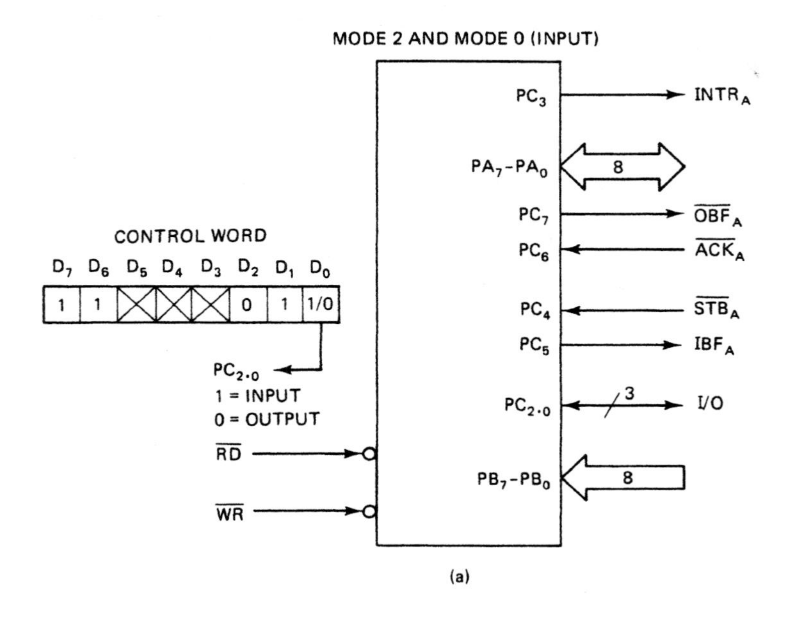

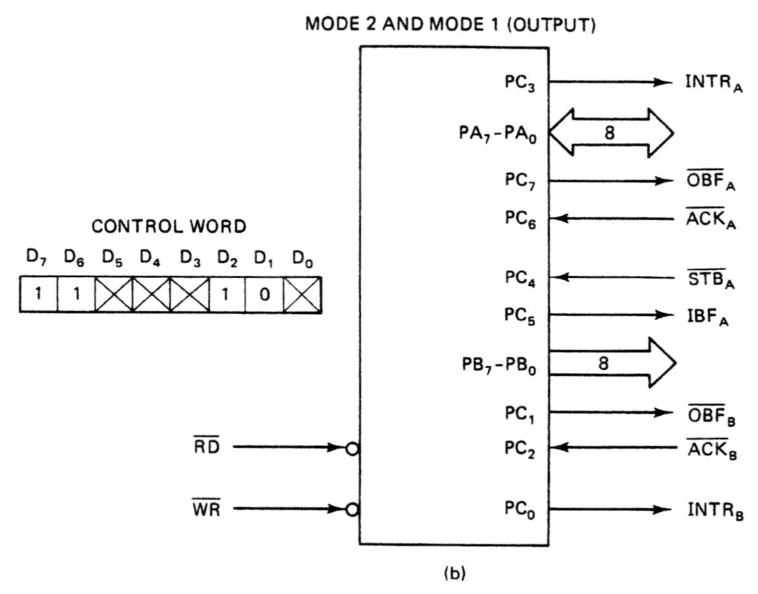

Figure 10–19 (a) Combined mode 2 and mode 0 (input) control word and I/O configuration. (Reprinted by permission of Intel Corporation. Copyright/Intel Corp. 1980) (b) Combined mode 2 and mode 1 (output) control word and I/O configuration. (Reprinted by permission of Intel Corporation. Copyright/Intel Corp. 1980)

configuration of an 82C55A set up for bidirectional mode 2 operation of port A and input mode 0 operation of port B. It should also be noted that in all modes, unused pins of port C are still available as general-purpose inputs or outputs.

EXAMPLE 10.10

What control word must be written into the control register of the 82C55A such that port A is configured for bidirectional operation and port B is set up with mode 1 outputs?

Solution

To configure the operating mode of the ports of the 82C55A, D_7 must be 1:

$$D_7 = 1$$

Port A is set up for bidirectional operation by making D_6 logic 1. In this case, D_5 through D_3 are don't-care states:

$$D_6 = 1$$

$$D_5D_4D_3 = XXX$$

Mode 1 is selected for port B by logic 1 in bit D_2 and output operation by logic 0 in D_1. Since mode 1 operation has been selected, D_0 is a don't-care state:

$$D_2 = 1$$

$$D_1 = 0$$

$$D_0 = X$$

This gives the control word

$$D_7D_6D_5D_4D_3D_2D_1D_0 = 11XXX10X_2$$

Assuming logic 0 for the don't-care states, we get

$$D_7D_6D_5D_4D_3D_2D_1D_0 = 11000100_2 = C4_{16}$$

This configuration is shown in Fig. 10–19(b).

EXAMPLE 10.11

Write the sequence of instructions needed to load the control register of an 82C55A with the control word formed in Example 10.10. Assume that the control register of the 82C55A resides at address $0F_{16}$ of the I/O address space.

Solution

First we must load AL with $C4_{16}$. This is the value of the control word that is to be written to the control register at address $0F_{16}$. The move instruction used to load AL is

```
MOV  AL, 0C4H
```

These data are output to the control register with the OUT instruction

```
OUT  0FH, AL
```

Because the I/O address of the control register is less than FF_{16}, this instruction uses direct I/O.

In our descriptions of mode 1 and mode 2 operations, we found that when the 82C55A is configured for either of these modes, most of the pins of port C perform I/O control functions. For instance, Fig. 10–12 shows that in mode 1 PC_3 works as the $INTR_A$ output. The MPU can be programmed to read the control information from port C through software. This is known as reading the status of port C. The format of the status information input by reading port C of an 82C55A operating in mode 1 is shown in Fig. 10–20(a). Note that if the ports are configured for input operation, the status byte contains the values of the IBF and INTR outputs and the INTE flag for both ports. Once read by the MPU, these bits can be tested with other software to control the flow of the program.

Input configuration

D_7	D_6	D_5	D_4	D_3	D_2	D_1	D_0
I/O	I/O	IBF_A	$INTE_A$	$INTR_A$	$INTE_B$	IBF_B	$INTR_B$
Group A					Group B		

Output configurations

D_7	D_6	D_5	D_4	D_3	D_2	D_1	D_0
$\overline{OBF}_A$	$INTE_A$	I/O	I/O	$INTR_A$	$INTE_B$	$\overline{OBF}_B$	$INTR_B$
Group A					Group B		

(a)

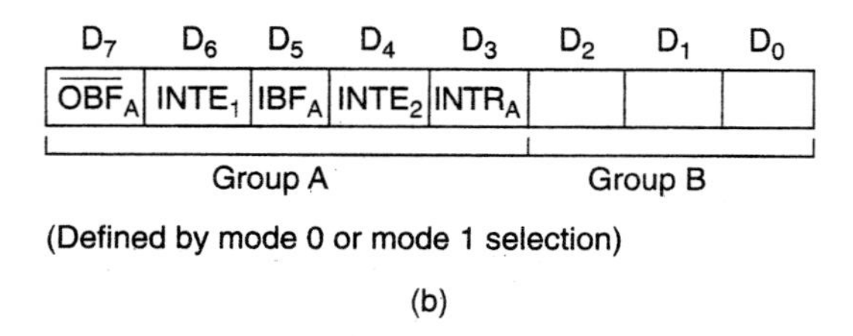

Figure 10–20 (a) Mode 1 status information for port C. (Reprinted by permission of Intel Corporation. Copyright/Intel Corp. 1980) (b) Mode 2 status information for port C. (Reprinted by permission of Intel Corporation. Copyright/Intel Corp. 1980)

By using a software handshake sequence that tests the status bits to change the program sequence, hardware signals such as interrupts can be saved. For instance, if port A is used as an input in mode 1, the processor can read (poll) the status register and check bit D_3 for $INTR_A$. If D_3 is 1, $INTR_A$ is active and the processor is signaled to read the data from port A of the 82C55A. In this way, the $INTR_A$ output may not be connected to the processor and one interrupt request input is saved.

If the ports are configured as outputs, the status byte contains the values of outputs $\overline{OBF}$ and INTR and the INTE flag for each port. Figure 10–20(b) shows the status byte format for an 82C55A configured for mode 2 operation.

▲ 10.6 82C55A IMPLEMENTATION OF PARALLEL INPUT/OUTPUT PORTS

In Sections 10.2 and 10.3, we showed how parallel input and output ports can be implemented for the 8088- and 8086-microcomputer systems using logic devices such as the 74F244 octal buffer and 74F373 octal latch, respectively. Even though logic ICs can be used to implement parallel input and output ports, the 82C55A PPI can be used to design a more versatile parallel I/O interface. This is because its ports can be configured either as inputs or outputs under software control. Here we will show how the 82C55A is used to design isolated parallel I/O interfaces for 8088- and 8086-based microcomputers.

The circuit in Fig. 10–21 shows how PPI devices can be connected to the bus of the 8088 to implement parallel input/output ports. This circuit configuration is for a minimum-mode 8088 microcomputer. Here we find a group of eight 82C55A devices connected to the data bus. A 74F138 address decoder is used to select one of the devices at a time for input and output data transfers. The ports are located at even-address boundaries. Each of these PPI devices provides up to three byte-wide ports. In the circuit, they are labeled port A, port B, and port C. These ports can be individually configured as inputs or outputs through software. Therefore, this circuit is capable of implementing up to 192 I/O lines.

Let us look more closely at the connection of the 82C55As. Starting with the inputs of the 74F138 address decoder, we see that its enable inputs are $\overline{G_{2B}} = A_0$ and $\overline{G_{2A}} = IO/\overline{M}$. A_0 is logic 0 whenever the 8088 outputs an even address on the bus. Moreover, $IO/\overline{M}$ is switched to logic 1 whenever an I/O bus cycle is in progress. This logic level is inverted and applies logic 0 to the $\overline{G_{2A}}$ input. For this reason, the decoder is enabled for all I/O bus cycles to an even address.

When the 74F138 decoder is enabled, the code at its A_0 through A_2 inputs causes one of the eight 82C55A PPIs attached to its outputs to get enabled for operation. Bits A_5 through A_3 of the I/O address are applied to these inputs of the decoder. It responds by switching the output corresponding to this 3-bit code to the 0 logic level. Decoder outputs O_0 through O_7 are applied to the chip select ($\overline{CS}$) inputs of the PPIs. For instance, $A_5A_4A_3$ = 000 switches output O_0 to logic 0. This enables the first 82C55A, numbered 0 in Fig. 10–21.

At the same time a PPI chip is selected, the 2-bit code A_2A_1 at its inputs A_1A_0 selects the port for which data are input or output. For example, A_2A_1 = 00 indicates that port A is to be accessed. Input/output data transfers take place over data bus lines D_0 through D_7. The timing of these read/write transfers is controlled by signals $\overline{RD}$ and $\overline{WR}$.

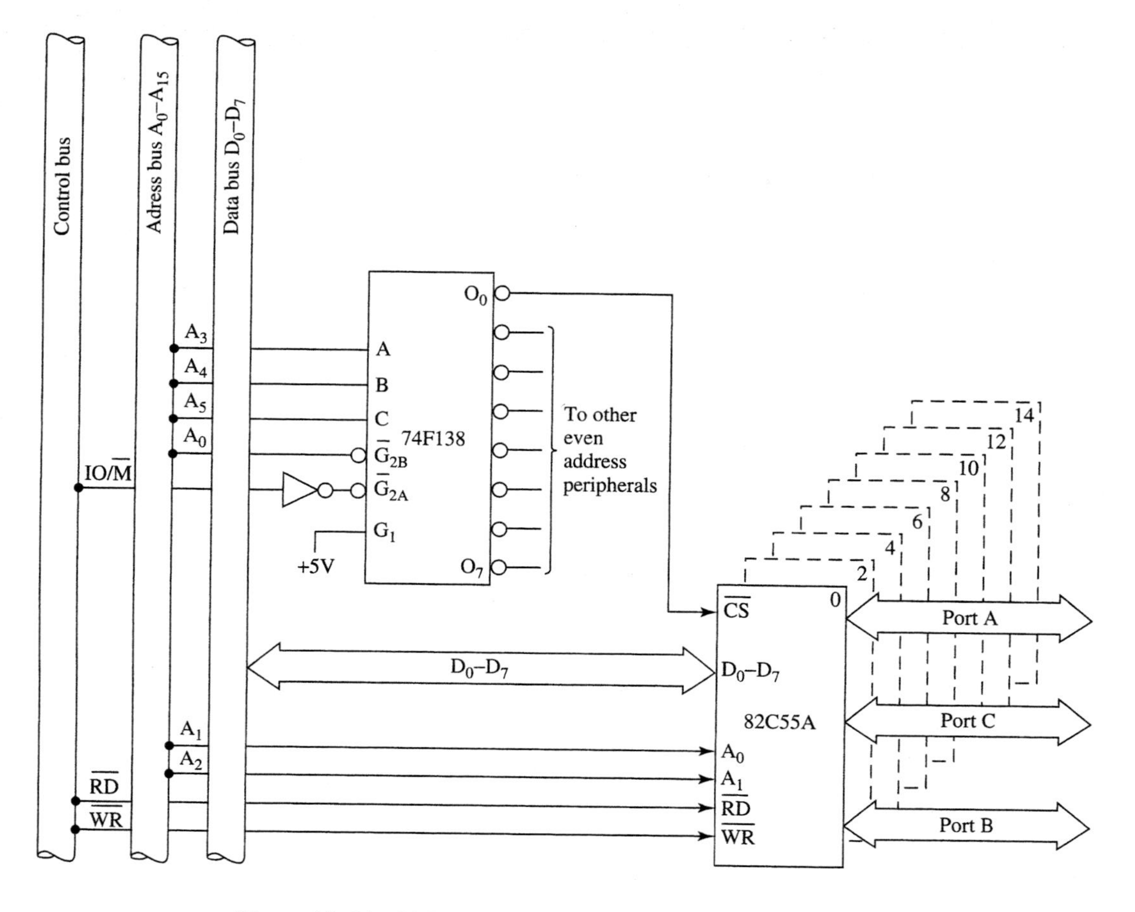

Figure 10–21 82C55A parallel I/O ports in an 8088-based microcomputer.

EXAMPLE 10.12

What must be the address bus inputs of the circuit in Fig. 10–21 if port C of PPI 14 is to be accessed?

Solution

To enable PPI 14, the 74F138 must be enabled for operation and its O_7 output switched to logic 0. This requires enable input $\overline{G}_{2B} = 0$ and chip select code CBA = 111. This in turn requires from the bus that

$$A_0 = 0 \text{ to enable 74F138}$$

and

$$A_5A_4A_3 = 111 \text{ to select PPI 14}$$

Port C of PPI 14 is selected with $A_1A_0 = 10$, which from the bus requires that

$$A_2A_1 = 10$$

The rest of the address bits are don't-care states.

EXAMPLE 10.13

Assume that, in Fig. 10–21, PPI 14 is configured so that port A is an output port, both ports B and C are input ports, and all three ports are set up for mode 0 operation. Write a program that will input the data at ports B and C, find the difference (port C) − (port B), and output this difference to port A.

Solution

From the circuit diagram in Fig. 10–21, we find that the addresses of the three I/O ports of PPI 14 are

$$\text{Port A address} = 00111000_2 = 38_{16}$$

$$\text{Port B address} = 00111010_2 = 3A_{16}$$

$$\text{Port C address} = 00111100_2 = 3C_{16}$$

The data at ports B and C can be input with the instruction sequence

```
IN   AL, 3AH     ;Read port B
MOV  BL, AL      ;Save data from port B
IN   AL, 3CH     ;Read port C
```

Now the data from port B are subtracted from the data from port C with the instruction

```
SUB  AL, BL    ;Subtract B from C
```

Finally, the difference is output to port A with the instruction

```
OUT  38H, AL   ;Write to port A
```

Figure 10–22 gives a similar circuit that implements parallel input/output ports for a minimum-mode 8086-based microcomputer system. Let us now look at the differences between this circuit and the 8088 microcomputer circuit shown in Fig. 10–21. In Fig. 10–22, we find that the I/O circuit has two groups of eight 82C55A devices; one connected to the lower eight data bus lines, and the other to the upper eight data bus lines. Each of these groups is capable of implementing up to 192 I/O lines to give a total I/O capability of 384 I/O lines.

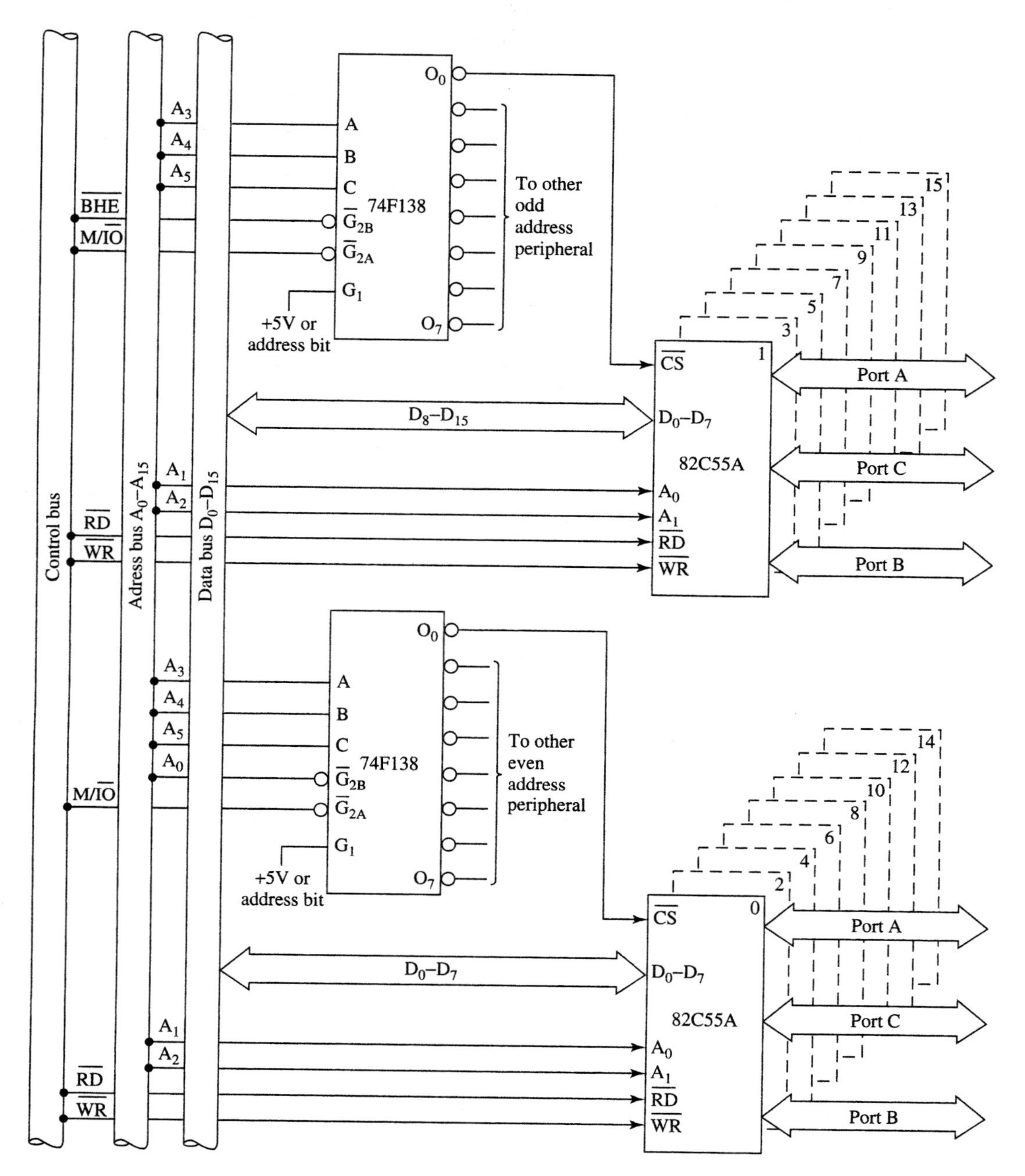

Figure 10–22 82C55A parallel I/O ports at even- and odd-address boundaries in an 8086-based microcomputer.

Each of the groups of 82C55As has its own 74F138 I/O address decoder. As in the 8088 microcomputer circuit, the address decoder is used to select devices in a group one at a time. The ports in the upper group are connected at odd-address boundaries and those in the lower group are at even-address boundaries. Let us first look more closely at the connection of the upper group of the 82C55As. Starting with the inputs of the 74F138 decoder, we see that its $\overline{G_{2B}}$ input is driven by control signal $\overline{BHE}$, the $\overline{G_{2A}}$ input is supplied by control signal $M/\overline{IO}$, and the G_1 input is permanently enabled by fixing it at the 1 logic level. $\overline{BHE}$ is logic 0 whenever the 8086 outputs an odd address on the bus. Moreover, $M/\overline{IO}$ is switched to logic 0 whenever an I/O bus cycle is in progress. In this way, we see that the upper decoder is enabled for I/O bus cycles that access a byte of data at an odd I/O address. Actually, it is also enabled during all word-wide I/O data accesses.

The code on address lines A_3 through A_5 selects one of the eight 82C55As for operation. When the upper 74F138 is enabled, the address code applied at the CBA inputs causes the corresponding output to switch to logic 0. This output is used as a chip-select ($\overline{CS}$) input to one of the 82C55As and enables it for input/output operation. The port that is accessed in the enabled PPI is selected by the code on lines A_1 and A_2 of the I/O address. Finally, the I/O data transfer takes place over data bus lines D_8 through D_{15}.

The connection of the lower group of PPIs in Fig. 10–22 is similar to that shown in Fig. 10–21. The only difference is that no inverter is required in the connection of the $M/\overline{IO}$ signal to the $\overline{G_{2A}}$ input of the 74F138 decoder. This bank is enabled for all byte-wide data accesses to an even address as well as for all word-wide data accesses.

▲ 10.7 MEMORY-MAPPED INPUT/OUTPUT PORTS

The *memory-mapped I/O interface* of a minimum-mode 8088 system is essentially the same as that employed in the accumulator I/O circuit of Fig. 10–21. Figure 10–23 shows the equivalent memory-mapped circuit. Ports are still selected by decoding an address on the address bus, and data are transferred between the 8088 and I/O device over the data bus. One difference is that now the full 20-bit address is available for addressing I/O. Therefore, memory-mapped I/O devices can reside anywhere in the 1Mbyte memory address space of the 8088.

Another difference is that during I/O operations memory read and write bus cycles are initiated instead of I/O bus cycles. This is because memory instructions, not input/output instructions, are used to perform the data transfers. Furthermore, $IO/\overline{M}$ stays at the 0 logic level throughout the bus cycle. This indicates that a memory operation, not an I/O operation, is in progress.

Since memory-mapped I/O devices reside in the memory address space and are accessed with read and write cycles, additional I/O address latch, address buffer, data bus transceiver, and address decoder circuitry are not needed. The circuitry provided for the memory interface can be used to access memory-mapped ports.

The key difference between the circuits in Figs. 10–21 and 10–23 is that $IO/\overline{M}$ is no longer inverted. Instead, it is applied directly to the $\overline{G_{2A}}$ input of the decoder. Another difference is that the G_1 input of the decoder is not fixed at the 1 logic level; instead, it is supplied by address line A_{10}. The I/O circuits are accessed whenever $IO/\overline{M}$ is equal to logic 0, A_{10} is equal to logic 1, and A_0 equals 0.

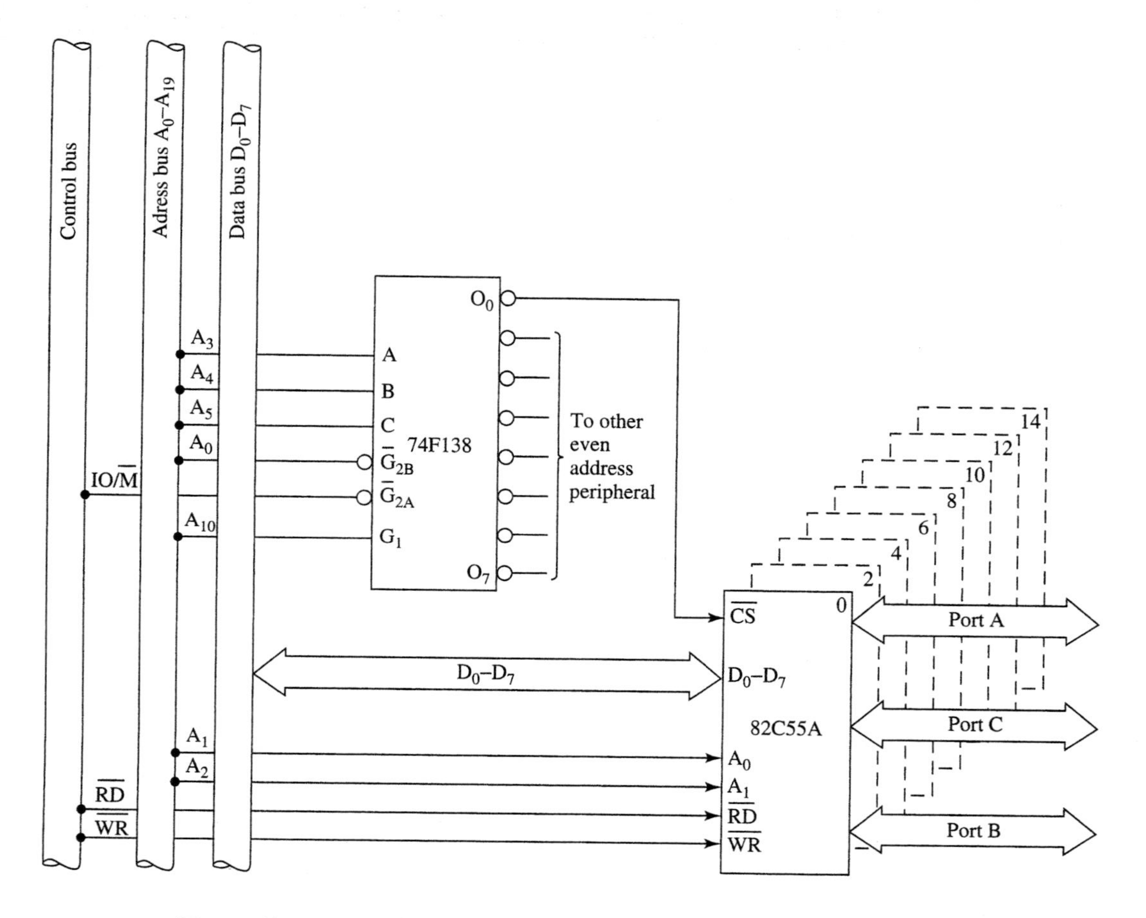

Figure 10–23 Memory-mapped 82C55A parallel I/O ports in an 8088-based microcomputer.

EXAMPLE 10.14

Which I/O port in Fig. 10–23 is selected for operation when the memory address output on the bus is 00402_{16}?

Solution

We begin by converting the address to binary form. This gives

$$A_{19} \ldots A_1A_0 = 00000000010000000010_2$$

In this address, bits $A_{10} = 1$ and $A_0 = 0$. Therefore, the 74F138 address decoder is enabled whenever IO/$\overline{M}$ = 0, which is the case during memory operations.

A memory-mapped I/O operation takes place at the port selected by $A_5A_4A_3 = 000$. This input code switches decoder output O_0 to logic 0 and chip selects PPI 0 for operation. That is,

$$A_5A_4A_3 = 000$$

makes

$$O_0 = 0$$

and selects PPI 0.

The address bits applied to the port select inputs of the PPI are $A_2A_1 = 01$. These inputs cause port B to be accessed. Thus, the address 00402_{16} selects port B on PPI 0 for memory-mapped I/O.

EXAMPLE 10.15

Write the sequence of instructions needed to initialize the control register of PPI 0 in the circuit of Fig. 10–23 so that port A is an output port, ports B and C are input ports, and all three ports are configured for mode 0 operation.

Solution

Referring to Fig. 10–9, we find that the control byte required to provide this configuration is

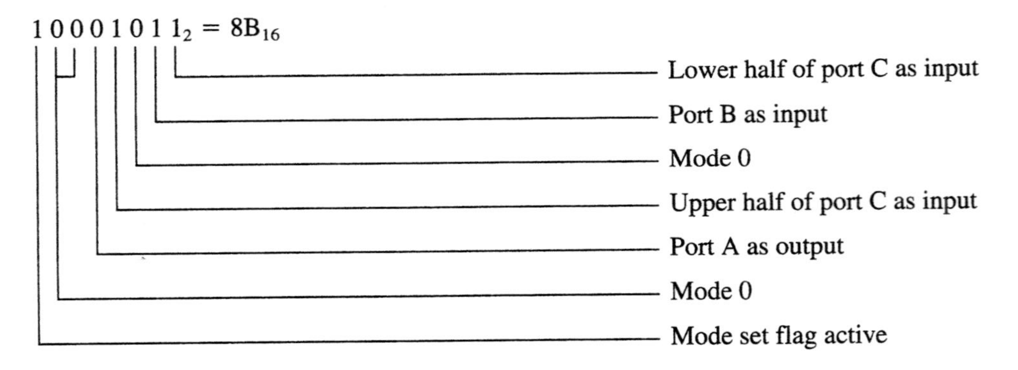

From the circuit diagram, the memory address of the control register for PPI 0 is found to be $00000000010000000110_2 = 00406_{16}$. Since PPI 0 is memory mapped, the following move instructions can be used to initialize the control register:

```
MOV  AX, 0          ;Create data segment at 00000H
MOV  DS, AX
MOV  AL, 8BH        ;Load AL with control byte
MOV  [406H], AL     ;Write control byte to PPI 0 control register
```

EXAMPLE 10.16

Assume that PPI 0 in Fig. 10–23 is configured as described in Example 10.15. Write a program that will input the contents of ports B and C, AND them together, and output the results to port A.

Solution

From the circuit diagram, we find that the addresses of the three I/O ports on PPI 0 are

$$\text{Port A address} = 00400_{16}$$
$$\text{Port B address} = 00402_{16}$$
$$\text{Port C address} = 00404_{16}$$

Now we set up a data segment at 00000_{16} and input the data from ports B and C:

```
MOV  AX, 0           ;Create data segment at 00000H
MOV  DS, AX
MOV  BL, [402H]      ;Read port B
MOV  AL, [404H]      ;Read port C
```

Next, the contents of AL and BL must be ANDed and the result output to port A. This is done with the instructions

```
AND  AL, BL          ;AND data at ports B and C
MOV  [400H], AL      ;Write to port A
```

Figure 10–24 shows a memory-mapped parallel I/O interface circuit for an 8086-based microcomputer system. Just like the accumulator-mapped circuit in Fig. 10–22, this circuit is capable of implementing up to 384 parallel I/O lines.

▲ 10.8 82C54 PROGRAMMABLE INTERVAL TIMER

The 82C54 is an LSI peripheral designed to permit easy implementation of *timer* and *counter functions* in a microcomputer system. It contains three independent 16-bit counters that can be programmed to operate in a variety of ways to implement timing functions. For instance, they can be set up to work as a one-shot pulse generator, square-wave generator, or rate generator.

Block Diagram of the 82C54

Let us begin our study of the 82C54 by looking at the signal interfaces shown in its block diagram of Fig. 10–25(a). The actual pin location for each of these signals is given

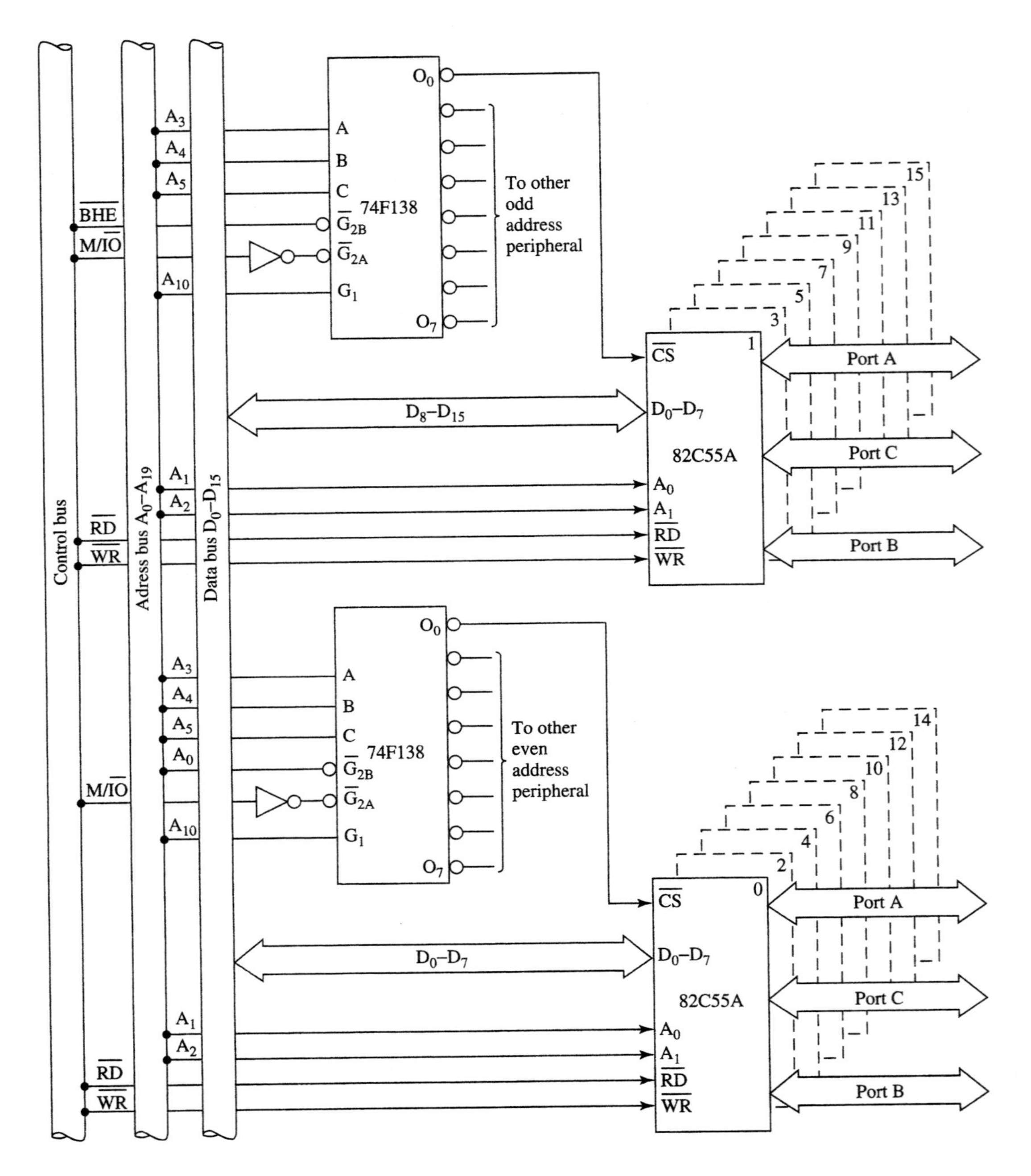

Figure 10–24 Memory-mapped 82C55A parallel I/O ports in an 8086-based microcomputer.

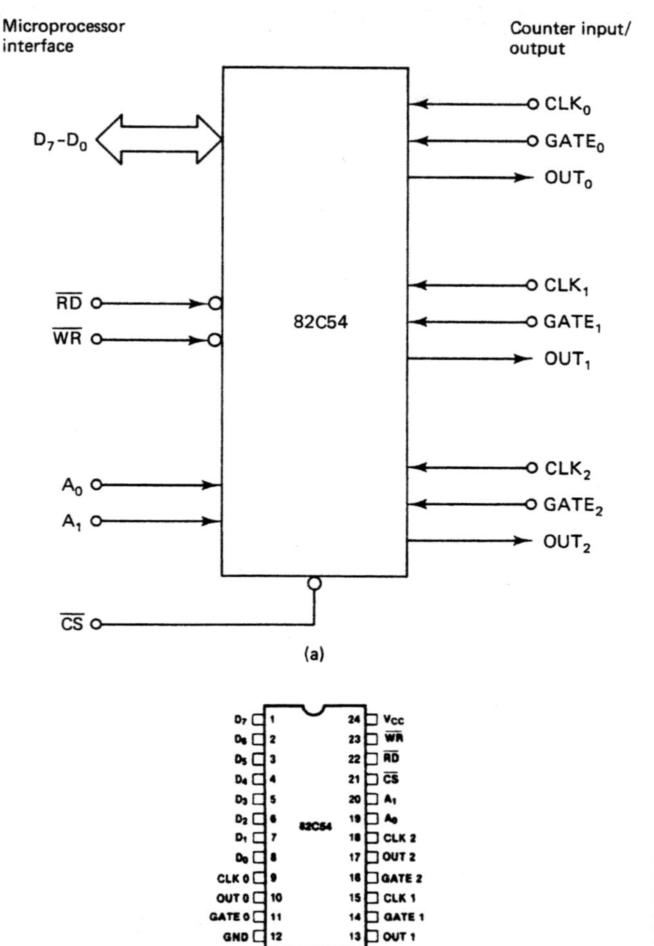

Figure 10–25 (a) Block diagram of the 82C54 interval timer. (b) Pin layout. (Reprinted by permission of Intel Corporation. Copyright/Intel Corp. 1987)

in Fig. 10–25(b). In a microcomputer system, the 82C54 is treated as a peripheral device. Moreover, it can be memory-mapped into the memory address space or I/O-mapped into the I/O address space. The microprocessor interface of the 82C54 allows the MPU to read from or write into its internal registers. In this way, it can be configured in various modes of operation.

Now we will look at the signals of the microprocessor interface. The microprocessor interface includes an 8-bit bidirectional data bus, D_0 through D_7. It is over these lines that data are transferred between the MPU and 82C54. Register address inputs A_0 and A_1 are used to select the register to be accessed, and control signals read ($\overline{RD}$) and write ($\overline{WR}$) indicate whether it is to be read from or written into, respectively. A chip-select ($\overline{CS}$) input is also provided to enable the 82C54's microprocessor interface. This input allows the designer to locate the device at a specific memory or I/O address.

At the other side of the block in Fig. 10–25(a), we find three signals for each counter. For instance, counter 0 has two inputs that are labeled CLK_0 and $GATE_0$. Pulses applied to the clock input are used to decrement counter 0. The gate input is used to enable or disable the counter. $GATE_0$ must be switched to logic 1 to enable counter 0 for operation. For example, in the square-wave mode of operation, the counter is to run continuously; therefore, $GATE_0$ is fixed at the 1 logic level, and a continuous clock signal is applied to CLK_0. The 82C54 is rated for a maximum clock frequency of 10 MHz. Counter 0 also has an output line that is labeled OUT_0. The counter produces either a clock or a pulse at OUT_0, depending on the mode of operation selected. For instance, when configured for the square-wave mode of operation, this output is a clock signal.

Architecture of the 82C54

Figure 10–26 shows the internal architecture of the 82C54. Here we find the *data bus buffer, read/write logic, control word register,* and three *counters.* The data bus buffer and read/write control logic represent the microprocessor interface we just described.

The control word register section actually contains three 8-bit registers used to configure the operation of counters 0, 1, and 2. The format of a *control word* is shown in Fig. 10–27. Here we find that the two most significant bits are a code that assigns the control word to a counter. For instance, making these bits 01 selects counter 1. Bits D_1 through D_3 are a 3-bit mode-select code, $M_2M_1M_0$, which selects one of six modes of counter operation. The least significant bit D_0 is labeled BCD and selects either binary or BCD mode of counting. For instance, if this bit is set to logic 0, the counter acts as a 16-bit

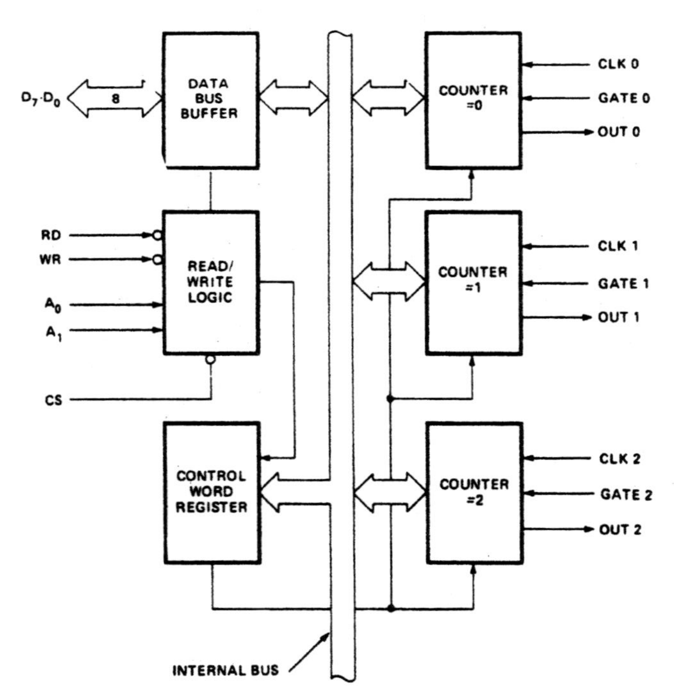

Figure 10–26 Internal architecture of the 82C54. (Reprinted by permission of Intel Corporation. Copyright/Intel Corp. 1987)

Control word format

D_7	D_6	D_5	D_4	D_3	D_2	D_1	D_0
SC_1	SC_0	RW/W_1	RW/W_0	M_2	M_1	M_0	BCD

Definition of control

SC-select counter:

SC_1	SC_0	
0	0	Select counter 0
0	1	Select counter 1
1	0	Select counter 2
1	1	Read back command

RW-read/write:

RW/W_1	RW/W_0	
0	0	Counter latch command
1	0	Read/write most significant byte only
0	1	Read/write least significant byte only
1	1	Read/write least significant byte first, then most significant byte

M-mode:

M_2	M_1	M_0	
0	0	0	Mode 0
0	0	1	Mode 1
X	1	0	Mode 2
X	1	1	Mode 3
1	0	0	Mode 4
1	0	1	Mode 5

BCD:

0	Binary counter 16-bits
1	Binary coded decimal (BCD) counter (4 decades)

Figure 10–27 Control word format of the 82C54. (Reprinted by permission of Intel Corporation. Copyright/Intel Corp. 1987)

binary counter. Finally, the 2-bit code RW/W_1 RW/W_0 is used to set the sequence in which bytes are read from or loaded into the 16-bit count registers.

EXAMPLE 10.17

An 82C54 receives the control word 10010000_2. What configuration is set up for the timer?

Solution

Since the SC bits are 10, the rest of the bits are for setting up the configuration of counter 2. Following the format in Fig. 10–27, we find that 01 in the RW/W bits sets counter 2 for the read/write sequence identified as the least significant byte only. This means that the next write operation performed to counter 2 will load the data into the least significant byte of its count register. Next the mode code is 000, and this selects mode 0 operation for this counter. The last bit, BCD, is also set to 0 and selects binary counting.

The three counters shown in Fig. 10–26 are each 16 bits in length and operate as *down counters*. That is, when enabled by an active gate input, the clock decrements the count. Each counter contains a 16-bit *count register* that must be loaded as part of the initialization cycle. The value held in the count register can be read at any time through software.

To read from or write to the counters of the 82C54 or load its control word register, the microprocessor needs to execute instructions. Figure 10–28 shows the bus-control information needed to access each register. For example, to write to the control register, the register address lines must be $A_1A_0 = 11$ and the control lines must be $\overline{WR} = 0$, $\overline{RD} = 1$, and $\overline{CS} = 0$.

EXAMPLE 10.18

Write an instruction sequence to set up the three counters of the 82C54 in Fig. 10–29 as follows:

Counter 0: Binary counter operating in mode 0 with an initial value of 1234H.

Counter 1: BCD counter operating in mode 2 with an initial value of 0100H.

Counter 2: Binary counter operating in mode 4 with an initial value of 1FFFH.

Solution

First, we need to determine the base address of the 82C54. The base address, which is also the address of counter 0, is determined with A_1A_0 set to 00. In Fig. 10–29 we find that to select the 82C54, $\overline{CS}$ must be logic 0. This requires that

$$A_{15}A_{14}\ldots A_7A_6A_5\ldots A_2 = 00000000010000_2$$

$\overline{CS}$	$\overline{RD}$	$\overline{WR}$	A_1	A_0	
0	1	0	0	0	Write into Counter 0
0	1	0	0	1	Write into Counter 1
0	1	0	1	0	Write into Counter 2
0	1	0	1	1	Write Control Word
0	0	1	0	0	Read from Counter 0
0	0	1	0	1	Read from Counter 1
0	0	1	1	0	Read from Counter 2
0	0	1	1	1	No-Operation (3-State)
1	X	X	X	X	No-Operation (3-State)
0	1	1	X	X	No-Operation (3-State)

Figure 10–28 Accessing the registers of the 82C54. (Reprinted by permission of Intel Corporation. Copyright/Intel Corp. 1987)

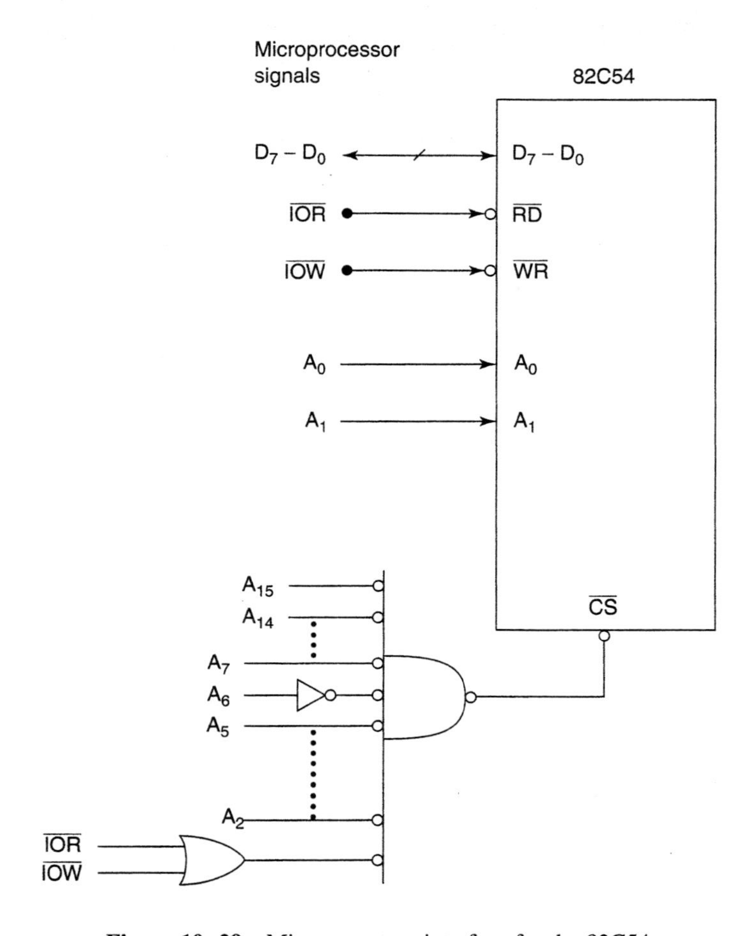

Figure 10–29 Microprocessor interface for the 82C54.

Combining this part of the address with the 00 at A_1A_0, gives the base address as

$$0000000001000000_2 = 40\text{H}$$

Since the base address of the 82C54 is 40H, and to select the mode register requires $A_1A_0 = 11$, its address is 43H. Similarly, the three counters 0, 1, and 2 are at addresses 40H, 41H, and 42H, respectively. Let us first determine the mode words for the three counters. Following the bit definitions in Fig. 10–27, we get

$$\text{Mode word for counter } 0 = 00110000_2 = 30_{16}$$

$$\text{Mode word for counter } 1 = 01010101_2 = 55_{16}$$

$$\text{Mode word for counter } 2 = 10111000_2 = \text{B}8_{16}$$

The following instruction sequence can be used to set up the 82C54 with the desired mode words and counts:

```
MOV   AL, 30H       ;Set up counter 0 mode
OUT   43H, AL
MOV   AL, 55H       ;Set up counter 1 mode
OUT   43H, AL
MOV   AL, 0B8H      ;Set up counter 2 mode
OUT   43H, AL
MOV   AL, 1234H     ;Initialize counter 0 with 1234H
OUT   40H, AL
MOV   AL, 12H
OUT   40H, AL
MOV   AL, 0100H     ;Initialize counter 1 with 0100H
OUT   41H, AL
MOV   AL, 01H
OUT   41H, AL
MOV   AL, 1FFFH     ;Initialize counter 2 with 1FFFH
OUT   42H, AL
MOV   AL, 1FH
OUT   42H, AL
```

Earlier we pointed out that the contents of a count register could be read at any time. Let us now look at how this is done in software. One approach is simply to read the contents of the corresponding register with an input instruction. Figure 10–28 shows that to read the contents of count register 0, the control inputs must be $\overline{CS} = 0$, $\overline{RD} = 0$, and $\overline{WR} = 1$, and the register address code must be $A_1A_0 = 00$. To ensure that a valid count is read out of count register 0, the counter must be inhibited before the read operation takes place. The easiest way to do this is to switch the $GATE_0$ input to logic 0 before performing the read operation. The count is read as two separate bytes, low byte followed by the high byte.

The contents of the count registers can also be read without first inhibiting the counter. That is, the count can be read on the fly. To do this in software, a command must first be issued to the mode register to capture the current value of the counter into a temporary internal storage register. Figure 10–27 shows that setting bits D_5 and D_4 of the mode byte to 00 specifies the latch mode of operation. Once this mode byte has been written to the 82C54, the contents of the temporary storage register for the counter can be read just as before.

EXAMPLE 10.19

Write an instruction sequence to read the contents of counter 2 on the fly. The count is to be loaded into the AX register. Assume that the 82C54 is located at I/O address 40H.

Solution

First, we latch the contents of counter 2 and then read this value from the temporary storage register. This is done with the following sequence of instructions:

```
MOV  AL, 1000XXXXB      ;Latch counter 2,
                        ;XXXX must be as per the mode and
                        ;counter type
OUT  43H, AL
IN   AL, 42H            ;Read the low byte
MOV  BL, AL
IN   AL, 42H            ;Read the high byte
MOV  AH, AL
MOV  AL, BL             ;(AX) = counter 2 value
```

Another mode of operation, *read-back mode,* permits a programmer to capture the current count values and status information of all three counters with a single command. In Fig. 10–27 we see that a read-back command has bits D_6 and D_7 both set to 1. The read-back command format is shown in more detail in Fig. 10–30. Note that bits D_1 (CNT 0), D_2 (CNT 1), and D_3 (CNT 2) are made logic 1 to select the counters, logic 0 in bit D_4 means that status information will be latched, and logic 0 in D_5 means that the counts will be latched. For instance, to capture the values in all three counters, the read-back command is $11011110_2 = DE_{16}$. This command must be written into the control word register of the 82C54. Figure 10–31 shows some other examples of read-back commands. Note that both count and status information can be latched with a single command.

Our read-back command example, DE_{16}, only latches the values of the three counters. The programmer must next read these values by issuing read commands for the individual counters. Once the value of a counter or status is latched, it must be read before a new value can be captured.

Figure 10–31 gives an example of a command that latches only the status for counters 1 and 2. This command is coded as

$$11101100_2 = EC_{16}$$

Figure 10–32 shows the format of the status information latched with this command. Here we find that bits D_0 through D_5 contain the mode-control information that was written into the counter. These bits are identical to the six least significant bits of the control

A0, A1 = 11 $\overline{CS}$ = 0 $\overline{RD}$ = 1 $\overline{WR}$ = 0

D_7	D_6	D_5	D_4	D_3	D_2	D_1	D_0
1	1	$\overline{COUNT}$	$\overline{STATUS}$	CNT 2	CNT 1	CNT 0	0

D_5: 0 = Latch count of selected counter(s)
D_4: 0 = Latch status of selected counter(s)
D_3: 1 = Select counter 2
D_2: 1 = Select counter 1
D_1: 1 = Select counter 0
D_0: Reserved for future expansion; must be 0

Figure 10–30 Read-back command format. (Reprinted by permission of Intel Corporation. Copyright/Intel Corp. 1990)

D_7	D_6	D_5	D_4	D_3	D_2	D_1	D_0	Description	Results
Command									
1	1	0	0	0	0	1	0	Read back count and status of Counter 0	Count and status latched for Counter 0
1	1	1	0	0	1	0	0	Read back status of Counter 1	Status latched for Counter 1
1	1	1	0	1	1	0	0	Read back status of Counters 2, 1	Status latched for Counters 1 and 2
1	1	0	1	1	0	0	0	Read back count of Counter 2	Count latched for Counter 2
1	1	0	0	0	1	0	0	Read back count and status of Counter 1	Count and status latched for Counter 1
1	1	1	0	0	0	1	0	Read back status of Counter 0	Status latched for Counter 0

Figure 10–31 Read-back command examples. (Reprinted by permission of Intel Corporation. Copyright/Intel Corp. 1990)

word in Fig. 10–27. In addition to this information, the status byte contains the logic state of the counter's output pin in bit position D_7 and the value of the null count flip-flop in bit position D_6. The programmer reads latched status information by issuing a read-counter command to the 82C54.

The first command in Fig. 10–31, $11000010_2 = C2_{16}$, captures both the count and status information for counter 0. When both count and status information is captured with a read-back command, two read-counter commands are required to return the information to the MPU. During the first read operation, the value of the count is read, and the status information is transferred during the second read operation.

Operating Modes of 82C54 Counters

As indicated earlier, each of the 82C54's counters can be configured to operate in one of six modes. Figure 10–33 shows waveforms that summarize operation for each mode. Note that mode 0 operation is known as interrupt on terminal count and mode 1 is called programmable one-shot. The GATE input of a counter takes on different functions, depending on which mode of operation is selected. Figure 10–34 summarizes the effect of the gate input. For instance, in mode 0, GATE disables counting when set to logic 0 and enables counting when set to 1. Let us now discuss each of these modes of operation in more detail.

The *interrupt on terminal count* mode of operation is used to generate an interrupt to the microprocessor after a certain interval of time has elapsed. As shown in the waveforms for mode 0 operation in Fig. 10–33, a count of $n = 4$ is written into the count register synchronously with the pulse at $\overline{WR}$. After the write operation is complete, the count

D_7	D_6	D_5	D_4	D_3	D_2	D_1	D_0
OUTPUT	NULL COUNT	RW1	RW0	M2	M1	M0	BCD

D_7 1 = Out Pin is 1
0 = Out Pin is 0
D_6 1 = Null count
0 = Count available for reading
D_5-D_0 Counter Programmed Mode

Figure 10–32 Status byte format. (Reprinted by permission of Intel Corporation. Copyright/Intel Corp. 1990)

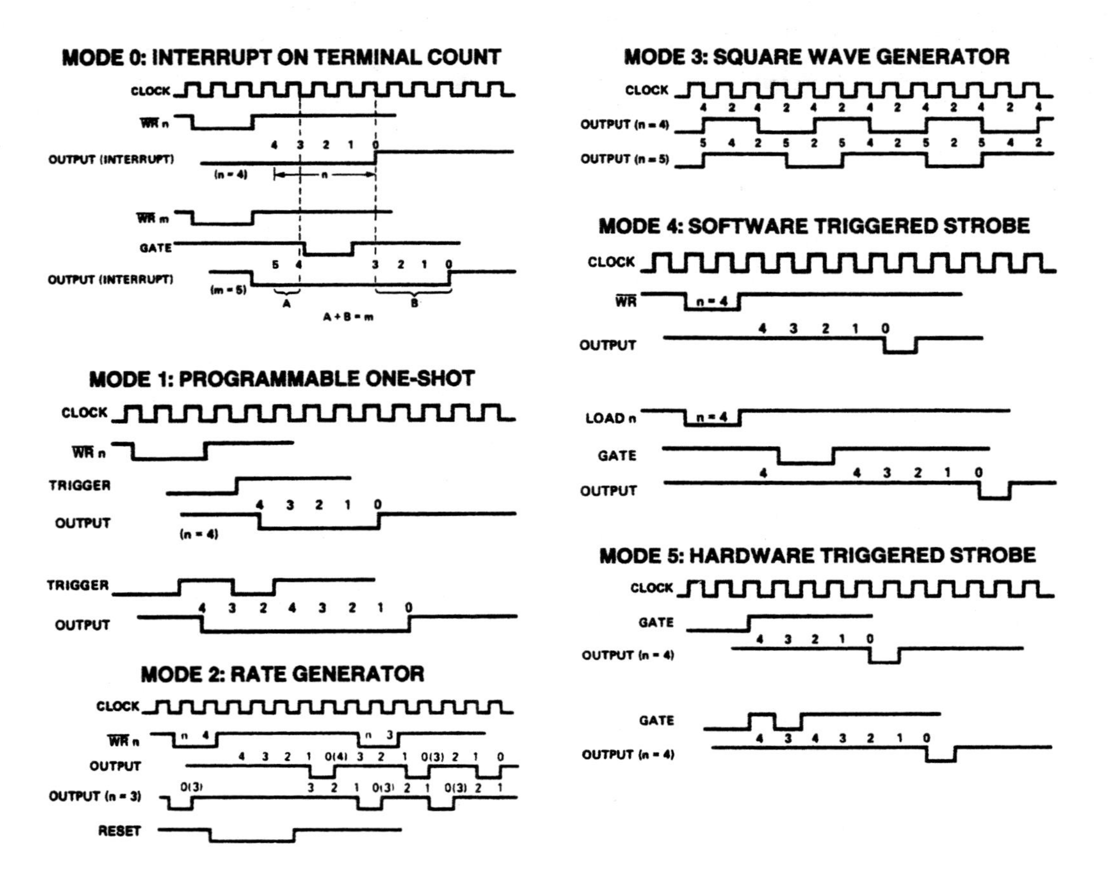

Figure 10–33 Operating modes of the 82C54. (Reprinted by permission of Intel Corporation. Copyright/Intel Corp. 1987)

is loaded into the counter on the next clock pulse and the count is decremented by 1 for each clock pulse that follows. When the count reaches 0, the terminal count, a 0-to-1 transition occurs at OUTPUT. This occurs after $n + 1$ (five) clock pulses. This signal can be used as the interrupt input to the microprocessor.

Earlier we found in Fig. 10–34 that GATE must be at logic 1 to enable the counter for interrupt on terminal count mode of operation. Figure 10–33 also shows waveforms for the case in which GATE is switched to logic 0. Here we see that the value of the count is 4 when GATE is switched to logic 0. It holds at this value until GATE returns to 1.

EXAMPLE 10.20

The counter of Fig. 10–35 is programmed to operate in mode 0. Assuming that the decimal value 100 is written into the counter, compute the time delay (T_D) that occurs until the positive transition takes place at the counter 0 output. The counter is configured for BCD counting. Assume the relationship between the $GATE_0$ and the CLK_0 signal as shown in the figure.

Signal Status Modes	Low Or Going Low	Rising	High
0	Disables counting	—	Enables counting
1	—	1) Initiates counting 2) Resets output after next clock	—
2	1) Disables counting 2) Sets output immediately high	Initiates counting	Enables counting
3	1) Disables counting 2) Sets output immediately high	Initiates counting	Enables counting
4	Disables counting	—	Enables counting
5	—	Initiates counting	—

Figure 10–34 Effect of the GATE input for each mode. (Reprinted by permission of Intel Corporation. Copyright/Intel Corp. 1987)

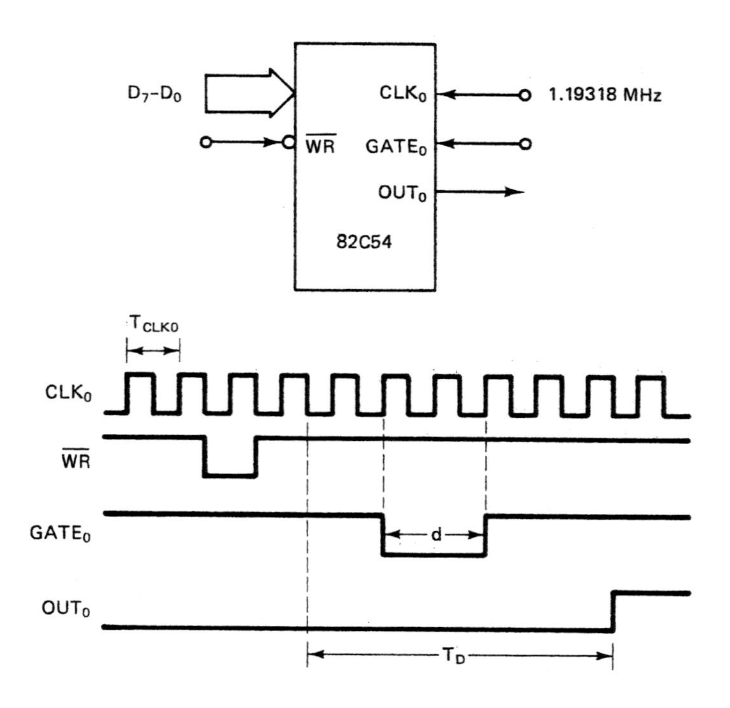

Figure 10–35 Mode 0 configuration.

Solution

Once loaded, counter 0 needs to count down for 100 pulses at the clock input. During this period, the counter is disabled by logic 0 at the $GATE_0$ input for two clock periods. Therefore, the time delay is calculated as

$$
\begin{aligned}
T_D &= (n + 1 + d)(T_{CLK0}) \\
&= (100 + 1 + 2)(1/1.19318)\ \mu s \\
&= 86.3\ \mu s
\end{aligned}
$$

Mode 1 operation implements what is known as a *programmable one-shot.* As Fig. 10–33 shows, when set for this mode of operation, the counter produces a single pulse at its output. The waveforms show that an initial count, which in this example is the number 4, is written into the counter synchronous with a pulse at $\overline{WR}$. When GATE, called TRIGGER in the waveshapes, switches from logic 0 to 1, OUTPUT switches to logic 0 on the next pulse at CLOCK and the count begins to decrement with each successive clock pulse. The pulse is completed as OUTPUT returns to logic 1 when the terminal count, 0, is reached. In this way, we see that the duration of the pulse is determined by the value loaded into the counter.

The pulse generator produced with an 82C54 counter is what is called a *retriggerable one-shot.* By retriggerable we mean that if, after an output pulse has been started, another rising edge is experienced at TRIGGER, the count is reloaded and restarting the count operation extends the pulse width. The lower one-shot waveform in Fig. 10–33 shows this type of operation. Note that after the count is decremented to 2, a second rising edge occurs at TRIGGER. On the next clock pulse, the value 4 is reloaded into the counter to extend the pulse width to 7 clock cycles.

EXAMPLE 10.21

Counter 1 of an 82C54 is programmed to operate in mode 1 and is loaded with the decimal value 10. The gate and clock inputs are as shown in Fig. 10–36. How long is the output pulse? Assume that the counter is configured for BCD counting.

Solution

The $GATE_1$ input in Fig. 10–36 shows that the counter is operated as a nonretriggerable one-shot. Therefore, the pulse width is given by

$$
\begin{aligned}
T &= \text{(counter contents) (clock period)} \\
&= (10)(1/1.19318)\ \text{MHz} \\
&= 8.38\ \mu s
\end{aligned}
$$

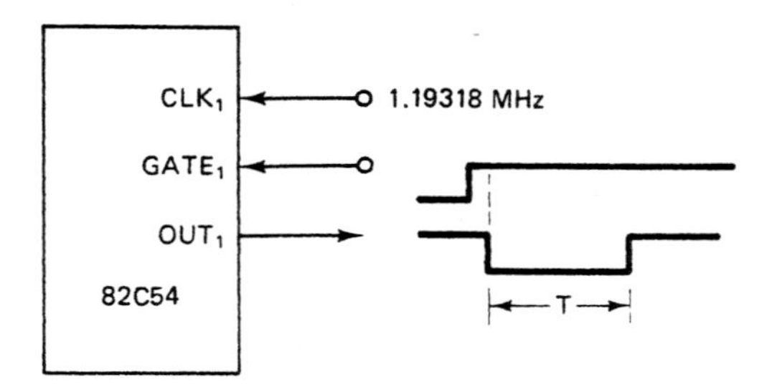

Figure 10–36 Mode 1 configuration.

When set for mode 2, *rate generator* operation, the counter within the 82C54 is set to operate as a divide-by-*N* counter. Here *N* stands for the value of the count loaded into the counter. Figure 10–37 shows counter 1 of an 82C54 set up in this way. Note that the gate input is fixed at the 1 logic level. As the table in Fig. 10–34 shows, this enables the counting operation. Looking at the waveforms for mode 2 operation in Fig. 10–33, we see that OUTPUT is at logic 1 until the count decrements to 1. Then OUTPUT switches to the active 0 logic level for just one clock pulse width. In this way, we see that there is one clock pulse at the output for every *N* clock pulses at the input. This is why it is called a divide-by-*N* counter.

EXAMPLE 10.22

Counter 1 of the 82C54, shown in Fig. 10–37, is programmed to operate in mode 2 and is loaded with decimal number 18. Describe the signal produced at OUT_1. Assume that the counter is configured for BCD counting.

Solution

In mode 2 the output goes low for one period of the input clock after the counter contents decrement to 0. Therefore,

$$T_2 = 1/1.19318 \text{ MHz} = 838 \text{ ns}$$

and

$$T = 18 \times T_2 = 15.094 \ \mu s$$

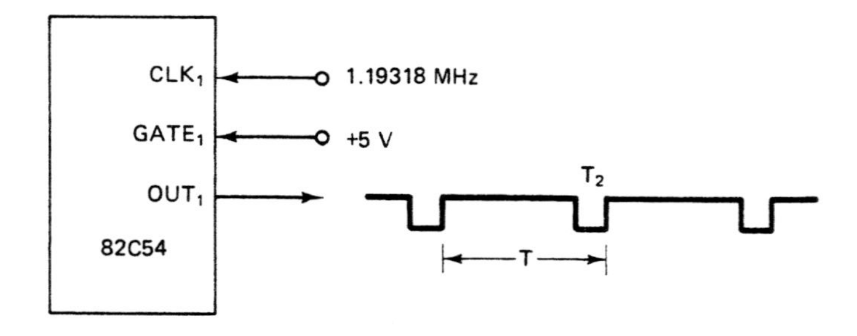

Figure 10–37 Mode 2 configuration.

Mode 3 sets the counter of the 82C54 to operate as a *square-wave rate generator.* In this mode, the output of the counter is a square wave with 50 percent duty cycle whenever the counter is loaded with an even number. That is, the output is at the 1 logic level for exactly the same amount of time that it is at the 0 logic level. As shown in Fig. 10–33, the count decrements by two with each pulse at the clock input. When the count reaches 0, the output switches logic levels, the original count ($n = 4$) is reloaded, and the count sequence repeats. Transitions of the output take place with respect to the negative edge of the input clock. The period of the symmetrical square wave at the output equals the number loaded into the counter multiplied by the period of the input clock.

If an odd number (N) is loaded into the counter instead of an even number, the time for which the output is high depends on $(N + 1)/2$, and the time for which the output is low depends on $(N - 1)/2$.

EXAMPLE 10.23

The counter in Fig. 10–38 is programmed to operate in mode 3 and is loaded with the decimal value 15. Determine the characteristics of the square wave at OUT_1. Assume that the counter is configured for BCD counting.

Solution

$$T_{CLK1} = 1/1.19318 \text{ MHz} = 838 \text{ ns}$$

$$T_1 = T_{CLK1}(N + 1)/2 = 838 \text{ ns} \times [(15 + 1)/2]$$

$$= 6.704\ \mu s$$

$$T_2 = T_{CLK1}(N - 1)/2 = 838 \text{ ns} \times [(15 - 1)/2]$$

$$= 5.866\ \mu s$$

$$T = T_1 + T_2 = 6.704\ \mu s + 5.866\ \mu s$$

$$= 12.57\ \mu s$$

Selecting mode 4 operation for a counter configures the counter to work as a *software-triggered strobed counter.* When in this mode, the counter automatically begins to decrement one clock pulse after it is loaded with the initial value through software. Again, it decrements at a rate set by the clock input signal. At the moment the terminal

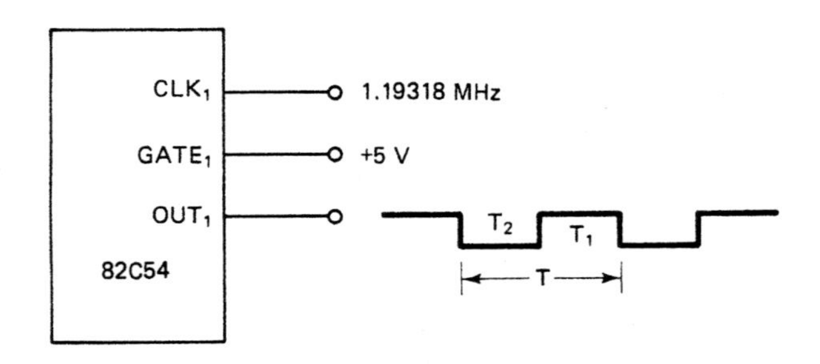

Figure 10–38 Mode 3 configuration.

count is reached, the counter generates a single strobe pulse with duration equal to one clock pulse at its output. That is, a strobe pulse is produced at the output after $n + 1$ clock pulses. Here n again stands for the value of the count loaded into the counter. This output pulse can be used to perform a timed operation. Figure 10–33 shows waveforms illustrating this mode of operation initiated by writing the value 4 into a counter. For instance, if CLOCK is 1.19318 MHz, the strobe occurs 4.19 μs after the count 4 is written into the counter. In the table of Fig. 10–34, we find that the gate input needs to be at logic 1 for the counter to operate.

This mode of operation can be used to implement a long-duration interval timer or a free-running timer. In either application, the strobe at the output can be used as an interrupt input to a microprocessor. In response to this pulse, an interrupt service routine can be used to reload the timer and restart the timing cycle. Frequently, the service routine also counts the strobes as they come in by decrementing the contents of a register. Software can test the value in this register to determine if the timer has timed out a certain number of times; for instance, to determine if the contents of the register have decremented to 0. When it reaches 0, a specific operation, such as a jump or call, can be initiated. In this way, we see that software has been used to extend the interval of time at which a function occurs beyond the maximum duration of the 16-bit counter within the 82C54.

EXAMPLE 10.24

Counter 1 of Fig. 10–39 is programmed to operate in mode 4. What value must be loaded into the counter to produce a strobe signal 10 μs after the counter is loaded?

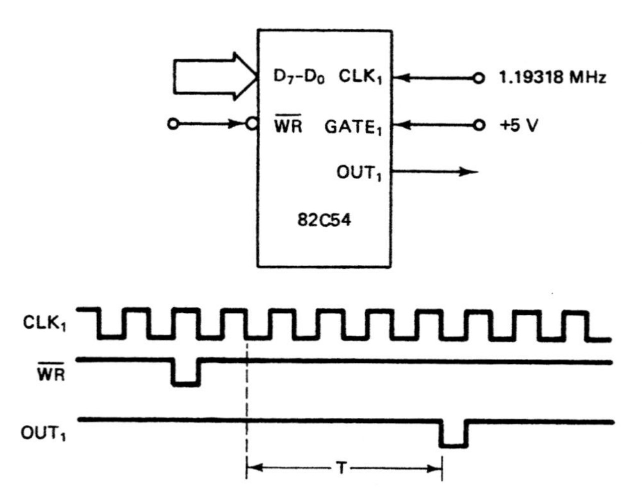

Figure 10–39 Mode 4 configuration.

Solution

The strobe pulse occurs after counting down the counter to zero. The number of input clock periods required for a period of 10 μs is given by

$$N = T/T_{CLK}$$
$$= 10\ \mu s/(1/1.19318\ MHz)$$
$$= 12_{10} = C_{16} = 00001100_2$$

Thus, the counter should be loaded with the number $n = 0B_{16}$ to produce a strobe pulse 10 μs after loading.

The last mode of 82C54 counter operation, mode 5, is called the *hardware-triggered strobe.* This mode is similar to mode 4 except that now counting is initiated by a signal at the gate input—that is, it is hardware triggered instead of software triggered. As shown in the waveforms of Fig. 10–33 and the table in Fig. 10–34, a rising edge at GATE starts the countdown process. Just as for software-triggered strobed operation, the strobe pulse is output after the count is decremented to 0. But in this case, OUTPUT switches to logic 0 N clock pulses after GATE becomes active.

▲ 10.9 82C37A PROGRAMMABLE DIRECT MEMORY ACCESS CONTROLLER

The 82C37A is the LSI controller IC that is widely used to implement the *direct memory access* (DMA) function in 8088- and 8086-based microcomputer systems. DMA capability permits devices, such as peripherals, to perform high-speed data transfers between either two sections of memory or between memory and an I/O device. In a microcomputer system, the memory or I/O bus cycles initiated as part of a DMA transfer are not performed by the MPU; instead, they are performed by a device known as a *DMA controller,* such as the 82C37A. The DMA mode of operation is frequently used when blocks or packets of data are to be transferred. For instance, disk controllers, local area network controllers, and communication controllers are devices that normally process data as blocks or packets. A single 82C37A supports up to four peripheral devices for DMA operation.

Microprocessor Interface of the 82C37A

A block diagram that shows the interface signals of the 82C37A DMA controller is given in Fig. 10–40(a). The pin layout in Fig. 10–40(b) identifies the pins at which these signals are available. Let us now look briefly at the operation of the microprocessor interface of the 82C37A.

In a microcomputer system, the 82C37A acts as a peripheral controller device, and its operation must be initialized through software. This is done by reading from or writ-

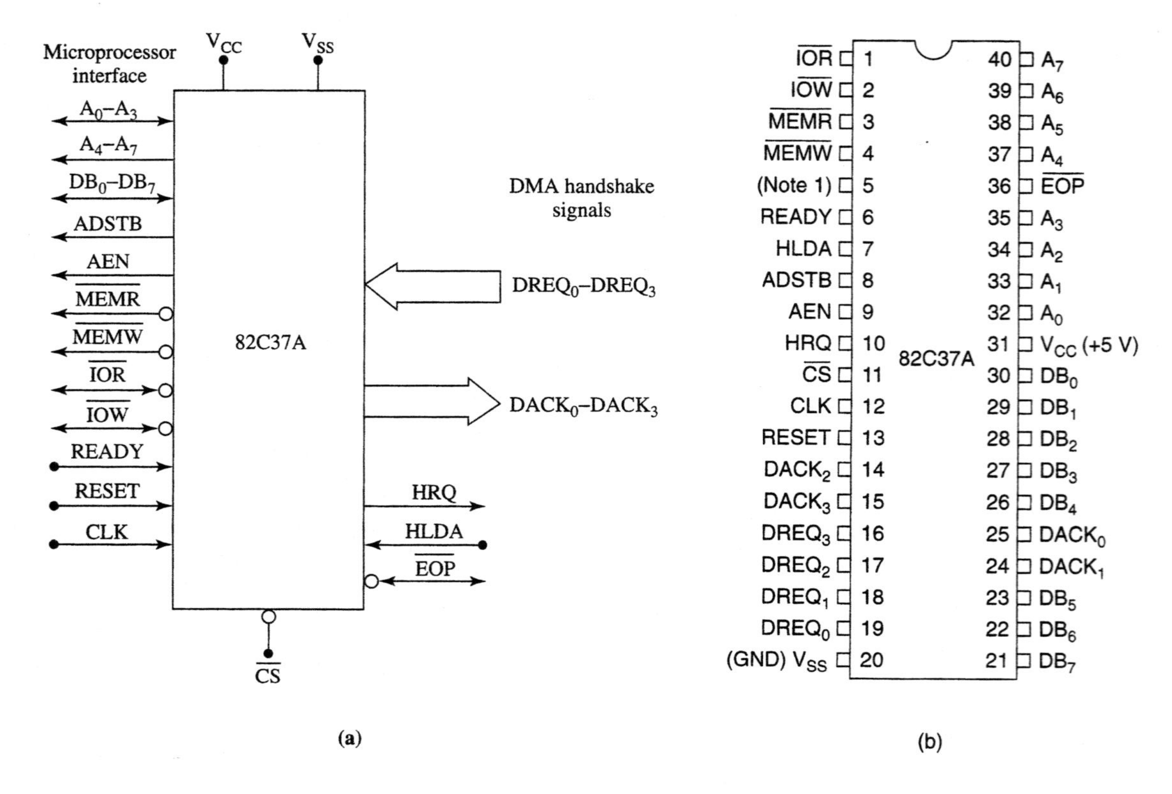

Figure 10–40 (a) Block diagram of the 82C37A DMA controller. (b) Pin layout. (Reprinted by permission of Intel Corporation. Copyright/Intel Corp. 1987)

ing into the bits of its internal registers. These data transfers take place through its microprocessor interface. Figure 10–41 shows how the 8088 connects to the 82C37A's microprocessor interface.

Whenever the 82C37A is not in use by a peripheral device for DMA operation, it is in a state known as the *idle state.* When in this state, the microprocessor can issue commands to the DMA controller and read from or write to its internal registers. Data bus lines DB_0 through DB_7 form the path over which these data transfers take place. Which register is accessed is determined by a 4-bit register address that is applied to address inputs A_0 through A_3. As Fig. 10–41 shows, address lines A_0 through A_3 of the microprocessor directly supply these inputs.

During the data-transfer bus cycle, other bits of the address are decoded in external circuitry to produce a chip-select ($\overline{CS}$) input for the 82C37A. When in the idle state, the 82C37A continuously samples this input, waiting for it to become active. Logic 0 at this input enables the microprocessor interface. The microprocessor tells the 82C37A whether an input or output bus cycle is in progress with the signal $\overline{IOR}$ or $\overline{IOW}$, respectively. In this way, we see that the 82C37A maps into the I/O address space of the 8088 microcomputer.

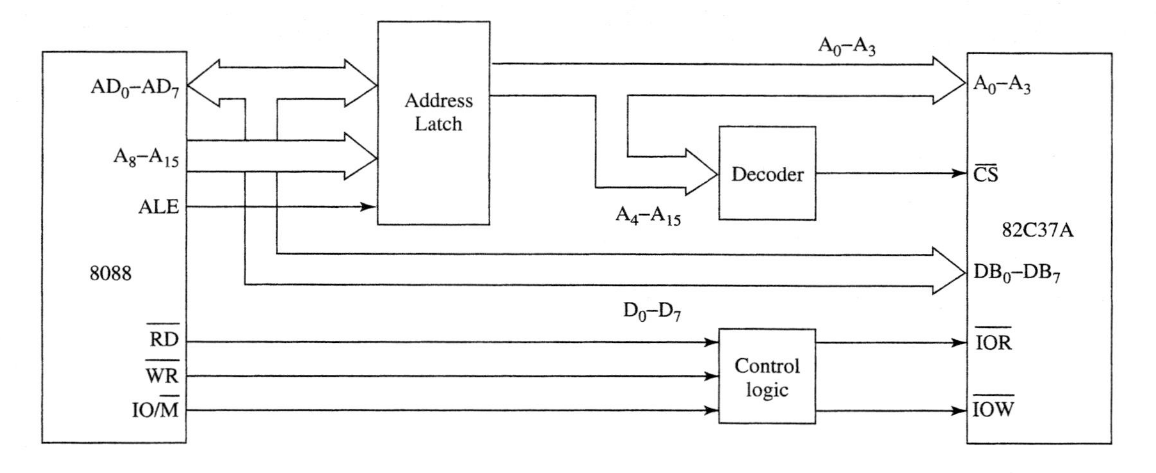

Figure 10–41 Microprocessor interface of 82C37A to the 8088.

DMA Interface of the 82C37A

Now that we have described how a microprocessor talks to the registers of the 82C37A, let us continue by looking at how peripheral devices initiate DMA service. The 82C37A contains four independent DMA channels, channels 0 through 3. Typically, each of these channels is dedicated to a specific peripheral device. Figure 10–42 shows that the device has four DMA request inputs, denoted as $DREQ_0$ through $DREQ_3$. These DREQ inputs correspond to channels 0 through 3, respectively. In the idle state, the 82C37A continuously tests these inputs to see if one is active. When a peripheral device wants to perform DMA operations, it makes a request for service at its DREQ input by switching it to logic 1.

In response to the active DMA request, the DMA controller switches the hold request (HRQ) output to logic 1. Normally, this output is supplied to the HOLD input of the 8088 and signals the microprocessor that the DMA controller needs to take control of the system bus. When the 8088 is ready to give up control of the bus, it puts its bus signals into the high-impedance state and signals this fact to the 82C37A by switching the HLDA (hold-acknowledge) output to logic 1. HLDA of the 8088 is applied to the HLDA input of the 82C37A and signals that the system bus is now available for use by the DMA controller.

When the 82C37A has control of the system bus, it tells the requesting peripheral device that it is ready by outputting a DMA-acknowledge (DACK) signal. Note in Fig. 10–42 that each of the four DMA request inputs, $DREQ_0$ through $DREQ_3$, has a corresponding DMA-acknowledge output, $DACK_0$ through $DACK_3$. Once this DMA-request/acknowledge handshake sequence is complete, the peripheral device gets direct access to the system bus and memory under control of the 82C37A.

During DMA bus cycles, the DMA controller, not the MPU, drives the system bus. The 82C37A generates the address and all control signals needed to perform the memory or I/O data transfers. At the beginning of all DMA bus cycles, a 16-bit address is output on lines A_0 through A_7 and DB_0 through DB_7. The upper 8 bits of the address, available

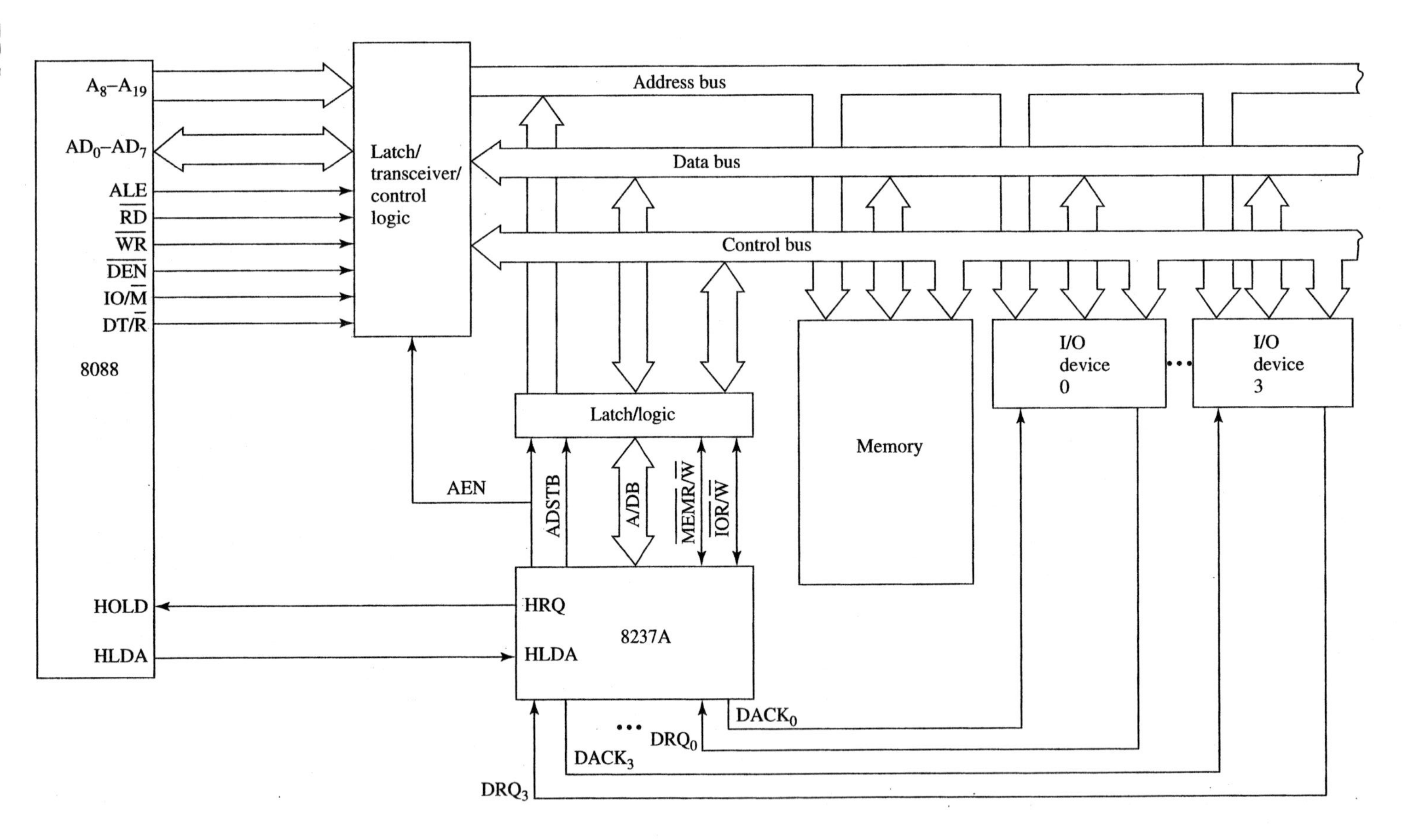

Figure 10–42 DMA interface to I/O devices.

on the data bus lines, appear at the same time that address strobe (ADSTB) becomes active. Thus, ADSTB is intended to be used to strobe the most significant byte of the address into an external address latch. This 16-bit address gives the 82C37A the ability to directly address up to 64Kbytes of storage locations. The address enable (AEN) output signal is active during the complete DMA bus cycle and can be used to both enable the address latch and disable other devices connected to the bus.

Let us assume for now that an I/O peripheral device is to transfer data to memory—that is, the I/O device wants to write data to memory. In this case, the 82C37A uses the $\overline{IOR}$ output to signal the I/O device to put the data onto data bus lines DB_0 through DB_7. At the same time, it asserts $\overline{MEMW}$ to signal that the data available on the bus are to be written into memory. In this case, the data are transferred directly from the I/O device to memory and do not go through the 82C37A.

In a similar way, DMA transfers of data can take place from memory to an I/O device. In this case, the I/O device reads data from memory and outputs it to the peripheral. For this data transfer, the 82C37A activates the $\overline{MEMR}$ and $\overline{IOW}$ control signals.

The 82C37A performs both the memory-to-I/O and I/O-to-memory DMA bus cycles in just four clock periods. The duration of these clock periods is determined by the frequency of the clock signal applied to the CLOCK input. For instance, at 5 MHz the clock period is 200 ns and the bus cycle takes 800 ns.

The 82C37A is also capable of performing memory-to-memory DMA transfers. In such a data transfer, both the $\overline{MEMR}$ and $\overline{MEMW}$ signals are utilized. Unlike the I/O-to-memory operation, this memory-to-memory data transfer takes eight clock cycles. This is because it is actually performed as a separate four-clock read bus cycle from the source memory location to a temporary register within the 82C37A and then another four-clock write bus cycle from the temporary register to the destination memory location. At 5 MHz, a memory-to-memory DMA cycle takes 1.6 μs.

The READY input is used to accommodate slow memory or I/O devices. READY must go active, logic 1, before the 82C37A will complete a memory or I/O bus cycle. As long as READY is at logic 0, wait states are inserted to extend the duration of the current bus cycle.

Internal Architecture of the 82C37A

Figure 10–43 is a block diagram of the internal architecture of the 82C37A DMA controller. Here we find the following functional blocks: the timing and control, the priority encoder and rotating priority logic, the command control, and 12 different types of registers. Let us now look briefly at the functions performed by each of these sections of circuitry and registers.

The timing and control part of the 82C37A generates the timing and control signals needed by the external bus interface. For instance, it accepts as inputs the READY and $\overline{CS}$ signals and produces output signals such as ADSTB and AEN. These signals are synchronized to the clock signal that is input to the controller. The highest-speed version of the 82C37A available today operates at a maximum clock rate of 5 MHz.

If multiple requests for DMA service are received by the 82C37A, they are accepted on a priority basis. One of two priority schemes can be selected for the 82C37A

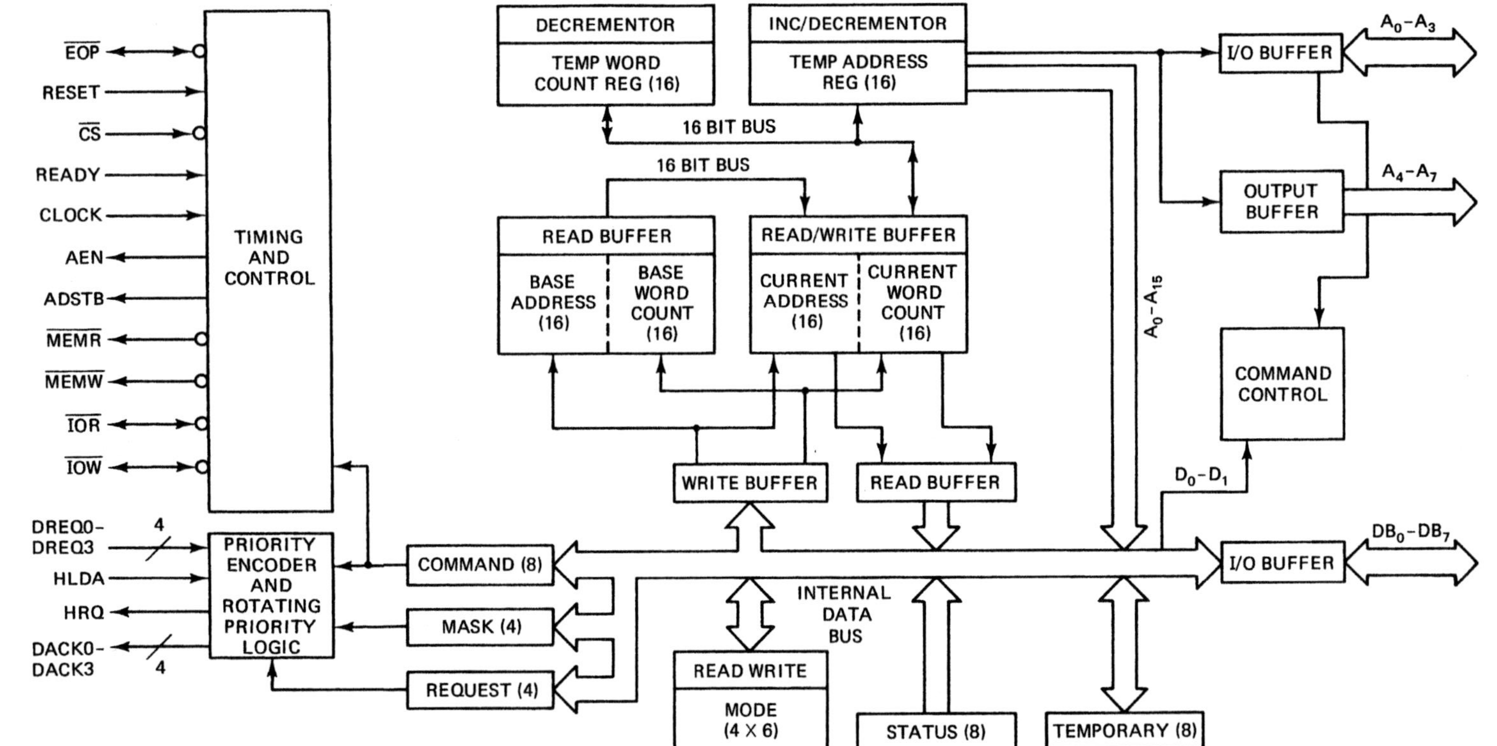

Figure 10–43 Internal architecture of the 82C37A. (Reprinted by permission of Intel Corporation. Copyright/Intel Corp. 1987)

under software control: *fixed priority* and *rotating priority.* The fixed-priority mode assigns priority to the channels in descending numeric order. That is, channel 0 has the highest priority and channel 3 the lowest priority. Rotating priority starts with the priority levels initially the same way as in fixed priority. However, after a DMA request for a specific level gets serviced, priority is rotated such that the previously active channel is reassigned to the lowest priority level. For instance, assuming that channel 1, which was initially at priority level 1, was just serviced, then $DREQ_2$ is now at the highest priority level and $DREQ_1$ rotates to the lowest level. The priority logic circuitry shown in Fig. 10–43 resolves priority for simultaneous DMA requests from peripheral devices based on the programmed priority scheme.

The command control circuit decodes the register commands applied to the 82C37A through the microprocessor interface. In this way it determines which register is to be accessed and what type of operation is to be performed. Moreover, it is used to decode the programmed operating modes of the device during DMA operation.

Looking at the block diagram in Fig. 10–43, we find that the 82C37A has 12 different types of internal registers. Some examples are the current address register, current count register, command register, mask register, and status register. Figure 10–44 lists the names for all the internal registers, along with their size and how many are provided in the 82C37A. Note that there are actually four current address registers and they are all 16 bits long. That is, there is one current address register for each of the four DMA channels. We will now describe the function served by each of these registers in terms of overall operation of the 82C37A DMA controller. Figure 10–45 summarizes the address information for the internal registers.

Each DMA channel has two address registers: the *base address register* and the *current address register.* The base address register holds the starting address for the DMA operation, and the current address register contains the address of the next storage location to be accessed. Writing a value to the base address register automatically loads the same value into the current address register. In this way, we see that initially the current address register points to the starting I/O or memory address.

These registers must be loaded with appropriate values prior to initiating a DMA cycle. To load a new 16-bit address into the base register, we must write two separate bytes, one after the other, to the address of the register. The 82C37A has an internal flip-flop called the *first/last flip-flop.* This flip-flop identifies which byte of the address is being written into the register. As the table in Fig. 10–45 shows, if the beginning state of the internal flip-flop (FF) is logic 0, then software must write the low byte of the address word to the register. On the other hand, if it is logic 1, the high byte must be written to

Name	Size	Number
Base Address Registers	16 bits	4
Base Word Count Registers	16 bits	4
Current Address Registers	16 bits	4
Current Word Count Registers	16 bits	4
Temporary Address Register	16 bits	1
Temporary Word Count Register	16 bits	1
Status Register	8 bits	1
Command Register	8 bits	1
Temporary Register	8 bits	1
Mode Registers	6 bits	4
Mask Register	4 bits	1
Request Register	4 bits	1

Figure 10–44 Internal registers of the 82C37A. (Reprinted by permission of Intel Corporation. Copyright/Intel Corp. 1987)

Channel(s)	Register	Operation	Register address	Internal FF	Data bus
0	Base and current address	Write	0_{16}	0 1	Low High
	Current address	Read	0_{16}	0 1	Low High
	Base and current count	Write	1_{16}	0 1	Low High
	Current count	Read	1_{16}	0 1	Low High
1	Base and current address	Write	2_{16}	0 1	Low High
	Current address	Read	2_{16}	0 1	Low High
	Base and current count	Write	3_{16}	0 1	Low High
	Current count	Read	3_{16}	0 1	Low High
2	Base and current address	Write	4_{16}	0 1	Low High
	Current address	Read	4_{16}	0 1	Low High
	Base and current count	Write	5_{16}	0 1	Low High
	Current count	Read	5_{16}	0 1	Low High
3	Base and current address	Write	6_{16}	0 1	Low High
	Current address	Read	6_{16}	0 1	Low High
	Base and current count	Write	7_{16}	0 1	Low High
	Current count	Read	7_{16}	0 1	Low High
All	Command register	Write	8_{16}	X	Low
All	Status register	Read	8_{16}	X	Low
All	Request register	Write	9_{16}	X	Low
All	Mask register	Write	A_{16}	X	Low
All	Mode register	Write	B_{16}	X	Low
All	Temporary register	Read	B_{16}	X	Low
All	Clear internal FF	Write	C_{16}	X	Low
All	Master clear	Write	D_{16}	X	Low
All	Clear mask register	Write	E_{16}	X	Low
All	Mask register	Write	F_{16}	X	Low

Figure 10–45 Accessing the registers of the 82C37A.

the register. For example, to write the address 1234_{16} into the base address register and the current address register for channel 0 of a DMA controller located at base I/O address DMA (where DMA is ≤ F0H and it is decided by how the $\overline{CS}$ for the 82C37A must be generated), the following instructions may be executed:

```
MOV  AL, 34H     ;Write low byte
OUT  DMA+0, AL
MOV  AL, 12H     ;Write high byte
OUT  DMA+0, AL
```

This routine assumes that the internal flip-flop was initially set to 0. Looking at Fig. 10–45, we find that a command can be issued to the 82C37A to clear the internal flip-flop. This is done by initiating an output bus cycle to address DMA + C_{16}.

If we read the contents of the register at address DMA + 0_{16}, the value obtained is the contents of the current address register for channel 0. Once loaded, the value in the base address register cannot be read out of the device.

The 82C37A also has two word-count registers for each of its DMA channels: the *base count register* and the *current count register.* In Fig. 10–44, we find that these registers are also 16 bits in length, and Fig. 10–45 identifies their address as 1_{16}, relative to the base address DMA for channel 0. The number of bytes of data to be transferred during a DMA operation is specified by the value in the base word-count register. Actually, the number of bytes transferred is always one more than the value programmed into this register. This is because the end of a DMA cycle is detected by the rollover of the current word count from 0000_{16} to $FFFF_{16}$. At any time during the DMA cycle, the value in the current word-count register tells how many bytes remain to be transferred.

The count registers are programmed in the same way as just described for the address registers. For instance, to program a count of $0FFF_{16}$ into the base and current count registers for channel 1 of a DMA controller located at address DMA (where DMA ≤ F0H), the following instructions can be executed:

```
MOV  AL, 0FFH    ;Write low byte
OUT  DMA+3, AL
MOV  AL, 0FH     ;Write high byte
OUT  DMA+3, AL
```

Again we have assumed that the internal flip-flop was initially cleared.

Figure 10–44 shows that the 82C37A has a single 8-bit command register. The bits in this register are used to control operating modes that apply to all channels of the DMA controller. Figure 10–46 identifies the function of each of its control bits. Note that the settings of the bits are used to select or deselect operating features such as memory-to-memory DMA transfer and the priority scheme. For instance, when bit 0 is set to logic 1, the memory-to-memory mode of DMA transfer is enabled, and when it is logic 0, DMA transfers take place between I/O and memory. Moreover, setting bit 4 to logic 0 selects the fixed priority scheme for all four channels or logic 1 in this location selects rotating priority. Looking at Fig. 10–45, we see that the command register is loaded by outputting the command code to the register at address 8_{16}, relative to the base address for the 82C37A.

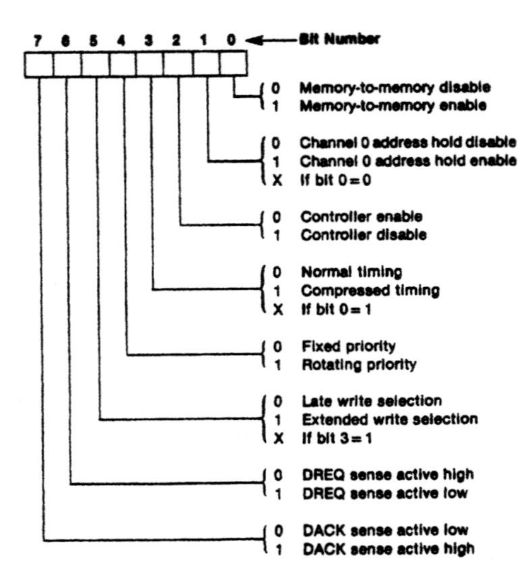

Figure 10–46 Command register format. (Reprinted by permission of Intel Corporation. Copyright/Intel Corp. 1987)

EXAMPLE 10.25

If the command register of an 82C37A is loaded with 01_{16}, how does the controller operate?

Solution

Representing the command word as a binary number, we get

$$01_{16} = 00000001_2$$

Referring to Fig. 10–46, we find that the DMA operation can be described as follows:

Bit 0 = 1 = Memory-to-memory transfers are disabled

Bit 1 = 0 = Channel 0 address increments/decrements normally

Bit 2 = 0 = 82C37A is enabled

Bit 3 = 0 = 82C37A operates with normal timing

Bit 4 = 0 = Channels have fixed priority, channel 0 having the highest priority and channel 3 the lowest priority

Bit 5 = 0 = Write operation occurs late in the DMA bus cycle

Bit 6 = 0 = DREQ is an active high (logic 1) signal

Bit 7 = 0 = DACK is an active low (logic 0) signal

The *mode registers* are also used to configure operational features of the 82C37A. Figure 10–44 shows that there is a separate mode register for each of the four DMA channels and that each is six bits in length. Their bits are used to select various operational features for the individual DMA channels. A mode register command, shown in Fig. 10–47,

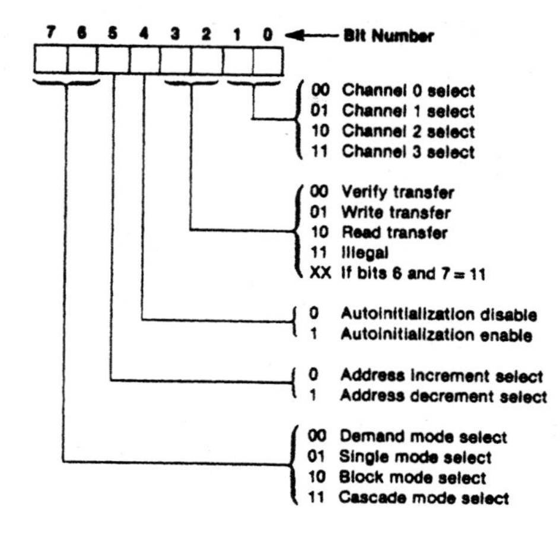

Figure 10–47 Mode register format. (Reprinted by permission of Intel Corporation. Copyright/Intel Corp. 1987)

has two least significant bits that are a 2-bit code, which identifies the channel to which the mode command byte applies. For instance, in a mode register command written for channel 1, these bits must be made 01. Bits 2 and 3 specify whether the channel is to perform data write, data read, or verify bus cycles. For example, if these bits are set to 01, the channel will only perform write-data transfers (DMA data transfers from an I/O device to memory).

The next two bits of the mode register affect how the values in the current address and current count registers are updated at the end of a DMA cycle and DMA data transfer, respectively. Bit 4 enables or disables the autoinitialization function. When autoinitialization is enabled, the current address and current count registers are automatically reloaded from the base address and base count registers, respectively, at the end of a DMA operation. In this way, the channel is prepared for the next DMA transfer. The setting of bit 5 determines whether the value in the current address register is automatically incremented or decremented at completion of each DMA data transfer.

The two most significant bits of the mode register select one of four possible modes of DMA operation for the channel: *demand mode, single mode, block mode,* and *cascade mode.* These modes allow for either one byte of data to be transferred at a time or a block of bytes. For example, when in the demand transfer mode, once the DMA cycle is initiated, bytes are continuously transferred as long as the DREQ signal remains active and the terminal count (TC) is not reached. By reaching the terminal count, we mean that the value in the current word-count register, which automatically decrements after each data transfer, rolls over from 0000_{16} to $FFFF_{16}$.

Block-transfer mode is similar to demand-transfer mode in that once the DMA cycle is initiated, data are continuously transferred until the terminal count is reached. However, they differ in that when in the demand mode, the return of DREQ to its inactive state halts the data transfer sequence. But when in block-transfer mode, DREQ can be released at any time after the DMA cycle begins, and the block transfer will still run to completion.

In the single-transfer mode, the channel is set up such that it performs just one data transfer at a time. At the completion of the transfer, the current word count is decremented and the current address either incremented or decremented (based on the selected option). Moreover, an autoinitialization, if enabled, will not occur unless the terminal count has been reached at the completion of the current data transfer. If the DREQ input becomes inactive before the completion of the current data transfer, another data transfer will not take place until DREQ once more becomes active. On the other hand, if DREQ remains active during the complete data transfer cycle, the HRQ output of the 82C37A is switched to its inactive 0 logic level to allow the microprocessor to gain control of the system bus for one bus cycle before another single transfer takes place. This mode of operation is typically used when it is necessary to not lock the microprocessor off the bus for the complete duration of the DMA operation.

EXAMPLE 10.26

Specify the mode byte for DMA channel 2 if it is to transfer data from an input peripheral device to a memory buffer starting at address $A000_{16}$ and ending at $AFFF_{16}$. Ensure that the microprocessor is not completely locked off the bus during the DMA cycle. Moreover, at the end of each DMA cycle, the channel is to be reinitialized so that the same buffer is filled when the next DMA operation is initiated.

Solution

For DMA channel 2, bit 1 and bit 0 must be loaded with 10_2:

$$B_1B_0 = 10$$

Transfer of data from an I/O device to memory represents a write bus cycle. Therefore, bit 3 and bit 2 must be set to 01:

$$B_3B_2 = 01$$

Selecting autoinitialization will set up the channel to automatically reset so that it points to the beginning of the memory buffer at completion of the current DMA cycle. Making bit 4 equal to 1 enables this feature:

$$B_4 = 1$$

The address that points to the memory buffer must increment after each data transfer. Therefore, bit 5 must be set to 0:

$$B_5 = 0$$

Finally, to ensure that the 8088 is not locked off the bus during the complete DMA cycle, we will select the single-transfer mode of operation. Making bits B7 and B6 equal to 01 does this:

$$B_7B_6 = 01$$

Thus, the mode register byte is

$$B_7B_6B_5B_4B_3B_2B_1B_0 = 01010110_2 = 56_{16}$$

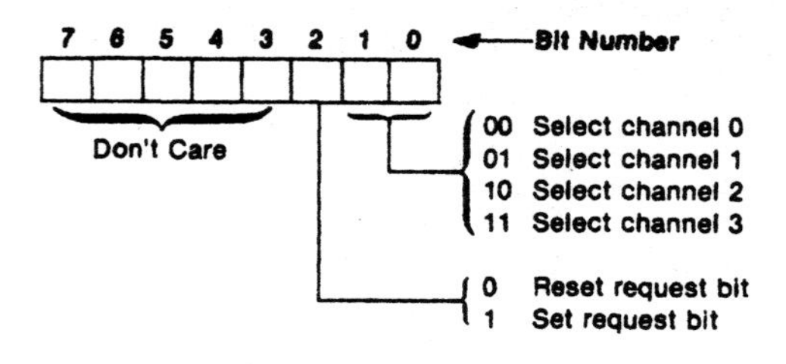

Figure 10–48 Request register format. (Reprinted by permission of Intel Corporation. Copyright/Intel Corp. 1987)

Up to now, we have discussed how DMA cycles can be initiated by a hardware request at a DREQ input. However, the 82C37A is also able to respond to software-initiated requests for DMA service. The *request register* has been provided for this purpose. Figure 10–44 shows that the request register has just four bits, one for each of the DMA channels. When the request bit for a channel is set, DMA operation is started, and when reset, the DMA cycle is stopped. Any channel used for software-initiated DMA must be programmed for block-transfer mode of operation.

The bits in the request register can be set or reset by issuing software commands to the 82C37A. The format of a request register command is shown in Fig. 10–48. For instance, if a command is issued to the address of the request register with bits 0 and 1 equal to 01 and with bit 2 at logic 1, a block-mode DMA cycle is initiated for channel 1. Figure 10–45 shows that the request register is located at register address 9_{16}, relative to the base address for the 82C37A.

A 4-bit *mask register* is also provided within the 82C37A. One bit is provided in this register for each of the DMA channels. When a mask bit is set, the DREQ input for the corresponding channel is disabled. Therefore, hardware requests to the channel are ignored. That is, the channel is masked out. On the other hand, if the mask bit is cleared, the DREQ input is enabled and an external device can activate its channel.

The format of a software command that can be used to set or reset a single bit in the mask register is shown in Fig. 10–49(a). For example, to enable the DREQ input for channel 2, the command is issued with bits 1 and 0 set to 10 to select channel 2, and with bit 2 equal to 0 to clear the mask bit. Therefore, the software command byte would be 02_{16}. The table in Fig. 10–45 shows that this command byte must be issued to the 82C37A with register address A_{16}, relative to the base address for the 82C37A.

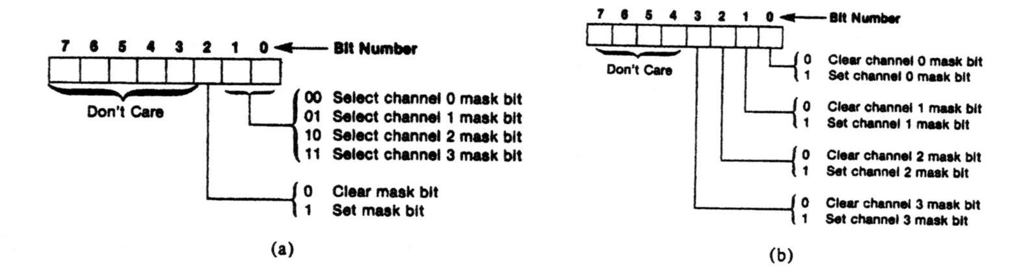

Figure 10–49 (a) Single-channel mask-register command format. (Reprinted by permission of Intel Corporation. Copyright/Intel Corp. 1987) (b) Four-channel mask-register command format. (Reprinted by permission of Intel Corporation. Copyright/Intel Corp. 1987)

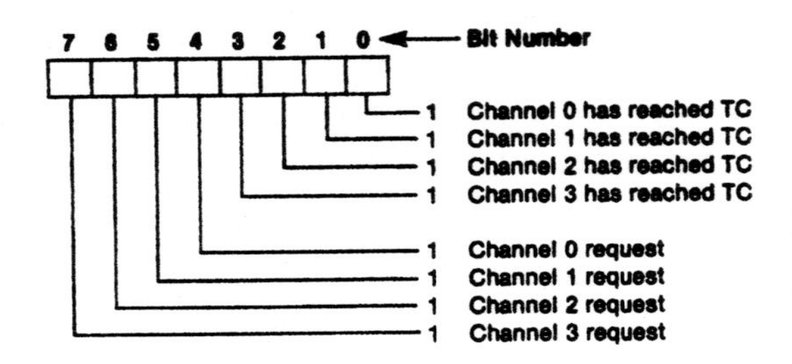

Figure 10–50 Status register. (Reprinted by permission of Intel Corporation. Copyright/Intel Corp. 1987)

A second mask register command is shown in Fig. 10–49(b). This command can be used to load all 4 bits of the register at once. In Fig. 10–45, we find that this command is issued to relative register address F_{16} instead of A_{16}. For instance, to mask out channel 2 while enabling channels 0, 1, and 3, the command code is 04_{16}. Either of these two methods can be used to mask or enable the DREQ input for a channel.

At system initialization, it is a common practice to clear the mask register. Looking at Fig. 10–45, we see that a special command is provided to perform this operation. Executing an output cycle to the register with relative address E_{16} clears the mask register.

The 82C37A has a *status register* that contains information about the operating state of its four DMA channels. Figure 10–50 shows the bits of the status register and defines their functions. Here we find that the four least significant bits identify whether or not channels 0 through 3 have reached their terminal count. When the DMA cycle for a channel reaches the terminal count, this fact is recorded by setting the corresponding TC bit to the 1 logic level. The four most significant bits of the register tell if a request is pending for the corresponding channel. For instance, if a DMA request has been issued for channel 0 either through hardware or software, bit 4 is set to 1. The 8088 can read the contents of the status register through software. This is done by initiating an input bus cycle for register address 8_{16}, relative to the base address for the 82C37A.

Earlier we pointed out that during memory-to-memory DMA transfers, the data read from the source address are held in a register known as the *temporary register*, and then a write cycle is initiated to write the data to the destination address. At the completion of the DMA cycle, this register contains the last byte that was transferred. The value in this register can be read by the microprocessor.

EXAMPLE 10.27

Write an instruction sequence to issue a master clear to the 82C37A and then enable all its DMA channels. Assume that the device is located at base I/O address DMA < F0H.

Solution

Figure 10–45 shows that a special software command is provided to perform a master reset of the 82C37A's registers. Since the contents of the data bus are a don't-care state when executing the master clear command, it is performed by simply writing into the register at relative address D_{16}. For instance, the instruction

```
OUT DMA+0DH, AL
```

can be used. To enable the DMA request inputs, all 4 bits of the mask register must be cleared. The clear-mask register command is issued by performing a write to the register at relative address E_{16}. Again, the data put on the bus during the write cycle are a don't-care state. Therefore, the command can be performed with the instruction

```
OUT   DMA+0EH, AL
```

DMA Interface for the 8088-Based Microcomputer Using the 82C37A

Figure 10–51 shows how the 82C37A is connected to the 8088 microprocessor to form a simplified DMA interface. Here we see that both the 8088 MPU and the 82C37A DMA controller drive the same three system buses, address bus, data bus, and control bus. Let us now look at how each of these devices attaches to the system bus. The 8088's multiplexed address/data bus is demultiplexed using three 74F373 latches to form independent system address and data buses. The address bus is 20 bits in length, and these lines are identified as A_0 through A_{19}. On the other hand, the data bus is byte-wide, with lines D_0 through D_7. Note that the ALE output of the 8088 is used as the CLK input to the latches.

Looking at the 82C37A, we find that the lower byte of its address, identified by A_0 through A_3 and A_4 through A_7, is supplied directly to the system address bus. On the other hand, the most significant byte of its address, A_8 through A_{15}, is demultiplexed from data bus lines DB_0 through DB_7 by another 74F373 latch. This latch is enabled by the AEN output of the DMA controller, and the address is loaded into the latch with the signal ADSTB. DB_0 through DB_7 are also directly attached to the system data bus.

Finally, let us look at how the system control bus signals are derived. The $IO/\overline{M}$, $\overline{RD}$, and $\overline{WR}$ control outputs of the microprocessor are gated together to produce the signals $\overline{MEMR}$, $\overline{MEMW}$, $\overline{IOR}$, and $\overline{IOW}$. These signals are combined to form the system control bus. Note that these same four signals are generated as outputs of the 82C37A and are also supplied to the control bus.

Now that we have shown how the independent address, data, and control signals of the 8088 and 82C37A are combined to form the system address, data, and control buses, let us continue by looking at how the DMA request/acknowledge interface is implemented. I/O devices request DMA service by activating one of the 82C37A's DMA request inputs, $DREQ_0$ through $DREQ_3$. When the 82C37A receives a valid DMA request on one of these lines, it sends a hold request to the HOLD input of the 8088. It does this by setting the HRQ output to logic 1. After the 8088 gives up control of the system buses, it acknowledges this fact to the 82C37A by switching its HLDA output to the 1 logic level. This signal is received by the DMA controller at its HLDA input and tells it that the system buses are available. The 82C37A is now ready to take over control of the system buses, and it signals this fact to the device that is requesting service by activating its DMA-acknowledge (DACK) line.

During the DMA operation, the 82C37A generates all of the bus signals that are needed to access I/O devices and the memory. It also generates the AEN signal, which is

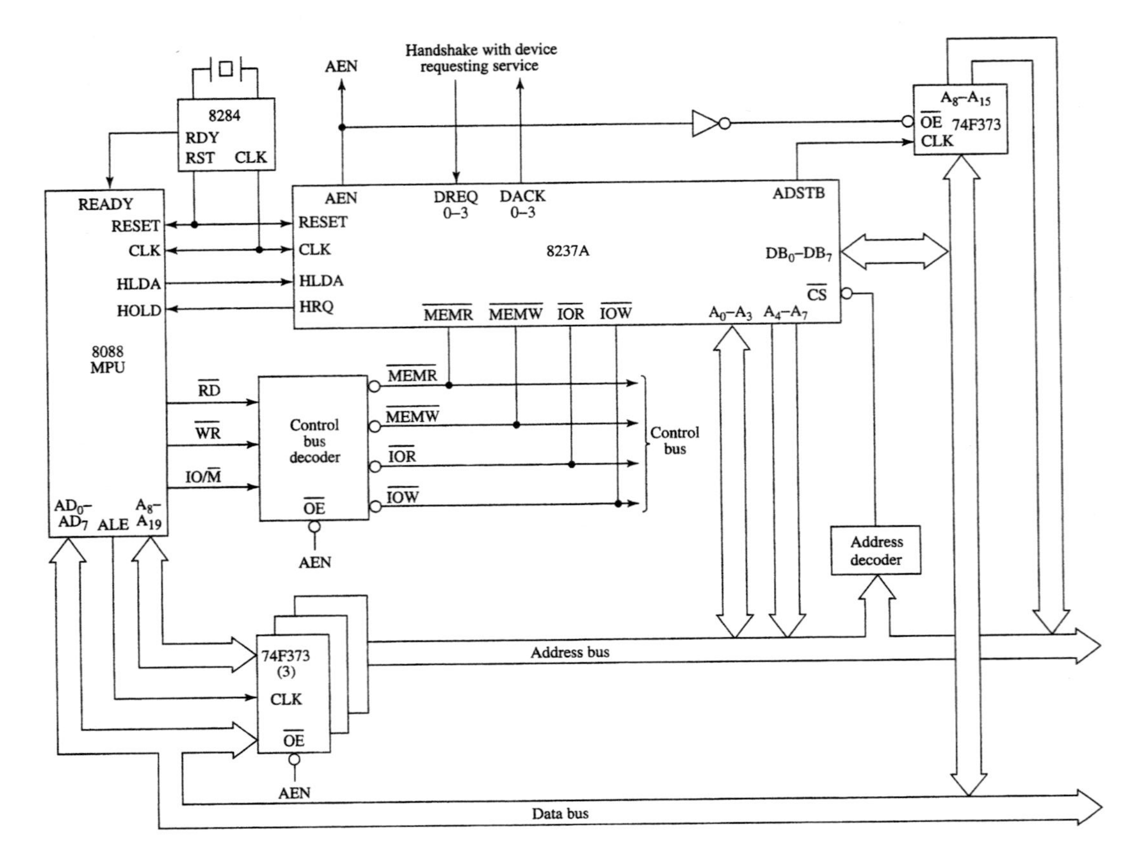

Figure 10–51 8088-based microcomputer with 82C37A DMA interface.

used to disable the microprocessor's connection to the system bus. AEN does this by disabling the control bus decoder and the latches for the address bus. The microprocessor's connection to the data bus is also disabled in response to the hold request received on its HOLD input. Remember that logic 1 at HOLD puts the data bus lines in the high-Z state. Thus, during a DMA operation, the 82C37A is in complete control of the address bus, control bus, and data bus.

▲ 10.10 SERIAL COMMUNICATIONS INTERFACE

Another type of I/O interface that is widely used in microcomputer systems is known as a *serial communication port.* This is the type of interface that is commonly used to connect peripheral units, such as CRT terminals, modems, and printers, to a microcomputer. It permits data to be transferred between two units using just two data lines. One line is used for transmitting data and the other for receiving data. For instance, data input at the keyboard of a terminal are passed to the MPU part of the microcomputer through this type of interface. Let us now look into the two different types of serial interfaces that are implemented in microcomputer systems.

Synchronous and Asynchronous Data Communications

Two types of *serial data communications* are widely used in microcomputer systems: *asynchronous communications* and *synchronous communications*. By synchronous, we mean that the receiver and transmitter sections of the two pieces of equipment communicating with each other must run synchronously. For this reason, as shown in Fig. 10–52(a), the interface includes a Clock line as well as Transmit data, Receive data, and Signal common lines. It is the clock signal that synchronizes both the transmission and reception of data.

The format used for synchronous communication of data is shown in Fig. 10–52(b). To initiate synchronous transmission, the transmitter first sends out synchronization characters to the receiver. The receiver reads the synchronization bit pattern and compares it to a known sync pattern. Once they are identified as being the same, the receiver begins to read character data off the data line. Transfer of data continues until the complete block of data is received. If large blocks of data are being sent, the synchronization characters

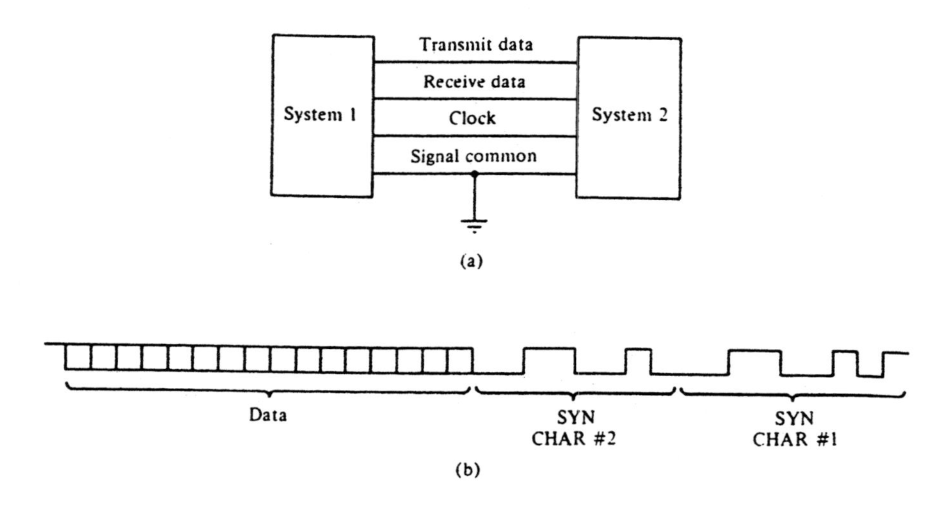

Figure 10–52 (a) Synchronous communications interface. (b) Synchronous data-transmission format.

may be periodically resent to assure that synchronization is maintained. The synchronous type of communications is typically used in applications where high-speed data transfer is required.

The asynchronous method of communications eliminates the need for the Clock signal. As shown in Fig. 10–53(a), the simplest form of an asynchronous communication interface could consist of a Receive data, Transmit data, and Signal common communication lines. In this case, the data to be transmitted are sent out one character at a time, and at the receiver examining synchronization bits that are included at the beginning and end of each character performs end of the communication line synchronization.

The format of a typical asynchronous character is shown in Fig. 10–53(b). Here we see that the synchronization bit at the beginning of the character is called the *start bit,* and that at the end of the character the *stop bit.* Depending on the communications scheme, 1, 1½, or 2 stop bits can be used. The bits of the character are embedded between the start and stop bits. Notice that the start bit is either input or output first. The LSB of the character, the rest of the character's bits, a parity bit, and the stop bits follow it in the serial bit stream. For instance, 7-bit ASCII can be used and parity added as an eighth bit for higher reliability in transmission. The duration of each bit in the format is called a *bit time.*

The fact that a 0 or 1 logic level is being transferred over the communication line is identified by whether the voltage level on the line corresponds to that of a *space* or a *mark,* respectively. The start bit is always to the mark level. It synchronizes the receiver to the transmitter and signals that the unit receiving data should start assembling the char-

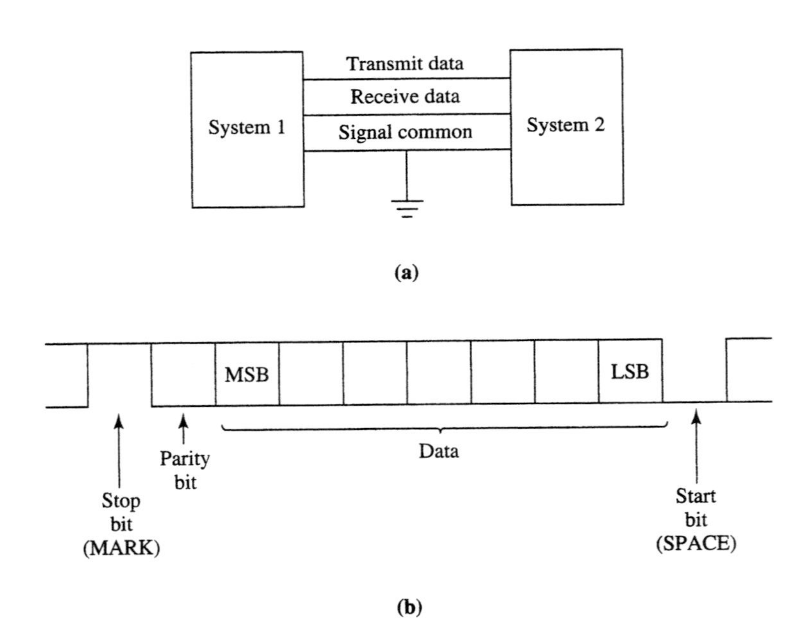

Figure 10–53 (a) Asynchronous communications interface. (b) Asynchronous data-transmission format.

acter. Stop bits are to the space level. The nontransmitting line is always at the space logic level. This scheme assures that the receiving unit sees a transition of logic level at the start bit of the next character.

Simplex, Half-Duplex, and Full-Duplex Communication Links

Applications require different types of asynchronous links to be implemented. For instance, the communication link needed to connect a printer to a microcomputer just needs to support communications in one direction. That is, the printer is an output-only device; therefore, the MPU needs only to transmit data to the printer. Data are not transmitted back. In this case, as shown in Fig. 10–54(a), a single unidirectional communication line can be used to connect the printer and microcomputer together. This type of connection is known as a *simplex communication link.*

Other devices, such as the CRT terminal with keyboard shown in Fig. 10–54(b), need to both transmit data to and receive data from the MPU. That is, they must both input and output data. Setting up a half-duplex communication link can also satisfy this requirement with a single communication line. In a half-duplex link, data are transmitted and received over the same line; therefore, transmission and reception of data cannot take place at the same time.

If higher-performance communication is required, separate transmit and receive lines can be used to connect the peripheral and microcomputer. When this is done, data can be transferred in both directions at the same time. This type of link, illustrated in Fig. 10–54(c), is called a *full-duplex communication link.*

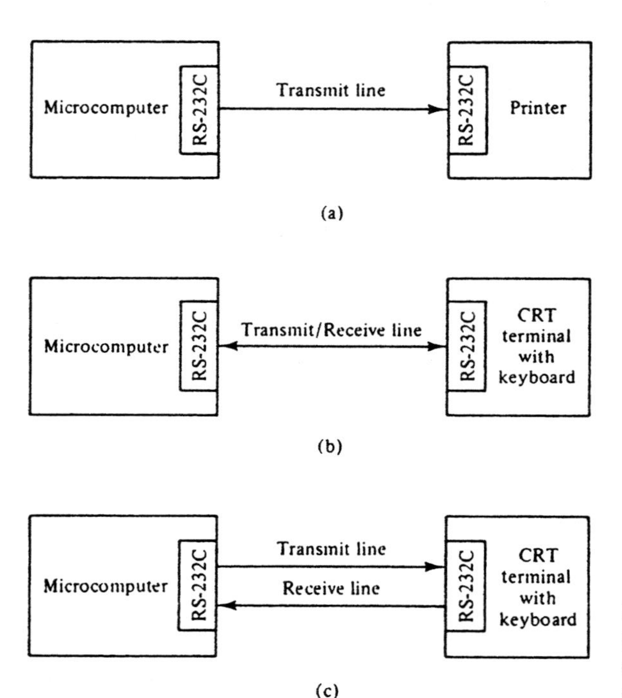

Figure 10–54 (a) Simplex communication link. (b) Half-duplex communication link. (c) Full-duplex communication link.

Baud Rate and the Baud-Rate Generator

The rate at which data transfers take place over the receive and transmit lines is known as the *baud rate.* By baud rate we mean the number of bits of data transferred per second. For instance, some of the common data transfer rates are 300 baud, 1200 baud, and 9600 baud. They correspond to 300 bits/second (bps), 1200 bps, and 9600 bps, respectively. Baud rate is set by a part of the serial communication interface called the *baud-rate generator.*

The baud rate at which data are transferred determines the bit time—that is, the amount of time each bit of data is on the communication line. At 300 baud, the bit time is found to be

$$t_{BT} = 1/300 \text{ bps} = 3.33 \text{ ms}$$

EXAMPLE 10.28

The data transfer across an asynchronous serial data communications line is observed and the bit time is measured as 0.833 ms. What is the baud rate?

Solution

Baud rate is calculated from the bit time as

$$\text{Baud rate} = 1/t_{BT} = 1/0.833 \text{ ms} = 1200 \text{ bps}$$

The RS-232C Interface

The *RS-232C interface* is a standard hardware interface for implementing asynchronous serial data communication ports on devices such as printers, CRT terminals, keyboards, and modems. The Electronic Industries Association (EIA) defines the pin definitions and electrical characteristics of this interface. The aim behind publishing standards, such as the RS-232C, is to assure compatibility between equipment made by different manufacturers.

Peripherals that connect to a microcomputer can be located within the systems or anywhere from several feet to many feet way. For instance, in large systems it is common to have the microcomputer part of the system in a separate room from the terminals and printers. This leads us to the main advantage of using a serial interface to connect peripherals to a microcomputer, which is that as few as three signal lines can be used to connect the peripheral to the MPU: a receive-data line, a transmit-data line, and signal common. This results in a large savings in wiring costs, and the small number of lines that need to be put in place also leads to higher reliability.

The RS-232C standard defines a 25-pin interface. Figure 10–55 lists each pin and its function. Note that the three signals that we mentioned earlier, transmit data (Tx_D), receive data (Rx_D), and signal ground, are located at pins 2, 3, and 7, respectively. Pins are also provided for additional control functions. For instance, pins 4 and 5 are the request-to-send and clear-to-send control signals.

Pin	Signal
1	Protective Ground
2	Transmitted Data
3	Received Data
4	Request to Send
5	Clear to Send
6	Data Set Ready
7	Signal Ground (Common Return)
8	Received Line Signal Detector
9	Reserved for Data Set Testing
10	Reserved for Data Set Testing
11	Unassigned
12	Secondary Received Line Signal Detector
13	Secondary Clear to Send
14	Secondary Transmitted Data
15	Transmission Signal Element Timing
16	Secondary Received Data
17	Receiver Signal Element Timing
18	Unassigned
19	Secondary Request to Send
20	Data Terminal Ready
21	Signal Quality Detector
22	Ring Indicator
23	Data Signal Rate Selector
24	Transmit Signal Element Timing
25	Unassigned

Figure 10–55 RS-232C interface pins and functions.

How the signals of the RS-232C interface are used in a device depends on whether it is configured as what is known as a *Data Terminal Equipment* (DTE) or a *Data Communications Equipment* (DCE). An example of a DTE is a PC, and that of a DCE is a modem. The direction for signals in a DTE and a DCE device are reversed. That is, signal lines that are outputs on a DTE device are inputs on a DCE and vice versa. This enables one to use a cable that makes direct connections between the pins of the DTE and a DCE. For instance, if pin 2 on a DTE is an output, it connects directly to pin 2 on the DCE, which acts as an input.

To make a DTE device communicate to a DTE device requires a cable that makes the pin-to-pin connections shown in Fig. 10–56. Note that when both devices are configured as DTEs, the data transmitted by one is received by the other and vice versa. It therefore requires a special cable in which the Tx_D pin of one is connected to the Rx_D pin of the other device and vice versa.

The control pins are provided to set up a handshake sequence for initiating communication between serial devices. These signals have the meanings expressed in their names; for instance, request to send (RTS) is used to send a request from a DTE device

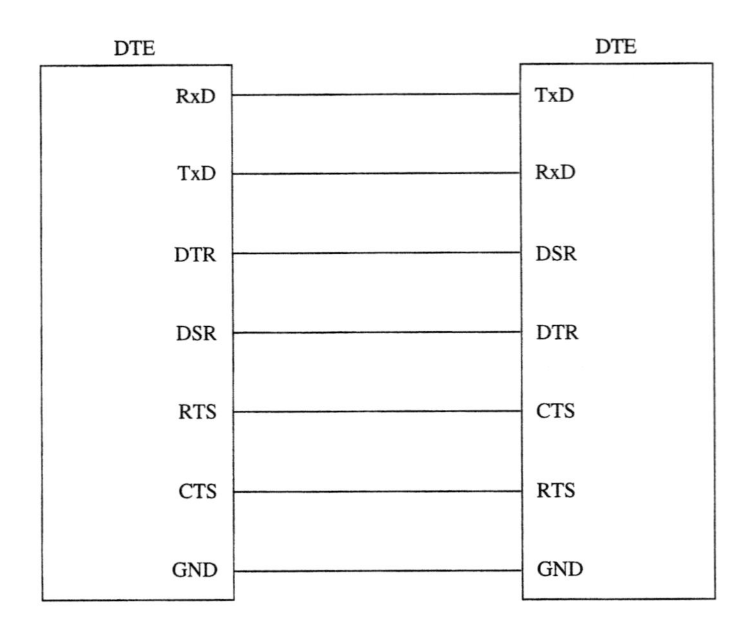

Figure 10–56 A DTE-to-DTE serial communication connection.

to a DCE or another DTE device to get clearance to send data. In many systems only three signals Tx_D, Rx_D, and common are used to provide serial communication. In such a set up no handshake sequence is used to initiate communication.

The RS-232C interface is specified to operate correctly over a distance of up to 100 feet. To satisfy this distance specification, a bus driver is used on the transmit line and a bus receiver is used on the receive line. RS-232C drivers and receivers are available as standard ICs. These buffers do both the voltage-level translation needed to convert the TTL-compatible signals to the mark (logic 1) and space (logic 0) voltage levels defined for the RS-232C interface. The voltage level for a mark can range from $-5V$ to $-15V$ at the transmitting end and $-3V$ or less at the receiving end. Similarly, a voltage level from $+5V$ to $+15V$ at the transmitting end and $+3V$ or more at the receiving end is considered a space.

The RS-232C data communication rate is specified as baud rate. Earlier we pointed out that this is a measure of the bits transferred through the communication interface per second (bps). For example, a common rate of 9600 baud means 9600 bits are transmitted or received in one second. In general, the receive and transmit baud rates do not need to be the same; however, in most simpler systems they are set to the same value.

▲ 10.11 PROGRAMMABLE COMMUNICATION INTERFACE CONTROLLERS

Because serial communication interfaces are so widely used in modern electronic equipment, special LSI peripheral devices have been developed to permit easy implementation of these types of interfaces. For instance, an RS-232C port is the type of interface needed to connect a CRT terminal or a modem to a microcomputer. To support connection of these two peripheral devices, the microcomputer would need two independent RS-232C I/O ports. This function is normally implemented with a programmable communication controller known as a *universal synchronous/asynchronous receiver transmitter* (USART). As the name implies, a USART is capable of implementing either an asynchronous or synchronous communication interface. Here we will concentrate on its use in implementing an asynchronous communication interface.

The programmability of the USART provides for a very flexible asynchronous communication interface. Typically, it contains a full-duplex receiver and transmitter, which can be configured through software for communication of data using formats with character lengths between 5 and 8 bits, with even or odd parity, and with 1, 1½, or 2 stop bits.

A USART has the ability to automatically check characters during data reception to detect the occurrence of parity, framing, and overrun errors. A framing error means that after the detection of the beginning of a character with a start bit the appropriate number of stop bits were not detected. This means that the character that was transmitted was not received correctly and should be resent. An overrun error means that the prior character that was received was not read out of the USART's receive data register by the microprocessor before another character was received. Therefore, the first character was lost and should be retransmitted.

8251A USART

A block diagram showing the internal architecture of the 8251A is shown in Fig. 10–57(a) and its pin layout in Fig. 10–57(b). From this diagram we find that it includes four key sections: the bus interface section, which consists of the data bus buffer and read/write control logic blocks; the transmit section, which consists of the transmit buffer and transmit control blocks; the receive section, which consists of the receive buffer and receive-control blocks; and the modem-control section. Let us now look at each of these sections in more detail.

A UART cannot stand alone in a communication system; its operation must typically be controlled by a microprocessor. The bus interface section is used to connect the 8251A to a microprocessor such as the 8086. Note that the interface includes an 8-bit bidirectional data bus D_0 through D_7 driven by the data bus buffer. It is over these lines that the microprocessor transfers commands to the 8251A, reads its status register, and inputs or outputs character data.

Data transfers over the bus are controlled by the signals $C/\overline{D}$ (control/data), $\overline{RD}$ (read), $\overline{WR}$ (write), and $\overline{CS}$ (chip select), all inputs to the read/write control logic section. Typically, the 8251A is located at a specific address in the microcomputer's I/O or memory address space. When the microprocessor is to access registers within the 8251A, it puts this address on the address bus. The address is decoded by external circuitry and must produce logic 0 at the $\overline{CS}$ input for a read or write bus cycle to take place to the 8251A.

The other three control signals, $C/\overline{D}$, $\overline{RD}$, and $\overline{WR}$, tell the 8251A what type of data transfer is to take place over the bus. Figure 10–58 shows the various types of read/write operations that can occur. For example, the first state in the table, $C/\overline{D} = 0$, $\overline{RD} = 0$, and $\overline{WR} = 1$, corresponds to a character data transfer from the 8251A to the microprocessor.

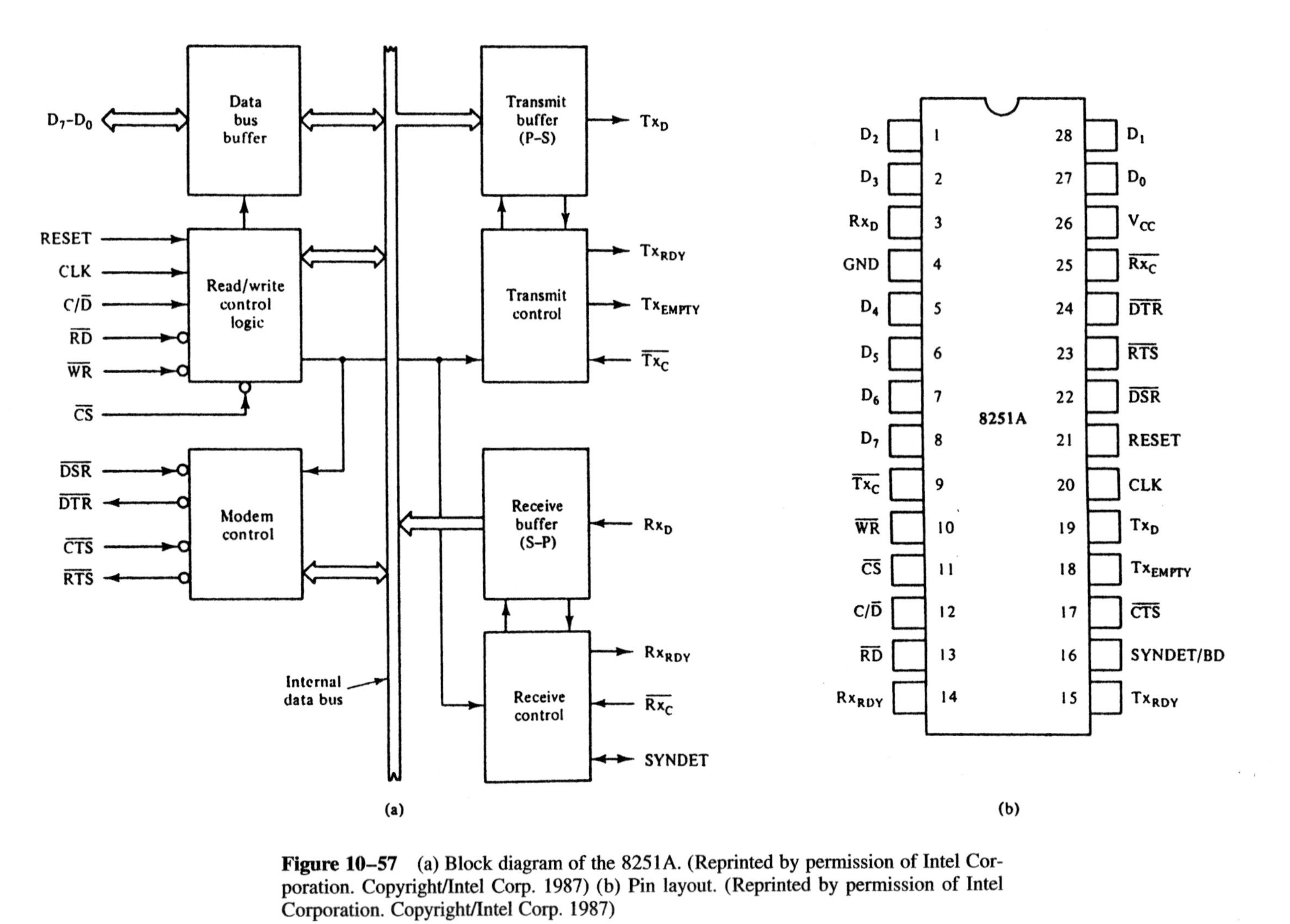

Figure 10–57 (a) Block diagram of the 8251A. (Reprinted by permission of Intel Corporation. Copyright/Intel Corp. 1987) (b) Pin layout. (Reprinted by permission of Intel Corporation. Copyright/Intel Corp. 1987)

$C/\overline{D}$	$\overline{RD}$	$\overline{WR}$	$\overline{CS}$	Operation
0	0	1	0	8251A Data → Data bus
0	1	0	0	Data bus → 8251A Data
1	0	1	0	Status → Data bus
1	1	0	0	Data bus → Control
X	1	1	0	Data bus → 3-State
X	X	X	1	Data bus → 3-State

Figure 10–58 Read/write operations. (Reprinted by permission of Intel Corporation. Copyright/Intel Corp. 1987)

Note that in general $\overline{RD} = 0$ signals that the microprocessor is reading data from the 8251A, $\overline{WR} = 0$ indicates that data are being written into the 8251A, and the logic level of $C/\overline{D}$ indicates whether character data, control information, or status information is on the data bus.

EXAMPLE 10.29

What type of data transfer is taking place over the bus if the control signals are at $\overline{CS} = 0$, $C/\overline{D} = 1$, $\overline{RD} = 0$, and $\overline{WR} = 1$?

Solution

Looking at the table in Fig. 10–58, we see that $\overline{CS} = 0$ means that the 8251A's data bus has been enabled for operation. Since $C/\overline{D}$ is 1 and $\overline{RD}$ is 0, status information is being read from the 8251A.

The receiver section is responsible for reading the serial bit stream of data at the Rx_D (receive-data) input and converting it to parallel form. When a mark voltage level is detected on this line, indicating a start bit, the receiver enables a counter. As the counter increments to a value equal to one-half a bit time, the logic level at the Rx_D line is sampled again. If it is still at the mark level, a valid start pulse has been detected. Then Rx_D is examined every time the counter increments through another bit time. This continues until a complete character is assembled and the stop bit is read. After this, the complete character is transferred into the receive-data register.

During reception of a character, the receiver automatically checks the character data for parity, framing, or overrun errors. If one of these conditions occurs, it is flagged by setting a bit in the status register. Then the Rx_{RDY} (receiver ready) output is switched to the 1 logic level. This signal is sent to the microprocessor to tell it that a character is available and should be read from the receive-data register. Rx_{RDY} is automatically reset to logic 0 when the MPU reads the contents of the receive-data register.

The 8251A does not have a built-in baud-rate generator. For this reason, the clock signal that is used to set the baud rate must be externally generated and applied to the Rx_C input of the receiver. Through software the 8251A can be set up to internally divide the clock signal input at Rx_C by 1, 16, or 64 to obtain the desired baud rate.

The transmitter does the opposite of the receiver section. It receives parallel character data from the MPU over the data bus. The character is then automatically framed with the start bit, appropriate parity bit, and the correct number of stop bits and put into the transmit data buffer register. Finally, it is shifted out of this register to produce a bit-serial output on the Tx_D line. When the transmit data buffer register is empty, the Tx_{RDY} output switches to logic 1. This signal can be returned to the MPU to tell it that another character should be output to the transmitter section. When the MPU writes another character out to the transmitter buffer register, the Tx_{RDY} output resets.

Data are output on the transmit line at the baud rate set by the external transmitter clock signal that is input at Tx_C. In most applications, the transmitter and receiver operate at the same baud rate. Therefore, the same baud-rate generator supplies both Rx_C and Tx_C. The circuit in Fig. 10–59 shows this type of system configuration.

The 8251A UART, just like the other peripheral ICs discussed earlier in the chapter, can be configured for various modes of operation through software. Its operation is controlled through the setting of bits in three internal control registers: the mode-control register, command register, and the status register. For instance, the way in which the 8251A's receiver and transmitter works is determined by the contents of the mode control register.

Figure 10–60 shows the organization of the mode control register and the function of each of its bits. Note that the two least significant bits B_1 and B_2 determine whether the device operates as an asynchronous or synchronous communication controller and in asynchronous mode how the external baud rate clock is divided within the 8251A. For example, if these two bits are 11, it is set for asynchronous operation with divide-by-64 for the baud-rate input. The two bits that follow these, L_1 and L_2, set the length of the character. For instance, when information is being transmitted and received as 7-bit ASCII characters, these bits should be loaded with 10.

The next two bits, PEN and EP, determine whether parity is in use and, if so, whether it is even parity or odd parity. Looking at Fig. 10–60, we see that PEN enables or disables parity. To enable parity, it is set to 1. Furthermore, when parity is enabled,

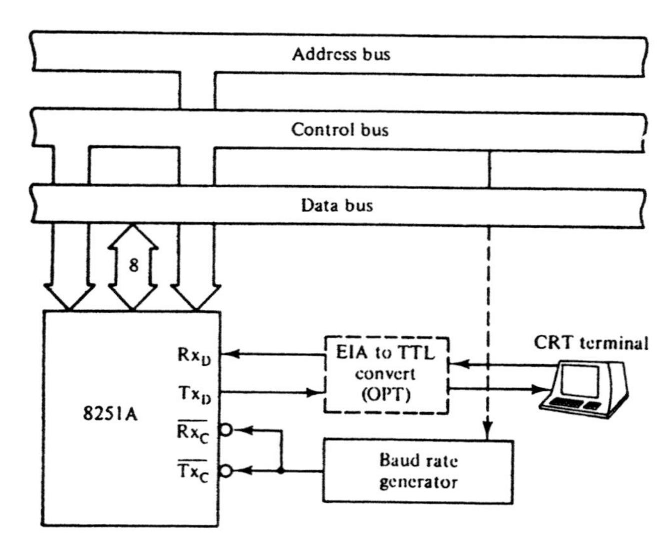

Figure 10–59 Receiver and transmitter driven at the same baud rate. (Reprinted by permission of Intel Corporation. Copyright/Intel Corp. 1987)

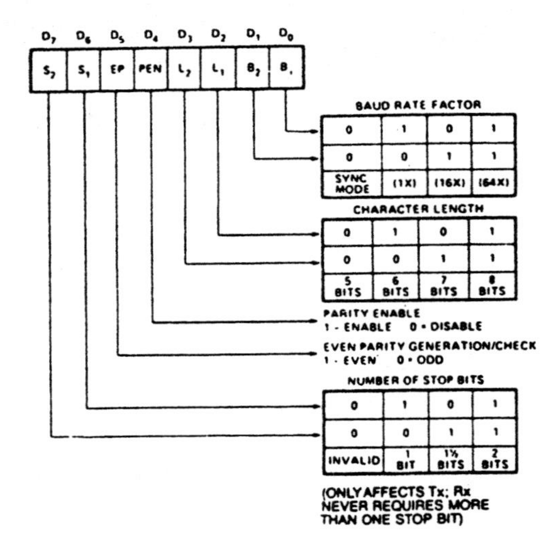

Figure 10–60 Mode instruction format. (Reprinted by permission of Intel Corporation. Copyright/Intel Corp. 1987)

logic 0 in EP selects odd parity, or logic 1 in this position selects even parity. To disable parity, all we need to do is reset PEN.

We will assume that the 8251A is working in the asynchronous mode; therefore, bits S_1 and S_2 determine the number of stop bits. Note that if 11 is loaded into these bit positions, the character is transmitted with 2 stop bits.

EXAMPLE 10.30

What value must be written into the mode-control register in order to configure the 8251A such that it works as an asynchronous communications controller with the baud rate clock internally divided by 16? Character size is to be 8 bits; parity is odd; and one stop bit is used.

Solution

From Fig. 10–60, we find that B_2B_1 must be set to 10 in order to select asynchronous operation with divide-by-16 for the external baud clock input.

$$B_2B_1 = 10$$

To select a character length of 8 bits, the next 2 bits are both made logic 1. This gives

$$L_2L_1 = 11$$

To set up odd parity, EP and PEN must be made equal to 0 and 1, respectively.

$$\text{EP PEN} = 01$$

Finally, S_2S_1 are set to 01 for one stop bit.

$$S_2S_1 = 01$$

Therefore, the complete control word is

$$D_7D_6 \ldots\ldots D_0 = 01011110_2$$
$$= 5E_{16}$$

Once the configuration for asynchronous communications has been set up in the mode control register, the microprocessor controls the operation of the serial interface by issuing commands to the command register within the 8251A. The format of the command instruction byte and the function of each of its bits is shown in Fig. 10–61. Let us look at the function of just a few of its bits.

Tx_{EN} and Rx_{EN} are enable bits for the transmitter and receiver. Since both the receiver and transmitter can operate simultaneously, these two bits can both be set. Rx_{EN} is actually an enable signal to the Rx_{RDY} signal. It does not turn the receiver section on and off. The receiver runs at all times, but if Rx_{EN} is set to 0, the 8251A does not signal the MPU that a character has been received by switching Rx_{RDY} to logic 1. The same is true for Tx_{EN}. It enables the Tx_{RDY} signal.

The 8251A USART has a status register that contains information related to its current state. The status register of the 8251A is shown in Fig. 10–62. Bits parity error (PE), overflow error (OE), and frame error (FE) are error flags for the receiver. If the incoming character is found to have a parity error, the PE (parity error) bit gets set. On the other hand, if an overrun or framing error condition occurs, the OE (overrun error) or FE (fram-

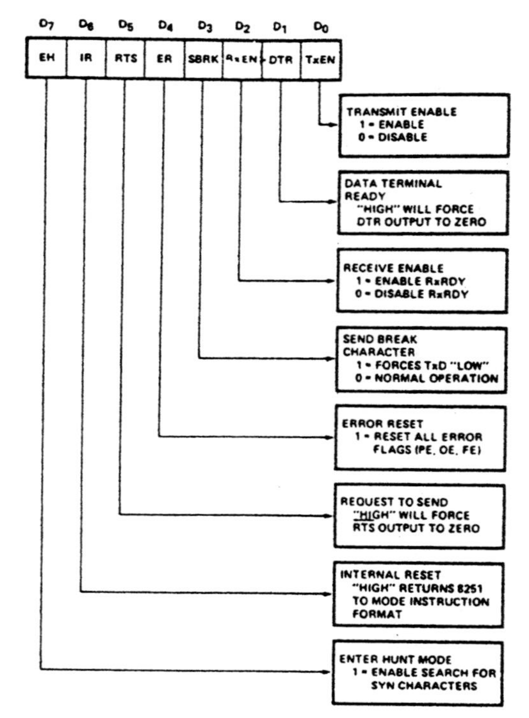

Figure 10–61 Command instruction format. (Reprinted by permission of Intel Corporation. Copyright/Intel Corp. 1987)

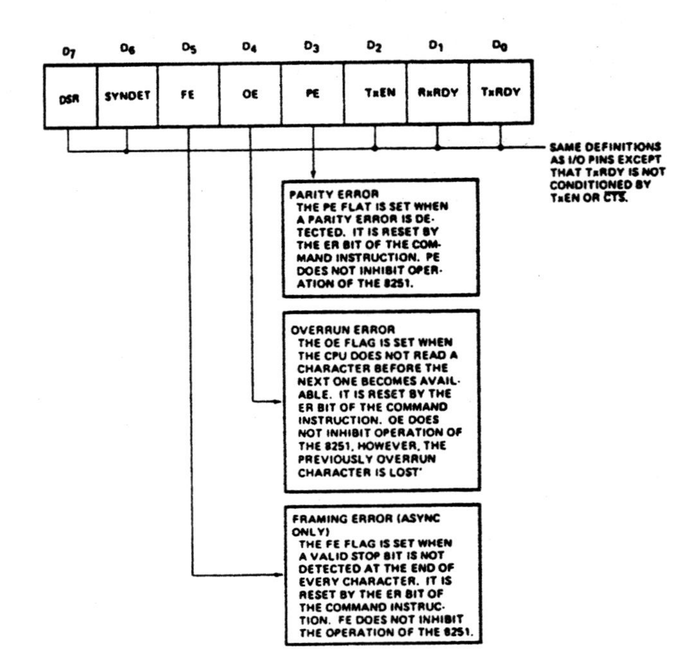

Figure 10–62 Status register. (Reprinted by permission of Intel Corporation. Copyright/Intel Corp. 1987)

ing error) flag is set, respectively. Before reading a character from the receive data register, the MPU should always verify that valid data has been received by examining these error bits. If an error is identified, a command can be issued to the command register to write a 1 into the ER bit. This causes all three of the error flags in the status register to be reset. Then a software routine can be initiated to cause the character to be retransmitted.

Let us look at just one more bit of the command register. The IR bit, which stands for internal reset, allows the 8251A to be initialized under software control. To initialize the device, the MPU simply writes a 1 into the IR bit.

Before the 8251A can be used to receive or transmit characters, its mode control and command registers must be initialized. The flowchart in Fig. 10–63 shows the sequence that must be followed when initializing the device. Let us just briefly trace through the sequence of events needed to set up the controller for asynchronous operation.

As the microcomputer powers up, it should issue a hardware reset to the 8251A. Switching its RESET input to logic 1 does this. After this, a load-mode instruction must be issued to write the new configuration byte into the mode-control register. Assuming that the 8251A is in the I/O address space of the 8088, the command byte formed in Example 10.30 can be written to the command register with the instruction sequence

```
MOV  DX, MODE_REG_ADDR
MOV  AL, 5EH
OUT  DX, AL
```

where MODE_REG_ADDR is a variable equal to the address of the mode register of the 8251A.

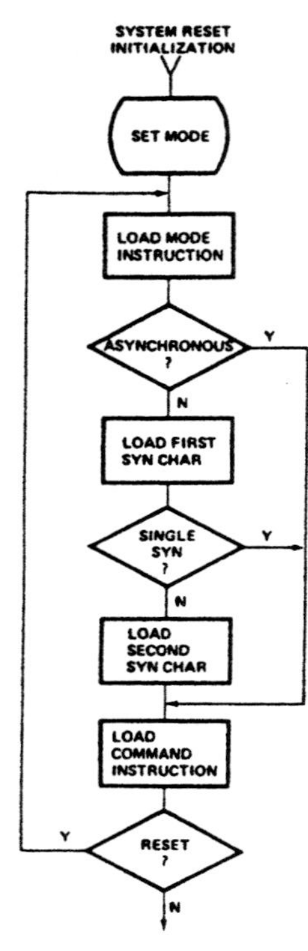

Figure 10–63 8251A initialization flowchart. (Reprinted by permission of Intel Corporation. Copyright/Intel Corp. 1987)

Since bits B_2B_1 of this register are not 00, asynchronous mode of operation is selected. Therefore, we go down the branch in the flowchart to the load command instruction. Execution of another OUT instruction can load the command register with its initial value. For instance, this command could enable the transmitter and receiver by setting the Tx_{EN} and Rx_{EN} bits, respectively. During its operation the status register can be read by the microprocessor to determine if the device has received the next byte, if it is ready to send the next byte, or if any problem occurred in the transmission such as a parity error.

EXAMPLE 10.31

The circuit in Fig. 10–64(a) implements serial I/O for the 8088 microprocessor using an 8251A. Write a program that continuously reads serial characters from the RS-232C interface, complements the received characters with software, and sends them back through the RS-232C interface. Each character is received and transmitted as an 8-bit character using 2 stop bits and no parity.

Solution

We must first determine the addresses for the registers in the 8251A that can be accessed from the microprocessor interface. Chip select ($\overline{CS}$) is enabled for I/O read or write operations to addresses for which

$$A_7A_6A_5A_4A_3A_2A_1 = 1000000$$

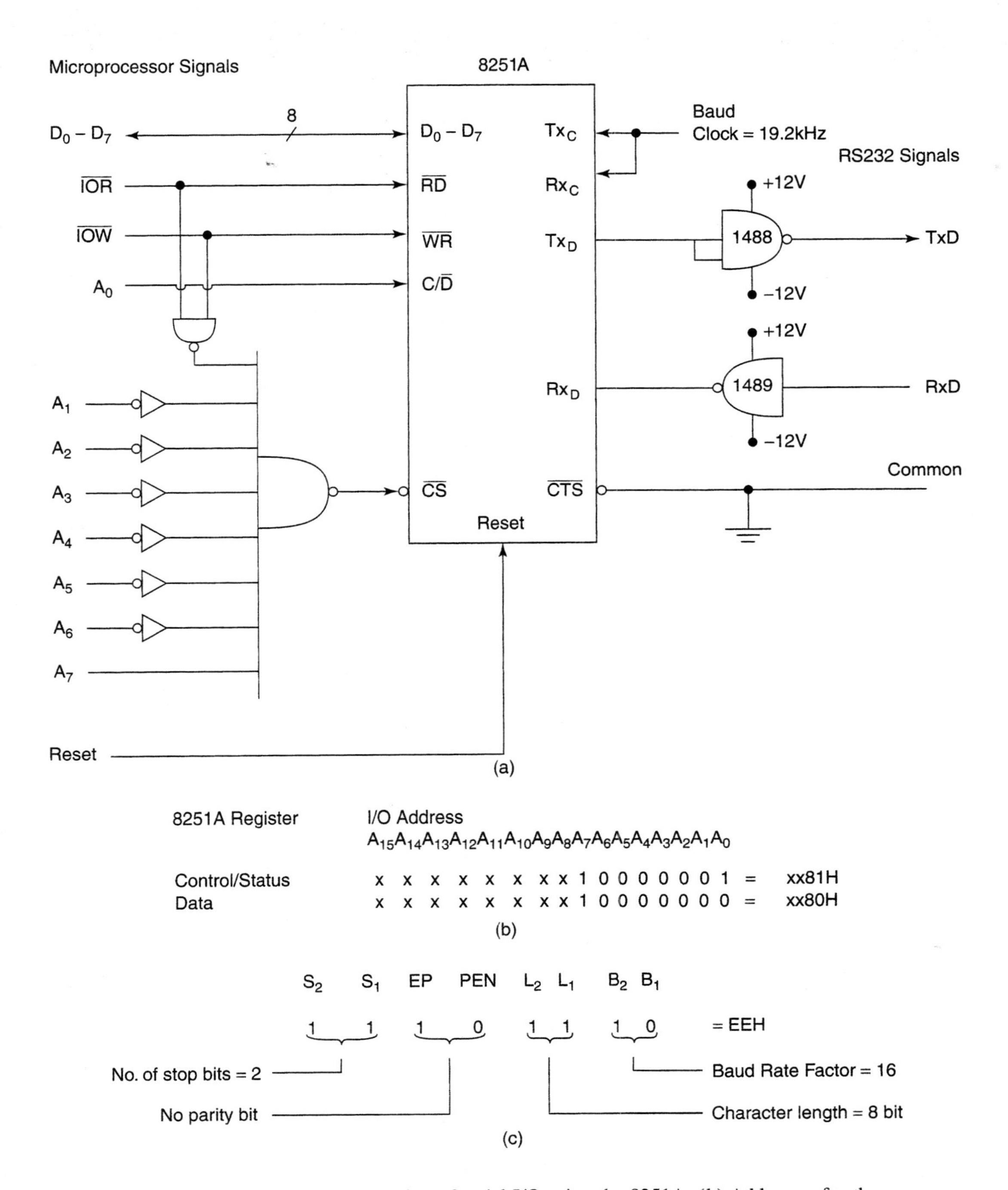

Figure 10–64 (a) Implementation of serial I/O using the 8251A. (b) Addresses for the 8251A registers. (c) Mode word. (d) Command word. (e) Flowchart for initialization, receive operation, and transmit operation. (f) Program for the implementation of initialization, receive operation, and transmit operation.

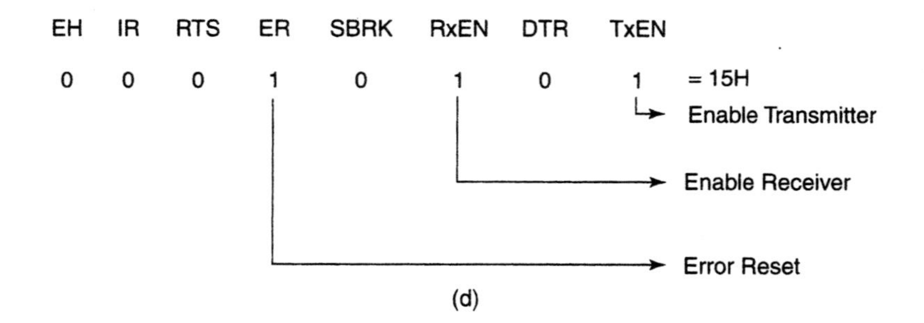

Figure 10–64 (continued)

Bit A_0 of the address bus is used to select between the data and control (or status) registers. As shown in Fig. 10–64(b), the addresses for the data and the control (or status) register are XX80H and XX81H, respectively

Next we must determine the mode word to select an 8-bit character with 2 stop bits and no parity. As shown in Fig. 10–64(c), the mode word is EEH. Here we have used a baud-rate factor of 16, which means that the baud rate is given as

$$\text{Baud rate} = \text{Baud-rate clock}/16 = 19{,}200/16 = 1200 \text{ bps}$$

To enable the transmitter as well as receiver operation of the 8251A, the required command word as shown in Fig. 10–64(d) is equal to 15H. Note that error reset has also been implemented by making the ER bit equal 1.

The flowchart of Fig. 10–64(e) shows how we can write software to implement initialization, the receive operation, and transmit operation. The program written to perform this sequence is shown in Fig. 10–64(f).

Initialization involves writing the mode word followed by the command word to the control register of the 8251A. It is important to note that this is done after the device has been reset. Since the control register's I/O address is 81H, the two words are output to this address using appropriate instructions.

The receive operation starts by reading the contents of the status register at address 81H and checking if the LSB, Rx_{RDY}, is at logic 1. If it is not 1, the routine keeps reading and checking until it does become 1. Next we read the data register at 80H for the receive data. The byte of data received is complemented and then saved for transmission.

The transmit operation also starts by reading the status register at address 81H and checking if bit 1, Tx_{RDY}, is logic 1. If it is not, we again keep reading and checking until it becomes 1. Next, the byte of data that was saved for transmission is written to the data register at address 81H. This causes it to be transmitted at the serial interface. The receive and transmit operations are repeated by jumping back to the point where the receive operation begins.

8250/16450 UART

The 8250 and 16450 are pin-for-pin and functionally equivalent universal asynchronous receiver transmitter ICs. These devices are newer than the 8251A UART and implement

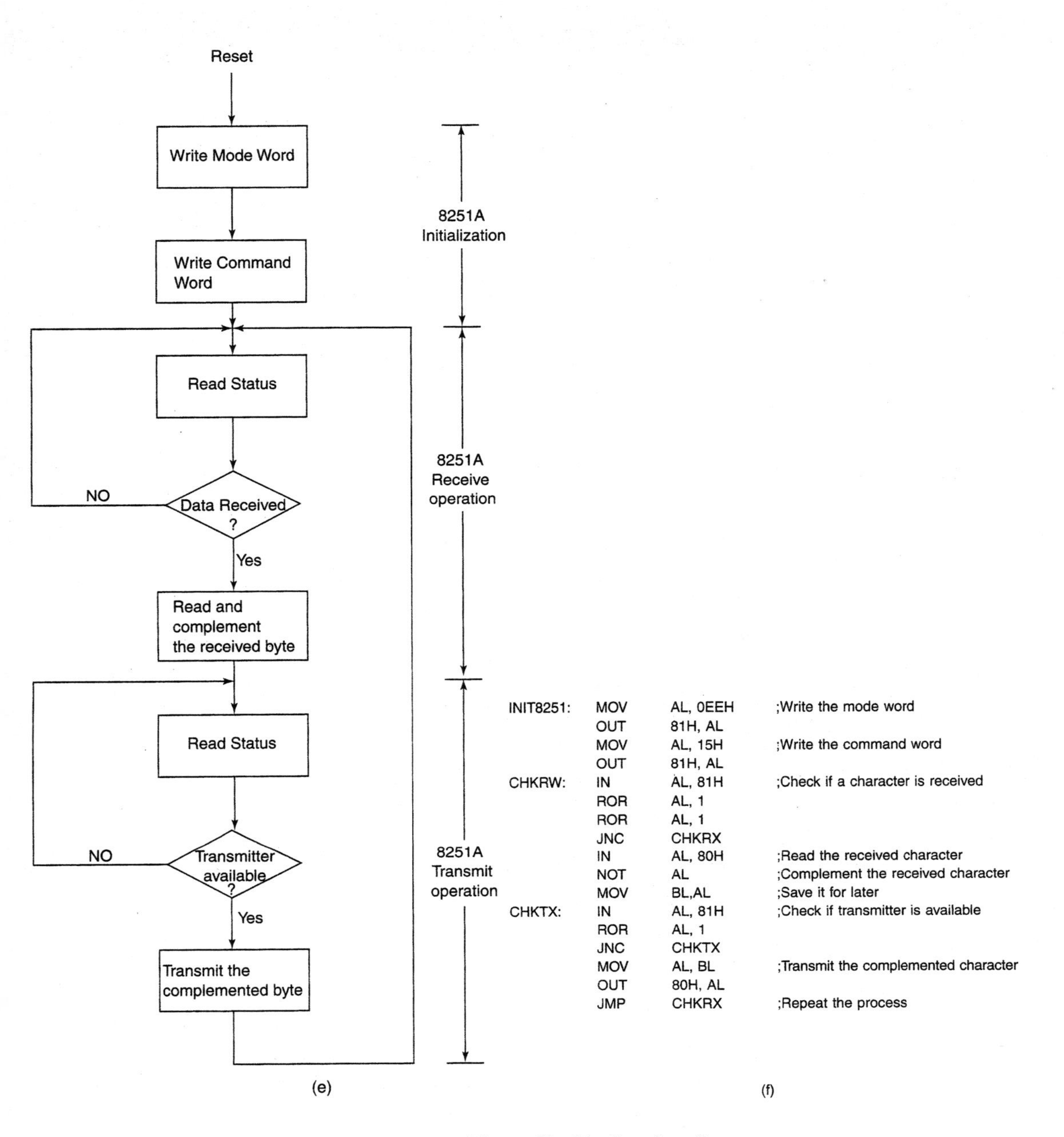

Figure 10–64 (continued)

a more versatile serial I/O operation. For instance, they have a built-in programmable baud-rate generator, double buffering on communication data registers, and enhanced status and interrupt signaling. The common pin layout for these devices is shown in Fig. 10–65(a).

The connection of the 8250/16450 to implement a simple RS-232C serial communications interface is shown in Fig. 10–65(b). Looking at the microprocessor interface, we

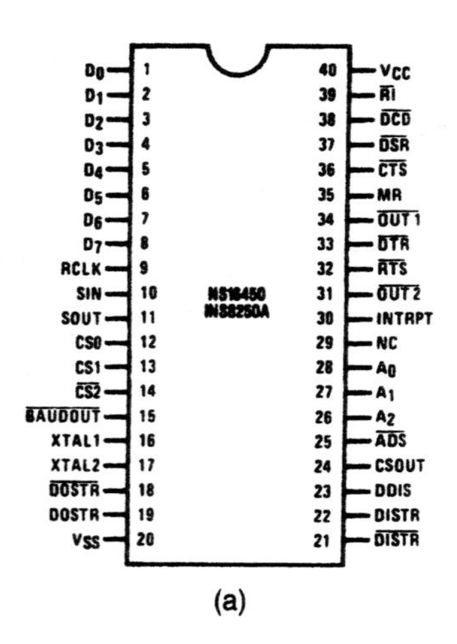

(a)

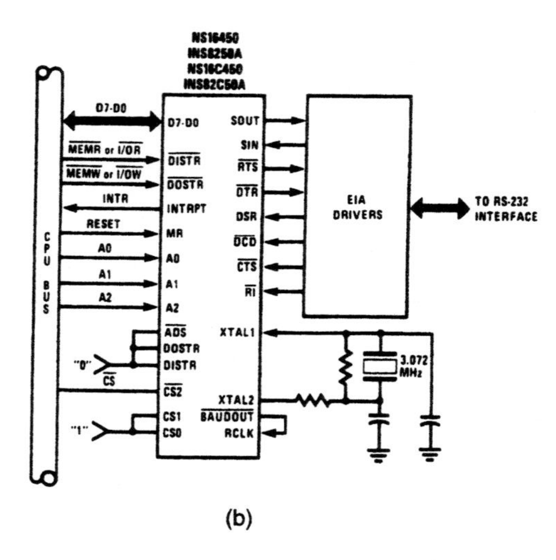

(b)

Figure 10–65 (a) Pin layout of the 8250/16450 UART. (Courtesy of National Semiconductor Corporation) (b) 8250/16450 RS-232C interface. (Courtesy of National Semiconductor Corporation)

find chip-select inputs CS_0, CS_1, and $\overline{CS}_2$. To enable the interface, these inputs must be at logic 1, 1, and 0, respectively, at the same time that address strobe ($\overline{ADS}$) is logic 0. Therefore, the interface in Fig. 10–65(b) is enabled whenever logic 0 is applied to $\overline{CS}_2$ from the MPU's bus.

Let us next look at how data are read from or written into the registers of the 8250/16450. Data transfers between the MPU and communication controller take place over data bus lines D_0 through D_7. The MPU signals the peripheral whether a data input or output operation is to occur with the logic level at the data-input strobe ($\overline{DISTR}$) and data-output strobe ($\overline{DOSTR}$) inputs. Note that when data are output during a memory write or output bus cycle, the MPU notifies the 8250/16450 with logic 0 on the $\overline{MEMW}$ or $\overline{I/OW}$ signal line, which is applied to the $\overline{DOSTR}$ input.

During the read or write bus cycle, the register that is accessed is determined by the code at register select inputs A_0, A_1, and A_2. Figure 10–65(b) shows that these inputs are attached to address lines A_0 through A_2, respectively. The registers selected by the various register-select codes are shown in Fig. 10–66. Note that the setting of the divisor latch bit (DLAB), which is in the line-control register, is also involved in the selection of the register. For example, to write to the line-control register, the code at $A_2A_1A_0$ must be 011_2. Moreover, to read the receive buffer register, the DLAB bit in the line-control register must first be set to 0 and then a read performed with register-select code $A_2A_1A_0$ equal to 000_2.

The function of the various bits of the 8250/16450's registers is summarized in the table of Fig. 10–67(a). Note that the receive buffer register (RBR) and transmitter hold register (THR) correspond to the read and write functions of register 0. However, as mentioned earlier, to perform these read or write operations the divisor latch bit (DLAB), which is bit 7 of the line control register (LCR), must have already been set to 0. From the table, we find that other bits of LCR are used to define the serial character data structure. For instance, Fig. 10–67(b) shows how the *word-length select bits,* bit 0 (WLS_0) and bit 1 (WLS_1) of LCR, select the number of bits in the serial character. Bit 2, *number of stop bits* (STB), selects the number of stop bits. If it is set to logic 0, one stop bit is generated for all transmitted data. On the other hand, if bit 2 is set to 1, one and a half stop bits are produced if character length is set to 5 bits and 2 stop bits are supplied if character length is 6 or more bits. The next 2 bits, bit 3 *parity enable* (PEN) and bit 4 *even parity select* (EPS), are used to select parity. First parity is enabled by making bit 3 logic 1 and then even or odd parity is selected by making bit 4 logic 1 or 0, respectively. The LCR can be loaded with the appropriate configuration information under software control.

Figure 10–65(b) shows that the baud-rate generator is operated off a 3.072-MHz crystal. This crystal frequency can be divided within the 8250/16450 to produce a variety of data communication baud rates. The divisor values that produce standard baud rates are shown in Fig. 10–68. For example, to set the asynchronous data communication rate to 300 baud, a divisor equal to 640 must be used. The 16-bit divider must be loaded under software control into the divisor latch registers, DLL and DLM. Figure 10–67(a) shows that the eight least significant bits of the divisor are in DLL and the eight most significant bits in DLM.

DLAB	A_2	A_1	A_0	Register
0	0	0	0	Receiver Buffer (read), Transmitter Holding Register (write)
0	0	0	1	Interrupt Enable
X	0	1	0	Interrupt Identification (read only)
X	0	1	1	Line Control
X	1	0	0	MODEM Control
X	1	0	1	Line Status
X	1	1	0	MODEM Status
X	1	1	1	Scratch
1	0	0	0	Divisor Latch (least significant byte)
1	0	0	1	Divisor Latch (most significant byte)

Figure 10–66 Register-select codes. (Courtesy of National Semiconductor Corporation)

Bit No.	Register Address										
	0 DLAB=0	0 DLAB=0	1 DLAB=0	2	3	4	5	6	7	0 DLAB=1	1 DLAB=1
	Receiver Buffer Register (Read Only)	Transmitter Holding Register (Write Only)	Interrupt Enable Register	Interrupt Ident. Register (Read Only)	Line Control Register	MODEM Control Register	Line Status Register	MODEM Status Register	Scratch Reg-ister	Divisor Latch (LS)	Latch (MS)
	RBR	THR	IER	IIR	LCR	MCR	LSR	MSR	SCR	DLL	DLM
0	Data Bit 0*	Data Bit 0	Enable Received Data Available Interrupt (ERBFI)	"0" if Interrupt Pending	Word Length Select Bit 0 (WLS0)	Data Terminal Ready (DTR)	Data Ready (DR)	Delta Clear to Send (DCTS)	Bit 0	Bit 0	Bit 8
1	Data Bit 1	Data Bit 1	Enable Transmitter Holding Register Empty Interrupt (ETBEI)	Interrupt ID Bit (0)	Word Length Select Bit 1 (WLS1)	Request to Send (RTS)	Overrun Error (OE)	Delta Data Set Ready (DDSR)	Bit 1	Bit 1	Bit 9
2	Data Bit 2	Data Bit 2	Enable Receiver Line Status Interrupt (ELSI)	Interrupt ID Bit (1)	Number of Stop Bits (STB)	Out 1	Parity Error (PE)	Trailing Edge Ring Indicator (TERI)	Bit 2	Bit 2	Bit 10
3	Data Bit 3	Data Bit 3	Enable MODEM Status Interrupt (EDSSI)	0	Parity Enable (PEN)	Out 2	Framing Error (FE)	Delta Data Carrier Detect (DDCD)	Bit 3	Bit 3	Bit 11
4	Data Bit 4	Data Bit 4	0	0	Even Parity Select (EPS)	Loop	Break Interrupt (BI)	Clear to Send (CTS)	Bit 4	Bit 4	Bit 12
5	Data Bit 5	Data Bit 5	0	0	Stick Parity	0	Transmitter Holding Register (THRE)	Data Set Ready (DSR)	Bit 5	Bit 5	Bit 13
6	Data Bit 6	Data Bit 6	0	0	Set Break	0	Transmitter Empty (TEMT)	Ring Indicator (RI)	Bit 6	Bit 6	Bit 14
7	Data Bit 7	Data Bit 7	0	0	Divisor Latch Access Bit (DLAB)	0	0	Data Carrier Detect (DCD)	Bit 7	Bit 7	Bit 15

*Bit 0 is the least significant bit. It is the first bit serially transmitted or received.

(a)

(b)

Bit 1	Bit 0	Word Length
0	0	5 Bits
0	1	6 Bits
1	0	7 Bits
1	1	8 Bits

Figure 10–67 (a) Register bit functions. (Courtesy of National Semiconductor Corporation) (b) Word-length select bits. (Courtesy of National Semiconductor Corporation)

Desired Baud Rate	Divisor Used to Generate 16 x Clock	Percent Error Difference Between Desired and Actual
50	3840	—
75	2560	—
110	1745	0.026
134.5	1428	0.034
150	1280	—
300	640	—
600	320	—
1200	160	—
1800	107	0.312
2000	96	—
2400	80	—
3600	53	0.628
4800	40	—
7200	27	1.23
9600	20	—
19200	10	—
38400	5	—

Figure 10–68 Baud rates and corresponding divisors. (Courtesy of National Semiconductor Corporation)

EXAMPLE 10.32

What count must be loaded into the divisor latch registers to set the data communication rate to 2400 baud? What register-select code must be applied to the 8250/16450 when writing the bytes of the divider count into the DLL and DLM registers?

Solution

Looking at Fig. 10–68, we find that the divisor for 2400 baud is 80. When writing the byte into DLL, the address must make

$$A_2A_1A_0 = 000_2 \text{ with DLAB} = 1$$

and the value that is written is

$$\text{DLL} = 80 = 50\text{H}$$

For DLM, the address must make

$$A_2A_1A_0 = 001_2 \text{ with DLAB} = 1$$

and the value is

$$\text{DLM} = 0 = 00\text{H}$$

Let us now turn our attention to the right side of the 8250/16450 in Fig. 10–65(b). Here the RS-232C serial communication interface is implemented. We find that the transmit data are output in serial form over the serial output (S_{OUT}) line, and receive data are input over the serial input (S_{IN}) line. Handshaking for the asynchronous serial interface is implemented with the request to send ($\overline{\text{RTS}}$) and data terminal ready ($\overline{\text{DTR}}$) outputs and the data set ready ($\overline{\text{DSR}}$), data carrier detect ($\overline{\text{DCD}}$), clear to send ($\overline{\text{CTS}}$), and ring indicator (RI) inputs.

The serial interface input/output signals are buffered by EIA drivers for compatibility with RS-232C voltage levels and drive currents. For example, a MC1488 driver IC can be used to buffer the output lines. It contains four TTL level to RS-232C drivers, each of which is actually a NAND gate. The MC1488 requires +12V, −12V, and ground supply connections to provide the mark and space transmission-voltage levels. The gates of a MC1489 RS-232C to TTL level driver can buffer the input lines of the interface. This IC contains four inverting buffers with tristate outputs and is operated from a single +5V supply. Figure 10–69 shows an RS-232C interface including the EIA driver circuitry.

▲ 10.12 KEYBOARD AND DISPLAY INTERFACE

The keyboard and display are important input and output devices in microcomputer systems such as the PC. Different types of keyboards and displays are used in many other

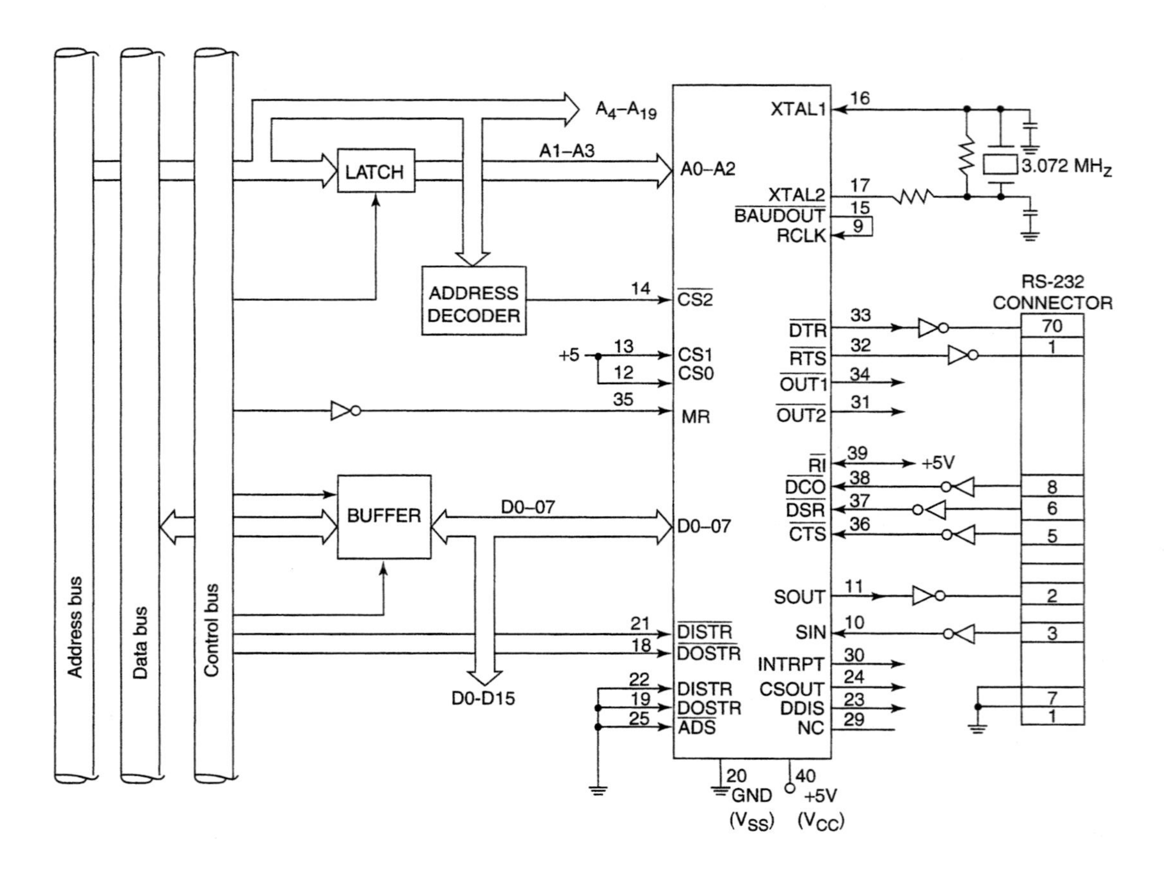

Figure 10–69 RS-232C interface with EIA drivers.

types of digital electronic systems. For instance, all calculators and hand-held computers have both a keyboard and a display; also many electronic test instruments have a display.

The circuit diagram in Fig. 10–70 shows how a keyboard is most frequently interfaced to a microcomputer. Note that the switches in the keyboard are arranged in an *array.* The size of the array is described in terms of the number of rows and the number of columns. In our example, the keyboard array has four rows, labeled R_0 through R_3, and four columns, labeled C_0 through C_3. A row and a column uniquely define the location of the switch for any key in the array. For instance, the 0 key is located at the junction of R_0 and C_0, whereas the 1 key is located at R_0 and C_1.

Now that we know how the keys of the keyboard are arranged, let us look at how the microcomputer services them. In most applications, the microcomputer scans the keyboard array. That is, it strobes one row of the keyboard after the other by sending out a short-duration pulse, to the 0 logic level, on the row line. During each row strobe, all column lines are examined by reading them in parallel. Typically, the column lines are pulled up to the 1 logic level; therefore, if a switch is closed, a logic 0 will be read on the corresponding column line. If no switches are closed, all 1s will be read when the column lines are examined.

For instance, if the 2 key is depressed when the microcomputer is scanning R_0, the column code read-back will be $C_3C_2C_1C_0 = 1011$. Since the microcomputer knows which row it is scanning (R_0) and which column the strobe was returned on (C_2), it can determine that the number 2 key was depressed. The microcomputer does not necessarily store the row and column codes in the form that we have shown. It is more common to just maintain the binary equivalent of the row or column. In our example, the microcomputer would internally store the row number as $R_0 = 00$ and the column number as $C_2 = 10$. This is a more compact representation of the row and column information.

Several other issues arise when designing keyboards for microcomputer systems. One is that when a key in the keyboard is depressed, its contacts bounce for a short period of time. Depending on the keyboard sampling method, this could result in incorrect reading of the keyboard input. This problem is overcome by a technique known as *keyboard debouncing.* Debouncing is achieved by resampling the column lines a second time, about 10 ms later, to assure that the same column line is at the 0 logic level. If so, it is then accepted as a valid input. This technique can be implemented either in hardware or software.

Another problem occurs in keyboard sampling when more than one key is depressed at a time. In this case, the column code read by the microcomputer would have more than 1 bit that is logic 0. For instance, if the 0 and 2 keys were depressed, the column code read back during the scan of R_0 would be $C_3C_2C_1C_0 = 1010$. Typically, two keys are not actually depressed at the same time. It is more common that the second key is depressed while the first one is still being held down and that the column code showing two key closures would show up in the second test that is made for debouncing.

Several different techniques are used to overcome this problem. One is called *two-key lockout.* With this method, the occurrence of a second key during the debounce scan causes both keys to be locked out, and neither is accepted by the microcomputer. If the second key that was depressed is released before the first key is released, the first key entry is accepted and the second key is ignored. On the other hand, if the first key is released before the second key, only the second key is accepted.

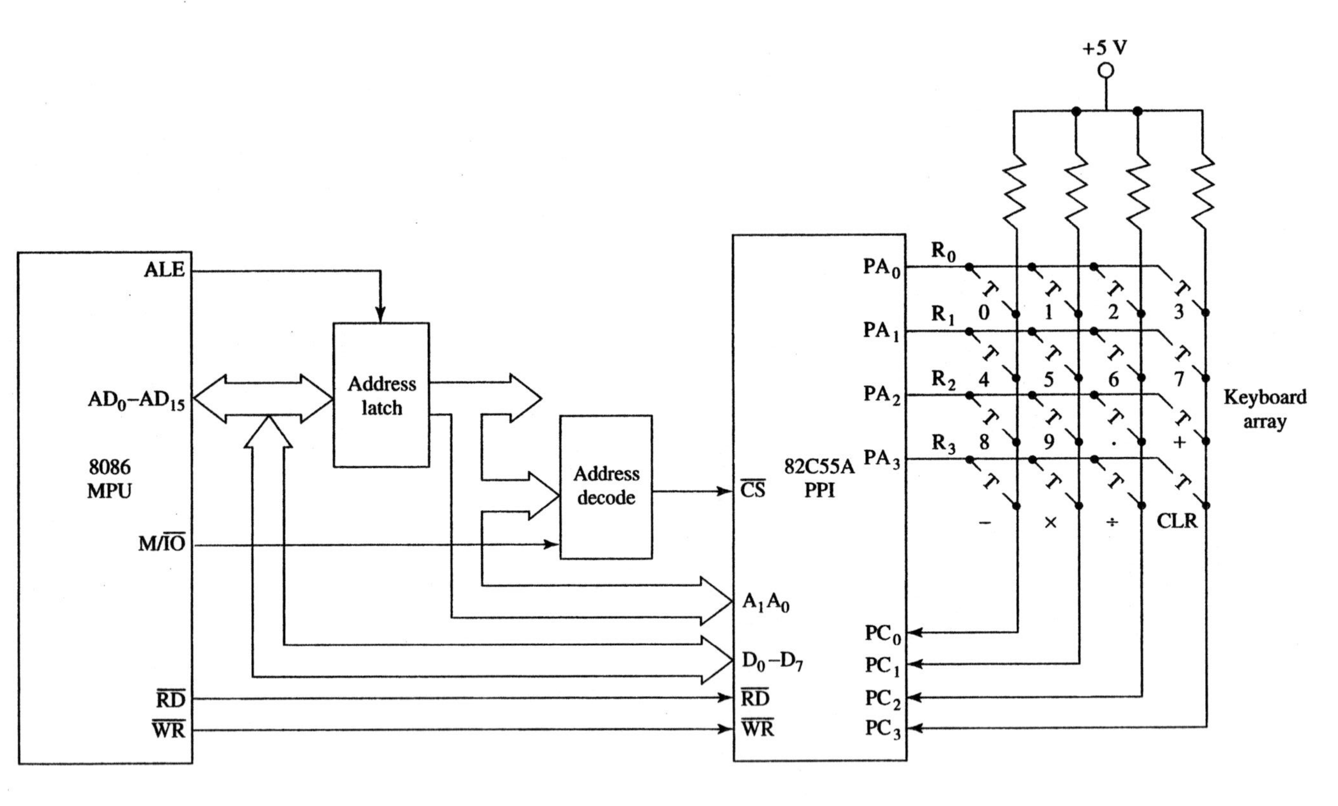

Figure 10–70 Keyboard interfaced to a microcomputer.

A second method of solving this problem is that known as *N-key rollover.* In this case, more than one key can be depressed at a time and be accepted by the microcomputer. The microcomputer keeps track of the order in which they are depressed and as long as the switch closures are still present at another keyboard scan 10 ms later, they are accepted. That is, in the case of multiple key depressions, the key entries are accepted in the order in which their switches are closed.

Figure 10–71 shows a display interface used in many microcomputer systems. Here we are using a four-digit, seven-segment numeric display. Note that segment lines a through g of all digits of the display are driven in parallel by outputs of the microcomputer. It is over these lines that the microcomputer outputs signals to tell the display which segments are to be lighted to form numbers in its digits. The way in which the segments of a seven-segment display digit are labeled is shown in Fig. 10–72. For instance, to form the number 1, a code is output to light only segments b and c.

The other set of lines in the display interface correspond to the digits of the display. These lines, labeled D_0 through D_3, correspond to digits 0, 1, 2, and 3, respectively. It is with these signals that the microcomputer tells the display in which digit the number corresponding to the code on lines a through g should be displayed.

The way in which the display is driven by the microcomputer is said to be *multiplexed.* That is, data are not permanently displayed; instead, they are output to one digit after the other in time. This scanning sequence is repeated frequently so the user cannot recognize the fact that the display is not permanently lighted.

The scanning of the digits of the display is similar to the scanning we have just described for the rows of the keyboard. A digit-drive signal is output to one digit of the display after the other in time and during each digit-drive pulse the seven-segment code for the number to be displayed in that digit is output on segment lines a through g. In fact, in most systems the digit-drive signals for the display and row-drive signals of the keyboard are supplied by the same set of outputs.

▲ 10.13 8279 PROGRAMMABLE KEYBOARD/DISPLAY CONTROLLER

Here we will introduce an LSI device, the *8279 programmable keyboard/display interface,* which can be used to implement keyboard and display interfaces similar to those described in the previous section. Use of the 8279 makes the design of a keyboard/display interface circuit quite simple. This device can drive an 8 × 8 keyboard switch array and a 16-digit, eight-segment display. Moreover, it can be configured through software to support key debouncing, two-key lockout, or *N*-key rollover modes of operation, and either left or right data entry to the display.

A block diagram of the device is shown in Fig. 10–73(a) and its pin layout in Fig. 10–73(b). From this diagram we see that there are four signal sections: the MPU interface, the key data inputs, the display data outputs, and the scan lines used by both the keyboard and display. Let us first look at the function of each of these interfaces.

The bus interface of the 8279 is similar to that found on the other peripherals we have considered up to this point. It consists of the eight data bus lines DB_0 through DB_7. These are the lines over which the MPU outputs data to the display, inputs key codes, issues commands to the controller, and reads status information from the controller. Other

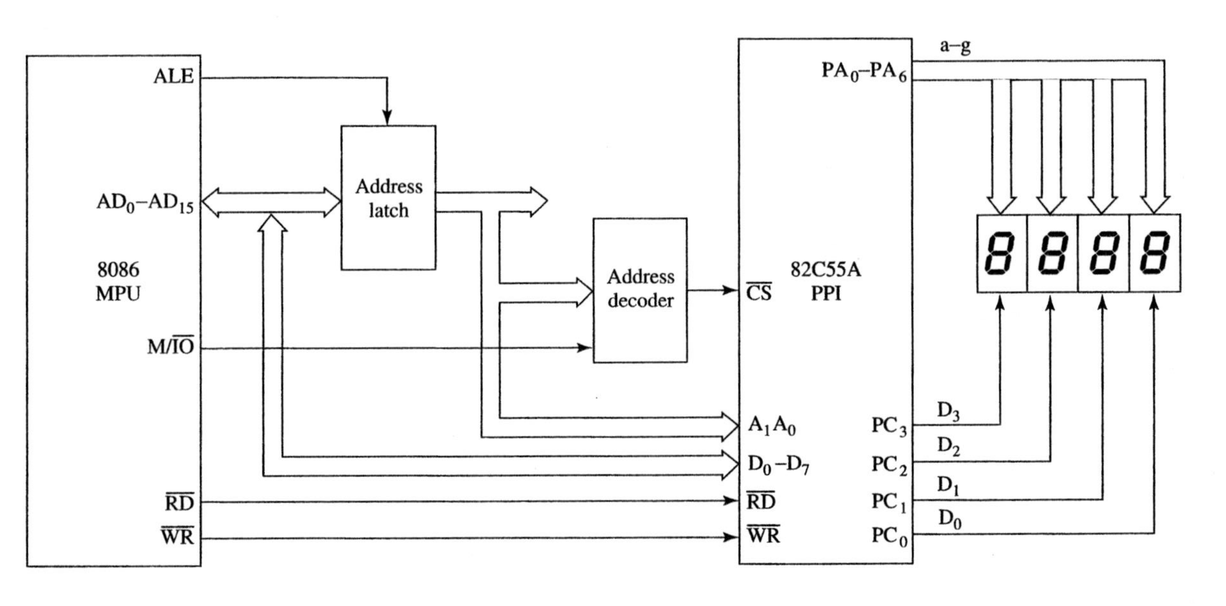

Figure 10–71 Display interfaced to a microcomputer.

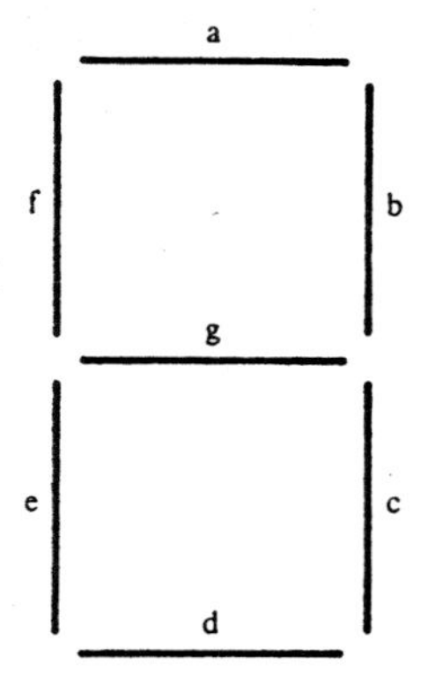

Figure 10–72 Seven-segment display labeling.

signals found at the interface are the read ($\overline{RD}$), write ($\overline{WR}$), chip-select ($\overline{CS}$), and address buffer (A_0) control signals. They are the signals that control the data bus transfers taking place between the microprocessor and 8279.

A new signal introduced with this interface is *interrupt request* (IRQ), an output that gets returned to an interrupt input of the microcomputer. This signal is provided so that the 8279 can tell the MPU it contains key codes that should be read.

The scan lines are used as row-drive signals for the keyboard and digit-drive signals for the display. There are just four of these lines, SL_0 through SL_3. However, they can be configured for two different modes of operation through software. In applications that require a small keyboard and display (four or less rows and digits), they can be used in what is known as the *decoded mode.* Scan output waveforms for this mode of operation are shown in Fig. 10–74(a). Note that a pulse to the 0 logic level is produced at one output after the other in time.

The second mode of operation, called *encoded mode,* allows use of a keyboard matrix with up to eight rows and a display with up to 16 digits. When this mode of operation is enabled through software, the binary-coded waveforms shown in Fig. 10–74(b) are output on the SL lines. These signals must be decoded with an external decoder circuit to produce the digit and column drive signals.

Even though 16 digit-drive signals are produced, only 8 row-drive signals can be used for the keyboard because the key code that is stored when a key depression has been sensed has just 3 bits allocated to identify the row. Figure 10–75 shows this kind of circuit configuration.

The key data lines include the eight return lines, RL_0 through RL_7. These lines receive inputs from the column outputs of the keyboard array. They are not tested all at once as we described earlier. Looking at the waveforms in Fig. 10–76, we see that the RL lines are examined one after the other during each 640-μs row pulse.

If logic 0 is detected at a return line, the number of the column is coded as a 3-bit binary number and combined with the 3-bit row number to make a 6-bit key code. This key code input is first debounced and then loaded into an 8×8 key code FIFO within the 8279. Once the FIFO contains a key code, the IRQ output is automatically set to logic 1. This signal can be used to tell the MPU that a keyboard input should be read from the 8279. There are two other signal inputs in this section: shift (SHIFT) and control/strobed

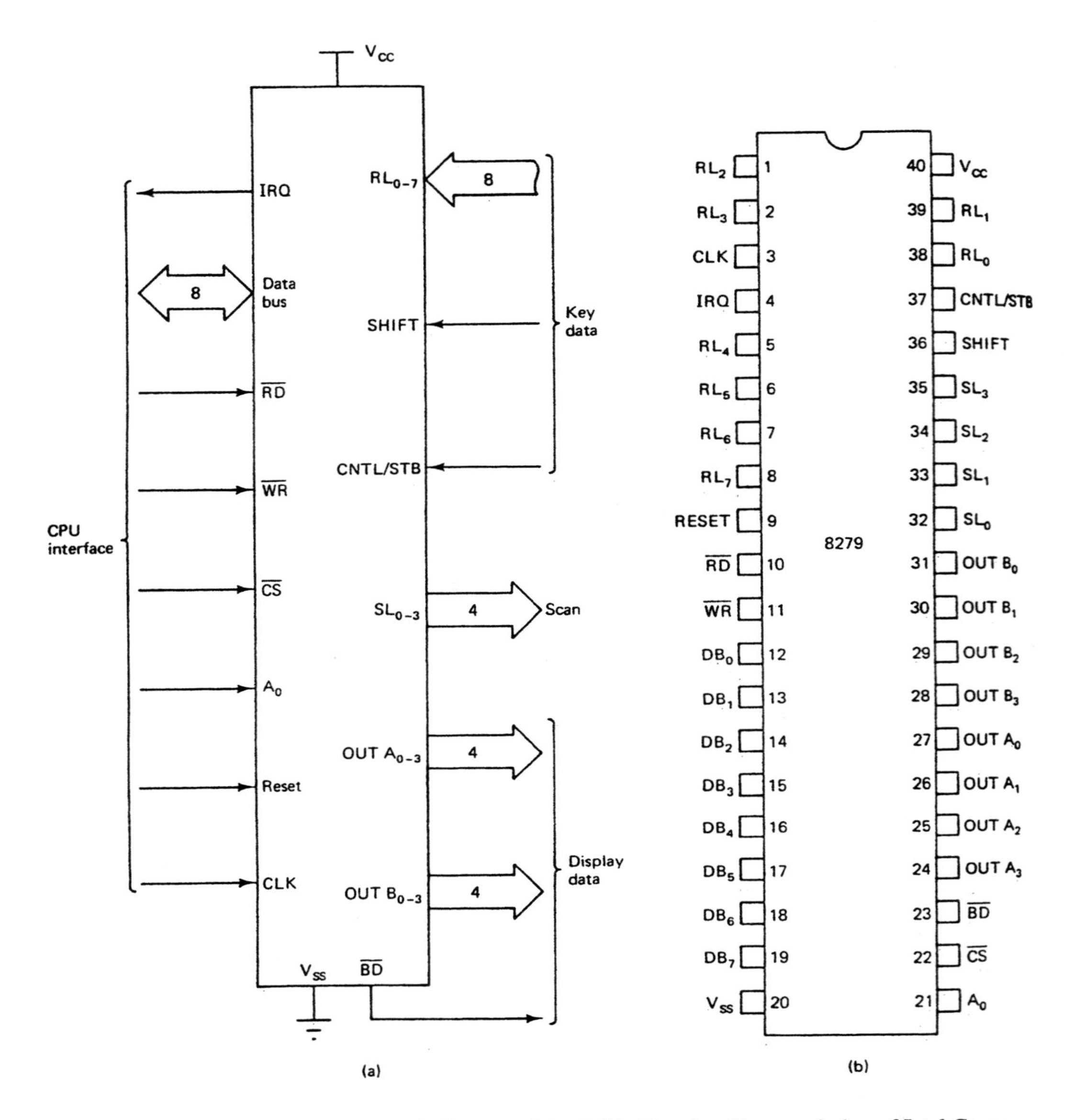

Figure 10–73 (a) Block diagram of the 8279. (Reprinted by permission of Intel Corporation. Copyright/Intel Corp. 1987) (b) Pin layout. (Reprinted by permission of Intel Corporation. Copyright/Intel Corp. 1987)

(CNTL/STB). The logic levels at these two inputs are also stored as part of the key code when a switch closure is detected. The format of the complete key code byte stored in FIFO is shown in Fig. 10–77.

A status register is provided within the 8279 that contains flags indicating the status of the *key code FIFO*. The bits of the status register and their meanings are shown in Fig. 10–78. Note that the three least significant bits, labeled NNN, identify the number of

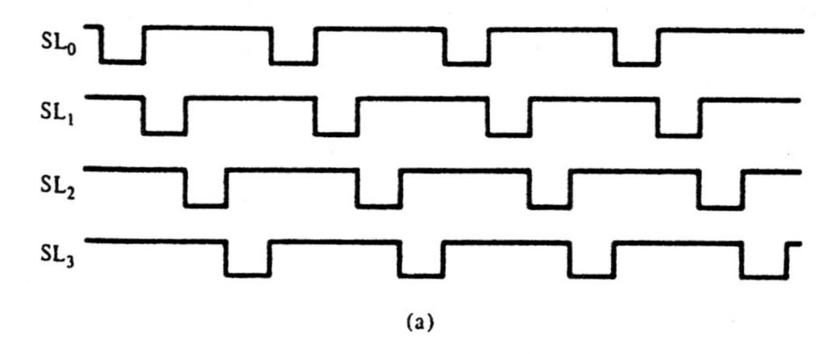

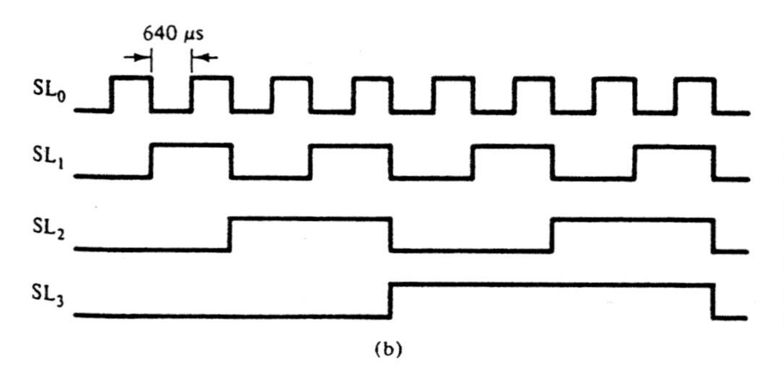

Figure 10–74 (a) Decoded-mode scan line signals. (Reprinted by permission of Intel Corporation. Copyright/Intel Corp. 1987) (b) Encoded-mode scan line signals. (Reprinted by permission of Intel Corporation. Copyright/Intel Corp. 1987)

key codes currently held in the FIFO. The next bit, F, indicates whether or not the FIFO is full. The two bits that follow it, U and O, represent two FIFO error conditions. O, which stands for overrun, indicates that an attempt was made to enter another key code into the FIFO, but it was already full. This condition could occur if the microprocessor does not respond quickly enough to the IRQ signal by reading key codes out of the FIFO. The other error condition, underrun (U), means that the microprocessor attempted to read the contents of the FIFO when it was empty. The microprocessor can read the contents of the status register under software control.

The display data lines include two 4-bit output ports, A_0 through A_3 and B_0 through B_3, that are used as display segment drive lines. Segment data that are output on these lines are held in a dedicated display RAM area within the 8279. This RAM is organized 16×8 and must be loaded with segment data by the microprocessor. Figure 10–76 shows that during each 640-μs scan time the segment data for one of the digits are output at the A and B ports.

The operation of the 8279 must be configured through software. Eight command words are provided for this purpose. These control words are loaded into the device by performing write (output) operations to the device with buffer address bit A_0 set to logic 1. Let us now look briefly at the function of each of these control words.

The first command (*command word 0*) is used to set the mode of operation for the keyboard and display. The general format of this word is shown in Fig. 10–79(a). Here we see that the three most significant bits are always reset. These 3 bits are a code by which the 8279 identifies which command is being issued by the microprocessor. The next 2 bits, labeled DD, are used to set the mode of operation for the display. The table in Fig. 10–79(b) shows the options available. After power-up reset, these bits are set to 01. From

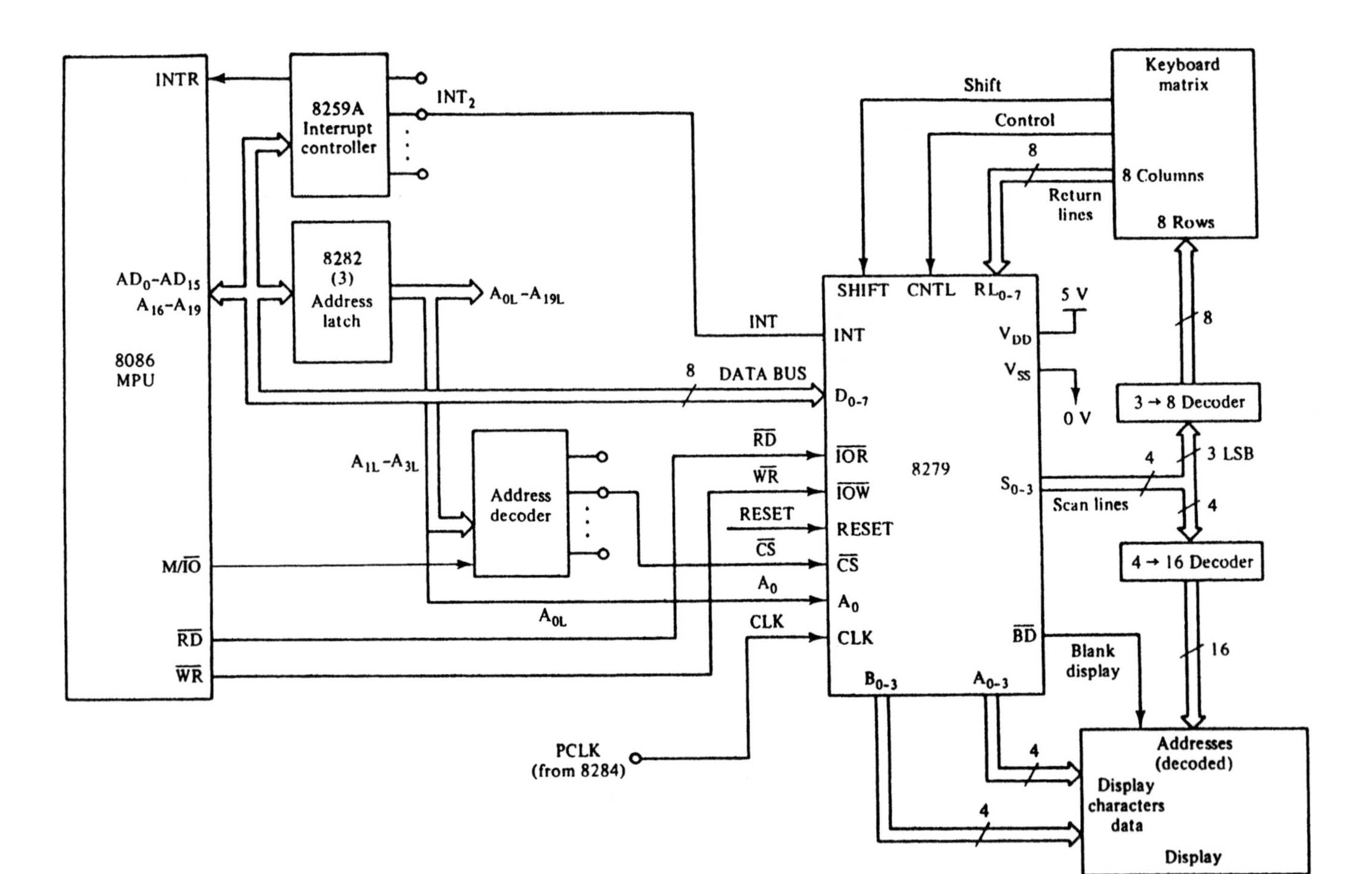

Figure 10–75 System configuration using the 8086 and 8279. (Reprinted by permission of Intel Corporation. Copyright/Intel Corp. 1987)

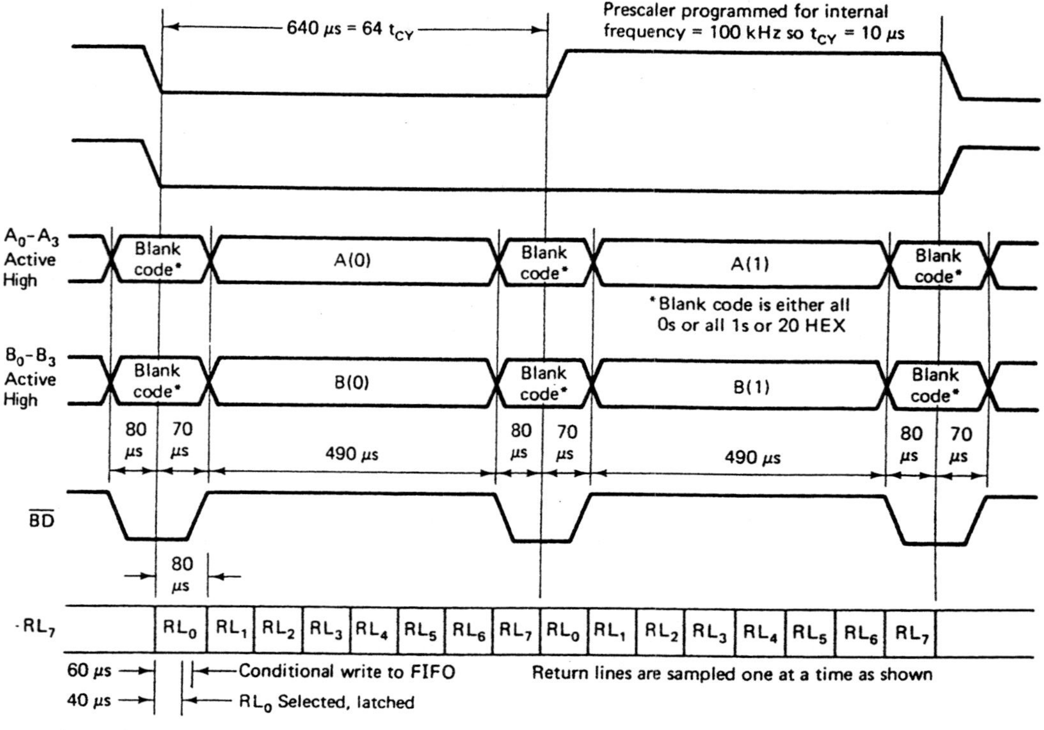

Figure 10–76 Keyboard and display signal timing. (Reprinted by permission of Intel Corporation. Copyright/Intel Corp. 1987)

the table we see that this configures the display for 16 digits with left entry. By left entry we mean that characters are entered into the display starting from the left.

The three least significant bits of the command word (KKK) set the scan mode of the keyboard and display. They are used to configure the operation of the keyboard according to the table in Fig. 10–79(c). The default code at power-up is 000 and selects encoded scan operation with two-key lockout.

EXAMPLE 10.33

What should be the value of command word 0 if the display is to be set for eight 8-segment digits with right entry and the keyboard for decoded scan with *N*-key rollover?

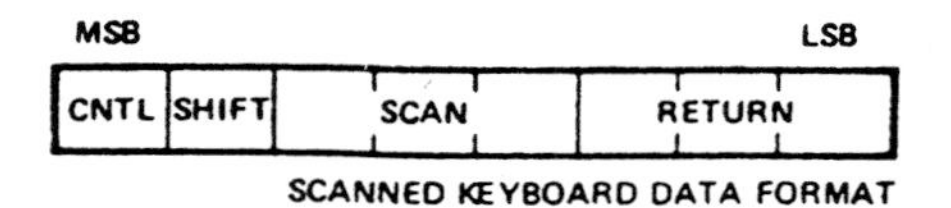

Figure 10–77 Key code byte format. (Reprinted by permission of Intel Corporation. Copyright/Intel Corp. 1987)

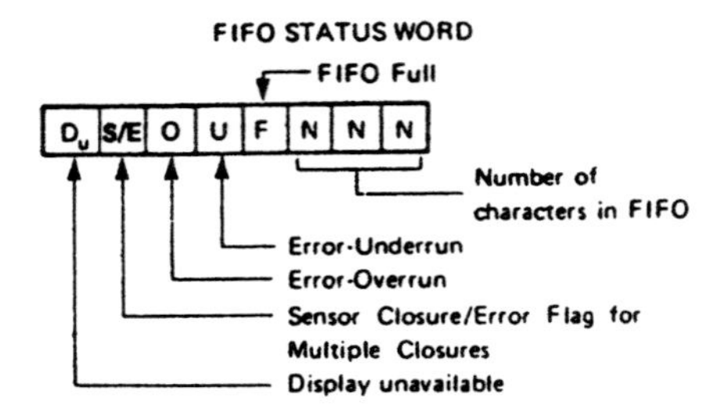

Figure 10–78 Status register. (Reprinted by permission of Intel Corporation. Copyright/Intel Corp. 1987)

MSB							LSB
0	0	0	D	D	K	K	K

(a)

D	D	Display operation
0	0	8 8-bit character display – Left entry
0	1	16 8-bit character display – Left entry
1	0	8 8-bit character display – Right entry
1	1	16 8-bit character display – Right entry

(b)

K	K	K	Keyboard operation
0	0	0	Encoded Scan Keyboard – 2-Key Lockout
0	0	1	Decoded Scan Keyboard – 2-Key Lockout
0	1	0	Encoded Scan Keyboard – N-Key Rollover
0	1	1	Decoded Scan Keyboard – N-Key Rollover
1	0	0	Encoded Scan Sensor Matrix
1	0	1	Decoded Scan Sensor Matrix
1	1	0	Strobed Input, Encoded Display Scan
1	1	1	Strobed Input, Decoded Display Scan

(c)

Figure 10–79 (a) Command word 0 format. (Reprinted by permission of Intel Corporation. Copyright/Intel Corp. 1987) (b) Display mode select codes. (Reprinted by permission of Intel Corporation. Copyright/Intel Corp. 1987) (c) Keyboard select codes. (Reprinted by permission of Intel Corporation. Copyright/Intel Corp. 1987)

Figure 10–80 Command word 1 format. (Reprinted by permission of Intel Corporation. Copyright/Intel Corp. 1987)

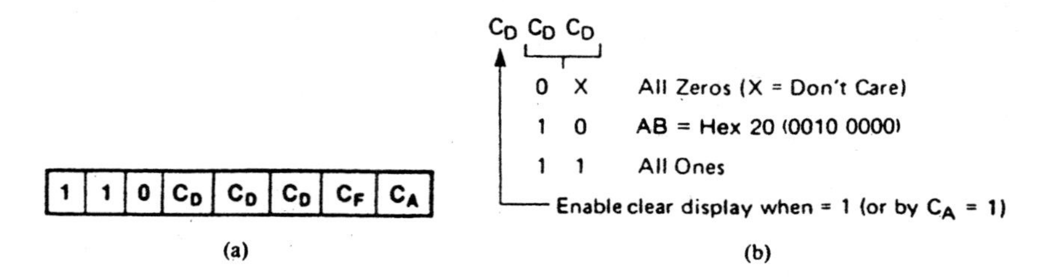

Figure 10–81 (a) Command word 6 format. (Reprinted by permission of Intel Corporation. Copyright/Intel Corp. 1987) (b) C_D coding. (Reprinted by permission of Intel Corporation. Copyright/Intel Corp. 1987)

Solution

The three MSBs of the command word are always 0. The next 2 bits, DD, must be set to 10 for eight 8-segment digits with right entry. Finally, the three LSBs are set to 011 for decoded keyboard scan with *N*-key rollover. This gives

$$\text{Command word 0} = 000\text{DDKKK}$$
$$= 00010011_2$$
$$= 13_{16}$$

Command word 1 is used to set the frequency of operation of the 8279. It is designed to run at 100 kHz; however, in most applications a much higher frequency signal is available to supply its CLK input. For this reason, a *5-bit programmable prescaler* is provided within the 8279 to divide down the input frequency. The format of this command word is shown in Fig. 10–80 (p. 568).

For instance, in a 5-MHz 8086-based microcomputer system, the PCLK output of the 8284 clock driver IC can be used for the 8279's clock input. PCLK is one-half the frequency of the oscillator, or 2.5 MHz. In this case the divider P must be

$$P = 2.5 \text{ MHz}/100 \text{ kHz} = 25$$

Twenty-five expressed as a 5-bit binary number is

$$P = 11001_2$$

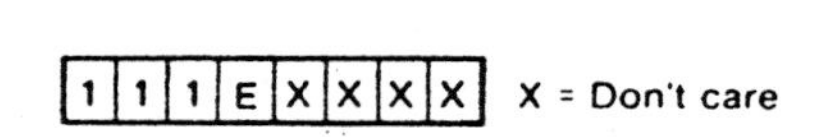

Figure 10–82 Command word 7 format. (Reprinted by permission of Intel Corporation. Copyright/Intel Corp. 1987)

Figure 10–83 Command word 2 format. (Reprinted by permission of Intel Corporation. Copyright/Intel Corp. 1987)

Therefore, the value of command word 1 written to the 8279 is

$$\text{Command word 1} = 001\text{PPPPP}_2 = 00111001_2$$
$$= 39_{16}$$

Let us skip now to *command word 6* because it is also used for initialization of the 8279. It is used to initialize the complete display memory, the FIFO status, and the interrupt-request output line. The format of this word is given in Fig. 10–81(a) (p. 569). The three C_D bits are used to control initialization of the display RAM. Figure 10–81(b) shows what values can be used in these locations. The C_F bit is provided for clearing the FIFO status and resetting the IRQ line. To perform the reset operation, a 1 must be written to C_F. The last bit, clear all (C_A), can be used to initiate both the C_D and C_F functions.

EXAMPLE 10.34

What clear operations are performed if the value of command word 6 written to the 8279 is $D2_{16}$?

Solution

First, we express the command word in binary form. This gives

$$\text{Command word 6} = D2_{16} = 11010010_2$$

Note that the three C_D bits are 100. This combination causes display memory to be cleared. The C_F bit is also set, and this causes the FIFO status and IRQ output to be reset.

As Fig. 10–82 (p. 569) shows, only one bit of *command word 7* is functional. This bit is labeled E and is an enable signal for what is called the special-error mode. When this mode is enabled and the keyboard has *N*-key rollover selected, a multiple-key depression causes the S/E flag of the FIFO status register to be set. This flag can be read by the microprocessor through software.

1	0	0	AI	A	A	A	A

Figure 10–84 Command word 4 format. (Reprinted by permission of Intel Corporation. Copyright/Intel Corp. 1987)

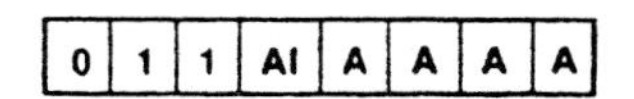

Figure 10–85 Command word 3 format. (Reprinted by permission of Intel Corporation. Copyright/Intel Corp. 1987)

The rest of the command codes are related to accessing the key code FIFO and display RAM. The key code FIFO is read only. However, before the microprocessor can access it, a read FIFO command must be issued to the 8279. This is *command word 2* and has the format shown in Fig. 10–83. When the 8279 is set up for keyboard scanning, the AI and AAA bits are don't-care states. Then all that needs to be done is to issue the command $01000000_2 = 40_{16}$ to the 8279 and initiate read (input) cycles to the address of the 8279. For each read bus cycle, the key code at the top of the FIFO is read into the MPU.

The display RAM can be either read from or written into by the MPU. Just like for the FIFO, a command must be sent to the 8279 before reading or writing can be initiated. For instance, when the microprocessor wants to send new data to the display, it must first issue *command word 4*. This command has the format shown in Fig. 10–84. Here the AAAA in the four least significant bit locations is the address of the first location to be accessed. For instance, if 0000_2 is put into these bits of the command, the first write operation will be to the first location in display RAM. Moreover, if the AI bit is set in the command, autoincrement addressing is enabled. In this way, the display RAM address pointer is automatically incremented after the write operation is complete and a write cycle can be initiated to address 0001_2 of display RAM without first issuing another write command.

The MPU can also read the contents of the display RAM in a similar way. This requires that *command word 3* be issued to the 8279. Figure 10–85 shows the format of this read display RAM command.

REVIEW PROBLEMS

Section 10.1

1. Give three examples of the special-purpose input/output interfaces of a microcomputer.
2. List three core input/output interfaces commonly used in microcomputer systems.

Section 10.2

3. What is the address of port 7 in the circuit shown in Fig. 10–1(a)?
4. What are the inputs of the I/O address decoder in Fig. 10–1(a) when the I/O address on the bus is $800A_{16}$? Which output is active? Which output port does this enable?
5. What operation does the instruction sequence that follows perform for the circuit in Fig. 10–1(a)?

```
MOV  AL, 0FFH
MOV  DX, 8004H
OUT  DX, AL
```

6. Write a sequence of instructions to output the word contents of the memory location called DATA to output ports 0 and 1 in the circuit shown in Fig. 10–1(a).

Section 10.3

7. Which input port in the circuit shown in Fig. 10–3 is selected for operation if the I/O address output on the bus is 8008_{16} ?
8. What operation is performed to the circuit in Fig. 10–3 when the instruction sequence that follows is executed?

```
MOV  DX, 8000H
IN   AL, DX
AND  AL, 0FH
MOV  [LOW_NIBBLE], AL
```

9. Write a sequence of instructions to read in the contents of ports 1 and 2 in the circuit shown in Fig. 10–3 and save them at consecutive memory addresses $A0000_{16}$ and $A0001_{16}$ in memory.
10. Write an instruction sequence that polls input I_{63} in the circuit shown in Fig. 10–3, checking for it to switch to logic 0.

Section 10.4

11. Name a method that can be used to synchronize the input or output of information to a peripheral device.
12. List the control signals in the parallel printer interface circuit shown in Fig. 10–6(a). Identify whether they are an input or output of the printer and briefly describe their functions.
13. What type of device provides the data lines for the printer interface circuit shown in Fig. 10–6(d)?
14. Give an overview of what happens in the circuit shown in Fig. 10–6(d) when a write bus cycle of byte-wide data is performed to I/O address 8000_{16}.
15. Show what push and pop instructions are needed in the program written in Example 10.6 to preserve the contents of registers used by it so that it can be used as a subroutine.

Section 10.5

16. What kind of input/output interface does a PPI implement?
17. How many I/O lines are available on the 82C55A?
18. What are the signal names of the I/O port lines of the 82C55A?
19. Describe the mode 0, mode 1, and mode 2 I/O operations of the 82C55A.
20. What function do the lines of port B of the 82C55A serve when port A is configured for mode 2 operation?
21. How is an 82C55A configured if its control register contains 9BH?
22. If the value $A4_{16}$ is written to the control register of an 82C55A, what is the mode and I/O configuration of port A? Port B?

23. If ports A, B, and C of an 82C55A are to be configured for mode 0 operation, where the A and B ports are inputs and C is an output port, what is the control word?
24. What value must be written to the control register of the 82C55A to configure the device such that both port A and port B are configured for mode 1 input operation?
25. If the control register of the 82C55A in problem 23 is at I/O address 1000_{16}, write an instruction sequence that will load the control word.
26. Assume that the control register of an 82C55A resides at memory address 00100_{16}. Write an instruction sequence to load it with the control word formed in problem 23.
27. What control word must be written to the control register of an 82C55A shown in Fig. 10–15(a) to enable the $INTR_B$ output? $INTE_B$ corresponds to bit PC_4 of port C.
28. If the value 03_{16} is written to the control register of an 82C55A set for mode 2 operation, what bit at port C is affected by the bit set/reset operation? Is it set to 1 or cleared to 0?
29. Assume that the control register of an 82C55A is at I/O address 0100_{16}. Write an instruction sequence to load it with the bit set/reset value given in problem 28.

Section 10.6

30. If I/O address $003E_{16}$ is applied to the circuit in Fig. 10–21 during a byte-write cycle and the data output on the bus is 98_{16}, which 82C55A is being accessed? Are data being written into port A, port B, port C, or the control register of this device?
31. If the instruction that follows is executed, what operation to the I/O interface circuit in Fig. 10–21 is performed?

```
IN AL, 08H
```

32. What are the addresses of the A, B, and C ports of PPI 2 in the circuit shown in Fig. 10–22?
33. Assume that PPI 2 in Fig. 10–22 is configured as defined in problem 23. Write a program that will input the data at ports A and B, add these values together, and output the sum to port C.

Section 10.7

34. Distinguish between memory-mapped I/O and isolated I/O.
35. What address inputs must be applied to the circuit in Fig. 10–23 in order to access port B of device 4? Assuming that all unused bits are 0, what would be the memory address?
36. Write an instruction that will load the control register of the port identified in problem 35 with the value 98_{16}.
37. Repeat problem 33 for the circuit in Fig. 10–24.

Section 10.8

38. What are the inputs and outputs of counter 2 of an 82C54?
39. Write a control word for counter 1 that selects the following options: load least significant byte only, mode 5 of operation, and binary counting.

40. What are the logic levels of inputs $\overline{CS}$, $\overline{RD}$, $\overline{WR}$, A_1, and A_0 when the byte in problem 39 is written to an 82C54?
41. Write an instruction sequence that loads the control word in problem 39 into an 82C54, starting at address 01000_{16} of the memory address space. A_1A_0 of the microprocessor are directly connected to A_1A_0 of the 82C54.
42. Write an instruction sequence that loads the value 12_{16} into the least significant byte of the count register for counter 2 of an 82C54, starting at memory address 01000_{16}. A_1A_0 of the microprocessor are directly connected to A_1A_0 of the 82C54.
43. Repeat Example 10.19 for the 82C54 located at memory address 01000_{16}, but this time just read the least significant byte of the counter. A_1A_0 of the microprocessor are directly connected to A_1A_0 of the 82C54.
44. What is the maximum time delay that can be generated with the timer in Fig. 10–35? What would be the maximum time delay if the clock frequency were increased to 2 MHz? Assume that it is configured for binary counting.
45. What is the resolution of pulses generated with the 82C54 in Fig. 10–35? What will be the resolution if the clock frequency is increased to 2 MHz?
46. Find the pulse width of the one-shot in Fig. 10–36 if the counter is loaded with the value 1000_{16}. Assume that the counter is configured for binary count operation.
47. What count must be loaded into the square-wave generator in Fig. 10–38 in order to produce a 25-KHz output?
48. If the counter in Fig. 10–39 is loaded with the value 120_{16}, how long of a delay occurs before the strobe pulse is output?

Section 10.9

49. Are signal lines $\overline{MEMR}$ and $\overline{MEMW}$ of the 82C37A used in the microprocessor interface?
50. Summarize the 82C37A's DMA request/acknowledge handshake sequence.
51. What is the total number of user-accessible registers in the 82C37A?
52. Write an instruction sequence that reads the value of the address from the current address register for channel 0 into the AX register. Assume that the 82C37A has the base address 10H.
53. Assuming that an 82C37A is located at I/O address 1000H, write an instruction sequence to perform a master clear operation.
54. Write an instruction sequence that writes the command word 00_{16} into the command register of an 82C37A located at address 2000H in the I/O address space.
55. Write an instruction sequence that loads the mode register for channel 2 with the mode byte obtained in Example 10.26. Assume that the 82C37A is located at I/O address F0H.
56. What must be output to the mask register in order to disable all of the DRQ inputs?
57. Write an instruction sequence that reads the contents of the status register into the AL register. Assume the 82C37A is located at I/O address 5000H.

Section 10.10

58. Name a signal line that distinguishes an asynchronous communication interface from that of a synchronous communication interface.

59. Define a simplex, a half-duplex, and a full-duplex communication link.

60. For an RS-232C interface, what voltage range defines a mark at the transmit end of a serial communication line?

Section 10.11

61. To write a byte of data to the 8251A, what logic levels must the microprocessor apply to control inputs $C/\overline{D}$, $\overline{RD}$, $\overline{WR}$, and $\overline{CS}$?

62. The mode-control register of an 8251A contains 11111111_2. What are the asynchronous character length, type of parity, and the number of stop bits?

63. Write an instruction sequence to load the control word obtained in problem 62 into a memory-mapped 8251A with the control register located at address MODE.

64. Describe the difference between a mode instruction and a command instruction used in 8251A initialization.

Section 10.12

65. Referring to Fig. 10–70, what is the maximum number of keys that can be supported using all 24 I/O lines of an appropriately configured 82C55A?

66. In the circuit shown in Fig. 10–70, what row and column code would identify the 9 key?

67. What codes would need to be output on the digital and segment lines of the circuit in Fig. 10–71 to display the number 7 in digit 1?

Section 10.13

68. Specify the mode of operation for the keyboard and display when an 8279 is configured with command word 0 equal to $3F_{16}$.

69. Determine the clock frequency applied to the input of an 8279 if it needs command word 1 equal to $1E_{16}$ to operate.

70. Summarize the function of each command word of the 8279.

Interrupt Interface of the 8088 and 8086 Microprocessors

▲ INTRODUCTION

In Chapters 8 and 10 we covered the input/output interface of the 8088- and 8086-based microcomputer systems. Here we continue with a special input interface, the *interrupt interface.* The topics presented in this chapter are as follows:

11.1 Interrupt Mechanism, Types, and Priority
11.2 Interrupt Vector Table
11.3 Interrupt Instructions
11.4 Enabling/Disabling of Interrupts
11.5 External Hardware-Interrupt Interface Signals
11.6 External Hardware-Interrupt Sequence
11.7 82C59A Programmable Interrupt Controller
11.8 Interrupt Interface Circuits Using the 82C59A
11.9 Software Interrupts
11.10 Nonmaskable Interrupt
11.11 Reset
11.12 Internal Interrupt Functions

▲ 11.1 INTERRUPT MECHANISM, TYPES, AND PRIORITY

Interrupts provide a mechanism for quickly changing program environment. Transfer of program control is initiated by the occurrence of either an event internal to the microprocessor or an event in its external hardware. For instance, when an interrupt signal occurs in external hardware indicating that an external device, such as a printer, requires service, the MPU must suspend what it is doing in the main part of the program and pass control to a special routine that performs the function required by the device. The section of program to which control is passed is called the *interrupt-service routine.* In the case of our example of a printer, the routine is usually called the *printer driver,* which is the piece of software that when executed drives the printer output interface.

As Fig. 11–1 shows, interrupts supply a well-defined context-switching mechanism for changing program environments. Here we see that interrupt 32 occurs as instruction N of the program is being executed. When the MPU terminates execution of the main program in response to interrupt 32, it first saves information that identifies the instruction following the one where the interrupt occurred, instruction $N + 1$, and then

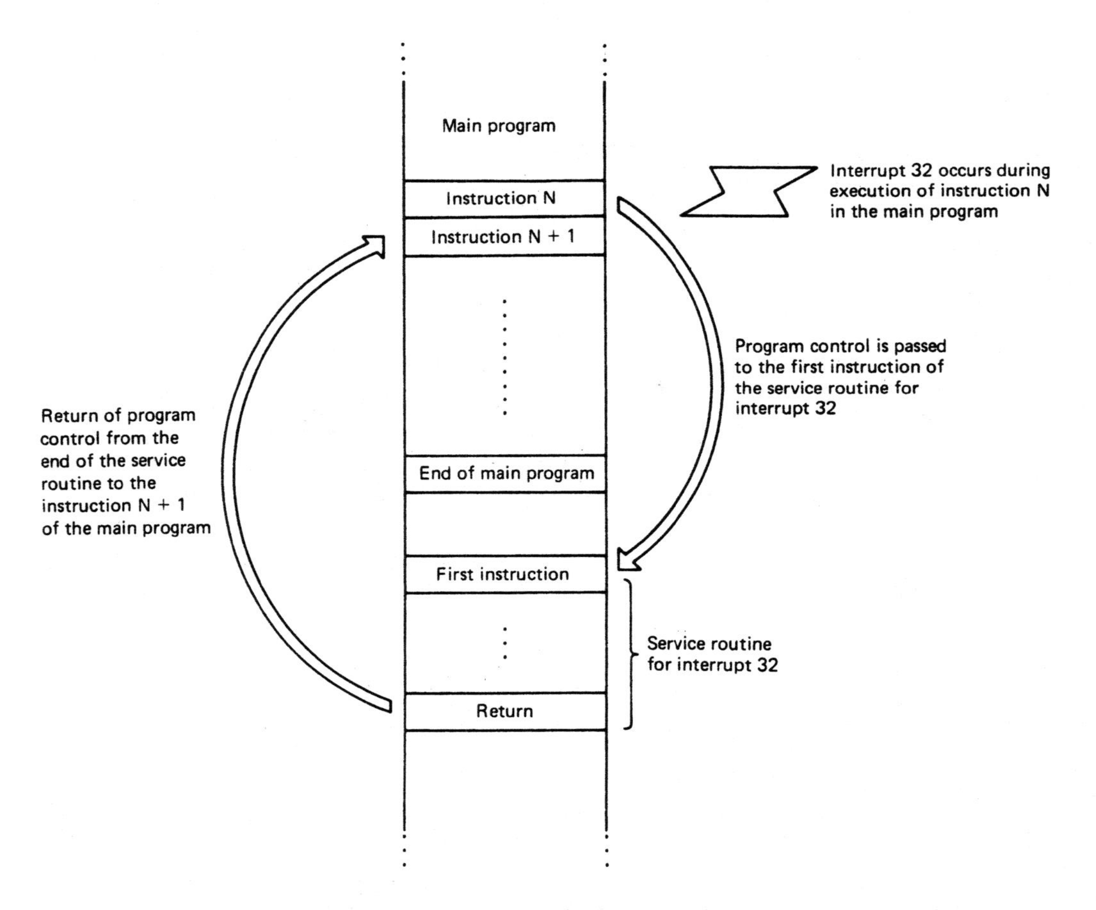

Figure 11–1 Interrupt program context switching mechanism.

picks up execution with the first instruction in the service routine. After this routine has run to completion, program control is returned to the point where the MPU originally left the main program, instruction $N + 1$, and then execution resumes.

The 8088 and 8086 microcomputers are capable of implementing any combination of up to 256 interrupts. As Fig. 11–2 shows, they are divided into five groups: *external hardware interrupts, nonmaskable interrupt, software interrupts, internal interrupts,* and *reset.* The user defines the function of the external hardware, software, and nonmaskable interrupts. For instance, hardware interrupts are often assigned to devices such as the keyboard, printer, and timers. On the other hand, the functions of the internal interrupts and reset are not user defined. They perform dedicated system functions.

Hardware, software, and internal interrupts are serviced on a *priority* basis. Priority is achieved in two ways. First, the interrupt-processing sequence implemented in the 8088/8086 tests for the occurrence of the various groups based on the hierarchy shown in Fig. 11–2. Thus, we see that internal interrupts are the highest-priority group, and the external hardware interrupts are the lowest-priority group.

Second, each of the interrupts is given a different priority level by assigning it a *type number.* Type 0 identifies the highest-priority interrupt, and type 255 identifies the lowest-priority interrupt. Actually, a few of the type numbers are not available for use with software or hardware interrupts. This is because they are reserved for special interrupt functions of the 8088/8086, such as internal interrupts. For instance, within the internal interrupt group, the interrupt known as *divide error* is assigned to type number 0. Therefore, it has the highest priority of the internal interrupts. Another internal interrupt, called *overflow,* is assigned the type number 4. Overflow is the lowest-priority internal interrupt.

The importance of priority lies in the fact that, if an interrupt-service routine has been initiated to perform a function assigned to a specific priority level, only devices with higher priority are allowed to interrupt the active service routine. Lower-priority devices will have to wait until the current routine is completed before their request for service can be acknowledged. For hardware interrupts, this priority scheme is implemented in external hardware. For this reason, the user normally assigns tasks that must not be interrupted frequently to higher-priority levels and those that can be interrupted to lower-priority levels.

An example of a high-priority service routine that should not be interrupted is that for a power failure. Once initiated, this routine should be quickly run to completion to assure that the microcomputer goes through an orderly power-down. A keyboard should also be assigned to a high-priority interrupt. This will assure that the keyboard buffer does not get full and lock out additional entries. On the other hand, devices such as the floppy disk or hard disk controller are typically assigned to a lower priority level.

We just pointed out that once an interrupt service routine is initiated, it could be interrupted only by a function that corresponds to a higher-priority level. For example, if a

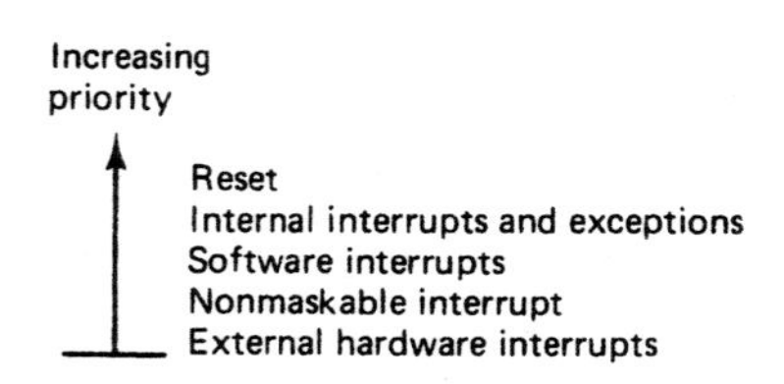

Figure 11–2 Types of interrupts and their priority.

type 50 external hardware interrupt is in progress, it can be interrupted by any software interrupt, the nonmaskable interrupt, all internal interrupts, or any external hardware interrupt with type number less than 50. That is, external hardware interrupts with priority levels equal to 50 or greater are *masked out.*

▲ 11.2 INTERRUPT VECTOR TABLE

An *address pointer table* is used to link the interrupt type numbers to the locations of their service routines in the program-storage memory. Figure 11–3 shows a map of the pointer table in the memory of the 8088 or 8086 microcomputer. Looking at this table, we see that it contains *256 address pointers (vectors),* which are identified as *vector 0* through *vector 255.* That is, one pointer corresponds to each of the interrupt types 0 through 255. These address pointers identify the starting location of their service routines in program memory. The contents of this table may be either held as firmware in EPROMs or loaded into RAM as part of the system initialization routine.

Note in Fig. 11–3 that the pointer table is located at the low-address end of the memory address space. It starts at address 00000_{16} and ends at $003FE_{16}$. This represents the first 1Kbytes of memory.

Each of the 256 pointers requires two words (4 bytes) of memory and is always stored at an even-address boundary. The higher-addressed word of the two-word vector is called the *base address.* It identifies the program memory segment in which the service routine resides.

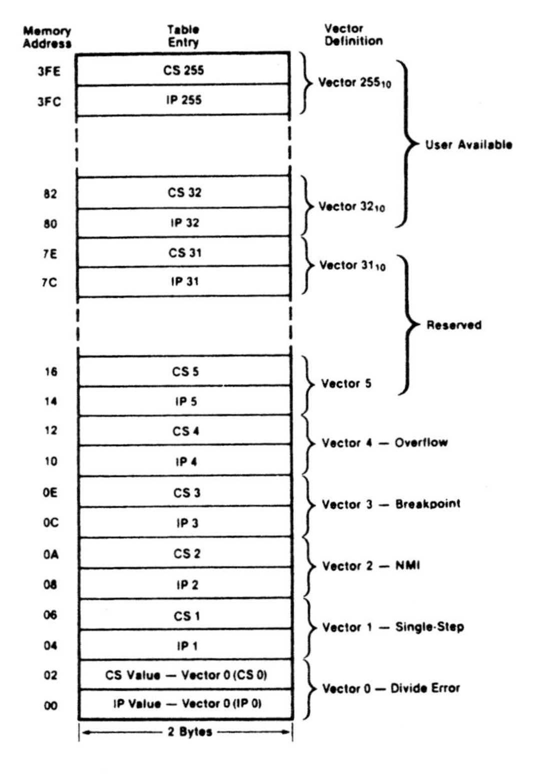

Figure 11–3 Interrupt vector table of the 8088/8086. (Reprinted by permission of Intel Corporation. Copyright/Intel Corp. 1979)

For this reason, it is loaded into the code segment (CS) register within the MPU. The lower-addressed word of the vector is the *offset* of the first instruction of the service routine from the beginning of the code segment defined by the base address loaded into CS. This offset is loaded into the instruction pointer (IP) register. For example, the offset and base address for type number 255, IP_{255} and CS_{255}, are stored at word addresses $003FC_{16}$ and $003FE_{16}$, respectively. When loaded into the MPU, it points to the instruction at $CS_{255}:IP_{255}$.

Looking more closely at the table in Fig. 11–3, we find that the first 31 pointers either have dedicated functions or are reserved. For instance, pointers 0, 1, 3, and 4 are used by the 8088's and 8086's internal interrupts: *divide error, single step, breakpoint,* and *overflow.* Pointer 2 is used to identify the starting location of the nonmaskable interrupt's service routine. The next 27 pointers, 5 through 31, represent a reserved portion of the pointer table and should not be used. The remainder of the table, the 224 pointers in the address range 00080_{16} through $003FF_{16}$, is available to the user for storage of software or hardware interrupt vectors. These pointers correspond to type numbers 32 through 255. In the case of external hardware interrupts, each type number (priority level) is associated with an interrupt input in the external interrupt interface circuitry.

EXAMPLE 11.1

At what address are CS_{50} and IP_{50} stored in memory?

Solution

Each vector requires four consecutive bytes of memory for storage. Therefore, its address can be found by multiplying the type number by 4. Since CS_{50} and IP_{50} represent the words of the type 50 interrupt pointer, we get

$$\text{Address} = 4 \times 50 = 200$$

Converting to binary form gives

$$\text{Address} = 11001000_2$$

and expressing it as a hexadecimal number results in

$$\text{Address} = C8_{16}$$

Therefore, IP_{50} is stored at $000C8_{16}$ and CS_{50} at $000CA_{16}$.

▲ 11.3 INTERRUPT INSTRUCTIONS

A number of instructions are provided in the instruction set of the 8088 and 8086 microprocessors for use with interrupt processing. Figure 11–4 lists these instructions, with brief descriptions of their functions.

For instance, the first two instructions, STI and CLI, permit manipulation of the interrupt flag through software. STI stands for *set interrupt enable flag.* Execution of this

Mnemonic	Meaning	Format	Operation	Flags Affected
CLI	Clear interrupt flag	CLI	0 → (IF)	IF
STI	Set interrupt flag	STI	1 → (IF)	IF
INT n	Type n software interrupt	INT n	(Flags) → ((SP) − 2) 0 → TF,IF (CS) → ((SP) − 4) (2 + 4 · n) → (CS) (IP) → ((SP) − 6) (4 · n) → (IP)	TF, IF
IRET	Interrupt return	IRET	((SP)) → (IP) ((SP) + 2) → (CS) ((SP) + 4) → (Flags) (SP) + 6 → (SP)	All
INTO	Interrupt on overflow	INTO	INT 4 steps	TF, IF
HLT	Halt	HLT	Wait for an external interrupt or reset to occur	None
WAIT	Wait	WAIT	Wait for $\overline{\text{TEST}}$ input to go active	None

Figure 11–4 Interrupt instructions.

instruction enables the external interrupt request (INTR) input for operation—that is, it sets interrupt flag (IF). On the other hand, execution of CLI (*clear interrupt enable flag*) disables the external interrupt input by resetting IF.

The next instruction listed in Fig. 11–4 is the *software-interrupt* instruction INT *n*. It is used to initiate a vectored call of a subroutine. Executing the instruction causes program control to be transferred to the subroutine pointed to by the vector for the number *n* specified in the instruction.

The operation outlined in Fig. 11–4 describes the effect of executing the INT instruction. For example, execution of the instruction INT 50 initiates execution of a subroutine whose starting point is identified by vector 50 in the pointer table of Fig. 11–3. First, the MPU saves the old flags on the stack, clears TF and IF, and saves the old program context, CS and IP, on the stack. Then it reads the values of IP_{50} and CS_{50} from addresses $000C8_{16}$ and $000CA_{16}$, respectively, in memory, loads them into the IP and CS registers, calculates the physical address CS_{50}:IP_{50}, and starts to fetch instruction from this new location in program memory.

An *interrupt-return* (IRET) instruction must be included at the end of each interrupt-service routine. It is required to pass control back to the point in the program where execution was terminated due to the occurrence of the interrupt. As shown in Fig. 11–4, when executed, IRET causes the old values of IP, CS, and flags to be popped from the stack back into the internal registers of the MPU. This restores the original program environment.

INTO is the *interrupt-on-overflow* instruction. This instruction must be included after arithmetic instructions that can result in an overflow condition, such as divide. It tests the overflow flag, and if the flag is found to be set, a type 4 internal interrupt is initiated. This condition causes program control to be passed to an overflow service routine located at the starting address identified by the vector IP_4 at 00010_{16} and CS_4 at 00012_{16} of the pointer table in Fig. 11–3.

The last two instructions associated with the interrupt interface are *halt* (HLT) and *wait* (WAIT). They produce similar responses by the 8088/8086 and permit the operation of the MPU to be synchronized to an event in external hardware. For instance, when HLT is executed, the MPU suspends operation and enters the idle state. It no longer executes instructions; instead, it remains idle waiting for the occurrence of an external hardware interrupt or reset interrupt. With the occurrence of either of these events, the MPU resumes execution with the corresponding service routine.

If the WAIT instruction is used instead of the HLT instruction, the MPU checks the logic level of the $\overline{\text{TEST}}$ input prior to going into the idle state. Only if $\overline{\text{TEST}}$ is at logic 1 will the MPU go into the idle state. While in the idle state, the MPU continues to check the logic level at $\overline{\text{TEST}}$, looking for its transition to the 0 logic level. As $\overline{\text{TEST}}$ switches to 0, execution resumes with the next sequential instruction in the program.

▲ 11.4 ENABLING/DISABLING OF INTERRUPTS

An *interrupt-enable flag* bit is provided within the 8088 and 8086 MPUs. Earlier we found that it is identified as IF. It affects only the external hardware-interrupt interface, not software interrupts, the nonmaskable interrupt, or internal interrupts. The ability to initiate an external hardware interrupt at the INTR input is enabled by setting IF or masked out by resetting it. Executing the STI instruction or the CLI instruction, respectively, does this through software.

During the initiation sequence of a service routine for an external hardware interrupt, the MPU automatically clears IF. This masks out the occurrence of any additional external hardware interrupts. In some applications, it may be necessary to permit other interrupts to interrupt the active service routine. If this is the case, the interrupt flag bit can be set with an STI instruction in the service routine to reenable the INTR input. Otherwise, the external hardware-interrupt interface is reenabled by the IRET instruction at the end of the service routine.

▲ 11.5 EXTERNAL HARDWARE-INTERRUPT INTERFACE SIGNALS

Up to this point, we have introduced the types of interrupts supported by the 8088/8086, its pointer table, interrupt instructions, and enabling/disabling interrupts. Let us now look at the signals of the *external hardware interrupt interface* of the 8088 and 8086 microcomputer systems.

Minimum-Mode Interrupt Interface

We will begin with an 8088 microcomputer configured for the minimum mode. The interrupt interface for this system is illustrated in Fig. 11–5(a). Here we see that it includes the multiplexed address/data bus and dedicated interrupt signal lines INTR and $\overline{\text{INTA}}$. We also see that external circuitry is required to interface the interrupt inputs,

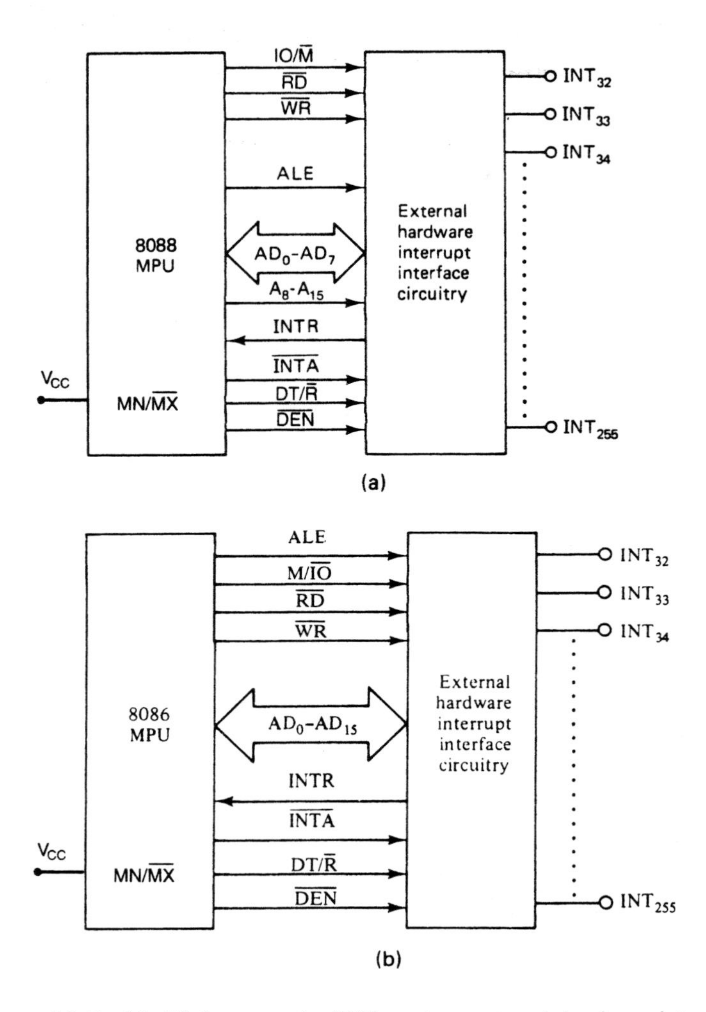

Figure 11–5 (a) Minimum-mode 8088 system external hardware-interrupt interface. (b) Minimum-mode 8086 system external hardware-interrupt interface.

INT_{32} through INT_{255}, to the 8088's interrupt interface. This interface circuitry must identify which of the pending active interrupts has the highest priority and then pass its type number to the microprocessor.

In this circuit we see that the key interrupt interface signals are *interrupt request* (INTR) and *interrupt acknowledge* ($\overline{INTA}$). The input at the INTR line signals the 8088 that an external device is requesting service. The 8088 samples this input during the last clock period of each instruction execution cycle. Logic 1 represents an active interrupt request. INTR is *level triggered;* therefore, its active 1 level must be maintained until tested by the 8088. If it is not maintained, the request for service may not be recognized. Moreover, the logic 1 at INTR must be removed before the service routine runs to completion; otherwise, the same interrupt may get acknowledged a second time.

When an interrupt request has been recognized by the 8088, it signals this fact to external circuitry. It does this with pulses to logic 0 at its $\overline{INTA}$ output. Actually, there are two pulses produced at $\overline{INTA}$ during the *interrupt acknowledge bus cycle.* The first pulse signals external circuitry that the interrupt request has been acknowledged and to prepare to send its type number to the 8088. The second pulse tells the external circuitry to put the type number on the data bus.

Note that the lower eight lines of the address/data bus, AD_0 through AD_7, are also part of the interrupt interface. During the second cycle in the interrupt acknowledge bus cycle, external circuitry must put an 8-bit type number on bus lines AD_0 through AD_7. The 8088 reads this number off the bus to identify which external device is requesting service. It uses the type number to generate the address of the interrupt's vector in the pointer table and to read the new values of CS and IP into the corresponding internal registers. CS and IP values from the interrupt vector table are transferred to the 8088 over the data bus. Before loading CS and IP with new values, their old values and the values of the internal flags are automatically pushed to the stack part of memory.

Figure 11–5(b) shows the interrupt interface of a minimum-mode 8086 microcomputer system. Comparing this diagram to Fig. 11–5(a), we find that the only difference is that the data path between the MPU and interrupt interface is now 16 bits in length.

Maximum-Mode Interrupt Interface

Figure 11–6(a) shows the maximum-mode interrupt interface of the 8088 microcomputer. The primary difference between this interrupt interface and that shown for the minimum mode in Fig. 11–5(a) is that the 8288 bus controller has been added. In the maximum-mode system, it is the bus controller that produces the $\overline{INTA}$ and ALE signals. Whenever the 8088 outputs an interrupt-acknowledge bus status code, the 8288 generates pulses at its $\overline{INTA}$ output to signal external circuitry that the 8088 has acknowledged an interrupt request. This interrupt-acknowledge bus status code, $\overline{S}_2\overline{S}_1\overline{S}_0 = 000$, is highlighted in Fig. 11–7.

A second change in Fig. 11–6(a) is that the 8088 provides a new signal for the interrupt interface. This output, labeled $\overline{LOCK}$, is called the *bus priority lock* signal. $\overline{LOCK}$ is applied as an input to a *bus arbiter.* In response to this signal, the arbitration logic ensures that no other device can take over control of the system bus until the interrupt-acknowledge bus cycle is complete.

Figure 11–6(b) illustrates the interrupt interface of a maximum-mode 8086 microcomputer system. Again, the only difference between this circuit and that of the 8088 microcomputer is that the complete 16-bit data bus is used to transfer data between the MPU and interrupt interface circuits.

▲ 11.6 EXTERNAL HARDWARE-INTERRUPT SEQUENCE

In the preceding section we showed the interrupt interfaces for the external hardware interrupts in minimum-mode and maximum-mode 8088 and 8086 microcomputer systems. We will continue by describing in detail the events that take place during the interrupt request, interrupt-acknowledge bus cycle, and device service routine. The events that

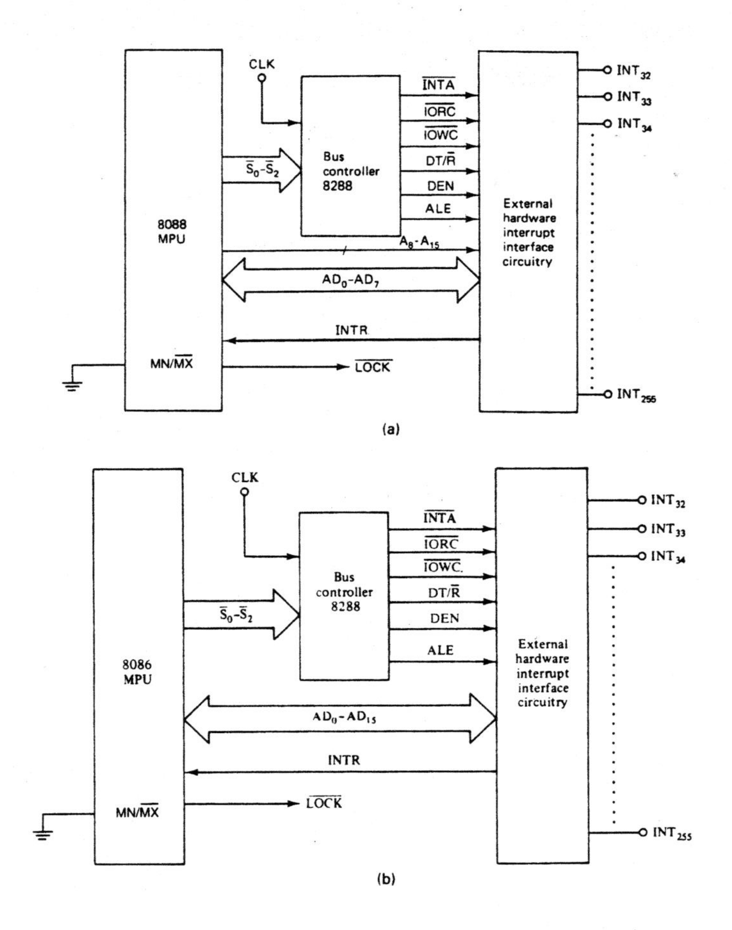

Figure 11–6 (a) Maximum-mode 8088 system external hardware interrupt interface. (b) Maximum-mode 8086 system external hardware interrupt interface.

Status inputs			CPU cycle	8288 command
$\overline{S_2}$	$\overline{S_1}$	$\overline{S_0}$		
0	0	0	Interrupt acknowledge	$\overline{\text{INTA}}$
0	0	1	Read I/O port	$\overline{\text{IORC}}$
0	1	0	Write I/O port	$\overline{\text{IOWC}}$, $\overline{\text{AIOWC}}$
0	1	1	Halt	None
1	0	0	Instruction fetch	$\overline{\text{MRDC}}$
1	0	1	Read memory	$\overline{\text{MRDC}}$
1	1	0	Write memory	$\overline{\text{MWTC}}$, $\overline{\text{AMWC}}$
1	1	1	Passive	None

Figure 11–7 Interrupt bus status code. (Reprinted by permission of Intel Corporation. Copyright/Intel Corp. 1979)

take place in the external hardware interrupt service sequence are identical for an 8088-based or 8086-based microcomputer system. Here we will describe this sequence for the 8088 microcomputer.

The interrupt sequence begins when an external device requests service by activating one of the interrupt inputs, INT_{32} through INT_{255}, of the 8088's external interrupt interface circuit in Fig. 11–5. For example, if the INT_{50} input is switched to the 1 logic level, it signals the microprocessor that the device associated with priority level 50 wants to be serviced.

The external interface circuitry evaluates the priority of this input. If there is no other interrupt already in progress or if this interrupt is of higher priority than the one presently active, the external circuitry issues a request for service to the MPU.

Let us assume that INT_{50} is the only active interrupt request input. In this case, the external circuitry switches INTR to logic 1. This tells the 8088 that an interrupt is pending for service. To ensure that it is recognized, the external circuitry must maintain INTR active until an interrupt-acknowledge pulse is issued by the 8088.

Figure 11–8 is a flow diagram that outlines the events taking place when the 8088 processes an interrupt. The 8088 tests for an active interrupt request during the last T state of the current instruction. Note that it tests first for the occurrence of an internal interrupt, then the occurrence of the nonmaskable interrupt, and finally checks the logic level of INTR to determine if an external hardware interrupt has occurred.

If INTR is logic 1, a request for service is recognized. Before the 8088 initiates the interrupt-acknowledge sequence, it checks the setting of IF. If IF is logic 0, external interrupts are masked out and the request is ignored. In this case, the next sequential instruction is executed. On the other hand, if IF is at logic 1, external hardware interrupts are enabled and the service routine is initiated.

Let us assume that IF is set to permit interrupts to occur when INTR is tested as 1. The 8088 responds by initiating the interrupt-acknowledge bus cycles. This bus cycle sequence is illustrated in Fig. 11–9. During T_1 of the first bus cycle, we see that a pulse is output on ALE along with putting the address/data bus in the high-Z state. The address/data bus stays at high-Z for the rest of the bus cycle. During periods T_2 and T_3, $\overline{INTA}$ is switched to logic 0. This signals external circuitry that the request for service is granted. In response to this pulse, the logic 1 at INTR can be removed.

The signal identified as $\overline{LOCK}$ is produced only in maximum-mode systems. Notice that $\overline{LOCK}$ is switched to logic 0 during T_2 of the first INTA bus cycle and is maintained at this level until T_2 of the second INTA bus cycle. During this time, the 8088 is prevented from accepting any HOLD request. The $\overline{LOCK}$ output is used in external logic to lock other devices off the system bus, thereby ensuring that the interrupt acknowledge sequence continues through to completion without interruption.

During the second interrupt-acknowledge bus cycle, a similar signal sequence occurs. However, this interrupt-acknowledge pulse tells the external circuitry to put the type number of the active interrupt on the data bus. External circuitry gates one of the interrupt codes $32 = 20_{16}$ through $255 = FF_{16}$, identified as vector type in Fig. 11–9, onto data bus lines AD_0 through AD_7. This code must be valid during periods T_3 and T_4 of the second interrupt-acknowledge bus cycle.

The 8088 sets up its bus control signals for an input data transfer to read the type number off the data bus. $DT/\overline{R}$ and $\overline{DEN}$ are set to logic 0 to enable the external data bus

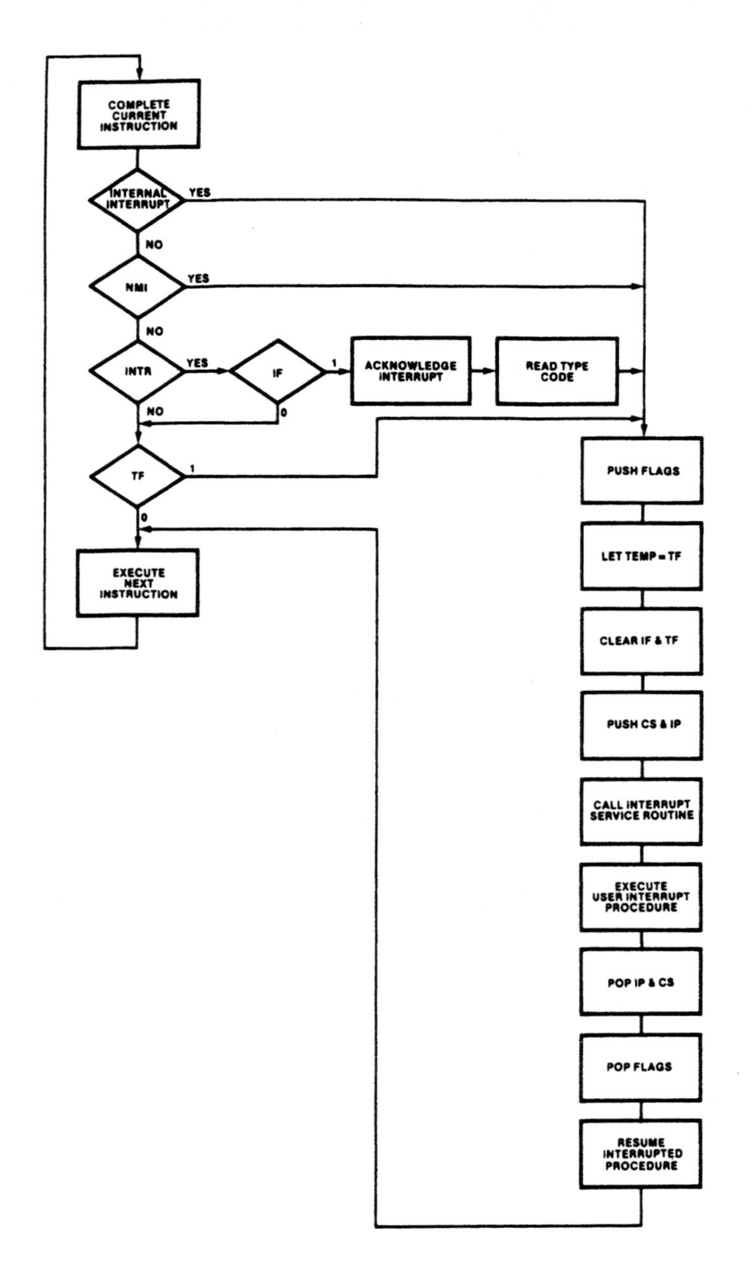

Figure 11–8 Interrupt processing sequence of the 8088 and 8086 microprocessors. (Reprinted by permission of Intel Corporation. Copyright/Intel Corp. 1979)

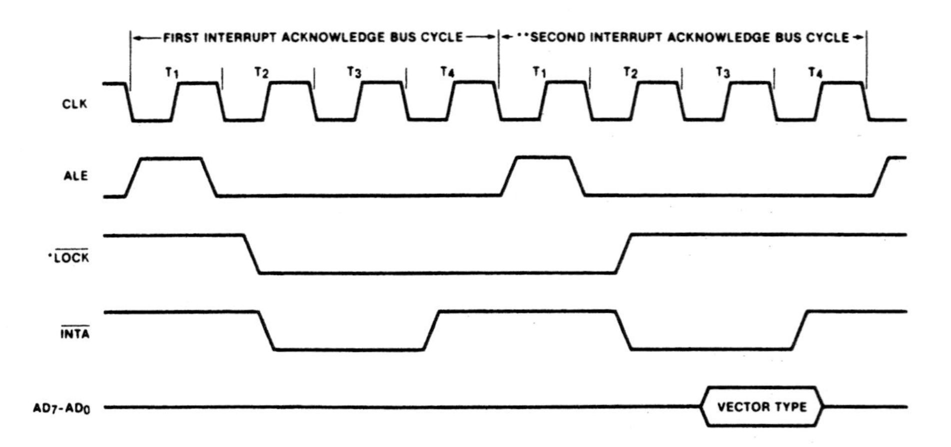

Figure 11–9 Interrupt-acknowledge bus cycle. (Reprinted by permission of Intel Corporation. Copyright/Intel Corp. 1979)

circuitry and set it for input of data. Also, IO/$\overline{M}$ is set to 1, indicating that data are to be input from the interrupt interface. During this input operation, the byte interrupt code is read off the data bus. For the case of INT_{50}, the code would be $00110010_2 = 32_{16}$. This completes the interrupt-request/acknowledge handshake part of the interrupt sequence.

Looking at Fig. 11–8, we see that the 8088 next saves the contents of the flag register by pushing it to the stack. This requires two write cycles and two bytes of stack. Then it clears IF. This disables external hardware interrupts from any other peripheral requesting service. Actually, the TF flag is also cleared. This disables the single-step mode of operation if it happens to be active. Now the 8088 automatically pushes the contents of CS and IP onto the stack. This requires four write bus cycles to take place over the data bus and uses four bytes of memory on the stack. The current value of the stack pointer is decremented by two as each of these values is placed onto the top of the stack.

Now the 8088 knows the type number associated with the external device that is requesting service. It must next call the service routine by fetching the interrupt vector that defines its starting point in the memory. The type number is internally multiplied by four, and this result is used as the address of the first word of the interrupt vector in the pointer table. Two-word read operations (four bus cycles) are performed to read the two-word vector from the memory. The first word, the lower-addressed word, is loaded into IP. The second, higher-addressed word is loaded into CS. For instance, the vector for INT_{50} would be read from addresses $000C8_{16}$ and $000CA_{16}$.

The service routine is now initiated. That is, execution resumes with the first instruction of the service routine. It is located at the address generated from the new value in CS and IP. Figure 11–10 shows the structure of a typical interrupt-service routine. The service routine includes PUSH instructions to save the contents of those internal registers that it will use. In this way, their original contents are saved in the stack during execution of the routine.

At the end of the service routine, the original program environment must be restored. This is done by first popping the contents of the appropriate registers from the stack by executing POP instructions. An IRET instruction must be executed as the last

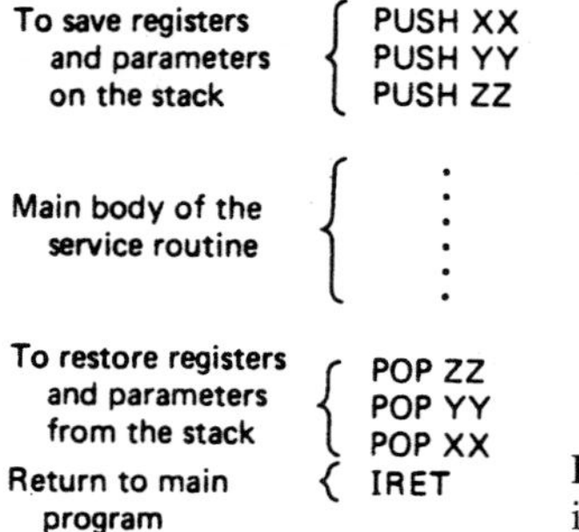

Figure 11–10 Structure of an interrupt-service routine.

instruction of the service routine. This instruction reenables the interrupt interface and causes the old contents of the flags, CS, and IP, to be popped from the stack back into the internal registers of the 8088. The original program environment has now been completely restored and execution resumes at the point in the program where it was interrupted.

Earlier we pointed out that the events that take place during the external hardware interrupt service sequence of the 8086 microcomputer are identical to those of the 8088 microcomputer. However, because the 8086 has a 16-bit data bus, slight changes are found in the external bus cycles that are produced as part of the program context switch sequence. For example, as the interrupt's program environment is initiated, three-word write cycles, instead of 6-byte write cycles, are required to save the old values of the flags, instruction pointer register, and code-segment register on the stack. Moreover, when the new values of CS and IP are fetched from the address pointer table in memory, just two bus cycles take place. Because five instead of ten bus cycles take place, the new program environment is entered faster by the 8086 microcomputer.

The same is true when the original program environment is restored at the completion of the service routine. Remember that this is done by popping the old flags, old CS, and old IP from the stack back into the MPU with an IRET instruction. This operation is performed with three-word read cycles by the 8086 and 6-byte read cycles by the 8088.

EXAMPLE 11.2

The circuit in Fig. 11–11(a) is used to count interrupt requests. The interrupting device interrupts the microprocessor each time the interrupt-request input signal transitions from 0 to 1. The corresponding interrupt type number generated by the 74LS244 is 60H.

a. Describe the operation of the hardware for an active request at the interrupt-request input.

b. What is the value of the type number sent to the microprocessor?

c. Assume that the original values in the segment registers are (CS) = (DS) = 1000H and (SS) = 4000H; the main program is located at offsets of 200H from the beginning of the original code segment; the count is held at an offset of 100H from the beginning of the current data segment; the interrupt-service routine starts at offset 1000H from the beginning of another code segment that begins at address 2000H:0000H; and the stack starts at an offset of 500H from the beginning of the current stack segment. Make a map showing the organization of the memory address space.

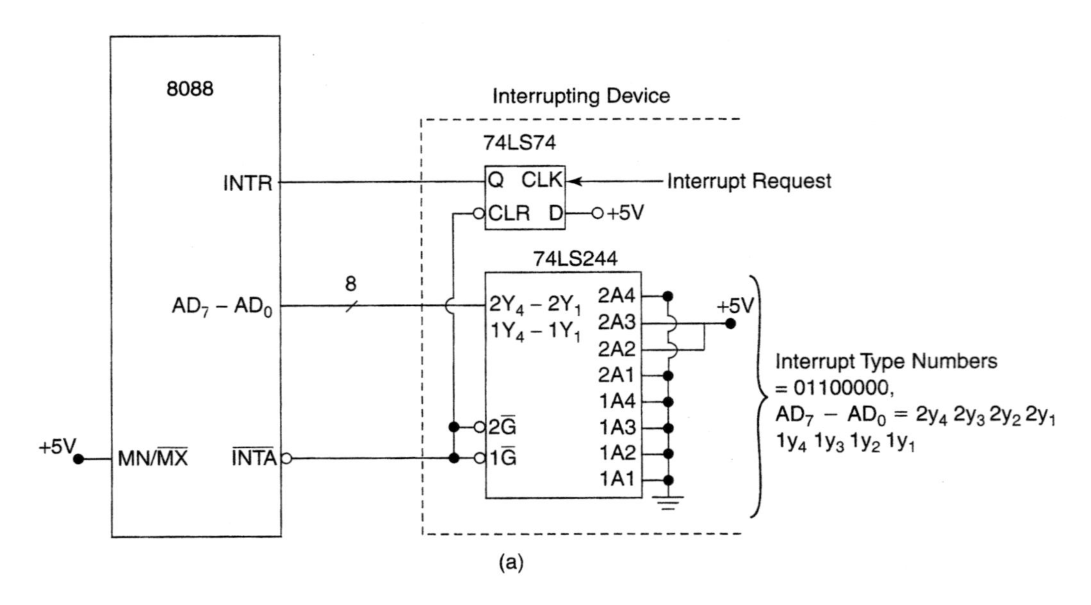

Figure 11–11 (a) Circuit for Example 11.2. (b) Memory organization. (c) Flowcharts for the main program and the interrupt-service routines. (d) Main program and interrupt-service routines.

d. Write the main program and the service routine for the circuit so that the positive transitions at INTR are counted as a decimal number.

Solution

a. Analysis of the circuit in Fig. 11–11(a) shows that a positive transition at the CLK input of the flip-flop (interrupt request) makes the Q output of the flip-flop logic 1 and presents a positive level signal at the INTR input of the 8088. When the 8088 recognizes this as an interrupt request, it responds by generating the $\overline{\text{INTA}}$ signal. The logic 0 output on this line clears the flip-flop and enables the 74LS244 buffer to present the type number to the 8088. This number is read off the data bus by the 8088 and is used to initiate the interrupt-service routine.

b. From the inputs and outputs of the 74LS244, we see that the type number is

$$AD_7 \ldots AD_1AD_0 = 2Y_42Y_32Y_22Y_1\ 1Y_41Y_31Y_21Y_1 = 01100000_2$$

$$AD_7 \ldots AD_1AD_0 = 60H$$

c. The memory organization in Fig. 11–11(b) shows where the various pieces of program and data are located. Here we see that the type 60H vector is located in the interrupt vector table at address 60H × 4 = 180H. Note that the byte-wide memory location used for Count is at address 1000H:0100H. This part of the memory address space is identified as the program data area in the memory map. The main part of the program, entered after reset, starts at address 2000H:1000H. On the other hand, the service routine is located at address 2000H:1000H in a separate code segment. For this reason, the vector held at 180H of the interrupt-vector table is

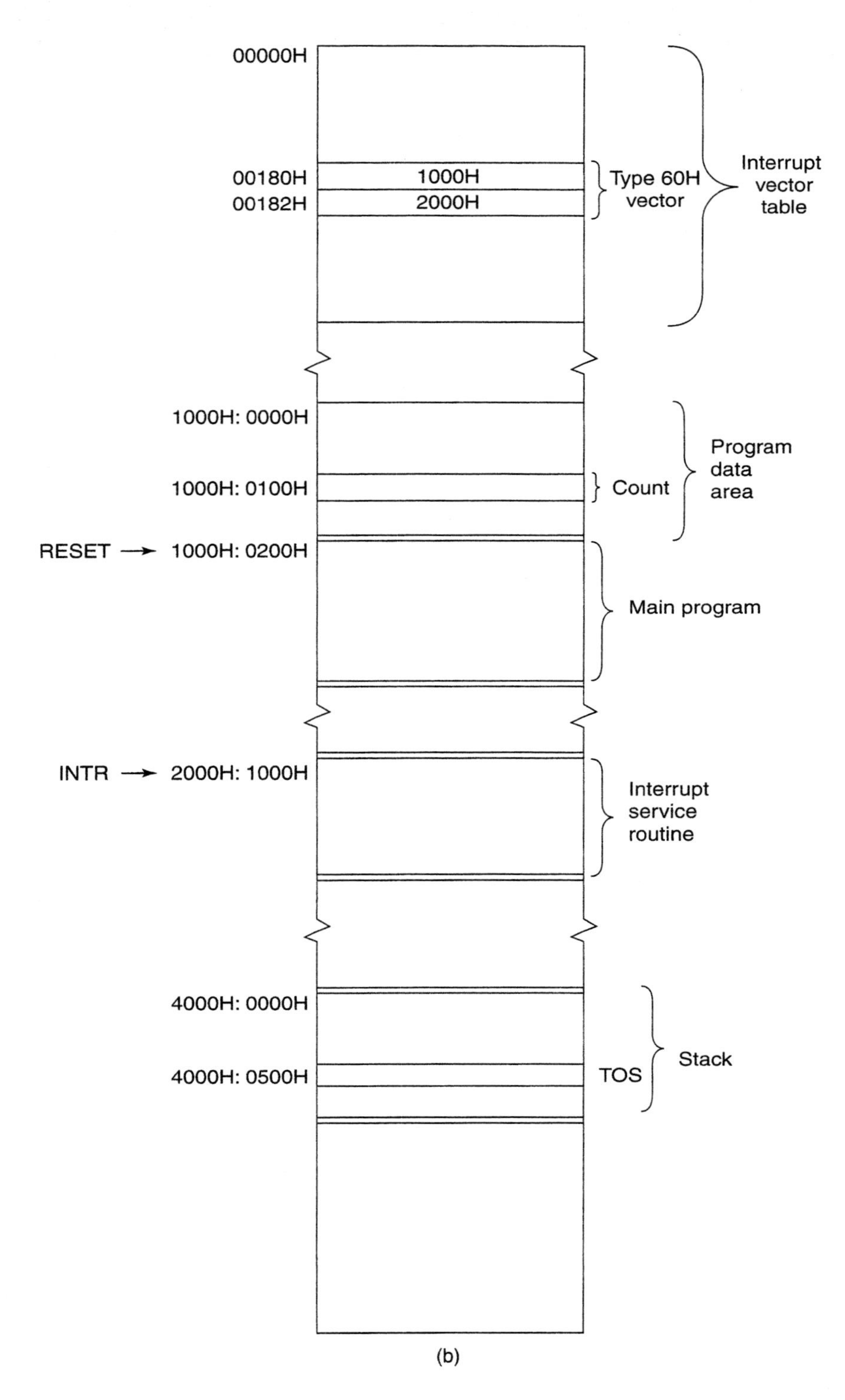

Figure 11–11 (continued)

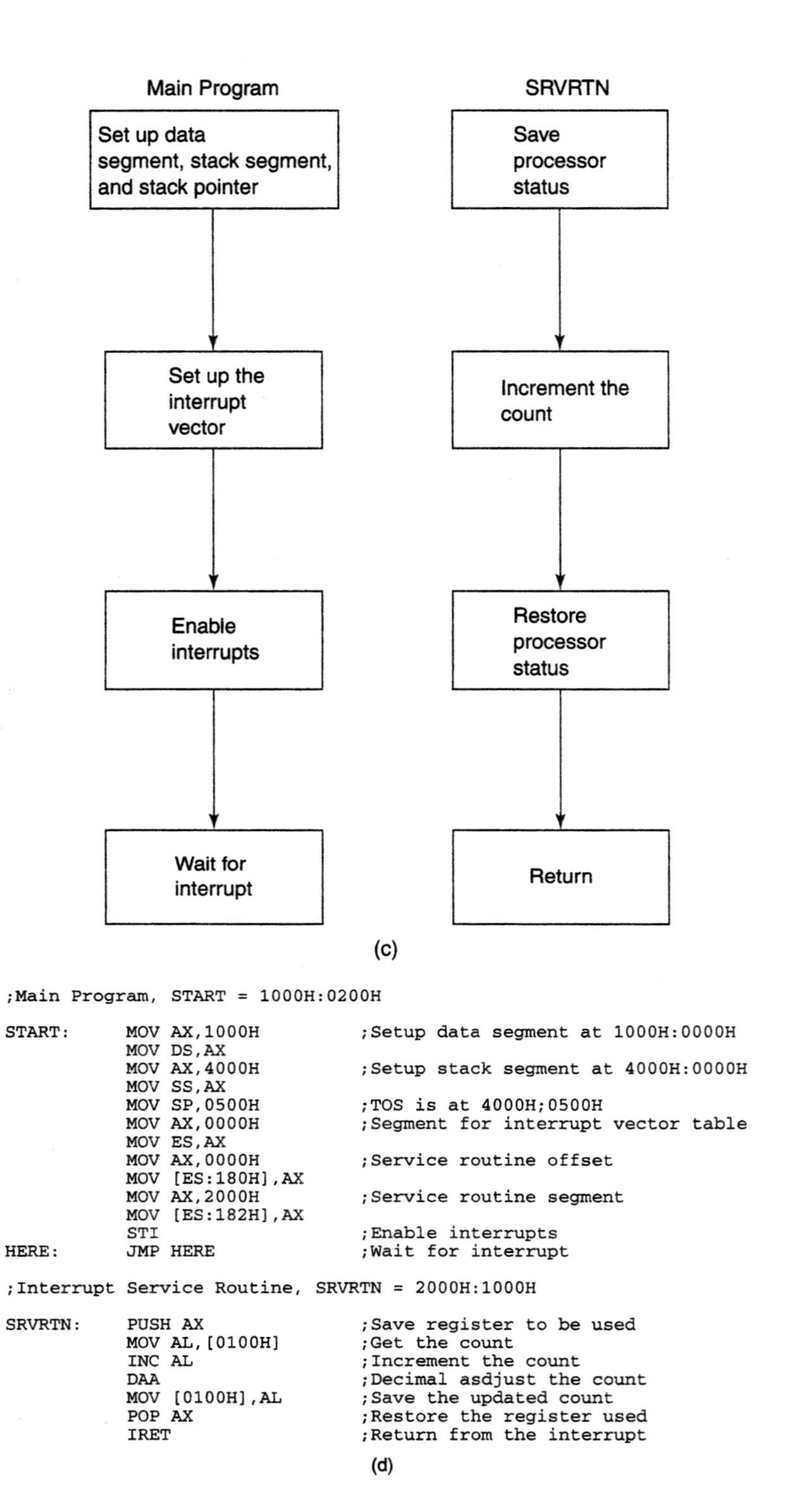

(c)

```
;Main Program, START = 1000H:0200H

START:      MOV AX,1000H           ;Setup data segment at 1000H:0000H
            MOV DS,AX
            MOV AX,4000H           ;Setup stack segment at 4000H:0000H
            MOV SS,AX
            MOV SP,0500H           ;TOS is at 4000H;0500H
            MOV AX,0000H           ;Segment for interrupt vector table
            MOV ES,AX
            MOV AX,0000H           ;Service routine offset
            MOV [ES:180H],AX
            MOV AX,2000H           ;Service routine segment
            MOV [ES:182H],AX
            STI                    ;Enable interrupts
HERE:       JMP HERE               ;Wait for interrupt

;Interrupt Service Routine, SRVRTN = 2000H:1000H

SRVRTN:     PUSH AX                ;Save register to be used
            MOV AL,[0100H]         ;Get the count
            INC AL                 ;Increment the count
            DAA                    ;Decimal asdjust the count
            MOV [0100H],AL         ;Save the updated count
            POP AX                 ;Restore the register used
            IRET                   ;Return from the interrupt
```

(d)

Figure 11–11 (continued)

(CS) = 2000H and (IP) = 1000H. Finally, the stack begins at 4000H:0000H with the current top of the stack located at 4000H:0500H.

d. The flowcharts in Fig. 11–11(c) show how the main program and interrupt-service routines are to function. The corresponding software is given in Fig. 11–11(d).

▲ 11.7 82C59A PROGRAMMABLE INTERRUPT CONTROLLER

The 82C59A is an LSI peripheral IC that is designed to simplify the implementation of the interrupt interface in the 8088- and 8086-based microcomputer systems. This device is known as a *programmable interrupt controller* or *PIC.* It is manufactured using the CMOS technology.

The operation of the PIC is programmable under software control, and it can be configured for a wide variety of applications. Some of its programmable features are the ability to accept level-sensitive or edge-triggered inputs, the ability to be easily cascaded to expand from 8 to 64 interrupt inputs, and the ability to be configured to implement a wide variety of priority schemes.

Block Diagram of the 82C59A

Let us begin our study of the PIC with its block diagram in Fig. 11–12(a). We just mentioned that the 82C59A is treated as a peripheral in the microcomputer. Therefore, its operation must be initialized by the microprocessor. The *host processor interface* is provided for this purpose. This interface consists of eight data bus lines, D_0 through D_7, and control signals read ($\overline{RD}$), write ($\overline{WR}$), and chip select ($\overline{CS}$). The data bus is the path over which data are transferred between the MPU and 82C59A. These data can be command words, status information, or interrupt-type numbers. Control input $\overline{CS}$ must be at logic 0 to enable the host processor interface. Moreover, $\overline{WR}$ and $\overline{RD}$ signal the 82C59A whether data are to be written into or read from its internal registers.

Two other signals, INT and $\overline{INTA}$, are identified as part of the host processor interface. Together, these two signals provide the handshake mechanism by which the 82C59A can signal the MPU of a request for service and receive an acknowledgment that the request has been accepted. INT is the interrupt request output of the 82C59A. It is applied directly to the INTR input of the 8088 or 8086. Logic 1 is produced at this output whenever the interrupt controller receives a valid request from an interrupting device.

On the other hand, $\overline{INTA}$ is an input of the 82C59A. It is connected to the $\overline{INTA}$ output of the 8088 or 8086. The MPU pulses this input of the 82C59A to logic 0 twice during the interrupt-acknowledge bus cycle, thereby signaling the 82C59A that the interrupt request has been acknowledged and that it should output the type number of the highest-priority active interrupt on data bus lines D_0 through D_7 so that it can be read by the MPU. The last signal line involved in the host processor interface is the A_0 input. An address line of the microprocessor, such as A_0 , normally supplies this input. The logic level at this input is involved in the selection of the internal register that is accessed during read and write operations.

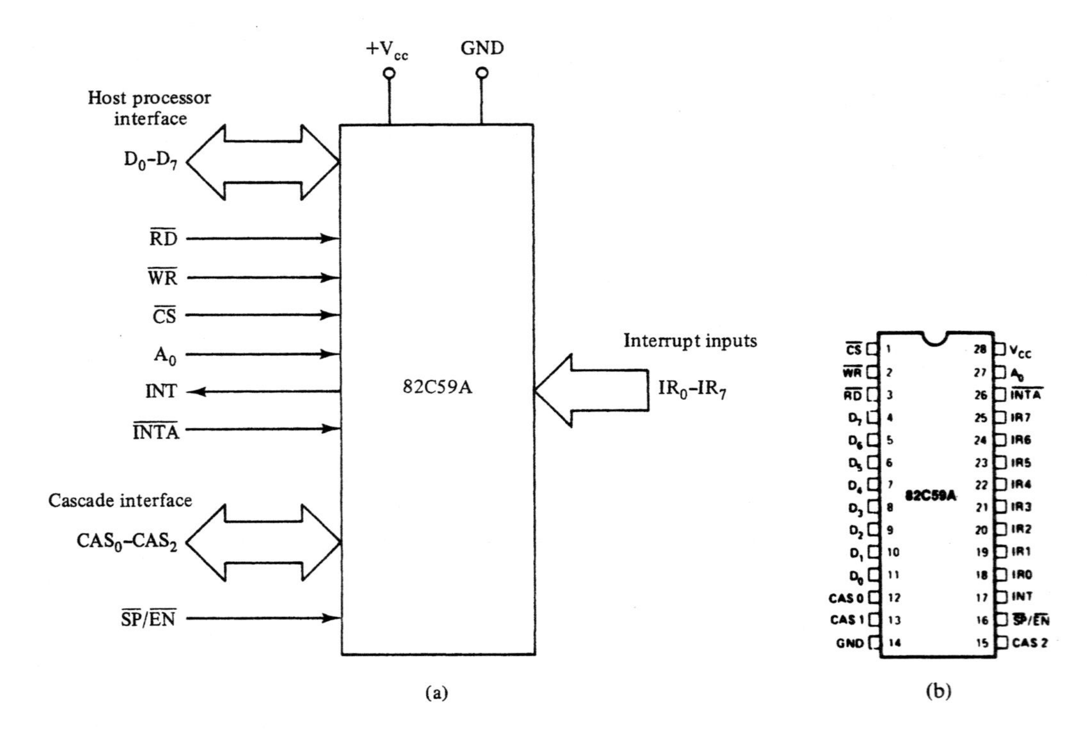

Figure 11–12 (a) Block diagram of the 82C59A. (b) Pin layout. (Reprinted by permission of Intel Corporation. Copyright/Intel Corp. 1979)

At the other side of the block in Fig. 11–12(a), we find the eight *interrupt inputs* of the PIC, labeled IR_0 through IR_7. It is through these inputs that external devices issue a request for service. One of the software options of the 82C59A permits these inputs to be configured for *level-sensitive* or *edge-triggered operation.* When configured for level-sensitive operation, logic 1 is the active level of the IR inputs. In this case, the request for service must be removed before the service routine runs to completion. Otherwise, the interrupt will be requested a second time and the service routine initiated again. Moreover, if the input returns to logic 0 before it is acknowledged by the MPU, the request for service will be missed.

Some external devices produce a short-duration pulse instead of a fixed logic level for use as an interrupt-request signal. If the MPU is busy servicing a higher-priority interrupt when the pulse is produced, the request for service could be completely missed if the 82C59A is in level-sensitive mode. To overcome this problem, the edge-triggered mode of operation is used.

Inputs of the 82C59A that are set up for edge-triggered operation become active on the transition from the inactive 0 logic level to the active 1 logic level. This represents what is known as a *positive edge-triggered input.* The fact that this transition has occurred at an IR line is latched internal to the 82C59A. If the IR input remains at the 1 logic level even after the service routine is completed, the interrupt is not reinitiated. Instead, it is locked out. To be recognized a second time, the input must first return to the 0 logic level and then be switched back to 1. The advantage of edge-triggered operation is that if the request at the IR input is removed before the MPU acknowledges service of the interrupt, its request is kept latched internal to the 82C59A until it can be serviced.

The last group of signals on the PIC implements what is known as the *cascade interface.* As shown in Fig. 11–12(a), it includes bidirectional cascading bus lines CAS_0 through CAS_2 and a multifunction control line labeled $\overline{SP}/\overline{EN}$. The primary use of these signals is in cascaded systems where a number of 82C59A ICs are interconnected in a *master/slave configuration* to expand the number of IR inputs from 8 to as high as 64. One of these 82C59A devices is configured as the *master* and all others are set up as *slaves.*

In a cascaded system, the CAS lines of all 82C59As are connected together to provide a private bus between the master and slave devices. In response to the first $\overline{INTA}$ pulse during the interrupt-acknowledge bus cycle, the master PIC outputs a 3-bit code on the CAS lines. This code identifies the highest-priority slave that is to be serviced. It is this device that is to be acknowledged for service. All slaves read this code off the *private cascading bus* and compare it to their internal ID code. A match condition at one slave tells the PIC that it has the highest-priority input. In response, it must put the type number of its highest-priority active input on the data bus during the second interrupt-acknowledge bus cycle.

When the PIC is configured through software for the cascaded mode, the $\overline{SP}/\overline{EN}$ line is used as an input. This corresponds to its $\overline{SP}$ (*slave program*) function. The logic level applied at $\overline{SP}$ tells the device whether it is to operate as a master or slave. Logic 1 at this input designates master mode, and logic 0 designates slave mode.

If the PIC is configured for single mode instead of cascade mode, $\overline{SP}/\overline{EN}$ takes on another function. In this case, it becomes an enable output that can be used to control the direction of data transfer through the bus transceiver that buffers the data bus.

Figure 11–12(b) presents a pin layout of the 82C59A.

Internal Architecture of the 82C59A

Now that we have introduced the input/output signals of the 82C59A, let us look at its internal architecture. Figure 11–13 is a block diagram of the PIC's internal circuitry. Here we find eight functional parts: the *data bus buffer, read/write logic, control logic, in-service register, interrupt-request register, priority resolver, interrupt-mask register,* and *cascade buffer/comparator.*

We will begin with the function of the data bus buffer and read/write logic sections. It is these parts of the 82C59A that let the MPU have access to the internal registers. Moreover, they provide the path over which interrupt-type numbers are passed to the microprocessor. The data bus buffer is an 8-bit bidirectional three-state buffer that interfaces the internal circuitry of the 82C59A to the data bus of the MPU. The direction, timing, and source or destination for data transfers through the buffer are under control of the outputs of the read/write logic block. These outputs are generated in response to control inputs $\overline{RD}$, $\overline{WR}$, A_0, and $\overline{CS}$.

The interrupt-request register, in-service register, priority resolver, and interrupt-mask register are the key internal blocks of the 82C59A. The interrupt-mask register (IMR) can be used to enable or mask out individually the interrupt request inputs. It contains eight bits, identified by M_0 through M_7. These bits correspond to interrupt-request inputs IR_0 through IR_7, respectively. Logic 0 in a mask register bit position enables the corresponding

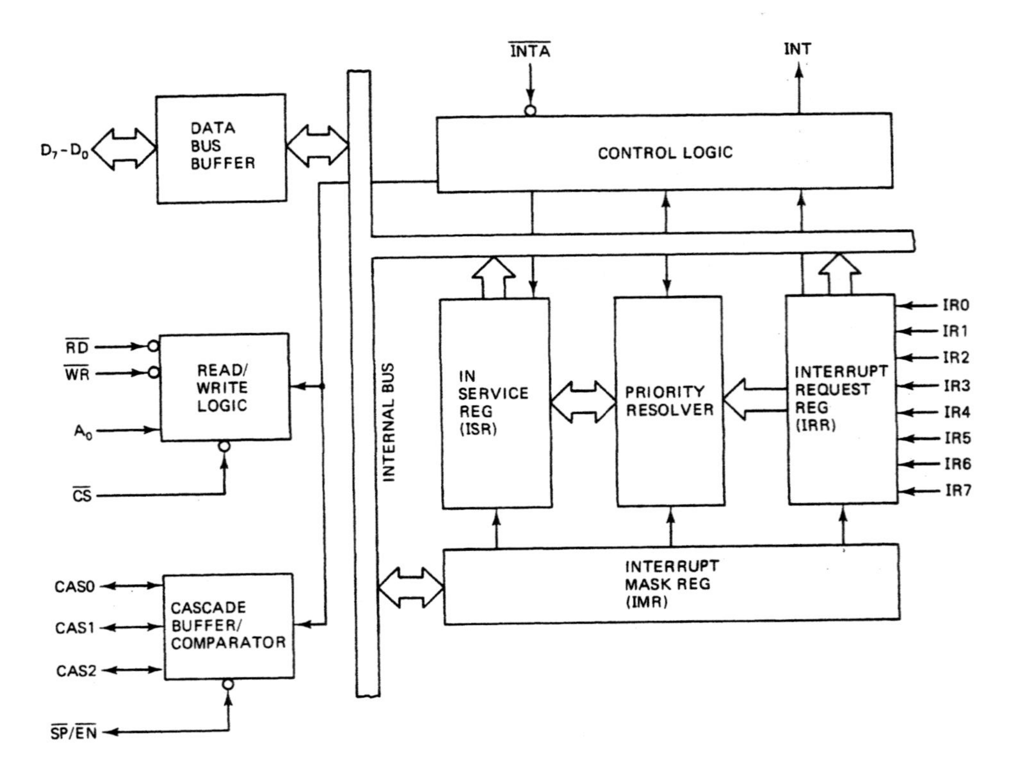

Figure 11–13 Internal architecture of the 82C59A. (Reprinted by permission of Intel Corporation. Copyright/Intel Corp. 1979)

interrupt input and logic 1 masks it out. The register can be read from or written into through software control.

On the other hand, the interrupt-request register (IRR) stores the current status of the interrupt-request inputs. It also contains one bit position for each of the IR inputs. The values in these bit positions reflect whether the interrupt inputs are active or inactive.

The priority resolver determines which of the active interrupt inputs has the highest priority. This section can be configured to work using a number of different priority schemes through software. Following the selected scheme, it identifies which of the active interrupts has the highest priority and signals the control logic that an interrupt is active. In response, the control logic causes the INT signal to be issued to the 8088 or 8086 microprocessor.

The in-service register differs in that it stores the interrupt level that is presently being serviced. During the first $\overline{\text{INTA}}$ pulse of an interrupt-acknowledge bus cycle, the level of the highest active interrupt is strobed into ISR. Loading of ISR occurs in response to output signals of the control logic section. This register cannot be written into by the microprocessor; however, its contents may be read as status.

The cascade buffer/comparator section provides the interface between master and slave 82C59As. As we mentioned earlier, this interface permits easy expansion of the interrupt interface using a master/slave configuration. Each slave has an *ID code* that is stored in this section.

Programming the 82C59A

The way in which the 82C59A operates is determined by how the device is programmed. Two types of command words are provided for this purpose: the *initialization command words* (ICW) and the *operational command words* (OCW). ICW commands are used to load the internal control registers of the 82C59A to define the basic configuration or mode in which it is used. There are four such command words, identified as ICW_1, ICW_2, ICW_3, and ICW_4. On the other hand, the three OCW commands, OCW_1, OCW_2, and OCW_3, permit the 8088 or 8086 microprocessor to initiate variations in the basic operating modes defined by the ICW commands.

Depending on whether the 82C59A is I/O-mapped or memory-mapped, the MPU issues commands to the 82C59A by initiating output or write cycles. Executing either the OUT instruction or MOV instruction, respectively, can do this. The address put on the system bus during the output bus cycle must be decoded with external circuitry to chip-select the peripheral. When an address assigned to the 82C59A is on the bus, the output of the decoder must produce logic 0 at the $\overline{CS}$ input. This signal enables the read/write logic within the PIC, and data applied at D_0 through D_7 are written into the command register within the control logic section synchronously with a write strobe at $\overline{WR}$.

The interrupt-request input (INTR) of the 8088 or 8086 must be disabled whenever commands are being issued to the 82C59A. Clearing the interrupt-enable flag by executing the CLI instruction can do this. After completion of the command sequence, the interrupt input must be reenabled. To do this, the microprocessor must execute the STI instruction.

The flow diagram in Fig. 11–14 shows the sequence of events that must take place to initialize the 82C59A with ICW commands. The cycle begins with the MPU outputting initialization command word ICW_1 to the address of the 82C59A.

The moment that ICW_1 is written into the control logic section of the 82C59A, certain internal setup conditions automatically occur. First, the internal sequence logic is set up so that the 82C59A will accept the remaining ICWs as designated by ICW_1. It turns out that if the least significant bit of ICW_1 is logic 1, command word ICW_4 is required in the initialization sequence. Moreover, if the next least significant bit of ICW_1 is logic 0, the command word ICW_3 is also required.

In addition to this, writing ICW_1 to the 82C59A clears ISR and IMR. Also, three operation command word bits, *special mask mode* (SMM) in OCW_3, *interrupt-request register* (IRR) in OCW_3, and *end of interrupt* (EOI) in OCW_2, are cleared to logic 0. Furthermore, the *fully nested masked mode* of interrupt operation is entered with an initial priority assignment so that IR_0 is the highest-priority input and IR_7 the lowest-priority input. Finally, the edge-sensitive latches associated with the IR inputs are all cleared.

If the LSB of ICW_1 was initialized to logic 0, one additional event occurs: all bits of the control register associated with ICW_4 are cleared.

Figure 11–14 shows that once the MPU starts initialization of the 82C59A by writing ICW_1 into the control register, it must continue the sequence by writing ICW_2 and then, optionally, ICW_3 and ICW_4 in that order. Note that it is not possible to modify just

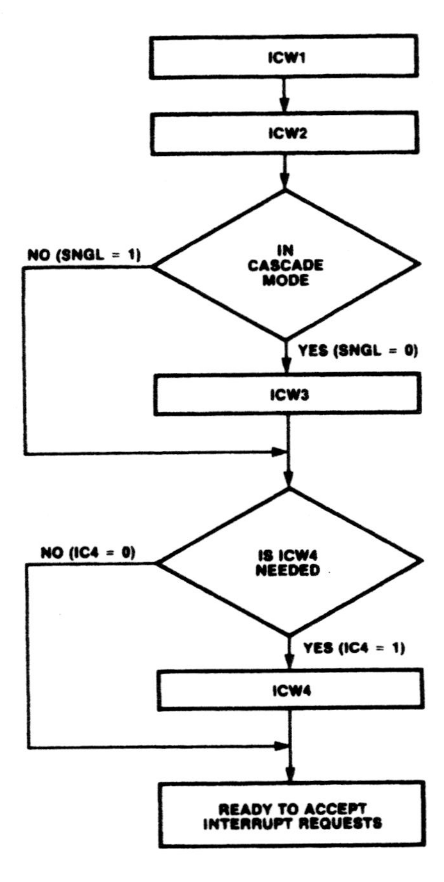

Figure 11–14 Initialization sequence of the 82C59A. (Reprinted by permission of Intel Corporation. Copyright/Intel Corp. 1979)

one of the initialization command registers. Instead, all words that are required to define the device's operating mode must be written into the 82C59A.

We found that all four words need not always be used to initialize the 82C59A. However, for its use in an 8088 or 8086 microcomputer system, words ICW_1, ICW_2, and ICW_4 are always required. ICW_3 is optional and is needed only if the 82C59A is to function in the cascade mode.

Initialization Command Words

Now that we have introduced the initialization sequence of the 82C59A, let us look more closely at the functions controlled by each of the initialization command words. We will begin with ICW_1. Its format and bit functions are identified in Fig. 11–15(a). Note that address bit A_0 is included as a ninth bit and it must be logic 0. This corresponds to an even address for writing ICW_1.

Here we find that the logic level of the LSB D_0 of the initialization word indicates to the 82C59A whether or not ICW_4 will be included in the programming sequence. As we mentioned earlier, logic 1 at D_0 (IC_4) specifies that it is needed. The next bit, D_1 (SNGL),

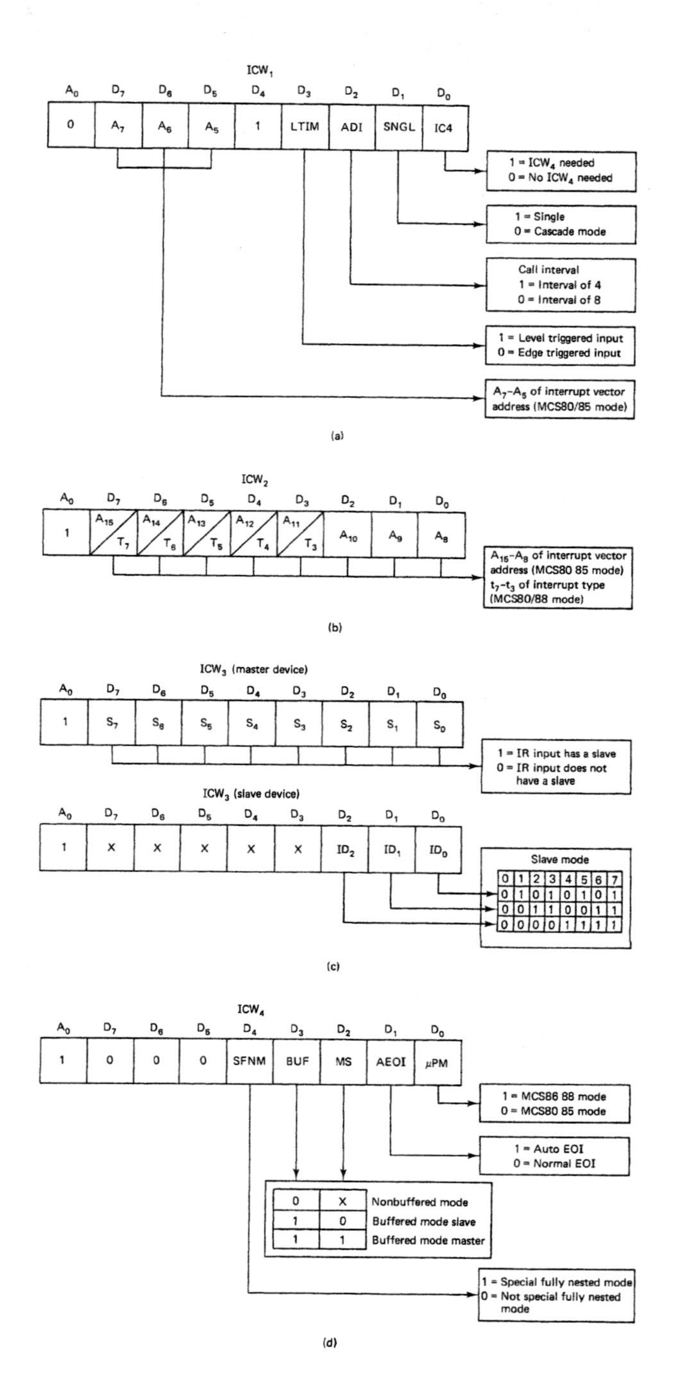

Figure 11–15 (a) ICW_1 format. (Reprinted by permission of Intel Corporation. Copyright/Intel Corp. 1979) (b) ICW_2 format. (Reprinted by permission of Intel Corporation. Copyright/Intel Corp. 1979) (c) ICW_3 format. (Reprinted by permission of Intel Corporation. Copyright/Intel Corp. 1979) (d) ICW_4 format. (Reprinted by permission of Intel Corporation. Copyright/Intel Corp. 1979)

selects between the *single device* and *multidevice cascaded mode* of operation. When D_1 is set to logic 0, the internal circuitry of the 82C59A is configured for cascaded mode. Selecting this state also sets up the initialization sequence such that ICW_3 must be issued as part of the initialization cycle. Bit D_2 has functions specified for it in Fig. 11–15(a); however, it can be ignored when the 82C59A is being connected to the 8088/8086 and is a don't-care state. D_3, labeled LTIM, defines whether the eight IR inputs operate in the level-sensitive or edge-triggered mode. Logic 1 in D_3 selects level-triggered operation, and logic 0 selects edge-triggered operation. Finally, bit D_4 is fixed at the 1 logic level and the three MSBs, D_5 through D_7, are not required in 8088- or 8086-based systems.

EXAMPLE 11.3

What value should be written into ICW_1 in order to configure the 82C59A so that ICW_4 is needed in the initialization sequence, the system is going to use multiple 82C59As, and its inputs are to be level sensitive? Assume that all unused bits are to be logic 0.

Solution

Since ICW_4 is to be initialized, D_0 must be logic 1.

$$D_0 = 1$$

For cascaded mode of operation, D_1 must be 0.

$$D_1 = 0$$

And for level-sensitive inputs, D_3 must be 1.

$$D_3 = 1$$

Bits D_2 and D_5 through D_7 are don't-care states and are all made logic 0.

$$D_2 = D_5 = D_6 = D_7 = 0$$

Moreover, D_4 must be fixed at the 1 logic level.

$$D_4 = 1$$

This gives the complete command word

$$D_7D_6D_5D_4D_3D_2D_1D_0 = 00011001_2 = 19_{16}$$

The second initialization word, ICW_2, has a single function in the 8088 or 8086 microcomputer. As Fig. 11–15(b) shows, its five most significant bits, D_7 through D_3, define a fixed binary code, T_7 through T_3, which is used as the most significant bits of its type number. Whenever the 82C59A puts the 3-bit interrupt type number corresponding to its active input onto the bus, it is automatically combined with the value T_7 through T_3 to form an 8-bit type number. The three least significant bits of ICW_2 are not used. Note that logic 1 must be applied to the A_0 input when this command word is put on the bus.

EXAMPLE 11.4

What should be programmed into register ICW_2 if the type numbers output on the bus by the device are to range from $F0_{16}$ through $F7_{16}$?

Solution

To set the 82C59A up so that type numbers are in the range of $F0_{16}$ through $F7_{16}$, its device code bits must be

$$D_7D_6D_5D_4D_3 = 11110_2$$

The lower three bits are don't-care states and all can be 0s. This gives the command word

$$D_7D_6D_5D_4D_3D_2D_1D_0 = 11110000_2 = F0_{16}$$

The information of initialization word ICW_3 is required by only those 82C59As configured for the cascaded mode of operation. Figure 11–15(c) shows its bits. Note that ICW_3 is used for different functions, depending on whether the device is a master or slave. If it is a master, bits D_0 through D_7 of the word are labeled S_0 through S_7. These bits correspond to IR inputs IR_0 through IR_7, respectively. They identify whether or not the corresponding IR input is supplied by either the INT output of a slave or directly by an external device. Logic 1 loaded in an S position indicates that a slave supplies the corresponding IR input.

On the other hand, ICW_3 for a slave is used to load the device with a 3-bit identification code $ID_2ID_1ID_0$. This number must correspond to the IR input of the master to which the slave's INT output is wired. The ID code is required within the slave so that it can be compared to the cascading code output by the master on CAS_0 through CAS_2.

EXAMPLE 11.5

Assume that a master PIC is to be configured so that its IR_0 through IR_3 inputs are to accept inputs directly from external devices, but IR_4 through IR_7 are to be supplied by the INT outputs of slaves. What code should be used for the initialization command word ICW_3?

Solution

For IR_0 through IR_3 to be configured to allow direct inputs from external devices, bits D_0 through D_3 of ICW_3 must be logic 0:

$$D_3D_2D_1D_0 = 0000_2$$

The other IR inputs of the master are to be supplied by INT outputs of slaves. Therefore, their control bits must be all 1:

$$D_7D_6D_5D_4 = 1111_2$$

This gives the complete command word

$$D_7D_6D_5D_4D_3D_2D_1D_0 = 11110000_2 = F0_{16}$$

The fourth control word, ICW_4, shown in Fig. 11–15(d), is used to configure the device for use with the 8088 or 8086 and selects various features that are available in its operation. The LSB D_0, called microprocessor mode (μPM), must be set to logic 1 whenever the device is connected to the 8088. The next bit, D_1, is labeled AEOI for *automatic end of interrupt*. If this mode is enabled by writing logic 1 into the bit location, the EOI (*end-of-interrupt*) command does not have to be issued as part of the service routine.

Of the next two bits in ICW_4, BUF is used to specify whether or not the 82C59A is to be used in a system where the data bus is buffered with a bidirectional bus transceiver. When buffered mode is selected, the $\overline{SP}/\overline{EN}$ line is configured as $\overline{EN}$. As indicated earlier, $\overline{EN}$ is a control output that can be used to control the direction of data transfer through the bus transceiver. It switches to logic 0 whenever data are transferred from the 82C59A to the MPU.

If buffered mode is not selected, the $\overline{SP}/\overline{EN}$ line is configured to work as the master/slave mode select input. In this case, logic 1 at the $\overline{SP}$ input selects master mode operation and logic 0 selects slave mode.

Assume that the buffered mode was selected; then the $\overline{SP}$ input is no longer available to select between the master and slave modes of operation. Instead, the MS bit of ICW_4 defines whether the 82C59A is a master or slave device.

Bit D_4 is used to enable or disable another operational option of the 82C59A. This option is known as the *special fully nested mode.* This function is used only in conjunction with the cascaded mode. Moreover, it is enabled only for the master 82C59A, not for the slaves. Setting the SFNM bit to logic 1 does this.

The 82C59A is put into the fully nested mode of operation as command word ICW_1 is loaded. When an interrupt is initiated in a cascaded system that is configured in this way, the occurrence of another interrupt at the slave corresponding to the original interrupt is masked out even if it is of higher priority. This is because the bit in ISR of the master 82C59A that corresponds to the slave is already set; therefore, the master 82C59A ignores all interrupts of equal or lower priority.

This problem is overcome by enabling the special fully nested mode of operation at the master. In this mode, the master will respond to those interrupts that are at lower or higher priority than the active level.

The last three bits of ICW_4, D_5 through D_7, must be logic 0.

Operational Command Words

Once the appropriate ICW commands have been issued to the 82C59A, it is ready to operate in the fully nested mode. Three operational command words are also provided for controlling the operation of the 82C59A. These commands permit further modifications to be made to the operation of the interrupt interface after it has been initialized. Unlike the initialization sequence, which requires that the ICWs be output in a special sequence after power-up, the OCWs can be issued under program control whenever needed and in any order.

The first operational command word, OCW_1, is used to access the contents of the interrupt-mask register (IMR). A read operation can be performed to the register to determine the present setting of the mask. Moreover, write operations can be performed to set or reset its bits. This permits selective masking of the interrupt inputs. Note in Fig. 11–16(a) that bits D_0 through D_7 of command word OCW_1 are identified as mask bits M_0 through M_7, respectively. In hardware, these bits correspond to interrupt inputs IR_0 through IR_7, respectively. Setting a bit to logic 1 masks out the associated interrupt input. On the other hand, clearing it to logic 0 enables the interrupt input.

For instance, writing $F0_{16} = 11110000_2$ into the register causes inputs IR_0 through IR_3 to be unmasked and IR_4 through IR_7 to be masked. Input A_0 must be logic 1 whenever the OCW_1 command is issued. In other words, the MPU address to access OCW_1 is an odd address.

EXAMPLE 11.6

What should be the OCW_1 code if interrupt inputs IR_0 through IR_3 are to be masked and IR_4 through IR_7 are to be unmasked?

Solution

For IR_0 through IR_3 to be masked, their corresponding bits in the mask register must be made logic 1:

$$D_3D_2D_1D_0 = 1111_2$$

On the other hand, for IR_4 through IR_7 to be unmasked, D_4 through D_7 must be logic 0:

$$D_7D_6D_5D_4 = 0000_2$$

Therefore, the complete word for OCW_1 is

$$D_7D_6D_5D_4D_3D_2D_1D_0 = 00001111_2 = 0F_{16}$$

The second operational command word, OCW_2, selects the appropriate priority scheme and assigns an IR level for those schemes that require a specific interrupt level. The format of OCW_2 is given in Fig. 11–16(b). Here we see that the three LSBs define the interrupt level. For example, using $L_2L_1L_0 = 000_2$ in these locations specifies interrupt level 0, which corresponds to input IR_0.

The other three active bits of the word D_7, D_6, and D_5 are called *rotation* (R), *specific level* (SL), and *end of interrupt* (EOI), respectively. They are used to select a priority scheme according to the table in Fig. 11–16(b). For instance, if these bits are all logic 1, the priority scheme known as *rotate on specific EOI* command is enabled. Since this scheme requires a specific interrupt, its value must be included in $L_2L_1L_0$. Input A_0 must be logic 0 whenever this command is issued to the 82C59A.

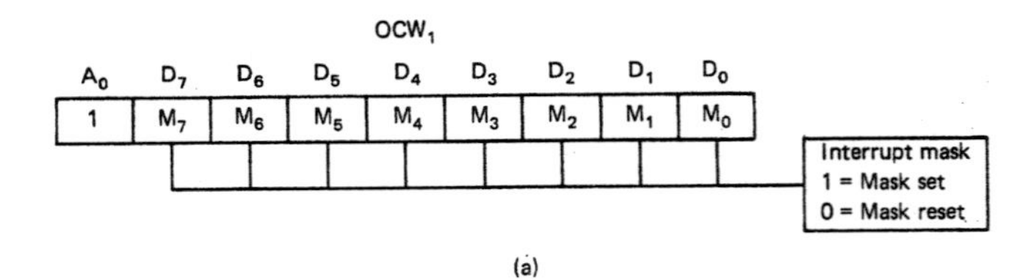

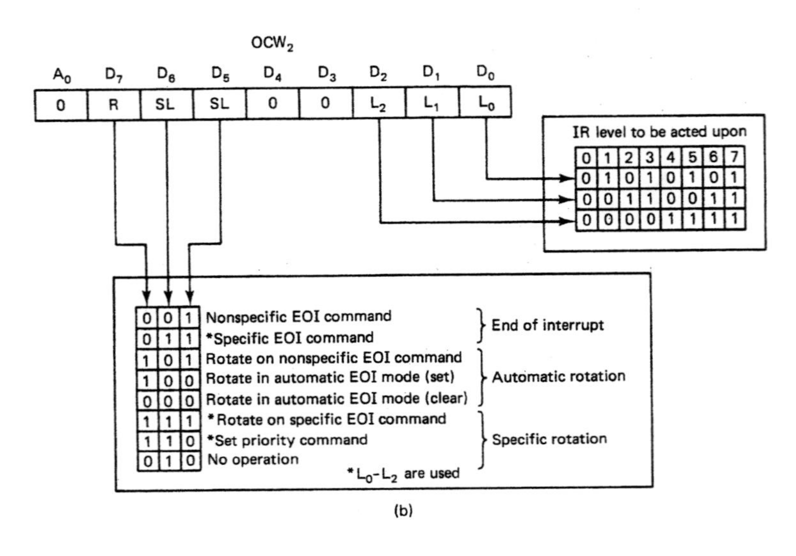

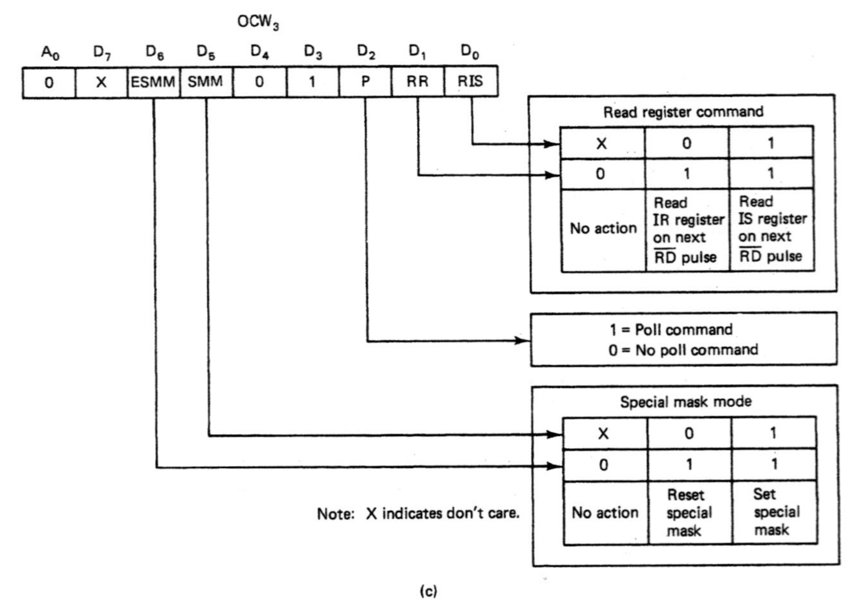

Figure 11–16 (a) OCW_1 format. (Reprinted by permission of Intel Corporation. Copyright/Intel Corp. 1979) (b) OCW_2 format. (Reprinted by permission of Intel Corporation. Copyright/Intel Corp. 1979) (c) OCW_3 format. (Reprinted by permission of Intel Corporation. Copyright/Intel Corp. 1979)

EXAMPLE 11.7

What OCW_2 must be issued to the 82C59A if the priority scheme rotate on nonspecific EOI command is to be selected?

Solution

To enable the rotate on nonspecific EOI command priority scheme, bits D_7 through D_5 must be set to 101. Since a specific level does not have to be specified, the rest of the bits in the command word can be 0. This gives OCW_2 as

$$D_7D_6D_5D_4D_3D_2D_1D_0 = 10100000_2 = A0_{16}$$

The last control word, OCW_3, which is shown in Fig. 11–16(c), permits reading of the contents of the ISR or IRR registers through software, issue of the poll command, and enable/disable of the special mask mode. Bit D_1, called *read register* (RR), is set to 1 to initiate reading of either the in-service register (ISR) or interrupt-request register (IRR). At the same time, bit D_0, labeled RIS, selects between ISR and IRR. Logic 0 in RIS selects IRR and logic 1 selects ISR. In response to this command, the 82C59A makes the contents of the selected register available on the data bus so that they can be read by the MPU.

If the next bit, D_2, in OCW_3 is logic 1, a *poll command* is issued to the 82C59A. The result of issuing a poll command is that the next $\overline{RD}$ pulse to the 82C59A is interpreted as an interrupt acknowledge. In turn, the 82C59A causes the ISR register to be loaded with the value of the highest-priority active interrupt. After this, a *poll word* is automatically put on the data bus. The MPU must read it off the bus.

Figure 11–17 illustrates the format of the poll word. Looking at this word, we see that the MSB is labeled I for interrupt. The logic level of this bit indicates to the MPU whether or not an interrupt input was active. Logic 1 means that an interrupt is active. The three LSBs, W_2, W_1, and W_0, identify the priority level of the highest-priority active interrupt input. This poll word can be decoded through software, and when an interrupt is found to be active, a branch is initiated to the starting point of its service routine. The poll command represents a software method of identifying whether or not an interrupt has occurred; therefore, the INTR input of the 8088 or 8086 should be disabled.

D_5 and D_6 are the remaining bits of OCW_3 for which functions are defined. They are used to enable or disable the special mask mode. ESMM (*enable special mask mode*) must be logic 1 to permit changing of the status of the special mask mode with the SMM (*special mask mode*) bit. Logic 1 at SMM enables the special mask mode of operation. If the 82C59A is initially configured for the fully nested mode of operation, only interrupts of higher priority are allowed to interrupt an active service routine. However, by enabling

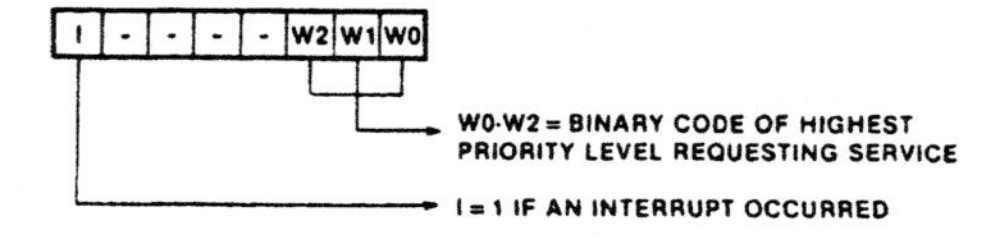

Figure 11–17 Poll word format. (Reprinted by permission of Intel Corporation. Copyright/Intel Corp. 1979)

the special mask mode, interrupts of higher or lower priority are enabled, but those of equal priority remain masked out.

EXAMPLE 11.8

Write a program that will initialize an 82C59A with the initialization command words ICW_1, ICW_2, and ICW_3 derived in Examples 11.3, 11.4, and 11.5, respectively and ICW_4 is equal to $1F_{16}$. Assume that the 82C59A resides at address $A000_{16}$ in the memory address space.

Solution

Since the 82C59A resides in the memory address space, we can use a series of move instructions to write the initialization command words into its registers. Note that the memory address for an ICW is $A000_{16}$ if $A_0 = 0$, and it is $A001_{16}$ if $A_0 = 1$. However, before doing this, we must first disable interrupts. This is done with the instruction

```
CLI                         ;Disable interrupts
```

Next we will create a data segment starting at address 00000_{16}:

```
MOV  AX, 0                  ;Create a data segment at 00000H
MOV  DS, AX
```

Now we are ready to write the command words to the 82C59A:

```
MOV  AL, 19H                ;Load ICW1
MOV  [0A000H], AL           ;Write ICW1 to 82C59A
MOV  AL, 0F0H               ;Load ICW2
MOV  [0A001H], AL           ;Write ICW2 to 82C59A
MOV  AL, 0F0H               ;Load ICW3
MOV  [0A001H], AL           ;Write ICW3 to 82C59A
MOV  AL, 1FH                ;Load ICW4
MOV  [0A001H], AL           ;Write ICW4 to 82C59A
```

Initialization is now complete and the interrupts can be enabled with the interrupt instruction

```
STI                         ;Enable interrupts
```

▲ 11.8 INTERRUPT INTERFACE CIRCUITS USING THE 82C59A

Now that we have introduced the 82C59A programmable interrupt controller, let us look at how it is used to implement the interrupt interface in 8088- and 8086-based microcomputer systems.

Figure 11–18(a) includes an interrupt interface circuit for a minimum-mode microcomputer system that is made with the 82C59A. Let us begin by looking at how the

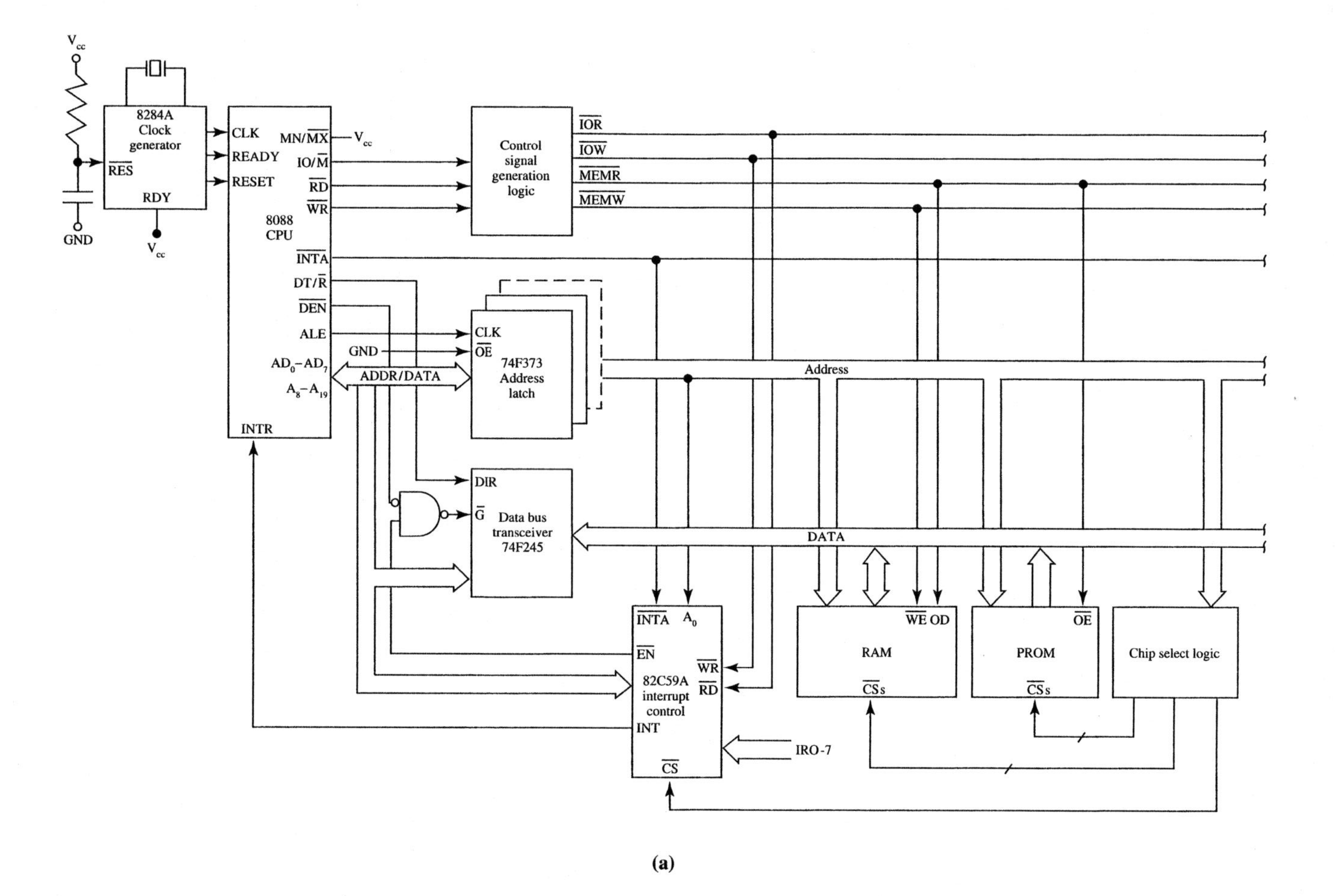

(a)

Figure 11–18 (a) Minimum-mode interrupt interface for the 8088 microcomputer using the 82C59A. (Reprinted by permission of Intel Corporation. Copyright/Intel Corp. 1979) (b) Minimum-mode interrupt interface for the 8086 microcomputer using cascaded 82C59As. (Reprinted by permission of Intel Corporation. Copyright/Intel Corp. 1979) (c) Master/slave connection. (Reprinted by permission of Intel Corporation. Copyright/Intel Corp. 1979)

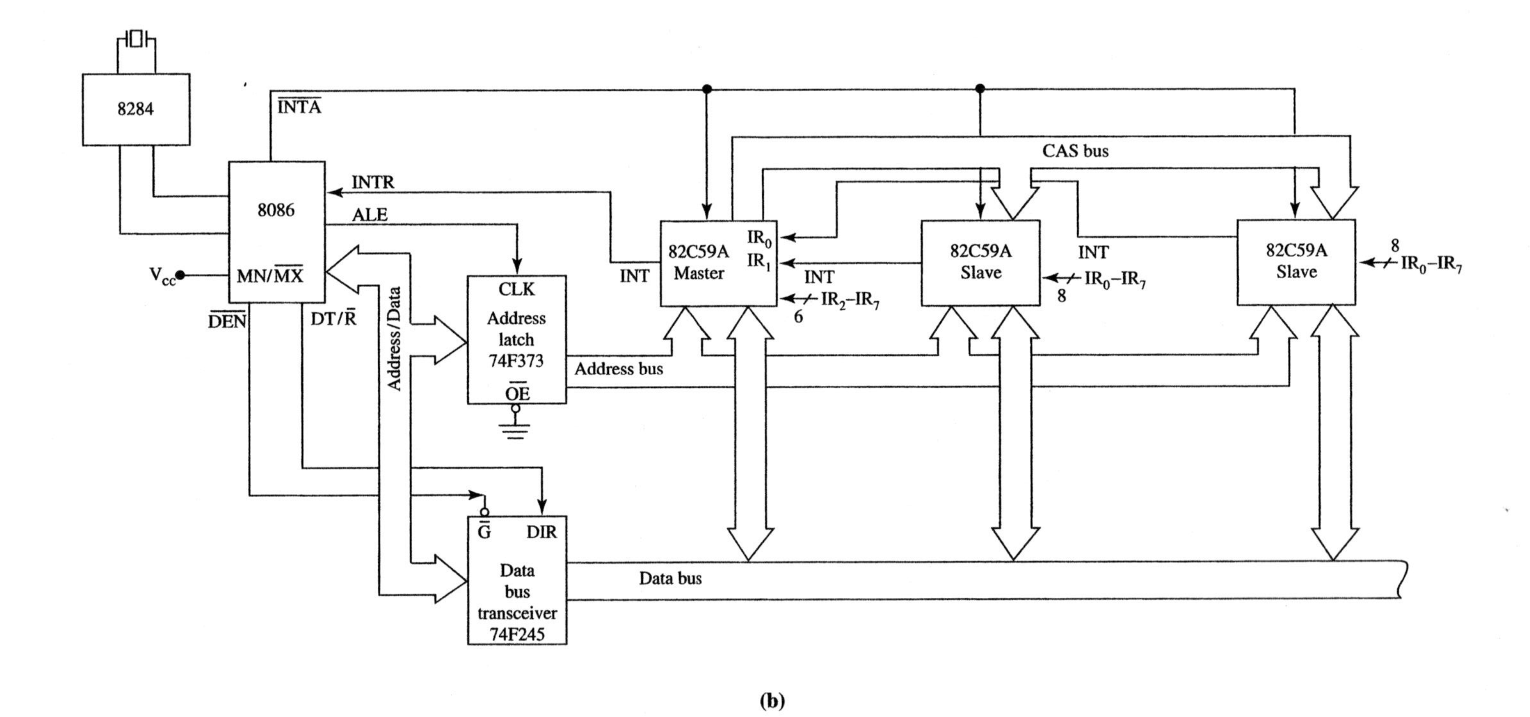

(b)

Figure 11–18 (continued)

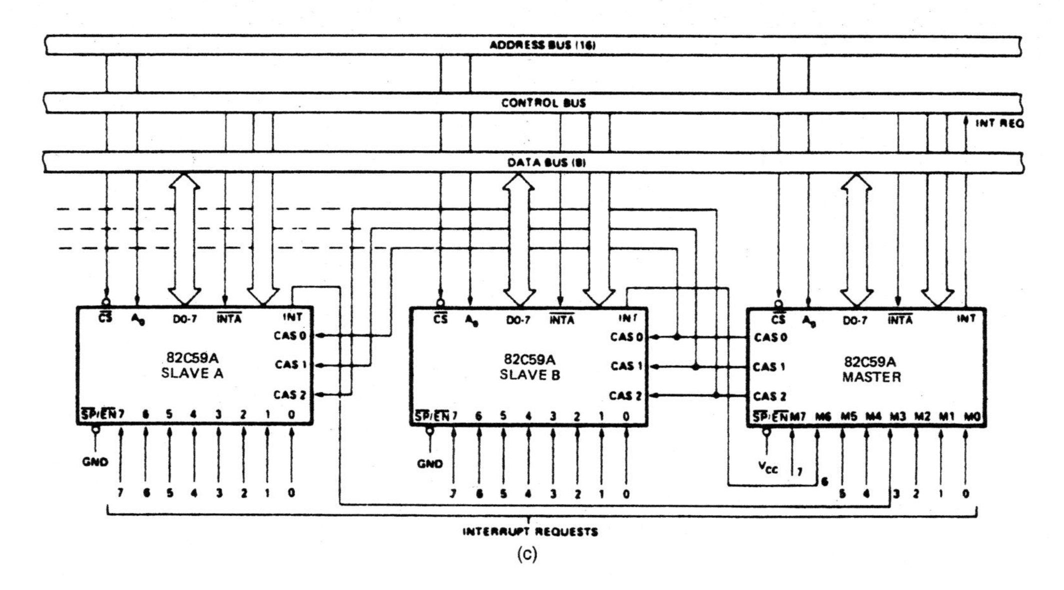

(c)

Figure 11–18 (continued)

82C59A interfaces to the MPU. Note that data bus lines D_0 through D_7 of the 82C59A are connected directly to the 8088's multiplexed address/data bus. It is over these lines that the 8088 initializes the internal registers of the 82C59A, reads the contents of these registers, and reads the type number of the active interrupt input during the interrupt-acknowledge bus cycle. In this circuit, the registers of the 82C59A are assigned to unique I/O addresses.

During input or output bus cycles to one of these addresses, bits of the demultiplexed address are decoded to generate the $\overline{CS}$ for the 82C59A. When this input is at its active, logic 0, level, the 82C59A's microprocessor interface is enabled for operation. The logic level of address bit A_0 selects a specific register within the PIC.

The control signal generation logic section produces the signals that identify whether an input or output data transfer is taking place. Note that the 8088's $\overline{RD}$, $\overline{WR}$, and $IO/\overline{M}$ control signals are decoded by this circuit to generate four control signals $\overline{IOR}$, $\overline{IOW}$, $\overline{MEMR}$, and $\overline{MEMW}$. Control signals $\overline{IOR}$ and $\overline{IOW}$ are supplied to the $\overline{RD}$ and $\overline{WR}$ inputs of the 82C59A, respectively. Logic 0 at one of these outputs tells the 82C59A whether an input or output bus operation is taking place.

Next we will trace the sequence of events that takes place as an external device requests service through the interrupt interface circuit. The external interrupt request inputs are identified as IR_0 through IR_7 in the circuit of Fig. 11–18(a). Whenever an interrupt input becomes active, and either no other interrupt is active or the priority level of the new interrupt is higher than that of the already active interrupt, the 82C59A switches its INT output to logic 1. This output is returned to the INTR input of the 8088. In this way, it signals the MPU that an external device needs to be serviced.

As long as the interrupt flag within the 8088 is set to 1, the interrupt interface is enabled. Assuming that IF is 1 when an IR input becomes active, the interrupt request is accepted and the interrupt-acknowledge bus cycle sequence is initiated. During the first interrupt-acknowledge bus cycle, the 8088 outputs a pulse at its $\overline{INTA}$ output. $\overline{INTA}$ is applied directly to the $\overline{INTA}$ input of the 82C59A and when logic 0, it signals that the active interrupt request will be serviced.

As the second interrupt-acknowledge bus cycle is executed, another pulse is output at $\overline{INTA}$. This pulse signals the 82C59A to output the type number of its highest-priority active interrupt onto the data bus. The 8088 reads this number off the bus and initiates a vectored transfer of program control to the starting point of the corresponding service routine in program memory.

For applications that require more than eight interrupt-request inputs, several 82C59As are connected into a *master-slave configuration.* The circuit in Fig. 11–18(b) shows how three devices are connected to construct a master-slave interrupt interface for a minimum-mode 8086 microcomputer system. Each of these devices must reside at unique addresses in the I/O or memory address space. In this way, during read or write bus cycles to the interrupt interface, the address output on the bus can be decoded to produce a chip-enable signal to select the appropriate device for operation.

Figure 11–18(c) shows a master/slave connection in more detail. Here we find that the rightmost device is identified as the master and the devices to the left as slave A and slave B. At the interrupt-request side of the devices, we find that slaves A and B are cascaded to the master 82C59A by attaching their INT outputs to the M_3 (IR_3) and M_6 (IR_6) inputs, respectively. This means that the identification code for slave A is 3 and that of slave B is 6. Moreover, the CAS lines on all three 82C59As are tied in parallel. Using the CAS lines, the master signals the slaves to tell them whose interrupt request has been acknowledged.

Whenever a slave signals the master that an interrupt input is active, the master determines whether or not its priority is higher than that of any already active interrupt. If the new interrupt is of higher priority, the master controller switches INTR to logic 1. This signals the MPU that an external device needs to be serviced. If the interrupt flag within the MPU is set to 1, the interrupt interface is enabled and the interrupt request will be accepted. Therefore, the MPU initiates the interrupt-acknowledge bus cycle sequence. As the first pulse is output at $\overline{INTA}$, the master 82C59A is signaled to output the 3-bit cascade code of the slave device whose interrupt request is being acknowledged on the CAS bus. All slaves read this code and compare it to their own internal code. In this way, the slave corresponding to the code is signaled to output the type number of its highest-priority active interrupt onto the data bus during the second interrupt-acknowledge bus cycle. The MPU reads this number off the bus and uses it to pass program control to the beginning of the corresponding interrupt service routine.

Figure 11–19 illustrates an interrupt interface implemented for a maximum-mode 8088 microcomputer system.

EXAMPLE 11.9

Analyze the circuit in Fig. 11–20(a) and write an appropriate main program and a service routine that counts as a decimal number the positive edges of the clock signal applied to the IR_0 input of the 82C59A.

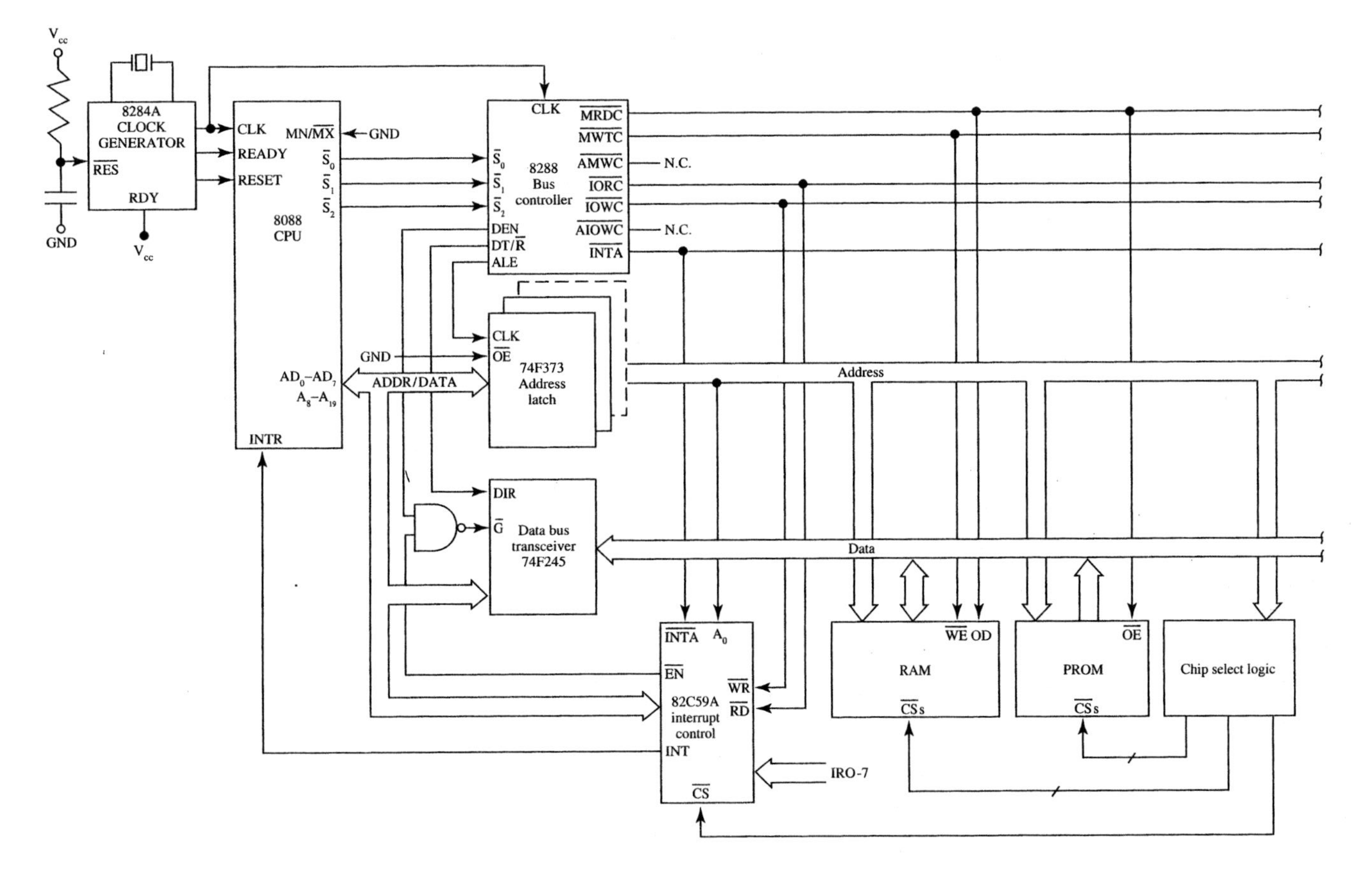

Figure 11–19 Maximum-mode interrupt interface for the 8088 microcomputer using the 82C59A. (Reprinted by permission of Intel Corporation. Copyright/Intel Corp. 1979)

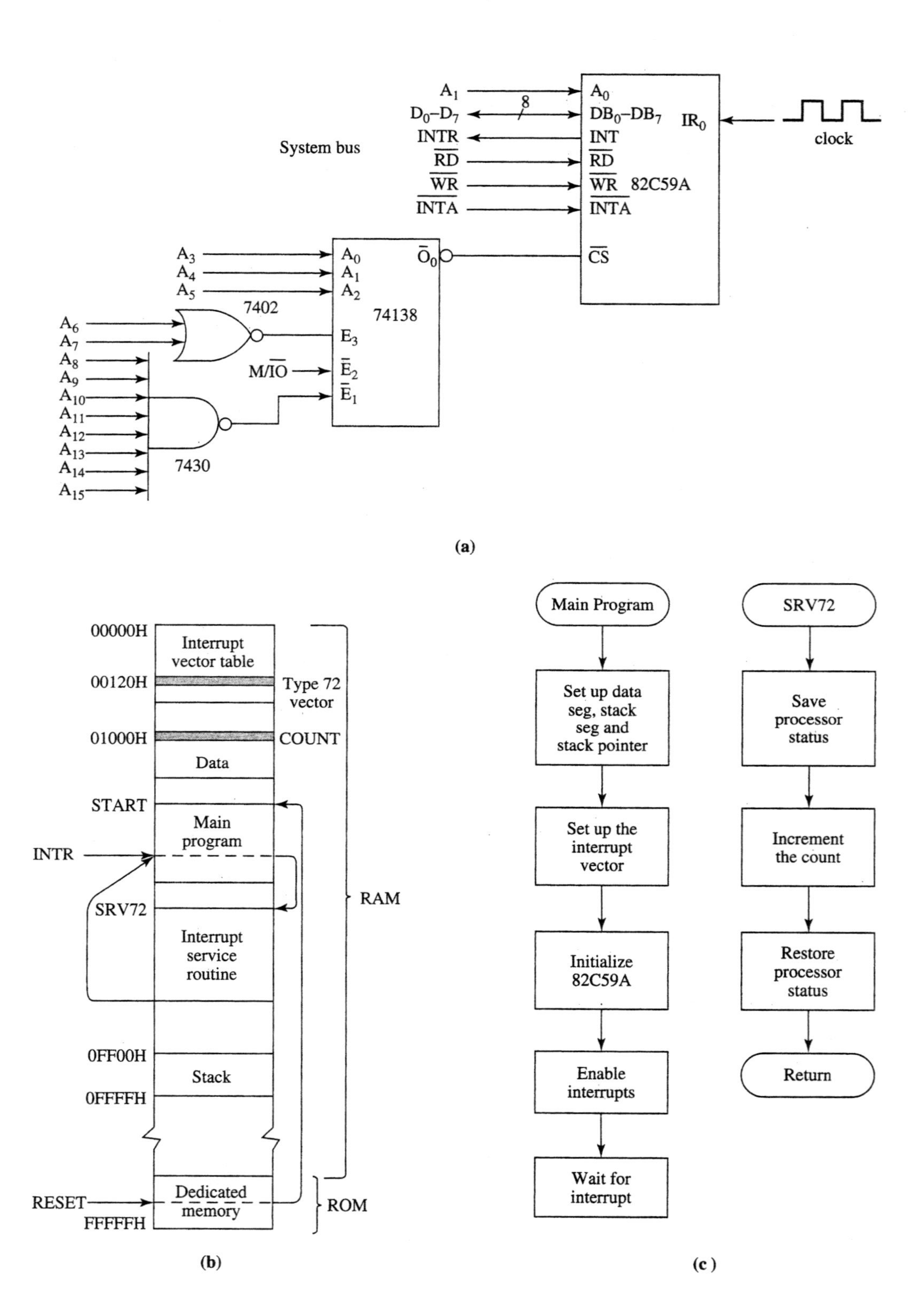

Figure 11–20 (a) Circuit for Example 11.9. (b) Software organization. (c) Flowcharts for the main program and service routine.

Solution

The microprocessor addresses to which the 82C59A in the circuit shown in Fig. 11–20(a) responds depend on how the $\overline{CS}$ signal for the 82C59A is generated as well as the logic level of address bit A_1, connected to input A_0. Note that the A_0 address line of the microprocessor is not used in the circuit and therefore it is a don't-care bit. Thus, if A_0 is taken as 0, the 82C59A responds to

$$A_{15}A_{14}A_{13}A_{12}A_{11}A_{10}A_9A_8A_7A_6A_5A_4A_3A_2A_1A_0$$

$$= 1111111100000000_2 \text{ for } A_1 = 0, M/\overline{IO} = 0 \text{ and}$$

$$= 1111111100000010_2 \text{ for } A_1 = 1, M/\overline{IO} = 0$$

These two I/O addresses are FF00H and FF02H, respectively. The address FF00H is for the ICW_1 and FF02H is for the ICW_2, ICW_3, ICW_4, and OCW_1 command words. Let us now determine the ICWs and OCWs for the 82C59A.

Because the $M/\overline{IO}$ signal is used in the circuit diagram, the 82C59A interface is for the 8086 microprocessor, there is only one 82C59A in the system, and the interrupt input is an edge, we are led to the following ICW_1:

$$ICW_1 = 00010011_2 = 13H$$

Let us assume that we will use interrupt type 72 to service an interrupt generated by an edge presented to the IR_0. This leads to the following ICW_2:

$$ICW_2 = 01001000_2 = 48H$$

For a single 82C59A, ICW_3 is not needed. To determine ICW_4, let us assume that we will use auto EOI and nonbuffered mode of operation. This leads to the following ICW_4:

$$ICW_4 = 00000011_2 = 03H$$

For OCWs, we will use only OCW_1 to mask all other interrupts but IR_0. This gives OCW_1 as

$$OCW_1 = 11111110_2 = FEH$$

Figure 11–20(b) shows the memory organization for the software. Let us understand the information presented in this memory organization. In the interrupt-vector table we need to set up the type 72 vector. The type 72 vector is located at $4 \times 72 = 288 = 120H$. At address 120H we need to place the offset of the service routine and at address 122H the code segment value of the service routine.

In the data area we need a location to keep a decimal count of the edges of the input clock. Let us assume that it is location 01000H. The stack segment starts at 0FF00H and ends at 0FFFFH. The start address of the main program is denoted as START, and that of the service routine is denoted as SRV72.

The flowcharts in Fig. 11–20(c) are for the main program and the service routine. The main program initializes the microprocessor and the 82C59A. First we establish various segments for data and stack. This can be done using the following instructions:

```
;MAIN PROGRAM
        CLI                     ;Start with interrupts disabled
START:  MOV  AX, 0              ;Extra segment at 00000H
        MOV  ES, AX
        MOV  AX, 1000H          ;Data segment at 01000H
        MOV  DS, AX
        MOV  AX, 0FF00H         ;Stack segment at 0FF00H
        MOV  SS, AX
        MOV  SP, 100H           ;Top of stack at 10000H
```

Next we can set up the IP and CS for the type 72 vector in the interrupt vector table. This can be accomplished using the following instructions:

```
MOV   AX, OFFSET SRV72     ;Get offset for the service routine
MOV   [ES:120H], AX        ;Set up the IP
MOV   AX, SEG SRV72        ;Get code segment for the service routine
MOV   [ES:122H], AX        ;Set up the CS
```

Having set up the interrupt-type vector, let us proceed now to initialize the 82C59A. Using the analyzed information, the following instructions can be executed:

```
MOV  DX, 0FF00H        ;ICW1 address
MOV  AL, 13H           ;Edge trig input, single 82C59A
OUT  DX, AL
MOV  DX, 0FF02H        ;ICW2,ICW4,OCW1 address
MOV  AL, 48H           ;ICW2, type 72
OUT  DX, AL
MOV  AL, 03H           ;ICW4, AEOI, nonbuff mode
OUT  DX, AL
MOV  AL, 0FEH          ;OCW1, mask all but IR0
OUT  DX, AL
STI                    ;Enable the interrupts
```

Now the processor is ready to accept interrupts. We can write an endless loop to wait for the interrupt to occur. In a real situation we may be doing some other operation in which the interrupt will be received and serviced. For simplicity let us use the following instruction to wait for the interrupt:

```
HERE:     JMP  HERE       ;Wait for an interrupt
```

Figure 11–20(c) shows the flowchart for the interrupt-service routine as well. The operations shown in the flowchart can be implemented using the following instructions:

```
SRV72:    PUSH  AX              ;Save register to be used
          MOV   AL, [COUNT]     ;Get the count
          INC   AL              ;Increment the count
          DAA                   ;Decimal adjust the count
          MOV   [COUNT], AL     ;Save the new count
          POP   AX              ;Restore the register used
          IRET                  ;Return from interrupt
```

▲ 11.9 SOFTWARE INTERRUPTS

The 8088 and 8086 microcomputer systems are capable of implementing up to 256 *software interrupts.* They differ from the external hardware interrupts in that their service routines are initiated in response to the execution of a software interrupt instruction, not an event in external hardware.

The INT *n* instruction is used to initiate a software interrupt. Earlier in this chapter we indicated that *n* represents the type number associated with the service routine. The software interrupt service routine vectors are also located in the memory locations in the vector table. These locations are shown in Fig. 11–3. Our earlier example was INT 50. It has a type number of 50, and causes a vector in program control to the service routine whose starting address is defined by the values of IP_{50} and CS_{50} stored at addresses $000C8_{16}$ and $000CA_{16}$, respectively.

The mechanism by which a software interrupt is initiated is similar to that described for the external hardware interrupts. However, no external interrupt-acknowledge bus cycles are initiated. Instead, control is passed to the start of the service routine immediately upon completion of execution of the interrupt instruction. As usual, first the old flags are automatically saved on the stack; then IF and TF are cleared; next the old CS and old IP are pushed onto the stack; now the new CS and new IP are read from memory and loaded into the MPU's registers; finally program execution resumes at $CS_{NEW}:IP_{NEW}$.

If necessary, the contents of other internal registers can be saved on the stack by including the appropriate PUSH instructions at the beginning of the service routine. Toward the end of the service routine, POP instructions are inserted to restore these registers. Finally, an IRET instruction is used at the end of the routine to return to the original program environment. In this way, we see that the program structure of a software-interrupt service routine is identical to that shown in Fig. 11–10.

Software interrupts are of higher priority than the external interrupts and are not masked out by IF. The software interrupts are actually *vectored subroutine calls.* A common use of these software routines is as *emulation routines* for more complex functions. For instance, INT 50 could define a *floating-point addition instruction* and INT 51 a *floating-point subtraction instruction.* These emulation routines are written using assembly language instructions, are assembled into machine code, and then are stored in the main memory of the 8088 microcomputer system. Other examples of their use are for *supervisor calls* from an operating system and for testing external hardware interrupt service routines.

▲ 11.10 NONMASKABLE INTERRUPT

The nonmaskable interrupt (NMI) is another interrupt that is initiated from external hardware. However, it differs from the other external hardware interrupts in several ways. First, as its name implies, it cannot be masked out with the interrupt flag. Second, requests for service by this interrupt are signaled to the 8088 or 8086 microprocessor by applying logic 1 at the NMI input, not the INTR input. Third, the NMI input is positive edge-triggered. Therefore, a request for service is automatically latched internal to the MPU.

On the 0 to 1 transition of the NMI input, the NMI flip-flop within the MPU is set. If the contents of the NMI latch are sampled as being active for two consecutive clock cycles, it is recognized and at completion of the current instruction the nonmaskable interrupt sequence is initiated. Just as with the other interrupts we have studied, initiation of NMI causes the current flags, current CS, and current IP to be pushed onto the stack. Moreover, the interrupt-enable flag is cleared to disable all external hardware interrupts, and the trap flag is cleared to disable the single-step mode of operation. Next the MPU fetches the words of the NMI vector from memory and loads them into IP and CS. Finally, execution resumes with the first instruction of the NMI service routine.

As Fig. 11–3 shows, NMI has a dedicated type number. It automatically vectors from the type 2 vector location in the pointer table. This vector is stored in memory at word addresses 0008_{16} and $000A_{16}$.

Typically, the NMI is assigned to hardware events that must be responded to immediately. Two examples are the detection of a power failure and detection of a memory-read error.

▲ 11.11 RESET

The RESET input of the 8088 and 8086 microprocessors provides a hardware means for initializing the microcomputer. This is typically done at power-up to provide an orderly startup of the system. However, some systems, such as a personal computer, also allow for a *warm start*—that is, a software-initiated reset.

Figure 11–21(a) shows that the reset interface of the 8088 includes part of the 8284 clock generator device. The 8284 contains circuitry that makes it easy to implement the hardware reset function. This circuit is used to detect an active reset input and synchronize the application and removal of the RESET signal with the clock. Note that the $\overline{\text{RES}}$ input (pin 11) of the clock generator is attached to an RC circuit. The signal at $\overline{\text{RES}}$ is applied to the input of an internal Schmitt trigger circuit. If the voltage across the capacitor is below the 1-logic-level threshold of the Schmitt trigger, the RESET output (pin 10) stays at logic 1. This output is supplied to the RESET input at pin 21 of the 8088. It can also be applied in parallel to reset inputs on LSI and VLSI peripheral devices in the microcomputer system. In this way, they are also initialized at power-on. Figure 11–21(a) provides the system $\overline{\text{reset}}$ signal line for this purpose.

At power-on, $\overline{\text{RES}}$ of the 8284 is shorted to ground through the capacitor. This represents logic 0 at the input of the Schmitt trigger and causes the RESET output to switch to its active 1 logic level. At the RESET input of the 8088, this signal is synchronized to

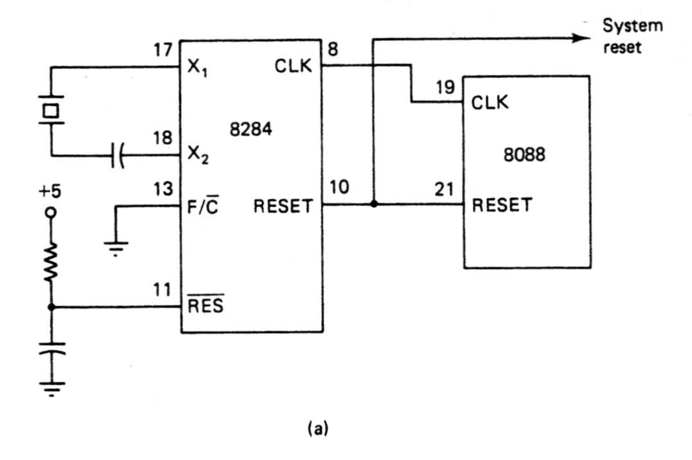

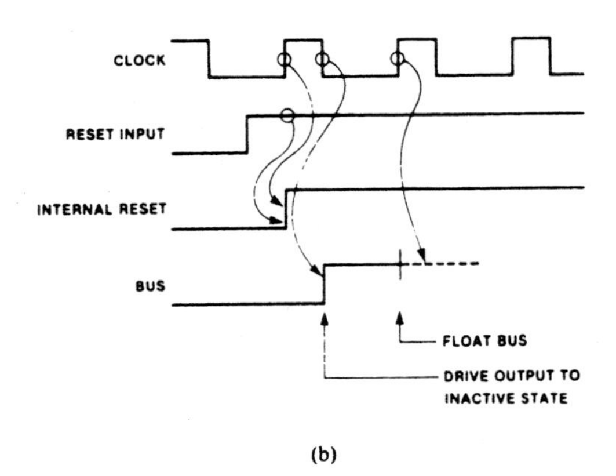

Figure 11–21 (a) Reset interface of the 8088. (Reprinted by permission of Intel Corporation. Copyright/Intel Corp. 1979) (b) Reset timing sequence. (Reprinted by permission of Intel Corporation. Copyright/Intel Corp. 1979)

the 0-to-1 edge of CLK to create an internal reset signal. This is shown in the waveforms of Fig. 11–21(b). RESET must be held at logic 1 for a minimum of four clock cycles; otherwise, it will not be recognized.

When RESET is recognized as active, the 8088 terminates operation, puts its buses in the high-Z state, and switches the control signals to their inactive states. Figure 11–22(a) summarizes these signal states. Here we see that in a minimum-mode system, signals AD_0 through AD_7, A_8 through A_{15}, and $A_{16}/\overline{S}_3$ through $A_{19}/\overline{S}_6$, are immediately put in the high-Z state. On the other hand, signal lines $\overline{SSO}$, $IO/\overline{M}$, $DT/\overline{R}$, $\overline{DEN}$, $\overline{WR}$, $\overline{RD}$, and $\overline{INTA}$ are first forced to logic 1 for one clock interval and then are put in the high-Z state synchronously with the positive edge of the next clock pulse. Moreover, signal lines ALE and HLDA are forced to their inactive 0 logic level. The 8088 remains in this state until the RESET input is returned to logic 0.

The hardware of the reset interface in an 8086 microcomputer system is identical to that just shown for the 8088 microprocessor. In fact, the reset and clock inputs, found at pins 21 and 19 of the 8088, respectively, in Fig. 11–21(a), are at these same pins on the 8086. Moreover, the waveforms given in Fig. 11–21(b) also describe the timing sequence

Signals	Condition
$AD_{7\text{-}0}$	Three-state
$A_{15\text{-}8}$	Three-state
$A_{19\text{-}16}/S_{6\text{-}3}$	Three-state
$\overline{SSO}$	Driven to 1, then three-state
$\overline{S_2}/(IO/\overline{M})$	Driven to 1, then three-state
$\overline{S_1}/(DT/\overline{R})$	Driven to 1, then three-state
$\overline{S_0}/\overline{DEN}$	Driven to 1, then three-state
$\overline{LOCK}/\overline{WR}$	Driven to 1, then three-state
$\overline{RD}$	Driven to 1, then three-state
$\overline{INTA}$	Driven to 1, then three-state
ALE	0
HLDA	0
$\overline{RQ}/\overline{GT_0}$	1
$\overline{RQ}/\overline{GT_1}$	1
QS_0	0
QS_1	0

(a)

Signals	Condition
$AD_{15\text{-}0}$	Three-state
$A_{19\text{-}16}/S_{6\text{-}3}$	Three-state
BHE/S_7	Three-state
$\overline{S_2}/(M/\overline{IO})$	Driven to "1" then three-state
$\overline{S_1}/(DT/\overline{R})$	Driven to "1" then three-state
$\overline{S_0}/\overline{DEN}$	Driven to "1" then three-state
$\overline{LOCK}/\overline{WR}$	Driven to "1" then three-state
$\overline{RD}$	Driven to "1" then three-state
$\overline{INTA}$	Driven to "1" then three-state
ALE	0
HLDA	0
$\overline{RQ}/\overline{GT_0}$	1
$\overline{RQ}/\overline{GT_1}$	1
QS_0	0
QS_1	0

(b)

Figure 11–22 (a) Bus and control signal status of the 8088 during system reset. (Reprinted by permission of Intel Corporation. Copyright/Intel Corp. 1979) (b) Bus and control signal status of the 8086 during system reset. (Reprinted by permission of Intel Corporation. Copyright/Intel Corp. 1979)

that occurs when the reset input of the 8086 is activated. Remember that the 8086 produces some different signals than the 8088. For example, it has a $\overline{BHE}$ output instead of an $\overline{SSO}$ output. Figure 11–22(b) shows the state of the 8086's bus and control signals during reset.

In the maximum-mode system, the 8088 and 8086 respond in a similar way to an active reset request. However, this time the $\overline{S_2}\overline{S_1}\overline{S_0}$ outputs, which are inputs to the 8288 bus controller, are also forced to logic 1 and then put into the high-Z state. These inputs

CPU COMPONENT	CONTENT
Flags	Clear
Instruction Pointer	0000H
CS Register	FFFFH
DS Register	0000H
SS Register	0000H
ES Register	0000H
Queue	Empty

Figure 11–23 Internal state of the 8088/8086 after reset. (Reprinted by permission of Intel Corporation. Copyright/Intel Corp. 1979)

of the 8288 have internal pullup resistors. Therefore, with the signal lines in the high-Z state, the input to the bus controller is $\overline{S_2}\overline{S_1}\overline{S_0} = 111$. In response, its control outputs are set to ALE = 0, DEN = 0, DT/$\overline{R}$ = 1, and all its command outputs are switched to the 1 logic level. Moreover, outputs QS_0 and QS_1 of the MPU are both held at logic 0 and the $\overline{RQ}/\overline{GT_0}$, and $\overline{RQ}/\overline{GT_1}$ lines are held at logic 1.

Return of the reset signal to logic 0 is also synchronized to CLK by the 8284. When the MPU recognizes the return to logic 0 at the RESET input, it initiates its internal initialization routine. At completion of initialization, the flags are all cleared, the instruction pointer is set to 0000_{16}, the CS register is set to $FFFF_{16}$, the DS, SS, and ES registers are set to 0000_{16}, and the instruction queue is emptied. The table in Fig. 11–23 summarizes this state.

Since the flags were all cleared as part of initialization, external hardware interrupts are disabled. Moreover, the code segment register contains $FFFF_{16}$ and the instruction pointer contains 0000_{16}. Therefore, program execution begins at address $FFFF0_{16}$ after reset. This storage location can contain an instruction that will cause a jump to the startup program that is used to initialize the rest of the microcomputer system's resources, such as I/O ports, the interrupt flag, and data memory. This startup program is known as a boot-strap program. After system-level initialization is complete, another jump can be performed to the starting point of the microcomputer's operating system or application program.

▲ 11.12 INTERNAL INTERRUPT FUNCTIONS

Earlier we indicated that four of the 256 interrupts of the 8088 and 8086 are dedicated to internal interrupt functions. Internal interrupts differ from external hardware interrupts in that they occur due to the result of executing an instruction, not an event that takes place in external hardware. That is, an internal interrupt is initiated because of a condition detected before, during, or after execution of an instruction. In this case, a routine must be initiated to service the internal condition before resuming execution of the same or next instruction of the program.

Looking at Fig. 11–8, we find that internal interrupts are not masked out with the interrupt-enable flag. For this reason, occurrence of any one of them is automatically detected by the MPU and causes an interrupt of program execution and a vectored transfer of program control to a corresponding service routine. During the control transfer sequence, no interrupt acknowledge bus cycles are produced.

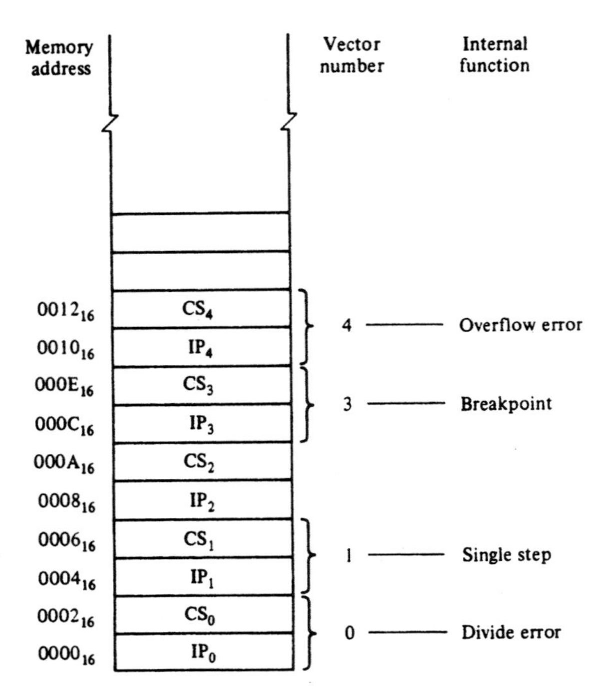

Figure 11–24 Internal interrupt vector locations.

Figure 11–24 identifies the internal interrupts of the 8088 and 8086 microprocessors. Here we find divide error, overflow error, single step, and breakpoint. Each of these functions is assigned a unique type number. Notice that they are the highest-priority type numbers.

Let us now look at each of these internal functions in more detail.

Divide Error

The *divide error* function represents an error condition that can occur in the execution of the division instructions. If the quotient that results from a DIV (divide) instruction or an IDIV (integer divide) instruction is larger than the specified destination, a divide error has occurred. This condition causes automatic initiation of a type 0 interrupt and passes control to a service routine whose starting point is defined by the values of IP_0 and CS_0 at addresses 00000_{16} and 00002_{16}, respectively, in the pointer table.

Overflow Error

The *overflow error* is an error condition similar to that of divide error. However, it can result from the execution of any arithmetic instruction. Whenever an overflow occurs, the overflow flag gets set. Unlike divide error, the transfer of program control to a service routine is not automatic at occurrence of the overflow condition. Instead, the INTO (interrupt on overflow) instruction must be executed to test the overflow flag (OF) and determine if the overflow service routine should be initiated. If the overflow flag is tested and found to be set, a type 4 interrupt service routine is initiated. Its vector consists of IP_4 and CS_4, which are stored at 00010_{16} and 00012_{16}, respectively, in memory. The routine pointed to by this vector is written to service the overflow condition. For instance, it could cause a message to be displayed to specify that an overflow has occurred.

Single Step

The *single-step* function relates to an operating option of the 8088 or 8086. If the trap flag (TF) bit in the flags register is set, the single-step mode of operation is enabled. This flag bit can be set or reset under software control.

When TF is set, the MPU initiates a type 1 interrupt to the service routine defined by IP_1 and CS_1 at addresses 00004_{16} and 00006_{16}, respectively, at the completion of execution of every instruction of the user program. This permits implementation of the single-step mode of operation so that the program can be executed one instruction at a time. For instance, the service routine could include a WAIT instruction. In this way, a transition to logic 0 at the $\overline{TEST}$ input of the 8088 or 8086 could be used as the mechanism for stepping through a program one instruction at a time. This single-step operation can be used as a valuable software debugging tool.

Breakpoint Interrupt

The *breakpoint* function can also be used to implement a software diagnostic tool. A *breakpoint interrupt* is initiated by execution of the breakpoint instruction (one-byte instruction with code = 11001100_2 = CC_{16}). This instruction can be inserted at strategic points in a program that is being debugged to cause execution to be stopped automatically. Breakpoint interrupt can be used in a way similar to that of the single-step option. The breakpoint service routine can stop execution of the main program, permit the programmer to examine the contents of registers and memory, and allow for the resumption of execution of the program down to the next breakpoint.

REVIEW PROBLEMS

Section 11.1

1. What are the five groups of interrupts supported on the 8088 and 8086 MPUs?
2. What name is given to the special software routine to which control is passed when an interrupt occurs?
3. List in order the interrupt groups; start with the lowest priority and end with the highest priority.
4. What is the range of type numbers assigned to the interrupts in the 8088 and 8086 microcomputer systems?
5. Is the interrupt assigned to type 21 at a higher or lower priority than the interrupt assigned to type 35?

Section 11.2

6. Where are the interrupt pointers held?
7. How many bytes of memory does an interrupt vector take up?

8. What two elements make up an interrupt vector?
9. Which interrupt function's service routine is specified by $CS_4:IP_4$?
10. The breakpoint routine in an 8086 microcomputer system starts at address $AA000_{16}$ in the code segment located at address $A0000_{16}$. Specify how the breakpoint vector will be stored in the interrupt-vector table.
11. At what addresses is the interrupt vector for type 40 stored in memory?

Section 11.3

12. What does STI stand for?
13. Which type of instruction does INTO normally follow? Which flag does it test?
14. What happens when the instruction HLT is executed?

Section 11.4

15. Explain how the CLI and STI instructions can be used to mask out external hardware interrupts during the execution of an uninterruptable subroutine.
16. How can the interrupt interface be reenabled during the execution of an interrupt service routine?

Section 11.5

17. What does $\overline{\text{INTA}}$ stand for?
18. Is the INTR input of the 8088 edge triggered or level triggered?
19. Explain the function of the INTR and $\overline{\text{INTA}}$ signals in the circuit diagram shown in Fig. 11–6(a).
20. Which device produces $\overline{\text{INTA}}$ in a minimum-mode 8088 microcomputer system? In a maximum-mode 8088 microcomputer system?
21. Over which signal lines does external circuitry send the type number of the active interrupt to the 8086?
22. What bus status code is assigned to interrupt acknowledge?

Section 11.6

23. Give an overview of the events in the order they take place during the interrupt-request, interrupt-acknowledge, and interrupt-vector-fetch cycles of an 8088 microcomputer system.
24. If an 8086-based microcomputer is running at 10 MHz with two wait states, how long does it take to perform the interrupt-acknowledge bus cycle sequence?
25. How long does it take the 8086 in problem 24 to push the values of the old flags, old CS, and old IP to the stack? How much stack space does this information take?
26. How long does it take the 8086 in problem 24 to fetch its vector $CS_{NEW}:IP_{NEW}$ from memory?

Section 11.7

27. Specify the value of ICW_1 needed to configure an 82C59A as follows: ICW_4 not needed, single-device interface, and edge-triggered inputs.

28. Specify the value of ICW_2 if the type numbers produced by the 82C59A are to be in the range 70_{16} through 77_{16}.
29. Specify the value of ICW_4 such that the 82C59A is configured for use in an 8086 system, with normal EOI, buffered-mode master, and special fully nested-mode disabled.
30. Write a program that initializes an 82C59A with the initialization command words derived in problems 27, 28, and 29. Assume that the 82C59A resides at address $0A000_{16}$ in the memory address space and that the contents of DS are 0000_{16}.
31. Write an instruction that, when executed, reads the contents of OCW_1 and places it in the AL register. Assume that the software in problem 30 has configured the 82C59A.
32. What priority scheme is enabled if OCW_2 equals 67_{16}?
33. Write an instruction sequence that when executed toggles the state of the read register bit in OCW_3. Assume that the 82C59A is located at memory address $0A000_{16}$ and that the contents of DS are 0000_{16}.

Section 11.8

34. The circuit in Fig. 11–18(a) can accept how many interrupt inputs?
35. The circuit in Fig. 11–18(b) can accept how many interrupt inputs?
36. Summarize the interrupt-request/acknowledge handshake sequence for an interrupt initiated at an input to slave B in the circuit in Fig. 11–18(c).
37. What is the maximum number of interrupt inputs that can be achieved by expanding the number of slaves in the master-slave configuration in Fig. 11–18(c)?

Section 11.9

38. Give another name for a software interrupt.
39. If the instruction INT 80 is to pass control to a subroutine at address $A0100_{16}$ in the code segment starting at address $A0000_{16}$, what vector should be loaded into the interrupt vector table?
40. At what address would the vector for the instruction INT 80 be stored in memory?

Section 11.10

41. What type number and interrupt vector table addresses are assigned to NMI?
42. What are the key differences between NMI and the other external-hardware-initiated interrupts?
43. Give a common use of the NMI input.

Section 11.11

44. What is the active logic level of the RESET input of the 8088?
45. To which signal must the application of the RESET input be synchronized?
46. What device is normally used to generate the signal for the RESET input of the 8088?
47. List the states of the address/data bus lines and control signals $\overline{BHE}$, ALE, $\overline{DEN}$, DT/$\overline{R}$, $\overline{RD}$, and $\overline{WR}$ in a minimum-mode 8086 system when reset is at its active level.

48. What is the address from where the first instruction is fetched by the MPU after the reset has been applied?

49. Write a reset subroutine that initializes the block of memory locations from address $0A000_{16}$ to $0A0FF_{16}$ to 0H. The initialization routine is at address 01000_{16}.

Section 11.12

50. List the internal interrupts serviced by the 8088.

51. Which vector numbers are allocated to internal interrupts?

52. What mode of operation is enabled with the trap flag? Which pointer holds the entry point for this service routine?

53. If the starting point of the service routine for problem 52 is defined by CS:IP = A000H:0200H, at what addresses in memory are the values of CS and IP held? At what physical address does the service routine start?

12

Hardware of the Original IBM PC Microcomputer

▲ INTRODUCTION

Having learned about the 8088 and 8086 microprocessors, their memory, input/output, and interrupt interfaces, we now turn our attention to a microcomputer system designed using this hardware. The microcomputer we will study in this chapter is the one found in the original IBM PC, the first 8088-based personal computer manufactured by IBM Corporation. The material covered in this chapter is organized as follows:

- 12.1 Architecture of the Original IBM PC System Processor Board
- 12.2 System Processor Circuitry
- 12.3 Wait-State Logic and NMI Circuitry
- 12.4 Input/Output and Memory Chip-Select Circuitry
- 12.5 Memory Circuitry
- 12.6 Direct Memory Access Circuitry
- 12.7 Timer Circuitry
- 12.8 Input/Output Circuitry
- 12.9 Input/Output Channel Interface

▲ 12.1 ARCHITECTURE OF THE ORIGINAL IBM PC SYSTEM PROCESSOR BOARD

The IBM PC is a practical application of the 8088 microprocessor and its peripheral chip set as a general-purpose microcomputer. A block diagram of the system processor board (main circuit board) of the original PC is shown in Fig. 12–1(a). This diagram identifies the major functional elements of the PC: MPU, PIC, DMA, PIT, PPI, ROM, and RAM. We will describe the circuitry used in each of these blocks in detail in the sections of this chapter that follow; however, let us begin with an overview of the architecture of the PC's microcomputer system.

The heart of the PC's system processor board is the 8088 microprocessor unit (MPU). It is here that instructions of the program are fetched and executed. To interface to the peripherals and other circuitry such as memory, the 8088 microprocessor generates address, data, status, and control signals. Together these signals form what is called the *local bus* in Fig. 12–1(a). Note that the local address and data bus lines are both buffered and demultiplexed to provide a separate 20-bit *system address bus* and 8-bit *system data bus.*

At the same time, the bus controller decodes the status and control lines of the local bus to generate the *system control bus.* This control bus consists of memory and I/O read and write control signals. The bus controller also produces the signals that control the direction of data transfer through the data bus buffers—that is, the signals needed to make the data bus lines work as inputs to the microprocessor during memory and I/O read operations and as outputs during write operations.

The operation of the microprocessor and other devices in a microcomputer system must be synchronized. The circuitry in the clock generator block of the PC in Fig. 12–1(a) generates clock signals for this purpose. The clock generator section also produces a power-on reset signal needed for initialization of the microprocessor and peripherals at power-up. Moreover, the clock generator section works in conjunction with the wait-state logic to synchronize the MPU to slow peripheral devices. In Fig. 12–1(a), we see that the wait-state logic circuitry monitors the system control bus signals and generates a wait signal for input to the clock generator. In turn, the clock generator synchronizes the wait input to the system clock to produce a ready signal at its output. This ready signal is input to the 8088 MPU and provides the ability to automatically extend bus cycles that are performed to slow devices by inserting wait states.

The memory subsystem of the PC system processor board we are studying in this chapter has 256Kbytes of dynamic R/W memory (RAM) and 48Kbytes of read-only memory (ROM). Fig. 12–1(b) shows a memory map for the PC's memory. From the map, we find that the RAM address ranges from 00000_{16} through $3FFFF_{16}$. This part of the memory subsystem can be implemented using 64K × 1-bit or 256K × 1-bit dynamic RAMs and is used to store operating system routines, application programs, and data to be processed. These programs and data are typically loaded into RAM from a mass storage device such as a diskette or hard disk.

Furthermore, the memory map shows that ROM is located in the address range from $F4000_{16}$ to $FFFFF_{16}$. This part of the memory subsystem contains the basic system ROM of the PC. Included in these ROMs are fixed programs such as the *BASIC interpreter, power-on system procedures,* and *I/O device drivers,* or *BIOS* as they are better known.

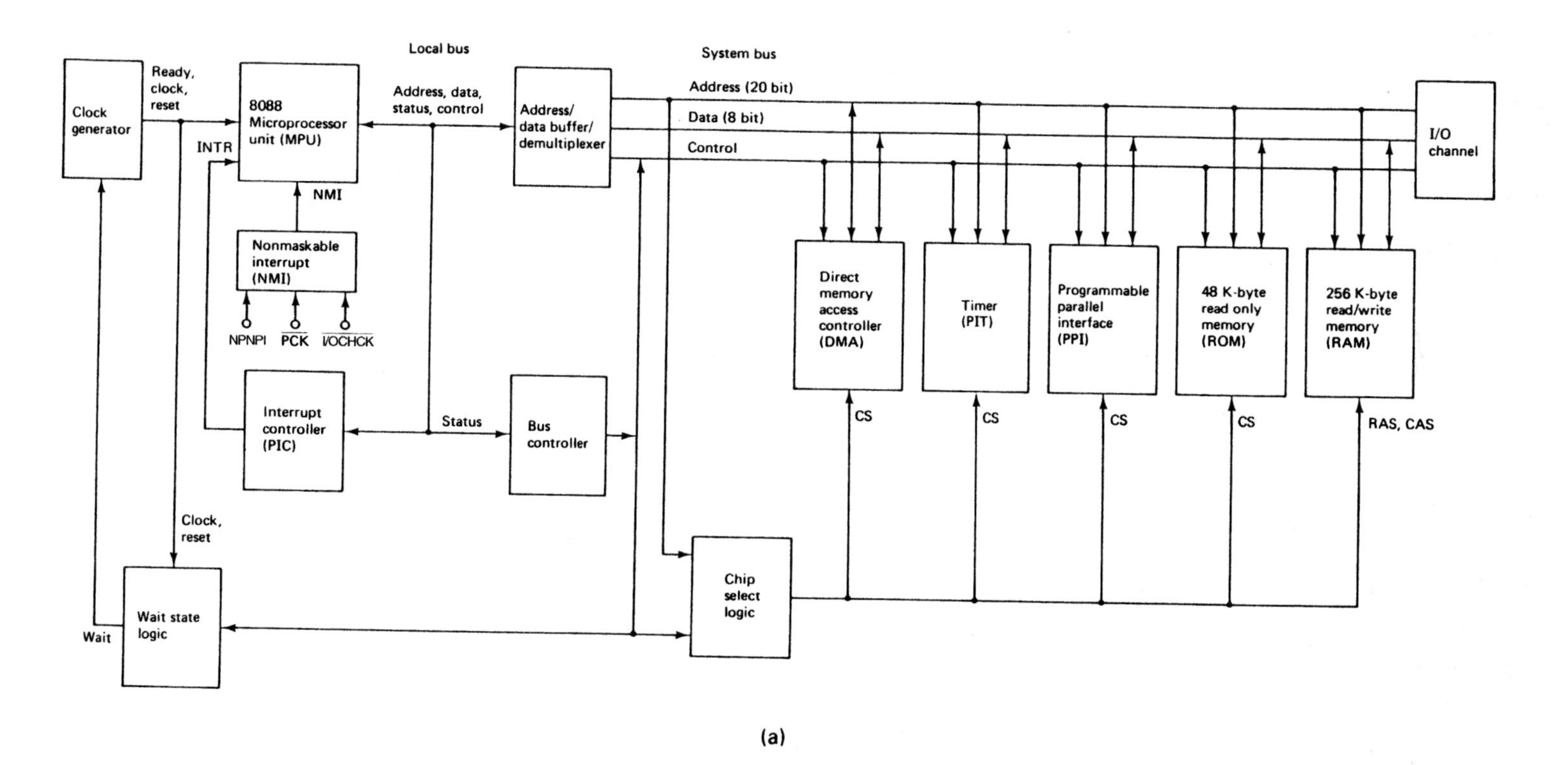

Figure 12–1 (a) Original IBM PC microcomputer block diagram. (b) Memory map. (c) PC system processor board peripheral addresses. (d) 8255A I/O map. (e) Interrupts. (Parts b, c, d, and e, courtesy of International Business Machines Corporation)

Start Address Decimal	Hex	Function
0	00000	64 to 256K Read/Write Memory on System Board
16K	04000	
32K	08000	
48K	0C000	
64K	10000	
80K	14000	
96K	18000	
112K	1C000	
128K	20000	
144K	24000	
160K	28000	
176K	2C000	
192K	30000	
208K	34000	
224K	38000	
240K	3C000	
256K	40000	Up to 384K Read/Write Memory in I/O Channel Up to 384K in I/O Channel
272K	44000	
288K	48000	
304K	4C000	
320K	50000	
336K	54000	
352K	58000	
368K	5C000	
384K	60000	
400K	64000	
416K	68000	
432K	6C000	
448K	70000	
464K	74000	
480K	78000	
496K	7C000	
512K	80000	
528K	84000	
544K	88000	
560K	8C000	
576K	90000	
592K	94000	
608K	98000	
624K	9C000	

(b)

Figure 12–1 (continued)

Start Address Decimal	Hex	Function
640K	A0000	128K Reserved
656K	A4000	
672K	A8000	
688K	AC000	
704K	B0000	Monochrome
720K	B4000	
736K	B8000	Color/Graphics
752K	BC000	
768K	C0000	192K Read Only Memory Expansion and Control
784K	C4000	
800K	C8000	Fixed Disk Control
816K	CC000	
832K	D0000	
848K	D4000	
864K	D8000	
880K	DC000	
896K	E0000	
912K	E4000	
928K	E8000	
944K	EC000	
960K	F0000	Reserved
976K	F4000	48K Base System ROM
992K	F8000	
1008K	FC000	

(b)

Figure 12–1 (continued)

The chip-select logic section, shown in the block diagram of Fig. 12–1(a), is used to select and enable the appropriate peripheral or memory devices whenever a bus cycle takes place over the system bus. To select a device in the I/O address space, such as the DMA controller, timer, or PPI, it decodes the address on the system bus to generate a chip-select (CS) signal for the corresponding I/O device. This chip-select signal is applied to the I/O device to enable it to operate. The memory chip selects are produced in a similar way by decoding the memory address on the system address bus.

The LSI peripherals included on the PC system processor board are the 8237A direct memory access (DMA) controller, 8253 programmable interval timer (PIT), 8255A programmable peripheral interface (PPI), and 8259A programmable interrupt controller (PIC). Note that each of these devices is identified with a separate block in Fig. 12–1(a). These peripherals are all located in the 8088's I/O address space, and their registers are accessed through software using the address ranges given in Fig. 12–1(c). For instance, the four registers within the PIT are located at addresses 0040_{16}, 0041_{16}, 0042_{16}, and 0043_{16}.

Hex Range	Usage
000-00F	DMA Chip 8237A-5
020-021	Interrupt 8259A
040-043	Timer 8253-5
060-063	PPI 8255A-5
080-083	DMA Page Registers
0Ax*	NMI Mask Register
0Cx	Reserved
0Ex	Reserved
100-1FF	Not Usable
200-20F	Game Control
210-217	Expansion Unit
220-24F	Reserved
278-27F	Reserved
2F0-2F7	Reserved
2F8-2FF	Asynchronous Communications (Secondary)
300-31F	Prototype Card
320-32F	Fixed Disk
378-37F	Printer
380-38C**	SDLC Communications
380-389**	Binary Synchronous Communications (Secondary)
3A0-3A9	Binary Synchronous Communications (Primary)
3B0-3BF	IBM Monochrome Display/Printer
3C0-3CF	Reserved
3D0-3DF	Color/Graphics
3E0-3F7	Reserved
3F0-3F7	Diskette
3F8-3FF	Asynchronous Communications (Primary)

* At power-on time, the Non Mask Interrupt into the 8088 is masked off. This mask bit can be set and reset through system software as follows:

Set mask: Write hex 80 to I/O Address hex A0 (enable NMI)

Clear mask: Write hex 00 to I/O Address hex A0 (disable NMI)

** SDLC Communications and Secondary Binary Synchronous Communications cannot be used together because their hex addresses overlap.

(c)

Figure 12–1 (continued)

To support high-speed memory and I/O data transfers, the 8237A direct memory access controller is provided on the PC system board. This DMA chip contains four DMA channels, *DMA channel 0* through *DMA channel 3*. One channel, DMA 0, is used to refresh the dynamic R/W memory (DRAM), and the other three channels are available for use with peripheral devices. For instance, DMA 2 is used to support floppy disk drive data transfers.

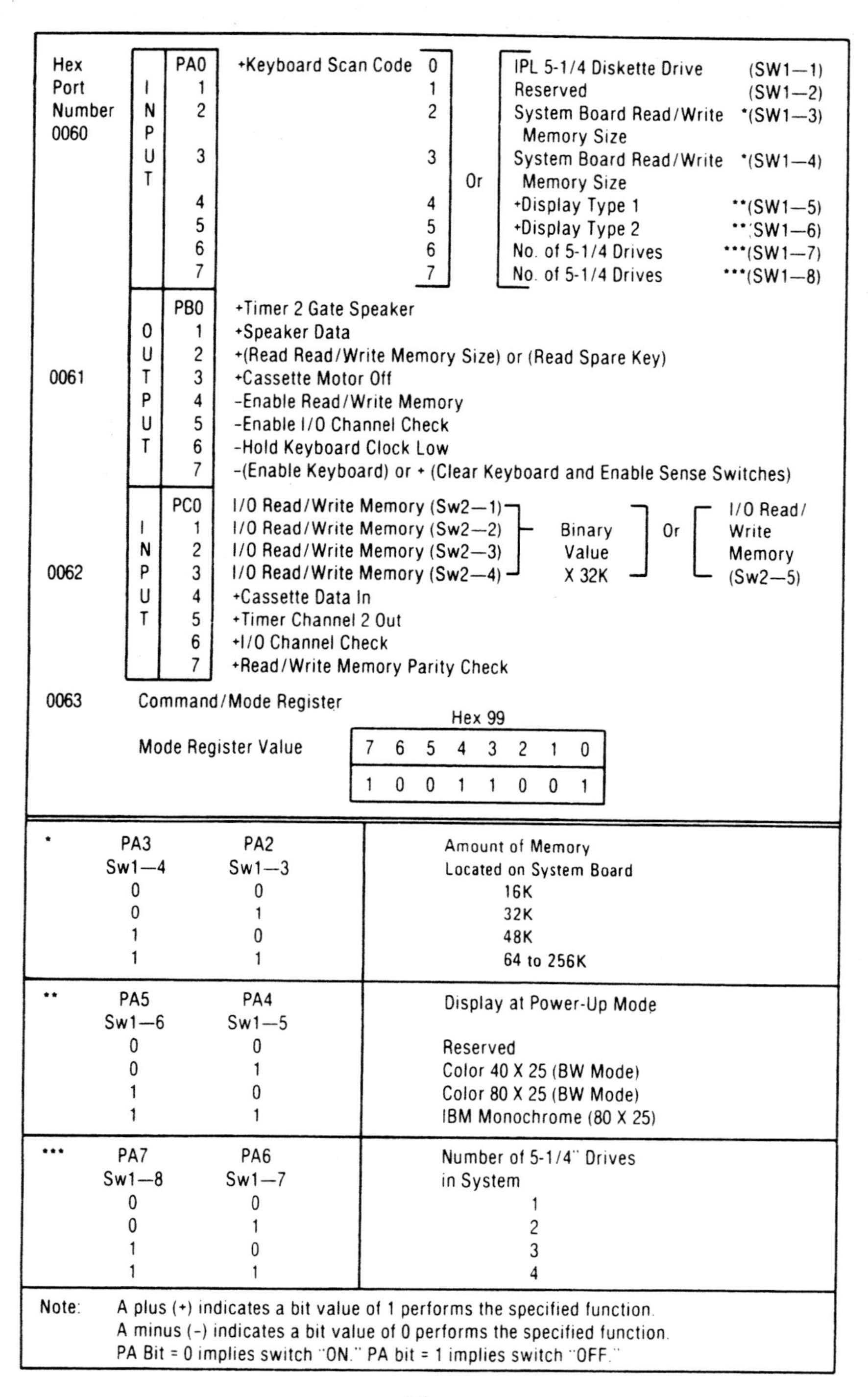

Hex Port Number		Bit	Function
0060	INPUT	PA0	+Keyboard Scan Code 0 — Or — IPL 5-1/4 Diskette Drive (SW1—1)
		1	1 — Or — Reserved (SW1—2)
		2	2 — Or — System Board Read/Write Memory Size *(SW1—3)
		3	3 — Or — System Board Read/Write Memory Size *(SW1—4)
		4	4 — Or — +Display Type 1 **(SW1—5)
		5	5 — Or — +Display Type 2 **(SW1—6)
		6	6 — Or — No. of 5-1/4 Drives ***(SW1—7)
		7	7 — Or — No. of 5-1/4 Drives ***(SW1—8)
0061	OUTPUT	PB0	+Timer 2 Gate Speaker
		1	+Speaker Data
		2	+(Read Read/Write Memory Size) or (Read Spare Key)
		3	+Cassette Motor Off
		4	-Enable Read/Write Memory
		5	-Enable I/O Channel Check
		6	-Hold Keyboard Clock Low
		7	-(Enable Keyboard) or + (Clear Keyboard and Enable Sense Switches)
0062	INPUT	PC0	I/O Read/Write Memory (Sw2—1) — Binary Value X 32K — Or — I/O Read/Write Memory (Sw2—5)
		1	I/O Read/Write Memory (Sw2—2)
		2	I/O Read/Write Memory (Sw2—3)
		3	I/O Read/Write Memory (Sw2—4)
		4	+Cassette Data In
		5	+Timer Channel 2 Out
		6	+I/O Channel Check
		7	+Read/Write Memory Parity Check
0063	Command/Mode Register		

Mode Register Value — Hex 99

7	6	5	4	3	2	1	0
1	0	0	1	1	0	0	1

*	PA3 Sw1—4	PA2 Sw1—3	Amount of Memory Located on System Board
	0	0	16K
	0	1	32K
	1	0	48K
	1	1	64 to 256K

**	PA5 Sw1—6	PA4 Sw1—5	Display at Power-Up Mode
	0	0	Reserved
	0	1	Color 40 X 25 (BW Mode)
	1	0	Color 80 X 25 (BW Mode)
	1	1	IBM Monochrome (80 X 25)

***	PA7 Sw1—8	PA6 Sw1—7	Number of 5-1/4" Drives in System
	0	0	1
	0	1	2
	1	0	3
	1	1	4

Note: A plus (+) indicates a bit value of 1 performs the specified function.
A minus (-) indicates a bit value of 0 performs the specified function.
PA Bit = 0 implies switch "ON." PA bit = 1 implies switch "OFF."

(d)

Figure 12–1 (continued)

Number	Usage
NMI	Parity
0	Timer
1	Keyboard
2	Reserved
3	Asynchronous Communications (Secondary)
	SDLC Communications
	BSC (Secondary)
4	Asynchronous Communications (Primary)
	SDLC Communications
	BSC (Primary)
5	Fixed Disk
6	Diskette
7	Printer

(e)

Figure 12–1 (continued)

The 8253-based timer circuitry is used to generate time-related functions and signals in the PC. There are three 16-bit counters in the 8253, and they are driven by a 1.19-MHz clock signal. Timer 0 is used to generate an interrupt to the microprocessor approximately every 55 ms. The system uses this timing function to keep track of time of the day. On the other hand, timer 1 is used to produce a DMA request every 15.12 μs to initiate refresh of the dynamic RAM. The last timer has multiple functions. It is used to generate programmable tones when driving the speaker and a record tone for use when sending data to the cassette for storage on tape.

The parallel I/O section of the PC's microcomputer, identified as PPI in Fig. 12–1(a), is implemented with the 8255A programmable peripheral interface controller. This device is configured through software to provide two 8-bit input ports and one 8-bit output port. Figure 12–1(d) gives the functions of the individual I/O lines of the PPI. Here we see that these I/O lines are used to input data from the keyboard (keyboard scan code), output tones to the speaker (speaker data), and read in the state of memory and system configuration switches (SW1-1 through SW1-8). Through the PPI's ports, the microcomputer also controls the cassette motor and enables or disables I/O channel check. Note in Fig. 12–1(d) that switch inputs SW1-3 and SW1-4 are used to tell the MPU how much RAM is implemented on the system processor board. In the lower part of this table, we find that for a system with 64Kbytes or 256Kbytes they are both set to the 1 position.

The circuitry in the nonmaskable interrupt (NMI) logic block allows nonmaskable interrupt requests derived from three sources to be applied to the microprocessor. As Fig. 12–1(a) shows, these interrupt sources are the *numeric coprocessor interrupt request* (N P NPI), *R/W memory parity check* ($\overline{\text{PCK}}$), and *I/O channel check* ($\overline{\text{I/O CH CK}}$). If any of these inputs are active, the NMI logic outputs a request for service to the 8088 over the NMI signal line.

In addition to the nonmaskable interrupt interface, the PC architecture provides for requests for service to the MPU by interrupts at another interrupt input called *interrupt request* (INTR). Note in Fig. 12–1(a) that this signal is supplied to the 8088 by the output of the interrupt controller (PIC) block. The 8259A LSI interrupt controller used in the PC provides for eight additional prioritized interrupt inputs. The inputs of the interrupt controller are supplied by peripherals such as the timer, keyboard, diskette drive, printer, and communication devices. Figure 12–1(e) gives interrupt priority assignments for these devices. For example, the timer (actually just timer 2 of the 8253) is at priority level 0.

The *I/O channel,* which is a collection of address, data, control, and power lines, is provided to support expansion of the PC system. The chassis of the PC has five 62-pin I/O channel card slots. In this way, the system configuration can be expanded by adding special function card slots, such as boards to control a monochrome or color display, floppy disk drives, a hard disk drive, expanded memory, or to attach a printer. Figure 12–1(b), shows that I/O channel expanded RAM resides in the part of the memory address space from 40000_{16} through $9FFFF_{16}$.

▲ 12.2 SYSTEM PROCESSOR CIRCUITRY

Figure 12–2 illustrates the system processor circuitry section of the IBM PC. It consists of the 8088 microprocessor, the 8284A clock generator, the 8288 bus controller, and the 8259A programmable interrupt controller. Here we will examine the operation of each of these sections of circuitry.

Clock Generator Circuitry

In Section 12.1, we pointed out that the clock generator circuitry serves three functions in terms of overall microcomputer system operation: *clock signal generation, reset signal generation,* and *ready signal generation.* Let us now explore the operation of the circuit for each of these functions in more depth.

The first function performed by the clock generator circuitry is the generation of the various clock signals needed to drive the 8088 microprocessor (U_3) and other circuits within the PC. As Fig. 12–2 shows, the 8284A clock generator/driver (U_{11}) has a 14.31818-MHz crystal (Y_1) connected between its X_1 and X_2 pins. This crystal causes the oscillator circuitry within the 8284A to run and generate three clock output signals: the oscillator clock (OSC) at 14.31818 MHz, the TTL peripheral clock (PCLK) at 2.385 MHz, and 8088 microprocessor clock (CLK88) at 4.77 MHz. Note in Fig. 12–2 that the CLK88 output at pin 8 of the 8284A is connected to the CLK input of the 8088 at pin 19. In this way, we see that the 8088 in the IBM PC runs at 4.77 MHz.

The second purpose served by the clock generator circuitry is to generate a power-on reset signal for the system. When power is first turned on, the power supply section of the PC tells the clock generator that power is not yet stable by setting its power good (PWR GOOD) output to logic 0. Looking at Fig. 12–2, we find that this signal is applied to the $\overline{\text{RES}}$ input at pin 11 of the 8284A. Logic 0 at $\overline{\text{RES}}$ represents an active input to the power-on reset circuit within the 8284A; therefore, the RESET output at pin 10 switches to its active level, logic 1, to signal that a reset operation is to take place. Note that the RESET output of the 8284A is applied directly to the RESET input at pin 21 of the 8088. When this input is at the 1 logic level, reset of the MPU is initiated.

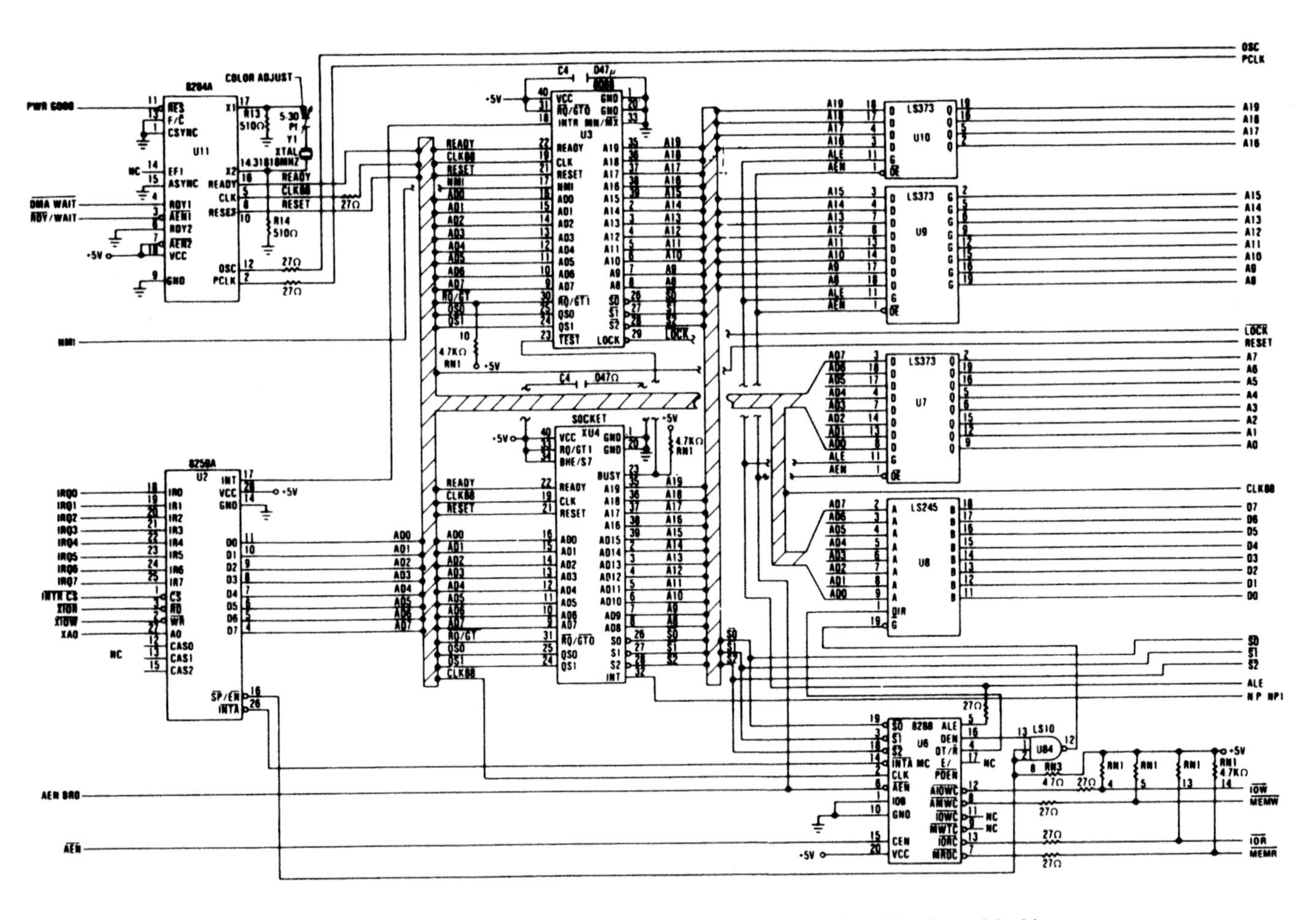

Figure 12–2 System processor circuitry. (Courtesy of International Business Machines Corporation)

As the voltage of the power supply builds up and becomes stable, the power supply switches PWR GOOD to logic 1. In response to this change in input, the RESET output of the 8284A returns to its inactive 0 logic level and the power-on reset is complete.

The last function served by the clock generator circuitry is to provide for synchronization of the 8088's bus operations with its memory and I/O peripherals. This synchronization is required to support the use of slow memory or peripheral devices on the system bus and is achieved by inserting wait states into the bus cycle to extend its duration. Let us now look at how and for what devices wait states are inserted into bus cycles of the IBM PC microcomputer.

The READY input at pin 22 of the 8088 is the signal that determines whether or not wait states are inserted into a bus cycle. If this input is at the 1 logic level when the processor samples it, bus cycles are run to completion without inserting wait states. However, if its logic level is 0 when sampled, wait states are inserted into the current bus cycle until READY returns to 1. In Fig. 12–2, we see that the READY input of the 8088 is directly supplied by the READY output (pin 5) of the 8284A. The logic level of this output is determined by inputs $\overline{\text{DMA WAIT}}$ and $\overline{\text{RDY}}$/WAIT. Whenever input $\overline{\text{DMA WAIT}}$ is logic 0, the READY output is switched to logic 0. This means that wait states are automatically inserted whenever DMA transfer bus cycles are performed. On the other hand, logic 1 at the $\overline{\text{RDY}}$/WAIT input also causes READY to switch to logic 0. This signals that a slow memory or I/O device is being accessed, and wait states are needed to extend the bus cycle. In the circuit under discussion, both I/O and DMA data transfers have one wait state inserted into each bus cycle.

Microprocessor, System Data Bus, and Bus Controller

The 8088 microprocessor (U_3) used in the PC is rated to operate at a maximum clock rate of 5 MHz. However, we just found that the CLK88 signal that is applied to its CLK input actually runs it at 4.77 MHz. At power-up, the RESET input of the 8088 is activated by the 8284A to initiate a power-on reset of the MPU. This reset operation causes the status, DS, SS, ES, and IP registers within the 8088 to be cleared, the instruction queue to be emptied, and the code segment register to be initialized to $FFFF_{16}$. When RESET returns to its inactive level, the 8088 begins to fetch instructions from program memory starting at address $FFFF0_{16}$. The instruction at this location passes control to the PC's power-up program, which causes the rest of the system resources to be initialized, diagnostic tests to be run on the hardware, and the operating system loaded from diskette or hard disk. At this point, the microcomputer is up and running. Let us now look at how it accesses memory and I/O devices.

Earlier we pointed out that the microcomputer of the IBM PC is architected to have both a multiplexed local bus and a demultiplexed system bus. In general, memory and I/O peripherals are attached to the 8088 microprocessor at the system bus. However, there are some exceptions; both the 8259A interrupt controller and 8087 numeric coprocessor are attached directly to the local bus.

Figure 12–2 shows that the local bus includes the 8088's multiplexed address data bus lines AD_0 through AD_7, address lines A_8 through A_{19}, and maximum-mode status lines $\overline{S}_0$ through $\overline{S}_2$. Note that the local bus lines are connected to the 8259A programmable interrupt controller (U_2) and the socket for the 8087 numeric coprocessor (XU_4).

Let us now turn our attention to how the local bus lines are demultiplexed and decoded to form the system bus. Looking at Fig. 12–2, we see that the upper address lines are latched using 74LS373 devices U_9 and U_{10} to give system address bus lines A_8 through A_{19}. Another 74LS373 latch (U_7) is used to demultiplex low address signals A_0 through A_7 from the data signals to complete the system address bus. Finally, the separate system data bus lines, D_0 through D_7, are implemented with the 74LS245 bus transceiver U_8. These latches and transceivers also buffer the address and data bus lines to increase the drive capability at the system bus.

The 8288 bus controller U_5 monitors the codes output on the 8088's status lines $\overline{S_0}$ through $\overline{S_2}$. Based on these codes, it produces appropriate system bus control signals. For example, in Fig. 12–2 we see that the address latch enable (ALE) signal is output at pin 5 of the 8288 and supplied to the system bus. To ensure that address information is latched at the appropriate time when demultiplexing the local bus, ALE is also applied to the enable input (G) of all three 74LS373 latches. The 8288 also produces the DEN and DT/$\overline{R}$ signals that are used to control operation of the 74LS245 system data bus transceiver. Logic 1 at DEN (pin 16) signals when a data transfer can take place over the data bus; therefore, it is inverted and applied to the enable input ($\overline{G}$) of the transceiver. On the other hand, the logic level of DT/$\overline{R}$ (pin 4) identifies whether data are to be input or output over the system bus. For this reason, it is applied to the direction (DIR) input of the 74LS245.

The 8288 also produces I/O and memory read and write control signals. The outputs $\overline{IOR}$ and $\overline{IOW}$ are used to identify I/O read and write operations, respectively. Moreover, $\overline{MEMR}$ or $\overline{MEMW}$ is output to tell that a memory read or write operation is in progress, respectively. These signals are made available on the system bus.

Address enable inputs AEN BRD and $\overline{AEN}$ are active during all DMA cycles. These signals are applied to enable inputs $\overline{AEN}$ and CEN, respectively, of the 8288 and disable it when DMA transfers are to take place over the system bus. When disabled, the 8288 stops producing the I/O and memory read/write control signals. Signal AEN BRD is also used to disable the address latches and data transceiver so that the system address lines float when the 8237A DMA controller is to use the system bus.

Interrupt Controller

As Fig. 12–2 shows, an external hardware interrupt interface is implemented for the IBM PC with the 8259A programmable interrupt controller device U_2. It monitors the state of interrupt request lines IRQ_0 through IRQ_7 to determine if any external device is requesting service. Figure 12–1(c) lists the functions of the priority 0 through priority 7 interrupts. For example, in this list we find that the IRQ_0 input is used to service the 8253 timer, and IRQ_1 is dedicated to servicing the keyboard.

If an interrupt request input becomes active, the PIC switches its interrupt request (INT) output to the 1 logic level. Note in Fig. 12–2 that the INT output at pin 17 of the 8259A is supplied to the INTR input at pin 18 of the 8088. At completion of execution of the current instruction, the 8088 samples the logic level of its INTR input. Assuming that it is active, the 8088 responds to the request for service by outputting the interrupt acknowledge status code to the 8288 bus controller. In turn, the 8288 outputs logic 0 on interrupt acknowledge ($\overline{INTA}$), pin 14 of U_5. This signal is sent to the $\overline{INTA}$ input at pin

26 of the interrupt controller. Upon receiving this signal, the 8259A generates an active 0 level at $\overline{\text{SP}}/\text{EN}$, which, in conjunction with the data enable (DEN) output of the 8288, is used to float the system data bus lines. Now the interrupt controller outputs the type number of the active interrupt over the local data bus to the 8088. The MPU uses the type number to fetch the vector of the service routine for the interrupt from memory, loads it into CS and IP, and then executes the service routine.

The operating configuration of the 8259A needs to be initialized at power-on of the system. This initialization is achieved by writing to the 8259A's internal registers over the local bus. Earlier we pointed out that the peripherals in the PC are located in the I/O address space. For this reason, I/O instructions are used to access the registers of the PIC. This is why its read ($\overline{\text{RD}}$) and write ($\overline{\text{WR}}$) inputs are supplied by the I/O read ($\overline{\text{XIOR}}$) and I/O write ($\overline{\text{XIOW}}$) control signals, respectively. Moreover, when inputting data from or outputting data to the 8259A, the address of the register, which is either 20_{16} or 21_{16}, is output on the address bus. This address is decoded in the chip-select logic circuit to produce chip-select signal $\overline{\text{INTR CS}}$. This signal is applied to the $\overline{\text{CS}}$ input of the 8259A and enables its microprocessor interface.

In the remaining sections of this chapter, we will trace the operation of each of these segments of circuitry in detail.

▲ 12.3 WAIT-STATE LOGIC AND NMI CIRCUITRY

The control logic circuitry shown in Fig. 12–3 provides several functions in terms of overall system operation. It consists of the wait-state control circuit needed to extend memory and I/O bus cycles, the wait-state and hold-acknowledge logic used to grant the 8237A DMA controller access to the system bus, and the circuitry that generates the nonmaskable interrupt request.

Wait-State Logic Circuitry

The wait-state logic circuitry is used to insert one wait state into all I/O channel, I/O, and DMA bus cycles. The circuit prouduces two wait-state control signals, $\overline{\text{RDY}}/\text{WAIT}$ and $\overline{\text{DMA WAIT}}$. $\overline{\text{RDY}}/\text{WAIT}$ is applied to the $\overline{\text{AEN}_1}$ input of the 8284A clock generator (see Fig. 12–2). Logic 1 at this input makes the READY output of the 8284A switch to logic 0. This output is applied to the READY input of the 8088 and initiates a wait state for the current bus cycle. On the other hand, signal $\overline{\text{DMA WAIT}}$ switches to logic 0 whenever a DMA bus cycle is initiated. It is applied to the RDY_1 input of the 8284A and, when at logic 0, it causes the READY output to switch to logic 0. In this way, it extends the DMA bus cycle by inserting wait states. Let us now examine just how the signal $\overline{\text{RDY}}/\text{WAIT}$ is produced.

$\overline{\text{I/O CH RDY}}$ (I/O channel ready) is one signal that can insert wait states into the processor's bus cycle. Cards located in the slots of the I/O channel interface use $\overline{\text{I/O CH RDY}}$. Figure 12–3 shows that this signal is applied to the preset (PR) input of the 74S74 flip-flop U_{82}. As long as $\overline{\text{I/O CH RDY}}$ is logic 0, the flip-flop is set and its Q output, signal $\overline{\text{RDY}}/\text{WAIT}$, is held at logic 1, and wait states are inserted into the current bus cycle.

Let us now look at how $\overline{\text{RDY}}/\text{WAIT}$ is produced for an I/O read, I/O write, or memory refresh cycle. Note in Fig. 12–3 that the CLR input at pin 1 of D-type flip-flop

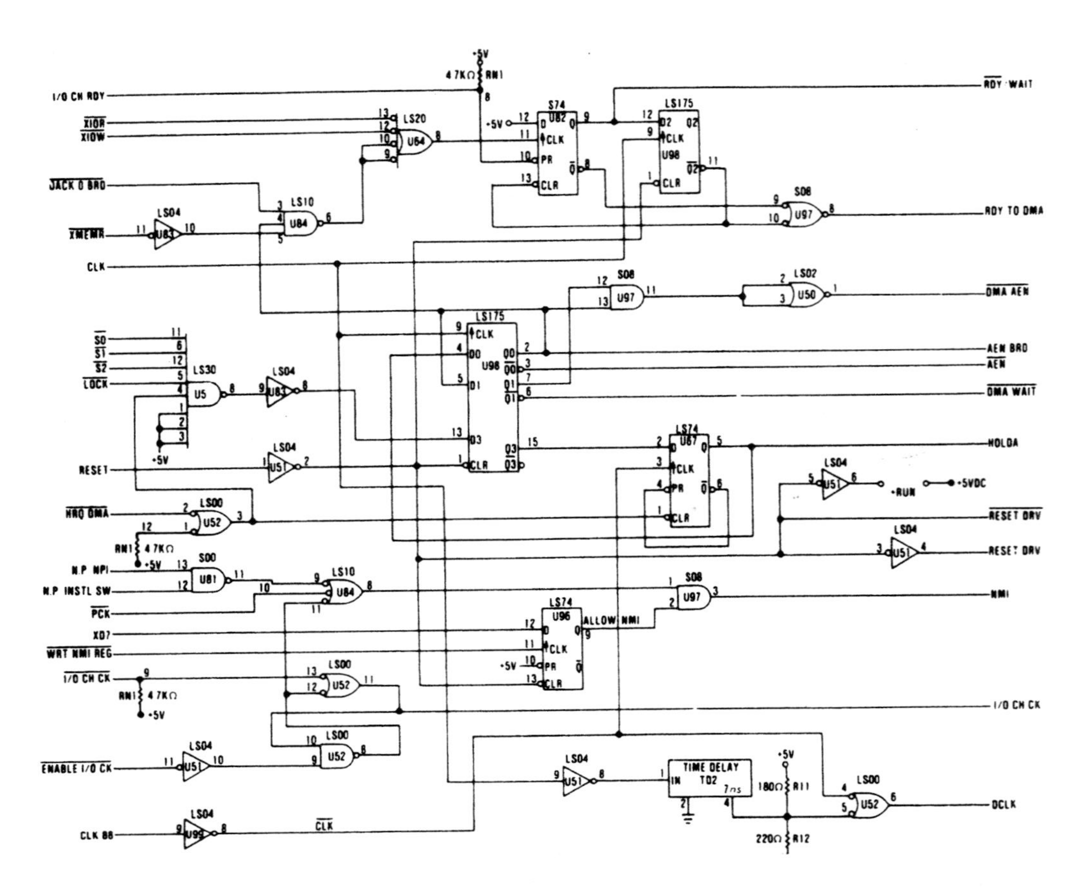

Figure 12–3 Wait-state logic and NMI circuitry. (Courtesy of International Business Machines Corporation)

U_{98} is tied to the RESET input through inverter U_{51}. This signal clears the flip-flop at power-up and initializes its Q_2 output to logic 1. As long as the I/O CH RDY input is logic 1, flip-flop U_{82} will set whenever a 0-to-1 transition occurs at its CLK input (pin 11). This causes its Q output to switch to logic 1 and its $\overline{\text{Q}}$ output to switch to logic 0. $\overline{\text{RDY/WAIT}}$ is now logic 1 and signals the 8088 that a wait state is to be inserted into the current bus cycle.

We will now look at what inputs cause an active transition at CLK. CLK is produced by the signals $\overline{\text{XIOR}}$ (I/O read), $\overline{\text{XIOW}}$ (I/O write), $\overline{\text{DACK 0 BRD}}$ (DMA acknowledge channel 0), $\overline{\text{XMEMR}}$ (memory read), and AEN BRD (DMA cycle in progress) with a logic circuit formed from gates U_{83}, U_{84}, and U_{64}. If any input of NAND gate U_{64} switches to logic 0, a 0-to-1 transition is produced at CLK and flip-flop U_{82} sets. In this way, we see that if either an I/O read ($\overline{\text{XIOR}} = 0$) or I/O write ($\overline{\text{XIOW}} = 0$) cycle is initiated, a wait state is generated. Moreover, a wait state is initiated if a memory read ($\overline{\text{XMEMR}} = 0$) occurs when a memory refresh is not in progress ($\overline{\text{DACK 0 BRD}} = 1$) and a DMA cycle is in progress (AEN BRD = 1).

Now that we see how the wait state is inserted, let us look at how it is terminated so that only one wait state is inserted into the bus cycle. Since the logic 1 at $\overline{\text{RDY}}$/WAIT is also the data input (pin 12) of the 74LS175 flip-flop U_{98}, the next pulse at the CLK input (pin 9) causes its outputs to set. Therefore, output $\overline{Q_2}$ switches to the 0 logic level. This logic 0 is returned to the CLR input at pin 13 of flip-flop U_{82} and causes it to reset. $\overline{\text{RDY}}$/WAIT returns to logic 0, signaling ready, and the bus cycle proceeds to completion after only one wait state.

Hold/Hold Acknowledge Circuitry

The 8088 in the PC is configured to operate in the maximum mode. When configured this way, there is no hold/hold acknowledge interface directly usable by the 8237A DMA controller. The $\overline{\text{DMA WAIT}}$ signal we mentioned earlier is coupled with a HOLDA signal produced in the control circuitry shown in Fig. 12–3 to implement a *simulated DMA interface* in the PC. Let us look at how DMA requests produce the $\overline{\text{DMA WAIT}}$ and HOLDA signals.

Peripheral devices issue a request for DMA service through the 8237A DMA controller. The 8237A signals the 8088 that it wants control of the system bus to perform DMA transfers by outputting the signal $\overline{\text{HRQ DMA}}$ (hold request DMA). In Fig. 12–3, we find that this signal is an input to NAND gate U_{52}. Whenever $\overline{\text{HRQ DMA}}$ is at its inactive 1 logic level, the output at pin 3 of U_{52} is logic 0. This signal is applied to the CLR input of the 74LS74 flip-flop U_{67} and holds it cleared. Therefore, HOLDA is at its inactive 0 logic level. The output at pin 3 of U_{52} is also applied to one input of NAND gate U_5. Here it is combined with status code $\overline{S_2}\overline{S_1}\overline{S_0}$. If the status output is 111_2 and $\overline{\text{LOCK}} = 1$, the output of U_5 switches to 0 and signals that the 8088's bus is in the passive state and DMA is permitted to take over control of the bus. On the next pulse at the CLK input, flip-flop 3 in latch U_{98} sets, and its Q_3 output switches to the 1 level. Q_3 is applied to the data input of the 74LS74 flip-flop U_{67}, and on the next pulse at CLK88, its Q output, HOLDA, becomes active. This output remains latched at the 1 logic level until the DMA request is removed. HOLDA is sent to the HLDA input of the 8237A (see Fig. 12–7) and signals that the 8088 has given up control of the system bus.

At the same time, the logic 1 at HOLDA is returned to the Q_0 input (pin 4) of 74LS175 latch U_{98}. On the next pulse at CLK, signals AEN BRD and $\overline{\text{AEN}}$ become active and signal that a DMA cycle is in progress. These signals are used to disable and tri-state the 8288 bus controller and system bus address latches (see Fig. 12–2), thereby isolating the 8088 microprocessor from the system bus. $\overline{\text{AEN}}$ also disables the decoder that generates peripheral chip selects for the I/O address space [see Fig. 12–4(a)]. AEN BRD is returned to the Q_1 input at pin 5 of latch U_{98}, and on the next pulse at CLK the $\overline{\text{DMA WAIT}}$ signal becomes active. The logic 0 at $\overline{\text{DMA WAIT}}$ is sent to the 8284A, where it deactivates the READY input to insert wait states. Finally, AEN BRD and the complement of $\overline{\text{DMA WAIT}}$ are gated together by the 74S08 AND gate U_{97} followed by a NOT to produce the signal $\overline{\text{DMA AEN}}$. (In Fig. 12–7, we will see how this signal is used to enable the DMA address circuitry. In the DMA circuitry section, we will find that it is used to enable the 8237A to produce its own address and I/O or memory read/write control signals.)

Nonmaskable Interrupt Circuitry

In Section 12.1, we indicated that there are three sources for applying a nonmaskable interrupt to the 8088 microprocessor: the 8087 numeric coprocessor, memory parity check, and I/O channel check. In Fig. 12–3, the signal mnemonics used to represent these three inputs are N P NPI, $\overline{\text{PCK}}$, and $\overline{\text{I/O CH CK}}$, respectively. These signals are combined in the NMI control logic circuitry to produce the nonmaskable interrupt request (NMI) signal. This output is applied directly to the NMI input at pin 17 of the 8088 (see Fig. 12–2). Let us now look at the operation of the NMI interrupt request control circuit.

The NMI control logic circuitry in Fig. 12–3 includes a *nonmaskable interrupt control register*. This register is implemented with the 74LS74 D-type flip-flop U_{96}. At reset of the PC, the NMI interface is automatically disabled. Note in Fig. 12–3 that the RESET signal is input to a 74LS04 inverter, and the output at pin 2 of this inverter is applied to the clear (CLR) input of the NMI control register flip-flop. Clearing the flip-flop causes its ALLOW NMI output to switch to logic 0. This output is used as the enable input of the 74S08 AND gate (U_{97}) that controls the NMI output. As long as ALLOW NMI is logic 0, the NMI output is held at its inactive 0 logic level, and the NMI interface is disabled.

We now look at how the NMI interface gets turned on. Looking at Fig. 12–3, we find that the data input (pin 12) of the 74LS74 flip-flop is supplied by XD_7 of the data bus, and its clock input (CLK) at pin 11 is supplied by a chip-select signal identified as $\overline{\text{WR NMI REG}}$ (write NMI register). As part of the initialization software of the PC, the NMI control register gets set by executing an instruction that writes a byte with its most significant bit (XD_7) set to logic 1 to any I/O address in the range $00A0_{16}$ through $00BF_{16}$. All these addresses decode to produce the chip-select signal $\overline{\text{WR NMI REG}}$ at CLK; therefore, the 1 at XD_7 is loaded into the flip-flop and ALLOW NMI switches to logic 1. This supplies the enable input for the 74S08 AND gate to the NMI output. The NMI interface is now enabled and waiting for one of the NMI interrupt functions to occur.

Now that the NMI interface is enabled, let us look at how the numeric coprocessor, parity check, or I/O channel check interrupt requests are handled. Figure 12–3 shows that the inputs for each of these three functions are combined with the 74LS10 NAND gate U_{84}. If any combination of the NAND gate inputs is logic 0, the output at pin 8 switches to logic 1. This represents an active NMI request. As long as the NMI interface is enabled, this logic 1 is passed to the NMI output and on to the NMI input of the 8088.

Actually, each NMI interrupt input also has an enable signal that allows it to be individually enabled or disabled. For instance, Fig. 12–3 shows that N P NPI is combined with the signal N P INSTL SW by the 74S00 NAND gate U_{81}. For the numeric coprocessor interrupt to be active, the N P INSTL SW input must be logic 1. N P INSTL SW stands for numerics processor install switch, which is the switch represented by the contacts marked 2-15 on SW1 in Fig. 12–9. Only when this switch is off (open) is the numeric coprocessor interrupt input enabled.

The parity check nonmaskable interrupt input can also be enabled or disabled; however, this part of the circuit is not shown in Fig. 12–3. To enable $\overline{\text{PCK}}$, a logic 0 must be written to bit 4 of output port PB of the 8255A U_{36} (see Fig. 12–9). This produces the signal $\overline{\text{ENB RAM PCK}}$ (enable RAM parity check), which is used to enable the parity check circuits that produce $\overline{\text{PCK}}$ in the RAM circuit (see Fig. 12–7).

Looking at Fig. 12–3, we see that to enable the NMI input for $\overline{\text{I/O CH CK}}$ (I/O channel check) logic 0 must be applied to the $\overline{\text{ENABLE I/O CK}}$ input. This signal is directly supplied by bit 5 of output port PB on the 8255A device U_{36} (see Fig. 12–9). This bit is set to logic 0 through software at power-up.

Up to this point, we have shown how the NMI input is enabled, disabled, or made active. However, since there are three possible sources for the NMI input, another question that must be answered is, how does the 8088 know which of the three interrupt inputs caused the request for service? It turns out that the signals PCK and I/O CH CK are returned to input ports on the 8255A device U_{36} (see Fig. 12–9). For instance, I/O CH CK is applied to input bit 6 on port PC of the 8255A. Therefore, the service routine for NMI can read these inputs through software, determine which has caused the request, and then branch to the part of the service routine that corresponds to the active input.

▲ 12.4 INPUT/OUTPUT AND MEMORY CHIP-SELECT CIRCUITRY

In the previous section, we found that the chip-select signal $\overline{\text{WRT NMI REG}}$ is used as an enable input for the NMI control register. Besides the NMI control register chip-select signal, chip selects are needed in the ROM, RAM, DMA, PPI, interval timer, and interrupt controller sections of the PC's system processor board circuitry. The I/O and memory chip-select circuit shown in Fig. 12–4(a) generates these chip selects. Two types of chip-select signals are produced, *I/O chip selects* and memory chip selects, and decoding of addresses generates them both. Let us now look at the operation of the circuits that produce these I/O and memory chip-select outputs.

I/O Chip Selects

Earlier we found that in the architecture of the PC, LSI peripheral devices, such as the DMA controller, interrupt controller, programmable interval timer, and programmable peripheral interface controller, are located in the I/O address space of the 8088 microprocessor. I/O chip select decoding for these devices takes place in the circuit shown in Fig. 12–4(a), which is formed from devices U_{66}, U_{50}, and U_{51}. Let us begin by looking at the operation of this segment of circuitry in detail.

To access a register within one of the peripheral devices, an I/O instruction must be executed to read from or write to the register. The address output on address lines A_0 through A_9 during the I/O bus cycle is used to both chip-select the peripheral device and select the appropriate register. Note in Fig. 12–4(a) that address bits XA_8 and XA_9 are applied to enable inputs G_{2B} and G_{2A}, respectively, of the 74LS138 three-line to eight-line decoder device (U_{66}). When these inputs are both at logic 0 and $\overline{\text{AEN}}$ is at logic 1, the decoder is enabled for operation. At the same time, address lines XA_5 through XA_7 apply a 3-bit code to the ABC inputs of the decoder. When U_{66} is enabled, the Y output corresponding to the code $XA_7XA_6XA_5$ is switched to its active 0 logic level. These Y signals produce I/O chip select outputs $\overline{\text{DMA CS}}$ (DMA chip select), $\overline{\text{INTR CS}}$ (interrupt request chip select), $\overline{\text{T/C CS}}$ (timer/counter chip select), $\overline{\text{PPI CS}}$ (parallel peripheral interface chip

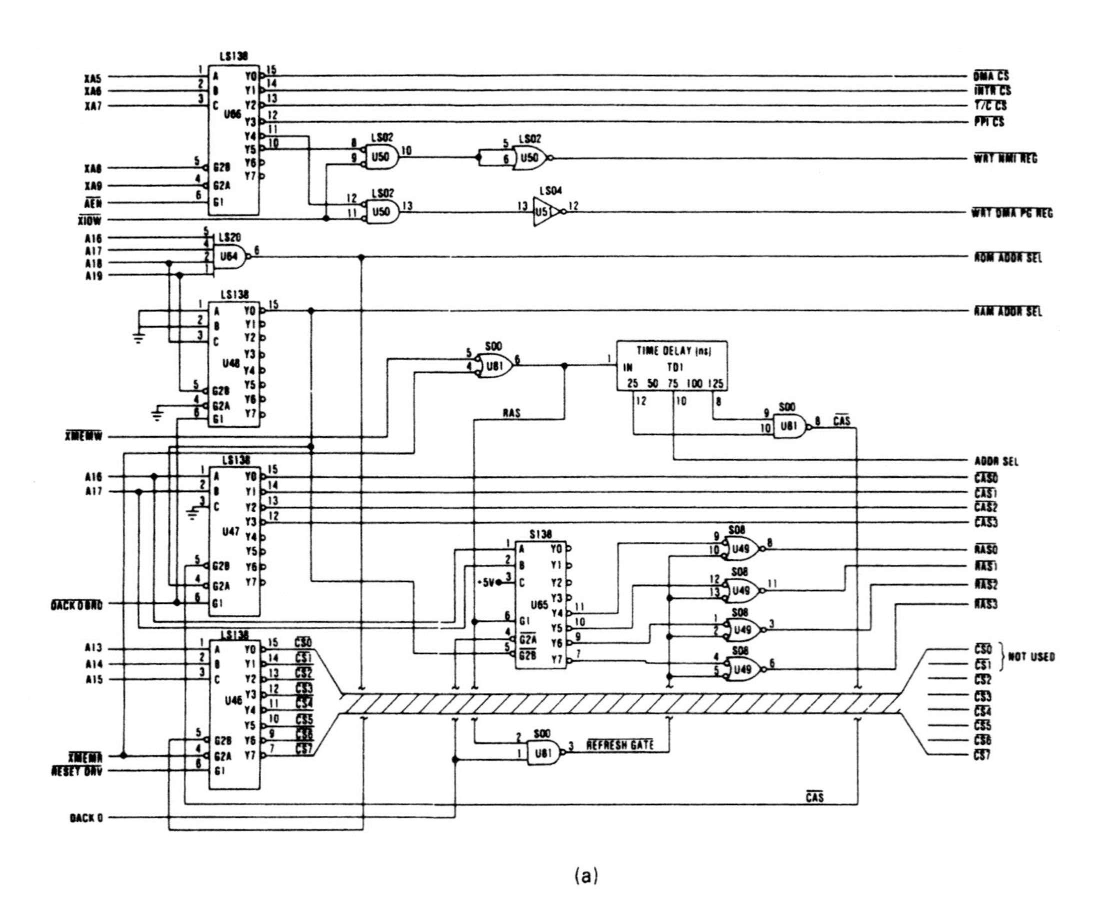

Address range	Signal	Function	Conditions
0-1F	$\overline{\text{DMA CS}}$	DMA controller	Non DMA bus cycle
20-3F	$\overline{\text{INTR CS}}$	Interrupt controller	Non DMA bus cycle
40-5F	$\overline{\text{T/C CS}}$	Interval timer	Non DMA bus cycle
60-7F	$\overline{\text{PPI CS}}$	Parallel peripheral interface	Non DMA bus cycle
80-9F	$\overline{\text{WRT DMA PG REG}}$	DMA page register	Non DMA bus cycle, XIOW active
A0-BF	$\overline{\text{WRT NMI REG}}$	NMI control register	Non DMA bus cycle, XIOW active

(b)

Figure 12–4 (a) Peripheral/memory chip-select circuitry. (Courtesy of International Business Machines Corporation) (b) Peripheral address decoding. (c) ROM address decoding. (d) RAM address decoding.

Address range	Chip select
F0000-F1FFF	$\overline{CS_0}$
F2000-F3FFF	$\overline{CS_1}$
F4000-F5FFF	$\overline{CS_2}$
F6000-F7FFF	$\overline{CS_3}$
F8000-F9FFF	$\overline{CS_4}$
FA000-FCFFF	$\overline{CS_5}$
FC000-FDFFF	$\overline{CS_6}$
FE000-FFFFF	$\overline{CS_7}$

(c)

Address range	Active signal	Condition
00000-3FFFF	$\overline{\text{RAM ADDR SEL}}$	Inactive DACK 0 BRD
00000-0FFFF	$\overline{RAS_0}$, $\overline{CAS_0}$	Active XMEMR or XMEMW
10000-1FFFF	$\overline{RAS_1}$, $\overline{CAS_1}$	Active XMEMR or XMEMW
20000-2FFFF	$\overline{RAS_2}$, $\overline{CAS_2}$	Active XMEMR or XMEMW
30000-3FFFF	$\overline{RAS_3}$, $\overline{CAS_3}$	Active XMEMR or XMEMW

(d)

Figure 12–4 (continued)

select), WRT NMI $\overline{\text{REG}}$ (NMI register chip select), and WRT DMA PG $\overline{\text{REG}}$ (DMA page register chip select).

For instance, if $XA_7XA_6XA_5 = 001$, output Y_1 switches to logic 0 and produces the chip-select output INTR $\overline{\text{CS}}$ at pin 14. In Fig. 12–2, we find that this signal is applied to the $\overline{\text{CS}}$ input at pin 1 of the 8259A interrupt controller and enables its microprocessor interface for operation. At the same time, appropriate lower-order address bits are applied directly to the register select inputs of the peripherals to select the register that is to be accessed. For the 8259A in Fig. 12–2, we find that only one address bit, XA_0 is used, and this signal is applied to register select input A_0 at pin 27.

To produce an I/O chip-select signal, address bits XA_0 to XA_4 are not used and, therefore, the individual I/O chip select signals produced actually correspond to a range of addresses. The address range for each chip-select output is shown in Fig. 12–4(b). For instance, any address in the range 0020_{16} through $003F_{16}$ decodes to produce the INTR $\overline{\text{CS}}$ chip-select signal.

The signal $\overline{\text{AEN}}$ is at its active 0 logic level only during DMA bus cycles. When $\overline{\text{AEN}}$ is at logic 0, decoder U_{66} is disabled. Thus, only the addresses output by the microprocessor will produce I/O chip-select signals. This is identified as a condition required for the occurrence of all chip selects in Fig. 12–4(b). Looking at the circuit in Fig. 12–4(a), we also find that the NMI control register and DMA page register chip selects are gated with the I/O write control signal $\overline{\text{XIOW}}$ by NOR gates in IC U_{50}. Therefore, as shown in Fig. 12–4(b), for these two chip selects to take place, an additional condition must be satisfied—that is, they are only produced if an I/O write (output) bus cycle is taking place.

Since the upper address lines XA_{10} through XA_{15} are not used in the I/O chip-select address decoder circuit, they represent don't-care states. Therefore, more than one range of addresses may be used to access each peripheral. For instance, any address in the ranges 0020_{16} through $FC20_{16}$, 0021_{16} through $FC21_{16}$, and 0022_{16} through $FC22_{16}$ will also decode to produce the signal $\overline{\text{INTR CS}}$.

Memory Chip Selects

The system processor board of the PC contains both read only memory (ROM) and random access read/write memory (RAM). The ROM part of memory is used to store embedded system software such as the BIOS, power-up diagnostics, and BASIC interpreter. On the other hand, programs that are typically loaded from disk, such as the operating system and application programs, are stored in the RAM. Here we will look only at the chip-select signals produced for enabling the memory devices. These chip-select signals are also generated in the I/O and memory chip-select circuit shown in Fig. 12–4(a).

Let us begin by examining the circuitry that produces the chip selects needed by ROM. The output signals produced for ROM in the circuit shown in Fig. 12–4(a) are ROM address select ($\overline{\text{ROM ADDR SEL}}$) and chip selects $\overline{CS_0}$ through $\overline{CS_7}$. Note that combining the upper four address bits, A16 through A19, with NAND gate U_{64} generates the signal $\overline{\text{ROM ADDR SEL}}$. If all three of these bits are at logic 1, the output at pin 6, $\overline{\text{ROM ADDR SEL}}$, switches to its active 0 logic level. This signal has two functions. First, it is used to enable the ROM chip select decoder U_{46} and, second, it is supplied to the *ROM array* (see Fig. 12–5) where it is used to control the direction of data transfer through the ROM data bus transceiver.

The chip-select outputs for the EPROMs, labeled $\overline{CS_0}$ through $\overline{CS_7}$ in Fig. 12–4(a), are produced by the 74LS138 three-line to eight-line decoder U_{46}. Note that $\overline{\text{ROM ADDR SEL}}$ is applied to the G_{2B} chip-enable input of the decoder. This enable signal ensures that the decoder decodes addresses in the range $F0000_{16}$ through $FFFFF_{16}$. Two other enable signals, $\overline{\text{XMEMR}}$ and RESET $\overline{\text{DRV}}$, are also applied to the decoder. $\overline{\text{XMEMR}}$ ensures that the decoder is enabled only during memory read operations.

Note that address lines A_{13} through A_{15} are applied to the ABC inputs of the decoder. This 3-bit code is decoded to generate the individual chip selects, $\overline{CS_0}$ through $\overline{CS_7}$. As Fig. 12–4(a) shows, chip selects $\overline{CS_0}$ and $\overline{CS_1}$ are not used. However, the other six, $\overline{CS_2}$ through $\overline{CS_7}$, are each used to enable an individual EPROM device in the ROM array (see Fig. 12–5).

For example, if the input to the ROM address decoder is $A_{14}A_{13}A_{12} = 010$, chip-select output $\overline{CS_2}$ is active, and a read operation is performed from one of the 8Kbyte storage locations in the EPROM device that is located in the address range $F4000_{16}$ through $F5FFF_{16}$. The actual storage location in the EPROM device accessed is selected by the lower 13 address bits, which are applied directly to all the EPROM devices in parallel. The memory address range that corresponds to each ROM chip-select output is given in Fig. 12–4(c). This chart shows that there are a total of 64K addresses decoded by the ROM address decoder circuitry.

We will now look at the circuitry used to produce the chip select, row address select, and column address select signals for the *RAM array* circuit. In Fig. 12–4(a), the chip-select outputs used to control the operation of RAM are $\overline{\text{RAM ADDR SEL}}$ (RAM address select) and ADDR SEL (address select). The 74LS138 decoder U_{48} produces $\overline{\text{RAM ADDR SEL}}$. Looking at the circuit diagram, we find that if A_{19} is logic 0 and $\overline{\text{DACK 0 BRD}}$ is logic 1, the decoder is enabled for operation. Moreover, as long as address bit A_{18} is also logic 0, output Y_0 of the decoder, which is the same as $\overline{\text{RAM ADDR SEL}}$, switches to logic 0. In this way, we see that $\overline{\text{RAM ADDR SEL}}$ goes active whenever the 8088 outputs an address in the range 00000_{16} through $3FFFF_{16}$. This is the full address range of RAM that resides on the system processor board. Note that this $\overline{\text{RAM ADDR SEL}}$ is applied to the G_{2A} input of the 74LS138 CAS decoder (U_{47}) and to input G_{2B} of the 74LS138 RAS decoder (U_{65}). It is also used in the RAM array circuit (see Fig. 12–6), where it controls the data bus transceiver.

The ADDR SEL signal is generated from $\overline{\text{XMEMW}}$ and $\overline{\text{XMEMR}}$ by NAND gate U_{81} and delay line TD_1. If either the memory read or write control input signal is at its active 0 logic level, the output at pin 6 of the NAND gate U_{81} switches to logic 1, and ADDR SEL becomes active after the time delay set by TD_1 elapses. ADDR SEL is supplied to the RAS/CAS address selector in the RAM array circuit (see Fig. 12–6), where it is used to select between the RAS and CAS parts of the address.

Note that the output at pin 6 of NAND gate U_{81}, which was used to produce ADDR SEL, is also the RAS (row address select) signal. RAS is applied to the G_1 enable input of the 74S138 RAS decoder (U_{65}). The other chip-select inputs of this decoder are supplied by the signals $\overline{\text{RAM ADDR SEL}}$ and DACK 0 and must be logic 0 and logic 1, respectively, to enable the device for operation.

Now when U_{65} is enabled, the code at the ABC input is decoded to produce the corresponding RAS output. Note that the C input of the decoder is fixed at the 1 logic level and the other two inputs, A and B, are supplied by address bits A_{16} and A_{17}, respectively. For instance, if these two address bits are both logic 0, the input code is 100, and output Y_4 switches to the 0 logic level and generates the signal $\overline{\text{RAS}_0}$. After a short delay, which is set by TD_1, the $\overline{\text{CAS}}$ signal is output at pin 8 of U_{81}. This signal is applied to the G_{2B} input of the CAS decoder U_{47}. Here the other decoder enable inputs are supplied by $\overline{\text{RAM ADDR SEL}}$ and $\overline{\text{DACK 0 BRD}}$. When enabled, the address at the AB inputs causes the corresponding column address select output to occur. Assuming that A_{16} and A_{17} are still both 0, $\overline{\text{CAS}_0}$ switches to its active 0 logic level. In this way, we see that each RAS chip select is followed after a short delay by the corresponding CAS chip select. Figure 12–4(d) summarizes the address decoding for the RAM address chip selects.

During DRAM refresh, DACK_0 becomes active, which along with the RAS signal is used to generate the $\overline{\text{RAS}_0}$, $\overline{\text{RAS}_1}$, $\overline{\text{RAS}_2}$, and $\overline{\text{RAS}_3}$ signals. These signals are generated independent of the RAS decoder outputs and are used to refresh the DRAM devices in the RAM array.

▲ 12.5 MEMORY CIRCUITRY

Earlier we found that the system processor board of the PC is equipped with 48Kbytes of ROM and either 64K or 256Kbytes of RAM. The ROM array is implemented using EPROM devices and provides for nonvolatile storage of fixed information, such as the BIOS of the PC. On the other hand, RAM is volatile and is used for temporary storage of information such as application programs. This part of the memory subsystem can be implemented with either 64K-bit or 256K-bit dynamic RAM chips. In the previous section, we showed how the ROM and RAM chip-select signals are generated. Here we will study how the EPROM devices are arranged to form the ROM array and how the DRAM devices are arranged to form the RAM array. We will also study how the memory arrays use the chip-select signals and interface to the system bus.

ROM Array Circuitry

Let us begin by briefly examining the architecture of the ROM array of the PC. Figure 12–5 illustrates the circuitry of the ROM array. Looking at this circuit diagram, we find that it is implemented with six 8K × 8-bit EPROMs. These devices are labeled XU_{28} through XU_{33}. Note that each of these EPROMs is enabled by one of the ROM chip-select signals, $\overline{CS_2}$ through $\overline{CS_7}$, which are generated by the ROM address decoder. For instance, $\overline{CS_2}$ enables EPROM XU_{28}. Figure 12–4(c) shows that this chip-select output is at its

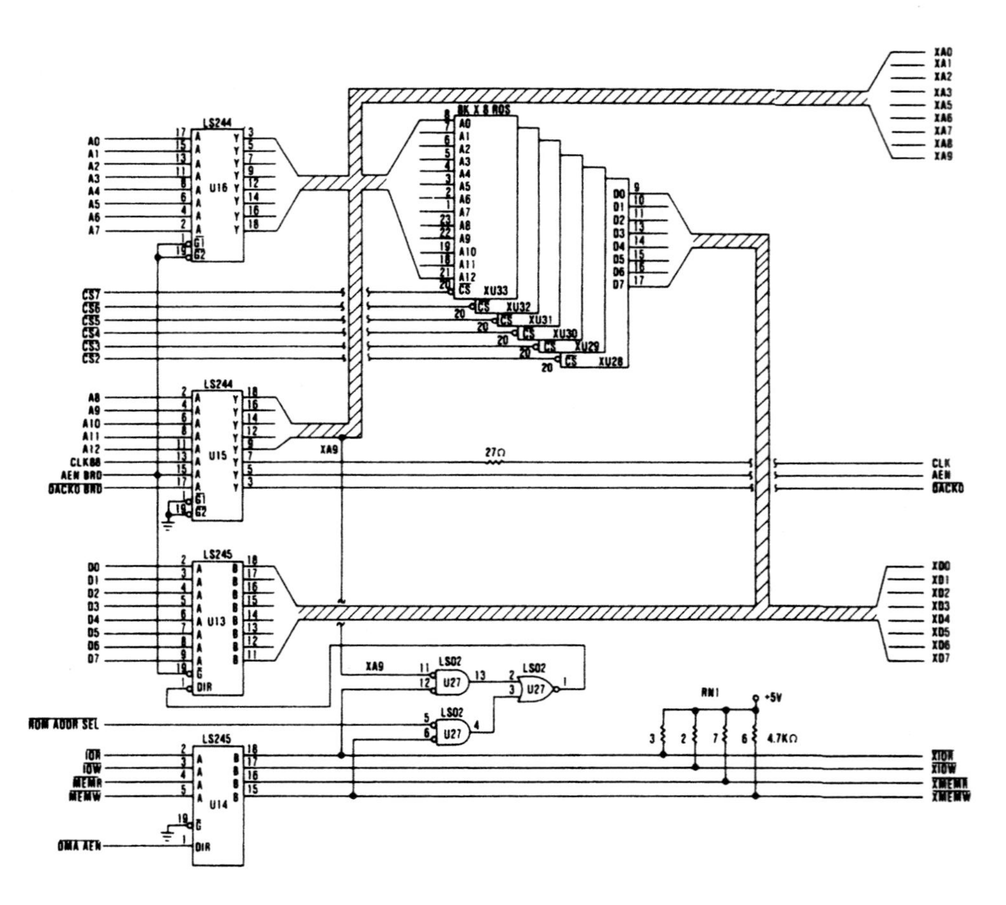

Figure 12–5 ROM circuitry. (Courtesy of International Business Machines Corporation)

active 0 logic level for all memory addresses in the range $F4000_{16}$ through $F5FFF_{16}$. Therefore, EPROM XU_{28} holds the information corresponding to these 8K addresses.

Now that we know how the individual EPROMs are selected, let us look at how a storage location within an EPROM is accessed and its data returned to the MPU. The address outputs on the lower 13 address lines of the system address bus, A_0 through A_{12}, are used to select the specific byte of data within an EPROM. These address inputs are first buffered with 74LS244 octal buffers U_{15} and U_{16} and then applied to the address inputs of all six EPROMs in parallel. Note that control signal AEN BRD must be at the 0 logic level for the address buffer to be enabled for operation.

The byte of code held at the addressed storage location in the chip-selected EPROM is output on data lines D_0 through D_7 for return to the MPU. These data outputs are interfaced to the 8088's system data bus by the 74LS245 bus transceiver U_{13}. During a read bus cycle, data must be transferred from the outputs of the ROM array to the system data bus lines D_0 through D_7.

The direction of data transfer through the data bus transceiver is set by the logic level at its data direction (DIR) input. Figure 12–5 shows that the logic level at DIR is determined by the operation of the control logic formed from transceiver U_{14} and three NOR gates of IC U_{27}. The $\overline{\text{IOR}}$ and $\overline{\text{MEMW}}$ outputs of U_{14}, along with chip-select signal $\overline{\text{ROM ADDR SEL}}$ and address bit XA_9, are inputs to the NOR gate circuit. In response to these inputs, the circuit switches DIR to logic 0 during all read bus cycles to storage locations in the address range of the ROM array and for all I/O read cycles from an address where XA_9 is logic 0. Logic 0 at DIR sets the direction of data transfer through U_{13} to be from memory to the 8088's system bus. That is, data are being read from the ROM array.

RAM Array Circuitry

Figure 12–6(a) shows the circuitry of the RAM array. This circuit shows only two of the four banks of RAM ICs provided for on the system processor board of the PC. These banks are identified as *bank 0* and *bank 1.* The circuitry for the other two banks, *bank 2* and *bank 3,* is shown in Fig. 12–6(b). In each bank, eight 64K $\times$ 1-bit dynamic RAMs (DRAMs) are used for data storage, and a ninth DRAM is included to hold parity bits for each of the 64K storage locations. Figure 12–6(a) shows that the DRAMs in bank 0 are labeled U_{37} through U_{45}. Device U_{37} is used to store the parity bit, and U_{38} through U_{45} store the bits of the byte of data. The data storage capacity of bank 0 is 64Kbytes, and all four banks together give the system processor board a maximum storage capacity of 256Kbytes.

Let us now examine how a byte of data is read from DRAMs in bank 0. Address lines A_0 through A_{15} are applied to inputs of the 74LS158 data selectors U_{62} and U_{79}. These devices are used to multiplex the 16-bit memory address into a byte-wide row address and a byte-wide column address. The multiplexed address outputs of the data selectors are called MA_0 through MA_7 and are applied to address inputs A_0 through A_7 of all DRAMs in parallel. The select signal ADDR SEL, which is applied to the select (S) input of both data selectors, is used to select whether the RAS or CAS byte of the address is output on the MA lines.

We have assumed that the storage location to be accessed is located in bank 0. In this case, the RAS and CAS address bytes are output from the address multiplexer

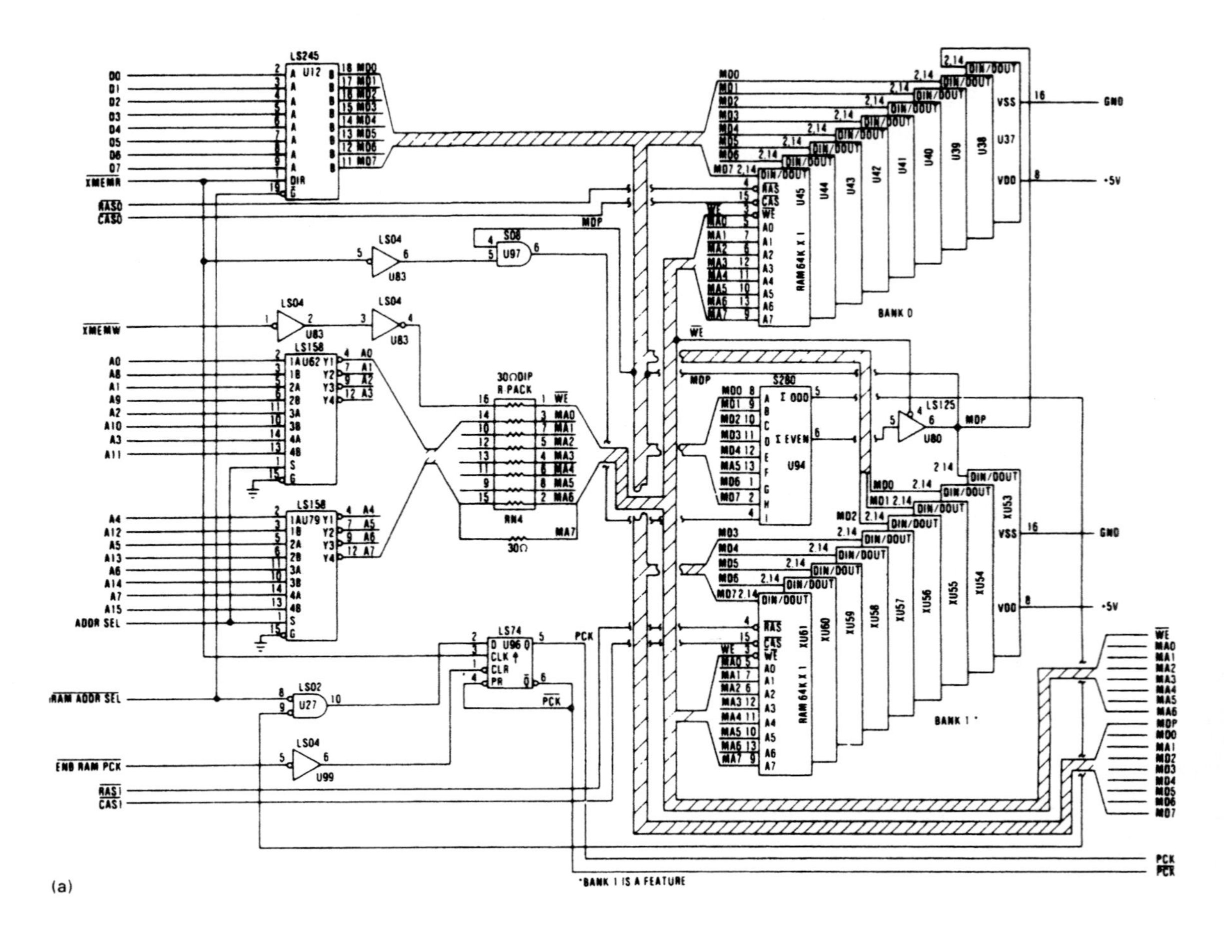

Figure 12–6 (a) RAM circuitry. (Courtesy of International Business Machines Corporation) (b) RAM banks 2 and 3. (Courtesy of International Business Machines Corporation)

synchronously with the occurrence of the active $\overline{RAS_0}$ and $\overline{CAS_0}$ strobe signals, respectively. ADDR SEL initially sets the multiplexer to output the RAS address byte on the MA line. When RAS_0 switches to logic 0, it signals all DRAMs in bank 0 to accept the row address off of the MA lines. Next, ADDR SEL switches the logic level and causes the column address to be output from the multiplexer. It is accompanied by $\overline{CAS_0}$, which is applied to the $\overline{CAS}$ inputs of all DRAMs in bank 0. Logic 0 at $\overline{CAS}$ causes them to accept the column address from the MA lines. At this point, the complete address of the storage location that is to be accessed has been supplied to RAM in bank 0.

We are also assuming that a read bus cycle is taking place. For this reason, the $\overline{XMEMW}$ input is logic 1 and signals all DRAMs that a read operation is to take place. Therefore, each device outputs a bit of data held in the storage location corresponding to the selected row and column address. The byte of data is passed over data lines MD_0 through MD_7 to the 74LS245 bus transceiver U_{12}. Here a 0 logic level at $\overline{XMEMR}$ sets the transceiver to pass data from the MA lines to system data bus lines D_0 through D_7 dur-

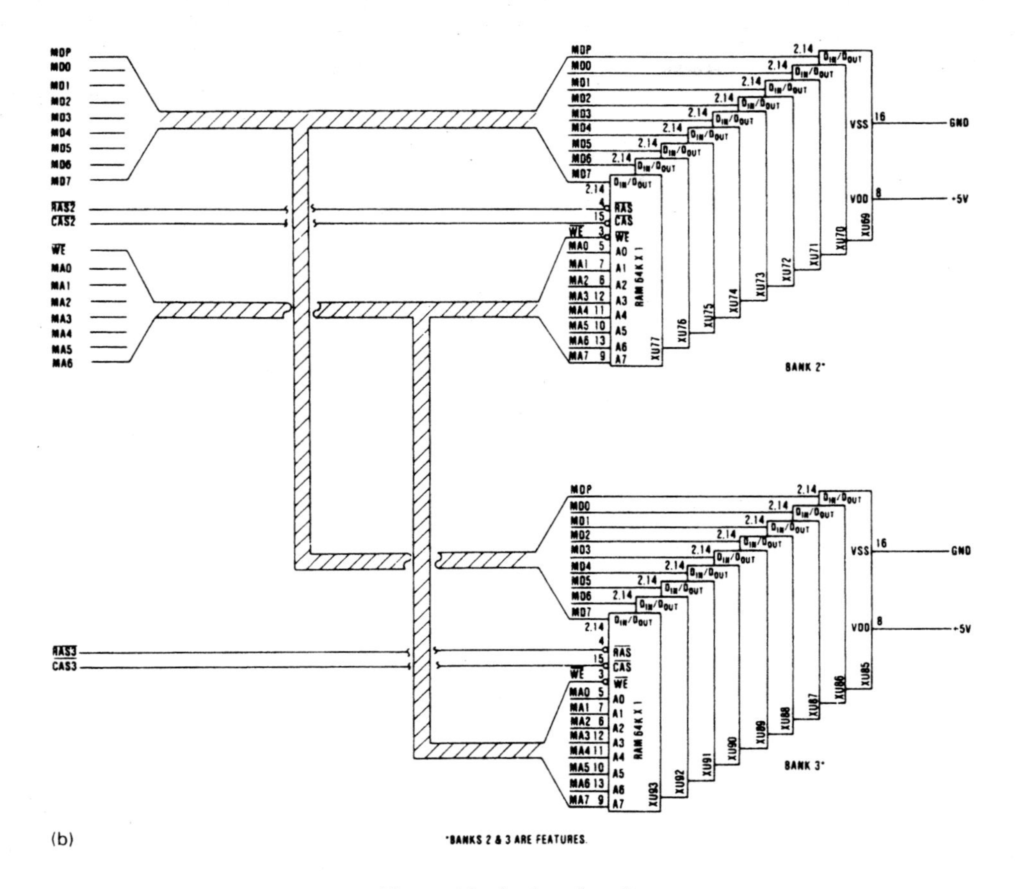

Figure 12–6 (continued)

ing all read cycles. Moreover, the signal $\overline{\text{RAM ADDR SEL}}$ enables the transceiver for operation during all bus cycles to the RAM array.

In our description of the read cycle, we did not consider the effect of the *parity generator/checker circuitry* that is included in the RAM array of the PC. *Parity* is a technique that is used to improve the reliability of data storage in a RAM subsystem. Whenever data are written into or read from the DRAMs in Fig. 12–6(a), the byte of data on lines MD_0 through MD_7 is also applied to inputs A through H of the 74S280 parity generator/checker device U_{94}. The I input during a write operation is at logic 0 as it is the inverse of $\overline{\text{XMEMR}}$, which is 1. Therefore, including a ninth bit at logic 0 does not change the parity of the byte being written. If the byte has *even parity* (contains an even number of bits at the 1 logic level), the Σ_{EVEN} output (pin 6) of U_{94} switches to logic 1. However, if parity is odd, Σ_{EVEN} switches to logic 0. During write bus cycles, this *parity bit* output is supplied to the DIN/DOUT pin of the *parity bit DRAM* over the MDP line and is stored in DRAM along with the byte of data.

On the other hand, during read operations the parity bit that is read out of the parity-bit DRAM on the MDP line and is gated by AND gate U_{97} to the ninth input (I) of the 74S280 parity generator/checker. If the 9-bit word read from memory has *odd parity* (an extra 1 is added to even parity words as the parity bit), the Σ_{ODD} output (pin 5) of U_{94} switches to logic 1 to indicate that parity is correct. Σ_{ODD} is sent through NOR gate U_{27} to the data input at pin 2 of the 74LS74 parity check interrupt latch U_{96}. As long as no *parity error* has occurred, the $\overline{\text{PCK}}$ output of the latch remains at its inactive 1 logic level. However, if Σ_{ODD} signals that a parity error has been detected by switching to logic 0, the parity error interrupt latch sets, and the logic 0 that results at $\overline{\text{PCK}}$ issues a nonmaskable interrupt request to the MPU. The NMI service routine must test the logic level of PCK through the 8255A I/O interface to determine if the source of the NMI is PCK. Moreover, at completion of the parity error interrupt service routine, issuing the signal $\overline{\text{ENB RAM PCK}}$ through the 8255A I/O interface clears the parity error interrupt latch.

▲ 12.6 DIRECT MEMORY ACCESS CIRCUITRY

The 8088-based system processor board of the IBM PC supports the direct memory access (DMA) mode of operation for both its memory and I/O address spaces. This DMA capability permits high-speed data transfers to take place between two sections of memory or an I/O device and memory. The bus cycles initiated for these DMA transfers are not under control of the 8088 MPU; instead, a special VLSI device known as a DMA controller performs them. The DMA circuitry in the PC implements this function using the 8237A-5 DMA controller IC. Looking at the circuit drawing in Fig. 12–7, we find that the 8237A is labeled U_{35}. This device provides four independent DMA channels for the PC.

Even though the 8237A performs the actual DMA bus cycles by itself, the 8088 controls overall operation of the device. There are 16 registers within the 8237A that determine how and when the four DMA channels work. Since the microprocessor interface of the 8237A is I/O mapped, the 8088 communicates with these registers by executing I/O instructions. For instance, the 8237A must be configured with operating features such as autoinitialization, address increment or decrement, and fixed or rotating channel priority. These options are selected by loading the command and mode registers within the 8237A through a software initialization routine. Moreover, before a DMA transfer can be performed, the 8088 must send the 8237A information related to the operation that is to take place. This information could include a source base address, destination base address, count of the words of data to be moved, and an operating mode. The modes of DMA operation available with the 8237A are demand transfer mode, single transfer mode, block transfer mode, and cascade mode. Finally, the 8088 can obtain status information about the current DMA bus cycle by reading the contents of registers. For example, it can read the values in the current address register and current count register to determine which data have been transferred.

Let us now look briefly at the signals and operation of the microprocessor interface of the 8237A. In Fig. 12–7, we find that the microprocessor interface of the 8237A is enabled by the signal $\overline{\text{DMA CS}}$, which is applied to its $\overline{\text{CS}}$ input at pin 11. Figure 12–4(b) shows that $\overline{\text{DMA CS}}$ is active whenever an I/O address in the range 0000_{16} through $001F_{16}$ is output on the system address bus. The specific register to be accessed is selected using the four least significant address lines, XA_0 through XA_3. Data are read from or written into the selected register over system data bus lines XD_0 through XD_7. The 8088

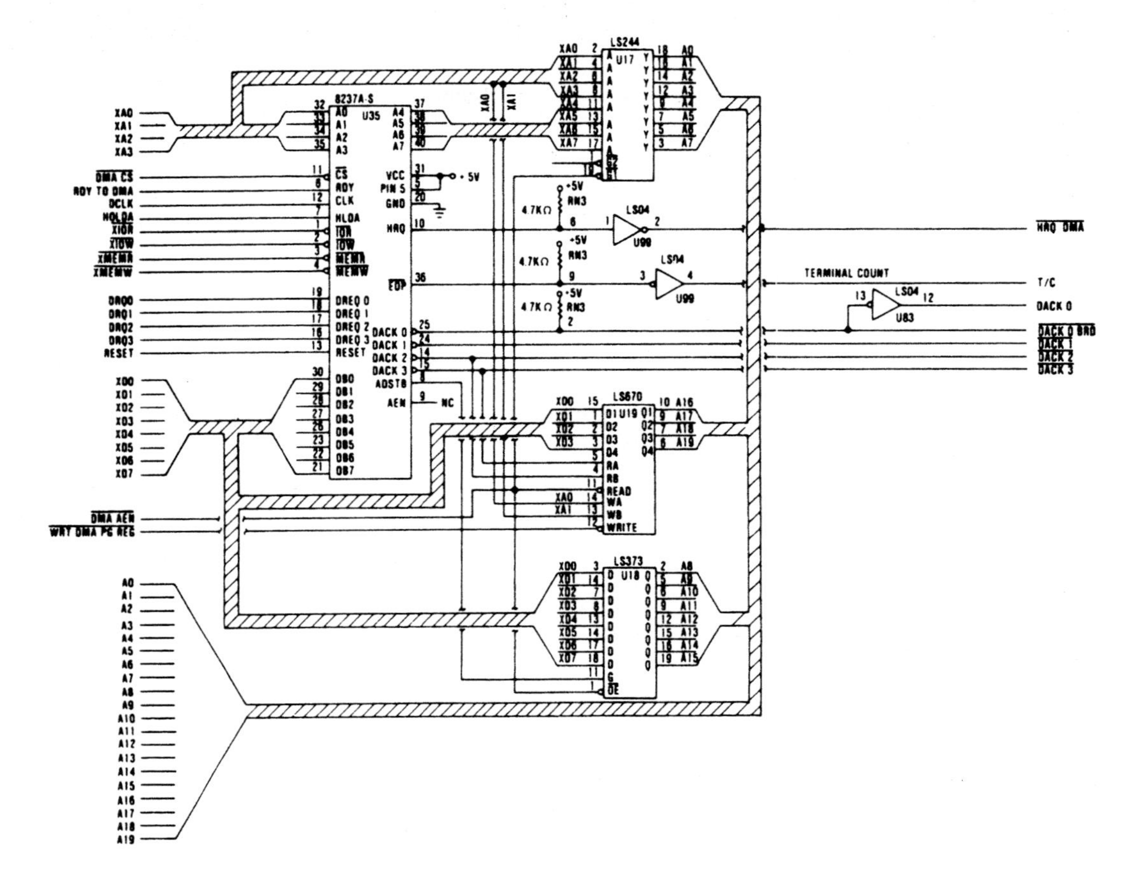

Figure 12–7 Direct memory access circuitry. (Courtesy of International Business Machines Corporation)

signals the 8237A whether data are to be input or output over the bus with the control signal $\overline{XIOR}$ or $\overline{XIOW}$, respectively.

Earlier we pointed out that the four DMA channels of the PC are identified as DMA channels 0 through 3. Moreover, we found that channel 0 is dedicated to RAM refresh and that channel 2 is used by the floppy disk subsystem. Use of a DMA channel is initiated by a request from hardware. In Fig. 12–7, the signals DRQ_0 through DRQ_3 are the hardware request inputs for DMA channels 0 through 3, respectively. DRQ_0 is generated by timer 1 of the 8253 programmable interval timer (see Fig. 12–8) and is used to initiate a DMA 0 refresh cycle for RAM every 15.12 μs. The other three DMA request lines are supplied from the I/O channel and are available for use by other I/O channel devices.

For a DMA request to be active, the corresponding DRQ input must be switched to the 1 logic level. Let us assume that a DRQ input has become active, the DMA request input for the active channel is not masked out within the 8237A, and a higher-priority channel is not already active. Then the response of the 8237A to the active DMA request is that it requests to take over control of the system bus by switching its hold request (HRQ) output to logic 1 and then waits in this state until the 8088 signals that it has given up the bus by returning a logic 1 on the hold acknowledge (HLDA) input of the 8237A.

The simulated hold/hold acknowledge handshake that takes place between the 8237A and 8088 is performed by the wait-state control logic circuitry that is shown in Fig. 12–3. The HRQ signal that is output at pin 10 of the 8237A is applied to the $\overline{\text{HRQ DMA}}$ input of the wait state control logic circuit. The operation of this circuitry was described in detail in Section 12.2. For this reason, here we will just give an overview of the events that take place in the hold acknowledge handshake sequence.

In response to logic 0 at the $\overline{\text{HRQ DMA}}$ input, the circuit first waits until the 8088 signals that its bus is in the passive state (no bus activity is taking place) and then switches the HOLDA output to logic 1. This signal is returned to the HLDA input at pin 7 of the 8237A, where it signals that the 8088 has given up control of the system bus.

Next, the control logic switches signal AEN BRD and $\overline{\text{AEN}}$ to their active logic levels. These signals are used to tri-state the outputs of the 8288 bus controller, data bus transceiver, and system bus address latches. With these outputs floating, the MPU is isolated from devices connected to the system bus. Additionally these signals disable the decoder that produces chip selects for the peripherals located on the system processor board.

One clock later, the signal $\overline{\text{DMA WAIT}}$ becomes active. This signal is returned through the ready/wait logic of the 8284A to the READY input of the 8088 and ensures that the 8088 does not initiate a new bus cycle. The signal $\overline{\text{DMA AEN}}$ is now produced by the control logic and sent to the DMA address logic (see Fig. 12–7). Logic 0 at this input enables the address buffers for operation. $\overline{\text{DMA AEN}}$ is also applied to the DIR input of transceiver U_{14} (see Fig. 12–5) and isolates the I/O and memory read/write control signals from the system bus so that the DMA controller itself can provide them.

At this point, the 8237A is free to take control of the system bus; therefore, it outputs the DMA acknowledge ($\overline{\text{DACK}_0}$ to $\overline{\text{DACK}_3}$) signal corresponding to the device requesting DMA service. $\overline{\text{DACK}_0}$ is output as $\overline{\text{DACK 0 BRD}}$ to the refresh control circuitry. Logic 0 on this line signals the wait-state circuit and RAM chip-select decoder that DMA refresh bus cycles are to be initiated. The other three DACK outputs are supplied to the I/O channel.

Now that the 8237A has taken control of the system data bus, let us look at how a block of data is transferred from memory to a device in the I/O address space. To perform this operation, the DMA controller first outputs a 16-bit address on address lines A_0 through A_7 and data lines DB_0 through DB_7. Address bit A_8 through A_{15}, output on the data lines, are output in conjunction with a pulse on the address strobe (ADSTB) line at pin 8 of the 8237A. This pulse is used to latch the address into the 74LS373 latch, U_{18}. The four most significant bits of the 20-bit address are not produced by the 8237A; instead, three DMA page registers within the 74LS670 register file device, U_{19}, generate them. The processor at power up initializes the page registers on time. Once initialized, they provide the upper four bits of the address. The device contains four registers; only three of them are used for channels 1, 2, and 3. To access a page register I/O addresses 81_{16} to 83_{16} can be used for channels 1, 2, and 3, respectively. These addresses activate the required $\overline{\text{WRT DMA PG REG}}$ signal along with the two bits XA_1 and XA_0 to select the desired page register. The $DACK_2$ and $DACK_3$ signals are used to read the appropriate four bits from a page register and feed to the four upper address lines of the address bus.

A valid 20-bit source address is now available on system address bus lines A_0 through A_{19}. Next, the memory read ($\overline{MEMR}$) and I/O write ($\overline{IOW}$) control signals become active, and the data held at the addressed storage location are read over system data bus lines XD_0 through XD_7 to the I/O device. This completes the first data transfer.

We will assume that during the DMA bus cycle the source or destination address is automatically incremented by the 8237A. In this way, its current value points to the next data element to be read from memory or written to memory. Moreover, at completion of the DMA bus cycle, the count in the current word register is decremented by 1. The new count stands for the number of data transfers that still remain to be performed.

This basic DMA transfer operation is automatically repeated by the 8237A until the current word register count rolls over from 0000_{16} to $FFFF_{16}$. At this moment, the DMA operation is complete and the end of process ($\overline{EOP}$) output is switched to logic 0. EOP is used to tell external circuitry that the DMA operation has run to completion. In Fig. 12–7, we see that $\overline{EOP}$ is inverted to produce the terminal count (T/C) signal for the I/O channel. In response to T/C, the requesting device removes its DMA request signal, and the 8237A responds by returning control of the system bus to the 8088.

▲ 12.7 TIMER CIRCUITRY

Figure 12–8 shows the timer circuitry of the IBM PC. This circuitry controls four basic system functions: *time-of-day clock, DRAM refresh, speaker,* and *cassette.* In the PC, the timers are implemented with the 8253-5 programmable interval timer IC. This device is labeled U_{34} in Fig. 12–8. The 8253 provides three independent, programmable, 16-bit counters for use in the microcomputer system. Here we will first look at how the 8253 is interfaced to the 8088 microprocessor and then at how it implements each of the four system functions.

Microprocessor Interface and Clock Inputs

The 8088 MPU communicates with the 8253's internal control registers through the microprocessor interface. Figure 12–1(c) shows that the control registers of the 8253 are located in the range 0040_{16} through 0043_{16} of the PC's I/O address space. Using I/O instructions, we can access the 8253's internal registers to configure the mode of operation for the individual timers and read or load their counters. For example, an input operation from I/O address 0040_{16} reads the current count in counter 0. On the other hand, an output operation to the same address loads an initial value into the count register for counter 0. The same type of operations can be performed to the registers for counters 1 and 2 by using address 0041_{16} or 0042_{16}, respectively. Moreover, the mode of operation for the counters is set up by writing a byte-wide control word to address 0043_{16}. However, the contents of the mode control register cannot be read through software.

Let us now look at how the 8088 performs data transfers to the 8253 over the system bus. The microprocessor interface of the 8253 is enabled by the signal $\overline{T/C\ CS}$, which

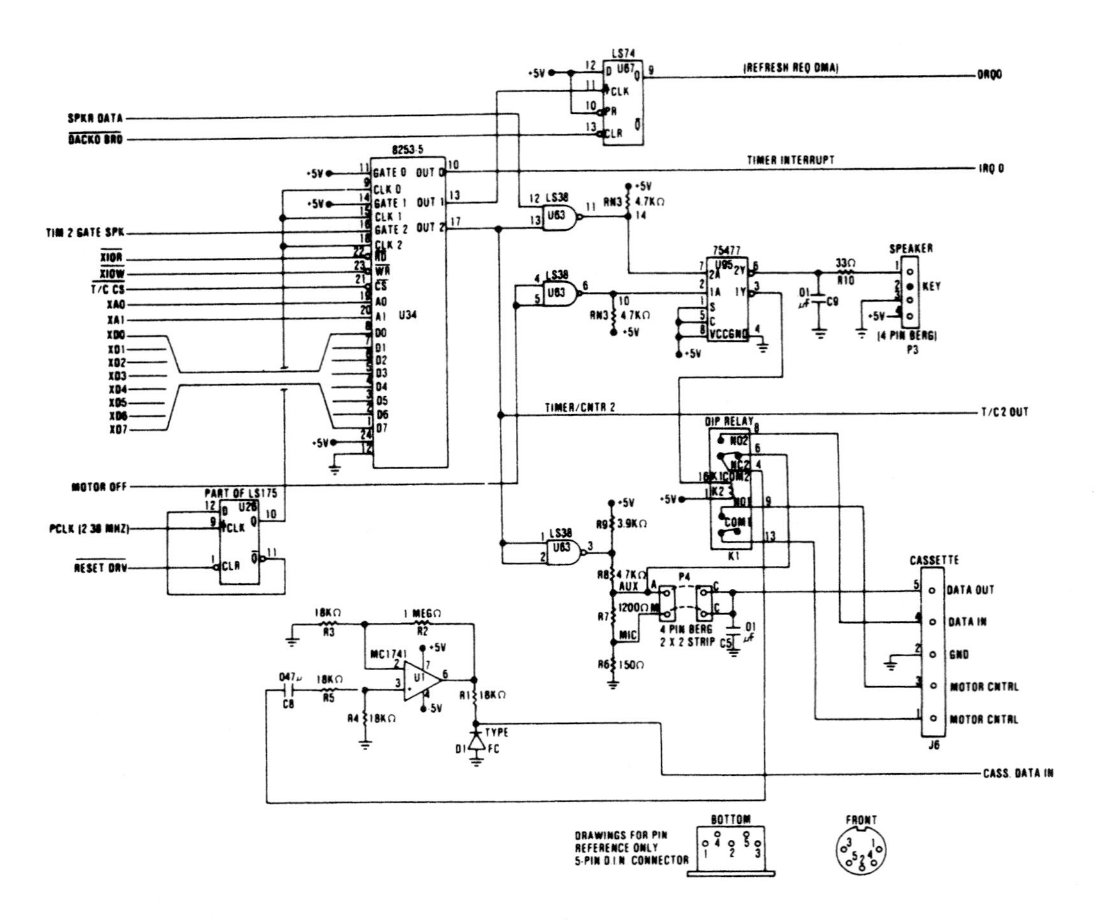

Figure 12–8 Timer circuitry. (Courtesy of International Business Machines Corporation)

is tied to its $\overline{CS}$ (chip-select) input at pin 21. Figure 12–4(b) shows that this signal is at its 0 active logic level whenever an I/O address in the range 0040_{16} through $005F_{16}$ is output on the system address bus. The internal control register that is to be accessed is selected by a code that is applied to register select inputs A_0 and A_1 over system address bus lines XA_0 and XA_1. Figure 12–8 shows that system data bus lines XD_0 through XD_7 connect to the data lines D_0 through D_7 of the 8253. The 8088 tells the PIT whether data are to be read from or written into the selected register over these lines with logic 0 at $\overline{XIOR}$ (I/O read) or $\overline{XIOW}$ (I/O write), respectively.

The signal that is applied to the CLK inputs of the timers is derived from the 2.38-MHz PCLK (peripheral clock) signal. Note that PCLK is first divided by 2 using the 74LS175 D-type flip-flop U_{26}. This generates a 1.19-MHz clock for input to the timers. This signal drives clock inputs CLK_0, CLK_1, and CLK_2 in parallel. Note in Fig. 12–8 that the first two of these clock signals are permanently enabled to the counter by having +5 V connected directly to the $GATE_0$ and $GATE_1$ inputs, respectively. However, CLK_2 is enabled to the counter with signal TIM 2 GATE SPK. This signal must be switched to logic 1 (see Fig. 12–9) under software control to enable the clock input for counter 2.

Outputs of the PIT

In Fig. 12–8, the three outputs of the 8253 timer are labeled OUT_0, OUT_1, and OUT_2. OUT_0 is produced by timer 0 and is set up to occur at a regular time interval equal to 54.936 ms. This output is applied to the timer interrupt request input (IRQ_0) of the 8259A interrupt controller, where it represents the time-of-day interrupt.

Timer output OUT_1 is generated by timer 1 and also occurs at a regular interval, every 15.12 μs. In Fig. 12–8, we find that this signal is applied to the CLK input (pin 11) of the 74LS74 flip-flop U_{67} and causes the DRQ_0 output to set. Logic 1 at this output sends a request for service to the 8237A DMA controller and asks it to perform a refresh operation for the dynamic RAM subsystem. When the DMA controller has taken control of the system bus and is ready to perform the refresh cycle, it acknowledges this fact by outputting the refresh acknowledge ($\overline{\text{DACK 0 BRD}}$) signal. Logic 0 on this line clears flip-flop U_{67}, thereby removing the refresh request.

The output of the third timer, OUT_2, is used three ways in the PC. First, it is sent as the signal T/C2 OUT to input 5 on port C of the 8255A PIC (see Fig. 12–9). In this way, its logic level can be read through software. Second, it is used as an enable signal for speaker data in the speaker interface. When the speaker is to be used, the 8088 must write logic 1 to bit 0 of port B on the 8255A PIC (see Fig. 12–9). This produces the signal TIM 2 GATE SPK, which enables the clock for timer 2. Pulses are now produced at OUT_2. When a tone is to be produced by the speaker, the 8088 outputs the signal SPKR DATA at pin 1 of port B of the 8255A (see Fig. 12–9). Logic 1 at input SPKR DATA enables the pulses output at OUT_2 to the 75477 driver U_{95}. The output of this driver is supplied to the speaker. Modifying the count in timer 2 changes the frequency of the tone produced by the speaker.

The last use of counter 2 is to supply the record tone for the cassette interface. As shown in Fig. 12–8, the PC's cassette interface is through connector J_6. Data that are to be recorded on the tape are output on the DATA OUT line at pin 5 of J_6. In Fig. 12–8, we find that the data to be recorded on the cassette are output from the OUT_2 pin of the 8253 timer and are supplied through inverter U_{63} to a voltage divider. Jumper P_4 is used to select the voltage level for the DATA OUT signal. For instance, if a jumper is installed from A to C, DATA OUT is set for a 0.68-V high signal level and 0 V as the low level.

Data played back from the cassette enter the microcomputer at the DATA IN input at pin 6 of connector J_6. DATA IN is passed through a set of contacts on DIP relay K_1 to the input of an amplifier made with the MC1741 device, U_1. Since it is a high-gain amplifier, the low-level signals read from the tape are saturated to produce a TTL-level signal at output CASS DATA IN. This signal is applied to input 4 at port B of the 8255A (see Fig. 12–9), where it can be read by the 8088 using IN instructions.

The motor of the cassette player is also turned on or off through circuitry shown in Fig. 12–8. When the MOTOR OFF input is switched to logic 0, DIP relay K_1 is activated. This connects the DATA IN signal to the input of the amplifier circuit formed from the MC1741 device U_1. At the same time, the motor control (MOTOR CNTRL) outputs at pins 1 and 3 of J_6 are connected through a relay contact and the motor turns on. MOTOR OFF is provided by output 3 at port B of the 8255A PIC (see Fig. 12–9).

▲ 12.8 INPUT/OUTPUT CIRCUITRY

Figure 12–9 shows the I/O circuitry of the IBM PC. Three basic types of functions are performed through this I/O interface. First, using this circuitry, the 8088 inputs data from the keyboard and outputs data to the cassette and speaker. Second, through this circuitry, the microprocessor reads the setting of DIP switches to determine system configuration information such as the size of the system memory, number of floppy-disk drives, type of monitor used on the system, and whether or not an 8087 numeric coprocessor is installed. Finally, certain I/O ports are used for special functions, such as clearing the parity check flip-flop and reading the state of the parity check flip-flop through software. The I/O circuitry of the PC system processor board is designed using the 8255A-5 programmable peripheral interface (PPI) IC. In this section, we will look at how the 8255A is interfaced to the 8088 MPU and at the different input/output operations that take place through its ports.

8255A Programmable Peripheral Interface

The 8255A PPI that implements the I/O circuitry is labeled U_{36} in Fig. 12–9. It has three 8-bit ports for implementing inputs or outputs. In the PC, ports PA and PC are configured to operate as inputs, and the lines of port PB are set up to work as outputs. Figure 12–1(d) shows that ports PA, PB, and PC reside at the I/O addresses 0060_{16}, 0061_{16}, and 0062_{16}, respectively.

Figure 12–1(d) also identifies the function of each pin at PA, PB, and PC. Here we find that input port PA is used to both read the configuration switches of SW1 and communicate with the keyboard. On the other hand, output port PB controls the cassette and speaker. It also supplies enable signals for RAM parity check, I/O channel check, and reading of the configuration switches or keyboard. Finally, we find that the input port PC is used to read the I/O channel RAM switches (SW2), parity check signal, I/O channel check signal, terminal count status from timer 2, and cassette data.

The operation of the ports of the 8255A are configurable under software control. Writing a configuration byte to the command/mode control register within the device does this. In Fig. 12–1(d) shows that the command/mode register is located at address 0063_{16}. When configured by the initialization software of the PC, it is loaded with the value 99_{16}. This configuration code selects mode 0 operation for all three ports.

Loading of the control register, as well as inputting of data from ports PA and PC or outputting of data to port PB, is performed through the 8255A's microprocessor interface. In Fig. 12–9, the microprocessor interface is activated by the $\overline{\text{PPI CS}}$ (PPI chip-select) signal, which is applied to the $\overline{\text{CS}}$ input at pin 6 of the 8255A. Figure 12–4(b) shows that this signal is at its active (logic 0) level whenever an I/O address in the range 0060_{16} through $007F_{16}$ is output on the system address bus. However, remember that just four of these addresses, 0060_{16} through 0063_{16}, are used by the 8255A interface. Note that the data bus inputs of the 8255A are tied to lines XD_0 through XD_7 of the system data

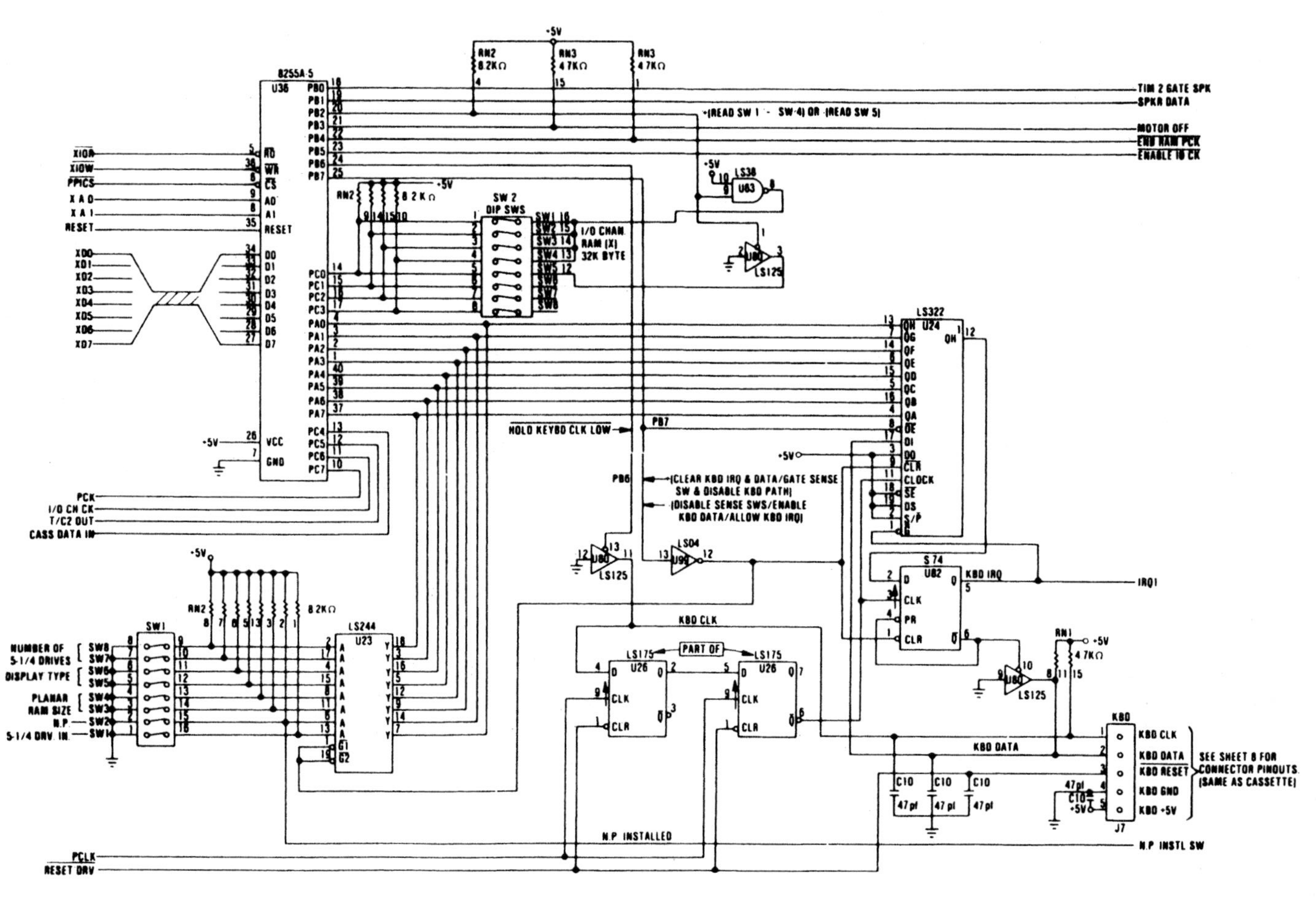

Figure 12–9 I/O circuitry. (Courtesy of International Business Machines Corporation)

bus. It is over these lines that the configuration information or input/output data are carried. The 8088 signals the PPI that data are to be read from or written into a register with signals $\overline{XIOR}$ and $\overline{XIOW}$, respectively, while the register to be accessed is determined by the register select code on address lines XA_0 and XA_1.

Inputting System Configuration DIP Switch Settings

Let us now look at how the settings of the system configuration DIP switches are input to the 8088 microprocessor. Looking at Fig. 12–9, we see that input port PA, at I/O address 0060_{16}, is connected to configuration switch SW1 through the 74LS244 buffer (U_{23}). To read the state of these switches, the keyboard data path must be disabled and the switch path enabled. Writing a 1 to bit PB_7 of the output port does this. This output is inverted and then applied to the enable inputs of buffer U_{23}. Logic 0 at these inputs enables the buffer and causes the switch setting to pass through to port PA. Now the instruction

```
IN  AL, 60H
```

can be used to read the contents of port PA. The byte of data read in can be decoded based on the table in Fig. 12–1(d) to determine the number of floppy disk drives, type of display, presence or absence of an 8087, and amount of RAM on the system board.

EXAMPLE 12.1

The system configuration byte read from input port PA is $7D_{16}$. Describe the PC configuration for these switch settings.

Solution

Expressing the switch setting byte in binary form, we get

$$PA_7PA_6PA_5PA_4PA_3PA_2PA_1PA_0 = 7D_{16} = 01111101_2$$

Referring to the table in Fig. 12–1(d), we find that

$PA_0 = 1$ indicates that the system has floppy-disk drive(s)

$PA_1 = 0$ indicates that an 8087 is not installed

$PA_3PA_2 = 11$ indicates that the system processor board has 256K of memory

$PA_5PA_4 = 11$ indicates that the system has a monochrome monitor

$PA_7PA_6 = 01$ indicates that the system has two floppy drives

Scanning the Keyboard

The keyboard of the PC is also interfaced to the 8088 through port PA of the 8255A. Figure 12–9 shows that the keyboard attaches to the system processor board at connector KB_0. The keyboard interface circuit includes devices U_{82}, U_{26}, and U_{24}. At completion of the power-on reset service routine, output PB_7 of the 8255A is switched to logic 0. This disables reading of configuration switch SW1 and enables the keyboard data path and interrupt.

We will now examine how the 8088 determines that a key on the keyboard has been depressed. The keyboard of the PC generates a *keyscan code* whenever one of its keys is depressed. Bits of the keyscan code are input to the system processor board in serial form at the KBD DATA pin of the keyboard connector synchronously with pulses at KBD CLK. Note in Fig. 12–9 that KBD DATA is applied directly to the data input (DI) of the 74LS322 serial-in, parallel-out shift register (U_{24}). On the other hand, KBD CLK is input to the data input at pin 4 of one of the two D-type flip-flops in the 74LS175 device, U_{26}. This flip-flop circuit divides the clock by 4 before outputting it at pin 6. The clock produced at pin 6 of U_{26} is applied to the clock input of the 74S74 keyboard interrupt request flip-flop U_{82}, as well as the CLOCK input of the 74LS322 shift register. CLOCK is used by the shift register to clock in bits of the serial keyscan code from DI. When a byte of data has been received, the Q_H output at pin 12 of the shift register switches to logic 1. Q_H is returned to the data input of the 74S74 flip-flop U_{82}, and when logic 1 is clocked into the device, the keyboard interrupt request signal KBD IRQ becomes active. At the same moment that the interrupt signal is generated, the KBD DATA line is driven to logic 0 by the output at pin 8 of buffer U_{80} and the shift register is disabled.

In response to the IRQ_1 interrupt request, the 8088 initiates a keyscan-code service routine. This routine reads the keyscan code by inputting the contents of the shift register through port PA. After reading the code, it drives output PB_7 to logic 1 to clear the keyboard interrupt request flip-flop and keyscan shift register. Next, the service routine drives PB_7 back to logic 0. This reenables the keyboard interface to accept another character from the keyboard.

Port C Input and Output Functions

The switch configuration identified as SW2 in Fig. 12–9 represents what is called the *I/O channel RAM switches.* The five connected switches are used to identify the amount of read/write memory provided through the I/O channel. The settings of these switches are also read through the 8255A PPI. Once the settings are read from the switches, the total amount of memory can be determined by multiplying the binary value of the switch settings by 32Kbytes.

Looking at the hardware in Fig. 12–9, we find that the settings of the five switches are returned to the 8088 over just four input lines, PC_0 through PC_3. To read the settings of switches SW2-1 through SW2-4, logic 1 must first be written to output PB_2 and then input the contents of port PC. The four least significant bits of this byte are the switch settings. Logic 1 in a bit position indicates that the corresponding switch is in the OFF position (open circuit). Switch SW2-5 is read by switching PB_2 to logic 0 and once again

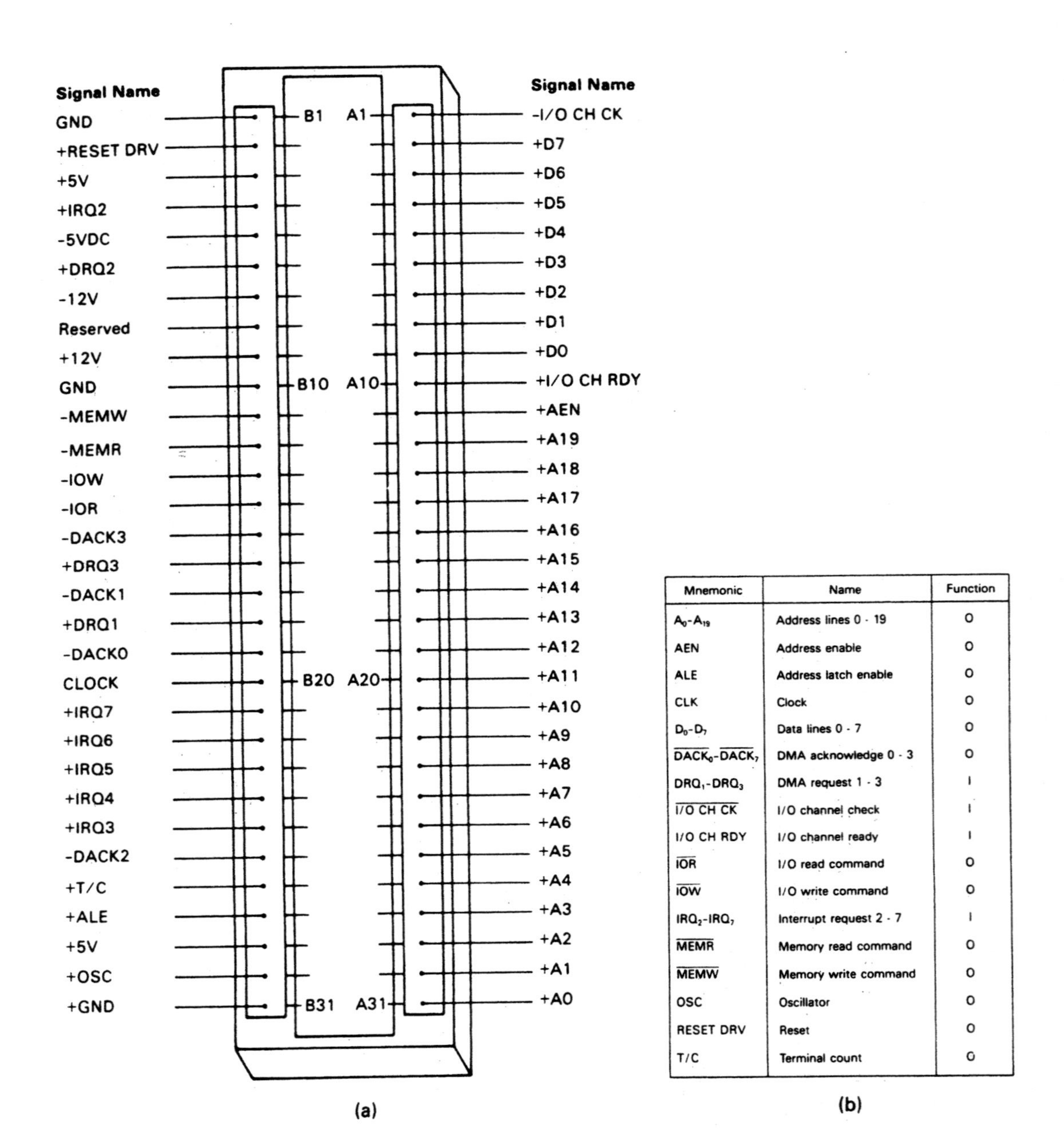

Mnemonic	Name	Function
A_0-A_{19}	Address lines 0 - 19	O
AEN	Address enable	O
ALE	Address latch enable	O
CLK	Clock	O
D_0-D_7	Data lines 0 - 7	O
$\overline{DACK_0}$-$\overline{DACK_7}$	DMA acknowledge 0 - 3	O
DRQ_1-DRQ_3	DMA request 1 - 3	I
$\overline{\text{I/O CH CK}}$	I/O channel check	I
I/O CH RDY	I/O channel ready	I
$\overline{\text{IOR}}$	I/O read command	O
$\overline{\text{IOW}}$	I/O write command	O
IRQ_2-IRQ_7	Interrupt request 2 - 7	I
$\overline{\text{MEMR}}$	Memory read command	O
$\overline{\text{MEMW}}$	Memory write command	O
OSC	Oscillator	O
RESET DRV	Reset	O
T/C	Terminal count	O

(b)

Figure 12–10 (a) I/O channel interface. (Courtesy of International Business Machines Corporation) (b) Signal mnemonics, names, and functions.

reading the contents of port PC. In this byte, the content of the least significant bit represents the setting of SW2-5. These two bytes can be combined through software to give a single byte that contains all five switch settings.

The four most significant bit lines of port PC are supplied by signals generated elsewhere on the system processor board. PC_5 through PC_7 allow the 8088 to read the state of the RAM parity check (PCK), I/O channel check (I/O CH CK), and timer terminal count (T/C2 OUT) signals through software. On the other hand, CASS DATA IN, which is available at PC_4, is the data input line from the cassette interface.

▲ 12.9 INPUT/OUTPUT CHANNEL INTERFACE

The input/output channel is the system expansion bus of the IBM personal computer. The chassis of the PC has five 62-pin I/O channel card slots. Earlier we pointed out that using these slots, special function adapter cards, such as boards to control a monochrome or color monitor, floppy disk drives, a hard disk drive, expanded memory, or a printer, can be added to the system to expand its configuration.

Figure 12–10(a) shows the electrical interface implemented with the I/O channel. In all, 62 signals are provided in each I/O channel slot. They include an 8-bit data bus, a 20-bit address bus, six interrupts, memory and I/O read/write controls, clock and timing signals, a channel check signal, and power and ground pins.

The table in Fig. 12–10(b) lists the mnemonic and name for each of the I/O channel signals. For instance, here we see that the signal AEN stands for address enable. This table also identifies whether the signal is an input (I), output (O), or input/output (I/O). Notice that input/output channel ready (I/O CH RDY) is an input signal; input/output write command ($\overline{IOW}$) is an example of an output; and data bus lines D_0 through D_7 are the only signals that are capable of operating as inputs or outputs. In the next chapter we will learn how to interface to the system board using the I/O channel signals.

REVIEW PROBLEMS

Section 12.1

1. Name the three system buses of the original PC.
2. What three functions are performed by the clock generator block shown in Fig. 12–1?
3. What I/O addresses are dedicated to the PPI?
4. What I/O addresses are assigned to the registers of the DMA controller? To the DMA page registers?
5. What functions are assigned to timer 0? Timer 1? Timer 2?
6. Is port PA of the PPI configured for the input or output mode of operation? Port PB? Port PC?
7. Over which PPI lines are the state of the memory and system configuration switches input to the microprocessor?
8. Which output port of the PPI is used to turn ON/OFF the cassette motor?
9. Which output port of the PPI is used to output data to the speaker?
10. What are the three sources of the NMI signal?

11. What is assigned to the lowest-priority interrupt request?
12. What I/O device is assigned to priority level 5?
13. How much I/O channel expansion RAM is supported in the PC?

Section 12.2

14. What is the frequency of CLK88? PCLK?
15. At what frequency does the 8087 in the PC run?
16. What are the input and output signals of the 8284A's reset circuitry?
17. To what pin of the 8087 is RESET applied?
18. What are the input and output signals of the 8284A's wait-state logic?
19. What does logic 0 at $\overline{\text{DMA WAIT}}$ mean? Logic 0 at $\overline{\text{RDY}}$/WAIT?
20. Which devices are attached to the local bus of the 8088?
21. What devices are used to demultiplex the local bus into the system address bus and system data bus?
22. What device is used to produce the system control bus signals?
23. At what pins of the 8288 are signals $\overline{\text{MEMW}}$ and $\overline{\text{MEMR}}$ output?
24. Give an overview of the interrupt request/acknowledge cycle that takes place between the 8259A and 8088.

Section 12.3

25. What is the source of the signal I/O CH RDY? To what logic level must it be switched to initiate a wait state?
26. What types of bus cycles cause the $\overline{\text{RDY}}$/WAIT output to switch to the 1 logic level? What input signal represents each of the bus cycles?
27. Give an overview of the operation of the hold/hold acknowledge circuitry.
28. Give an overview of how the NMI interface is enabled for operation.
29. Write an instruction sequence to disable NMI.
30. Can the parity check interrupt request be individually enabled/disabled? Explain.
31. How does the 8088 determine which of the NMI sources has initiated the request for service?

Section 12.4

32. Trace through the operation of the I/O chip-select circuitry when an I/O write takes place to address $A0_{16}$.
33. Which I/O chip selects can occur during either an input or output bus cycle to an address in the range 0000_{16} through $007F_{16}$?
34. What are the outputs of the ROM chip-select circuitry?
35. Trace through the operation of the ROM address decoder as address $FA000_{16}$ is applied to the input.
36. At what logic level must address bits A_{18} and A_{19} be for the $\overline{\text{RAM ADDR SEL}}$ output to switch to its active 0 logic level?
37. What $\overline{\text{RAS}}$ output is produced by U_{65} if the address input is 10100_{16}?
38. What $\overline{\text{CAS}}$ output is produced by U_{47} if the address input is 20200_{16}?

Section 12.5

39. Trace through the operation of the ROM circuitry in Fig. 12–5 as a read cycle is performed to address $F4000_{16}$.

40. Give an overview of the operation of the RAM circuitry in Fig. 12–6(a) as a byte of data is written to the DRAMs in bank 0.

Section 12.6

41. What are the sources of DMA requests for channels 1, 2, and 3?

42. Give an overview of the DMA request/acknowledge handshake sequence.

43. What must be loaded into the DMA page registers to implement DMA operation as follows: channel 1 DMA memory address range $A0000_{16}$ through $AFFFF_{16}$, channel 2 DMA address range $B0000_{16}$ through $BFFFF_{16}$, and channel 3 DMA address range $C0000_{16}$ through $CFFFF_{16}$? Write an instruction sequence to initialize the 74LS670 device.

Section 12.7

44. What is the frequency of the timer interrupt produced by the 8253 timer? The refresh request signal?

45. What is the divisor loaded into counter 1?

46. Give an overview of how timer 2 is used to drive the speaker.

47. Draw the waveform of the signal applied to the speaker if the signal at OUT 2 is a square waveform of 3 kHz and SPKR DATA is a square waveform of 100 Hz.

Section 12.8

48. Write an instruction sequence to read SW1 through the 8255A in Fig. 12–9.

49. What is the function of signal KBD IRQ?

50. Write a simple keyboard interrupt service routine.

Section 12.9

51. What is the purpose of the I/O channel slots in the system processor board? How many are provided?

52. Which I/O channel connector pin is used to supply the signal I/O CH RDY to the system processor board? Is it active low or active high?

13

PC Bus Interfacing, Circuit Construction, Testing, and Troubleshooting

▲ INTRODUCTION

In the previous chapter we learned about the electronics of the original IBM PC's main processor board. Here we continue our study of microcomputer electronics with circuits built using the PC's I/O channel bus signals. This study includes the analysis, design, building, testing, and troubleshooting of a variety of bus interface, input/output, and peripheral circuits. The following subjects are covered in this chapter:

13.1 PC Bus-Based Interfacing
13.2 The PCμLAB Laboratory Test Unit
13.3 Experimenting with the On-Board Circuitry of the PCμLAB
13.4 Building, Testing, and Troubleshooting Interface Circuits
13.5 Observing Microcomputer Bus Activity with a Digital Logic Analyzer

▲ 13.1 PC BUS-BASED INTERFACING

In this section, we examine some of the hardware that can be used to experiment with external circuitry in the PC bus-based laboratory environment. That is, we will now work with circuitry that is not already available as part of the PC's main processor board; instead, the circuits will be constructed external to the PC. This includes prebuilt circuits readily available on PC add-on boards, such as a serial communication interface, a parallel I/O expansion module, an analog-to-digital (A-to-D) converter, and a digital-to-analog (D-to-A) converter, or custom circuits that are hand-built on special prototyping boards. We call an experimental circuit that is built to test out a function a *prototype.*

The I/O channel expansion bus is where additional circuits are added into the PC's microcomputer. A variety of methods can be used to implement these circuits. That is, a wide range of hardware is available for building experimental circuits. Figures 13–1(a) and (b) show two examples. These cards are known as *breadboards*—that is, a card meant for prototyping circuits. The breadboard card in Fig. 13–1(a) requires the circuit to be constructed on the board by inserting the leads of the devices through the holes; then the leads of the devices are soldered to permanently connect them. Similar boards are available where the devices are interconnected by wire wrapping instead of with solder. On the other hand, the module shown in Fig. 13–1(b) is what is known as a *solderless breadboard.* Here the components are plugged in and interconnected with jumper wires. Therefore, it is more practical in that the breadboard can be reused many times.

Prototyping modules are plugged directly into the PC's bus slots. However, this does not permit easy access to the circuits on the board for testing. One solution to this problem is the *board-extender card* shown in Fig. 13–2. The board extender is plugged into the slot in the PC's main processor board, and the card with the experimental circuitry is plugged into the top of the extender card. In this way, the circuitry to be tested becomes more accessible because it is located above the PC's case.

The PC add-on prototyping cards we just discussed are widely used in industry; however, they require the PC's cover to be left off and the testing of circuits to take place in close contact to the other circuitry within the PC. In an educational environment, it is beneficial to have the complete experimental environment external to the PC. Moving the breadboard outside permits easier access to the circuits for testing and modification and limits the risk of damage to the PC.

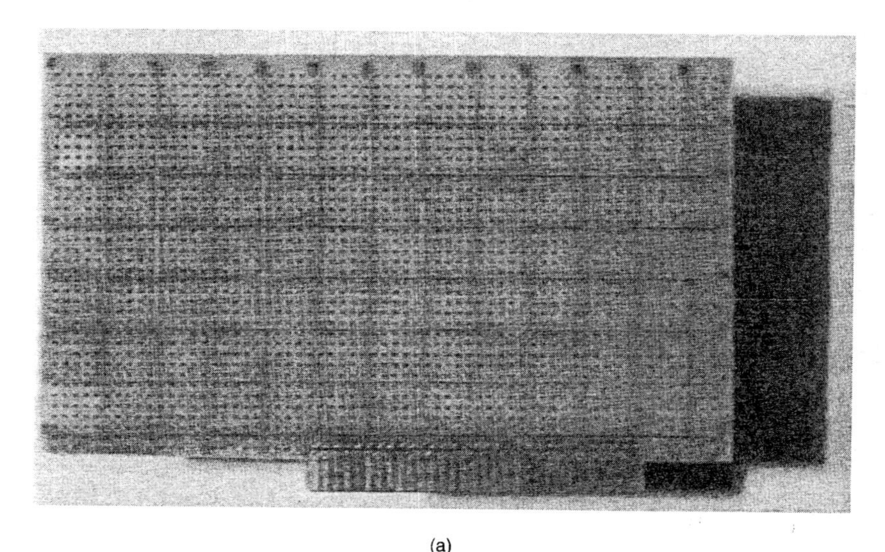

(a)

Figure 13–1 (a) Breadboard card. (b) Solderless breadboard card.

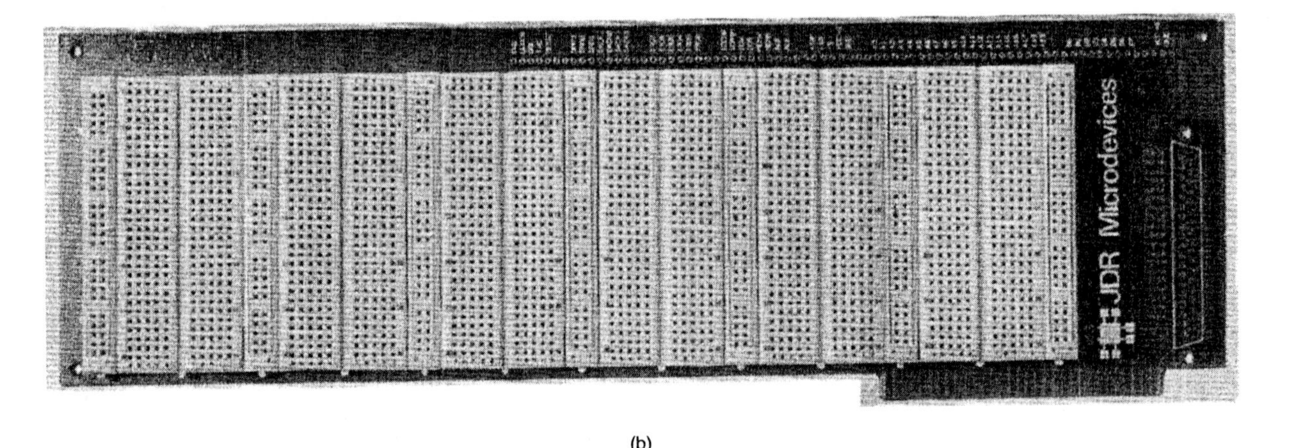

(b)

Figure 13–1 (continued)

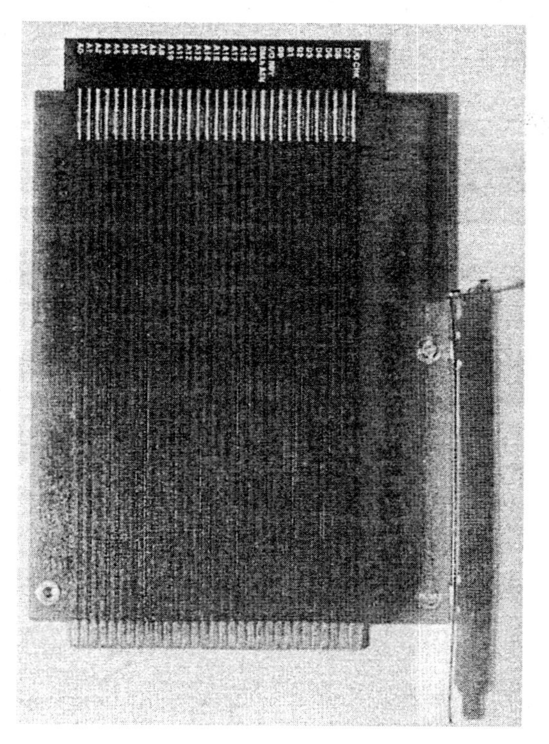

Figure 13–2 Extender card.

The PCμLAB shown in Fig.13–3 implements this type of laboratory environment. It is a bench-top laboratory test unit. The illustration in Fig. 13–4 shows that the PCμLAB uses a bus interface module that is installed inside the PC. This interface board buffers all the bus signals. Cables carry the signals of the I/O channel expansion bus over to the PCμLAB breadboard unit. If the PCμLAB is installed on a PC/AT compatible microcomputer, the signals of the complete ISA bus are available for use in breadboarding circuits. This unit has a large solderless breadboarding area for easy construction of circuits and connectors of a single I/O channel slot (ISA slot when installed on a PC/AT) for using prebuilt add-on boards. This type of system offers a better solution for an educational microcomputer laboratory and will be used here for discussion.

▲ 13.2 THE PCμLAB LABORATORY TEST UNIT

In the previous section we showed how the PCμLAB attaches to the personal computer. Here we examine the features it offers for experimentation in the laboratory. Earlier we indicated that it has a breadboard area for building circuitry and an I/O channel slot for plugging in a prebuilt board. It also has basic I/O devices such as *switches, LEDs,* a *speaker,* and some *internal I/O interface circuitry.* This built-in I/O circuitry permits exploration of simple parallel I/O techniques, such as reading switches as inputs, lighting LEDs as outputs, polling a switch input, and generating tones at the speaker, without having to build any circuitry.

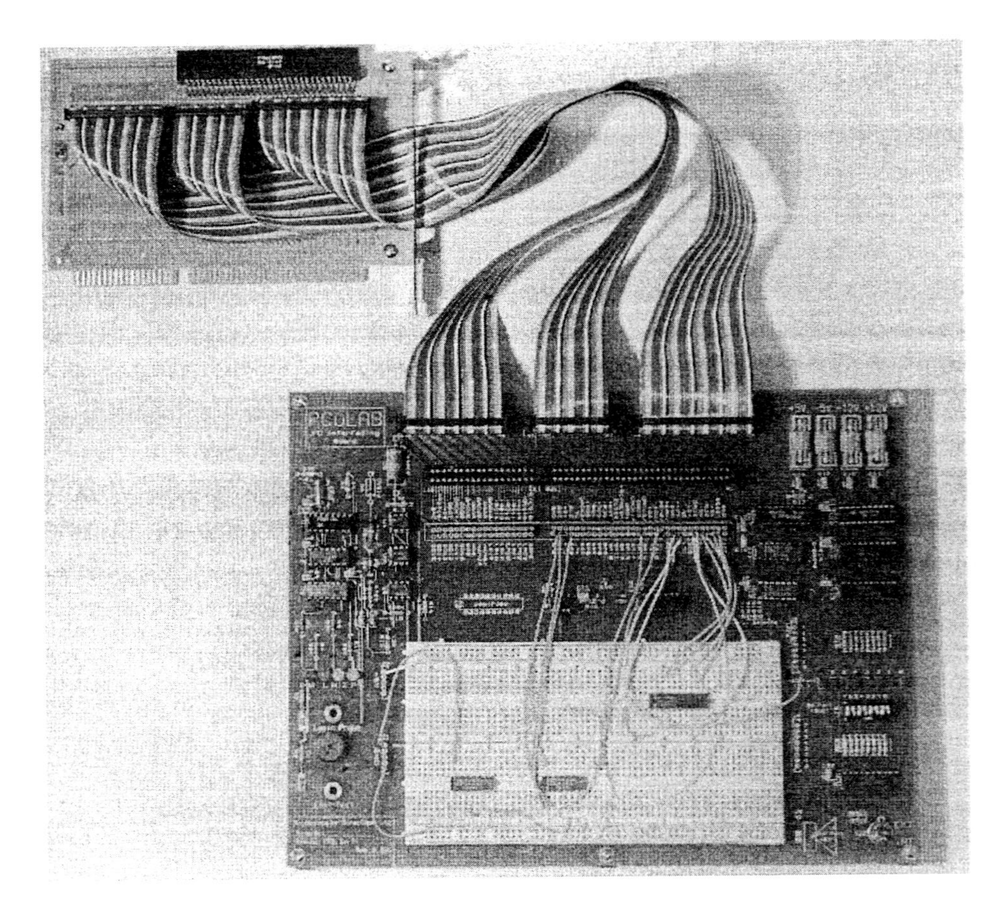

Figure 13–3 PCμLAB. (Reprinted with the permission of Microcomputer Directions Inc. P.O. Box 15127, Fremont, CA 94539)

The layout of the PCμLAB is shown in detail in Fig. 13–5. We will begin by identifying the input/output devices. On the right side we find both a block of eight switches, labeled 0 through 7, and a row of eight red LEDs, 0 through 7. The switches can be used to supply inputs, and the LEDs can be used to produce outputs for either the built-in circuits or circuitry constructed on the breadboard area. The INT/EXT switch determines the

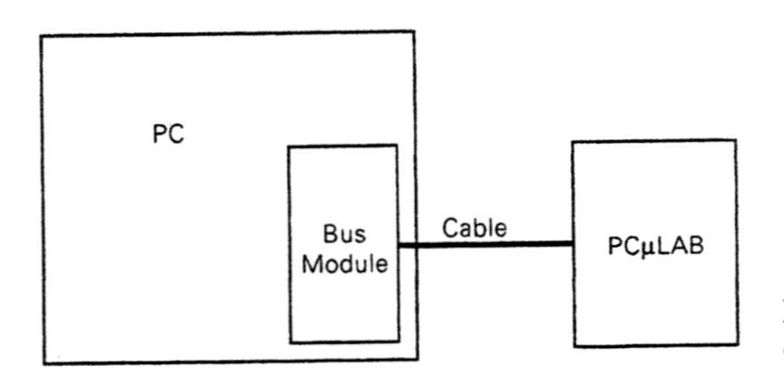

Figure 13–4 PCμLAB system configuration.

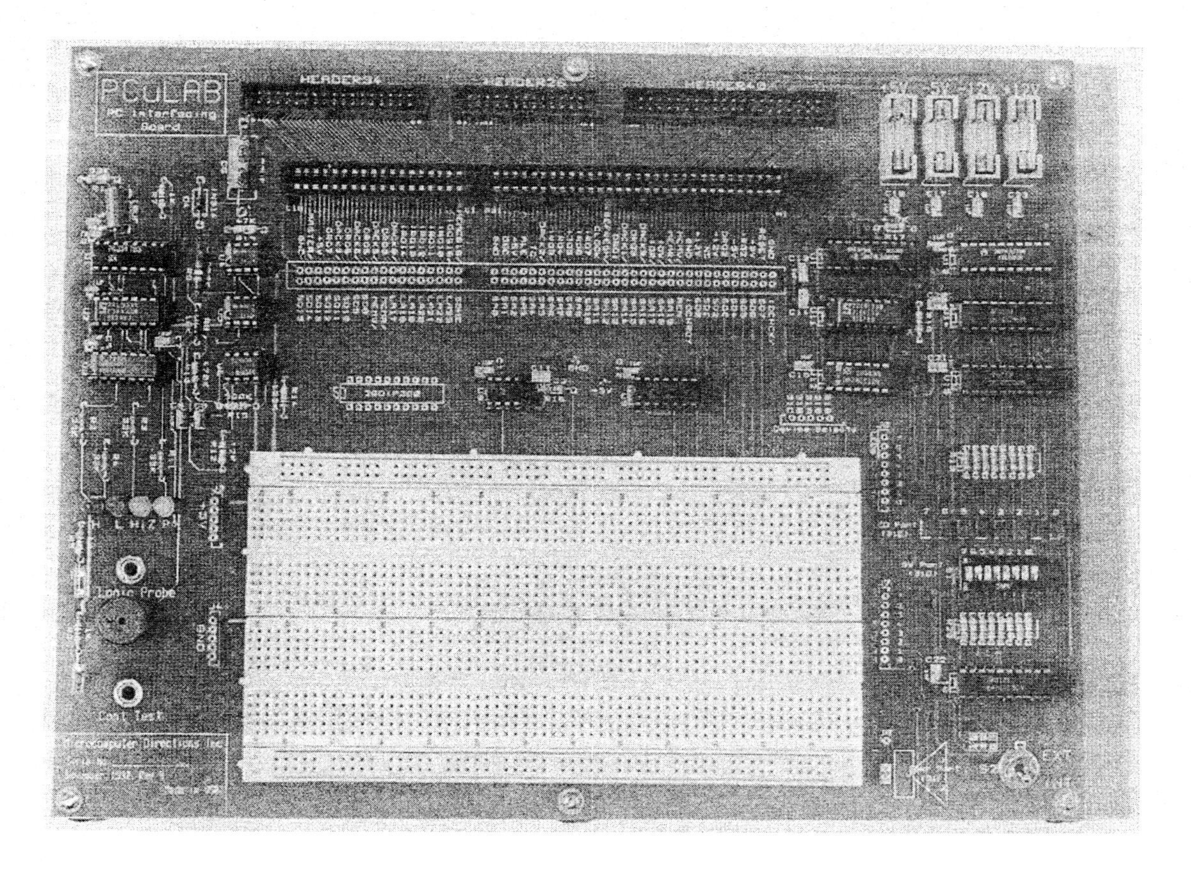

Figure 13–5 PCμLAB layout. (Reprinted with the permission of Microcomputer Directions Inc. P.O. Box 15127, Fremont, CA 94539)

use of these I/O devices. When it is in the INT (internal) position, they are connected directly to the on-board circuits. However, moving the switch to the EXT (external) position makes them available for connection to circuits implemented on the breadboard area.

All the signals of the PC's I/O channel expansion bus are made available at the connectors at the top of the front panel. Here we find that the signals are supplied at the PC slot and a receptacle connector. The slot is the connectors into which prebuilt boards are inserted. Figure 13–6 shows the PCμLAB with an add-on card inserted for testing. The receptacle connector is provided to permit connection of the bus signals to circuits built on the breadboard. For ease of use, mnemonics for all signals are labeled next to the connector. The table in Fig. 13–7 identifies the signal name for each of these mnemonics and whether it is an input or output. Remember that when the PCμLAB is attached to a PC/AT compatible microcomputer, the signals for the complete ISA bus are available. The signals listed at C and D pins in Fig. 13–7 are only active when the PCμLAB is installed on a PC/AT to implement an ISA bus environment. Those at A and B pins are active for both the I/O channel bus of the PC and ISA bus of the PC/AT.

Figure 13–6 PCμLAB with add-on card. (Reprinted with the permission of Microcomputer Directions Inc. P.O. Box 15127, Fremont, CA 94539)

Let us next look at the breadboarding area. This area permits the experimenter to build and test custom circuits. Figure 13–8 shows a circuit constructed on the breadboard area of the PCμLAB.

Looking at Fig. 13–5, we see that the breadboard area is implemented with two solderless breadboards. For this reason, it permits installation of two rows of ICs. A drawing of the electrical connection of the wire insertion clips is shown in the PCμLAB circuit layout master of Fig. 13–9. Notice that the column of five vertical clips from a device pin is internally attached. One is used up when the IC is inserted and the other four are for use in making jumper wire connections to other circuits. The jumpers must be made with the appropriate rated wire (26 AWG—American wire gauge).

At both the top and bottom of the board are two horizontal rows of attached wire insertion clips. These four rows of clips are provided for power supply distribution. Two rows are intended to implement the +5V power supply bus, and the other two are used as the common ground bus. The power supply for the circuit can be picked up with jumpers from the I/O channel connector (ISA bus connector) or at the separate power supply terminal strip. Notice in Fig. 13–7 that +5 V is available at contacts B_3 and B_{29} of the I/O channel connectors.

Pin	Name	Type
A1	I/O CH CK	I
A2	D7	I/O
A3	D6	I/O
A4	D5	I/O
A5	D4	I/O
A6	D3	I/O
A7	D2	I/O
A8	D1	I/O
A9	D0	I/O
A10	I/O CH RDY	I
A11	AEN	O
A12	A19	O
A13	A18	O
A14	A17	O
A15	A16	O
A16	A15	O
A17	A14	O
A18	A13	O
A19	A12	O
A20	A11	O
A21	A10	O
A22	A9	O
A23	A8	O
A24	A7	O
A25	A6	O
A26	A5	O
A27	A4	O
A28	A3	O
A29	A2	O
A30	A1	O
A31	A0	O

Pin	Name	Type
B1	GND	
B2	RESET DRV	O
B3	+5 V	
B4	IRQ2	I
B5	−5 V	
B6	DRQ2	I
B7	−12 V	
B8	RESERVED	
B9	+12 V	
B10	GND	
B11	SMEMW	O
B12	SMEMR	O
B13	IOW	O
B14	IOR	O
B15	DACK3	O
B16	DRQ3	I
B17	DACK1	O
B18	DRQ1	I
B19	REFRESH	O
B20	CLOCK	O
B21	IRQ7	I
B22	IRQ6	I
B23	IRQ5	I
B24	IRQ4	I
B25	IRQ3	I
B26	DACK2	O
B27	T/C	O
B28	ALE	O
B29	+5 V	
B30	OSC	O
B31	GND	

Pin	Name	Type
C1	SBHE	O
C2	LA23	O
C3	LA22	O
C4	LA21	O
C5	LA20	O
C6	LA19	O
C7	LA18	O
C8	LA17	O
C9	MEMR	O
C10	MEMW	O
C11	SD08	I/O
C12	SD09	I/O
C13	SD10	I/O
C14	SD11	I/O
C15	SD12	I/O
C16	SD13	I/O
C17	SD14	I/O
C18	SD15	I/O

Pin	Name	Type
D1	MEM CS 16	I
D2	IO CS 16	I
D3	IRQ10	I
D4	IRQ11	I
D5	IRQ12	I
D6	IRQ13	I
D7	IRQ14	I
D8	DACK0	O
D9	DRQ0	I
D10	DACK5	O
D11	DRQ5	I
D12	DACK6	O
D13	DRQ6	I
D14	DACK7	O
D15	DRQ7	I
D16	+5 V	
D17	MASTER	I
D18	GND	

Figure 13–7 ISA bus interface signals.

Figure 13–8 Breadboard circuit. (Reprinted with the permission of Microcomputer Directions Inc. P.O. Box 15127, Fremont, CA 94539)

EXAMPLE 13.1

Which contacts of the I/O channel interface connectors can be used as ground points?

Solution

Figure 13–7 shows that ground (GND) is provided by contacts B_1, B_{10}, and B_{31} of the I/O channel interface connectors.

Earlier we pointed out that the switches, LEDs, and speaker supply inputs and outputs for the on-board circuitry and can also be connected to circuits built on the breadboard. In both cases, I/O addresses are output over the address bus part of the I/O channel interface, A_0 through A_{15}, and data are input or output over the data bus lines, D_0 through D_{15}.

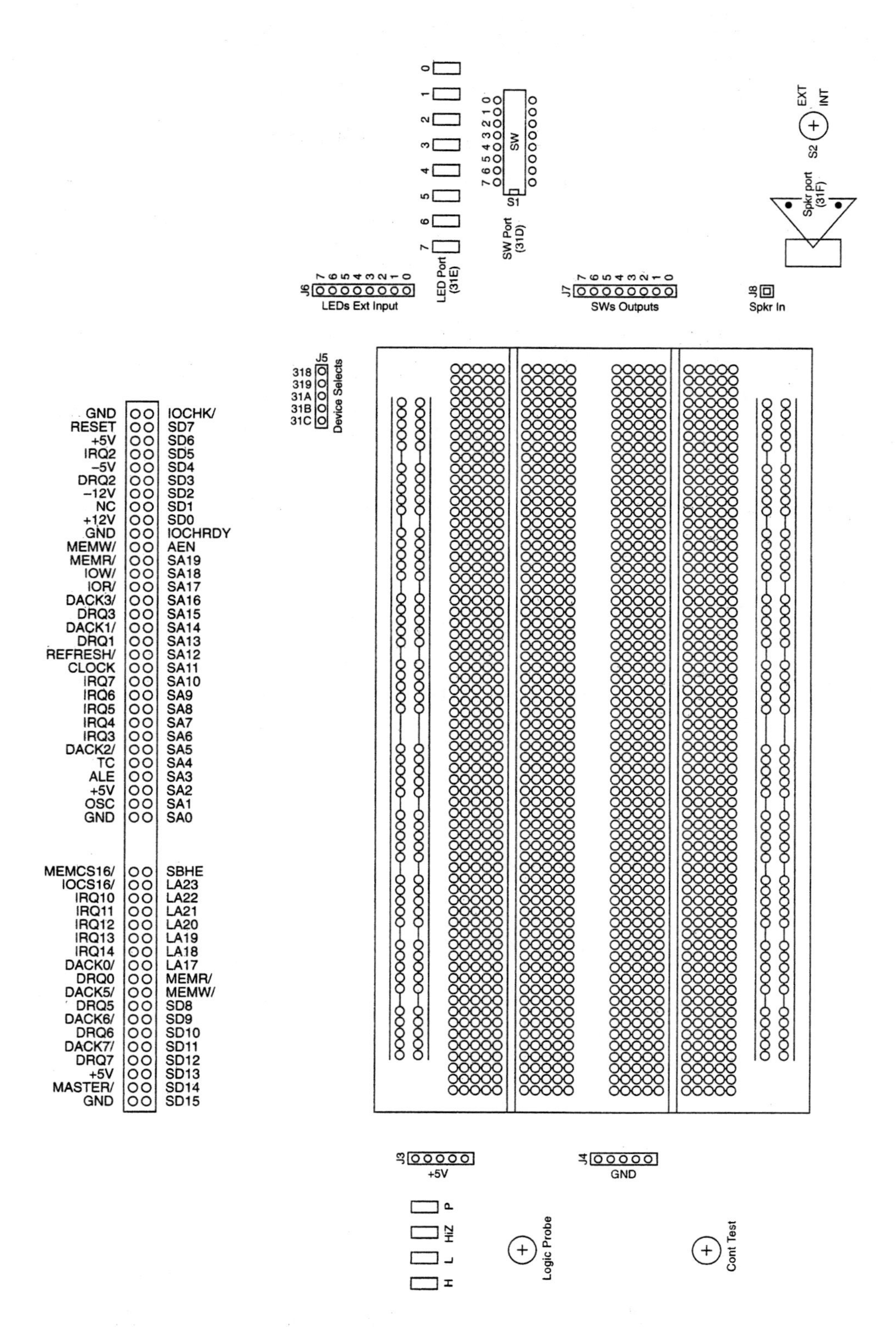

Figure 13–9 Circuit layout master. (Reprinted with the permission of Microcomputer Directions Inc. P.O. Box 15127, Fremont, CA 94539)

EXAMPLE 13.2

At what contacts of the I/O channel connector are data bus lines D_0 through D_7 available?

Solution

Figure 13–7 shows that the data bus lines are supplied at contact A_2 through A_9 of the connector.

The PCμLAB also has built-in circuit test capability. It has both a continuity tester and a logic probe. Looking at Fig. 13–5, we see that the probes for the continuity tester are inserted into the female connectors identified as *CT.* The continuity tester is useful for debugging circuit connections—that is, it can be used to verify whether or not two points of a circuit are wired together. This is done by attaching one of the probes to the first point in the circuit and then touching the other probe to the second point. If they are connected, the buzzer sounds. Care must be taken to assure that the power is not applied to the circuit under test while continuity tests are being made; otherwise, the tester circuits may be damaged.

The purpose of the logic probe is not to verify circuit connections; instead, it is used to determine the logic level of signals at various test points in a circuit. The probe used to input the signal from the circuit is inserted into the LP connector. Then, the probe is touched to the test point in the circuit. Based on the logic level of this signal, the red, green, or amber LED lights. Here red stands for logic 1, green is logic 0, and amber is the high-Z state. The table in Fig. 13–10 shows the voltage levels corresponding to these three states. The second red LED, marked P, identifies that the signal at the test point is pulsing. By pulsing, we mean a signal, such as a square wave, that is repeatedly switching back and forth between the 0 and 1 logic levels.

▲ 13.3 EXPERIMENTING WITH THE ON-BOARD CIRCUITRY OF THE PCμLAB

The on-board circuitry of the PCμLAB implements simple parallel input/output interfaces. Having this circuitry built into the experimental board allows us to examine some basic I/O techniques without having to take the time to construct the circuitry. These interface circuits provide users with the ability to input the settings of the switches, light the LEDs, or sound a tone at the speaker. Earlier we pointed out that this circuitry becomes active when the INT/EXT switch is set to the INT position. In this section, we describe the design and operation of the internal (on-board) circuits. The operation of these circuits is illustrated using several input and output examples.

Logic level	LED	Voltage
1	Red	$V > 2.4$ V
Hi–Z	Amber	$0.4\text{ V} < V < 2.4$ V
0	Green	$V < 0.4$ V

Figure 13–10 Logic state voltage levels.

I/O Address Decoding

Let us begin our study of the on-board circuitry with the address decoder circuit shown in Fig. 13–11(a). Here we find that a 74LS688 parity generator/checker IC (U_{10}), a 74LS138 3-line to 8-line decoder IC (U_{11}), and a 74LS32 quad-OR gate IC (U_{12}) perform the address decode function. They accept as inputs address bits A_0 through A_9 and the address-enable (AEN) control signal. These inputs are directly picked up from the I/O channel connector of the ISA expansion bus. As Fig. 13–11(b) shows, they correspond to the signals available at contacts A_{22} through A_{30} and A_{11} of the on-board ISA bus connector.

At the other side of the circuit, we find three I/O-select outputs. They are labeled $\overline{\text{IORX31D}}$, $\overline{\text{IOWX31E}}$, and $\overline{\text{IOWX31F}}$ and stand for *I/O read address X31DH, I/O write address X31EH,* and *I/O write address X31FH,* respectively. These signals are used to select between the on-board I/O devices: switches, LEDs, or speaker. For instance, Fig. 13–11(c) shows that output $\overline{\text{IORX31D}}$ is used to enable input of the state of the switch settings.

Figure 13–11(b) shows that the address bits available at the I/O channel connectors of the ISA bus interface are A_0 through A_{23}. However, just the lower 16 address lines A_0 through A_{15} are used in I/O addressing, and many of these address bits are not used in the on-board address decoder circuit. For this reason, the unused address bits are considered don't-care states. Therefore, the decoded address is

$$A_{15} \ldots A_0 = XXXXXXA_9A_8A_7A_6A_5A_4A_3A_2A_1A_0$$

Since many address bits are don't-care states, the outputs of the decoder do not correspond to unique addresses. Instead, a large number of I/O addresses decode to produce each chip select output.

Next we will look at how the higher-order address bits are decoded by the 74LS688 comparator. Looking at Fig. 13–11(a), we see that inputs P_0 through P_7 of the comparator are supplied by address signals A_3 through A_9 and AEN. This gives

$$P_7P_6 \ldots P_0 = AENA_9A_8A_7A_6A_5A_4A_3$$

On the other hand, the Q inputs are set at fixed logic levels and represent

$$Q_7Q_6 \ldots Q_0 = 01100011_2$$

The circuit within the 74LS688 compares the address information at the P inputs to the fixed code at the Q inputs, and if they are equal, it switches the P = Q output to logic 0. This means that the address on the bus corresponds to an on-board I/O device. This output is applied to the G_{2A} input of the 74LS138 decoder and enables it for operation. In this way, we see that all addresses with

$$A_{15} \ldots A_0 = XXXXXX1100011A_2A_1A_0 \text{ along with AEN} = 0$$

map to the PCμLAB's on-board I/O address space.

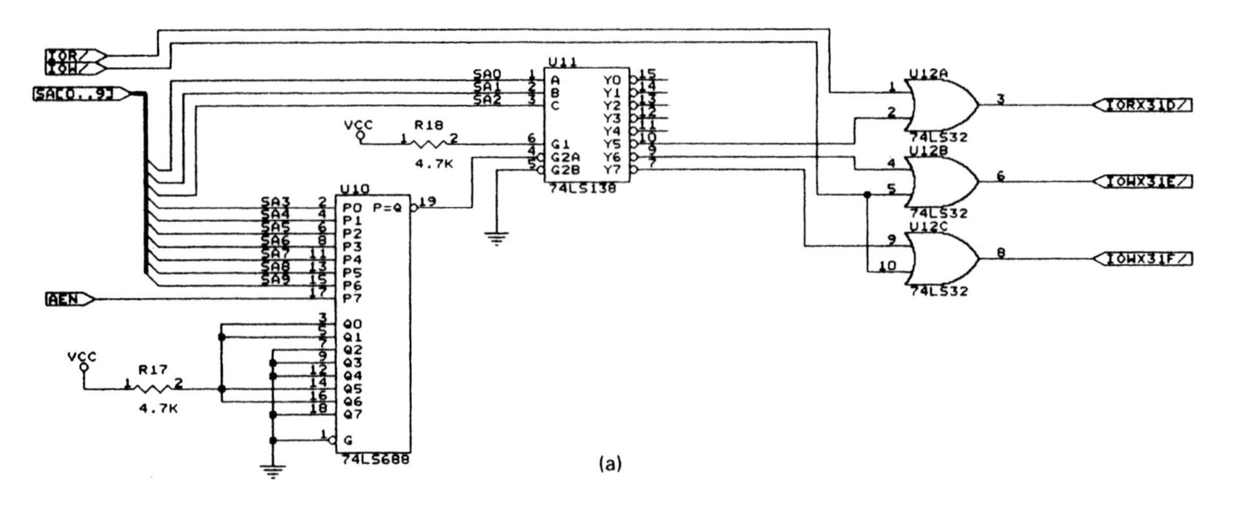

Figure 13–11 (a) Address decoder circuit. (Reprinted with the permission of Microcomputer Directions Inc. P.O. Box 15127, Fremont, CA 94539) (b) ISA bus interface signals. (c) Output signals.

Signal	Pin	Pin	Signal
$\overline{\text{I/O CH CK}}$	A1	B1	GND
D7	A2	B2	RESET DRV
D6	A3	B3	+5 V
D5	A4	B4	IRQ2
D4	A5	B5	–5 V
D3	A6	B6	DRQ2
D2	A7	B7	–12 V
D1	A8	B8	RESERVED
D0	A9	B9	+12 V
I/O CH RDY	A10	B10	GND
AEN	A11	B11	$\overline{\text{SMEMW}}$
A19	A12	B12	$\overline{\text{SMEMR}}$
A18	A13	B13	$\overline{\text{IOW}}$
A17	A14	B14	$\overline{\text{IOR}}$
A16	A15	B15	$\overline{\text{DACK3}}$
A15	A16	B16	DRQ3
A14	A17	B17	$\overline{\text{DACK1}}$
A13	A18	B18	DRQ1
A12	A19	B19	$\overline{\text{REFRESH}}$
A11	A20	B20	CLOCK
A10	A21	B21	IRQ7
A9	A22	B22	IRQ6
A8	A23	B23	IRQ5
A7	A24	B24	IRQ4
A6	A25	B25	IRQ3
A5	A26	B26	$\overline{\text{DACK2}}$
A4	A27	B27	T/C
A3	A28	B25	ALE
A2	A29	B29	+5 V
A1	A30	B30	OSC
A0	A31	B31	GND

Signal	Pin	Pin	Signal
$\overline{\text{SBHE}}$	C1	D1	$\overline{\text{MEM CS16}}$
LA23	C2	D2	$\overline{\text{IO CS16}}$
LA22	C3	D3	IRQ10
LA21	C4	D4	IRQ11
LA20	C5	D5	IRQ12
LA19	C6	D6	IRQ13
LA18	C7	D7	IRQ14
LA17	C8	D8	$\overline{\text{DACK0}}$
$\overline{\text{MEMR}}$	C9	D9	DRQ0
$\overline{\text{MEMW}}$	C10	D10	$\overline{\text{DACK5}}$
SD08	C11	D11	DRQ5
SD09	C12	D12	$\overline{\text{DACK6}}$
SD10	C13	D13	DRQ6
SD11	C14	D14	$\overline{\text{DACK7}}$
SD12	C15	D15	DRQ7
SD13	C16	D16	+5 V
SD14	C17	D17	$\overline{\text{MASTER}}$
SD15	C18	D18	GND

(b)

I/O device	Type	Device-Select Signal	Address
Switches	Input	$\overline{\text{IORX31D}}$	031DH
LEDs	Output	$\overline{\text{IOWX31E}}$	031EH
Speaker	Output	$\overline{\text{IOWX31F}}$	031FH

(c)

Figure 13–11 (continued)

Once the 74LS138 decoder is enabled, the code on address lines A_0 through A_2 is used to produce the appropriate output. The circuit diagram in Fig. 13–11(a) shows that the codes that produce the enable signals for the switches, LEDs, and speaker are

$$A_2A_1A_0 = 101_2 \text{ and active } \overline{IOR} \text{ produces } \overline{IORX31D}$$

$$A_2A_1A_0 = 110_2 \text{ and active } \overline{IOW} \text{ produces } \overline{IOWX31E}$$

$$A_2A_1A_0 = 111_2 \text{ and active } \overline{IOW} \text{ produces } \overline{IOWX31F}$$

This results in the device addresses as listed in Fig. 13–11(c). For example, reading of the switches is enabled by any address that is of the form

$$A_{15} \ldots A_0 = XXXXXX1100011101_2 \text{ provided AEN} = 0.$$

Some examples of valid addresses are $031D_{16}$, $F31D_{16}$, $FF1D_{16}$, and $0F1D_{16}$. All these addresses make the Y_5 output of the decoder circuit switch to logic 0. Notice that this output is gated with the I/O channel expansion bus signal $\overline{IOR}$ by OR gate U_{12A}. In this way, the $\overline{IORX31D}$ output can be active only during input (I/O read) bus cycles.

EXAMPLE 13.3

Which output chip select does the I/O address $F71F_{16}$ produce when applied to the input of the circuit in Fig. 13–11(a)? What type of bus cycle must be in progress to produce this chip-select output?

Solution

First the address expressed in binary form is

$$F71F_{16} = 1111011100011111_2$$

Considering the lower 10 bits, we get

$$A_9 \ldots A_0 = 1100011111_2 = 31F_{16}$$

Tracing the circuit, we find that this address-bit combination makes the Y_7 output of U_{11} equal to logic 0. Y_7 is gated with bus signal $\overline{IOW}$ to produce the $\overline{IOWX31F}$ output. Therefore, $\overline{IOWX31F}$ is at its active 0 logic level as long as an output bus cycle is taking place.

Switch Input Circuit

Figure 13–12 shows the interface of switches S_0 through S_7 to the data bus. Note that one contact from each of the eight switches is connected to ground (0 V). The other contact on each switch is supplied to one of the resistors in resistor pack R_{20}, and the other end of each resistor connects to V_{cc} (+5 V). The connections between the resistors

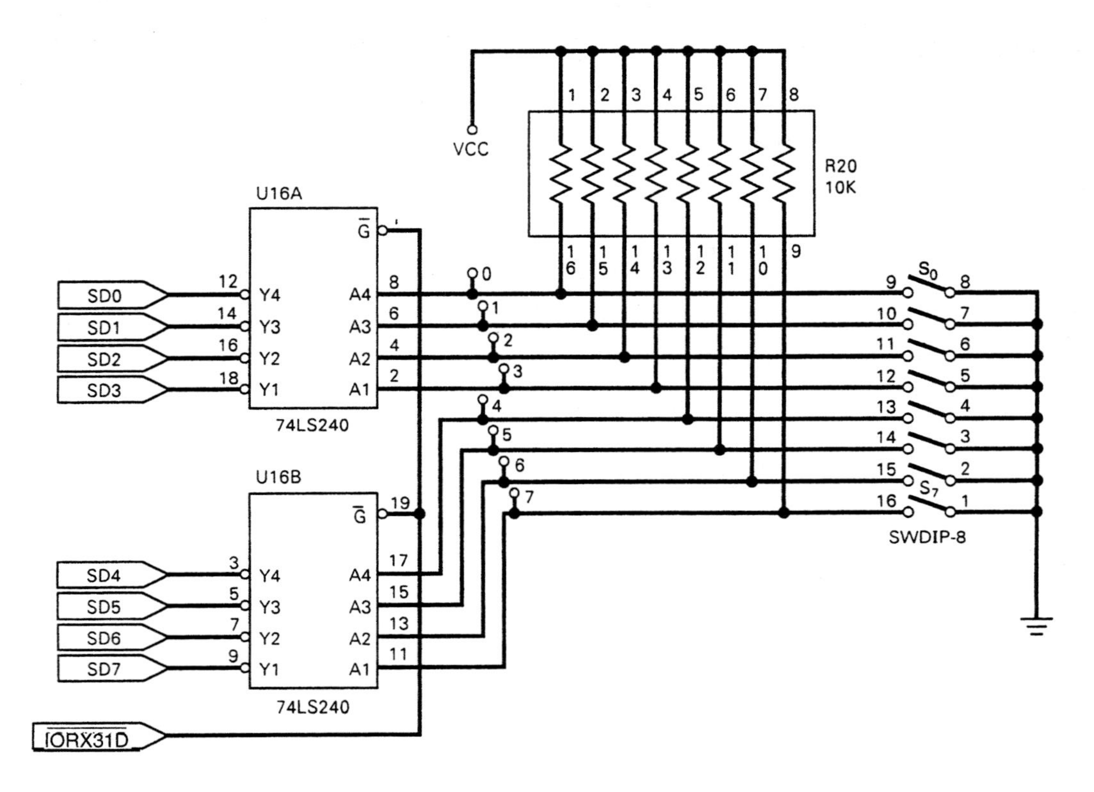

Figure 13–12 Switch input interface circuit. (Reprinted with the permission of Microcomputer Directions Inc. P.O. Box 15127, Fremont, CA 94539)

and the switch contacts are supplied as inputs to the data bus through the 74LS240 inverting buffers of IC U_{16}. For instance, S_0 is supplied from input A_4 of buffer U_{16A} to output Y_4 and onto data bus line D_0. Similarly, the state of switch 7 is passed through the inverter at input A_1 of U_{16B} to data bus line D_7. Because inverting buffers are used in the circuit, logic 0 is applied to the data bus whenever a switch is open, and logic 1 is put on the data bus if a switch is closed.

The circuit diagram shows that the switch input buffer is enabled by the signal $\overline{\text{IORX31D}}$. Earlier we showed that this signal is at its active 0 logic level whenever an input bus cycle is performed to address $031D_{16}$. But, remember that the complete I/O address is not decoded; therefore, many other addresses also decode to enable this buffer.

The state of the switches can be read with an INPUT command. For instance, the DEBUG command

```
I 31D (↵)
```

causes the settings of all eight switches to be displayed as a hexadecimal byte. In this byte, the most significant bit represents the S_7 state and the least significant bit S_0. Remember that a bit at logic 1 means a closed switch and logic 0 an open switch.

Let us now look at how to read the status of the switches into the accumulator of the MPU. This is done by simply executing an IN instruction. Therefore, after executing the instruction sequence

```
MOV  DX, 31DH
IN   AL, DX
```

the switch setting is held in AL.

In practical applications, it is common to want to determine the setting of a single switch. Additional processing of the byte in AL does this. For instance, to find the setting of S_7, we can use the instruction

```
AND  AL, 80H
```

Execution of this instruction ANDs the contents of AL with the value 80_{16}. Therefore, the result in AL will be 10000000_2 if switch 7 is closed or 00000000_2 if it is open. That is, the zero flag (ZF) is 1 if S_7 is open and 0 if it is closed.

EXAMPLE 13.4

Write a program that polls S_0 waiting for it to be closed. Use a shift instruction to isolate and determine the setting of switch 0. Use a valid address other than $31D_{16}$ to read the setting of the switches.

Solution

The settings of the switches are input to AL with the instructions

```
       MOV  DX, 0FF1DH
POLL:  IN   AL, DX
```

Here we have used $FF1D_{16}$ as the I/O address for the switch port. Now the setting of switch 0, which is in the bit 0 position, is shifted into the carry flag (CF) with the instruction

```
SHR  AL, 1
```

Finally, the setting of the switch is tested for 0 (open) by checking the carry flag with the instruction

```
JNC  POLL
```

If CF is 0, the switch is open and the poll loop is repeated. But if the switch is closed, CF is 1, the poll loop is complete, and the instruction following JNC is executed.

LED Output Circuit

Let us next look at the output circuit that drives the LEDs. Figure 13–13 shows the drive circuitry for LEDs 0 through 7. Here we see that the anode side of the individual LEDs are all connected in parallel and supplied by +5 V. On the other hand, the cathode sides of the LEDs are wired through separate resistors of resistor pack R_{19} to the outputs of the 74LS240 LED drive buffer (U_{14}). For example, the cathode of LED 0 connects through the uppermost 330-Ω resistor to the Y_1 output of IC U_{14A}. The inputs to the inverting buffer are supplied by the outputs of the 74LS374 LED port latch (U_{13}).

To light an LED, an output bus cycle must be performed to load logic 1 into the corresponding bit of the LED port latch. As identified earlier, the I/O address accompanying this data and $\overline{IOW}$ must decode to produce logic 0 at $\overline{IOWX31E}$. Note that this signal is used to clock the data on data bus lines D_0 through D_7 into the 74LS374 latch. The 74LS240 buffer inverts the bits at the output of this latch. Logic 0 at any output of the buffer provides a path to ground for the corresponding LED, and thus turns it on.

To try out the LEDs on the PCμLAB, we can turn them all on by issuing a single OUT command from DEBUG:

```
O  31E FF (↵)
```

They can be turned off with the command

```
O  31E 00 (↵)
```

EXAMPLE 13.5

Write a program that blinks LED 7 on the LED port of the PCμLAB.

Solution

To turn on LED 7, 80_{16} must be output to the LED port latch, and it is turned off by outputting 00_{16}. Therefore, we begin with the instructions

```
          MOV   DX, 31EH
          MOV   AL, 80H
ON_OFF:   OUT   DX, AL
```

Next we need to delay for a period of time before turning the LED off. To do this, a count is loaded into CX and a time delay is implemented using a LOOP instruction.

```
          MOV CX, 0FFFFH
HERE:     LOOP HERE
```

The duration of the time delay can be adjusted by simply changing the value loaded into CX.

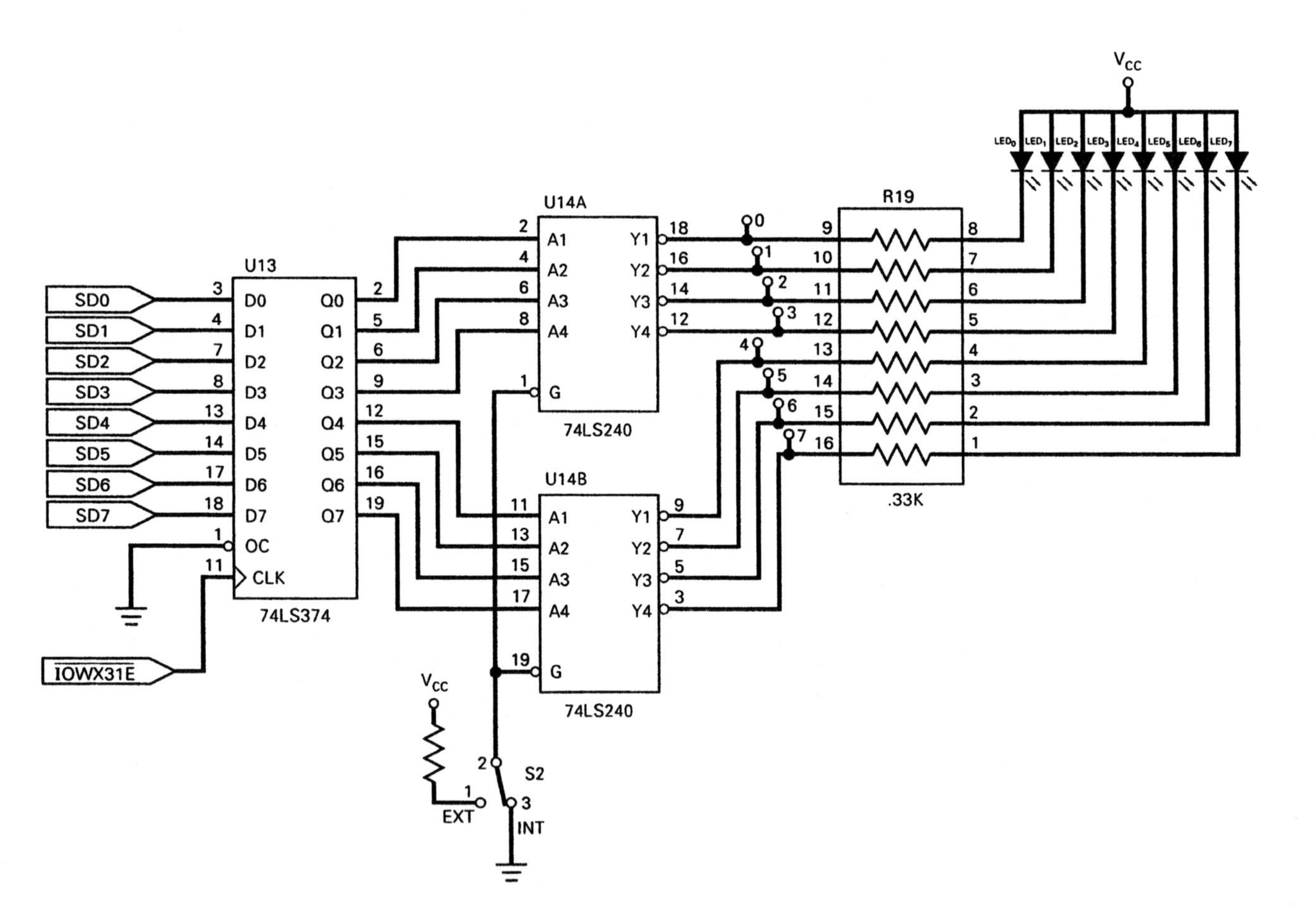

Figure 13–13 LED output interface circuit. (Reprinted with the permission of Microcomputer Directions Inc. P.O. Box 15127, Fremont, CA 94539)

```
            MOV   DX,31EH
            MOV   AL,80H
ON_OFF:     OUT   DX,AL
            MOV   CX,0FFFFH
HERE:       LOOP HERE
            XOR    AL,80H
            JMP   ON_OFF
```

Figure 13–14 LED blink program.

After the time delay has elapsed, the value in bit 7 of AL is inverted with the instruction

```
XOR  AL, 80H
```

The new contents of AL equal 00_{16}. This value will cause LED 7 to turn off. Finally, a JMP instruction returns program control to ON_OFF.

```
JMP  ON_OFF
```

and the loop repeats. Figure 13–14 illustrates the complete program.

Speaker-Drive Circuit

The speaker-drive circuit of Fig. 13–15 is an output interface. It is implemented with 74LS74 data latch device, U_9, and the 75477 speaker driver, U_5. The tone to be sounded at the speaker can be generated under software control.

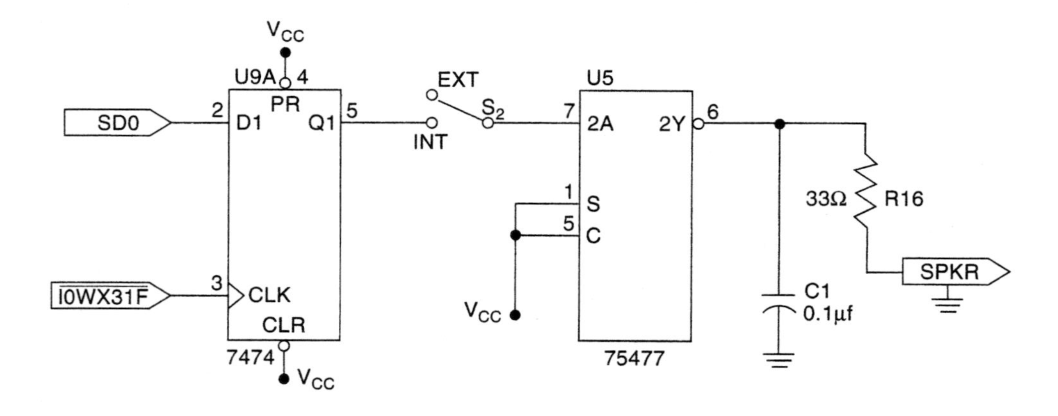

Figure 13–15 Speaker-drive circuit. (Reprinted with the permission of Microcomputer Directions Inc. P.O. Box 15127, Fremont, CA 94539)

Let us begin by studying how a tone signal can be generated by the MPU and sent to the speaker. Applying a square wave to the speaker produces a tone. This signal is output over data bus line D_0 to the D_1 input at pin 2 of U_{9A}. Looking at the circuit diagram, we find that the tone is passed from the Q_1 output at pin 5 of U_9 through the EXT/INT switch to the 2A input at pin 7 of U_5. It is then output at pin 6 (2Y) of U_5 and sent through resistor R_{16} to the speaker. The other end of the speaker's coil is connected to $+V_{cc}$.

The square wave can be generated with a program similar to the one we used to blink LED 7. However, the data is output on data bus line D_0 rather than D_7, and it must be accompanied by address $31F_{16}$ instead of $31E_{16}$. This gives the program

```
          MOV   DX, 31FH
          MOV   AL, 01H
ON_OFF:   OUT   DX, AL
          MOV   CX, 0FFFFH
HERE:     LOOP  HERE
          XOR   AL, 01H
          JMP   ON_OFF
```

Varying the frequency of the square wave can change the pitch of the tone. Adjusting the duration of the time delay does this—that is, changing the count that is loaded into CX. The lower the count, the higher is the pitch.

▲ 13.4 BUILDING, TESTING, AND TROUBLESHOOTING INTERFACE CIRCUITS

In the previous section, we examined the on-board circuits of the PCμLAB. Here we turn our attention to circuits that can be built on the breadboard area of the PCμLAB. First, we look at how circuits are constructed, then we consider how their operation is tested, and finally, we explore troubleshooting techniques that can be used if the circuit does not work.

Building a Circuit

Earlier we showed that the breadboard area is where custom circuits can be built and that it allows for mounting two rows of ICs. Assuming that a schematic diagram of the circuit to be built is already available, the first step in the process of building the circuit is to make a layout diagram to show how the circuit will be constructed on the breadboard. This drawing can be made on a circuit layout master similar to the one shown in Fig. 13–9.

Figure 13–16(a) is the diagram for a circuit that implements a parallel output port to drive LED 0. This is the circuit we will use to illustrate the method used to breadboard a circuit. Note that the circuit uses a 74LS138 3-line to 8-line decoder, a 7400 quad 2-input NAND gate, and a 74LS374 octal latch. The 74LS240 inverting buffer, 330-Ω resistor, and LED 0 are supplied by the on-board circuitry of the PCμLAB by using input 0 of receptacle J_6. We begin by marking the IC pin numbers for each of the inputs and outputs onto the circuit diagram. For instance, from the pin layout of the 74LS138 in Fig. 13–17, we find that its A, B, and C inputs are at pins 1, 2, and 3, respectively. Moreover, the Y_7 output is identified as pin 7. This is done for each IC to give the circuit shown in Fig. 13–16(b).

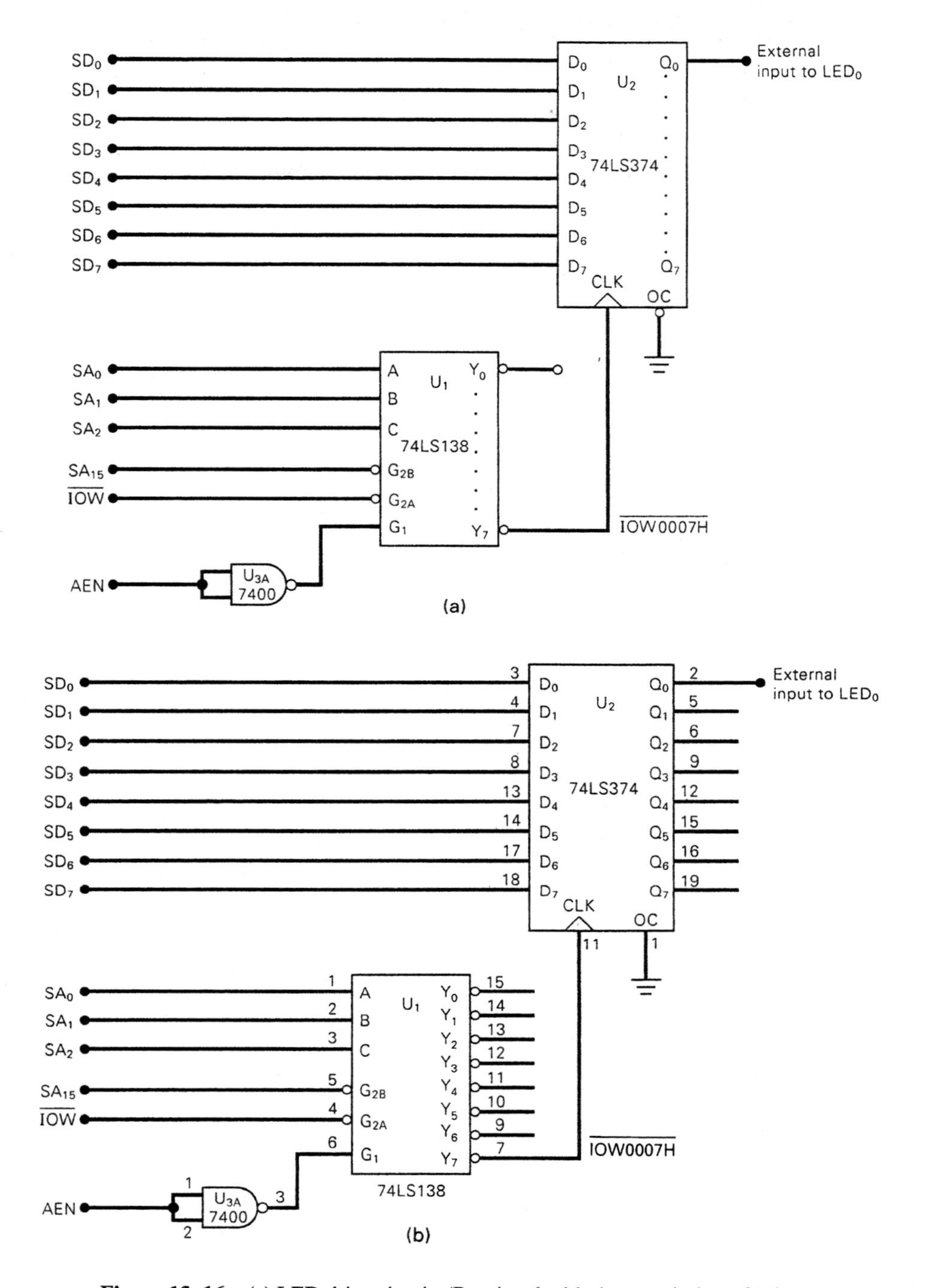

Figure 13–16 (a) LED drive circuit. (Reprinted with the permission of Microcomputer Directions Inc. P.O. Box 15127, Fremont, CA 94539) (b) Schematic with pin numbers marked. (Reprinted with the permission of Microcomputer Directions Inc. P.O. Box 15127, Fremont, CA 94539) (c) Completed layout master. (Reprinted with the permission of Microcomputer Directions Inc. P.O. Box 15127, Fremont, CA 94539)

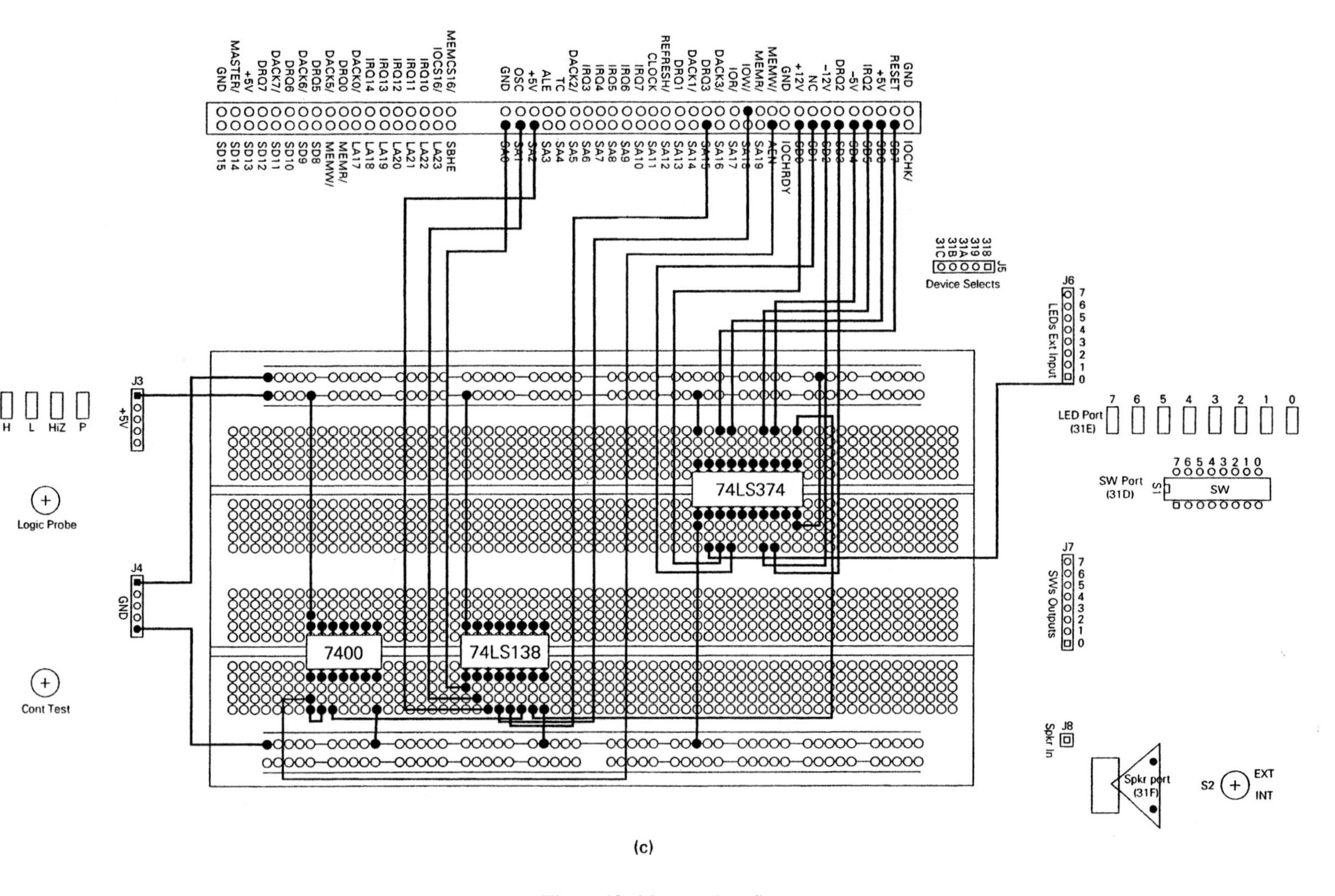

(c)

Figure 13–16 (continued)

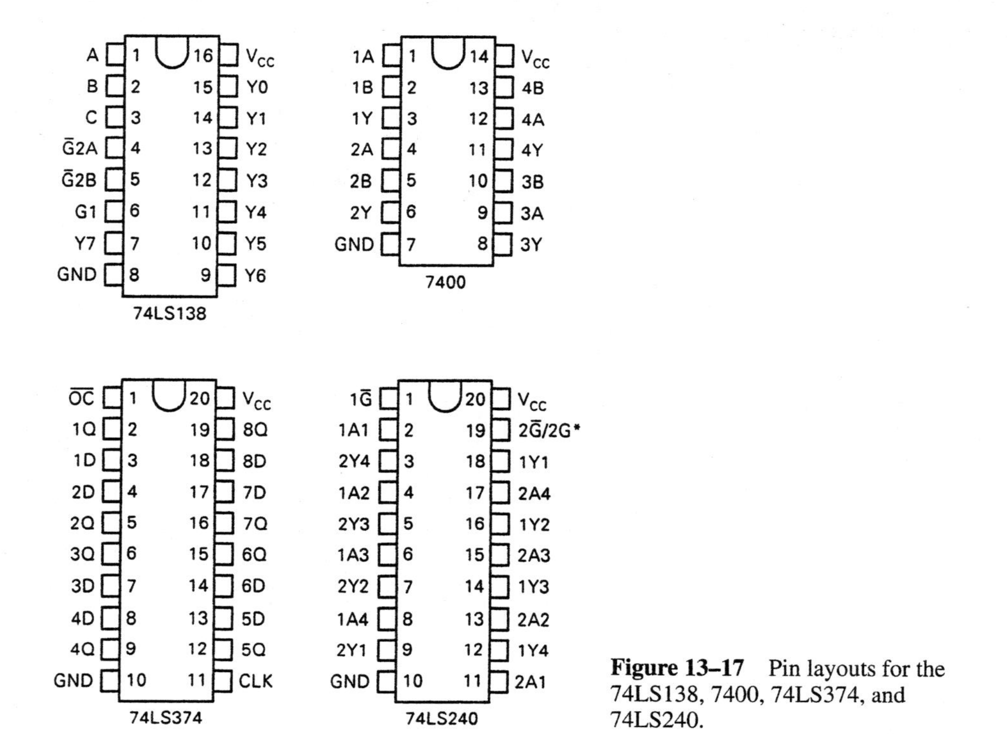

Figure 13–17 Pin layouts for the 74LS138, 7400, 74LS374, and 74LS240.

Now we are ready to make the drawing that shows how the circuits will be laid out on the PCμLAB's breadboard area. To do this, we simply draw the ICs and pin connections onto one of the circuit layout masters. Figure 13–16(c) shows the layout for our test circuit. Looking at this diagram, we see that +5 V is picked up by inserting a jumper between connector J_3 and one of the power bus lines of the breadboard. Ground is supplied in a similar way, from connector J_4 to another power bus line. Then +5 V and GND are jumpered from IC to IC. For instance, pin 14 of the 7400 is connected to +5 V and pin 7 to GND. This completes the power distribution for the circuit.

Let us next look at how the inputs and outputs of the circuit are provided. At the output side, a jumper is used to connect the Q_0 output of the 74LS374 to the 0 input of connector J_6. The data input at pin 3 of the 74LS374 latch is picked up with a jumper to data bus line D_0 at pin A_9 of the I/O channel connector of the ISA bus. The output of the 74LS138 decoder (pin 7) that supplies the clock to the latch is jumpered to pin 11 of the 74LS374 IC. Notice that the AEN input is inverted with a NAND gate. Connecting pins 1 and 2 of the 7400 IC and then applying AEN to pin 1 forms the inverter. The inverted output at pin 3 of the 7400 IC is supplied to the G_1 input at pin 6 of the 74LS138 decoder IC.

EXAMPLE 13.6

Use the circuit diagram layout in Fig. 13–16(c) to identify which pins of the I/O channel connector are used to supply the address signals to the A, B, and C inputs of the 74LS138 decoder.

Solution

In the circuit diagram, we find that the A, B, and C inputs of the decoder are attached to address lines A_0, A_1, and A_2, respectively. These signals are picked up at pins A_{31}, A_{30}, and A_{29} of the I/O channel connector.

The layout drawing in Fig. 13–16(c) is our plan for constructing the circuit. This drawing and the schematic diagram marked with pin numbers serve as valuable tools when testing and troubleshooting circuits. Figure 13–18 shows the breadboard of this example circuit.

Testing the Operation of a Circuit

Now that the circuit has been constructed, we are ready to check out its operation. The process of checking out how an electronic circuit works is called *testing*. To test a circuit, we must first know the events that should take place when it is functioning correctly. Usually this means that it will produce certain outputs. These outputs may be a visual event, such as lighting an LED; an audible event, such as sounding a tone; a mechanical event, such as positioning a mechanism; or simply a signal waveshape that can be observed with an *instrument*. For instance, the function of our example circuit in Fig. 13–16(a) is to light an LED.

To test the operation of our example circuit, we can simply turn on the LED or even better make it blink. However, to do this, the LED output interface must be driven with software. In this way, we see that to test microcomputer interface circuits, they must be driven with software. In fact, they are normally driven by a special piece of software that is specifically written to exercise the interface. This segment of program is sometimes referred to as a *diagnostic program.*

The diagnostic routine does not have to be complex. For instance, to turn on LED 0 in our breadboard circuit, we can simply execute the instructions

```
MOV  DX, 0007H
MOV  AL, 01H
OUT  DX, AL
```

Note that the I/O address 0007_{16} along with active $\overline{IOW}$ and inactive AEN produce an active low pulse at output Y_7 of the decoder. This pulse is used to clock the data into the octal latch to make the Q_0 output become logic 1. A software routine that will blink LED 0 is as follows:

```
          MOV  DX, 0007H
          MOV  AL, 01H
ON_OFF:   OUT  DX, AL
          MOV  CX, 0FFFFH
HERE:     LOOP HERE
          XOR  AL, 01H
          JMP  ON_OFF
```

The operation of a circuit that produces an electrical output can be tested with instrumentation. For instance, if the circuit we just constructed did not include an LED at the output, we would need to observe the signal at the Q_0 output (pin 2) of the 74LS374 latch. Accessories, such as *IC test clips,* are available to provide easy attachment of instruments to the pins of an IC. Figure 13–18 shows some IC test clips. This type of clip is spring loaded and snaps tightly over the top of the IC. Instruments are connected to its pins instead of to those of the IC. For our example of the 74LS374 IC, a 20-pin IC test clip would be attached and then the probe of the instrument clipped onto pin 2 at the top of the test clip.

Various instruments are available to test the electrical signals in a circuit. Figures 13–19(a), (b), and (c) show three examples, the *logic probe, multimeter,* and *oscilloscope,* respectively. The logic probe is a hand-held instrument that can be used to observe the logic level at a test point in a microcomputer interface circuit. As we pointed out earlier, a logic probe is built into the PCμLAB. This instrument has the ability to tell whether the signal tested is in the 0, 1, or high-Z logic state, or if it is pulsating. The probe is simply touched to the point in the circuit where the signal is to be observed and the logic level is signaled by one of the LEDs.

In microcomputer interface circuits, the multimeter is useful for measuring static voltage levels. For instance, it can be used to verify that +5 V is applied to each of the ICs. A multimeter can also be used to measure the logic levels at inputs and outputs, but they must be stable voltages, not pulsating signals. In this case, the meter displays the amount of voltage at the test point and from this value we can determine whether the signal is at logic 0, at logic 1, or in the high-Z state.

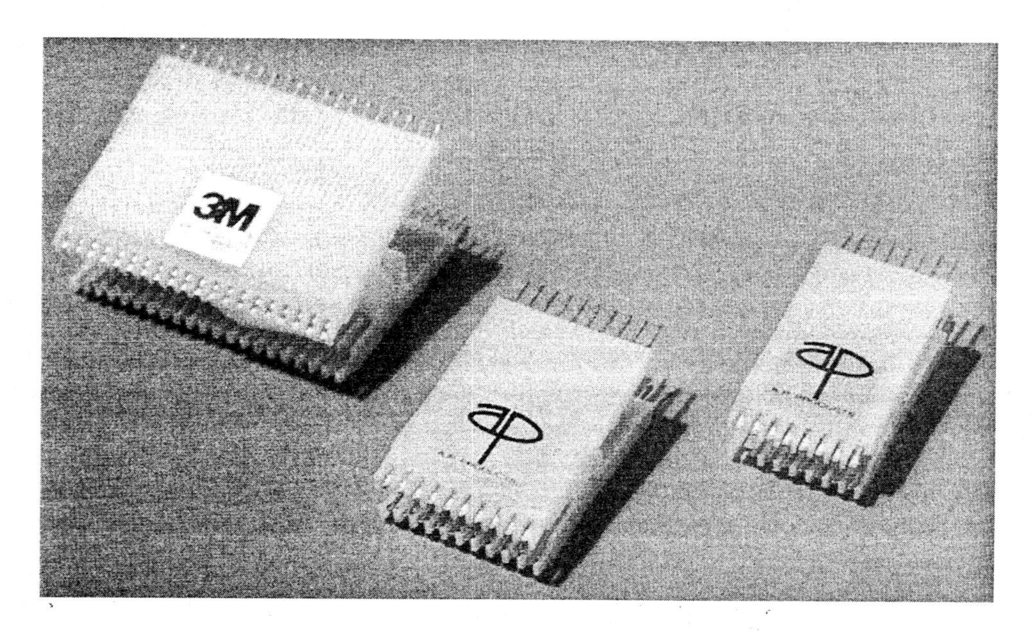

Figure 13–18 IC test clips.

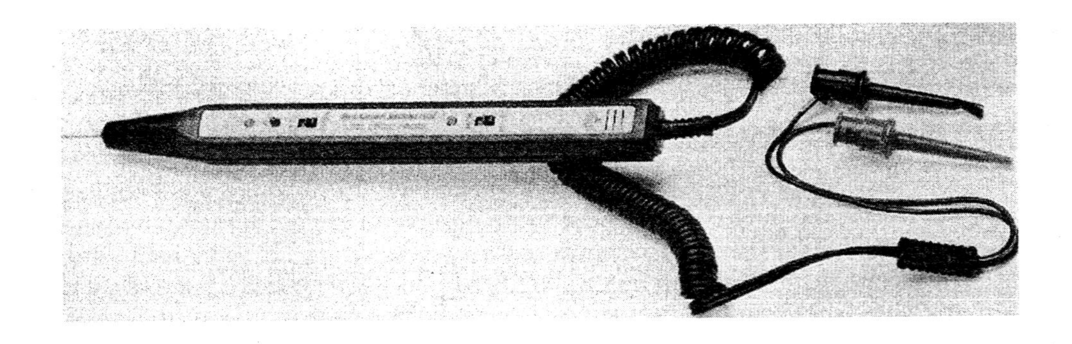

(a)

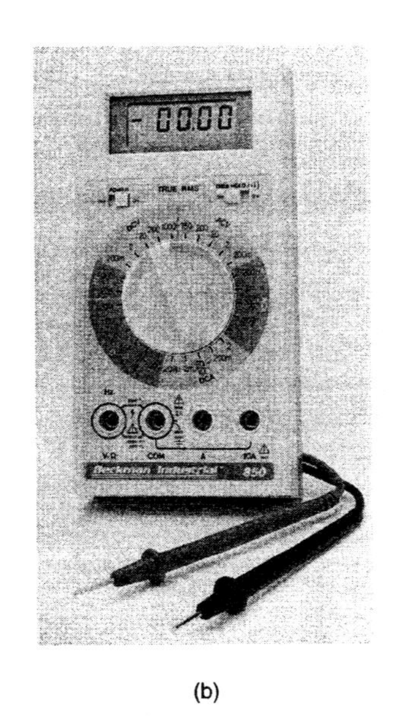

(b)

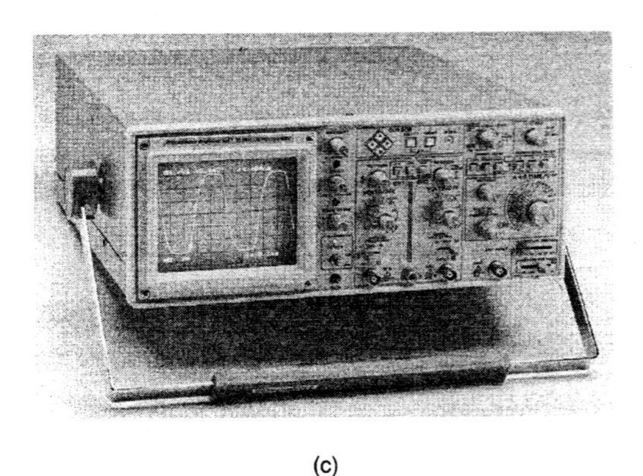

(c)

Figure 13–19 (a) Logic probe. (b) Digital multimeter. (c) Oscilloscope.

Most multimeters also have the ability to measure resistance, AC voltage, and AC and DC current. For instance, it could be used to find the amount of current the circuit draws from the V_{cc} supply.

We just mentioned that the logic probe could tell if the signal at a point under test is pulsating. However, in this case, a better instrument for observing the operation of the circuit is an oscilloscope (or *scope* as it is better known). The scope is the most widely used instrument for observing *periodic signals.* That is, signals, such as a square wave, that have a repeating pattern.

The scope displays the exact waveform of the pulsating signal on its screen. This type of representation gives us much more information. For instance, we can find the value of the high-voltage (logic 1), the value of the low-voltage (logic 0), how long the signal is at the 0 and 1 logic levels, and the shape of the signal as it transitions back and forth between 0 and 1.

Figure 13–20(a) shows the shape of the signal produced at the Q_0 output at pin 2 of the 74LS374 when the diagnostic program that blinks LED 0 is running. The display of the waveform allows us to measure the period (T) of the square wave and calculate its frequency using

$$f = 1/T$$

Most scopes have the ability to display several signals on the screen at the same time—that is, they have several *signal channels.* The most common scope in use is the *dual-trace scope,* which has the ability to display two signals simultaneously. Figure 13–20(b) shows both the output square-wave and clock input signals of our test circuit. In this example, the scope has been set up to synchronize the display of the square-wave output, applied to channel 1, to the clock input at channel 2. For this reason, the waveforms represent their true relationship in time. Note that a clock pulse is associated with the loading of each logic 0 and logic 1 into the Q_0 output of the latch. This mode of operation is known as using an *external sync.* That is, the sweep of the scope is initiated by an external signal, which in this case is the clock input.

Troubleshooting Microcomputer Interface Circuitry

In the testing of the operation of a circuit, we may find that it does not work. That is, it does not perform the function for which it was designed. In this case, we must identify the cause of the malfunction and then correct the problem. The process of finding the cause of a malfunction is called *troubleshooting,* and the process of correcting the problem is known as *repair.* Here we look at some causes of malfunctions in circuits and then outline methods that can be used to troubleshoot microcomputer interface circuits.

The cause of problems found in malfunctioning circuits depends on the type of circuit being tested. In general, electronic circuits fall into several categories. A first example is a breadboard of a new circuit design. In this case, the circuit may be working correctly, but not perform the function for which it was designed. That is, the malfunction may simply be due to the fact that a mistake was made in the design. This may be the most complicated type of failure to find and resolve.

The breadboard of a circuit we use in our laboratory exercises for this book is a second example. Here the circuit is known to operate correctly. For this reason, the most common causes for a circuit not to work are that a wiring mistake was made when the breadboard was built, or the software that was written to exercise the hardware has a bug.

A third example is a circuit in an existing electronic system, such as the PC, that has failed. In this case, we know that the system worked correctly in the past, but now malfunctions. Therefore, the cause of the problem may not be a mistake in the design, a wiring error, or incorrect software; instead, it is likely due to the failure of a component.

The final example is a circuit board, such as the main processor board of the PC that has just been built on a manufacturing line. Here a wide variety of potential causes exits for the malfunction. For instance, a lead of a component may not be correctly soldered, a lead of a device may be short-circuited to a pin on another device with excess solder, the wrong component may have been inserted, or a component may have been installed in reverse orientation.

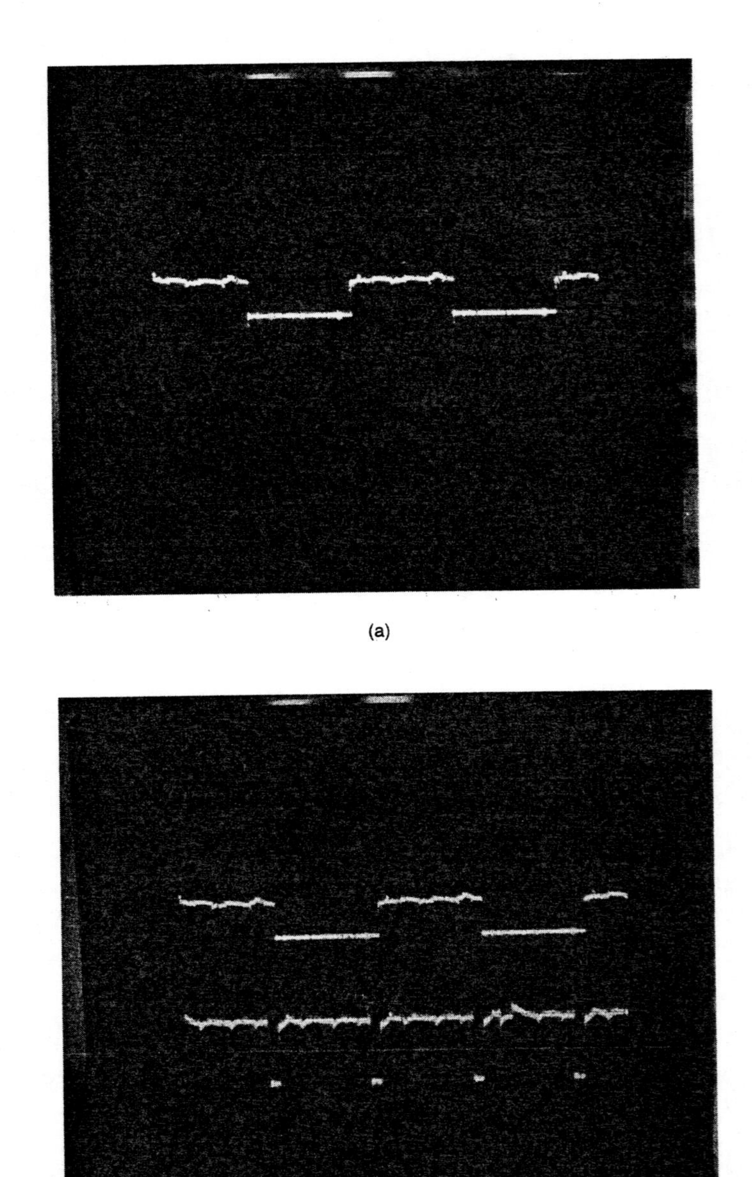

Figure 13–20 (a) Q_0 output waveform. (b) Clock input and square wave output.

Thus, there are many causes for a circuit to malfunction. In fact, most of the causes we just stated can affect any of the circuits. For instance, a short circuit could occur between the pins of two devices on the PC's main processor board. This short may have been accidentally created when the system was opened to install an interface board, change the system-configuration DIP switches, or replace another failing subsystem, such as a disk drive. As another example, it is also possible that a bad IC gets installed when building a breadboard of a circuit or even during the manufacturing of a printed circuit board. Finally, an open circuit may occur in a copper trace on the main processor circuit board of the PC, even though the board had been working correctly for a long time. For instance, using too much force when inserting an interface board into the expansion bus connector may have made a crack.

Having looked at some of the causes of circuit failures, let us continue by exploring troubleshooting methods. We assume that the circuit is known to have worked previously. This would be the case in troubleshooting a circuit built for one of our laboratory exercises or when repairing an electronic system such as a PC. We will begin with a general procedure that can be used to troubleshoot microcomputer interface circuits.

Figure 13–21 shows a general flowchart for testing and troubleshooting a microcomputer interface circuit. Here we will assume that a circuit breadboard and diagnostic software exist. Therefore, the first step is to test the operation of the interface. This is done by exercising the hardware by running the program and observing its operation visually or with instrumentation. If the hardware correctly performs its intended operation, the flowchart's Y path shows that we are done. On the other hand, if it does not work, the N path is taken. That is, we need to troubleshoot the circuit. It is important to remember that circuits may not work due to problems in software or hardware or both software and hardware.

Figure 13–21 shows that the first step in the troubleshooting part of the process is to identify and describe the symptoms of the failure. It is important to make a clear and concise description of the problem before beginning to examine the software or hardware. For example, in the case of the test circuit in Fig. 13–16(a), running the diagnostic program blinks LED 0. The failure symptom may be that the LED just remains off or it may turn on, but not blink.

Now we must decide whether or not software can be ruled out as a cause of the problem. For example, in a laboratory exercise where the program is given, software should not be the cause of the malfunction. Also, in the case of an application program running on a PC, which is known to have no bugs, software may be ruled out. In cases where correct software operation cannot be assumed, the operation of the program should be analyzed before testing the circuitry. That is, as shown in Fig. 13–21, software debugging is the next step. If the program is found to be correct, the N path is followed in the flowchart and then hardware troubleshooting begins.

When bugs are found in the program, they must be corrected; then the Y path is taken. Here we see that the interface circuit is retested to verify whether the software fixes make it work correctly. If the interface circuit does operate correctly, troubleshooting is complete.

Let us assume that the interface still does not function correctly. Then Fig. 13–21 shows that hardware troubleshooting must begin. After the hardware problems are identified and corrected, the interface circuit is once again tested.

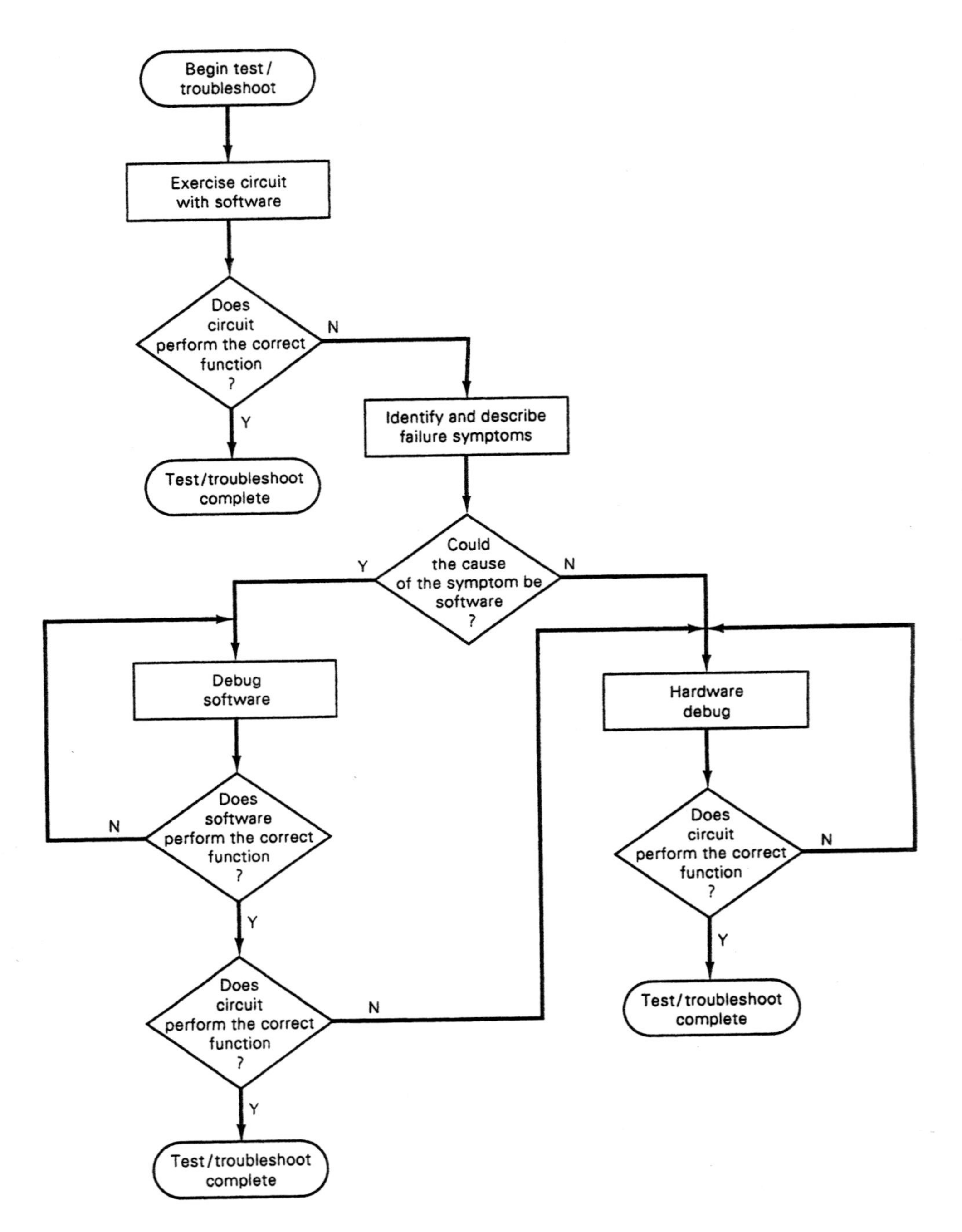

Figure 13–21 General test/troubleshoot flowchart.

Now that we have covered the general test and troubleshooting procedure, we continue by looking more closely at the software-debug part of the process. The flowchart in Fig. 13–22 identifies the steps in the software-debug process. Note that first the programming of any VLSI peripheral ICs in the circuit must be verified to be correct. The programming sequence and command values can be reviewed if they are programmed as part of the program. In fact, software can be added to read back the contents of the registers (if possible) after programming to verify that they have been updated. Our example circuit in Fig. 13–16(a) has no peripheral ICs, so this step is not required.

Next, the addresses corresponding to I/O devices or memory locations and data that are to be transferred over the bus must be checked. This will verify that the correct data are transferred and that they will go to the correct place in the circuit. In the square-wave program we wrote earlier to blink LED 0, the address of the LED latch, 0007_{16}, is loaded into DX and the initial data to be output to the latch, 01_{16}, are loaded into AL. Both of these values are correct for the circuit under test.

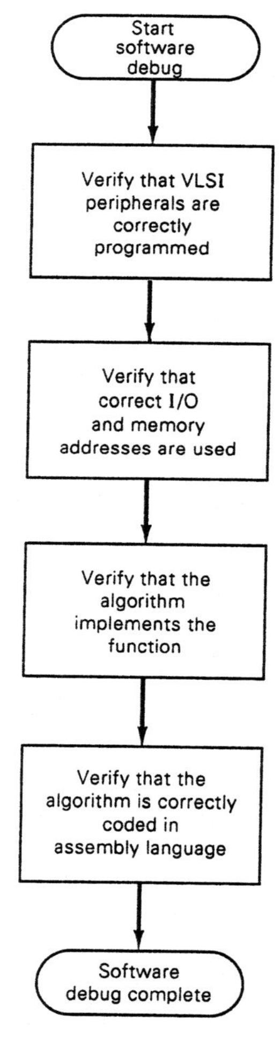

Figure 13–22 Software-debug flowchart.

Finally, the flowchart in Fig. 13–22 shows that the last step in the software-debug process is to check the algorithm and its software implementation. This can be done by rechecking the flowchart to confirm that it provides a valid solution to the problem, and then comparing the instruction sequence against the flowchart to assure that it implements this algorithm. The algorithm is next *desk checked.* That is, the operations of the instructions of the program are traced through to verify correct operation for a known test case.

If software is not the cause of the problem, attention must be turned to the hardware. Figure 13–23 outlines a general hardware-troubleshooting procedure. For now we will assume that the circuit undergoing troubleshooting is a breadboard built on the PCμLAB.

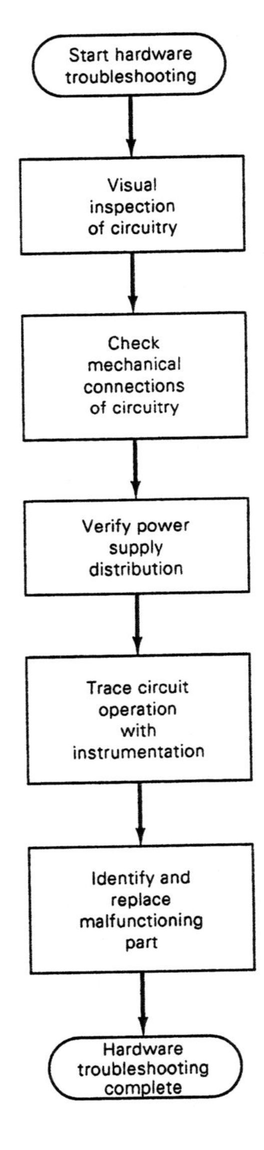

Figure 13–23 Hardware-debug flowchart.

The first step identified in the flowchart is to make a thorough visual inspection of the circuit to assure that it is correctly constructed. This includes verifying that the correct IC pin numbers are marked into the schematic diagram, the circuit diagram layout does correctly implement the circuit in the schematic, and that all jumper connections are consistent with those identified in the layout diagram. Second, the mechanical connections of the circuit should be checked to verify that they make good electrical connections—that is, that the pins of the ICs are making contact with the contacts of the solderless breadboard and that the wire connections provide continuity between the pins of the various ICs. The continuity tester of the PCμLAB can be used to check out these connections.

If the circuit connections are correct, the flowchart shows that the next step is to check out the power supply. That is, we should verify that +5 V is applied between the V_{cc} and GND pins of each IC. Here we are normally interested in knowing the exact amount of voltage. For this reason, the voltage measurements are usually taken with a multimeter. Since the $+V_{cc}$ supply of TTL ICs is rated at +5 V ±5%, power supply measurements between +5.25 V and +4.75 V are satisfactory.

After the circuit connections and power supply have been ruled out as the source of the malfunction, we are ready to begin checking the operation of the circuit. To be successful at this, we must understand the operation, signal flow through the circuit, and wave shapes expected at select test points. Typically, operation is traced by observing signals starting from the output and working back toward the input in an attempt to identify the point in the circuit up to which correct signals exist. For instance, in our breadboard circuit, we can begin by examining the waveform applied to the LED with an oscilloscope. This spot is identified as test point 1 (TP1) in the schematic shown in Fig. 13–24 and corresponds to pin 8 of resistor pack R_{19}. Assuming a symmetrical square wave is not observed, the probe of the scope can be moved to pin 18 of the 74LS240 IC, test point 2 (TP2). If a square wave is not present there either, the next test point should be the input of the inverter at pin 2 of the 74LS240 (TP3). Assuming that the square-wave signal is again not found, the output at pin 2 of the 74LS374 data latch, TP4, should be checked. This completes tracing of the data path from LED 0 to the data bus.

If a square wave is not observed at any test point in the data path from data bus line SD_0 to LED 0, the problem may be in the chip-select decoder circuit. Earlier we found that this circuit produces the clock that loads the data into the 74LS374 latch. Notice that the clock is applied to pin 11 on U_2, identified as TP5. This signal is not a symmetrical square wave; instead, it is a repeating pulse that would appear as an asymmetrical square wave on the screen of a scope. Assuming that a pulse is not observed, this would be the reason data are not being loaded from the data bus into the latch. In this case, the signal path of the latching pulse must be traced back to pin 7 of the 74LS138, TP6, in an attempt to locate the pulse.

Let us assume that no clock pulse is observed at pin 7 of the 74LS138 IC. Then we should continue by checking for signals $\overline{IOW}$ and $\overline{AEN}$ at inputs $\overline{G}_{2A}$ and G_1, respectively. There should be active low pulses on both $\overline{IOW}$ and $\overline{AEN}$, identified as TP7 and TP8 in the circuit diagram shown in Fig. 13–24. If that is the case, the only signals that remain to be checked are address lines SA_0, SA_1, SA_2, and SA_{15} connected to A, B, C, and $\overline{G}_{2B}$ and denoted as TP9, TP10, TP11, and TP12, respectively. To do this, the sweep of the scope can be synchronized with $\overline{IOW}$ at $\overline{G}_{2A}$ and the logic levels that exist at inputs A, B, C, and $\overline{G}_{2B}$ observed during this pulse. Assuming that code CBA equals 110 and $\overline{G}_{2A} = 0$, output

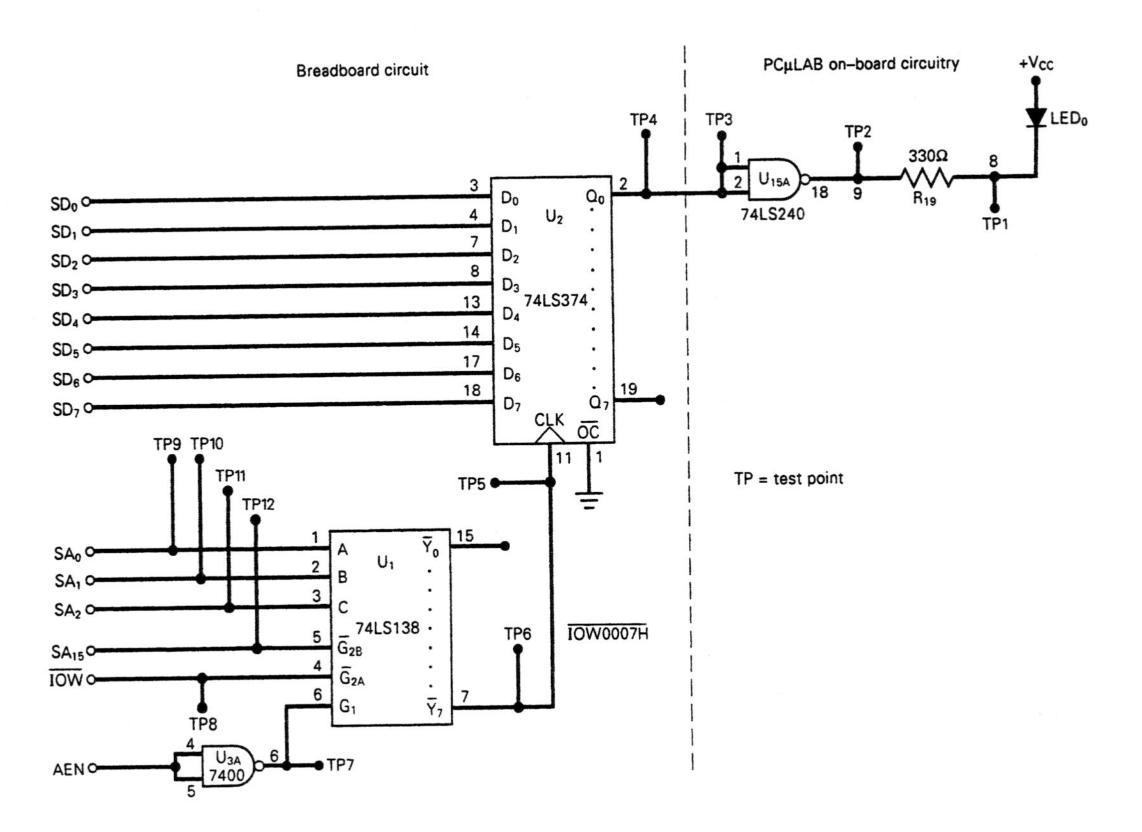

Figure 13–24 Breadboarded circuit schematic with test points.

$\overline{Y_7}$ should be a pulse similar to that at $\overline{G_{2A}}$. Since no pulse was found at pin 7, we have found the source of the problem. The 74LS138 IC is bad and must be replaced.

After the bad IC is replaced, the operation is again observed. If LED 0 blinks, troubleshooting is complete. Otherwise, troubleshooting resumes by verifying that the clock pulse is produced at pin 7 of the 74LS138 and is passed to the clock input of the 74LS374 data latch.

If an oscilloscope is not available, many of the test measurements during the troubleshooting process we just outlined can be made with the logic probe of the PCμLAB. For instance, the signals at inputs G_1, $\overline{G_{2A}}$, and $\overline{G_{2B}}$ of the 74LS138 address decoder can be tested. However, the logic probe is not as versatile as the oscilloscope. With the scope, we could verify the logic level of the address inputs to the address decoder. This type of synchronous measurement at the time when $\overline{IOW}$ is 0 cannot be made with a logic probe.

Another useful instrument in troubleshooting microcomputer interface circuits is a *logic pulser.* The logic pulser can be set to output either a one-shot pulse or a square wave. This instrument can be used to inject a pulse or square wave into the input of a device in the circuit. Figure 13–25 shows a typical logic pulser.

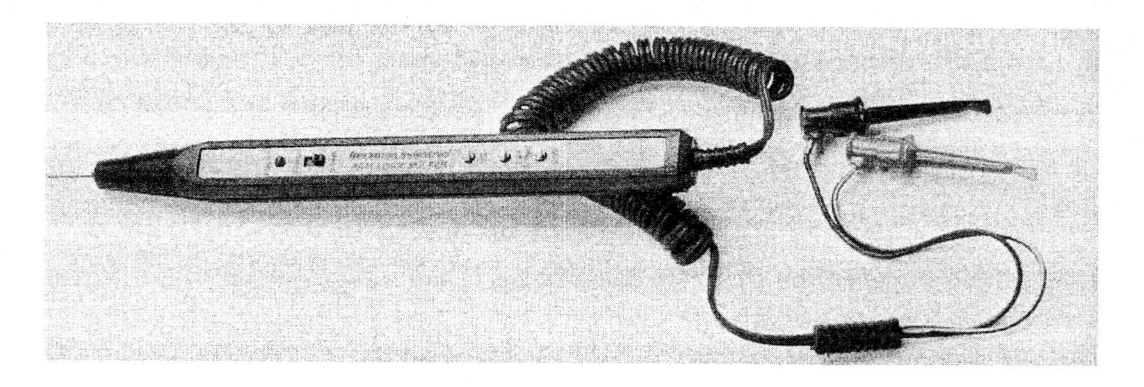

Figure 13–25 Logic pulser.

Let us now look briefly at how a logic pulser can be used when troubleshooting the circuit in Fig. 13–24. The pulser would be set to pulse mode of operation and then the pulse injected at test point 2. As long as the connection through the resistor pack is good, LED 0 should blink. Assuming that this part of the circuit works correctly, the pulser's probe can next be touched to test point 3. Again, the LED should blink. This verifies whether or not the 74LS240 inverter operates correctly. Just as with the oscilloscope, the logic pulser is used to check out the circuit step by step.

Hardware troubleshooting of circuit boards in a manufacturing environment can be quite different. A visual inspection is still performed, but it should look for different things. For instance, the quality of solder joints is checked to determine whether they have enough solder, are of the correct shape, and that there are no solder shorts between pins of ICs and other components.

After a board has passed visual inspection, it is ready for circuit test. In this case, the board is not tested with instruments circuit by circuit as we just described; instead, it is checked out with an automatic test system. The tester is programmed to perform a series of tests on the circuit board. The system provides information on tests that have passed and failed to guide the repair process. In this way, the board is tested and repaired step by step until it is completely functional.

Troubleshooting an electronic system, such as a PC, that has been working is also different. In this case, we will assume that the symptoms of the problem have been identified, that the software is functional, and that we are investigating a hardware problem. Actually, the servicing of a PC is normally a system-level repair—that is, the failing subassembly (main processor board, add-on card, power supply, floppy-disk drive, keyboard, or monitor) is identified and replaced.

The hardware troubleshooting procedure outlined in Fig. 13–23 still applies to a system-level repair. Therefore, the first step is a visual inspection; however, the inspection performed is different than that used when examining a breadboard or circuit board that just came off of the manufacturing line and includes a number of mechanical checks. For example, the system should be checked to verify that all cables are securely connected, the setting of the DIP switches and jumpers should be checked to assure that the system configuration is correct, and the surface of the circuit boards can be examined for overheated or burned components. An example of a mechanical check is to touch the components to see if any are abnormally hot (but be careful with this because some devices can get very hot). Moreover, to assure that the problem is not due to dirty connector contacts, the cables and add-on boards are removed, their contacts are cleaned, and then are reseated into the connector.

Next we must begin to check the circuits and subassemblies. If the PC does not come up at all when the power switch is turned on, the first step should be to check the power supply voltages. However, if it boots up and loads the DOS operating system, diagnostic software, such as QAPlus™ by DiagSoft, Inc., can be used to analyze the function of the system. The diagnostic disk is inserted into the floppy-disk drive and the diagnostic program is initiated from the keyboard. This program can exercise each of the PC's subassemblies, and it displays information indicating whether they have passed or failed the diagnostic tests.

Let us assume that the hard disk controller in a PC has failed the diagnostic test. The normal system-level repair procedure is to replace the complete controller with another one to quickly get the system back up running. The bad board is usually returned to a circuit-board repair location for IC-level troubleshooting and repair. Some diagnostic programs permit IC-level troubleshooting of the dynamic memory subsystem. In the case of a memory failure, the diagnostic program can identify the bad IC. Since the DRAMs in some PCs are socket mounted, the repair may be made by replacing the failing device.

▲ 13.5 OBSERVING MICROCOMPUTER BUS ACTIVITY WITH A DIGITAL LOGIC ANALYZER

Up to this point, we observed signal waveforms in the microcomputer with an oscilloscope. However, this instrument permits viewing of only a limited number of periodic signals at a time. The address, data, and control buses in a microcomputer have many lines. For instance, the data bus alone is eight bits wide in an 8088-based PC. When the microcomputer is running, the data bus could be returning read data or instruction code to the MPU, sending write data to memory, or be in the high-Z state if no bus activity is taking place. Moreover, the data being transferred is rarely the same; therefore, data bus activity is not periodic.

To observe the operation of the data bus signals, we need to see the logic states of all eight data bits and some of the read/write control signals at the same time. This is not possible with a scope. It is for this type of measurement that an instrument known as a *digital logic analyzer* was developed. Let us now look briefly at what a logic analyzer is and what it is used for in testing microcomputer systems.

The logic analyzer is a modern digital test instrument that is very useful for testing and troubleshooting microcomputer systems. With it, nonperiodic signals, such as those of the address bus, data bus, and control bus, can be measured and their waveforms viewed. Figure 13–26 shows a typical logic analyzer. Today, this type of instrument is available with 8, 16, or 32 channels. This means that they are capable of simultaneously sampling and displaying the waveshapes of up to 8, 16, or 32 signals. The probe is a pod that has a clip for input to each channel. They are attached to the signals that are to be observed. For example, the data lines D_0 through D_7 and control signals, such as $\overline{RD}$, $\overline{WR}$, DT/$\overline{R}$, IO/$\overline{M}$, $\overline{DEN}$, READY, S_3, and S_4, can be sampled to monitor transfers over the data bus.

Figure 13–26 Digital logic analyzer. (Hewlett-Packard Co.)

The logic analyzer operates differently than an oscilloscope. The oscilloscope immediately displays the voltage of the signal applied at its input. On the other hand, the logic analyzer samples the voltage at all inputs at a very high rate. This information is stored in memory as a logic 0 or logic 1, and not as a specific voltage level. The waveform of the signal can then be displayed on the screen using the stored data. The user of the instrument has the ability to start the sampling based on the occurrence of a specific event or events indicated by a combination of the logic values of the signals being monitored and continue until the trace buffer memory is full. Moreover, most logic analyzers permit the stored information to be displayed in a variety of ways.

Sample waveforms taken from our test circuit in Fig. 13–16(a) are shown in Fig. 13–27(a). This display shows the address, data transfer, and address decoder output produced when the LED blink diagnostic routine runs. Here we see a timing diagram that clearly illustrates the relationship between the address, decoder output, and the data transfer to the latch. Notice that whenever address decoder output $\overline{\text{IOW0007H}}$ switches from logic 0 to logic 1, the byte of data on the data bus, 00000001_2, is latched into the 74LS374 device and makes Q_0 switch to logic 1. Figure 13–27(b) shows the signals when the Q_0 output switches from logic 1 back to 0. Both transitions at Q_0 are shown with a single timing diagram in Fig. 13–27(c). Remember that the logic analyzer cannot display signals the way they really would look if observed with an oscilloscope. Since it saves only the logic level (0 or 1) of the signals in memory, waveshapes are shown with sharp transitions between these logic levels. When observed with a scope, the waveshapes may show rise and fall times between the 0 and 1 levels, and possibly *ringing, overshoots,* and *undershoots,* around the 0 and 1 logic levels.

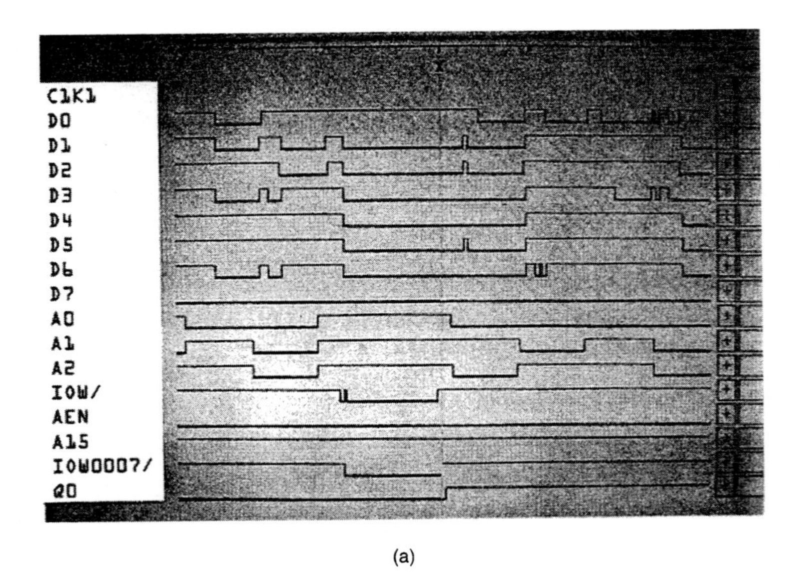

(a)

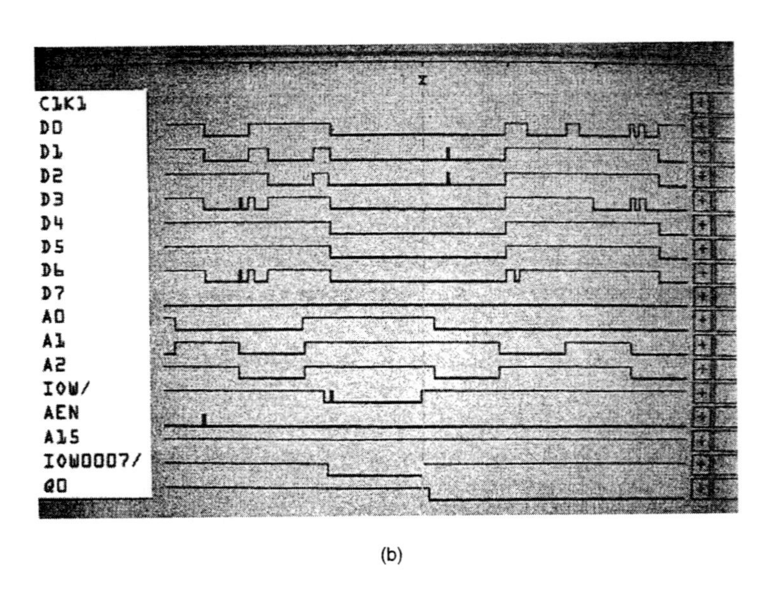

(b)

Figure 13–27 (a) Timing diagram for 0 to 1 transition at Q_0. (b) Timing diagram for 1 to 0 transition at Q_0. (c) Timing diagram showing both transitions at Q_0.

A logic analyzer can also be set up to collect only code or data transfers over the data bus. Once this information is stored in memory, it can be *disassembled* into assembly language instructions and displayed on the screen. For example, the logic analyzer was used to disassemble a series of instructions shown in Fig. 13–28. This capability permits us to monitor the execution of instructions by the MPU and compare the instruction execution sequence to events observed in the hardware.

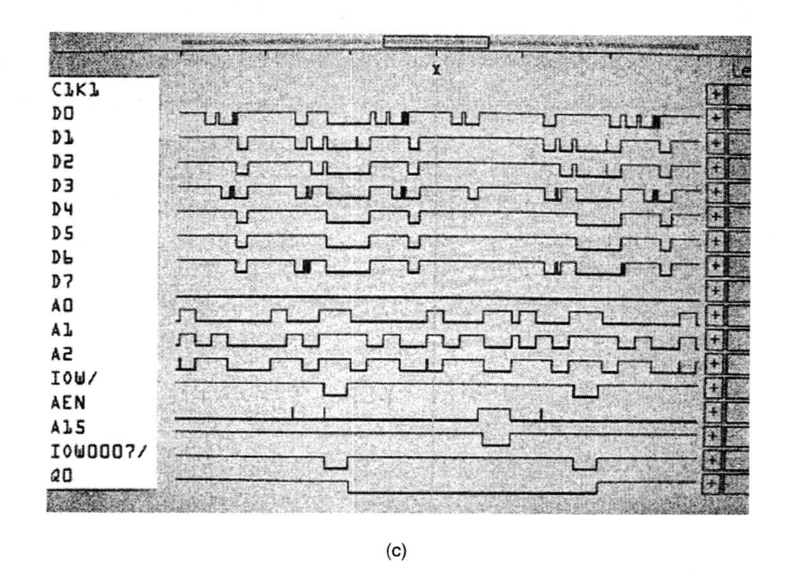

(c)

Figure 13–27 (continued)

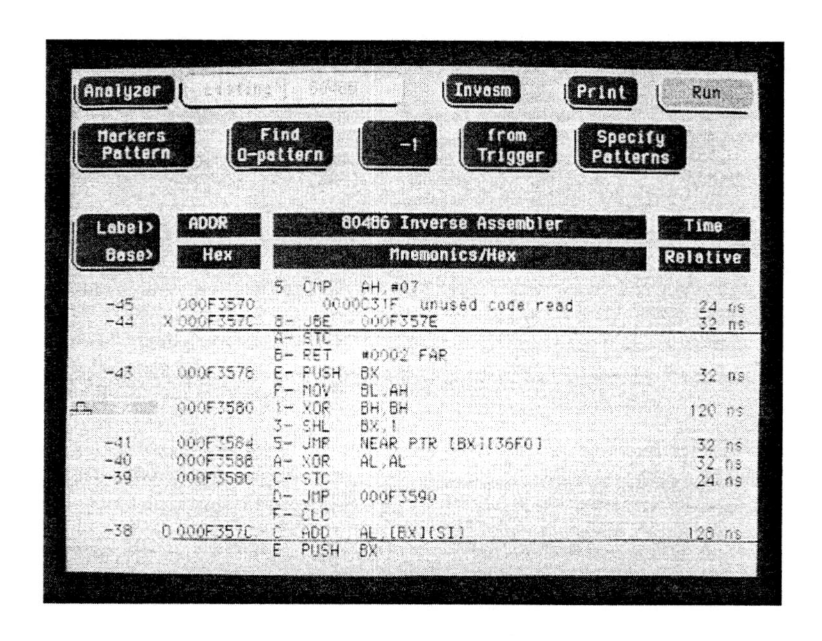

Figure 13–28 Disassembling code with a logic analyzer. (Hewlett-Packard Co.)

REVIEW PROBLEMS

Section 13.1

1. What is an experimental circuit board built to test an electronic function called?
2. What is a breadboard card?
3. What kind of breadboard does not require devices to be soldered in place but rather that they are plugged in?
4. What is the purpose of an extender card?
5. List the parts of the PCμLAB.

Section 13.2

6. List three types of I/O devices that are built into the PCμLAB.
7. How are the on-board I/O devices set up for use with circuits built on the breadboard area?
8. How many ISA expansion slots are provided on the PCμLAB?
9. What size wire jumpers must be used to interconnect circuits on the breadboard area?
10. Which wire insertion clips are intended for use as the +5 V and GND for power distribution on the solderless breadboard?
11. At which contacts of the I/O channel connector are address lines A_0 through A_{19} available?
12. What is the purpose of the PCμLAB's continuity tester? How does it signal continuity?
13. Identify how the PCμLAB's logic probe signals the 0, 1, and high-Z logic states.
14. How does the logic probe identify that the signal at a test point is switching between the 0 and 1 logic levels?

Section 13.3

15. Which ICs are used in the I/O address decoder circuit of the PCμLAB?
16. Which output of the I/O address decoder circuit is used to enable data output to the LEDs?
17. Which I/O address bits are don't-care states?
18. Does the I/O address $771E_{16}$ activate an output of the I/O address decoder circuit? If so, which output signal is activated?
19. Why are the I/O select outputs of the I/O address decoder circuit produced only during I/O bus cycles?
20. If switches S_0 through S_3 are closed and switches S_4 through S_7 are open, what value will be transferred over the data bus when the switches are read with an IN instruction?
21. Does the command

```
I FF1D (↵)
```

read the state of the on-board switches?

22. Write an instruction sequence that will read the state of the switches and mask off all switch settings but S_0 and S_1. If both switches are read as closed, a jump is initiated to a service routine called SERVE_3.
23. What operation is performed by the instruction sequence that follows?

```
        MOV     DX, 31DH
POLL:   IN      AL, DX
        MOV     CL, 8
        SHR     AL, CL
        JC      POLL
```

24. Which LED is driven with data from data bus line D_0? From D_7?
25. What does the command O FF1E 0F accomplish?
26. Write a program that will scan the LEDs on the PCμLAB. That is, first light LED 0 for a period of time, next turn off LED 0 and turn on LED 1, and so on until LED 7 is lighted. The scan sequence should repeat continuously.
27. Describe the operation performed by the following instruction sequence.

```
        MOV     DX, 31EH
        MOV     AL, 0H
BIN:    OUT     DX, AL
        MOV     CX, 0FFFFH
DELAY:  DEC     CX
        JNZ     DELAY
        INC     AL
        JMP     BIN
```

28. What IC is used to drive the speaker?
29. Which instruction in the tone-generation program given in the section on the speaker-drive circuit needs to be changed to double the frequency of the tone? Write the new instruction.

Section 13.4

30. Write an instruction that reads the switch setting into AL in Fig. 13–29.
31. Mark the pin numbers into the circuit shown in Fig. 13–29, and then make a layout drawing on a circuit diagram master.
32. What is a program written specifically to test the operation of a microcomputer circuit called?
33. What accessory is attached to the top of an IC to make it easier to connect the probe of an instrument?
34. Name three instruments that can be used to test the operation of a microcomputer interface circuit.
35. What information does a logic probe provide about the signal at a test point in a circuit?

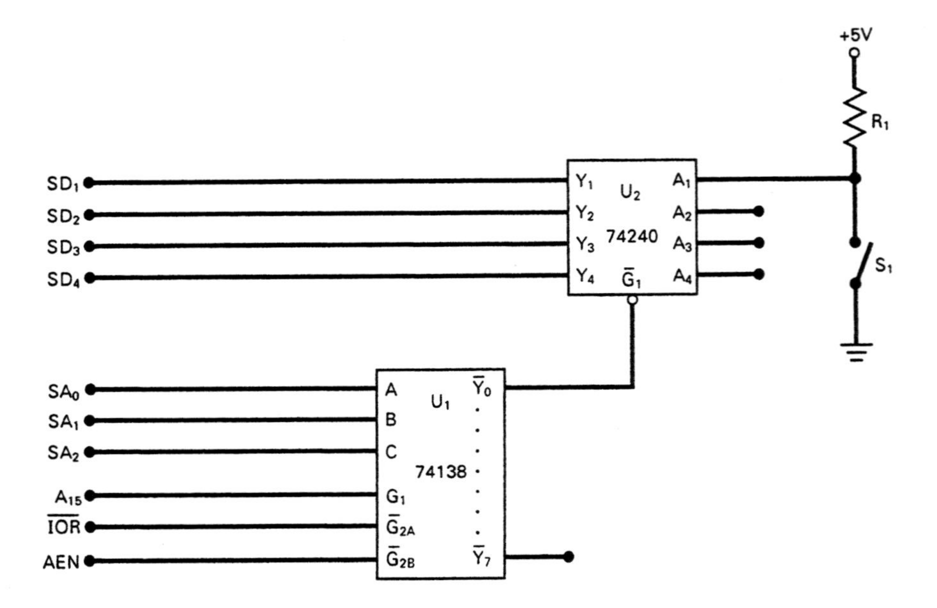

Figure 13–29 Switch input interface circuit.

36. When used to measure voltage, what information does a multimeter tell about the signal at a test point?

37. What information does an oscilloscope provide about the signal at a test point?

38. What is meant by *periodic signal?*

39. What process is used for determining the cause of a hardware malfunction in a circuit?

40. If a breadboard of a circuit does not work when tested with a diagnostic program that was just written and never checked on working hardware, should hardware troubleshooting or software debug take place next?

41. Assume that a diagnostic program is available that has been checked out and verified to correctly test the interface. If a new PC interface module is tested and fails when checked out with this diagnostic program, is the next step software debug or hardware troubleshooting?

42. List three items that should be checked as part of the software-debug process.

43. List three visual inspections that can be made as part of the hardware troubleshooting process used to determine why a circuit built on the breadboard area of the PCμLAB does not work.

44. During the hardware troubleshooting process of a breadboard circuit, what should be checked next if the circuit is found to be correctly constructed?

45. Assume that the circuit in Fig. 13–29 is being driven by a software routine that polls the state of switch 0 and that this program is known to operate correctly. What type of signal would you expect to see at test point 1 when the switch is closed? When open? At test point 2? At test point 3?

46. The results found at test points 1 and 2 of problem 45 when troubleshooting the circuit are as follows:

Test Point	*Switch Open*	*Switch Closed*
1	1	0
2	1	1

What do you think is the problem with the circuit?

Section 13.5

47. List three key groups of signals of the microcomputer system that are, in general, nonperiodic.

48. What instrument is usually used to observe nonperiodic signals in a microcomputer?

49. Compare an oscilloscope and a logic analyzer.

Real-Mode Software and Hardware Architecture of the 80286 Microprocessor

▲ INTRODUCTION

The 80286 microprocessor is the MPU used to design IBM's original PC/AT. This microprocessor can operate in either of two modes, the *real-address mode* (real mode) or the *protected-address mode* (protected mode). In the PC/AT, it is used to implement a real mode microcomputer. This chapter focuses on the real mode software and hardware architecture of the 80286. Here we will first examine the real-address mode software architecture and its extended instruction set. Then we study the hardware architecture of the 80286-based microcomputer system: the signal interfaces of the 80286, its memory interface, I/O interface, and interrupts and exception processing. For this purpose, we discuss the following topics in this chapter:

14.1 80286 Microprocessor
14.2 Internal Architecture
14.3 Real-Mode Software Model
14.4 Real-Mode Extended Instruction Set
14.5 Interfaces of the 80286
14.6 82C288 Bus Controller
14.7 System Clock

▲ 14.1 80286 MICROPROCESSOR

The 80286, first announced in 1982, was the fifth member of Intel Corporation's 8086 microprocessor family. We already mentioned that the 80286 offers two modes of operation: real mode, for compatibility with the existing 8086/8088 software base, and protected mode, which offers enhanced system-level features such as memory management, multitasking, and protection. A number of changes have been made to both the software and the hardware architecture of the 80286, primarily to improve its performance. For example, additional pipelining is provided within the 80286 to provide higher performance, the instruction set is enhanced with new instructions, the address and data buses are demultiplexed to simplify system design, and the bus is designed to support interleaved memory subsystems.

The original 80286 was manufactured using the high-performance metal-oxide-semiconductor III (HMOSIII) process, and its circuitry is equivalent to approximately 125,000 transistors. It is available in *plastic leaded chip carrier* (PLCC), *ceramic leadless chip carrier* (LCC), and *pin grid array* (PGA) packages. Figure 14–1 shows an 80286 in the LCC package. The PLCC is the lowest-cost package type and is the most widely used in commercial applications such as personal computers. The LCC and PGA are more rugged and are typically used in applications that require higher reliability.

Each of these packages has 68 leads. Figure 14–2 illustrates the signal available on each lead. Note that unlike the earlier 8086 and 8088 devices, none of the signal lines of the 80286 are multiplexed with another signal. This is intended to simplify the microcomputer circuit design.

▲ 14.2 INTERNAL ARCHITECTURE

Figure 14–3 shows the internal architecture of the 80286 microprocessor. The 8086 and 8088 each have only two processing units, the bus interface unit (BIU) and execution unit (EU). Figure 14–3 shows that the 80286 is internally partitioned into four independent

Figure 14–1 80286 IC (Courtesy of Intel Corporation)

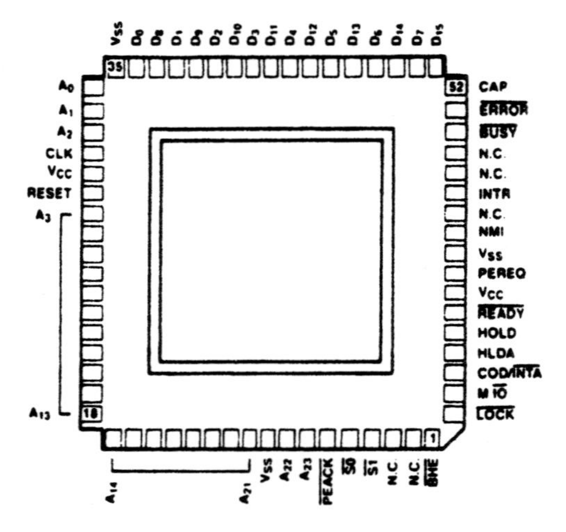

Figure 14–2 Pin layout of the 80286. (Reprinted by permission of Intel Corporation. Copyright/Intel Corp. 1987)

processing units: the *bus unit* (BU), the *instruction unit* (IU), the *execution unit* (EU), and the *address unit* (AU). This additional parallel processing provides an important contribution to the higher level of performance achieved with the 80286 architecture.

The bus unit is the 80286's interface with the outside world. It provides a 16-bit data bus, a 24-bit address bus, and the signals needed to control bus transfers. These buses are demultiplexed instead of multiplexed as in the 8086/8088 hardware architecture—that is, the 80286 has separate pins for its address and data lines. Demultiplexing of these buses improves the performance of the 80286's hardware architecture.

The bus unit is responsible for performing all external bus operations. Figure 14–3 shows that this processing unit contains the latches and drivers for the address bus, transceivers for the data bus, and control logic for generating the control signals needed to perform memory and I/O bus cycles.

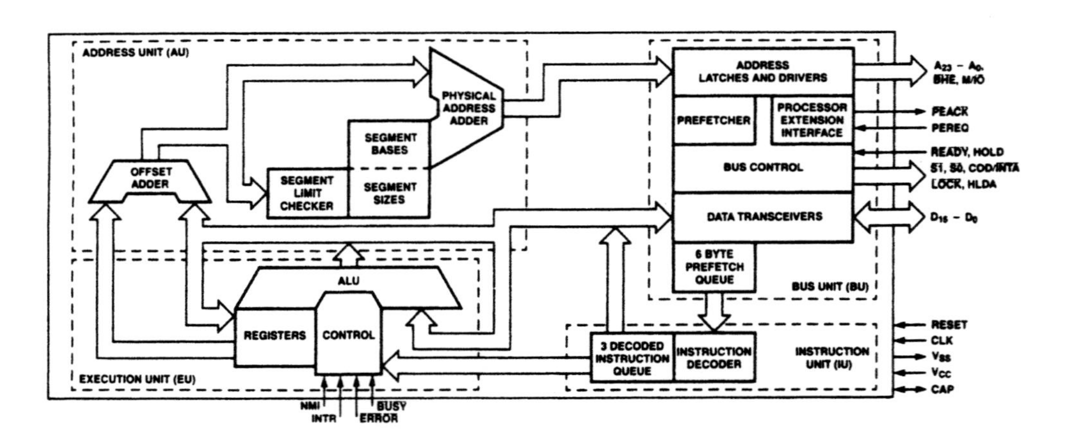

Figure 14–3 Internal architecture of the 80286. (Reprinted by permission of Intel Corporation. Copyright/Intel Corp. 1987)

Note in Fig. 14–3 that the bus unit also contains elements called the *prefetcher* and *prefetch queue.* Together these elements implement a mechanism known as an *instruction stream queue.* This queue permits the 80286 to prefetch up to six bytes of instruction code. Whenever the queue is not full—that is, whenever it has room for at least two more bytes and, at the same time, the execution unit is not asking it to read or write operands from memory—the bus unit is free to look ahead in the program by fetching the next sequential instructions. These prefetched instructions are held in a first-in, first-out (FIFO) queue. With its 16-bit data bus, the 80286 fetches two bytes of instruction code in a single memory cycle. When a byte is loaded at the input end of the queue, all other bytes shift up through the FIFO to the empty locations nearer the output. In this way, the fetch time for many instructions is hidden.

The prefetcher provided in the 80286's bus unit has more intelligence than the one implemented in the 8086/8088 architecture. It has the ability to determine if an instruction that has been prefetched will cause a transfer in program control—for instance, a jump. When this type of instruction is detected, the queue is flushed and prefetch resumes at the point to which control is to be passed. The reset of the queue happens prior to the actual execution of the instruction. This feature results in an additional performance improvement for the 80286.

The address unit provides the memory management and protection services for the 80286. It off-loads the responsibility for address generation, translations, and checking from the bus unit and thereby further boosts the performance of the MPU. It contains dedicated hardware for performing high-speed address calculations, virtual-to-physical address translations, and limit and access rights attribute checks. For instance, in the real mode, the address unit calculates the address of the next instruction to be fetched. This is done by shifting the current contents of the code segment (CS) register left by four bit positions, filling the least significant bits with 0s, and adding the value in the instruction pointer (IP) register. This gives the 20-bit physical address that is output on the address bus. For protected mode, the address unit performs more functions. For instance, it performs the various address translations and protection checks needed when performing protected mode bus cycles.

Looking at Fig. 14–3, we see that the instruction unit accesses the output end of the prefetch queue. It reads one instruction byte after the other from the output of the queue and decodes them into the 69-bit instruction format used by the 80286's execution unit—that is, it off-loads the responsibility for instruction decoding from the execution unit. The instruction queue within the instruction unit permits three fully decoded instructions to be held waiting for execution by the execution unit. Once again the result is improved performance for the MPU.

Notice in Fig. 14–3 that the execution unit includes the *arithmetic logic unit* (ALU), the 80286's registers, and a control ROM. The block labeled registers represents all the user-accessible registers, such as the general-purpose data registers and segment registers. The control ROM contains the microcode sequences that define the operations performed by the 80286's instructions. The execution unit reads decoded instructions from the instruction queue and performs the operations that they specify. If necessary during the execution of an instruction, it requests that the address unit generate operand addresses and that the bus unit perform read or write bus cycles to access data in memory or at an I/O device.

▲ 14.3 REAL-MODE SOFTWARE MODEL

We will begin our study of the 80286 microprocessor by exploring its real-address mode software model and operation. Whenever the 80286 is powered on or reset, it comes up in the real mode. The 80286 will remain in the real mode unless it is switched to protected mode under software control. In fact, in many applications the 80286 is simply used in real mode.

In real mode, the 80286 operates like a high-performance 8086. For example, the standard 8-MHz 80286 provides more than five times higher performance than the standard 5-MHz 8086. Furthermore, the 10-MHz 80286 outperforms the 10-MHz 8086 by a factor of 7.

When in the real mode, the 80286 can be used to execute the base instruction set of the 8086/8088 architecture. The object code for the base instructions of the 80286 is identical to that of the 8086/8088. This means that operating systems and application programs written for the 8086 and 8088 can be run on the 80286 without modification. An example is the disk operating system (DOS) of the original IBM PC. This operating system also runs on the 80286 in the original PC/AT. For this reason, we can say that the 80286 is *object code compatible* with the 8086 and 8088 microprocessors.

A number of new instructions have been added in the instruction set of the 80286 to enhance performance and functionality. For example, instructions have been added to push or pop the complete register set, perform string I/O, and check the boundaries of data array accesses. We also say that object code is *upward compatible* within the 8086 architecture. By this we mean that 8086/8088 object code will run on the 80286, but the reverse is not true if any of the new instructions are in use.

Figure 14–4 shows the real-mode software model of the 80286. Here we see that it has 15 internal registers. Fourteen of them, the instruction pointer (IP), data registers (AX, BX, CX, and DX), pointer registers (BP and SP), index registers (SI and DI), segment registers (CS, DS, SS, and ES), and the flag register (F) are identical to the corresponding registers in the 8086's software model, and they serve the same functions. For instance, CS:IP points to the next instruction that is to be fetched.

A new register called the *machine status word register* (MSW) is added in the 80286 model. The only bit in the MSW that is active in the real mode is the *protected-mode enable* (PE) bit. This is the bit used to switch the 80286 from real to protected mode.

Looking at Fig. 14–4, we find that the 80286 microcomputer's real-mode address space is also identical to that of the 8086 microcomputer. It is partitioned into a 1Mbyte memory address space and a separate 64Kbyte input/output address space. The memory address space resides in the range from address 00000_{16} to $FFFFF_{16}$. I/O addresses span 64Kbytes of the range 0000_{16} to $FFFF_{16}$. Moreover, in the memory address space, only four 64Kbyte segments are active at a time for a total of 256Kbytes of active memory. Again, 64Kbytes of the active memory are allocated for code storage, 64Kbytes for stack, and 128Kbytes for data storage.

Figures 14–5(a) and (b) show that the real-mode 80286 memory and I/O address spaces are partitioned into general-use and reserved areas in the same way as for the 8086.

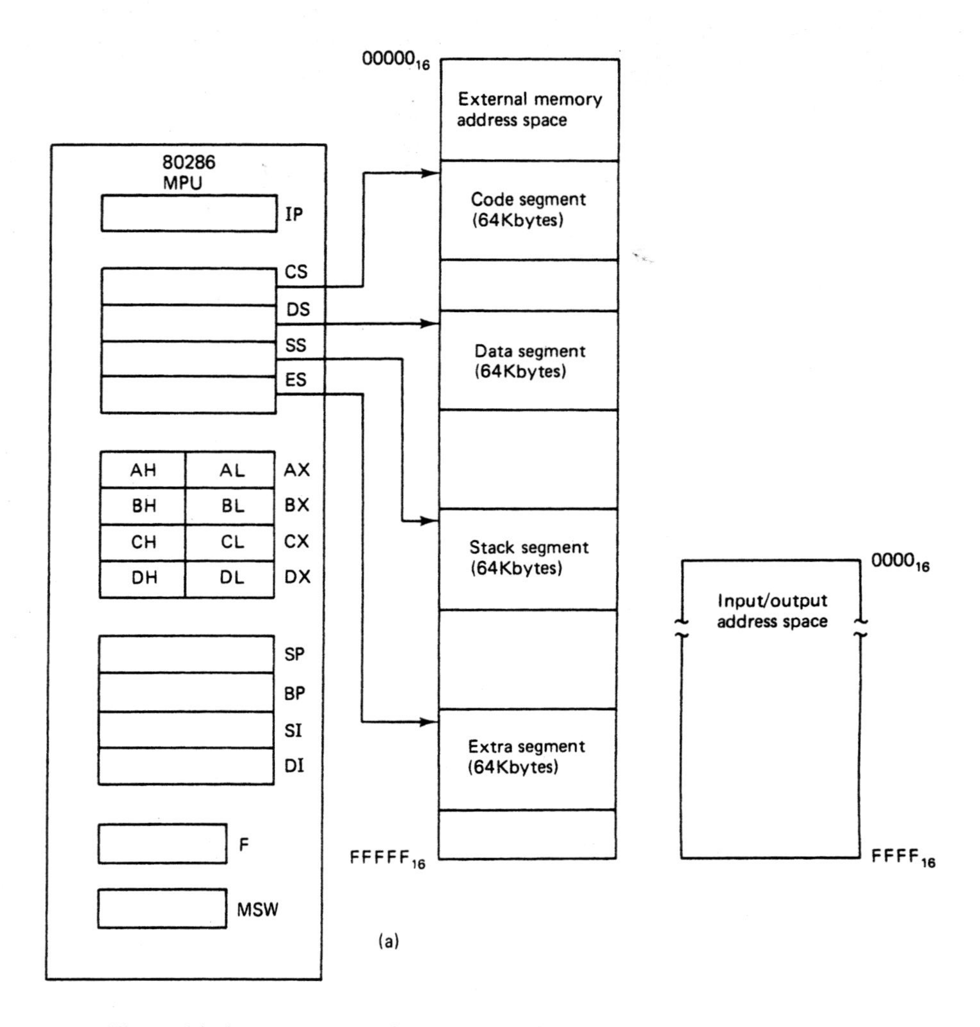

Figure 14–4 Real-mode software model of the 80286 microprocessor.

For instance, in Fig. 14–5(a) we find that the first 1Kbytes of the memory address space, addresses 0_{16} through $3FF_{16}$, are reserved and are used for storage of the interrupt vector table. Figure 14–5(b) shows that the first 256-byte I/O addresses are identified as page 0. Page 0 is the I/O addresses that are directly accessible with an IN or OUT instruction.

Finally, the real-mode 80286 generates physical addresses in the same manner as the 8086, as illustrated in Fig. 14–6. Note that the 16-bit contents of a segment register, such as CS, are shifted left by four bit positions, the four least significant bits are filled with 0s, and the result is added to a 16-bit logical address, such as the value in IP, to form the 20-bit physical memory address.

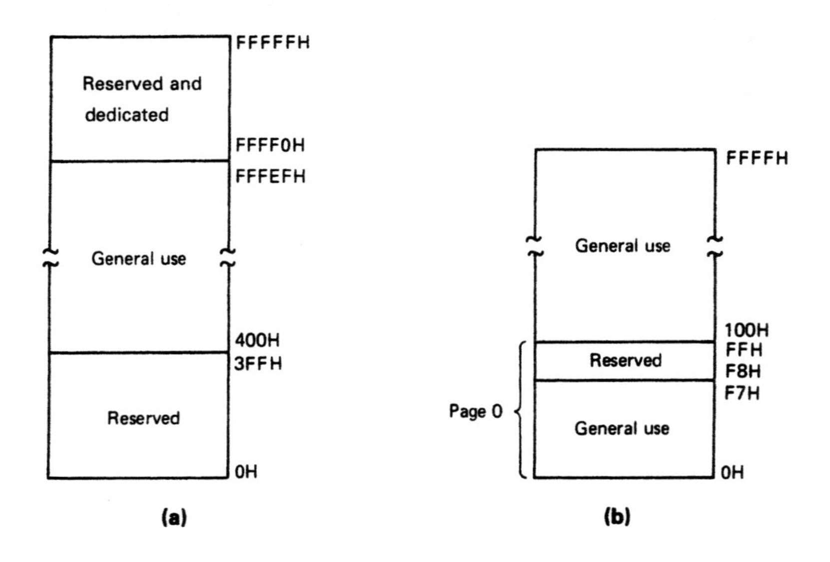

Figure 14–5 (a) Dedicated and general use of memory in real-mode. (Reprinted by permission of Intel Corporation. Copyright/Intel Corp. 1987) (b) I/O address space. (Reprinted by permission of Intel Corporation. Copyright/Intel Corp. 1987)

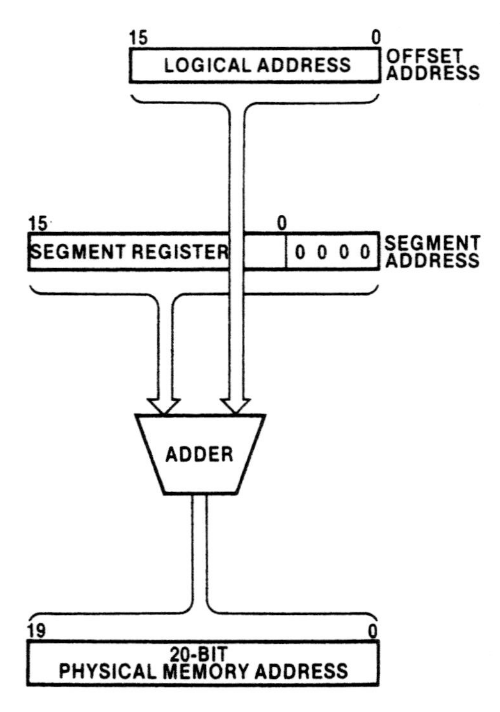

Figure 14–6 Real-address mode physical address generation. (Reprinted by permission of Intel Corporation. Copyright/Intel Corp. 1987)

▲ 14.4 REAL-MODE EXTENDED INSTRUCTION SET

Figure 14–7 shows the evolution of the instruction set for the 8086 architecture. The instruction set of the 8086 and 8088 microprocessors, the *base instruction set,* was enhanced in the 80286 microprocessor to implement what is called the *extended instruction set.* This extended instruction set includes several new instructions and implements additional addressing modes. For example, two instructions added as extensions to the basic instruction set are *push all* (PUSHA) and *pop all* (POPA). Figure 14–8 shows that the PUSH and IMUL instructions have been enhanced to permit the use of immediate operand addressing. In this way, we see that the 80286's real-mode instruction set is a superset of the base instruction set. These instructions are all executable by the 80286 in the real mode. Let us now look at these and the other new instructions in more detail.

Push-All and Pop-All Instructions—PUSHA and POPA

When writing the compiler for a high-level language such as C, it is very common to push the contents of all of the 80286 general registers to the stack before calling a subroutine. If we use the PUSH instruction to perform this operation, many instructions are needed. To simplify this operation, special instructions are provided in the instruction set of the 80286. They are called *push all* (PUSHA) and *pop all* (POPA).

Looking at Fig. 14–8 we see that execution of PUSHA causes the values in AX, CX, DX, BX, SP, BP, SI, and DI to be pushed, in that order, onto the top of the stack. Figure 14–9 shows the state of the stack before and after execution of the instruction. As shown in Fig. 14–10, executing a POPA instruction at the end of the subroutine restores the old state of the 80286.

Stack Frame Instructions—ENTER and LEAVE

Before the main program calls a subroutine, quite often it is necessary for the calling program to pass the values of some *variables* (parameters) to the subroutine. It is a common practice to push these variables onto the stack before calling the routine. Then during the execution of the subroutine, they are accessed by reading from the stack and used in computations. Two instructions are provided in the extended instruction set of the 80286 to allocate and de-allocate a data area called a *stack frame.* This data area, located

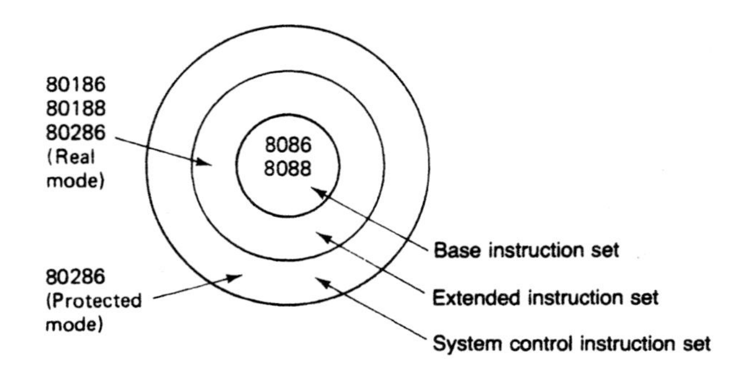

Figure 14–7 Evolution of the instruction set for the 8086 microprocessor family.

Mnemonic	Meaning	Format	Operation
PUSH	Push	PUSH dw/db	Push the specified data word (dw) or sign extended data byte (db) onto the stack.
PUSHA	Push all	PUSHA	Push the contents of registers AX, CX, DX, BX, original SP, BP, SI, and DI onto the stack.
POPA	Pop all	POPA	Pop the stack contents into the registers DI, SI, BP, SP, BX, DX, CX, and AX.
IMUL	Integer multiply	IMUL rw, ew, dw/db	Perform the signed multiplication as follows: rw = ew*dw/db where rw is the word size register, ew is the effective word size operand, and the third operand is the immediate data word (dw) or a byte (db).
Logic instructions		Instruction db	Perform the logic instruction using the specified byte (db) as the count.
INS	Input string	INSB, INSW	Input the byte or the word size element of the string from the port specified by DX to the location ES: [DI].
OUTS	Output string	OUTSB, OUTSW	Output the byte or the word size element of the string from ES: [SI] to port specified by DX.
ENTER	Enter procedure	ENTER dw, 0/1/db	Make stack frame for procedure parameters.
LEAVE	Leave procedure	LEAVE	Release the stack space used by the procedure.
BOUND	Check array index against bounds	BOUND rw, md	Interrupt 5 occurs if the register word (rw) is not greater than or equal to the memory word at md and not less than or equal to the second memory word at md + 1.

Figure 14–8 Extended instruction set.

in the stack part of memory, is used for local storage of parameters and other data used by the subroutine.

Normally, high-level languages allocate a stack frame for each procedure in a program. The stack frame provides a dynamically allocated local storage space for the procedure and contains data such as variables, pointers to the stack frames of the previous procedures from which the current procedure was called, and a return address for linkage to the stack frame of the calling procedure. This mechanism also permits access to the data in stack frames of the calling procedures.

The instructions used for allocation and de-allocation of stack frames are given in Fig. 14–8 as *enter* (ENTER) and *leave* (LEAVE). Execution of an ENTER instruction allocates a stack frame; it is de-allocated by executing the LEAVE instruction. For this reason, as shown in Fig. 14–11, the ENTER instruction is used at the beginning of a subroutine, and LEAVE at the end, just before the return instruction.

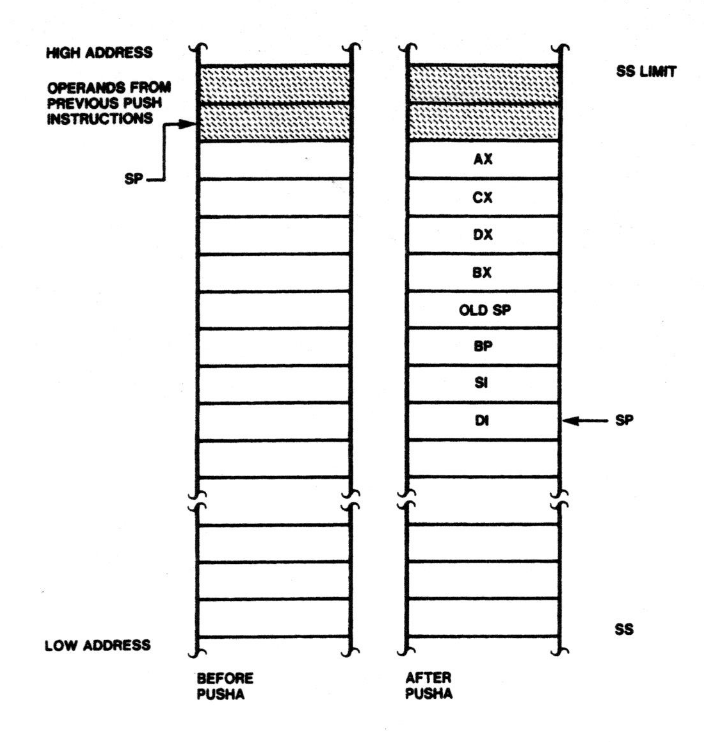

Figure 14–9 State of the stack before and after executing PUSHA. (Reprinted by permission of Intel Corporation. Copyright/Intel Corp. 1987)

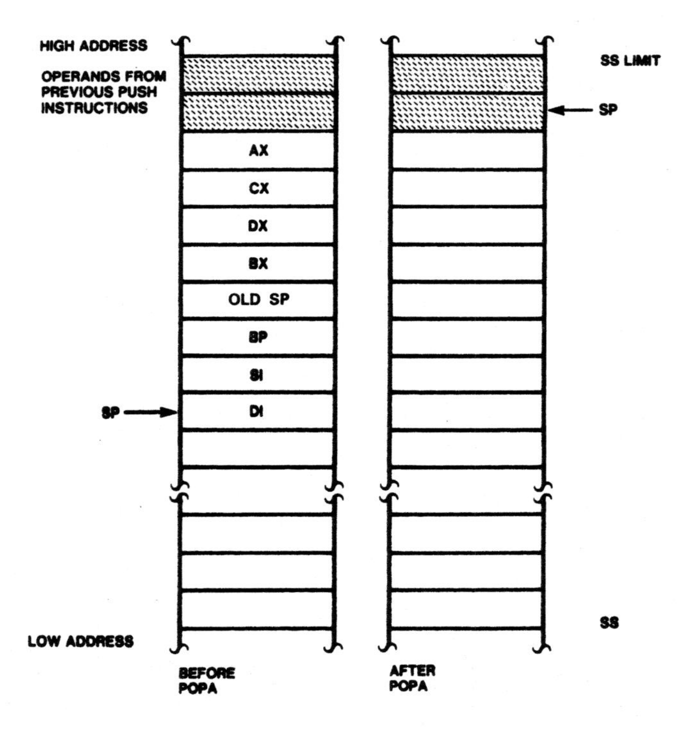

Figure 14–10 State of the stack before and after executing POPA. (Reprinted by permission of Intel Corporation. Copyright/Intel Corp. 1987)

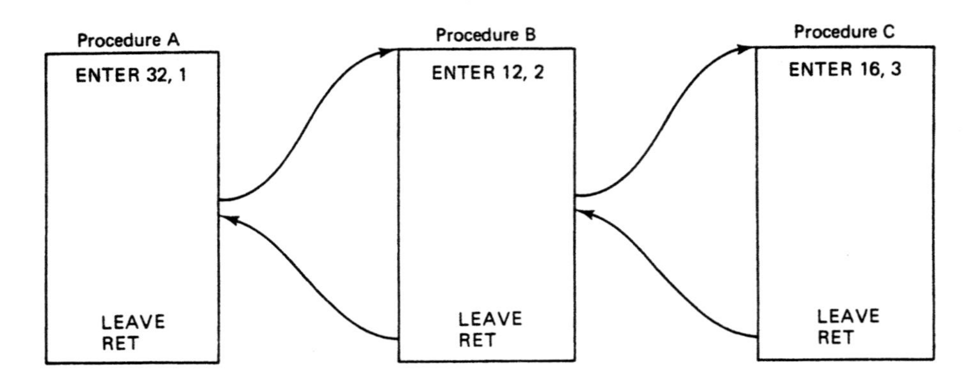

Figure 14–11 Enter/leave example.

Looking at Fig. 14–8, we find that the ENTER instruction has two operands. The first operand, identified as dw, is a word-size immediate operand. This operand specifies the number of bytes to be allocated on the stack for local data storage of the procedure. The second operand, a byte-size immediate operand, specifies what is called the *lexical nesting level* of the routine. This lexical level defines how many pointers to previous stack frames can be stored in the current stack frame. This list of previous stack frame pointers is called a *display.* The value of the lexical level byte must be limited to a maximum of 32 in 80286 programs.

An example of an ENTER instruction is

```
ENTER  12, 2
```

Execution of this instruction allocates 12 bytes of local storage on the stack for use as a stack frame. It does this by decrementing SP by 12. This defines a new top of stack at the address formed with SS and SP–12. Also the *base pointer* (BP) that identifies the beginning of the previous stack frame is copied into the stack frame created by the ENTER instruction. This value is called the *dynamic link* and is held in the first storage location of the stack frame. The number of stack frame pointers that can be saved in a stack frame is equal to the value of the byte that specifies the lexical level of the procedure. Therefore, in our example, just two levels of nesting are specified.

The BP register is used as a pointer into the stack segment of memory. When a procedure is called, the value in BP points to the stack frame location that contains the previous stack frame pointer (dynamic link). Therefore, based-indexed addressing can be used to access variables in the stack frame by referencing the BP register.

The LEAVE instruction reverses the process of an ENTER instruction—that is, its execution de-allocates the stack frame. This is done by first automatically loading SP from the BP register. This returns the storage locations of the current stack frame to the stack. Now SP points to the location where the dynamic link (pointer to the previous stack frame) is stored. Next, popping the contents of the stack into BP returns the pointer to the stack frame of the previous procedure.

To illustrate the operation of the stack frame instructions, let us consider the example of Fig. 14–11. Here we find three procedures. Procedure A is used to call procedure B, which in turn calls procedure C. It is assumed that the lexical levels for these procedures are 1, 2, and 3, respectively. The ENTER, LEAVE, and RET instructions for each procedure are shown in the diagram. Note that the ENTER instructions specify the lexical levels for the procedures.

The stack frames created by executing the ENTER instructions in the three procedures are shown in Fig. 14–12. As the ENTER instruction in procedure A is executed, the old BP from the procedure that called procedure A is pushed onto the stack. BP is loaded from SP to point to the location of the old BP. Since the second operand is 1, only the current BP that is the BP for procedure A is pushed onto the stack. Finally, to allocate 32 bytes for the stack frame, 32 is subtracted from the current value in SP.

After entering procedure B, a second ENTER instruction is encountered. This time the lexical level is 2. Therefore, the instruction first pushes the old BP—that is, the BP for procedure A—onto the stack, then pushes the BP previously stored on the stack frame for A to the stack, and lastly pushes the current BP for procedure B to the stack. This

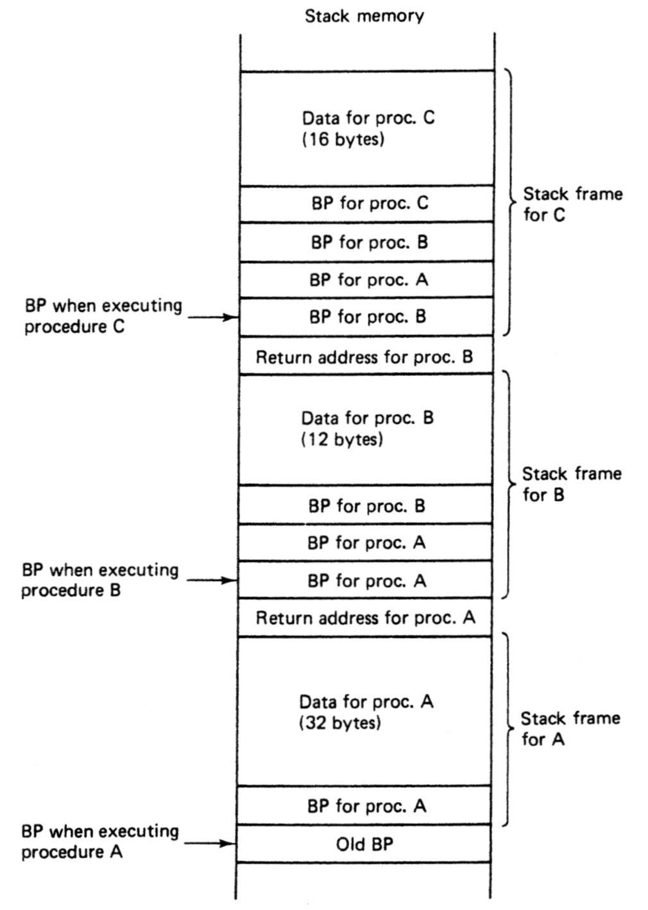

Figure 14–12 Stack after execution of ENTER instructions for procedures A, B, and C.

mechanism provides access to the stack frame for procedure A from procedure B. Next, 12 bytes, as specified by the instruction, are allocated for local storage.

The ENTER instruction in procedure C has the lexical level of 3; therefore it pushes the BPs for the two previous procedures—that is, B and A—to the stack in addition to the BP for C and the BP for the calling procedure B.

Input String and Output String Instructions—INS and OUTS

Figure 14–8 shows that there are also input and output string instructions in the extended instruction set of the 80286. Using these string instructions, a programmer can either input data from an input port to a storage location directly in memory or output data from a memory location to an output port.

The first instruction, called *input string,* can be written in one of two ways, INSB or INSW. INSB stands for *input string byte,* and INSW means *input string word.* Let us now look at the operation performed by the INSB instruction. INSB assumes that the address of the input port to be accessed is in the DX register. This value must be loaded prior to executing the string instruction. The address of the memory storage location into which the bytes of data are input is identified by the current values in ES and DI—that is, when executed, the input operation performed is

$$(ES{:}DI) \leftarrow ((DX))$$

Just as for the other byte string instructions, the value in DI is either incremented or decremented by 1 after the data transfer takes place:

$$(DI) \leftarrow (DI) \pm 1$$

In this way, it points to the next byte-wide storage location in memory to be accessed. Whether the value in DI is incremented or decremented depends on the setting of the DF flag. Note in Fig. 14–8 that INSW performs the same data transfer operation except that since the contents of a word-wide I/O port are input, the value in DI is incremented or decremented by 2.

The INSB instruction performs the operation we just described on one data element, not on an array of elements. However, this basic operation can be repeated to handle a block input operation. Block operations are done by inserting a repeat (REP) prefix in front of the string instruction. For example, the instruction

```
REPINSW
```

will cause the contents of the word-wide port pointed to by the I/O address in DX to be input and saved in the memory location at the address specified by ES and DI. Then the value in DX is incremented by 2 (assuming that the DF flag equals 0), the count in CX is decremented by 1, and the value in CX is tested to determine if it is 0. As long as the value in CX is not 0, the input operation is repeated. When CX equals 0, all elements of the array have been input, and the input string operation is complete. Remember that the count representing the number of times the string operation is to be repeated must be loaded into the CX register prior to executing the repeat input string instruction.

Figure 14–8 shows that OUTSB and OUTSW are the two forms of the *output string* instruction. These instructions operate in a way similar to the input string instructions; however, they perform an output operation. For instance, executing OUTSW causes the operation

$$(ES:SI) \rightarrow ((DX))$$

$$SI \pm 2 \rightarrow SI$$

That is, the word of data held at the memory location pointed to by the address specified by ES and SI is output to the word-wide port pointed to by the I/O address in DX. After the output data transfer is complete, the value in SI is either incremented or decremented by 2.

An example of an output string instruction that can be used to output an array of data is

```
REPOUTSB
```

When executed, this instruction causes the data elements of the array of byte-wide data pointed to by ES and SI to be output one after the other to the output port located at the I/O address specified by DX. Again, the count in CX defines the size of the array.

Check Array Index against Bounds Instruction—BOUND

The *check array index against bounds* (BOUND) instruction, as its name implies, can determine if the contents of a register, known as the *array index,* lie within a set of minimum and maximum values, called the *upper bound* and *lower bound.* This type of operation is important when accessing elements of an array of data in memory.

Figure 14–8 presents the format of the BOUND instruction. An example is the instruction

```
BOUND  SI, LIMITS
```

Note that the instruction contains two operands. The first operand represents the register whose word contents are to be tested to verify whether or not it lies within the boundaries. In our example, this is the source index register (SI). The second operand is the effective relative address of the first of two word-storage locations in memory that contain the values of the lower and upper boundaries. In the example, the word of data starting at address LIMITS is the value of the lower bound and that stored at address LIMITS + 2 is the value of the upper bound.

When this BOUND instruction is executed, the contents of SI are compared with both the value of the lower bound at LIMITS and the upper bound at LIMITS + 2. If it is found to be either less than the lower bound or more than the upper bound, an exception occurs and control is passed to a service routine through the vector for type 5. Otherwise, the next sequential instruction is executed.

▲ 14.5 INTERFACES OF THE 80286

Earlier we found that the 80286 microprocessor can be configured to work in either of two modes: the real-address mode and the protected-address mode. Let us now look at the signals produced at each of the 80286's interfaces when in these modes.

Figure 14–13 presents a block diagram of the 80286 microprocessor. Here we have grouped its signal lines into four interfaces: the *memory/IO interface, the interrupt interface, the DMA interface,* and *the processor extension interface.* Figure 14–14 lists each of the signals at the 80286's interfaces. Included in this table are a mnemonic, function, and type for each signal. For instance, the memory/IO control signal with the mnemonic $M/\overline{IO}$ stands for memory/IO select. This signal is an output produced by the 80286 that is used to signal external circuitry whether the current address available on the address bus is for memory or an I/O device. On the other hand, the signal INTR at the interrupt interface is the interrupt request input of the 80286. Using this input, external devices can signal the 80286 that they need to be serviced.

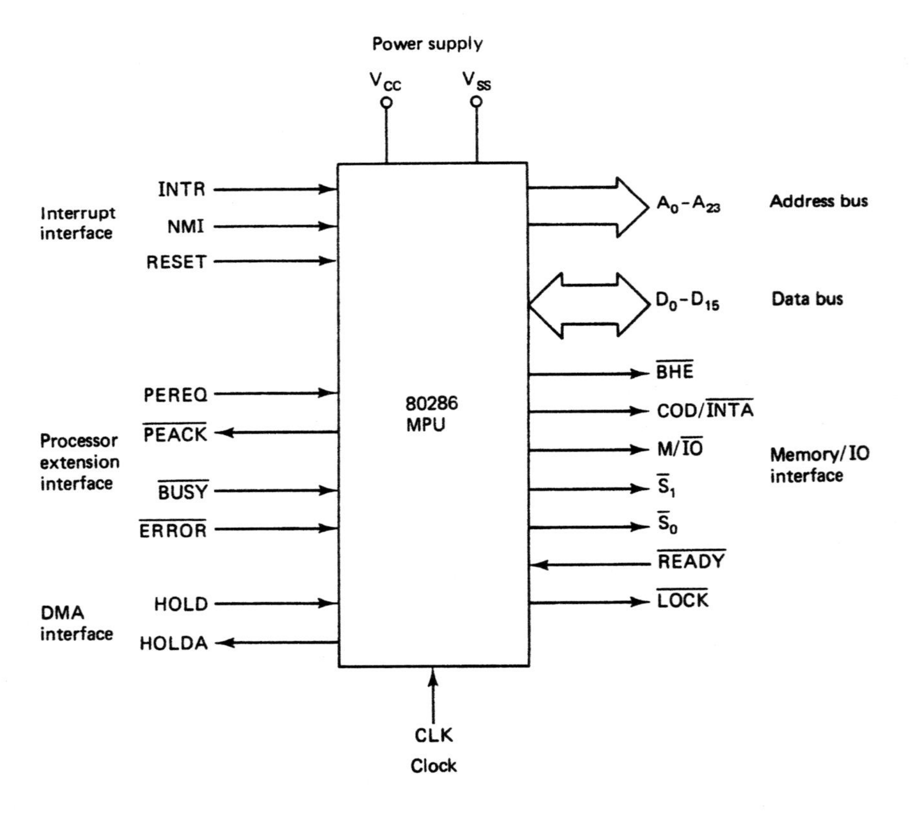

Figure 14–13 Block diagram of the 80286 MPU.

Mnemonic	Function	Type
CLK	System clock	I
$D_{15}-D_0$	Data bus	I/O
$A_{23}-A_0$	Address bus	O
$\overline{BHE}$	Bus high enable	O
$\overline{S}_1, \overline{S}_0$	Bus cycle status	O
$M/\overline{IO}$	Memory I/O select	O
$COD/\overline{INTA}$	Code/interrupt acknowledge	O
$\overline{LOCK}$	Bus lock	O
$\overline{READY}$	Bus ready	I
HOLD	Bus hold	I
HLDA	Hold acknowledge	O
INTR	Interrupt request	I
NMI	Nonmaskable interrupt request	I
PEREQ	Processor extension request	I
$\overline{PEACK}$	Processor extension acknowledge	O
$\overline{BUSY}$	Processor extension busy	I
$\overline{ERROR}$	Processor extension error	I
RESET	System reset	I
V_{SS}	System ground	I
V_{CC}	System power	I

Figure 14–14 Signal mnemonics, functions, and types.

Memory/IO Interface

In a microcomputer system, the address bus and data bus signal lines form the path over which the MPU talks with its memory and I/O subsystems. Unlike the older 8086 and 8088 microprocessors, the 80286 address/data bus is demultiplexed. Notice in Fig. 14–2 that the address bus and data bus lines are located at different pins of the IC.

From an external hardware point of view, there is only one difference between an 80286 configured for the real-address mode or one configured for the protected-address mode. This difference is the size of the address bus. When in real mode, just the lower 20 address lines, A_0 through A_{19}, are active; in the protected mode, all 24 lines, A_0 through A_{23} , are used. Of these, A_{23} is the most-significant address bit and A_0 the least significant bit. As Fig. 14–14 shows, the address lines are outputs. They are used to carry address information from the 80286 to memory and I/O ports. In real-address mode, the 20-bit address gives the 80286 the ability to address a 1Mbyte physical memory address space. On the other hand, in protected mode the extended 24-bit address results in a 16Mbyte physical memory address space; however, in protected mode, virtual addressing is provided through software, and this results in a 1Gbyte virtual memory address space.

In both the real and protected modes, the 80286 microcomputer has an independent I/O address space 64Kbytes in length. Therefore, only address lines A_0 through A_{15} are used when addressing I/O devices.

Since the 80286 is a 16-bit microprocessor, its data bus is the 16 data lines D_0 through D_{15}. Data line D_{15} is the MSB and D_0 the LSB. These lines are identified as bidirectional in Fig. 14–14 because they have the ability to carry data either in or out of the MPU. The kinds of data transferred over these lines are read/write data for memory, I/O data for I/O devices, and interrupt-type codes from an interrupt controller.

Control signals are required to support data transfers over the 80286's address/data bus. They are needed to signal when a valid address is on the address bus, in which direction data are to be transferred over the data bus, when valid write data are on the data bus, and when an external device can put read data on the data bus. The 80286 does not produce these signals directly. Instead, just like the 8086 in maximum mode, it outputs a 4-bit *bus status code* prior to the initiation of each bus cycle. This code identifies which type of bus cycle is to follow and must be decoded in external circuitry to produce the needed memory and I/O control signals.

Figure 14–15 shows that the bus status code is output on four of the 80286's signal lines: COD/$\overline{\text{INTA}}$, M/$\overline{\text{IO}}$, $\overline{S}_1$, and $\overline{S}_0$. The logic level of *code/interrupt acknowledge* (COD/$\overline{\text{INTA}}$) identifies whether the current bus cycle is for an instruction fetch or interrupt-acknowledge operation. From the table we find that COD/$\overline{\text{INTA}}$ is logic 1 whenever an instruction-fetch is taking place and logic 0 when an interrupt is being acknowledged. The next signal, *memory/IO* (M/$\overline{\text{IO}}$), tells whether a memory or I/O cycle is to take place over the bus. Logic 1 at this output signals a memory operation, and logic 0 signals an I/O operation. Looking at the table in Fig. 14–15 more closely, we find that if the code on these two lines is 00, an interrupt is to be acknowledged; if it is 01, a data memory read or write is taking place; if it is 10, an I/O operation is in progress; and, finally, if it is 11, instruction code is being fetched.

The last two signals in Fig. 14–15 are called *status lines* and they identify the specific type of memory or I/O operation that will occur during a bus cycle. For example, when COD/$\overline{\text{INTA}}$ M/$\overline{\text{IO}}$ is 01, a status code of 01 indicates that data are to be read from memory. On the other hand, 10 at $\overline{S}_1\overline{S}_0$ indicates that data are being written into memory.

COD/$\overline{\text{INTA}}$	M/$\overline{\text{IO}}$	$\overline{\text{S1}}$	$\overline{\text{S0}}$	Bus cycle initiated
0 (LOW)	0	0	0	Interrupt acknowledge
0	0	0	1	Will not occur
0	0	1	0	Will not occur
0	0	1	1	None; not a status cycle
0	1	0	0	IF A1 = 1 then halt; else shutdown
0	1	0	1	Memory data read
0	1	1	0	Memory data write
0	1	1	1	None; not a status cycle
1 (HIGH)	0	0	0	Will not occur
1	0	0	1	I/O read
1	0	1	0	I/O write
1	0	1	1	None; not a status cycle
1	1	0	0	Will not occur
1	1	0	1	Memory instruction read
1	1	1	0	Will not occur
1	1	1	1	None; not a status cycle

Figure 14–15 Bus status codes. (Reprinted by permission of Intel Corporation. Copyright/Intel Corp. 1987)

As another example, we find that COD/$\overline{\text{INTA}}$ M/$\overline{\text{IO}}$ equals 11 and $\overline{S}_1\overline{S}_0$ equals 01 whenever the 80286 reads instruction code from memory.

EXAMPLE 14.1

What type of bus cycle is taking place if the bus status code COD/$\overline{\text{INTA}}$ M/$\overline{\text{IO}}$ $\overline{S}_1\overline{S}_0$ is 1001?

Solution

Looking at the table in Fig. 14–15, we see that the status code 1001 identifies an I/O read (input) bus cycle.

A bus control signal directly produced by the 80286 is *bus high enable* ($\overline{\text{BHE}}$). The logic levels of $\overline{\text{BHE}}$ and address bit A_0 identify whether a word or byte of data will be transferred during the current bus cycle. Moreover, if a byte transfer is to take place, the 2-bit code output at $\overline{\text{BHE}}\ A_0$ tells whether the byte will be transferred over the upper or lower eight lines of the data bus. Figure 14–16 summarizes the state of these two signals for each type of data transfer. Notice that whenever a word of data is transferred over the bus, both $\overline{\text{BHE}}$ and A_0 are set to logic 0.

EXAMPLE 14.2

If a byte of data is being written to memory over data bus lines D_0 through D_7, what code is output at $\overline{\text{BHE}}\ A_0$?

Solution

Figure 14–16 shows that all transfers of byte data over the lower part of the address bus are accompanied by the code

$$\overline{\text{BHE}}\ A_0 = 10$$

$\overline{\text{BHE}}$	A0	Function
0	0	Word transfer
0	1	Byte transfer on upper half of data bus (D_{15}–D_8)
1	0	Byte transfer on lower half of data bus (D_7–D_0)
1	1	Will never occur

Figure 14–16 $\overline{\text{BHE}}$ and A_0 encoding. (Reprinted by permission of Intel Corporation. Copyright/Intel Corp. 1987)

The $\overline{\text{READY}}$ signal is used to insert wait states into the current bus cycle to extend it by a number of clock periods. This signal is an input of the 80286. Normally it is produced by the microcomputer's memory or I/O subsystem and supplied to the 80286 by way of the clock generator device. By signaling $\overline{\text{READY}}$, slow memory or I/O devices can tell the 80286 when they are ready to permit a data transfer to be completed.

The 80286 supplies one other control signal, the *bus lock* ($\overline{\text{LOCK}}$) output, which supports multiple-processor system architectures. In multiprocessor systems that employ shared resources, such as global memory, this signal can be employed to assure that the 80286 can have control of the system bus to use the shared resource—that is, by switching the $\overline{\text{LOCK}}$ output to logic 0, an MPU can lock up the shared resource for its exclusive use.

Interrupt Interface

Looking at Fig. 14–13, we find that the key interrupt interface signals are *interrupt request* (INTR), *nonmaskable interrupt* (NMI), and *system reset* (RESET). INTR is an input to the 80286 that can be used by external devices to signal that they need to be serviced. This input is sampled at the beginning of each processor cycle. Logic 1 on INTR represents an active interrupt request. After the 80286 recognizes an interrupt request, it initiates interrupt acknowledge bus cycles. In Fig. 14–15, the occurrence of an interrupt acknowledge bus cycle is signaled to external circuitry with the bus status code $\text{COD}/\overline{\text{INTA}}\ \text{M}/\overline{\text{IO}}\ \overline{S}_1\overline{S}_0$ equal to 0000. This status code is decoded in external circuitry to produce an interrupt acknowledge signal.

The INTR input is maskable—that is, its operation can be enabled or disabled with the interrupt enable flag (IF) within the 80286's flag register. On the other hand, the NMI input, as its name implies, is a nonmaskable interrupt input. On the 0-to-1 transition of NMI, a request for service is latched within the 80286. Independent of the setting of the IF flag, at completion of execution of the current instruction control is passed to the beginning of the nonmaskable interrupt service routine.

Finally, the RESET input is used to provide hardware reset to the 80286 microprocessor. Switching RESET to logic 1 initializes the internal registers of the 80286. When it is returned to logic 0, program control is automatically passed to a reset service routine.

DMA Interface

Now that we have examined the signals of the 80286's interrupt interface, let us turn to the *direct memory access* (DMA) interface. Figure 14–13 shows that the DMA interface is implemented with just two signals: *bus hold request* (HOLD) and *hold acknowledge* (HLDA). When an external device, such as a *DMA controller,* wants to take control of the system bus, it signals this fact to the 80286 by switching the HOLD input to logic 1. At completion of the current bus cycle, the 80286 enters the hold state. When in this state, the local bus signals are in the high-Z state. The 80286 signals external devices that it has given up control of the bus by switching its HLDA output to the 1 logic level. This completes the hold/hold acknowledge handshake.

Processor Extension Interface

A processor extension interface is provided on the 80286 microprocessor to permit it to easily interface with the *80287 numeric coprocessor.* Whenever the 80287 needs the 80286 to read or write operands from memory, it signals this fact to the 80286. The 80287 does this by switching the *processor extension operand request* (PEREQ) input of the 80286 to logic 1. The processor extension handshake is completed when the 80286 signals the 80287 that it can have access to the bus. It signals this to the 80287 by switching the *processor extension acknowledge* ($\overline{PEACK}$) output to logic 0.

The other two signals included in the external coprocessor interface are $\overline{BUSY}$ and $\overline{ERROR}$. *Processor extension busy* ($\overline{BUSY}$) is an input to the 80286. Whenever the 80287 is executing a numeric instruction, it signals this fact to the 80286 by switching the $\overline{BUSY}$ input to logic 0. In this way, the 80286 knows not to request the numeric coprocessor to perform another calculation until $\overline{BUSY}$ returns to 1. If an error occurs in a calculation performed by the numeric coprocessor, it signals this condition to the 80286 by switching the *bus extension error* ($\overline{ERROR}$) input to the 0 logic level.

▲ 14.6 82C288 BUS CONTROLLER

In the previous section we pointed out that the 80286 does not directly produce all the signals required to control the memory, I/O, and interrupt interfaces. Instead, it outputs a status code prior to the initiation of each bus cycle that identifies which type of bus cycle is to follow. In an 80286-based microcomputer, three of these status lines, $M/\overline{IO}$, $\overline{S}_1$, and $\overline{S}_0$, are supplied as inputs to an external *bus controller* device, the 82C288. Just as in a maximum-mode 8086 system, the bus controller decodes them to identify the type of MPU bus cycle and then generates appropriately timed command and control signals at its outputs.

Let us now look at how the 82C288 interfaces with the 80286 MPU. A block diagram for the 82C288 IC is shown in Fig. 14–17(a), and its pin layout is given in Fig. 14–17(b). Here we can find each of the signals that it accepts as inputs and produces as outputs. For instance, status inputs $M/\overline{IO}$, $\overline{S}_1$, and $\overline{S}_0$ are located at pins 18, 3, and 19, respectively.

The 82C288 connects to the 80286 as shown in Fig. 14–18. Note that the status outputs of the 80286 are simply connected directly to the corresponding input of the 82C288. Moreover, the clock (CLK) and ready ($\overline{READY}$) inputs of the 82C288 are wired to the respective inputs of the 80286 and both are driven by outputs of the 82C284 clock generator.

At the other side of the bus controller, we find the bus command and control signal outputs. The command outputs are *memory read command* ($\overline{MRDC}$), *memory write command* ($\overline{MWTC}$), *I/O read command* ($\overline{IORC}$), *I/O write command* ($\overline{IOWC}$), and *interrupt acknowledge command* ($\overline{INTA}$). Figure 14–19 shows that only one of these command outputs becomes active for a given bus status code. For instance, when the 80286 outputs the code $M/\overline{IO}\,\overline{S}_1\overline{S}_0$ equal to 001, it indicates that an I/O read cycle is to be performed. In turn, the 82C288 makes its $\overline{IORC}$ output switch to logic 0. On the other hand, if the code 111 is output by the 80286, it is signaling the bus controller that no bus activity is to take place.

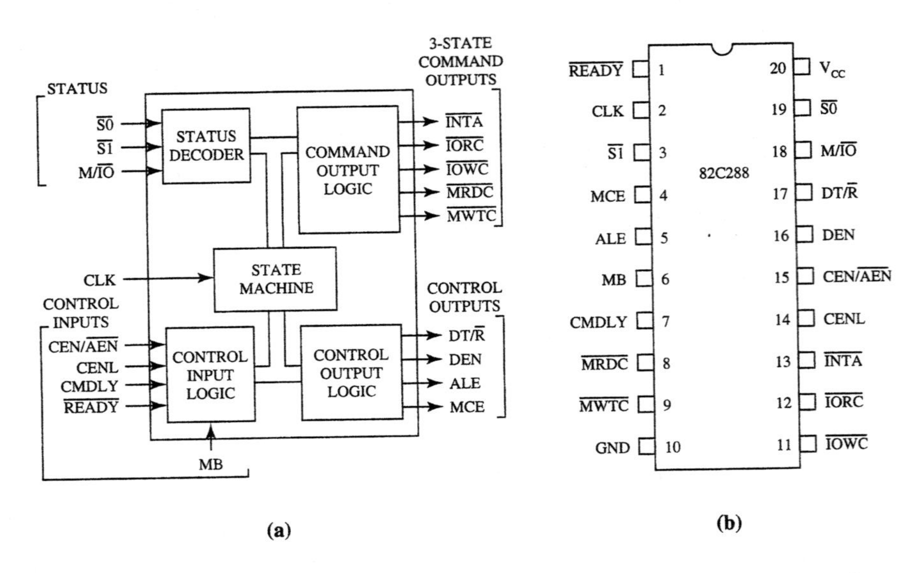

Figure 14–17 (a) Block diagram of the 82C288. (Reprinted by permission of Intel Corporation. Copyright/Intel Corp. 1987) (b) Pin layout of the 82C288. (Reprinted by permission of Intel Corporation. Copyright/Intel Corp. 1987)

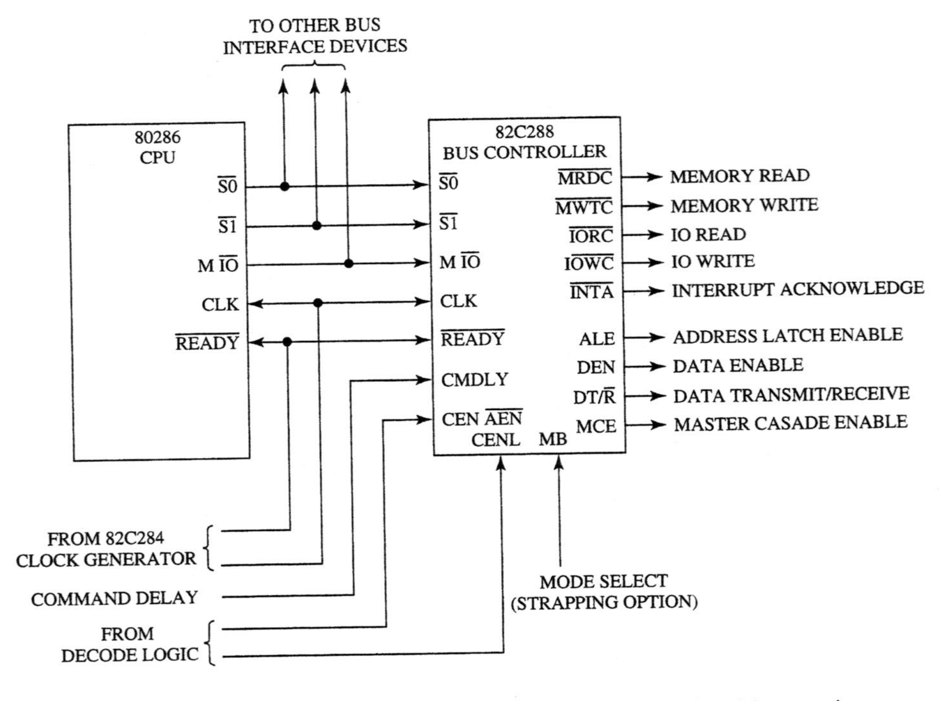

Figure 14–18 Decoding bus status with the 82C288. (Reprinted by permission of Intel Corporation. Copyright/Intel Corp. 1987)

Type of Bus Cycle	$M/\overline{IO}$	$\overline{S1}$	$\overline{S0}$	Command Activated	$DT/\overline{R}$ State	ALE, DEN Issued?	MCE Issued?
Interrupt Acknowledge	0	0	0	$\overline{INTA}$	LOW	YES	YES
I/O Read	0	0	1	$\overline{IORC}$	LOW	YES	NO
I/O Write	0	1	0	$\overline{IOWC}$	HIGH	YES	NO
None; Idle	0	1	1	None	HIGH	NO	NO
Halt/Shutdown	1	0	0	None	HIGH	NO	NO
Memory Read	1	0	1	$\overline{MRDC}$	LOW	YES	NO
Memory Write	1	1	0	$\overline{MWTC}$	HIGH	YES	NO
None; Idle	1	1	1	None	HIGH	NO	NO

Figure 14–19 Control signals for each bus cycle. (Reprinted by permission of Intel Corporation. Copyright/Intel Corp. 1987)

EXAMPLE 14.3

If the $\overline{MRDC}$ output of the 82C288 has just switched to logic 0, what bus status code was output by the 80286?

Solution

Looking at Fig. 14–19, we find that the status code input to the 82C288 that makes $\overline{MRDC}$ active is

$$M/\overline{IO}\ \overline{S}_1\overline{S}_0 = 101$$

The 82C288 also produces the control signals *data transmit/receive* ($DT/\overline{R}$), *address latch enable* (ALE), *data enable* (DEN), and *master cascade enable* (MCE). Figure 14.19 shows the state of these control signals during each bus cycle. For example, during an I/O read cycle $DT/\overline{R}$ is set to logic 0, and both ALE and DEN are issued.

These command and control signals are needed to support memory and I/O data transfers over the address and data buses. Earlier we pointed out that they need to identify when the bus is carrying a valid address, in which direction data are to be transferred over the bus, when valid write data are on the bus, and when to put read data on the bus. For example, the ALE line signals external circuitry when a valid address word is on the bus. This address can be latched in external circuitry on the 0-to-1 edge of the pulse at ALE.

The direction in which data are to be transferred over the bus is signaled to external circuitry with the $DT/\overline{R}$ output. When this line is at logic 1 during the data transfer part of a bus cycle, the bus is in the transmit mode. Therefore, data are either written into memory or output to an I/O device. On the other hand, logic 0 at $DT/\overline{R}$ signals that the bus is in the receive mode. This corresponds to reading data from memory or input of data from an input device.

The signals $\overline{MRDC}$ and $\overline{MWTC}$ indicate a memory read bus cycle and a memory write bus cycle, respectively. The 82C288 switches $\overline{MWTC}$ to logic 0 to signal external devices that valid write data are on the bus. On the other hand, logic 0 on $\overline{MRDC}$ indicates that the 80286 is performing a read of data off the bus. During read operations, one other control signal is supplied. This is DEN, and it is the signal used to enable the data path buffers. Input and output bus cycles are performed in the same way; however, in this case $\overline{IORC}$ and $\overline{IOWC}$ are activated by the 82C288 instead of $\overline{MRDC}$ and $\overline{MWTC}$.

The 82C288 also has several control inputs. In Fig. 14–18 we find a *Multibus mode select* (MB) input. When the system is not using *Multibus* as the system bus, this input is connected to ground. Figure 14–17 includes three other signal lines that were not shown in the 80286/82C288 interface circuit in Fig. 14–18. For instance, the *command enable/address enable* (CEN/$\overline{AEN}$) input can be used to enable or disable the command and DEN outputs of the 82C288. To enable the 82C288 for operation, this input must be at the 1 logic level. Designs that use either multiple 82C288s or require the command output signals to be delayed are implemented by the last two inputs, CENL and CMDLY, respectively.

▲ 14.7 SYSTEM CLOCK

The time base for synchronization of the internal and external operations of the 80286 microprocessor is provided by the *clock* (CLK) input signal. The 80286 is available with three different clock speeds. The standard 80286 MPU operates at 8 MHz, and its two faster versions, the 80286-10 and 80286-12, operate at 10 MHz and 12.5 MHz, respectively. CLK is generated externally by the 82C284 clock generator/driver IC. Figure 14–20(a) is a block diagram of this device, and Fig. 14–20(b) shows its pin layout.

The normal way in which the clock chip is used is by connecting a crystal with twice the microprocessor clock frequency between the X_1 and X_2 inputs, as shown in Fig. 14–21. For instance, to run the 80286 at 8 MHz, a 16-MHz crystal is needed. Note that loading capacitors C_1 and C_2 are required from X_1 and X_2, respectively, to ground. Typical capacitance values (device capacitance plus stray capacitance) for these capacitors are 25 pF and 15 pF, respectively. The fundamental crystal frequency produced by the oscillator within the 82C284 is divided by 2 to give the CLK output. As Fig. 14–21 shows, the CLK output of the 82C284 is applied directly to the CLK input of the 80286.

The waveform of the CLK input of the 80286 is given in Fig. 14–22. Here we see that the signal is specified at MOS-compatible voltage levels and not TTL levels. Its minimum and maximum low logic levels are $V_{Lmin} = -0.5$ V and $V_{Lmax} = 0.6$ V, respectively. Moreover, the minimum and maximum high logic levels are $V_{Hmin} = 3.8$ V and $V_{Hmax} = -5.5$ V, respectively. The period of the 8-MHz signal is a minimum of 62.5 ns, and the maximum rise and fall times of its edges are equal to 10 ns.

Figure 14–20(a) illustrates another clock output on the 82C284, the *processor clock* (PCLK) signal, which is provided to drive peripheral ICs. The clock produced at PCLK is half the frequency of CLK—that is, if the 80286 is to operate at 8 MHz, PCLK will be 4 MHz. Figure 14–23 illustrates this relationship. Also, it is at TTL-compatible voltage levels rather than at CMOS levels.

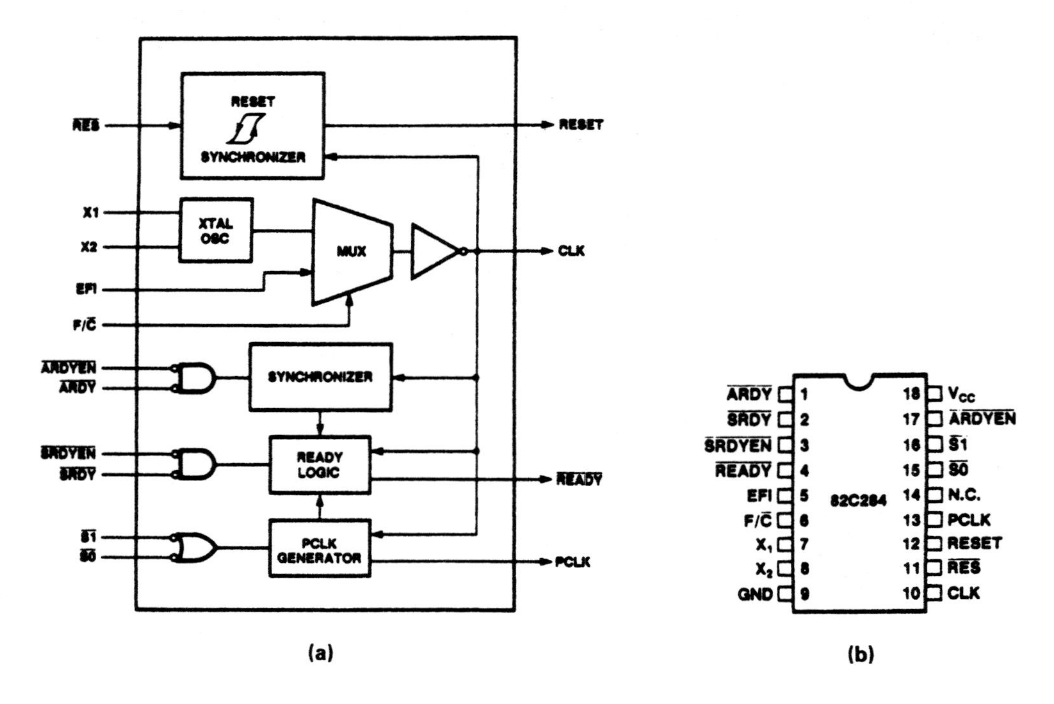

Figure 14–20 (a) Block diagram of the 82C284 clock generator. (Reprinted by permission of Intel Corporation. Copyright/Intel Corp. 1987) (b) Pin layout. (Reprinted by permission of Intel Corporation. Copyright/Intel Corp. 1987)

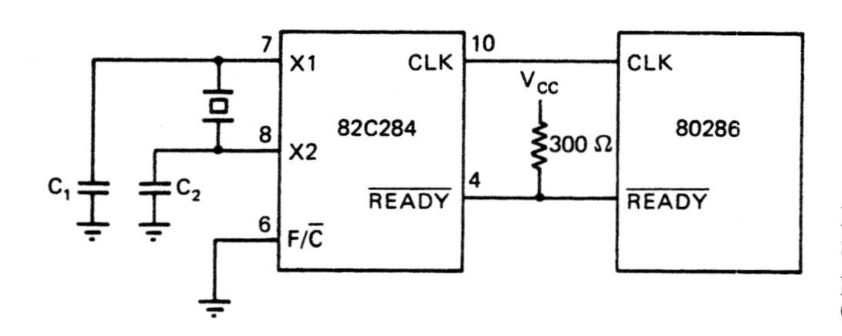

Figure 14–21 Connecting the 82C284 to the 80286. (Reprinted by permission of Intel Corporation. Copyright/Intel Corp. 1987)

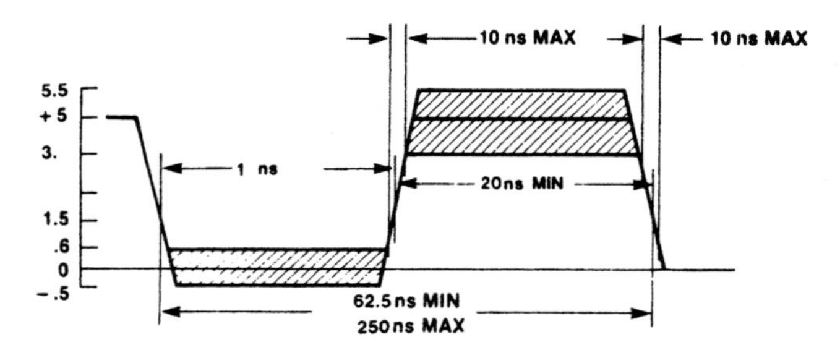

Figure 14–22 CLK voltage and timing characteristics. (Reprinted by permission of Intel Corporation. Copyright/Intel Corp. 1987)

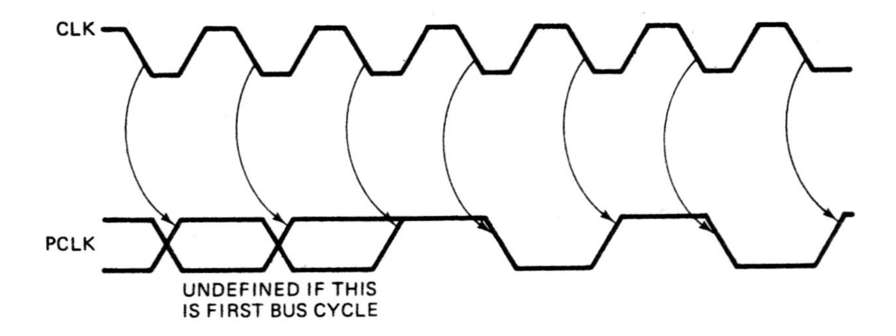

Figure 14–23 Timing relationship between CLK and PCLK. (Reprinted by permission of Intel Corporation. Copyright/Intel Corp. 1987)

The 82C284 can also be driven from an external clock source. In this case, the external clock signal is applied to the *external frequency input* (EFI). Input F/$\overline{C}$ (*frequency/crystal select*) is provided for clock source selection. When strapped to the 0 logic level, as in Fig. 14–21, the crystal between X_1 and X_2 is used. On the other hand, applying logic 1 to F/$\overline{C}$ selects EFI as the source of the clock.

▲ 14.8 BUS CYCLE AND BUS STATES

A *bus cycle* is the activity performed whenever a microprocessor accesses information in program memory, data memory, or an I/O device. Figure 14–24 shows a traditional microprocessor bus cycle. It represents a sequence of events that starts with an address, denoted as *n*, being output on the address bus during clock state T_1. Later in the bus cycle, while the address is still available on the address bus, a read or write data transfer takes place over the data bus. Note that the data transfer for address *n* is shown to occur during clock state T_3. The interval denoted as *address access time* in Fig. 14–24 represents the amount of time that the address is stable prior to the actual read or write of data. During the bus cycle, a series of control signals is also produced by the MPU to control the direction and timing of the bus operation.

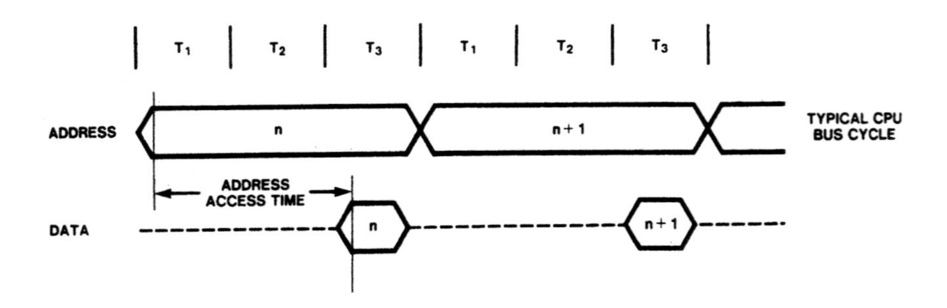

Figure 14–24 Traditional bus cycle. (Reprinted by permission of Intel Corporation. Copyright/Intel Corp. 1987)

Before looking at the bus cycle of the 80286, let us examine the relationship between the timing of the 80286's CLK input and bus cycle states. In the previous section we pointed out that the processor clock (PCLK) signal is at half the frequency of CLK. Therefore, as Fig. 14–25 shows, one processor clock cycle corresponds to two CLK cycles. These CLK cycles are labeled as *phase 1* (ϕ_1) and *phase 2* (ϕ_2). In an 8-MHz 80286 microcomputer system, each system clock cycle has a duration of 125 ns. Therefore, a processor clock cycle is 250 ns long.

Figure 14–25 shows that the two phases ($\phi_1 + \phi_2$) of a processor cycle are identified as one bus *T state.* Figure 14–26 (a) shows a typical 80286 bus cycle. Notice that here the clock T states are labeled T_s and T_c. T_s, the first state, stands for the *send-status state.* During this part of the bus cycle, the 80286 outputs a bus status code to the 82C288 bus controller, and in the case of a write cycle, write data are also output on the data bus. The second state, T_c, is called the *perform-command state.* It is during this part of the bus cycle that external devices are to accept write data from the bus, or in the case of a read cycle, put data on the bus. Also, the address for the next bus cycle is output on the address bus during T_c. Since each bus cycle has a minimum of two T states, the minimum bus cycle duration in an 8-MHz system is 250 ns.

The bus cycle of the 80286 employs a technique known as *pipelining.* By pipelining we mean that addressing for the next bus cycle is overlapped with the data transfer of the prior bus cycle. In Fig. 14–26(a), address *n* becomes valid in the T_c state of the prior bus cycle and then the data transfer for address *n* takes place in the next T_c state. Moreover, note that at the same time that data transfer *n* occurs, address $n + 1$ is output on the address bus. In this way, we see that the 80286 begins addressing the next storage location to be accessed while it is still performing the read or write of data for the previously addressed storage location. Because of the address/data pipelining, the memory or I/O subsystem actually has five CLK periods to complete the data transfer, even though the effective duration of every bus cycle is just four CLK periods.

Figure 14–26(a) shows that the effective address access time equals the duration of a complete bus cycle. This leads us to the benefit of the pipelined bus operation of the 80286 not found in the traditional bus operation shown in Fig. 14–24. That is, for a fixed address access time (equal-speed memory design), the 80286 will have a shorter duration bus cycle than a processor that uses a nonpipelined bus cycle. This results in improved bus performance for the 80286. Another way of looking at this is to say that when using equal-speed memory designs, a pipelined microprocessor can be operated at a higher

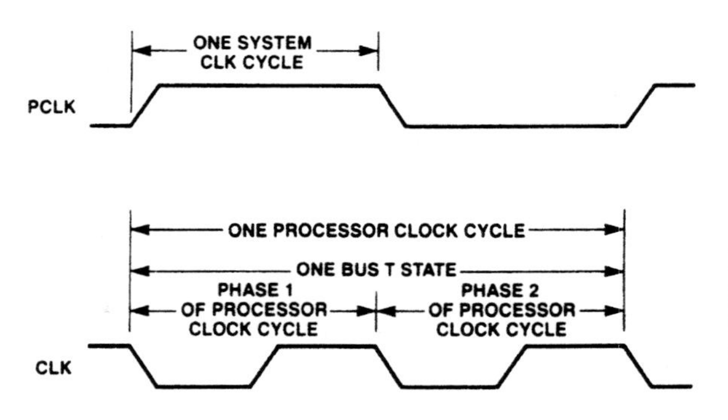

Figure 14–25 Send-status and perform-command bus states. (Reprinted by permission of Intel Corporation. Copyright/Intel Corp. 1987)

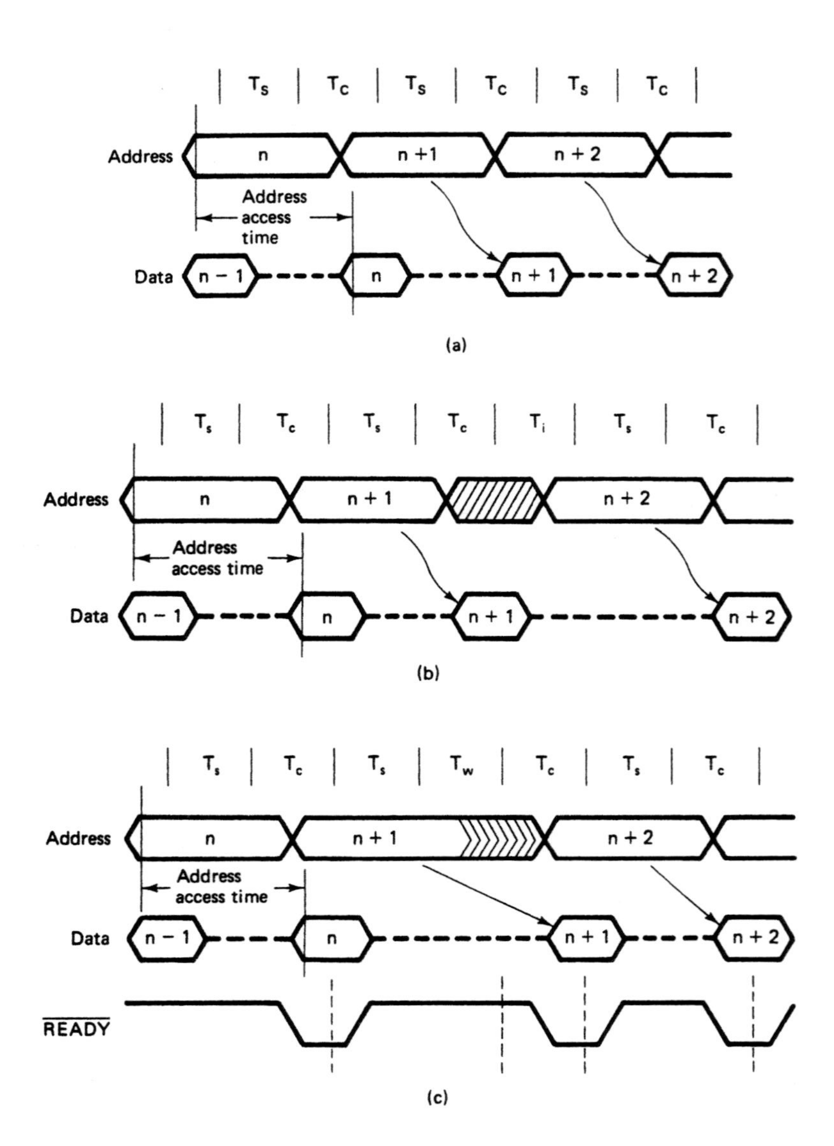

Figure 14–26 (a) Pipelined 80286 bus cycle. (Reprinted by permission of Intel Corporation. Copyright/Intel Corp. 1987) (b) Bus cycle with idle states. (Reprinted by permission of Intel Corporation. Copyright/Intel Corp. 1987) (c) Bus cycle with wait states. (Reprinted by permission of Intel Corp. Copyright/Intel Corp. 1987)

clock rate than a processor that executes a nonpipelined bus cycle. Once again, the result is higher system performance. A pipelined memory subsystem design is much more complex and requires more hardware than that of a nonpipelined memory subsystem.

In Fig. 14–26(a) we see that at completion of the bus cycle for address *n*, another bus cycle is initiated immediately for address $n + 1$. Sometimes another bus cycle will not be initiated immediately. For instance, if the 80286's prefetch queue is already full and the instruction that is currently being executed does not need to access operands in memory, no bus activity will take place. In this case, the bus goes into a mode of operation known as an *idle state* and no bus activity occurs. Figure 14–26(b) shows a sequence of bus activity in which several idle states exist between the bus cycles for addresses n and $n + 1$. The duration of a single idle state is equal to one processor cycle (two clock cycles).

Wait states can also be inserted into the 80286's bus cycle. This is done in response to a request by an event in external hardware instead of an internal event such as a full queue. In fact the $\overline{READY}$ input of the 80286 is provided specifically for this purpose. This input is sampled by the processor in the later part of the T_c state of every bus cycle to determine if the data transfer should be completed. Figure 14–26(c) shows that logic 1 at this input indicates that the current bus cycle should not be completed. As long as $\overline{READY}$ is held at the 1 level, the read or write data transfer does not take place and the current T_s state becomes a wait state (T_w) to extend the bus cycle. The bus cycle is not completed until external hardware returns $\overline{READY}$ back to logic 0. This ability to extend the duration of a bus cycle permits the use of slower memory or I/O devices in the microcomputer system.

▲ 14.9 MEMORY INTERFACE

In the preceding sections we studied the 80286 microprocessor, its internal architecture, the extended instruction set, the 82C288 bus controller, the 82C284 clock generator, and read/write bus cycles. Here we continue by introducing its memory interface, hardware organization of the address space, data transfers through the memory interface, and the read and write bus cycle timing.

Memory Interface Circuit

A memory interface circuit diagram for a real-mode 80286-based microcomputer system is shown in Fig. 14–27. Here we find that the interface includes the 82C288 bus controller, an address decoder and address latches, data bus transceiver/buffers, and bank select logic. Status signals, $M/\overline{IO}$, $\overline{S}_1$, and $\overline{S}_0$, which are output by the 80286, are supplied directly to the 82C288 bus controller. Here they are decoded to produce the command and control signals needed to control data transfers over the bus. Figure 14–28 highlights the two status codes that relate to the memory interface. The code $M/\overline{IO}\,\overline{S}_1\overline{S}_0 = 101$ indicates that a memory read bus cycle is in progress. This code makes the $\overline{MRDC}$ output switch to logic 0. Note in Fig. 14–27 that $\overline{MRDC}$ is applied to the output enable ($\overline{OE}$) input of the memory subsystem.

Next let us look at how the address bus is decoded, latched, and buffered. Address lines A_0 through A_{17} and $\overline{BHE}$ are sent directly to inputs of the address latches. On the other hand, bits A_{18} and A_{19} are first decoded and then passed on to the latch. Finally, the pulse at ALE is used to strobe address bits A_0 through A_{17}, $\overline{BHE}$, and the output of the decoder, chip enable ($\overline{CE}$), into the latches. The address latches buffer the address lines, $\overline{BHE}$, and $\overline{CE}$ before passing them on to the memory subsystem.

This part of the memory interface demonstrates one of the benefits of the 80286's pipelined bus. Remember that the 80286 actually outputs the address in the T_c state of the prior bus cycle. Therefore, by putting the address decoder before the address latches

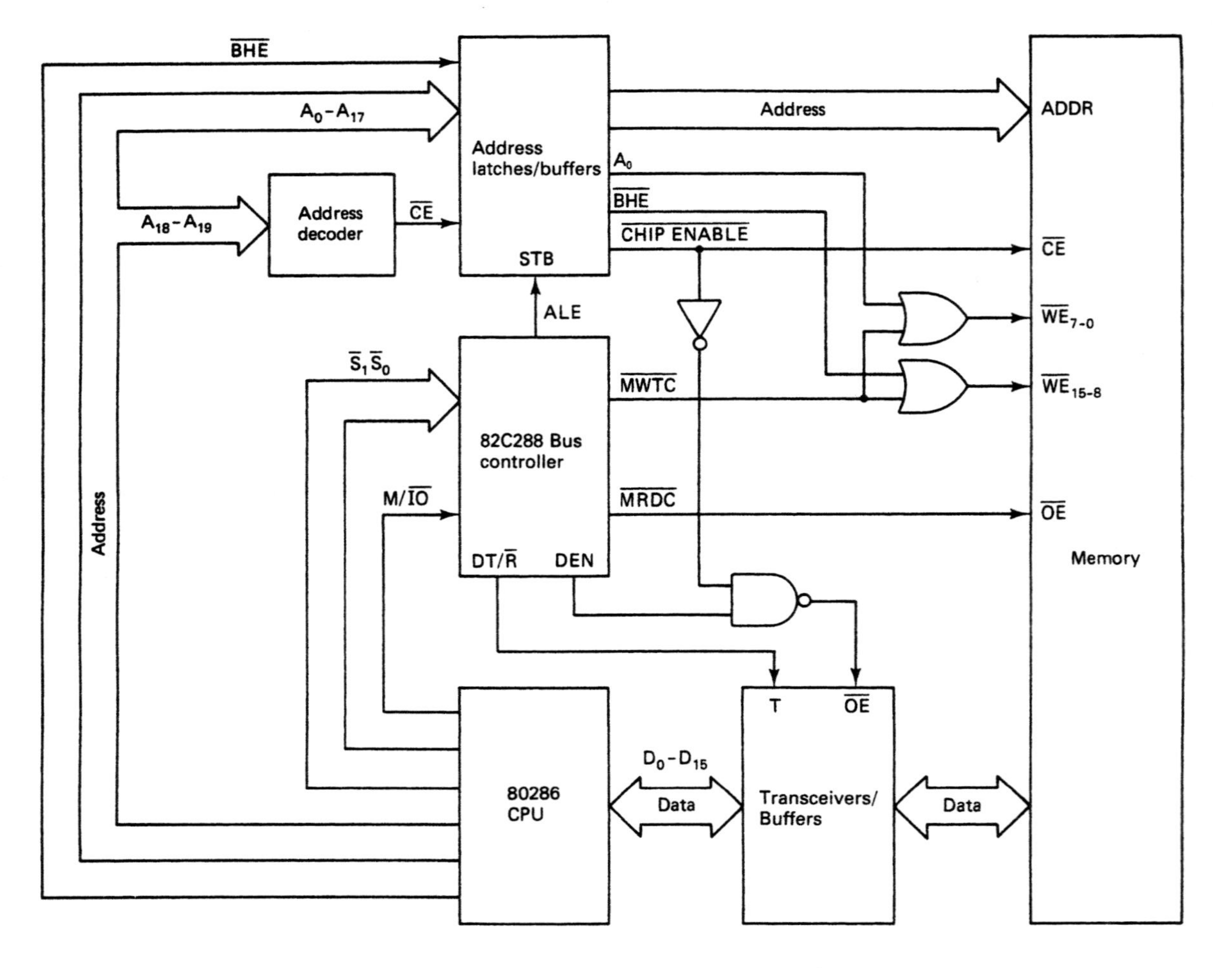

Figure 14–27 80286 system memory interface.

M/$\overline{IO}$	$\overline{S}_1$	$\overline{S}_0$	Type of bus cycle
0	0	0	Interrupt acknowledge
0	0	1	I/O Read
0	1	0	I/O Write
0	1	1	None; idle
1	0	0	Halt or shutdown
1	0	1	Memory read
1	1	0	Memory write
1	1	1	None; idle

Figure 14–28 Memory bus cycle status codes. (Reprinted by permission of Intel Corporation. Copyright/Intel Corp. 1987)

instead of after, the code at address lines A_{18} and A_{19} can be fully decoded and stable prior to the occurrence of ALE in the T_s state of the current bus cycle—that is, since the address can be decoded prior to ALE, the decode propagation delay is transparent and does not decrease the overall access time of the memory subsystem.

The circuit in Fig. 14–27 allows only word reads from the memory, but either byte or word writes to the memory.

During read bus cycles, the $\overline{\text{MRDC}}$ output of the bus controller enables the data at the outputs of the memory subsystem onto data bus lines D_0 through D_{15}. On the other hand, during write operations to memory, control logic must determine whether data are written into the lower memory bank, the upper memory bank, or both banks. This depends on whether an even-byte, odd-byte, or word-data transfer is taking place over the bus. The bank-select logic performs this function. Note in Fig. 14–27 that $\overline{\text{MWTC}}$ is gated with address bit A_0 to produce write enable $\overline{\text{WE}_{7\text{-}0}}$ for the low memory bank. For $\overline{\text{WE}_{7\text{-}0}}$ to be at its active logic 0 level, both A_0 and $\overline{\text{MWTC}}$ must be logic 0. This happens only when either an even-addressed byte of data or a word of data is written to memory. Furthermore, $\overline{\text{MWTC}}$ is gated with $\overline{\text{BHE}}$ to produce the high memory bank write enable signal $\overline{\text{WE}_{15\text{-}8}}$. For this signal to switch to its active 0 logic level, $\overline{\text{BHE}}$ and $\overline{\text{MWTC}}$ must both be logic 0. This occurs when either an odd-addressed byte of data or a word of data is written to memory.

The data bus transceiver/buffers control the direction of data transfers between the MPU and memory subsystem. Switching the output enable ($\overline{\text{OE}}$) inputs of the transceivers to logic 0 enables them. Signals data bus enable (DEN) and chip enable ($\overline{\text{CE}}$) are combined with an inverter and NAND gate to produce $\overline{\text{OE}}$. When DEN is logic 1 and $\overline{\text{CE}}$ is logic 0, the output of the NAND gate switches to logic 0 and the transceivers are enabled. This happens during all read and write bus cycles to memory.

The bus controller controls the direction in which data are passed through the transceiver/buffers. The bus controller sets its data transmit/receive (DT/$\overline{\text{R}}$) output to logic 0 during all read cycles and logic 1 during all write cycles. This signal is applied to the DIR input of the transceiver to select the direction in which data are transferred.

Hardware Organization of the Memory Address Space

From a hardware point of view, the memory address space of the 80286 is implemented exactly the same as that of the 8086—that is, it is organized as two independent banks called the *low (even) bank* and the *high (odd) bank.* Data bytes associated with an even address (000000_{16}, 000002_{16}, etc.) reside in the low bank, and those with odd addresses (000001_{16}, 000003_{16}, etc.) reside in the high bank. When in the protected mode, each of these banks can be as large as 8Mbytes.

Looking at Fig. 14–29, we find that the 80286's real-mode physical address space is partitioned into a 512Kbyte low bank and a 512Kbyte high bank. In the real mode, address bits A_1 through A_{19} select the storage location to be accessed. Therefore, they are applied to both banks in parallel. Just as for the 8086, A_0 and bank high enable ($\overline{\text{BHE}}$) are used as bank-select signals. Logic 0 at A_0 identifies an even-addressed byte of data and causes the low bank of memory to be enabled. On the other hand, $\overline{\text{BHE}}$ equal to 0 enables the high bank for access of an odd-addressed byte of data. Each of the memory banks supplies half

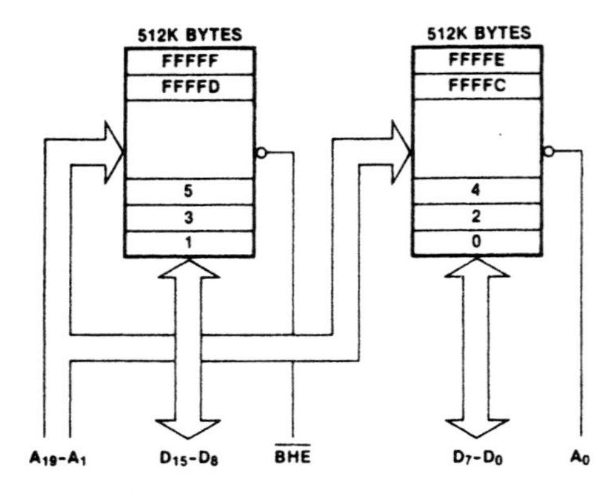

Figure 14–29 High and low memory banks. (Reprinted by permission of Intel Corporation. Copyright/Intel Corp. 1987)

the 80286's 16-bit data bus. Note that the lower bank transfers bytes of data over data lines D_0 through D_7; data transfers for the high bank use D_8 through D_{15}. Figures 14–30(a) through (d) show how an even-addressed byte of data, odd-addressed byte of data, even-addressed word of data, and odd-addressed word of data, respectively, are accessed. These accesses are the same as those in the 8086-based system described in an earlier chapter.

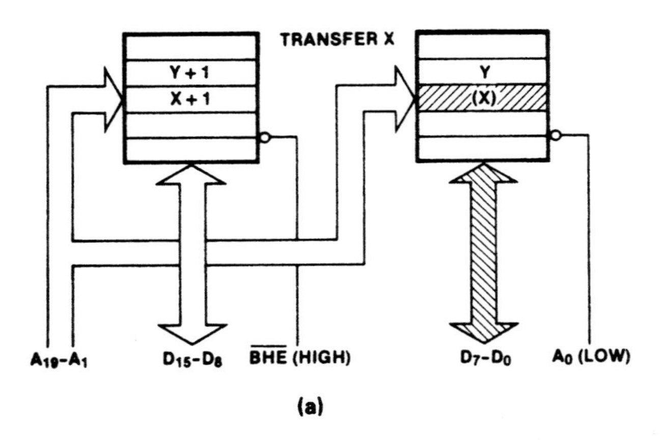

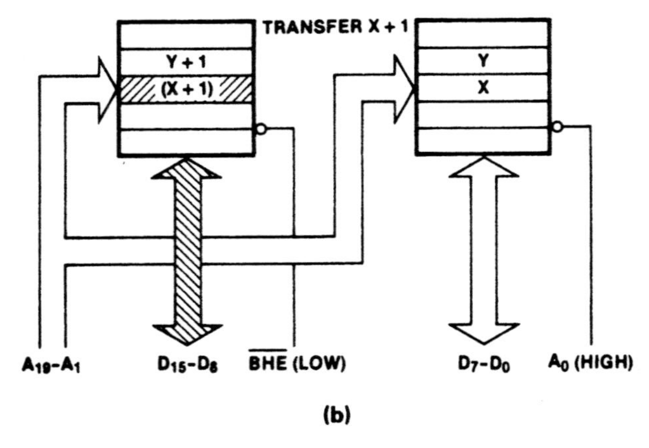

Figure 14–30 (a) Even-address byte transfer. (Reprinted by permission of Intel Corporation. Copyright/Intel Corp. 1987) (b) Odd-address byte transfer. (Reprinted by permission of Intel Corporation. Copyright/Intel Corp. 1987) (c) Even-address word transfer. (Reprinted by permission of Intel Corporation. Copyright/Intel Corp. 1987) (d) Odd-address word transfer. (Reprinted by permission of Intel Corporation. Copyright/Intel Corp. 1987)

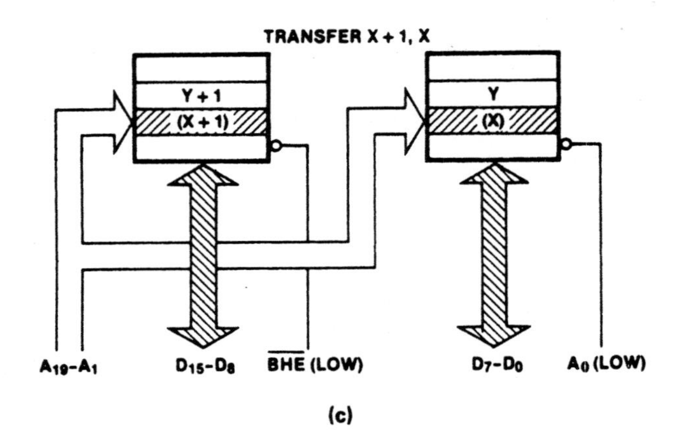

(c)

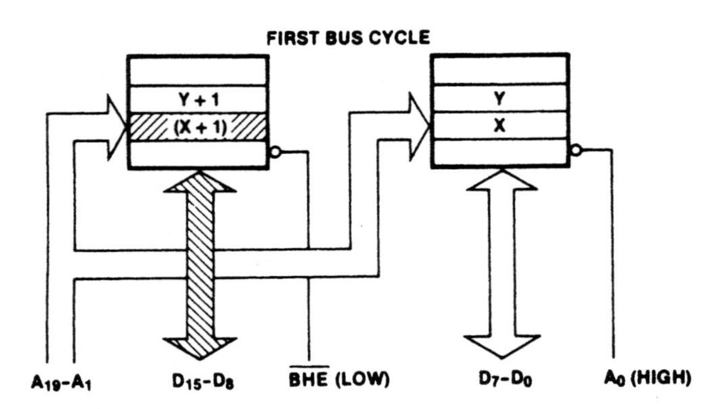

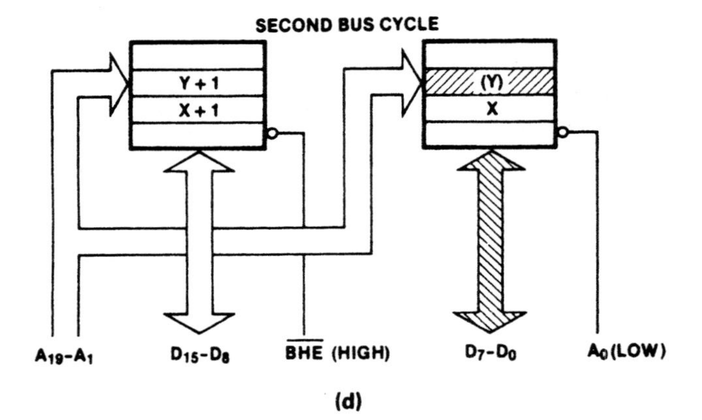

(d)

Figure 14–30 (continued)

Read Cycle Timing

Figure 14–31 shows the memory interface signal sequence that occurs when the 80286 reads data from memory. Since the pipelined bus of the 80286 overlaps bus cycles, we have shown parts of three cycles in the timing diagram: the previous cycle, the current read cycle, and the next cycle. Let us now trace through the events that take place as data or instructions are read from memory.

The occurrence of all signals in the read bus cycle timing diagram is illustrated relative to the two timing states, T_s (send-status state) and T_c (perform-command state), of the 80286's bus cycle. The read cycle begins at phase 2 (ϕ_2) in the T_c state of the previous bus cycle. During this period, the 80286 outputs the 24-bit address of the memory location to be accessed on address bus lines A_0 through A_{23}. Also we see in Fig. 14–31 that at the start of ϕ_2, signal M/$\overline{\text{IO}}$ is set to logic 1 to indicate to the circuitry in the memory interface that a memory bus cycle is in progress, and COD/$\overline{\text{INTA}}$ is set to logic 1 if the bus cycle is being performed to fetch code data. In our description of the memory

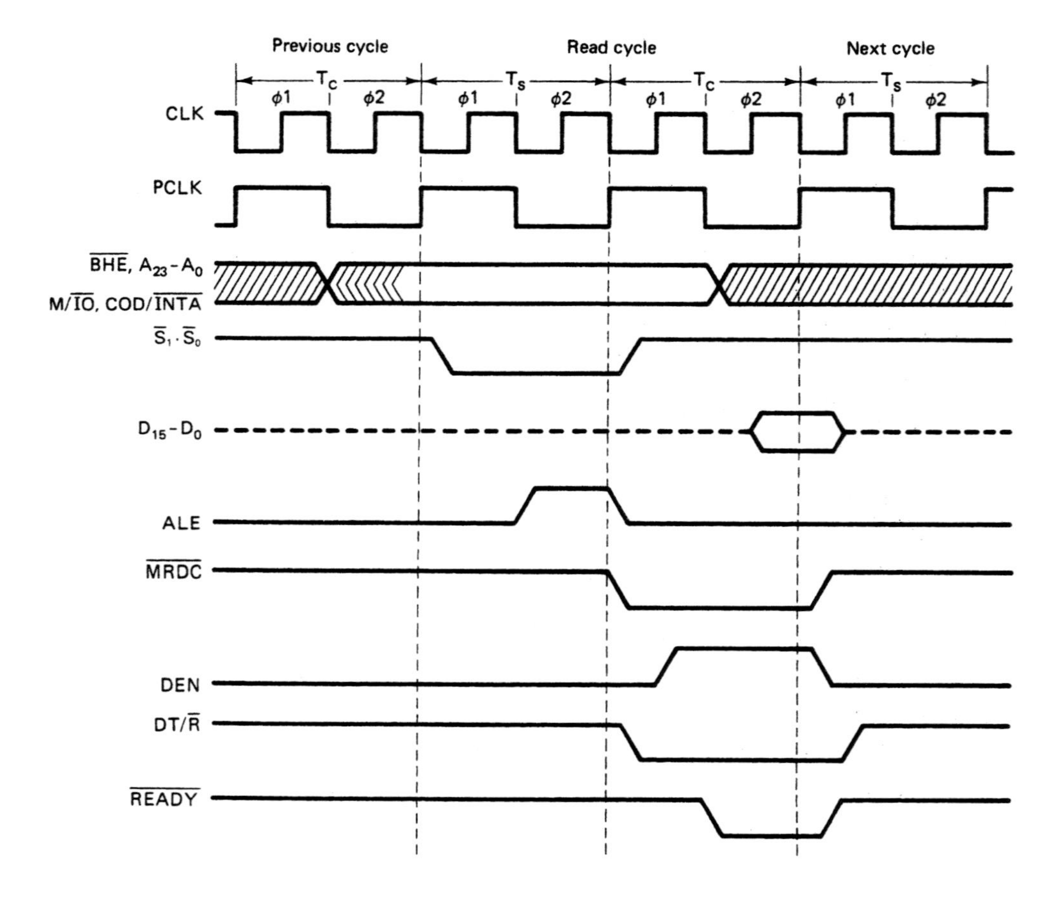

Figure 14–31 Memory read bus cycle timing. (Reprinted by permission of Intel Corporation. Copyright/Intel Corp. 1987)

interface circuit in Fig. 14–27, we pointed out that these signals are stable during the T_c state of the previous bus cycle; therefore, the address decode logic circuitry can begin to decode them immediately. This will assure that a maximum amount of time will be available to access the memory subsystem. Note that the status signals, M/$\overline{\text{IO}}$ and COD/$\overline{\text{INTA}}$, are maintained valid through ϕ_1 of state T_c of the read cycle. In ϕ_2 of T_c they are replaced with new values in preparation for the next bus cycle.

At the beginning of ϕ_1 of the read cycle, the 80286 outputs status code $\overline{S}_1\overline{S}_0$ equal to 01 on the status bus. Remember that this code identifies that a read bus transfer is to take place. This status information is maintained through the T_s state. At the falling edge of CLK in the middle of T_s, the 82C288 bus controller samples the status lines and the read cycle bus control sequence is started. At the same time the status information is output, $\overline{\text{BHE}}$ is also switched to logic 0 or 1, depending on whether or not the high memory bank is to be enabled during the read cycle.

With ϕ_2 of T_s, the bus controller takes over the timing control for the bus. It begins by producing a pulse to logic 1 at ALE during ϕ_2. The leading edge of this pulse can be used to latch the address and $\overline{\text{BHE}}$ into external circuitry.

At the start of the T_c state of the read cycle, the bus controller switches both DT/$\overline{\text{R}}$ and $\overline{\text{MRDC}}$ to logic 0. Logic 0 at DT/$\overline{\text{R}}$ is used to set the direction of the data bus transceivers so that they pass data from the memory subsystem to the 80286. On the other hand, the logic 0 at $\overline{\text{MRDC}}$ enables the output buffers of the memory subsystem so that they output data to the inputs of the data bus transceivers. Later in ϕ_1, DEN is set to 1. This active logic level enables the data bus transceivers and lets the data available at its inputs pass through to the outputs attached to the data bus of the 80286.

The 80286 and 82C288 sample the logic level of their $\overline{\text{READY}}$ inputs at the end of ϕ_2 in the T_c state of the read cycle. Assuming that $\overline{\text{READY}}$ is at its active 0 logic level, the MPU reads the data off the bus. The read cycle is now completed as the bus controller returns $\overline{\text{MRDC}}$, DEN, and DT/$\overline{\text{R}}$ to their inactive logic levels.

Write Cycle Timing

The write bus cycle timing diagram, shown in Fig. 14–32, is similar to that given for a read cycle in Fig. 14–31. Looking at the write cycle waveforms, we find that the address, M/$\overline{\text{IO}}$, and the COD/$\overline{\text{INTA}}$ signals are output at the beginning of ϕ_2 of the T_c state of the previous bus cycle. All these signals can be latched in external circuitry with the ALE pulse during ϕ_2 of the T_s state in the write cycle. Up to this point, the bus cycle is identical to that for the read cycle.

The $\overline{\text{BHE}}$ signal and status code $\overline{S}_1\overline{S}_0$ are output at the beginning of the T_s state of the write cycle. This also happened in the read cycle, but this time the status code is 10 instead of 01. This code is sampled by the 82C288 later in the T_s state to determine if it is a memory write cycle.

Let us now look at the control signal sequence produced by the bus controller for the write bus cycle. Note that the 80286 outputs the data to be written to memory on the data bus at the beginning of ϕ_2 in the T_s state. After identifying the status code as that for a write cycle, the 82C288 switches DT/$\overline{\text{R}}$ (DT/$\overline{\text{R}}$ is normally in the 1 state) and DEN to logic 1. This sets the data bus transceivers to pass data from the MPU to the memory subsystem and enables it for operation. Finally, the memory write ($\overline{\text{MWTC}}$) output is switched

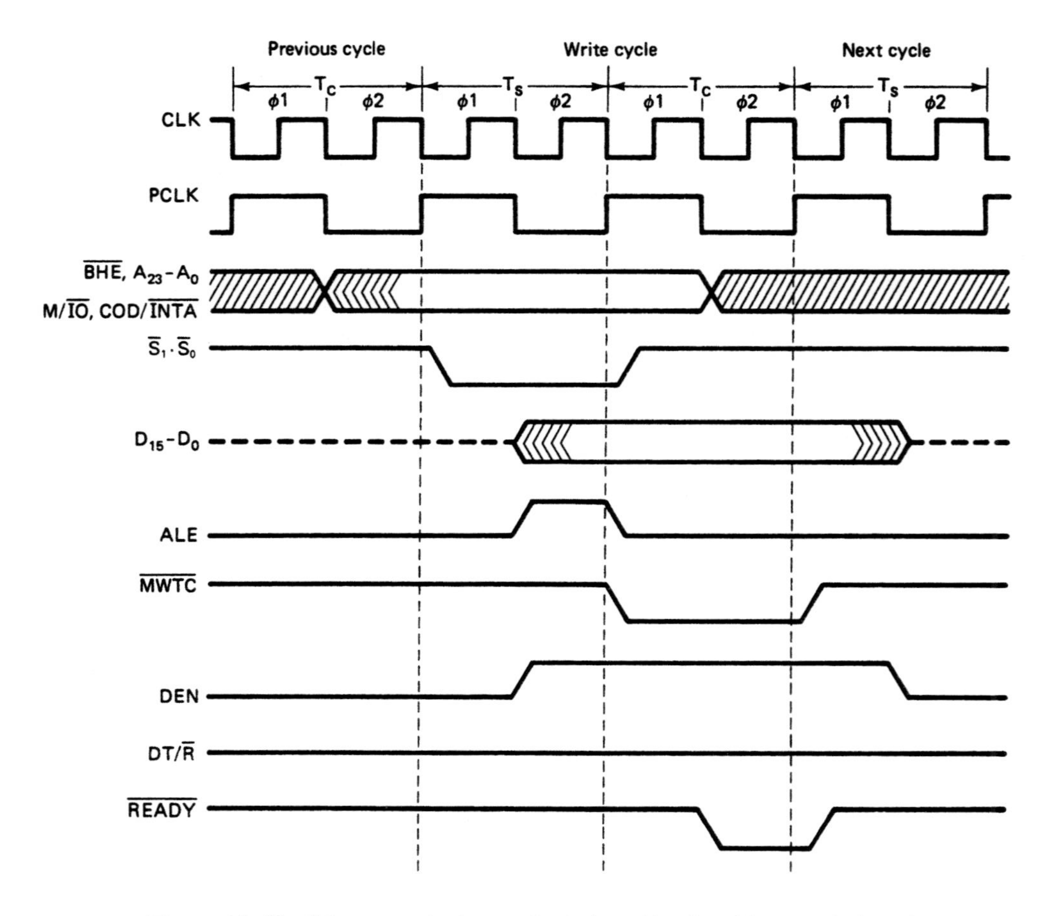

Figure 14–32 Memory write bus cycle timing. (Reprinted by permission of Intel Corporation. Copyright/Intel Corp. 1987)

to logic 0 at the beginning of the T_c state of the write cycle. $\overline{MWTC}$ signals the memory subsystem that valid write data are on the bus. Finally, late in the T_c state, the 80286 and 82C288 test the logic level of $\overline{READY}$. If $\overline{READY}$ is logic 0, the write cycle is completed and the buses and control signals are prepared for the next read or write cycle.

Wait States in the Memory Bus Cycle

Wait states can be inserted to lengthen the memory bus cycle of the 80286. This is done with the $\overline{READY}$ input signal. Upon request from an event in external hardware, for instance, slow memory, the $\overline{READY}$ input is switched to logic 1. This signals the 80286 and the 82C288 that the current bus cycle should not be completed. Instead, it extends the T_c state by the duration of one wait state (T_w), 125 ns for 8-MHz clock operation. Figure 14–33 shows a read cycle extended by one wait state. Note that the control signals $DT/\overline{R}$, DEN, and $\overline{MRDC}$ are maintained throughout the wait-state period. In this

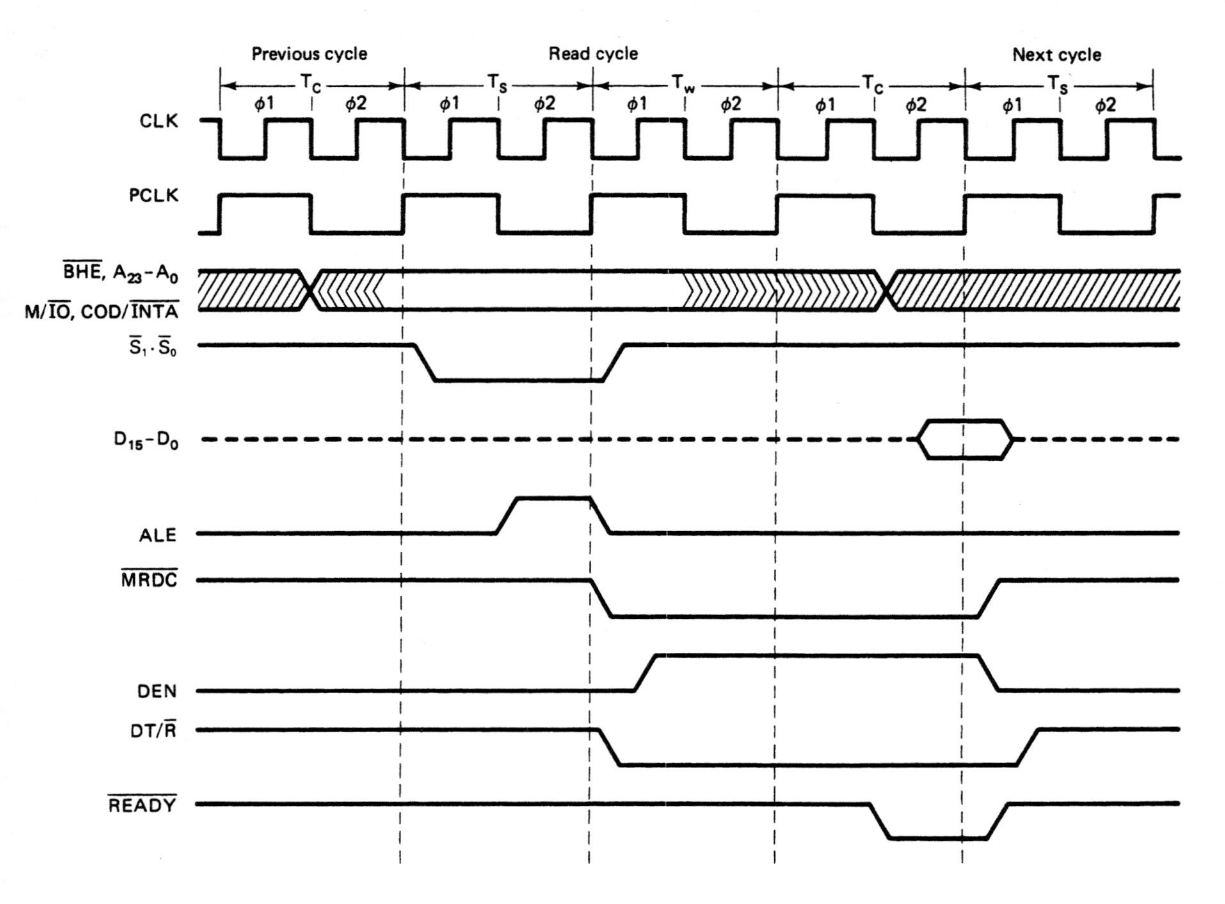

Figure 14–33 Read bus cycle with wait states. (Reprinted by permission of Intel Corporation. Copyright/Intel Corp. 1987)

way, the read cycle is not completed until $\overline{READY}$ switches to logic 0 in the T_c state that follows.

▲ 14.10 INPUT/OUTPUT INTERFACE

In Section 14.9, we studied the memory interface of the 80286 microprocessor. Here we will examine another important interface of the 80286-based microcomputer system, the I/O interface.

I/O Interface Circuit

The I/O interface of the 80286 microcomputer permits it to communicate with the outside world. The way in which the 80286 deals with I/O circuitry is similar to the way in which it interfaces with memory circuitry—that is, the transfer of I/O data also takes place over the data bus. The I/O interface allows easy interface to LSI peripheral devices such as parallel I/O expanders, interval timers, and serial communication controllers. Let us continue by looking at how the 80286 interfaces to its I/O subsystem.

Figure 14–34 shows a typical I/O interface. Here we see that the way in which the 80286 interfaces with I/O devices is identical to that of the maximum-mode 8086 microcomputer system. Note that the interface includes the 82C288 bus controller, an address decoder, address latches/buffers, data bus transceiver/buffers, and I/O devices. An example of an I/O device is a programmable peripheral interface IC, such as the 82C55A, which can be used to implement parallel I/O ports.

The I/O device that is accessed for input or output of data is selected by an *I/O address.* This address is specified as part of the instruction that performs the I/O operation. Just as for the 8086 architecture, the 80286's I/O address is 16 bits in length. As Figure 14–34 shows, they are output to the I/O interface over address bus lines A_0 through

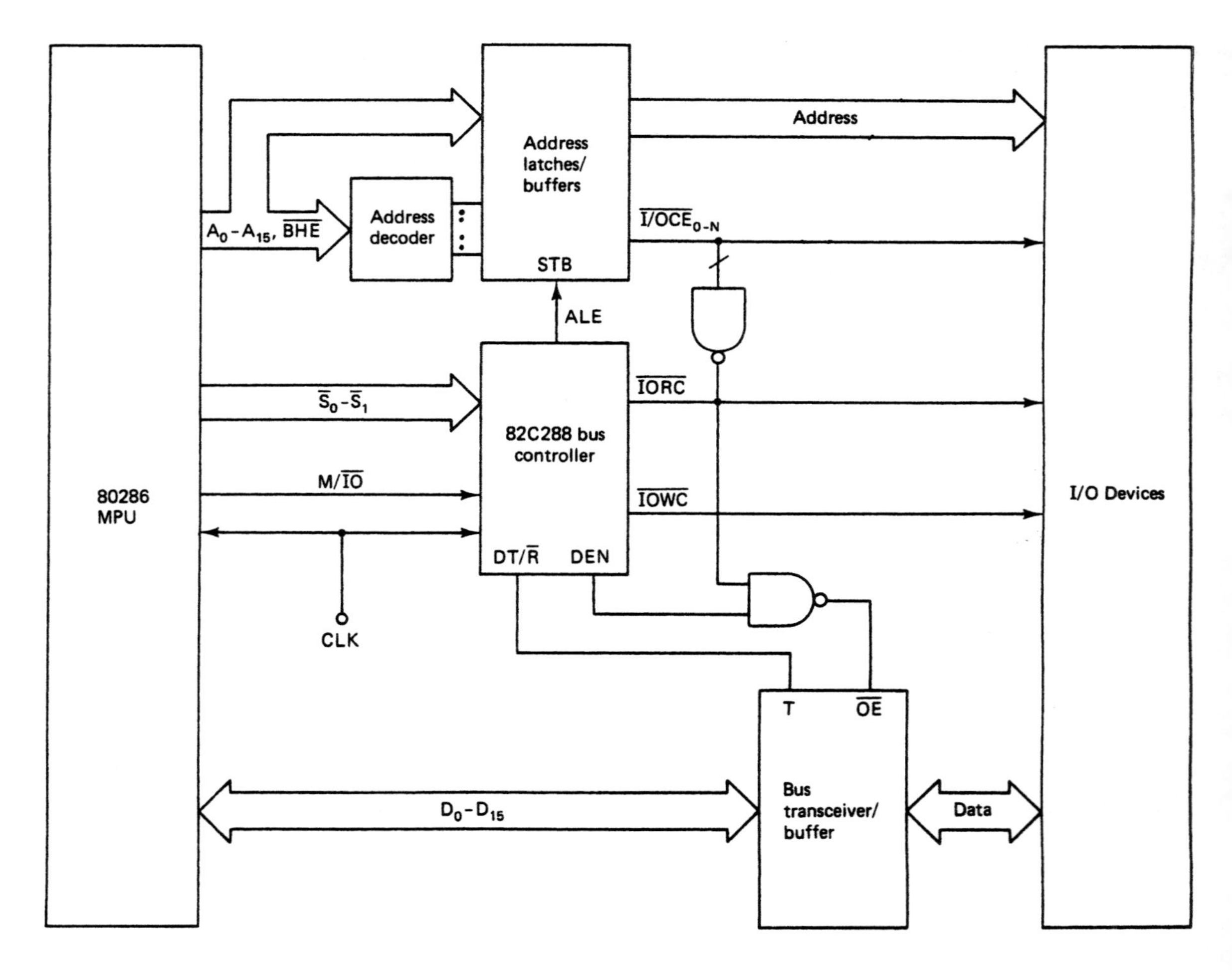

Figure 14–34 I/O interface circuit.

A_{15}. The more significant address bits, A_{16} through A_{23}, are held at the 0 logic level during the address period of all I/O bus cycles.

In the circuit shown in Fig. 14–34, some of the address bits on address lines A_0 through A_{15} of the 80286 are decoded by the address decoder to produce I/O chip-enable signals, $I/\overline{OCE}_0$ through $I/\overline{OCE}_N$, for the I/O devices. For instance, with three address bits, we can produce enough chip-enable outputs to select up to eight I/O devices. The outputs of the address decoder are latched along with the rest of the address bits into the address latches. Latching of the address is achieved with the pulse output on the ALE line of the 82C288. The address bits sent directly to the I/O devices are typically used to select the register within the peripheral device to be accessed. For example, with four address lines, we can select one of 16 I/O device registers.

During I/O bus cycles, data are passed between the 80286 and the selected register within the enabled I/O device over data bus lines D_0 through D_{15}. The 80286 can input or output data in either byte-wide or word-wide format; however, most LSI peripherals used as I/O controllers in the 80286 microcomputer system are designed to interface with an 8-bit bus and are attached to either the upper or lower part of the bus. For this reason, I/O operations usually involve byte-wide data transfers.

Just as for the memory interface, the signals A_0 and $\overline{BHE}$ are used to signal whether an even- or odd-addressed byte of data is being transferred over the bus. Again logic 0 at A_0 and $\overline{BHE}$ equal to 1 identifies an even byte, and the byte of data is input or output over data bus lines D_0 through D_7. Moreover, logic 1 at A_0 and logic 0 at $\overline{BHE}$ identifies an odd byte, and the data transfer occurs over bus lines D_8 through D_{15}.

As in the memory interface, the 82C288 bus controller produces the control signals for the I/O interface shown in Fig. 14–34. The 82C288 decodes bus cycle status codes that are output by the 80286 on $M/\overline{IO}$, $\overline{S}_1$, and $\overline{S}_0$. The table in Fig. 14–35 shows the bus command status codes together with the type of bus cycle they produce. Those for I/O bus cycles have been highlighted. If the code corresponds to an I/O read bus cycle ($M/\overline{IO}\,\overline{S}_1\overline{S}_0$ = 001), the 82C288 generates the *I/O read command output* ($\overline{IORC}$), and for I/O write bus cycles ($M/\overline{IO}\,\overline{S}_1\overline{S}_0$ = 010), it generates the *I/O write command output* ($\overline{IOWC}$). Looking at Fig. 14–34, we see that $\overline{IORC}$ and $\overline{IOWC}$ are applied directly to the I/O devices and tell them whether data are to be input from or output to the enabled I/O device.

Note in Fig. 14–34 that the 82C288 also produces control signals ALE, $DT/\overline{R}$, and DEN. These signals are used to set up the I/O interface circuitry for the input or output data transfer. The data bus transceiver/buffers control the direction of data transfers

$M/\overline{IO}$	$\overline{S}_1$	$\overline{S}_0$	Type of bus cycle
0	0	0	Interrupt acknowledge
0	0	1	I/O Read
0	1	0	I/O Write
0	1	1	None; idle
1	0	0	Halt or shutdown
1	0	1	Memory read
1	1	0	Memory write
1	1	1	None; idle

Figure 14–35 I/O bus status codes. (Reprinted by permission of Intel Corporation. Copyright/Intel Corp. 1987)

between the 80286 and I/O devices. The transceivers get enabled for operation whenever their output enable ($\overline{OE}$) inputs are switched to logic 0. Notice that the signals data bus enable DEN and I/$\overline{OCE}_{0-N}$ are combined with NAND gates to produce $\overline{OE}$. To enable the transceiver, DEN must be at logic 1 and at least one of the I/$\overline{OCE}_{0-N}$ lines must be at logic 0. These conditions occur during all I/O bus cycles.

The logic level of the DIR input determines the direction in which data are passed through the transceivers. This input is supplied by the DT/$\overline{R}$ output of the 82C288. During all input cycles, DT/$\overline{R}$ is logic 0, and the transceivers are set to pass data from the selected I/O device to the 80286. On the other hand, during output cycles, DT/$\overline{R}$ is switched to logic 1, and data passes from the 80286 to the I/O device.

Input and Output Bus Cycle Timing

We just found that the I/O interface signals of the 80286 microcomputer are essentially the same as those involved in the memory interface. In fact, the function, logic levels, and timing of all signals other than the M/$\overline{IO}$ are identical to those already described for the memory interface in Section 14.9.

Timing diagrams for the 80286's *input bus cycle* and *output bus cycle* are shown in Figs. 14–36 and 14–37, respectively. Looking at the input bus cycle waveforms, we see that address A_0 through A_{15} along with $\overline{BHE}$ and M/$\overline{IO}$ are output during the T_c state of the previous bus cycle. This time the 80286 switches M/$\overline{IO}$ to logic 0 to signal that an I/O bus cycle is taking place. At the beginning of T_s of the input cycle, $\overline{S}_1\overline{S}_0$ is set to 01 to signal that an input operation is in progress. This status information is input to the 82C288 and initiates an I/O read bus control sequence.

Let us continue with the sequence of events that takes place in external circuitry during the read bus cycle. First, the 82C288 outputs a pulse to the 1 logic level on ALE. As shown in the circuit in Fig. 14–34, this pulse is used to latch the address information into the external address latch devices. At the beginning of T_c, $\overline{IORC}$ is switched to logic 0 to signal the enabled I/O device that data are to be input to the MPU, and DT/$\overline{R}$ is switched to logic 0 to set the data bus transceivers to the input direction. A short time later, the transceivers are enabled as DEN switches to logic 1, and the data from the I/O device are passed onto the 80286's data bus. As Fig. 14–36 shows, $\overline{READY}$ is at its active 0 logic level when sampled at the end of the T_c state; therefore, the data are read by the 80286. Finally, the 82C288 returns $\overline{IORC}$, DEN, and DT/$\overline{R}$ to their inactive logic level, and the input cycle is complete.

Looking at the output bus cycle timing diagram in Fig. 14–37, we see that the 80286 puts the data to be output onto the data bus at the beginning of ϕ_2 in the T_s state of the output cycle. At this same time, the 82C288 switches DEN to logic 1 and DT/$\overline{R}$ is maintained at the 1 level for transmit mode. From Fig. 14–34 we find that since DEN is logic 1, as soon as one of the I/O $\overline{CE}_{0-N}$ lines switches to logic 0, the data bus transceivers are enabled and set up to pass data from the 80286 to the I/O devices. Therefore, the data output on the bus is available on the data inputs of the enabled I/O device. Finally, the signal $\overline{IOWC}$ switches to logic 0 during the T_c state and tells the I/O device that valid output data is on the bus. Now the I/O device must read the data off the bus before the bus controller terminates the bus cycle. If the device cannot read data at this rate, it can hold $\overline{READY}$ at the 1 logic level to extend the bus cycle with wait states.

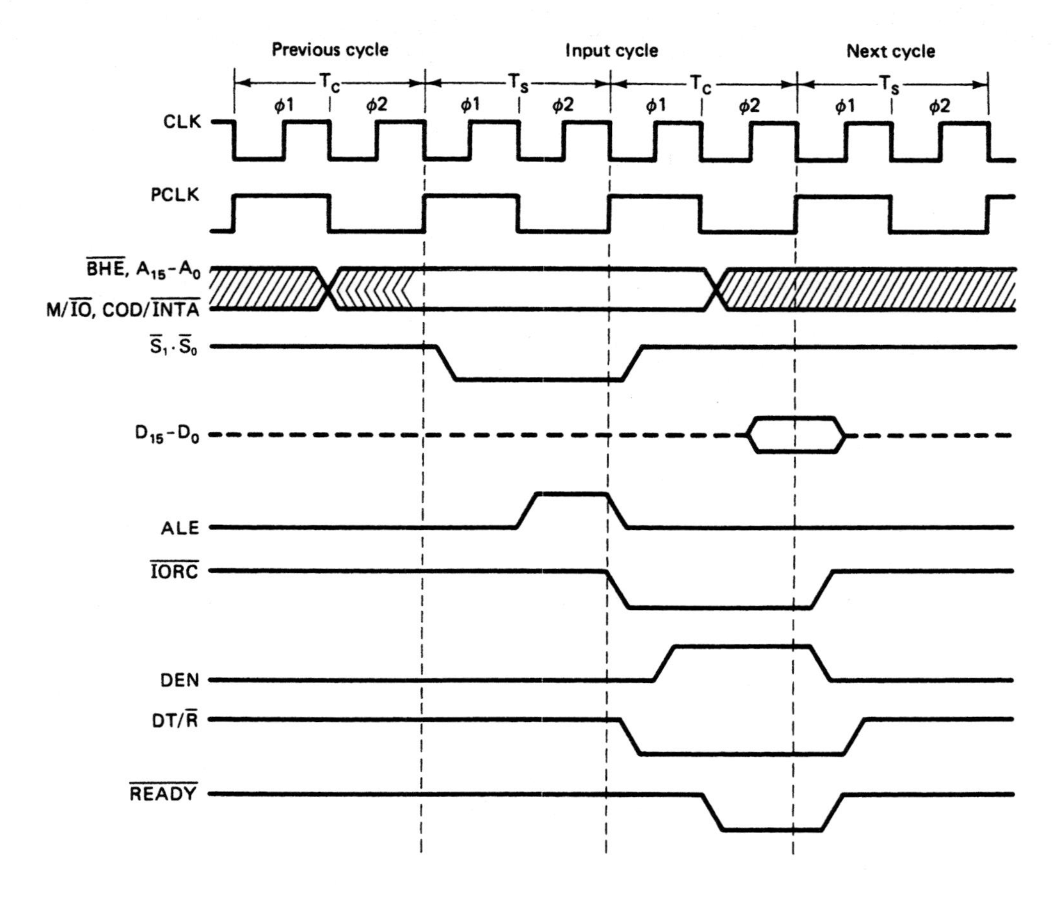

Figure 14–36 Input bus cycle timing. (Reprinted by permission of Intel Corporation. Copyright/Intel Corp. 1987)

▲ 14.11 INTERRUPT AND EXCEPTION PROCESSING

In our study of the 8086 microprocessor, we found that interrupts provide a mechanism for quickly changing program environments. We also found that the transfer of program control to the interrupt service routine can be initiated by either an event internal to the microprocessor or an event in its external hardware.

Just as for the 8086-based microcomputer, the 80286 is capable of implementing up to 256 prioritized interrupts, which are divided into five groups: *external hardware interrupts, software interrupts, internal interrupts and exceptions,* the *nonmaskable interrupt,* and *reset.* The functions of the external hardware, software, and nonmaskable interrupts are identical to those in the 8086 microcomputer and are again defined by the user. On the other hand, the internal interrupt and exception processing capability of the 80286 has been greatly expanded. These internal interrupts and reset perform dedicated system functions in the 80286 microcomputer. The priority scheme by which the 80286 services interrupts and exceptions is identical to that described earlier for the 8086.

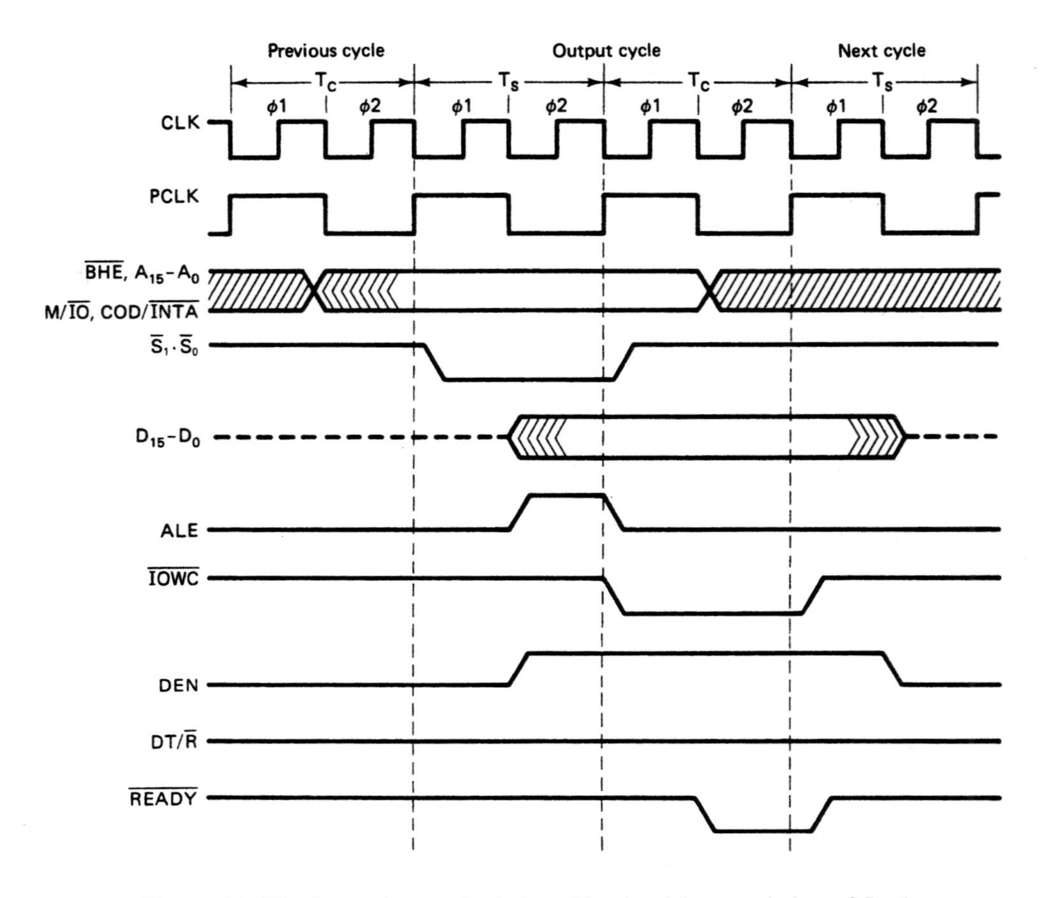

Figure 14–37 Output bus cycle timing. (Reprinted by permission of Intel Corporation. Copyright/Intel Corp. 1987)

In the real mode, the 80286 processes interrupts in exactly the same way as the 8086—that is, the same events that were described in Chapter 11 take place during the interrupt request, interrupt acknowledge bus cycle, and device service routine.

Interrupt Vector and Interrupt Descriptor Tables

An address pointer table is used to link the interrupt type numbers to the locations of their service routines in program storage memory. In a real-mode 80286-based microcomputer system, this table is called the *interrupt vector table.* On the other hand, in a protected-mode system, the table is referred to as the *interrupt descriptor table.* Figure 14–38 shows a map of the interrupt vector table in the memory of a real-mode 80286 microcomputer. Looking at the table, we see that it is identical to the interrupt vector table of the 8086 microcomputer. It contains 256 address pointers (vectors), one for each of the interrupt type numbers 0 through 255. These address pointers identify the starting locations of their service routines in program memory. The content of this table is either

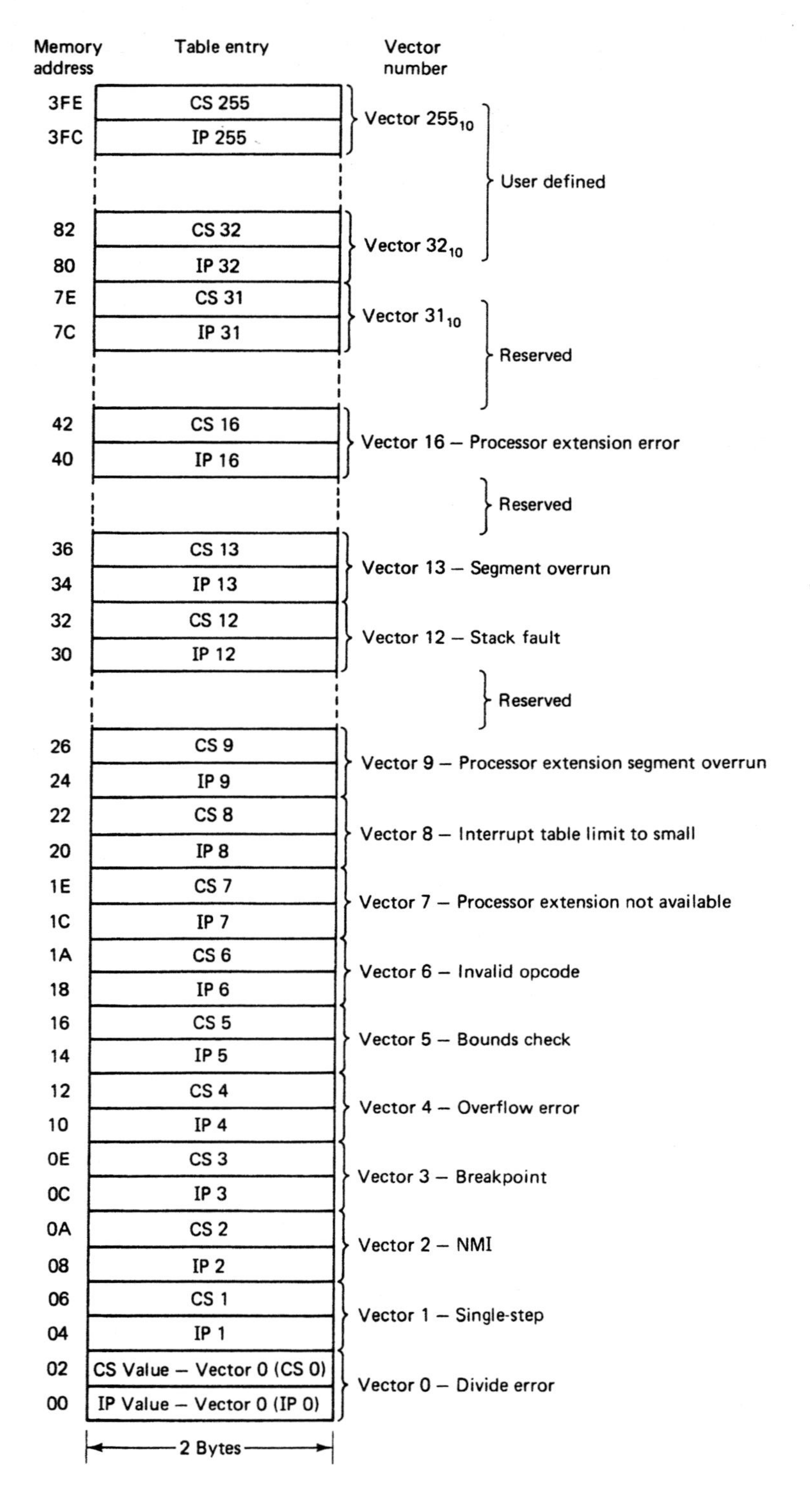

Figure 14–38 Real-mode interrupt vector table. (Reprinted by permission of Intel Corporation. Copyright/Intel Corp. 1987)

held as firmware in EPROMs or loaded into RAM as part of the system initialization procedure.

Note that in Fig. 14–38 the interrupt vector table is again located at the low-address end of the memory address space. It starts at address 00000_{16} and ends at word address $003FE_{16}$. Each of the 256 vectors requires two words (base address and offset) and is stored at an even-address boundary. Unlike the 8086, the interrupt vector table or interrupt descriptor table in an 80286-based microcomputer can be located anywhere in the memory address space. Its starting location and size are actually identified by the contents of a register within the 80286 called the *interrupt descriptor table register* (IDTR). When the 80286 is initialized at power-on, it comes up in the real mode with the bits of the base address in IDTR all equal to zero and the limit set to $03FF_{16}$. This positions the interrupt vector table as shown in Fig. 14–38. In the real mode, the value in IDTR should be left at this initial value to maintain compatibility with 8086/8088-based microcomputer software.

External Hardware Interrupt Interface Circuit

Up to this point in the section, we have introduced the types of interrupts supported by the 80286 and the real-mode interrupt vector table. Let us now look at the external hardware interrupt interface of the 80286 microcomputer.

Figure 14–39(a) illustrates a general interrupt interface circuit for an 80286-based microcomputer system. Looking at this diagram, we see that it is similar to the interrupt interface of a maximum-mode 8086 microcomputer system. Note that it includes the address and data buses, status signals $\overline{S}_0$, $\overline{S}_1$, and $M/\overline{IO}$, and dedicated interrupt signals INTR and $\overline{INTA}$. The external circuitry is required to interface interrupt inputs, INT_{32} through INT_{255}, to the 80286's interrupt interface. Just as in an 8086 system, this interface circuitry must identify which of the pending active interrupts has the highest priority, perform an interrupt request/acknowledge handshake, and then pass a type number to the 80286.

In this circuit we see that the key interrupt interface signals are *interrupt request* (INTR) and *interrupt acknowledge* ($\overline{INTA}$). The logic level input at the INTR line signals the 80286 that an external device is requesting service. The 80286 samples this input during the last clock period of each instruction execution cycle—that is, at instruction boundaries. Logic 1 at INTR represents an active interrupt request. INTR is *level triggered;* therefore, its active level must be maintained until tested by the 80286. If it is not maintained, the request for service may not be recognized. For this reason, inputs INT_{32} through INT_{255} are normally latched. The 1 at INTR must be removed before the service routine runs to completion; otherwise, the same interrupt may be acknowledged a second time.

When the 80286 recognizes an interrupt request, it signals this to external circuitry by outputting the interrupt acknowledge bus cycle status code on $M/\overline{IO}\ \overline{S}_1\overline{S}_0$. This code, which equals 000, is highlighted in Fig. 14–40. Note in Fig. 14–39(a) that this code is input to the 82C288 bus controller where it is decoded to produce a pulse to logic 0 at the $\overline{INTA}$ output. Actually, there are two pulses produced at $\overline{INTA}$ during the *interrupt acknowledge bus cycle.* The first pulse signals external circuitry that the interrupt request has been acknowledged and to prepare to send its type number to the 80286. The second pulse tells the external circuitry to put the interrupt type number on the data bus.

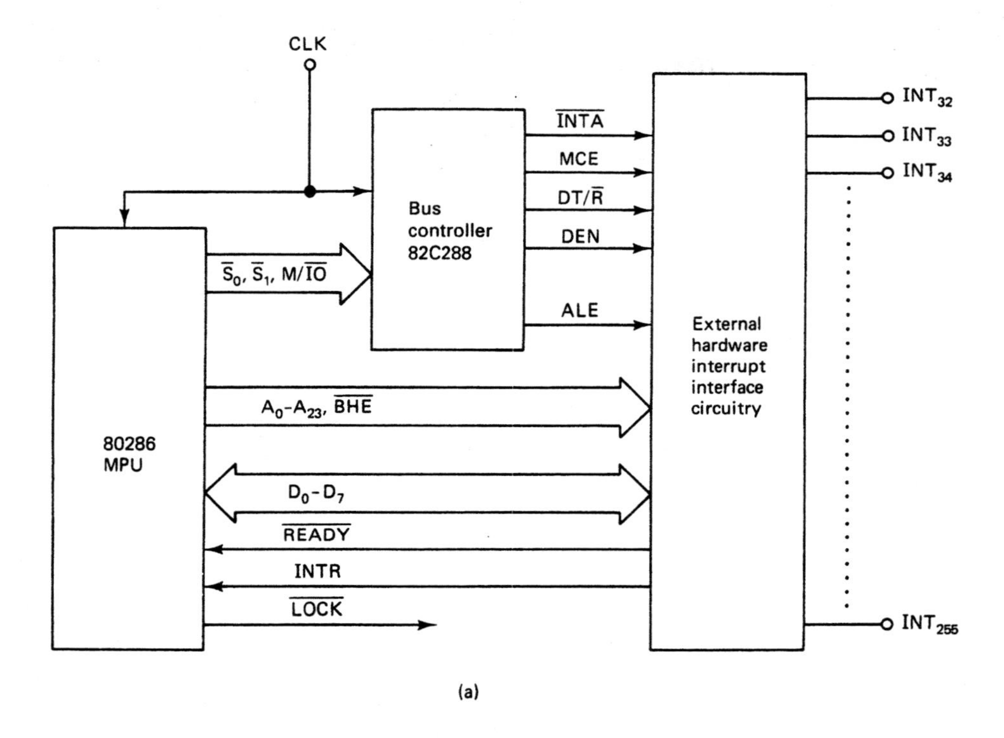

Figure 14–39 (a) 80286 microcomputer system external hardware interrupt interface. (b) Interrupt interface circuit using an 82C59A. (Reprinted by permission of Intel Corporation. Copyright/Intel Corp. 1987)

Notice that only the lower eight lines of the data bus, D_0 through D_7, are part of the interrupt interface. During the second cycle in the interrupt acknowledge bus sequence, external circuitry must put an 8-bit type number of the highest priority active interrupt request onto this part of the data bus. The 80286 reads this number off the bus to identify which external device is requesting service. Then it uses the type number to generate the address of the interrupt's vector in the interrupt vector table or gate in the interrupt descriptor table.

Address lines A_0 through A_{23} are also shown in the interrupt interface circuit in Fig. 14–39(a). LSI interrupt controller devices are typically used to implement most of the external circuitry. When a read or write bus cycle is performed to the controller, for example, to initialize its internal registers after system reset, some of the address bits are decoded to produce a chip select to enable the controller device, and other address bits are used to select the internal register that is to be accessed. Of course, the interrupt controller could be I/O mapped instead of memory mapped; in this case, just address lines A_0 through A_{15} are used in the interface.

Figure 14–39(b) shows an interrupt interface circuit that uses a single 82C59A programmable interrupt controller. This circuit implements eight interrupt request inputs. For applications that require more than eight interrupt request inputs, 82C59A devices are cascaded into a master/slave configuration similar to the one for the 8088 microprocessor discussed in an earlier chapter.

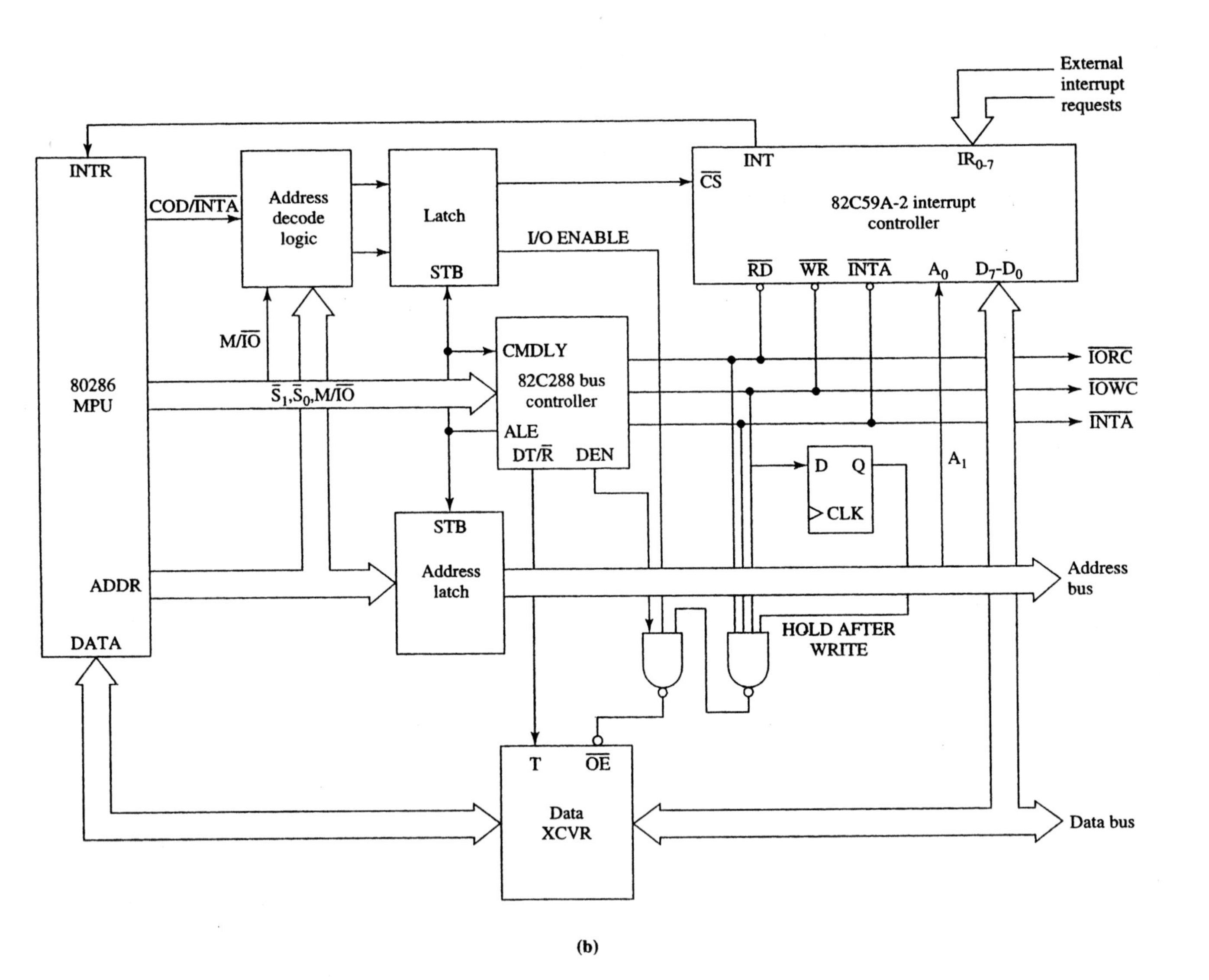

(b)

Figure 14–39 (continued)

$M/\overline{IO}$	$\overline{S}_1$	$\overline{S}_0$	Type of bus cycle
0	0	0	Interrupt acknowledge
0	0	1	I/O Read
0	1	0	I/O Write
0	1	1	None; idle
1	0	0	Halt or shutdown
1	0	1	Memory read
1	1	0	Memory write
1	1	1	None; idle

Figure 14–40 Interrupt bus status code. (Reprinted by permission of Intel Corporation. Copyright/Intel Corp. 1987)

Internal Interrupt and Exception Functions

Earlier we indicated that some of the 256 interrupt vectors of the 80286 are dedicated to internal interrupt and exception functions. Internal interrupts and exceptions differ from external hardware interrupts in that they result from the execution of an instruction, not an event, that takes place in external hardware. That is, an internal interrupt or exception may be initiated because a fault condition was detected either during or after execution of an instruction. In this case, a routine is provided to service the internal condition.

Figure 14–41 identifies the internal interrupts and exceptions that are active in real mode. Here we find internal interrupts such as single step and breakpoint and exception functions such as divide error and overflow error that were also detected by the 8086; however, the 80286 also implements several new real-mode exceptions. For example, invalid opcode, bounds check, and processor-extension error are all new with the 80286. Let us now look at these new real-mode internal functions in more detail.

Bounds Check Exception. Earlier we pointed out that the BOUND (check array index against bounds) instruction could be used to test an operand, which is used as the index into an array, to verify that it is within a predefined range. If the index is less than the lower bound (minimum value) or greater than the upper bound (maximum value), a *bounds check exception* has occurred and control is passed to the exception handler pointed to by the values of CS_5 and IP_5.

Invalid Opcode Exception. The exception processing capability of the 80286 permits detection of undefined opcodes. This feature of the 80286 allows it to detect automatically whether or not the opcode to be executed as an instruction corresponds to one of the instructions in the instruction set. If it does not, execution is not attempted; instead, the opcode is identified as being undefined and the *invalid opcode exception* is initiated. In turn, control is passed to the exception handler identified by IP_6 and CS_6. This *undefined opcode detection* mechanism permits the 80286 to detect errors in its instruction stream.

Processor Extension Not Available Exception. When the 80286 comes up in the real mode, both the EM (emulate) and MP (math present) bits of its machine status word are reset. This mode of operation corresponds to that of the 8088 or 8086 microprocessors. When set in this way, the *processor extension not available exception* cannot occur; how-

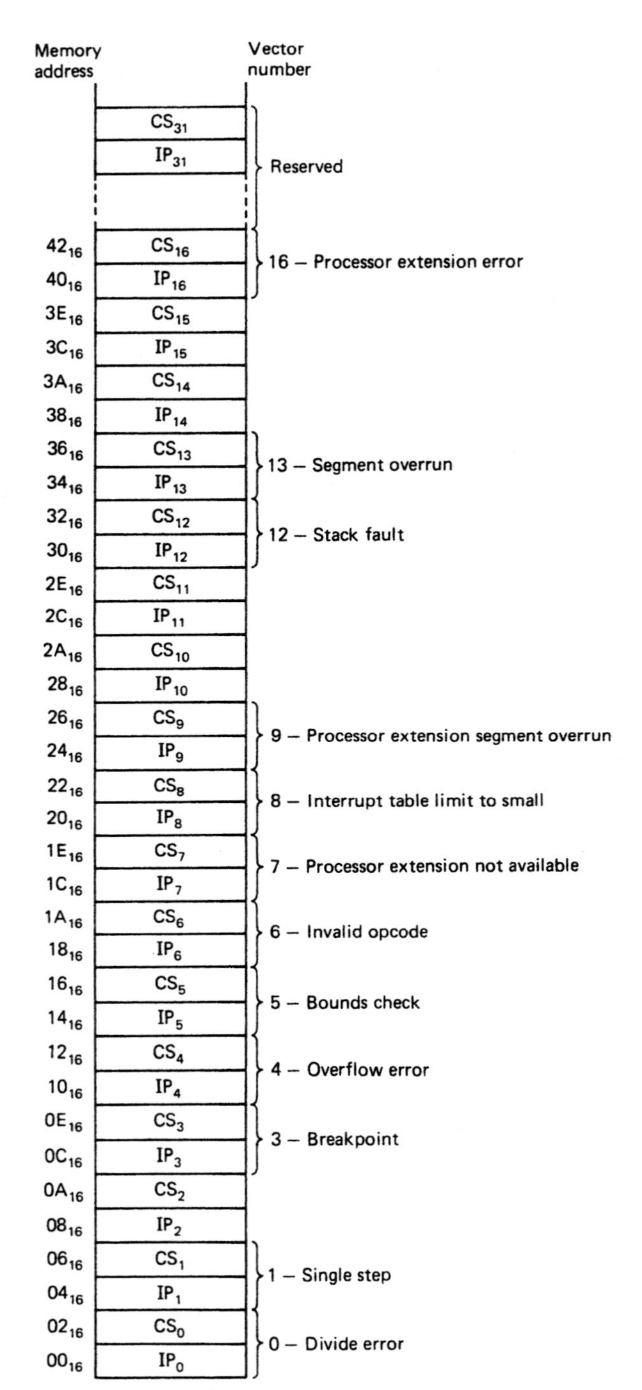

Figure 14–41 Real-mode internal exception vector locations. (Reprinted by permission of Intel Corporation. Copyright/Intel Corp. 1987)

ever, if the EM bit has been set to 1 under software control (no math coprocessor is present) and the 80286 executes an ESC (escape) instruction for the 80287 math coprocessor, a processor extension not present exception is initiated through the vector specified by CS_7 and IP_7. This service routine could pass control to a software emulation routine for the floating-point arithmetic operation. If the MP (math present) bit and TS bit are both set (meaning that a math coprocessor is available in the system and a task is in progress) when the ESC or wait instruction is executed, an exception also takes place.

Interrupt Table Limit Too Small Exception. Earlier we pointed out that the IDTR could be used to relocate the interrupt vector table in memory. If the location of the real-mode table has been changed such that the table extends beyond address $003FF_{16}$ and an interrupt is invoked that attempts to access a vector stored at an address higher than the limit, the *interrupt table limit too small exception* occurs. In this case, control is passed to the service routine by the vector CS_8:IP_8.

Processor Extension Segment Overrun Exception. The *processor extension segment overrun exception* signals that the 80287 numeric coprocessor has overrun the limit of a segment while attempting to read or write its operand. This event is detected by the processor extension data channel within the 80286 and passes control to the service routine through interrupt vector 9. This exception handler can clear the exception, reset the 80287, determine the cause of the exception by examining the registers within the 80287, and then initiate a corrective action.

Stack Fault Exception. In the real-mode, if the address of an operand access for the stack segment crosses the boundaries of the stack, a *stack fault exception* is produced. This causes control to be transferred to the service routine defined by CS_{12} and IP_{12}.

Segment Overrun Exception. This exception occurs in the real mode if an instruction attempts to access an operand that extends beyond the end of a segment. For instance, if a word access is made to the address CS:FFFFH, DS:FFFFH, or ES:FFFFH, a *segment overrun exception* occurs.

Processor Extension Exception. As part of the handshake sequence between the 80286 microprocessor and 80287 math coprocessor, the 80286 checks the status of its $\overline{\text{ERROR}}$ input. If the 80287 encounters a problem performing a numeric operation, it signals this to the 80286 by switching its $\overline{\text{ERROR}}$ output to logic 0. This signal is normally applied directly to the $\overline{\text{ERROR}}$ input of the 80286. Logic 0 at this input signals that an error condition has occurred and causes a *processor extension exception* through vector 16.

REVIEW PROBLEMS

Section 14.1

1. Name the technology used to fabricate the 80286 microprocessor.
2. What is the approximate transistor count of the 80286?
3. In what three packages is the 80286 manufactured?

Section 14.2

4. Name the four internal processing units of the 80286.
5. What are the word lengths of the 80286's address bus and data bus?
6. Does the 80286 have a multiplexed address/data bus or demultiplexed address and data buses?
7. How large is the 80286's instruction queue?
8. List three functions performed by the address unit.
9. What is the function of the instruction queue?

Section 14.3

10. How does the performance of an 8-MHz 80286 in real mode compare to that of a 5-MHz 8086?
11. What is meant when we say that the 80286 is object code compatible with the 8086?
12. What additional register has been provided in the 80286 operating in real-mode as compared to the 8086?

Section 14.4

13. Describe the operation performed by the instruction PUSHA.
14. Which registers and in what order does the instruction POPA pop from the stack?
15. What is a stack frame?
16. How much stack does the instruction ENTER 20H, 4 allocate for the stack frame? What is the lexical level?
17. If (DS) = (ES) = 1075_{16}, (DI) = 100_{16}, (DF) = 0, and (DX) = 1000_{16}, what happens when the instruction INSW is executed?
18. If (DS) = (ES) = 1075_{16}, (SI) = 100_{16}, (DF) = 1, (CX) = F_{16}, and (DX) = 2000_{16}, what happens when the instruction REPOUTSB is executed?
19. Explain the function of the bound instruction in the sequence

```
MOV     DI, VALUE
BOUND   DI, [LIMITS]
```

Assume that address LIMITS contains the value 0000_{16} and LIMITS + 2 holds the value $00FF_{16}$.

Section 14.5

20. How large are the real-address mode address bus and physical address space? How large is the protected-address mode address bus and physical address space? How large is the protected-mode virtual address space?
21. What type of bus cycle is in progress when the bus status code COD/$\overline{\text{INTA}}$ M/$\overline{\text{IO}}$ $\overline{S}_1\overline{S}_0$ equals 1010?
22. If the code output as $\overline{\text{BHE}}$ A_0 equals 01, what type of data transfer is taking place over the bus?
23. Is the logic level output on COD/$\overline{\text{INTA}}$ intended to be used directly as the interrupt acknowledge signal to the external hardware interrupt interface circuitry?

24. Which signals implement the DMA interface?
25. What numeric processor can be attached to the processor extension interface?

Section 14.6

26. Which of the 80286's status outputs are input to the 82C288?
27. If the input status code to the 82C288 is 010, what command output is active?
28. What bus control signals are produced by the 82C288?
29. To what logic level must CEN/$\overline{\text{AEN}}$ be set to enable the command outputs of the 82C288?

Section 14.7

30. What speed 80286 ICs are available from Intel Corporation? How are these speeds denoted in the part number?
31. What frequency crystal must be connected between the X_1 and X_2 inputs of the 80C284-12 to run the device at full speed?
32. What clock outputs are produced by the 82C284? What would be their frequencies if a 20-MHz crystal were used?

Section 14.8

33. How many clock states are in an 80286 bus cycle that has no wait states? What would be the duration of this bus cycle for the 80286 operating at 10 MHz?
34. What does T_s stand for? What happens in this part of the processor cycle?
35. What does T_c stand for? What happens in this part of the processor cycle?
36. Explain what pipelining of the 80286's bus means.
37. What is an idle state?
38. What is a wait state? If an 80286 running at 10 MHz performs a bus cycle with two wait states, what is the duration of the bus cycle?

Section 14.9

39. Summarize the function of each of the blocks in the memory interface diagram shown in Fig. 14–27.
40. When the instruction PUSH AX is executed, what bus status code is output by the 80286 and what read/write control signal is produced by the 82C288?
41. What four types of data transfers can take place over the data bus? How many bus cycles are required for each type of data transfer?
42. If an 80286 is running at 10 MHz and all memory accesses involve one wait state, how long will it take to fetch the word of data starting at address $0FF1A_{16}$? At address $0FF1D_{16}$?
43. During a bus cycle that involves an odd-addressed word transfer, which byte of data is transferred over the bus during the first bus cycle?
44. Describe the bus activity that takes place as the 80286 writes a byte of data into memory address $B0010_{16}$.
45. If the write cycle in Fig. 14–32 is for an 80286 running at 12.5 MHz, what is the duration of the bus cycle?

46. If the read cycle in Fig. 14–33 is for an 80286 running at 12.5 MHz, what is the duration of the bus cycle?

Section 14.10

47. Which signal indicates to the bus controller and external circuitry that the current bus cycle is for the I/O interface and not for the memory interface?
48. Which device produces the input (read), output (write), and bus control signals for the I/O interface?
49. Briefly describe the function of each block in the I/O interface circuit in Fig. 14–34.
50. If an 80286 running at 10 MHz inserts two wait states into all I/O bus cycles, what is the duration of a bus cycle in which a byte of data is being output?
51. If the 80286 in problem 50 were outputting a word of data to a word-wide port at I/O address $1A1_{16}$, what would be the duration of the bus cycle?
52. Summarize the sequence of events that take place at the I/O interface in Fig. 14–34 as a word of I/O data is output over data bus lines D_0 through D_{15} to a port at an even address.

Section 14.11

53. What five groups of interrupts are supported by the 80286 MPU?
54. What is the range of type numbers assigned to the interrupts in the 80286 microcomputer system?
55. What is the real-mode interrupt address pointer table called? The protected-mode address pointer table?
56. What is the size of a real-mode interrupt vector?
57. The contents of which register determines the location of the interrupt address pointer table? To what value is this register initialized at reset?
58. The breakpoint routine in a real-mode 80286 microcomputer system starts at address $AA000_{16}$ in the code segment located at address $A0000_{16}$. Specify how the breakpoint vector will be stored in the interrupt vector table.
59. Explain the function of INTR and $\overline{\text{INTA}}$ in the circuit diagram shown in Fig. 14–39(a).
60. How many interrupt inputs can be directly accepted by the circuit shown in Fig. 14–39(b)?
61. List the real-mode internal interrupts serviced by the 80286.
62. Which real-mode vector numbers are reserved for internal interrupts and exceptions?

The 80386, 80486, and Pentium Processor Families: Software Architecture

▲ INTRODUCTION

In this chapter, we study the software architecture of the *80386, 80486,* and *Pentium processor* families. The 80386 supports three modes of software operation: the *real-address mode* (real mode), the *protected-address mode* (protected mode), and *virtual 8086 mode.* We will explore each of these modes in the chapter. After first introducing the 80386 microprocessor, we examine its real-address-mode software architecture and assembly language instruction set. This is followed by a detailed study of the 80386's protected-address mode of operation and system control instruction set; the material on the 80386 completes with a description of the virtual 8086 mode. The chapter closes with sections that describe the differences between the software architecture of the 80386, 80486, and Pentium processor families. This includes introductions to the floating-point and multimedia architectures and instruction sets. The following topics are covered in the chapter:

▲ 15.1 80386 MICROPROCESSOR FAMILY

The 80386 family of microprocessors, announced by Intel Corporation in 1985, represent the first 32-bit members of Intel's popular microprocessor architecture. The 80386DX MPU, the first entry in the 80386 family, was the sixth member of the 8086 family of microprocessors. This device is the high-performance member of the 80386 family of MPUs. The 80386DX is a full 32-bit processor—that is, it has both 32-bit internal registers and a 32-bit external data bus. Several years later Intel brought the 80386SX microprocessor to market. This device, with its 32-bit internal registers and 16-bit data bus, provides a lower performance MPU for the 80386-based microcomputer system. In this chapter, we will primarily study the 80386DX MPU. However, both the 80386DX and 80386SX devices operate essentially the same way from a software point of view.

The chart in Fig. 15–1 is called the *iCOMP™ index*. A bar is used in this chart to represent a measure of the performance for each of Intel's MPUs. Intel Corporation provides this index so that the relative performance of its microprocessors can be compared. Note that the members of the 80386 family offer low performance when compared to the newer 80486 and Pentium processor families. In fact, the slowest 80386SX MPU shown in Fig. 15–1, the -20, has a performance rating of 32, while the fastest 80386DX, the -33, is rated at 68. In this way, we see that by selecting between the various members of the 80386 family, we can achieve a wide range of system performance levels.

We already learned that from the software point of view the 80386DX offers several modes of operation: real-address mode, protected-address mode, and virtual 8086 mode. The real mode is for compatibility with the large existing 8086/8088 software base; the protected mode offers an advanced software architecture with enhanced system-level features such as memory management, multitasking, and protection; and the virtual 8086 mode provides 8086 real-mode compatibility while operating in the protected mode. The virtual 8086 mode was not supported by the protected mode of the 80286 microprocessor.

▲ 15.2 INTERNAL ARCHITECTURE OF THE 80386DX MICROPROCESSOR

As part of the evolutionary process from the original 8086 to the 80386, the internal architecture of the 8086 family of microprocessors has changed considerably. All members of the 8086 family employ what is called *parallel processing*. That is, they are implemented with multiple, simultaneously operating processing units. Each unit has a dedicated function and each operates at the same time. The more parallel processing, the higher is the microprocessor's performance.

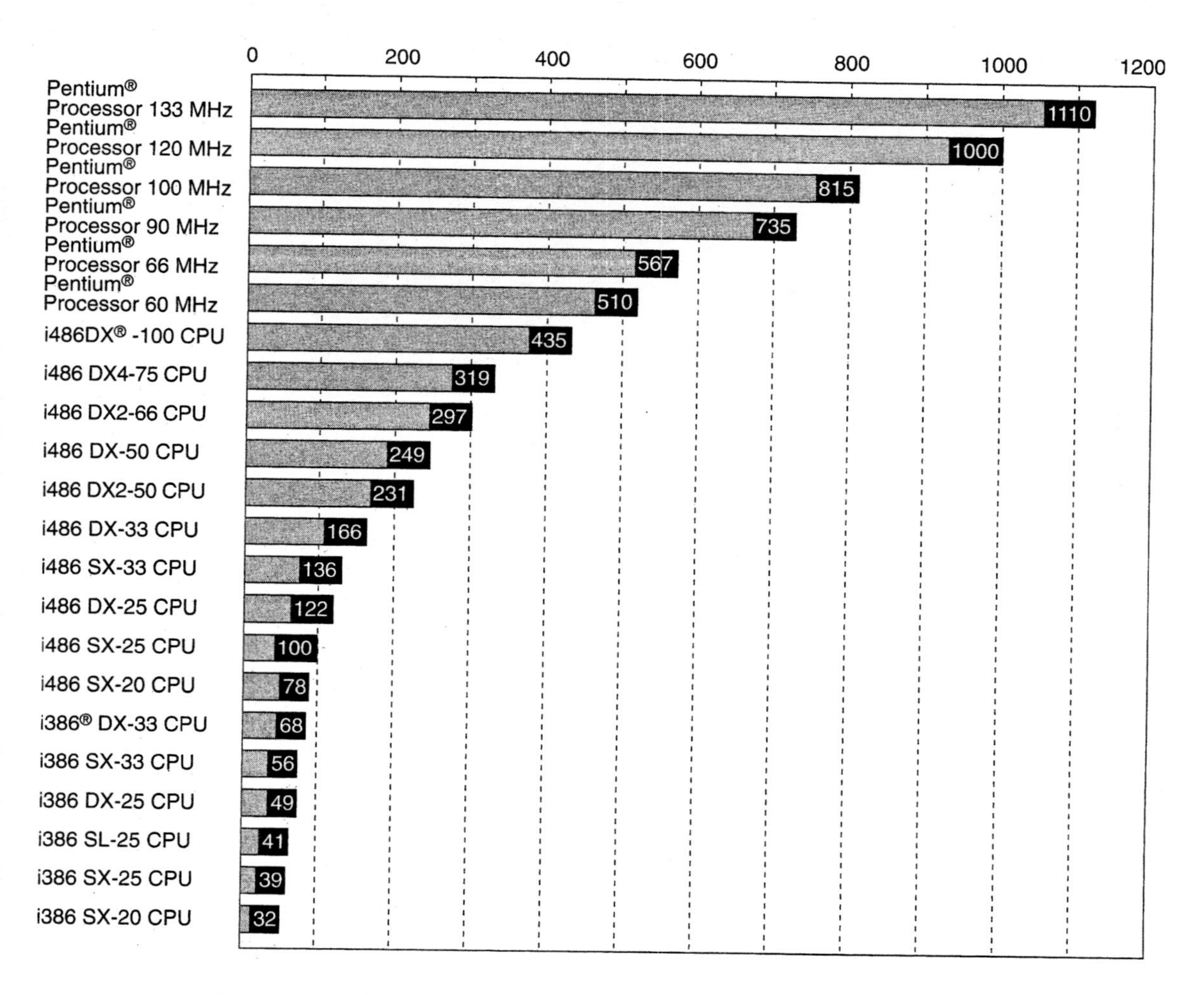

Figure 15–1 iCOMP™ Index rating chart. (Reprinted by permission of Intel Corporation. Copyright/Intel Corp. 1993)

In earlier chapters, we examined the internal architecture of the 8086 and 80286 microprocessors. We found that the 8086 microprocessor contains just two processing units: the bus interface unit and execution unit. In the 80286 microprocessor, the internal architecture was further partitioned into four independent processing elements: the bus unit, the instruction unit, the execution unit, and the address unit. This additional parallel processing provided an important contribution to the higher level of performance achieved with the 80286 architecture.

Figure 15–2 illustrates the 80386DX's internal architecture. Here we see that to enhance performance, more parallel processing elements are provided. Note that now there are six functional units: the *execution unit,* the *segment unit,* the *page unit,* the *bus unit,* the *prefetch unit,* and the *decode unit.* Let us now look more closely at each of the processing units of the 80386DX.

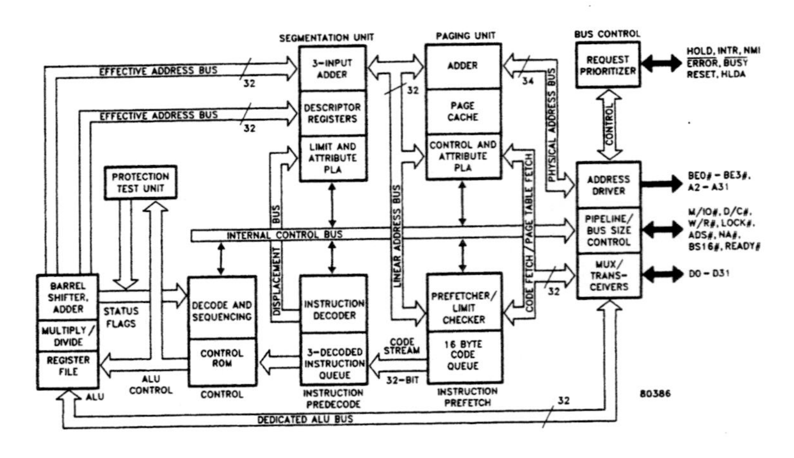

Figure 15–2 Internal architecture of the 80386DX MPU. (Reprinted by permission of Intel Corporation. Copyright/Intel Corp. 1987)

The bus unit is the 80386DX's interface to the outside world. By interface, we mean the path by which it connects to external devices. The bus interface provides a 32-bit data bus, a 32-bit address bus, and the signals needed to control transfers over the bus. In fact, 8-bit, 16-bit, and 32-bit data transfers are supported. These buses are demultiplexed like those of the 80286. That is, the 80386DX has separate pins for its address and data bus lines. This demultiplexing of address and data results in higher performance and easier hardware design. Expanding the data bus width to 32 bits further improves the performance of the 80386DX's hardware architecture as compared to that of either the 8086 or 80286.

The bus unit is responsible for performing all external bus operations. This processing unit contains the latches and drivers for the address bus, transceivers for the data bus, and control logic for signaling whether a memory, input/output, or interrupt-acknowledge bus cycle is being performed. Looking at Fig. 15–2, we find that for data accesses, the address of the storage location to be accessed is input from the paging unit, and for code accesses, the prefetch unit provides the address.

The prefetch unit implements a mechanism known as an *instruction stream queue.* This queue permits the 80386DX to prefetch up to 16 bytes of instruction code. Whenever the queue is not full—that is, it has room for at least four more bytes, and at the same time, the execution unit is not asking it to read or write data from memory—the prefetch unit supplies addresses to the bus interface unit and signals it to look ahead in the program by fetching the next sequential instructions. Prefetched instructions are held in the FIFO queue for use by the instruction decoder. Whenever bytes are loaded at the input end of the queue, they are automatically shifted up through the FIFO to the empty locations near the output. With its 32-bit data bus, the 80386DX fetches 4 bytes of instruction code in a single memory cycle. Through this prefetch mechanism, the fetch time for most instructions is hidden.

If the queue in the prefetch unit is full and the execution unit is not requesting access to data in memory, the bus interface unit does not need to perform any bus cycle.

These intervals of no bus activity, which occur between bus cycles, are known as *idle states.*

The prefetch unit prioritizes bus activity. Highest priority is given to operand accesses for the execution unit. However, if the bus unit is already in the process of fetching instruction code when the execution unit requests it to read or write operands from memory or I/O, the current instruction fetch is first completed before the operand read/write cycle is initiated.

Figure 15–2 shows that the decode unit accesses the output end of the prefetch unit's instruction queue. It reads machine-code instructions from the output side of the prefetch queue and decodes them into the microcode instruction format used by the execution unit. That is, it off-loads the responsibility for instruction decoding from the execution unit. The *instruction queue* within the 80386DX's instruction unit permits three fully decoded instructions to be held waiting for use by the execution unit. Once again, the result is improved performance for the MPU.

The execution unit includes the arithmetic/logic unit (ALU), the 80386DX's registers, special multiply, divide, and shift hardware, and a control ROM. By registers, we mean the 80386DX's general-purpose registers, such as EAX, EBX, and ECX. The control ROM contains the microcode sequences that define the operation performed by each of the 80386DX's machine-code instructions. The execution unit reads decoded instructions from the instruction queue and performs the operations that they specify. It is the ALU that performs the arithmetic, logic, and shift operations required by an instruction. If necessary, during the execution of an instruction, it requests the segment and page units to generate operand addresses and the bus interface unit to perform read or write bus cycles to access data in memory or I/O devices. The extra hardware provided to perform multiply, divide, shift, and rotate operations improves the performance of instructions that employ these operations.

The segment and page units provide the memory management and protection services for the 80386DX. They off-load the responsibility for address generation, address translation, and segment checking from the bus interface unit, thereby further boosting the performance of the MPU. The segment unit implements the segmentation model of the 80386DX's memory management. That is, it contains dedicated hardware for performing high-speed address calculations, logical-to-linear address translation, and protection checks. For instance, when in the real mode, the execution unit requests the segment unit to obtain the address of the next instruction to be fetched by adding an appended version of the current contents of the code segment (CS) register with the value in the instruction pointer (IP) register to obtain the 20-bit physical address to be output on the address bus. This address is passed on to the bus unit.

For protected mode, the segment unit performs the logical-to-linear address translation and various protection checks needed when performing bus cycles. It contains the segment registers and the *6-word × 64-bit cache* that is used to hold the current descriptors within the 80386DX.

The page unit implements the protected mode paging model of the 80386DX's memory management. It contains the *translation lookaside buffer* that stores recently used page directory and page table entries. When paging is enabled, the linear address produced by the segment unit is used as the input of the page unit. Here the linear address is translated into the physical address of the memory or I/O location to be accessed. This physical memory or I/O address is output to the bus interface unit.

▲ 15.3 REAL-ADDRESS-MODE SOFTWARE MODEL OF THE 80386DX

Let us begin our study of the 80386DX microprocessor with its real-address-mode software model and operation. Just like the 80286 microprocessor, 80386DX comes up in the real mode after it is reset. The 80386DX will remain in this mode unless it is switched to protected mode by the software. In real mode, the 80386DX operates as a very high performance 8086. For instance, the original 16-MHz 80386DX provides more than 10 times higher performance than the standard 5-MHz 8086.

When in the real mode, the 80386DX can be used to execute the base instruction set of the 8086/8088 architecture. Similar to the 80286, object code for the base instructions of the 80386DX is identical to that of the 8086/8088. In this way, we see that object code compatibility is maintained between the 8086/8088 and 80386 family of microprocessors. This means that the operating systems and programs written for the 8086 and 8088 can be run directly on the 80386DX without modification.

As for the 80286, a number of new instructions have been added to the instruction set of the 80386DX to enhance its performance and functionality. In fact, the exact same instructions that were added to the 80286 to make the 80286's extended instruction set are also available in the 80386DX's instruction set. For instance, instructions have been added to push or pop the complete register set, perform string input/output, and check the boundaries of data array accesses. However, the real-mode instruction set has been further enhanced in the 80386DX. For example, it has a group of instructions that are provided to perform bit test and set operations. The object code of the 80386DX is also upward compatible within the 8086 architecture—that is, the 8086/8088's object code will run on the real-mode 80386DX. But, the reverse is not true. For instance, if the bit test and set instructions are employed in the writing of a program, it will not run on the 8086 or the 8088 MPU.

The real-mode software model of the 80386DX is shown in Fig. 15–3. This register model is very different from those of the 8088, 8086, and 80286. Here we have highlighted the 17 internal registers that are used in real-mode application programming. Nine of them—the data registers (EAX, EBX, ECX, and EDX), the pointer registers (EBP and ESP), the index registers (ESI and EDI), and the flag register (FLAGS)—are identical to the corresponding registers in the 8086's software model except that they are now all 32 bits in length. On the other hand, the segment registers (CS, DS, SS, and ES) and instruction pointer (IP) are both identical and still 16 bits in length. From a software point of view, all these registers serve functions similar to those they performed in the 8088/8086. For instance, CS and IP together point to the next instruction to be fetched.

Several new registers are found in the real-mode 80386DX's software model. For instance, it has two more data segment registers, FS and GS. These registers are not implemented in either the 8088/8086 or the 80286 microprocessors. Another new register is called *control register zero* (CR_0). The five least significant bits of this register are called the *machine status word* (MSW) and are identical to the MSW of the 80286. The

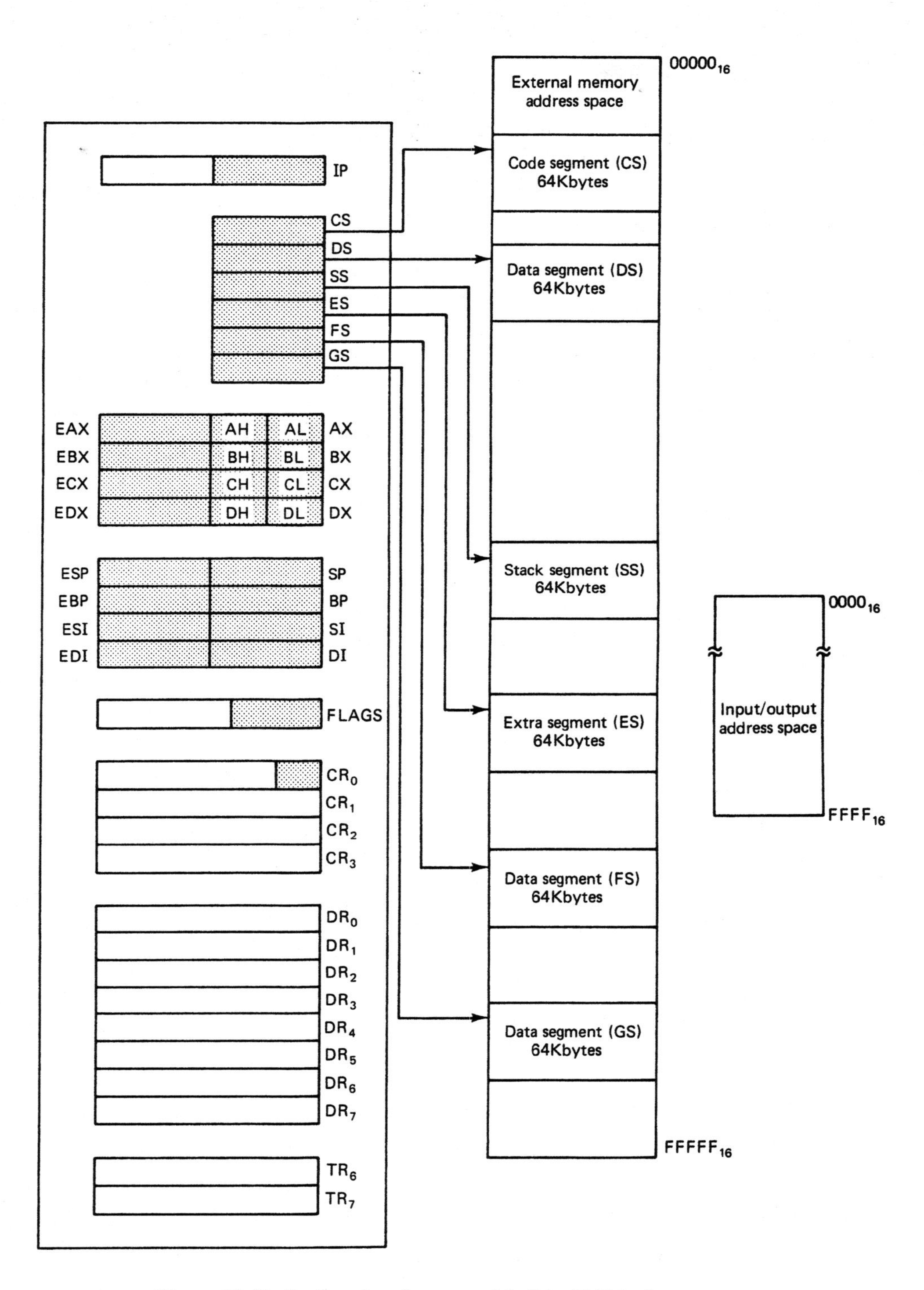

Figure 15–3 Real-mode software model of the 80386 microprocessor.

only bit in CR_0 that is active in the real mode is bit 0, which is the *protection enable* (PE) *bit.* PE is the bit used to switch the 80386DX from real to protected mode. At reset, PE is set to 0 and selects real-mode operation. The software model of the 80386SX is exactly the same as that shown in Fig. 15–3.

Looking at the software model in Fig. 15–3, we see that the 80386DX microcomputer's real-mode address space is identical to that of the 8086 and 80286 microcomputer. Again, it is partitioned into a 1Mbyte memory address space and a separate 64Kbyte input/output address space. The memory address space is from address 00000_{16} to $FFFFF_{16}$ and the I/0 addresses space is from address 0000_{16} to $FFFF_{16}$. Since the 80386DX has six segment registers, not four as in the 8086 and 80286, six 64Kbyte segments of the memory address space are active at a time and give a maximum of 384Kbytes of active memory; 64Kbytes of the active memory are allocated for code, 64Kbytes for stack, and 256Kbytes for data storage. Figures 15–4(a) and (b) show that the real-mode 80386DX memory and I/O address spaces are partitioned into general-use and reserved areas in the same way as for the 8086 or 80286 microcomputer. Memory in

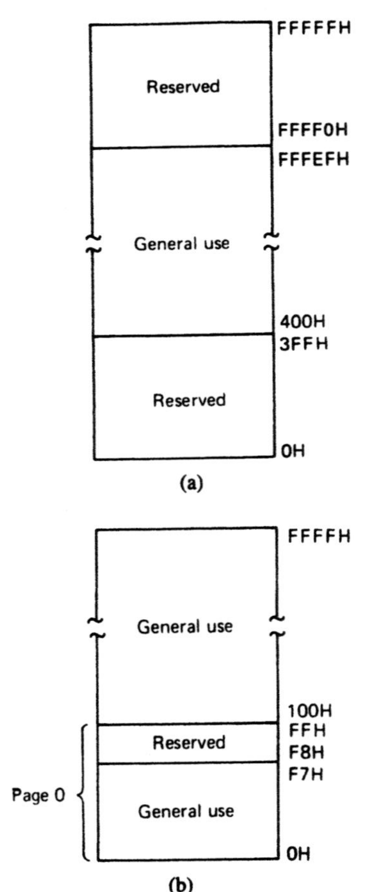

Figure 15–4 (a) Dedicated and general use of memory in the real mode. (Reprinted by permission of Intel Corporation. Copyright/Intel Corp. 1979) (b) I/O address space. (Reprinted by permission of Intel Corporation. Copyright/Intel Corp. 1979)

an 80386SX-based microcomputer system is organized in the same way from a software point of view.

Finally, the real-mode 80386DX generates physical addresses in the same manner as the 8086 or 80286. That is, the 16-bit contents of a segment register, such as CS, are shifted left by four bit positions, the four least significant bits are filled with 0s, and then it is added to a 16-bit offset, such as the value in IP, to form the 20-bit physical memory address. Note that the IP register is shown to be larger that 16 bits in the software model, but only the lower 16 bits are active when the 80386DX is in real mode.

▲ 15.4 REAL-ADDRESS-MODE INSTRUCTION SET OF THE 80386DX

Figure 15–5 shows the evolution of the instruction set for the 8086 architecture. The instruction set of the 8086 and 8088 microprocessors, called the *base instruction set,* is a subset of the 80386DX's real-address-mode instruction set. The instructions of the base instruction set were covered in detail in Chapters 5 and 6.

This base instruction set was enhanced in the 80286 microprocessor with a group of instructions known as the *extended instruction set.* All these instructions are also available in the real-mode instruction set of the 80386DX. We studied the operation of these instructions in Chapter 14.

The last group of instructions, identified in Fig. 15–5 as the *80386 specific instruction set,* was first implemented with the 80386DX microprocessor. In this way, we see that the 80386DX's real-mode instruction set is a superset of the 8086 and 80286 microprocessor instruction sets. The 80386SX microprocessor executes the exact same instruc-

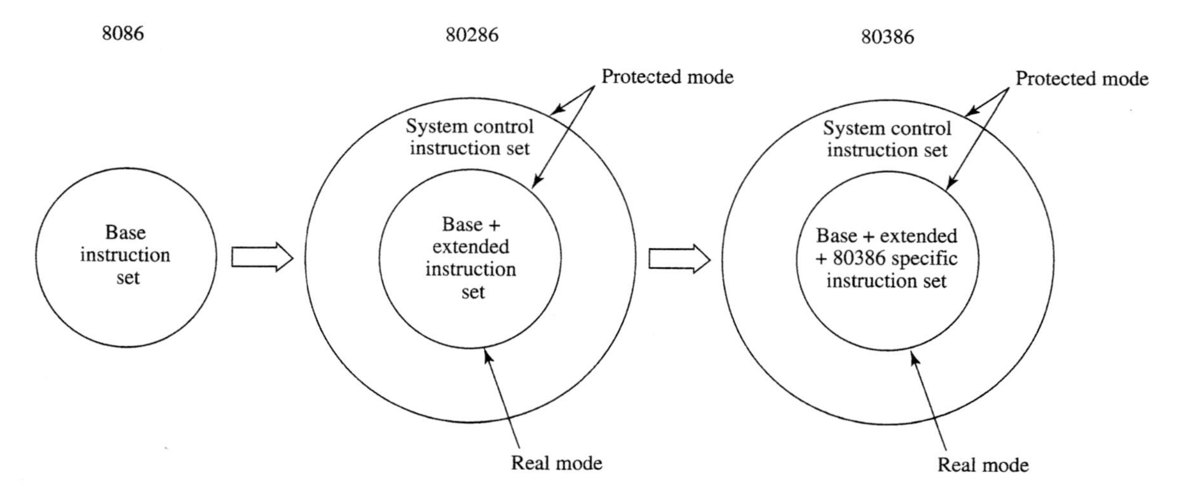

Figure 15–5 Evolution of the 8086 family instruction set.

tion set as the 80386DX. We will now continue by examining the instructions of the 80386 specific instruction set.

80386 Specific Instruction Set

The enhancements to the 80386DX's real-mode instruction set, highlighted in Fig. 15–6(a), represent the 80386 specific instruction set. For example, it includes instructions to directly load a pointer into the FS, GS, and SS registers. Moreover, it contains additional forms of existing instructions that have been added to perform the identical operations in a more general way, on data in special registers, or on a double word of data. First, we will look briefly at some of the instructions with expanded functions.

Figure 15–6(b) shows that the MOV instructions can be written with a control register (CR), debug register (DR), or test register (TR) as its source or destination operand. As an example, let us look at what function the instruction

```
MOV EAX, CR0
```

performs. Execution of this instruction causes the value of the flags in CR_0 to be copied into the EAX register. Looking at Fig. 15–6(b), we see that the string instructions have been expanded to support double-word (32-bit) operands. The instruction mnemonics for the double-word string operations are MOVSD, CMPSD, SCASD, LODSD, STOSD, INSD, and OUTSD. In all seven cases, the basic operation performed by the instruction is the same as described earlier for the 8088 processor; however, a double-word data transfer takes place. The same is true for the shift, convert, compare, jump, push, and pop instructions in Fig. 15–6(b)—they simply perform their normal operation on double-word operands. One exception is that the 80386DX limits the count for shift instructions to a count of 32, instead of 256 as on the 8086 or 80286. Let us next look at the instructions that are implemented for the first time on the 80386DX MPU.

Sign-Extend and Zero-Extend Move Instructions: MOVSX and MOVZX

In Fig. 15–6(b), we find that a number of special-purpose move instructions have been added in the instruction set of the 80386DX. The first two instructions, *move with sign-extend* (MOVSX) and *move with zero-extend* (MOVZX), are used to sign extend or zero extend, respectively, a source operand as it is moved to the destination operand location. The source operand is either a byte or a word of data in a register or a storage location in memory, whereas the destination operand is either a 16- or 32-bit register.

For example, the instruction

```
MOVSX EBX, AX
```

is used to copy the 16-bit value in AX into EBX. As the copy is performed, the value in the sign bit, bit 15 of AX, is extended into the 16 higher-order bits of EBX. For example, if AX contains $FFFF_{16}$, the sign bit is logic 1. Therefore, after execution of the MOVSX

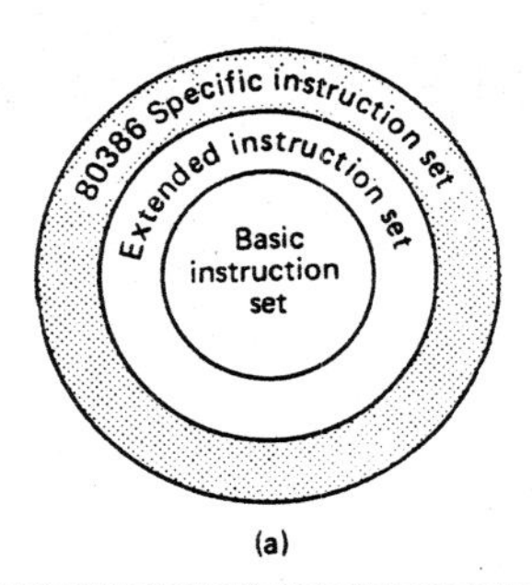

(a)

Mnemonic	Meaning	Format	Operation
MOV	Move	MOV D, S	Moves the contents of a special register to/from a general purpose register. Special registers are CR0, CR2, CR3, DR0–DR3, DR6, DR7, TR6, and TR7.
Double word operand instructions	–	–	Operation extended to double word, operands. The instructions include: MOVSD, CMPSD, SCASD, LODSD, STOSD, INSD, OUTSD, JMP rel/32, JMP r/m32, CMPSD, CWD/CDQ, POPAD, POPFD, PUSHAD, PUSHFD, SHLD, AND SHRD.
MOVSX	Move with sign-extend	MOVSX D, S	Sign extend the S and place in D. Allows for 8 to 16 or 16 to 32 bit sign extension from a source (r/m8, or r/m16) to a destination (r16 or r32).
MOVZX	Move with zero-extend	MOVZX D, S	Zero extend the S and place in D. Allows for 8 to 16 or 16 to 32 bit zero extension from a source (r/m8, or r/m16) to a destination (r16 or r32).
LSS	Load register and SS	LSS r16, m16:16 LSS r32, m16:32	Load stack segment register and the specified register with the pointer from memory.
LFS	Load register and FS	LFS r16, m16:16 LFS r32, m16:32	Load data segment register FS and the specified register with the pointer from memory.
LGS	Load register and GS	LGS r16, m16:16 LGS r32, m16:32	Load data segment register GS and the specified register with the pointer from memory.
BT	Bit test	BT opr1, opr2	CF ← BIT (opr1, opr2)
BTR	Bit test and reset	BTR opr1, opr2	CF ← BIT (opr1, opr2), BIT(opr1, opr2) ← 0
BTS	Bit test and set	BTR opr1, opr2	CF ← BIT (opr1, opr2), BIT(opr1, opr2) ← 1
BTC	Bit test and complement	BTC opr1, opr2	CF ← BIT (opr1, opr2), BIT(opr1, opr2) ← NOT BIT(opr1, opr2)
BSF	Bit scan forward	BSF r16, r/m16 BSF r32, r/m32	Scan the second operand starting from bit 0. ZF = 0 if all bits are 0, else ZF = 1 and the register (r16 or r32) contains the bit index of the first set bit.
BSR	Bit scan reverse	BSR r16, r/m16 BSR r32, r/m32	Scan the second operand starting from the MSB. ZF = 0 if all bits are 0, else ZF = 1 and the register (r16 or r32) contains the bit index of the first set bit.
SETcc	Set byte on condition	SETcc r/m8	r/m8 ← 1 if cc is true, r/m8 ← 0 if cc is not true

(b)

Figure 15–6 (a) 80386 specific instructions set. (b) Instructions of the 80386 specific instruction set.

instruction, the value that results in EBX is $FFFFFFFF_{16}$. The MOVZX instruction performs a similar function to the MOVSX instruction except that it extends the value moved to the destination operand location with 0s.

EXAMPLE 15.1

Explain the operation performed by the instruction

```
MOVZX CX, BYTE PTR [DATA_BYTE]
```

if the value of data at memory address DATA_BYTE is FF_{16}.

Solution

When the MOVZX instruction is executed, the value FF_{16} is copied into the lower byte of CX and the upper 8 bits are filled with 0s. This gives

$$(CX) = 00FF_{16}$$

Load Full Pointer Instructions: LSS, LFS, and LGS

The base instruction set of the 8086 includes two *load full pointer instructions,* LDS and LES. Three additional instructions of this type are performed by the 80386DX. Looking at Fig. 15–6(b), we find that they are LSS, LFS, and LGS. Note that executing the *load register and SS* (LSS) instruction causes both the register specified in the instruction and the stack segment register to be loaded from the source operand. For example, the instruction

```
LSS ESP, [STACK_POINTER]
```

causes the first 32 bits starting at memory address STACK_POINTER to be loaded into the 32-bit register ESP and the next 16 bits into the SS register. The other two instructions, *load register and FS* (LFS) and *load register and GS* (LGS), perform a function similar to LSS. However, they load the specified register and the FS or GS register, respectively.

EXAMPLE 15.2

Write an instruction that will load the 48-bit pointer starting at memory address DATA_G_ADDRESS into the ESI and GS registers.

Solution

This operation is performed with the instruction

```
LGS ESI, [DATA_G_ADDRESS]
```

Bit Test and Bit Scan Instructions: BT, BTR, BTS, BTC, BSF, and BSR

The *bit test and bit scan instructions* of the 80386DX enable a programmer to test the logic value of a bit in either a register or a storage location in memory. Let us begin by examining the bit test instructions. They are used to test the state of a single bit in a register or memory location. When the instruction is executed, the value of the tested bit is saved in the carry flag. Instructions are provided that can also reset, set, or complement the contents of the tested bit during the execution of the instruction.

In Fig. 15.6(b), we see that the *bit test* (BT) instruction has two operands. The first operand identifies the register or memory location that contains the bit that is to be tested. The second operand contains an index that selects the bit that is to be tested. Note that the index may be either an immediate operand or the value in a register. When this instruction is executed, the state of the tested bit is simply copied into the carry flag.

Once the state of the bit is saved in CF, it can be tested through software. For instance, a conditional jump instruction could be used to test the value in CF, and if CF equals 1, program control could be passed to a service routine. On the other hand, if CF equals 0, the value of the index could be incremented, a jump performed back to the BT instruction, and the next bit in the operand tested.

Another example is the instruction

```
BTR EAX, EDI
```

Execution of this instruction causes the bit in 32-bit register EAX that is selected by the index in EDI to be tested. The value of the tested bit is first saved in the carry flag and then it is reset in the register EAX.

EXAMPLE 15.3

Describe the operation that is performed by the instruction

```
BTC BX, 7
```

Assume that register BX contains the value $03F0_{16}$.

Solution

Let us first express the value in BX in binary form. This gives

$$(BX) = 0000001111110000_2$$

Execution of the *bit test and complement* instruction causes the value of bit 7 to be first tested and then complemented. Since this bit is logic 1, CF is set to 1. This gives

$$(CF) = 1$$

$$(BX) = 0000001101110000_2 = 0370_{16}$$

The *bit scan forward* (BSF) and *bit scan reverse* (BSR) instructions are used to scan through the bits of a register or storage location in memory to determine whether or not they are all 0. For example, by executing the instruction

```
BSF ESI, EDX
```

the bits of 32-bit register EDX are tested one after the other, starting from bit 0. If all bits are found to be 0, the ZF is cleared. On the other hand, if the contents of EDX are not zero, ZF is set to 1 and the index value of the first bit tested as 1 is copied into ESI.

Byte Set on Condition: SETcc

The *byte set on condition* (SETcc) instruction can be used to test for various states of the flags. In Fig. 15–6(b), we see that the general form of the instruction is denoted as

```
SETcc D
```

Here the cc part of the mnemonic stands for a general-flag relationship and must be replaced with a specific relationship when writing the instruction. Figure 15–7 is a list of the mnemonics that can be used to replace cc and their corresponding flag relationship.

Instruction	Meaning	Conditions code relationship
SETA *r/m8*	Set byte if above	CF = 0 · ZF = 0
SETAE *r/m8*	Set byte if above or equal	CF = 0
SETB *r/m8*	Set byte if below	CF = 1
SETBE *r/m8*	Set byte if below or equal	CF = 1 + ZF = 1
SETC *r/m8*	Set if carry	CF = 1
SETE *r/m8*	Set byte if equal	ZF = 1
SETG *r/m8*	Set byte if greater	ZF = 0 + SF = OF
SETGE *r/m8*	Set byte if greater	SF = OF
SETL *r/m8*	Set byte if less	SF <> OF
SETLE *r/m8*	Set byte if less or equal	ZF = 1 · SF <> OF
SETNA *r/m8*	Set byte if not above	CF = 1
SETNAE *r/m8*	Set byte if not above	CF = 1
SETNB *r/m8*	Set byte if not below	CF = 0
SETNBE *r/m8*	Set byte if not below	CF = 0 · ZF = 0
SETNC *r/m8*	Set byte if not carry	CF = 0
SETNE *r/m8*	Set byte if not equal	ZF = 0
SETNG *r/m8*	Set byte if not greater	ZF = 1 + SF <> OF
SETNGE *r/m8*	Set if not greater or equal	SF <> OF
SETNL *r/m8*	Set byte if not less	SF = OF
SETNLE *r/m8*	Set byte if not less or equal	ZF = 1 · SF <> OF
SETNO *r/m8*	Set byte if not overflow	OF = 0
SETNP *r/m8*	Set byte if not parity	PF = 0
SETNS *r/m8*	Set byte if not sign	SF = 0
SETNZ *r/m8*	Set byte if not zero	ZF = 0
SETO *r/m8*	Set byte if overflow	OF = 1
SETP *r/m8*	Set byte if parity	PF = 1
SETPE *r/m8*	Set byte if parity even	PF = 1
SETPO *r/m8*	Set byte if parity odd	PF = 0
SETS *r/m8*	Set byte if sign	SF = 1
SETZ *r/m8*	Set byte if zero	ZF = 1

Figure 15–7 SET instruction conditions.

For instance, replacing cc by A gives the mnemonic SETA. This stands for *set byte if above* and tests the flags to determine if

$$(CF) = 0 \cdot (ZF) = 0$$

If these conditions are satisfied, a byte of 1s is written to the register or memory location specified as the destination operand. On the other hand, if the conditions are not valid, a byte of 0s is written to the destination operand.

An example is the instruction

```
SETE AL
```

Looking at Fig. 15–7, we find that execution of this instruction causes the ZF to be tested. If ZF equals 1, 11111111_2 is written into AL; otherwise, it is loaded with 00000000_2.

EXAMPLE 15.4

Write an instruction that will load memory location EVEN_PARITY with the value FF_{16} if the result produced by the last instruction had even parity.

Solution

As shown in Fig. 15–7, the instruction that tests for PF equal to 1 and sets the byte destination operand at memory address EVEN_PARITY to FF_{16} is

```
SETPE  BYTE PTR [EVEN_PARITY]
```

▲ 15.5 PROTECTED-ADDRESS-MODE SOFTWARE ARCHITECTURE OF THE 80386DX

Having completed our study of the real-mode software operation and instruction set of the 80386DX microprocessor, we are now ready to turn our attention to its protected-address mode (protected mode) of operation. Earlier we indicated that whenever the 80386DX microprocessor is reset, it comes up in real mode. Moreover, we indicated that the PE bit of control register zero (CR_0) is used to switch the 80386DX into the protected mode under software control. When configured for protected-mode operation, the 80386DX provides an advanced software architecture that supports memory management, virtual addressing, paging, protection, and multitasking. In this section we examine the 80386DX's protected-mode register model, virtual-memory address space, and memory management.

Protected-Mode Register Model

Figure 15–8 shows the protected-mode register set of the 80386DX microprocessor. Looking at this diagram, we see that its application register model is a superset of the real-mode register set shown in Fig. 15–3. Comparing these two diagrams, we find four

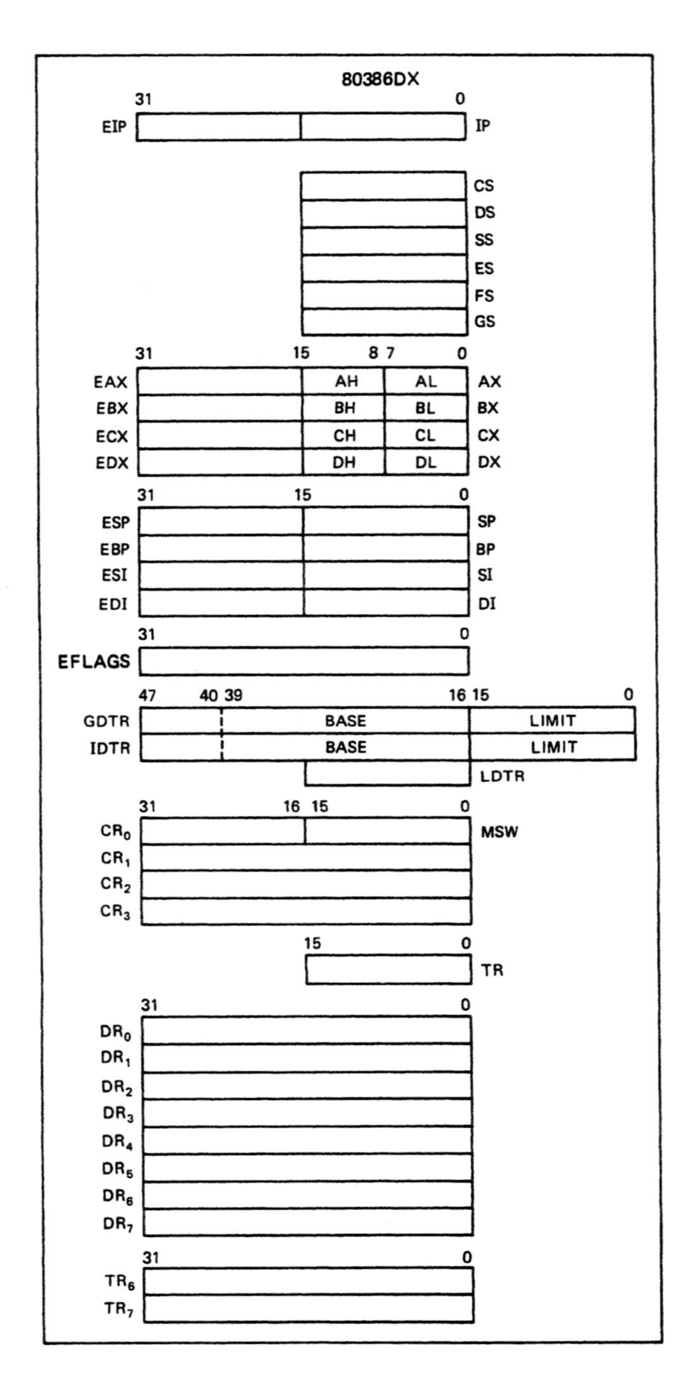

Figure 15–8 Protected-mode register model.

new registers in the protected-mode model: the *global descriptor table register* (GDTR), *interrupt descriptor table register* (IDTR), *local descriptor table register* (LDTR), and *task register* (TR). Furthermore, the functions of a few registers have been extended. For example, the instruction pointer, now called EIP, is 32 bits in length; more bits of the flag register (EFLAGS) are active; and all four control registers, CR_0 through CR_3, are functional. Let us next discuss the purpose of each new and extended register and how they are used in the segmented memory-protected-mode operation of the microprocessor.

Global Descriptor Table Register. As shown in Fig. 15–9, the contents of the *global descriptor table register* define a table in the 80386DX's physical memory address space called the *global descriptor table* (GDT). This global descriptor table is one important element of the 80386DX's memory management system.

GDTR is a 48-bit register located inside the 80386DX. The lower 2 bytes of this register, identified as LIMIT in Fig. 15–9, specify the size in bytes of the GDT. The value of LIMIT is one less than the actual size of the table. For instance, if LIMIT equals $00FF_{16}$, the table is 256 bytes in length. Since LIMIT has 16 bits, the GDT can be up to 65,536 bytes long. The upper 4 bytes of the GDTR, labeled BASE in Fig. 15–9, locate the beginning of the GDT in physical memory. This 32-bit base address allows the table to be positioned anywhere in the 80386DX's 4Gbyte linear address space.

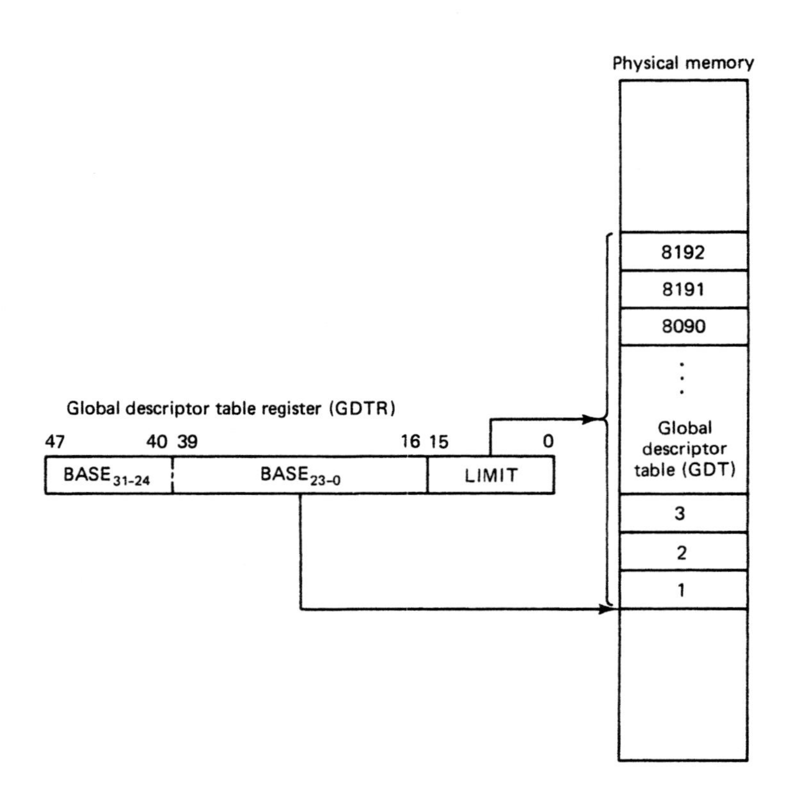

Figure 15–9 Global descriptor table mechanism.

EXAMPLE 15.5

If the limit and base in the global descriptor table register are $0FFF_{16}$ and 00100000_{16}, respectively, what is the beginning address of the descriptor table, size of the table in bytes, and the ending address of the table?

Solution

The starting address of the global descriptor table in physical memory is given by the BASE. Therefore,

$$GDT_{START} = 00100000_{16}$$

The limit is the offset to the end of the table. This gives

$$GDT_{END} = 00100000_{16} + 0FFF_{16} = 00100FFF_{16}$$

Finally, the size of the table is equal to the decimal value of LIMIT plus 1:

$$GDT_{SIZE} = FFF_{16} + 1_2 = 4096 \text{ bytes}$$

The GDT provides a mechanism for defining the characteristics of the 80386DX's global memory address space. *Global memory* is a general system resource shared by many or all software tasks. That is, storage locations in global memory are accessible by any task that runs on the microprocessor.

This table contains what are called *system segment descriptors*. These descriptors identify the characteristics of the segments of global memory. For instance, a segment descriptor provides information about the size, starting point, and access rights of a global memory segment. Each descriptor is 8 bytes long; thus, our earlier example of a 256-byte table provides storage space for just 32 descriptors. Remember that the size of the global descriptor table can be expanded by simply changing the value of LIMIT in the GDTR under software control. If the table is increased to its maximum size of 65,536 bytes, it can hold up to 8192 descriptors.

EXAMPLE 15.6

How many descriptors can be stored in the global descriptor table defined in Example 15.5?

Solution

Each descriptor takes up 8 bytes; therefore, a 4096-byte table can hold

$$4096/8 = 512 \text{ descriptors}$$

The value of the BASE and LIMIT must be loaded into the GDTR before the 80386DX is switched from the real mode of operation to the protected mode. Special instructions are provided for this purpose in the system control instruction set of the 80386DX. These instructions will be introduced later in this chapter. Once the 80386DX is in protected mode, the location of the table is typically not changed.

Interrupt Descriptor Table Register. Just like the global descriptor table register, the *interrupt descriptor table register* (IDTR) defines a table in physical memory. However, this table contains what are called *interrupt descriptors,* not segment descriptors. For this reason, it is known as the *interrupt descriptor table* (IDT). This register and table of descriptors provide the mechanism by which the microprocessor passes program control to interrupt and exception service routines.

As shown in Fig. 15–10, just like the GDTR, the IDTR is 48 bits in length. Again, the lower two bytes of the register (LIMIT) define the table size. That is, the size of the table equals LIMIT+1 bytes. Since two bytes define the size, the IDT can also be up to 65,536 bytes long. But the 80386DX only supports up to 256 interrupts and exceptions; therefore, the size of the IDT should not be set to support more than 256 interrupts. The upper 4 bytes of IDTR (BASE) identify the starting address of the IDT in physical memory.

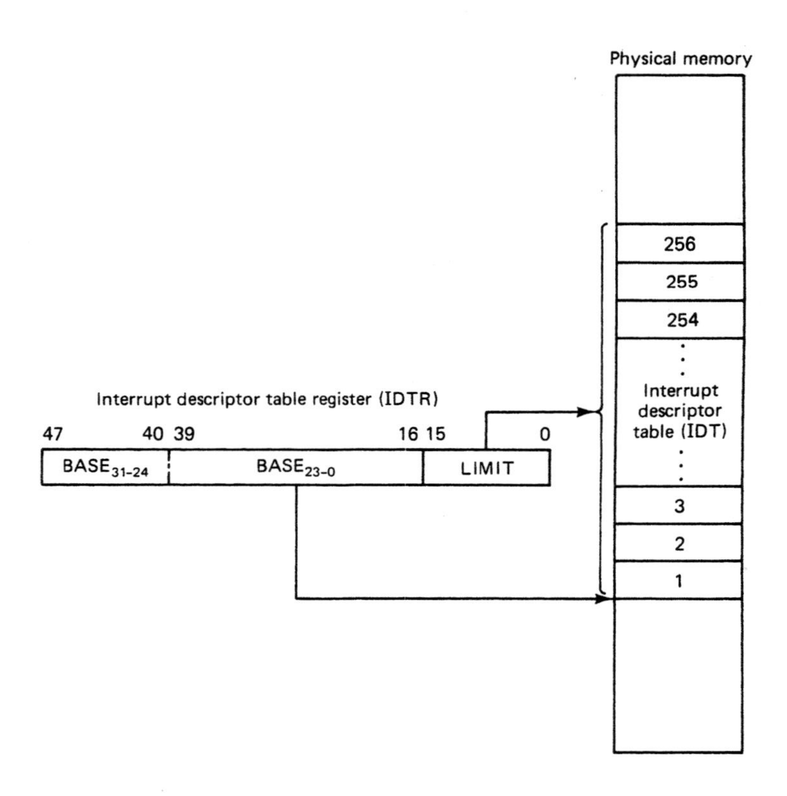

Figure 15–10 Interrupt descriptor table mechanism.

The types of descriptors used in the IDT are called *interrupt gates.* These gates provide a means for passing program control to the beginning of an interrupt service routine. Each gate is 8 bytes long and contains both attributes and a starting address for the service routine.

EXAMPLE 15.7

What is the maximum value that should be assigned to LIMIT in the IDTR of 80386DX?

Solution

The maximum number of interrupt descriptors that can be used in an 80386DX microcomputer system is 256. Therefore, the maximum table size in bytes is

$$IDT_{SIZE} = 8_{10} \times 256 = 4096 \text{ bytes}$$

Thus,

$$LIMIT = 0FFF_{16}$$

This table can also be located anywhere in the linear address space addressable with the 80386DX's 32-bit address. Just like the GDTR, the IDTR needs to be loaded before the 80386DX enters protected mode. Special instructions are provided for loading and saving the contents of the IDTR. Once the location of the table is set, it is typically not changed after entering protected mode.

EXAMPLE 15.8

What is the address range of the last descriptor in the interrupt descriptor table defined by base address 00011000_{16} and limit $01FF_{16}$?

Solution

From the values of the base and limit, we find that the table is located in the address range defined by

$$IDT_{START} = 00011000_{16}$$

and

$$IDT_{END} = 000111FF_{16}$$

The last descriptor in this table takes up the 8 bytes of memory from address $000111F8_{16}$ through $000111FF_{16}$.

Local Descriptor Table Register. The *local descriptor table register* (LDTR) is also part of the 80386DX's memory management support mechanism. As Fig. 15–11(a) shows, each task can have access to its own private descriptor table in addition to the global descriptor table. This private table is called the *local descriptor table* (LDT) and defines a *local memory* address space for use by the task. The LDT holds segment descriptors that provide access to code and data in segments of memory that are reserved for the current task. Since each task can have its own segment of local memory, the protected-mode software system may contain many local descriptor tables. For this reason, we have identified LDT_0 through LDT_N in Fig. 15–11(a).

Figure 15–11(b) shows us that the contents of the 16-bit LDTR do not directly define the local descriptor table. Instead, it holds a selector that points to an *LDT descriptor* in the GDT. Whenever a selector is loaded into the LDTR, the corresponding descriptor is transparently read from global memory and loaded into the *local descriptor table cache* within the 80386DX. It is this descriptor that defines the local descriptor table. As shown in Fig. 15–11(b), the 32-bit BASE value identifies the starting point of the table in physical memory, and the value of the 16-bit LIMIT determines the size of the table. Loading of this descriptor into the cache creates the LDT for the current task. That is, every time a selector is loaded into the LDTR, a local descriptor-table descriptor is cached and a new LDT is activated.

Control Registers. The protected-mode model includes the four system-control registers, identified as CR_0 through CR_3 in Fig. 15–8. Figure 15–12 shows these registers in more detail. Note that the lower 5 bits of CR_0 are system-control flags. These bits make up what is known as the *machine status word* (MSW). The most significant bit of CR_0 and registers CR_2 and CR_3 are used by the 80386DX's paging mechanism.

Let us continue by examining the machine status word bits of CR_0. They contain information about the 80386DX's protected-mode configuration and status. The four bits labeled PE, MP, EM, and R are control bits that define the protected-mode system configuration. The fifth bit, TS, is a status bit. These bits can be examined or modified through software.

The *protected-mode enable* (PE) bit determines if the 80386DX is in the real or protected mode. At reset, PE is cleared. This enables real-mode of operation. To enter protected mode, we simply switch PE to 1 through software. Once in protected mode, the 80386DX cannot be switched back to real mode under software control by clearing the PE bit. The only way to return to real mode is by initiating hardware reset.

The *math present* (MP) bit is set to 1 to indicate that a numeric coprocessor is present in the microcomputer system. On the other hand, if the system is to be configured so that a software emulator is used to perform numeric operations, the *emulate* (EM) bit is set to 1. Only one of these two bits can be set at a time. Finally, the *extension-type* (R) bit is used to indicate whether an 80287 or 80387 numeric coprocessor is in use. Logic 1 in R indicates that an 80387 is installed. The last bit in the MSW, *task switch* (TS), automatically gets set whenever the 80386DX switches from one task to another. It can be cleared under software control.

The protected-mode software architecture of the 80386DX also supports paged memory operation. Switching the PG bit in CR_0 to logic 1 turns on paging. Now addressing of physical memory is implemented with an address translation mechanism that con-

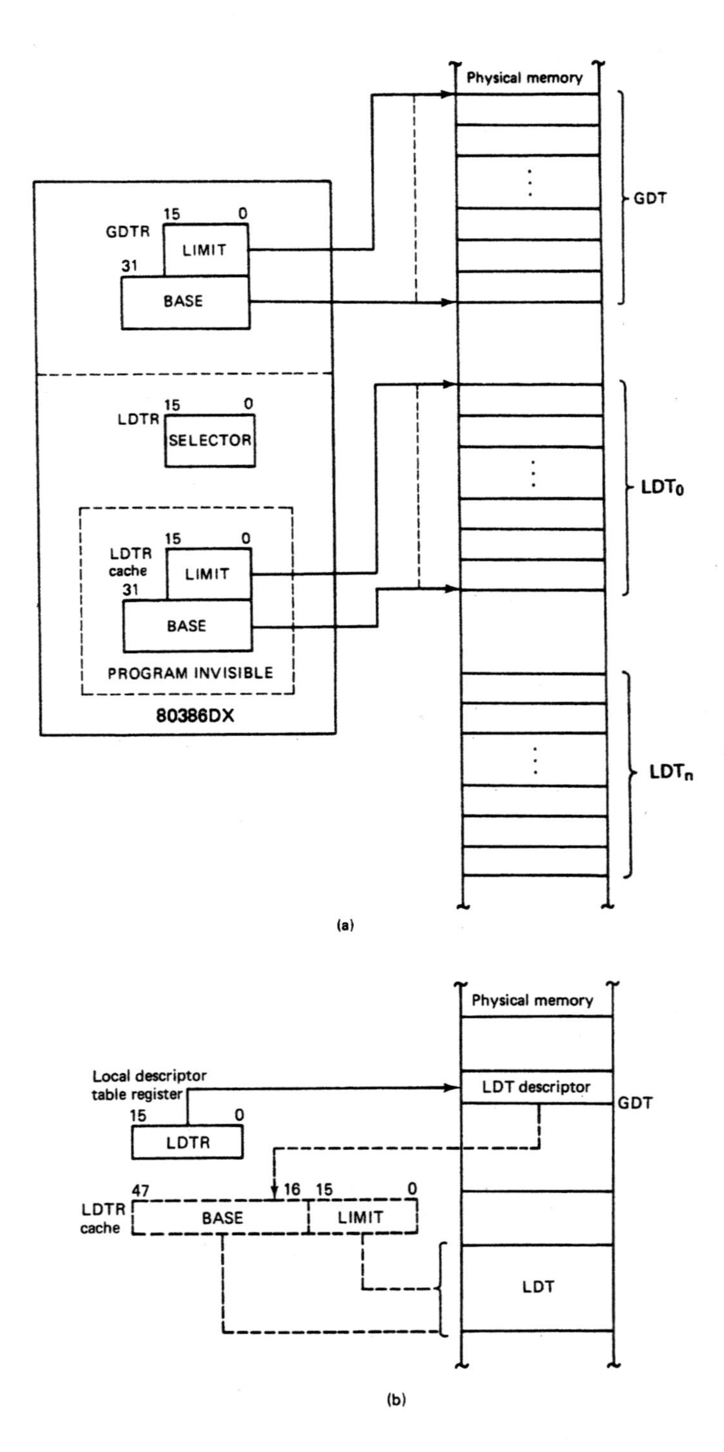

Figure 15–11 (a) Task with global and local descriptor table. (b) Loading the local descriptor table register to define a local descriptor table.

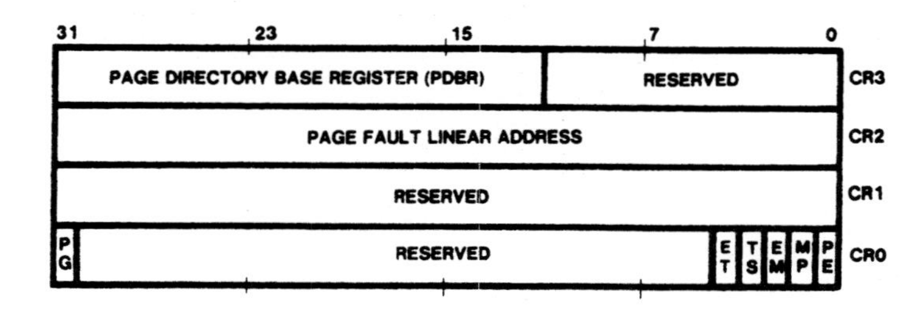

Figure 15–12 Control registers. (Reprinted by permission of Intel Corporation. Copyright/Intel Corp. 1986)

sists of a page directory and page table, which are both held in physical memory. Figure 15–12 shows that CR_3 contains the *page directory base register* (PDBR). This register holds a 20-bit *page directory base address* that points to the beginning of the page directory. A page-fault error occurs during the page-translation process if the page is not present in memory. In this case, the 80386DX saves the address at which the page fault occurred in register CR_2. This address is denoted as *page-fault linear address* in Fig. 15–12.

Task Register. The *task register* (TR) is a key element in the protected-mode task switching mechanism of the 80386DX microprocessor. This register holds a 16-bit index value called a *selector.* The initial selector must be loaded into TR under software control. This starts the initial task. After this is done, the selector is changed automatically whenever the 80386DX executes an instruction that performs a task switch.

As shown in Fig. 15–13, the selector in TR is used to locate a descriptor in the global descriptor table. Note that when a selector is loaded into TR, the corresponding *task state segment* (TSS) *descriptor* automatically gets read from memory and loaded into

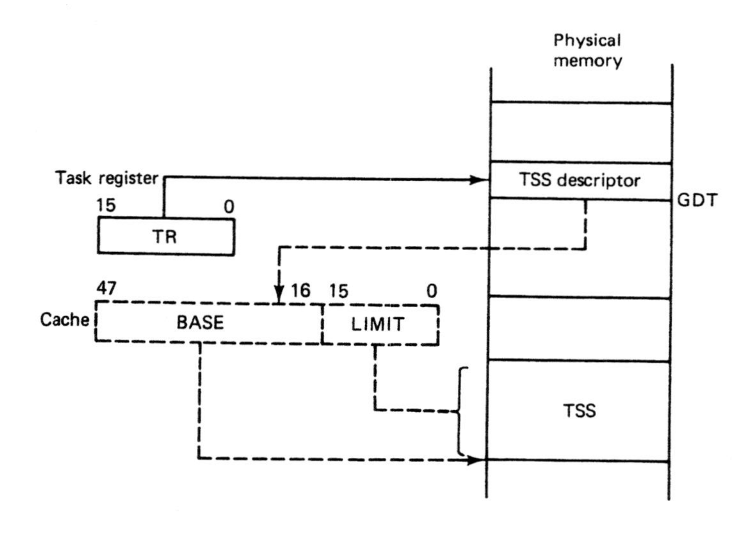

Figure 15–13 Task register and the task-switching mechanism.

the on-chip *task descriptor cache.* This descriptor defines a block of memory called the *task state segment* (TSS). It does this by providing the starting address (BASE) and the size (LIMIT) of the segment. Every task has its own TSS. The TSS holds the information needed to initiate the task, such as initial values for the user-accessible registers.

EXAMPLE 15.9

What is the maximum size of a TSS? Where can it be located in the linear address space?

Solution

Since the value of LIMIT is 16 bits in length, the TSS can be as long as 64Kbytes. Moreover, the base is 32 bits in length. Therefore, the TSS can be located anywhere in the 80386DX's 4Gbyte address space.

EXAMPLE 15.10

Assume that the base address of the global descriptor table is 00011000_{16} and the selector in the task register is 2108_{16}. In what address range is the TSS descriptor stored?

Solution

The beginning address of the TSS descriptor is

$$\begin{aligned} \text{TSS\_DESCRIPTOR}_{\text{START}} &= 00011000_{16} + 2108_{16} \\ &= 00013108_{16} \end{aligned}$$

Since the descriptor is 8 bytes long, it ends at

$$\text{TSS\_DESCRIPTOR}_{\text{END}} = 0001310F_{16}$$

Registers with Changed Functionality. Earlier we pointed out that the function of a few of the registers common to both the real-mode and protected-mode register models changes as the 80386DX is switched into the protected mode of operation. For instance, the segment registers are now called the *segment selector registers,* and instead of holding a base address they are loaded with what is known as a *selector.* The selector does not directly specify a storage location in memory. Instead, it selects a descriptor that defines the size and characteristics of a segment of memory.

The format of a selector is shown in Fig. 15–14. Here we see that the two least significant bits are labeled RPL, which stands for *requested privilege level.* These bits contain 00 = 0, 01 = 1, 10 = 2, or 11 = 3 and assign a request protection level to the selector. The next bit, identified as *task indicator* (TI) in Fig. 15–14, selects the table to be used when accessing a segment descriptor. Remember that in protected mode two descriptor tables are active at a time, the global descriptor table and a local descriptor table. Looking at Fig. 15–14, we find that if TI is 0, the selector corresponds to a descrip-

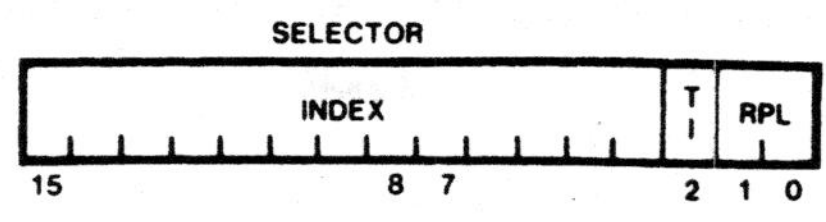

BITS	NAME	FUNCTION
1-0	REQUESTED PRIVILEGE LEVEL (RPL)	INDICATES SELECTOR PRIVILEGE LEVEL DESIRED
2	TABLE INDICATOR (TI)	TI 0 USE GLOBAL DESCRIPTOR TABLE (GDT) TI 1 USE LOCAL DESCRIPTOR TABLE (LDT)
15-3	INDEX	SELECT DESCRIPTOR ENTRY IN TABLE

Figure 15–14 Selector format. (Reprinted by permission of Intel Corporation. Copyright/Intel Corp. 1986)

tor in the global descriptor table. Finally, the 13 most significant bits contain an *index* that is used as a pointer to a specific descriptor entry in the table selected by the TI bit.

EXAMPLE 15.11

Assume that the base address of the LDT is 00120000_{16} and the GDT base address is 00100000_{16}. If the value of the selector loaded into the CS register is 1007_{16}, what is the request privilege level? Is the segment descriptor in the GDT or LDT? What is the address of the segment descriptor?

Solution

Expressing the selector value in binary form, we get

$$(CS) = 0001000000000111_2$$

Since the two least significant bits are both 1,

$$RPL = 3$$

The next bit, bit 2, is also 1. This means that the segment descriptor is in the LDT. Finally, the value in the 13 most significant bits must be scaled by 8 to give the offset of the descriptor from the base address of the table. Therefore,

$$\begin{aligned} OFFSET &= 0001000000000_2 \times 8_{10} = 512_{10} \times 8_{10} = 4096_{10} \\ &= 1000_{16} \end{aligned}$$

and the address of the segment descriptor is

$$\begin{aligned} DESCRIPTOR_{ADDRESS} &= 00120000_{16} + 1000_{16} \\ &= 00121000_{16} \end{aligned}$$

Another register whose function changes when the 80386DX is switched to protected mode is the flag register. As Fig. 15–8 shows, the flag register is now identified as EFLAGS and expands to 32 bits in length. The functions of the bits in EFLAGS are given in Fig. 15–15. Comparing this illustration to the 8086 flag register in Fig. 2–17, we see that five additional bits are implemented. These bits are only active when the 80386DX is in protected mode. They are the 2-bit *input/output privilege level* (IOPL) code, the *nested task* (NT) flag, the *resume* (RF) flag, and the *virtual 8086 mode* (VM) flag.

Note in Fig. 15–15 that each of these flags is identified as a system flag. That is, they represent protected-mode system operations. For example, the IOPL bits are used to assign a maximum privilege level to input/output. If 00 is loaded into IOPL, I/O can only be performed when the 80386DX is in the highest privilege level, level 0. On the other hand, if IOPL is 11, I/O is assigned to the lowest privilege level, level 3.

The NT flag identifies whether or not the current task is a nested task—that is, if it was called from another task. This bit is automatically set whenever a nested task is initiated and can only be reset through software.

Protected-Mode Memory Management and Address Translation

Up to this point in the section, we have introduced the register set of the protected-mode software model for the 80386DX microprocessor. However, the software model of a microprocessor also includes its memory structure. Because of the memory-management capability of the 80386DX, the organization of protected-mode memory appears quite complex. Here we will examine how the *memory-management unit* (MMU) of the 80386DX implements the address space and how it translates virtual (logical) addresses to physical addresses. We begin here with what are called the *segmented-* and *paged-models* of memory.

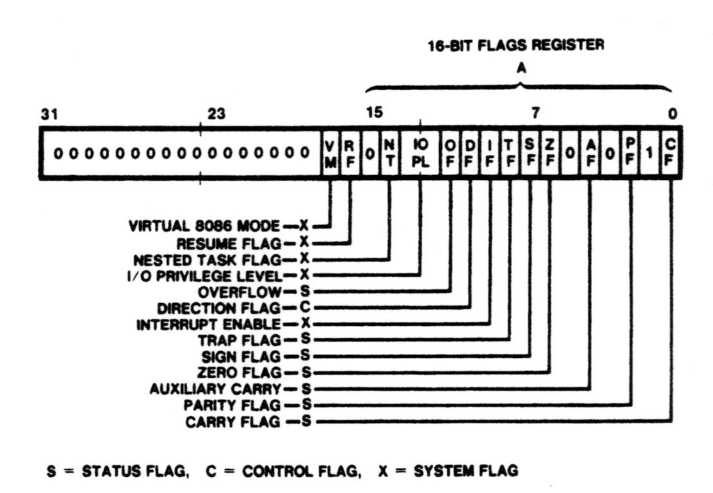

Figure 15–15 Protected-mode flag register. (Reprinted by permission of Intel Corp. Copyright/Intel Corp. 1986)

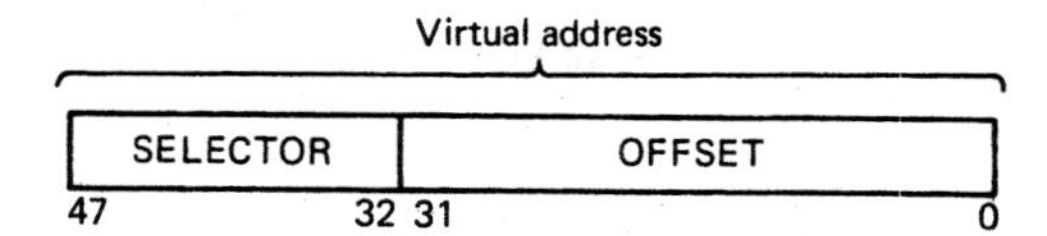

Figure 15–16 Protected-mode memory pointer.

Virtual Address and Virtual Address Space. The protected-mode memory management unit employs memory pointers that are 48 bits in length and consist of two parts, the *selector* and the *offset.* This 48-bit memory pointer is called a *virtual address* and is used to specify the memory locations of instructions or data. As shown in Fig. 15–16, the selector is 16 bits in length and the offset is 32 bits long. Earlier we pointed out that one source of selectors is the segment selector registers within the 80386DX. For instance, if code is being accessed in memory, the active segment selector is held in CS. This part of the pointer selects a unique segment of the 80386DX's *virtual address space.*

The offset is held in one of the 80386DX's other user-accessible registers. For our example of a code access, the offset would be in the EIP register. This part of the pointer is the displacement of the memory location that is to be accessed within the selected segment of memory. In our example, it points to the first byte of the double word of instruction code to be fetched for execution. Since the offset is 32 bits in length, segment size can be as large as 4Gbytes. We say as large as 4Gbytes because segment size is actually variable and can be defined to be as small as 1 byte to as large as 4Gbytes.

Figure 15.17 shows that the 16-bit selector breaks down into a *13-bit index, table select bit,* and two bits used for a *request privilege level.* The two RPL bits are not used in the selection of the memory segment. That is, just 14 of its 16 bits are employed in addressing memory. Therefore, the virtual address space consists of 2^{14} (16,384 = 16K) unique segments of memory, each of which has a maximum size of 4Gbytes. These segments are the basic elements into which the memory management unit of the 80386DX organizes the virtual address space.

Another way of looking at the size of the virtual address space is that by combining the 14-bit segment selector with the 32-bit offset we get a 46-bit virtual address. Therefore, the 80386DX's virtual address space contains 2^{46} equals 64Tbytes (terabytes) unique addresses.

Segmented Partitioning of the Virtual Address Space. The memory-management unit of the 80386DX implements both a segmented model and a paged model of virtual memory. In the segmented model, the 80386DX's 64Tbyte virtual address space is partitioned into a 32Tbyte *global memory address space* and a 32Tbyte *local memory address*

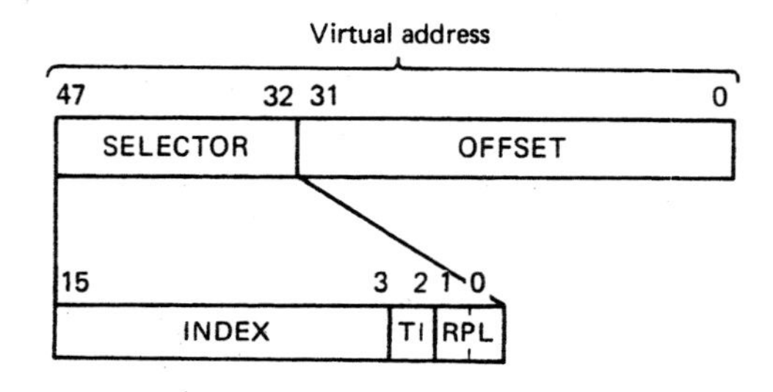

Figure 15–17 Segment selector format.

space. This partitioning is illustrated in Fig. 15–18. The TI bit of the selector shown in Fig. 15–17 is used to select between the global or local descriptor tables that define the virtual address space. Each of these address spaces may contain as many as 8,192 segments of memory. This assumes that every descriptor in both the global descriptor table and local descriptor table is in use and set for maximum size. These descriptors define the attributes of the corresponding segments. However, in practical system applications not all the descriptors are normally in use. Let us now look briefly at how software uses global and local segments of memory.

In the multiprocessing software environment of the 80386DX, an application is expressed as a collection of tasks. By *task* we mean a group of program routines that together perform a specific function. When the 80386DX initiates a task, it can activate both global and local segments of memory. Figure 15–19 illustrates this idea. Note that tasks 1, 2, and 3 each have a reserved segment of the local address space. This part of memory stores data or code that can only be accessed by the corresponding task. That is, task 2 cannot access any of the information in the local address space of task 1. On the other hand, all the tasks are shown to share the same segment of the global address space. This segment typically contains operating system resources and data that are to be shared by all or many tasks.

Physical Address Space and Virtual-to-Physical Address Translation. We have just found that the virtual address space available to the programmer is 64Tbytes in length. However, the 32-bit protected-mode address bus of the 80386DX supports only a 4Gbyte

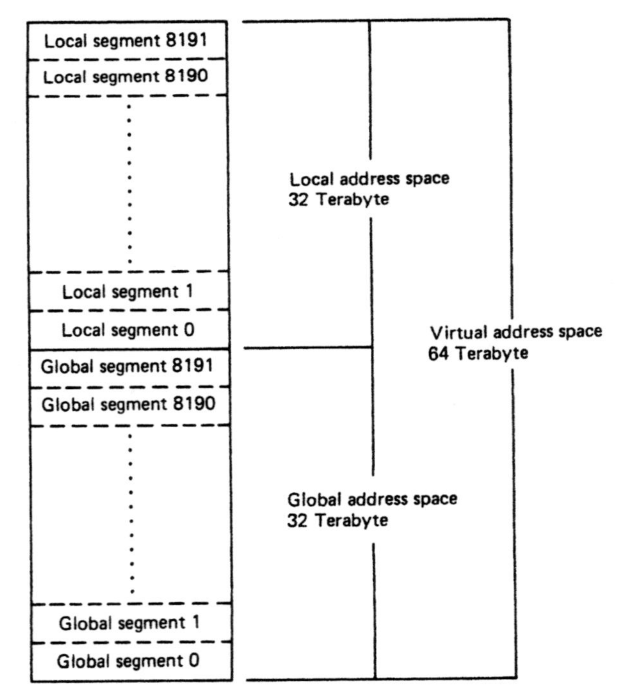

Figure 15–18 Partitioning the virtual address space.

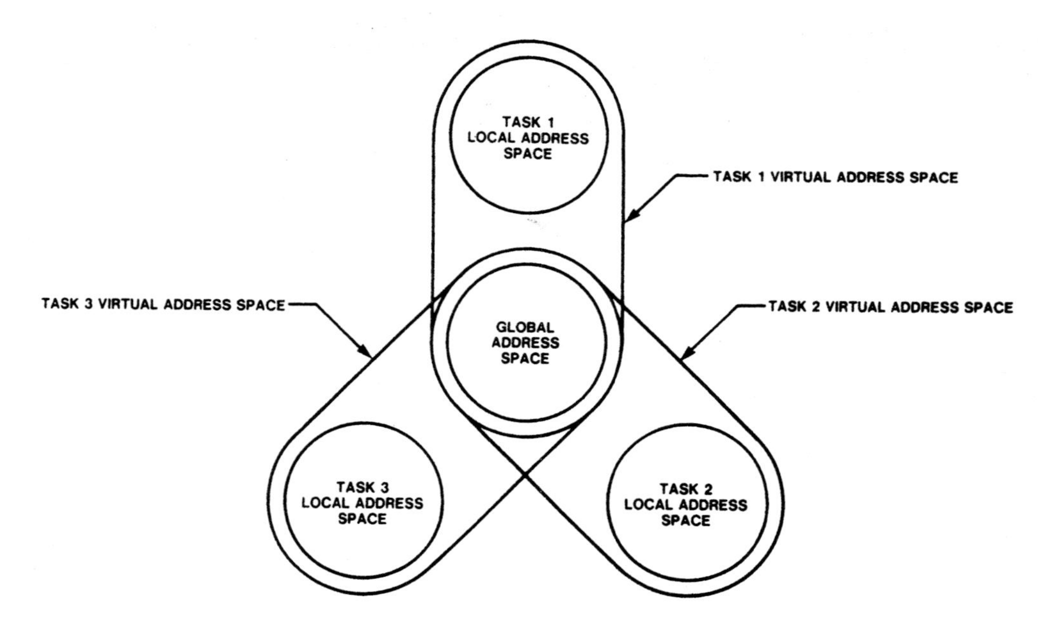

Figure 15–19 Global and local memory for a task. (Reprinted by permission of Intel Corporation. Copyright/Intel Corp. 1987)

physical address space. Just a small amount of the information in virtual memory can reside in physical memory at a time. For this reason systems that employ a virtual address space that is larger than the implemented physical memory are equipped with a secondary storage device such as a hard disk. The segments not currently in use are stored on disk.

If a program accesses a segment of memory not present in physical memory and space is available in physical memory, the segment is simply read from the hard disk and copied in physical memory. On the other hand, if the physical memory address space is full, another segment must first be sent out to the hard disk to make room for the new information. The memory-manager part of the operating system controls the allocation and deallocation of physical memory and the swapping of data between the hard disk and physical memory of the microcomputer. In this way, the memory address space of the microcomputer appears much larger than the physical memory in the microcomputer.

The segmentation and paging memory management units of the 80386DX provide the mechanism by which 48-bit virtual addresses are mapped into the 32-bit physical addresses needed by hardware. They employ a memory-based lookup table address-translation process. The diagram in Fig. 15–20 illustrates this address translation in general. Note that first a *segment translation* is performed on the virtual (logical) address. Then, if paging is disabled, the *linear address* produced is equal to the physical address. However, if paging is enabled, the linear address goes through a second translation process, known as *page translation,* to produce the physical address.

As part of the translation process, the MMU determines whether or not the corresponding segment or page of the virtual address space currently exists in physical memory. If the segment or page already resides in memory, the operation is performed on the

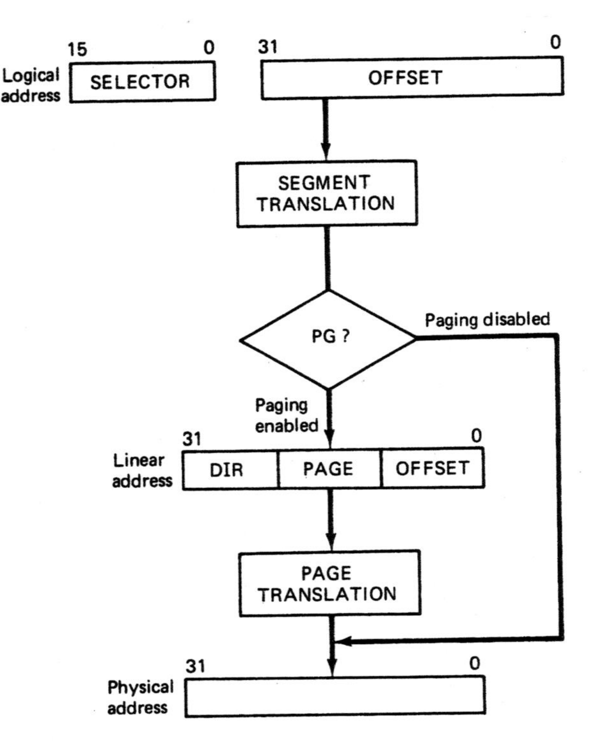

Figure 15–20 Virtual-to-physical address translation. (Reprinted by permission of Intel Corporation. Copyright/Intel Corp. 1986)

information. However, if the segment or page is not present, it signals this condition as an error. Once this condition is identified, the memory-manager software initiates loading of the segment or page from the external storage device to physical memory. This operation is called a *swap*. That is, an old segment or page gets swapped out to disk to make room in physical memory, and then the new segment or page is swapped into this space. Even though a swap has taken place, it appears to the program that all segments or pages are available in physical memory.

Segmentation Virtual-to-Physical Address Translation. Let us now look more closely at the address-translation process. We begin by assuming that paging is turned off. In this case, the address-translation sequence that takes place is the one highlighted in Fig. 15–21(a). Figure 15–21(b) describes the operations that take place during the segment translation process. Earlier we found that the 80386DX's segment selector registers, CS, DS, ES, FS, GS, and SS, provide the segment selectors used to index into either the global descriptor table or the local descriptor table. Whenever a selector value is loaded into a segment register, the descriptor pointed to by the index in the table selected by the TI bit is automatically fetched from memory and loaded into the corresponding *segment descriptor cache register.* It is the contents of this descriptor, not the selector that defines the location, size and characteristics of the segment of memory.

Note in Fig. 15–22 that the 80386DX has one 64-bit internal segment descriptor cache register for each segment selector register. These cache registers are not accessible by the programmer. Instead, they are transparently loaded with a complete descriptor

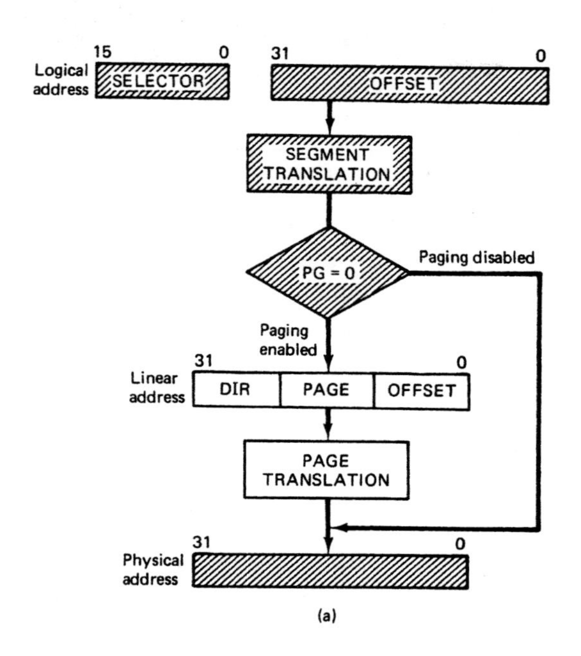

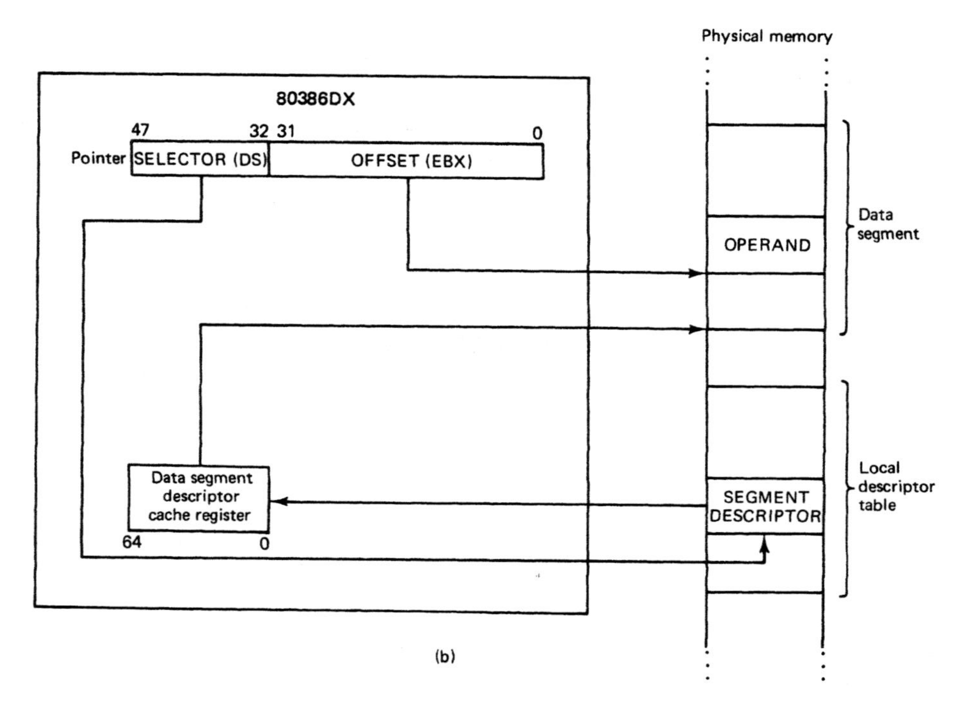

Figure 15–21 (a) Virtual-to-linear address translation. (Reprinted by permission of Intel Corporation. Copyright/Intel Corp. 1986) (b) Translating a virtual address into a physical (linear) address.

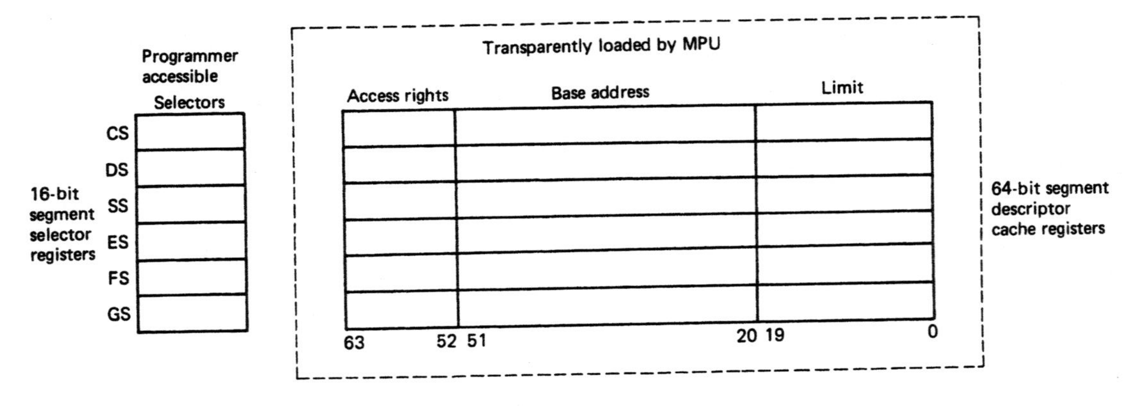

Figure 15–22 Segment selector registers and the segment descriptor cache registers.

whenever an instruction is executed that loads a new selector into a segment register. For instance, if an operand were to be accessed from a new data segment, a local memory data segment selector would be first loaded into DS with the instruction

```
MOV DS, AX
```

As this instruction is executed, the selector in AX is loaded into DS and then the corresponding descriptor in the local descriptor table is read from memory and loaded into the data segment descriptor cache register. The MMU looks at the information in the descriptor and performs checks to determine whether or not it is valid.

In this way, we see that the segment descriptors held in the cache dynamically change as a task is performed. At any one time, the memory-management unit permits just six segments of memory to be active. These segments correspond to the six segment selector registers, CS, DS, ES, FS, GS, and SS, and can reside in either local or global memory. Once the descriptors are cached, subsequent references to them are performed without any overhead for loading of the descriptor.

In Fig. 15–22, we find that this data segment descriptor has three parts: 12 bits of *access rights* information, a 32-bit *segment base address,* and a 20-bit *segment limit.* The value of the 32-bit base address identifies the beginning of the data segment that is to be accessed. The loading of the data segment descriptor cache completes the table lookup that maps the 16-bit selector to its equivalent 32-bit data segment base address.

The location of the operand in this data segment is determined by the offset part of the virtual address. For example, let us assume that the next instruction to be executed needs to access an operand in this data segment and that the instruction uses based-addressing mode to specify the operand. Then the EBX register holds the offset of the operand from the base address of the data segment. Figure 15–21(a) shows that the base address is directly added to the offset to produce the 32-bit physical address of the operand. This addition completes the translation of the 48-bit virtual address into the 32-bit linear address. As Fig. 15–21(a) shows, when paging is disabled, PG = 0, the linear address is the physical address of the storage location to be accessed in memory.

EXAMPLE 15.12

Assume that, in Fig. 15–21(b), the virtual address is made up of a segment selector equal to 0100_{16}, offset equal to 00002000_{16}, and that paging is disabled. If the segment base address read in from the descriptor is 00030000_{16}, what is the physical address of the operand?

Solution

The virtual address is given as

$$\text{Virtual address} = 0100{:}00002000_{16}$$

This virtual address translates to the physical address

$$\begin{aligned}\text{Linear address} &= \text{Base address} + \text{Offset}\\ &= 00030000_{16} + 00002000_{16}\\ &= 00032000_{16}\end{aligned}$$

Paged Partitioning of the Virtual Address Space and Virtual-to-Physical Address Translation. The paging memory management unit works beneath the segmentation memory management unit and when enabled it organizes the 80386DX's address space in a different way. Earlier we pointed out that when paging is not in use, the 4Gbyte physical address space is organized into segments that can be any size from 1 byte to 4Gbytes. However, when paging is turned on, the paging unit arranges the physical address space into 1,048,496 pages that are each 4,096 bytes long. Figure 15–23 shows how the physical address space may be organized in this way. The fixed-size blocks of paged memory are a disadvantage in that 4K addresses are allocated by the memory manager even though it may not all be used. This creation of unused sections of memory is called *fragmentation.* Fragmentation results in less efficient use of memory. However, paging greatly simplifies the implementation of the memory-manager software. Let us continue by looking at what happens to the address translation process when paging is enabled.

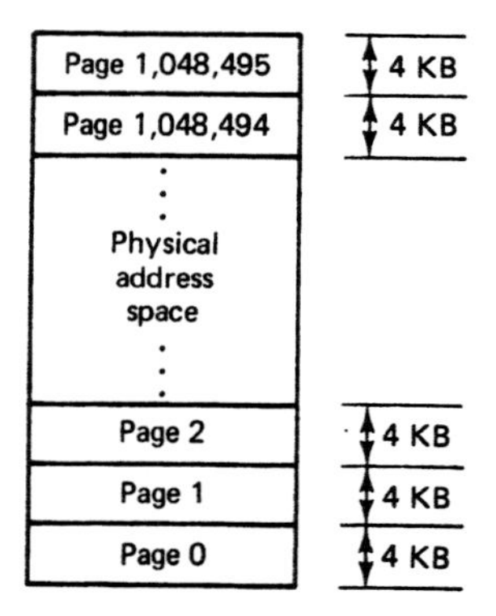

Figure 15–23 Paged organization of the physical address space.

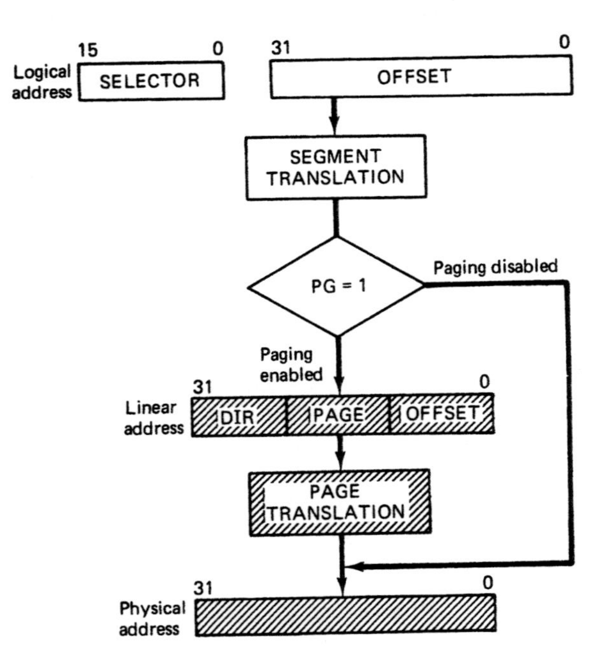

Figure 15–24 Paged translation of a linear address to a physical address. (Reprinted by permission of Intel Corporation. Copyright/Intel Corp. 1986)

In Fig. 15–24, we see that the linear address produced by the segment-translation process is no longer used as the physical address. Instead, it undergoes a second translation called the *page translation.* Figure 15–25 shows the format of a linear address. Note that it is composed of three elements: a 12-bit offset field, a 10-bit page field, and a 10-bit directory field.

The diagram in Fig. 15–26 illustrates how a linear address is translated into its equivalent physical address. Earlier we found that the location of the *page directory table* in memory is identified by the address in the page directory base register (PDBR) in CR_3. These 20 bits are actually the MSBs of the base address. The 12 lower bits are assumed to start at 000_{16} at the beginning of the directory and range to FFF_{16} at its end. Therefore, the page directory contains 4Kbyte memory locations and is organized as 1K, 32-bit addresses. These addresses each point to a separate page table, which is also in physical memory.

Note that the 10-bit directory field of the linear address is the offset from the value in PDBR that selects one of the 1K, 32-bit *page directory entries* in the page directory table. This pointer is cached inside the 80386DX in what is called the *translation look-*

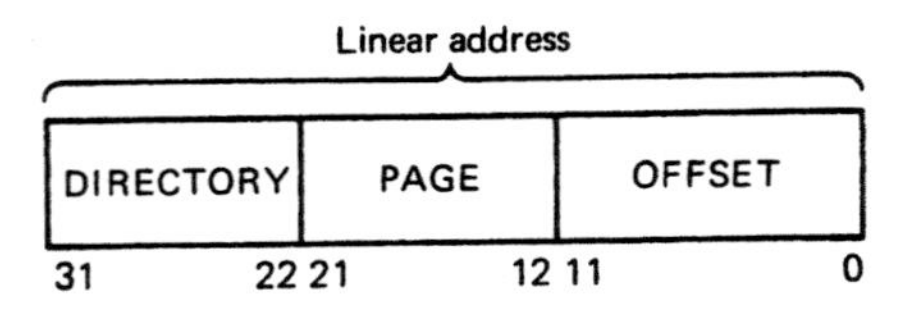

Figure 15–25 Linear address format.

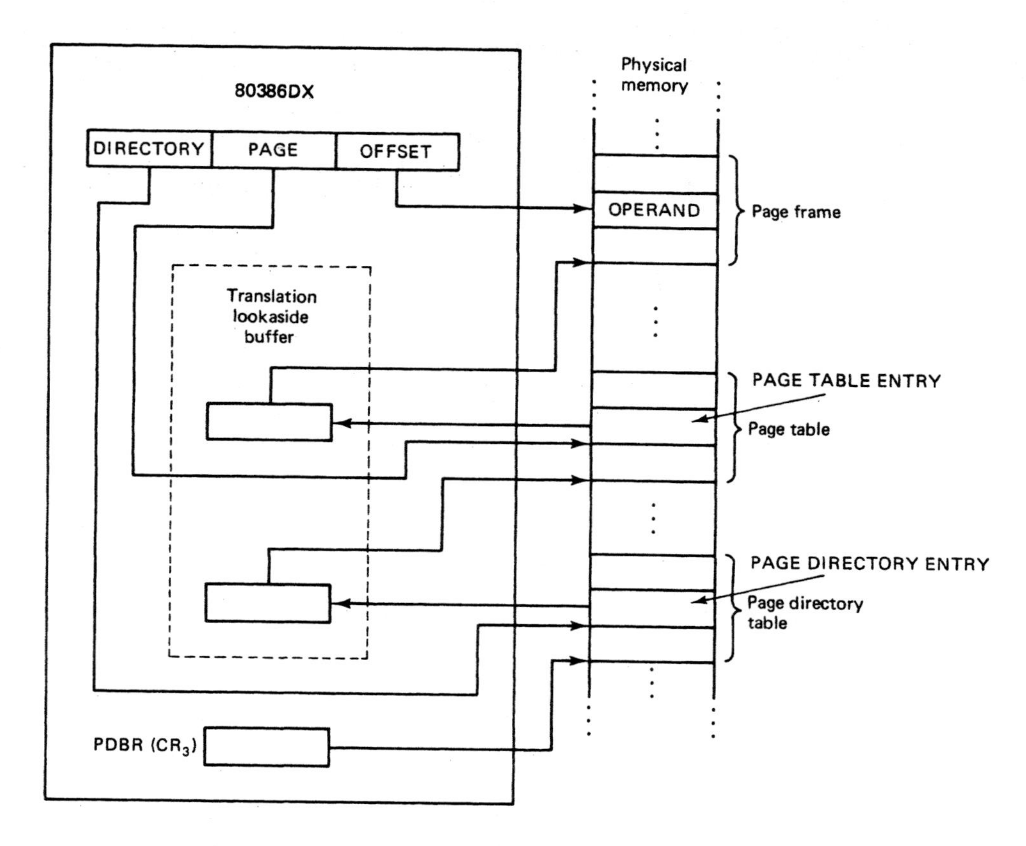

Figure 15–26 Translating a linear address to a physical address.

aside buffer. Its value is used as the base address of a *page table* in memory. As with the page directory, each page table is also 4Kbytes long and contains 1K, 32-bit addresses. However, these addresses are called *page frame addresses.* Each page frame address points to a 4K frame of data storage locations in physical memory.

Next, the 10-bit page field of the linear address selects one of the 1K, 32-bit *page table entries* from the page table. This table entry is also cached in the translation lookaside buffer. In Fig. 15–26, we see that it is another base address and selects a 4Kbyte *page frame* in memory. This frame of memory locations is used for storage of data. The 12-bit offset part of the linear address identifies the location of the operand in the active page frame.

The 80386DX's translation lookaside buffer is actually capable of maintaining 32 sets of table entries. In this way, we see that 128K-bytes of paged memory is always directly accessible. Operands in this part of memory can be accessed without first reading new entries from the page tables. If an operand to be accessed is not in one of these pages, overhead is required to first read the page table entry into the translation lookaside buffer.

▲ 15.6 DESCRIPTOR AND PAGE TABLE ENTRIES OF THE 80386DX

In the previous section of this chapter, we frequently used the terms *descriptor* and *page table entry.* We talked about the descriptor as an element of the global descriptor, local descriptor, and interrupt descriptor tables. Actually, the 80386DX supports several kinds of descriptors, and they all serve different functions relative to overall system operation. Some examples are the *segment descriptor, system segment descriptor, local descriptor table descriptor, call gate descriptor, task state segment descriptor,* and *task gate descriptor.* We also discussed page table entries in our description of the 80386DX's page translation of virtual addresses. There are only two types of page table entries: the *page directory entry* and the *page table entry.* Let us now explore the structure of descriptors and page table entries.

Descriptors are the elements by which the on-chip memory manager hardware manages the segmentation of the 80386DX's 64Tbyte virtual memory address space. One descriptor exists for each segment of memory in the virtual address space. Descriptors are assigned to the local descriptor table, global descriptor table, task state segment, call gate, task gate, and interrupts. The contents of a descriptor provide mapping from virtual addresses to linear addresses for code, data, stack, and the task state segments and then assign attributes to the segment.

Each descriptor is 8 bytes long and contains three kinds of information. Earlier we identified the 20-bit *LIMIT* field and showed that its value defines the size of the segment or the table. Moreover, we found that the 32-bit *BASE* value provides the beginning address for the segment or the table in the 64Gbyte linear address space. The third element of a descriptor, called the *access rights byte,* is different for each type of descriptor. Let us now look at the format of two types of descriptors: the segment descriptor and system segment descriptor.

The segment descriptor is the type of descriptor used to describe code, data, and stack segments. Figure 15–27(a) shows the general structure of a segment descriptor. Here we see that the two lowest-addressed bytes, bytes 0 and 1, hold the 16 least significant bits of the limit; the next three bytes contain the 24 least significant bits of the base address; byte 5 is the access rights byte; the lower 4 bits of byte 6 are the four most significant bits of the limit; the upper 4 bits include the *granularity* (G) and the *programmer available* (AVL) bits; and byte 7 contains the eight most significant bits of the 32-bit base. Segment descriptors are only found in the local and global descriptor tables.

Figure 15–28 shows how a descriptor is loaded from the local descriptor table in global memory to define a code segment in local memory. Note that the LDTR descriptor defines a local descriptor table between addresses 00900000_{16} and $0090FFFF_{16}$. The value 1005_{16}, held in the code segment selector register, causes the descriptor at offset 1000_{16} in the local descriptor table to be cached into the code segment descriptor cache. In this way, a 1Mbyte code segment is activated starting at address 00600000_{16} in local memory.

The bits of the access rights byte define the operating characteristics of a segment. For example, they contain information about a segment such as whether the descriptor has been accessed, if it is a code or data segment descriptor, its privilege level, if it is read-

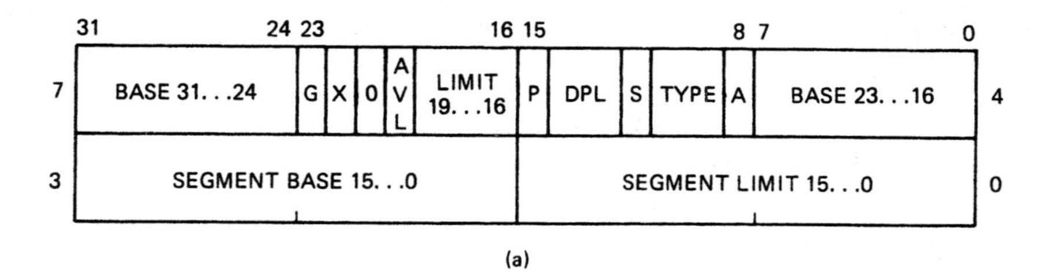

(a)

	Bit Position	Name	Function		
	7	Present (P)	P = 1 P = 0	Segment is mapped into physical memory. No mapping to physical memory exists, base and limit are not used.	
	6–5	Descriptor Privilege Level (DPL)		Segment privilege attribute used in privilege tests.	
	4	Segment Descriptor (S)	S = 1 S = 0	Code or Data (includes stacks) segment descriptor System Segment Descriptor or Gate Descriptor	
Type Field Definition	3 2 1	Executable (E) Expansion Direction (ED) Writeable (W)	E = 0 ED = 0 ED = 1 W = 0 W = 1	Data segment descriptor type is: Expand up segment, offsets must be ≤ limit. Expand down segment, offsets must be > limit. Data segment may not be written into. Data segment may be written into.	If Data Segment (S = 1, E = 0)
Type Field Definition	3 2 1	Executable (E) Conforming (C) Readable (R)	E = 1 C = 1 R = 0 R = 1	Code Segment Descriptor type is: Code segment may only be executed when CPL ≥ DPL and CPL remains unchanged. Code segment may not be read Code segment may be read.	If Code Segment (S = 1, E = 1)
	0	Accessed (A)	A = 0 A = 1	Segment has not been accessed. Segment selector has been loaded into segment register or used by selector test instructions.	

(b)

Figure 15–27 (a) Segment descriptor format. (b) Access byte bit definitions. (Reprinted by permission of Intel Corporation. Copyright/Intel Corp. 1987)

able or writeable, and if it is currently loaded into internal memory. Let us next look at the function of each of these bits in detail.

Figure 15–27(b) lists the function of each bit in the access rights byte. Note that if bit 0 is logic 1, the descriptor has been accessed. A descriptor is marked this way to indicate that its value has been cached on the 80386DX. The memory-manager software checks this information to find out if the segment is already in physical memory. Bit 4 identifies whether the descriptor represents a code/data segment or is a control descriptor. Let us assume that this bit is 1 to identify a segment descriptor. Then, the type bits, bits 1 through 3, determine whether the descriptor describes a code segment or a data segment. For instance, 000 means that it is a read/write data segment that grows upward from the base to the limit. The DPL bits, bits 5 and 6, assign a descriptor privilege level to the segment. For example, 00 selects the most privileged level, level 0. Finally, the present bit indicates whether or not the segment is currently loaded into physical memory. The oper-

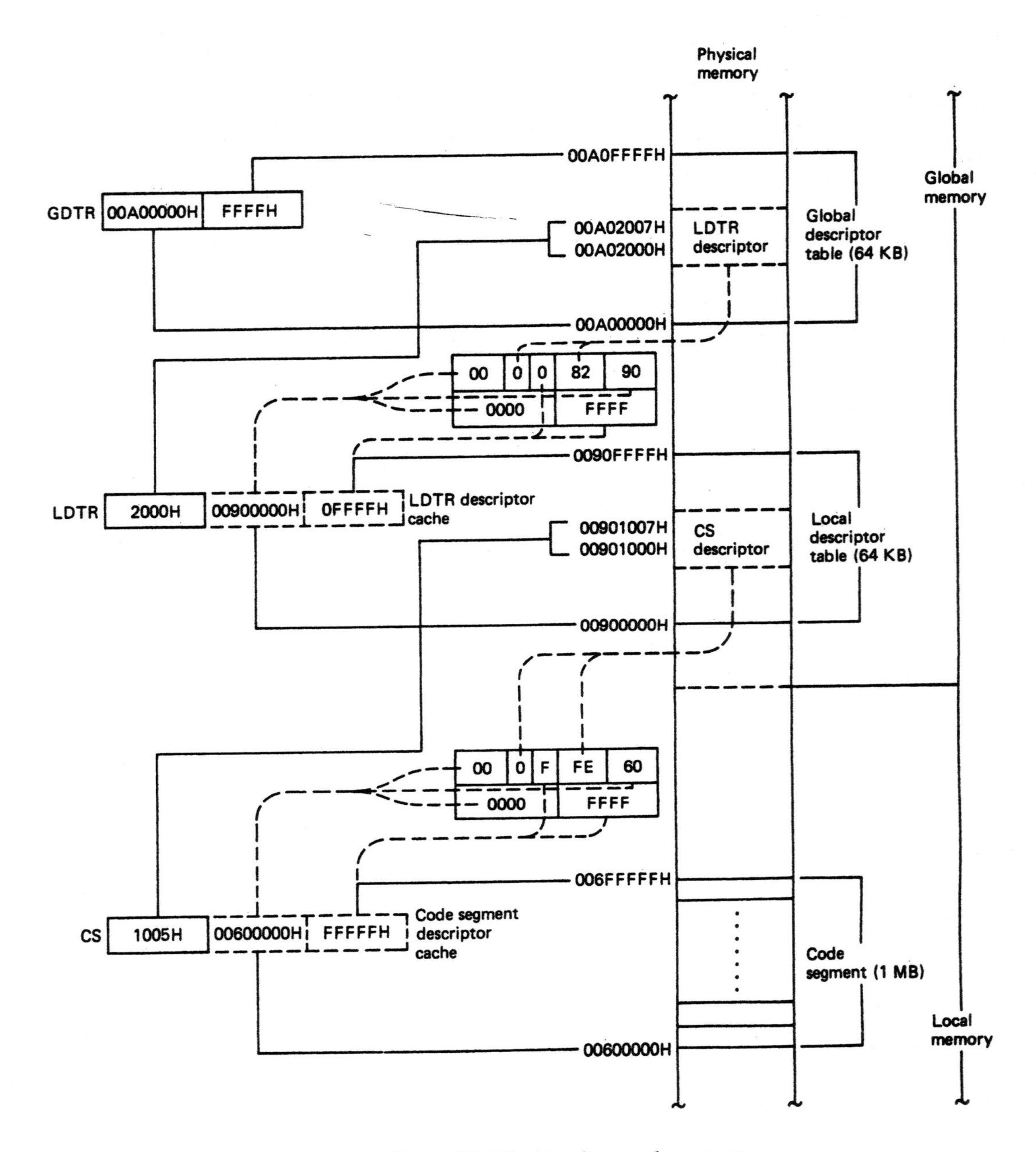

Figure 15–28 Creating a code segment.

ating system software to determine if the segment should be loaded from a secondary storage device such as a hard disk can test this bit. For example, if the access rights byte has logic 1 in bit 7, the data segment is already available in physical memory and does not have to be loaded from an external device. Figure 15–29(a) shows the general form of a code segment descriptor, and Fig. 15–29(b) shows a general data/stack segment descriptor.

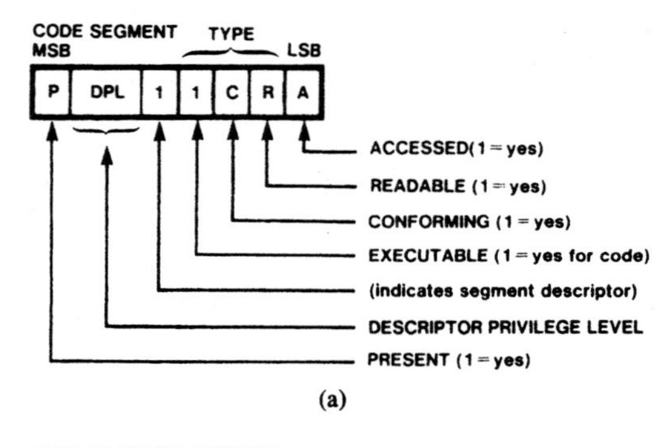

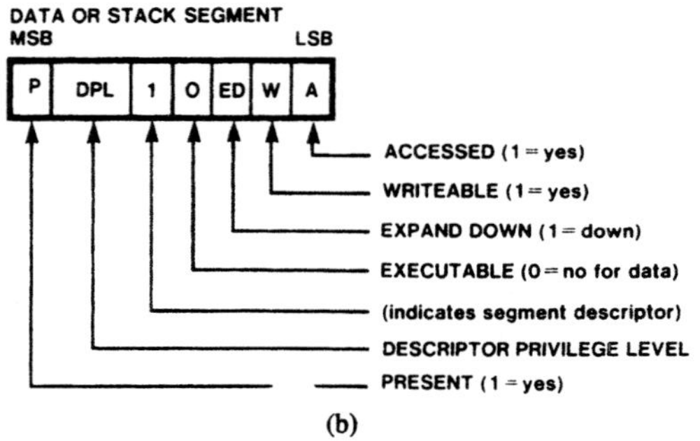

Figure 15–29 (a) Code segment descriptor access byte configuration. (Reprinted by permission of Intel Corporation. Copyright/Intel Corp. 1987) (b) Data or stack segment access byte configuration. (Reprinted by permission of Intel Corporation. Copyright/Intel Corp. 1987)

EXAMPLE 15.13

The access rights byte of a segment descriptor contains FE_{16}. What type of segment descriptor does it describe, and what are its characteristics?

Solution

Expressing the access rights byte in binary form, we get

$$FE_{16} = 11111110_2$$

Since bit 4 is 1, the access rights byte is for a code/data segment descriptor. This segment has the characteristics that follow:

$$P = 1 \quad = \text{segment is mapped into physical memory}$$

$$DPL = 11 = \text{privilege level 3}$$

$$E = 1 \quad = \text{executable code segment}$$

$$C = 1 \quad = \text{conforming code segment}$$

$$R = 1 \quad = \text{readable code segment}$$

$$A = 0 \quad = \text{segment has not been accessed}$$

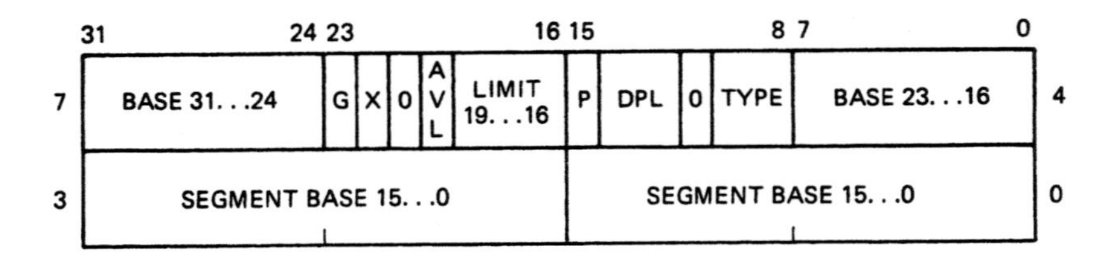

Name	Value	Description
TYPE	0	Reserved by Intel
	1	Available 80286 TSS
	2	LDT
	3	Busy 80286 TSS
	4	Call gate
	5	Task gate
	6	80286 interrupt gate
	7	80286 trap gate
	8	Reserved by Intel
	9	Available 80386 TSS
	A	Reserved
	B	Busy 80386 TSS
	C	80386 call gate
	D	Reserved by Intel
	E	80386 interrupt gate
	F	80386 trap gate
P	0	Descriptor contents are not valid
	1	Descriptor contents are valid
DPL	0-3	Descriptor privilege level 0, 1, 2 or 3
BASE	32-bit number	Base address of special system data segment in memory
LIMIT	20-bit number	Offset of last byte in segment from the base

Figure 15–30 System segment descriptor format and field definitions.

An example of a system segment descriptor is the descriptor used to define the local descriptor table. This descriptor is located in the GDT. Looking at Fig. 15–30, we find that the format of a system segment descriptor is similar to the segment descriptor we just discussed. However, the type field of the access rights byte takes on new functions.

EXAMPLE 15.14

If a system segment descriptor has an access rights byte equal to 82_{16}, what type of descriptor does it represent? What is its privilege level? Is the descriptor present?

Solution

First, we express the access rights byte in binary form. This gives

$$82_{16} = 10000010_2$$

Now we see that the bits that describe the type of the descriptor are given as

$$\text{TYPE} = 0010 = \text{local descriptor table descriptor}$$

The privilege level is given by

$$\text{DPL} = 00 = \text{privilege level 0}$$

and since

$$\text{P} = 1$$

the descriptor is present in physical memory.

Now that we have discussed the format and use of descriptors, let us continue with page table entries. Figure 15–31 shows the format of either a page directory or page table entry. Here we see that the 20 most significant bits are either the base address of the page table if the entry is in the page directory table or the base address of the page frame if the entry is in the page table. Note that only bits 12 through 31 of the base address are supplied by the entry. The 12 least significant bits are assumed to be equal to 0. In this way, we see that page tables and page frames are always located on a 4Kbyte address boundary. In Fig. 15–26, we found that these entries are cached into the translation lookaside buffer.

The 12 lower bits of the entry supply protection characteristics or statistical information about the use of the page table or page frame. For example, the *user/supervisor* (U/S) and *read/write* (R/W) bits implement a two-level page protection mechanism. Setting U/S to 1 selects user-level protection. User is the low privilege level and is the same as protection level 3 of the segmentation model. That is, user is the protection level assigned to pages of memory that are accessible by application software. On the other hand, making U/S equal to 0 assigns supervisor-level protection to the table or frame. Supervisor corresponds to levels 0, 1, and 2 of the segmentation model and is the level assigned to operating system resources. The *read/write* (R/W) bit is used to make a user-

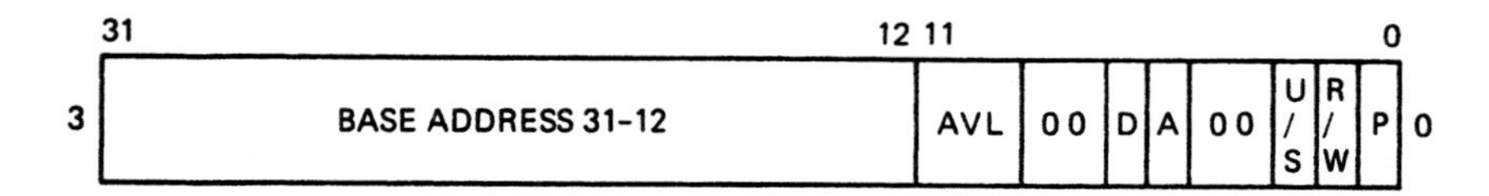

Figure 15–31 Directory or page table entry format.

U/S	R/W	User	Supervisor
0	0	None	Read/write
0	1	None	Read/write
1	0	Read-only	Read/write
1	1	Read/write	Read/write

Figure 15–32 User- and supervisor-level access rights.

level table or frame read-only or read/write. Logic 1 in R/W selects read-only operation. Figure 15–32 summarizes the access characteristics for each setting of U/S and R/W.

Protection characteristics assigned by a page directory entry are applied to all page frames defined by the entries in the page table. On the other hand, the attributes assigned to a page table entry apply only to the page frame that it defines. Since two sets of protection characteristics exist for all page frames, the page-protection mechanism of the 80386DX is designed always to enforce the higher-privileged (more restricting) of the two protection rights.

EXAMPLE 15.15

If the page directory entry for the active page frame is $F1000007_{16}$ and its page table entry is 01000005_{16}, is the frame assigned to the user or supervisor? What access is permitted to the frame from user mode and from supervisor mode?

Solution

First, the page directory entry is expressed in binary form. This gives

$$F1000007_{16} = 11110001000000000000000000000111_2$$

Therefore, the page protection bits are

$$U/S\ R/W = 11$$

This assigns user-mode and read/write accesses to the complete page frame.

Next, the page table entry for the frame is expressed in binary form as

$$01000005_{16} = 00000001000000000000000000000101_2$$

Here we find that

$$U/S\ R/W = 10$$

This defines the page frame as a user-mode, read-only page. Since the page frame attributes are the more restrictive, they apply. Figure 15–32 shows that user software (application software) can only read data in this frame. On the other hand, supervisor software (operating system software) can either read data from or write data into the frame.

The other implemented bits in the directory and page table entry shown in Fig. 15–31 provide statistical information about the table or frame usage. For instance, the *present* (P) bit identifies whether or not the entry can be used for page address translation. P equal to logic 1 indicates that the entry is valid and is available for use in address translation. On the other hand, if P equals 0, the entry is either undefined or not present in physical memory. If an attempt is made to access a page table or page frame that has its P bit marked 0, a page fault results. This page fault needs to be serviced by the operating system.

The 80386DX also records the fact that a page table or page frame has been accessed. Just before a read or write is performed to any address in a table or frame, the *accessed* (A) bit of the entry is set to 1. This marks it as having been accessed. For page frame accesses, it also records whether the access was for a read or write operation. The *dirty* (D) bit is defined only for a page table entry, and it gets set if a write is performed to any address in the corresponding page frame. In a virtual demand paged memory system, the operating system can check the state of these bits to determine if a page in physical memory needs to be updated on the virtual storage device (hard disk) when a new page is swapped into its physical memory address space. The last 3 bits are labeled AVL and are available for use by the programmer.

▲ 15.7 PROTECTED-MODE SYSTEM-CONTROL INSTRUCTION SET OF THE 80386DX

In Chapters 5 and 6, we studied the instruction set of the 8086/8088 microprocessor. The part of the instruction set introduced in those chapters represents the base instruction set of the 8086 architecture. In Sections 14.4 and 15.4, we introduced a number of new instructions that are available in the real-mode instruction set of the 80386 family. In protected mode, the 80386DX executes all the instructions available in the real mode. Moreover, it is enhanced with a number of additional instructions that either apply only to protected-mode operation or are used in the real mode to prepare the 80386DX for entry into the protected mode. As Fig. 15–33 shows, these instructions are known as the *system control instruction set.*

Figure 15–34 lists the instructions of the system control instruction set. Here we find the format of each instruction along with a description of its operation. Moreover, the mode or modes in which the instruction is available are identified. Let us now look at the operation of some of these instructions in detail.

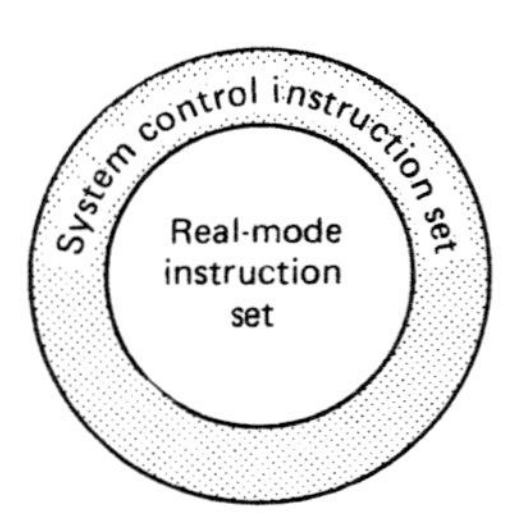

Figure 15–33 Protected-mode instruction set.

Instruction	Description	Mode
LGDT S	Load the global descriptor table register. S specifies the memory location that contains the first byte of the 6 bytes to be loaded into the GDTR.	Both
SGDT D	Store the global descriptor table register. D specifies the memory location that gets the first byte of the 6 bytes to be stored from the GDTR.	Both
LIDT S	Load the interrupt descriptor table register. S specifies the memory location that contains the first byte of the 6 bytes to be loaded into the IDTR.	Both
SIDT D	Store the interrupt descriptor table register. D specifies the memory location that gets the first byte of the 6 bytes to be stored from the IDTR.	Both
LMSW S	Load the machine status word. S is an operand to specify the word to be loaded into the MSW.	Both
SMSW D	Store the machine status word. D is an operand to specify the word location or register where the MSW is to be stored.	Both
LLDT S	Load the local descriptor table register. S specifies the operand to specify a word to be loaded into the LDTR.	Protected
SLDT D	Store the local descriptor table register. D is an operand to specify the word location where the LDTR is to be saved.	Protected
LTR S	Load the task register. S is an operand to specify a word to be loaded into the TR.	Protected
STR D	Store the task register. D is an operand to specify the word location where the TR is to be stored.	Protected
LAR D, S	Load access rights byte. S specifies the selector for the descriptor whose access byte is loaded into the upper byte of the D operand. The low byte specified by D is cleared. The zero flag is set if the loading completes successfully; otherwise it is cleared.	Protected
LSL R16, S	Load segment limit. S specifies the selector for the descriptor whose limit word is loaded into the word register operand R16. The zero flag is set if the loading completes successfully; otherwise it is cleared.	Protected
ARPL D, R16	Adjust RPL field of the selector. D specifies the selector whose RPL field is increased to match the PRL field in the register. The zero flag is set if successful; otherwise it is cleared.	Protected
VERR S	Verify read access. S specifies the selector for the segment to be verified for read operation. If successful the zero flag is set; otherwise it is reset.	Protected
VERW S	Verify write access. S specifies the selector for the segment to be verified for write operation. If successful the zero flaq is set; otherwise it is reset.	Protected
CLTS	Clear task switched flag.	Protected

Figure 15–34 Protected-mode system control instruction set.

Figure 15–34 shows that the first six instructions can be executed in both the real and protected mode. They provide programmers with the ability to load (L) or store (S) the contents of the global descriptor table (GDT) register, interrupt descriptor table (IDT) register, and machine status word (MSW) part of CR_0. Note that the instruction *load global descriptor table register* (LGDT) is used to load the GDTR from memory. Operand S specifies the location of the 6 bytes of memory that hold the limit and base values needed to specify the size and beginning address of the GDT. The first word of memory contains the limit, and the next 4 bytes contain the base. For instance, executing the instruction

```
LGDT [INIT_GDTR]
```

loads the GDTR with the base and limit stored at address INIT_GDTR to create a global descriptor table in memory. This instruction is meant to be used during system initialization and before switching the 80386DX to the protected mode.

Once loaded, the current contents of the GDTR can be saved in memory by executing the *store global descriptor table* (SGDT) instruction. An example is the instruction

```
SGDT [SAVE_GDTR]
```

The instructions LIDT and SIDT perform similar operations for the interrupt descriptor table register. The IDTR is also set up during initialization.

The instructions *load machine status word* (LMSW) and *store machine status word* (SMSW) allow the contents of the machine status word (MSW) to be loaded and stored respectively. These are the instructions that are used to switch the 80386DX from real to protected mode. To do this, we must set the least significant bit in the MSW to 1. This can be done by first reading the contents of the machine status word, modifying the LSB (PE), and then writing the modified value back into the MSW part of CR_0. The instruction sequence that follows will switch an 80386DX operating in real mode to the protected mode:

```
SMSW  AX        ;Read from the MSW
OR    AX, 1     ;Set the PE bit
LMSW  AX        ;Write to the MSW
```

The next four instructions in Fig. 15–34 are also used to initialize or save the contents of protected-mode registers. However, they can be used only when the 80386DX is in the protected mode. To load and to save the contents of the LDTR, we have the instructions LLDT and SLDT, respectively. Moreover, for loading and saving the contents of the TR, the equivalent instructions are LTR and STR.

The rest of the instructions in Fig. 15–34 are for accessing the contents of descriptors. For instance, to read a descriptor's access rights byte, the *load access rights byte* (LAR) instruction is executed. An example is the instruction

```
LAR AX, [LDIS_1]
```

Execution of this instruction causes the access rights byte of the specified local descriptor to be loaded into AH. To read the segment limit of a descriptor, we use the *load segment limit* (LSL) instruction. For instance, to copy the segment limit for the specified local descriptor into register EBX, the instruction

```
LSL EBX, [LDIS_1]
```

is executed. In both cases, ZF is set to 1 if the operation is performed correctly.

The instruction *adjust RPL field of selector* (ARPL) can be used to increase the RPL field of a selector in memory or a register, destination (D), to match the protection level of the selector in a register, source (S). If an RPL-level increase takes place, ZF is set to 1. Finally, the instructions VERR and VERW are provided to test the accessibility of a segment for a read or write operation, respectively. If the descriptor permits the type of access tested for by executing the instruction, ZF is set to 1.

▲ 15.8 MULTITASKING AND PROTECTION

We say that the 80386DX microprocessor implements a *multitasking* software architecture. By this we mean that it contains on-chip hardware that both permits multiple tasks to exist in a software system and allows them to be scheduled for execution in a time-shared manner. That is, program control is switched from one task to another after a fixed interval of time elapses. For instance, the tasks can be executed in a round-robin fashion. This means that the most recently executed task is returned to the end of the list of tasks being executed. Even though the processes are executed in a time-shared fashion, an 80386DX microcomputer has the performance to make it appear to the user that they are all running simultaneously.

Earlier we defined a task as a collection of program routines that performs a specific function. This function is also called a *process*. Software systems typically need to perform many processes. In the protected-mode 80386DX-based microcomputer, each process is identified as an independent task. The 80386DX provides an efficient mechanism, called the *task-switching mechanism,* for switching between tasks. For instance, an 80386DX running at 16 MHz can perform a task switch operation in just 19μs.

We also indicated earlier that when a task is called into operation it could have both global and local memory resources. The local memory address space is divided between tasks. This means that each task normally has its own private segments of local memory. Segments in global memory can be shared by all tasks. Therefore, a task can have access to any of the segments in global memory. As Fig. 15–35 shows, task A has both a private address space and a global address space available for its use.

Protection and the Protection Model

Safeguards can be built into the protected-mode software system to deny unauthorized or incorrect accesses of a task's memory resources. The concept of safeguarding memory is known as *protection.* The 80386DX includes on-chip hardware that implements a *protection mechanism.* This mechanism is designed to put restrictions on the access of local and system resources by a task and to isolate tasks from each other in a multitasking environment.

Segmentation, paging, and descriptors are the key elements of the 80386DX's protection mechanism. We already identified that when using a segmented memory model, a segment is the smallest element of the virtual memory address space that has unique protection attributes. The access rights information and limit fields in the segment's descriptor define these attributes. As Fig. 15–36(a) shows, the on-chip protection hardware performs a number of checks during all memory accesses. Figure 15–36(b) is a list

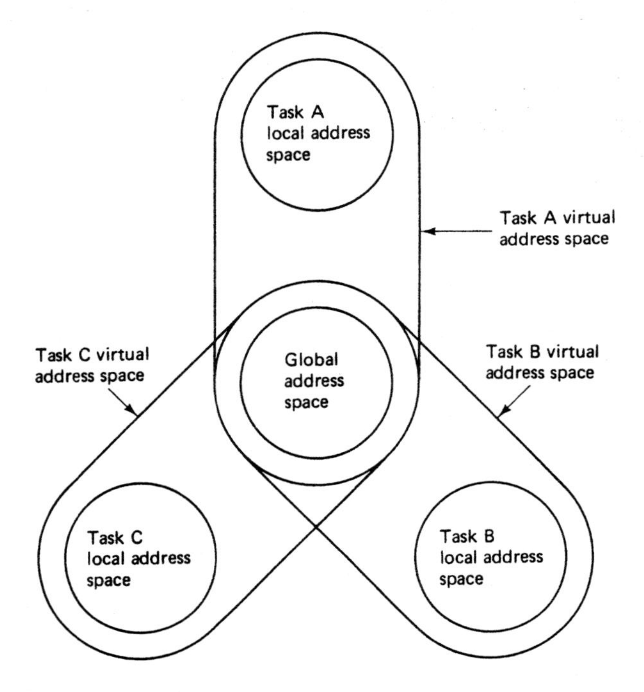

Figure 15–35 Virtual address space of a task. (Reprinted by permission of Intel Corporation. Copyright/Intel Corp. 1987)

of the protection checks and restrictions imposed on software by the 80386DX. For example, when a data-storage location in memory is written to, the type field in the access rights byte of the segment is tested to assure that its attributes are consistent with the register cache being loaded, and the offset is checked to verify that it is within the limit of the segment.

Let us just review the attributes that can be assigned to a segment with the access rights information in its descriptor. Figure 15–37 shows the format of a data segment descriptor and an executable (code) segment descriptor. The P bit defines whether a segment of memory is present in physical memory. Assuming that a segment is present, bit 4 of the type field makes it either a code segment or data segment. Note that this bit is 0 if the descriptor is for a data segment and it is 1 for code segments. Segment attributes such as readable, writeable, conforming, expand up or down, and accessed are assigned by other bits in the type field. Finally, a privilege level is assigned with the DPL field.

Earlier we showed that whenever a segment is accessed, the base address and limit are cached inside the 80386DX. In Fig. 15–36(a), we find that the access rights information is also loaded into the cache register. However, before loading the descriptor the MMU verifies that the selected segment is currently present in physical memory, it is at a privilege level accessible from the privilege level of the current program, the type is consistent with the target segment selector register (CS = code segment, DS, ES, FS, GS, or SS = data segment), and the reference into the segment does not exceed the address limit of the segment. If a violation is detected, an error condition is signaled. The memory-manager software can determine the cause, correct the problem, and then reinitiate the operation.

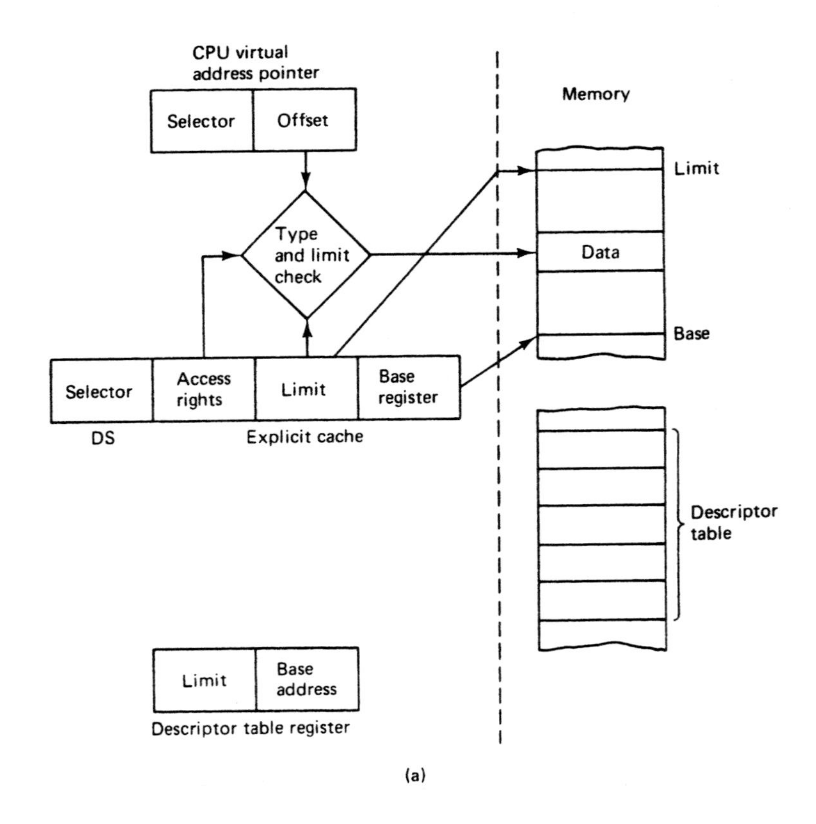

Type check
Limit check
Restriction of addressable domain
Restriction of procedure entry point
Restriction of instruction set

(b)

Figure 15–36 (a) Testing the access rights of a descriptor. (Reprinted by permission of Intel Corporation. Copyright/Intel Corp. 1982) (b) Protection checks and restrictions.

Let us now look at some examples of memory accesses that result in protection violations. For example, if the selector loaded into the CS register points to a descriptor that defines a data segment, the type check leads to a protection violation. Another example of an invalid memory access is an attempt to read an operand from a code segment that is not marked as readable. Finally, any attempt to access a byte of data at an offset greater than LIMIT, a word at an offset equal to or greater than LIMIT, or a double word at an offset equal to or greater than LIMIT−2 extends beyond the end of the data segment and results in a protection violation.

The 80386DX's protection model provides four possible privilege levels for each task—*levels 0, 1, 2,* and *3* — which are illustrated by concentric circles as in

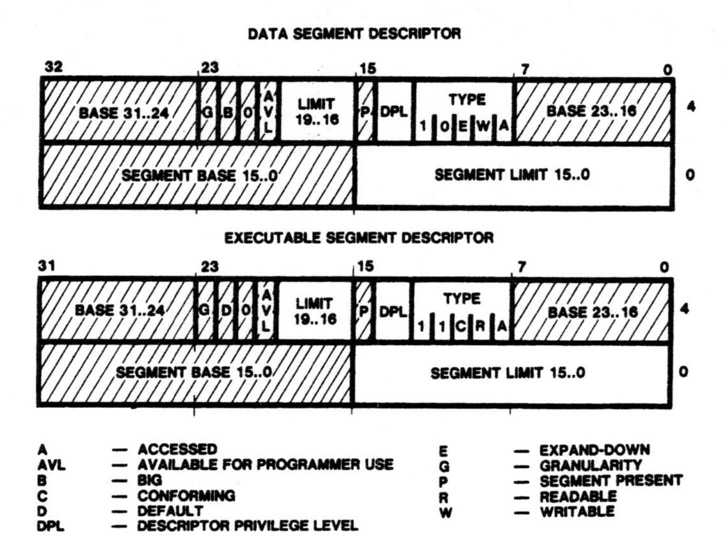

Figure 15–37 Data segment and executable (code) segment descriptors. (Reprinted by permission of Intel Corporation. Copyright/Intel Corp. 1986)

Fig. 15–38. Here level 0 is the most privileged level and level 3 is the least privileged level.

System and application software are typically partitioned in the manner shown in Fig. 15–38. The kernel represents application-independent software that provides microprocessor-oriented functions such as I/O control, task sequencing, and memory management. For this reason, it is kept at the most privileged level, level 0. Level 1 contains processes that provide system services such as file accessing. Level 2 is used to implement custom routines to support special-purpose system operations. Finally, the least privileged level, level 3, is the level at which user applications are run. This example also demonstrates how privilege levels are used to isolate system-level software (operating system software in levels 0 through 2) from the user's application software (level 3). Tasks at a less privileged level can use programs from the more privileged levels but cannot modify the contents of these routines in any way. In this way, applications are permitted to use system software routines from the three higher privilege levels without affecting their integrity.

Earlier we indicated that protection restrictions are put on the instruction set. An example of this is that the system control instructions can only be executed in a code segment at protection level 0. We also pointed out that each task is assigned its own local descriptor table. Therefore, as long as none of the descriptors in a task's local-descriptor-table references code or data available to another task, it is isolated from all other tasks. That is, it has been assigned a unique part of the virtual address space. For example, in Fig. 15–38 multiple applications running at level 3 are isolated from each other by assigning them different local resources. This shows that segments, privilege levels, and the local descriptor table provide protection for both code and data within a task. These types

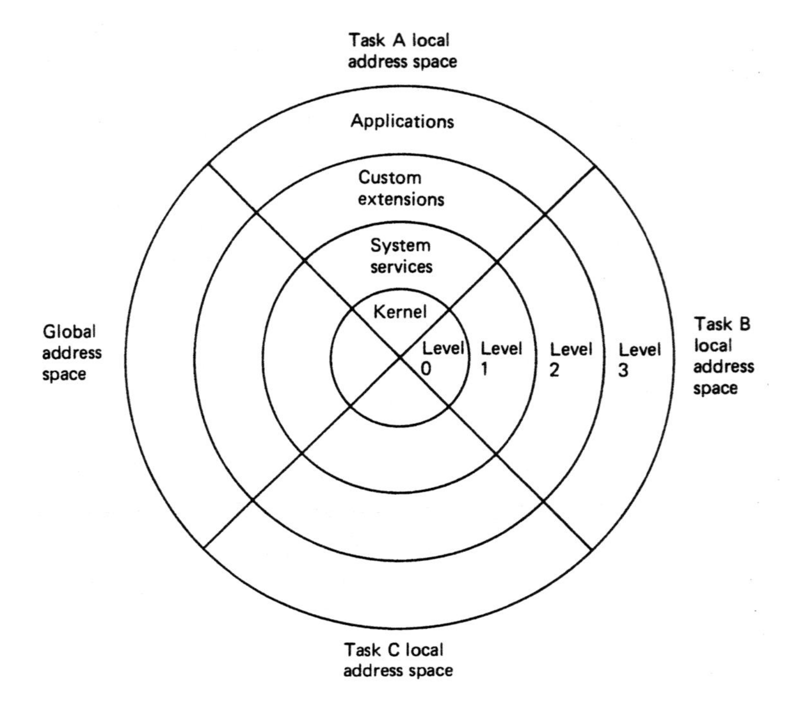

Figure 15–38 Protection model.

of protection result in improved software reliability because errors in one application will not affect the operating system or other applications.

Let us now look more closely at how the privilege level is assigned to a code or data segment. Remember that when a task is running, it has access to both local and global code segments, local and global data segments, and stack segments. A privilege level is assigned to each of these segments through the access rights information in its descriptor. A segment may be assigned to any privilege level simply by entering the number for the level into the DPL bits.

To provide more flexibility, input/output has two levels of privilege. First, the I/O drivers, which are normally system resources, are assigned to a privilege level. For the software system of Fig. 15–38, we indicate that the I/O control routines are part of the kernel and are at level 0.

The IN, INS, OUT, OUTS, CLI, and STI instructions are what are called *trusted instructions* because the protection model of the 80386DX puts additional restrictions on their use in protected mode. They can only be executed at a privilege level equal to or more privileged than the *input/output privilege level* (IOPL) code. IOPL supplies the second level of I/O privilege. Remember that the IOPL bits exist in the protected-mode flag register. These bits must be loaded with the value of the privilege level to be assigned to input/output instructions through software. The value of IOPL may change from task to task. Assigning the I/O instructions to a level higher than 3 restricts applications from directly performing I/O. Therefore, to perform an I/O operation, the application must request service by an I/O driver through the operating system.

Accessing Code and Data through the Protection Model

During the running of a task, the 80386DX may need to either pass control to program routines at another privilege level or access data in a segment at a different privilege level. Accesses to code or data in segments at a different privilege level are governed by strict rules. These rules are designed to protect the code or data at the more privileged level from contamination by the less privileged routine.

Before looking at how accesses are made for routines or data at the same or different privilege levels, let us first look at some terminology used to identify privilege levels. We have already been using the terms *descriptor privilege level* (DPL) and *I/O privilege level* (IOPL). However, when discussing the protection mechanisms by which processes access data or code, two new terms come into play: *current privilege level* (CPL) and *requested privilege level* (RPL). CPL is defined as the privilege level of the code or data segment currently being accessed by a task. For example, the CPL of an executing task is the DPL of the access rights byte in the descriptor cache for the CS register. This value normally equals the DPL of the code segment. RPL is the privilege level of the new selector loaded into a segment register. For instance, in the case of code, it is the privilege level of the code segment that contains the routine being called. That is, RPL is the DPL of the code segment to which control is to be passed.

As a task in an application runs, it may require access to program routines that reside in segments at any of the four privilege levels. Therefore, the current privilege level of the task changes dynamically with the programs it executes because the CPL of the task is normally switched to the DPL of the code segment currently being accessed.

The protection rules of the 80386DX determine what code or data a program can access. Before looking at how control is passed to code at different protection levels, let us first look at how data segments are accessed by code at the current privilege level. Figure 15–39 illustrates the protection-level checks that are made for a data access. The general rule is that code can access only data that are at the same or a less privileged level. For instance, if the current privilege level of a task is 1, it can access operands that are in data segments with DPL equal to 1, 2, or 3. Whenever a new selector is loaded into the DS, ES, FS, or GS register, the DPL of the target data segment is checked to make sure that it is equal to or less privileged than the most privileged of either CPL or RPL. As long as DPL satisfies this condition, the descriptor is cached inside the 80386DX, and the data access takes place.

One exception to this rule is when the SS register is loaded. In this case, the DPL must always equal the CPL. That is, the active stack (one is required for each privilege level) is always at the CPL.

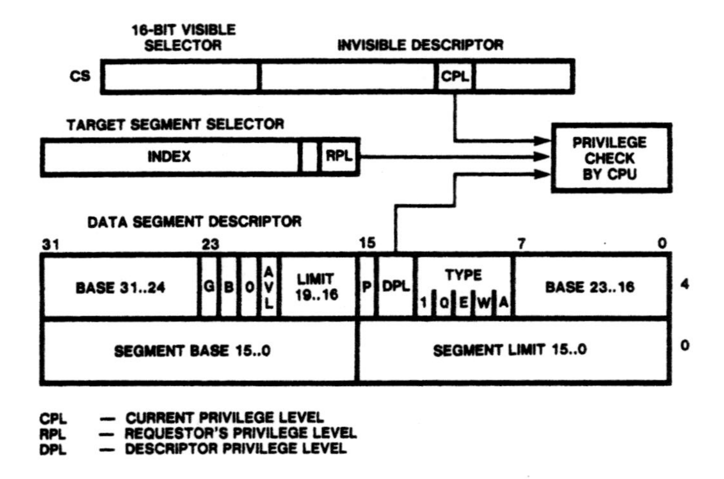

Figure 15–39 Privilege-level checks for a data access. (Reprinted by permission of Intel Corporation. Copyright/Intel Corp. 1986)

EXAMPLE 15.16

Assuming that, in Fig. 15–39, DPL = 2, CPL = 0, and RPL = 2, will the data access take place?

Solution

DPL of the target segment is 2, and this value is less privileged than CPL = 0, which is the more privileged of CPL and RPL. Therefore, the protection criterion is satisfied and the access will take place.

Different rules apply to how control is passed between code at the same privilege level and between code at different privilege levels. To transfer program control to another instruction in the same code segment, we can simply use a near jump or call instruction. In either case, only a limit check is made to ensure that the destination of the jump or call does not exceed the limit of the current code segment.

To pass control to code in another segment that is at the same or a different privilege level, a far jump or far call instruction is used. For this transfer of program control, both type and limit checks are performed and privilege-level rules are applied. Figure 15–40 shows the privilege checks made by the 80386DX. There are two conditions under which the transfer in program control will take place. First, if CPL equals the DPL, the two segments are at the same protection level and the transfer occurs. Second, if the CPL represents a more privileged level than DPL but the conforming code (C) bit in the type field of the new segment is set, the routine is executed at the CPL.

The general rule that applies when control is passed to code in a segment that is at a different privilege level is that the new code segment must be at a more privileged level.

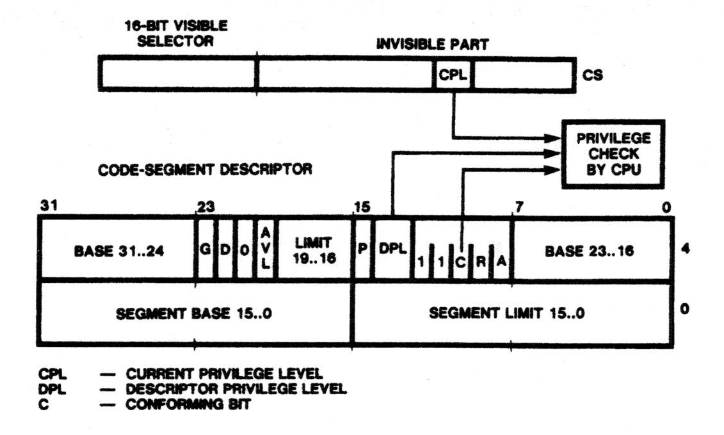

Figure 15–40 Privilege-level checks when directly passing program control to a program in another segment. (Reprinted by permission of Intel Corporation. Copyright/Intel Corp. 1986)

A special kind of descriptor, a *gate descriptor,* comes into play to implement the change in privilege level. An attempt to transfer control to a routine in a code segment at a higher privilege level is still initiated with either a far call or far jump instruction. This time the instruction does not directly specify the location of the destination code; instead, it references a gate descriptor. In this case the 80386DX goes through a much more complex program control-transfer mechanism.

The structure of a gate descriptor is shown in Fig. 15–41. Note that there are four types of gate descriptors: the *call gate, task gate, interrupt gate,* and *trap gate.* The call gate implements an indirect transfer of control within a task from code at the CPL to code at a higher privilege level. It does this by defining a valid entry point into the more privileged segment. The contents of a call gate are the virtual address of the entry point: the *destination selector* and the *destination offset.* In Fig. 15–41, we see that the destination selector identifies the code segment that contains the program to which control is to be redirected. The destination offset points to the instruction in this segment where execution is to resume. Call gates can reside in either the GDT or a LDT.

Figure 15–42 illustrates the operation of the call gate mechanism. Here we see that the call instruction includes an offset and a selector. When the instruction is executed, this selector is loaded into CS and points to the call gate. In turn, the call gate causes its destination selector to be loaded into CS. This leads to the caching of the descriptor for the called code segment (executable segment descriptor). The executable segment descriptor provides the base address for the executable segment (code segment) of memory. Note that the offset in the call gate descriptor locates the entry point of the procedure in the executable segment.

Whenever the task's current privilege level is changed, a new stack is activated. As part of the program context-switching sequence, the old ESP and SS are saved on the new stack along with any parameters and the old EIP and CS. This information is needed to preserve linkage for return to the old program environment.

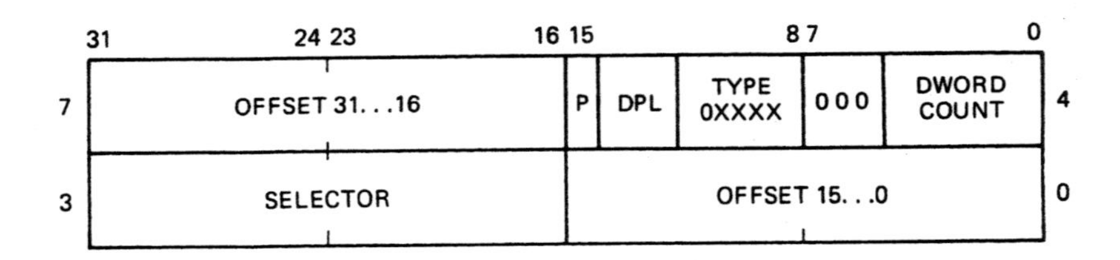

Name	Value	Description
TYPE	4 5 6 7	– Call gate – Task gate – Interrupt gate – Trap gate
P	0 1	– Descriptor contents are not valid – Descriptor contents are valid
DPL	0–3	Descriptor privilege level
WORD COUNT	0–31	Number of double words to copy from callers stack to called procedures stack. Only used with call gate
DESTINATION SELECTOR	16-Bit selector	Selector to the target code segment (call interrupt or trap gate) Selector to the target task state segment (task gate)
DESTINATION OFFSET	32-Bit offset	Entry point within the target code segment

Figure 15–41 Gate descriptor format. (Reprinted by permission of Intel Corporation. Copyright/Intel Corp. 1987)

Now the procedure at the higher privilege level begins to execute. At the end of the routine, a RET instruction must be included to return program control back to the calling program. Execution of RET causes the old values of EIP, CS, the parameters, ESP, and SS to be popped from the stack. This restores the original program environment. Now program execution resumes with the instruction following the call instruction in the lower privileged code segment. Figure 15–43 shows the privilege checks performed when program control transfer is initiated through a call gate. For the call to be successful, the DPL of the gate must be the same as the CPL, and the RPL of the called code must be higher than the CPL.

Task Switching and the Task State Segment Table

Earlier we identified the task as the key program element of the 80386DX's multitasking software architecture and indicated that another important feature of this architecture is the high-performance task-switching mechanism. A task can be invoked either

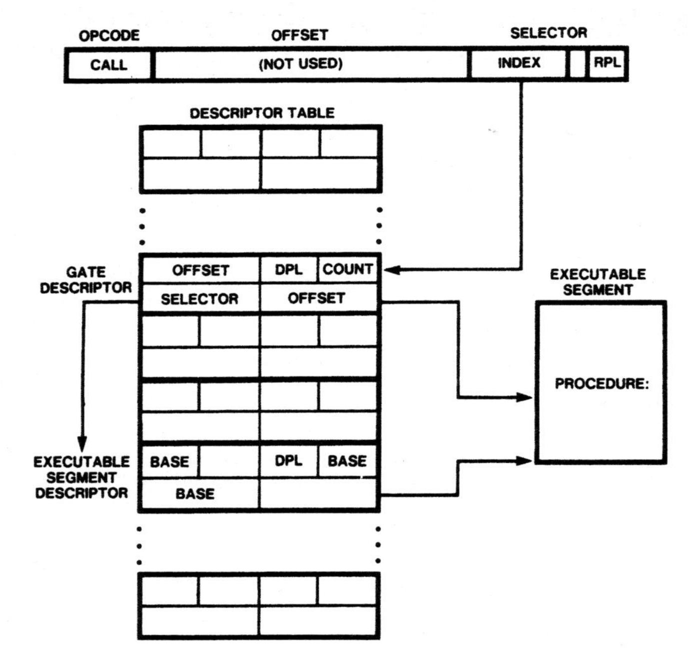

Figure 15–42 Call gate operation. (Reprinted by permission of Intel Corporation. Copyright/Intel Corp. 1986)

directly or indirectly by executing either the intersegment jump or intersegment call instruction. When a jump instruction is used to initiate a task switch, no return linkage to the prior task is supported. On the other hand, if a call is used to switch to the new task instead of a jump, back linkage information is saved automatically. This information permits a return to be performed to the instruction that follows the calling instruction in the old task at completion of the new task.

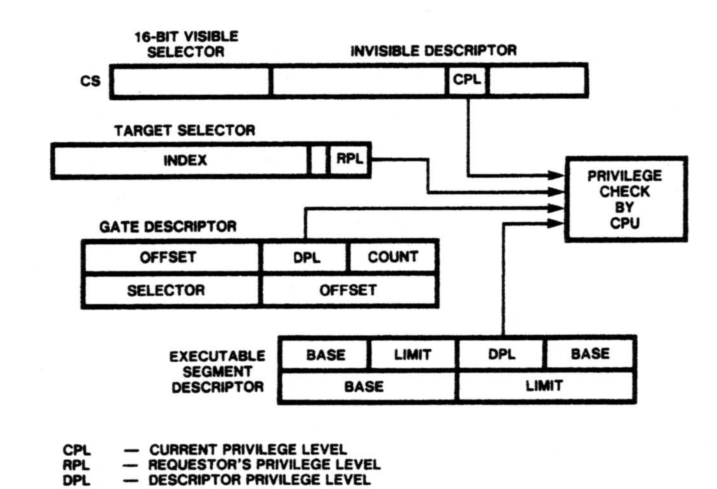

Figure 15–43 Privilege-level checks for program control transfer with a call gate. (Reprinted by permission of Intel Corporation. Copyright/Intel Corp. 1986)

Each task that is to be performed by the 80386DX is assigned a unique selector called a *task state selector*. This selector is an index to a corresponding *task state segment descriptor* (TSS) in the global descriptor table. The format of a task state segment descriptor is given in Fig. 15–44.

If a jump or call instruction has a task state selector as its operand, a direct entry is performed to the task. As Fig. 15–45 shows, when a call instruction is executed, the selector is loaded into the 80386DX's task register (TR). Then the corresponding task state segment descriptor is read from the GDT and loaded into the task register cache. This only happens if the criteria specified by the access rights information of the descriptor are satisfied. That is, the descriptor is present (P = 1), the task is not busy (B = 0), and protection is not violated (CPL must be equal to DPL). Looking at Fig. 15–45, we see that, once loaded, the base address and limit specified in the descriptor define the starting point and size of the task's *task state segment* (TSS). This TSS contains all the information needed to either start or stop a task.

Before explaining the rest of the task-switch sequence, let us first look more closely at what is contained in the task state segment. Figure 15–45 show a typical TSS. Its minimum size is 104 bytes. For this reason, the minimum limit that can be specified in a TSS descriptor is 00067_{16}. Note that the segment contains information such as the state of the microprocessor (general registers, segment selectors, instruction pointer, and flags) needed to initiate the task, a back-link selector to the TSS of the task that was active when this task was called, the local descriptor table register selector, a stack selector and pointer for privilege levels 0, 1, and 2, and an I/O permission bit map.

Now we will continue with the procedure that invokes a task. Let us assume that a task was already active when a new task was called. Then the new task is what is called a *nested task* and it causes the NT bit of the flag word to be set to 1. In this case, the current task is first suspended and the state of the 80386DX's user-accessible registers is saved in the old TSS. Next, the B bit in the new task's descriptor is marked busy; the TS bit in the machine status word is set to indicate that a task is active; the state information from the new task's TSS is loaded into the MPU; and the selector for the old TSS is saved as the back-link selector in the new task state segment. The task switch operation is now complete and execution resumes with the instruction identified by the new contents of the code segment selector (CS) and instruction pointer (IP).

The old program context is preserved by saving the selector for the old TSS as the back-link selector in the new TSS. By executing a return instruction at the end of the new

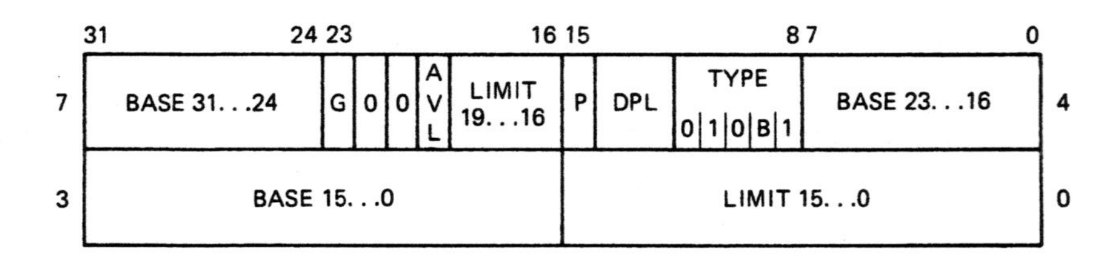

Figure 15–44 TSS descriptor format. (Reprinted by permission of Intel Corporation. Copyright/Intel Corp. 1986)

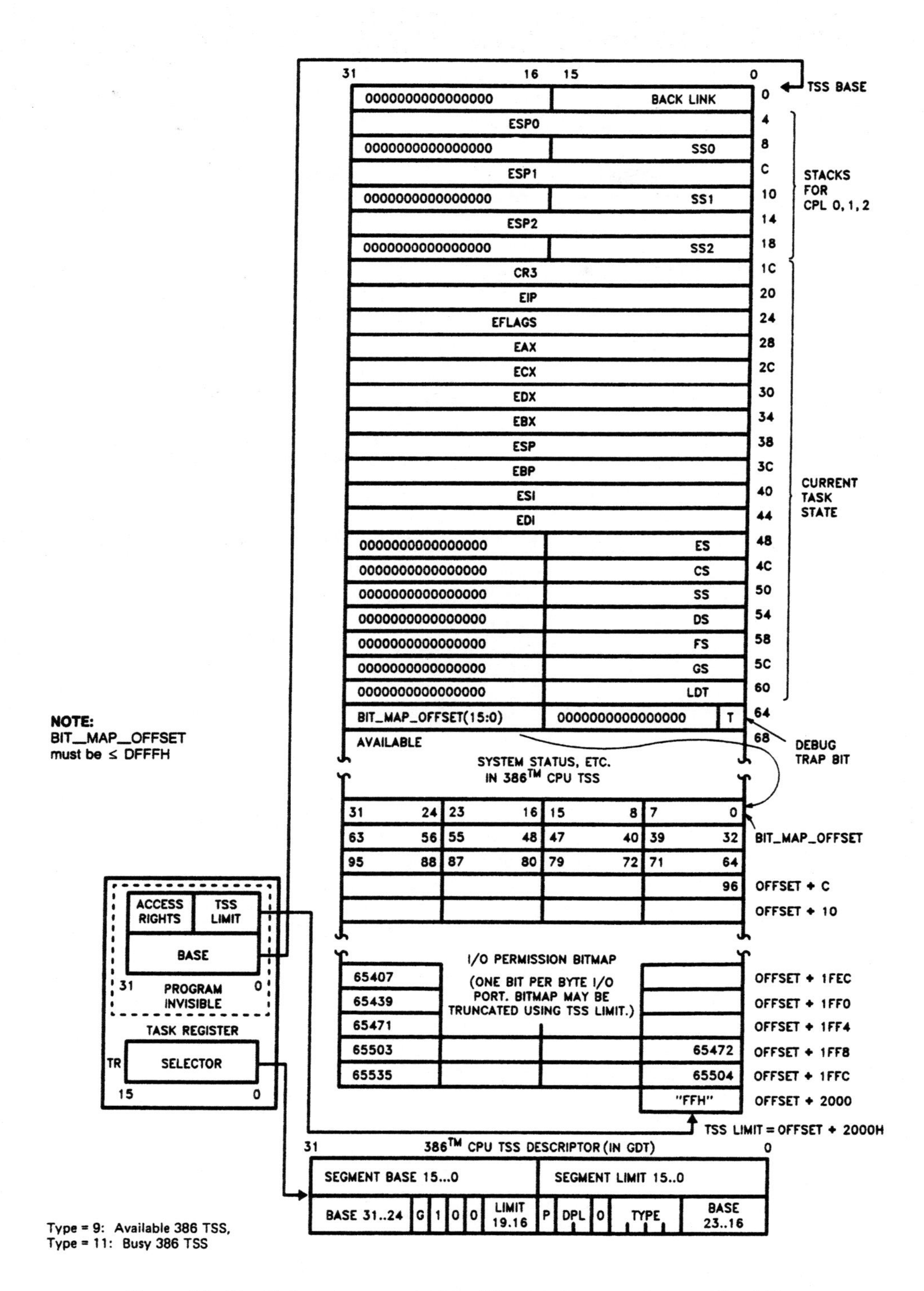

Figure 15–45 Task state segment table. (Reprinted by permission of Intel Corporation. Copyright/Intel Corp. 1987)

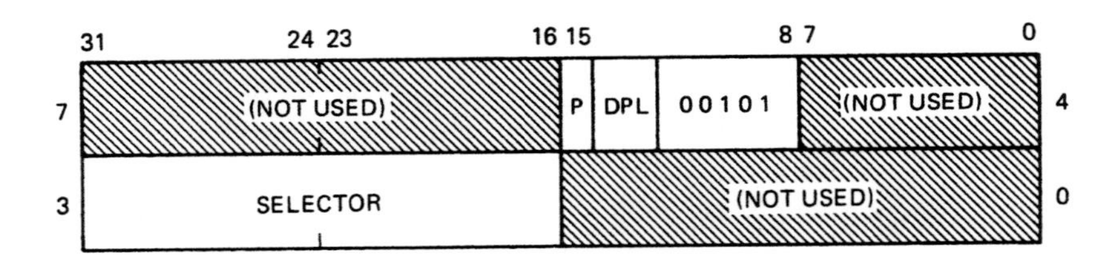

Figure 15–46 Task gate descriptor format. (Reprinted by permission of Intel Corporation. Copyright/Intel Corp. 1986)

task, the back-link selector for the old TSS is automatically reloaded into TR. This activates the old TSS and restores the prior program environment. Now program execution resumes at the point where it left off in the old task.

The indirect method of invoking a task is by jumping to or calling a *task gate.* This is the method used to transfer control to a task at an RPL that is higher than the CPL. Figure 15–46 shows the format of a task gate. This time the instruction includes a selector that points to the task gate, which is in either the LDT or the GDT, instead of a task state selector. The TSS selector held in this gate is loaded into TR to select the TSS and initiate the task. Figure 15–47 illustrates a task initiated through a task gate.

Let us consider an example to illustrate the principle of task switching. The table in Fig. 15–48 contains TSS descriptors SELECT0 through SELECT3. These descriptors contain access rights and selectors for tasks 0 through 3, respectively. To invoke the task corresponding to selector SELECT2 in the data segment where these selectors are stored, we can use the following procedure. First, the data segment register is loaded with the

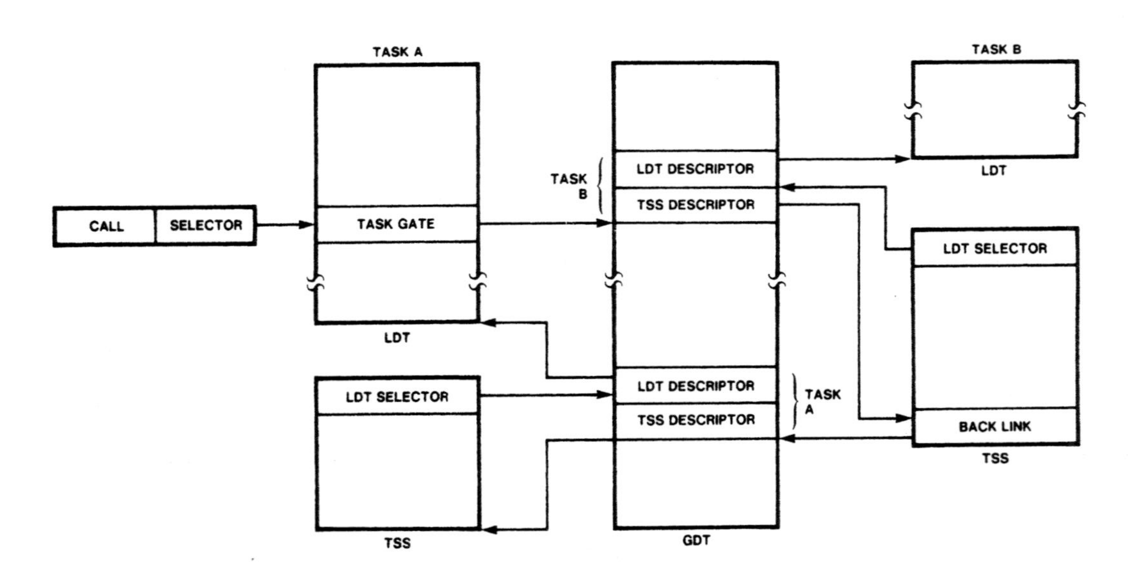

Figure 15–47 Task switch through a task gate. (Reprinted by permission of Intel Corporation. Copyright/Intel Corp. 1987)

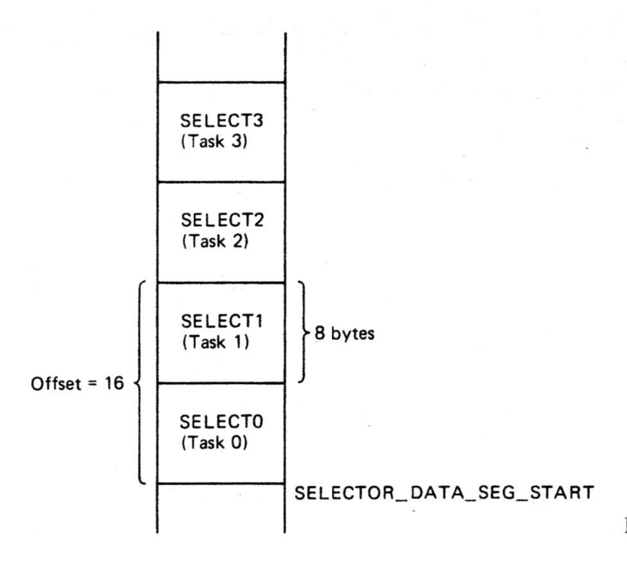

Figure 15–48 Task gate descriptors.

address SELECTOR_DATA_SEG_START to point to the segment that contains the selectors. This is done with the instructions

```
MOV AX, SELECTOR_DATA_SEG_START
MOV DS, AX
```

Since each selector is 8 bytes long, SELECT2 is offset from the beginning of the segment by 16 bytes. Let us load the offset into register EBX:

```
MOV EBX, 0FH
```

At this point we can use SELECT2 to implement an intersegment jump with the instruction

```
JMP DWORD PTR [EBX]
```

Execution of this instruction switches program control to the task specified by the selector in descriptor SELECT2. In this case, no program linkage is preserved. On the other hand, by calling the task with the instruction

```
CALL DWORD PTR [EBX]
```

the linkage is maintained.

▲ 15.9 VIRTUAL 8086 MODE

8086 and 8088 application programs, such as those written for the PC's DOS operating system, can be directly run on the 80386DX in real mode. A protected-mode operating system, such as UNIX, can also run DOS applications without change. This is done through what is called *virtual 8086 mode.* When in this mode, the 80386DX supports an 8086 microprocessor programming model and can directly run programs written for the 8086. That is, it creates a virtual 8086 machine for executing programs.

In this kind of application, the 80386DX is switched back and forth between protected mode and virtual 8086 mode. The UNIX operating system and UNIX applications are run in protected mode, and when the DOS operating system and a DOS application are to run, the 80386DX is switched to the virtual 8086 mode. A program known as a virtual 8086 monitor controls this mode switching.

The *virtual mode* (VM) bit in the extended flag register selects virtual 8086 mode of operation. VM must be switched to 1 to enable virtual 8086 mode of operation. Actually, the VM bit in EFLAGS is not directly switched to 1 by software because virtual 8086 mode is normally entered as a protected-mode task. Therefore, the copy of EFLAGS in the TSS for the task would include VM equal to 1. These EFLAGS are loaded as part of the task switching process. In turn, the virtual 8086 mode of operation is initiated. The virtual 8086 program is run at privilege level 3. The virtual 8086 monitor is responsible for setting and resetting the VM bit in the task's copy of EFLAGS and permits both protected-mode tasks and virtual-8086-mode tasks to coexist in a multitasking program environment.

Another way of initiating virtual 8086 mode is through an interrupt return. In this case, the EFLAGS are reloaded from the stack. Again the copy of EFLAGS must have the VM bit set to 1 to enter the virtual 8086 mode of operation.

▲ 15.10 80486 MICROPROCESSOR FAMILY

Intel's second generation of 32-bit microprocessors, the 80486 family, was introduced in 1989. The first product offered in this family was the *80486DX* MPU. This device is a full 32-bit microprocessor—that is, its internal registers and external data paths are both 32 bits wide. This device offers a number of advanced software and hardware architecture features as compared to the 80386DX. Two major changes that greatly improved performance were the addition of an on-chip *floating-point math coprocessor* and an on-chip *code and data cache memory.* The 80386 family supports both a math coprocessor and cache memory, but they need to be implemented external to the device.

The 80486 family maintains real-mode and protected-mode software compatibility with the 80386 architecture. However, important changes have been made in the instruction set. First, and most important, is that the execution speed of most instructions of the instruction set has been improved for the 80486 family. This was done by changing the way in which they are performed by the MPU so that now most of the basic instructions execute in just one clock cycle. For instance, the move, add, subtract, and logic operations can all be performed in a single clock cycle. With the 80386DX, these same operations took two or more clock cycles to be completed. Finally, a number of new instructions have been added to make the instruction set even more versatile. The result of these architectural changes is an improvement of more than 2× in the overall performance for the 80486 family.

Similar to the 80386DX, the 80486DX was followed by an *80486SX* device. However, this time the SX version did not have a 16-bit external architecture. It also is a full 32-bit MPU, although it does not include the on-chip floating-point coprocessor unit. The

80486DX and 80486SX were followed by several new generations of 80486 MPUs, which introduced additional architectural features that further enhanced the performance of the family. For instance, the *80486DX2* MPU, which is both hardware and software compatible with the 80486DX, has increased performance by a technique known as *clock doubling*. In the *80486DX4*, this internal clock scaling was expanded to permit multiplication by 2×, 2.5×, or 3×. Here we will focus on the 80486SX, not the 80486DX.

Let us next briefly compare the levels of performance offered by the 80386 family and 80486 family MPUs. Referring back to the iCOMP index chart in Fig. 15–1, we see that the 80486SX-20 has an iCOMP rating of 78 compared to a rating of 32 for the 80386SX-20. Moreover, the 80486DX-33 is rated at a level of 166, whereas the 80386DX-33 is at 68. In this way, we see that comparable 80486 family members do deliver more than twice the performance. Also, newer members of the 80486 family, such as the 80486DX2-66 (rated at 297 in the iCOMP chart), have widened this performance advantage to more than 4×.

EXAMPLE 15.17

What is the iCOMP rating of the 80486DX4-100?

Solution

Looking at Fig. 15–1, we find that the iCOMP rating of the 80486DX4-100 is 435.

Internal Architecture of the 80486

We already mentioned that the internal architecture of the 80486 family is an improvement over that of the 80386 family. For instance, we said that a floating-point coprocessor and cache memory are now on-chip, but these are not the only changes that have been made to improve the performance of the 80486 family. Here we will explore the functional elements within the 80486DX microprocessor's architecture and how they have changed from that of the 80386DX.

Figure 15–49 shows a block diagram of the internal architecture of the 80486 family. Similar to the 80386 architecture, we find the execution unit, segmentation unit, paging unit, bus interface unit, prefetch unit, and decode unit. However, the function of many of these elements has been enhanced for the 80486 family. For instance, we already mentioned that the coding of instructions in the control ROM had been changed to permit instructions to be performed in fewer clock cycles. Some other changes are that the code queue in the prefetch unit has been doubled in size to 32 bytes. This permits more instructions to be held on-chip ready for decode and execution. Also, the translation lookaside buffer in the paging unit now uses an improved algorithm. Finally, the bus interface unit has been modified to give the 80486 architecture a much faster and more versatile processor bus.

The 8086 family microprocessors available before the 80486DX are what are known as *complex instruction set computers* or *CISC* processors. That is, they have a large versatile instruction set that supports many complex addressing modes. In general,

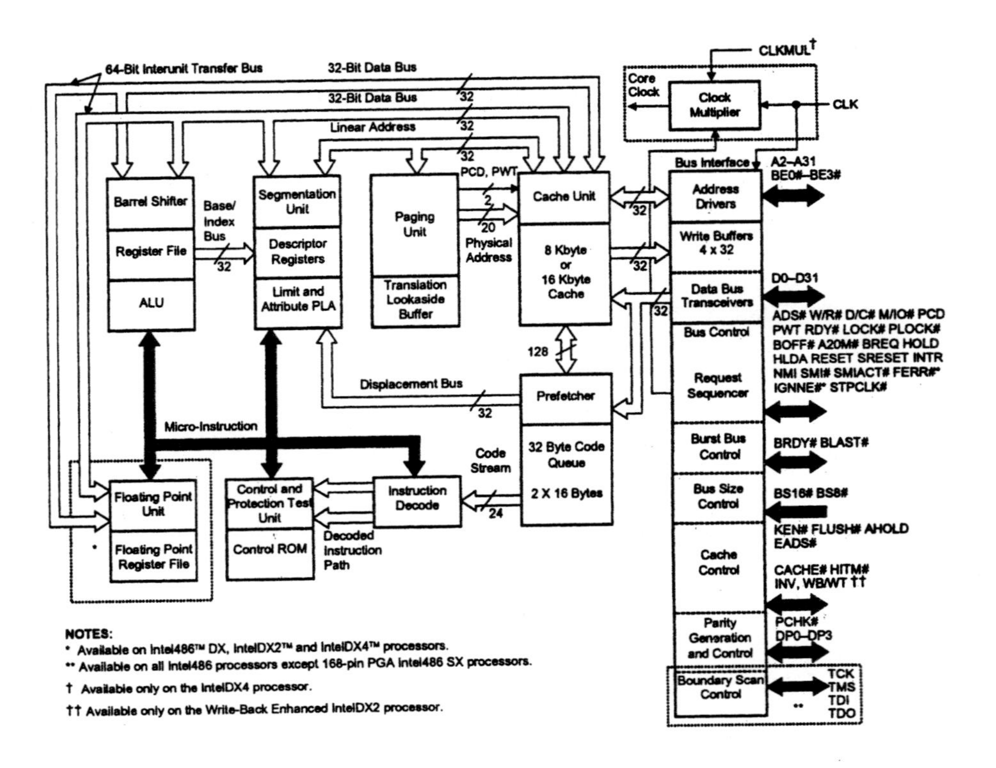

Figure 15–49 Internal architecture of the 80486DX MPU. (Reprinted by permission of Intel Corporation. Copyright/Intel Corp. 1994)

the instructions execute in two to many clock cycles. For instance, the 80386's register-to-register ADD instruction takes two clock cycles, but the same instruction when adding the content of a storage location in memory takes seven clock cycles. Other instructions, for instance, the integer multiply (IMUL) and integer divide (IDIV) take even more clock cycles.

The 80486DX is the first member of this family with a high-performance *reduced-instruction-set computer* (RISC) integer core. A RISC processor is typically characterized as having a small instruction set, limited addressing modes, and single-clock execution for instructions. The MPUs of the 80486 family are best described as *complex reduced-instruction-set computers* (CRISC). This is because a core group of instructions in the 80486's instruction set executes in a single clock cycle. For example, the register-to-register ADD is performed in a single cycle. At the same time, it retains the many complex instructions and addressing modes that make the instruction set more versatile. However, the number of clock cycles needed to execute many of these complex instructions has also been reduced in the 80486DX. For instance, the register-to-memory ADD is reduced from seven to three clock cycles.

Let us now look more closely at the new elements of the 80486DX's internal architecture. Traditionally, an externally attached floating-point coprocessor has performed the floating-point operations of the microcomputer. With the 80486DX, this function is integrated into the MPU. This *floating-point math unit* supports the processing of the 32-bit, 64-bit, and 80-bit number formats specified in the *IEEE 754 standard* for floating-point numbers. At the same time, it is upward software compatible with the older 8087, 80287, and 80387 numeric coprocessors. The result of this on-chip implementation is higher-performance floating-point operation. Remember that this unit is not provided in the 80486SX MPU.

Addition of a high-speed *cache memory* to a microcomputer system provides a way of improving overall system performance while permitting the use of low-cost, slow-speed memory devices in the main system memory. During system operation, the cache memory contains recently used instructions, data, or both. The objective is that the MPU accesses code and data in the cache most of the time, instead of from the main memory. Since less time is required to access the information from the cache memory, the result is a higher level of system performance. The internal *cache memory unit* of the 80486DX is 8Kbytes in size and caches both code and data.

EXAMPLE 15.18

In Fig. 15–49, identify an architectural element other than the floating-point unit that is not implemented in the PGA-packaged 80486SX MPU.

Solution

Looking at Fig. 15–49, we find that the *boundary scan control* block of the bus interface unit is not available on the 80486SX MPU.

Real-Mode Software Model and Instruction Set of the 80486SX

At this point, we will turn our attention to the 80486SX MPU and its real-mode software model. Figure 15–50 shows the registers in the software model of the 80486SX. These registers are essentially the same as those shown for the 80386DX in Fig. 15–3. The organization and functionality are also the same. One exception is control register 0 (CR_0). In the 80386DX, this register has just one bit that is active in the real mode. For the 80486SX, two other bits, caches disable (CD) and not write-through (NW), are active. They are used to enable and configure the operation of the on-chip cache memory. Of course, the 80486DX has more registers in its real-mode model. The new registers are needed to support the operation of the floating-point coprocessor.

The real-mode instruction set has been enhanced with new instructions for the 80486 family. Figure 15–51(a) shows that the 80486SX's instruction set is simply a superset of that of the 80386DX. A group of new instructions called the *80486 specific instruction set* has been added. Figure 15–51(b) summarizes the instructions of the 80486 specific instruction set. Since all the earlier instructions are retained in all 80486 family processors and their object code is compatible with the 8086, 8088, 80286, and 80386 processors, upward software compatibility is maintained. Let us now look at the operation of each of the new instructions.

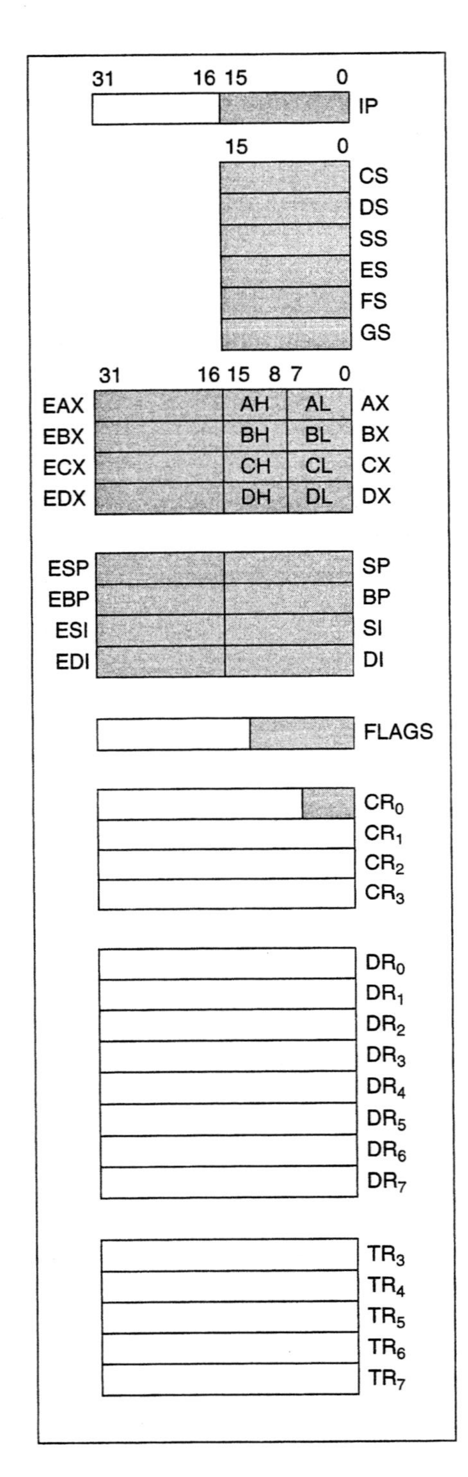

Figure 15–50 Real-mode software model of the 80486SX microprocessor.

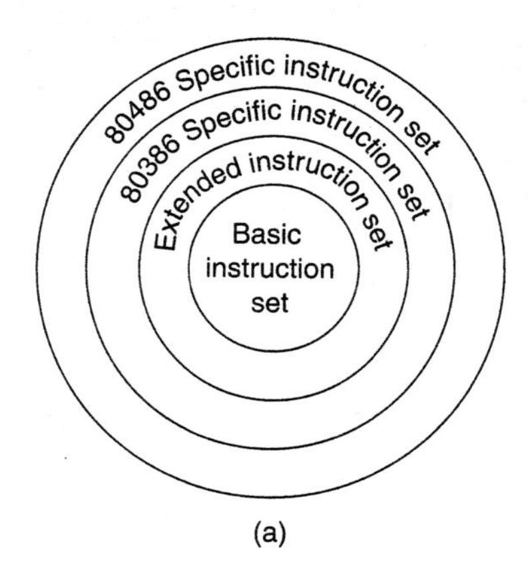

(a)

Mnemonic	Meaning	Format	Operation
BSWAP	Byte swap	BSWAP r32	Reverse the bye order of the 32-bit register.
XADD	Exchange and add	XADD D,S	(D) ↔ (S), (D) ← (S) + (D)
CMPXCHG	Compare and exchange	CMPXCHG D,S	if (ACC) = (D) (ZF) ← 1, (D) ← (S) Else (ZF) ← 0, (ACC) ← (D)

(b)

Figure 15–51 (a) 80486SX instruction set. (b) 80486 specific instructions set.

Byte-Swap Instruction: BSWAP. When studying the 8086 microprocessor, we showed how the bytes of a double word of data were stored in memory. As Fig. 15–52(a) shows, the least significant byte is stored at the lowest-value byte address, which is identified as address *m*. The next more significant bytes are held at address m + 1 and m + 2. Finally, the most significant byte is saved at the highest-value byte address, m + 3. This method of storing information in memory is known as *little endian* organization.

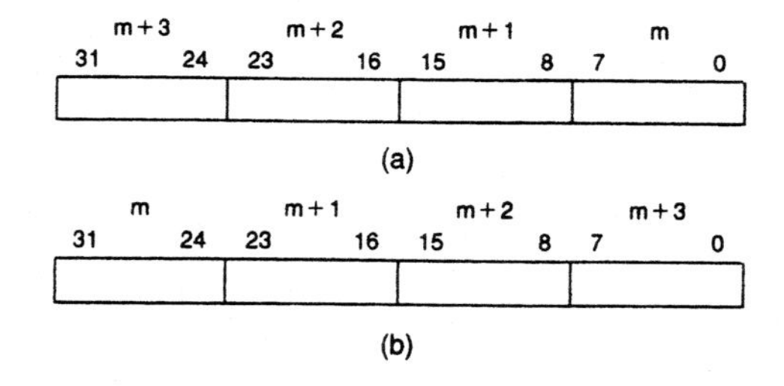

Figure 15–52 (a) Little endian memory format. (Reprinted by permission of Intel Corporation. Copyright/Intel Corp. 1994) (b) Big endian memory format. (Reprinted by permission of Intel Corporation. Copyright/Intel Corp. 1994)

Other microprocessor architectures employ another method of double-word data organization, called *big endian.* For instance, Motorola's 68000 family of microprocessors stores data in this way. Figure 15–52(b) shows how the bytes of a 32-bit word are arranged in big endian format. Note that they are stored in the opposite order.

To make it easier for the 80486SX to process data that had been initially created in the big endian format, a special instruction was added to the real-mode instruction set. This *byte-swap* (BSWAP) instruction in provided to convert the organization of the bytes of a double word of data between the big endian format and little endian format. Figure 15–51(b) shows that the instruction has a single 32-bit register as its destination. Therefore, the double word that is to be converted must first be loaded into an internal register of the 80486SX.

An example is the instruction

```
BSWAP EAX
```

Let us assume that the contents of EAX are in big endian format. Then, when this instruction is executed, the bytes of the double word of data in register EAX are rearranged into the little endian format. For instance, if the big endian contents of the register are

$$(EAX) = 01234567H = 00000001001000110100010101100111_2$$

the new contents of EAX after the byte swap has taken place will be

$$(EAX) = 67452301H = 01100111010001010010001100000001_2$$

Actually, BSWAP will convert the format of data either way. If the bytes of data in EAX are in little endian form when the instruction is executed, it will be changed to big endian form.

EXAMPLE 15.19

Write an instruction sequence that reads the double-word contents of storage location DS:1000H in memory, rearrange the bytes from big endian to little endian organization, and then return the new value to the original storage location in memory.

Solution

The big endian-format double-word data is read from memory with the instruction

```
MOV EAX, [1000H]
```

Then, the double word is converted to little endian form by

```
BSWAP EAX
```

Finally, the double word is returned to memory with the instruction

```
MOV [1000H], EAX
```

Exchange and Add Instruction: XADD. The second instruction added to the real-mode instruction set of the 80486SX is the *exchange and add* (XADD) instruction. Looking at Fig. 15–51(b), we find that this instruction performs both an add and exchange operation on the contents of the source and destination operands. The source operand must be an internal register, whereas the destination can be either another register or a storage location in memory.

For example, let us determine the operation of the register-to-register exchange and add instruction

```
XADD AX, BX
```

We will assume that the contents of registers AX and BX are 1234_{16} and 1111_{16}, respectively. After the exchange and add operation takes place, the sum of these two values ends up in destination register AX:

$$(AX) = 0001001000110100_2 + 0001000100010001_2 = 0010001101000101_2$$
$$= 2345_{16}$$

and the original contents of AX are in BX

$$(BX) = 1234_{16}$$

EXAMPLE 15.20

If the value in EAX is 00000001H and the double word at memory address DS:1000H is 00000002H, what results are produced by executing the instruction

```
XADD EAX, [1000H]
```

Solution

Execution of the instruction causes the addition

$$(EAX) + (DS{:}1000H) = 00000001_{16} + 00000002_{16}$$
$$= 00000003_{16}$$

The value that results in the source location is

$$(DS{:}1000H) = 00000001_{16}$$

and that in the destination is

$$(EAX) = 00000003_{16}$$

Compare and Exchange Instruction: CMPXCHG. The last of the new real-mode instructions is *compare and exchange* (CMPXCHG). This instruction performs a compare operation and an exchange operation that depends on the result of the compare. As Fig. 15.51(b) shows, the compare that takes place is not between the values of the source and destination operand. It is between the content of the accumulator register (AL, AX, EAX) and the corresponding size destination. If the accumulator and destination contain the same value, the zero flag is set to 1 and the content of the source register is loaded into the destination location. Otherwise, ZF is cleared to 0 and the content of the destination is loaded into the accumulator. The destination can be either a register or storage location in memory. The value in the accumulator must be loaded prior to execution of the CMPXCHG instruction.

As an example, let us consider the instruction

```
CMPXCHG [2000H], BL
```

and assume that register AL contains 11_{16}, register BL contains 22_{16}, and the byte memory location at address 2000H contains 12_{16}. When the instruction is executed, the value in AL (11_{16}) is compared to that at address 2000H in memory (12_{16}). Since they are not equal, ZF is made logic 0 and the value 12_{16} is copied from memory into AL. Therefore, after execution of the instruction, the results are

$$(AL) = 12_{16}$$

$$(BL) = 22_{16}$$

$$(DS{:}2000H) = 12_{16}$$

EXAMPLE 15.21

Assume that the values in registers AL and BL are 12_{16} and 22_{16}, respectively, and that the byte-wide memory location DS:2000H contains 12_{16}. What results are produced by executing the instruction

```
CMPXCHG  [2000H], BL
```

Solution

Since (AL) = (DS:2000H) = 12_{16}, the zero flag is set and the value of the source operand is loaded into the destination. This gives

$$(AL) = 12_{16}$$

$$(BL) = 22_{16}$$

$$(DS{:}2000H) = 22_{16}$$

Protected-Mode Software Architecture of the 80486SX

Now that we have examined the differences between the software architectures of the 80386DX and 80486SX in the real mode, let us turn our attention to protected-mode operation. The protected-mode operation of the 80486SX is essentially the same as that of the 80386DX. Enhancements have been made to the register set, system control instruction set, and the page tables. Here we will focus on the differences between the two MPUs.

Protected-Mode Software Model. Similar to the real mode, the protected-mode software model of the 80486SX is essentially the same as that shown for the 80386DX in Fig. 15–8. That is, the same set of registers exists in both processors, and they serve the same functions relative to the protected-mode operation. However, the 80486SX has three additional test registers and new bits defined in the flags register and control registers. We will begin by examining these new register functions.

Figure 15–53 shows the protected-mode flags (EFLAGS) register of the 80486SX. In this illustration, the system-control flags, bits that affect protected-mode operation, are identified. Comparing these bits to those of the 80386DX in Fig, 15–15, we find that just one new bit has been added, the *alignment-check* (AC) flag, which is located in bit position 18. When this bit is set to 1, an alignment check is performed during all memory access operations that are performed at privilege level 3. A double word of data that is not stored at an address that is a multiple of four is said to be unaligned. If an unaligned double-word storage location is accessed, two memory bus cycles must be performed. The extra bus cycle that is introduced because the data is unaligned reduces overall system performance. The alignment-check feature of the 80486SX can be used to identify when unaligned elements of data are accessed. If an unaligned access takes place, an alignment-check exception, which is exception 17, occurs.

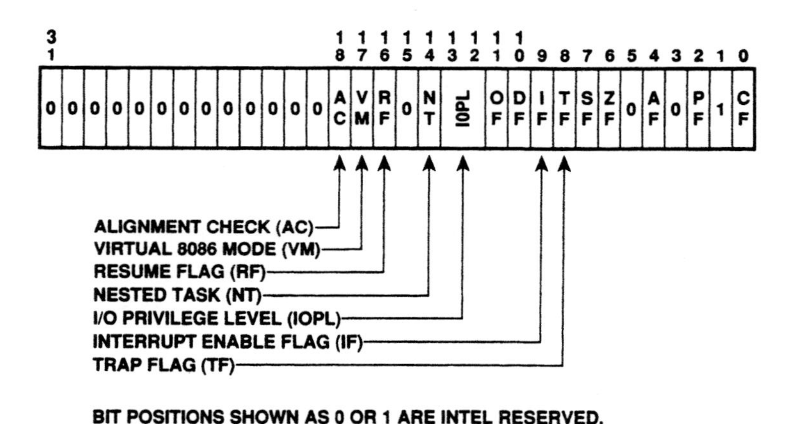

Figure 15–53 Protected-mode flags register. (Reprinted by permission of Intel Corporation. Copyright/Intel Corp. 1992)

The 80486SX has four control registers just like the 80386DX; however, a number of new bits are now active. The control registers of the 80486SX are shown in Fig. 15–54, and those of the 80386DX are in Fig. 15–12. Note that five additional bits have been activated in CR_0 of the 80486SX: the *alignment mask* (AM), *numeric error* (NE), *write protect* (WP), *cache disable* (CD), and *not write-through* (NW). Let us look at a few of these bits in detail.

We just said that the AC system-control flag enabled memory data alignment checks. Actually it takes more than just setting AC to 1 to enable this mode of operation. The AM bit, which is bit 18 in CR_0, must also be set to 1. If AM is switched to 0, the alignment check operation is masked out.

Two other bits in CR_0, CD (bit 30) and NW (bit 29), are used to enable and control the operation of the on-chip cache memory. To enable the cache for operation, CD must be cleared to 0. The NW bit enables write through and cache validation cycles to take place when it is set to 0. Therefore, to permit normal cache operation, both of these bits should be cleared to 0.

Some more changes are found in CR_3. Two new bits, *page-level cache disable* (PCD) and *page-level writes transparent* (PWT), have been defined. The state of these bits is output on signal lines PCD and PWT, respectively, during all bus cycles that are not paged. They are used as input signals to the control circuitry for an external cache memory subsystem.

System-Control Instruction Set. The system-control instruction set has been expanded by three instructions for the 80486SX microprocessor: *invalidate cache* (INVD), *write-back and invalidate data cache* (WBINVD), and *invalidate translation lookaside buffer entry* (INVLPG). Figure 15–55 shows the format of these instructions and briefly describes their operation.

The first two instructions, INVD and WBINVD, support management of the on-chip and external cache memories. When an INVD instruction is executed, the on-chip cache is flushed. That is, all of the data that it holds is made invalid. In addition to invalidating the content of the on-chip cache, execution of this instruction also initiates a spe-

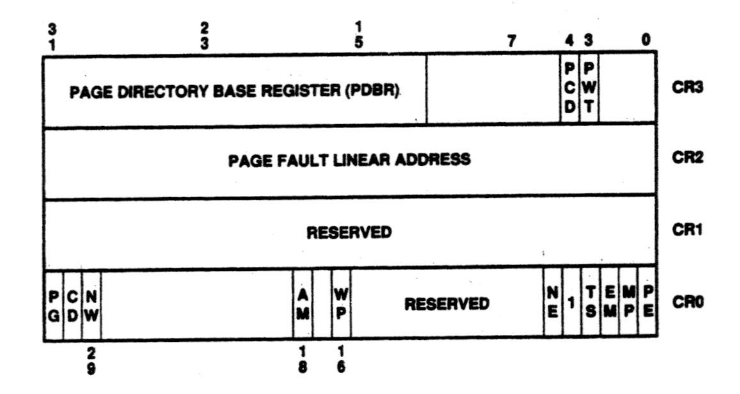

Figure 15–54 Control registers of the 80486SX. (Reprinted by permission of Intel Corporation. Copyright/Intel Corp. 1992)

Mnemonic	Meaning	Format	Operation
INVD	Invalidate cache	INVD	Flush internal cache and signal external cache to flush.
WBINVD	Write back and invalidate cache	WBINVD	Flush internal cache, signal external cache to write-back and flush.
INVLPG m	Invalidate TLB entry	INVLPG	Invalidate the signal TLB entry.

Figure 15–55 80486SX specific system-control instructions.

cial bus cycle known as a *flush bus cycle.* External circuitry must detect the occurrence of this cycle and initiate a flush of the data held in the external cache memory subsystem. WBINVD is similar to INVD in that it initiates a flush of the on-chip cache memory; however, it initiates a different special bus cycle, a *write-back bus cycle.* External circuitry must again identify that a write back cycle has taken place and tell the external cache to write back its content to the main memory.

The INVLPG instruction is used to invalidate a single entry in the 80486SX's internal translation lookaside table. Notice that the instruction has an operand *m* that identifies which of the 32 table entries is to be marked invalid.

Page Directory and Page Table Entries. The page directory and page tables of the 80486SX are the same size and serve the same function as they did in the 80386DX's protected-mode software architecture. However, a change has been made in the format of the page directory and page table entries. Two additional bits of the 32-bit entry are defined. Let us now look at the function of these two bits.

Figure 15–31 gives the format of a page directory/page table entry for the 80386DX. Comparing it to the format of the 80486SX's entry in Fig. 15–56, we find that the two new bits are *page cache disable* (PCD) and *page write transparent* (PWT). These

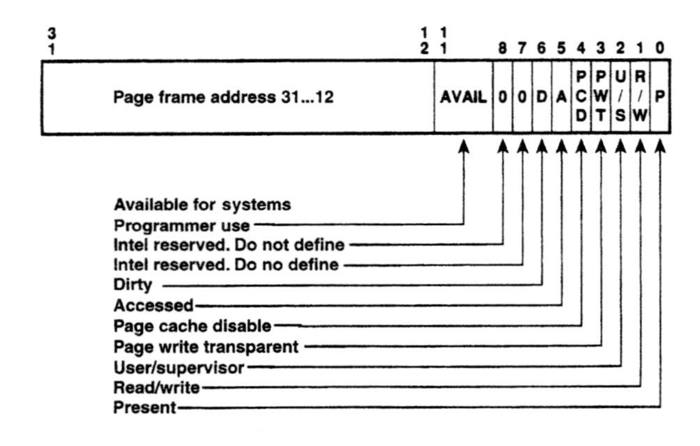

Figure 15–56 80486SX directory and page table entry format. (Reprinted by permission of Intel Corporation. Copyright/Intel Corp. 1992)

two bits are used for page-level control of the internal and external caches. To enable caching of a page, the PCD bit in the page table entry must be set to 0. Logic 1 in PWT selects page-level write through operation of the cache for the corresponding page.

When paging is in use, the logic levels of the PCD and PWT bits in the page table entry are output at the PCD and PWT pins of the MPU. This permits control of an external cache memory subsystem.

▲ 15.11 80486DX FLOATING-POINT ARCHITECTURE AND INSTRUCTIONS

Up to this point, the instructions that we examined all performed operations on integer data. This limits computations to the use of whole numbers. Many practical applications require the processing of real numbers—requiring a much larger range of numbers that includes signed integers, fractions, and mixed numbers. Some examples of floating-point numbers are

$$1\tfrac{1}{2}, -201.75, 2.5 \times 10^{-6}$$

In reference to microprocessors, these types of numbers are referred to as floating-point numbers. The instructions of the on-chip floating-point unit of the 80486DX microprocessor perform computations using these types of numbers. The floating-point unit also has the ability to operate on integer and packed BCD data. However, this section focuses on the processing of data expressed as real numbers.

Organization of Floating-Point Data

Let us first look at how floating-point numbers are expressed in the floating-point registers of the 80486DX. This microprocessor has eight registers that are used to hold floating-point operands for processing, and they are each 80 bits wide. Binary floating-point numbers are expressed in normalized scientific notation form for processing. A normalized binary floating-point number is expressed in general as

$$FP_{Norm} = \pm \text{Significand} \times 2^{\pm \text{Exponent}} = \pm \ 1.\text{bbbb} \times 2^{\pm \text{Exp}}$$

Note that the significand, also called the mantissa, is expressed in normalized form as 1.bbbb.

Data, such as decimal real numbers, must be expressed in this form for processing. The process for converting a decimal number to this form is as follows:

1. Convert the integer part of the decimal number to binary form.
2. Convert the fractional part of the decimal number to binary form.
3. Form the binary equivalent of the number by combining the integer and decimal parts.
4. Express the number as a normalized binary number.

As an example, let us apply this process to convert the number 20.75 to its equivalent normalized binary number. First, the integer and fractional parts are converted to binary form as

Integer	*Fraction*
$20 \div 2 \rightarrow 0$	$2 \times .75$
$10 \div 2 \rightarrow 0$	$2 \times 1.5 \rightarrow 1$
$5 \div 2 \rightarrow 1$	$2 \times 1.0 \rightarrow 1$
$2 \div 2 \rightarrow 0$	
$1 \div 2 \rightarrow 1$	$.75 = .11_2$
$20 = 10100_2$	

Combining the integer and fractional parts gives the binary number

$$+20.75 = +10100.11_2$$

Finally, shifting the point 4 bit positions left and expressing as a power of 2 normalizes the binary number:

$$+20.75 = +10100.11_2 = +1.010011 \times 2^{+4}$$

Data in a floating-point register is coded according to IEEE Standard 754. This standard defines three different size floating-point numbers: single precision, double precision, and extended double precision. Figure 15–57 shows how these different types of numbers are placed in the 80-bit floating-point registers. Note that the single precision number is coded into the 32 least significant bits of the 80-bit register. The most significant bit, bit position 31, represents the sign and is made 0 for positive numbers and 1 for negative numbers. Our earlier example is a positive number; therefore, the sign bit is made 0. The 23 least significant bits, bits 0 through 22, contain the fractional part of the signficand in the normalized number expressed with 23 bits of precision. For our earlier example, these bits contain

$$01001100000000000000000$$

Note that unused less significant bits are simply filled with 0s. The 8 bits, bit 23 through 30, are an 8-bit biased exponent. This is not the exponent in the normalized number. It is biased by adding 127_{10} (1111111_2 = 7FH). Therefore, the exponent for our example is

$$+4_{10} + 127_{10} = 131_{10}$$

or

$$00000100_2 + 01111111_2 = 10000011_2$$

Coding as a 32-bit value and expressing as a hexadecimal number gives

Sign = 0

Bias exponent = 10000011

Fractional significand = 01001100000000000000000

+20.75 = 0 10000011 01001100000000000000000 = 41A60000H

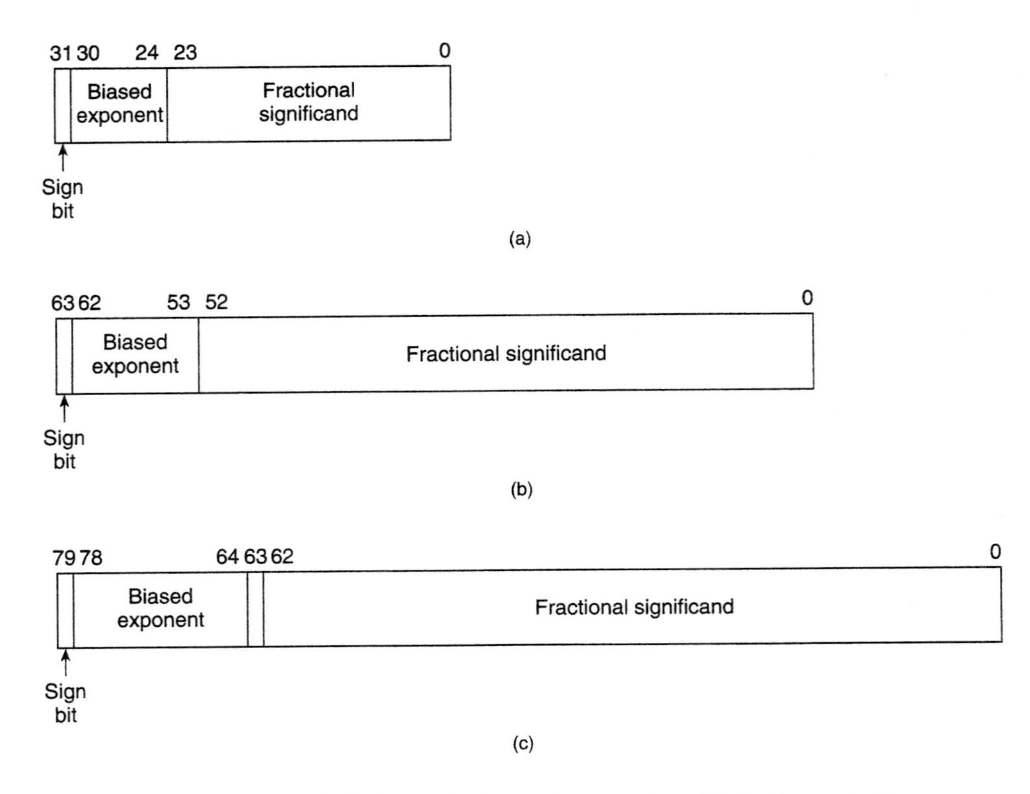

Figure 15–57 (a) Coding a single precision number. (b) Coding a double precision number. (c) Coding an extended double precision number.

The 32 bit number is placed in the 80-bit floating-point register left justified, with unused less significant bits of the register filled with 0s.

Numbers are coded as double precision and extended precision numbers in a similar way. However, when coding a number in double precision, the size of the fractional significand is expanded to 52 bits, the exponent to 11 bits, and bias added to the exponent increases to 3FFH. The earlier example is expressed as a double precision number as

Sign = 0

Biased exponent = 00000000100 + 01111111111 = 10000000011

Fractional significand = 0100110000000000000000000000000000000000000000000000

+20.75 = 0 10000000011 0100110000000000000000000000000000000000000000000000
= 4034C00000000000

The larger exponent of the double precision notation permits it to represent a larger range of numbers and the additional bits of the significand result in higher precision. In this way, we see that double precision notation represents a wider range of numbers and expresses them more accurately than the single precision notation.

Finally, extended precision uses a 63-bit fractional significand, a 15-bit biased exponent, and bias value of 3FFFH. Note that for extended precision numbers the integer bit of the significand is placed in bit position 64. Again the larger exponent and additional significand bits expand the range of numbers and increase their accuracy, respectively.

Floating-Point Register Stack

Earlier we pointed out that the 80486DX has eight floating-point registers. Figure 15–58 shows that these registers are denoted R_0 through R_7. From a software point of view, they are organized as a register stack and accessed by floating-point instructions using the notation ST(n) where n represents the nth register from the current top of stack. Figure 15–59(a) shows the organization of the stack when it is full. Note that register ST(7) represents the bottom of the stack and ST(0), or just ST, corresponds to the top of the stack. As floating-point values are pushed onto the stack, the stack grows from the bottom of the stack R_7 toward register R_0.

The stack in Fig. 15.59(b) illustrates the state of the stack after just two values are loaded onto an empty floating-point stack. The first value is placed in R_7 and is denoted in software as ST(1), and the second is placed into R_6; therefore, ST(0) corresponds to register R_6 and represents the current top of stack (ST). As additional floating-point numbers are loaded onto the stack, the software references are adjusted so that ST(0) references the new register associated with the top of the stack. For example, if two more values are loaded onto the stack, ST(0) now corresponds to registers R_4 and the bottom of the stack (still R_7) is identified from a software point of view as ST(3). When information is removed from the stack, the stack adjusts in the opposite direction. For instance, if one value is now popped from the stack, the new top of stack ST(0) points to the data in register R_5 and the bottom of stack R_7 represents ST(2).

Our description of stack operation raises the question of how the floating-point unit keeps track of the current top of this stack. The answer is that there is a 3-bit code TOP in bits 11 through 13 of a floating-point status register that identifies which register is currently at the top of stack. Figure 15–60 shows the format of the status word in the 80486DX's floating-point status register. This value decrements or increments, respectively, as values are pushed onto or popped off the stack. For our earlier example, TOP equals 100_2 meaning R_4 after the fourth value is placed on the stack and 101_2 after the value is popped from the stack.

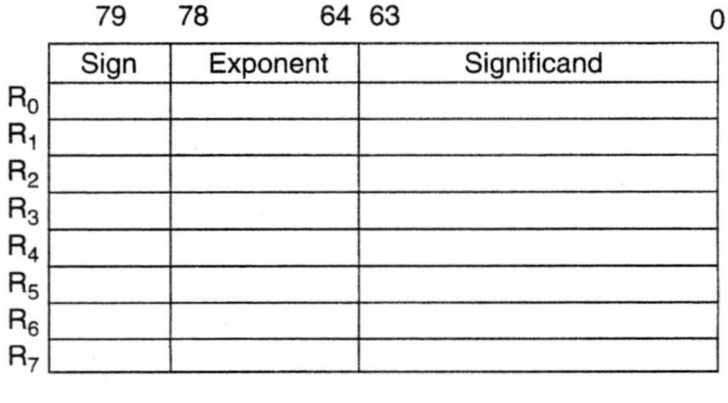

	79	78 — 64	63 — 0
	Sign	Exponent	Significand
R_0			
R_1			
R_2			
R_3			
R_4			
R_5			
R_6			
R_7			

Figure 15–58 Organization of the floating-point registers of the 80486DX.

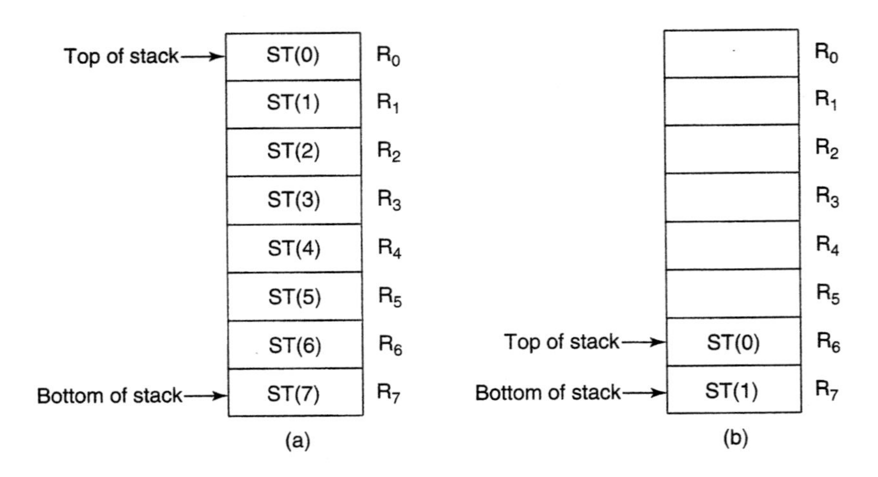

Figure 15–59 (a) Software notation for a full stack. (b) Partially filled stack.

Let us look briefly at the purpose of some of the other bits in the floating-point status register. Control bit B tells whether or not the floating-point unit is busy executing an instruction. Flag bits, such as underflow error (UE), overflow error (OE), and zero divide error (ZE), mark the occurrence of error conditions that result in a floating-point exception. The content of this register can be copied to either memory or the AX register. Then individual bits can be examined and, through software, an appropriate response initiated.

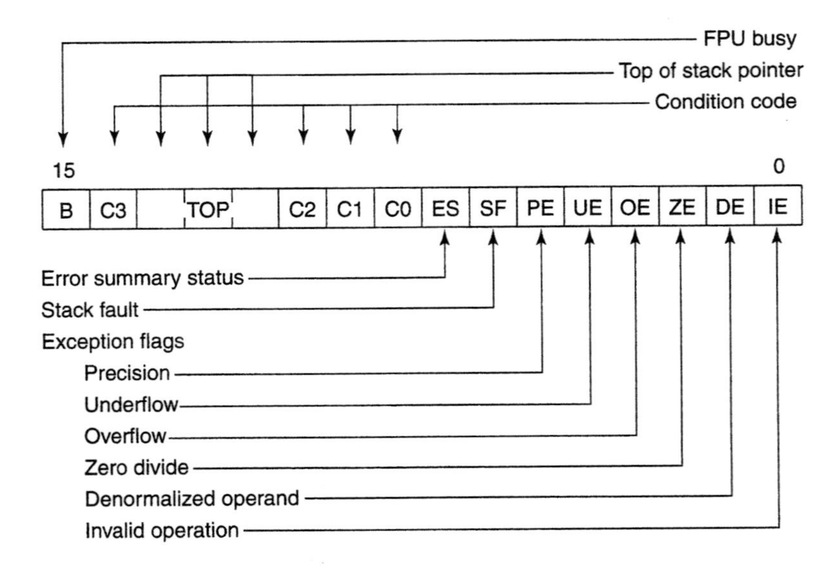

Figure 15–60 Format of the floating-point status register.

Floating-Point Instructions

The floating-point instruction set of the 80486DX microprocessor consists of 68 instructions. Instructions are provided for loading information to and storing information from the stack, performing numeric computations and comparisons of information of the stack, and controlling operations of the floating-point unit. These instructions are categorized into the following groups: data transfer instructions, arithmetic instructions, compare instructions, transcendental instructions, constant instructions, and control instructions

A number of the commonly used instructions of the 80486DX's floating-point instruction set are listed in Fig. 15–61. The table lists their mnemonic, name, and format. Let us briefly look at the operation of a few of these instructions.

The load and store instructions from the data transfer group are used to load values onto the floating-point stack for processing or to save values on the stack that result from computation by storing them in memory. For instance, the source operand can load a value from memory to the current top of the stack or from a register within the stack to the top of the stack. Figure 15–62 presents the notation used to identify memory and reg-

Group	Mnemonic	Name	Format	
Data transfer	FLD	Load real number	FLD	S
	FST	Store real number	FST	D
	FSTP	Store real number and pop	FSTP	D
	FXCH	Exchange registers	FXCH	D
Arithmetic	FADD	Add real numbers	FADD	D, S
	FADDP	Add real numbers and pop	FADDP	D, S
	FSUB	Subtract real numbers	FSUB	D, S
	FSUBP	Subtract real numbers and pop	FSUBP	D, S
	FMUL	Multiply real numbers	FMUL	D, S
	FMULP	Multiply real numbers and pop	FMULP	D, S
	FDIV	Divide real numbers	FDIV	D, S
	FDIVP	Divide real numbers and pop	FDIVP	D, S
	FQRT	Compute square root	FSQRT	D, S
	FABS	Compute absolute value	FABS	D, S
	FCHS	Change sign	FCHS	D, S
Comparison	FCOM	Compare real numbers	FCOM	S
	FCOMP	Compare real numbers and pop	FCOMP	S
	FCOMPP	Compare real numbers and pop twice	FCOMPP	S
Transcendental	FSIN	Compute sine function	FSIN	
	FCOS	Compute cosine function	FCOS	
	FTAN	Compute tangent function	FTAN	
	FATAN	Compute arctangent function	FATAN	
Constant	FLDZ	Load 0.0	FLDZ	
	FLD1	Load 1.0	FLD1	
	FLDPI	Load PI=3.14.159...	FLDPI	
Control	FINIT	Reset the floating point unit	FINIT	
	FINCSTP	Increment the stack pointer	FINCSTP	
	FDECSTP	Decrement the stack pointer	FDECSTP	
	FSTSW	Store status word	FSTSW	D
	FSTSW	Store status word to AX	FSTSW	AX

Figure 15–61 Commonly used floating-point instructions.

Operand	Location of operand
None (ST or ST (0))	Current top of stack
ST(n)	Stack register offset by n from the current top of stack
Mem32/64/80real	Single precision, double precision, or extended precision memory location

Figure 15–62 Floating-point instruction operand notations.

ister operands. For instance, to load a single precision real number from memory, the instruction is written as

```
FLD Mem32real
```

Here Mem32real stands in general for an address pointer to a single precision floating point number stored in memory. When the instruction executes, the top of stack pointer (TOP) is decremented by 1 and then the value at this memory location is read and loaded onto the new top of the stack. An example of this instruction is

```
FLD DATA1_32B
```

To copy a value from a register within the stack to the top of the stack, the operand is expressed in general as

```
FLD ST(n)
```

For example, to copy the value in the third register below the current top of the stack to the new top of stack, the instruction is written as

```
FLD ST(3)
```

Note that the 3 in ST(3) does not stand for R_3. It is the offset of the source register relative to the current top of the stack. For instance, if the top of the stack before executing the instruction is R_2, ST(3) stands for R_5. The value in this register is copied to register R_2, which is the new top of stack.

The FST instruction is used to copy the current top of stack to a destination that is either another register or a storage location in memory. For example, execution of the instruction

```
FST DATA1_32B
```

stores the value at the current top of stack in memory at the address pointed to by DATA1_32B. Likewise, the instruction

```
FST ST(3)
```

copies the value in the register at the top of the stack into the third register below the TOS. Neither of these stored instructions affects the value in TOP. To pop the value off the stack after it is saved, use the FSTP instruction. Execution of the instruction

```
FSTP DATA1_32B
```

saves the current value at the top of stack and then pops this value from the stack by incrementing TOP by 1.

Now that we know how to place data on the floating-point stack and copy it from register to register, let us use the FADD instruction to perform a computation. This instruction has the ability to add a value in memory to the value at the current top of the stack, add the value in another stack register to the top of the stack, or add the value currently at the top of the stack to another that is in another stack register. For example, executing the instruction

```
FADD DATA2_32B
```

adds the single precision value held at the memory location pointed to by address DATA2_32B to the value at the current top of the stack and places the result in the register corresponding to TOP. Execution of this instruction does not affect the new value in TOP.

EXAMPLE 15.22

The instruction that follows adds the value in the source register to the value of the destination register and places their sum in the destination registers:

```
FADD ST(2), ST(0)
```

If the values in ST(0) and ST(2) are −2.5 and +10.75, respectively, find

a. The destination registers.

b. The decimal value of SUM produced in the destination register.

c. Express SUM as a single precision binary number.

d. Express SUM in hexadecimal notation.

Solution

a. This instruction performs this addition:

$$(ST(2)) + (ST(0)) \rightarrow (ST(2))$$

Therefore, the SUM is produced in destination floating-point register ST(2).

b. The decimal computation is as follows:

$$SUM = +10.75 + (-2.5) = +7.25$$

c. Converting the integer and fractional parts to binary form gives

Integer	*Fraction*
$7 \div 2 \rightarrow 1$	$2 \times .25$
$3 \div 2 \rightarrow 1$	$2 \times .5 \rightarrow 0$
$1 \div 2 \rightarrow 1$	$2 \times 1.0 \rightarrow 1$
$7 = 111_2$	$.25 = .01_2$

Combining the integer and fractional parts gives the binary number

$$+7.25 = +111.01_2$$

Now, normalizing the binary number gives

$$+111.01_2 = +1.1101 \times 2^{+2}$$

and prepares it to be expressed in 32-bit single precision form

Sign = 0

Bias exponent = $+2_{10} + 127_{10} = 00000010_2 + 01111111_2 = 10000001_2$

Fractional significand = 11010000000000000000000

+7.25 = 0 10000001 11010000000000000000000

$= 01000000111010000000000000000000_2$

d. Finally, bits are grouped to form hexadecimal digits to give

$$+7.25 = 40E80000H$$

The FSUB instruction supports the same operand combinations as the FADD instruction. An example of a floating-point subtraction of real numbers in two registers is

```
FSUB ST(0), ST(3)
```

This instruction performs the subtraction

$$(ST(3)) - (ST(0)) \rightarrow (ST(0))$$

Note that the result is placed at the current top of the stack. Again, the value in TOP is unaffected. The FADDP and FSUBP instructions perform add and subtract operations, but also pop the result at the current top of the stack by incrementing TOP.

In many floating-point instructions, no operands are directly specified. In these cases, the current top of stack, ST(0), is normally assumed as the source and destination if there is a single operand. If the instruction requires two operands, ST(1) is assumed as the destination and ST(0) the source. An example is the instruction

```
FADD
```

When executed, it performs the computation

$$(ST(0)) + (ST(1)) \rightarrow (ST(1))$$

$$TOP = TOP + 1$$

Another example is the instruction

```
FABS
```

Execution of this instruction computes the absolute value of the value at the current top of the stack and places this result in the same register. It does this by clearing the sign bit of the value in this register equal to 0.

EXAMPLE 15.23

Describe what the following sequence of instructions does?

```
FLD DATA3_64B
FLD DATA4_64B
FSUBP
FABS
FSTP DATA5_64B
```

Solution

The two instruction decrement TOP and then load the double precision real number from memory location DATA3_64B to the new top of the stack. The next instruction performs the same operation for the real number held at address DATA4_64B. This makes

$$(ST(0)) = (DATA4\_64B)$$

$$(ST(1)) = (DATA3\_64B)$$

Once the data are loaded, the FADD instruction performs the addition

$$(ST(0)) - (ST(1)) \rightarrow (ST(1))$$

and then increments the stack pointer to pop the current top of stack. Therefore, the new top of stack contents is

$$(ST(0)) = (DATA4\_64B) - (DATA3\_64B)$$

The next instruction computes the absolute value of the double precision real number at the current top of stack. This gives

$$(ST(0)) = |(DATA4\_64B) - (DATA3\_64B)|$$

Finally, the absolute value from the current top of stack is saved in the memory location pointed to by the address DATA5_64B:

$$|(DATA4\_64B) - (DATA3\_64B)| \rightarrow (DATA5\_64B)$$

and the result is popped from the current top of stack by incrementing TOP.

▲ 15.12 PENTIUM PROCESSOR FAMILY

The Pentium processor family, introduced in 1993, represents the high-performance end of Intel's 8086 architecture. Like the 80486 MPUs, Pentium processors are 32-bit MPUs—that is, they have a 32-bit register set and the instructions can process words of data as large as 32 bits in length. However, the Pentium processor's 32-bit architecture is enhanced with a 64-bit external data bus and a variety of internal data paths that are 64 bits, 128 bits, or 256 bits wide. These large internal and external data paths result in an increased level of performance.

The Pentium processors' internal and software architectures have been enhanced in many other ways. For instance, they employ an advanced *superscaler* pipelined internal architecture. The parallel processing provided by this superscaler pipelining gives the Pentium processors the ability to execute more than one instruction per clock cycle. From a software point of view, the instruction set has been enhanced with new instructions and, more important, the performance of the floating-point unit has been greatly increased.

Figure 15–1 shows the performance of the Pentium processor family members relative to those of the 80386 and 80486 families. Note that the iCOMP index of the original Pentium processor (60-MHz) MPU, 510, is about 20 percent higher than the fastest 80486 MPU, the 80486DX4-100. With the introduction of faster family members, the Pentium processor's performance edge has increased to more than 2. For instance, the 133-MHz device has an iCOMP rating of 1110.

EXAMPLE 15.24

Which speed Pentium processor is the first device to offer more than a 2× performance increase over the 80486DX4-100?

Solution

Figure 15–1 shows that the iCOMP rating of the 80486DX4-100 is 435. Therefore, the Pentium 120 MHz, which has an iCOMP rating of 1000, is the first member of the Pentium processor family that offers more than a 2× performance increase over the 80486DX4-100.

Internal Architecture

Figure 15–63 presents a block diagram of the Pentium processor's internal architecture. Earlier we pointed out a number of important architectural advances introduced with the Pentium processor. Three of these are its *superscaler* pipelined architecture, independent code and data caches, and high-performance floating-point unit. Let us next look briefly at each of these architectural features.

Intel uses the term *superscaler* to describe an architecture that has more than one execution unit. In the case of the Pentium processor, there are two execution units. These execution units, or *pipelines* as they are also known, process the instructions of the microcomputer program. Each of these execution units has its own ALU, address generation

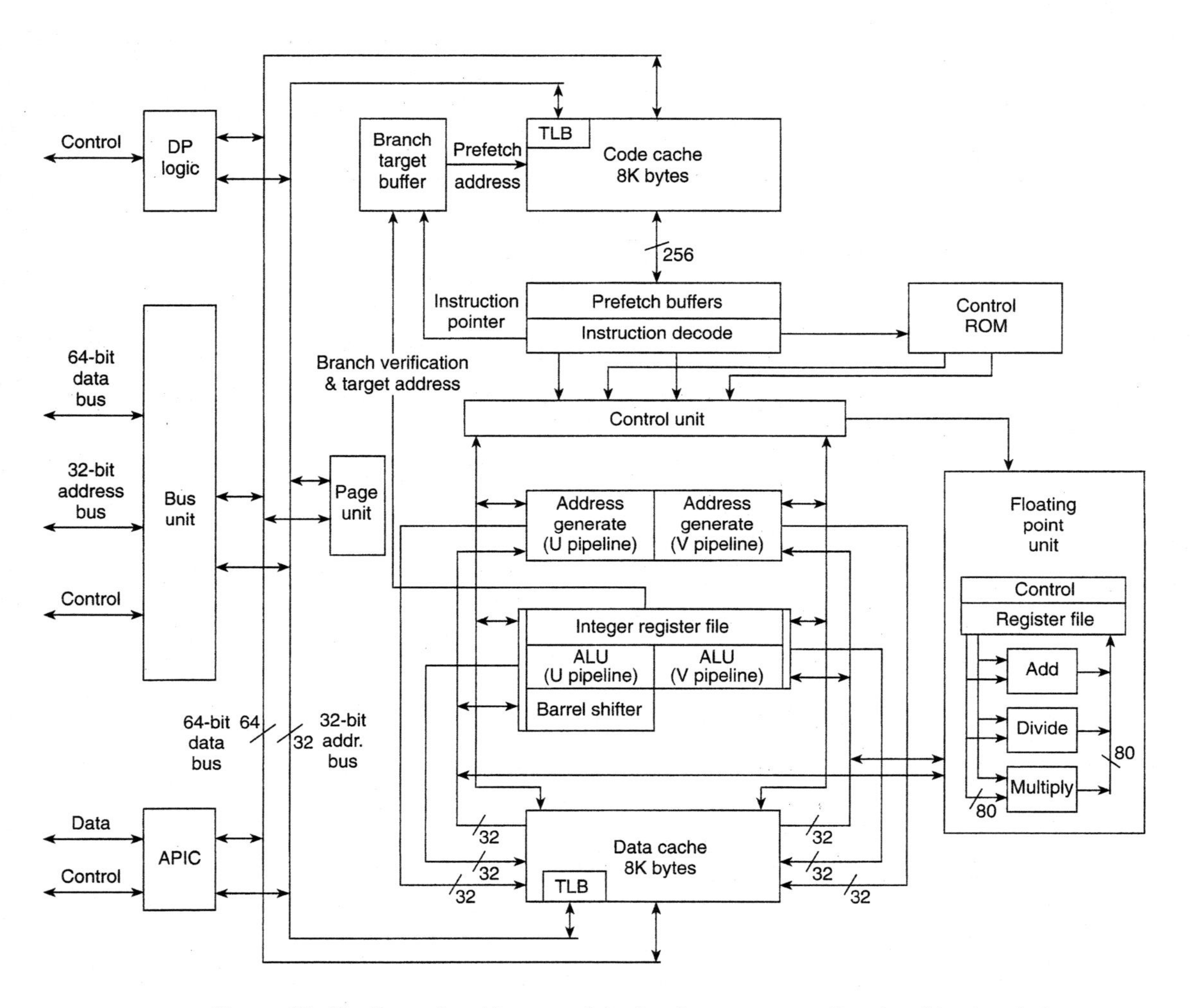

Figure 15–63 Internal architecture of the Pentium processor. (Reprinted by permission of Intel Corporation. Copyright/Intel Corp. 1994)

circuitry, and data cache interface. They are identified in Fig. 15–63 as the *U pipeline* and the *V pipeline.*

In Section 15.10 we found that the execution time of a core group of instructions were reduced to one clock cycle in the 80486 MPU family. With the Pentium processor's dual-pipeline architecture, two instructions can be processed at the same time. Therefore, they have the capability of executing as many as two instructions per clock cycle. For this reason, pipelining makes a significant contribution to the higher level of performance achieved with the Pentium processor family.

The 80486 family of MPUs has an on-chip cache memory that is used to cache both code and data. With the Pentium processor, the on-chip cache memory subsystem has been further enhanced. Its cache memory, like that of the 80486DX4 MPU, has been expanded to 16Kbytes, but it is also partitioned into separate 8Kbyte code and 8Kbyte data caches. Also, like the 80486DX4, the write update method can be configured for either the write-through or the write-back mode of operation. Note in Fig. 15–63 that the U pipe and V pipe of the ALU accesses the data cache independently. This is known as a *dual-port interface* and permits both ALUs to access data in the cache at the same time. These independent caches result in more frequent use of the cache memory and lead to a higher level of performance for the Pentium processor-based microcomputer system.

All members of the Pentium processor family have a built-in floating-point unit. This floating-point math unit has been further enhanced from that used in the 80486 family. For instance, it employs faster hardwired, instead of microcoded, implementations of the floating-point add, multiply, and divide operations. The result is higher-performance floating-point operation for the Pentium processor. In fact, it can execute floating-point instructions 5 to 10 times as fast as the 80486DX-33 MPU.

Software Architecture of the Pentium Processor

The MPUs of the Pentium processor family remain fully software compatible with the 80486 architecture. Just like the 80386 and 80486 MPUs, the Pentium processor comes up in real mode after reset and can be switched to the protected mode by executing a single instruction. Its real- and protected-mode software models and instruction sets are both supersets of those of the 80486 family MPU. They have all of the same instructions, functional registers, and register bit definitions. However, just like for the 80486 MPU and 80386 MPU before that, a number of new instructions, flags, and control bits have been defined.

Real-Mode and Protected-Mode Register Sets. The real-mode and protected-mode application register sets of the Pentium processor are essentially the same as those of the 80486SX microprocessor. However, some changes are found in the functionality of both the flags register and the control registers. Figure 15–64 shows the EFLAGS register of the Pentium processor. Three additional flag bits, *ID flag* (ID), *virtual interrupt pending* (VIP), and *virtual interrupt flag* (VIF), have been activated. The ID flag can be used to determine if the MPU supports a new instruction called CPUID. If this bit can be set or reset under software control, CPUID is included in the instruction set. The VIP and VIF flag bits are used to implement a virtualized system-interrupt flag for protected-mode multitasking software environments.

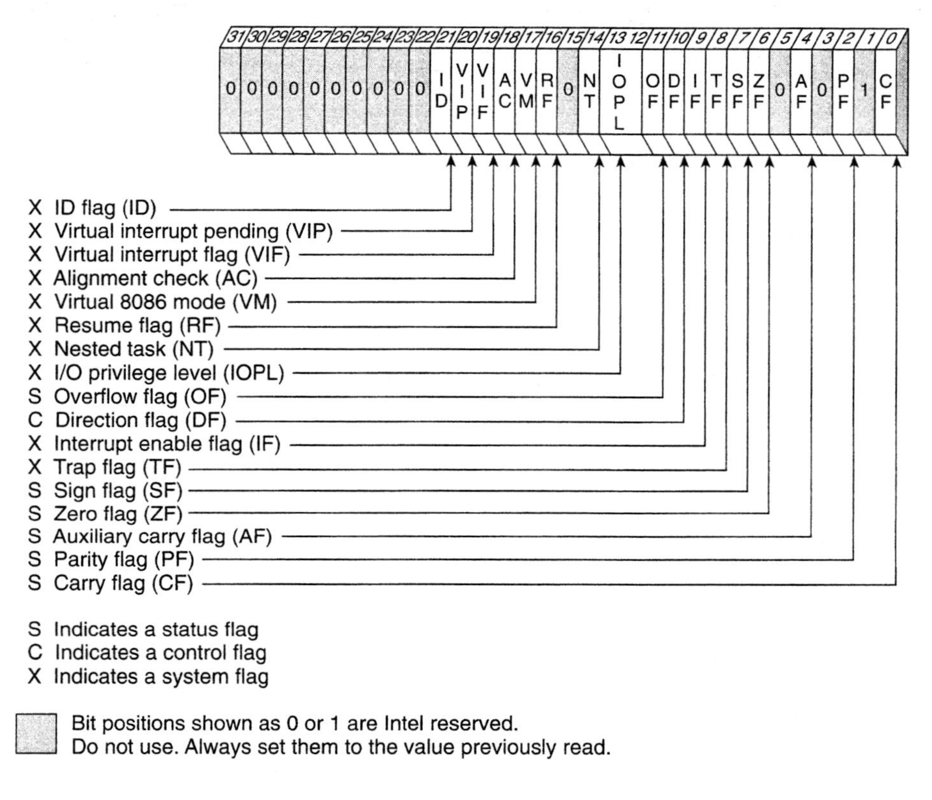

Figure 15–64 EFLAGS register of the Pentium processor. (Reprinted by permission of Intel Corporation. Copyright/Intel Corp. 1995)

The function of the control registers has been expanded in the Pentium processor. Figure 15–65(a) shows that a fifth control register, CR_4, has been added, and the six new control bits are all in this register. The meaning of each bit and a brief description of its function is given in Fig. 15–65(b). Looking at the functional descriptions for these control bits, we see that they are all used to enable or disable new capabilities of the Pentium processor. For instance, the *virtual-8086 mode extensions* (VME) and *protected-mode virtual interrupts* (PVI) bits are used to enable the virtualized system interrupt capability implemented with the VIP and VIF flag bits in the virtual-8086 mode and protected-mode, respectively.

EXAMPLE 15.25

What is the protected-mode page size when the PSE bit in CR_4 is logic 0? Logic 1?

Solution

When PSE is logic 0, the page size is 4Kbytes, and when it is switched to logic 1, the size is increased to 4Mbytes.

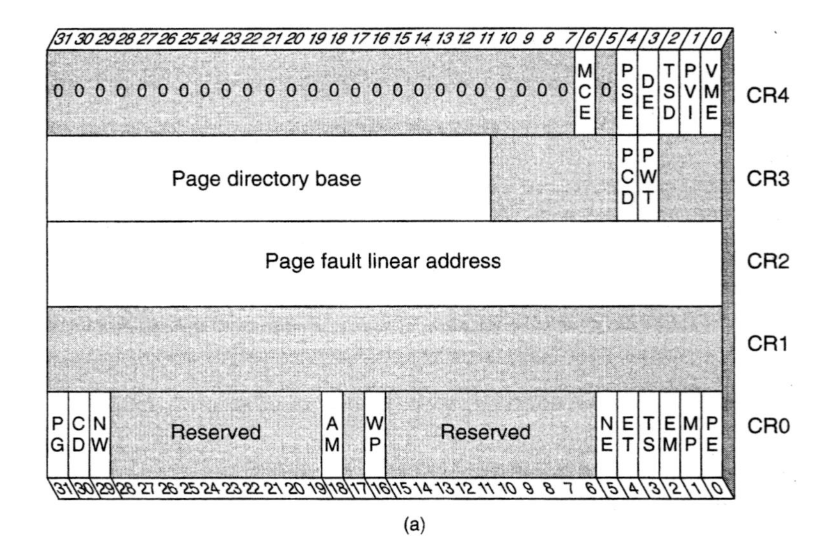

(a)

Bit	Name	Function
0	Virtual-8086 mode Extensions (VME)	Logic 1 enables support for a virtual interrupt flag in virtual-8086 mode.
1	Protected mode Virtual interrupts (PVI)	Logic 1 enables support for a virtual interrupt flag in protected mode.
2	Time-date stamp Disable (TSD)	Logic 1 makes the read from time stamp counter (RDTSC) instruction a privileged instruction.
3	Debugging Extensions (DE)	Logic 1 enables I/O breakpoints.
4	Page size Extensions (PSE)	Logic 1 enables 4M-byte page size.
6	Machine check enable (MCE)	Logic 1 enables the machine-check exceptions.

(b)

Figure 15–65 (a) Control registers of the Pentium processor. (Reprinted by permission of Intel Corporation. Copyright/Intel Corp. 1995) (b) Functions of the CR_4 control bits.

Both 80386 and 80486 MPUs also contained debug and test registers in their real- and protected-mode software models. However, these registers are not used in application programming. Another difference in the register models of the Pentium processor is that debug registers DR_4 and DR_5 are now implemented. Also, the test registers no longer exist; instead, their functionality is implemented in a new group of registers called the *model-specific registers.*

For simplicity, we have ignored the registers that are provided for the floating-point math unit in the application software models.

Enhancements to the Instruction Set. Upwards software compatibility is maintained in the instruction set of the Pentium processor family. That is, it retains all of the instructions of the 8086, 8088, 80286, 80386, and 80486 microprocessor families. This means that all software written for a microcomputer constructed with one of these earlier MPUs can be directly run on a Pentium processor-based microcomputer.

The instruction set of the Pentium processor is enhanced with three new Pentium processor specific-instructions and four additional system-control instructions. The chart in Fig. 15–66 summarizes the function of each of the new instructions. For instance, the instruction *compare and exchange 8 bytes* (CMPXCHG8B) is an enhancement in the Pentium processor's specific instruction set. CMPXCHG8B is a variation of the CMPXCHG instruction, which was first introduced with the 80486 instruction set. It enables a 64-bit compare and exchange operation to be performed. Note in Fig. 15–66 that when the instruction is executed, the 64-bit value formed from EDX and EAX, denoted [EDX:EAX] (where EDX is the more significant double word), is compared to the value of a 64-bit data word (quad-word) in memory. If the two values are equal, ZF is set to 1 and the quad-word value formed from [ECX:EBX] (where ECX is the more significant double

Mnemonic	Meaning	Format	Operation
CMPXCH8B	Compare and exchange 8 bytes	CMPXCH8B D	if [EDX:EAX] = D (ZF)←1, (D)←[ECX:EBX] else (ZF)←0, [EDX:EAX]←(D)

CPUID	CPU identification	CPUID	if (EAX) = 0H
			[EAX,EBX,ECX,EDX]←
			Vendor information
			if (EAX) = 1H
			[EAX,EBX,ECX,EDX]←
			MPU information

RDTSC	Read from time stamp counter	RDTSC	[EDX:EAX]← Time stamp counter
RDMSR	Read from model specific register	RDMSR	[EDX:EAX]←MSR(ECX) (ECX) = 0H selects MCA (ECX) = 1H selects MCT
WRMSR	Write to model specific registers	WRMSR	MSR(ECX)←[EDX:EAX] (ECX) = 0H selects MCA (ECX) = 1H selects MCT
RSM	Resume from system management mode	RSM	Resume operation from SMM
MOV CR4	Move to/from CR4	MOV CR4,r32 MOV r32, CR4	(CR4)←(r32) (r32)←(CR4)

Figure 15–66 Additions to the instruction set in the Pentium processor.

word) is written into the quad-word storage location in memory. On the other hand, if the value [EDX:EAX] does not match that in memory, ZF is cleared to 0 and the quad-word held in memory is read into [EDX:EAX].

EXAMPLE 15.26

If the contents of EAX, EBX, ECX, and EDX are 11111111_{16}, 22222222_{16}, 33333333_{16}, and $FFFFFFFF_{16}$, respectively, and the content of the quad-word memory storage location pointed to by the address TABLE is $FFFFFFFF11111111_{16}$, what is the result produced by executing the instruction

```
CMPXCHG8B [TABLE]
```

Solution

When the instruction is executed, the value of

$$(EDX:EAX) = FFFFFFFF11111111_{16}$$

is compared to the quad-word pointed to by address TABLE. The contents of this memory location are given as

$$(DS:TABLE) = FFFFFFFF11111111_{16}$$

Since these two values are equal, the results produced are

$$(ZF) \leftarrow 1$$

$$(DS:TABLE) \leftarrow (ECX:EBX) = 3333333322222222_{16}$$

Another one of the new Pentium processor's specific instructions, *CPU identification* (CPUID), permits software to identify the type and feature set of the Pentium processor that is in use in the microcomputer system. The EAX register must be set to either 0 or 1 before executing CPUID. Depending on which value is selected for EAX, different identification information is made available to software. By executing this instruction with EAX equal to 0, the processor identification information provided is as follows:

$$EAX = 1 \leftarrow \text{Pentium processor}$$

$$EBX = \text{vendor identification string} \leftarrow \text{Genu}$$

$$ECX = \text{vendor identification string} \leftarrow \text{inel}$$

$$EDX = \text{vendor identification string} \leftarrow \text{ntel}$$

Note that the MPU is identified as a Pentium processor and that it is a Genuine Intel device. Now, executing the instruction again with EAX equal to 1 provides more information about the MPU. The results produced in EAX, EBX, ECX, and EDX are as follows:

$$
\begin{aligned}
EAX(3{:}0) &\leftarrow \text{stepping ID}\\
EAX(7{:}4) &\leftarrow \text{model}\\
EAX(11{:}8) &\leftarrow \text{family}\\
EAX(31{:}12) &\leftarrow \text{reserved bits}\\
EBX &\leftarrow \text{reserved bits}\\
ECX &\leftarrow \text{reserved bits}\\
EDX(0{:}0) &\leftarrow \text{FPU on-chip}\\
EDX(2{:}2) &\leftarrow \text{I/O breakpoints}\\
EDX(4{:}4) &\leftarrow \text{time-stamp counter}\\
EDX(5{:}5) &\leftarrow \text{Pentium CPU style model-specific registers}\\
EDX(7{:}7) &\leftarrow \text{machine-check exception}\\
EDX(8{:}8) &\leftarrow \text{CMPXCHG8B instruction}\\
EDX(31{:}9) &\leftarrow \text{reserved}
\end{aligned}
$$

EXAMPLE 15.27

Figure 15–67 shows the contents of the EAX register after executing the CPUID instruction. What are the family, model, and stepping?

Solution

In Fig. 15–67 we find

$$
\begin{aligned}
&\text{family} = 0101 = 5 = \text{Pentium processor}\\
&\text{model} = 0000 = 0\\
&\text{stepping} = 0000 = 0
\end{aligned}
$$

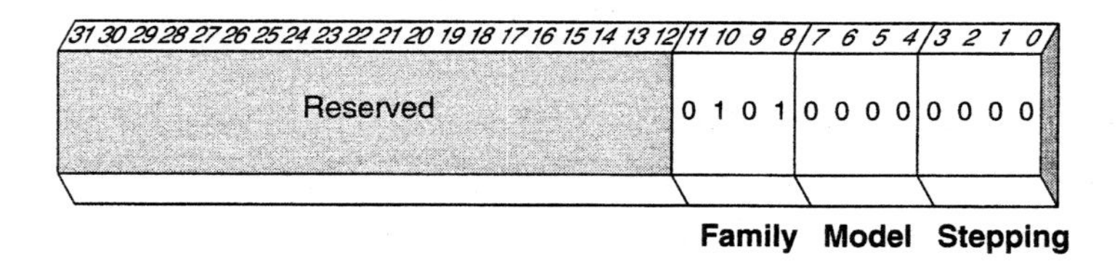

Figure 15–67 EAX after executing the CPUID instruction. (Reprinted by permission of Intel Corporation. Copyright/Intel Corp. 1995)

The last of the new Pentium processor-specific instructions is *read from time-stamp counter* (RDTSC). The Pentium processor has an on-chip 64-bit counter called the *time-stamp counter.* The value in this counter is incremented during every clock cycle. Executing the RDTSC instruction reads the value in this counter into the register set [EDX:EAX]. EDX holds the upper 32 bits of the 64-bit count, and EAX holds the lower 32 bits. In some software applications, it may be necessary to determine how many clock cycles have elapsed during an event. Reading the value of the time-stamp counter before and after the event being measured and then determining the number of elapsed clock cycles by forming the difference of the two count readings can do this.

Looking at Fig. 15–66, we find that the four new system-control instructions are *read from model-specific register* (RDMSR), *write to model-specific register* (WRMSR), *resume from system management* (RSM), and a new form of the move instruction (MOV CR4, r32 and MOV r32, CR4) that permits data to be moved directly between control register 4 (CR_4) and an internal register. The instructions RDMSR and WRMSR are used to permit software access to the contents of the model-specific registers of the Pentium processor. The two model-specific registers that can be accessed with these instructions are the *machine check address register* (MCA) and the *machine check type register* (MCT). To access MCA, the value 0H must be loaded into ECX prior to executing the instruction, or to select MCT, ECX must be loaded with the value 1H. Execution of the instructions cause either a 64-bit read- or 64-bit write-data transfer to take place between the selected model-specific register and the register set [EDX:EAX]. The RSM instruction is used by the Pentium processor's system management mode.

System-Management Mode. The Pentium processor also has a mode of operation known as *system-management mode* (SMM). This mode is used primarily to manage the system's power consumption. SMM is entered by an interrupt request from external hardware, and the return to real or protected mode is initiated by executing the RSM instruction.

▲ 15.13 MULTIMEDIA ARCHITECTURE AND INSTRUCTIONS

Multimedia applications, such as audio, graphics, and speech recognition, require the MPU to perform the same operation on multiple elements of data. Digital signal processors (DSP) have traditionally performed these types of applications. The architecture of the *Pentium processor with MMX technology* has been enhanced to produce high performance for these types of applications in a PC, without the need for a DSP. This is the first Pentium family processor to implement the *MMX technology*—new multimedia data types and special instructions for processing of these types of data. In this section, we examine the new multimedia data types, the registers in which multimedia information is processed, and instructions that process multimedia data.

Organization of SIMD Data

Many multimedia applications process large arrays of 8-bit or 16-bit data elements and frequently require that the same operation be performed repeatedly on groups of this data. For this reason, the MMX technology packs multiple elements of data into a single operand and performs an operation, such as arithmetic computations, repositioning of data, and alignment of data, with them in parallel. That is, the operation is independently performed on each element of data at the same time. Data types that represent more than one piece of data that are to be processed at the same time are known as *single instruction multiple data* (SIMD). Four new 64-bit SIMD data types are implemented in the Pentium processor with MMX technology.

Figure 15–68 shows the new packed SIMD data types—the *packed byte, packed word, packed double word,* and *packed quad-word.* The packed byte format represents eight independent elements of 8-bit integer data. This format is also referred to as an 8 × 8-bit word of data. Note that the packed word and packed double word forms correspond to four 16-bit integers (4 × 16-bit) and two 32-bit double words (2 × 32-bit) of integer data, respectively. Finally, the packed quad-word (1 × 64-bit) form is simply a single 64-bit element of integer data. In this way, we see that the source and destination operands of multimedia instructions are always 64 bits in width.

Multimedia Register File

Eight 64-bit registers are provided for storage of operands processed by MMX instructions. These are not new registers; they are aliases with the existing floating-point register stack. That is, as shown in Fig. 15–69 MMX registers MM_0 through MM_7 are mapped on top of the subtrahend part (lower 64 bits) of the registers in the floating-point stack. Dual use of these registers means that applications must not intermix instructions that process floating-point data and MMX data. Therefore, it is important to clear these registers after using them for either floating-point or MMX computations. The MMX instruction set has an instruction, the EMMS (empty MMX technology state instruction), specifically for this purpose. Execution of this instruction invalidates the data in all MMX registers.

The MMX registers are only used to hold source and destination operands that represent MMX data. Data are placed in the register using one of the data formats shown in Fig. 15–68. MMX instructions that access memory use an integer register, such as SI, for the address pointer. When a value of MMX is loaded into a register, the unused, more significant bits of the floating-point register are all set to 1. Unlike the floating-point registers, the MMX registers can be randomly accessed.

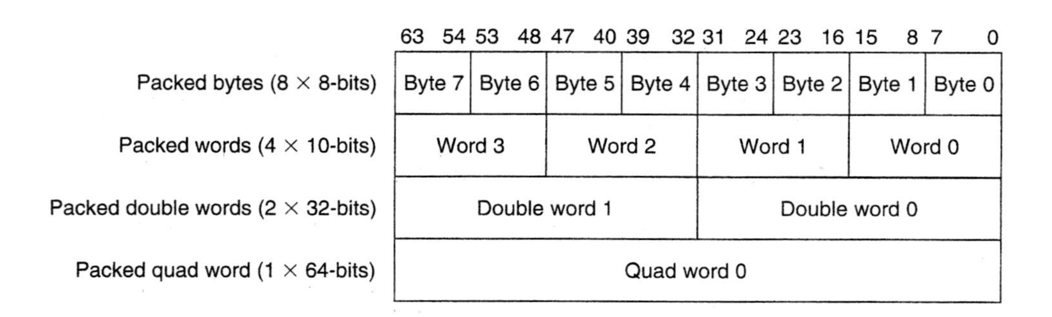

Figure 15–68 MMX technology packed data types.

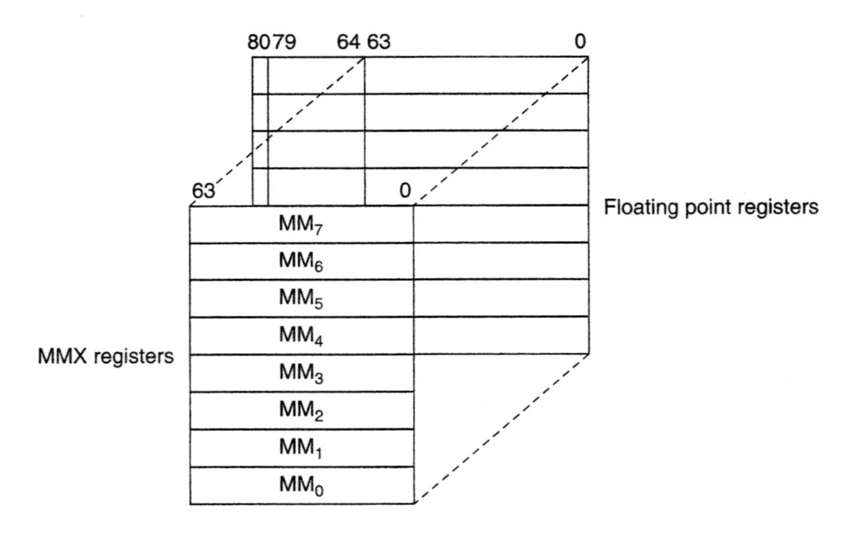

Figure 15–69 MMX technology register set.

Multimedia Instruction Set

By adding the U and V pipes, performance of the Pentium processor increased by enabling it to execute more than one instruction at a time. Performance was further improved in the Pentium processor with MMX technology by extending the instruction set with 57 new instructions to perform operations on SIMD data. Introduction of these new MMX instructions expands parallelism by allowing a single instruction to perform an operation on multiple elements of data at the same time. That is, MMX technology exploits parallelism in data to speed up applications.

Here we introduce the MMX instructions groups and examine the operation of a few of them. Categorizing the MMX instructions based on their function gives the following groups: data transfer instructions, arithmetic instructions, comparison instructions, logical instructions, and conversion instructions. The table in Fig. 15–70 lists the mnemonic, name, and format for the instructions in each of these groups.

The move quad word (MOVQ) instruction allows an MMX register to be initialized with data from memory or to copy data from one MMX register to another. Figure 15–71 shows the allowed operands from the MMX instructions. Note that all instructions except MOVD require the destination operand to be in an MMX register (MM). Moreover, all instruction, with the exception of MOVD and the shift instructions, allow the source operand to be either a quad word storage location in memory (M64) or another MMX register (MM). The shift instructions also permit use of an 8-bit (IMM8) source operand. In the case of a source operand in memory, the address that points to this operand is held in one of the MPUs integer registers, for instance, SI or BX. An example of an instruction that performs a register-to-register quad word move is

```
MOV MM7, MM0
```

Group	Mnemonic	Name	Format	
Data transfer	MOVQ	Move quad word	MOVQ	D, S
	MOVD	Move double word	MOVQ	D, S
Arithmetic	PADDB	Packed add bytes	PADDB	D, S
	PADDW	Packed add words	PADDW	D, S
	PADDD	Packed add double words	PADDD	D, S
	PADDSB	Packed add signed bytes	PADDSB	D, S
	PADDSW	Packed add signed words	PADDSW	D, S
	PADDUSB	Packed add unsigned bytes	PADDSB	D, S
	PADDUSW	Packed add unsigned words	PADDSW	D, S
	PSUBB	Packed subtract bytes	PSUBB	D, S
	PSUBW	Packed subtract words	PSUBW	D, S
	PSUBD	Packed subtract double words	PSUBD	D, S
	PSUBSB	Packed subtract signed bytes	PSUBSB	D, S
	PSUBSW	Packed subtract signed words	PSUBSW	D, S
	PSUBSD	Packed subtract signed double words	PSUBSD	D, S
	PSUBUSB	Packed subtract unsigned bytes	PSUBUSB	D, S
	PSUBUSW	Packed subtract unsigned words	PSUBSW	D, S
	PMULLW	Packed multiply low words	PMULLW	D, S
	PMULHW	Packed multiply high words	PMULHW	D, S
	PMADDWD	Packed multiply and add words	PMULHW	D, S
	PSRAW	Packed shift right arithmetic words	PSRAW	D, S
	PSRAD	Packed shift right arithmetic double words	PSRAD	D, S
Comparison	PCMPEQB	Packed compare for equal bytes	PCMPEQB	D, S
	PCMPEQW	Packed compare for equal words	PCMPEQB	D, S
	PCMPEQD	Packed compare for equal double words	PCMPEQB	D, S
	PCMPGTB	Packed compare for greater than bytes	PCMPEQB	D, S
	PCMPGTW	Packed compare for greater than words	PCMPEQB	D, S
	PCMPGTD	Packed compare for greater than double words	PCMPEQB	D, S
Logical	PAND	Bitwise logical AND	PAND	D, S
	PANDN	Bitwise logical AND NOT	PANDN	D, S
	POR	Bitwise logical OR	POR	D, S
	PXOR	Bitwise logical exclusive-OR	PXOR	D, S
	PSLLW	Packed shift left logical word	PSLLW	D, S
	PSLLD	Packed shift left logical double word	PSLLD	D, S
	PSLLQ	Packed shift left logical quad word	PSLLQ	D, S
	PSRLW	Packed shift right logical word	PSRLW	D, S
	PSRLD	Packed shift right logical double word	PSRLW	D, S
	PSRLQ	Packed shift right logical quad word	PSRLQ	D, S
Conversion	PACKUSWB	Pack words to bytes with unsigned saturation	PACKUSWB	D, S
	PACKSSWB	Pack words to bytes with signed saturation	PACKSSWB	D, S
	PACKSSDW	Pack double words to words with signed saturation	PACKSSDW	D, S
	PUNPCKHBW	Unpack high packed data bytes to words	PUNPCKHBW	D, S
	PUNPCKHWD	Unpack high packed data words to double words	PUNPCKHWD	D, S
	PUNPCKHDQ	Unpack high packed data double words to quad word	PUNPCKHDQ	D, S
	PUNPCKLBW	Unpack low packed data bytes to words	PUNPCKLBW	D, S
	PUNPCKLWD	Unpack low packed data words to double words	PUNPCKLWD	D, S
	PUNPCKLDQ	Unpack low packed data double words to quad word	PUNPCKLDQ	D, S

Figure 15–70 MMX instruction set.

Operand	Allowed operands	Location of operand
Destination	MM	MMX register
	R32*	Integer register
	M32*	Memory
Source	MM	MMX register
	M64	Memory
	R32*	Integer register
	M32*	Memory
	IMM8+	Immediate data

* MOVD only
+ PSRAW/D, PSLLW/D/Q, and PSRLW/D/Q only

Figure 15–71 MMX instruction operand notations.

When this instruction is executed, the SIMD data in MM_0 is copied into MM_7. The instruction that follows is used to initialize register MM_3 with data from a storage location pointed to by the address in the source index register:

```
MOV MM3, SI
```

Now that we know how data are loaded from memory into the MMX registers, let us examine the operation of some instructions that process SIMD data. The PADDW instruction independently adds the four word-wide data elements of the source to their corresponding elements in the destination and places the four 16-bit sums that result in the destination register. An example is the instruction

```
PADDW MM0, MM1
```

When executed, this instruction performs the addition

MM_1	Word 3	Word 2	Word 1	Word 0
	+	+	+	+
MM_0	Word 3	Word 2	Word 1	Word 0

That is, it processes each element of the MMX operands in parallel. Word 0 of register MM_1 is added to Word 0 of MM_0 and the sum of these two values is placed in the least significant word location of MM_0, Word 1 of register MM_1 is added to Word 1 of MM_0, and the sum of these two values is placed in the next least significant word position in MM_0, and so on. If the sum is larger than 16 bits it cannot fit in the register. The PADD instruction resolves this overflow with a method known as wraparound. It simple truncates any bits above the 16th bit.

EXAMPLE 15.28

If registers MM_3 and MM_4 contain the values FFFFFFFF12345678H and 000100028765-4321H respectively, what result does the instruction execution produce?

```
PADDUSW MM4, MM3
```

Solution

The PADDUSW instruction treats the word-wide operands as unsigned integers when it performs the add operation. If the result of the addition is greater that 16 bits, an overflow condition has occurred. In this case, the PADDUSB/W instructions resolve the overflow by applying saturation. Using saturation, the result is simply replaced with the largest value that can be represented with 16 bits, which is FFFFH for an unsigned word. Therefore, the instruction performs the four additions that follow:

Word 3	FFFFH + 0001H = 10000H → FFFFH due to saturation
Word 2	FFFFH + 0002H = 10001H → FFFFH due to saturation
Word 1	1234H + 8765H = 9999H
Word 0	5678H + 4321H = 9999H

The result produced in the MMX register is

$$MM_4 = \text{FFFFFFFF99999999H}$$

If an underflow occurs in an unsigned, parallel subtraction (PSUBUSB/W), the result is replaced by the smallest value that can be represented with the number of bits, either 00H or 0000H. The maximum and minimum saturation limits for signed byte and word overflows and underflows are 7FH for a byte (7FFFH for a word) and 80H for a byte (8000H for a word), respectively.

The compare instructions permit parallel comparison operations to be performed on parallel byte-wide, word-wide, and double-word-wide elements of data. For example, the PCMPEQ instruction independently compares the corresponding elements of the source and destination operand to see if they are equal. If the corresponding elements of the source and destination operand are equal, the comparison is true and all bits of the destination element are set to 1; otherwise, the comparison is false and all bits of the destination element are cleared to 0. The PCMPGT performs the same operation except it checks to see if the individual values in the destination operand are greater than their corresponding elements in the source operand. An example is the instruction

```
PCMPGTB MM4, MM5
```

Execution of this instruction compares the source to the destination for each of the independent signed byte elements of data. This is, it performs the eight comparisons that follow:

MM_4	Byte 7	Byte 6	Byte 5	Byte 4	Byte 3	Byte 2	Byte 1	Byte 0
	$>$	$>$	$>$	$>$	$>$	$>$	$>$	$>$
MM_5	Byte 7	Byte 6	Byte 5	Byte 4	Byte 3	Byte 2	Byte 1	Byte 0

If the value of a specific byte in MM_4 is greater than its corresponding byte in MM_5, the comparison is true and the byte in MM_4 is set to FFH; otherwise it is false and made 00H.

EXAMPLE 15.29

Determine the results produced in the destination register after executing the instruction

```
PCMPEQD MM3, MM4
```

Assume that MM_3 and MM_4 contain the values FFFF000012345678H and FFFF000087654321H. What result does the instruction execution produce?

Solution

This instruction compares the individual double-word destination elements to their corresponding source elements and determines whether or not they are equal. This gives the following result:

Double word 1 FFFF0000H = FFFF0000H = True → FFFFFFFFH

Double word 0 12345678H ≠ 87654321H = False → 00000000H

Therefore, the result in the destination register is

$$MM_3 = \text{FFFFFFFF00000000H}$$

The pack and unpack conversion instructions are used for storing data after converting the size of elements according to the destination size. For example, the instruction

```
PACKUSDW MM0, MM1
```

converts the four double-word elements of source and destination registers MM_0 and MM_1 into four words and packs them into destination register MM_0. Figure 15–72 shows how the elements are merged together in the destination. Just like the PADD instruction, if an element results in an overflow or underflow when it is compressed during the exe-

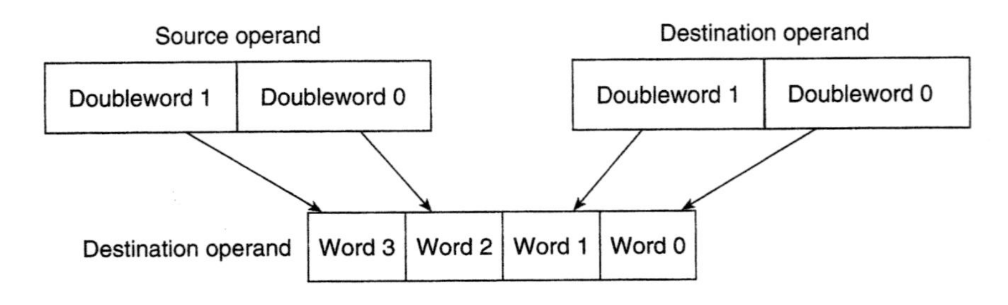

Figure 15–72 Pack operation of the PACKUSDW instruction.

cution of the PACKUSDW instruction, FFFFH and 0000H replace it, respectively. For the instruction PACKUSDW, the signed number overflow and underflow maximum and minimum values are 7FFFH and 8000H, respectively.

EXAMPLE 15.30

What results are produced in the destination register by executing the instruction that follows?

```
PACKUSDW MM6, MM5
```

Assume that MM_5 and MM_6 contain the values 000100FF80000123H and 0000AA00000-04321H.

Solution

The word-wide elements produced in destination MM_6 are as follows:

MM_5	Word 3	000100FFH → FFFFH Unsigned saturation overflow
	Word 2	80000123H → 0000H Unsigned saturation underflow
MM_6	Word 1	0000AA00H → AA00H
	Word 0	00004321H → 4321H

Therefore, the result in the destination register is

$$MM_6 = \text{FFFF0000AA004321H}$$

Let us now briefly look at how the unpack instruction is used to realign data. Figure 15–73 shows how the PUNPCKLBW MM_4, MM_3 merges the four low bytes of byte-wide source and destination operands to form four words in the destination operand. For instance, if the original data in registers MM_3 and MM_4 are 0F0F0F0F07050301H and FFFF000006040200H, respectively, executing the instruction merges the bytes into the destination as

$$MM_4 = \text{0706050403020100H}$$

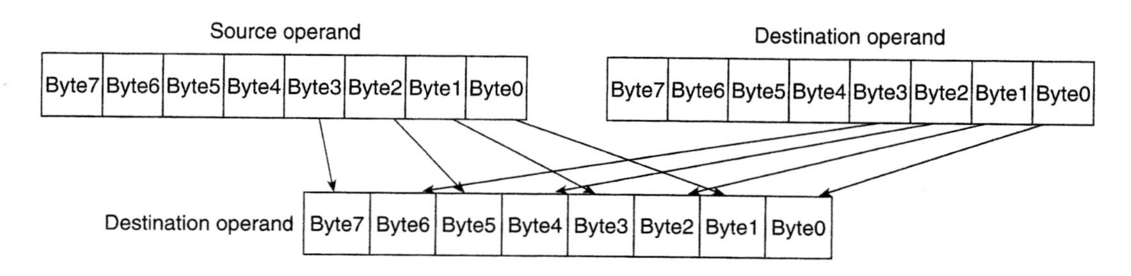

Figure 15–73 Unpack operation of the PUNPCKLBW instruction.

REVIEW PROBLEMS

Section 15.1

1. Name two MPUs in the 80386 family.
2. What size are the registers and the data bus of the 80386DX? The 80386SX?
3. What is the iCOMP rating of an 80386SX-25 MPU? An 80386DX-25 MPU?
4. List the three modes of software operation supported by the 80386DX.

Section 15.2

5. Name the six internal processing units of the 80386DX.
6. What are the word lengths of the 80386DX's address bus and data bus?
7. Does the 80386DX have a multiplexed address/data bus or separate address and data buses?
8. How large is the 80386DX's instruction stream queue?
9. In which unit is the instruction stream queue located?
10. How large is the descriptor cache?
11. Where are recently used page directory and page table entries stored?

Section 15.3

12. How does the performance of a 16-MHz 80386DX compare to that of a 5-MHz 8086?
13. What is meant when we say that the 80386DX is object code compatible with the 8086?
14. In the real mode, is the accumulator register 16 bits or 32 bits in length? The DS register?
15. What new registers are found in the 80386DX's real-mode software model?

Section 15.4

16. Write an instruction that will move the contents of control register 1 to the extended base register.
17. What instruction does the mnemonic SHLD stand for?
18. Describe the operation performed by the instruction MOVSX EAX, BL.
19. Write an instruction that will zero-extend the word of data at address DATA_WORD and copy it into register EAX.
20. What operation is performed when the instruction LFS EDI, DATA_F_ADDRESS is executed?
21. If the values in AX and CL are $F0F0_{16}$ and 04_{16}, respectively, what is the result in AX and CF after execution of each of the instructions that follow:
    (a) BT AX, CL
    (b) BTR AX, CL
    (c) BTC AX, CL
22. What does the mnemonic SETNC stand for? What flag condition does it test for?

Section 15.5

23. List the protected-mode registers that are not part of the real-mode model.
24. What are the two parts of the GDTR called?

25. What function is served by the GDTR?
26. If the contents of the GDTR are $0021000001FF_{16}$, what are the starting and ending addresses of the table? How large is the table? How many descriptors can be stored in the table?
27. What is stored in the GDT?
28. What do IDTR and IDT stand for?
29. What is the maximum limit that should be used in the IDTR?
30. What is stored in the IDT?
31. What descriptor table defines the local memory address space?
32. What gets loaded into the LDTR? What happens when it gets loaded?
33. Which control register contains the MSW?
34. Which bit is used to switch the 80386DX from real-address mode to protected-address mode?
35. What MSW bit settings identify that floating-point operations are to be performed by an 80387 coprocessor?
36. What does TS stand for?
37. What must be done to turn on paging?
38. Where is the page directory base register located?
39. How large is the page directory?
40. What is held in the page table?
41. What gets loaded into TR? What is its function?
42. What is the function of the task descriptor cache?
43. What determines the location and size of a task state segment?
44. What is the name of the CS register in the protected mode? The DS register?
45. What are the names and sizes of the three fields in a selector?
46. What does TI equals 1 mean?
47. If the GDT register contains $0013000000FF_{16}$ and the selector loaded into the LDTR is 0040_{16}, what is the starting address of the LDT descriptor that is to be loaded into the cache?
48. What does NT stand for? RF?
49. If the IOPL bits of the flag register contain 10, what is the privilege level of the I/O instructions?
50. What size is the 80386DX's virtual address?
51. What are the two parts of a virtual address called?
52. How large can a data segment be? How small?
53. How large is the 80386DX's virtual address space? What is the maximum number of segments that can exist in the virtual address space?
54. How large is the global memory address space? How many segments can it contain?
55. In Fig. 15–19, which segments of memory does task 3 have access to? Which segments does it not have access to?
56. What part of the 80386DX is used to translate virtual addresses to physical addresses?

57. What happens when the instruction sequence that follows is executed?

```
MOV AX, [SI]
MOV CS, AX
```

58. If the descriptor accessed in problem 57 has the value $00200000FFFF_{16}$ and IP contains 0100_{16}, what is the physical address of the next instruction to be fetched?
59. Into how many pages is the 80386DX's address space mapped when paging is turned on? What is the size of a page?
60. What are the three elements of the linear address produced by page translation? Give the size of each element.
61. What is the purpose of the translation lookaside buffer?
62. How large is a page frame? What selects the specific storage location in the page frame?

Section 15.6

63. How many bytes are in a descriptor? Name each of its fields and give their sizes.
64. Which registers are segment descriptors associated with? System segment descriptors?
65. The selector 0224_{16} is loaded into the data segment register. This value points to a segment descriptor starting at address 00100220_{16} in the local descriptor table. If the words of the descriptor are

$$(00100220_{16}) = 0110_{16}$$
$$(00100222_{16}) = 0000_{16}$$
$$(00100224_{16}) = 1A20_{16}$$
$$(00100226_{16}) = 0000_{16}$$

what are the LIMIT and BASE?

66. Is the segment of memory identified by the descriptor in problem 65 already loaded into physical memory? Is it a code segment or a data segment?
67. If the current value of IP is 00000226_{16}, what is the physical address of the next instruction to be fetched from the code segment of problem 65?
68. What do the 20 most significant bits of a page directory or page table entry stand for?
69. The page mode protection of a page frame is to provide no access from the user protection level and read/write operation at the supervisor protection level. What are the settings of R/W and U/S?
70. What happens when an attempt is made to access a page frame that has $P = 0$ in its page table entry?
71. What does the D bit in a page directory entry stand for?

Section 15.7

72. If the instruction LGDT [INIT_GDTR] is to load the limit $FFFF_{16}$ and base 00300000_{16}, show how the descriptor must be stored in memory.
73. Write an instruction sequence that can be used to clear the task-switched bit of the MSW.

74. Write an instruction sequence that will load the local descriptor table register with the selector $02F0_{16}$ from register BX.

Section 15.8

75. Define the term *multitasking.*
76. What is a task?
77. What two safeguards are implemented by the 80386DX's protection mechanism?
78. What happens if either the segment limit check or segment attributes check fails?
79. What is the highest privilege level of the 80386DX protection model called? What is the lowest level called?
80. At what protection level are applications run?
81. What protection mechanism is used to isolate local and global resources?
82. What protection mechanism is used to isolate tasks?
83. What is the privilege level of the segment defined by the descriptor in problem 72?
84. What does CPL stand for? RPL?
85. State the data access protection rule.
86. Which privilege-level data segments can be accessed by an application running at level 3?
87. Summarize the code access protection rules.
88. If an application is running at privilege level 3, what privilege-level operating system software is available to it?
89. What purpose does a call gate serve?
90. Explain what happens when the instruction CALL [NEW_ROUTINE] is executed within a task. Assume that NEW_ROUTINE is at a privilege level that is higher than the CPL.
91. What is the purpose of the task state descriptor?
92. What is the function of a task state segment?
93. Where is the state of the prior task saved? Where is the linkage to the prior task saved?
94. Into which register is the TSS selector loaded to initiate a task?
95. Give an overview of the task switch sequence illustrated in Fig. 15–47.

Section 15.9

96. Which bit position in EFLAGS is VM?
97. Is 80386DX protection active or inactive in virtual 8086 mode? If active, what is the privilege level of a virtual 8086 program?
98. Can both protected mode and virtual 8086 tasks coexist in an 80386DX multitasking environment?
99. Can multiple virtual 8086 tasks be active in an 80386DX multitasking environment?

Section 15.10

100. Give two on-chip additions of the 80486DX MPU that result in greater performance.
101. What is the key difference between the 80486DX and 80486SX MPUs?

102. What is the iCOMP rating of the 80486SX-33 MPU? The 80486DX-50 MPU?

103. What is the size of the 80486SX's instruction code queue?

104. What does CISC stand for? RISC? CRISC?

105. List three characteristics of a RISC processor.

106. Is the 80486SX best categorized as that of a CISC, RISC, or CRISC?

107. What new bits are active in the CR_0 of the real-mode 80486SX?

108. Does the 8086 architecture use little endian or big endian organization for data stored in memory?

109. If the contents of EAX are 0F0F0F0FH, what is the result in the register after executing the instruction SWAP EAX?

110. Write an instruction sequence that will read the big endian double-word elements of a table starting at address BIG_E_TABLE and convert them to little endian format in a table starting at address LIT_E_TABLE. Assume that the number of double-word elements in the table equals COUNT.

111. Write an instruction that performs an exchange and add operation on the double word storage location SUM and register EBX. If the original value in SUM is 00000001H and that in EBX is 00000000H, what result is produced in destination SUM by executing the instruction five times?

112. If the instruction CMPXCHG [DATA], BL is executed when the content of AL is 11_{16}, BL is 22_{16}, and storage location DATA is 11_{16}, what results are produced?

113. What new flag is active in the 80486SX's EFLAGS register, and in which bit position is it found?

114. Which bits in the protected mode CR_0 are used to control the operation of the on-chip cache memory?

115. What instruction should be executed to flush the on-chip cache and initiate a flush bus cycle?

116. What is the difference between the operations performed by the INVD and WBINVD instructions?

117. Name the two new active bits in the 80486SX's page table entry.

Section 15.11

118. Convert the floating-point number -9.5 to its equivalent normalized binary number.

119. What are the three sizes of floating-point numbers defined by IEEE Standard 754? How many bits are required to code each of these types of floating point-numbers?

120. What are the three elements of a double precision floating-point number and how many bits wide is each?

121. Express the normalized floating-point number found in problem 118 as a binary single precision floating-point number. What is the hexadecimal value of this number?

122. How are the registers of the floating-point stack labeled from a hardware point of view? How are the registers identified from a software point of view?

123. If the value of TOP in the floating-point status register is 010, which hardware register is currently the top of the stack? How many valid values are currently held in the stack? Which hardware register currently corresponds to ST(4)?

124. If the current top of stack corresponds to floating-point register R_5, what operation does the instruction FLD ST(2) perform?

125. If the values corresponding to DATA3_64B and DATA4_64B are −2.5 and −10.75, respectively, what binary result does the program in Example 15.23 store in the memory location corresponding to DATA4_64B? Hexadecimal result?

Section 15.12

126. How wide is the Pentium processor's external data bus?

127. What does the term *superscaler* mean?

128. What is the iCOMP rating of the 90-MHz Pentium processor MPU?

129. How many pipelines are in a Pentium processor MPU? What are they called?

130. List three improvements made in the Pentium processor's on-chip cache memory over that provided in the 80486SX MPU.

131. How many times faster is the floating-point unit of the Pentium processor compared to that of the 80486DX-33?

132. Give the mnemonics for the three new flags activated in the Pentium processor's EFLAGS register.

133. What does PSE stand for? How many 32-bit entries are in a page when PSE is set to 1?

134. What capability does the MCE bit of CR_4 enable/disable?

135. List three new real-mode instructions supported by the Pentium processor MPU.

136. If the contents of EAX, EBX, ECX, and EDX are 11111111_{16}, 22222222_{16}, 33333333_{16}, and $FFFFFFFF_{16}$, respectively, and the content of the quad-word memory storage location pointed to by the address TABLE is $11111111FFFFFFFF_{16}$, what is the result produced by executing the instruction CMPXCHG8B [TABLE]?

137. Which model-specific register is accessed when the RDMSR system control instruction is executed when ECX contains the value 1_{16}?

Section 15.13

138. What does SIMD stand for?

139. What four types of SIMD data can be processed by the Pentium processor with MMX technology?

140. How many bits wide are the MMX registers? How are they labeled?

141. If an MMX register contains the value FF0012345678ABCDH

(a) What are the values of the individual elements of 8 × 8 data?

(b) 4 × 16 data?

(c) 2 × 32 data?

142. When the instruction PADDUSB MM3, MM4 is executed and the contents of registers MM_3 and MM_4 are FFFFFFFF12345678H and 0001000287654321H, respectively, what results are produced in the destination register?

143. If the operands for the instruction PCMPGTD M3, M4 process the operand values given in Problem 142, what results are produced in the destination?

144. Assume that MM_5 and MM_6 contain the values 000100FF80000123H and 0000AA-0000004321H. What results are produced in the destination register by executing the instruction PACKUSDW MM6, MM5.

16

The 80386, 80486, and Pentium Processor Families: Hardware Architecture

▲ INTRODUCTION

In Chapter 15, we studied the software architecture of the 80386, 80486, and Pentium processor families. We covered their real- and protected-mode software architectures, the 80386 and 80486 specific instruction sets, and the system control instruction set. Now we will turn our attention to the hardware architecture of the 80386, 80486, and Pentium processors. In this chapter, we examine the signal interfaces of the 80386DX MPU, its memory interface, input/output interface, and interrupts/exception processing. Next we will introduce the 80486SX and Pentium processor families. Here we focus on the hardware architecture difference between these newer processors and the 80386DX. For this purpose, we have included the following topics in this chapter:

16.1 80386 Microprocessor Family
16.2 Signal Interfaces of the 80386DX
16.3 System Clock of the 80386DX
16.4 80386DX Bus States and Pipelined and Nonpipelined Bus Cycles
16.5 Memory Organization and Interface Circuits
16.6 Input/Output Interface Circuits and Bus Cycles
16.7 Interrupt and Exception Processing
16.8 80486SX and 80486DX Microprocessors
16.9 Other 80486 Family Microprocessors—80486DX2 and 80486DX4

▲ 16.1 80386 MICROPROCESSOR FAMILY

Hardware compatibility of the 80386 family of microprocessor with either the 8086 or 80286 microprocessor is much less of a concern than software compatibility. In fact, a number of changes have been made to the hardware architecture of the 80386DX to improve both its versatility and its performance. For example, additional pipelining has been provided within the 80386DX, and the address and data buses have both been made 32 bits in length. These two changes in the hardware result in increased performance for 80386DX-based microcomputers. Another feature, *dynamic bus sizing* for the data bus, provides more versatility in system hardware design.

The original 80386DX was manufactured using Intel's complementary high-performance metal-oxide-semiconductor III (CHMOSIII) process. Its circuitry is equivalent to approximately 275,000 transistors, more than twice those used in the design of the 80286 MPU and almost 10 times that of the 8086.

The 80386DX is housed in a 132-pin ceramic *pin grid array* (PGA) package. An 80386DX in this package is shown in Fig. 16–1. This package can be mounted in a socket that is soldered to the circuit board or have its leads inserted through holes in the board and soldered. The signal at each pin is shown in Fig. 16–2(a). Note that all the 80386DX's signals are supplied at separate pins on the package. This is intended to simplify the microcomputer circuit design.

Figure 16–2(a) shows that the rows of pins on the package are identified by row numbers 1 through 14 and the columns of pins are labeled A through P. Therefore, the location of the pin for each signal is uniquely defined by a column and row coordinate. For example, in Fig. 16–2(a) address line A_{31} is at the junction of column N and row 2—that is, it is at pin N2. Figure 16–2(b) lists the pin locations for all of the 80386DX's signals.

EXAMPLE 16.1

At what pin location is the signal D_0?

Solution

Looking at Fig. 16–2(a), we find that the pin for D_0 is located in column H at row 12. Therefore, its pin is identified as H12 in Fig. 16–2(b).

The 80386SX MPU is not packaged in a PGA. It is available in a 100-lead *plastic quad flat package* (PQFP). A plastic package is used to permit a lower cost for the device. This type of package is meant for *surface-mount* installation. That is, its pins do not go

Figure 16–1 80386 IC. (Courtesy of Intel Corporation)

through the board; instead, the device is laid on top of the circuit board and then soldered in place.

▲ 16.2 SIGNAL INTERFACES OF THE 80386DX

Figure 16–3 presents a block diagram of the 80386DX microprocessor. Here we have grouped its signal lines into four interfaces: the *memory/IO interface,* the *interrupt interface,* the *DMA interface,* and the *coprocessor interface.* Figure 16–4 lists each of the signals at the 80386DX's interfaces. Included in this table are a mnemonic, function, type, and active level for each signal. For instance, we find that the signal with the mnemonic M/$\overline{IO}$ stands for memory/IO indication. This signal is an output produced by the 80386DX that is used to tell external circuitry whether the current address available on the address bus is for memory or an input/output device. Its active level is listed as 1/0, which means that logic 1 on this line identifies a memory bus cycle and logic 0 an input/output bus cycle. On the other hand, the signal INTR at the interrupt interface is the maskable interrupt request input of the 80386DX. This input is active when at logic 1. By using this input, external devices can signal the 80386DX that they need to be serviced.

Memory/IO Interface

In a microcomputer system, the address bus and data bus signal lines form a parallel path over which the MPU talks with its memory and I/O subsystems. Like the 80286 microprocessor, but unlike the older 8086 and 8088, the 80386DX has a demultiplexed address/data bus. Note in Fig. 16–2(b) that the address bus and data bus lines are located at different pins of the IC.

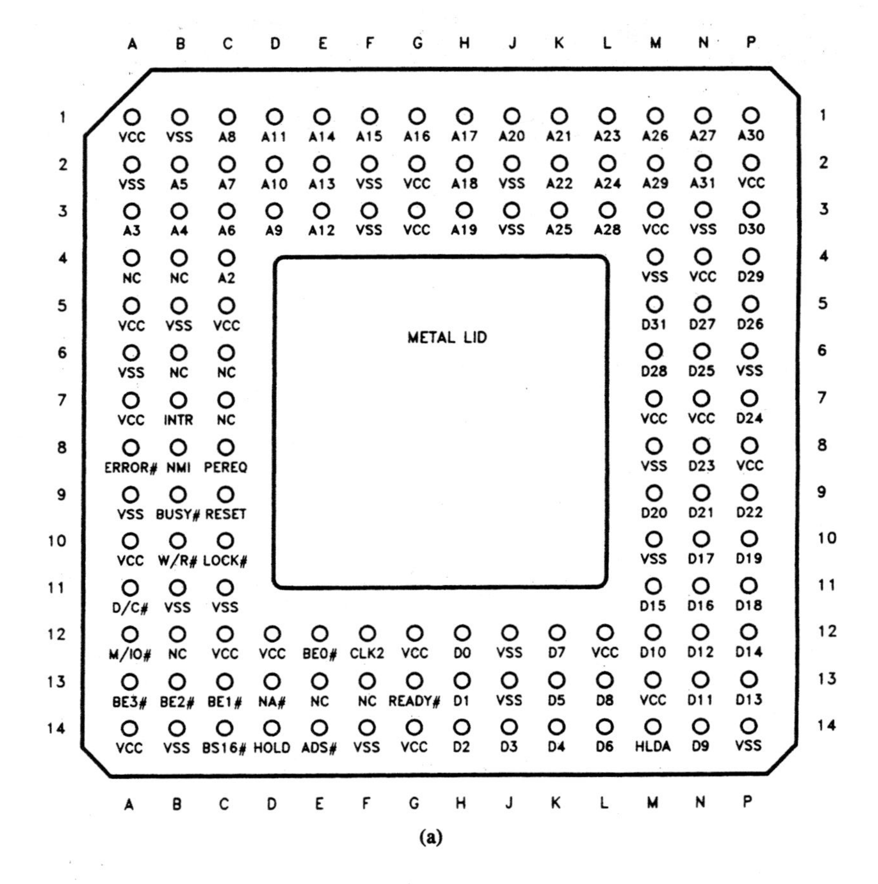

(a)

Figure 16–2 (a) Pin layout of the 80386DX. (Reprinted by permission of Intel Corporation. Copyright/Intel Corp. 1987) (b) Signal pin numbering. (Reprinted by permission of Intel Corporation. Copyright/Intel Corp. 1987)

From a hardware point of view, there is only one difference between an 80386DX configured for the real-address mode and one configured for protected-address mode. This difference is the size of the address bus. When in real mode, only the lower 18 address lines, A_2 through A_{19}, are active, whereas in the protected mode all 30 lines, A_2 through A_{31}, are functional. Of these, A_{19} and A_{31} are the most significant address bits, respectively. Actually, real-mode addresses are 20 bits long and protected-mode addresses are 32 bits long. The other two bits, A_0 and A_1, are decoded inside the 80386DX, along with information about the size of the data to be transferred, to produce *byte-enable* outputs, $\overline{BE}_0$, $\overline{BE}_1$, $\overline{BE}_2$, and $\overline{BE}_3$.

As Fig. 16–4 shows, the address lines are outputs. They are used to carry address information from the 80386DX to memory and input/output ports. In real-address mode, the 20-bit address gives the 80386DX the ability to address a 1Mbyte physical memory address space. On the other hand, in protected mode the extended 32-bit address results in a 4Gbyte physical memory address space. Moreover, when in protected mode, virtual

Pin / Signal	Pin / Signal	Pin / Signal	Pin / Signal
N2 A31	M5 D31	A1 V_{CC}	A2 V_{SS}
P1 A30	P3 D30	A5 V_{CC}	A6 V_{SS}
M2 A29	P4 D29	A7 V_{CC}	A9 V_{SS}
L3 A28	M6 D28	A10 V_{CC}	B1 V_{SS}
N1 A27	N5 D27	A14 V_{CC}	B5 V_{SS}
M1 A26	P5 D26	C5 V_{CC}	B11 V_{SS}
K3 A25	N6 D25	C12 V_{CC}	B14 V_{SS}
L2 A24	P7 D24	D12 V_{CC}	C11 V_{SS}
L1 A23	N8 D23	G2 V_{CC}	F2 V_{SS}
K2 A22	P9 D22	G3 V_{CC}	F3 V_{SS}
K1 A21	N9 D21	G12 V_{CC}	F14 V_{SS}
J1 A20	M9 D20	G14 V_{CC}	J2 V_{SS}
H3 A19	P10 D19	L12 V_{CC}	J3 V_{SS}
H2 A18	P11 D18	M3 V_{CC}	J12 V_{SS}
H1 A17	N10 D17	M7 V_{CC}	J13 V_{SS}
G1 A16	N11 D16	M13 V_{CC}	M4 V_{SS}
F1 A15	M11 D15	N4 V_{CC}	M8 V_{SS}
E1 A14	P12 D14	N7 V_{CC}	M10 V_{SS}
E2 A13	P13 D13	P2 V_{CC}	N3 V_{SS}
E3 A12	N12 D12	P8 V_{CC}	P6 V_{SS}
D1 A11	N13 D11		P14 V_{SS}
D2 A10	M12 D10		
D3 A9	N14 D9	F12 CLK2	A4 N.C.
C1 A8	L13 D8		B4 N.C.
C2 A7	K12 D7	E14 ADS#	B6 N.C.
C3 A6	L14 D6		B12 N.C.
B2 A5	K13 D5	B10 W/R#	C6 N.C.
B3 A4	K14 D4	A11 D/C#	C7 N.C.
A3 A3	J14 D3	A12 M/IO#	E13 N.C.
C4 A2	H14 D2	C10 LOCK#	F13 N.C.
A13 BE3#	H13 D1		
B13 BE2#	H12 D0	D13 NA#	C8 PEREQ
C13 BE1#		C14 BS16#	B9 BUSY#
E12 BE0#		G13 READY#	A8 ERROR#
	D14 HOLD		
C9 RESET	M14 HLDA	B7 INTR	B8 NMI

(b)

Figure 16–2 (continued)

addressing is provided through software. This results in a 64Tbyte virtual memory address space.

In both the real and protected modes, the 80386DX microcomputer has an independent I/O address space. This I/O address space is 64Kbytes in length. Therefore, only address lines A_2 through A_{15} and the $\overline{BE}$ outputs are used when addressing I/O devices.

The 80386SX has the same number of active address lines as the 80386DX in the real mode, but fewer address lines in the protected mode. Its protected-mode address is only 25 bits long; therefore, the address space is limited to 16Mbytes. This is the first key difference between the 80386SX and 80386DX MPUs. The 80386SX accesses the I/O address space in exactly the same way as the 80386DX.

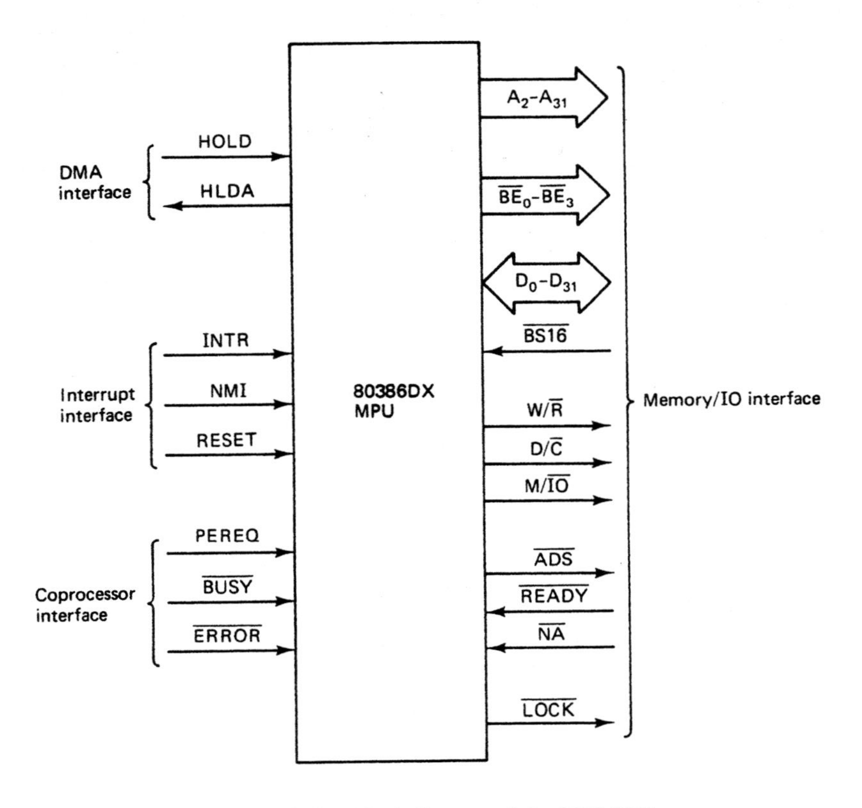

Figure 16–3 Block diagram of the 80386DX.

Since the 80386DX is a 32-bit microprocessor, its data bus is formed from the 32 data lines D_0 through D_{31}. Data line D_{31} is the most significant bit and D_0 the least significant bit. These lines are identified as bidirectional in Fig. 16–4 because they have the ability to carry data either in or out of the MPU. The kinds of data transferred over these lines are read/write data and instructions for memory, input/output data for input/output devices, and interrupt type codes from an interrupt controller.

The 80386SX has a 16-bit data bus instead of a 32-bit data bus. This is the second important hardware difference between the 80386SX and 80386DX. This results in lower performance, but has the advantage of enabling the design of a lower-cost memory subsystem for the microcomputer system.

Earlier we indicated that the 80386DX supports dynamic bus sizing. Even though the 80386DX has 32 data lines, the size of the bus can be dynamically switched to 16 bits. Simply switching the *bus size 16* ($\overline{\text{BS 16}}$) input to logic 0 does this. When in this mode, 32-bit data transfers are performed as two successive 16-bit data transfers over bus lines D_0 through D_{15}. Since the 80386SX has a 16-bit data bus, it does not have this input.

Remember that the 80386DX supports byte, word, and double-word data transfers over its data bus during a single bus cycle. Therefore, it must signal to external circuitry what type of data transfer is taking place and over which part of the data bus the data will

Name	Function	Type	Level
CLK2	System clock	I	–
A_{31}-A_2	Address bus	O	1
$\overline{BE_3}$-$\overline{BE_0}$	Byte enables	O	0
D_{31}-D_0	Data bus	I/O	1
$\overline{BS_{16}}$	Bus size 16	I	0
W/$\overline{R}$	Write/read indication	O	1/0
D/$\overline{C}$	Data/control indication	O	1/0
M/$\overline{IO}$	Memory I/O indication	O	1/0
$\overline{ADS}$	Address status	O	0
$\overline{READY}$	Transfer acknowledge	I	0
$\overline{NA}$	Next address request	I	0
$\overline{LOCK}$	Bus lock indication	O	0
INTR	Interrupt request	I	1
NMI	Nonmaskable interrupt request	I	1
RESET	System reset	I	1
HOLD	Bus hold request	I	1
HLDA	Bus hold acknowledge	O	1
PEREQ	Coprocessor request	I	1
$\overline{BUSY}$	Coprocessor busy	I	0
$\overline{ERROR}$	Coprocessor error	I	0

Figure 16–4 Signals of the 80386DX.

be carried. The bus unit does this by activating the appropriate byte enable ($\overline{BE}$) output signals.

Figure 16–5 lists each byte-enable output signal and the part of the data bus it is intended to enable. For instance, here we see that $\overline{BE_0}$ corresponds to data bus lines D_0 through D_7. If a byte of data is being read from memory, only one of the $\overline{BE}$ outputs is made active. For instance, if the most significant byte of an aligned double word is read from memory, $\overline{BE_3}$ is switched to logic 0 and the data moves on data lines D_{24} through D_{31}. On the other hand, if a word of data is being read, two outputs become active. An example would be to read the most significant word of an aligned double word from

Byte enable	Data bus lines
$\overline{BE_0}$	D_0-D_7
$\overline{BE_1}$	D_8-D_{15}
$\overline{BE_2}$	D_{16}-D_{23}
$\overline{BE_3}$	D_{24}-D_{31}

Figure 16–5 Byte enable outputs and data bus lines.

memory. In this case, $\overline{BE_2}$ and $\overline{BE_3}$ are both switched to logic 0 and the data is carried by lines D_{16} through D_{31}. Finally, if an aligned double-word read is taking place, all four $\overline{BE}$ outputs are made active and all 32 data lines are used to transfer the data.

EXAMPLE 16.2

What code is output on the byte-enable lines whenever the address on the bus is for an instruction-acquisition bus cycle?

Solution

Since code is always fetched as 32-bit words (aligned double words), all the byte-enable outputs are made active. Therefore,

$$\overline{BE_3}\,\overline{BE_2}\,\overline{BE_1}\,\overline{BE_0} = 0000_2$$

The byte-enable lines work exactly the same way when write data transfers are performed over the bus. Figure 16–6(a) identifies what type of data transfer takes place for all possible variations of the byte-enable outputs. For instance, we find that $\overline{BE_3}\,\overline{BE_2}\,\overline{BE_1}\,\overline{BE_0} = 1110_2$ means that a byte of data is written over data bus lines D_0 through D_7.

EXAMPLE 16.3

What type of data transfer takes place and over which data bus lines are data transferred if the byte-enable code output is

$$\overline{BE_3}\,\overline{BE_2}\,\overline{BE_1}\,\overline{BE_0} = 1100_2$$

Solution

In Fig. 16–6(a), we see that a word of data is transferred over data bus lines D_0 through D_{15}.

With its 16-bit data bus, the 80386SX can transfer only a byte or word of data over the bus during a single bus cycle. For this reason, the four byte-enable signals of the 80386DX are replaced by just two signals: *byte high enable* ($\overline{BHE}$) and *byte low enable* ($\overline{BLE}$) on the 80386SX. Logic 0 at $\overline{BLE}$ tells that a byte of data is being transferred over data bus line D_0 through D_7, and logic 0 at $\overline{BHE}$ means a byte transfer is taking place over data bus lines D_8 through D_{15}. When a word of data is transferred, both of these signals are at their active 0 logic level.

The 80386DX performs what is called *data duplication* during certain types of write cycles. Data duplication is provided in the 80386DX to optimize the performance of the data bus when it is set for 16-bit mode. Note that whenever a write cycle is per-

$\overline{BE_3}$	$\overline{BE_2}$	$\overline{BE_1}$	$\overline{BE_0}$	D_{31}-D_{24}	D_{23}-D_{16}	D_{15}-D_8	D_7-D_0
1	1	1	0				XXXXXXXX
1	1	0	1			XXXXXXXX	
1	0	1	1		XXXXXXXX		
0	1	1	1	XXXXXXXX			
1	1	0	0			XXXXXXXX	XXXXXXXX
1	0	0	1		XXXXXXXX	XXXXXXXX	
0	0	1	1	XXXXXXXX	XXXXXXXX		
1	0	0	0		XXXXXXXX	XXXXXXXX	XXXXXXXX
0	0	0	1	XXXXXXXX	XXXXXXXX	XXXXXXXX	
0	0	0	0	XXXXXXXX	XXXXXXXX	XXXXXXXX	XXXXXXXX

(a)

$\overline{BE_3}$	$\overline{BE_2}$	$\overline{BE_1}$	$\overline{BE_0}$	D_{31}-D_{24}	D_{23}-D_{16}	D_{15}-D_8	D_7-D_0
1	1	1	0				XXXXXXXX
1	1	0	1			XXXXXXXX	
1	0	1	1		XXXXXXXX		DDDDDDDD
0	1	1	1	XXXXXXXX		DDDDDDDD	
1	1	0	0			XXXXXXXX	XXXXXXXX
1	0	0	1		XXXXXXXX	XXXXXXXX	
0	0	1	1	XXXXXXXX	XXXXXXXX	DDDDDDDD	DDDDDDDD
1	0	0	0		XXXXXXXX	XXXXXXXX	XXXXXXXX
0	0	0	1	XXXXXXXX	XXXXXXXX	XXXXXXXX	
0	0	0	0	XXXXXXXX	XXXXXXXX	XXXXXXXX	XXXXXXXX

(b)

Figure 16–6 (a) Types of data transfers for the various byte-enable combinations. (b) Data transfers that include duplication.

formed in which data are transferred only over the upper part of the 32-bit data bus, the data are duplicated on the corresponding lines of the lower part of the bus. For example, looking at Fig. 16–6(b), we see that when $\overline{BE_3}\overline{BE_2}\overline{BE_1}\overline{BE_0} = 1011_2$, data (denoted as XXXXXXXX) are actually being written over data bus lines D_{16} through D_{23}. However, at the same time, the data (denoted as DDDDDDDD in Fig. 16–6(b)) are automatically duplicated on data bus lines D_0 through D_7. Despite the fact that the byte is available on the lower eight data bus lines, $\overline{BE_0}$ stays inactive. The same thing happens when a word of data is transferred over D_{16} through D_{31}. In this example, $\overline{BE_3}\overline{BE_2}\overline{BE_1}\overline{BE_0} = 0011_2$, and Fig. 16–6(b) shows that the word is duplicated on data lines D_0 through D_{15}.

EXAMPLE 16.4

If a word of data that is being written to memory is accompanied by the byte-enable code 1001_2, over which data bus lines are the data carried? Is data duplication performed for this data transfer?

Solution

The tables in Fig. 16–6 show that for the byte-enable code 1001_2, the word of data is transferred over data bus lines D_8 through D_{23}. For this transfer, data duplication does not occur.

Control signals are required to support information transfers over the 80386DX's address and data buses. They are needed to signal when a valid address is on the address bus, in which direction data are to be transferred over the data bus, when valid write data are on the data bus, and when an external device can put read data on the data bus. The 80386DX does not directly produce signals for all these functions. Instead, it outputs bus cycle definition and control signals at the beginning of each bus cycle. These bus cycle indication signals must be decoded in external circuitry to produce the needed memory and I/O control signals.

Three signals are used to identify the type of 80386DX bus cycle that is in progress. In Figs. 16–3 and 16–4, they are labeled *write/read indication* ($W/\overline{R}$), *data/control indication* ($D/\overline{C}$), and *memory/input-output indication* ($M/\overline{IO}$). The table in Fig. 16–7 lists all possible combinations of the bus cycle indication signals and the corresponding type of bus cycle. Here we find that the logic level of memory/input-output ($M/\overline{IO}$) tells whether a memory or input/output cycle is to take place over the bus. Logic 1 at this output signals a memory operation, and logic 0 signals an I/O operation. The next signal in Fig. 16–7, data/control indication ($D/\overline{C}$), identifies whether the current bus cycle is a data or control cycle. In the table we see that it signals *control cycle* (logic 0) for instruction fetch, interrupt acknowledge, and halt/shutdown operations and *data cycle* (logic 1) for memory and I/O read and write operations. Looking more closely at the table in Fig. 16–7, we find that if the code on these two lines, $M/\overline{IO}$ $D/\overline{C}$, is 00, an interrupt is to be acknowledged; if it is 01, an input/output operation is in progress; if it is 10, instruction code is being fetched; and finally, if it is 11, a data memory read or write is taking place.

The last signal noted in Fig. 16–7, write/read indication ($W/\overline{R}$), identifies the specific type of memory or input/output operation that will occur during a bus cycle. For example, when $W/\overline{R}$ is logic 0, data are to be read from memory or an I/O port. On the other hand, logic 1 at $W/\overline{R}$ says that data are to be written into memory or an I/O device. For example, all bus cycles that read instruction code from memory are accompanied by logic 0 on the $W/\overline{R}$ line.

$M/\overline{IO}$	$D/\overline{C}$	$W/\overline{R}$	Type of bus cycle
0	0	0	Interrupt acknowledge
0	0	1	Idle
0	1	0	I/O data read
0	1	1	I/O data write
1	0	0	Memory code read
1	0	1	Halt/shutdown
1	1	0	Memory data read
1	1	1	Memory data write

Figure 16–7 Bus cycle indication signals and types of bus cycles.

EXAMPLE 16.5

If the bus cycle indication code $M/\overline{IO}$ $D/\overline{C}$ $W/\overline{R}$ equals 010, what type of bus cycle is taking place?

Solution

Looking at the table in Fig. 16–7, we see that bus cycle indication code 010 identifies an I/O read (input) bus cycle.

Three bus cycle-control signals are produced directly by the 80386DX. They are identified in Figs. 16–3 and 16–4 as *address status* ($\overline{ADS}$), *transfer acknowledge* ($\overline{READY}$), and *next-address request* ($\overline{NA}$). The $\overline{ADS}$ output is switched to logic 0 to indicate that the bus cycle indication code ($M/\overline{IO}$ $D/\overline{C}$ $W/\overline{R}$), byte-enable code ($\overline{BE}_3\overline{BE}_2$ $\overline{BE}_1\overline{BE}_0$), and address ($A_2$ through A_{31}) signals are all stable. Therefore, it is normally applied to an input of the external bus-control logic circuit and tells it that a valid bus cycle indication code and address are available. In Fig. 16–7 the bus cycle indication code $M/\overline{IO}$ $D/\overline{C}$ $W/\overline{R}$ = 001 is identified as *idle.* That is, it is the code that is output whenever no bus cycle is being performed.

$\overline{READY}$ can be used to insert wait states into the current bus cycle such that it is extended by a number of clock periods. In Fig. 16–4, we find that this signal is an input to the 80386DX. Normally, it is produced by the microcomputer's memory or input/output subsystem and supplied to the 80386DX by way of external bus control logic circuitry. By switching $\overline{READY}$ to logic 0, slow memory or I/O devices can tell the 80386DX when they are ready to permit a data transfer to be completed.

Earlier we pointed out that the 80386DX supports address pipelining at its bus interface. By address pipelining, we mean that the address and bus cycle indication code for the next bus cycle is output before $\overline{READY}$ becomes active to signal that the prior bus cycle can be completed. This mode of operation is optional. The external bus-control logic circuitry activates pipelining by switching the next-address request ($\overline{NA}$) input to logic 0. By using pipelining, delays introduced by the decode logic can be made transparent and the address to data access time is increased. In this way, the same level of performance can be obtained with slower, lower-cost memory devices.

One other bus interface control output that is supplied by the 80386DX is *bus lock indication* ($\overline{LOCK}$). This signal is needed to support multiple-processor architectures. In multiprocessor systems that employ shared resources, such as global memory, this signal can be employed to assure that the 80386DX has uninterrupted control of the system bus and the shared resource. That is, by switching its $\overline{LOCK}$ output to logic 0, the MPU can lock up the shared resource for exclusive use.

The 80386SX has the same bus cycle indication and control signals as the 80386DX. The signals that identify the type of bus cycle are write/read indication ($W/\overline{R}$), data/control indication ($D/\overline{C}$), and memory/input-output indication ($M/\overline{IO}$), whereas those that control the bus cycle are address status ($\overline{ADS}$), transfer acknowledge ($\overline{READY}$), next address request ($\overline{NA}$), and bus lock indication ($\overline{LOCK}$). Each of these signals serves the exact same function as they do for the 80386DX MPU.

Interrupt Interface

Looking at Figs. 16–3 and 16–4, we find that the key interrupt interface signals are *interrupt request* (INTR), *nonmaskable interrupt request* (NMI), and *system reset* (RESET). INTR is an input to the 80386DX that can be used by external devices to signal that they need to be serviced. The 80386DX samples this input at the beginning of each instruction. Logic 1 on INTR represents an active interrupt request.

When the 80386DX recognizes an active interrupt request, it signals this fact to external circuitry and initiates an interrupt-acknowledge bus cycle sequence. In Fig. 16–7, the occurrence of an interrupt-acknowledge bus cycle is signaled to external circuitry with the bus cycle definition $M/\overline{IO}$ $D/\overline{C}$ $W/\overline{R}$ equal to 000. This bus cycle indication code can be decoded in the external bus control logic circuitry to produce an interrupt-acknowledge signal. With this interrupt-acknowledge signal, the 80386DX tells the external device that its request for service has been granted. This completes the interrupt request/acknowledge handshake. At this point, program control is passed to the interrupt's service routine.

The INTR input is maskable. That is, its operation can be enabled or disabled with the interrupt flag (IF) within the 80386DX's flag register. On the other hand, the NMI input, as its name implies, is a nonmaskable interrupt input. On any 0-to-1 transition of NMI, a request for service is latched within the 80386DX. Independent of the setting of the IF flag, control is passed to the beginning of the nonmaskable interrupt service routine at the completion of execution of the current instruction.

Finally, the RESET input provides a method of implementing a hardware reset for the 80386DX microprocessor. For instance, using this input can reset the microcomputer at power on. Switching RESET to logic 1 initializes the internal registers of the 80386DX. When it is returned to logic 0, program control is passed to the beginning of a reset service routine. This routine is used to initialize the rest of the system's resources, such as I/O ports, the interrupt flag, and data memory. A diagnostic routine that tests the 80386DX microprocessor can also be initiated as part of the reset sequence. This assures an orderly startup of the microcomputer system.

The 80386SX's interrupt interface is exactly the same as that of the 80386DX. It is implemented with the same signals, interrupt request (INTR), nonmaskable interrupt request (NMI), and system reset (RESET), and it is identified to external circuitry with the same bus cycle code.

DMA Interface

Now that we have examined the signals of the 80386DX's interrupt interface, let us turn our attention to the *direct memory access* (DMA) interface. Figures 16–3 and 16–4 show that the DMA interface is implemented with just two signals: *bus hold request* (HOLD) and *bus hold acknowledge* (HLDA). When an external device, such as a *DMA controller,* wants to take over control of the local address and data buses, it signals this fact to the 80386DX by switching the HOLD input to logic 1. At completion of the current bus cycle, the 80386DX enters the hold state. When in the hold state, its local bus signals are in the high-impedance state. Next, the 80386DX signals external devices that it has given up control of the bus by switching its HLDA output to the 1 logic level. This completes the hold/hold acknowledge handshake sequence. The 80386DX remains in this state until the hold request is removed. The 80386SX's DMA interface is exactly the same as that of the 80386DX.

Coprocessor Interface

Figure 16–3 shows that a coprocessor interface is provided on the 80386DX microprocessor to permit it to easily interface to the *80387DX numeric coprocessor.* The 80387DX cannot perform transfers over the data bus by itself. Whenever the 80387DX needs to read or write operands from memory, it must signal the 80386DX to initiate the data transfers. The 80387DX does this by switching the *coprocessor request* (PEREQ) input of the 80386DX to logic 1.

The other two signals included in the external coprocessor interface are $\overline{\text{BUSY}}$ and $\overline{\text{ERROR}}$. *Coprocessor busy* ($\overline{\text{BUSY}}$) is an input of the 80386DX. Whenever the 80387DX is executing a numeric instruction, it signals this fact to the 80386DX by switching the $\overline{\text{BUSY}}$ input to logic 0. In this way, the 80386DX knows not to request the numeric coprocessor to perform another calculation until $\overline{\text{BUSY}}$ returns to 1. Moreover, if an error occurs in a calculation performed by the numeric coprocessor, this condition is signaled to the 80386DX by switching the *coprocessor error* ($\overline{\text{ERROR}}$) input to the 0 logic level. This interface is implemented the same way on the 80386SX MPU.

▲ 16.3 SYSTEM CLOCK OF THE 80386DX

The time base for synchronization of the internal and external operations of the 80386DX microprocessor is provided by the *clock* (CLK2) input signal. At present, the 80386DX has been available with four different clock speeds. The original 80386DX-16 MPU operates at 16 MHz and its three faster versions, the 80386DX-20, 80386DX-25, and 80386DX-33, operate at 20, 25, and 33 MHz, respectively. The clock signal applied to the CLK2 input of the 80386DX is twice the frequency rating of the microprocessor. Therefore, CLK2 of an 80386DX-16 is driven by a 32-MHz signal. This signal must be generated in external circuitry. The 80386SX is also available in each of these four speeds.

Figure 16–8 illustrates the waveform of the CLK2 input of the 80386DX. This signal is specified at CMOS-compatible voltage levels and not TTL levels. Its minimum and

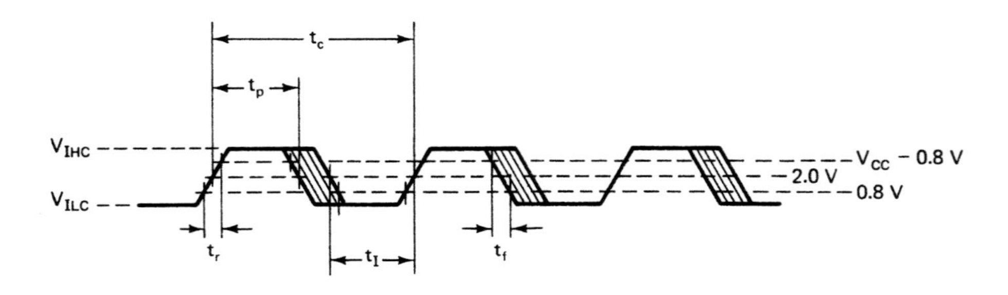

Figure 16–8 System clock (CLK2) waveform.

maximum low logic levels are $V_{ILCmin} = -0.3$ V and $V_{ILCmax} = 0.8$ V, respectively. Moreover, the minimum and maximum high logic levels are $V_{IHCmin} = V_{cc} - 0.8$ V and $V_{IHCmax} = V_{cc} + 0.3$ V, respectively. The minimum period of the 16-MHz clock signal is $t_{cmin} = 31$ ns (measured at the 2.0V level); its minimum high time t_{pmin} and low time t_{lmin} (measured at the 2.0V level) are both equal to 9 ns; and the maximum rise time t_{rmax} and fall time t_{fmax} of its edges (measured between the $V_{cc} - 0.8$V and 0.8V levels) are equal to 8 ns.

▲ 16.4 80386DX BUS STATES AND PIPELINED AND NONPIPELINED BUS CYCLES

Before looking at the bus cycles of the 80386DX, let us first examine the relationship between the timing of the 80386DX's CLK2 input and its bus cycle states. The *internal processor clock* (PCLK) signal is at half the frequency of the external clock input signal. Therefore, as shown in Fig. 16–9, one processor clock cycle corresponds to two CLK2 cycles. Note that these CLK2 cycles are labeled as *phase 1* (ϕ_1) and *phase 2* (ϕ_2). In a 20-MHz 80386DX microprocessor, CLK2 equals 40 MHz and each clock cycle has a duration of 25 ns. In Fig. 16–9, the two phases ($\phi_1 + \phi_2$) of a processor cycle are identified as one processor clock period. A processor clock period is also called a *T state.* Therefore, the minimum length of an internal processor clock cycle is 50 ns.

Nonpipelined and Pipelined Bus Cycles

A *bus cycle* is the activity performed whenever a microprocessor accesses information in program memory, data memory, or an input/output device. The 80386DX can perform bus cycles with either of two types of timing: *nonpipelined* and *pipelined.* Here we will examine the difference between these two types of bus cycles.

Figure 16–10 shows a typical nonpipelined microprocessor bus cycle. Note that the bus cycle contains two T states, T_1 and T_2. During the T_1 part of the bus cycle, the

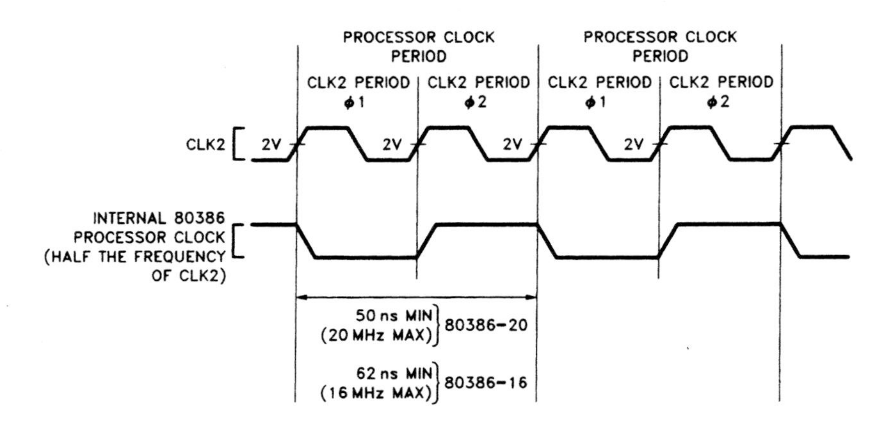

Figure 16–9 Processor clock cycles. (Reprinted by permission of Intel Corporation. Copyright/Intel Corp. 1987)

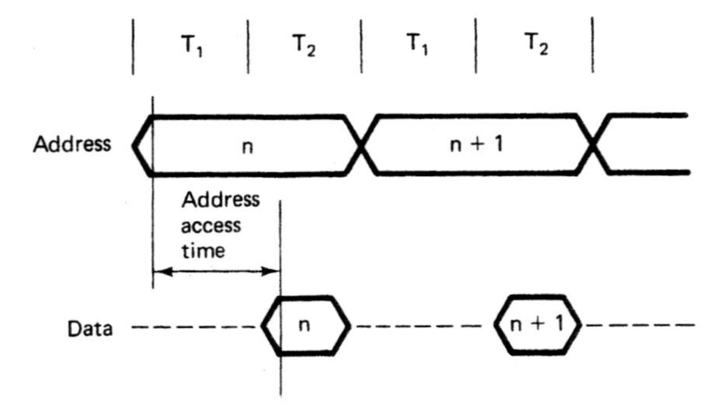

Figure 16–10 Typical nonpipelined read/write bus cycle.

80386DX outputs the address of the storage location that is to be accessed on the address bus, a bus cycle indication code, and control signals. In the case of a write cycle, write data are also output on the data bus during T_1. The second state, T_2, is the part of the bus cycle during which external devices are to accept write data from the data bus or, in the case of a read cycle, put data on the data bus.

For instance, in Fig. 16–10, the sequence of events starts with an address, denoted as *n,* being output on the address bus in clock state T_1. Later in the bus cycle, while the address is still available on the address bus, a read or write data transfer takes place over the data bus. Note that the data transfer for address *n* is shown to occur in clock state T_2. Since each bus cycle has a minimum of two T states (four CLK2 cycles), the minimum bus cycle duration for an 80386DX-20 is 100 ns.

Let us now look at a microprocessor bus cycle that employs *pipelining.* By pipelining we mean that addressing for the next bus cycle is overlapped with the data transfer of the prior bus cycle. When address pipelining is in use, the address, bus cycle indication code, and control signals for the next bus cycle are output during T_2 of the prior cycle, instead of the T_1 that follows.

In Fig. 16–11, address *n* becomes valid in the T_2 state of the prior bus cycle, and then the data transfer for address *n* takes place in the next T_2 state. Moreover, note that at the same time that data transfer *n* occurs, address *n* + 1 is output on the address bus. In this way we see that the microprocessor begins addressing the next storage location to be

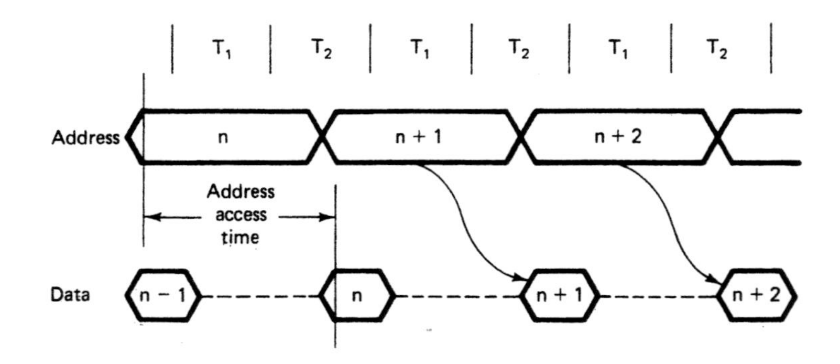

Figure 16–11 Pipelined bus cyle. (Reprinted by permission of Intel Corporation. Copyright/Intel Corp. 1987)

accessed while it is still performing the read or write of data for the previously addressed storage location. Due to the address/data pipelining, the memory or I/O subsystem actually has five CLK2 cycles (125 ns for an 80386DX-20 running at full speed) to perform the data transfer, even though the duration of every bus cycle is just four CLK2 cycles (100 ns).

The interval of time denoted as *address-access time* in Fig. 16–10 represents the amount of time that the address must be stable prior to the read or write of data actually taking place. Note that this duration is less than the four CLK2 cycles in a nonpipelined bus cycle. Figure 16–11 shows that in a pipelined bus cycle the *effective address-access time* equals the duration of a complete bus cycle. This leads us to the benefit of the 80386DX's pipelined mode of bus operation over the nonpipelined mode of operation—that is, for a fixed address-access time (equal speed memory design), the 80386DX pipelined bus cycle will have a shorter duration than its nonpipelined bus cycle. This results in improved bus performance.

Another way of looking at this is to say that when using equal-speed memory designs, an 80386DX that uses a pipelined bus can be operated at a higher clock rate than a design that executes a nonpipelined bus cycle. Once again, the result is higher system performance.

Figure 16–11 shows that at completion of the bus cycle for address n, another bus cycle is initiated immediately for address $n + 1$. Sometimes another bus cycle will not be initiated immediately. For instance, if the 80386DX's prefetch queue is already full and the instruction that is currently being executed does not need to access operands in memory, no bus activity will take place. In this case, the bus goes into a mode of operation known as an *idle state* and no bus activity occurs. Figure 16–12 shows a sequence of bus activity in which an idle state exists between the bus cycles for addresses $n + 1$ and $n + 2$. The duration of a single idle state is equal to two CLK2 cycles.

Wait states can be inserted to extend the duration of the 80386DX's bus cycle. This is done in response to a request by an event in external hardware instead of an internal event such as a full queue. In fact, the $\overline{\text{READY}}$ input of the 80386DX is provided specifically for this purpose. This input is sampled in the later part of the T_2 state of every bus cycle to determine if the data transfer should be completed. Figure 16–13 shows that logic 1 at this input indicates that the current bus cycle should not be completed. As long as

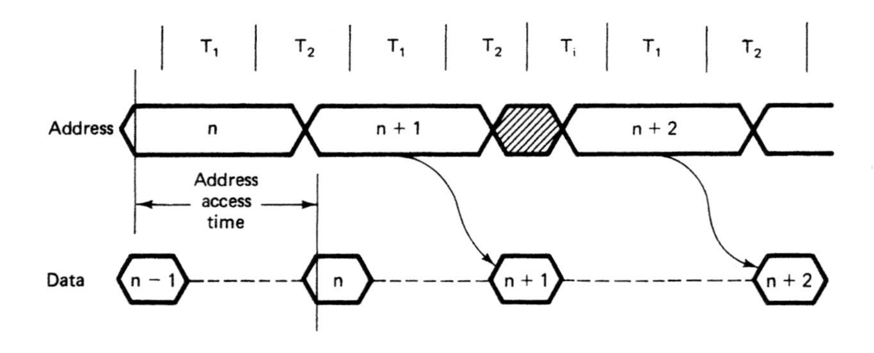

Figure 16–12 Idle states in bus activity.

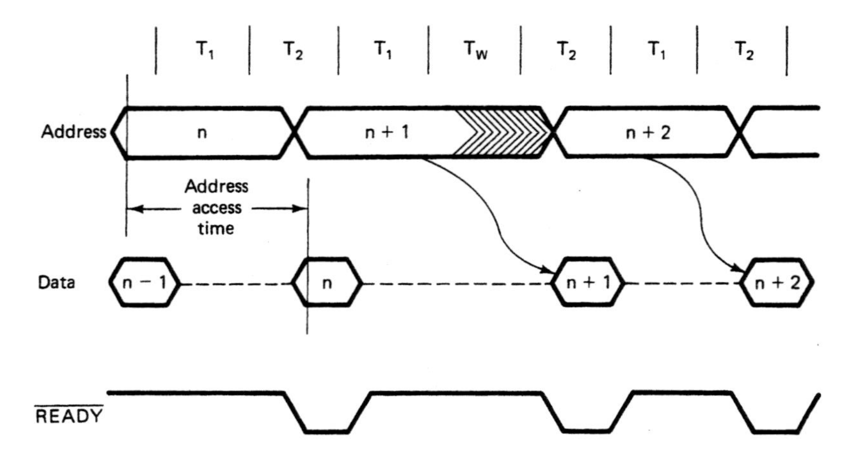

Figure 16–13 Bus cycle with wait states.

$\overline{READY}$ is held at the 1 level, the read or write data transfer does not take place and the current T_2 state becomes a wait state (T_w) to extend the bus cycle. The bus cycle is not completed until external hardware returns $\overline{READY}$ back to logic 0. This ability to extend the duration of a bus cycle permits the use of slow memory or I/O devices in the microcomputer system.

Nonpipelined Read Cycle Timing

The memory interface signals that occur when the 80386DX reads data from memory are shown in Fig. 16–14. This diagram shows two separate nonpipelined read cycles. They are *cycle 1,* which is performed without wait states, and *cycle 2,* which includes one wait state. Let us now trace through the events that take place in cycle 1 as data or instructions are read from memory.

The occurrence of all the signals in the read bus cycle timing diagram are illustrated relative to the two timing states, T_1 and T_2, of the 80386DX's bus cycle. The read operation starts at the beginning of phase 1 (ϕ_1) in the T_1 state of the bus cycle. At this moment, the 80386DX outputs the address of the double-word memory location to be accessed on address bus lines A_2 through A_{31}, outputs the byte-enable signals $\overline{BE}_0$ through $\overline{BE}_3$ that identify the bytes of the double word that are to be fetched, and switches address strobe ($\overline{ADS}$) to logic 0 to signal that a valid address is on the address bus. Looking at Fig. 16–14, we see that the address and bus cycle indication signals are maintained stable during the complete bus cycle; however, they must be latched into the external bus control logic circuitry synchronously with the pulse to logic 0 on $\overline{ADS}$. At the end of ϕ_2 of T_1, $\overline{ADS}$ is returned to its inactive 1 logic level.

Note in Fig. 16–14 that the bus cycle indication signals, $M/\overline{IO}$, $D/\overline{C}$, and $W/\overline{R}$, are also made valid at the beginning of ϕ_1 of state T_1. Figure 16–15 highlights the bus cycle indication codes that apply to a memory read cycle. Here we see that if code is being read from memory, $M/\overline{IO}$ $D/\overline{C}$ $W/\overline{R}$ equals 100. That is, signal $M/\overline{IO}$ is set to logic 1 to indi-

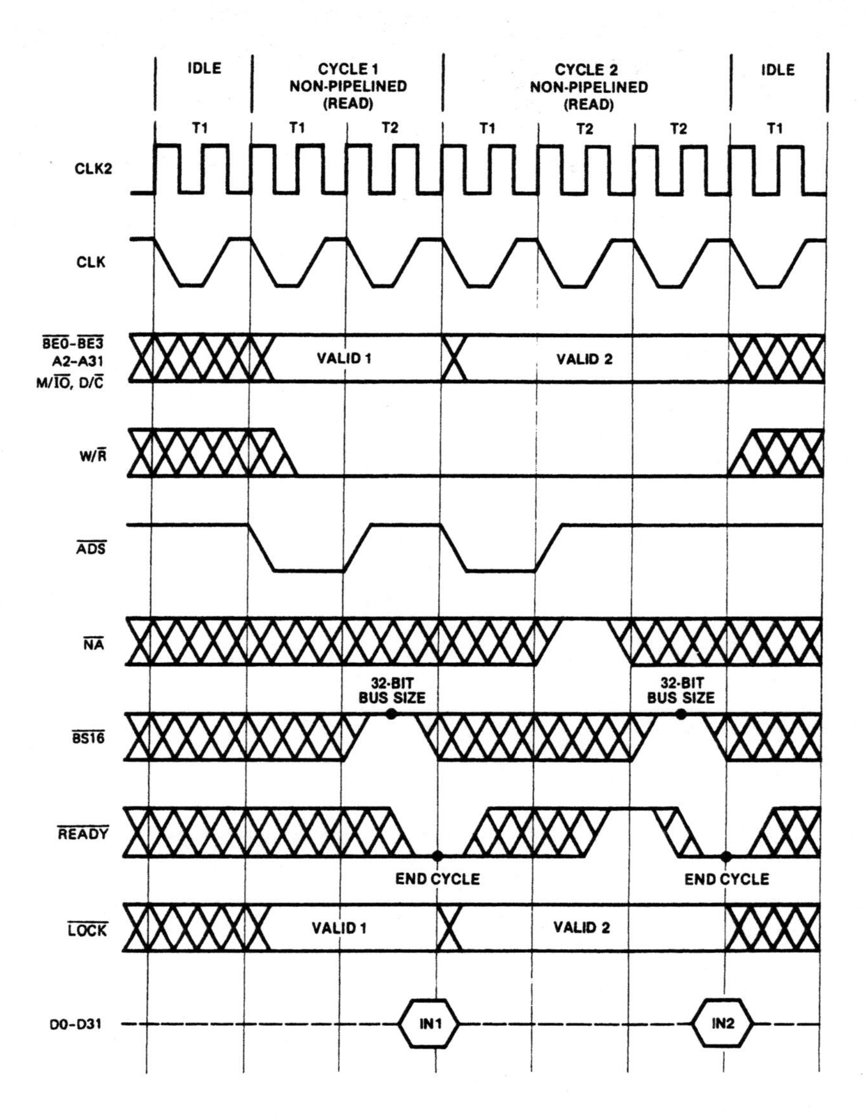

Figure 16–14 Nonpipelined read cycle timing. (Reprinted by permission of Intel Corporation. Copyright/Intel Corp. 1987)

cate to the circuitry in the memory interface that a memory bus cycle is in progress, D/$\overline{\text{C}}$ is set to 0 to indicate that code memory is to be accessed, and W/$\overline{\text{R}}$ is set to 0 to indicate that data are being read from memory.

At the beginning of ϕ_1 in T_2 of the read cycle, external circuitry must signal the 80386DX whether the bus is to operate in the 16- or 32-bit mode. In Fig. 16–14, we see that it does this with the $\overline{\text{BS16}}$ signal. The 80386DX samples this input in the middle of the T_2 bus cycle state. The 1 logic level shown in the timing diagram indicates that a 32-bit data transfer is to take place.

$M/\overline{IO}$	$D/\overline{C}$	$W/\overline{R}$	Type of Bus Cycle
0	0	0	Interrupt acknowledge
0	0	1	Idle
0	1	0	I/O data read
0	1	1	I/O data write
1	0	0	Memory code read
1	0	1	Halt/shutdown
1	1	0	Memory data read
1	1	1	Memory data write

Figure 16–15 Memory read bus cycle indication codes.

Note in Fig. 16–14 that at the end of T_2 the $\overline{READY}$ input is tested by the 80386DX. The logic level at this input signals whether the current bus cycle is to be completed or extended with wait states. The logic 0 at this input means that the bus cycle is to run to completion. For this reason we see that data available on data bus lines D_0 through D_{31} are read into the 80386DX at the end of T_2.

Nonpipelined Write Cycle Timing

The nonpipelined write bus cycle timing diagram, shown in Fig. 16–16, is similar to that given for a nonpipelined read cycle in Fig. 16–14. It includes waveforms for both a no-wait-state write operation (cycle 1) and a one-wait-state write operation (cycle 2). Looking at the write cycle waveforms, we find that the address, byte-enable, and bus cycle indication signals are output at the beginning of ϕ_1 of the T_1 state. All these signals are to be latched in external circuitry with the pulse at $\overline{ADS}$. The one difference here is that $W/\overline{R}$ is at the 1 logic level instead of 0. In fact, as shown in Fig. 16–17, the bus cycle indication code for a memory data write is $M/\overline{IO}$ $D/\overline{C}$ $W/\overline{R}$ equals 111; therefore, $M/\overline{IO}$ and $D/\overline{C}$ are also at the logic 1 level.

Let us now look at what happens on the data bus during a write bus cycle. Note in Fig. 16–16 that the 80386DX outputs the data to be written to memory onto the data bus at the beginning of ϕ_2 in the T_1 state. These data are maintained valid until the end of the bus cycle. In the middle of the T_2 state, the logic level of the $\overline{BS16}$ input is tested by the 80386DX and indicates that the bus is to be used in the 32-bit mode. Finally, at the end of T_2, $\overline{READY}$ is tested and found to be at its active 0 logic level. Since the memory subsystem has made $\overline{READY}$ logic 0, the write cycle is complete and the buses and control signal lines are prepared for the next write cycle.

Wait States in a Nonpipelined Memory Bus Cycle

Earlier we showed how wait states are used to lengthen the duration of the memory bus cycle of microprocessors. Wait states are inserted with the $\overline{READY}$ input signal. Upon request from an event in external hardware, for instance, slow memory, the $\overline{READY}$ input is switched to logic 1. This signals the 80386DX that the memory subsystem is not

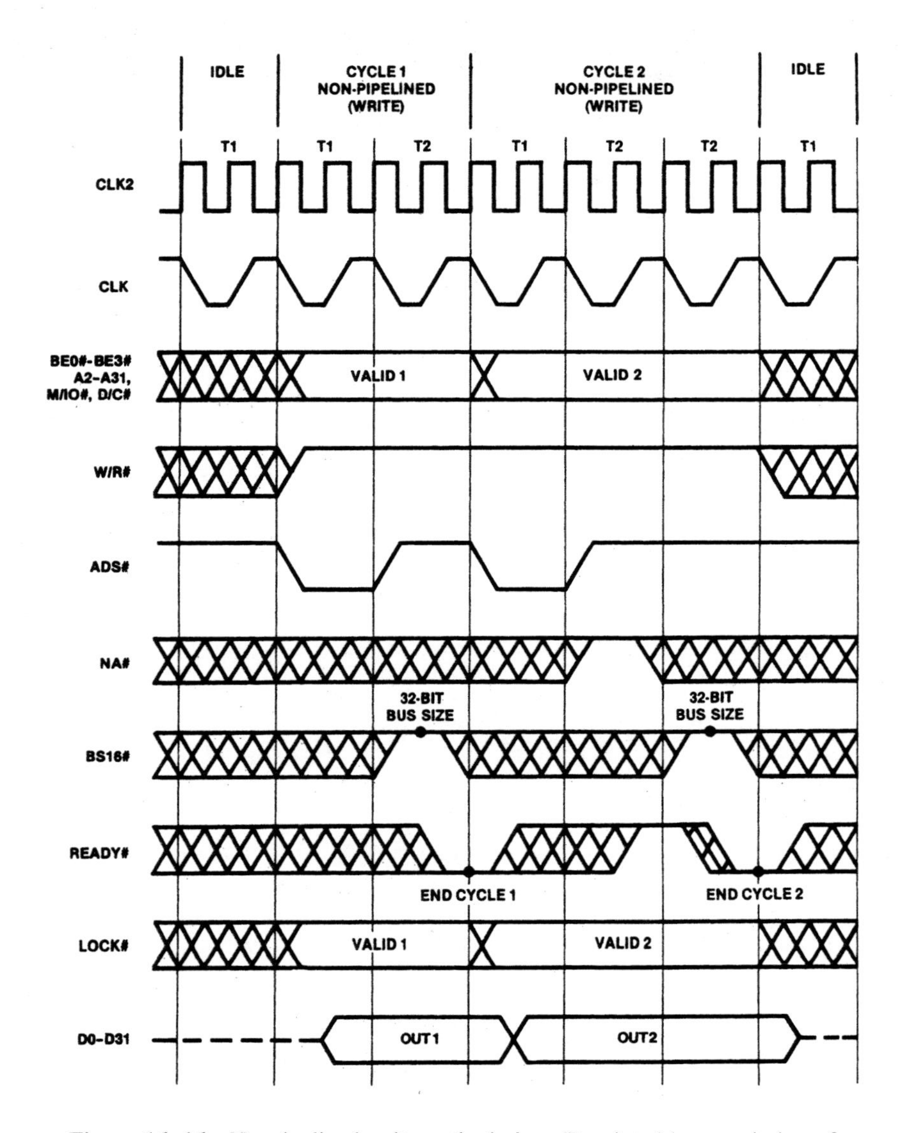

Figure 16–16 Nonpipelined write cycle timing. (Reprinted by permission of Intel Corporation. Corporation/Intel Corp. 1987)

ready and that the current bus cycle should not be completed. Instead, repeating the T_2 state extends it. Therefore, the duration of one wait state ($T_w = T_2$) equals 50 ns for 20-MHz clock operation.

Cycle 2 in Fig. 16–14 shows a read cycle extended by one wait state. Note that the address, byte-enable, and bus cycle indication signals are maintained throughout the wait-state period. In this way, the read cycle is not completed until $\overline{\text{READY}}$ is switched to logic 0 in the second T_2 state.

$M/\overline{IO}$	$D/\overline{C}$	$W/\overline{R}$	Type of bus cycle
0	0	0	Interrupt acknowledge
0	0	1	Idle
0	1	0	I/O data read
0	1	1	I/O data write
1	0	0	Memory code read
1	0	1	Halt/shutdown
1	1	0	Memory data read
1	1	1	Memory data write

Figure 16–17 Memory write bus cycle indication code.

EXAMPLE 16.6

If cycle 2 in Fig. 16–16 is for an 80386DX-20 running at full speed, what is the duration of the write bus cycle?

Solution

Each T state in the bus cycle of an 80386DX running at 20 MHz is 50 ns. Since the write cycle is extended by one wait state, the write cycle takes 150 ns.

Pipelined Read/Write Cycle Timing

Timing diagrams for both nonpipelined and pipelined read and write bus cycles are shown in Fig. 16–18. Here we find that the cycle identified as *cycle 3* is an example of a pipelined write bus cycle. Let us now look more closely at this bus cycle.

Remember that when pipelined addressing is in use, the 80386DX outputs the address information for the next bus cycle during the T_2 state of the current cycle. The signal next address ($\overline{NA}$) is used to signal the 80386DX that a pipelined bus cycle is to be initiated. This input is sampled by the 80386DX during any bus state when $\overline{ADS}$ is not active. In Fig. 16–18, we see that ($\overline{NA}$) is first tested as 0 (active) during T_2 of cycle 2. This nonpipelined read cycle is also extended with period T_{2P} because $\overline{READY}$ is not active. Note that the address, byte-enable, and bus cycle indication signals for cycle 3 become valid (identified as VALID 3 in Fig. 16–18) during this period and a pulse is produced at $\overline{ADS}$. This information is externally latched synchronously with $\overline{ADS}$ and decoded to produce bus enable and control signals. In this way, the memory access time for a zero-wait-state memory cycle has been increased.

Bus cycle 3 represents a pipelined write cycle. The data to be written to memory are output on D_0 through D_{31} at ϕ_2 of T_{1P} and remain valid for the rest of the cycle. Logic 0 on $\overline{READY}$ at the end of T_{2P} indicates that the write cycle is to be completed without wait states.

Looking at Fig. 16–18, we find that $\overline{NA}$ is also active during T_{1P} of cycle 3. This means that cycle 4 will also be performed with pipelined timing. Cycle 4 is an example of a zero-wait-state pipelined read cycle. In this case, the address information, bus cycle indication code, and address strobe are output during T_{2P} of cycle 3 (the previous cycle), and memory data are read into the MPU at the end of T_{2P} of cycle 4.

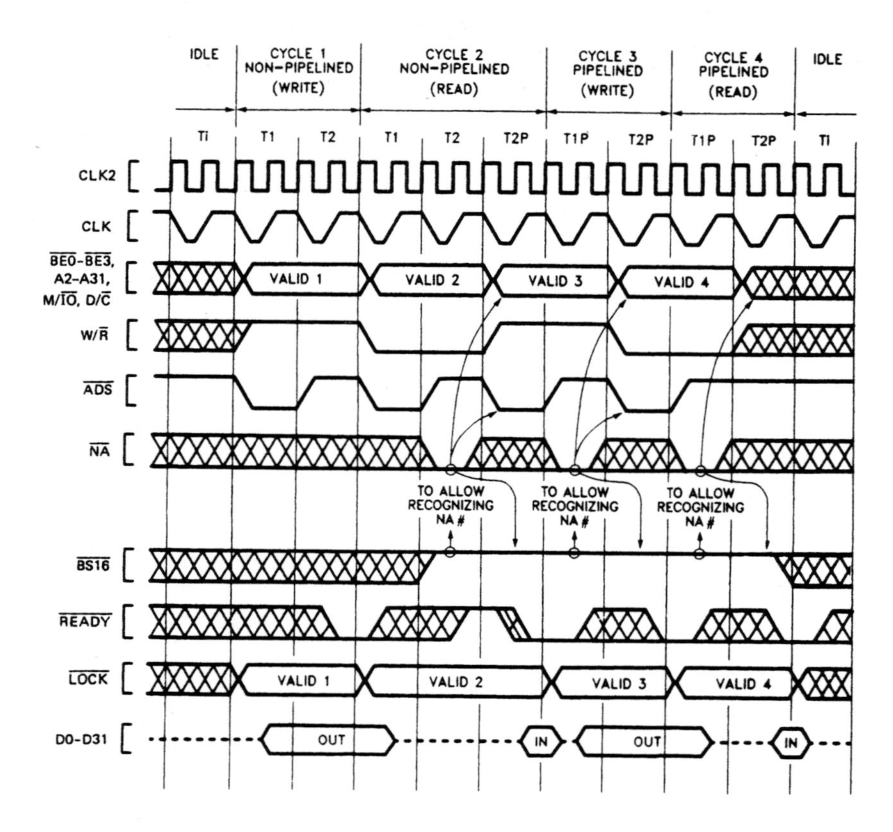

Figure 16–18 Pipelined read and write cycle timing. (Reprinted by permission of Intel Corporation. Copyright/Intel Corp. 1987)

▲ 16.5 MEMORY ORGANIZATION AND INTERFACE CIRCUITS

Earlier we indicated that in protected mode the 32-bit address bus of the 80386DX results in a 4Gbyte physical memory address space. As Fig. 16–19 shows, from a software point of view, this memory is organized as individual bytes over the address range from 00000000_{16} through $FFFFFFFF_{16}$. The 80386DX can also access data in this memory as words or double words.

Hardware Organization of the Memory Address Space

From a hardware point of view, the physical address space is implemented as four independent byte-wide banks, and each of these banks is 1Gbyte in size. In Fig. 16–20 the banks are identified as *bank 0, bank 1, bank 2,* and *bank 3.* Note that they correspond to addresses that produce byte-enable signals $\overline{BE}_0$, $\overline{BE}_1$, $\overline{BE}_2$, and $\overline{BE}_3$, respectively. Logic 0 at a byte-enable input selects the bank for operation. Looking at Fig. 16–20, we see that address bits A_2 through A_{31} are applied to all four banks in parallel. On the other

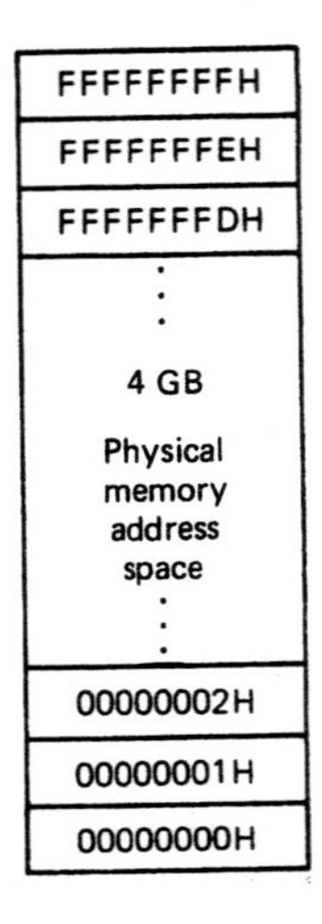

Figure 16–19 Physical address space.

hand, each memory bank supplies just eight lines of the 80386DX's 32-bit data bus. For example, byte data transfers for bank 0 take place over data bus lines D_0 through D_7, whereas byte data transfers for bank 3 are carried over data bus lines D_{24} through D_{31}.

When the 80386DX is operated in real mode, only the value on address lines A_2 through A_{19} and the $\overline{BE}$ signals are used to select the storage location to be accessed. For this reason, the physical address space is 1Mbyte in length, not 4Gbyte. The memory subsystem is once again partitioned into four banks, as shown in Fig. 16–20, but this time each bank is 256Kbyte in size.

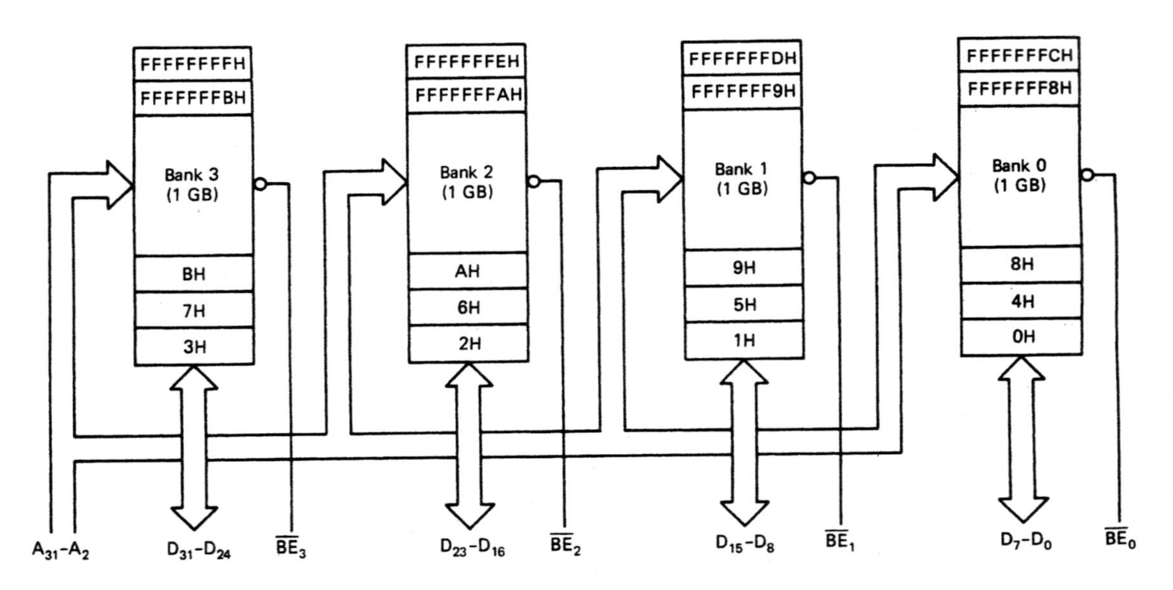

Figure 16–20 Hardware organization of the physical address space.

Figure 16–20 shows that in hardware the memory address space is physically organized as a sequence of double words. The address on lines A_2 through A_{31} selects the double-word storage location. Therefore, each aligned double word starts at a physical address that is a multiple of 4. For instance, in Fig. 16–20 aligned double words start at addresses 00000000_{16}, 00000004_{16}, 00000008_{16}, up through $FFFFFFFC_{16}$.

Each of the four bytes of a double word corresponds to one of the byte-enable signals. For this reason, they are each stored in a different bank of memory. In Fig. 16–20, we have identified the range of byte addresses that correspond to the storage locations in each bank of memory. For example, byte data accesses to addresses such as 00000000_{16}, 00000004_{16}, and 00000008_{16} all produce $\overline{BE_0}$, which enables memory bank 0, and the read or write data transfer takes place over data bus lines D_0 through D_7. Figure 16–21(a) illustrates how the byte at double-word aligned memory address X is accessed.

On the other hand, in Fig. 16–20 byte addresses 00000001_{16}, 00000005_{16}, and 00000009_{16} correspond to data held in memory bank 1. Figure 16–21(b) shows how the byte of data at address X + 1 is accessed. Notice that $\overline{BE_1}$ is made active to enable bank 1 of memory.

Most memory accesses produce more than one byte-enable signal. For instance, if the word of data beginning at aligned address X is read from memory, both $\overline{BE_0}$ and $\overline{BE_1}$ are generated. In this way, bank 0 and bank 1 of memory are enabled for operation. As Fig. 16–21(c) shows, the word of data is transferred to the MPU over data bus lines D_0 through D_{15}.

Let us now look at what happens when a double word of data is written to aligned double-word address X. As Fig. 16–21(d) shows, $\overline{BE_0}$, $\overline{BE_1}$, $\overline{BE_2}$, and $\overline{BE_3}$ are made 0 to enable all four banks of memory, and the MPU writes the data to memory over the complete data bus, D_0 through D_{31}.

All the data transfers we have described so far have been for what are called *double-word aligned data.* For each of the pieces of data, all the bytes existed within the same double word, that is, a double word that is on an address boundary equal to a multiple of four. The diagram in Fig. 16–22 illustrates a number of aligned words and double words of data. Byte, aligned word, and aligned double-word data transfers are all performed by the 80386DX in a single bus cycle.

It is not always possible to have all words or double words of data aligned at double-word boundaries. Figure 16–23 shows some examples of misaligned words and double words of data that can be accessed by the 80386DX. Note that word 3 consists of byte 3 that is in aligned double word 0 and byte 4 that is in aligned double word 4. Let us now look at how misaligned data are transferred over the bus.

The diagram in Fig. 16–24 illustrates a misaligned double-word data transfer. Here the double word of data starting at address X + 2 is to be accessed. However, this word consists of bytes X + 2 and X + 3 of the aligned double word at physical address X and bytes Y and Y + 1 of the aligned double word at physical address Y. Looking at the diagram, we see that $\overline{BE_0}$ and $\overline{BE_1}$ are active during the first bus cycle, and the word at address Y is transferred over D_0 through D_{15}. A second bus cycle automatically follows in which $\overline{BE_3}$ and $\overline{BE_4}$ are active, address X is put on the address bus, and the second word of data, X + 2 and X + 3, is carried over D_{16} through D_{31}. In this way we see that data transfers of misaligned words or double words take two bus cycles.

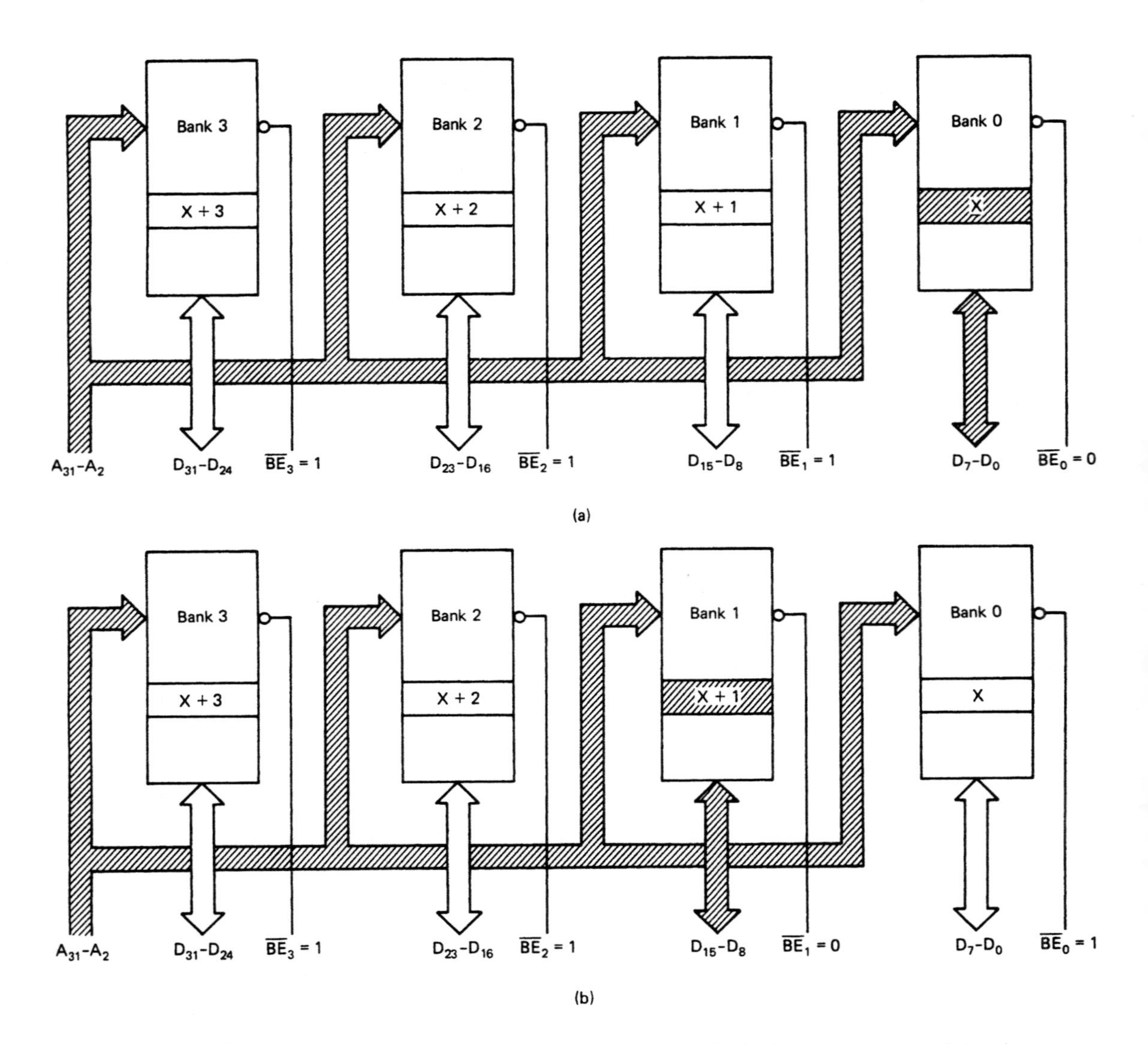

Figure 16–21 (a) Accessing a byte of data in bank 0. (b) Accessing a byte of data in bank 1. (c) Accessing a word of data in memory. (d) Accessing an aligned double word in memory.

EXAMPLE 16.7

Is the word at address $0000123F_{16}$ aligned or misaligned? How many bus cycles are required to read it from memory?

Solution

The first byte of the word is the fourth byte at aligned double-word address $0000123C_{16}$ and the second byte of the word is the first byte of the aligned double word at address

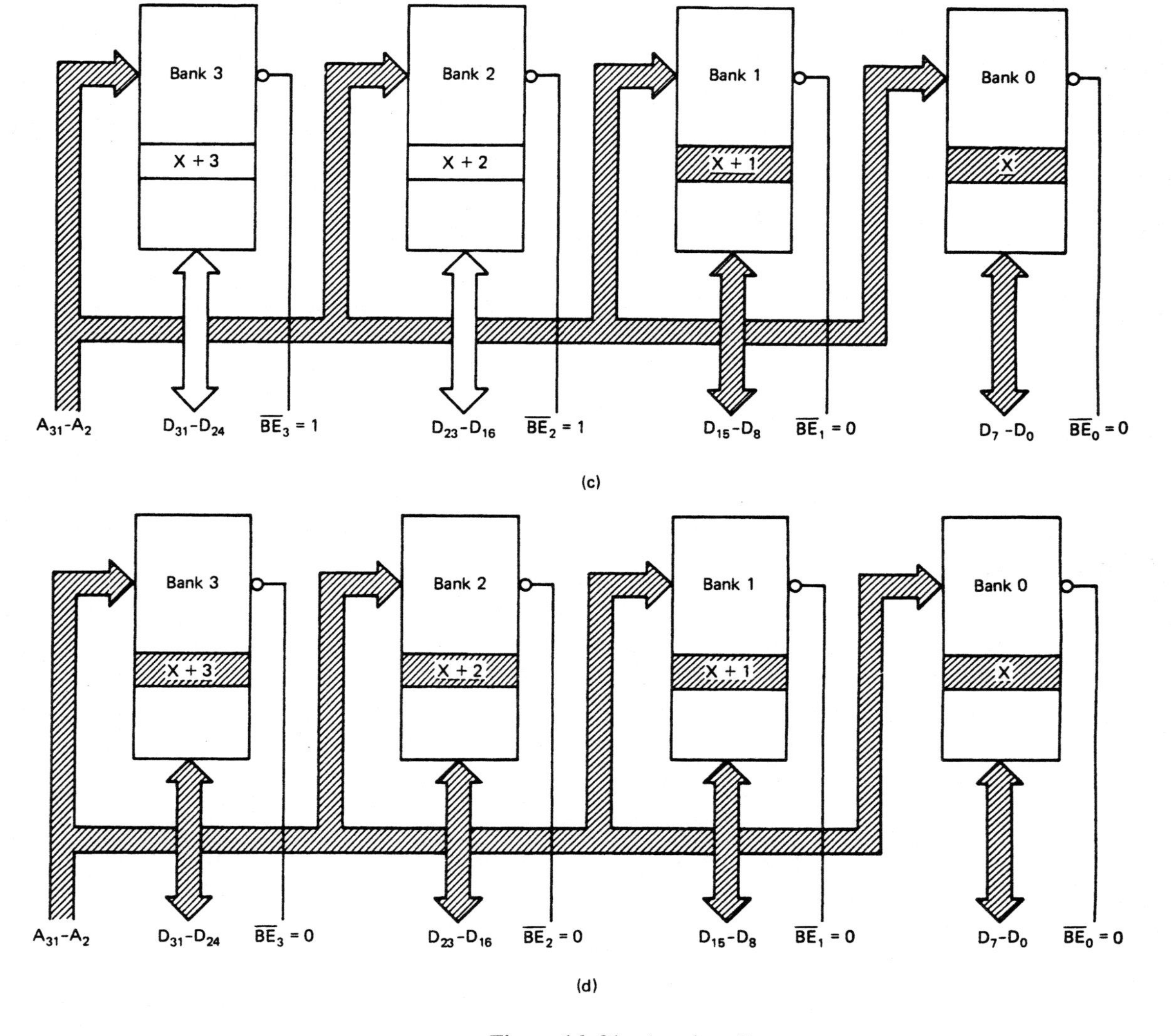

Figure 16–21 (continued)

00001240_{16}. Therefore, the word is misaligned and requires two bus cycles to be read from memory.

Memory Interface Circuitry

Figure 16–25 presents a memory interface diagram for a protected-mode, 80386DX-based microcomputer system. Here we find that the interface includes bus control logic, address bus latches and an address decoder, data bus transceiver/buffers, and

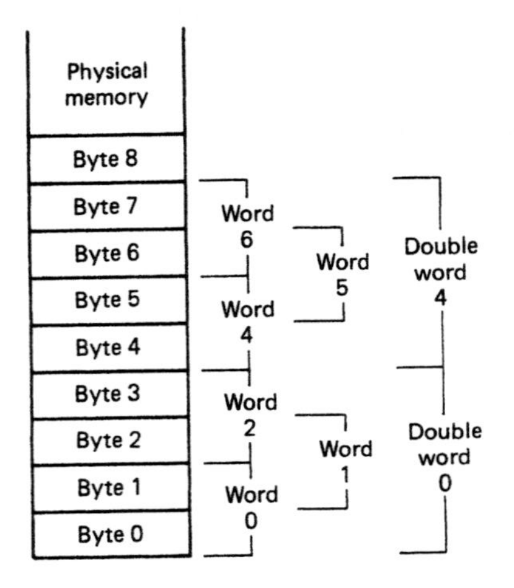

Figure 16–22 Examples of aligned data words and double words.

bank write control logic. The bus cycle indication signals, $M/\overline{IO}$, $D/\overline{C}$, and $W/\overline{R}$, which are output by the 80386DX, are supplied directly to the bus control logic. Here they are decoded to produce the command and control signals needed to control data transfers over the bus. Figs. 16–15 and 16–17 highlight the status codes that relate to the memory interface. For example, the code $M/\overline{IO}$ $D/\overline{C}$ $W/\overline{R}$ equal to 110 indicates that a data memory-read bus cycle is in progress. This code switches the $\overline{MRDC}$ command output of the bus control logic to logic 0. Note in Fig. 16–25 that $\overline{MRDC}$ is applied directly to the $\overline{OE}$ input of the memory subsystem.

Next let us look at how the address bus is decoded, buffered, and latched. Looking at Fig. 16–25, we see that address lines A_{29} through A_{31} are decoded to produce chip-enable outputs $\overline{CE}_0$ through $\overline{CE}_7$. These chip-enable signals are latched along with address bits A_2 through A_{28} and byte-enable lines $\overline{BE}_0$ through $\overline{BE}_3$ into the address latches. Note that the bus-control logic receives $\overline{ADS}$ and the bus cycle indication code as inputs and produces the address latch enable (ALE), memory read command ($\overline{MRDC}$),

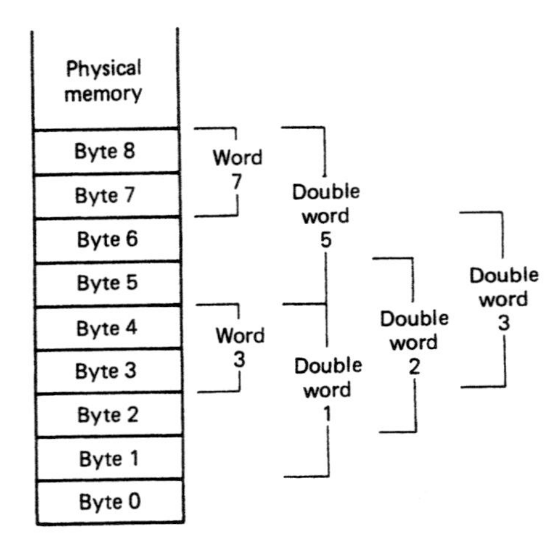

Figure 16–23 Examples of misaligned data words and double words.

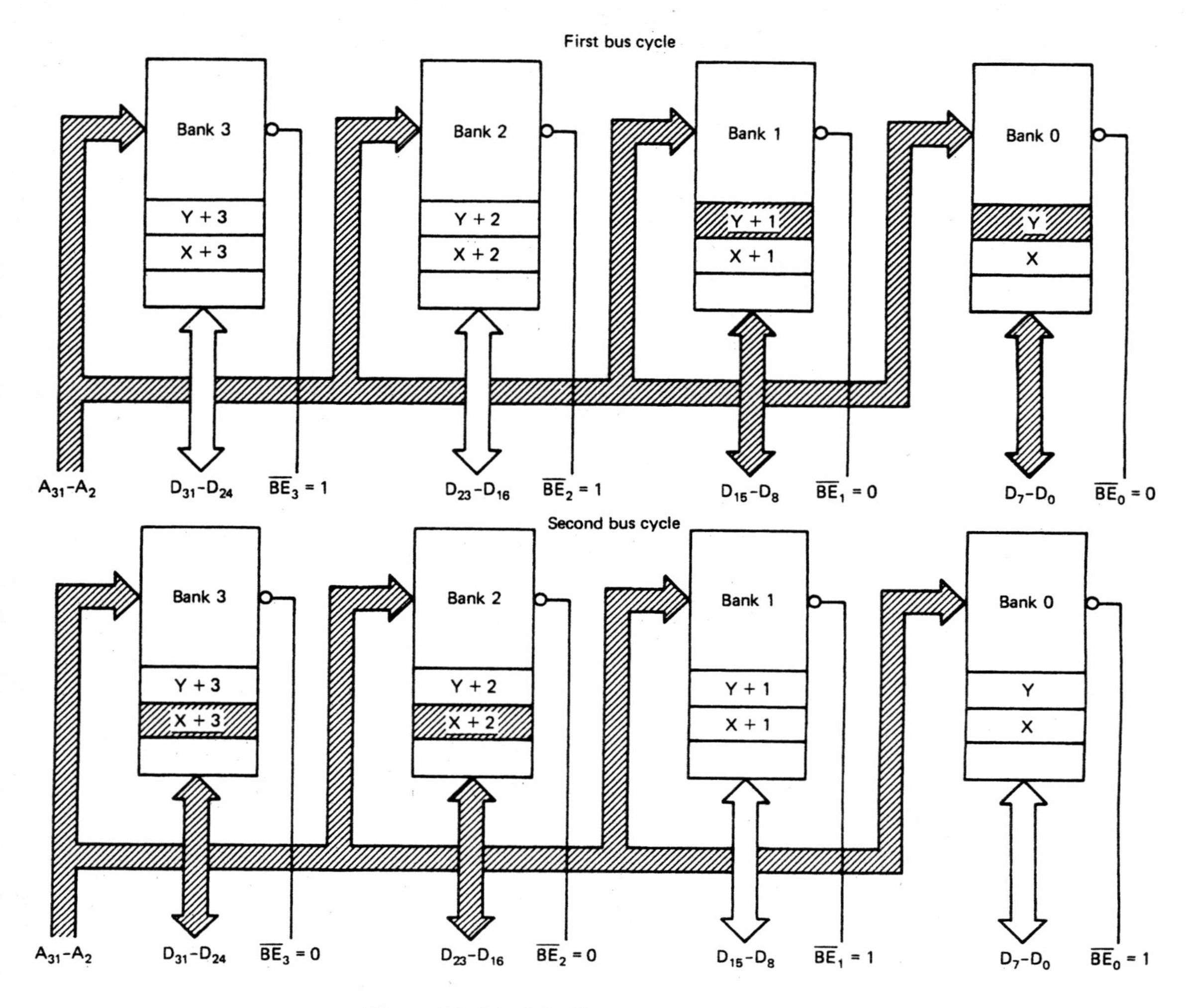

Figure 16–24 Misaligned double-word data transfers.

and memory write command ($\overline{MWTC}$) signals at its output. ALE is applied to the CLK input of the latches and strobes the bits of the address, byte-enable, and chip-enable signals into the address latches. These signals are buffered by the address latch devices; the address and chip enables are output directly to the memory subsystem, and the byte enables are supplied as inputs to the bank write control logic circuit.

This part of the memory interface demonstrates one of the benefits of the 80386DX's pipelined bus mode. When working in the pipelined mode, the 80386DX actually outputs the address in the T_2 state of the prior bus cycle. Therefore, by putting the address decoder before the address latches instead of after, the code at address lines A_{28} through A_{31} can be fully decoded and stable prior to the T_1 state of the next bus cycle. In this way, the access time of the memory subsystem is reduced.

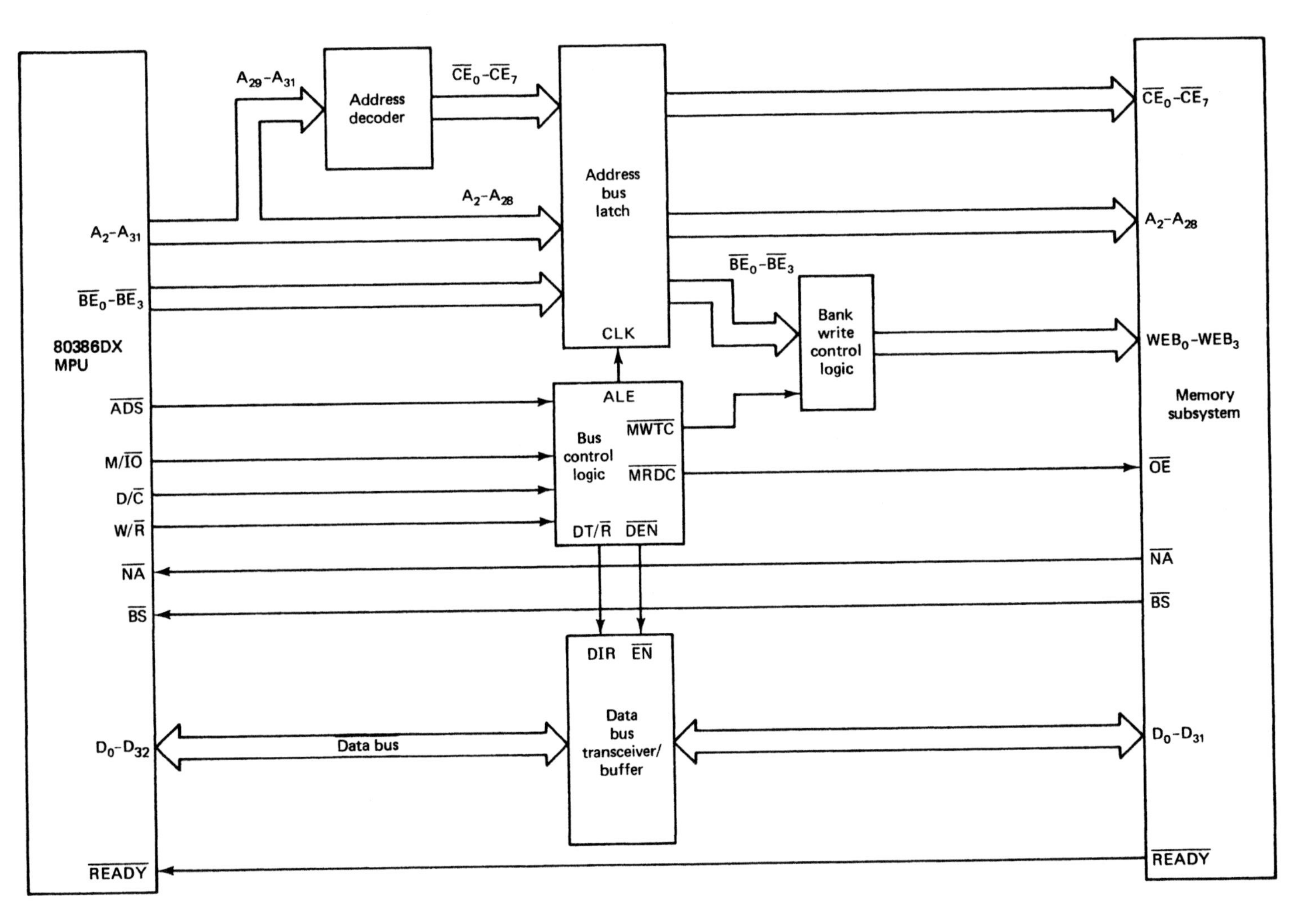

Figure 16–25 Memory interface block diagram.

During read bus cycles, the $\overline{MRDC}$ output of the bus control logic enables the data at the outputs of the memory subsystem onto data bus lines D_0 through D_{31}. The 80386DX will read the appropriate byte, word, or double word of data. On the other hand, during write operations to memory, the bank write control logic determines into which of the four memory banks the data are written. This depends on whether a byte, word, or double-word data transfer is taking place over the bus.

Note in Fig. 16–25 that the latched byte-enable signals $\overline{BE_0}$ through $\overline{BE_3}$ are gated with the memory-write command signal $\overline{MWTC}$ to produce a separate write-enable signal for each of the four banks of memory. These signals are denoted as $\overline{WEB_0}$ through $\overline{WEB_3}$ in Fig. 16–25. For example, if a word of data is to be written to memory over data bus lines D_0 through D_{15}, $\overline{WEB_0}$ and $\overline{WEB_1}$ are switched to their active 0 logic level.

The bus transceivers control the direction of data transfer between the MPU and memory subsystem. In Fig. 16–25, the operation of the transceivers is controlled by the data-transmit/receive (DT/$\overline{R}$) and data bus-enable ($\overline{DEN}$) outputs of the bus control logic. $\overline{DEN}$ is applied to the enable ($\overline{EN}$) input of the transceivers and enables them for operation. This happens during all read and write bus cycles. DT/$\overline{R}$ selects the direction of data transfer through the transceivers. Note that it is supplied to the DIR input of the devices. When a read cycle is in process, DT/$\overline{R}$ is set to 0 and data are passed from the memory subsystem to the MPU. On the other hand, when a write cycle is taking place, DT/$\overline{R}$ is switched to logic 1 and data are carried from the MPU to the memory subsystem.

▲ 16.6 INPUT/OUTPUT INTERFACE CIRCUITS AND BUS CYCLES

In Section 16.5 we studied the memory interface of the 80386DX microprocessor. Here we will examine another important interface of the 80386DX microcomputer system, the input/output interface.

Input/Output Interface and I/O Address Space

The input/output interface of the 80386DX-based microcomputer permits it to communicate with the outside world. The way in which the 80386DX and 80386SX MPUs deal with input/output circuitry is similar to the way in which they interface with memory circuitry. That is, input/output data transfers also take place over the data bus. This parallel bus permits easy interface to LSI peripheral devices such as parallel I/O expanders, interval timers, and serial communication controllers. Let us continue by looking at how the 80386DX interfaces to its I/O subsystem.

Figure 16–26 shows a typical I/O interface circuit for an 80386DX-based microcomputer system. Note that the interface between the microprocessor and I/O subsystem includes the bus controller logic, an I/O address decoder, I/O address latches, I/O data bus transceiver/buffers, I/O bank write control logic, and the I/O subsystem. An example of a typical device used in the I/O subsystem is a programmable peripheral interface (PPI) IC, such as the 82C55A. This type of device can be used to implement an interface that employs parallel input and output ports. Some examples are a keyboard/display interface circuit and a parallel printer interface circuit. Let us now look at the function of each of the blocks in this circuit more closely.

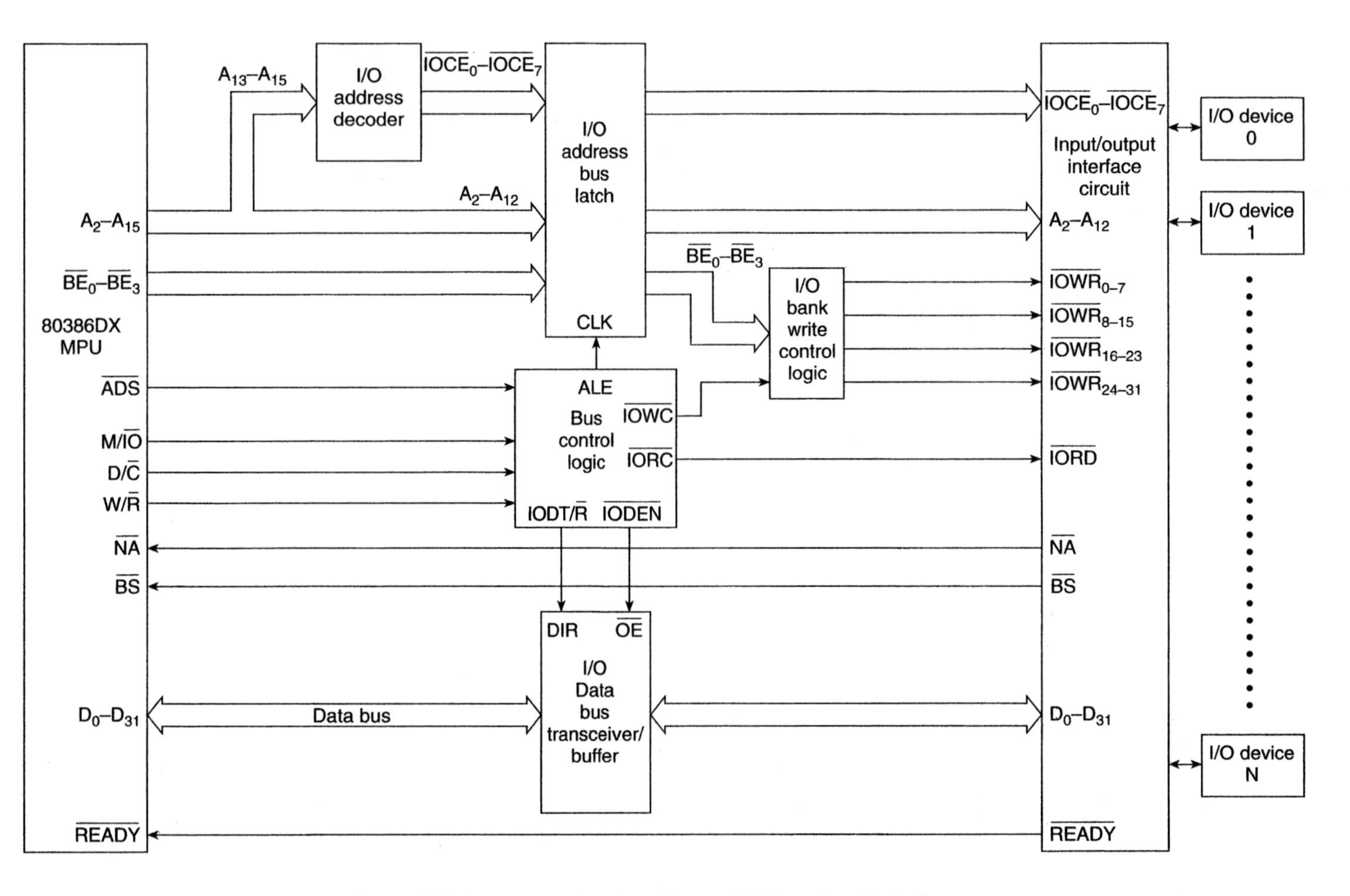

Figure 16–26 Byte, word, and double-word I/O interface block diagram.

The I/O interface shown in Fig. 16–26 is designed to support 8-, 16-, and 32-bit I/O data transfers. The I/O device that is accessed for input or output of data is selected by an *I/O address.* This address is specified as part of the instruction that performs the I/O operation. Just as for the 8086 architecture, the 80386DX's I/O addresses are 16 bits in length and support 64K independent byte-wide I/O ports. The address on lines A_2 through A_{15} is used to specify the double-word I/O port that is to be accessed. When data are output to output ports, the logic levels of $\overline{BE}_0$ through $\overline{BE}_3$ determine which byte-wide port or ports are enabled for operation. The more significant address bits, A_{16} through A_{31}, are held at the 0 logic level during the address period of all I/O bus cycles. The 80386DX signals external circuitry that an I/O address is on the bus by switching its $M/\overline{IO}$ output to logic 0.

Note in the circuit diagram that the I/O address decoder decodes the part of the I/O address that is output on address lines A_2 through A_{15} of the 80386DX. The bits of the address that are decoded produce I/O chip-enable signals for the individual I/O devices. For instance, Fig. 16–26 shows that with three address bits, A_{13} through A_{15}, enough chip-enable outputs are produced to select up to eight I/O devices. Note that the outputs of the I/O address decoder are labeled $\overline{IOCE}_0$ through $\overline{IOCE}_7$. The I/O chip-enable signals are latched along with the address and byte-enable signals in the I/O address latches. Latching of this information is achieved with a pulse at the ALE output of the bus control logic. If a microcomputer employs a very simple I/O subsystem, it may be possible to eliminate the address decoder and simply use some of the latched high-order address bits as I/O enable signals.

In Fig 16–26, all the low-order address bits, A_2 through A_{12}, are latched and sent directly to the I/O devices. Typically, these address bits are used to select the register within the peripheral device that is to be accessed. For example, with just four of these address lines, we can select any one of 16 registers.

EXAMPLE 16.8

If address bits A_7 through A_{15} are used directly as chip-enable signals and address lines A_2 through A_6 are used as register-select inputs for the I/O devices, how many I/O devices can be used, and what is the maximum number of registers that each device can contain?

Solution

When latched into the address latch, the nine address lines produce the nine I/O chip-enable signals, $\overline{IOCE}_0$ through $\overline{IOCE}_8$, for I/O devices 0 through 8. The lower five address bits are able to select one of $2^5 = 32$ registers for each peripheral IC.

During input and output bus cycles, data are passed between the selected register in the enabled I/O device and the 80386DX over data bus lines D_0 through D_{31}. Earlier we pointed out that the 80386DX can input or output data in byte-wide, word-wide, or double-word-wide format. Just as for the memory interface, the signals $\overline{BE}_0$ through $\overline{BE}_3$ are used to signal which byte or bytes of data are being transferred over the bus. Again logic

0 at $\overline{BE}_0$ identifies that a byte of data is input or output over data bus lines D_0 through D_7. On the other hand, logic 0 at $\overline{BE}_3$ means that a byte-data transfer is taking place over bus lines D_{24} through D_{31}.

Many of the peripheral ICs used in the 80386DX microcomputer have a byte-wide data bus. For this reason they are normally attached to the lower part of the data bus. That is, they are connected to data bus lines D_0 through D_7. If this is done in the circuit of Fig. 16–26, all I/O addresses must be scaled by 4. This is because the first byte I/O address that corresponds to a byte transfer across the lower eight data bus lines is 0000_{16}, the next byte address, which represents a byte transfer over D_0 through D_7, is 0004_{16}, the third I/O address is 0008_{16}, and so on. In fact, if only 8-bit peripherals are used in the 80386DX microcomputer system and they are all attached to I/O data bus lines D_0 through D_7, the byte-enable signals are not needed in the I/O interface. In this case, address bit A_2 is used as the least significant bit of the I/O address and A_{15} the most significant bit. Therefore, from a hardware point of view, the I/O address space appears as 16K contiguous byte-wide storage locations over the address range from

$$A_{15} \ldots A_3A_2 = 00000000000000_2$$

to

$$A_{15} \ldots A_3A_2 = 11111111111111_2$$

This puts the burden on software to assure that bytes of data are input or output only for addresses that are a multiple of 4 and correspond to a data transfer over data bus lines D_0 through D_7.

As in the memory interface, bus control logic is needed to produce the control signals for the I/O interface. Figure 16–26 shows that the inputs of the bus control logic are the bus cycle indication signals that are output by the 80386DX on $M/\overline{IO}$, $D/\overline{C}$, and $W/\overline{R}$. They are decoded to produce the I/O read command ($\overline{IORC}$) and I/O write command ($\overline{IOWC}$) outputs. $\overline{IORC}$ is applied directly to the input/output read ($\overline{IORD}$) input of the I/O devices and tells them when data are to be input to the MPU. In this case, 32-bit data are always put on the data bus. However, the 80386DX inputs only the appropriate byte, word, or double word. On the other hand, $\overline{IOWC}$ is gated with the $\overline{BE}$ signals to produce a separate write-enable signal for each byte of the data bus. They are labeled $\overline{IOWR}_{0\text{-}7}$, $\overline{IOWR}_{8\text{-}15}$, $\overline{IOWR}_{16\text{-}23}$, and $\overline{IOWR}_{24\text{-}31}$. These signals are needed to support writing of 8-, 16-, or 32-bit data through the interface.

The bus-control logic section also produces the signals needed to latch the address and set up the data bus for an input and output data transfer. The data bus transceiver/buffers control the direction of data transfers between the 80386DX and I/O devices. They are enabled for operation when their output-enable ($\overline{OE}$) input is switched to logic 0. Note that the signal *I/O data bus enable* ($\overline{IODEN}$) is applied to the $\overline{OE}$ inputs.

The logic level of the DIR input determines the direction in which data are passed through the transceivers. This input is supplied by the *I/O data-transmit/receive* ($IODT/\overline{R}$) output of the bus-control logic. During all input cycles, $IODT/\overline{R}$ is logic 0 and the transceivers are set to pass data from the selected I/O device to the 80386DX. On the other hand, during output cycles, $IODT/\overline{R}$ is switched to logic 1 and data passes from the 80386DX to the I/O device.

Figure 16–27 presents another input/output interface diagram. This circuit includes an I/O bank select decoder in the data bus interface. This circuit is used to multiplex the 32-bit data bus of the 80386DX to an 8-bit I/O data bus for connection to 8-bit peripheral devices. By using this circuit configuration, data can be input from or output to all 64K contiguous byte addresses in the I/O address space. In this case, hardware, instead of software, assures that byte data transfers to consecutive byte I/O addresses are performed to contiguous byte-wide I/O ports. The I/O bank-select decoder circuit maps bytes of data from the 32-bit data bus to the 8-bit I/O data bus. It does this by assuring that only one byte-enable ($\overline{BE}$) output of the 80386DX is active. That is, it checks to assure that a byte-input or byte-output operation is in progress. If more than one of the $\overline{BE}$ inputs of the decoder is active, none of the $\overline{OE}$ outputs of the decoder is produced, and the data transfer does not take place. Now the I/O address 0000_{16} corresponds to a I/O cycle over data bus lines D_0 through D_7 to an 8-bit peripheral attached to I/O data bus lines, IOD_0 through IOD_7; 0001_{16} corresponds to an input or output of a byte of data for the peripheral over lines D_8 through D_{15}; 0002_{16} represents a byte I/O transfer over D_{16} through D_{23}; and finally, 0003_{16} accompanies a byte transfer over data bus lines D_{24} through D_{31}. That is, even though the byte of data for addresses 0000_{16} through 0003_{16} are output by the 80386DX on different parts of its data bus, they are all multiplexed in external hardware to the same 8-bit I/O data bus, IOD_0 through IOD_7. In this way, the addresses of the peripheral's registers no longer need to be scaled by four in software.

Input and Output Bus Cycle Timing

We just found that the I/O interface signals of the 80386DX microcomputer are essentially the same as those involved in the memory interface. In fact, the function, logic levels, and timing of all signals other than the $M/\overline{IO}$ are identical to those already described for the memory interface in Section 16.5.

The timing diagram in Fig. 16–28 shows some *nonpipelined input* and *output bus cycles.* Looking at the waveforms for the first input/output bus cycle, cycle 1, we see that it represents a zero-wait-state input bus cycle. Note that the byte-enable signals $\overline{BE}_0$ through $\overline{BE}_3$, the address lines A_2 through A_{15}, the bus cycle indication signals $M/\overline{IO}$, $D/\overline{C}$, and $W/\overline{R}$, and address status ($\overline{ADS}$) signal are all output at the beginning of the T_1 state. This time the 80386DX switches $M/\overline{IO}$ to logic 0, $D/\overline{C}$ to logic 1, and $W/\overline{R}$ to logic 0 to signal external circuitry that an I/O data input bus cycle is in progress.

As shown in the block diagram in Fig. 16–27, the bus cycle indication code is input to the bus-control logic. An input of $M/\overline{IO}$ $D/\overline{C}$ $W/\overline{R}$ equals 010 initiates an I/O input bus-control sequence. Let us continue with the sequence of events that takes place in external circuitry during the input cycle. First, the bus-control logic outputs a pulse to the 1 logic level on ALE. As the circuit in Fig. 16–27 shows, this pulse is used to latch the address information into the I/O address latch devices. The decoded part of the latched address ($\overline{IOCE}_0$ through $\overline{IOCE}_7$) selects the I/O device to be accessed, and the code on

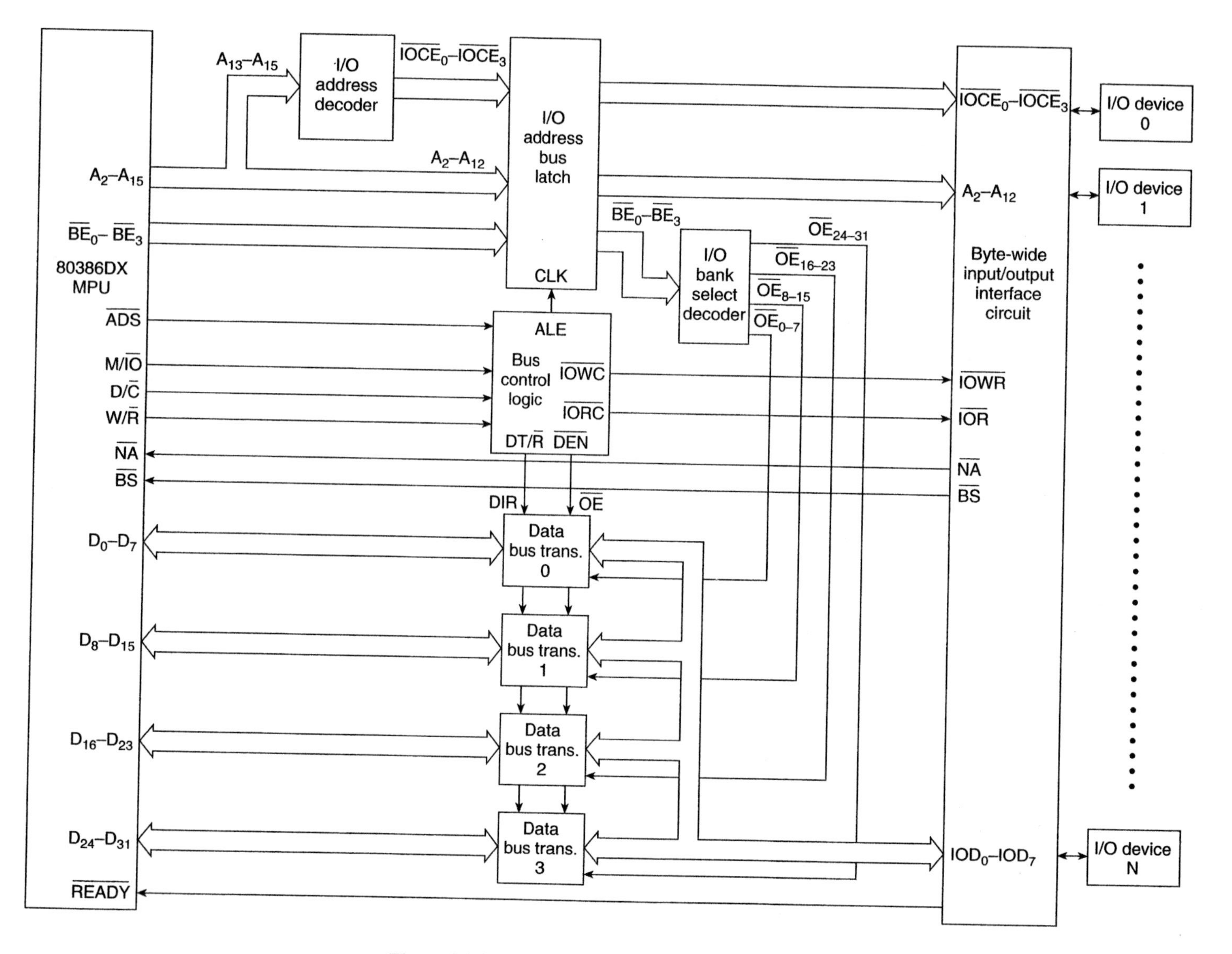

Figure 16–27 Byte-wide I/O interface block diagram.

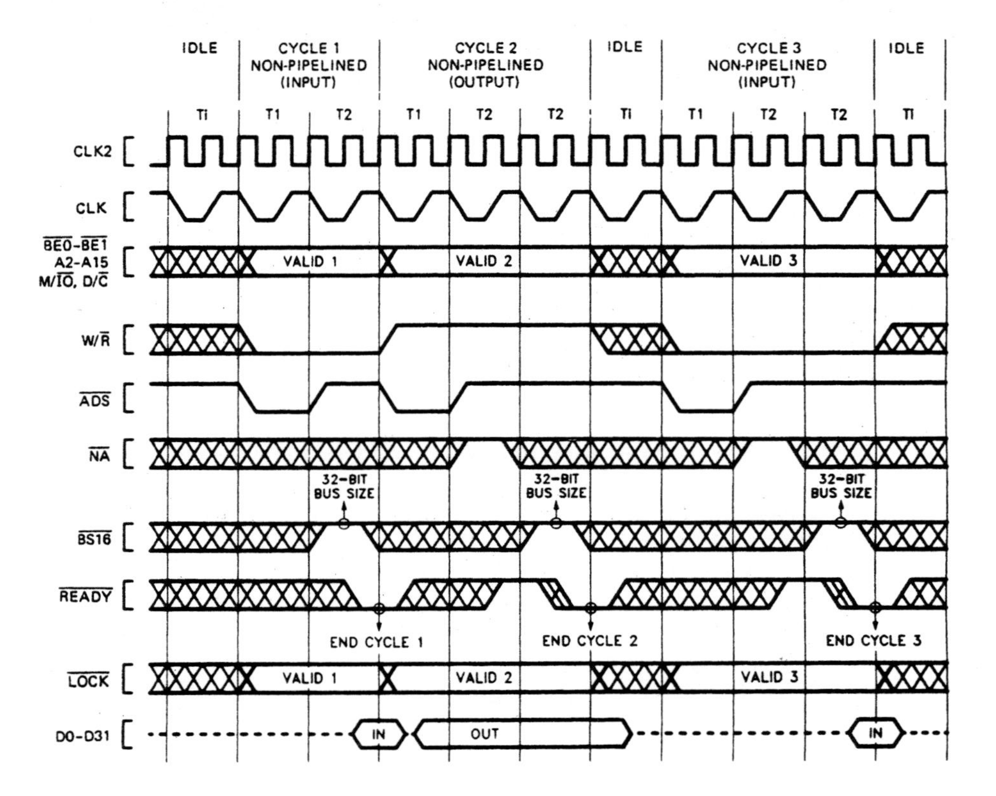

Figure 16–28 I/O read and I/O write bus cycles. (Reprinted by permission of Intel Corporation. Copyright/Intel Corp. 1987)

the lower address lines selects the register that is to be accessed. Later in the bus cycle, $\overline{\text{IORC}}$ is switched to logic 0 to signal the enabled I/O device that data are to be input to the MPU. In response to $\overline{\text{IORC}}$, the enabled input device puts the data from the addressed register onto the data bus. A short time later, IODT/$\overline{\text{R}}$ is switched to logic 0 to set the data bus transceivers to the input direction, and then the transceivers are enabled as $\overline{\text{IODEN}}$ is switched to logic 0. At this point the data from the I/O device are available on the 80386DX's data bus.

In the waveforms shown in Fig. 16–28, we see that at the end of the T_2 state the 80386DX tests the logic level at its ready input to determine if the I/O bus cycle should be completed or extended with wait states. As Fig. 16–28 shows, $\overline{\text{READY}}$ is at its active 0 logic level when sampled. Therefore, the 80386DX inputs the data off the bus. Finally, the bus-control logic returns $\overline{\text{IORC}}$, $\overline{\text{IODEN}}$, and IODT/$\overline{\text{R}}$ to their inactive logic levels, and the input bus cycle is finished.

Cycle 3 in Fig. 16–28 is also an input bus cycle. However, looking at the $\overline{\text{READY}}$ waveform, we find that this time it is not logic 0 at the end of the first T_2 state. Therefore, the input cycle is extended with a second T_2 state (wait state). Since some of the peripheral devices used with the 80386DX are older, slower devices, it is common to have several wait states in I/O bus cycles.

EXAMPLE 16.9

If the 80386DX that is executing cycle 3 in Fig. 16–28 is running at 20 MHz, what is the duration of this input cycle?

Solution

An 80386DX that is running at 20 MHz has a T state equal to 50 ns. Since the input bus cycle takes three T states, its duration is 150 ns.

Looking at the output bus cycle, cycle 2 in the timing diagram in Fig. 16–28, we see that the 80386DX puts the data to be output onto the data bus at the beginning of ϕ_2 in the T_1 state. This time the bus-control logic switches $\overline{IODEN}$ to logic 0 and maintains $IODT/\overline{R}$ at the 1 level for transmit mode. From Fig. 16–26, we find that since $\overline{IODEN}$ is logic 0 and $IODT/\overline{R}$ is 1, the transceivers are enabled and set up to pass data from the 80386DX to the I/O devices. Therefore, the data output on the bus are available on the data inputs of the enabled I/O device. Finally, the signal $\overline{IOWC}$ is switched to logic 0. It is gated with $\overline{BE_0}$ through $\overline{BE_3}$ in the I/O bank write-control logic to produce the needed bank write-enable signals. These signals tell the I/O device that valid output data are on the bus. Now the I/O device must read the data off the bus before the bus control logic terminates the bus cycle. If the device cannot read data at this rate, it can hold $\overline{READY}$ at the 1 logic level to extend the bus cycle.

EXAMPLE 16.10

If the output bus cycles performed to byte-wide ports by an 80386DX running at 20 MHz are to be completed in a minimum of 250 ns, how many wait states are needed?

Solution

Since each T state is 50 ns in duration, the bus cycle must last at least

$$\text{number of T states} = 250\text{ ns}/50\text{ ns} = 5$$

A zero-wait-state output cycle lasts just two T states; therefore, all output cycles must include three wait states.

Similar to memory, the I/O bus cycle requirements exist for data transfers for aligned and unaligned I/O ports. That is, all word and double-word data transfers to aligned port addresses take place in only one bus cycle. However, two bus cycles are required to perform data transfers for unaligned 16- or 32-bit I/O ports.

Protected-Mode Input/Output

When the 80386DX is in the protected-address mode, the input/output instructions can be executed only if the current privilege level is greater than or equal to the I/O privilege level (IOPL). That is, the numerical value of CPL must be lower than or equal to the numerical value of IOPL. Remember that IOPL is defined by the code in bits 12 and 13 of the flags register. If the current privilege level is less than IOPL, the instruction is not executed; instead, a general protection fault occurs. The general protection fault is an example of an 80386DX exception and is examined in more detail in the next section.

In Chapter 15 we indicated that the task state segment (TSS) of a task includes a section known as the *I/O permission bit map.* This I/O permission bit map provides a second protection mechanism for the protected-mode I/O address space. Remember that the size of the TSS segment is variable. Its size is specified by the limit in the TSS descriptor. Figure 16–29 shows a typical task state segment. Here we see that the 16-bit *I/O map base* offset, which is held at word offset 66_{16} in the TSS, identifies the beginning of the I/O permission bit map. The limit field in the descriptor for the TSS sets the upper end of the bit map. Let us now look at what the bits in the I/O permission bit map stand for.

Figure 16–30 shows a more detailed representation of the I/O permission bit map. Note that it contains one bit position for each of the 65,536 byte-wide I/O ports in the 80386DX's I/O address space. In the bit map we find that the bit position that corresponds to I/O port 0 (I/O address 0000_{16}) is the least significant bit at the address defined with the I/O bit map base offset. The remaining bits in this first double word in the map represent I/O ports 1 through 31. Finally, the last bit in the table, which corresponds to port 65,535 and I/O address $FFFF_{16}$, is the most significant bit in the double word located at an offset of $1FFC_{16}$ from the I/O bit map base. Figure 16–30 shows that the byte address that follows the map must always contain FF_{16}. This is the least significant byte in the last double word of the TSS. The value of the I/O map base offset must be less than $DFFF_{16}$; otherwise the complete map may not fit within the TSS.

Using this bit map, restrictions can be put on input/output operations to each of the 80386DX's 65,536 I/O port addresses. In protected mode, the bit for the I/O port in the I/O permission map is checked only if the CPL, when the I/O instruction is executed, is less privileged than the IOPL. If logic 0 is found in a bit position, it means that an I/O

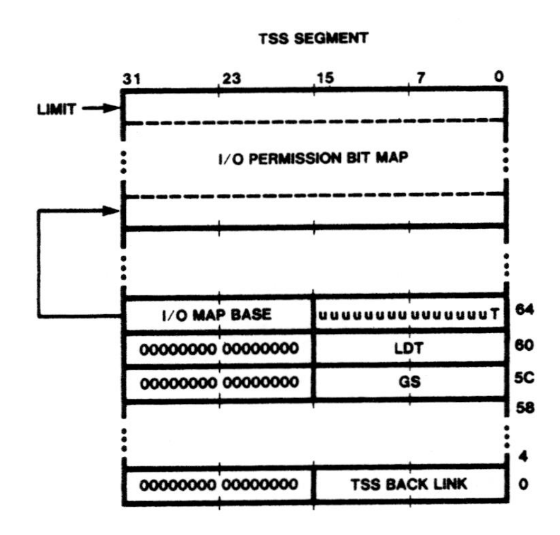

Figure 16–29 Location of the I/O permission bit map in the TSS. (Reprinted by permission of Intel Corporation. Copyright/Intel Corp. 1986)

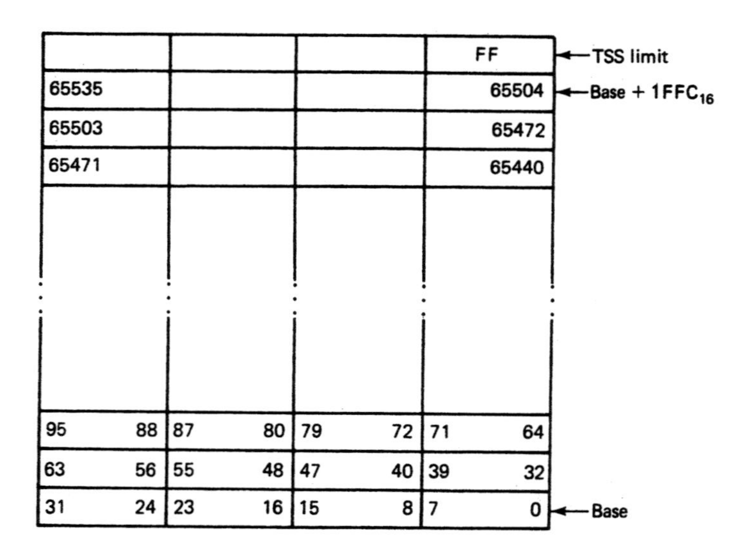

Figure 16–30 Contents of the I/O permission bit map.

operation can be performed to the port address. On the other hand, logic 1 inhibits the I/O operation. Any attempt to input or output data for an I/O address marked with a 1 in the I/O permission bit map by code with a CPL that is less privileged than the IOPL results in a general protection exception. In this way an operating system can detect attempts to access certain I/O devices and trap to special service routines for the devices through the general protection exception. In virtual 8086 mode, all I/O accesses reference the I/O permission bit map.

The I/O permission configuration defined by a bit map applies only to the task that uses the TSS. For this reason, many different I/O configurations can exist within a protected-mode software system. Actually, a different bit map could be defined for every task.

In practical applications, most tasks would use the same I/O permission bit map configuration. In fact, in some applications not all I/O addresses need to be protected with the I/O permission bit map. It turns out that any bit map position located beyond the limit of the TSS is interpreted as containing a 1. Therefore, all accesses to an I/O address that corresponds to a bit position beyond the limit of the TSS will produce a general protection exception. For instance, a protected-mode I/O address space may be set up with a small block of I/O addresses to which access is permitted at the low end of the I/O address space and with access to the rest of the I/O address space restricted. A smaller table can be set up to specify this configuration. By setting the values of the bit map base and TSS limit such that the bit positions for all the restricted addresses fall beyond the end of the TSS segment, they are caused to result in an exception. On the other hand, the bit positions located within the table are all made 0 to permit I/O accesses to their corresponding ports. Moreover, if the complete I/O address space is to be restricted for a task, the I/O permission map base address can simply be set to a value greater than the TSS limit.

▲ 16.7 INTERRUPT AND EXCEPTION PROCESSING

In our study of the 8086 microprocessor, we found that interrupts provide a mechanism for quickly changing program environments. Moreover, we identified that the transfer of program control is initiated by the occurrence of either an event internal to the microprocessor or an event in its external hardware. Finally, we determined that the interrupts employ a well-defined context-switching mechanism for changing the program environment.

Figure 16–31 illustrates the program context-switching mechanism. Here we see that interrupt 32 occurs as instruction N of the main program is being executed. When the MPU terminates execution of the main program in response to interrupt 32, it first saves information that identifies the instruction following the one where the interrupt occurred, instruction $N + 1$, and then picks up execution with the first instruction in the service routine. After this routine has run to completion, program control is returned to the point where the MPU originally left the main program, instruction $N + 1$, and then execution resumes.

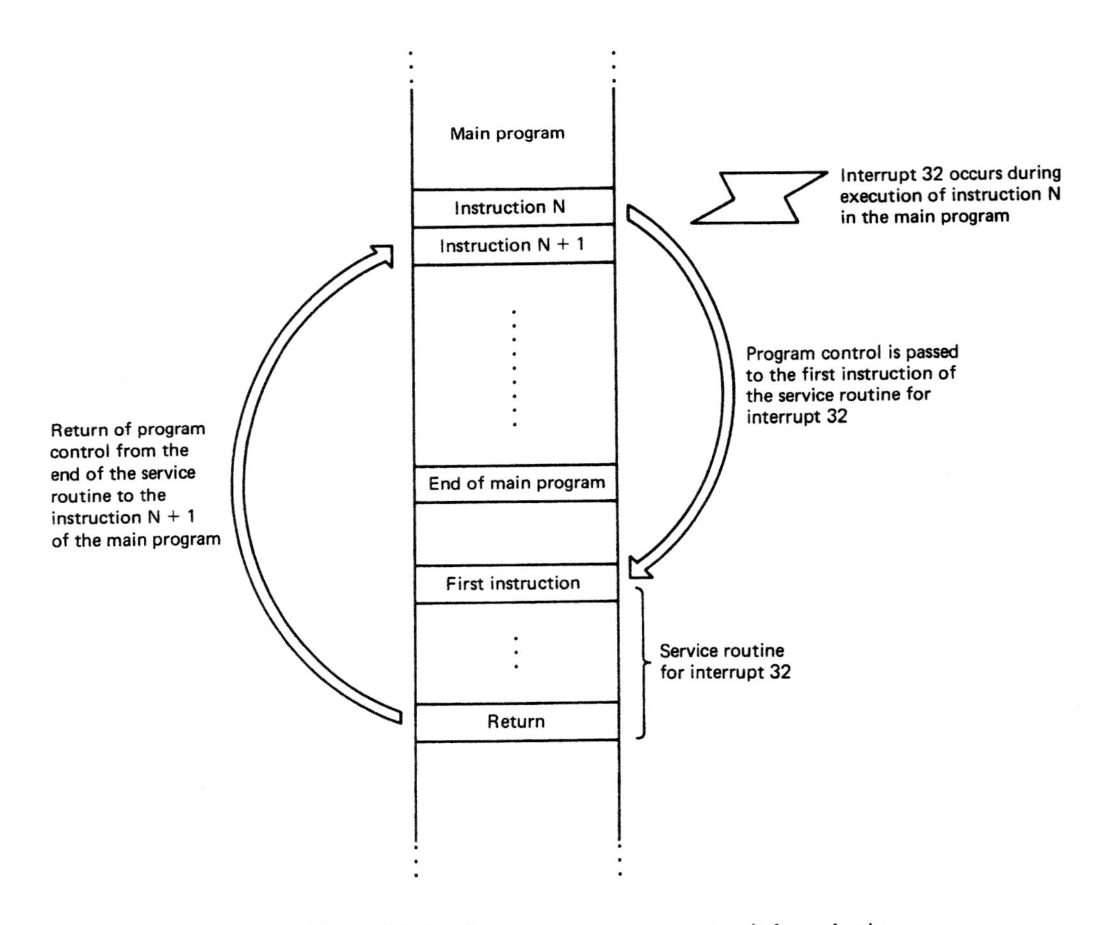

Figure 16–31 Interrupt program context-switch mechanism.

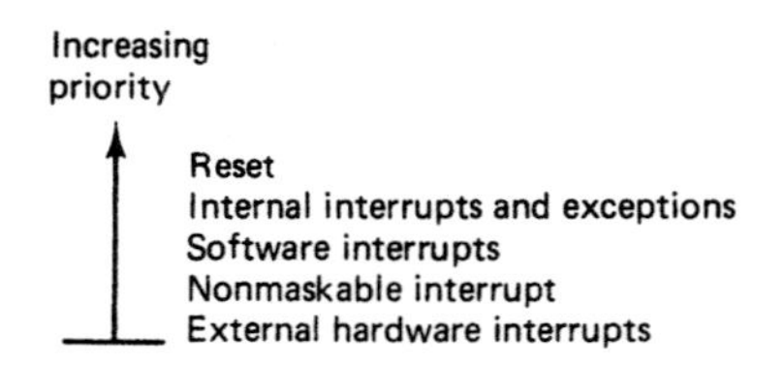

Figure 16–32 Types of interrupts and their priority.

The 80386DX's interrupt mechanism is essentially the same as that of the 8086. Just like the 8086-based microcomputer, the 80386DX is capable of implementing any combination of up to 256 interrupts. As Fig. 16–32 shows, they are divided into five groups: *external hardware interrupts, nonmaskable interrupt, software interrupts, internal interrupts and exceptions,* and *reset.* The functions of the external hardware, software, and nonmaskable interrupts are identical to those in the 8086 microcomputer and are defined by the user. On the other hand, the internal interrupts and exception processing capability of the 80386DX has been greatly enhanced. These internal interrupts and exceptions and reset perform dedicated system functions. Figure 16–32 identifies the priority by which the 80386DX services interrupts and exceptions.

Interrupt Vector and Interrupt Descriptor Tables

An address pointer table is used to link interrupt type numbers to the locations of their service routines in program-storage memory. In a real-mode 80386DX-based microcomputer system, this table is called the *interrupt-vector table.* On the other hand, in a protected-mode system, the table is referred to as the *interrupt-descriptor table.* Figure 16–33 shows a map of the interrupt-vector table in the memory of a real-mode 80386DX microcomputer. The table contains 256 *address pointers,* identified as *vector 0* through *vector 255.* That is, one pointer corresponds to each of the interrupt types 0 through 255. As in the 8086 microcomputer system, these address pointers identify the starting locations of their service routines in program memory. The contents of these tables are either held as firmware in EPROMs or loaded into RAM as part of the system initialization routine.

Note in Fig. 16–33 that the interrupt vector table is located at the low-address end of the memory address space. It starts at address 00000_{16} and ends at $003FE_{16}$. Unlike the 8086 microcomputer system, the interrupt vector table or interrupt descriptor table in an 80386DX microcomputer can be located anywhere in the memory address space. Its starting location and size are identified by the contents of a register within the 80386DX called the *interrupt-descriptor table register* (IDTR). When the 80386DX is reset at power on, it comes up in the real mode with the bits of the base address in IDTR all equal to zero and the limit set to $03FF_{16}$. This positions the interrupt vector table, as shown in Fig. 16–33. Moreover, when in the real mode, the value in IDTR is normally left at this initial value to maintain compatibility with 8086/8088-based microcomputer software.

The protected-mode interrupt-descriptor table can reside anywhere in the 80386DX's physical address space. The location and size of this table are again defined by the contents of the IDTR. Figure 16–34 shows that the IDTR contains a 32-bit *base address* and a 16-bit *limit.* The base address identifies the starting point of the table in memory. On the other hand, the limit determines the number of bytes in the table.

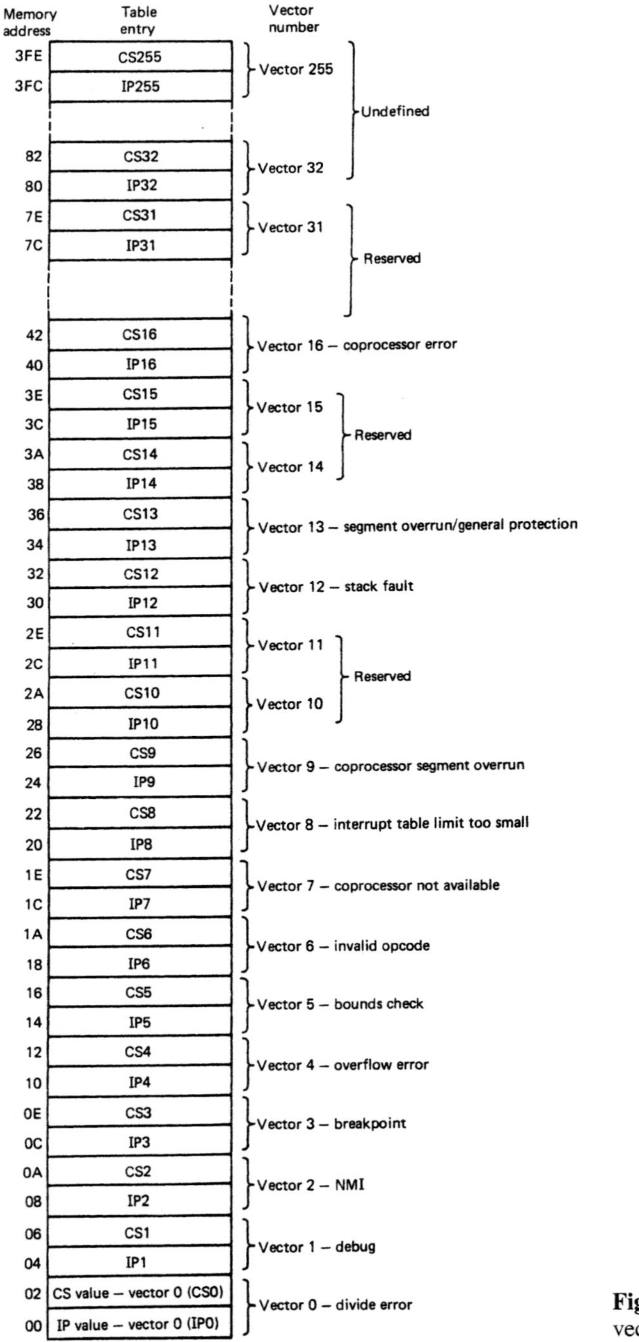

Figure 16–33 Real-mode interrupt vector table.

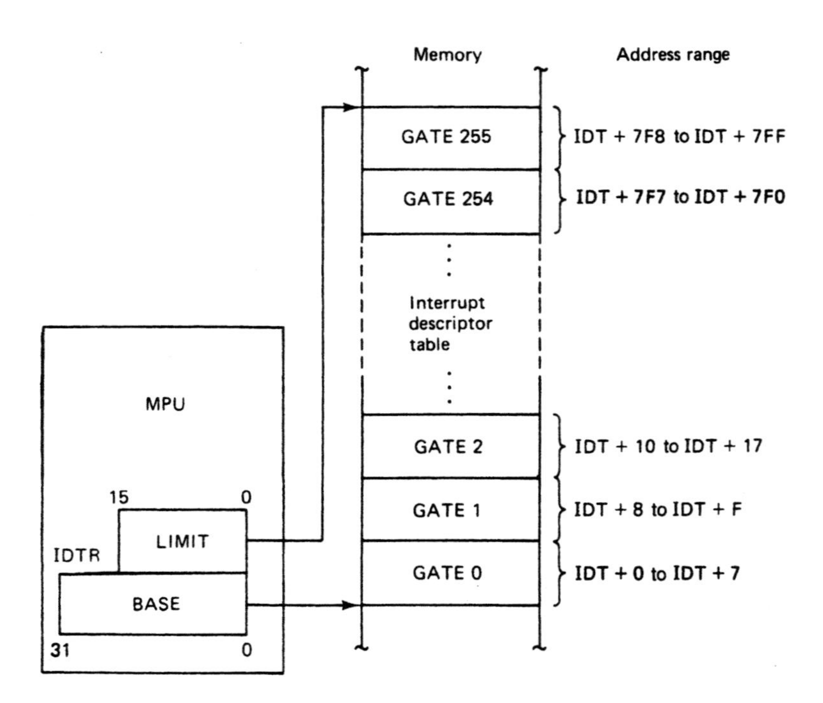

Figure 16–34 Accessing a gate in the protected-mode interrupt-descriptor table.

The interrupt-descriptor table contains gate descriptors, not vectors. The table in Fig. 16–34 contains a maximum of 256 gate descriptors. These descriptors are identified as *gate 0* through *gate 255.* Each gate descriptor can be defined as a *trap gate, interrupt gate,* or *task gate.* Interrupt and trap gates permit control to be passed to a service routine that is located within the current task. On the other hand, the task gate permits program control to be passed to a different task.

Just like a real-mode interrupt vector, a protected-mode gate acts as a pointer that is used to direct program execution to the starting point of a service routine. However, unlike an interrupt vector, a gate descriptor takes up 8 bytes of memory. For instance, in Fig. 16–34, we see that gate 0 is located at addresses IDT + 0H through IDT + 7H and gate 255 is at addresses IDT + 7F8H through IDT + 7FFH. If all 256 gates are not needed for an application, limit in the IDTR can be set to a value lower than $07FF_{16}$ to minimize the amount of memory reserved for the table.

Figure 16–35 illustrates the format of a typical interrupt or trap gate descriptor. Here we see that the two lower-addressed words, 0 and 1, are the interrupt's *code offset 0 through 15* and *segment selector,* respectively. The highest-addressed word, word 3, is the interrupt's *code offset 16 through 31.* These three words identify the starting point of the service routine. The upper byte of word 2 of the descriptor is called the *access rights byte.* The settings of the bits in this byte identify whether or not this gate descriptor is valid, the privilege level of the service routine, and the type of gate. For example, the *present*

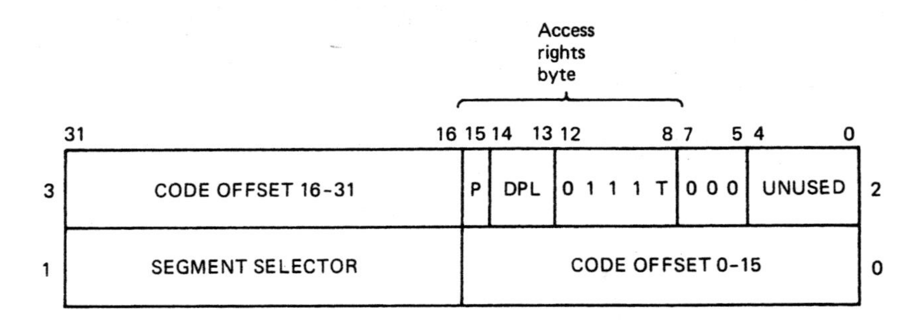

Figure 16–35 Format of a trap or interrupt gate descriptor.

bit (P) needs to be set to logic 1 if the gate descriptor is to be active. The next 2 bits, identified as DPL in Fig. 16–35, are used to assign a privilege level to the service routine. If these bits are made 00, level 0, which is the most privileged level, is assigned to the gate. Finally, the setting of the *type bit* (T) determines if the descriptor works as a trap gate or an interrupt gate. T equal to 0 selects the interrupt-gate mode of operation. The only difference between the operations of these two types of gates is that when a trap gate context switch is performed IF is not cleared to disable external hardware interrupts.

Normally, external hardware interrupts are configured with interrupt-gate descriptors. Once an interrupt request has been acknowledged for service, the external hardware-interrupt interface is disabled with IF. In this way, additional external interrupts cannot be accepted unless the interface is reenabled under software control. On the other hand, internal interrupts, such as software interrupts, usually use trap gate descriptors. In this case, the hardware-interrupt interface in not affected when the service routine for the software interrupt is initiated. Sometimes low-priority hardware interrupts are assigned trap gates instead of an interrupt gate. This will permit higher-priority external events to easily interrupt their service routine.

External Hardware-Interrupt Interface

Up to this point in the section, we have introduced the types of interrupts supported by the 80386DX, its interrupt descriptor table, and interrupt descriptor format. Let us now look at the external hardware-interrupt interface of the 80386DX microcomputer system.

Figure 16–36 shows a general interrupt interface for an 80386DX-based microcomputer system. Here we see that it is similar to the interrupt interface of the maximum-mode 8086 microcomputer system. Note that it includes the address and data buses, byte-enable signals, bus cycle indication signals, lock output, and the ready and interrupt-request inputs. Moreover, external circuitry is required to interface interrupt inputs, INT_{32} through INT_{255}, to the 80386DX's interrupt interface. This interface circuit must identify which of the pending active interrupts has the highest priority, perform an interrupt-request/acknowledge handshake, and then set up the bus to pass an interrupt-type number to the MPU.

In this circuit we see that the key interrupt interface signals are *interrupt request* (INTR) and *interrupt acknowledge* ($\overline{\text{INTA}}$). The logic-level input at the INTR line signals

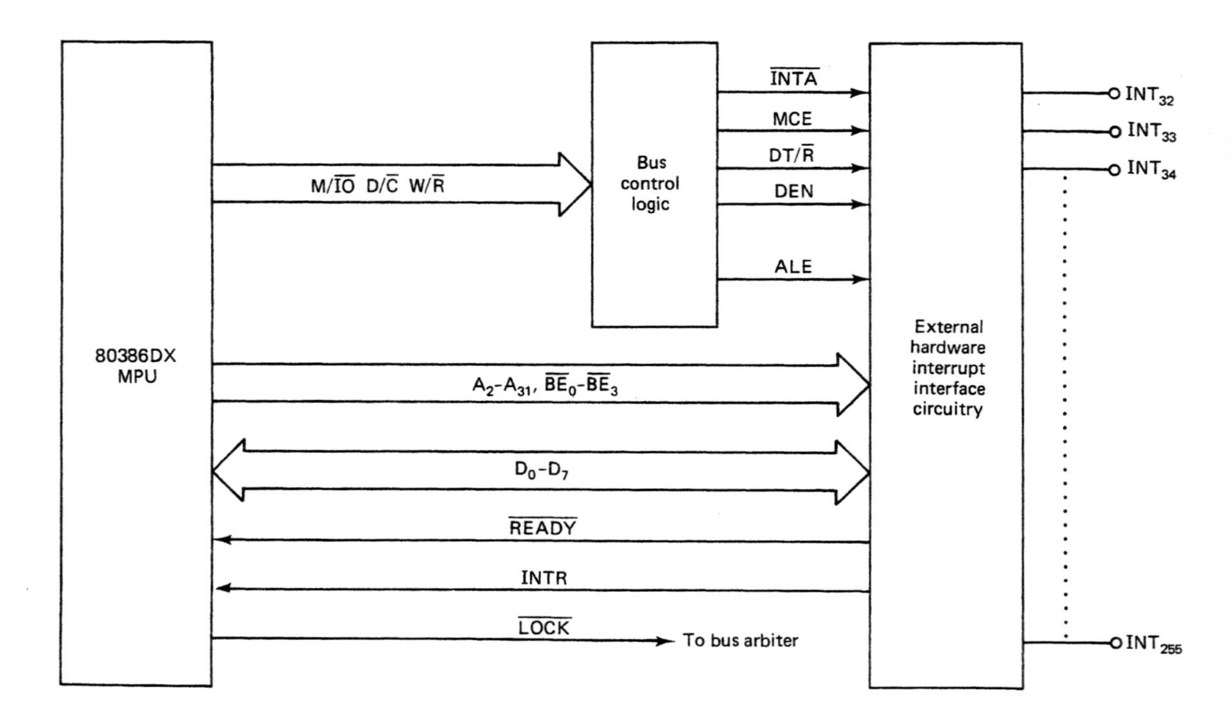

Figure 16–36 80386DX microcomputer system external hardware interrupt interface.

the 80386DX that an external device is requesting service. The 80386DX samples this input at the beginning of each instruction execution cycle—that is, at instruction boundaries. Logic 1 at INTR represents an active interrupt request. INTR is *level triggered;* therefore, the external hardware must maintain the active level until it is tested by the MPU. If it is not maintained, the request for service may not be recognized. For this reason, inputs INT_{32} through INT_{255} are normally latched. Moreover, the 1 at INTR must be removed before the service routine runs to completion; otherwise, the same interrupt may be acknowledged a second time.

When the 80386DX has recognized an interrupt request, it signals this fact to external circuitry by outputting the interrupt-acknowledge bus cycle indication code on M/$\overline{IO}$ C/$\overline{D}$ W/$\overline{R}$. This code, which equals 000_2, is highlighted in Fig. 16–37. Note in Fig. 16–36 that this code is input to the bus controller logic, where it is decoded to produce a pulse to logic 0 at the $\overline{INTA}$ output. Actually, two pulses are produced at $\overline{INTA}$ during the *interrupt-acknowledge bus cycle* sequence. The first pulse, which is output during cycle 1, signals external circuitry that the interrupt request has been acknowledged and to prepare to send its type number to the MPU. The second pulse, which occurs during cycle 2, tells the external circuitry to put the type number on the data bus. The ready ($\overline{READY}$) input can be used to insert wait states into these bus cycles.

Note that the lower eight lines of the data bus, D_0 through D_7, are also part of the interrupt interface. During the second cycle in the interrupt-acknowledge bus cycle, external circuitry must put the 8-bit type number of the highest-priority active interrupt-request

$M/\overline{IO}$	$D/\overline{C}$	$W/\overline{R}$	Type of Bus Cycle
0	0	0	Interrupt acknowledge
0	0	1	Idle
0	1	0	I/O data read
0	1	1	I/O data write
1	0	0	Memory code read
1	0	1	Halt/shutdown
1	1	0	Memory data read
1	1	1	Memory data write

Figure 16–37 Interrupt-acknowledge bus cycle indication code.

input onto this part of the data bus. The 80386DX reads the type number off the bus to identify which external device is requesting service. It uses the type number to generate the address of the interrupt's vector or gate in the interrupt vector or descriptor table, respectively, and to read the new values of IP and CS into the corresponding internal registers. IP and CS values from the interrupt vector table are transferred to the MPU over the data bus. Before loading IP and CS with new values, their old values and the values of the internal flags are automatically written to the stack part of memory.

Address lines A_2 through A_{31} and byte-enable line $\overline{BE}_0$ through $\overline{BE}_3$ are also shown in the interrupt interface circuit in Fig. 16–36. This is because LSI interrupt-controller devices are typically used to implement most of the external circuitry. When a read or write bus cycle is performed to the controller, for example, to initialize its internal registers after system reset, some of the address bits are decoded to produce a chip select to enable the controller device, and other address bits are used to select the internal register to be accessed. The interrupt controller could be I/O mapped instead of memory mapped; in this case only address lines A_2 through A_{15} are used in the interface. Addresses are also output on A_2 through A_{31} during the write cycles that save the old program context in the stack and the read cycles used to load the new program context from the vector table in program memory.

Another signal shown in the interrupt interface of Fig. 16–36 is the *bus lock indication* ($\overline{LOCK}$) output of the 80386DX. $\overline{LOCK}$ is used as an input to the bus arbiter circuit in multiprocessor systems. The 80386DX switches this output to its active 0 logic level and maintains it at this level throughout the complete interrupt-acknowledge bus cycle. In response to this signal, the arbitration logic assures that no other device can take over control of the system bus until the interrupt-acknowledge bus cycle sequence is completed.

External Hardware-Interrupt Sequence

In the real mode, the 80386DX processes interrupts in exactly the same way as the 8086. That is, the same events that we described in Chapter 11 take place during the interrupt request, interrupt-acknowledge bus cycle, and device service routine. On the other hand, in protected mode a more complex processing sequence is performed. When the 80386DX-based microcomputer is configured for the protected mode of operation, the interrupt request/acknowledge handshake sequence appears to take place exactly the same way in the external hardware; however, a number of changes do occur in the internal-processing sequence of the MPU. Let us now look at how the protected mode 80386DX reacts to an interrupt request.

When processing interrupts in protected mode, the general protection mechanism of the 80386DX comes into play. The general protection rules dictate that program control can be directly passed only to a service routine that is in a segment with equal or higher privilege—that is, a segment with an equal- or lower-numbered descriptor privilege level. Any attempt to transfer program control to a routine in a segment with lower privilege (higher-numbered descriptor privilege level) results in an exception unless the transition is made through a gate.

Typically, interrupt drivers are in code segments at a high privilege level, possibly level 0. Moreover, interrupts occur randomly; therefore, there is a good chance that the microprocessor will be executing application code at a low privilege level. In the case of interrupts, the current privilege level (CPL) is the privilege level assigned by the descriptor of the software that was executing when the interrupt occurred. This could be any of the 80386DX's valid privilege levels. The privilege level of the service routine is that defined in the interrupt or trap gate descriptor for the type number. That is, it is the descriptor privilege level (DPL).

When a service routine is initiated, the current privilege level may change. This depends on whether the software that was interrupted was in a code segment that was configured as *conforming* or *nonconforming*. If the interrupted code is in a conforming code segment, CPL does not change when the service routine is initiated. In this case, the contents of the stack after the context switch is as illustrated in Fig. 16–38(a). Since the privilege level does not change, the current stack (OLD SS:ESP) is used. Note that as part of the interrupt initiation sequence, the OLD EFLAGS, OLD CS, and OLD EIP are

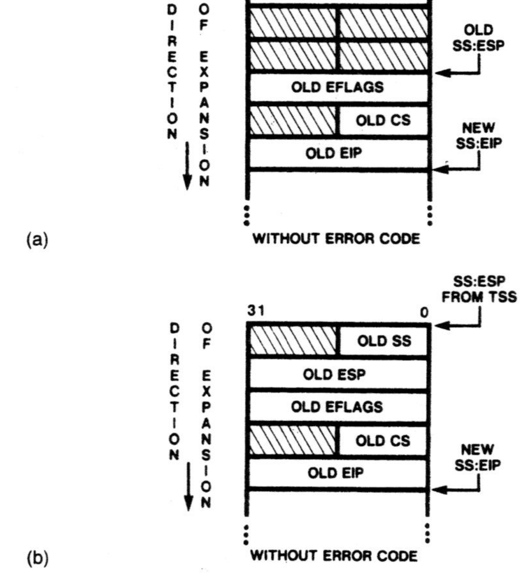

Figure 16–38 (a) Stack after context switch with no privilege-level transition. (Reprinted by permission of Intel Corporation. Copyright/Intel Corp. 1986) (b) Stack after context switch with a privilege-level transition. (Reprinted by permission of Intel Corporation. Copyright/Intel Corp. 1986)

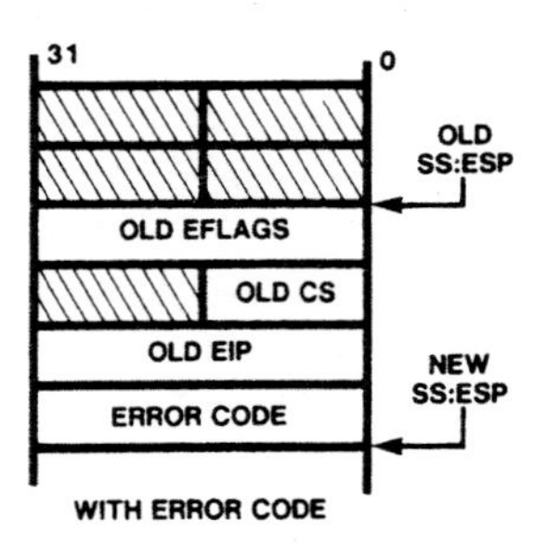

Figure 16–39 Stack contents after interrupt with an error. (Reprinted by permission of Intel Corporation. Copyright/Intel Corp. 1986)

automatically saved on the stack. Actually, the *requested privilege level* (RPL) code is also saved on the stack. This is because it is part of OLD CS. Code RPL identifies the protection level of the interrupted routine.

However, if the segment is nonconforming, the value of DPL is assigned to CPL as long as the service routine is active. As Fig. 16–38(b) shows, this time the stack is changed to that for the new privilege level. The MPU is loaded with a new SS and new ESP from TSS, and the old stack pointer, OLD SS, and OLD ESP are saved on the stack followed by the OLD EFLAGS, OLD CS, and OLD EIP. Remember that for an interrupt gate, IF is cleared as part of the context switch, but for a trap gate, IF remains unchanged. In both cases, the TF flag is reset after the contents of the flag register are pushed to the stack.

Figure 16–39 shows the stack as it exists after an attempt to initiate an interrupt-service routine that did not involve a privilege-level transition failed. Note that the context switch to the exception service routine caused an *error code* to be pushed onto the stack following the values of OLD EFLAGS, OLD CS, and OLD EIP.

One format of the error code is given in Fig. 16–40. This type error code is known as an *IDT error code.* Here we see that the least significant bit, EXT, indicates whether the error was for an externally or internally initiated interrupt. For external interrupts, such as the hardware interrupts, the EXT bit is always set to logic 1. The next bit, IDT, is set to 1 if the error is produced as a result of an interrupt. That is, it is the result of a reference to a descriptor in the IDT. If IDT is not set, the third bit indicates whether the descriptor is in the GDT (TI = 0) or the LDT (TI = 1). The next 14 bits contain the segment selector that produced the error condition. With this information available on the stack, the exception service routine can determine which interrupt attempt failed and whether it was internally or externally initiated. A second format is used for errors that result from a protected-mode page fault. Figure 16–41 illustrates this error code and the function of its bits.

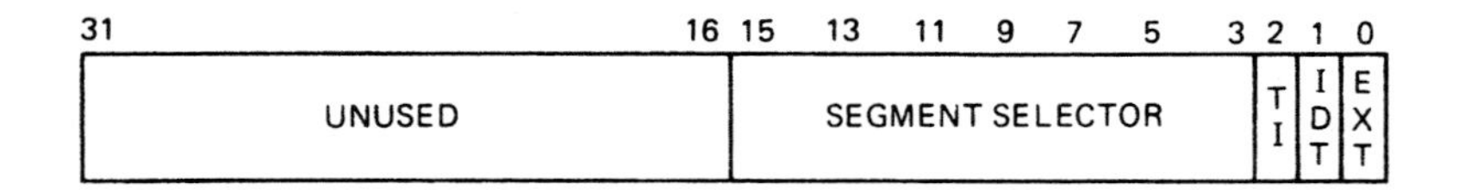

Figure 16–40 IDT error code format.

Field	Value	Description
U/S	0	The access causing the fault originated when the processor was executing in supervisor mode.
	1	The access causing the fault originated when the processor was executing in user mode.
W/R	0	The access causing the fault was a read.
	1	The access causing the fault was a write.
P	0	The fault was caused by a not-present page.
	1	The fault was caused by a page-level protection violation.

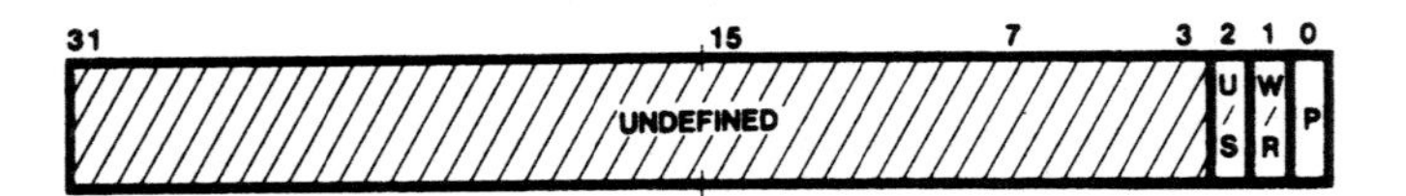

Figure 16–41 Page fault error code format and bit functions. (Reprinted by permission of Intel Corporation. Copyright/Intel Corp. 1986)

Just as in real-mode, the IRET instruction is used to return from a protected-mode interrupt-service routine. For service routines using an interrupt gate or trap gate, IRET is restricted to the return from a higher privilege level to a lower privilege level, for instance, from level 1 to level 3. Once the flags, OLD CS, and OLD EIP are returned to the 80386DX, the RPL bits of OLD CS are tested to see if they equal CPL. If RPL = CPL, an intralevel return is in progress. In this case, the return is complete and program execution resumes at the point in the program where execution had stopped.

If RPL is greater than CPL, an interlevel return, not an intralevel return, is taking place. During an interlevel return, checks are performed to determine if a protection violation will occur due to the protection-level transition. Assuming that no violation occurs, the OLD SS and OLD ESP are popped from the stack into the MPU and then program execution resumes.

Internal Interrupt and Exception Functions

Earlier we indicated that some of the 256 interrupt vectors of the 80386DX are dedicated to internal interrupt and exception functions. Internal interrupts and exceptions differ from external hardware interrupts in that they occur as the result of executing an instruction, not an event that takes place in external hardware. That is, an internal interrupt or exception is initiated because an error condition was detected before, during, or after execution of an instruction. In this case, a routine must be initiated to service the internal condition before resuming execution of the same or next instruction of the program.

Internal interrupts and exceptions are not masked out with the interrupt enable flag. For this reason, the 80386DX automatically detects the occurrence of any one of these internal conditions and causes an interrupt of program execution and a vectored transfer of control to a corresponding service routine. During the control-transfer sequence, no interrupt-acknowledge bus cycles are produced.

Figure 16–42 identifies the internal interrupts and exceptions that are active in real mode. Here we find internal interrupts such as breakpoint and exception functions such as divide error and overflow error that were also detected by the 8086. However, the 80386DX also implements several new real-mode exceptions. Examples of exceptions that are not implemented on the 8086 are invalid opcode, bounds check, and interrupt table limit too small.

Internal interrupts and exceptions are further categorized as a *fault, trap,* or *abort* based on how the failing function is reported. In the case of an exception that causes a fault, the values of CS and IP saved on the stack point to the instruction that resulted in the fault. Therefore, after servicing the exception, the faulting instruction can be reexecuted. On the other hand, for those exceptions that result in a trap, the values of CS and IP pushed to the stack point to the next instruction to be executed, instead of the instruction that caused the trap. Therefore, upon completion of the service routine, program execution resumes with the instruction that follows the instruction that produced the trap. Finally, exceptions that produce an abort do not preserve any information that identifies the location that caused the error. In this case the system may need to be restarted. Let us now look at the new 80386DX real-mode internal interrupts and exceptions in more detail.

Debug Exception. The *debug exception* relates to the debug mode of operation of the 80386DX. The 80386DX has a set of eight on-chip debug registers. Using these registers, the programmer can specify up to four breakpoint addresses and specify conditions under which they are to be active. For instance, the activating condition could be an instruction fetch from the address, a data write to the address, or either a data read or write for the address, but not an instruction fetch. Moreover, for data accesses, the size of the data element can be specified as a byte, word, or double word. Finally, the individual addresses can be locally or globally enabled or disabled. If an access that matches any of these debug conditions is attempted, a debug exception occurs and control is passed to the service routine defined by IP_1 and CS_1 at word addresses 00004_{16} and 00006_{16}, respectively. The service routine could include a mechanism that allows the programmer to view the contents of the 80386DX's internal registers and its external memory.

If the trap flag (TF) bit in the flags register is set, the single-step mode of operation is enabled. This flag bit can be set or reset under software control. When TF is set, the 80386DX initiates a type 1 interrupt to the service routine defined by IP_1 and CS_1 at addresses 00004_{16} and 00006_{16}, respectively, at the completion of execution of every instruction. This permits implementation of the single-step mode of operation so that the program can be executed one instruction at a time.

Bounds Check Exception. Earlier we pointed out that the BOUND (check array index against bounds) instruction can be used to test an operand that is used as the index into an array to verify that it is within a predefined range. If the index is less than the lower bound (minimum value) or greater than the upper bound (maximum value), a *bound check exception* has occurred and control is passed to the exception handler pointed to by

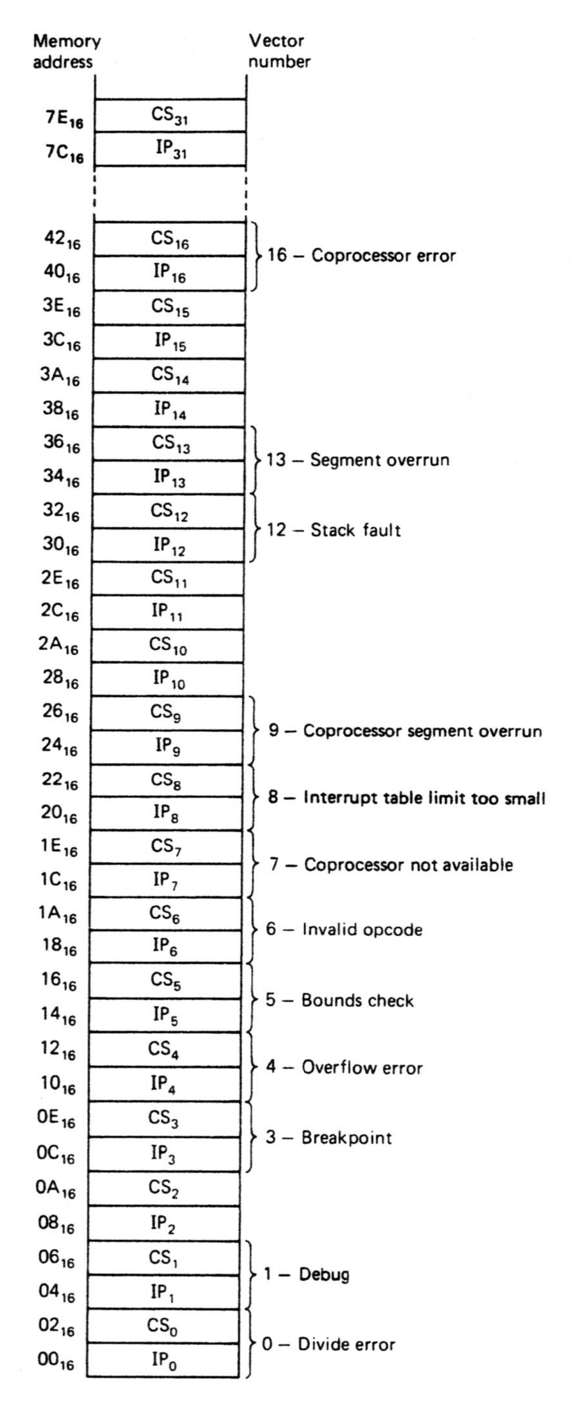

Figure 16–42 Real-mode internal interrupt and exception vector table.

$CS_5:IP_5$. The exception produced by the BOUND instruction is an example of a fault. Therefore, the values of CS and IP pushed to the stack represent the address of the instruction that produced the exception.

Invalid Opcode Exception. The exception-processing capability of the 80386DX permits detection of undefined opcodes. This feature of the 80386DX allows it to detect automatically whether or not the opcode to be executed as an instruction corresponds to one of the instructions in the instruction set. If it does not, execution is not attempted; instead, the opcode is identified as being undefined and the *invalid opcode exception* is initiated. In turn, control is passed to the exception handler identified by IP_6 and CS_6. This *undefined opcode-detection mechanism* permits the 80386DX to detect errors in its instruction stream. Invalid opcode is an example of an exception that produces a fault.

Coprocessor Extension Not Available Exception. When the 80386DX comes up in the real mode, both the EM (emulate coprocessor) and MP (math present) bits of its machine status word are reset. This mode of operation corresponds to that of the 8088 or 8086 microprocessor. When set in this way, the *coprocessor not available exception* cannot occur. However, if the EM bit has been set to 1 under software control (do not monitor coprocessor) and the 80386DX executes an ESC (escape) instruction for the math coprocessor, a processor extension not present exception is initiated through the vector at CS_7: IP_7. This service routine could pass control to a software emulation routine for the floating-point arithmetic operation. Moreover, if the MP and TS bits are set (meaning that a math coprocessor is available in the system and a task is in progress), when an ESC or WAIT instruction is executed, an exception also takes place.

Interrupt Table Limit Too Small Exception. Earlier we pointed out that the LIDT instruction can be used to relocate or change the limit of the interrupt vector table in memory. If the real-mode table has been changed, for example, its limit is set lower than address $003FF_{16}$ and an interrupt is invoked that attempts to access a vector stored at an address higher than the new limit, the *interrupt table limit too small exception* occurs. In this case, control is passed to the service routine by the vector $CS_8:IP_8$. This exception is a fault; therefore, the address of the instruction that exceeded the limit is saved on the stack.

Coprocessor Segment-Overrun Exception. The *coprocessor segment-overrun exception* signals that the 80387DX numeric coprocessor has overrun the limit of a segment while attempting to read or write its operand. This event is detected by the coprocessor data channel within the 80386DX and passes control to the service routine through interrupt vector 9. This exception handler can clear the exception, reset the 80387DX, determine the cause of the exception by examining the registers within the 80387DX, and then initiate a corrective action.

Stack Fault Exception. In the real mode, if the address of an operand access for the stack segment crosses the boundaries of the stack, a stack fault exception is produced. This causes control to be transferred to the service routine defined by CS_{12} and IP_{12}.

Segment Overrun Exception. This exception occurs in the real mode if an instruction attempts to access an operand that extends beyond the end of a segment. For instance, if a word access is made to the address CS:FFFFH, DS:FFFFH, or ES:FFFFH, a fault occurs to the *segment overrun exception* service routine.

Coprocessor Error Exception. As part of the handshake sequence between the 80386DX microprocessor and 80387DX numeric coprocessor, the 80386DX checks the status of its $\overline{\text{ERROR}}$ input. If the 80387DX encounters a problem performing a numeric operation, it signals this fact to the 80386DX by switching its $\overline{\text{ERROR}}$ output to logic 0. This signal is normally applied directly to the $\overline{\text{ERROR}}$ input of the 80386DX and signals that an error condition has occurred. Logic 0 at this input causes a *coprocessor error exception* through vector 16.

Protected-Mode Internal Interrupts and Exceptions. In protected mode, more internal conditions can initiate an internal interrupt or exception. Figure 16–43 identifies each of these functions and its corresponding type number.

▲ 16.8 80486SX AND 80486DX MICROPROCESSORS

In the previous chapter we pointed out that the 80486 family of microprocessors is Intel Corporation's second generation of 32-bit processors. It brought a higher level of performance and more versatility to the 8086 architecture. Just like for the 80386 family, maintaining compatibility of the 80486 family's hardware architecture to earlier 8086 family MPUs was also less important. This does not mean that the signal interfaces were purposely changed. In fact, many of the 80486's interface signals are the same as those provided on the 80386DX. However, a number of enhancements have been made to the 80486 family that are directed at improving its performance and making the 80486-based microcomputer more versatile. Much of the focus of these enhancements was on the memory interface. For instance, it is now enabled to do dynamic bus sizing down to eight bits, high-speed burst data transfers over the bus, and write operations are buffered. The addition of these new capabilities has expanded the number of interface signals.

A number of hardware elements that were normally implemented in external circuitry are for the first time added into the MPU with the 80486 family. Examples of these new on-chip hardware functions are parity generation/checking, code/data cache memory, and—in the case of the 80486DX—a floating-point mathematics unit. Addition of these capabilities called for further expansion of the number of interface signals.

The more advanced processes used to manufacture the 80486 family of MPUs permitted the integration of many more transistors into a single IC. The circuitry of the 80486DX is equivalent to approximately 1.2M transistors, four times more than the 80386DX. Actually, the 80486DX was the first IC made by Intel Corporation that contained more than 1 million transistors.

Originally the 80486DX and 80486SX were both manufactured in a 168-lead pin grid array package. Figure 16–44 shows a device in this package. The layout of the pins

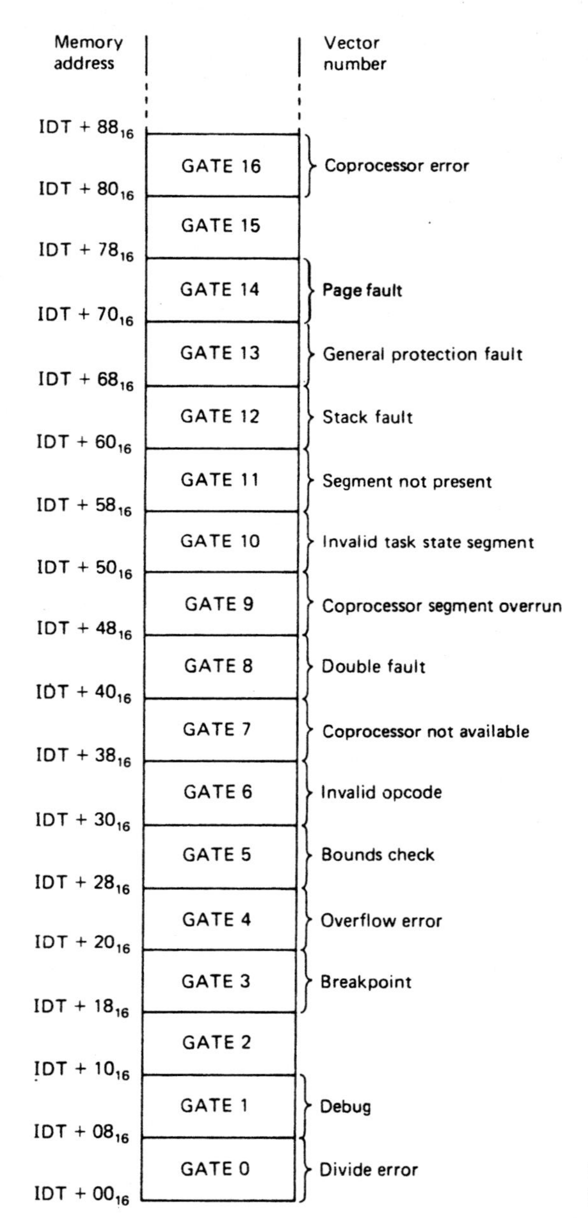

Figure 16–43 Protected-mode internal exception gate locations.

and signals on this package are shown in Fig. 16–45(a) and (b). For example, the pin located at the uppermost left corner, which corresponds to row S and column 1, is address bit A_{27}. Another example is the data bus line D_{20}, which is located at the junction of row A and column 1. Later both devices were made available in a lower-cost 196-lead plastic quad flat package.

Figure 16–44 80486DX IC. (Reprinted by permission of Intel Corporation.)

EXAMPLE 16.11

What signal is located at pin S17 of the 80486SX's PGA package?

Solution

Looking at Fig. 16–45(a), we find that the signal corresponding to this pin is $\overline{\text{ADS}}$.

Signal Interfaces of the 80486SX MPU

Let us continue our study of the 80486 family of microprocessors by exploring its signal interfaces. Figure 16–46 is a block diagram showing the signal interfaces of the 80486SX MPU. Many of the 80486SX's interface signals are identical in name, mnemonic, and function to those on the 80386DX. For instance, we find that the 80486SX's 30 address bus lines are labeled A_2 through A_{31}, and its four byte-enable signals are labeled $\overline{\text{BE}}_0$ through $\overline{\text{BE}}_3$. However, most of the interfaces have some new signals that are provided to implement enhanced functions. In fact, the interrupt interface is the only interface that is completely unchanged. Here we will focus on the new signals at each of the 80486SX's interfaces.

Memory/IO Interface. Earlier we pointed out that many of the hardware enhancements of the 80486 family are in the memory interface. For this reason, most of the new signal lines of the 80486SX are located at this interface. Let us look at the function of these new signals.

The 80386DX MPU had the ability to configure the data bus as 16 bits instead of 32 bits by activating the $\overline{\text{BS16}}$ input. The 80486SX also has this capability; however, another input *bus size 8* ($\overline{\text{BS8}}$) also gives designers the ability to configure the data bus 8 bits wide. If $\overline{\text{BS8}}$ is at the active 0 logic level, data are transferred one byte at a time over data bus lines D_0 through D_7.

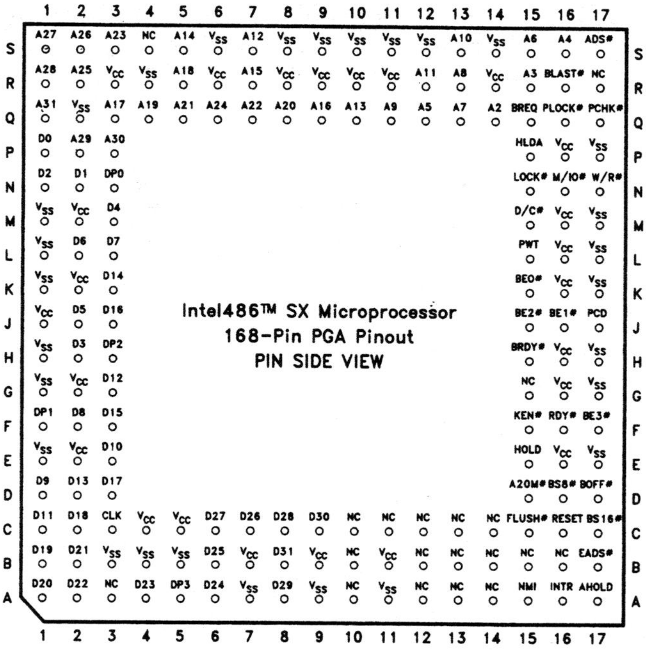

	1	2	3	4	5	6	7	8	9	10	11	12	13	14	15	16	17	
S	A27	A26	A23	NC	A14	V_{SS}	A12	V_{SS}	V_{SS}	V_{SS}	V_{SS}	V_{SS}	A10	V_{SS}	A6	A4	ADS#	S
R	A28	A25	V_{CC}	V_{SS}	A18	V_{CC}	A15	V_{CC}	V_{CC}	V_{CC}	V_{CC}	A11	A8	V_{CC}	A3	BLAST#	NC	R
Q	A31	V_{SS}	A17	A19	A21	A24	A22	A20	A16	A13	A9	A5	A7	A2	BREQ	PLOCK#	PCHK#	Q
P	D0	A29	A30												HLDA	V_{CC}	V_{SS}	P
N	D2	D1	DP0												LOCK#	M/IO#	W/R#	N
M	V_{SS}	V_{CC}	D4												D/C#	V_{CC}	V_{SS}	M
L	V_{SS}	D6	D7												PWT	V_{CC}	V_{SS}	L
K	V_{SS}	V_{CC}	D14												BE0#	V_{CC}	V_{SS}	K
J	V_{CC}	D5	D16												BE2#	BE1#	PCD	J
H	V_{SS}	D3	DP2												BRDY#	V_{CC}	V_{SS}	H
G	V_{SS}	V_{CC}	D12												NC	V_{CC}	V_{SS}	G
F	DP1	D8	D15												KEN#	RDY#	BE3#	F
E	V_{SS}	V_{CC}	D10												HOLD	V_{CC}	V_{SS}	E
D	D9	D13	D17												A20M#	BS8#	BOFF#	D
C	D11	D18	CLK	V_{CC}	V_{CC}	D27	D26	D28	D30	NC	NC	NC	NC	NC	FLUSH#	RESET	BS16#	C
B	D19	D21	V_{SS}	V_{SS}	V_{SS}	D25	V_{CC}	D31	V_{CC}	NC	V_{CC}	NC	NC	NC	NC	NC	EADS#	B
A	D20	D22	NC	D23	DP3	D24	V_{SS}	D29	V_{SS}	NC	V_{SS}	NC	NC	NC	NMI	INTR	AHOLD	A
	1	2	3	4	5	6	7	8	9	10	11	12	13	14	15	16	17	

Intel486™ SX Microprocessor
168-Pin PGA Pinout
PIN SIDE VIEW

(a)

Address		Data		Control		N/C	V_{CC}	V_{SS}
A_2	Q14	D_0	P1	A20M#	D15	A3	B7	A7
A_3	R15	D_1	N2	ADS#	S17	A10	B9	A9
A_4	S16	D_2	N1	AHOLD	A17	A12	B11	A11
A_5	Q12	D_3	H2	BE0#	K15	A13	C4	B3
A_6	S15	D_4	M3	BE1#	J16	A14	C5	B4
A_7	Q13	D_5	J2	BE2#	J15	B10	E2	B5
A_8	R13	D_6	L2	BE3#	F17	B12	E16	E1
A_9	Q11	D_7	L3	BLAST#	R16	B13	G2	E17
A_{10}	S13	D_8	F2	BOFF#	D17	B14	G16	G1
A_{11}	R12	D_9	D1	BRDY#	H15	B15	H16	G17
A_{12}	S7	D_{10}	E3	BREQ	Q15	B16	J1	H1
A_{13}	Q10	D_{11}	C1	BS8#	D16	C10	K2	H17
A_{14}	S5	D_{12}	G3	BS16#	C17	C11	K16	K1
A_{15}	R7	D_{13}	D2	CLK	C3	C12	L16	K17
A_{16}	Q9	D_{14}	K3	D/C#	M15	C13	M2	L1
A_{17}	Q3	D_{15}	F3	DP0	N3	C14	M16	L17
A_{18}	R5	D_{16}	J3	DP1	F1	G15	P16	M1
A_{19}	Q4	D_{17}	D3	DP2	H3	R17	R3	M17
A_{20}	Q8	D_{18}	C2	DP3	A5	S4	R6	P17
A_{21}	Q5	D_{19}	B1	EADS#	B17		R8	Q2
A_{22}	Q7	D_{20}	A1	FLUSH#	C15		R9	R4
A_{23}	S3	D_{21}	B2	HLDA	P15		R10	S6
A_{24}	Q6	D_{22}	A2	HOLD	E15		R11	S8
A_{25}	R2	D_{23}	A4	INTR	A16		R14	S9
A_{26}	S2	D_{24}	A6	KEN#	F15			S10
A_{27}	S1	D_{25}	B6	LOCK#	N15			S11
A_{28}	R1	D_{26}	C7	M/IO#	N16			S12
A_{29}	P2	D_{27}	C6	NMI	A15			S14
A_{30}	P3	D_{28}	C8	PCD	J17			
A_{31}	Q1	D_{29}	A8	PCHK#	Q17			
		D_{30}	C9	PWT	L15			
		D_{31}	B8	PLOCK#	Q16			
				RDY#	F16			
				RESET	C16			
				W/R#	N17			

(b)

Figure 16–45 (a) Pin layout of the 80486SX PGA. (Reprinted by permission of Intel Corporation. Copyright/Intel Corp. 1992) (b) Signal pin numbering. (Reprinted by permission of Intel Corporation. Copyright/Intel Corp. 1992)

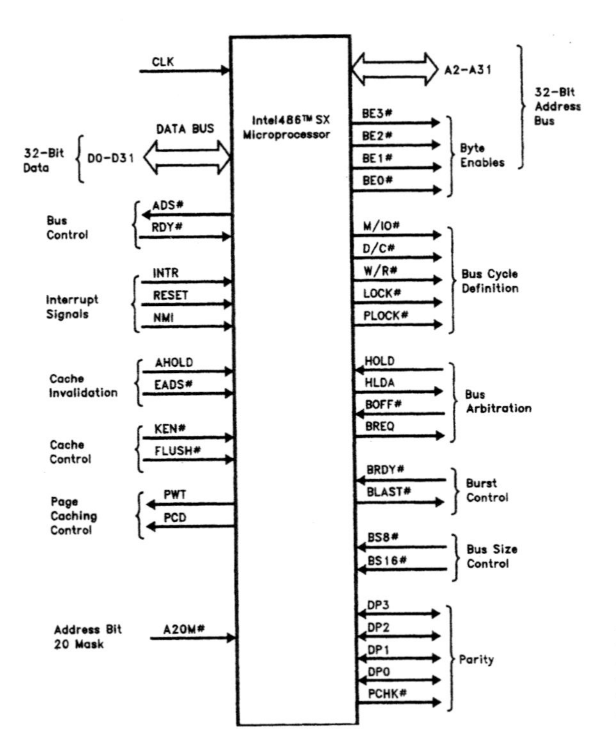

Figure 16–46 Block diagram of the 80486SX. (Reprinted by permission of Intel Corporation. Copyright/Intel Corp. 1992)

The bus cycle indication signals $M/\overline{IO}$, $D/\overline{C}$, and $W/\overline{R}$ of the 80486SX are the same as those on the 80386DX, except a change has been made in the coding of the bus cycles. For the 80486SX, the halt/shutdown bus cycle is identified by the code 001_2, instead of 101_2. The code 101_2 is now reserved. Codes for all other types of bus cycles are unchanged.

An important difference between the memory interface of the 80386DX MPU and that of the 80486SX is that automatic *parity generation and checking* has been added. Parity has been added to each byte of the 80486SX's 32-bit data bus. An even parity bit is automatically generated for each byte of the data written to memory, and each byte of data on read operations is checked for even parity. For this reason, four bidirectional *data parity* (DP_0 through DP_3) lines and a *parity status* ($\overline{PCHK}$) output have been added into the memory interface. The data parity lines are additional data bus lines used to carry parity data to and from memory. On the other hand, the parity status output is used to signal external circuitry whether or not a parity error has occurred on a read operation. Logic 0 at this output identifies a parity error condition.

The 80486SX automatically performs an operation known as *address bit 20 mask*. This is an operation that must be performed in all ISA bus-compatible computers. In an 80386DX-based microcomputer, masking of A_{20} is accomplished with external circuitry. An extra input, *address bit 20 mask* ($\overline{A20M}$), has been added on the 80486SX to perform this function. Whenever $\overline{A20M}$ is logic 0, address bit A_{20} is masked out for bus cycles that access internal cache memory or external memory.

Another enhancement in the 80486 family of MPUs is the ability to perform what is known as *burst bus cycles.* A burst bus cycle is a special bus cycle that permits faster reads and writes of data. During a burst bus cycle, the transfer of the first element of data takes place in two clock cycles, and each additional data element is transferred in a single clock cycle. Nonburst bus cycles transfer one data element at a time and require a minimum of two clock cycles for each data transfer. Whenever the 80486SX requires data, it can perform the transfer with normal or burst bus cycles. If the external device can perform burst data transfers, it signals this fact to the MPU. Switching the control signal *burst ready* ($\overline{\text{BRDY}}$), instead of $\overline{\text{RDY}}$, to logic 0 does this.

During all memory bus cycles, the $\overline{\text{BLAST}}$ output signals when the last data transfer takes place. In the case of a normal bus cycle, only one data transfer takes place. This data transfer is marked by $\overline{\text{BLAST}}$ switching to logic 0. For a burst cycle, multiple data transfers are performed, but $\overline{\text{BLAST}}$ is active only for the last one.

The 80486SX has a second type of lock signal. This signal, *pseudo-lock* ($\overline{\text{PLOCK}}$), differs from $\overline{\text{LOCK}}$ in that it locks out access to the bus by other devices for more than one bus cycle.

Cache Memory-Control Interface. A new group of interface signals is provided on the 80486 family MPUs to support internal and external cache memory subsystems. They include the *cache-enable* ($\overline{\text{KEN}}$) input, *cache-flush* ($\overline{\text{FLUSH}}$) input, the *page cache disable* (PCD) output, the *page write-through* (PWT) output, the *address hold* (AHOLD) input, and the *valid external address* ($\overline{\text{EADS}}$) input. Let us next look briefly at the function of each of these signals.

The external memory subsystem has the ability to tell the 80486SX whether or not a bus cycle is cacheable. It does this by switching the $\overline{\text{KEN}}$ input to logic 0 or 1. Whenever $\overline{\text{KEN}}$ is set to logic 0 during a memory-read bus cycle, the information carried over the bus is copied into the on-chip cache. If $\overline{\text{KEN}}$ remains at its inactive 1 logic level, a noncacheable bus cycle takes place.

External circuitry also has the ability to invalidate all of the data in the on-chip cache memory of the 80486SX. This operation is known as a *cache flush.* To flush the on-chip cache, the $\overline{\text{FLUSH}}$ input is simply switched to its active 0 logic level for one clock cycle.

PCD and PWT are outputs of the MPU and are used to control external cache. Earlier we pointed out that the programmed logic levels of the page attribute bits in the page entry table, page directory table, or control register 3 are output on these lines when caching is enabled. The logic level of PCD signals the external cache memory subsystem whether or not the page of memory being accessed is configured as cacheable. Logic 1 at PWT indicates that write operations to the external cache are performed in a write-through fashion.

The last two signals, AHOLD and $\overline{\text{EADS}}$, are used to perform what is known as a *cache-invalidate cycle.* This type of bus cycle is used to maintain consistency between data in the internal cache and external main memory. For instance, if another bus master device modifies data in main memory, a check must be made immediately to see if the contents of this storage location are currently held in the 80486SX's internal cache memory. If they are and the 80486SX would read the value at this address, the wrong value would be accessed from the cache. For this reason, the value in the cache must be invalidated.

The signal AHOLD is used to tristate the address bus during a cache-invalidate cycle. The first step in this cycle is for an external device to apply logic 1 to the AHOLD input of the 80486SX. In response to this input, the address lines are immediately put into the high-Z state. Next, the external device puts the address of the main memory storage location that was modified onto the 80486SX's address bus and then switches $\overline{\text{EADS}}$ to logic 0 to signal the MPU that a valid address is on the address bus. In this case, the address lines are inputs to the MPU. If the internal cache subsystem identifies that the content of this memory location is stored in the cache, the value held in the cache is invalidated. Since the cache entry is no longer valid, consistency is restored between cache memory and main memory.

Bus Arbitration Interface. The DMA interface of the 80386DX is expanded in the 80486SX MPU to make it into what is called the *bus arbitration interface.* Two new signals, *backoff input* ($\overline{\text{BOFF}}$) and *bus request output* (BREQ), have been added to the interface. Let us look at the function of each of these signals.

The BREQ output of the 80486SX signals whether or not the MPU is generating a request to use the external bus. When a bus cycle is to be performed, BREQ is switched to logic 0 and remains at this level until the bus cycle is completed. This signal can be used by external circuitry to tell other bus masters that the MPU has a bus access pending.

$\overline{\text{BOFF}}$ is similar to the HOLD input of the MPU in that its active logic level tristates the bus interface signals. However, there are two differences between the operation of $\overline{\text{BOFF}}$ and HOLD. First, a bus backoff operation is initiated at the completion of the current clock cycle, not at the end of the current bus cycle. Moreover, no hold-acknowledge response is made to external circuitry. When $\overline{\text{BOFF}}$ returns to its inactive logic level, the interrupted bus cycle is restarted. External bus masters can use this input to quickly take over control of the system bus.

Memory Interface That Employs Dynamic Bus Sizing

The 80386DX has the ability to dynamically adjust the physical width of the data bus as either 32 bits or 16 bits. This capability has been expanded further in the 80486SX to enable sizing of the data bus as 32 bits wide, 16 bits wide, or 8 bits wide. The memory interface signals that select the size of the bus are *bus size 16* ($\overline{\text{BS16}}$) and *bus size 8* ($\overline{\text{BS8}}$). If both of these signals are at their active 0 logic level, the bus defaults to 8-bit-wide operation.

EXAMPLE 16.12

How would the data bus of the 80486SX be configured if both $\overline{\text{BS16}}$ and $\overline{\text{BS8}}$ are wired to +5V?

Solution

The 80486SX's data bus is permanently set for 32-bit-wide operation.

Figure 16–47 shows a typical application where the memory address space is partitioned in this way. Note that a single-boot EPROM is used in an 8-bit memory configuration, a memory-mapped peripheral that has a 16-bit data bus in its microprocessor interface is treated as a 16-bit segment of memory, and the main program and data storage-memories are both 32 bits wide.

If the data bus is sized at 16 bits or 8 bits, additional bus cycles may need to be performed to complete a read- or write-data transfer. For instance, if $\overline{BS16}$ is logic 0 when an instruction is executed that initiates an aligned 32-bit read from the memory-mapped peripheral in Fig. 16–47, the data bus is set for 16-bit width, and two consecutive 16-bit data-read bus cycles will take place. Moreover, when an instruction fetch occurs from the boot ROM, $\overline{BS8}$ is active during this memory operation, and the data transfer is performed with four 8-bit data-read bus cycles.

EXAMPLE 16.13

How many bus cycles are required to read the double word of data from address $F1001_{16}$ in the address space of the memory-mapped peripheral in Fig. 16–47?

Solution

Since the 32-bit word at address $F1001_{16}$ is misaligned, three 16-bit read-data transfer bus cycles must take place.

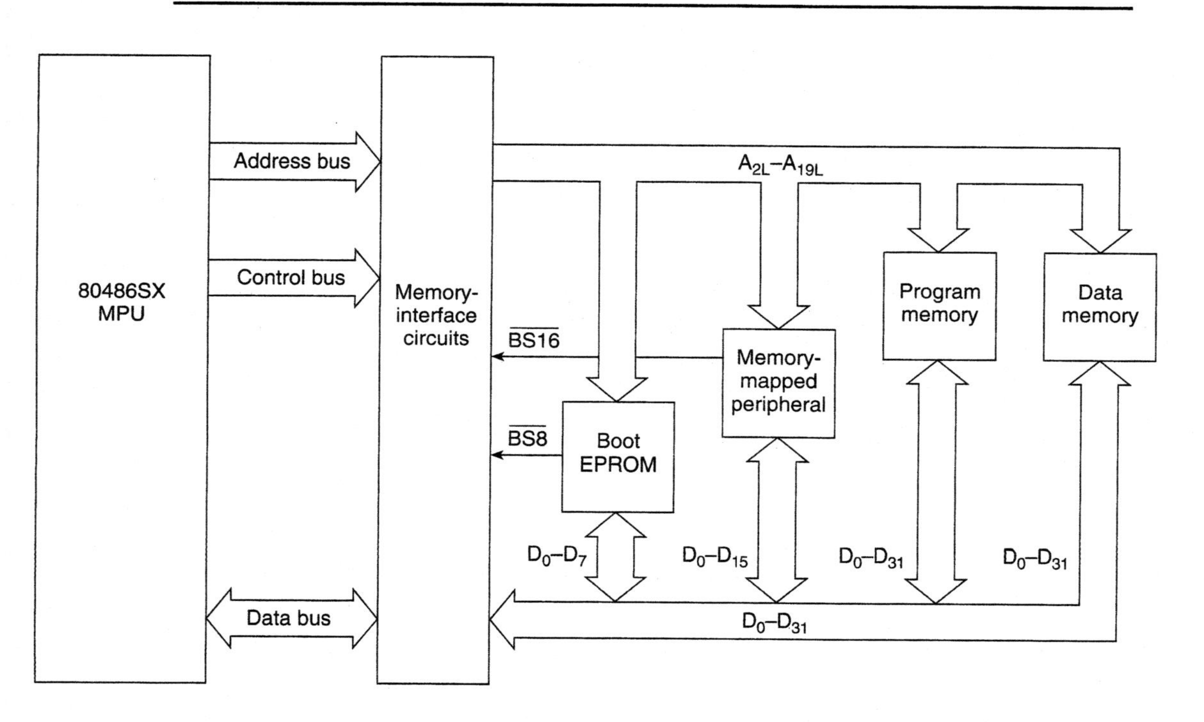

Figure 16–47 8-bit, 16-bit, and 32-bit memory subsystems.

Nonburst and Burst Bus Cycles

In our description of the memory interface signals, we found that the 80486SX's data bus can be dynamically configured for an 8-bit, a 16-bit, or a 32-bit mode of operation. For each of these modes, the 80486SX can perform either of two bus cycles, a *nonburst bus cycle* or a *burst bus cycle.* Both of these cycles can be made cacheable or noncacheable. Here we look briefly at the bus activity for each of these bus cycles.

Nonburst, Noncacheable Bus Cycle. The timing diagram in Fig. 16–48 shows the sequence of bus activity that takes place as the 80486SX reads or writes data to memory or an I/O device with a *nonburst, noncacheable bus cycle.* Note that the minimum duration of a bus cycle is two clock cycles. They are identified as T_1 and T_2 in the bus cycle timing diagram.

Looking at Fig. 16–48, we see that early in clock cycle T_1 the address (A_2 through A_{31}), byte enables ($\overline{BE}_0$ through $\overline{BE}_3$), and memory indication signals ($M/\overline{IO}$, $D/\overline{C}$, and $W/\overline{R}$) are made available and latched into external circuitry with the transition of $\overline{ADS}$. Assuming that the data transfer is to take place in a single bus cycle, $\overline{BLAST}$ is switched to the 0 logic level during clock cycle T_2. This tells the external circuitry that the data transfer is to be complete at the end of the current bus cycle. Therefore, at the end of T_2, external circuitry switches $\overline{RDY}$ to logic 0 to tell the MPU that the data transfer is to take place.

A bus cycle can be extended by any number of clock cycles by holding $\overline{RDY}$ at logic 1 during T_2.

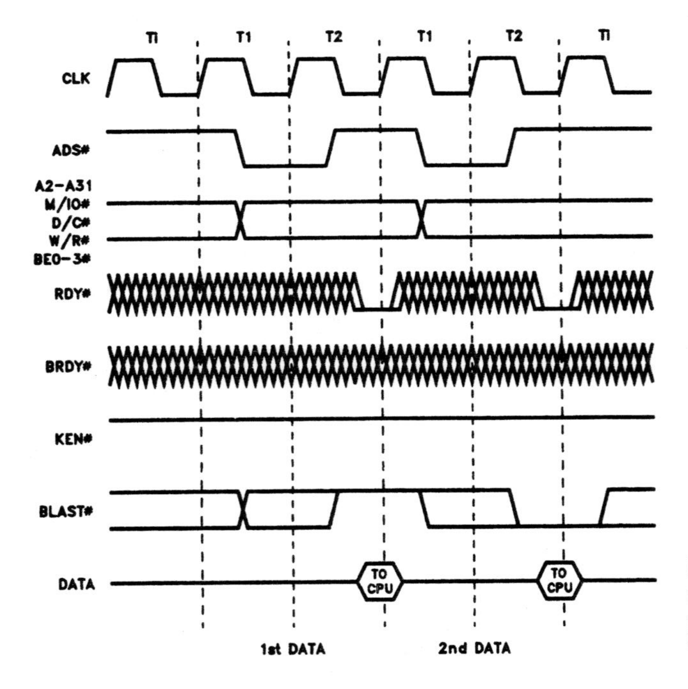

Figure 16–48 Nonburst, noncacheable bus cycle. (Reprinted by permission of Intel Corporation. Copyright/Intel Corp. 1992)

Nonburst, Cacheable Bus Cycle. Earlier we pointed out that the $\overline{KEN}$ input determines whether or not a bus cycle is cacheable. Figure 16–49 shows the 80486SX's *nonburst, cacheable bus cycle.* Here we see that the bus cycle starts the same way, but later in T_1 the external circuitry switches the $\overline{KEN}$ input to logic 0. This indicates that the cycle is a cacheable bus cycle. Only memory-read bus cycles are cacheable; therefore, the MPU ignores $\overline{KEN}$ during all write and I/O bus cycles.

Information is stored in the internal cache memory as *lines,* which are 16 bytes wide. Whenever a cacheable read cycle is performed, a complete line of code or data (four double words), instead of a single 32-bit word, is read from memory. Looking at the timing diagram in Fig. 16–49, we see that four read-data transfers take place. The address is automatically adjusted to point to the appropriate double word after each read operation and then $\overline{RDY}$ is made active. For this reason, the complete bus cycle takes a total of eight clock states. Note that $\overline{BLAST}$ is not switched to the active 0 logic level until T_2 of the fourth and last double word of data is read.

Burst, Cacheable Bus Cycle. The *burst, cacheable bus cycle* of Fig. 16–50 is similar to the nonburst, cacheable bus cycle we just described in that four read operations are performed. One difference is that during a burst bus cycle external circuitry signals that burst data transfers are to take place by replying with logic 0 on $\overline{BRDY}$ instead of $\overline{RDY}$. A second and very important difference is that only the first data transfer takes two clock

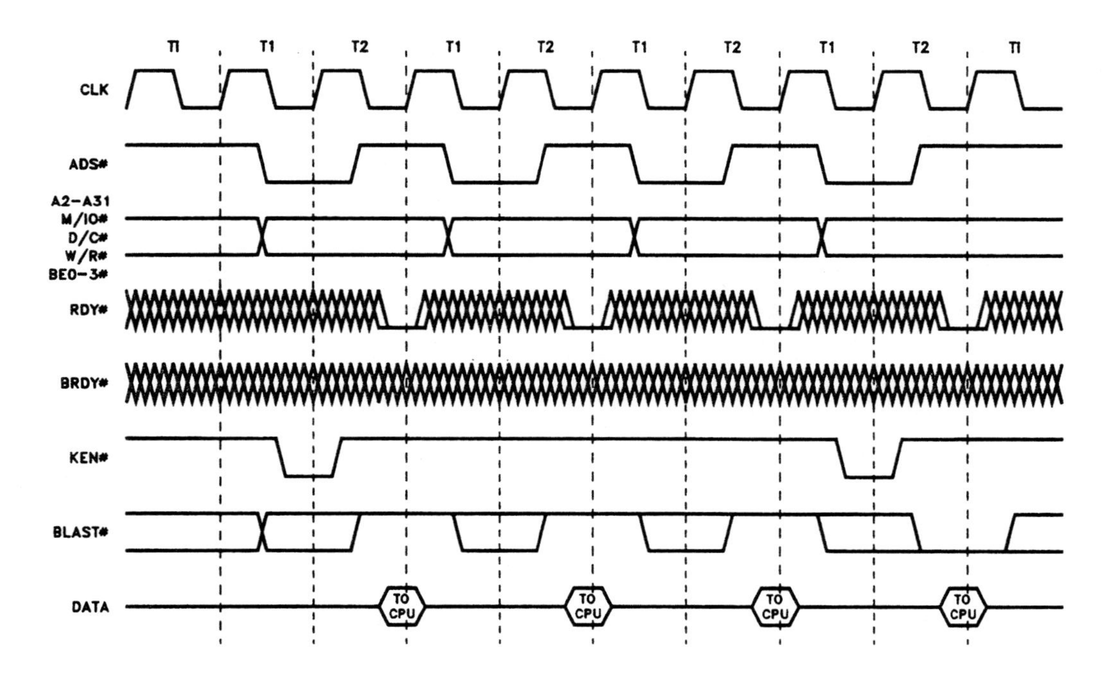

Figure 16–49 Nonburst, cacheable bus cycle. (Reprinted by permission of Intel Corporation. Copyright/Intel Corp. 1992)

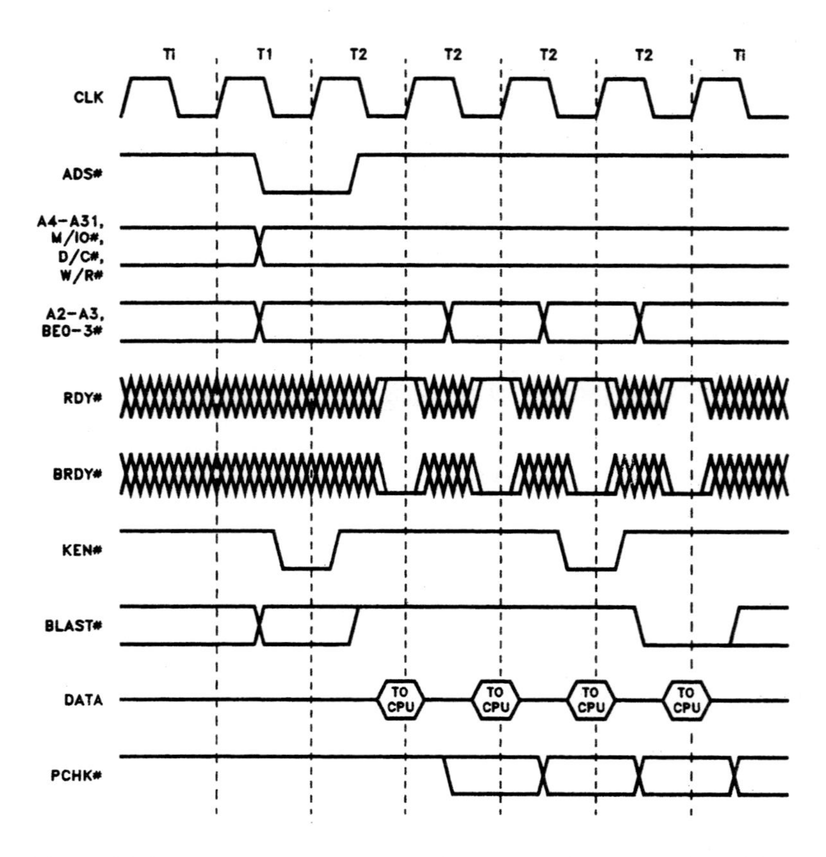

Figure 16–50 Burst, cacheable bus cycle. (Reprinted by permission of Intel Corporation. Copyright/Intel Corp. 1992)

cycles. Note that all four data transfers of the burst bus cycle are completed in just five clock states.

Cache Memory

When a microcomputer system employs a large main memory subsystem of several megabytes, it is normally made with high-capacity but relative slow-speed dynamic RAMs, EPROMs, and FLASH memory. Even though DRAMs are available with access times as short as 60 ns and EPROMs as fast as 70 ns, these high-speed versions of the devices are expensive and still too slow to work in a microcomputer system that is running with zero wait states. For example, an 80486SX-25 microprocessor running at 25 MHz requires DRAMs with a 40-ns access time to implement a zero-wait state memory design. For this reason, wait states are introduced in all bus cycles to data and program memory. These waits states degrade the overall performance of the microcomputer system.

Addition of a cache memory subsystem to the microcomputer provides a means for improving overall system performance while still permitting the use of low-cost, slow-speed memory devices in main memory. In a microcomputer system with cache, a second smaller, but very fast memory section is added between the MPU and main memory subsystem. Figure 16–51(a) illustrates this type of system architecture. This small, high-speed memory section is known as the *cache memory.* The cache is designed with fast, more expensive static RAMs and can be accessed without wait states. During system operation, the cache memory contains recently used instructions and data. The objective is that most of the time the MPU accesses code and data in the cache rather than from main memory. This results in close to zero-wait-state memory system operation even though accesses of the main memory require one or more wait states, thus resulting in higher performance for the microcomputer. As Fig. 16–51(b) shows, the 80486SX's cache differs in that it is on-chip—that is, it is internal to the MPU.

External cache memories are widely used in high-performance 80486-based microcomputer systems today. Notice in Fig. 16–51(a) that at one side the external cache memory subsystem attaches to the local bus of the MPU, and at the other side it drives the system bus of the microcomputer's main memory subsystem. External caches typically range in size from 32Kbytes to 512Kbytes and can be used to cache both data and code. The 80486SX's internal cache also stores both code and data, but is smaller (8Kbytes) in size. The internal cache of the 80486SX is called the *first-level cache* and the external cache a *second-level cache.*

Function of a Cache. Let us continue our examination of cache for the microcomputer system by looking at how it affects the execution of a program. The first time the MPU executes a segment of program, one instruction after the other is fetched from main memory and executed. The most recently fetched instructions are automatically saved in the cache memory. That is, a copy of these instructions is held within the cache. For example, a segment of program that implements a loop operation could be fetched, executed, and placed in the cache. In this way, we see that the cache always holds some of the most recently executed instructions.

Now that we know what the cache holds, let us look at how the cached instructions are used during program execution. Many software operations involve repeated execution of the same sequence of instructions. A loop is a good example of this type of program structure. In Fig. 16–52, we find that the first execution of the loop references code held in the slow main program memory. During this access, the routine is copied into the cache. When the instructions of the loop are repeated, the MPU reaccesses the routine by using the instructions held in the cache instead of refetching them from the main memory. Accesses to code in the cache are performed with no wait states, whereas those of code in main memory normally require multiple wait states. In this way, we see that the use of cache has reduced the number of accesses made from the slower main memory. The more frequent instructions held in the cache are used, the closer to zero-wait-state operation is achieved, the more the overall execution time of the program is decreased, and the higher is the performance for the microcomputer system.

During execution of the loop routine, data operands that are accessed can also be cached in the internal cache of the 80486SX. If these operands are reaccessed during the

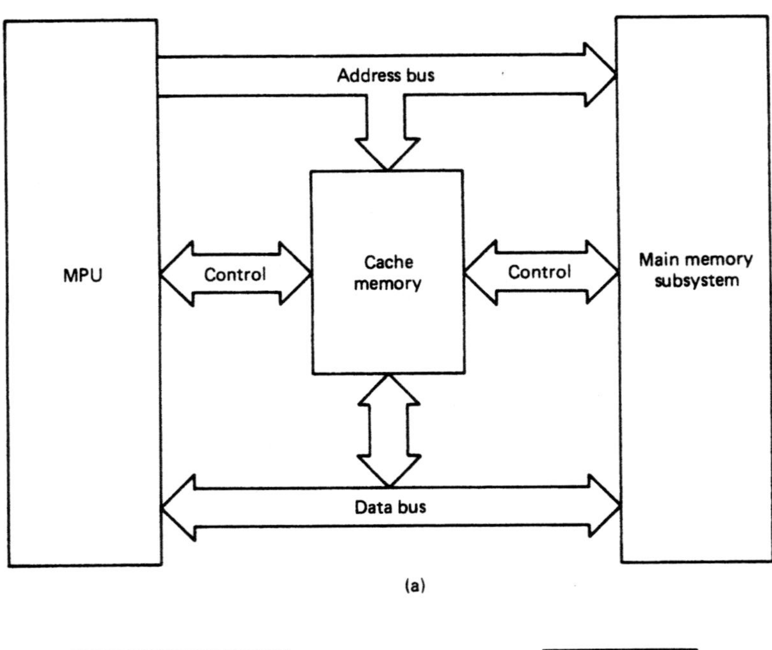

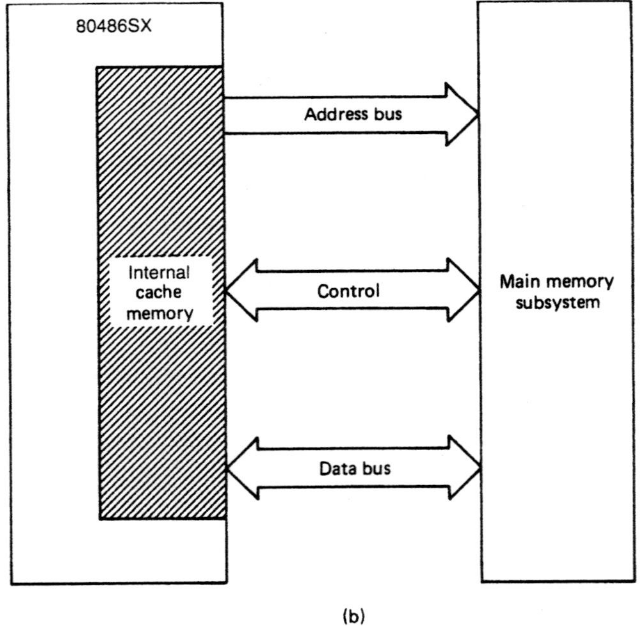

Figure 16–51 (a) Microcomputer system with cache memory. (b) Internal cache of the 80486SX.

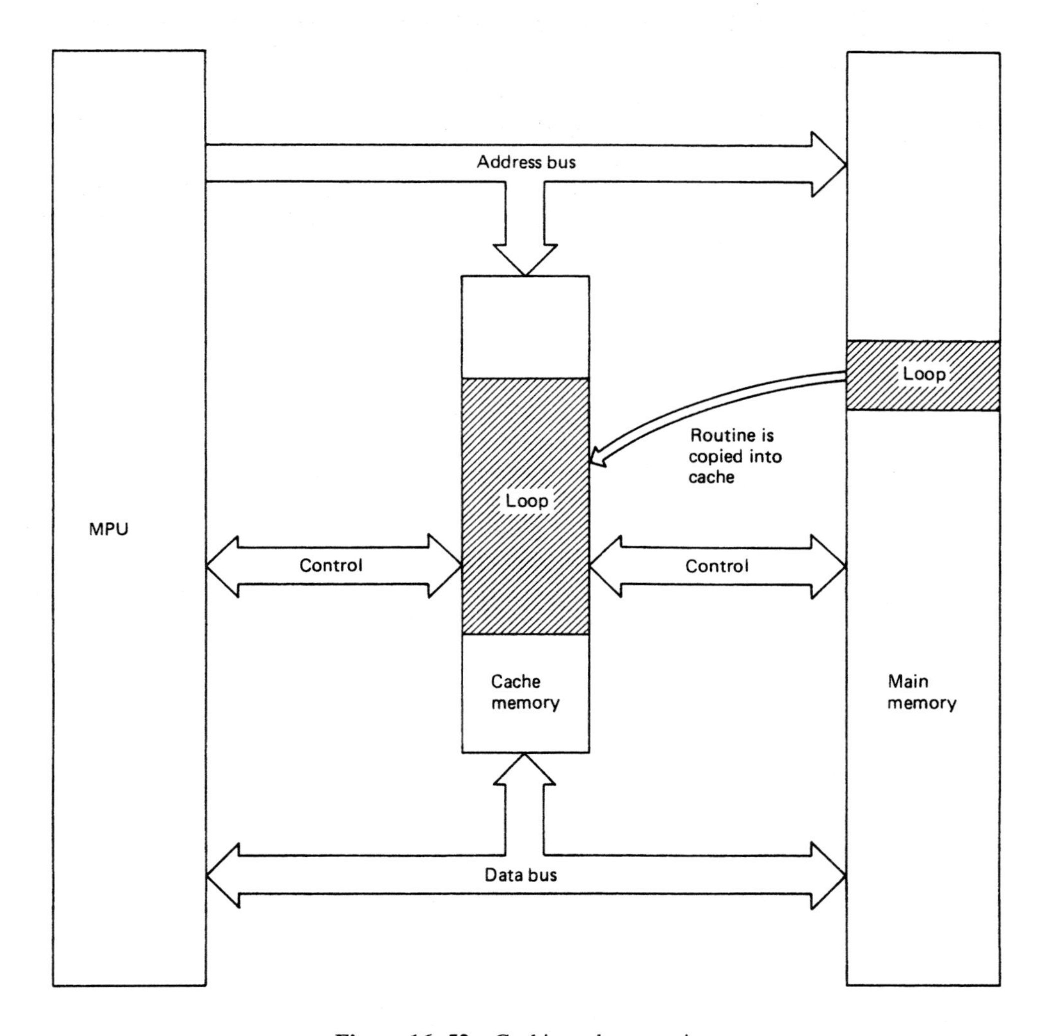

Figure 16–52 Caching a loop routine.

repeated execution of the loop, they are read from the cache instead of from main data memory. This further reduces execution time of the segment of program.

Cache Hit, Cache Miss, and Hit Rate. We just found that the concept behind cache memory is that it stores recently used code and data and that if this information is to be reaccessed, it may be read from the cache with zero-wait-state bus cycles rather than from main memory. When the address of a code or data-storage location to be read is output on the local bus, the cache subsystem must determine whether or not the information to be accessed resides in both main memory and cache memory. If it does, the memory cycle is considered a *cache hit* condition. In this case, a bus cycle is not initiated to the main memory subsystem; instead, the copy of the information in the cache is accessed.

On the other hand, if the address output on the local bus does not correspond to information that is already cached, the condition represents what is called a *miss.* This time the MPU reads the code or data from main memory and writes it into a corresponding location in cache.

Hit rate is a measure of how effective the cache subsystem operates and is defined as the ratio of the number of cache hits to the total number of memory accesses, expressed as a percentage. That is, hit rate equals

$$\text{Hit rate} = \frac{\text{Number of hits}}{\text{Number of bus cycles}} \times 100\%$$

The higher the value of hit rate, the better the cache memory design. For instance, a cache may have a hit rate of 85 percent. This means that the MPU reads code or data from the cache memory for 85 percent of its memory bus cycles. In other words, just 15 percent of the memory accesses are from the main memory subsystem. The hit rate is not a fixed value for a cache design. It depends on the code being executed and data used. That is, a hit rate may be one value for a specific application program and a totally different value for another.

Cache design also affects the hit rate. For instance, the size, organization, and the update method of the cache memory subsystem all determine the maximum hit rate that may be achieved by a cache. Practical cache memories for microcomputer systems range in size from as small as 8Kbytes to as large as 512Kbytes. In general, the larger the size of the cache, the higher the hit rate. This is because a larger cache can contain more data and code, which yields a greater chance that the information to be accessed resides in the cache. However, the improvement in hit rate decreases if we keep increasing cache size, whereas the cost of the cache subsystem may increase substantially.

Types of Cache Memory Organizations. Three widely used cache memory organizations are those known as the *direct-mapped cache, two-way set associative cache,* and *four-way set associative cache.* The direct-mapped cache is also called a *one-way set associative cache.* Figure 16–53 illustrates the organization of a 64Kbyte direct-mapped cache memory. Note that the cache memory array is arranged as a single 64Kbyte bank of memory, and the main memory is viewed as a series of 64Kbyte pages, denoted page 0 through page *n.* Note that the data-storage location at the same offset (X) in all pages of main memory, X(0) through X(n) in Fig. 16–53, map to a single storage location, marked X, in the cache memory array. That is, each location in a 64Kbyte page of main memory maps to a different location in the cache memory array.

On the other hand, the 64Kbyte memory array of a two-way set associative cache memory is organized into two 32Kbyte banks. That is, the cache array is divided two ways, *BANK A* and *BANK B.* This cache memory subsystem configuration is shown in Fig. 16–54. Again main memory is mapped into pages equal to the size of a bank in the cache array. But because a bank is now 32Kbyte, there are twice as many main-memory pages as in the direct-mapped organization. In this case, the storage location at a specific offset in every page of main memory can map to the same storage location in either the A or B

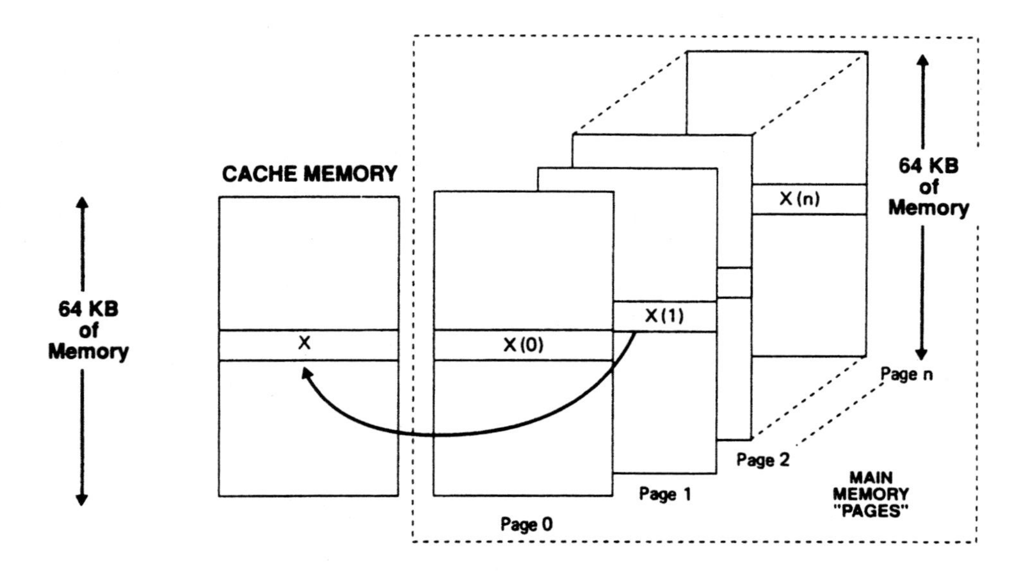

Figure 16–53 Organization of a direct-mapped memory subsystem. (Reprinted by permission of Intel Corporation. Copyright/Intel Corp. 1990)

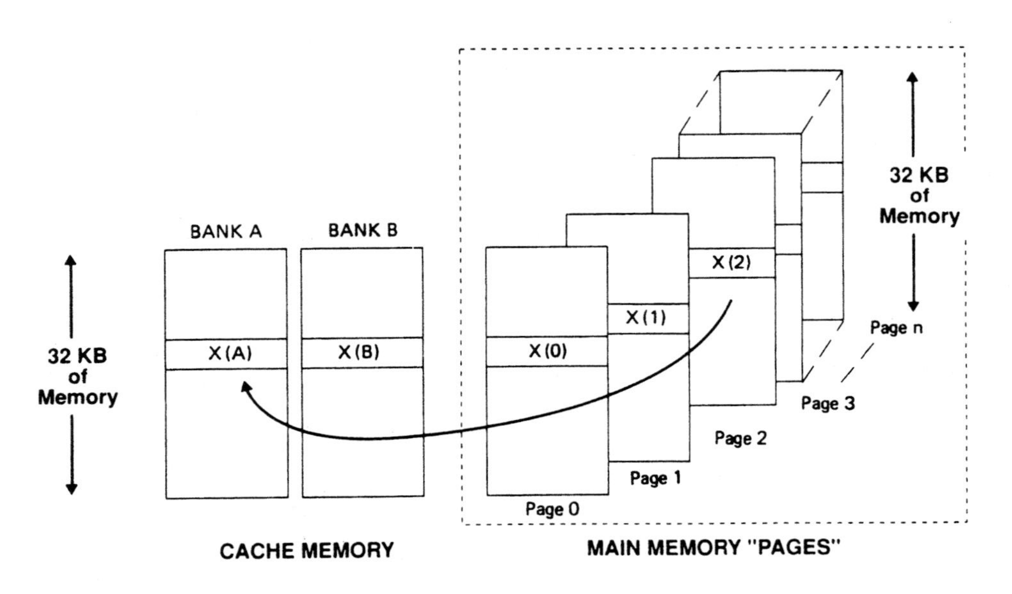

Figure 16–54 Organization of a two-way set associative memory subsystem. (Reprinted by permission of Intel Corporation. Copyright/Intel Corp. 1990)

bank. For example, the contents of storage location X(2) can be cached into either X(A) or X(B). The two-way set associative organization results in higher hit rate operation.

The 80486SX's internal 8Kbyte cache memory uses a four-way set associative memory. Therefore, its configuration is similar to that shown in Fig. 16–53 except that it is arranged into four 2Kbyte banks.

An example of a memory update method that affects hit rate is the information replacement algorithm. Replacement methods are based on the fact that there is a higher chance that more recently used information will be reused. For instance, the two-way set associative cache organization permits the use of a *least recently used* (LRU) replacement algorithm. In this method, the cache subsystem hardware keeps track of whether information X(A) in the BANK A storage location or X(B) in BANK B is most recently used. For example, let us assume that the value at storage location X(A) in BANK A of Fig. 16–54 is X(0) from page 0 and that it was just loaded into the cache. On the other hand, X(B) in BANK B is from page 1 and has not been accessed for a long time. Therefore, the value of X(B) in BANK B is tagged as the least recently used information. When a new value of code or data, for instance, from offset X(3) in page 3 is accessed, it must replace the value of X(A) in BANK A or X(B) in BANK B. Therefore, the cache replacement algorithm automatically selects the cache storage location corresponding to the least recently accessed bank for storage of X(3). For our example, this would be storage location X(B) in BANK B. This shows that the replacement algorithm retains more recently used information in the cache memory array. The four-way set associative cache memory of the 80486SX MPU uses the LRU replacement algorithm. This results in a higher maximum hit rate for the internal cache and a higher level of performance for the microcomputer system.

We found earlier that use of a cache reduced the number of accesses of main memory over the system bus. In our example of a loop routine, we saw that repeated accesses of the instructions that perform the loop operation were made from the internal cache, not from main memory. Therefore, fewer code and data accesses are performed across the system bus. That is, availability of the bus has been increased for external devices and is another advantage of using cache in a microcomputer system. The freed-up bus bandwidth is available to other bus masters, such as DMA controllers or other processors in a multiprocessor system.

Organization and Operation of the 80486SX's Internal Cache. Let us next examine the organization and operation of the 80486SX's internal 8Kbyte four-way set associative cache memory. We begin by determining how data are stored in the cache memory array.

Since the internal cache uses a four-way set associative organization, the data-storage array is partitioned into four separate 2Kbyte areas. Figure 16–55 illustrates how this memory is organized. We will refer to these areas of memory as *SET 0* through *SET 3*. Data are loaded, stored, validated, and invalidated in 16-byte-wide elements called a *line of data*. The 80486SX's cache does not support filling of partial lines. Therefore, if a single double word of data is to be read from memory and copied into the cache, the MPU must fetch from memory the complete line in which this double word is contained. This is the reason that cacheable read bus cycles initiate four double-word data transfers. That is, they always access code or data one line (16 bytes) at a time so that a complete

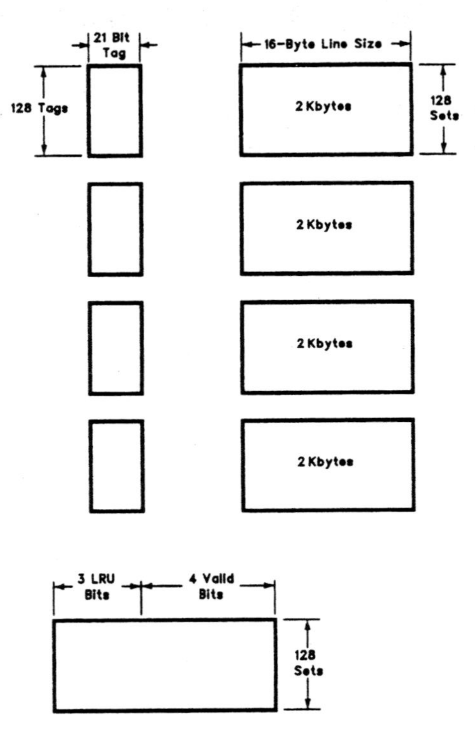

Figure 16–55 Organization of the on-chip cache of the 80486SX. (Reprinted by permission of Intel Corporation. Copyright/Intel Corp. 1992)

line of the cache gets filled. In this way, we see that the contents of each 2Kbyte bank in the cache's storage array is further arranged into 128 lines of data.

Associated with each data-storage set is a separate tag directory. The tag directory contains 128 21-bit tag entries; one corresponds to each line of data in the set. Each tag entry includes bits of information about the use of the line and whether or not it currently holds valid information.

When the MPU initiates a read operation and the information that is to be accessed is already in the internal cache, the information is obtained without performing external bus cycles. Instead, the information is simply read from the cache memory. However, if a cache miss occurs, a line of code or data must be read from main memory and copied into the cache. The on-chip cache circuitry must determine if there is room in the cache and, if not, which of the current valid lines of information is to be replaced. It does this by checking the information in the tag directory. If the cache is found to contain invalid lines of data or code, one of them can be simply replaced with the new information. On the other hand, if there are no vacant line-storage locations, the least recently used mechanism of the cache automatically checks the use information in the tag to determine which valid line of information to replace.

Let us next look at what happens when a data-write operation is performed. Whenever a new value of data is to be written to a storage location in memory, the internal cache circuitry must first be checked to confirm whether or not the contents of this storage location also exist within the cache. If they do, the value must be either invalidated or updated as part of the write operation. Otherwise, a cache data consistency problem is created.

The 80486SX's on-chip cache is implemented with a write-update method known as *write-through.* With this method, all write bus cycles that result in a cache hit automatically update both the corresponding storage location in the internal cache and external memory. That is, write operations to main memory can be viewed as going through the cache. On the other hand, for a cache miss, the data are written only to main memory. Remember that a cache write-through operation can be enabled or disabled with the no-write-through (NW) bit of CR_0. However, write-through would not be disabled during normal operation.

Enabling and Disabling Internal Caching. The 80486SX is equipped with a variety of methods for controlling the operation of the internal cache. For instance, the complete cache memory can be turned on or off, the memory-address space can be mapped with cacheable and noncacheable areas, and external circuitry can define any bus cycle as cacheable or noncacheable. Let us briefly review how each of these cache memory controls are implemented on the 80486SX.

Remember that the operation of the cache memory can be enabled or disabled under software control. Logic 1 in the cache fill disable (CD) bit of control register CR_0 can be used to turn off filling of the cache. However, this does not completely disable the cache; it just stops it from being refilled. To completely disable the cache, the no-write-through (NW) bit in CR_0 must also be set to 1 and then the cache must be flushed. The flush operation is needed to remove the stale data that were left in the cache when cache fill was turned off.

Mapping of parts of the memory address space as cacheable or noncacheable can be achieved either through software or hardware. Under software control, each page of the memory address space can be configured as cacheable or noncacheable with the page-level cache-disable (PCD) bit in its page table entry. For instance, to make a page of memory noncacheable the PCD bit is made 0. The memory address space can be mapped cacheable or noncacheable on a byte-wide basis by external circuitry. The *cache-enable* ($\overline{KEN}$) input can be used to indicate whether or not the data for the current bus cycle should be cached. By decoding addresses in external circuitry and returning logic 1 at $\overline{KEN}$, a part of the address space can be designated as noncacheable.

Flushing the Cache. When the internal cache is flushed, all of the line valid bits in the tag directory are invalidated. Therefore, after a *flush* occurs the cache is empty and will need to be refilled. The cache is flushed whenever the MPU is reset; it can be flushed under software control by executing the invalidate cache (INVD) instruction or with external circuitry by activating the $\overline{FLUSH}$ input.

Cache Line Invalidations. We just described how the complete contents of the cache can be invalidated with a flush operation. It is also possible to invalidate individual lines of information within the cache. This is known as a *cache line invalidation* and is normally done to make the contents of the internal cache consistent with that of external memory.

If an external device changes the contents of a storage location in external memory and the value of this storage location is currently held in the 80486SX's internal cache memory, a cache consistency problem can occur. That is, the corresponding storage location in cache and external memory no longer contain the same value and the value in cache is no longer valid. If the 80486SX were to read this storage location, the incorrect value in the cache would be accessed, and if this data were processed and written back to memory, the results in external memory would then also be wrong.

To protect against inconsistency problems between the contents of the internal cache and external memory, external circuitry needs to initiate a cache line invalidation operation each time an external device modifies the content of a storage location in external memory. This is done by initiating an invalidate bus cycle. The first step in this process is to switch *address hold* (AHOLD) to logic 1. This puts the address bus lines into the high-Z state. Next, the external circuitry puts the address of the external memory storage location whose contents were changed onto the address lines and signals the 80486SX that a valid external address is available by switching $\overline{\text{EADS}}$ (*external address*) to logic 0. Now the address lines of the MPU act as inputs, instead of outputs. If this address corresponds to an element of data that is currently held in cache, the corresponding cache entry is invalidated. In this way, cache consistency is restored.

Internal Exceptions

Earlier we indicated that the external hardware interrupt interface of the 80486SX MPU is exactly the same as that of the 80386DX processor. In fact, the only differences between the interrupt/exception processing capability of the 80486SX and 80386DX are that one additional internal exception function is defined for the 80486SX and one that was performed by the 80386DX is no longer supported in the 80486SX. Next we look briefly at these two changes in exception processing.

The new exception that is activated for the first time in 80486 family MPUs is the *alignment check exception.* In our description of the software model of the 80486SX in Chapter 15, we identified two new control bits, the alignment-check (AC) flag in EFLAGS and the alignment-mask (AM) control bit in CR_0, that are used to enable address alignment checking for memory-access operations. At that time, we indicated that this function gives the 80486SX the ability to detect an attempt to access an unaligned element of data. This exception is detected only for memory accesses initiated while in protected mode and executing user-code privilege level 3. When this option is enabled, any attempt by the program to access an unaligned operand in memory results in an alignment-check exception through exception gate 17.

The 80386DX exception function not supported by 80486 family MPUs is exception 9, coprocessor segment overrun.

▲ 16.9 OTHER 80486 FAMILY MICROPROCESSORS—80486DX2 AND 80486DX4

The 80486DX2 and 80486DX4 are hardware- and software-compatible upgrades of the 80486DX MPU. They contain a number of architectural enhancements that result in higher performance for the 80486-based microcomputer system. Note in Fig. 15–1 that the 80486DX2-50 has an iCOMP rating of 231. This value is close to twice the rating of the 80486DX-25 and very close to that of the 80486DX-50.

Two architectural enhancements made in the 80486DX2 are *clock doubling* and *write-back enhanced cache.* 80486DX2 MPUs are driven by what is called a ½ × *clock,* instead of a 1 × clock like the 80486SX and 80486DX. That is, the signal applied to the clock input of a 50-MHz 80486DX2 MPU (80486DX2-50) is actually a 25-MHz signal, or half the clock speed rating of the device. Similarly, the 80486DX2-66 is driven by a 33-MHz clock signal. On-chip clock multiplying circuitry doubles the input clock frequency to produce a 66-MHz clock that runs the internal circuitry of the MPU. Only the core of the 80486DX2-66 MPU operates at 66MHz, not the external bus interface. Like the 80486DX-33, the external bus interface of the 80486DX2-66 is operated at 33MHz. In this way, we see that the 80486DX2 achieves a higher level of performance by running its internal circuitry twice as fast. At the same time, it permits simpler external interface circuit designs by maintaining the clock speed of the external local bus interface at 33 MHz.

The 80486DX2's 8Kbyte on-chip cache memory is configurable to operate with either the write-through or write-back methods for updating external memory. Actually, the lines of the cache memory array can be individually configured as write-through or write-back. The write-though mode of operation is compatible with that of the 80486SX and 80486DX MPUs, whereas write-back is an improvement that offers higher system-level performance. Unlike the write-through operation, write-back updates of external memory are not performed at the same time the information is written into the internal cache. Instead, the updates are accumulated in the on-chip cache memory subsystem and written to memory at a later time. This reduces the bus activity and therefore enhances the microcomputer's performance.

When in write-back mode, the 80486DX2's cache also supports snooping. The ability to snoop is needed to assure that data coherency is maintained between the data in the on-chip cache and the external main memory. A data coherency problem can occur whenever another bus master attempts to read information from or write information into the main memory. For example, if the storage location being read in main memory corresponds to an address whose data is cached within the 80486DX2 and modified, but not yet written back to main memory, the information that would be read from main memory is incorrect. In this case, a write-back must be performed to update main memory before the bus master can complete the read cycle. On the other hand, if the bus master writes to a storage location in main memory whose data are also cached, the value held in the cache must be invalidated. In this way, we see that whenever a bus master accesses main memory, a snoop bus cycle must be performed to determine if the content of the storage location to be accessed is cached within the MPU and, if so, initiate a write-back. Otherwise, cache coherency may be lost. The 80486DX2 employs the *MESI* (modify/exclusive/shared/invalid) write-back cache-consistency protocol.

Figure 16–56 is a block diagram that shows the new pin functions defined for the 80486DX2 and 80486DX4 MPUs. Added are four signals, $\overline{\text{CACHE}}$, $\overline{\text{HITM}}$, INV, and $\overline{\text{WBWT}}$, for use in control of the 80486DX2's write-back cache memory. The cacheability ($\overline{\text{CACHE}}$) output switches to its active level for cacheable data reads, instruction code fetches, and data write-backs. Logic 0 at $\overline{\text{CACHE}}$ signals external circuitry that a cacheable read cycle or a burst write-back cycle is taking place.

The hit/miss ($\overline{\text{HITM}}$) output and invalidate request (INV) input are used during snoop bus cycles of the on-chip cache. The $\overline{\text{HITM}}$ output is activated by the cache-

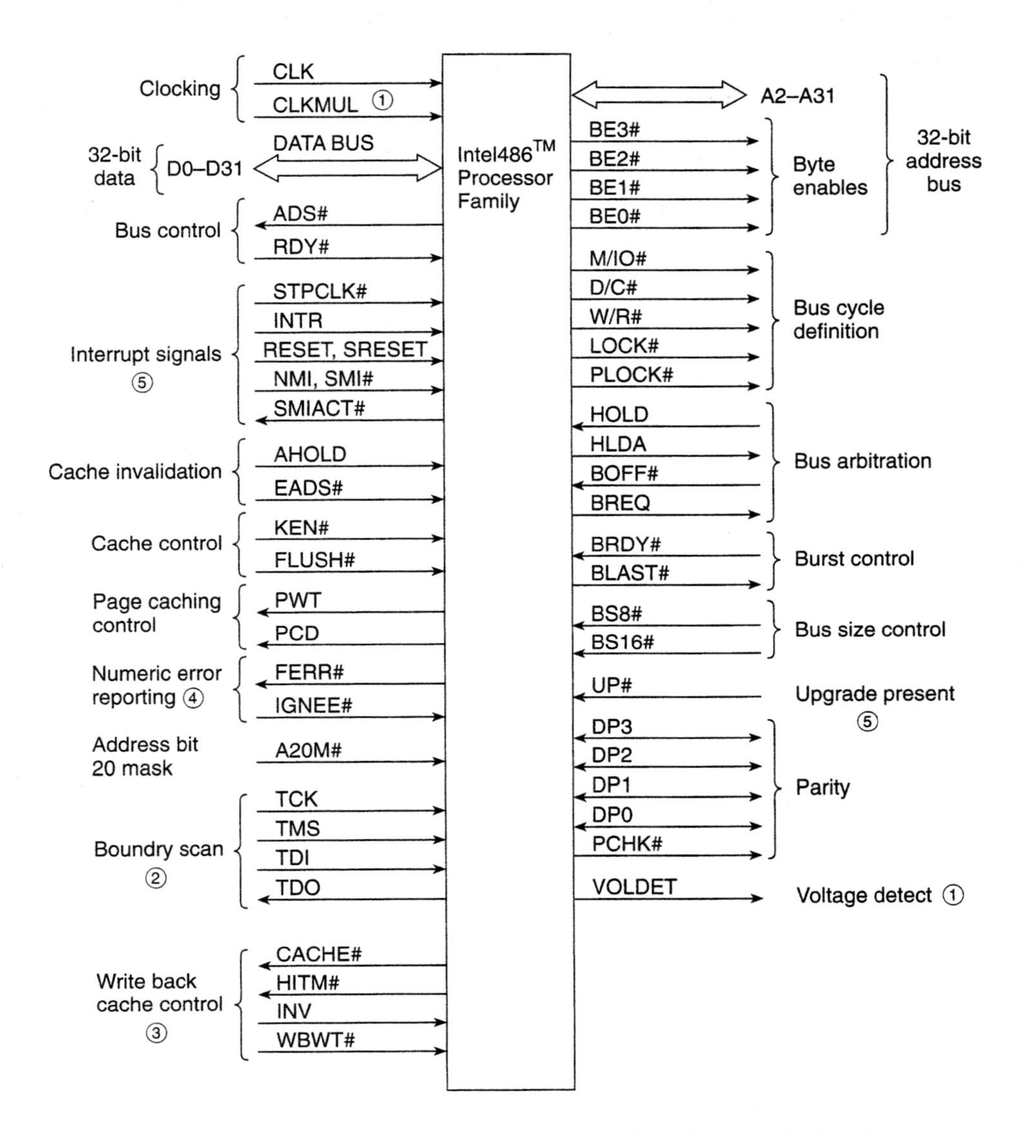

Figure 16–56 Block diagram of the 80486 MPU, including new pin functions for the 80486DX2 and 80486DX4. (Reprinted by permission of Intel Corporation. Copyright/Intel Corp. 1994)

coherency protocol of the 80486DX2. If during a snoop bus cycle, the line of information checked for is found to be cached and modified but not yet written back to main memory, this fact is signaled to the external circuitry by logic 1 at $\overline{\text{HITM}}$. If INV is at its active 1 logic level during a snoop cycle for a write to main memory that identifies a line of data currently cached on-chip, this line of data is invalidated whether the cache is configured for write-through or write-back operation. However, if a line of data is found in the cache and it has been modified, it is first written back to main memory and then invalidated. INV should be held at logic 0 during snoop cycles for a read of main memory. In this case, the snoop cycle initiates a write-back to main memory but does not invalidate the data in the cache. In this way, cache invalidations are minimized.

The 80486DX2's cache memory is configured for the write-back mode of operation as part of the hardware-reset process. If the write-back/write-through ($\overline{\text{WBWT}}$) input is held at logic 1 for at least two clock periods before and after the falling edge of RESET, the write-back configuration is enabled for the cache. The function of the $\overline{\text{FLUSH}}$ input changes when the cache is set up for write-back mode. Now logic 0 at $\overline{\text{FLUSH}}$ initiates a write-back of all lines of data in the cache that are modified before the contents of the cache are invalidated.

With the 80486DX4 came additional architectural enhancements and even higher performance. This device is designed to operate off a 3.3-V dc power supply instead of 5V dc; the clock-scaling circuitry is enhanced to permit multiplication of the clock by 2, 2.5, or 3; and the size of the cache is increased from 8Kbytes to 16Kbytes. The result is an increase of the iCOMP performance rating for the 80486DX4-100 to 435, almost 50 percent higher than the 80486DX2-66.

The V_{cc} power supply of the 80486DX4 MPU is rated at +3.3 V dc ± 0.3V. This does not mean that the 80486DX4 can only be used in 3.3-V system designs. Its inputs are 5-V tolerant and outputs are TTL compatible. Therefore, the 80486DX4 can be interfaced to 5-V logic components even though the internal circuitry is operating at 3.3 V. This is known as a *mixed voltage system.* Applying 5V dc to the V_{cc5} pin enables mixed voltage system operation.

Earlier we pointed out that the clock-multiplier circuitry in the 80486DX4 MPU has been enhanced to permit scaling of the clock by 2, 2.5 or 3. The logic level of a new input pin, clock multiplier (CLKMUL), determines the value of the multiplier. The logic level of this input is sampled during hardware reset of the MPU to select the clock multiplier. Logic 1 at this input selects the clock-tripled mode of operation. For instance, if the 80486DX4-100 is run by a 33-MHz clock signal, the internal operation of the MPU is 99MHz, and the speed of the local bus is maintained at 33 MHz. This increase in internal speed is the primary cause of higher performance for the 80486DX4-based microcomputer system.

The last architectural change in the 80486DX4 is that the size of the on-chip cache memory has been expanded to 16Kbytes. However, unlike the 80486DX2, this cache is write-though only and does not support the write-back mode of operation. Figure 16–57 shows the organization of the cache memory. Expanding the size of the cache results in a higher hit rate. This is another cause for the improved performance obtained with the 80486DX4 MPU.

▲ 16.10 PENTIUM PROCESSOR FAMILY

Many enhancements have been made to the hardware architecture of the MPUs in the Pentium processor family. Some examples of important improvements are that the data bus has been expanded to 64 bits, parity has been provided for the address bus, the on-chip cache

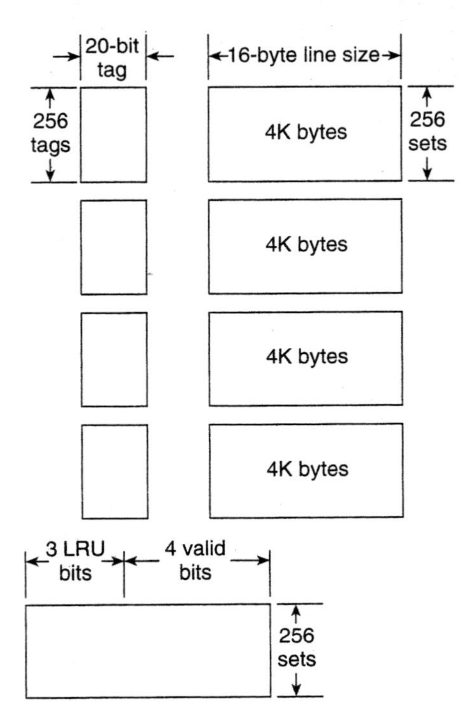

Figure 16–57 Organization of the on-chip cache of the 80486DX4. (Reprinted by permission of Intel Corporation. Copyright/Intel Corp. 1994)

memory is partitioned into separate code and data caches, the internal caches support either the write-through or write-back cache-update methods, and pipelined bus cycles have been implemented. These hardware changes play a key role in giving the Pentium processor its higher level of performance and simplifying the design of Pentium processor-based multiprocessor microcomputer systems.

We have seen that continued advances in processor technology have enabled Intel Corporation to integrate more and more transistors into its MPUs. The Pentium family of microprocessors were originally built on a 0.6-μm manufacturing process but have been moved to an even smaller geometry process. The reduction in transistor size achieved with these more advanced processes has permitted the integration of more than 3 million transistors into the Pentium processor.

Because of the process used to manufacture the Pentium processor, it must be powered by a 3.3-V power supply, V_{cc}. Its inputs and outputs are also rated at 3.3 V. Output signal lines are TTL-compatible in that they do meet the minimum high-logic level (V_{IHmin}) for a TTL input. For this reason, they may directly drive external interface circuits made with either 3.3-V or 5-V TTL logic devices. On the other hand, the inputs of the Pentium processor cannot tolerate more than 3.3 V as the V_{IHmax}. Therefore, inputs signal lines must be driven from outputs of 3.3-V logic devices or 5-V devices with open-collector outputs that are set up to convert between 5-V and 3.3-V logic levels.

The Pentium processor and its pin layout are shown in Figs. 16–58(a) and (b), respectively. Here we see that it is housed in a 296-pin staggered-pin grid array (SPGA) package.

EXAMPLE 16.14

What are the pin locations of the signals D_0 and D_{64}?

Solution

The pin layout diagram in Fig. 16–58(b) shows that signal D_0 is at pin D3 and signal D_{63} at pin H18.

In this section, we will examine some of the architectural advancements that were first introduced in the Pentium processor family.

Signal Interfaces of the Pentium Processor

Figure 16–59 presents a block diagram of the Pentium processor. Although a lot of changes have been made in the hardware architecture of the Pentium processor, many of the interface signals remain the same as those used on 80386 and 80486 family MPUs. For example, similar to the 80386DX and 80486SX, the type of bus cycle is defined by the code output on $M/\overline{IO}$, $D/\overline{C}$, and $W/\overline{R}$, and the interrupt interface consists of INTR, NMI, and RESET inputs. Most of the interface signals first introduced on the 80486 family of MPUs are also provided on Pentium processors. For instance, the signal lines $\overline{A20M}$, $\overline{BOFF}$, $\overline{BRDY}$, $\overline{FLUSH}$, $\overline{KEN}$, PWT, and PCD are all part of the Pentium processor's memory/IO and cache memory interfaces. Here we will focus on the interface signals first introduced with the Pentium processor family.

Memory/IO Interface. Earlier we pointed out that the memory/IO interface has been improved by making the data bus 64 bits wide and by adding parity on the address bus. The data bus now consists of bidirectional data lines D_0 through D_{63}. Because of the larger data bus, the number of byte-enable and data parity lines have been increased to eight. In Fig. 16–59, they are labeled $\overline{BE}_0$ through $\overline{BE}_7$ and DP_0 through DP_7, respectively. Extending this bus to 64 bits results in increased data-transfer rate between the MPU and its memory and I/O subsystem and higher-performance operation.

During all write cycles, the bus interface unit generates a code at DP_0 through DP_7 that produces even parity for each byte of data on D_0 through D_{63}. When a read bus cycle takes place, the data on D_0 through D_{63} and DP_0 through DP_7 are tested for even parity on a byte-wide basis. If a data parity error is detected in any byte of the quad-data word, this fact is signaled to external circuitry with logic 0 at the $\overline{PCHK}$ output. The *parity-enable* ($\overline{PEN}$) input is used to determine whether or not an exception is initiated when a data-read parity error occurs. Logic 0 at this input configures the MPU to initiate an exception automatically whenever a read parity error is detected.

In the Pentium processor, parity generation and checking has been added to the address bus. Whenever an address is output on A_3 through A_{31}, an even parity bit is generated and output at pin *address parity* (AP). In this way, the memory subsystem can

(a)

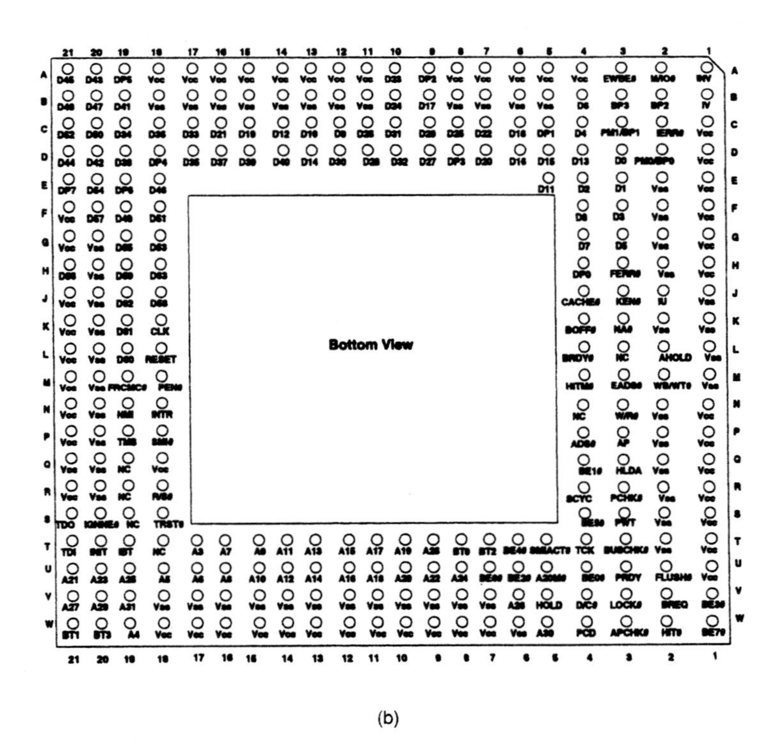

(b)

Figure 16–58 (a) Pentium processor IC. (Reprinted by permission of Intel Corporation. Copyright/Intel Corp. 1995) (b) Pin layout of the Pentium processor. (Reprinted by permission of Intel Corporation. Copyright/Intel Corp. 1995)

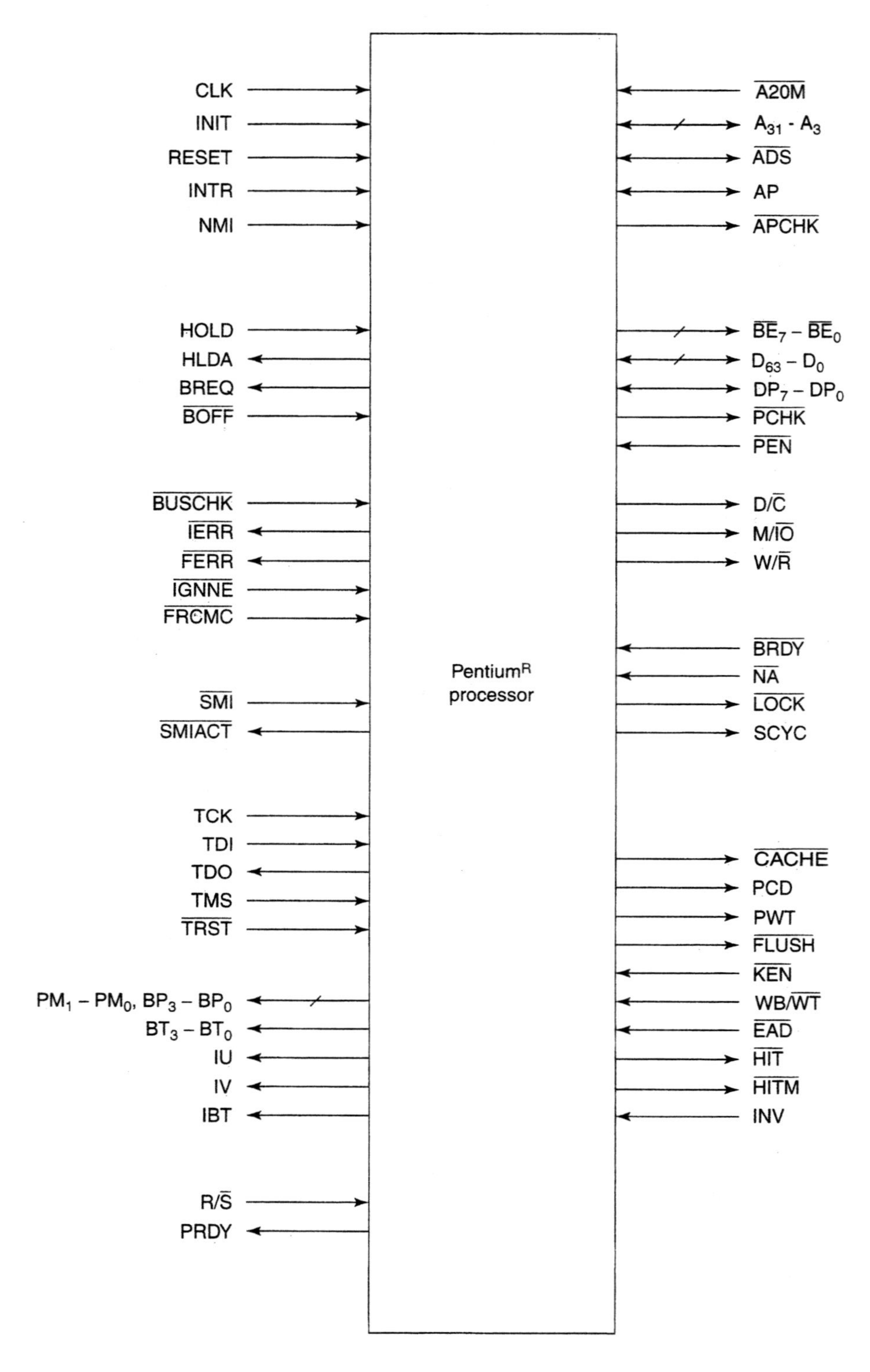

Figure 16–59 Pentium processor block diagram. (Reprinted by permission of Intel Corporation. Copyright/Intel Corp. 1995)

perform parity checks on both the data and the address. Adding address parity checking and detection to the memory/IO subsystem results in an increased level of data integrity for the microcomputer system.

The address bus is actually bidirectional because the Pentium processor, like the 80486SX, permits external devices to examine the contents of its internal caches. The operation is enhanced on the Pentium processor with address parity checking. The external system applies what is known as an *inquire address* to the processor on A_5 through A_{31}. This is the address of the cache-storage location to be accessed during the inquire cycle. Logic 0 at the valid external address ($\overline{EADS}$) input signals the MPU that the address is available. As part of the read operation, a parity-check operation is performed on the inquire address, and if a parity error is detected, it is identified by logic 0 at the address parity check ($\overline{APCHK}$) output.

The Pentium processor's memory/IO interface can also detect whether or not a bus cycle has run to completion correctly. This is known as a *bus error* condition. The *bus check* ($\overline{BUSCHK}$) input is used for this purpose. External circuitry must determine whether or not the current bus cycle is not successfully completed. If the bus cycle is not completed, it switches $\overline{BUSCHK}$ to logic 0. This input is sampled during read and write bus cycles. If an active 0 logic level is detected, the address and type of the failing bus cycle are latched within the MPU. An exception can also be automatically initiated to transfer program control to a service routine for the bus error problem.

Cache Memory-Control Interface. The number of control signals at the cache memory-control interface of the Pentium processor have also be expanded. Comparing the block diagram shown in Fig. 16–59 to that of the 80486SX in Fig. 16–46, we find that five new signals are provided: cacheability ($\overline{CACHE}$), write-back/write-through (WB/$\overline{WT}$), inquire cycle hit/miss indication ($\overline{HIT}$), hit/miss to a modified line ($\overline{HITM}$), and invalidation request (INV). Here we look briefly at how they function in support of the internal cache memory.

The logic level of the $\overline{CACHE}$ output has a different meaning, depending on whether a read or write operation is taking place. This output is switched to logic 0 during bus cycles where the data read from external memory can be cached. That is, it signals external circuitry that a cacheable data read or cacheable code fetch is taking place. $\overline{CACHE}$ is also made active during write cycles that represent a write-back to external memory of data updated in the internal cache.

WB/$\overline{WT}$ is an input that can be used to define the individual storage locations of the internal cache memory as write-back or write-through. By applying logic 0 or 1 to this input, external circuitry can dynamically decide whether the external memory-update method for the addressed storage location is write-back or write-through.

The other three signals, the $\overline{HIT}$ output, $\overline{HITM}$ output, and INV input, are involved in the inquire address operation.

Interrupt Interface. Comparing the interrupt interface of the Pentium processor to that of the 80486SX, we find that just one new signal has been added. The function of this input, initialization (INIT), is similar to that of RESET in that it is used to initialize the MPU to a known state. However, in the case of INIT, the content of the internal cache and a number of other registers remain unchanged.

Bus Cycles: Nonpipelined, Pipelined, and Burst

The bus interface of the Pentium processor has been designed to be very versatile and permit a number of different types of data-transfer bus cycles to be performed. Similar to the 80486 family of MPUs, the Pentium processor can perform bus cycles with either a single data transfer or burst of data transfers, and these bus cycles can be made either noncacheable or cacheable. Moreover, like the 80386 family, it can perform nonpipelined and pipelined bus cycles. The table in Fig. 16–60 shows the relationship between the bus cycle indication signals and corresponding bus activity. For instance, if

$M/\overline{IO}$	$D/\overline{C}$	$W/\overline{R}$	$\overline{CACHE}$*	$\overline{KEN}$	Cycle description	No. of Transfers
0	0	0	1	x	Interrupt acknowledge (2 locked cycles)	1 transfer each cycle
0	0	1	1	x	Special cycle	1
0	1	0	1	x	I/O read, 32 bits or less, noncacheable	1
0	1	1	1	x	I/O write, 32 bits or less, noncacheable	1
1	0	0	1	x	Code read, 64 bits, noncacheable	1
1	0	0	x	1	Code read, 64 bits, noncacheable	1
1	0	0	0	0	Code read, 256-bit burst line fill	4
1	0	1	x	x	Intel reserved (will not be driven by the Pentium® processor)	n/a
1	1	0	1	x	Memory read, 64 bits or less, noncacheable	1
1	1	0	x	1	Memory read, 64 bits or less, noncacheable	1
1	1	0	0	0	Memory read, 256-bit burst line fill	4
1	1	1	1	x	Memory write, 64 bits or less, noncacheable	1
1	1	1	0	x	256-bit burst writeback	4

* $\overline{CACHE}$ will not be asserted by any cycle in which $M/\overline{IO}$ is driven low or for any cycle in which PCD is driven high.

Figure 16–60 Types of bus cycles. (Reprinted by permission of Intel Corporation. Copyright/Intel Corp. 1995)

$M/\overline{IO}$ $D/\overline{C}$ $W/\overline{R}$ = 010_2 and $\overline{CACHE}$ = 1, a noncacheable 32-bit single data-transfer I/O read-bus cycle is taking place. Moreover, if $M/\overline{IO}$ $D/\overline{C}$ $W/\overline{R}$ = 100_2 and $\overline{CACHE}$ = 1, a noncacheable 64-bit code read is in process. They are both examples of single-data-transfer bus cycles. On the other hand, if $M/\overline{IO}$ $D/\overline{C}$ $W/\overline{R}$ = 100_2, $\overline{CACHE}$ = 0, and $\overline{KEN}$ = 0, code is read into the on-chip code cache memory with a 256-bit burst line-fill bus cycle. Let us next look more closely at the different types of read and write bus cycles that can be performed by the Pentium processor.

Nonpipelined Read and Write Cycles. Figure 16–61(a) shows the waveforms for the nonpipelined single-data read or write bus cycle. Note that a single-data-transfer read or write takes a minimum of two clock cycles, denoted T_1 and T_2 in the timing diagram. The read bus cycle starts with an address being output on the address bus accompanied by an address strobe pulse at $\overline{ADS}$. At the same time, $W/\overline{R}$ is switched to logic 0 to identify a read-data transfer. Note that $\overline{NA}$ and $\overline{CACHE}$ are both left at logic 1 throughout the bus

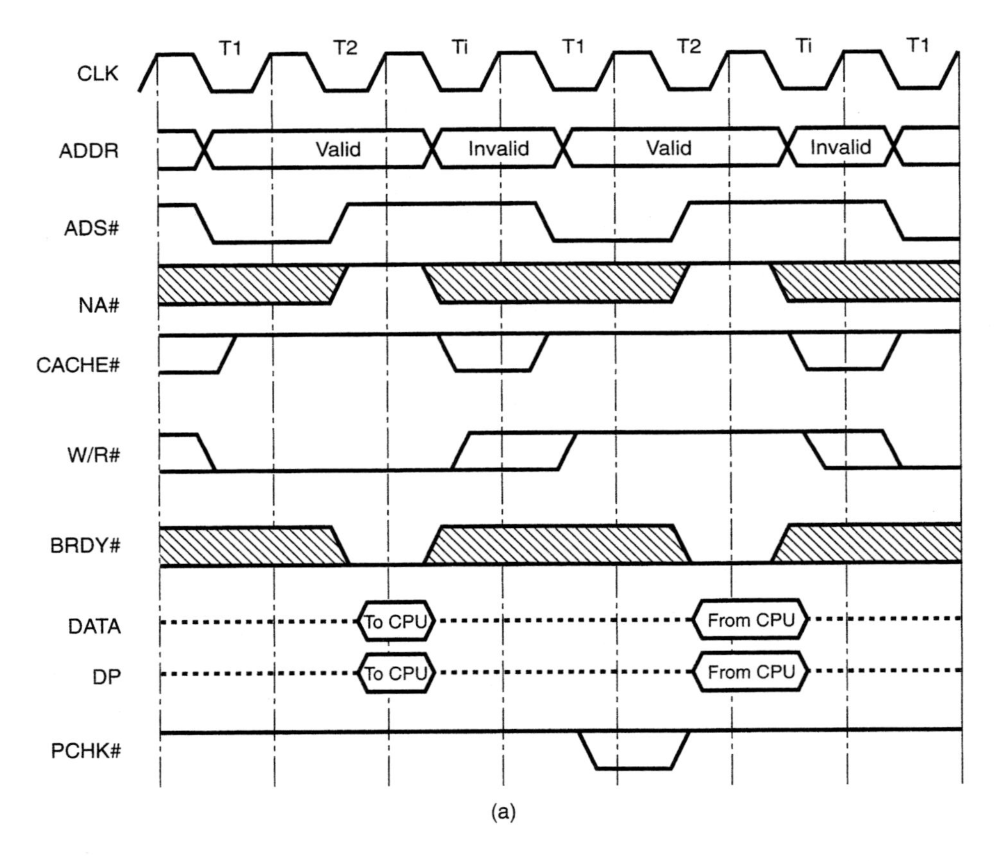

Figure 16–61 (a) Nonpipelined read and write bus cycles. (Reprinted by permission of Intel Corporation. Copyright/Intel Corp. 1995) (b) Nonpipelined read and write bus cycle with wait states. (Reprinted by permission of Intel Corporation. Copyright/Intel Corp. 1995)

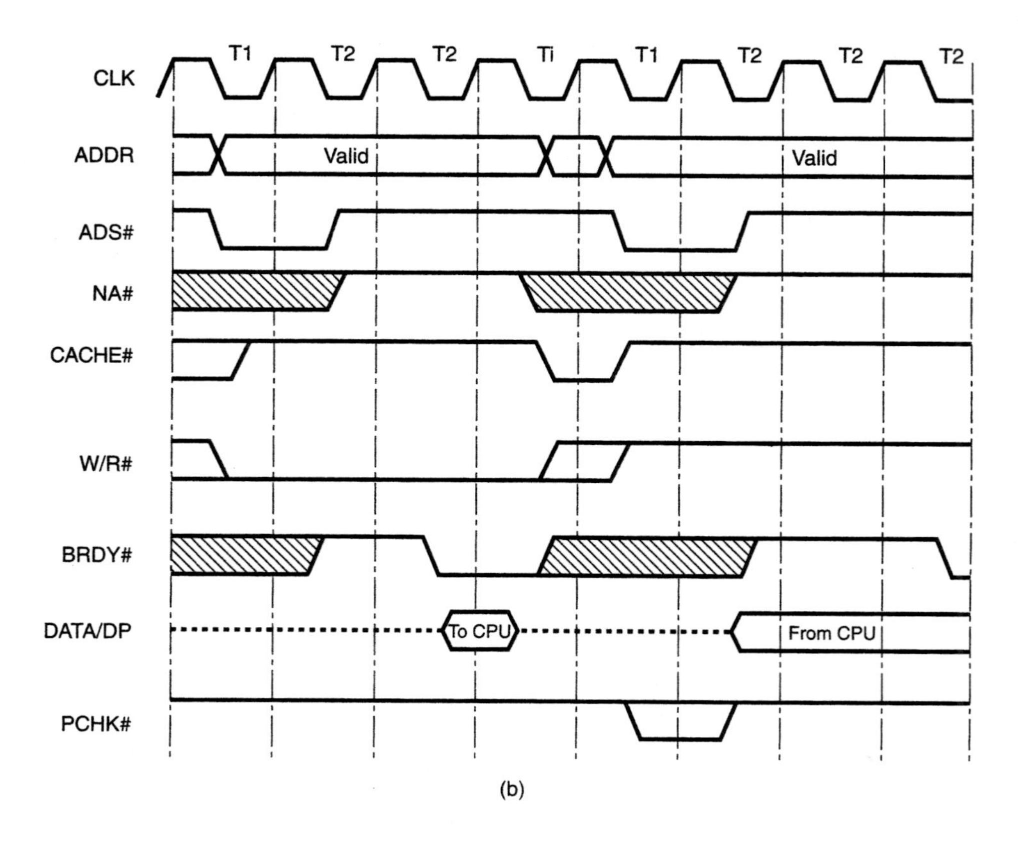

(b)

Figure 16–61 (continued)

cycle. This means that the bus cycle is nonpipelined and noncacheable. If the Pentium processor samples the $\overline{BRDY}$ input late in T_2 and finds that it is at its active 0 logic level as shown, the read-data transfer takes place to the MPU and the bus cycle is complete. Otherwise, the bus cycle is extended with additional clock periods until a logic 0 is detected at $\overline{BRDY}$. The bus cycle diagram in Fig. 16–61(b) shows a read extended with one wait state and a write bus cycle extended with two wait states.

EXAMPLE 16.15

What bus cycle indication code is output when a nonpipelined, noncacheable 64-bit data-write bus cycle is in progress?

Solution

From the table in Fig. 16–60, we find that

$$M/\overline{IO}\ D/\overline{C}\ W/\overline{R} = 111_2$$

Burst Read and Write Bus Cycles. In Fig. 16–60 we find that the Pentium processor performs only three types of burst bus cycles: a *code-read burst line fill, data-read line fill,* and a *burst write-back.* Each of these cycles represents an update of the cache memory. Note that a burst bus cycle involves 256 bits of data—that is, transfer of four quad data words.

Figures 16–62(a) and (b) show typical burst read and burst write bus cycles, respectively. Since all cacheable bus cycles are performed as burst cycles, we see that the $\overline{\text{CACHE}}$ output is held at its active 0 logic level throughout the bus cycle. For the burst read bus cycle, logic 0 must be returned to the $\overline{\text{KEN}}$ input of the MPU in clock 2 of the first data transfer. This signals that the memory subsystem will support the current read bus cycle as a burst line fill. $\overline{\text{KEN}}$ is not active during burst write bus cycles. The address and byte enables of the first quad-word of data to be accessed are output by the MPU along with a pulse at $\overline{\text{ADS}}$ at the beginning of the bus cycle. This original address is maintained valid throughout the bus cycle. For this reason, the address must be incremented in external hardware to point to the storage locations for each of the other three quad-word data transfers that follow. Note that the first 64-bit data transfer of a burst read or write takes place in two clock cycles; however, just one additional cycle is needed for each of the other three data transfers.

EXAMPLE 16.16

What bus cycle indication code is output when a burst write-back bus cycle is performed? What are the values of $\overline{\text{CACHE}}$ and $\overline{\text{KEN}}$ during this bus cycle?

Solution

The table in Fig. 16–60 shows that

$$M/\overline{IO}\ D/\overline{C}\ W/\overline{R} = 111_2$$

$$\overline{\text{CACHE}} = 0$$

$$\overline{\text{KEN}} = X$$

Pipelined Read and Write Bus Cycles. The Pentium processor is equipped with a *next address* ($\overline{\text{NA}}$) input to enable it to perform pipelined bus cycles. Remember that in a pipelined-mode bus cycle, the address for the next bus cycle is output overlapping with the data transfer for the prior bus cycle. Both single data-transfer and burst data-transfer bus cycles can be performed in a pipelined manner.

Figure 16–63(a) illustrates the bus activity for back-to-back pipelined burst read cycles. First a cacheable burst read is initiated to address a. When $\overline{\text{BRDY}}$ switches to logic 0, $\overline{\text{NA}}$ also becomes active. Logic 0 at $\overline{\text{NA}}$ signals the MPU that the next address, identified as address b, can be output on the address bus. As expected in a pipelined bus cycle, the address for the second cacheable burst read cycle is available to the external memory subsystem prior to the completion of the current burst read bus cycle. Therefore,

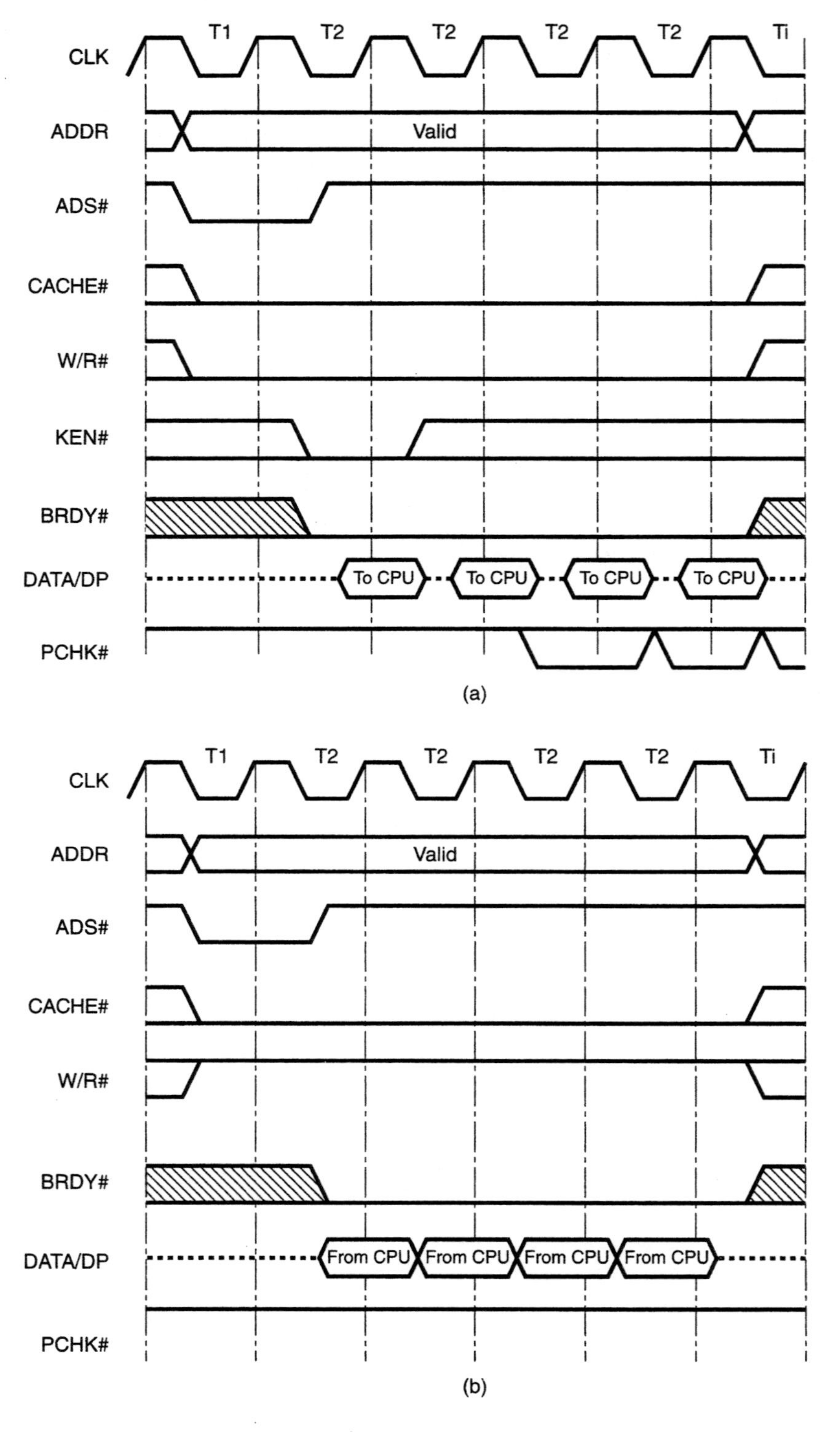

Figure 16–62 (a) Burst read bus cycle. (Reprinted by permission of Intel Corporation. Copyright/Intel Corp. 1995) (b) Burst write bus cycle. (Reprinted by permission of Intel Corporation. Copyright/Intel Corp. 1995)

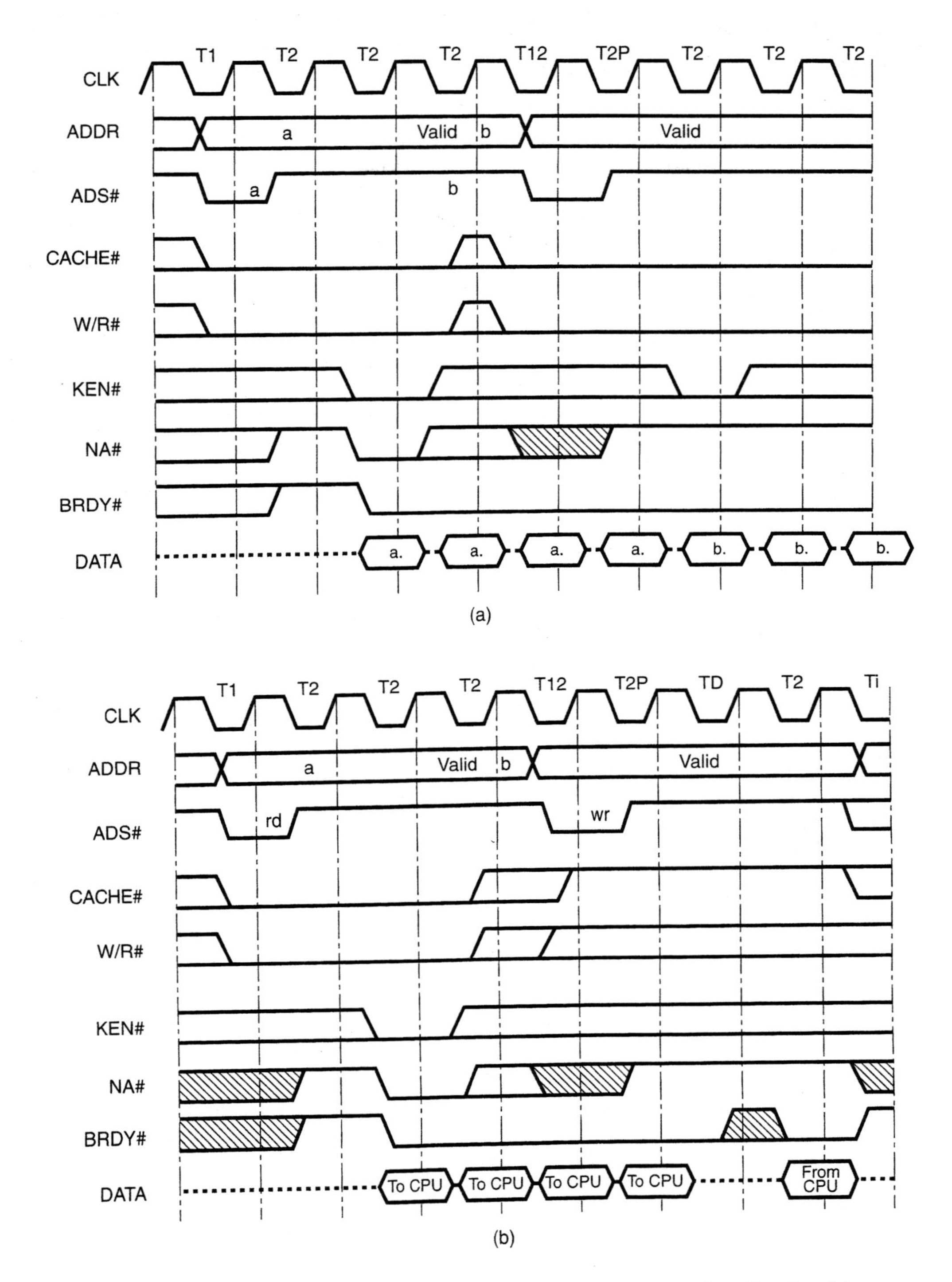

Figure 16–63 (a) Pipelined back-to-back burst read bus cycles. (Reprinted by permission of Intel Corporation. Copyright/Intel Corp. 1995) (b) Pipelined back-to-back read and write bus cycles. (Reprinted by permission of Intel Corporation. Copyright/Intel Corp. 1995)

the valid address is available for a longer period of time, and the memory subsystem can be designed with slower-access-time memory devices.

The waveforms in Fig. 16–63(b) represent pipelined back-to-back read and write cycles. The first cycle is a cacheable burst read and the second is a noncacheable single data-transfer write. Note that there is one dead clock period between the read and write cycles. It is identified as T_D in the timing diagram. The Pentium processor needs this period of time to turn around the bus from input for the read cycle to output for the write cycle.

Cache Memory of the Pentium Processor

The architecture of the Pentium processor-based microcomputer system supports both an internal and an external cache memory subsystem. Here we focus on the architecture, organization, and operation of the internal cache memory of the Pentium processor. Its on-chip cache memory differs from that of the 80486SX in several ways. For instance, there are separate cache memories for storage of data and code; they employ a two-way set associative organization instead of a four-way set associative organization, and two write-update methods—write-through and write-back—are supported for the data cache. The separate caches and write-back capability lead to higher performance for the Pentium processor-based microcomputer.

Organization and Operation of the Internal Cache Memory. Let us begin by examining the organization of the Pentium processor's on-chip data and code cache memories. They both are 8Kbytes in size. Since the two-way set associative organization is used, the storage array in each cache memory is organized into two separate 4Kbyte areas called *Way 0* and *Way 1,* as shown in Fig. 16–64. Just as with the cache of the 80486SX MPU, updates to the data or code cache of the Pentium processor are always done a line of data at a time; however, its line width is 256 bits (32 bytes) instead of 128 bits (16 bytes). Therefore, Way 0 and Way 1 can each hold 256 lines (sets) of data.

WAY 0 and WAY 1 each have a separate tag directory. The tag directory contains one tag for each of the 256 set entries. As Fig. 16–64 shows, a data cache tag entry is formed from a tag address and two MESI (modified-exclusive-shared-invalid) state bits. The MESI bits are used to maintain consistency between the Pentium processor's on-chip data cache and external caches. Note that the code cache tag only supports one bit for storage of MESI state information.

When a data or code read is initiated by the MPU to a storage location whose information is already held in an on-chip cache, a cache hit has occurred, and the information is read from the internal cache memory. On the other hand, if a cache miss results from the read operation, a line-fill cache read takes place from external memory. All line-fill cache reads involve four quad-word (64-bit) data transfers and are performed with burst bus cycles. Depending on whether the data read is for the code cache or data cache, the data transfer is performed as a code-read or memory-read burst line-fill bus cycle, respectively. If an invalid line exists within the cache, this quad-word of information is loaded into it; otherwise, the LRU algorithm is used to decide which of the valid lines of data will get replaced.

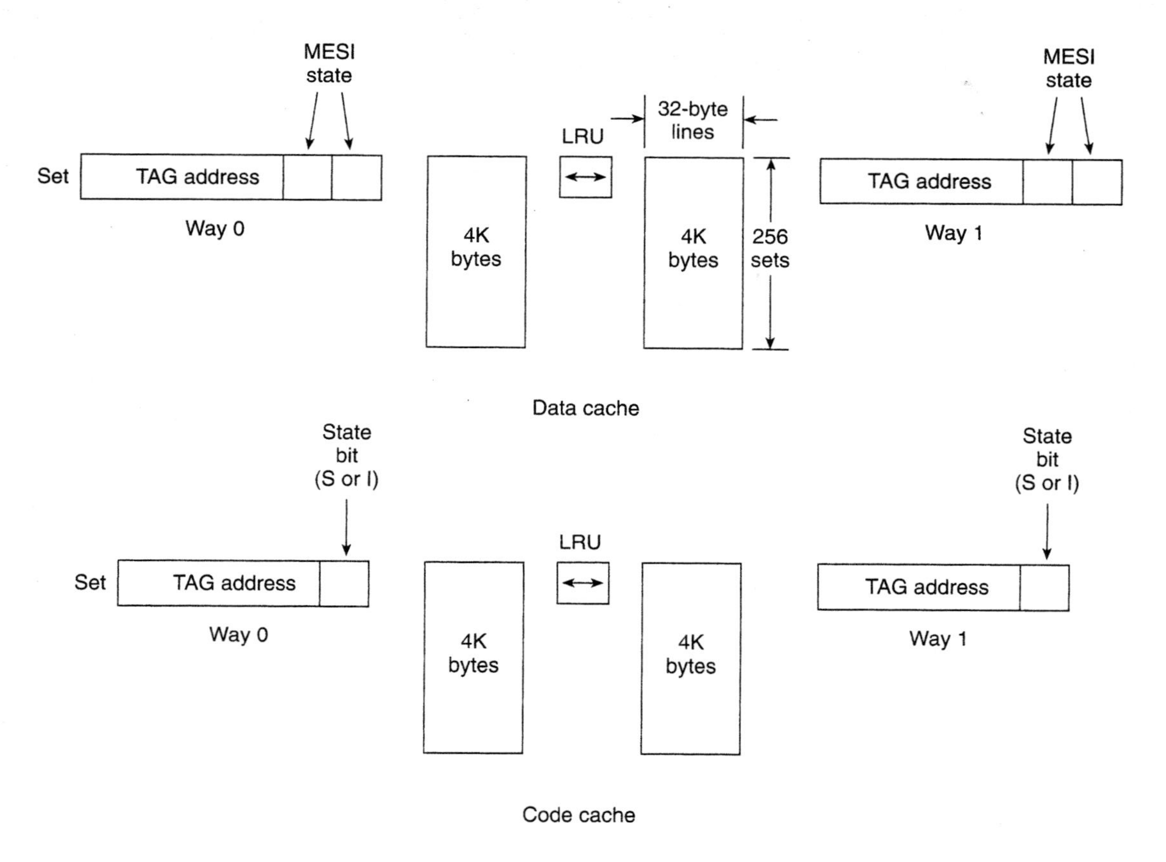

Figure 16–64 Organization of the on-chip cache of the Pentium processor. (Reprinted by permission of Intel Corporation. Copyright/Intel Corp. 1993)

During data write operations, cache-data consistency must be maintained between the on-chip data cache and external memory. The internal cache circuitry must determine whether or not the data held at the storage location being accessed also exist within the data cache. If a cache hit occurs during a write operation, the corresponding storage location in both the cache and the external memory must be updated.

As identified earlier, the Pentium processor's data cache supports both the write-through and write-back update methods. The 80486SX's on-chip cache provided only the write-through method. With this method, both the line in the internal data cache and its corresponding storage locations in external memory are updated as the write operation is performed. Unlike write-through operations, write-back updates of external memory are not performed at the same time the information is written into the cache. Instead, they are accumulated in the cache memory subsystem and written to memory at a later time. This reduces the bus activity and therefore enhances the microcomputer's performance. The data transfers that take place during a write-back operation to external memory are performed with 256-bit burst write-back bus cycles.

Different areas of the memory address space can be defined as either write-through or write-back and this can be done through software or hardware. For instance, logic 1 at the write-back/write-through (WB/$\overline{\text{WT}}$) input selects the write-back operation for the current write update.

The memory address space can also be partitioned into noncacheable and cacheable sections through software or hardware. The way in which this is done on the Pentium processor is identical to how it is performed on the 80486SX MPU. As discussed earlier for the 80486SX MPU, individual pages of memory are made cacheable or noncacheable under software control with the page-level cache disable (PCD) bit in its page table entry, and the memory address space can be hardware mapped as cacheable or noncacheable on a line-by-line basis with the cache-enable ($\overline{\text{KEN}}$) input.

Enabling, Disabling, and Flushing the On-Chip Cache. Just like the cache of the 80486SX, the operation of the Pentium processor's cache memories can be controlled with software and hardware. The table in Fig. 16–65 shows how the cache disable (CD)

CD	NW	Purpose/Description
0	0	**Normal highest performance cache operation.** Read hits access the cache. Read misses may cause replecement. Write hits update the cache. Only writes to shared lines and write misses appear externally. Write hits can change shared lines to exclusive under control of WB/WT#. Invalidation is allowed.
0	1	**Invalid setting.** A general-protection exception with an error code of zero is generated.
1	0	**Cache disabled. Memory consistency maintained. Existing contents locked in cache.** Read hits access the cache. Read misses do not cause replacement. Write hits update cache. Only write hits to shared lines and write misses update memory. Write hits can change shared lines to exclusive under control of WB/WT#. Invalidation is allowed.
1	1	**Cache disabled. Memory consistency not maintained.** Read hits access the cache. Read misses do not cause replacement. Write hits update cache but not memory. Write hits change exclusive lines to be modified. Shared lines remain shared lines after write hit. Write misses access memory. Invalidation is inhibited.

Figure 16–65 On-chip cache operating modes. (Reprinted by permission of Intel Corporation. Copyright/Intel Corp. 1995)

and not-write-through (NW) bits of control registers CR_0 affect the operation of the caches. The values of CD and NW can be changed under software control. For instance, the cache memories are enabled for operation by setting both CD and NW to logic 0. Note that making both CD and NW logic 1 does not totally disable the caches. Instead, as shown in the table for this state, read hits still access valid information held in the cache memory, but misses do not result in an update of the corresponding storage locations in the cache. To completely disable the cache, it must also be flushed after performing the software disable. This can be done with the cache flush ($\overline{\text{FLUSH}}$) input. Switching $\overline{\text{FLUSH}}$ to logic 0 initiates a write-back to external memory of the contents of all modified lines in the data cache, and when this is completed, all of the data held in the caches are invalidated.

The contents of the cache memories can also be invalidated under software control. This can be done with the write-back and invalidate cache (WBINVD) instruction. Execution of WBINVD causes a write-back to take place to external memory of any modified lines in the data cache and then the contents of both the code and data caches are invalidated. The table in Fig. 16–65 shows that a software invalidation can be performed only when the NW control bit is logic 0.

Applying logic 1 to the RESET input resets the MPU, sets both CD and NW to 1, and flushes the internal caches. Therefore, a hardware reset also completely disables the cache.

Interrupts and Internal Exceptions of the Pentium Processor

Interrupt and exception-processing capability has undergone very little change as part of the evolutionary path from the 80386 family of microprocessor to the Pentium processor family. In our study of the 80486SX MPU, we found that the only changes made in the exception processing of the 80486SX are that one new exception, alignment check exception, has been added and that one of the 80386DX's exceptions, coprocessor segment overrun, is no longer supported. The extensions made to the exception handling of the Pentium processor are also very small. In fact, just one new exception, *machine-check exception,* has been defined. Here we will look at this change.

Machine check is a new exception that was first implemented in the Pentium processor. This exception is enabled for operation by making the machine-check-enable (MCE) control bit, identified as bit position 6 of CR_4 in Fig. 15–59(b), to logic 1 under software control. Two events that can occur in external hardware will cause this type of exception: the detection of a parity error in a data read or the unsuccessful completion of a bus cycle. The data-read parity error condition is signaled to the MPU by external circuitry with logic 0 at the $\overline{\text{PEN}}$ input and the occurrence of a bus cycle error is identified by logic 0 at the $\overline{\text{BUSCHK}}$ input. If either of these events occurs, the address and type information for the bus cycle is latched in the machine check-address (MCA) and machine check-type (MCT) registers, respectively. Figure 16–66 shows the bus cycle-type information saved in MCT. The CHK bit is set to 1 when data are latched into MTC and automatically cleared when the contents are read through software. The exception service routine, initiated through exception gate 18, can access this information in an effort to determine the cause of the exception. The read from model-specific register (RDMSR) instruction is used to read the contents of MCA and MCT.

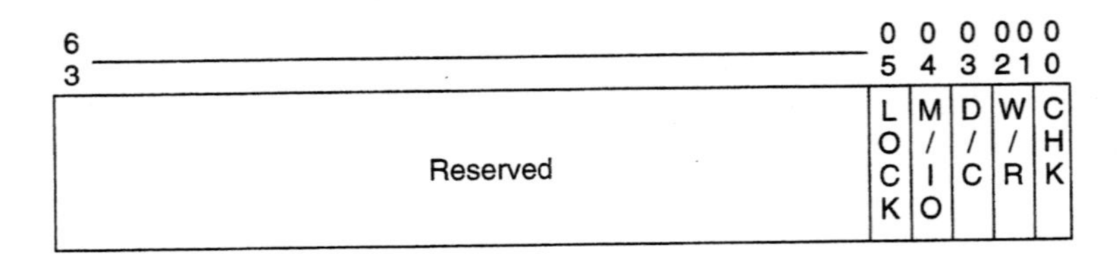

Figure 16–66 Machine check-type register contents. Reprinted by permission of Intel Corporation. Copyright/Intel Corp. 1995

▲ 16.11 PENTIUM PRO PROCESSOR AND PENTIUM PROCESSOR WITH MMX TECHNOLOGY

Intel Corporation has continued to extend the hardware and software capabilities of the Pentium processor family through new members. For instance, a second generation of devices, the *Pentium Pro processor,* was introduced in 1995. This MPU contains advanced features needed by high-performance personal computers, workstations, and servers. Some examples of these capabilities are support for easy, low-cost implementation of multiprocessor systems, and data integrity and reliability functions such as error checking and correction (ECC), fault analysis/recovery, and functional redundancy checking (FRC). The architecture of another family member, the *Pentium processor with MMX technology,* has been enhanced to provide higher performance for multimedia and communication applications. The intended use for this device, which was introduced in January 1997, is in desktop and laptop personal computers.

Let us begin by looking more closely at the Pentium Pro processor. As Fig. 16–67 shows, this device is actually two separate die that are housed in a single 387-pin dual-cavity staggered-pin grid-array package (SPGA). One die is the Pentium Pro processor's MPU and the other is a custom second-level cache. These die are manufactured using Intel's 0.35 μm BiCMOS process. This process technology uses bipolar transistors to implement high-speed circuitry and CMOS transistors for low-power, high-density circuitry. The circuitry of the Pentium Pro processor is equivalent to 5.5 million transistors.

The Pentium Pro processor design implements a new higher-performance microarchitecture—the P6 microarchitecture. P6 microarchitecture employs what is known as *dynamic execution.* Unlike earlier members of the Pentium processor family, the Pentium Pro processor does not execute instructions in order. In the dynamic execution architecture, a larger group of instructions are prefetched and decoded and made available for execution. They are identified as the *instruction pool* in Fig. 16–68.

Figure 16–68 shows that the traditional instruction-execution phase is replaced in the Pentium Pro processor by dispatch/execute and retire phases. This execution architecture permits instructions to be executed out of order, but assures that they are put back in their original order when completed. *Data flow analysis* is performed to determine the best order for execution of instructions. That is, instructions are executed based on whether they are ready to be executed, not based on their order in the program. For

Figure 16–67 Pentium Pro processor IC. (Reprinted by permission of Intel Corporation. Copyright/ Intel Corp. 1996)

instance, if an instruction that is being executed cannot be completed because it is waiting for additional data, the Pentium Pro processor looks ahead in the instruction pool and begins working on the execution of other instructions. This is known as *speculative execution.*

Several branch instructions exist in the large pool of decoded instructions. The *multiple branch prediction* capability of the Pentium Pro processor gives it the ability to analyze these branch conditions and predict the flow of the program through several levels of branching. Adjusting the instruction execution sequence based on this information, achieves more efficient instruction execution and a higher level of performance.

Some hardware architectural elements that in past family members were implemented with external circuitry are now implemented within the Pentium Pro processor. For instance, the device is available with either a 256Kbyte or 512Kbyte level-2 cache. Unlike an external second-level cache, this internal level-2 cache runs at full speed and results in a higher level of performance. Another hardware element that is now implemented internal to the processor is the advanced programmable interrupt control (APIC).

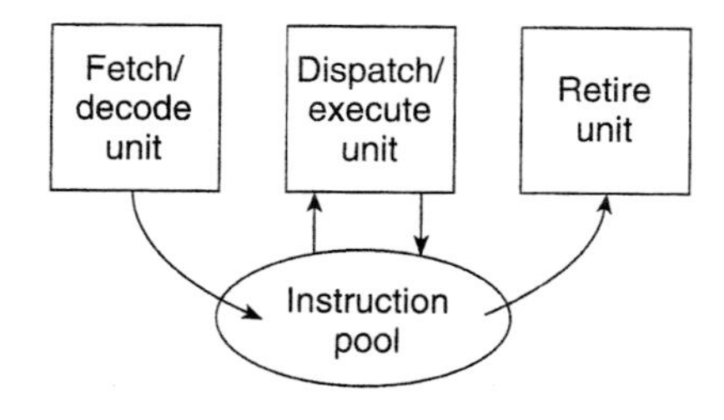

Figure 16–68 Instruction execution of the Pentium Pro processor. (Reprinted by permission of Intel Corporation. Copyright/ Intel Corp. 1995)

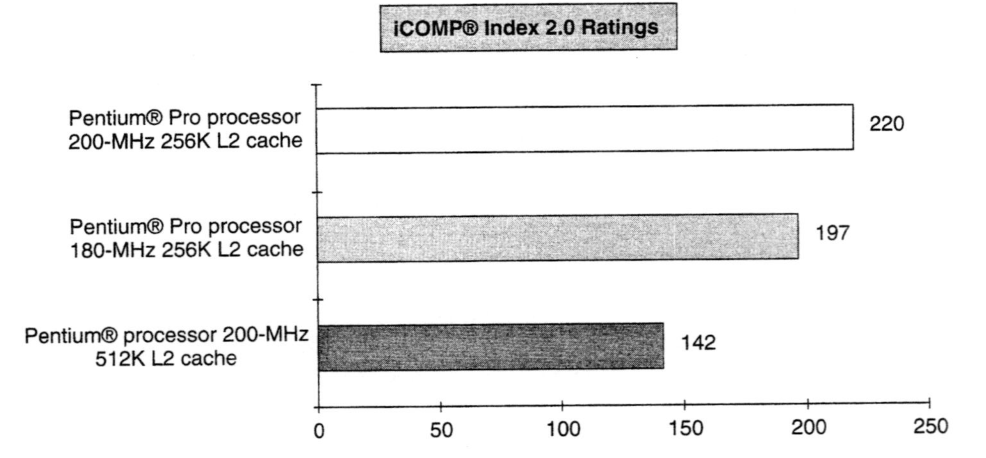

Figure 16–69 Pentium Pro processor iCOMP Index 2.0 ratings. (Reprinted by permission of Intel Corporation. Copyright/Intel Corp. 1996)

The performance of Pentium processor family devices is compared with the iCOMP Index 2.0. This new version of the index cannot be compared to the original iCOMP index introduced in Chapter 1 because it is formulated based on a different set of benchmarks. The chart in Fig. 16–69 compares the 180-MHz and 200-MHz Pentium Pro processors with an internal 256Kbyte level-2 cache to a 200-MHz Pentium processor with an external 512Kbyte level-2 cache. Note that the 200-MHz Pentium Pro processor has a rating of 220.

Figure 16–70 shows a Pentium processor with MMX technology. This device is both software and pin-for-pin compatible with the earlier Pentium processor family members. Like the Pentium Pro processor, it is manufactured on Intel's 0.35 μm process technology, but it is made with a single die that contains 4.5 million transistors.

Unlike the Pentium Pro, the Pentium processor with MMX technology is implemented with the microarchitecture of the original Pentium processor; however, the internal architecture has been enhanced in a number of ways. First, the instruction set has been expanded with a new group of *MMX technology* instructions and data types. They include 57 new instructions and four new 64-bit data types. An additional pipeline stage, the fetch stage, has been added into the instruction execution sequence, and its performance is further increased by improvements made in the dynamic branch-prediction capability. Finally, the size of the internal first-level code and data caches has been doubled to 16Kbytes, the caches are four-way set associative instead of two-way set associative, and the write buffers that are used to improve memory-write performance have been expanded from two to four buffers. These enhancements lead to the higher level of performance achieved with the Pentium processor with MMX technology. The iCOMP Index 2.0 ratings of the 166-MHz and 200-MHz devices are shown in Fig. 16–71.

Figure 16–70 Pentium processor with MMX technology IC. (Reprinted by permission of Intel Corporation. Copyright/Intel Corp. 1997)

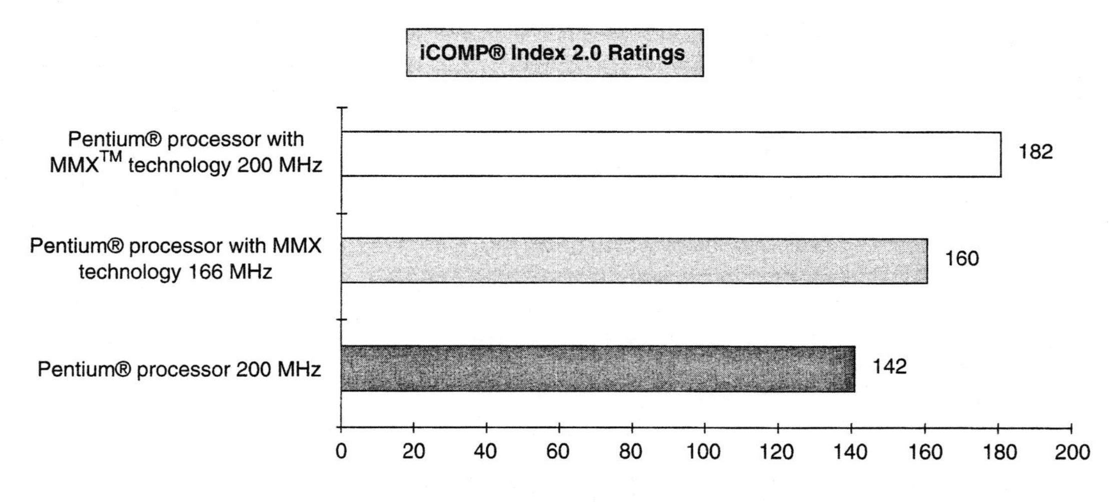

Figure 16–71 Pentium processor with MMX technology iCOMP Index 2.0 ratings. (Reprinted by permission of Intel Corporation. Copyright/Intel Corp. 1997)

▲ 16.12 PENTIUM II PROCESSOR, CELERON PROCESSOR, AND PENTIUM II XEON PROCESSOR

Intel Corporation introduced a third generation of Pentium processor class MPUs in 1998. The first member, the *Pentium II processor* is shown in Fig. 16–72. The intended use of this new processor is in the high-performance desktop personal computer. Two other MPUs, the Celeron processor and the Pentium II Xeon processor have followed it. The Celeron processor enables a new lower cost personal computer known as the Basic PC. On the other hand, the Pentium II Xeon processor offers higher performance than the standard Pentium II processor and is for use in workstations, servers, and other multiprocessor computer systems.

The architecture of the Pentium II processor builds on the P6 dynamic execution architecture that was first introduced in the Pentium Pro processor by merging into it the MMX technology instruction set. Additional improvements made in the Pentium II processor family include *.25 μm manufacturing process technology, dual independent bus architecture,* and a *100-MHz front-side bus.* These enhancements produce the higher level of performance achieved with the Pentium II processor.

Like the Pentium processor, the original Pentium II processor had independent 8Kbyte L1 code and data caches; however, the instruction cache is implemented with a four-way set associative organization. The original L2 cache was 256Kbytes and four-way set associative.

The original 266 MHz Pentium II processor provided a 40 percent improvement in performance, measured with the iCOMP Index 2.0 rating, over the 200-MHz Pentium processor with MMX technology. Figure 16–73 shows that the iCOMP rating of the 266-MHz MPU is 303 and that this rating increases to 440 at a clock frequency of 400 MHz, which represents more than a 2 × improvement in performance.

Dual independent bus architecture was actually first introduced in the Pentium Pro processor, but its implementation in the Pentium II processor is different. Figure 16–74 illustrates the dual independent bus architecture. This shows that in the Pentium II processor, two separate buses are provided: the level 2 (L2) cache bus (backside bus) and the

Figure 16–72 Pentium II processor. (Reprinted by permission of Intel Corporation. Copyright/Intel Corp. 1998)

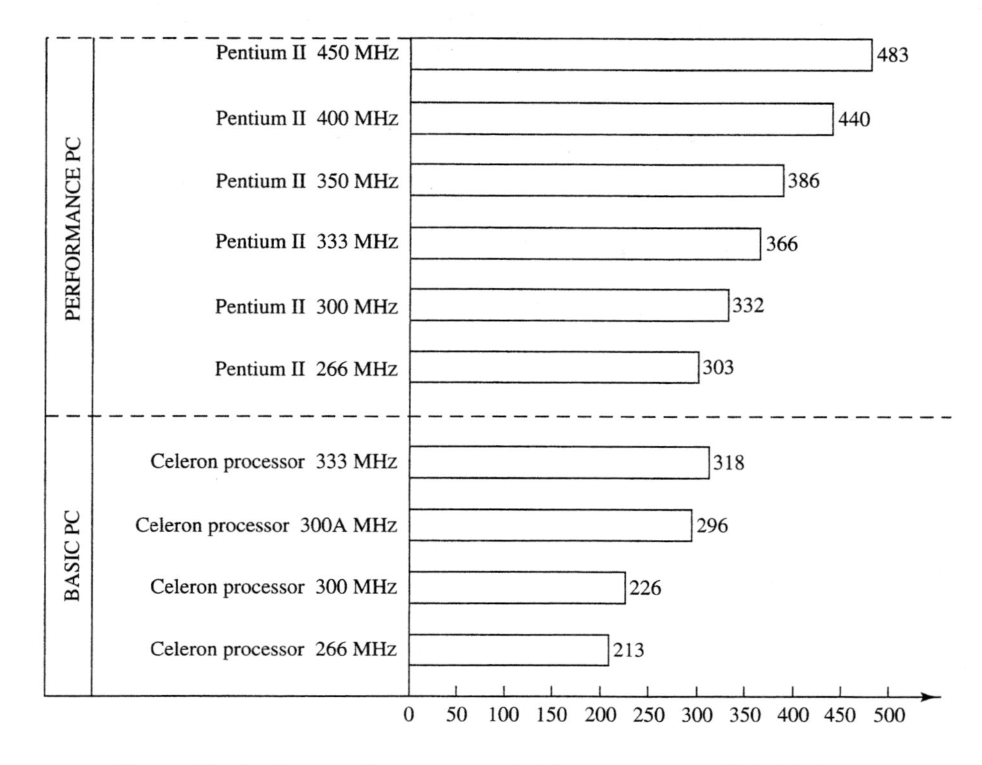

Figure 16–73 Pentium II processor and Celeron processor iCOMP Index 2.0 ratings. (Reprinted by permission of Intel Corporation. Copyright/Intel Corp. 1998)

system memory bus (front side bus). The front-side bus is fixed at 66.67 MHz on all earlier members of the Pentium processor family.

Peak bandwidth of a bus is calculated as the product of the bus frequency and the number of bytes transferred over the bus per clock cycle. That is,

$$BW_{Peak} = f \times \text{\# bytes/clock}$$

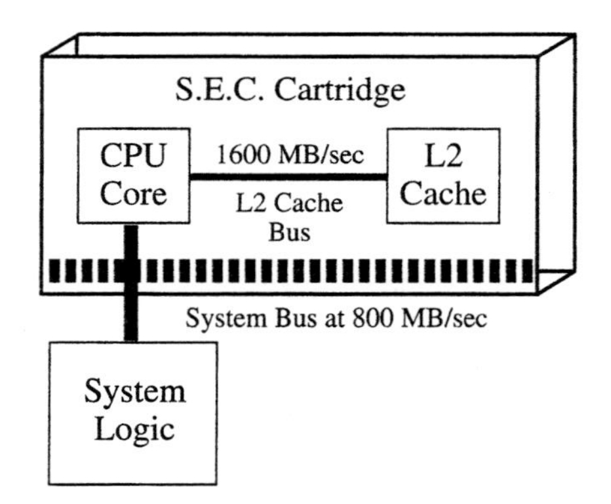

Figure 16–74 Dual independent bus architecture. (Reprinted by permission of Intel Corporation. Copyright/Intel Corp. 1998)

The system bus of both the Pentium and Pentium II processors carry 8 bytes per clock. With the Pentium II processor, the maximum speed of this bus is increased to 100 MHz. This results in higher bandwidth for the system memory bus—that is, it is capable of transferring more bytes of information. For instance, at 100 MHz, the peak bandwidth of the Pentium II processor is 800 Mbytes/second. Actually, the front-side bus of the Pentium II processor can be set to run at either 66.67 MHz or 100 MHz.

The backside bus between the L2 cache and MPU is higher speed. The speed of this bus is half that of the processor frequency. For instance, this bus is 200 MHz for the 400-MHz Pentium II processor. Therefore, its peak bus bandwidth is 1600 Mbytes/second. The total peak bus bandwidth of the 400-MHz Pentium II processor is the sum of the bus bandwidth of the system bus and the L2 cache bus. In this way, we find that the total peak bandwidth at 400 MHz is 2400 Mbytes/second.

The cache implementation in the Pentium II processor's dual independent bus architecture differs from that in the Pentium Pro processor. Like the Pentium Pro processor, the L2 cache memory of the Pentium II processor is not implemented on the same die as the MPU; however, in this case standard ICs are used to implement the tag RAM and cache storage array. The MPU and cache chips are separately packaged ICs. These devices are mounted on a standard printed circuit board and housed in a *single edge cartridge connector cartridge* (SECC). The signal leads of the processor are brought out at an edge connector on the SECC cartridge. The Pentium II processor is mounted on the PC's main processor board by inserting it into a connector known as the *slot 1 connector.*

The Celeron processor and Pentium II processor are similar in many ways. For instance, they both employ the dynamic execution architecture and MMX technology instructions. Let us just briefly look at the key differences between them. The Celeron processor runs at a lower clock rate, 266 MHz, its front side bus runs at the standard 66.67 MHz, and it does not have any built-in level 2 cache memory.

On the other hand, the architecture of the Pentium II Xeon processor has been enhanced to provide higher performance and to support multiprocessor applications. For instance, increasing the maximum size of the L2 cache to 1Mbyte and making the clock rate of the L2 cache bus equal to that of the processor improve performance. Therefore, the 400-MHz Pentium II Xeon processor has a 400-MHz backside bus and 100-MHz front-side bus.

Because workstations and servers are normally implemented with multiple processors, advanced system management and multiprocessing support capabilities have been added in the Pentium II Xeon processor. For instance, data integrity and reliability features, such as error checking and correction (ECC) and functional redundancy checking (FRC), which were first introduced in the Pentium Pro processor, are also implemented in the Pentium II Xeon processor. However, within the Pentium II Xeon processor, this system management capability has been enhanced with a dedicated system management bus interface, processor information ROM, and a scratch EEPROM for storage of user-defined system management information.

The design of multiprocessor systems has also been made easier. The Pentium II Xeon processor permits up to eight processors to be directly connected together in a multiprocessor system architecture. This capability enables low-cost implementation of very high performance multiprocessing computer systems.

▲ 16.13 PENTIUM III PROCESSOR AND PENTIUM IV PROCESSOR

The Pentium II processor was followed by two new generations of Pentium class MPUs for the high-performance desktop market—the Pentium III in 1999 and Pentium IV in 2000. The internal architecture and feature set have continued to evolve to make the Pentium family of processors more versatile and higher performance. For example, in the Pentium III, system bus bandwidth is improved by increasing the maximum speed of the front-side bus to 133 MHz, the L1 code and data caches are both 16Kbytes in size, and the L2 cache is now accessed at full clock speed. These architectural enhancements and the 0.18 process technology used to manufacture the Pentium III enabled Intel Corporation to offer their first GHz range microprocessor—maximum clock rate of 1.13 GHz. The Pentium IV contains 42 million transistors and runs at clock speeds in excess of 2 GHz.

In the Pentium III, the P6 microarchitecture is enriched with an additional 70 instructions. This new instruction group, known as the Internet Streaming SIMD extension, implements single instruction-multiple data operations similar to those available in the MMX technology instruction set extension, but performs these operations on floating-point numbers. These types of operations are important for applications such as high-resolution graphics, high-quality audio, and fast speech recognition.

The Pentium IV processor is based on a new high-performance microarchitecture called the Intel NetBurst architecture. This architecture vastly improves performance through the use of a deeper pipeline, expanded out-of-order, and increased parallel instruction execution. Special logic is provided for reordering instructions for parallel execution, and instructions can be dispatched to separate execution units for integer, floating point, load, and store operations. In fact, the high-speed integer ALU is able to perform two operations per main clock cycle. Availability of these extensive execution resources increased parallel instruction execution.

Other architectural enhancements that contribute to the high performance attained with the Pentium IV processor are the increase of the front-side bus speed to 400 MHz, a new level of internal cache—the execution trace cache—that stores decoded instructions, and the full-speed L2 cache is now eight-way associative.

Again, the MMX technology Internet SSE instruction set extensions are enhanced with 14 new instructions in the Pentium IV processors. These instructions permit arithmetic on 128-bit integer values and double precision floating-point numbers. This is known as the SSE2 instruction set extension.

REVIEW PROBLEMS

Section 16.1

1. Name the technology used to fabricate the 80386DX microprocessor.
2. What is the transistor count of the 80386DX?
3. Which signal is located at pin B7?

Section 16.2

4. How large is the real-mode address and physical address space of the 80386DX MPU? How large is the protected-mode address and physical address space? How large is the protected-mode virtual address space?
5. If the byte-enable code output during a data-write bus cycle is $\overline{BE_3}\overline{BE_2}\overline{BE_1}\overline{BE_0} = 1110_2$, is a byte, word, or double-word data transfer taking place? Over which data bus lines are the data transferred? Does data duplication occur?
6. For which byte-enable codes does data duplication take place?
7. What type of bus cycle is in progress when the bus status code $M/\overline{IO}\ D/\overline{C}\ W/\overline{R} = 010_2$?
8. Which signals implement the DMA interface?
9. What processor is most frequently attached to the processor extension interface?

Section 16.3

10. What are the speeds of the 80386DX ICs available from Intel Corporation? How are these speeds denoted in the part number?
11. At what pin is the CLK2 input applied?
12. What frequency clock signal must be applied to the CLK2 input of an 80386DX-25 if it is to run at full speed?

Section 16.4

13. What is the duration of PCLK for an 80386DX that is driven by CLK2 = 50 MHz?
14. What two types of bus cycles can be performed by the 80386DX?
15. Explain what pipelining the 80386DX's bus means.
16. What is an idle state?
17. What is a wait state?
18. What are the two T states of the 80386DX's bus cycle called?
19. If an 80386DX-25 is executing a nonpipelined write bus cycle that has no wait states, what would be the duration of this bus cycle if the 80386DX is operating at full speed?
20. If an 80386DX-25 that is running at full speed performs a read bus cycle with two wait states, what is the duration of the bus cycle?

Section 16.5

21. How is memory organized from a hardware point of view in a protected-mode 80386DX microcomputer system? Real-mode 80386DX microcomputer system?
22. What five types of data transfers can take place over the data bus? How many bus cycles are required for each type of data transfer?
23. If an 80386DX-25 is running at full speed and all memory accesses involve one wait state, how long will it take to fetch the word of data starting at address $0FF1A_{16}$? At address $0FF1F_{16}$?
24. During a bus cycle that involves a misaligned word transfer, which byte of data is transferred over the bus during the first bus cycle?

25. Give an overview of the function of each block in the memory interface diagram shown in Fig. 16–25.
26. When the instruction PUSH AX is executed, what bus status code is output by the 80386DX, which byte enable signals are active, and what read/write control signal is produced by the bus control logic?

Section 16.6

27. Which signal can be used to distinguish an I/O bus cycle and a memory bus cycle?
28. Which block produces the input (read), output (write), and bus control signals for the I/O interface?
29. Describe briefly the function of each block in the I/O interface circuit in Fig. 16–27.
30. If an 80386DX-25 running at full speed inserts two wait states into all I/O bus cycles, what is the duration of a nonpipelined bus cycle in which a byte of data is being output?
31. If the 80386DX in problem 30 was outputting a word of data to a word-wide port at I/O address $1A3_{16}$, what would be the duration of the bus cycle?
32. What parameter identifies the beginning of the I/O permission bit map in a TSS? At what address of the TSS is this parameter held?
33. At what double-word address in the I/O permission bit map is the bit for I/O port 64 held? Which bit of this double word corresponds to port 64?
34. To what logic level should the bit in problem 33 be set if I/O operations are to be inhibited to the port in protected mode?

Section 16.7

35. What is the real-mode interrupt address pointer table called? Protected-mode address pointer table?
36. What is the size of a real-mode interrupt vector? Protected-mode gate?
37. The contents of which register determines the location of the interrupt address pointer table? To what value is this register initialized at reset?
38. At what addresses is the protected-mode gate for type number 20 stored in memory? Assume that the table starts at address 00000_{16}.
39. Assume that gate 3 consists of the four words that follow:

$$(\text{IDT} + 8\text{H}) = 1000_{16}$$

$$(\text{IDT} + \text{AH}) = \text{B}000_{16}$$

$$(\text{IDT} + \text{CH}) = \text{AE}00_{16}$$

$$(\text{IDT} + \text{EH}) = 0000_{16}$$

(a) Is the gate descriptor active?

(b) What is the privilege level?

(c) Is the gate a trap gate or an interrupt gate?

(d) What is the starting address of the service routine?

40. Values stored in memory locations are as follows,

$$(\text{IDT\_TABLE}) = 01FF_{16}$$

$$(\text{IDT\_TABLE} + 2H) = 0000_{16}$$

$$(\text{IDT\_TABLE} + 4H) = 0001_{16}$$

What address is loaded into the interrupt-descriptor table register when the instruction LIDT [IDT_TABLE] is executed? What is the maximum size of the table? How many gates are provided for in this table?

41. What is the primary difference between the real-mode and protected-mode interrupt-request/acknowledge handshake sequence for the 80386DX microprocessor?
42. List the real-mode internal interrupts serviced by the 80386DX.
43. Internal interrupts and exceptions are categorized into groups based on how the failing function is reported. List the names of these three groups.
44. Which real-mode vector numbers are allocated to internal interrupts and exceptions?
45. Into which reporting group is the invalid opcode exception classified?
46. What is the cause of a stack fault exception?
47. Which exceptions take on a new meaning or are only active in the protected mode?

Section 16.8

48. What signal is located at pin A16 of the 80486SX's package?
49. Which lines of the 80486SX's memory/IO interface carry parity information? What type of parity is automatically generated?
50. What input signal and logic level does the external circuitry use to tell the 80486SX MPU that it can perform a burst bus cycle?
51. What does $\overline{\text{KEN}}$ stand for?
52. What signal permits an external device to take control of the 80486SX's bus interface at the completion of the current clock cycle?
53. What is the maximum number of bytes that can be transferred with a single-burst bus cycle? What signal must be supplied to the MPU by the memory subsystem to initiate a burst bus cycle? How many clock cycles does it take for completion?
54. What is meant when a read cycle is said to be cacheable? What signal must be supplied to the MPU by the memory subsystem to initiate a cacheable bus cycle? How many bytes of data are transferred during a cacheable bus cycle?
55. What is the result obtained by using a cache memory in a microcomputer system?
56. Is the internal cache of the 80486SX a first-level or second-level cache?
57. Define the term *cache hit.*
58. When an application program is tested on an 80486SX-based microcomputer system, it is found that 1340 instruction and data accesses are from the internal cache memory and 97 are from main memory. What is the hit rate?

59. If the cache memory in problem 58 operates with zero wait states and main memory bus cycles are performed with three wait states, what is the average number of wait states experienced executing the application?
60. What type of cache organization is used for the 80486SX's internal cache?
61. How large is the 80486SX's internal cache? What is the smallest element of data that can be loaded into the cache?
62. What write update method is implemented for the internal cache of the 80486SX?
63. What happens when the $\overline{\text{FLUSH}}$ input of the 80486SX is switched to logic 0 by external circuitry?
64. What new internal exception is implemented in the 80486SX MPU? What protected mode exception gate is assigned to this exception?

Section 16.9

65. Name two enhancements made in the architecture of the 80486DX2 to improve its performance?
66. What method is employed by the 80486DX2 to assure coherency between data in the on-chip cache and that in external memory?
67. What cache-coherency protocol is employed by the 80486DX2?
68. What output signal and logic level indicates that the line of data checked during a snoop operation is held in the 80486SX's on-chip cache and has been modified but has not yet been written back to main memory?
69. How is the 80486SX's cache put into the write-back mode?
70. Name two enhancements made to the architecture of the 80486DX2 to improve its performance.
71. How many times greater is the performance of the 80486DX4-100 compared to that of the 80386DX-33?
72. What value voltage supply is needed to power the 80486DX4?

Section 16.10

73. Approximately how many transistors are used to implement the Pentium processor?
74. What is the nominal value of the Pentium processor power supply?
75. At what pin of the Pentium's package is data bus line D_{45} located? Address line A_3?
76. How many byte-enable, data parity, address parity lines are provided in the memory/IO interface of the Pentium processor?
77. What kinds of parity are supported on the Pentium processor's bus interface? How are parity errors identified?
78. What does logic 0 at the $\overline{\text{BUSCHK}}$ input mean?
79. What type of bus cycle is in progress if the bus cycle indication information output by the processor is M/$\overline{\text{IO}}$ D/$\overline{\text{C}}$ W/$\overline{\text{R}}$ $\overline{\text{CACHE}}$ $\overline{\text{KEN}}$ = 0111X?
80. What does logic 0 at $\overline{\text{NA}}$ mean about the current bus cycle?
81. What type of organization is implemented with the Pentium processor's caches?

82. How much cache memory is provided on the Pentium processor?
83. How many bits are in the Pentium processor's cache line?
84. What does logic 0 on $WB/\overline{WT}$ mean about the current write bus cycle?
85. Describe the operation of the cache memories enable by making CD = 0 and NW = 0.

Section 16.11

86. Which of the new Pentium processor family's microprocessors is intended for use in workstations and servers?
87. How many transistors are in the Pentium Pro processor? The Pentium processor with MMX technology?
88. With what size level-2 cache is the Pentium Pro processor available?
89. How large are the internal caches of the Pentium processor with MMX technology? How are they organized?
90. What is the iCOMP Index 2.0 rating of the 200-MHz Pentium processor with MMX technology?

Section 16.12

91. Which processor is intended for use in the Basic PC?
92. What is the iCOMP rating of the 350 MHz Pentium II processor? 266 MHz Celeron processor?
93. What is the peak bandwidth of the front-side bus of the Pentium II processor if it is set to operate at 66 MHz?
94. What is the peak bandwidth of the backside bus of the L2 cache on a 300-MHz Pentium II processor?
95. What packaging technology is used to house the Pentium II processor?
96. What is the total bus bandwidth of the 400-MHz Pentium II Xeon processor?

Section 16.13

97. What is the name of the microarchitecture used to design the Pentium III processor? Pentium IV processor?
98. How many transistors does the Pentium IV processor have?
99. What is the speed of the front side bus of the Pentium III processor? The Pentium IV processor?
100. What is the name of the cache in the Pentium IV that holds decoded instructions?

Bibliography

Bistry, David, Carole Delong, Dr. Mickey Gutman, Michael Julier, Michael Keith, Lawrence M. Mennemeier, Millind Mittal, Alex D. Peleg, and Dr. Uri Weiser. *The Complete Guide to MMX™ Technology.* New York: McGraw-Hill, 1997.

Bradley, David J. *Assembly Language Programming for the IBM Personal Computer.* Upper Saddle River, NJ: Prentice-Hall, 1984.

Ciarcia, Steven. "The Intel 8086," *Byte,* November 1979.

Coffron, James W. *Programming the 8086/8088.* Berkeley, CA: Sybex, 1983.

Intel Corporation. *Components Data Catalog.* Santa Clara, CA: Intel Corporation, 1980.

———. *80286 Hardware Reference Manual.* Santa Clara, CA: Intel Corporation, 1987.

———. *80286 Operating Systems Writer's Guide.* Santa Clara, CA: Intel Corporation, 1986.

———. *80286 and 80287 Programmer's Reference Manual.* Santa Clara, CA: Intel Corporation, 1987.

———. *80386 Microprocessor Hardware Reference Manual.* Santa Clara, CA: Intel Corporation, 1987.

———. *80386 Programmer's Reference Manual.* Santa Clara, CA: Intel Corporation, 1987.

———. *80386 System Software Writer's Guide.* Santa Clara, CA: Intel Corporation, 1987.

———. *iAPX86,88 User's Manual.* Santa Clara, CA: Intel Corporation, July 1981.

———. *Introduction to the 80386.* Santa Clara, CA: Intel Corporation, September 1985.

———. *Memory.* Santa Clara, CA: Intel Corporation, 1989.

———. *i486™ Microprocessor Family Programmer's Reference Manual.* Santa Clara, CA: Intel Corporation, 1992.

———. *i486™ Microprocessor Hardware Reference Manual.* Santa Clara, CA: Intel Corporation, 1990.

———. *iAPX86,88 User's Manual.* Santa Clara, CA: Intel Corporation, July 1981.

———. *Intel486™ SX Microprocessor Data Book.* Santa Clara, CA: Intel Corporation, 1992.

———. *MCS-86™ User's Manual.* Santa Clara, CA: Intel Corporation, February 1979.

———. *Microprocessor.* Santa Clara, CA: Intel Corporation, 1989.

———. *Microprocessor and Peripheral Handbook,* vols. 1 and 2. Santa Clara, CA: Intel Corporation, 1989.

———. *Pentium® Processors and Related Products.* Santa Clara, CA: Intel Corporation, 1995.

———. *Pentium® Processor Family Developer's Manual,* vols. 1, 2, and 3. Santa Clara, CA: Intel Corporation, 1995.

———. *Pentium® II Processors Data Sheet.* Santa Clara, CA: Intel Corporation, 1996,1997.

———. *Peripheral.* Santa Clara, CA: Intel Corporation, 1989.

———. *Peripheral Design Handbook.* Santa Clara, CA: Intel Corporation, April 1978.

Morse, Stephen P. *The 8086 Primer.* Rochelle Park, N.J.: Hayden Book Company, 1978.

National Semiconductor Corporation, *Series 32000 Databook.* Santa Clara, CA: National Semiconductor Corporation, 1986.

Norton, Peter. *Inside the IBM PC.* Bowie, MD: Robert J. Brady, 1983.

Rector, Russell, and George Alexy. *The 8086 Book.* Berkeley, CA: Osborne/McGraw-Hill, 1980.

Scanlon, Leo J. *IBM PC Assembly Language.* Bowie, MD: Robert J. Brady, 1983.

Schneider, Al. *Fundamentals of IBM PC Assembly Language.* Blue Ridge Summit, PA: Tab Books, 1984.

Singh, Avtar, and Walter A. Triebel. *IBM PC/8088 Assembly Language Programming.* Upper Saddle River, NJ: Prentice-Hall, 1985.

Singh, Avtar, and Walter A. Triebel. *The 8088 Microprocessor: Programming, Interfacing, Software, Hardware, and Applications.* Upper Saddle River, NJ: Prentice-Hall, 1989.

Singh, Avtar, and Walter A. Triebel. *The 8086 and 80286 Microprocessors: Hardware, Software, and Interfacing.* Upper Saddle River, NJ: Prentice-Hall, 1990.

Strauss, Ed, *Inside the 80286.* New York: Brady Books, 1986.

Texas Instruments Incorporated, *Programmable Logic Data Book.* Dallas, TX: Texas Instruments Incorporated, 1990.

Triebel, Walter A. *Integrated Digital Electronics.* Upper Saddle River, NJ: Prentice-Hall, 1985.

Triebel, Walter A. *The 80386DX Microprocessor: Hardware, Software, and Interfacing.* Upper Saddle River, NJ: Prentice-Hall, 1992.

Triebel, Walter A. *The 80386, 80486, and Pentium® Processor: Hardware, Software, and Interfacing.* Upper Saddle River, NJ: Prentice-Hall, 1998.

Triebel, Walter A., and Alfred E. Chu. *Handbook of Semiconductor and Bubble Memories.* Upper Saddle River, NJ: Prentice-Hall, 1982.

Triebel, Walter A., and Avtar Singh. *The 8086 Microprocessor: Architecture, Software, and Interface Techniques.* Upper Saddle River, NJ: Prentice-Hall, 1985.

Triebel, Walter A., and Avtar Singh. *The 8088 and 8086 Microprocessors: Programming, Interfacing, Software, Hardware, and Applications,* 3rd edition. Upper Saddle River, NJ: Prentice-Hall, 2000.

Willen, David C., and Jeffrey I. Krantz. *8088 Assembly Language Programming: The IBM PC.* Indianapolis, IN: Howard W. Sams, 1983.

Answers to Selected Review Problems

CHAPTER 1

Section 1.1

1. Original IBM PC.

3. I/O channel.

5. Industry standard architecture.

7. A reprogrammable microcomputer is a general-purpose computer designed to run programs for a wide variety of applications, for instance, accounting, word processing, and languages such as BASIC.

9. The microcomputer is similar to the minicomputer in that it is designed to perform general-purpose data processing; however, it is smaller in size, has reduced capabilities, and costs less than a minicomputer.

Section 1.2

11. Input unit, output unit, microprocessing unit, and memory unit.

13. 16-bit.

15. Monitor and printer.

17. 360Kbytes; 10Mbytes.

19. 48Kbytes; 256Kbytes.

Section 1.3

21. 4-bit, 8-bit, 16-bit, 32-bit, and 64-bit.

23. 8086, 8088, 80186, 80188, 80286.

25. 27 MIPS.

27. 39; 49.

29. A special-purpose microcomputer that performs a dedicated control function.

31. A multichip microcomputer is constructed from separate MPU, memory, and I/O ICs. On the other hand, in a single-chip microcomputer, the MPU, memory, and I/O functions are all integrated into one IC.

33. Real mode and protected mode.

35. Memory management, protection, and multitasking.

Section 1.4

37. MSB and LSB.

39. 1 and $2^{+5} = 16_{10}$; 1 and $2^{-4} = 1/16$.

41. Min $= 00000000_2 = 0_{10}$, Max $= 11111111_2 = 255_{10}$.

43. 0000000111110100_2.

45. C and $16^{+2} = 256_{10}$.

47. **(a)** 39H, **(b)** E2H, **(c)** 03A0H.

49. C6H, 198_{10}.

51. 8005AH, $1{,}048{,}666_{10}$.

CHAPTER 2

Section 2.1

1. Bus interface unit and execution unit.

3. 20 bits; 16 bits.

5. General-purpose registers, temporary operand registers, arithmetic logic unit (ALU), and status and control flags.

Section 2.2

7. There purpose, function, operating capabilities, and limitations.

9. 1,048,576 (1M) bytes.

Section 2.3

11. $FFFFF_{16}$ and 00000_{16}.

13. $00FF_{16}$; aligned word.

15.

Address	*Contents*
0A003H	CDH
0A004H	ABH

aligned word.

Section 2.4

17. Unsigned integer, signed integer, unpacked BCD, packed BCD, and ASCII.

19. (0A000H) = F4H

(0A001H) = 01H

21. **(a)** 00000010, 00001001; 00101001

(b) 00001000, 00001000; 10001000

23. NEXT I.

Section 2.5

25. 64Kbytes.

27. CS.

29. Up to 128Kbytes.

Section 2.6

31. 80_{16} through $FFFEF_{16}$.

33. Control transfer to the reset power-up initialization software routine.

Section 2.7

35. The instruction is fetched from memory; decoded within the 8088; operands are read from memory or internal registers; the operation specified by the instruction is performed on the data; and results are written back to either memory or an internal register.

Section 2.8

37. Accumulator (A) register, base (B) register, count (C) register, and data (D) register.

39. DH and DL.

Section 2.9

41. Offset address of a memory location relative to a segment base address.

43. SS

45. Source index register; destination index register.

Section 2.10

47.

Flag	*Type*
CF	Status
PF	Status
AF	Status
ZF	Status
SF	Status
OF	Status
TF	Control
IF	Control
DF	Control

49. Instructions can be used to test the state of these flags and, based on their setting, modify the sequence in which instructions of the program are executed.

51. DF.

Section 2.11

53. 20 bits.

55. **(a)** 11234H
(b) 0BBCDH
(c) A32CFH
(d) C2612H

57. $021AC_{16}$.

59. 1234_{16}.

Section 2.12

61. $CFF00_{16}$.

63. FEFEH → (SP)
(AH) = EEH → (CFEFFH)
(AL) = 11H → (CFEFEH)

Section 2.13

65. 64 Kbytes.

CHAPTER 3

Section 3.1

1. Software.

3. Operating system.

5. Instructions encoded in machine language are coded in 0s and 1s, while assembly language instructions are written with alphanumeric symbols such as MOV, ADD, or SUB.

7. The data that is to be processed during execution of an instruction; source operand and destination operand.

9. An assembler is a program that is used to convert an assembly language source program to its equivalent program in machine code. A compiler is a program that converts a program written in a high-level language to equivalent machine code.

11. It takes up less memory and executes faster.

13. Floppy disk subsystem control and communications to a printer; code translation and table sort routines.

Section 3.2

15. Algorithm; software specification.

19. Assembler.

21. Linker.

23. **(a)** PROG_A.ASM
(b) PROG_A.LST and PROG_A.OBJ
(c) PROG_A.EXE and PROG_A.MAP

Section 3.3

25. Data transfer instructions, arithmetic instructions, logic instructions, string manipulation instructions, control transfer instructions, and processor control instructions.

Section 3.5

27. An addressing mode means the method by which an operand can be specified in a register or a memory location.

29. Base, index, and displacement.

31.

Instruction	*Destination*	*Source*
(a)	Register	Register
(b)	Register	Immediate
(c)	Register indirect	Register
(d)	Register	Register indirect
(e)	Based	Register
(f)	Indexed	Register
(g)	Based-indexed	Register

CHAPTER 4

Section 4.1

1. 6 bytes.

3. **(a)** 1000100100010101_2 = 8915H; **(b)** 1000100100011000_2 = 8918H; **(c)** $100010100101011100010000_2$ = 8A5710H

Section 4.2

5. 3 bytes.

Section 4.3

7. The DEBUG program allows us to enter a program into the PC's memory, execute it under control, view its operation, and modify it to fix errors.

9. Error.

11.
```
-R F           (↵)
NV UP EI PL NZ NA PO NC -PE (↵)
```

Section 4.4

13.
```
-D CS:0000 000F  (↵)
```

15.
```
-E CS:100 FF FF FF FF FF    (↵)
```

17.
```
-F CS:100 105 11   (↵)
-F CS:106 10B 22   (↵)
-F CS:10C 111 33   (↵)
-F CS:112 117 44   (↵)
-F CS:118 11D 55   (↵)
-E CS:105          (↵)
CS:0105 XX.FF      (↵)
-E CS:113          (↵)
-CS:0113 XX.FF     (↵)
-D CS:100 11D      (↵)
-S CS:100 11D FF   (↵)
```

Section 4.5

19. Contents of the byte-wide input port at address 0123_{16} is input and displayed on the screen.

Section 4.6

21. The sum and difference of two hexadecimal numbers.

23.
```
H FA 5A                 (↵)
```

Section 4.7

25.
```
-L CS:400 1 50 1        (↵)
-U CS:400 403           (↵)
1342:0400 320E3412   XOR  CL,[1234]
-
```

Section 4.8

27.
```
-A CS:200               (↵)
1342:0200 ROL BL,CL     (↵)
1342:0202               (↵)
-U CS:200 201           (↵)
1342:0200 D2C3 ROL BL,CL
-
```

Section 4.9

29.
```
-N A:BLK.EXE            (↵)
-L CS:200               (↵)
-R DS                   (↵)
DS 1342
:2000                   (↵)
-F DS:100 10F FF        (↵)
-F DS:120 12F 00        (↵)
-D DS:100 10F           (↵)
2000:0100 FF FF FF FF FF FF FF FF-FF FF FF FF FF FF FF FF
-D DS:120 12F           (↵)
2000:0120 00 00 00 00 00 00 00 00-00 00 00 00 00 00 00 00
-R DS                   (↵)
DS 2000
:1342                   (↵)
-R                      (↵)
AX=0000 BX=0000 CX=0000 DX=0000 SP=FFEE BP=0000 SI=0000 DI=0000
DS=1342 ES=1342 SS=1342 CS=1342 IP=0100 NV UP EI PL NZ NA PO NC
1342:0100 8915    MOV  [DI], DX   DS:0000=20CD
-U CS:200 217     (↵)
1342:0200 B80020 MOV  AX, 2000
1342:0203 8ED8   MOV  DS, AX
1342:0205 BE0001 MOV  SI, 0100
1342:0208 BF2001 MOV  DI, 0120
1342:020B B91000 MOV  CX, 0010
1342:020E 8A24   MOV  AH, [SI]
1342:0210 8825   MOV  [DI], AH
1342:0212 46     INC  SI
1342:0213 47     INC  DI
1342:0214 49     DEC  CX
1342:0215 75F7   JNZ  020E
1342:0217 90     NOP
```

```
-G =CS:200 217   (↵)
AX=FF00 BX=0000 CX=0000 DX=0000 SP=FFEE BP=0000 SI=0110 DI=0130
DS=2000 ES=1342 SS=1342 CS=1342 IP=0217 NV UP EI PL ZR NA PE NC
1342:0217 90   NOP
-D DS:100 10F (↵)
2000:0100 FF FF FF FF FF FF FF FF-FF FF FF FF FF FF FF FF
-D DS:120 12F     (↵)
2000:0120 FF FF FF FF FF FF FF FF-FF FF FF FF FF FF FF FF
```

Section 4.10

31. Bugs.

33.

```
-N A:BLK.EXE          (↵)
-L CS:200             (↵)
-U CS:200 217         (↵)
1342:0200 B80020   MOV AX, 2000
1342:0203 8ED8     MOV DS, AX
1342:0205 BE0001   MOV SI, 0100
1342:0208 BF2001   MOV DI, 0120
1342:020B B91000   MOV CX, 0010
1342:020E 8A24     MOV AH, [SI]
1342:0210 8825     MOV [DI], AH
1342:0212 46       INC SI
1342:0213 47       INC DI
1342:0214 49       DEC CX
1342:0215 75F7     JNZ 020E
1342:0217 90       NOP
-R DS                 (↵)
DS 1342
:2000                 (↵)
-F DS:100 10F FF      (↵)
-F DS:120 12F 00      (↵)
-R DS                 (↵)
DS 2000
:1342                 (↵)
-G =CS:200 20E        (↵)
AX=2000 BX=0000 CX=0010 DX=0000 SP=FFEE BP=0000 SI=0100 DI=0120
DS=2000 ES=1342 SS=1342 CS=1342 IP=020E NV UP EI PL NZ NA PO NC
1342:020E 8A24     MOV AH,[SI]    DS:0100=FF
-D DS:120 12F         (↵)
2000:0120 00 00 00 00 00 00 00 00-00 00 00 00 00 00 00 00
-G 212                (↵)
AX=FF00 BX=0000 CX=0010 DX=0000 SP=FFEE BP=0000 SI=0100 DI=0120
DS=2000 ES=1342 SS=1342 CS=1342 IP=0212 NV UP EI PL NZ NA PO NC
1342:0212 46       INC   SI
-D DS:120 12F         (↵)
2000:0120 FF 00 00 00 00 00 00 00-00 00 00 00 00 00 00 00
-G 215                (↵)
AX=FF00 BX=0000 CX=000F DX=0000 SP=FFEE BP=0000 SI=0101 DI=0121
DS=2000 ES=1342 SS=1342 CS=1342 IP=0215 NV UP EI PL NZ AC PE NC
1342:0215 75F7     JNZ   020E
```

```
-G 20E              (↵)
AX=FF00 BX=0000 CX=000F DX=0000 SP=FFEE BP=0000 SI=0101 DI=0121
DS=2000 ES=1342 SS=1342 CS=1342 IP=020E NV UP EI PL NZ AC PE NC
1342:020E 8A24      MOV AH,[SI]          DS:0101=FF
-G 215              (↵)
AX=FF00 BX=0000 CX=000E DX=0000 SP=FFEE BP=0000 SI=0102 DI=0122
DS=2000 ES=1342 SS=1342 CS=1342 IP=0215 NV UP EI PL NZ NA PO NC
1342:0215 75F7      JNZ    020E
-D DS:120 12F       (↵)
2000:0120 FF FF 00 00 00 00 00 00-00 00 00 00 00 00 00 00
-G 20E              (↵)
AX=FF00 BX=0000 CX=000E DX=0000 SP=FFEE BP=0000 SI=0102 DI=0122
DS=2000 ES=1342 SS=1342 CS=1342 IP=020E NV UP EI PL NZ NA PO NC
1342:020E 8A24      MOV AH,[SI]          DS:0102=FF
-G 217              (↵)
AX=FF00 BX=0000 CX=0000 DX=0000 SP=FFEE BP=0000 SI=0110 DI=0130
DS=2000 ES=1342 SS=1342 CS=1342 IP=0217 NV UP EI PL ZR NA PE NC
1342:0217 90        NOP
-D DS:120 12F       (↵)
2000:0120 FF FF FF FF FF FF FF FF-FF FF FF FF FF FF FF FF
```

CHAPTER 5

Section 5.1

1. **(a)** Value of immediate operand 0110H is moved into AX.
   (b) Contents of AX are copied into DI.
   (c) Contents of AL are copied into BL.
   (d) Contents of AX are copied into memory address DS:0100H.
   (e) Contents of AX are copied into the data segment memory location pointed to by (DS)0 + (BX) + (DI).
   (f) Contents of AX are copied into the data segment memory location pointed to by (DS)0 + (DI) + 4H.
   (g) Contents of AX are copied into the data segment memory location pointed to by (DS)0 + (BX) + (DI) + 4H.

3.
```
MOV AX, 1010H
MOV ES, AX
```

5. Destination operand CL is specified as a byte, and source operand AX is specified as a word. Both must be specified with the same size.

7. 10750H + 100H + 10H = 10860H.

9. LDS AX, [0200H].

Section 5.2

11. 110011000_2, 198H, 408_{10}.

13. **(a)** 00001001_2. **(b)** 01101001_2.

15. **(a)** 00FFH is added to the value in AX.
    (b) Contents of AX and CF are added to the contents of SI.
    (c) Contents of DS:100H are incremented by 1.

(d) Contents of BL are subtracted from the contents of DL.

(e) Contents of DS:200H and CF are subtracted from the contents of DL.

(f) Contents of the byte-wide data segment storage location pointed to by (DS)0 + (DI) + (BX) are decremented by 1.

(g) Contents of the byte-wide data segment storage location pointed to by (DS)0 + (DI) + 10H are replaced by its negative.

(h) Contents of word register DX are signed-multiplied by the word contents of AX. The double-word product that results is produced in DX, AX.

(i) Contents of the byte storage location pointed to by (DS)0 + (BX) + (SI) are multiplied by the contents of AL.

(j) Contents of AX are signed-divided by the byte contents of the data segment storage location pointed to by (DS)0 + (SI) + 30H.

(k) Contents of AX are signed-divided by the byte contents of the data segment storage location pointed to by (DS)0 + (BX) + (SI) + 30H.

17. `ADC DX, 111FH.`

19. `ADD SI, 2H,`

or

```
INC SI
INC SI
```

21. DAA.

23. (AX) = FFA0H.

25. Let us assume that the memory locations NUM1, NUM2, and NUM3 are in the same data segment.

```
MOV AX, DATA_SEG    ;Establish data segment
MOV DS, AX
MOV AL, [NUM2]      ;Get the second BCD number
SUB AL, [NUM1]      ;Subtract the binary way
DAS        ;Apply BCD adjustment
MOV [NUM3], AL      ;Save the result.
```

Note that storage locations NUM1, NUM2, and NUM3 are assumed to have been declared as byte locations.

Section 5.3

27. (a) 00011101_2. (b) 11011111_2.

29. 00011000_2, 18H.

31. (a) (DS:300H) = 0AH

(b) (DX) = A00AH

(c) (DS:210H) = FFFFH

(d) (DS:220H) = F5H

(e) (AX) = AA55H

(f) (DS:300H) = 55H

(g) (DS:210H) = 55H, (DS:211H) = 55H

33. AND WORD PTR [100H],0080H.

35. XOR AH, 80H.

37. The first instruction reads the byte of data from memory location CONTROL_FLAGS and loads it into BL. The AND instruction masks all bits but B_3 to 0; the XOR instruction toggles bit B_3 of this byte. That is, if the original value of B_3 equals logic 0, it is switched to 1, or if it is logic 1, it is switched to 0. Finally, the byte of flag information is written back to memory. This instruction sequence can be used to selectively complement one or more bits of the control flag byte.

Section 5.4

39. **(a)** (DX) = 2220H, (CF) = 0
(b) (DS:400H) = 40H, (CF) = 1
(c) (DS:200H) = 11H, (CF) = 0
(d) (DS:210H) = 02H, (CF) = 1
(e) (DS:210H,211H) = D52AH, (CF) = 1
(f) (DS:220H,221H) = 02ADH, (CF) = 0

41.
```
MOV  CL, 08H
SHL  WORD PTR [DI],CL
```

43. (AX) = F800H; CF = 1.

45.
```
MOV  AX, [ASCII_DATA]        ;Get the word into AX
MOV  BX, AX                  ;and BX
MOV  CL, 08H                 ;(CL) = bit count
SHR  BX, CL                  ;(BX) = higher character
AND  AX, 00FFH               ;(AX) = lower character
MOV  [ASCII_CHAR_L], AX      ;Save lower character
MOV  [ASCII_CHAR_H], BX      ;Save higher character
```

Section 5.5

47. **(a)** (DX) = 2222H, (CF) = 0
(b) (DS:400H) = 5AH, (CF) = 1
(c) (DS:200H) = 11H, (CF) = 0
(d) (DS:210H) = AAH, (CF) = 1
(e) (DS:210H,211H) = D52AH, (CF) = 1
(f) (DS:220H,221H) = AAADH, (CF) = 0

49.
```
MOV BL, AL ;Move bit 5 to bit 0 position
MOV CL, 5
SHR BX, CL
AND BX, 1   ;Mask the other bit
```

Advanced Problems

51.
```
MOV   AX, DATA_SEG    ;Establish the data segment
MOV   DS, AX
MOV   AL, [MEM1]      ;Get the given code at MEM1
MOV   BX, TABL1
XLAT                  ;Translate
MOV   [MEM1], AL      ;Save new code at MEM1
MOV   AL, [MEM2]      ;Repeat for the second code at MEM2
MOV   BX, TABL2
XLAT
MOV   [MEM2], AL
```

53.
```
;(RESULT) = (AL) • (NUM1) + (AL) • (NUM2) + (BL)
NOT [NUM2]     ;(NUM2) ← (NUM2)
MOV CL, AL
```

AND CL, [NUM2] ; $(CL) \leftarrow (AL) \bullet (\overline{NUM2})$

OR CL, BL ; $(CL) \leftarrow (AL) \bullet (\overline{NUM2}) + (BL)$

AND AL, [NUM1] ; $(AL) \leftarrow (AL) \bullet (\overline{NUM2})$

OR AL, CL

MOV [RESULT], AL ; $(RESULT) = (AL) \bullet (NUM1) + (AL) \bullet (\overline{NUM2}) + (BL)$

CHAPTER 6

Section 6.1

1. Executing the first instruction causes the contents of the status register to be copied into AH. The second instruction causes the value of the flags to be saved in memory location (DS)0 + (BX) + (DI).

3. STC; CLC.

5.
```
CLI                  ;Disable interrupts
MOV   AX, 0H         ;Establish data segment
MOV   DS, AX
MOV   BX, 0A000H     ;Establish destination pointer
LAHF                 ;Get flags
MOV   [BX], AH       ;and save at 0A000H
CLC                  ;Clear CF
```

Section 6.2

7. **(a)** The byte of data in AL is compared with the byte of data in memory at address DS:100H by subtraction, and the status flags are set or reset to reflect the result.

(b) The word contents of the data storage memory location pointed to by (DS)0 + (SI) are compared with the contents of AX by subtraction, and the status flags are set or reset to reflect the results.

(c) The immediate data 1234H are compared with the word contents of the memory location pointed to by (DS)0 + (DI) by subtraction, and the status flags are set or reset to reflect the results.

9.

Instruction	*(ZF)*	*(CF)*
Initial state	0	0
After MOV BX, 1111H	0	0
After MOV AX, 0BBBBH	0	0
After CMP BX, AX	0	1

Section 6.3

11. IP; CS and IP.

13. Intersegment.

15. **(a)** 1075H:10H
(b) 1075H:1000H
(c) 1075H:1000H

17. (SF) = 0.

19. **(a)** Intrasegment; short-label; if the carry flag is reset, a jump is performed by loading IP with 10H.

(b) Intrasegment; near-label; if PF is not set, a jump is performed by loading IP with 1000_{16}.

(c) Intersegment; memptr32; if the overflow flag is set, a jump is performed by loading the two words of the 32-bit pointer addressed by the value (DS)0 + (BX) into IP and CS, respectively.

21. **(a)** $1000_{16} = 2^{12} = 4096$ times.

(b)
```
;Implement the loop with the counter = 17

        MOV  CX, 11H
DLY:    DEC  CX
        JNZ  DLY
NXT:    ---  ---
```

(c)
```
;Set up a nested loop with 16-bit inner and 16-bit outer
;counters. Load these counters so that the JNZ
;instruction is encountered 2^32 times.
       MOV AX, 0FFFFH
DLY1:  MOV CX, 0H
DLY2:  DEC CX
       JNZ DLY2
       DEC AX
       JNZ DLY1
NXT:   ---
```

23.
```
            MOV CX, 64H        ;Set up array counter
            MOV SI, 0A000H     ;Set up source array pointer
            MOV DI, 0B000H     ;Set up destination array
                               ;pointer
GO_ON:      MOV  AX, [SI]
            CMP  AX, [DI]      ;Compare the next element
            JNE  MIS_MATCH     ;Skip on a mismatch
            ADD  SI, 2         ;Update pointers and counter
            ADD  DI, 2
            DEC  CX
            JNZ  GO_ON         ;Repeat for the next element
            MOV  [FOUND], 0H   ;If arrays are identical, save
                               ;a zero
            JMP  DONE
MIS_MATCH:  MOV  [FOUND], SI   ;Else, save the mismatch address
DONE:       ---  ---
```

Section 6.4

25. The call instruction saves the value in the instruction pointer, or in both the instruction pointer and code segment register, in addition to performing the jump operation.

27. IP; IP and CS.

29. **(a)** 1075H:1000H

(b) 1075H:0100H

(c) 1000H:0100H

31. **(a)** The value in the DS register is pushed onto the top of the stack, and the stack pointer is decremented by 2.

(b) The word of data in memory location (DS)0 + (SI) is pushed onto the top of the stack, and SP is decremented by 2.

(c) The word at the top of the stack is popped into the DI register, and SP is incremented by 2.

(d) The word at the top of the stack is popped into the memory location pointed to by (DS)0 + (BX) + (DI), and SP is incremented by 2.

(e) The word at the top of the stack is popped into the status register, and SP is incremented by 2.

33. When the contents of the flags must be preserved for the instruction that follows the subroutine.

Section 6.5

35. ZF.

37. Jump size = −126 to +129.

39.

```
        MOV    AL, 1H
        MOV    CL, N
        JCXZ   DONE          ;N = 0 case
        LOOPZ  DONE          ;N = 1 case
        INC    CL            ;Restore N
AGAIN:  MUL    CL
        LOOP   AGAIN
 DONE:  MOV    [FACT], AL
```

Section 6.6

41. DF.

43. **(a)** CLD

```
MOV ES, DS
MOVSB
```

(b)

```
CLD
LODSW
```

(c)

```
STD
CMPSB
```

Advanced Problems

45.

```
        MOV CX, 64H         ;Set up the counter
        MOV AX, 0H          ;Set up the data segment
        MOV DS, AX
        MOV BX, 0A000H      ;Pointer for the given array
        MOV SI, 0B000H      ;Pointer for the +ve array
        MOV DI, 0C000H      ;Pointer for the -ve array
AGAIN:  MOV AX, [BX]        ;Get the next source element
        CMP AX, 0H          ;Skip if positive
        JGE POSTV
NEGTV:  MOV [DI], AX        ;Else place in -ve array
        INC DI
        INC DI
        JMP NXT             ;Skip
```

```
POSTV:  MOV [SI], AX        ;Place in the +ve array
        INC SI
        INC SI
  NXT:  DEC CX              ;Repeat for all elements
        JNZ AGAIN
        HLT
```

47. ;Assume that all arrays are in the same data segment

```
        MOV  AX, DATASEG          ;Set up data segment
        MOV  DS, AX
        MOV  ES, AX
        MOV  SI, OFFSET_ARRAYA    ;Set up pointer to array A
        MOV  DI, OFFSET_ARRAYB    ;Set up pointer to array B
        MOV  CX, 62H
        MOV  AX, [SI]             ;Initialize A(I-2) and B(1)
        MOV  ARRAYC, AX
        MOV  [DI], AX
        ADD  SI, 2
        ADD  DI, 2
        MOV  AX, [SI]             ;Initialize A(I-1) and B(2)
        MOV  ARRAYC+1, AX
        MOV  [DI], AX
        ADD  SI, 2
        ADD  DI, 2
        MOV  AX, [SI]             ;Initialize A(I)
        MOV  ARRAYC+2, AX
        ADD  SI, 2
        MOV  AX, [SI]             ;Initialize A(I+1)
        MOV  ARRAYC+3, AX
        ADD  SI, 2
        MOV  AX, [SI]             ;Initialize A(I+2)
        MOV  ARRAYC+4, AX
        ADD  SI, 2
 NEXT:  CALL SORT                 ;Sort the 5 element array
        MOV  AX, ARRAYC+2         ;Save the median
        MOV  [DI], AX
        ADD  DI, 2
        MOV  AX, ARRAYC+1         ;Shift the old elements
        MOV  ARRAYC, AX
        MOV  AX, ARRAYC+2
        MOV  ARRAYC+1, AX
        MOV  AX, ARRAYC+3
        MOV  ARRAY+2, AX
        MOV  AX, ARRAYC+4
        MOV  ARRAYC+3, AX
        MOV  AX, [SI]             ;Add the new element
        MOV  ARRAY+4, AX
        ADD  SI, 2
        LOOP NEXT                 ;Repeat
        SUB  SI, 4                ;The last two elements of array B
        MOV  AX, [SI]
```

```
        MOV [DI], AX
        ADD SI, 2
        ADD DI, 2
        MOV AX, [SI]
        MOV [DI], AX
DONE:   --- ---
;SORT subroutine
SORT:  PUSHF                              ;Save registers and flags
       PUSH AX
       PUSH BX
       PUSH DX
       MOV  SI, OFFSET_ARRAYC
       MOV  BX, OFFSET_ARRAYC+4
  AA:  MOV  DI, SI
       ADD  DI, 02H
  BB:  MOV  AX, [SI]
       CMP  AX, [DI]
       JLE  CC
       MOV  DX, [DI]
       MOV  [SI], DX
       MOV  [DI], AX
  CC:  INC  DI
       INC  DI
       CMP  DI, BX
       JBE  BB
       INC  SI
       INC  SI
       CMP  SI, BX
       JB   AA
       POP  DX                            ;Restore registers and flags
       POP  BX
       POP  AX
       POPF
       RET
```

49.

```
;Assume that the offset of A[I] is AI1ADDR
;and the offset of B[I] is BI1ADDR
        MOV  AX, DATA_SEG    ;Initialize data segment
        MOV  DS, AX
        MOV  CX, 62H
        MOV  SI, AI1ADDR     ;Source array pointer
        MOV  DI, BI1ADDR     ;Destination array pointer
        MOV  AX, [SI]
        MOV  [DI], AX        ;B[1] = A[1]
MORE:   MOV  AX, [SI]        ;Store A[I] into AX
        ADD  SI, 2           ;Increment pointer
        MOV  BX, [SI]        ;Store A[I+1] into BX
        ADD  SI, 2
        MOV  CX, [SI]        ;Store A[I+2] into CX
        ADD  SI, 2
        CALL ARITH           ;Call arithmetic subroutine
```

```
        MOV  [DI], AX
        SUB  SI, 4
        ADD  DI, 2
        LOOP MORE               ;Loop back for next element
        ADD  SI, 4
DONE:   MOV  AX, [SI]           ;B[100] = A[100]
        MOV  [DI], AX
        HLT
;Subroutine for arithmetic
;(AX) ← [(AX) - 5(BX) + 9(CX)]/4
ARITH:  PUSHF                   ;Save flags and registers in stack
        PUSH BX
        PUSH CX
        PUSH DX
        PUSH DI
        MOV  DX, CX             ;(DX) ← (CX)
        MOV  DI, CX
        MOV  CL, 3
        SAL  DX, CL
        ADD  DX, DI
        MOV  CL, 2              ;(AX) ← 5(BX)
        MOV  DI, BX
        SAL  BX, CL
        ADD  BX, DI
        SUB  AX, BX             ;(AX) ← [(AX) - 5(BX) + 9(CX)]/4
        ADD  AX, DX
        SAR  AX, CL
        POP  DI                 ;Restore flags and registers
        POP  DX
        POP  CX
        POP  BX
        POPF
        RET                     ;Return
```

CHAPTER 7

Section 7.1

1. Macroassembler.
3. Assembly language instructions tell the MPU what operations to perform.
5. Label, opcode, operand(s), and comment(s).
7. **(a)** Fields must be separated by at least one blank space.
   (b) Statements that do not have a label must have at least one blank space before the opcode.
9. 31.
11. Operands tell where the data to be processed resides and how it is to be accessed.
13. Document what is done by the instruction or a group of instructions; the assembler ignores comments.

15. `MOV AX,[0111111111011000B]; MOV AX,[7FD8H].`

17. `MOV AX,0.`

Section 7.2

19. Data directives, conditional directives, macro directives, listing directives.

21. The symbol SRC_BLOCK is given 0100H as its value, and symbol DEST_BLOCK is given 0120H as its value.

23. The variable SEG_ADDR is allocated word-size memory and is assigned the value 1234_{16}.

25. `INIT_COUNT DW 0F000H.`

27.
```
SOURCE_BLOCK DW 0000H,1000H,2000H,3000H,4000H,5000H,
6000H,7000H,8000H,9000H,A000H,B000H,C000H,D000H,E000H,F000H.
```

29.
```
DATA_SEG SEGMENT WORD COMMON 'DATA'
        .
        .
DATA_SEG ENDS
```

31. A section of program that performs a specific function and can be called for execution from other modules.

33. An ORG statement specifies where the machine code generated by the assembler for subsequent instructions will reside in memory.

35.
```
PAGE 55 80
TITLE BLOCK-MOVE PROGRAM
```

Section 7.3

37. Move, copy, delete, find, and find and replace.

39. Use Save As operations to save the file under the filenames BLOCK.ASM and BLOCK.BAK. When the file BLOCK.ASM is edited at a later time, an original copy will be preserved during the edit process in the file BLOCK.BAK.

Section 7.4

41. Object module: machine language version of the source program.
    Source listing: listing that includes memory address, machine code, source statements, and a symbol table.

43. In Fig. 7–20, this error is in the comment and the cause is a missing ";" at the start of the statement.

45. **(a)** Since separate programmers can work on the individual modules, the complete program can be written in a shorter period of time.
    (b) Because of the smaller size of modules, they can be edited and assembled in less time.
    (c) It is easier to reuse old software.

47. Run module: executable machine code version of the source program.
    Link map: table showing the start address, stop address, and length of each memory segment employed by the program that was linked.

49. Object Modules[.OBJ]:A:MAIN.OBJ+A:SUB1.OBJ+A:SUB2.OBJ

CHAPTER 8

Section 8.1

1. HMOS.

3. 17.

5. 1Mbyte.

Section 8.2

7. The logic level of input $MN/\overline{MX}$ determines the mode. Logic 1 puts the MPU in minimum mode, and logic 0 puts it in maximum mode.

9. $\overline{WR}$, $\overline{LOCK}$.

11. $\overline{SSO}$.

Section 8.3

13. 20-bit, 8-bit; 20-bit, 16-bit.

15. A_0, D_7.

17. $\overline{BHE}$.

19. $\overline{WR}$.

21. HOLD, HLDA.

Section 8.4

23. HOLD, HLDA, $\overline{WR}$, $IO/\overline{M}$, $DT/\overline{R}$, $\overline{DEN}$, ALE, and $\overline{INTA}$ in minimum mode are $\overline{RQ}/\overline{GT}_{1,0}$, $\overline{LOCK}$, $\overline{S}_2$, $\overline{S}_1$, $\overline{S}_0$, QS_0, and QS_1, respectively, in the maximum mode.

25. $\overline{MRDC}$, $\overline{MWTC}$, $\overline{AMWC}$, $\overline{IORC}$, $\overline{IOWC}$, $\overline{AIORC}$, $\overline{INTA}$, MCE/PDEN, DEN, $DT/\overline{R}$, and ALE.

27. $\overline{S}_2\overline{S}_1\overline{S}_0 = 101_2$.

29. $QS_1QS_0 = 10_2$.

Section 8.5

31. +4.5 V to +5.5 V.

33. +2.0 V.

Section 8.6

35. 5 MHz and 8 MHz.

37. CLK, PCLK, and OSC; 10 MHz, 5 MHz, and 30 MHz.

Section 8.7

39. 4; T_1, T_2, T_3, and T_4.

41. An idle state is a period of no bus activity that occurs because the prefetch queue is full and the instruction currently being executed does not require bus activity.

43. 600 ns.

Section 8.8

45. Address $B0003_{16}$ is applied over the lines A_0 through A_{19} of the address bus, and a byte of data is fetched over data bus lines D_0 through D_7. Only one bus cycle is required to read a byte from memory. Control signals in minimum mode at the time of the read are $A_0 = 1$, $\overline{WR} = 1$, $\overline{RD} = 0$, $IO/\overline{M} = 0$, $DT/\overline{R} = 0$, and $\overline{DEN} = 0$.

47. High bank, $\overline{BHE}$.

49. $\overline{BHE} = 0$, $A_0 = 0$, $\overline{WR} = 0$, $M/\overline{IO} = 1$, $DT/\overline{R} = 1$, and $\overline{DEN} = 0$.

Section 8.9

51. $S_4S_3 = 10$.

Section 8.10

53. $IO/\overline{M}$.

55. $\overline{S}_2\overline{S}_1\overline{S}_0 = 100$; $\overline{MRDC}$.

57. $S_4S_3 = 01$ and $\overline{S}_2\overline{S}_1\overline{S}_0 = 110$; $\overline{MWTC}$ and $\overline{AMWC}$.

Section 8.11

59. Address is output on A_0 through A_{19}, ALE pulse is output, and $IO/\overline{M}$, $\overline{DEN}$, and $DT/\overline{R}$ are set to the appropriate logic levels.

61. $\overline{WR}$, $DT/\overline{R}$.

Section 8.12

63. The 8288 bus controller produces the appropriately timed command and control signals needed to coordinate transfers over the data bus.

The address bus latch is used to latch and buffer the address bits.

The address decoder decodes the higher order address bits to produce chip-enable signals.

The bank write control logic determines which memory bank is selected during a write bus cycle.

The bank read control logic determines which memory bank is selected during a read bus cycle.

The data bus transceiver/buffer controls the direction of data transfers between the MPU and memory subsystem and supplies buffering for the data bus lines.

65. D-type latches.

67.

Operation	$\overline{RD_U}$	$\overline{RD_L}$	$\overline{WR_U}$	$\overline{WR_L}$	$\overline{BHEL}$	$\overline{MRDC}$	$\overline{MWRC}$	$\overline{A}_{0L}$
(a) Byte read from address 01234H	1	0	1	1	1	0	1	0
(b) Byte write to address 01235H	1	1	0	1	0	1	0	1
(c) Word read from address 01234H	0	0	1	1	0	0	1	0
(d) Word write to address 01234H	0	0	1	1	0	1	0	0

69. $\overline{DEN} = 0$, $DT/\overline{R} = 0$.

71. Three address lines decode to generate eight chip selects. Therefore, three of them need not be used.

73. $Y_5 = 0$.

Section 8.13

75. Programmable logic array.

77. Fuse links.

81. 20 inputs; 8 outputs.

Section 8.14

85. Isolated I/O.

87. Isolated I/O.

Section 8.15

89. $IO/\overline{M}$.

91. $M/\overline{IO}$ is the complement of $IO/\overline{M}$.

93. The bus controller decodes I/O bus commands to produce the input/output and bus control signals for the I/O interface. The I/O interface circuitry provides for addressing of I/O devices as well as the path for data transfer between the MPU and the addressed I/O device.

95. $\overline{IORC} = 1$, $\overline{IOWC} = 0$, $\overline{AIOWC} = 0$.

Section 8.16

97. 0000_{16} through $FFFF_{16}$.

99. $A_0 = 0$ and $\overline{BHE} = 1$; $A_0 = 0$ and $\overline{BHE} = 0$.

Section 8.17

101. Execution of this input instruction causes accumulator AX to be loaded with the contents of the word-wide input port at address 1AH.

103. Execution of this output instruction causes the value in the lower byte of the accumulator (AL) to be loaded into the byte-wide output port at address 2AH.

105.
```
MOV  DX, 0A000H    ;Input data from port at A000H
IN   AL, DX
MOV  BL, AL        ;Save it in BL
MOV  DX, 0B000H    ;Input data from port at B000H
IN   AL, DX
ADD  BL, AL        ;Add the two pieces of data
MOV  [IO_SUM],BL   ;Save result in the memory location
```

Section 8.18

107. $IO/\overline{M}$ and ALE in T_1, and $\overline{RD}$ and $\overline{DEN}$ in T_2.

109. With zero wait states, the 8088 needs to perform two output bus cycles. They require 8 T-states, which at 5 MHz equals 1.6 μs.

111. To write a word of data to an odd address, the 8086 requires two bus cycles. Since each bus cycle has two wait states, it takes 12 T-states to perform the output operation. With a 10-MHz clock, the output operation takes 1200 ns.

CHAPTER 9

Section 9.1

1. Program-storage memory; data-storage memory.

3. Firmware.

Section 9.2

5. When the power supply for the memory device is turned off, its data contents are not lost.

7. Ultraviolet light.

9. We are assuming that external decode logic has already produced active signals for $\overline{CE}$ and $\overline{OE}$. Next, the address is applied to the A inputs of the EPROM and decoded within the device to select the storage location to be accessed. After a delay equal to t_{ACC}, the data at this storage location are available at the D outputs.

11. The access time of the 27C64 is 250 ns and that of the 27C64-1 is 150 ns. That is, the 27C64-1 is a faster device.

13. 1 ms.

Section 9.3

15. Volatile.

17. 32K $\times$ 32 bits (1MB).

19. Higher density and lower power.

Section 9.4

23. Parity-checker/generator circuit.

25. $\Sigma_{EVEN} = 0$; $\Sigma_{ODD} = 1$.

Section 9.5

29. The storage array in the bulk-erase device is a single block, whereas the memory array in both the boot block and FlashFile is organized as multiple independently erasable blocks.

31. Bulk erase.

33. 28F002 and 28F004.

35.

Type	*Quantity*	*Sizes*
Boot block	1	16Kbyte
Parameter block	2	8Kbyte
Main block	4	(1) 96Kbyte, (3) 128Kbyte

37. Logic 0 at the RY/$\overline{BY}$ output signals that the on-chip write state machine is busy performing an operation. Logic 1 means that it is ready to start another operation.

Section 9.6

39. READY.

41. Seven.

Section 9.7

43. Byte addresses 00000H through 03FFFH; word addresses 00000H through 03FFEH.

45. 6;3.

CHAPTER 10

Section 10.1

1. Keyboard interface, display interface, and parallel printer interface.

Section 10.2

3. $A_{15L}A_{14L}. \dots . A_{4L}A_{3L}A_{2L}A_{1L}A_{0L} = 1X. \dots .X1110_2 = 800E_{16}$ with $X = 0$.

5. Sets all outputs at port 2 (O_{16}–O_{23}) to logic 1.

Section 10.3

7. Port 4.

9.
```
MOV  AX, 0A000H      ;Set up the segment to start at A0000H
MOV  DS, AX
MOV  DX, 8002H       ;Input from port 1
IN   AL, DX
MOV  [0000H], AL     ;Save the input at A0000H
MOV  DX, 8004H       ;Input from port 2
IN   AL, DX
MOV  [0001H], AL     ;Save the input at A0001H
```

Section 10.4

11. Handshaking.

13. 74F373 octal latch.

15.
```
PUSH  DX    ;Save all registers to be used
PUSH  AX
PUSH  CX
PUSH  SI
PUSH  BX
  .         ;Program of Example 10.6 starts here
  .
  .
  .         ;Program of Example 10.6 ends here
POP   BX    ;Restore the saved registers
POP   SI
POP   CX
POP   AX
POP   DX
RET         ;Return from the subroutine
```

Section 10.5

17. 24.

19. Mode 0 selects simple I/O operation. This means that the lines of the port can be configured as level-sensitive inputs or latched outputs. Port A and port B can be configured as 8-bit input or output ports, and port C can be configured for operation as two independent 4-bit input or output ports.

Mode 1 operation represents what is known as strobed I/O. In this mode, ports A and B are configured as two independent byte-wide I/O ports, each of which has a 4-bit control port associated with it. The control ports are formed from port C's lower and upper nibbles, respectively. When configured in this way, data applied to an input port must be strobed in with a signal produced in external hardware. An output port is provided with handshake signals that indicate when new data are available at its outputs and when an external device has read these values.

Mode 2 represents strobed bidirectional I/O. The key difference is that now the port works as either input or output and control signals are provided for both functions. Only port A can be configured to work in this way.

21. D_0 = 1 Lower 4 lines of port C are inputs
D_1 = 1 Port B lines are inputs
D_2 = 0 Mode 0 operation for both port B and the lower 4 lines of port C
D_3 = 1 Upper 4 lines of port C are inputs
D_4 = 1 Port A lines are inputs
D_6D_5 = 00 Mode 0 operation for both port A and the upper 4 lines of port C
D_7 = 1 Mode being set

23. Control word bits = $D_7D_6D_5D_4D_3D_2D_1D_0 = 10010010_2$ = 92H.

25.
```
MOV  DX, 1000H       ;Load the control register with 92H
MOV  AL, 92H
OUT  DX, AL
```

27. To enable $INTR_B$, the INTE B bit must be set to 1. This is done with a bit set/reset operation that sets bit PC_4 to 1. This command is

$D_7 - D_0$ = 0XXX1001

29.
```
MOV  AL, 03H      ;Load the control register with 03H
MOV  DX, 100H
OUT  DX, AL
```

Section 10.6

31. The value at the inputs of port A of PPI 2 is read into AL.

33.
```
IN   AL, 08H     ;Read port A
MOV  BL, AL      ;Save in BL
IN   AL, 0AH     ;Read port B
ADD  AL, BL      ;Add the two numbers
OUT  0CH, AL     ;Output to port C
```

Section 10.7

35. To access port B on PPI 4

$A_0 = 0$, $A_2A_1 = 01$, and $A_5A_4A_3 = 010$

This gives the address = $XXXXXXXXXX010010_2 = 00012_{16}$ with Xs = 0.

37.
```
MOV  BL, [0408H]     ;Read port A
MOV  AL, [040AH]     ;Read port B
ADD  AL, BL          ;Add the two readings
MOV  [040CH], AL     ;Write to port C
```

Section 10.8

39. Control word $D_7D_6D_5D_4D_3D_2D_1D_0 = 01011010_2$ = 5AH.

41.
```
MOV  DX, 1003H     ;Select the I/O location
MOV  AL, 5AH       ;Get the control word
MOV  [DX], AL      ;Write it
```

43.
```
MOV  AL, 10000000B      ;Latch counter 2
MOV  DX, 1003H
MOV  [DX], AL
MOV  DX, 1002H
MOV  AL, [DX]           ;Read the least significant byte
```

45. 838 ns; 500 ns.

47. $N = 48_{10} = 30_{16}$.

Section 10.9

49. No.

51. 27.

53.
```
MOV  DX, 100DH   ;Master clear for 82C37A
OUT  DX, AL
```

55.
```
MOV  AL, 56H     ;Load channel 2 mode register
OUT  FBH, AL
```

57.
```
MOV  DX, 5008H   ;Read status register of 82C37A
IN   AL, DX
```

Section 10.10

59. Simplex: capability to transmit in one direction only.
Half-duplex: capability to transmit in both directions but at different times.
Full-duplex: capability to transmit in both directions at the same time.

Section 10.11

61. $C/\overline{D} = 0$, $\overline{RD} = 1$, $\overline{WR} = 0$, and $\overline{CS} = 0$.

63.
```
MOV AL, FFH
MOV MODE, AL
```

Section 10.12

65. 12 rows $\times$ 12 columns = 144 keys.

67. $D_3D_2D_1D_0 = 1101$, abcdefg = 1110000.

Section 10.13

69. P = 30.
CLK = (100 kHz) (30) = 3 MHz

CHAPTER 11

Section 11.1

1. External hardware interrupts, software interrupts, internal interrupts, nonmaskable interrupt, and reset.

3. External hardware interrupts, nonmaskable interrupt, software interrupts, internal interrupts, and reset.

5. Higher priority.

Section 11.2

7. 4 bytes.

9. Overflow.

11. (IP_{40}) = (Location A0H), and (CS_{40}) = (Location A2H).

Section 11.3

13. Arithmetic; overflow flag.

Section 11.4

15.
```
;This is an uninterruptible subroutine
CLI    ;Disable interrupts at entry point
 .
 .     ;Body of subroutine
 .
 .
STI    ;Enable interrupts
RET    ;Return to calling program
```

Section 11.5

17. Interrupt acknowledge.

19. INTR is the interrupt request signal that must be applied to the 8088 MPU by external interrupt interface circuitry to request service for an interrupt-driven device. When the MPU has acknowledged this request, it outputs an interrupt acknowledge bus status code on $\overline{S_2}\overline{S_1}\overline{S_0}$, and the 8288 bus controller decodes this code to produce the $\overline{INTA}$ signal. $\overline{INTA}$ is the signal used to tell the external device that its request for service has been granted.

21. D_0 through D_7.

Section 11.6

23. When the 8088 microprocessor recognizes an interrupt request, it checks whether the interrupts are enabled. It does this by checking the IF. If IF is set, an interrupt-acknowledge cycle is initiated. During this cycle, the $\overline{INTA}$ and $\overline{LOCK}$ signals are asserted. This tells the external interrupt hardware that the interrupt request has been accepted. Following the acknowledge bus cycle, the 8088 initiates a cycle to read the interrupt vector type. During this cycle the $\overline{INTA}$ signal is again asserted to get the vector type presented by the external interrupt hardware. Finally, the interrupt vector words corresponding to the type number are fetched from memory and loaded into IP and CS.

25. 1.8 μs for three write cycles; 6 bytes.

Section 11.7

27. $D_0 = 0$ ICW_4 not needed
$D_1 = 1$ Single-device
$D_3 = 0$ Edge-triggered
and assuming that all other bits are logic 0 gives
$ICW_1 = 00000010_2 = 02_{16}$.

29. $D_0 = 1$ Use with the 8086/8088
$D_1 = 0$ Normal end of interrupt
$D_3D_2 = 11$ Buffered mode master
$D_4 = 0$ Disable special fully nested mode
and assuming that the rest of the bits are logic 0, we get
$ICW_4 = 00001101_2 = 0D_{16}$.

31. `MOV AL, [0A001H]`

33.
```
MOV  AL, [0A001H]    ;Read OCW3
MOV  [OCW3], AL      ;Copy in memory
NOT  AL              ;Extract RR bit
AND  AL, 2H          ;Toggle RR bit
OR   [OCW3], AL      ;New OCW3
MOV  AL, [OCW3]      ;Prepare to output OCW3
MOV  [0A001H], AL    ;Update OCW3
```

Section 11.8

35. 22.

37. 64.

Section 11.9

39. CS_{80} = A000H and IP_{80} = 0100H.

Section 11.10

41. Type number 2; IP_2 is at location 08H and CS_2 is at location 0AH.

43. Initiate a power failure service routine.

Section 11.11

45. CLK.

47. AD_0 through AD_{15} = High-Z
A_{16} through A_{19} = High-Z
$\overline{BHE}$ = High-Z
ALE = 0
$\overline{DEN}$ = 1 then High-Z
DT/$\overline{R}$ = 1 then High-Z
$\overline{RD}$ = 1 then High-Z
$\overline{WR}$ = 1 then High-Z

49.
```
RESET:    MOV AX, 0         ;Set up the data segment
          MOV DS, AX
          MOV CX, 100H      ;Set up the count of bytes
          MOV DI, 0A000H    ;Point to the first byte
NXT:      MOV [DI], 0       ;Write 0 in the next byte
          INC DI            ;Update pointer, counter
          DEC CX
          JNZ NXT           ;Repeat for 100H bytes
          RET               ;Return
```

Section 11.12

51. Vectors 0 through 4.

53. CS_1 is held at 00006H and IP_1 is held at 00004H; A0200H.

CHAPTER 12

Section 12.1

1. System address bus, system data bus, and system control bus.

3. 060H, 061H, 062H, and 063H.

5. Timer 0—to keep track of the time of the day, generate an interrupt to the microprocessor every 55 ms.

Timer 1—to produce a DMA request every 15.12 μs to initiate a refresh cycle of DRAM.

Timer 2—has multiple functions, such as to generate programmable tones for the speaker and a record tone for the cassette.

7. PA_0 through PA_7 and PC_0 through PC_3.

9. Port B (PB_1).

11. Printer

13. 384Kbytes.

Section 12.2

15. 4.77 MHz.

17. Pin 21.

19. Logic 0 at $\overline{\text{DMA WAIT}}$ means wait states are required. Logic 0 at $\overline{\text{RDY}}$/WAIT means data are ready and CPU can complete the cycle, thus wait states are not required.

21. 74LS373 latches and 74LS245 bus transceiver.

23. $\overline{\text{MEMR}}$ = pin 7 and $\overline{\text{MEMW}}$ = pin 8.

Section 12.3

25. I/O channel cards; 0.

27. When a DMA request (DRQ_0 through DRQ_3) goes active (logic 1), 8237A outputs logic 0 at $\overline{\text{HRQ DMA}}$. This signal is input to NAND gate U_{52} in the wait state logic circuit and causes logic 1 at its output. This output drives the CLR input of 74LS74 flip-flop U_{67}, which produces HOLDA, and releases the cleared flip-flop for operation. The output of NAND gate U_{52} is also used as an input to NAND gate U_5. When the 8088 outputs the status code $\overline{S_2}\overline{S_1}\overline{S_0}$ = 111 (passive state) and $\overline{\text{LOCK}}$ = 1, the output of U_5 switches to 0. This output is inverted to logic 1 at pin 8 of U_{83}.

On the next pulse at CLK, the logic 1 applied to input D_3 of flip-flop U_{98} is latched at output Q_3. Next, this output is latched into the 74LS74 flip-flop U_{67} synchronously with a pulse at CLK88 to make HOLDA logic 1. HOLDA is sent to the HLDA input of 8237A and signals that the 8088 has given up control of the system bus.

29.
```
MOV AL, 80H    ;Disable NMI
OUT 0A0H, AL
```

31. Since the signals PCK and I/O CH CK are connected to PC_7 and PC_6 of the 8255A, respectively, the 8088 can read port C to determine which NMI source is requesting service.

PC_7	PC_6	*NMI source*
0	0	N P NPI
0	1	I/O CH CK
1	0	PCK

Section 12.4

33. DMA controller chip select ($\overline{\text{DMA CS}}$), interrupt controller chip select ($\overline{\text{INTR CS}}$), interval timer chip select ($\overline{\text{T/C CS}}$), and parallel peripheral interface chip select ($\overline{\text{PPI CS}}$).

35. Expressing the address in binary form, we get

$$A_{19}A_{18}A_{17}A_{16}A_{15}A_{14}A_{13}A_{12}A_{11}A_{10}A_9A_8A_7A_6A_5A_4A_3A_2A_1A_0 = 11111010000000000000$$

As the address FA000H is applied at the input of the ROM address decoder circuitry, address bits A_{16} through A_{19}, which are all 1, drive the inputs of NAND gate U_{64}. This input condition makes the $\overline{\text{ROM ADDR SEL}}$ output at pin 6 becomes logic 0. This signal, along with $\overline{\text{XMEMR}}$ (logic 0) and $\overline{\text{RESET DRV}}$ (logic 1), enables the 74LS138 three-line-to-eight-line decoder U_{46}. The inputs of this decoder are $A_{15}A_{14}A_{13}$ = 101. Therefore, output $\overline{CS_5}$ switches to its active 0 level. $\overline{CS_5}$ enables EPROM XU_{31} in the ROM array, and the signal $\overline{\text{ROM ADDR SEL}}$ controls the data direction through the 74LS245 bus transceiver U_{13}.

37. $\overline{RAS_1}$.

Section 12.5

39. For the address $F4000_{16}$ we have A_{16} through A_{19} = 1111, A_{14} = 1, and the rest of the address bits are 0. Since A_{16} through A_{19} = 1111, $\overline{\text{ROM ADDR SEL}}$ is at its active 0 logic level. This output enables the 74LS138 decoder, U_{46}. Since $A_{15}A_{14}A_{13}$ = 010, $\overline{CS_2}$ is active. $\overline{CS_2}$ selects XU_{28} in the ROM array and data are read from the first storage location of the EPROM chip. Also, $\overline{\text{ROM ADDR SEL}}$ directs the data from ROM to the 8088 via the 74LS245 bus transceiver U_{13}.

Section 12.6

41. DMA requests for channels 1, 2, and 3 are the I/O channel devices (boards plugged into the I/O channel).

43.

DMA Page Register	*Contents*
1	0AH
2	0BH
3	0CH

Instruction sequence:

```
MOV AL, 0AH     ;Init. channel 1 page register
OUT 81H, AL
MOV AL, 0BH     ;Init. channel 2 page register
OUT 82H, AL
MOV AL, 0CH     ;Init. channel 3 page register
OUT 83H, AL
```

Section 12.7

45. Counter 1 divisor = 1.1 M/18.2 = 60,440.

Section 12.8

49. KBD IRQ is an output that is used as an interrupt to the MPU and, when active, it signals that a keyscan code needs to be read.

Section 12.9

51. I/O channel slots provide the system interface to add-on cards. Five 62-pin card slots are provided on the system board.

CHAPTER 13

Section 13.1

1. Prototype circuit.

3. Solderless breadboard.

5. Bus interface module, I/O expansion bus cables, breadboard unit.

Section 13.2

7. The INT/EXT switch must be set to the EXT position.

9. 26 AWG.

11. A_{31} through A_{12}.

13. Logic 0 lights the green LED; logic 1 lights the red LED; and the high-Z level lights the amber LED.

Section 13.3

15. 74LS688, 74LS138, and 74LS32.

17. A_{10} through A_{15}.

19. The select outputs of the 74LS138 are gated with either $\overline{IOR}$ or $\overline{IOW}$ in 74LS32 OR gates. These signals are active only during an I/O cycle.

21. Yes.

23. The setting of switch 7 is polled waiting for it to close.

25. Lights LEDs 0 through 3.

27. The LEDs are lit in a binary counting pattern.

29. Change MOV CX, FFFFH to MOV CX, 7FFFH.

Section 13.4

33. IC test clip.

35. Whether the test point is at the 0, 1, or high-Z logic state, or if it is pulsating.

37. Amount of voltage, duration of the signal, and the signal waveshape.

39. Troubleshooting.

41. Hardware troubleshooting.

43. **(i)** Check to verify that correct pin numbers are marked into the schematic diagram.
(ii) Verify that the circuit layout diagram correctly implements the schematic.
(iii) Check that the ICs and jumpers are correctly installed to implement the circuit.

45.

Test Point	Switch Open	Switch Closed
1	1	0
2	1	0
3	Pulse	Pulse

Section 13.5

47. Address bus, data bus, and control-bus signals.

49.

Oscilloscope	Logic analyzer
(i) Requires periodic signal to display	**(i)** Can display periodic or nonperiodic signals
(ii) Small number of channels	**(ii)** Large number of channels
(iii) Displays actual voltage values	**(iii)** Displays logic values
(iv) Generally does not store the signals for display	**(iv)** Stores signals for display
(v) Simple trigger condition using a single signal	**(v)** Trigger signal can be a combination of a number of signals

CHAPTER 14

Section 14.1

1. HMOSIII.

3. PLCC, LCC, and PGA.

Section 14.2

5. 24 bits, 16 bits.

7. 6 bytes.

9. The queue holds the fetched instructions for the execution unit to decode and perform the operations that they specify.

Section 14.3

11. That an 8086 object code program can run on the 80286.

Section 14.4

13. Saves the contents of various registers of the processor such as AX, SP, and so forth on the stack.

15. Data area on the stack for a subroutine to provide space for the storage of local variables, linkage to the calling subroutine, and the return address.

17. A word of data from the word size port at address 1000H is input to the memory address 1075H:100H. The SI register is incremented to 102H, and CX is decremented by 2.

19. The instruction tests if VALUE lies between 0000H and 00FFH. If it is outside these bounds, interrupt 5 occurs.

Section 14.5

21. I/O write (output bus cycle).

23. No, it is one bit of a status code that must be decoded to produce an interrupt-acknowledge signal.

25. 80287.

Section 14.6

27. $\overline{IOWC}$.

29. 1.

Section 14.7

31. 25 MHz.

Section 14.8

33. Four clocks; 400 ns.

35. Perform-command state; external devices accept write data from the bus, or in the case of a read cycle, place data on the bus.

37. An idle state is a period of no bus activity that occurs because the prefetch queue is already full and the instruction currently being executed requires no bus activity.

Section 14.9

39. The bus controller produces the appropriately timed command and control signals needed to control transfers over the data bus. The decoder decodes the higher-order address bits to produce chip-enable signals. The address latch is used to latch and buffer the lower bits of the address and chip-enable signals. The data bus buffer/transceiver controls the direction of data transfers between the MPU and memory subsystem.

41. Odd-addressed byte, even-addressed byte, even-addressed word, and odd-addressed word. One bus cycle is required for all types of cycles, except the odd-addressed word cycle, which requires two bus cycles.

43. Odd-addressed byte.

45. 320 ns.

Section 14.10

47. $M/\overline{IO}$.

49. The decoder is used to decode several of the upper I/O address bits to produce the $\overline{I/OCE}$ signals. The latch is used to latch the lower-order address bits and $\overline{I/OCE}$ outputs of the decoder. The bus controller decodes the I/O bus commands to produce the I/O and bus control signals for the I/O interface. The bus transceivers control the direction of data transfer over the bus.

51. 1.2 μs.

Section 14.11

53. Hardware interrupts, software interrupts, internal interrupts and exceptions, software interrupts, and reset.

55. Interrupt vector table; interrupt descriptor table.

57. Interrupt descriptor table register, 0.

59. INTR is the interrupt request signal that must be applied to the 80286 MPU by the external interrupt interface circuitry to request service for an interrupt-driven device. When the MPU acknowledges this request, it outputs an interrupt acknowledge bus status code on $M/\overline{IO}\ \overline{S}_1\overline{S}_0$, and this code is decoded by the 82C288 bus controller to produce the $\overline{INTA}$ signal. $\overline{INTA}$ is the signal that is used to tell the external device that its request for service has been granted.

61. 22.

63. Vectors 0 through 16.

CHAPTER 15

Section 15.1

1. 80386DX and 80386SX.

3. 39; 49.

Section 15.2

5. Bus unit, prefetch unit, decode unit, execution unit, segment unit, and page unit.

7. Separate address and data buses.

9. Prefetch unit.

11. Translation lookaside buffer.

Section 15.3

13. *Object code compatible* means that programs and operating systems written for the 8088/8086 will run directly on the 80386DX and 80386SX in real-address mode.

15. FS, GS, and CR_0 registers.

Section 15.4

17. Double precision shift left.

19. `MOVZX EAX, [DATA_WORD]`.

21. **(a)** (AX) = F0F0H, (CF) = 1.
(b) (AX) = F0E0H, (CF) = 1.
(c) (AX) = F0E0H, (CF) = 1.

Section 15.5

23. Global descriptor table register, interrupt descriptor table register, task register, and local descriptor table register.

25. Defines the location and size of the global descriptor table.

27. System segment descriptors.

29. 0FFFH.

31. Local descriptor table.

33. CR_0.

35. (MP) = 1, (EM) = 0, and (ET) = 1.

37. Switch the PG bit in CR_0 to 1.

39. 4Kbyte.

41. Selector; selects a task state segment descriptor.

43. BASE and LIMIT of the TSS descriptor.

45. RPL = 2 bits
TI = 1 bit
INDEX = 13 bits

47. 00130020H.

49. Level 2.

51. Selector and offset.

53. 64Tbyte, 16,384 segments.

55. Task 3 has access to the global memory address space and the task 3 local address space, but it cannot access either the task 1 local address space or task 2 local address space.

57. The first instruction loads the AX register with the selector from the data storage location pointed to by SI. The second instruction loads the selector into the code segment selector register. This causes the descriptor pointed to by the selector in CS to be loaded into the code segment descriptor cache.

59. 1,048,496 pages; 4096 bytes long.

61. Cache page directory and page table pointers on-chip.

Section 15.6

63. 8, BASE = 32-bits, LIMIT = 20-bits, ACCESS RIGHTS BYTE = 8-bits, AVAILABLE = 1 bit, and GRANULARITY = 1 bit.

65. LIMIT = 00110H, BASE = 00200000H.

67. 00200226H.

69. R/W = 0 and U/S = 0 or R/W = 1 and U/S = 0.

71. Dirty bit.

Section 15.7

73.
```
LMSW  AX              ;Get MSW
AND   AX, 0FFF7H      ;Clear task-switched bit
SMSW  AX              ;Write new MSW
```

Section 15.8

75. The running of multiple processes in a time-shared manner.

77. Local memory resources are isolated from global memory resources, and tasks are isolated from each other.

79. Level 0, level 3.

81. LDT and GDT.

83. Level 0.

85. A task can access data in a data segment at the CPL and at all lower privilege levels, but it cannot access data in segments that are at a higher privilege level.

87. A task can access code in segments at the CPL or at higher privilege levels, but cannot modify the code at a higher privilege level.

89. The call gate is used to transfer control within a task from code at the CPL to a routine at a higher privilege level.

91. Identifies a task state segment.

93. The state of the prior task is saved in its own task state segment. The linkage to the prior task is saved as the back link selector in the first word of the new task state segment.

Section 15.9

97. Active, level 3.

99. Yes.

Section 15.10

101. The 80486SX does not have an on-chip floating-point math coprocessor.

103. 32 bytes.

105. Small instruction set, limited addressing modes, and single clock execution for instructions.

107. Cache disable (CD) and not write-through (NW).

109. F0F0H.

111. `XADD [SUM], EBX`

(EAX)	*(SUM)*	
01H	00H	
01H	01H	1st execution
01H	02H	2nd execution
02H	03H	3rd execution
03H	05H	4th execution

113. Alignment check (AC); bit 18.

115. INVD.

117. Page cache disable (PCD) and page write transparent (PWT).

Section 15.11

119. Single precision number = 32 bits, double precision number = 64 bits, and extended precision number = 80 bits.

121. Sign = 1

Biased exponent = $+3 + 127 = 00000011_2 + 01111111_2 = 10000010$
Fractional significand = 00110000000000000000000
$-1.0011 \times 2^{+3}$ = 1 10000010 00110000000000000000000;
$-1.0011 \times 2^{+3}$ = C1180000H

123. R_2, 5, R_5.

125.

|(DATA4_64B) − (DATA3_64B)| → DATA5_64B
$|-10.75 - (-2.5)| = |-10.75 + 2.5| = |-8.25| = 8.25$
$8.25 = 1000.01_2 = 1.00001 \times 2^{+3}$
Sign = 0
Biased exponent = $00000000011_2 + 01111111111_2 = 10000000010$
Fractional significand = 0000100000000000000000000000000000000000000000000000
8.25 = 0 10000000010 0000100000000000000000000000000000000000000000000000
8.25 = $0100000000100001000000000000000000000000000000000000000000000000_2$
8.25 = 4020800000000000H

Section 15.12

127. A microprocessor architecture that employs more than one execution unit.

129. 2; U pipe and V pipe.

131. 5 to 10 times faster.

133. Page size extensions, 1M 32-bit entries.

135. Compare and exchange 8 bytes (CMPXCHG8B), CPU identification (CPUID), and read from time stamp counter (RDTSC).

137. Machine check type (MCT).

Section 15.13

139. Packed byte, packed word, packed double word, and packed quad-word; 8 × 8-bit, 4 × 16-bit, 2 × 32-bit, and 1 × 64-bit.

141. **(a)** Byte 7 = FFH, Byte 6 = 00H, Byte 5 = 12H, Byte 4 = 34H, Byte 3 = 56H, Byte 2 = 78H, Byte 1 = ABH, Byte 0 = CDH.

(b) Word 3 = FF00H, Word 2 = 1234H, Word 1 = 5678H, Word 0 = ABCDH.

(c) Double word 1 = FF001234H Double word 0 = 5678ABCDH.

143. Data is compared as signed numbers. This gives

Double word 1 FFFFFFFFH > 00010002H = False → 00000000H

Double word 0 12345678H > 876554321H = True → FFFFFFFFH

MM_3 = 00000000FFFFFFFFH

CHAPTER 16

Section 16.1

1. CHMOSIII.

3. INTR.

Section 16.2

5. Byte, D_0 through D_7, no.

7. I/O data read.

9. 80387DX numeric coprocessor.

Section 16.3

11. F12.

Section 16.4

13. 40 ns.

15. In Fig. 16–11 address *n* becomes valid in the T_2 state of the prior bus cycle and then the data transfer takes place in the next T_2 state. Also, at the same time that data transfer *n* occurs, address *n* + 1 is output on the address bus. This shows that during pipelining, the 80386DX starts to address the next storage location to be accessed while still reading or writing data for the previously addressed storage location.

17. An extension of the current bus cycle by a period equal to one or more T states because the $\overline{\text{READY}}$ input was tested and found to be logic 1.

19. 80 ns.

Section 16.5

21. Four independent banks each organized as 1G × 8 bits; four banks each organized as 256K × 8 bits.

23. 120 ns; 240 ns.

25. The bus control logic produces the appropriately timed command and control signals needed to control transfers over the data bus.

The address decoder decodes the higher-order address bits to produce chip-enable signals.

The address bus latch is used to latch and buffer the lower bits of the address, byte-enable signals, and chip-enable signals.

The bank write control logic determines to which memory banks $\overline{MWTC}$ is applied during write bus cycles.

The data bus transceiver/buffer controls the direction of data transfers between the MPU and memory subsystem and supplies buffering for the data bus lines.

Section 16.6

27. $M/\overline{IO}$.

29. The I/O address decoder is used to decode several of the upper I/O address bits to produce the $\overline{I/OCE}$ signals.

The I/O address bus latch is used to latch lower-order address bits, byte-enable signals, and $\overline{I/OCE}$ outputs of the decoder.

The bus-control logic decodes I/O bus commands to produce the input/output and bus-control signals for the I/O interface.

The data bus transceivers control the direction of data transfer over the bus, multiplex data between the 32-bit microprocessor data bus and the 8-bit I/O data bus, and supplies buffering for the data bus lines.

The I/O bank-select decoder controls the enabling and multiplexing of the data bus transceivers.

31. 320 ns.

33. BASE + 8H; LSB (bit 0).

Section 16.7

35. Interrupt vector table; interrupt descriptor table.

37. Interrupt descriptor table register; $0000000003FF_{16}$.

39. **(a)** Active; **(b)** privilege level 2; **(c)** interrupt gate; **(d)** B000H:1000H.

41. In the protected mode, the 80386DX's protection mechanism comes into play and checks are made to confirm that the gate is present; the offset is within the limit of the interrupt descriptor table; access byte of the descriptor for the type number is for a trap, interrupt, or task gate; and to assure that a privilege level violation will not occur.

43. Faults, traps, and aborts.

45. Fault.

47. Double fault, invalid task state segment, segment not present, general protection fault, and page fault.

Section 16.8

49. DP_0, DP_1, DP_2, DP_3, and $\overline{PCHK}$; even parity.

51. Cache enable.

53. Four double words = 16 bytes; $\overline{BRDY}$ = 0; 5 clock cycles.

55. Near to zero-wait-state operation even though the system employs a main memory subsystem that operates with one or more wait states.

57. A bus cycle that reads code or data from the cache memory is called a cache hit.

59. 0.203 wait states/bus cycle.

61. 8Kbytes; line of data = 128 bits (16 bytes).

63. The contents of the internal cache are cleared. That is, the tag for each of the lines of information in the cache is marked as invalid.

Section 16.9

65. Clock doubling and write-back cache.

67. Modify/exclusive/shared/invalid (MESI) protocol.

69. The $\overline{\text{WBWT}}$ input must be held at logic 1 for at least two clock periods before and after a hardware reset.

71. 435/68 = 6.4.

Section 16.10

73. 3 million transistors.

75. D_{45} is at pin A21; A_3 is at pin T17.

77. Data parity and address parity; $\overline{\text{APCHK}}$ = 0 address parity error and $\overline{\text{PCHK}}$ = 0 data parity error.

79. I/O write.

81. Two-way set associative.

83. 256 bits (32 bytes).

85. Read hits access the cache, read misses may cause replacement, write hits update the cache, writes to shared lines and write misses appear externally, write hits can change shared lines to exclusive under control of WB/$\overline{\text{WT}}$, and invalidation is allowed.

Section 16.11

87. 5.5 million; 4.5 million.

89. Both the code and data caches are 16Kbytes in size; they are organized four-way set-associative.

Section 16.12

91. Celeron processor.

93. 533Mbytes/sec.

95. Single edge contact cartridge.

Section 16.13

97. P6 microarchitecture; NetBurst microarchitecture.

99. 133 MHz; 400 MHz.

Index